LONDON

A-1
DELUXE
STREET
ATLAS

LONDON

A1 DELUXE STREET ATLAS

ISBN : 0 09 202 2006

The maps in this Atlas are based upon the Ordnance Survey Maps with the sanction of the Controller of H.M. Stationery Office, with additions obtained from Local Authorities. The Ordnance Survey is not responsible for the accuracy of the National Grid on this Production.

The representation on these maps of a Road, Track or Footpath is no evidence of the existence of a right of way.

The contents of this publication are believed correct at the time of printing. Nevertheless the Publishers can accept no responsibility for errors or omissions or for changes in the details given.

GEOGRAPHIA

Geographia Ltd., 63 Fleet Street, London EC4Y 1PE
Telephone 01-353 2701/2

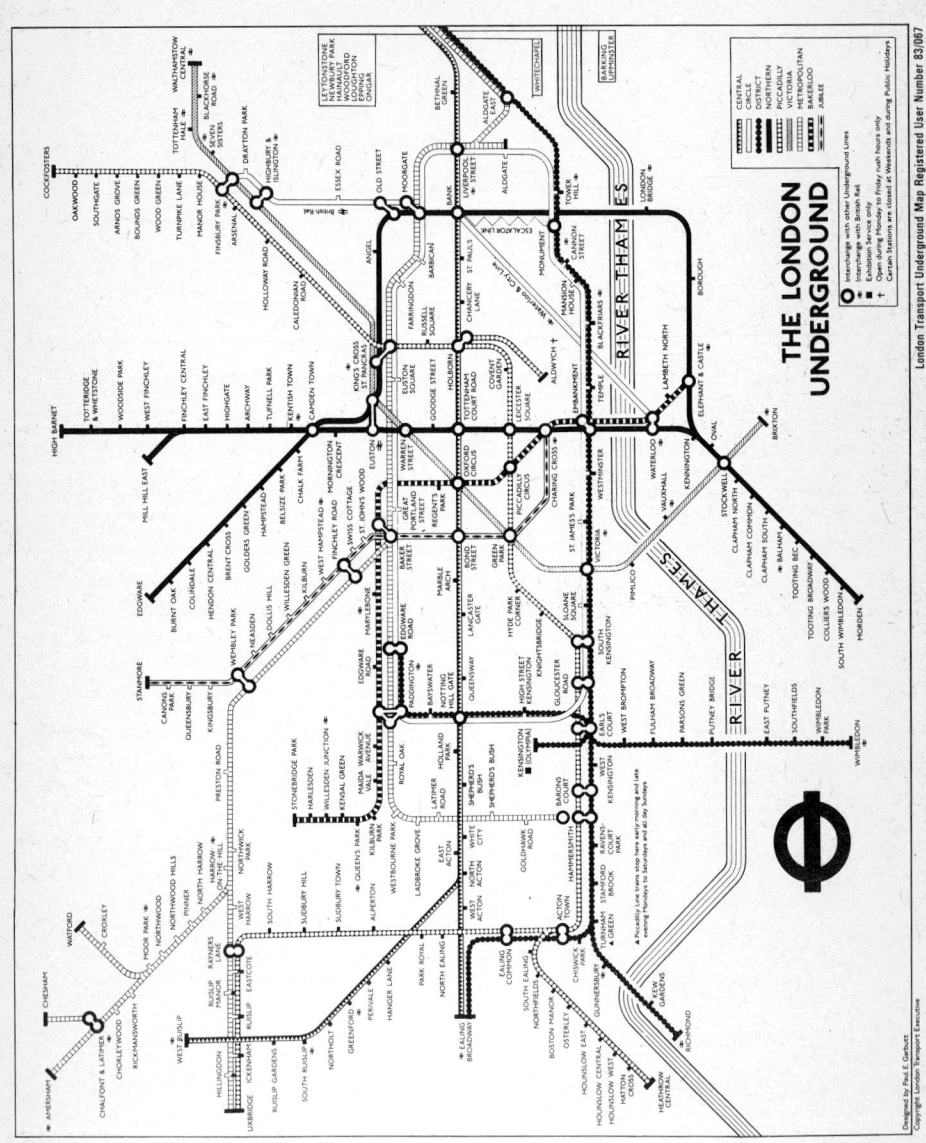

THE LONDON UNDERGROUND

CENTRAL
CIRCLE
DISTRICT
NORTHERN
PICCADILLY
VICTORIA
METROPOLITAN
BAKERLOO
JUBILEE

○ Interchange with other Underground Lines
● Interchange with British Rail
+ Exhibition: Service only
▪ Open during Monday to Friday rush hours only
Certain Stations are closed at Weekends and during Public Holidays

London Transport Underground Map Registered User Number 83/067

Designed by Paul E. Garbutt.
Copyright London Transport Executive

CONTENTS MATIÈRE INHALT CONTENUTO

Scale
Echelle 1:20,000 (3.17 inches to 1 mile—5cm to 1 km) Masstab
Scala

0	¼	½	¾	1Mile

0	½	1	1½Kms.

LEGEND   LÉGENDE   LEGENDE   LEGGENDA

Motorways with junctions Autoroutes avec les jonctions	**M1** ❶	Autobahn mit Anschlüssen Autostrade da congiunzione
Dual carriageway Route a deux voies		Zweibahnige Strasse Strade duale
House numbers Nombres des maisons	1 105 2 106	Hausnummern Numeri delle case
Railways and stations Chemins de Fers et les Gares		Eisenbahnen und Bahnhofen Ferrovie e Stazione
Underground stations Gares de Metro	WEST FINCHLEY ⊖	U-Bahnhofen Stazione delle Metro
Administrative Boundaries Limite Administrative	–·–·–·–	Bezirksgrenze Limite Amministrativo
Place of worship Eglise	+	Kirche Chiesa
Fire Stations Poste des Pompiers	F.S.	Feuerwachen Caserme dei Pompieri
Police Stations Commissariat	P.S.	Polizeiwachen Uffici dei Polizia
Post Offices Bureaux des Postes	P.O.	Postamt Poste
Map continuation numbers Nombres des cartes voisines	10	Anschlusskarten Numeri delle carte di seguito

GREEN SECTION

Scale
Echelle 1:11,520 (5.5 inches to 1 mile—8.7 cm to 1 km) Masstab
Scala

0	¼	½Mile

0	½	1Km

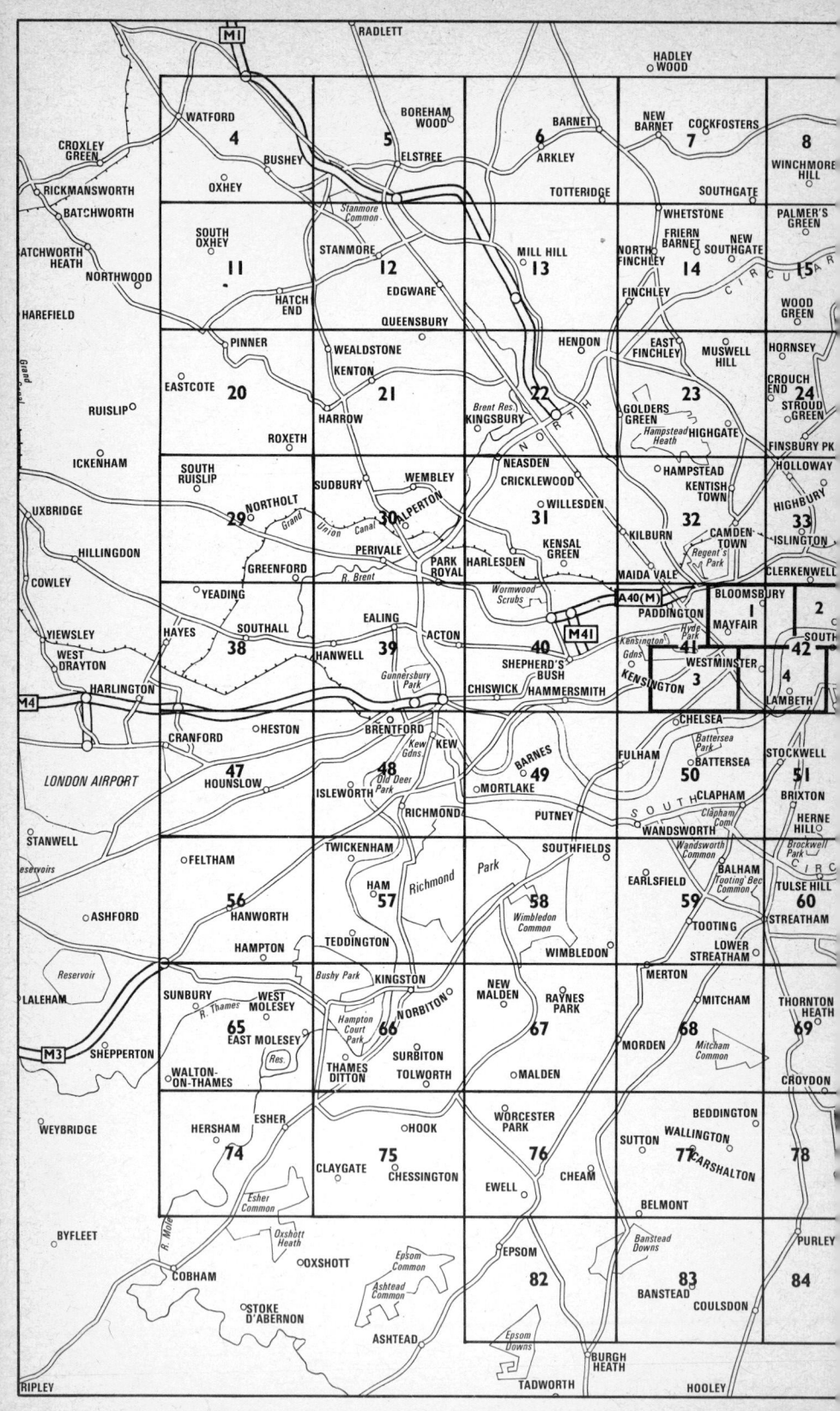

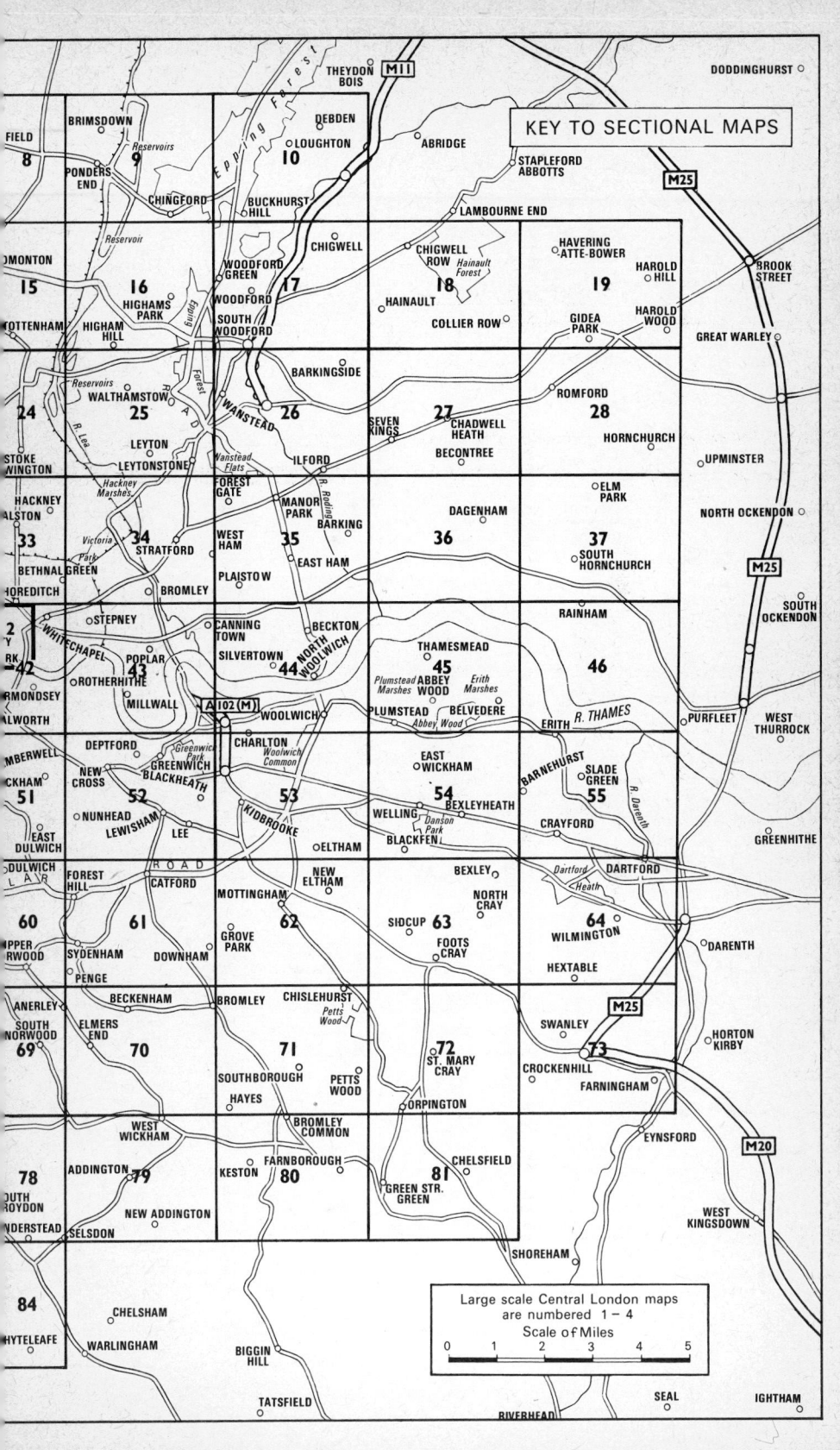

KEY TO SECTIONAL MAPS

Large scale Central London maps
are numbered 1 – 4
Scale of Miles

0   1   2   3   4   5

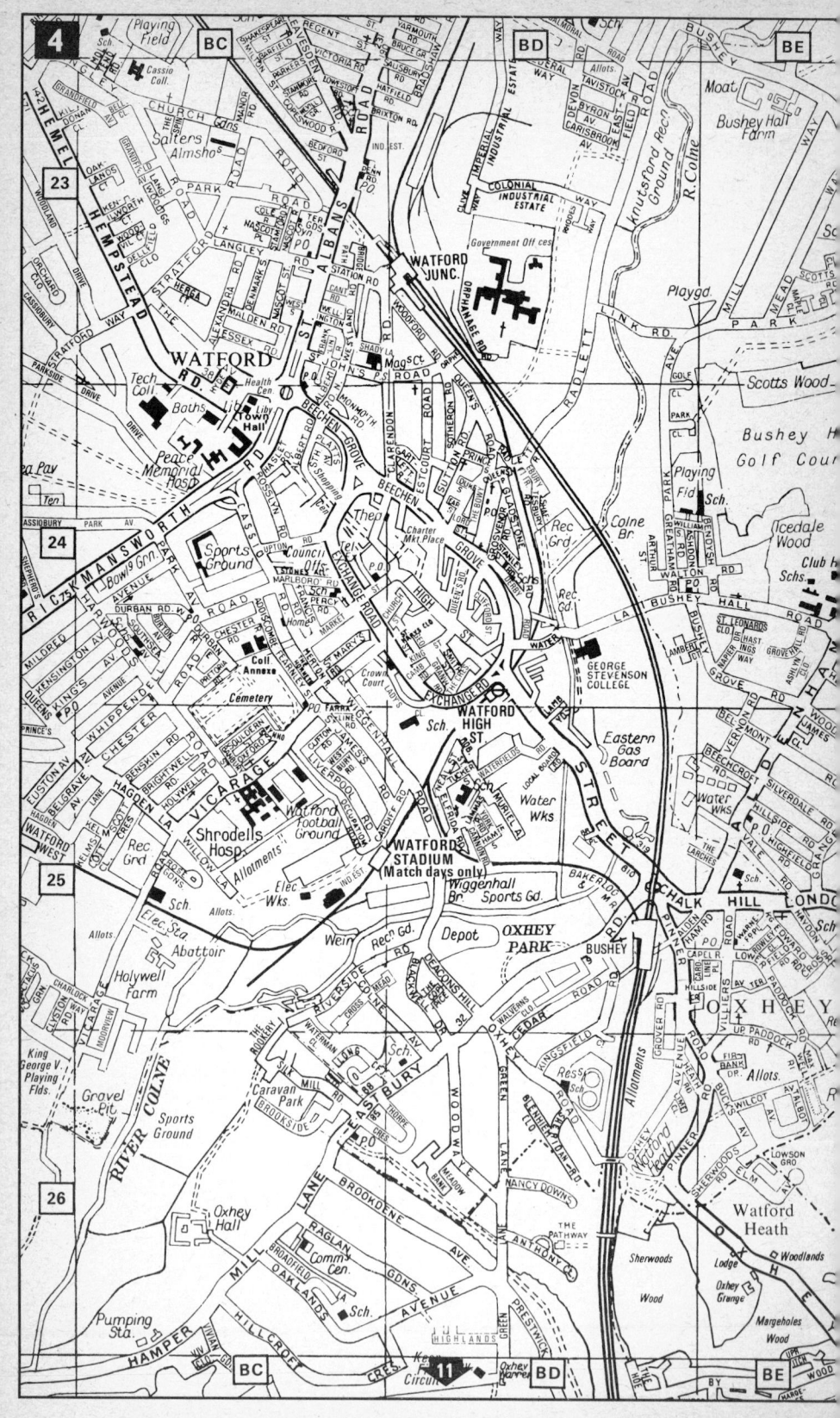

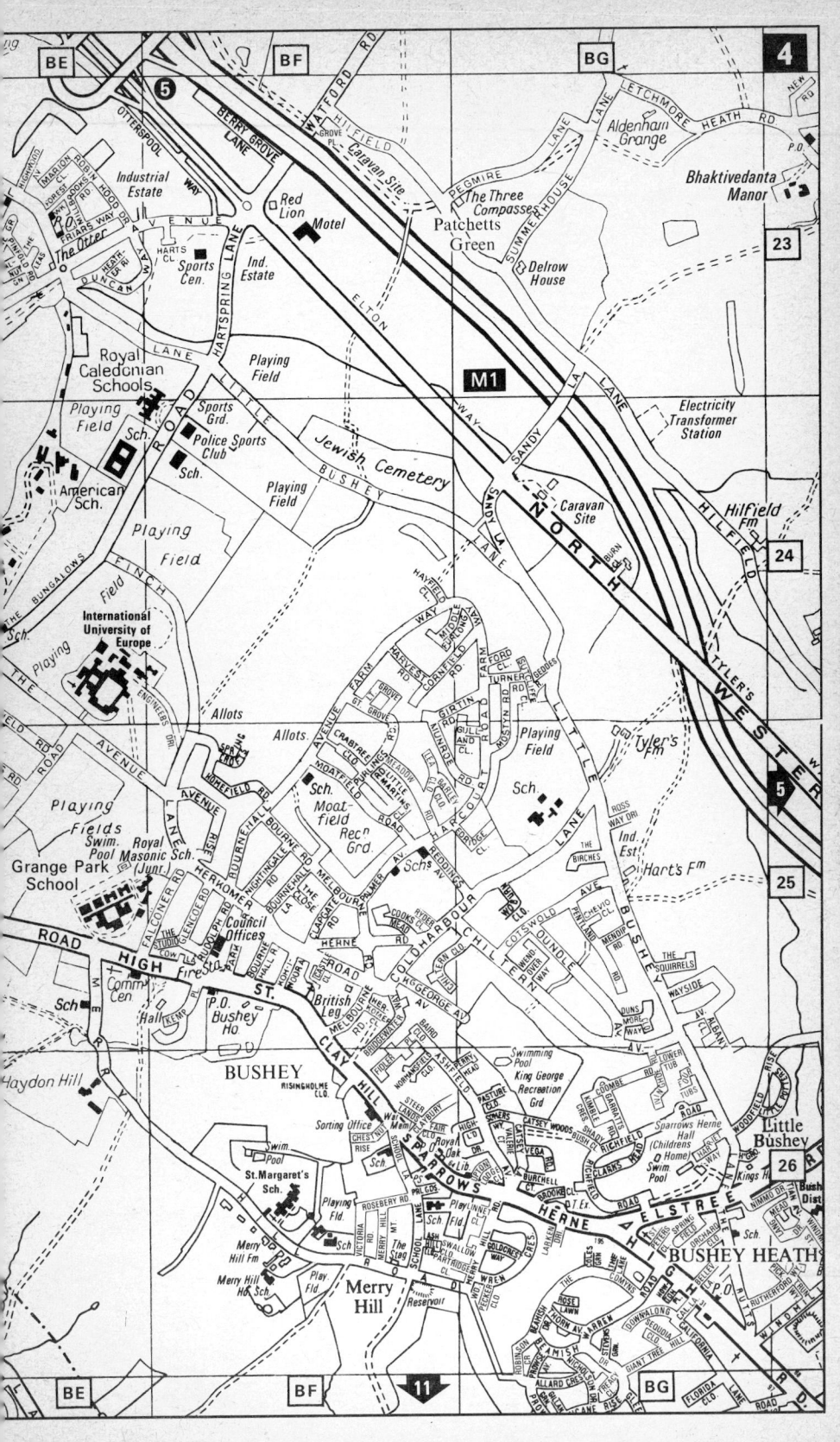

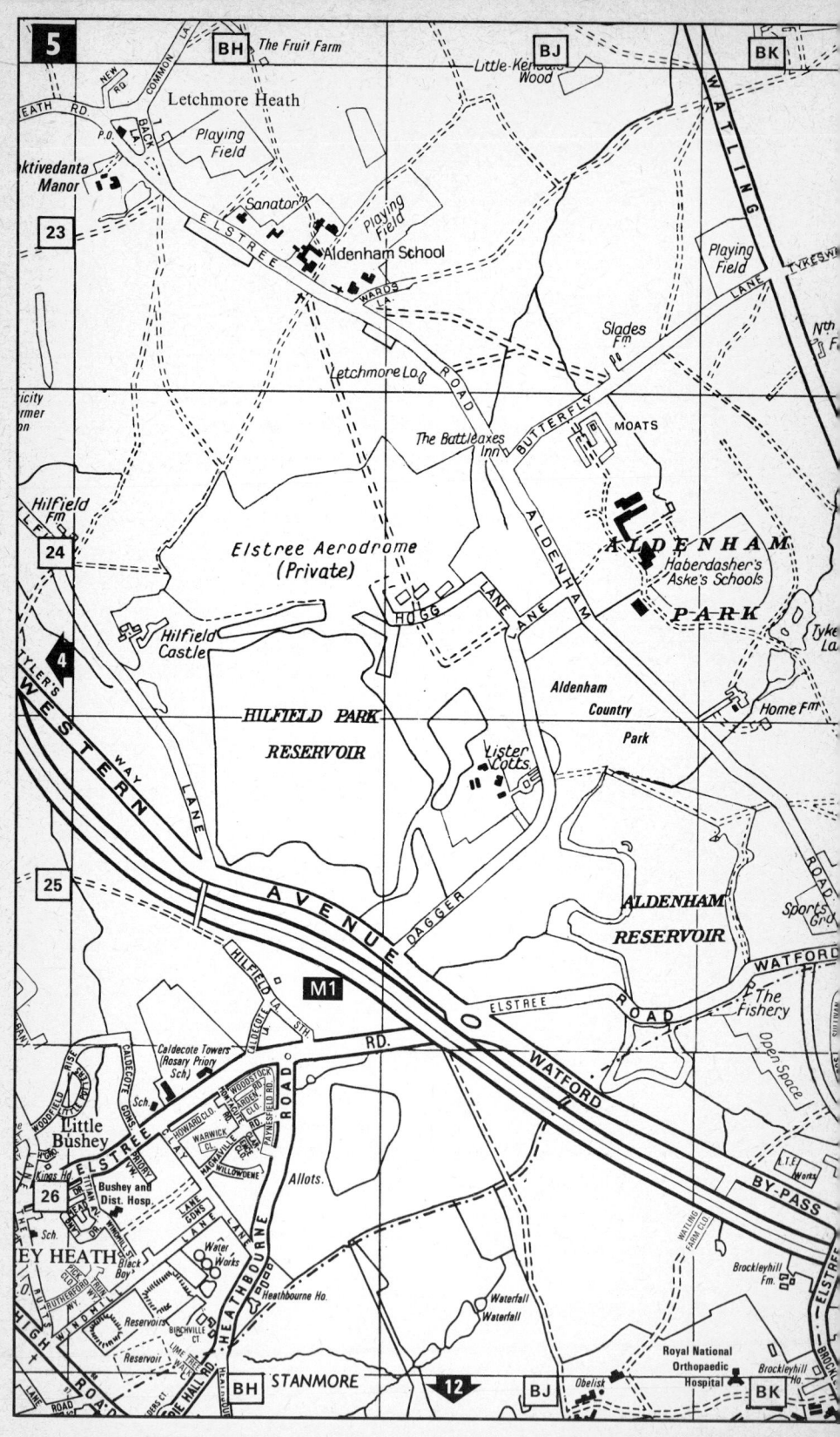

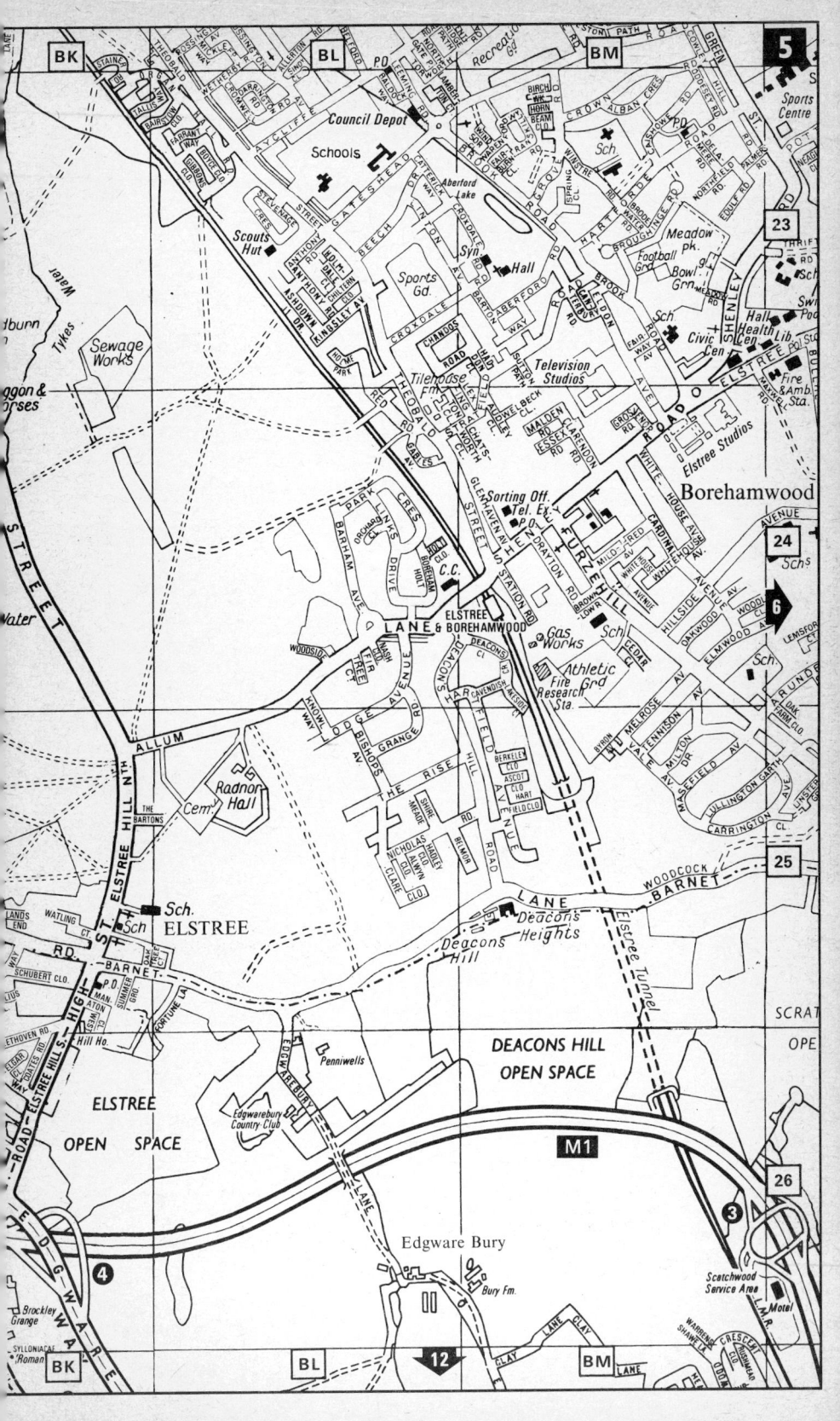

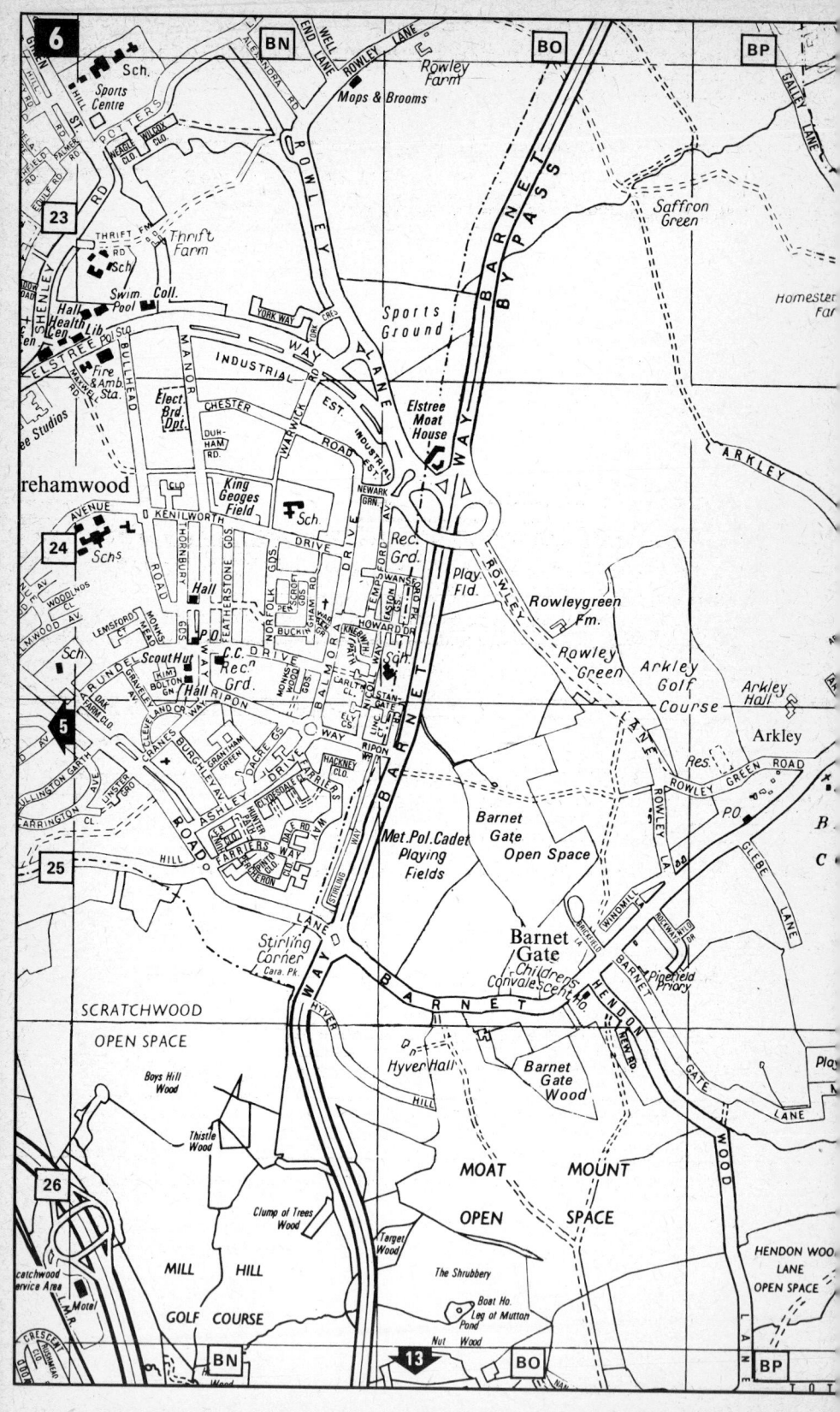

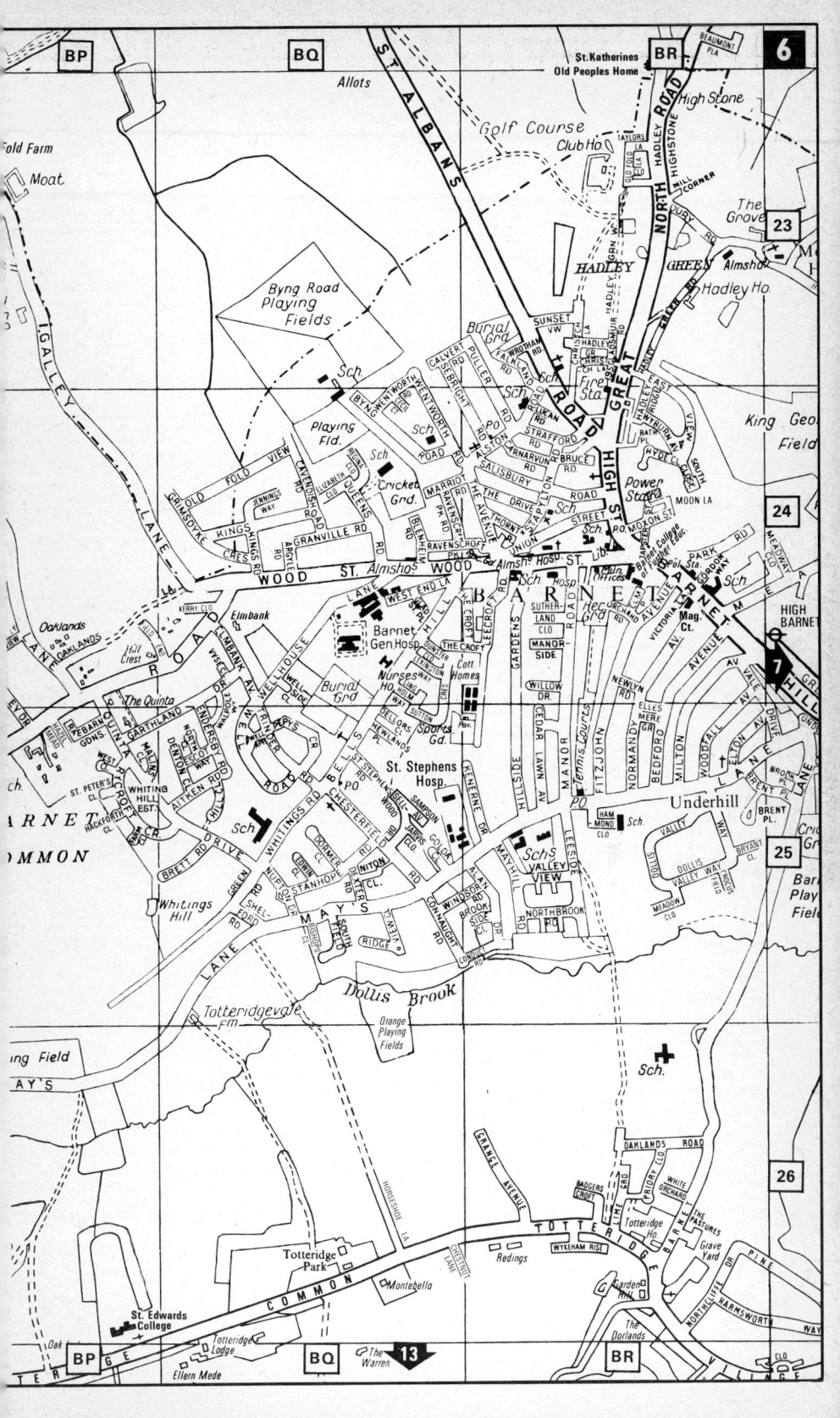

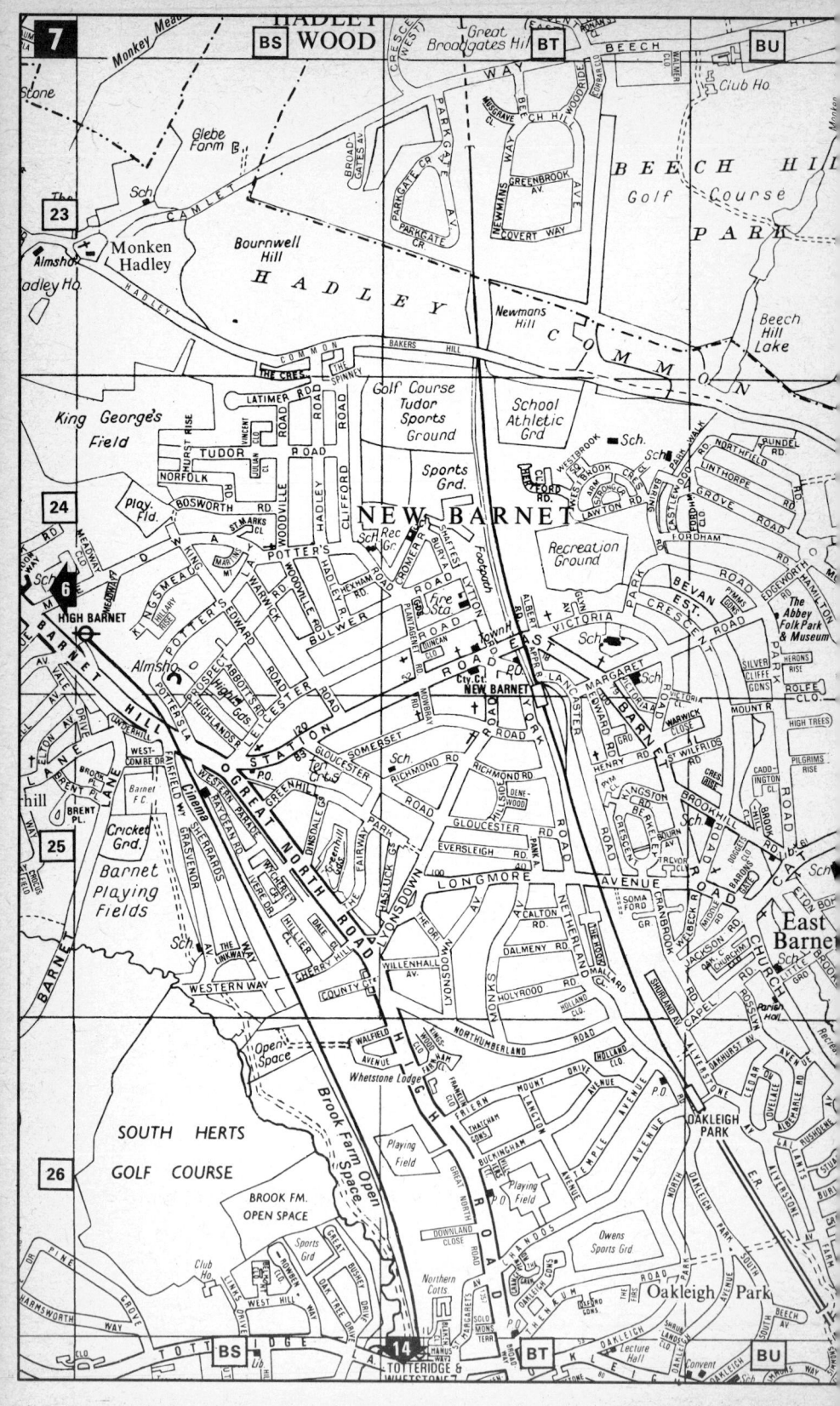

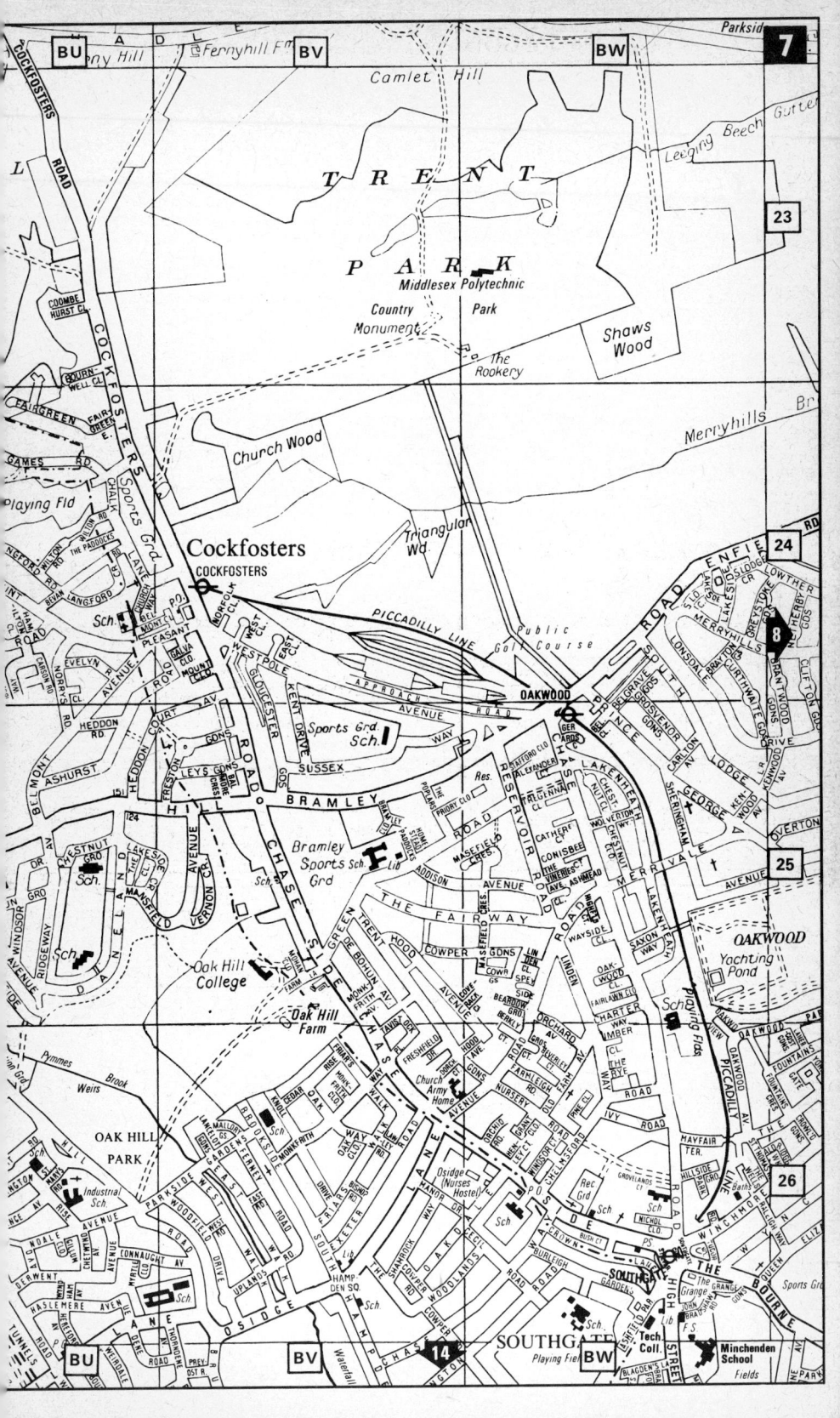

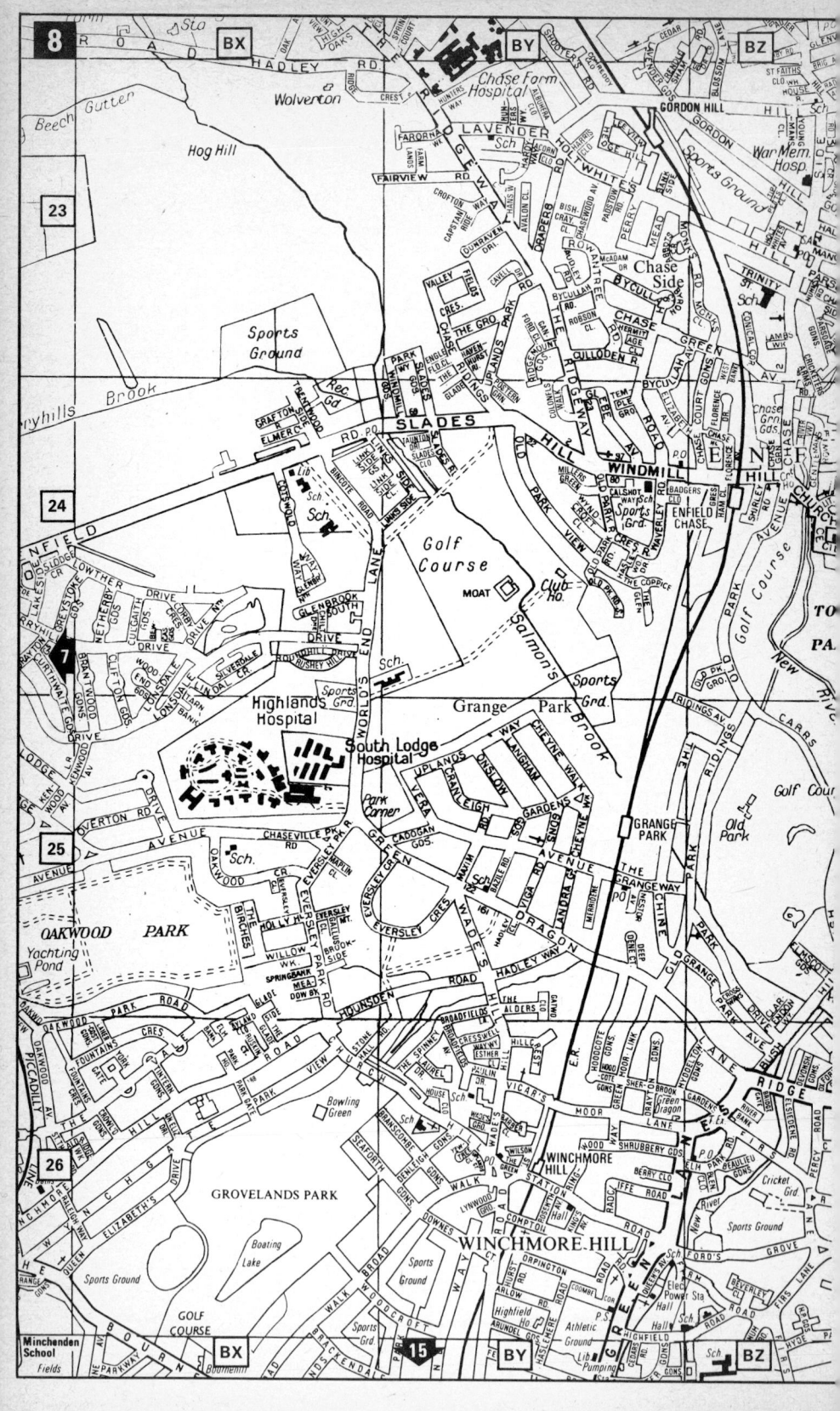

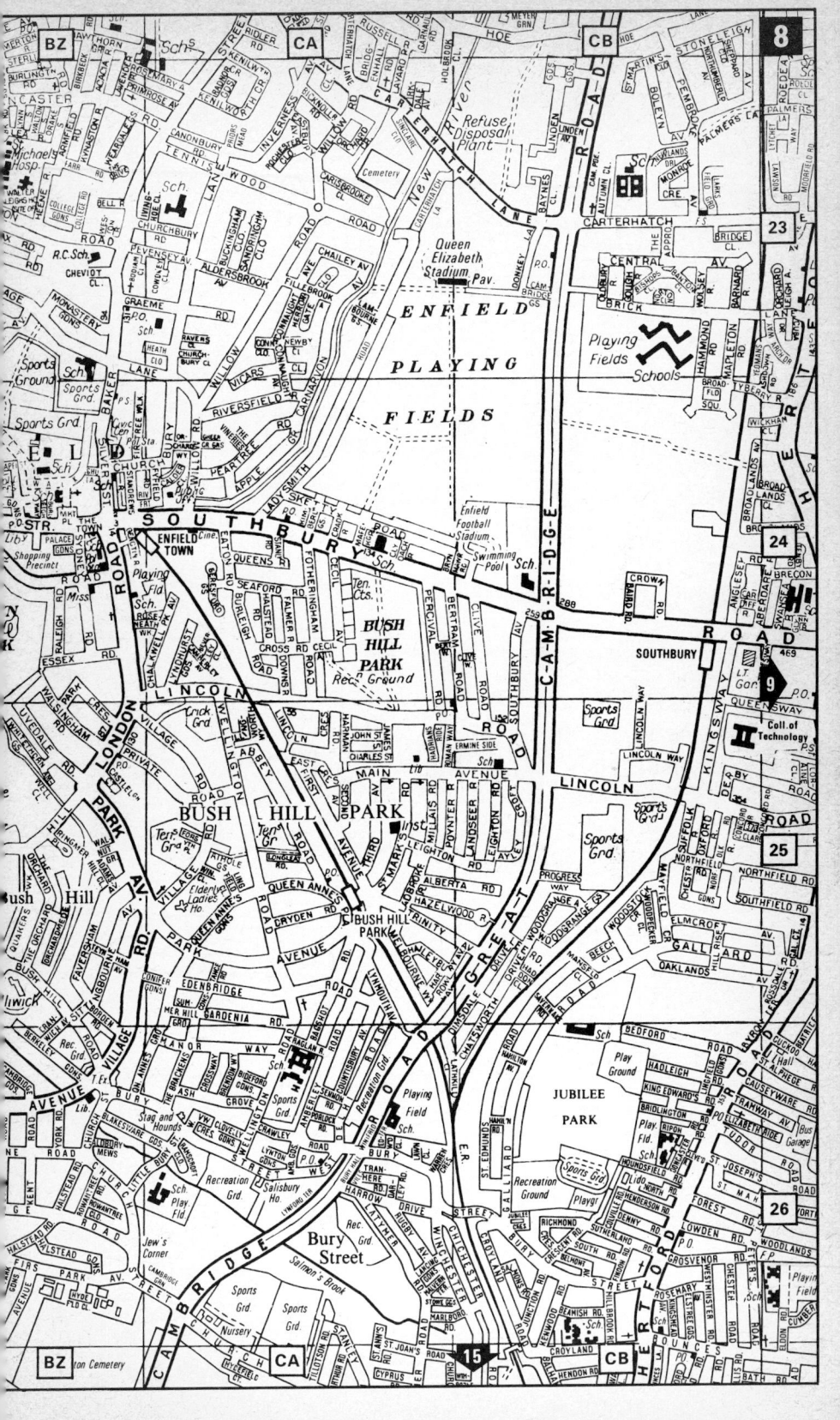

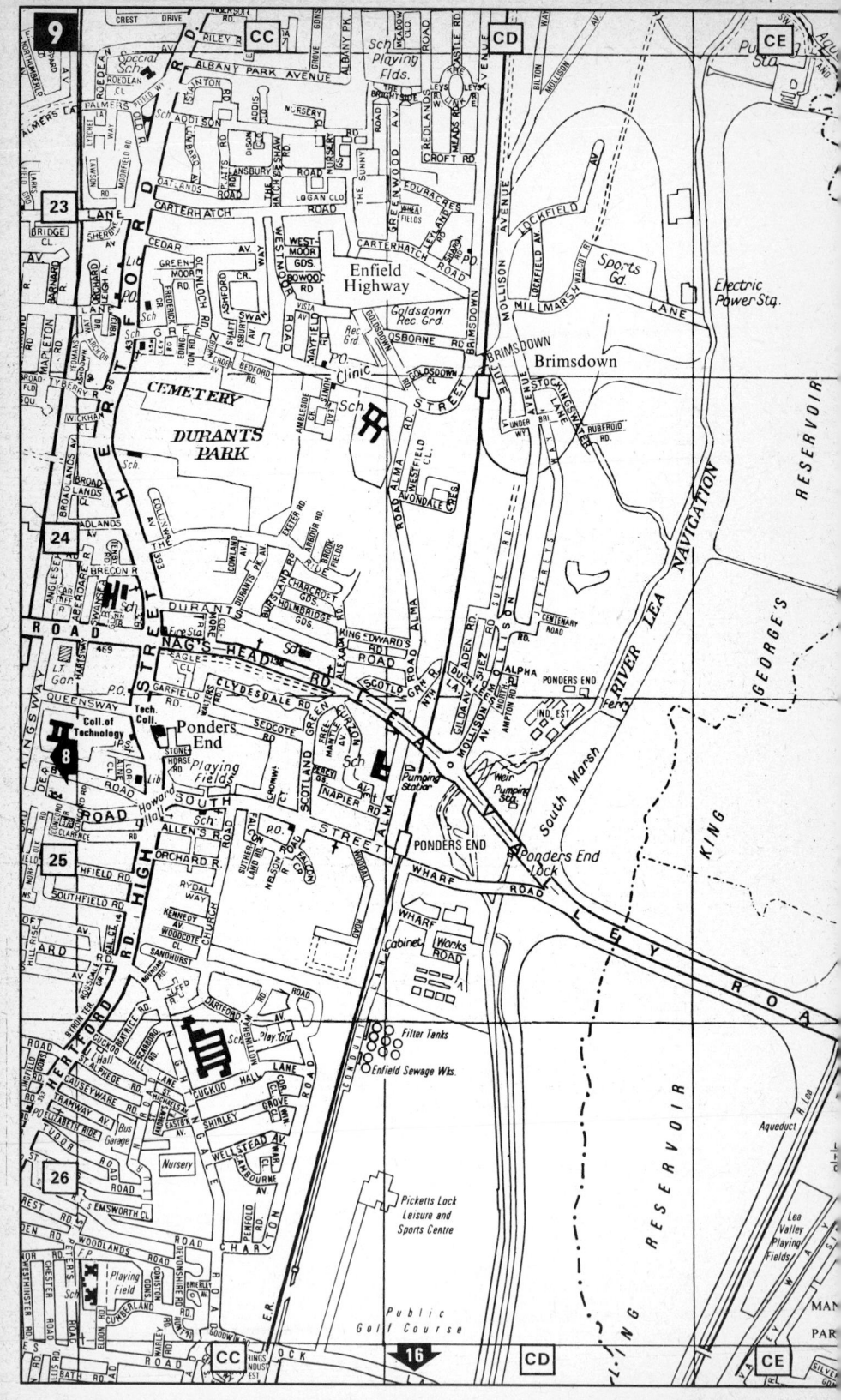

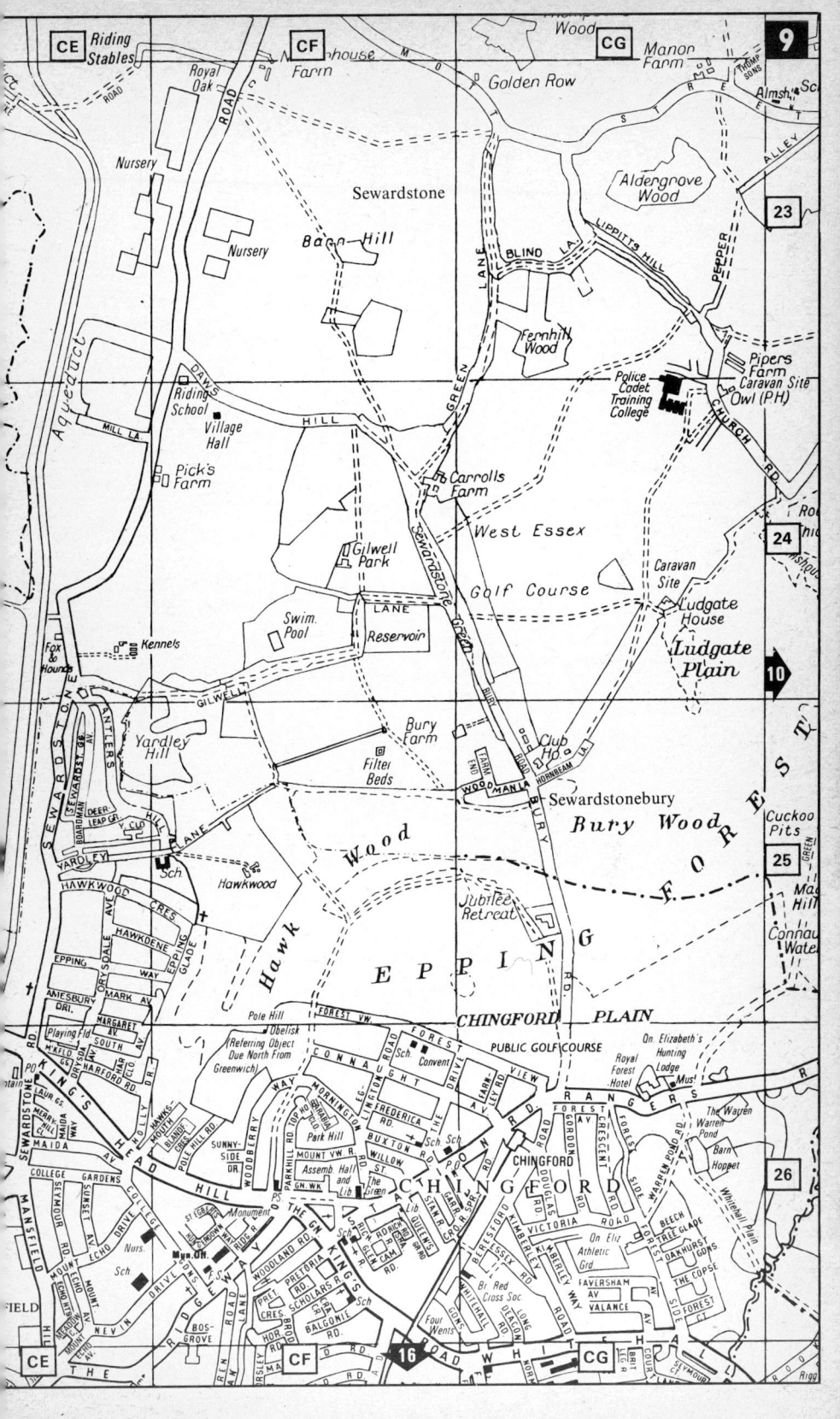

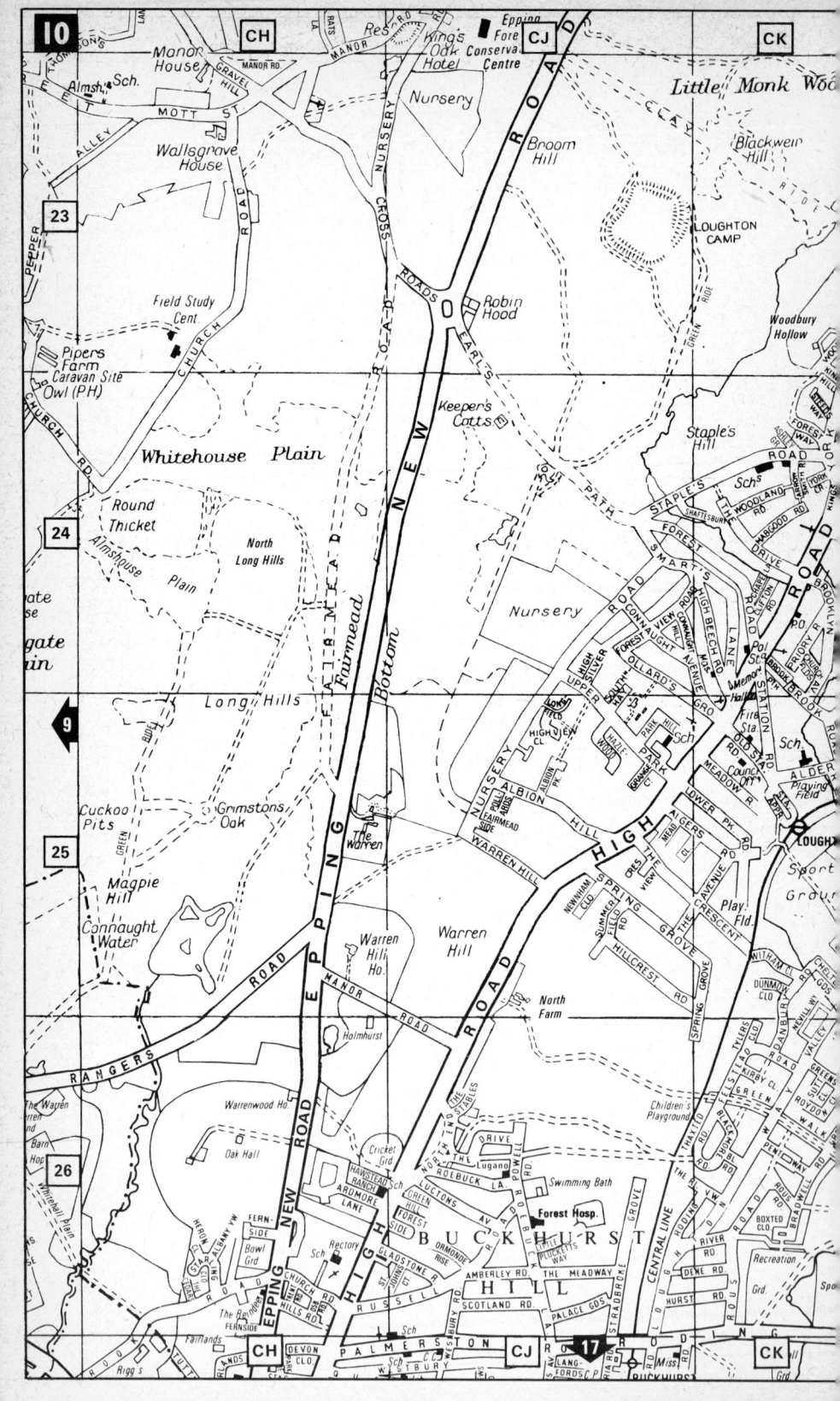

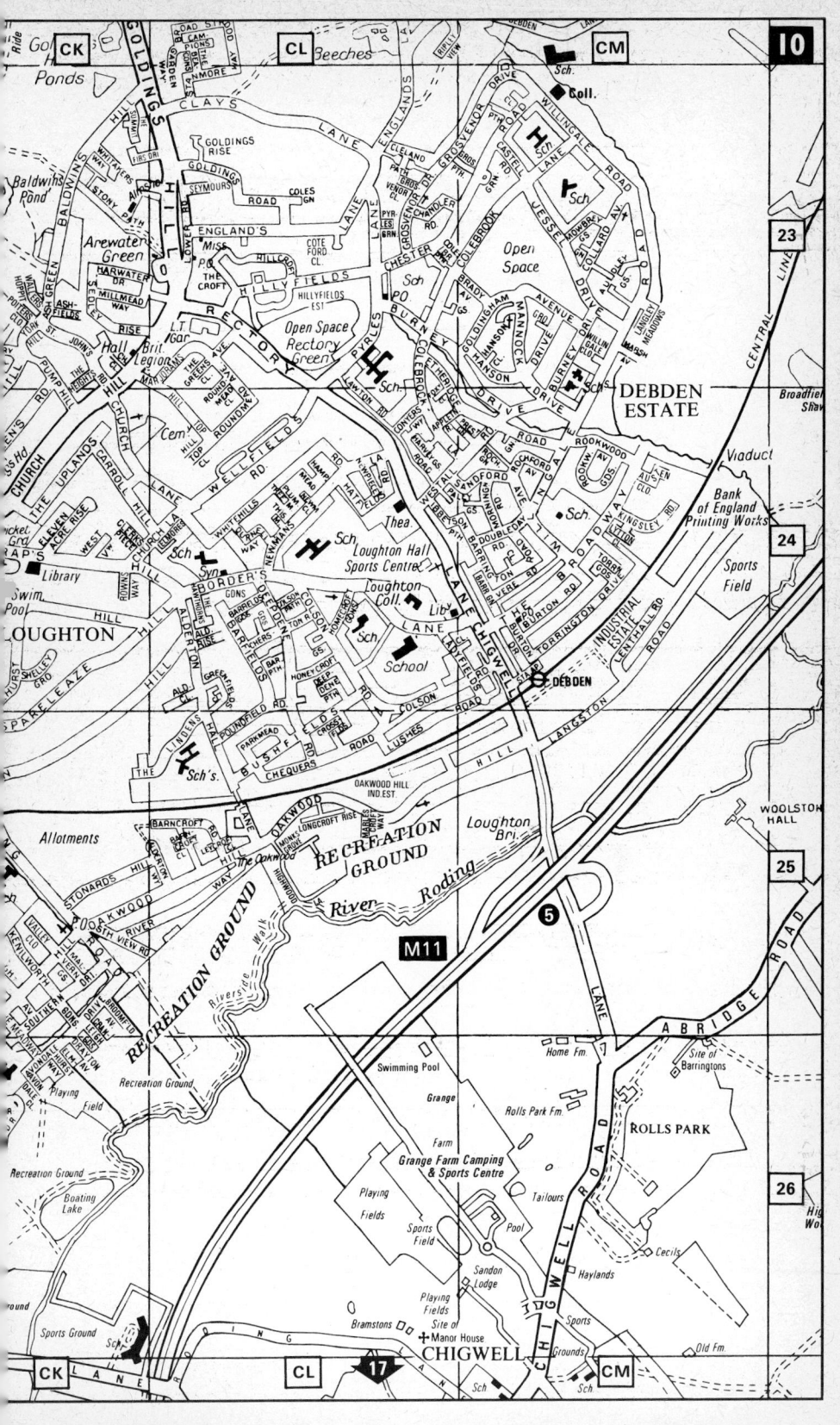

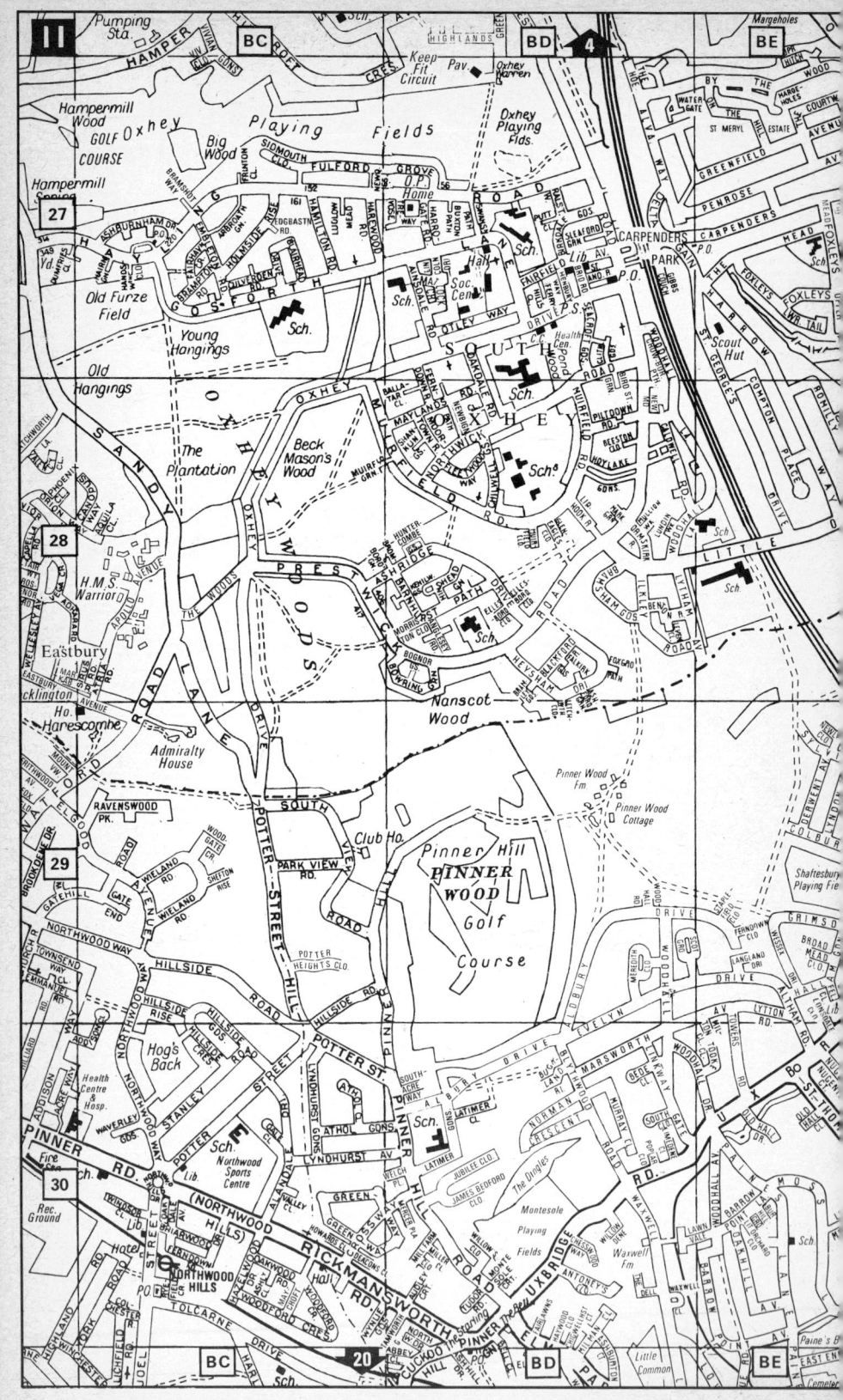

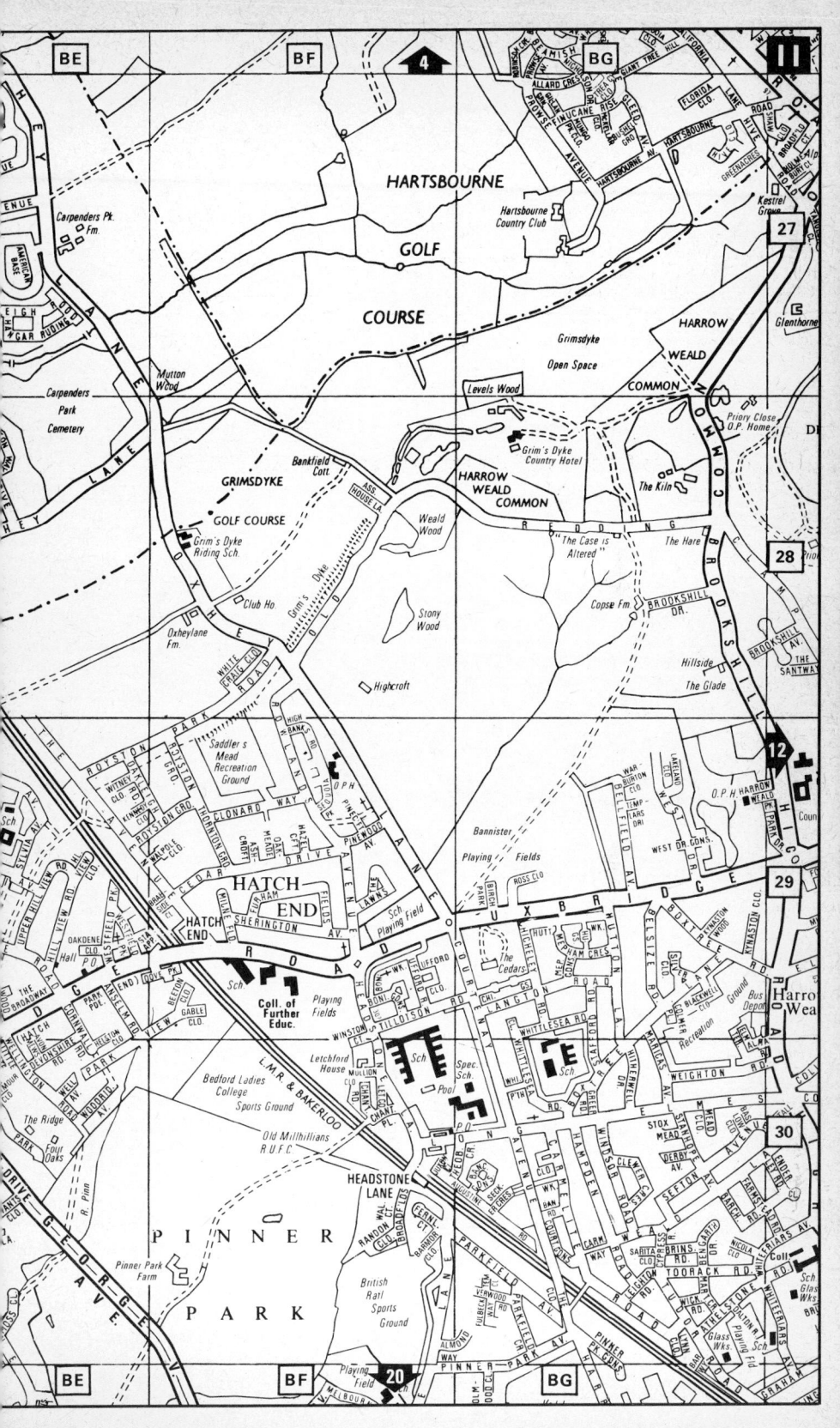

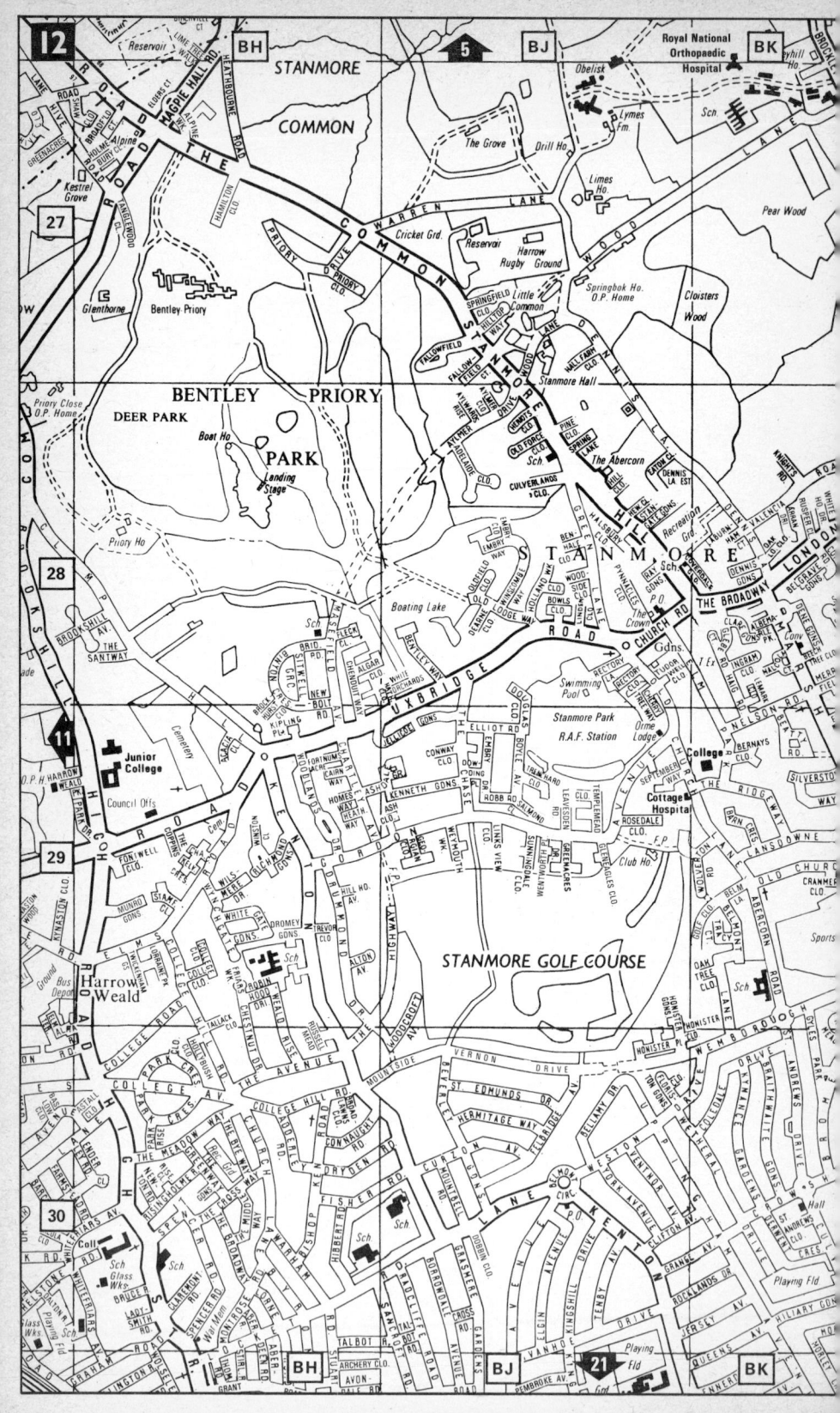

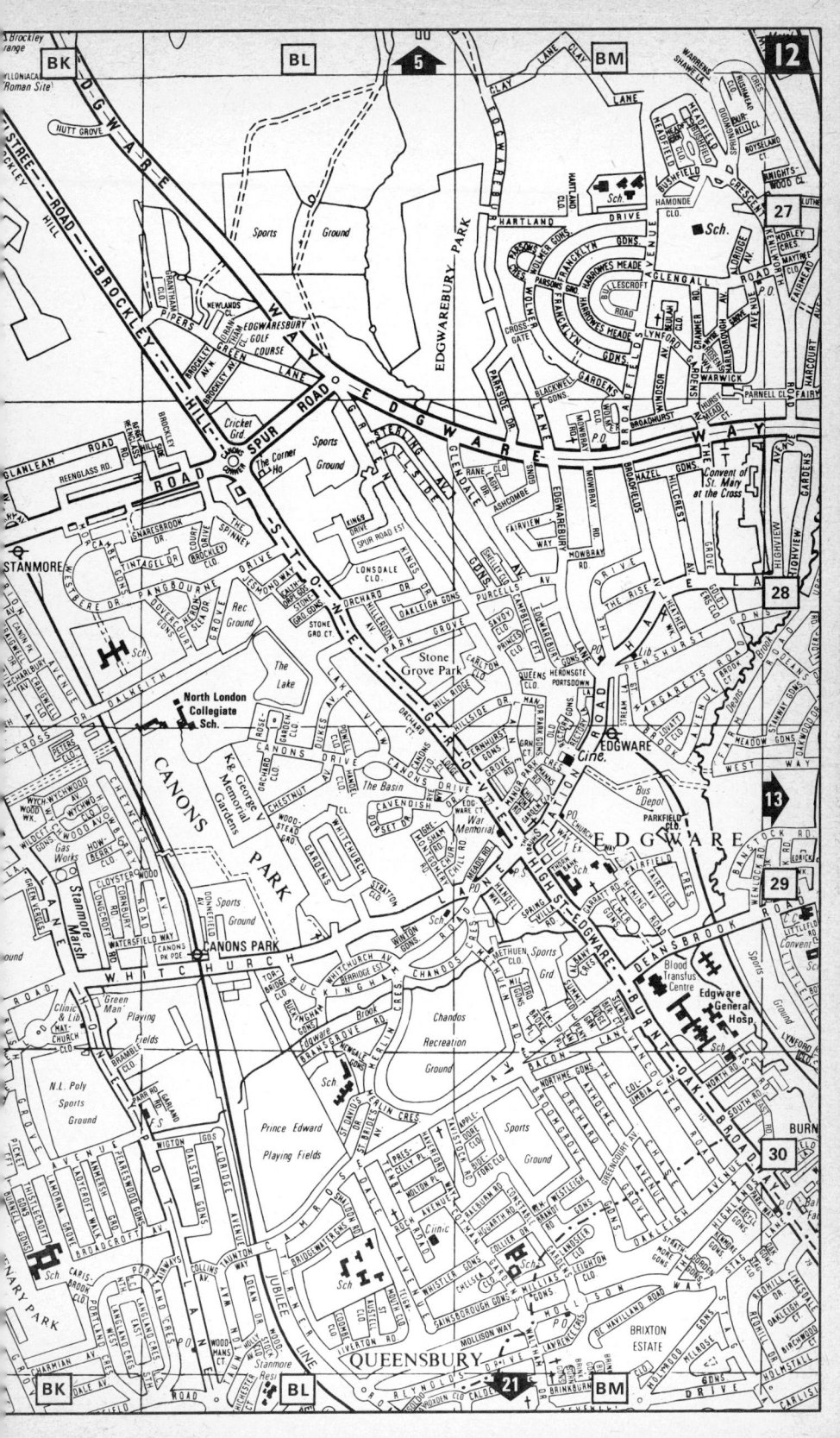

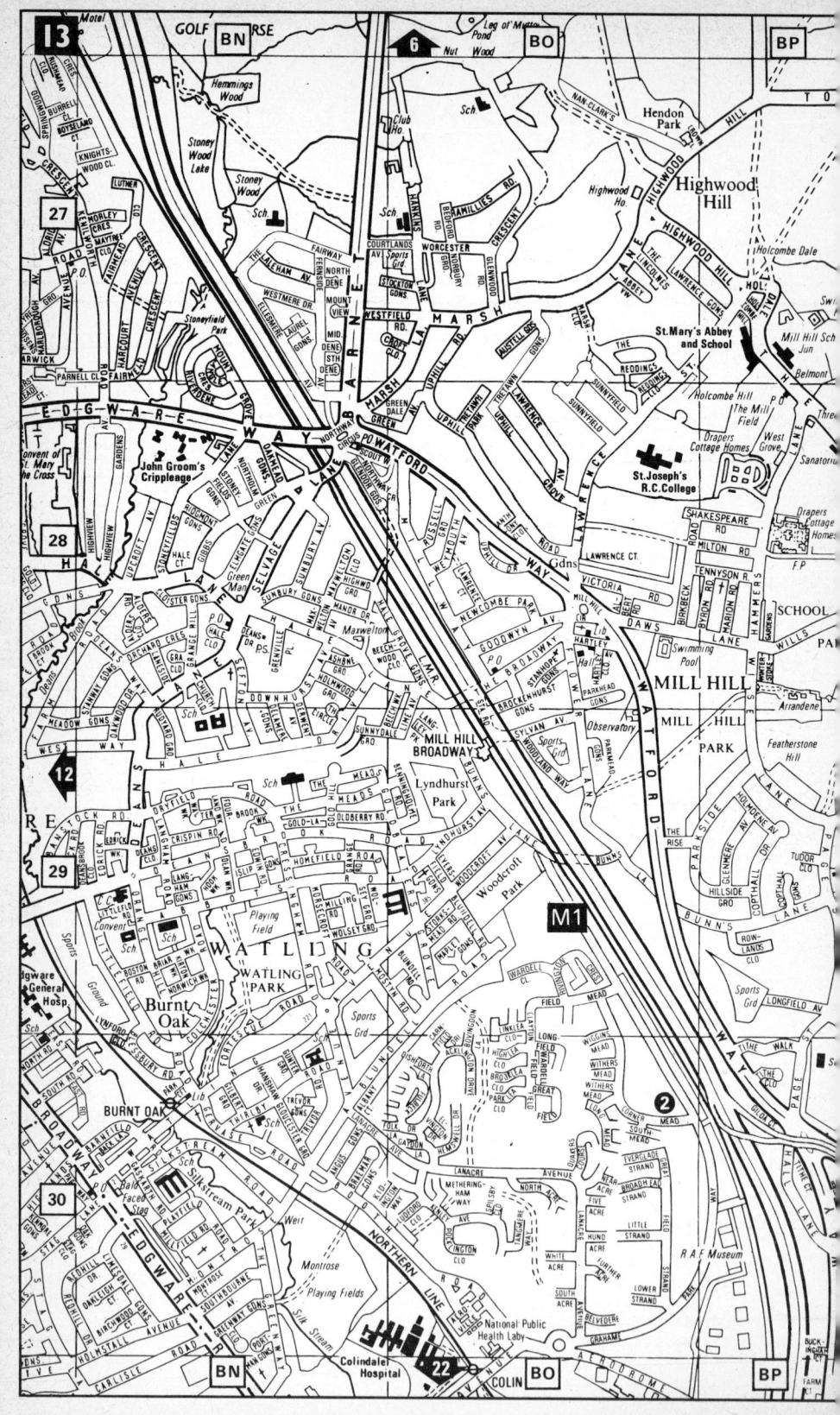

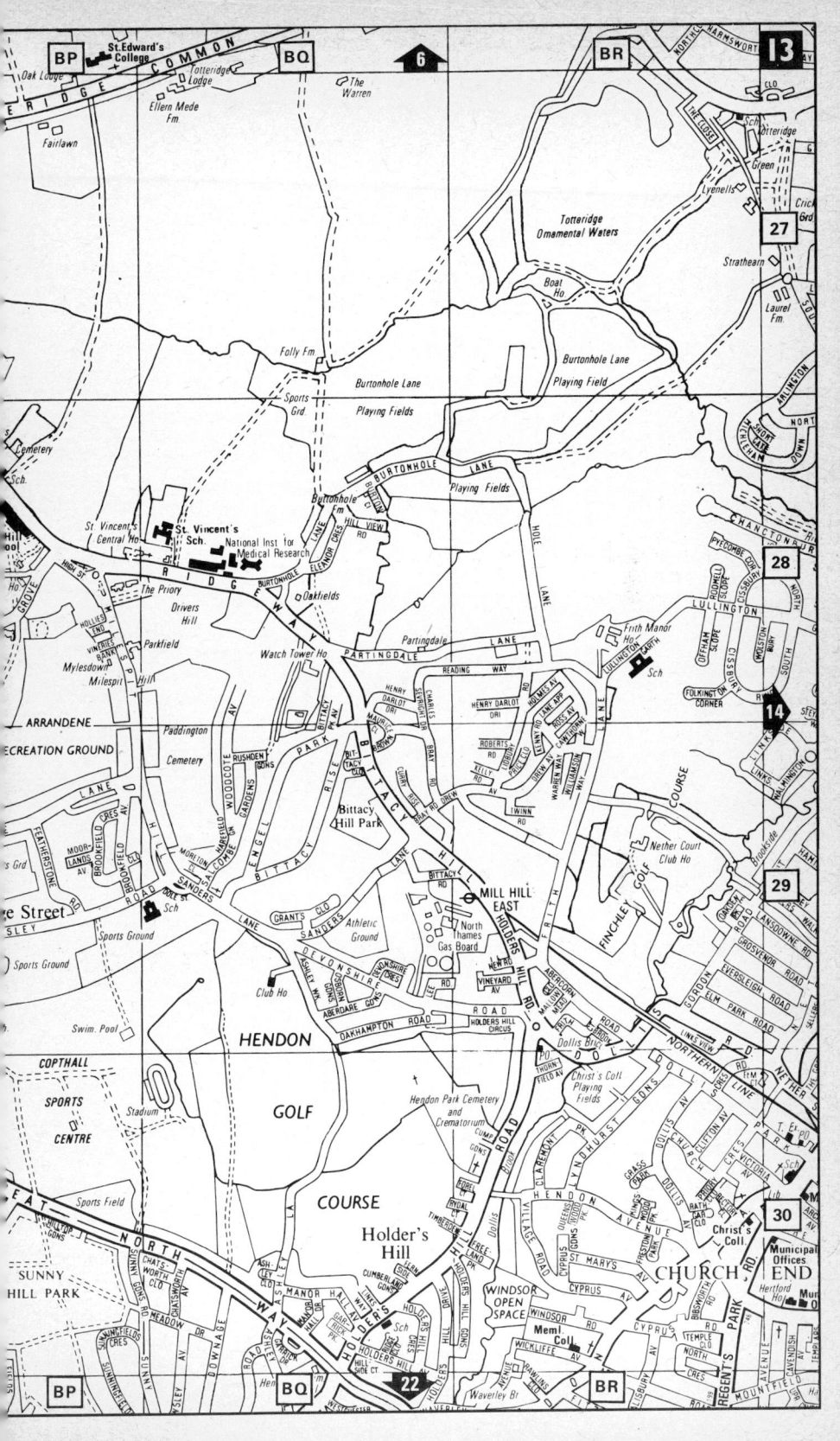

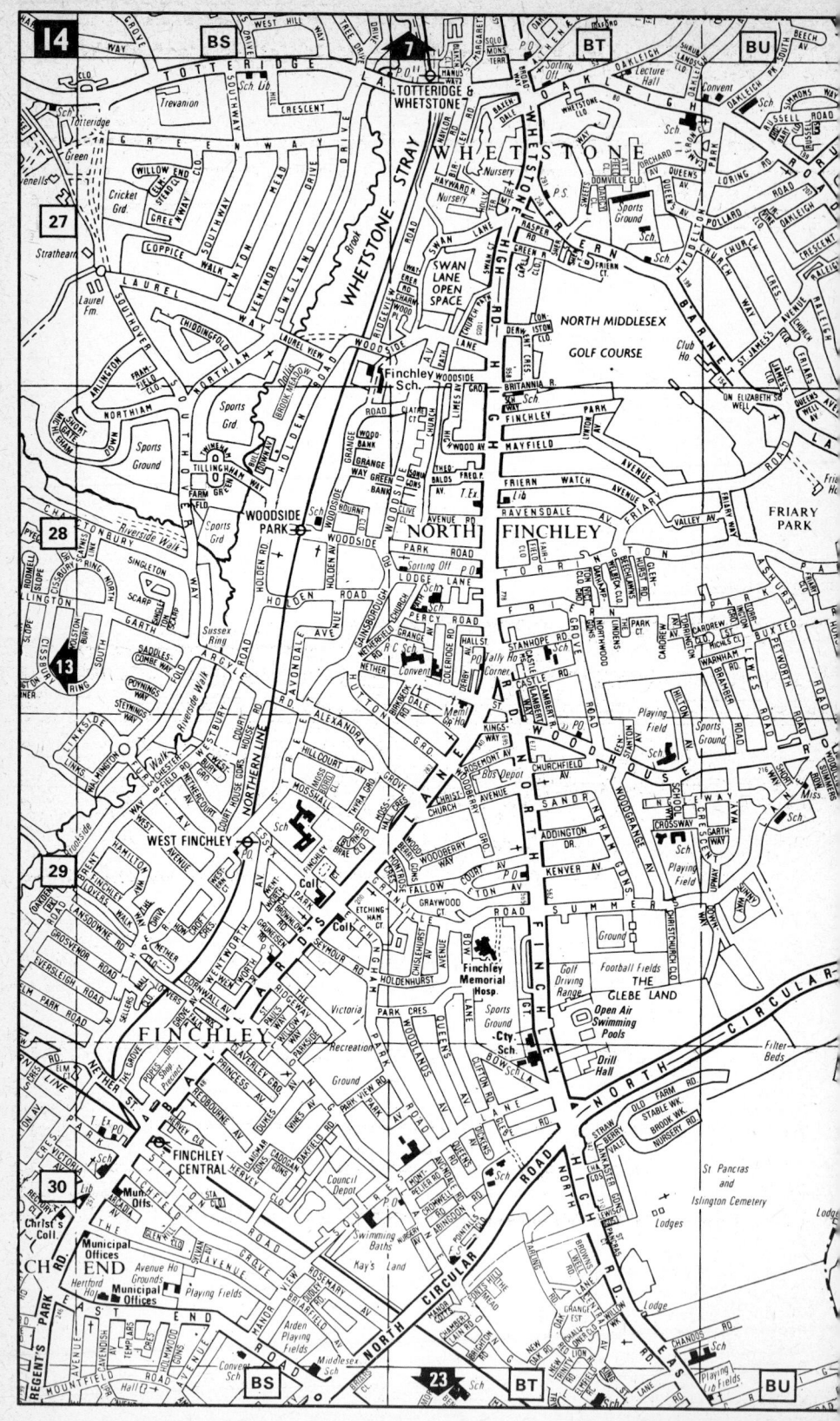

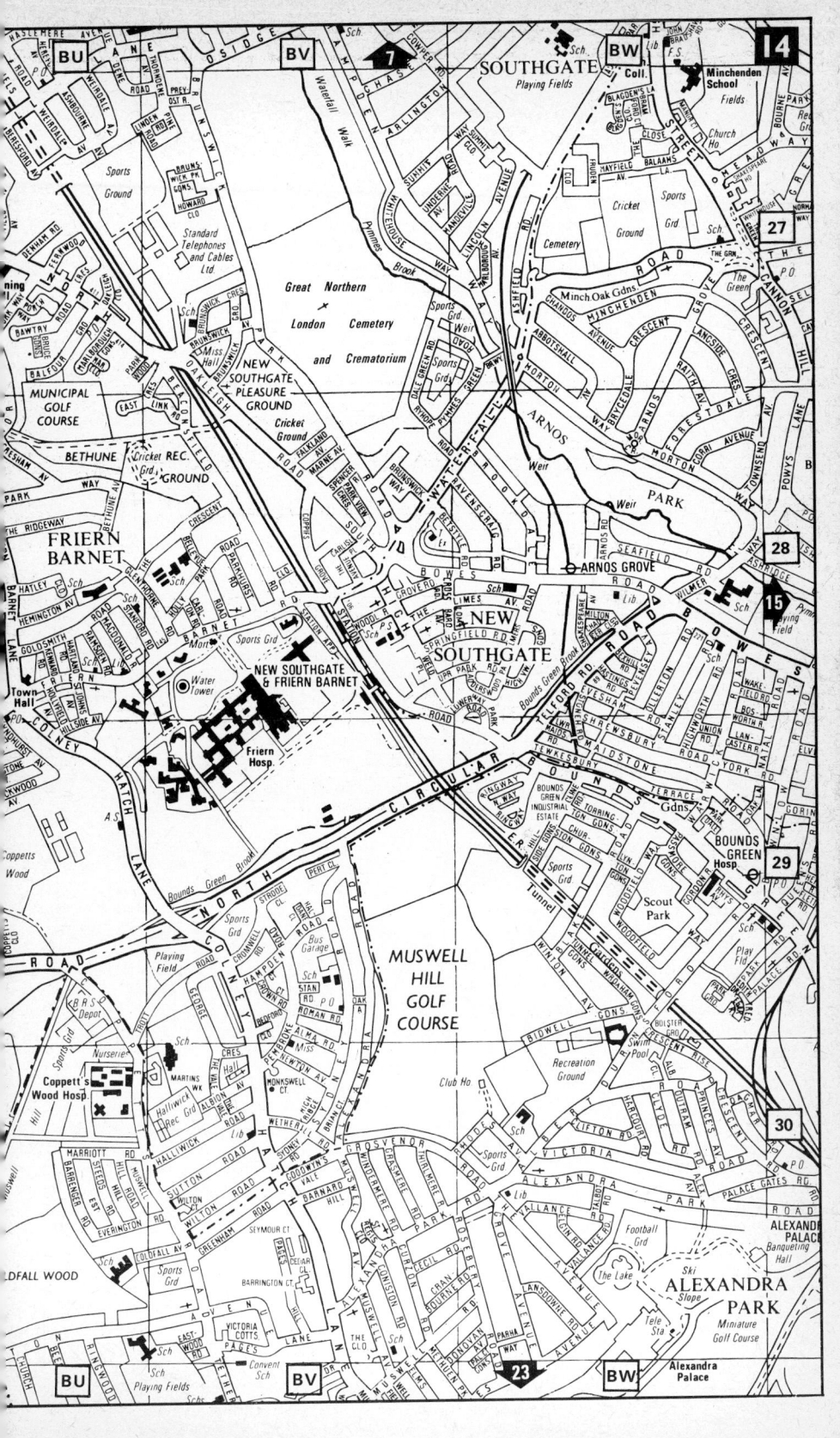

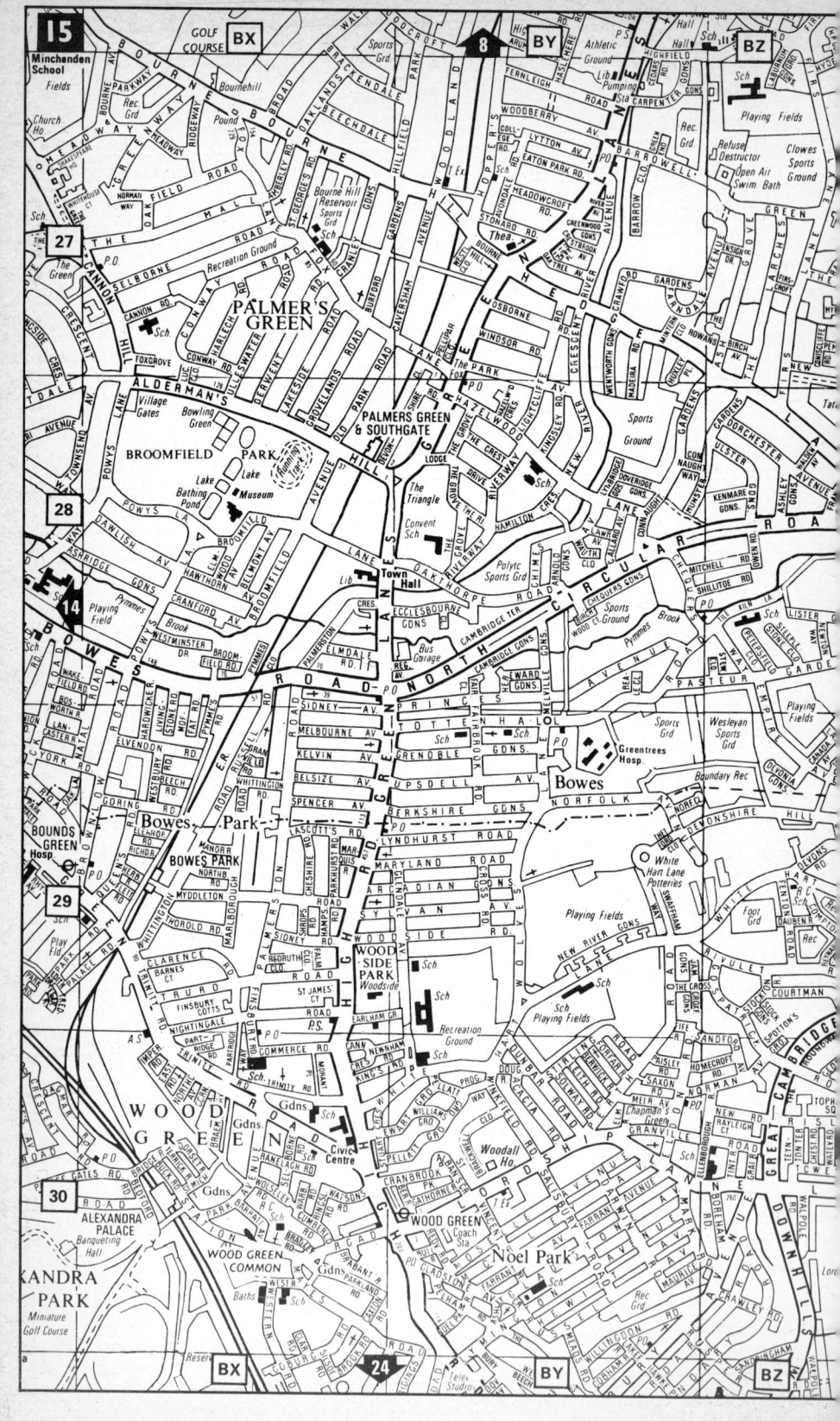

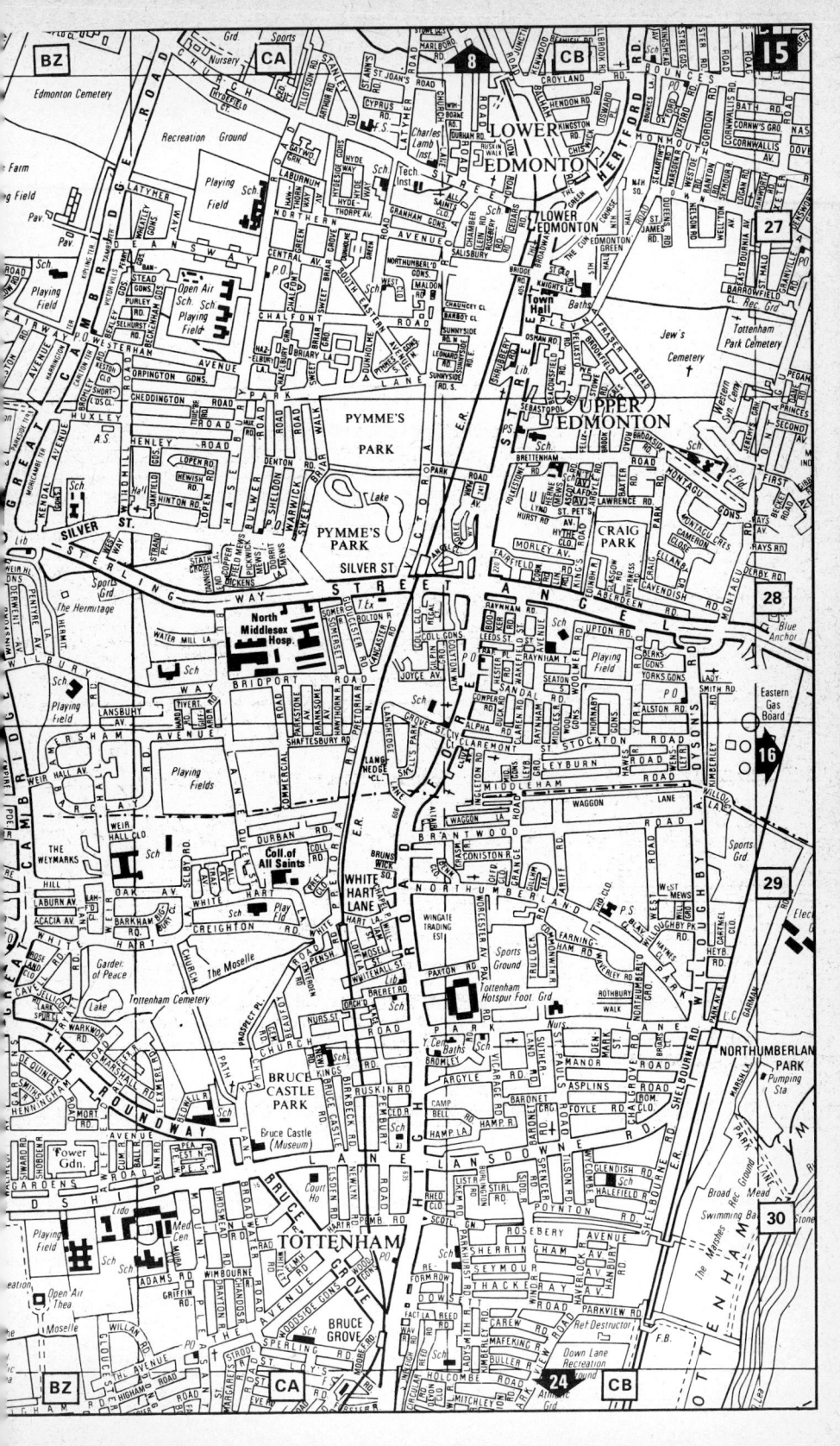

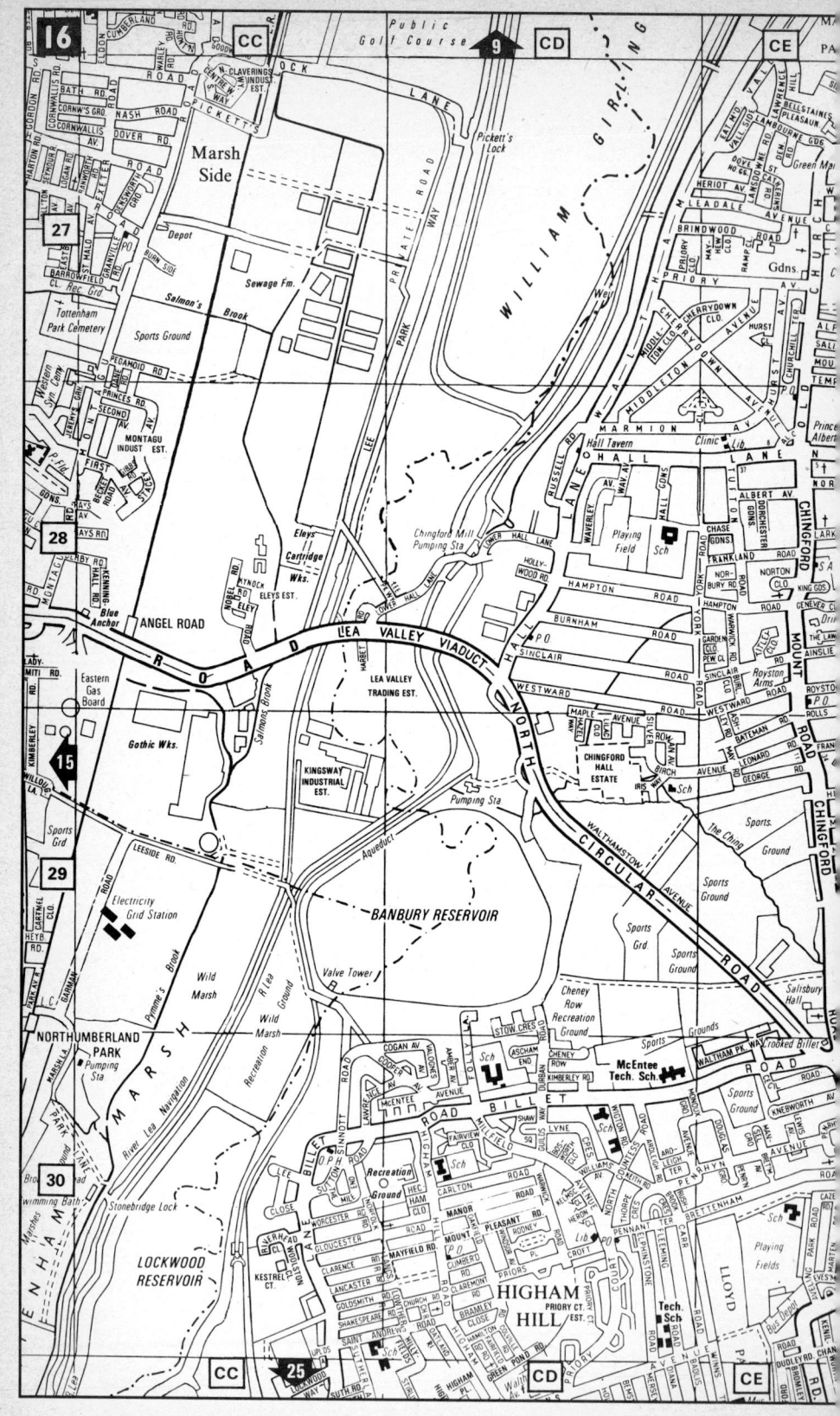

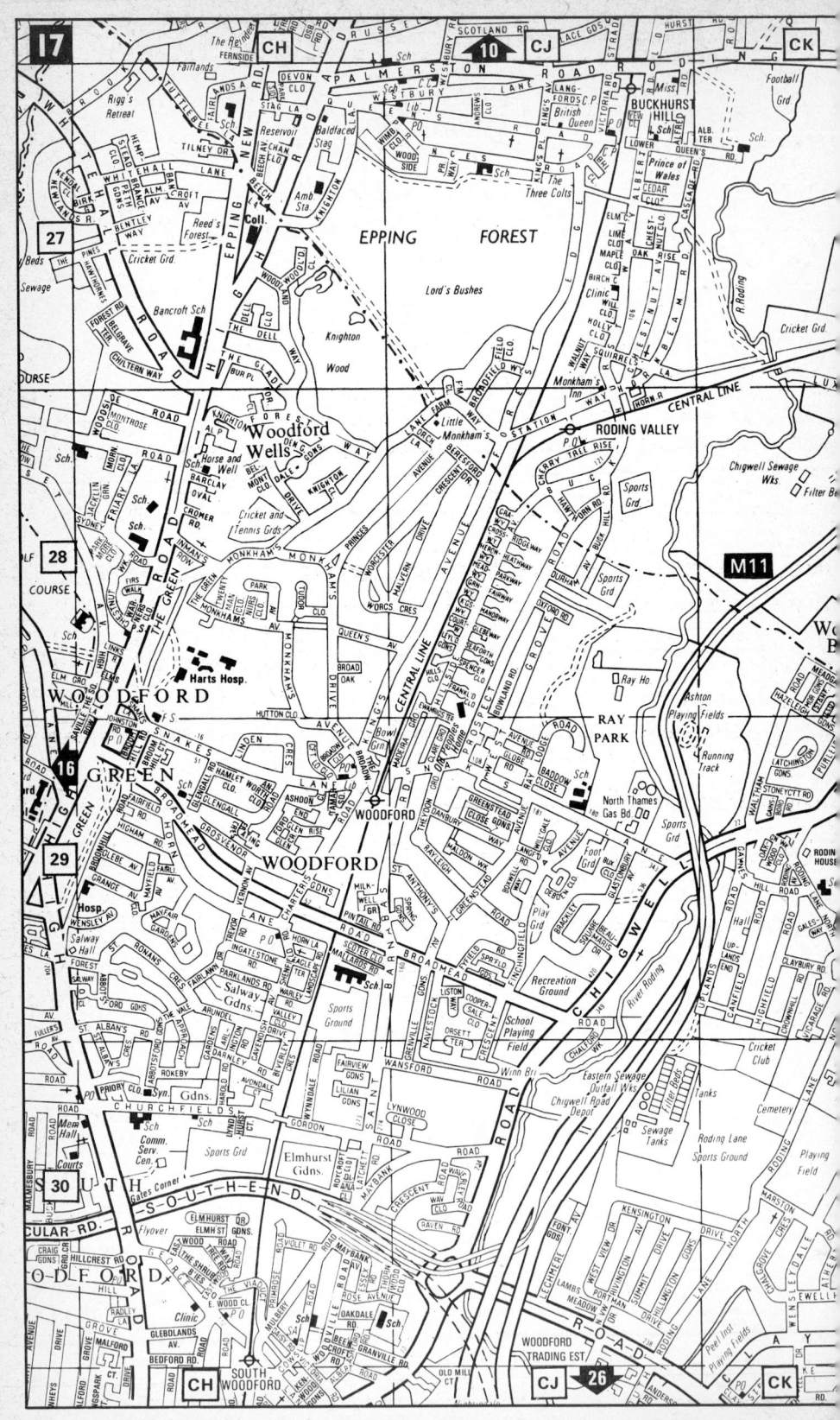

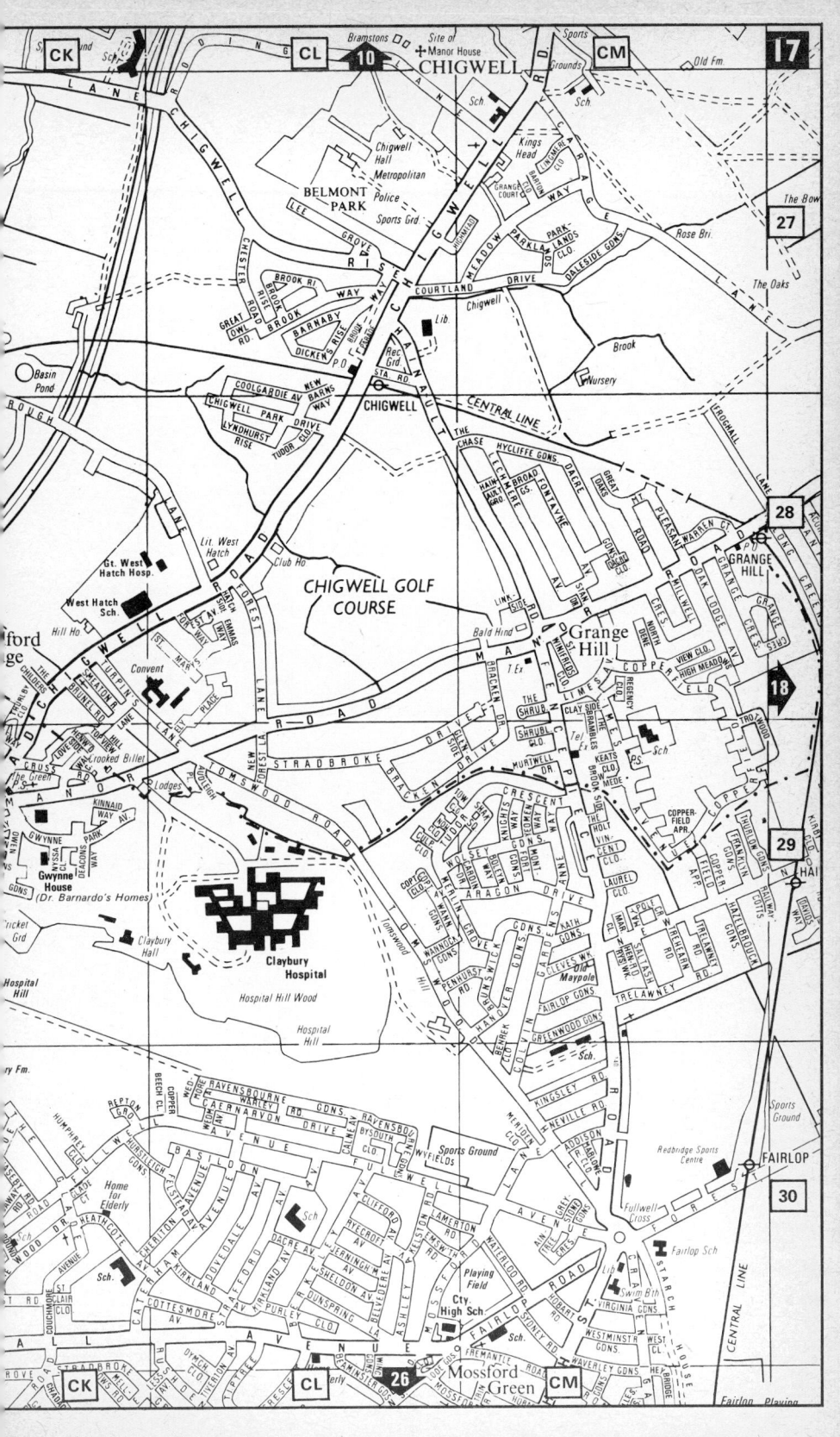

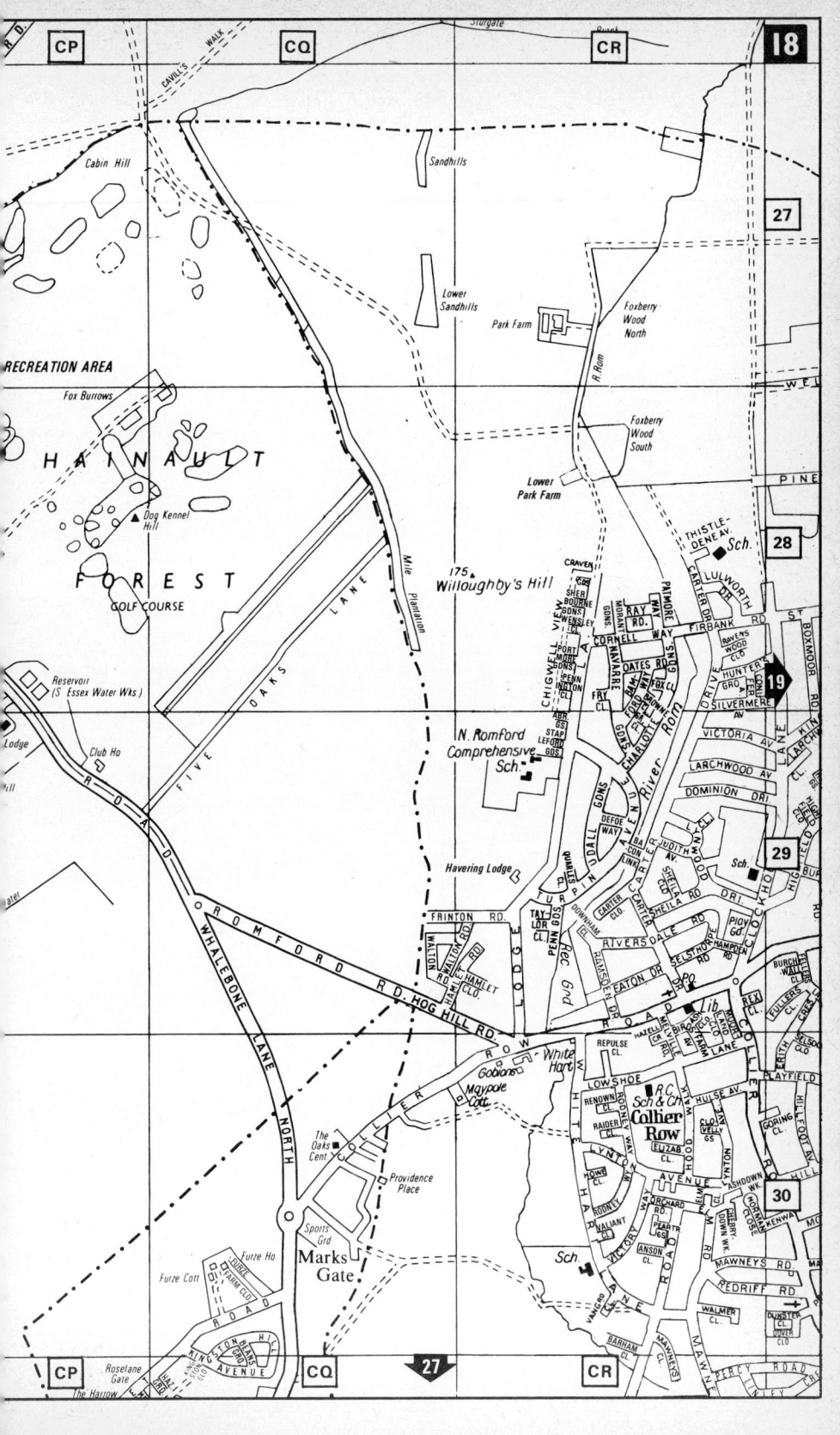

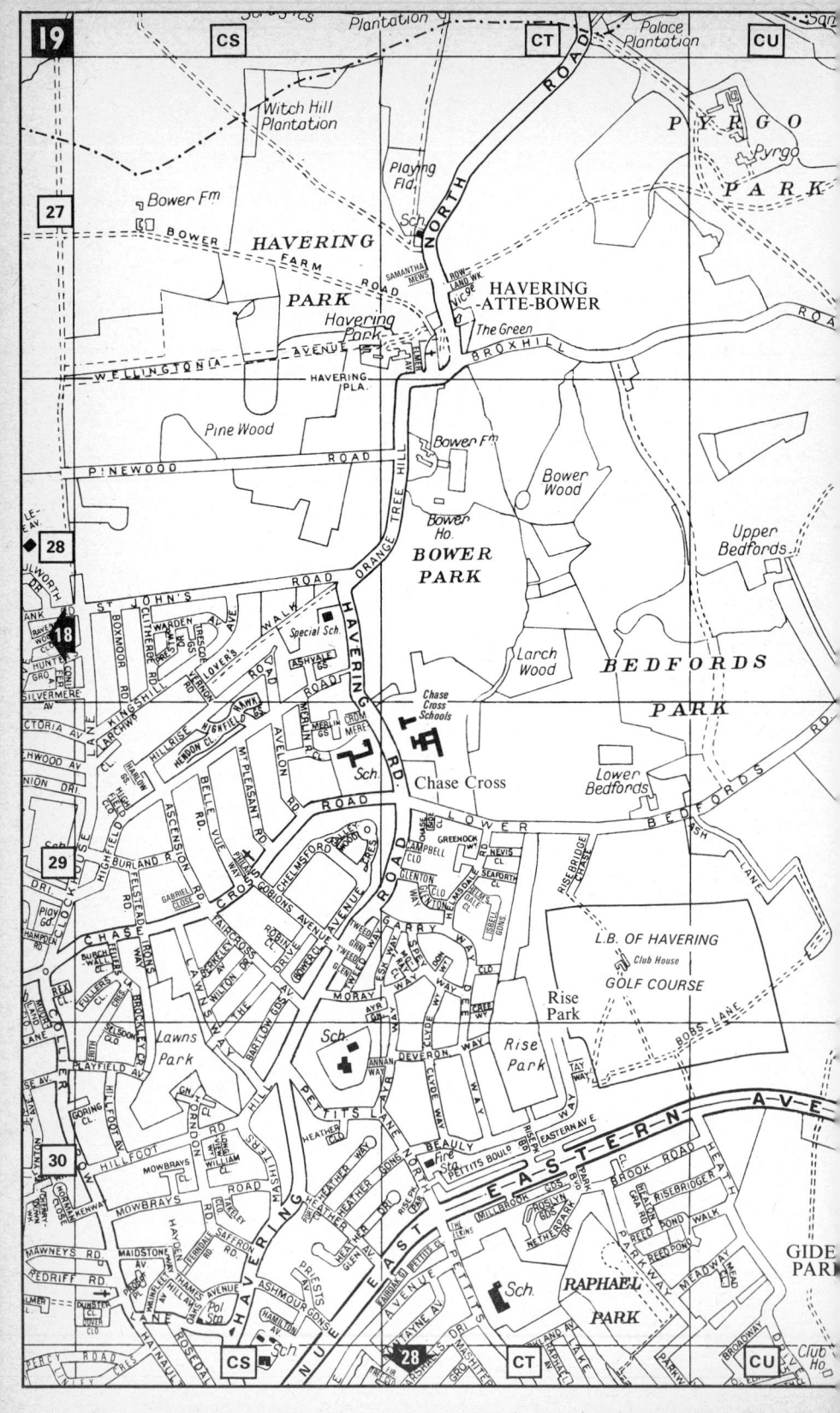

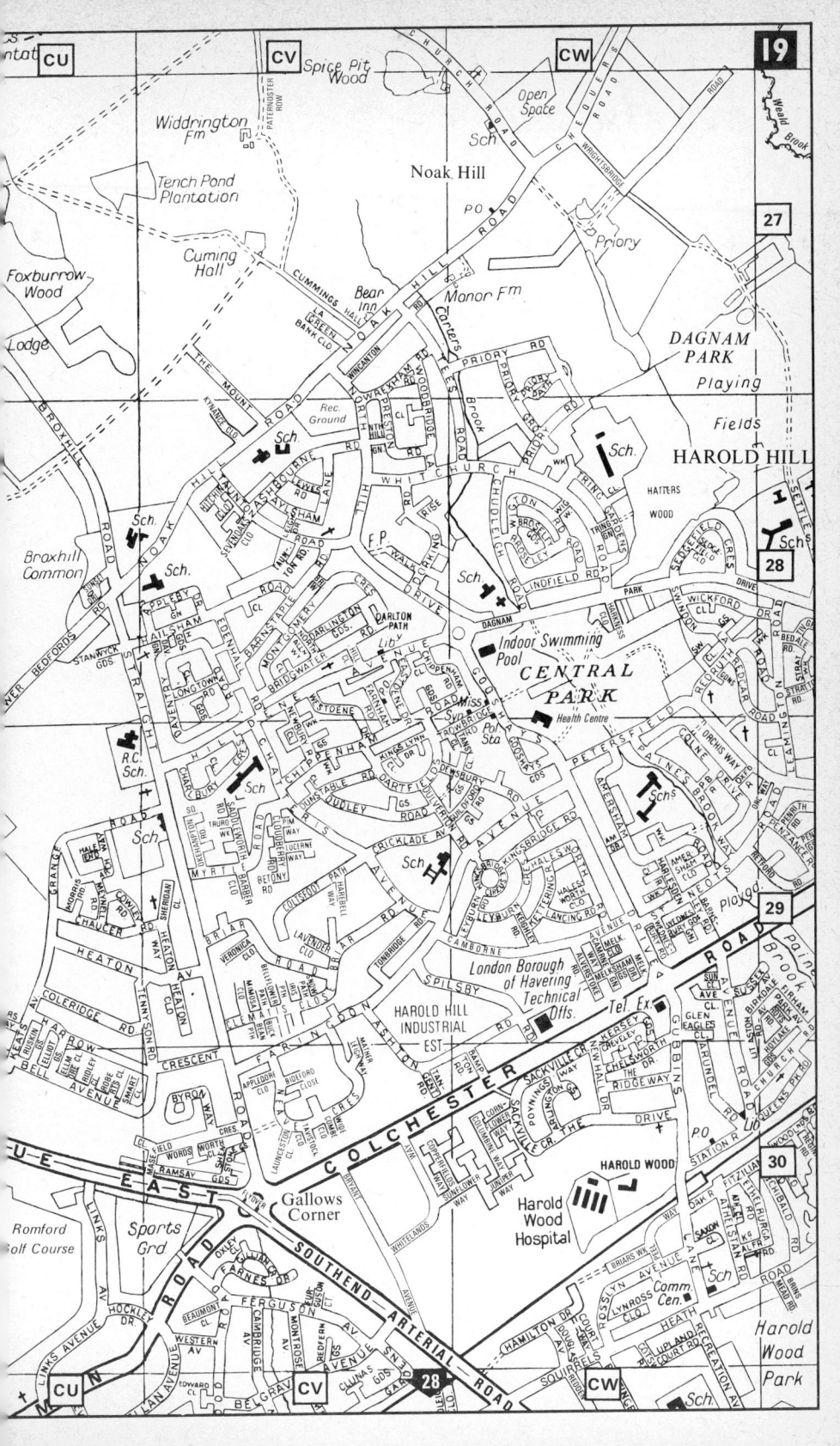

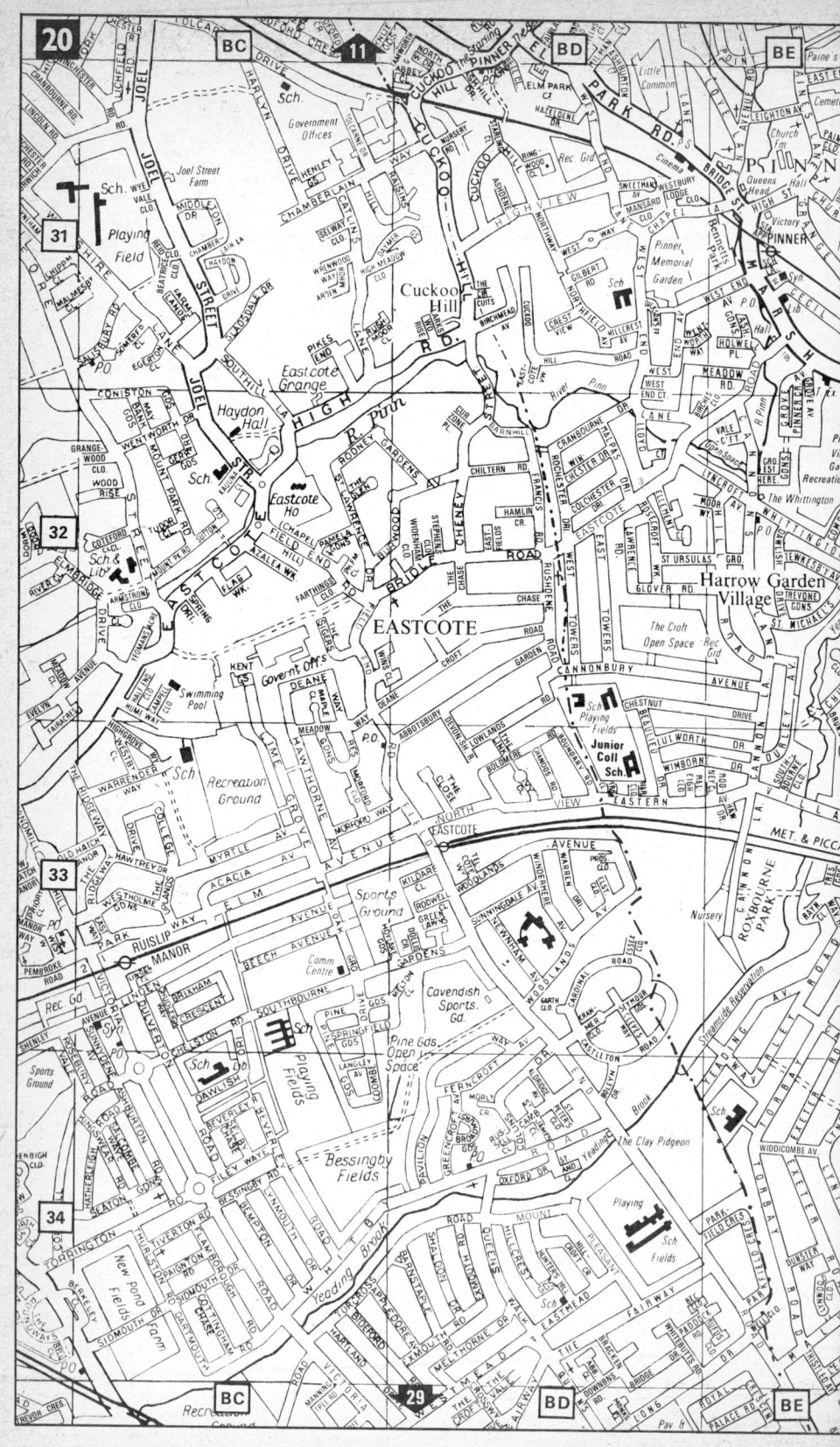

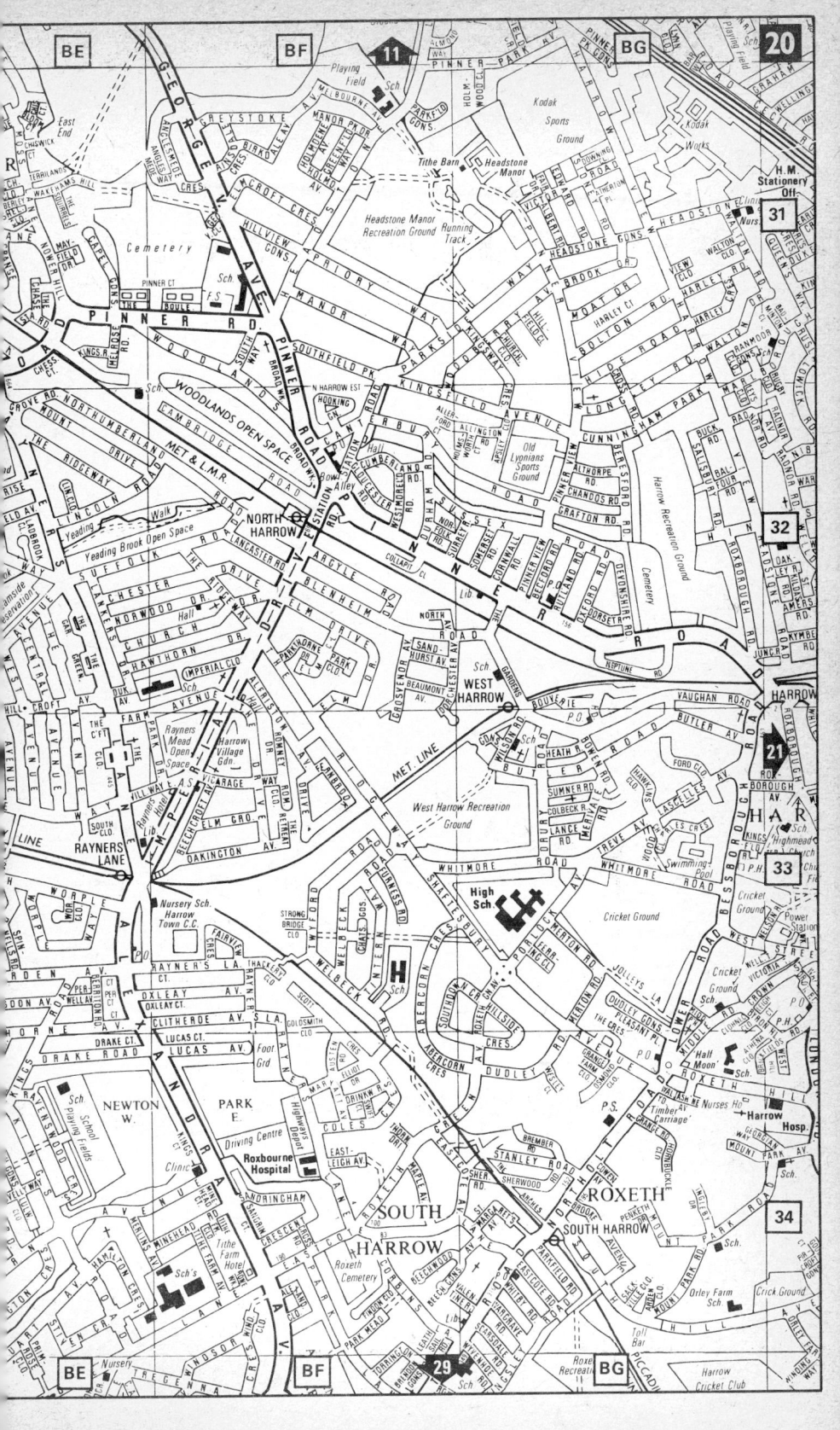

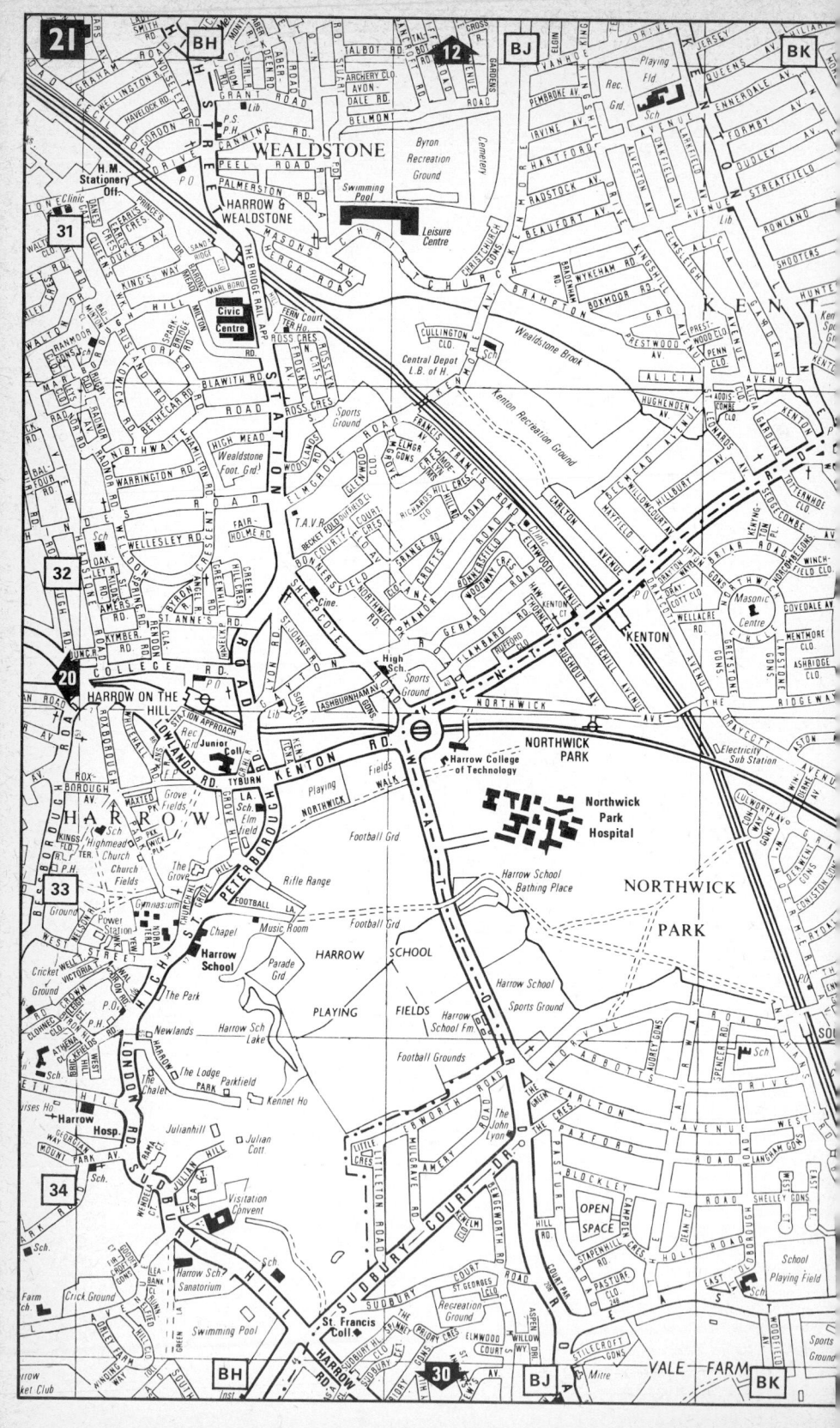

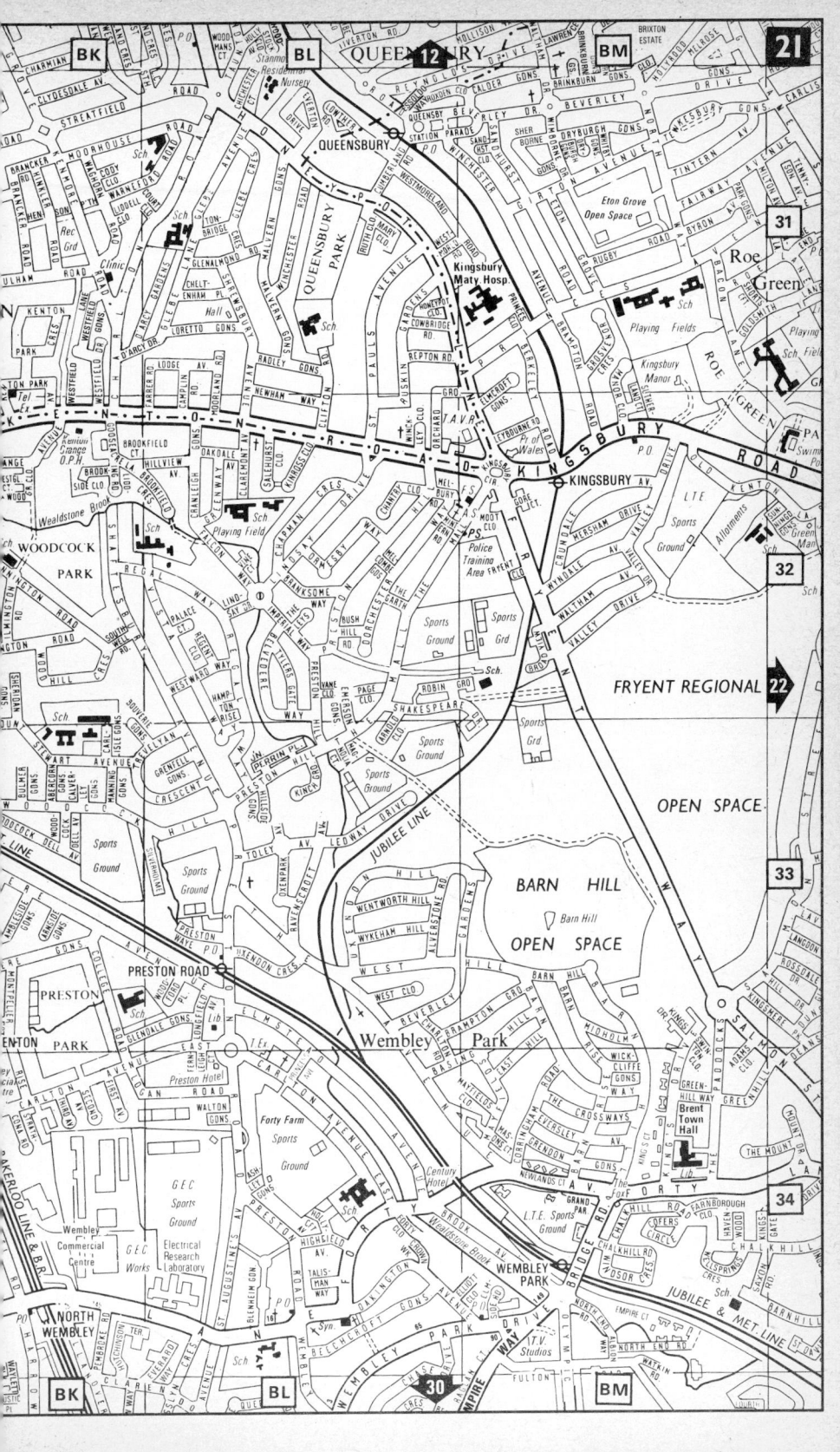

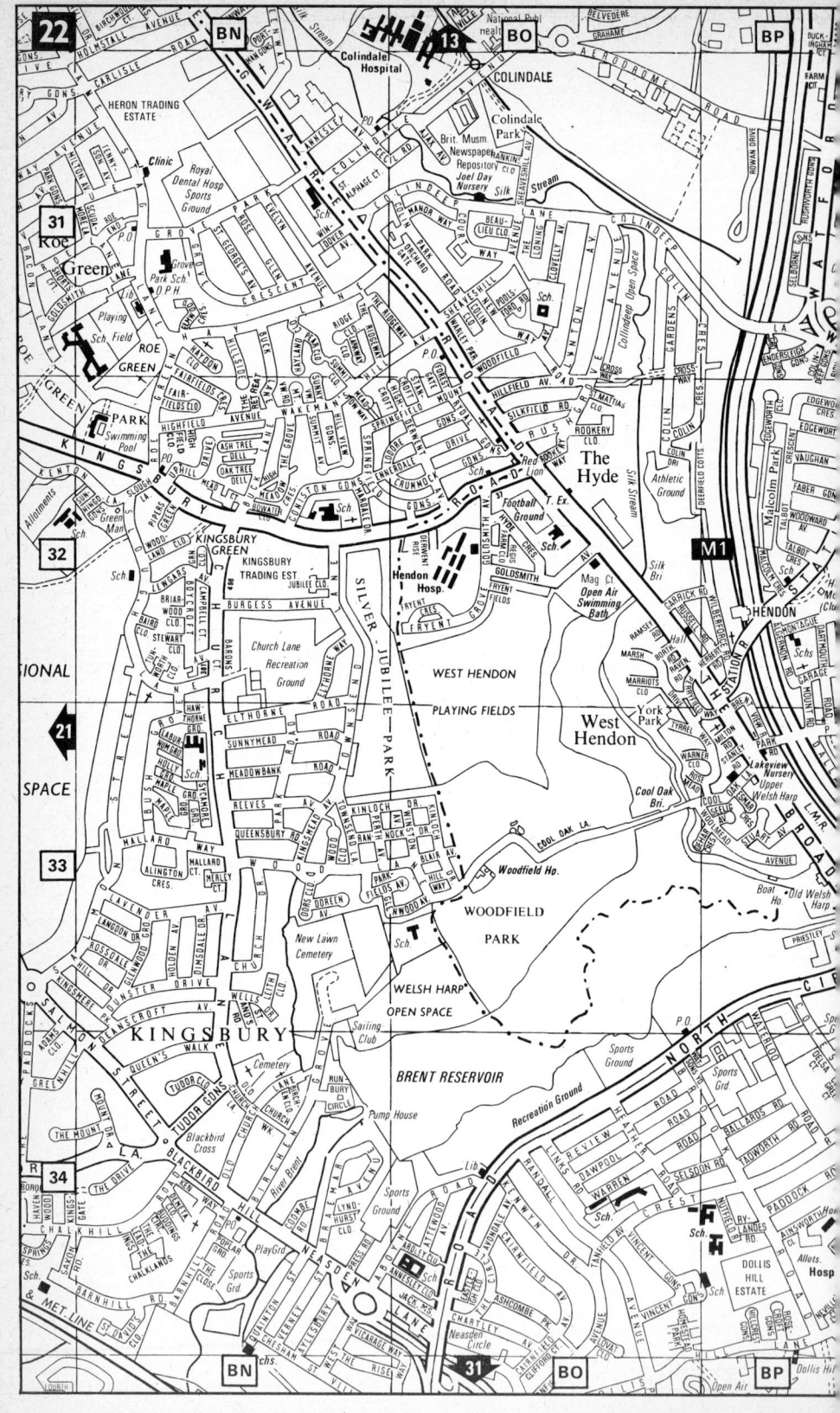

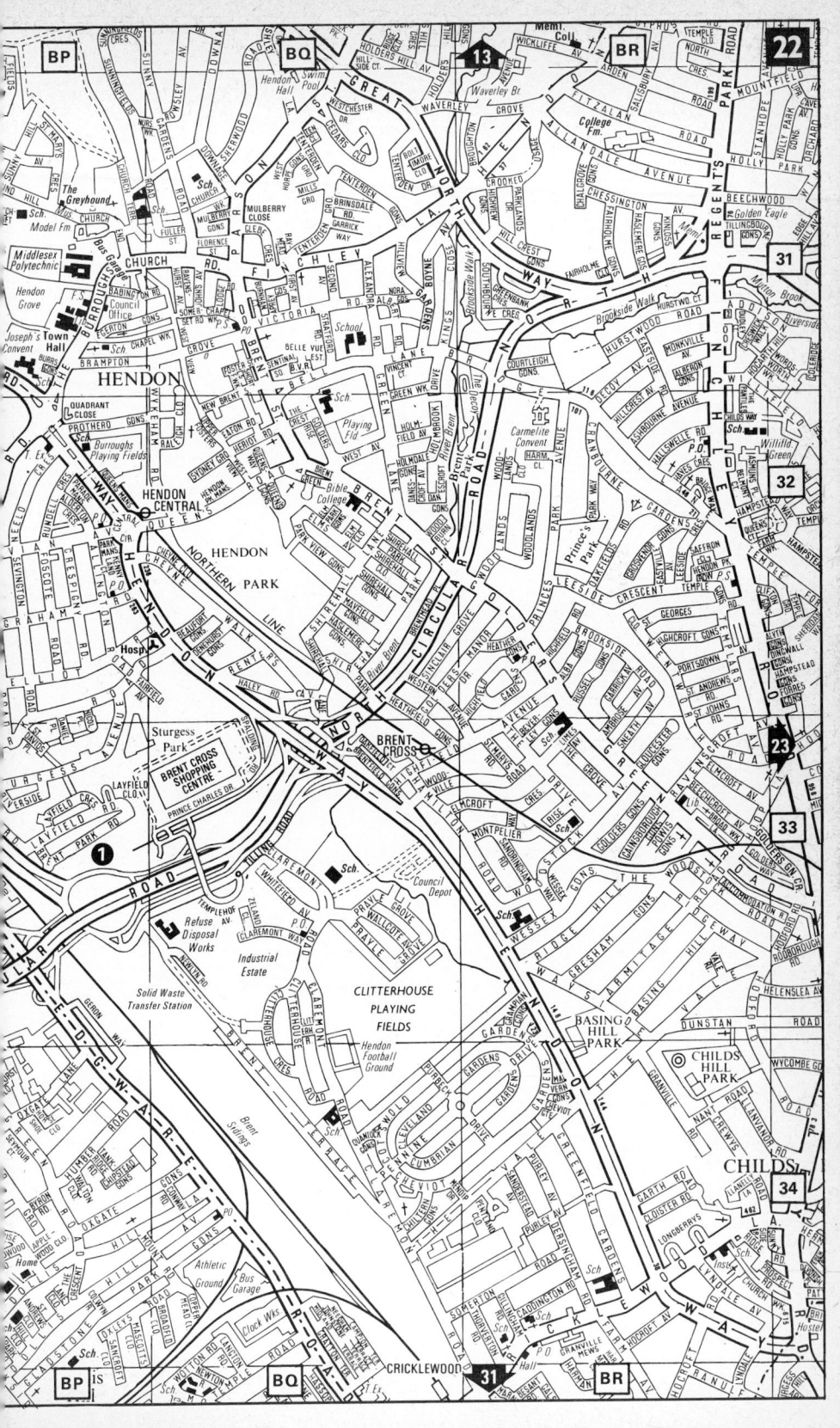

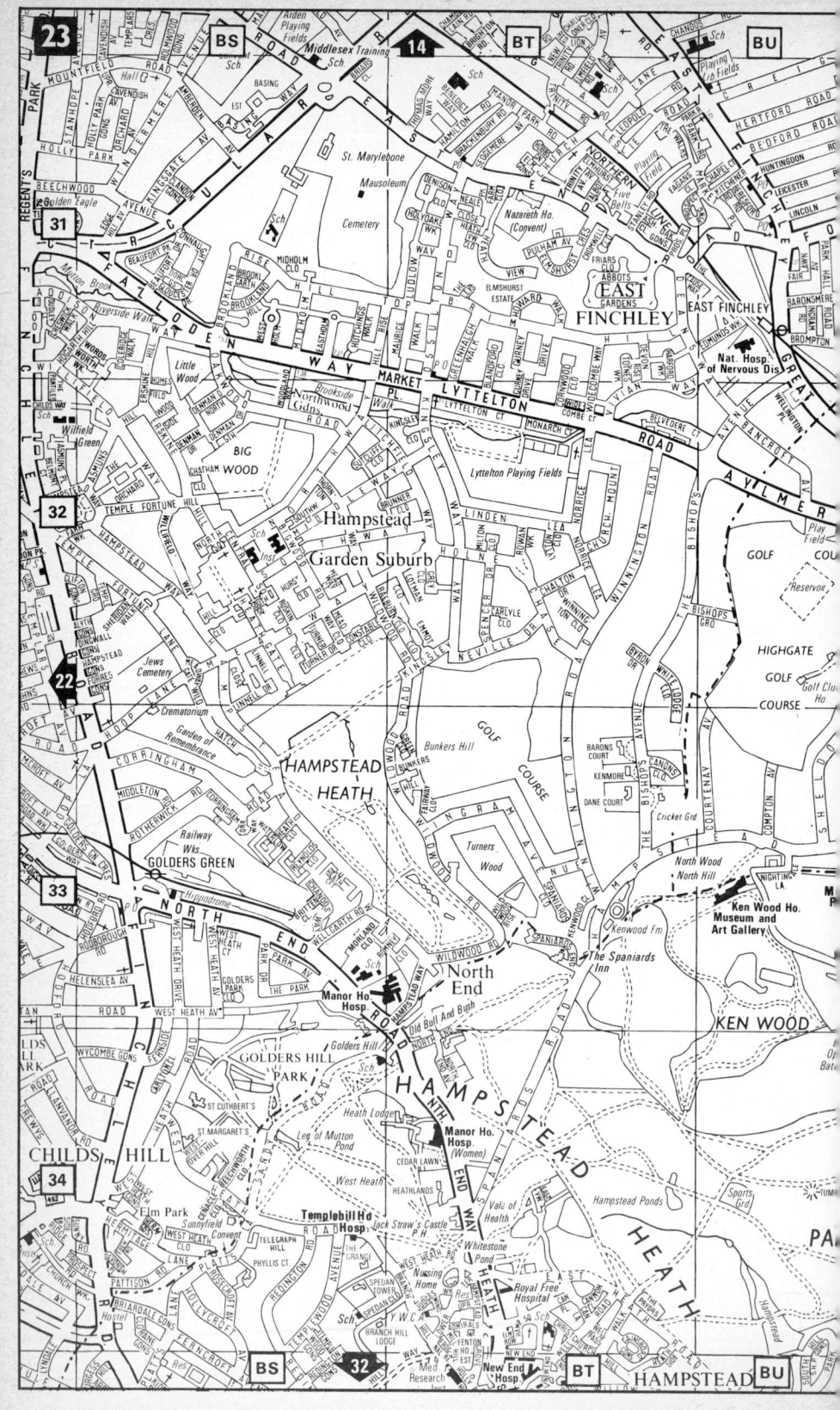

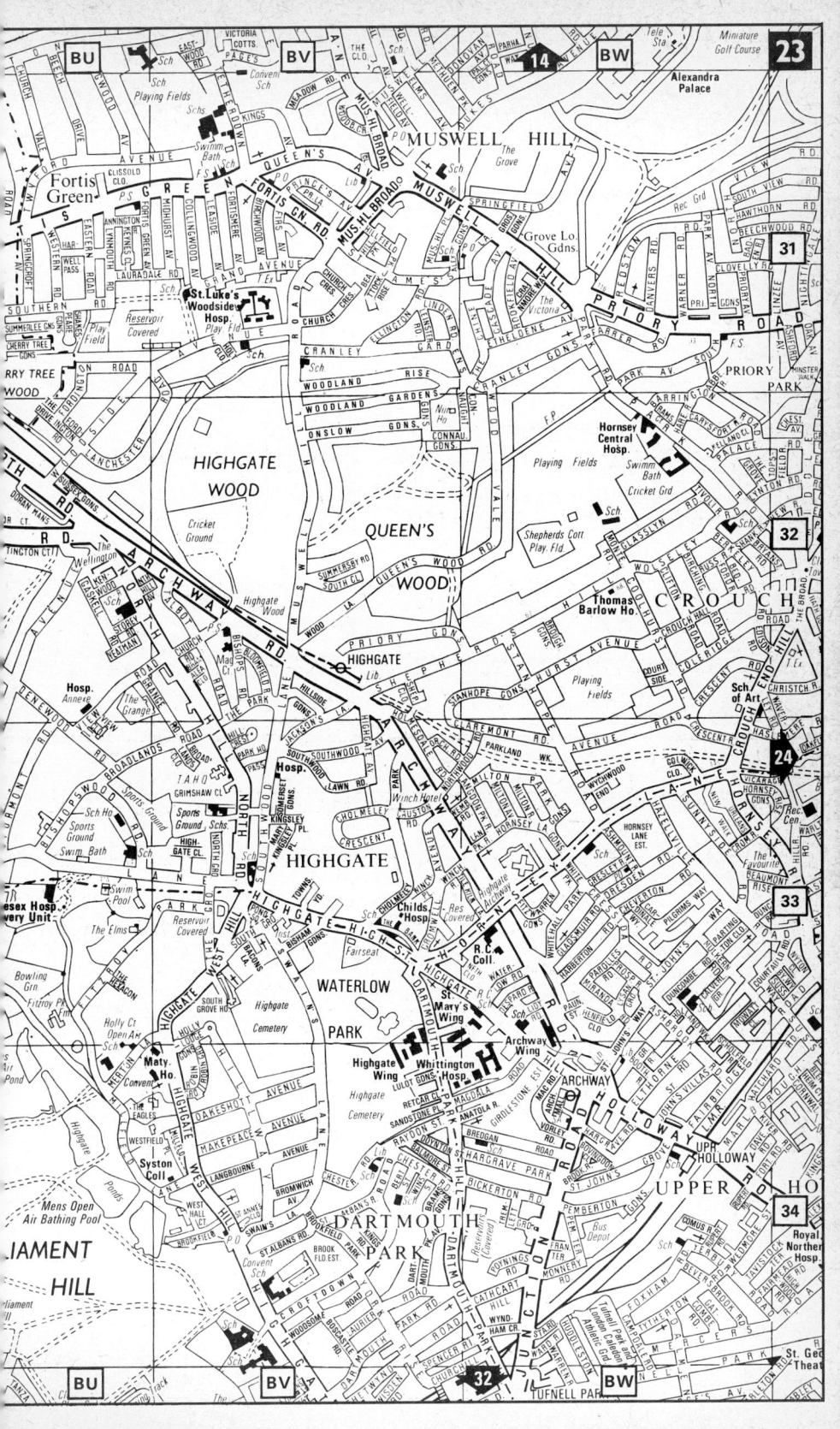

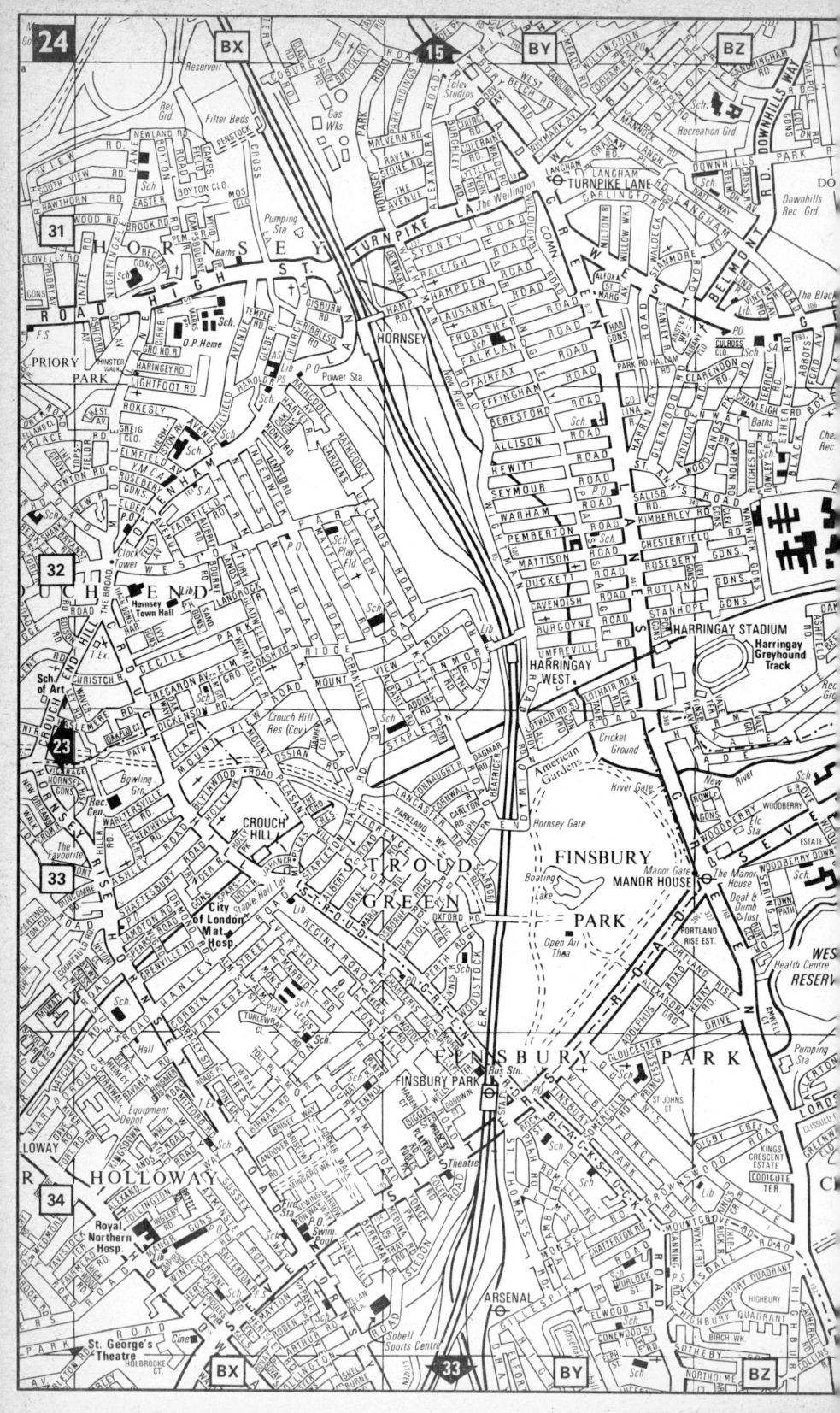

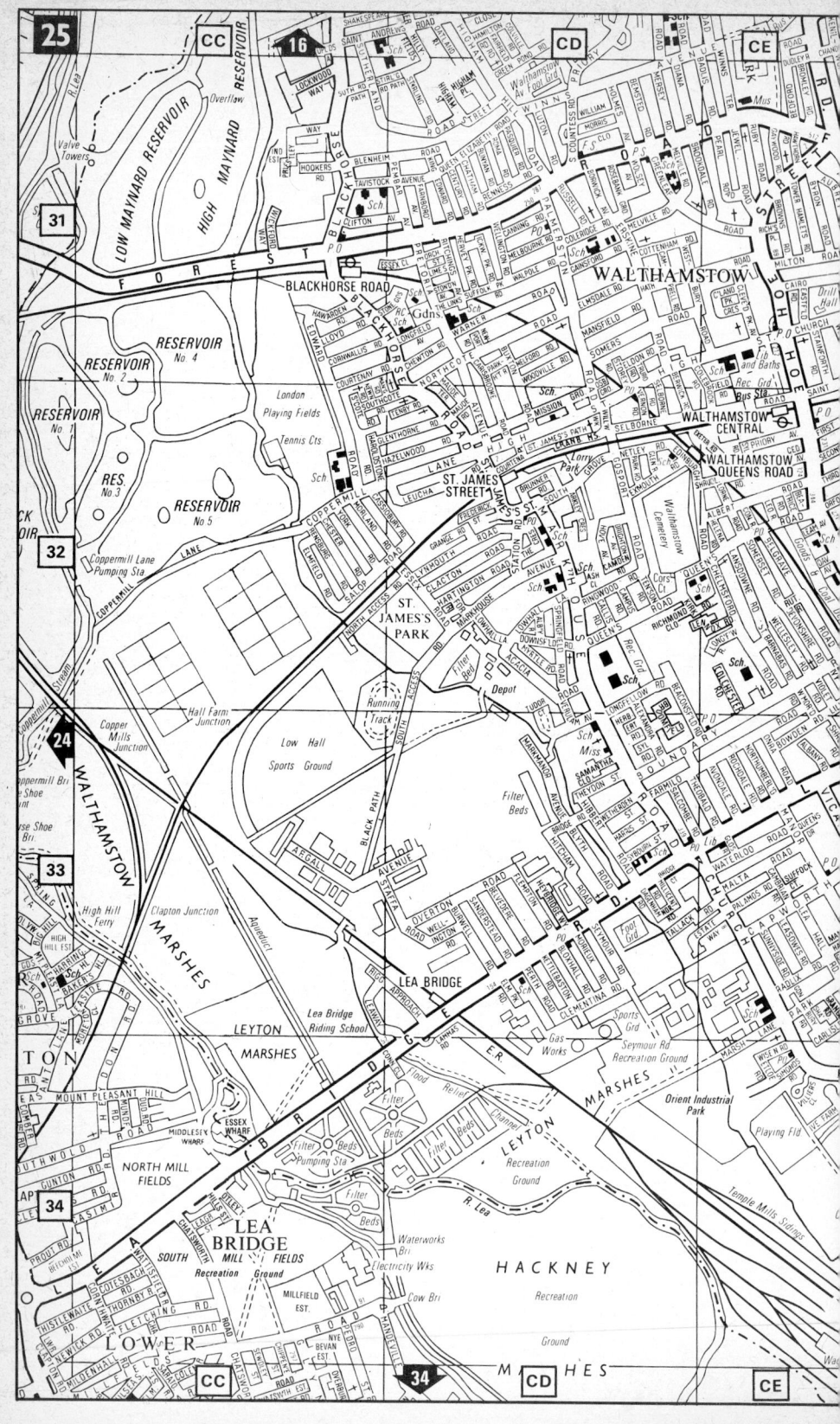

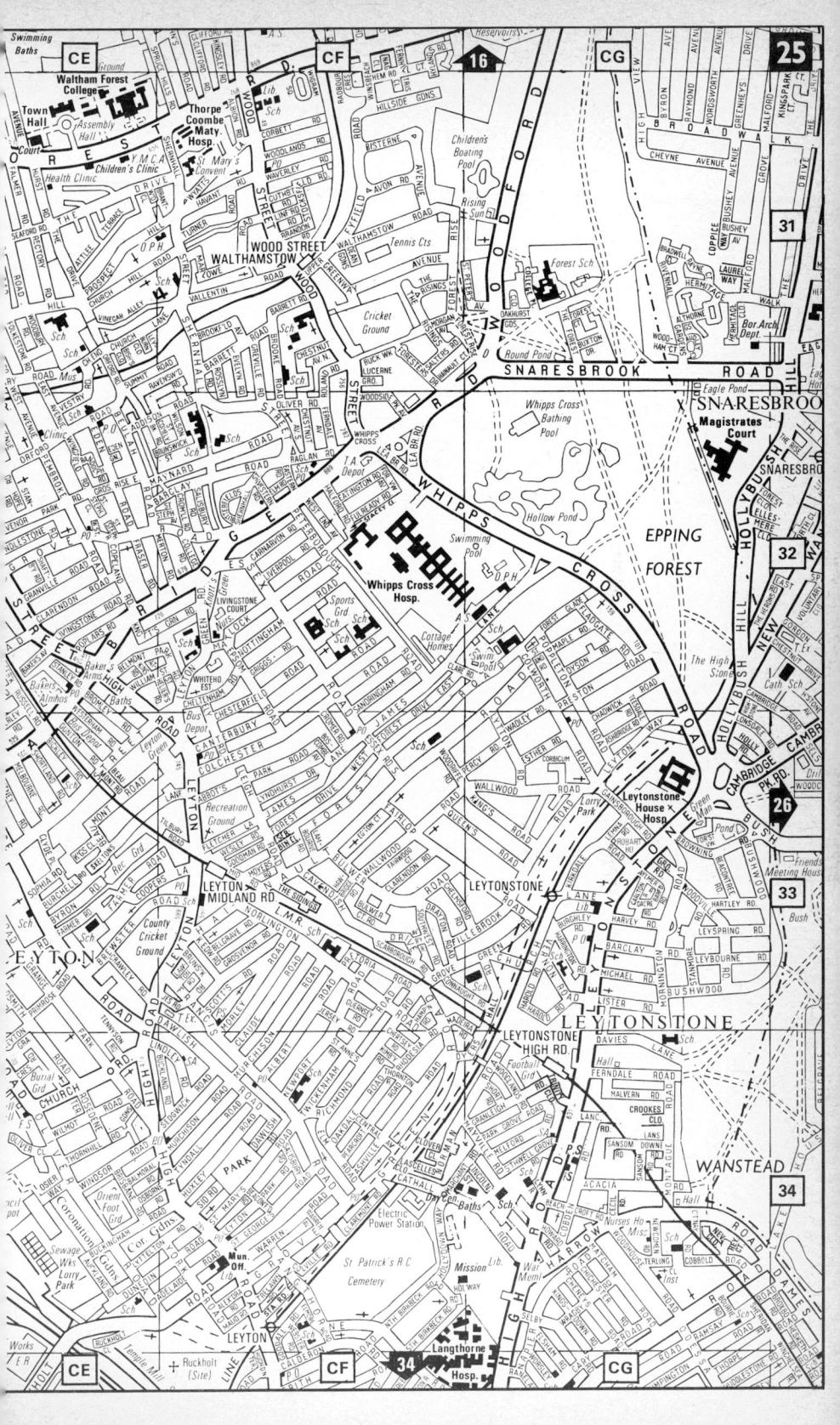

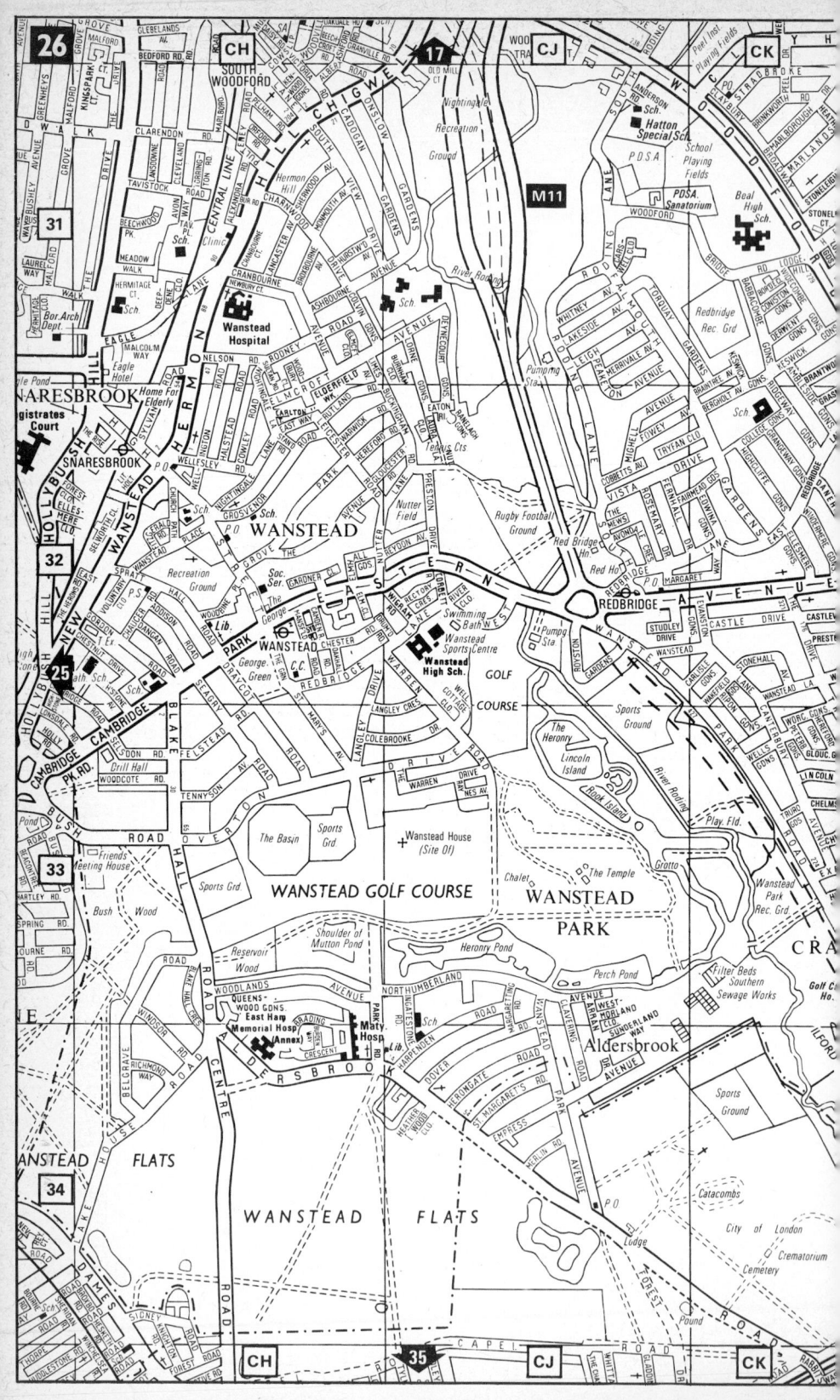

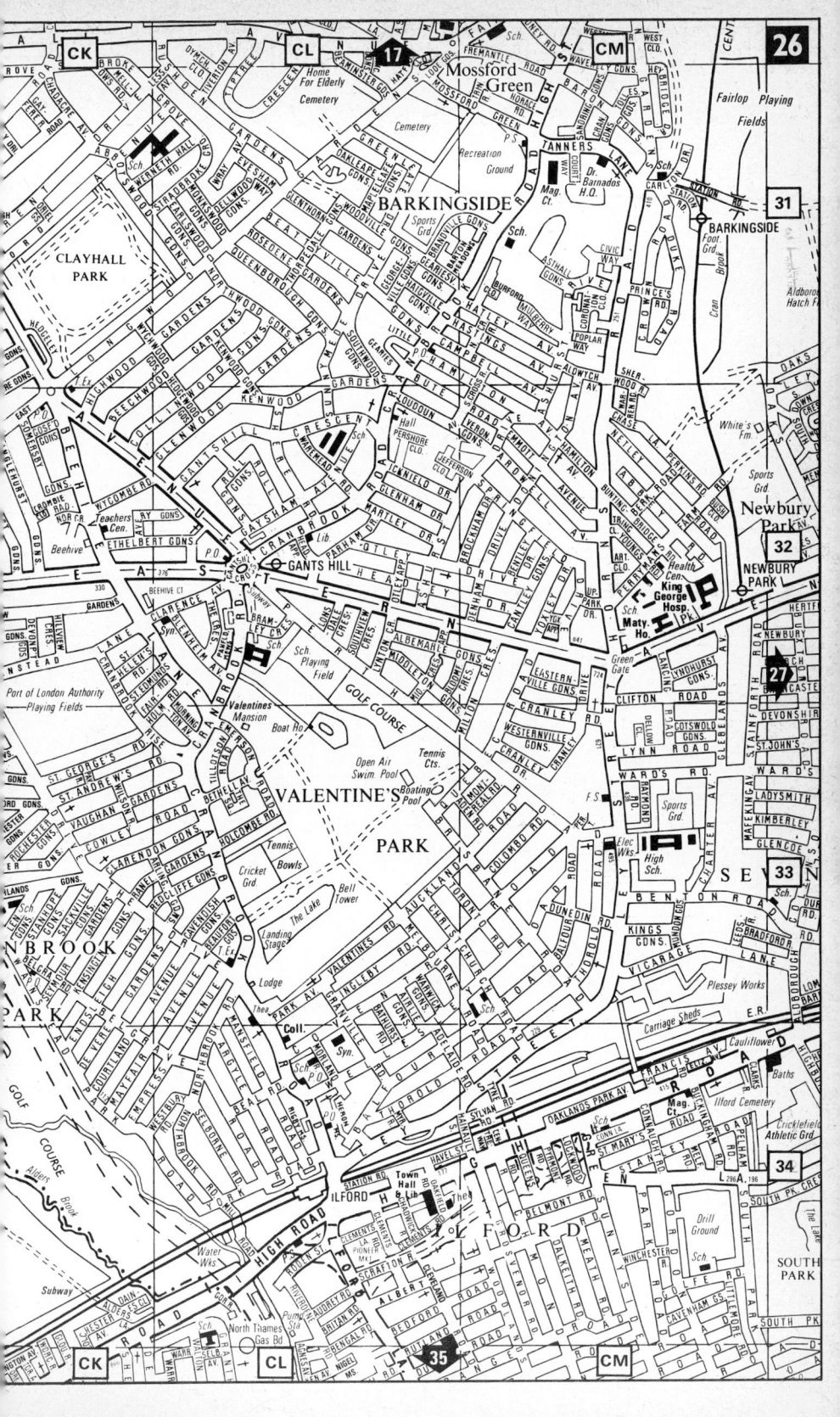

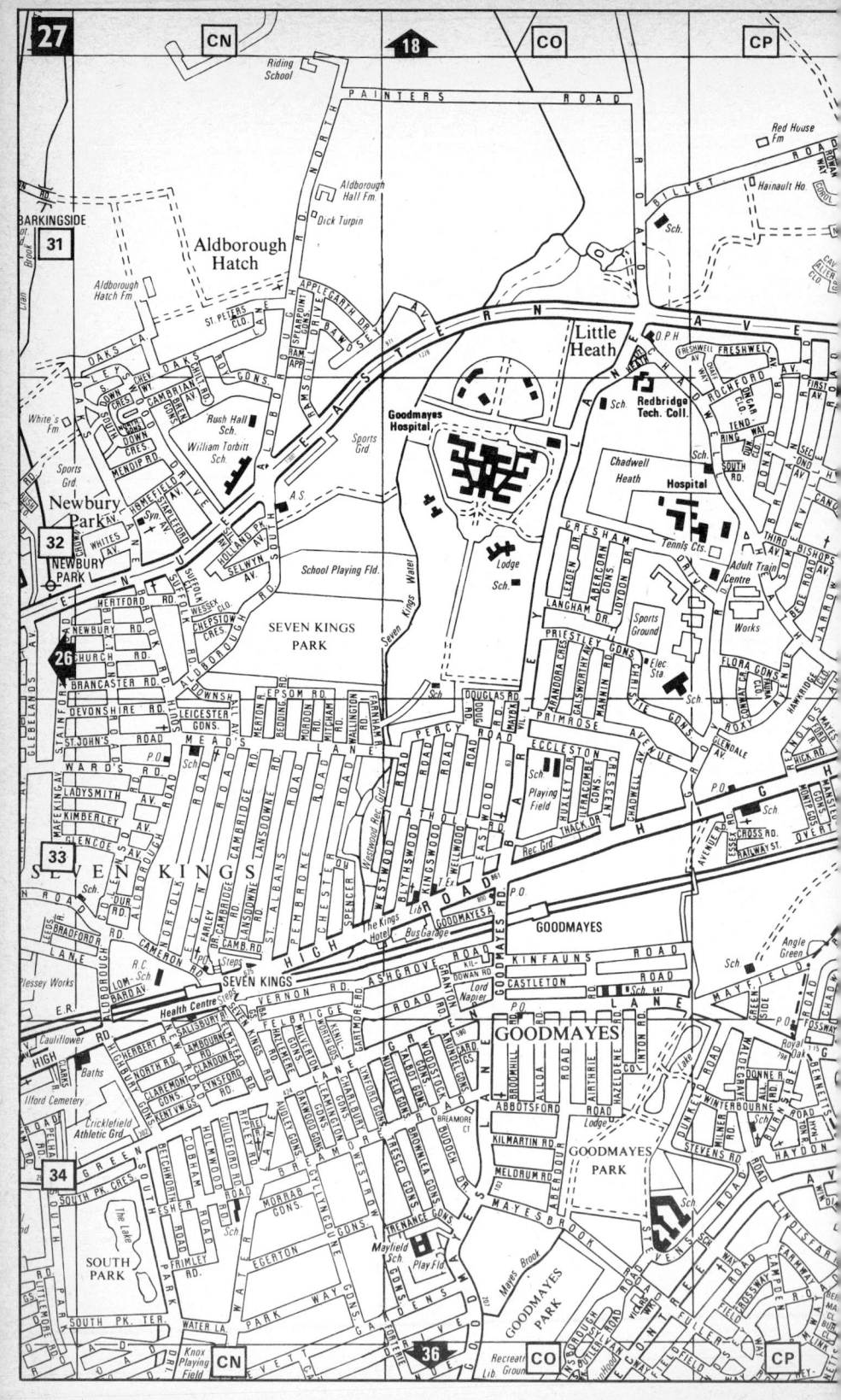

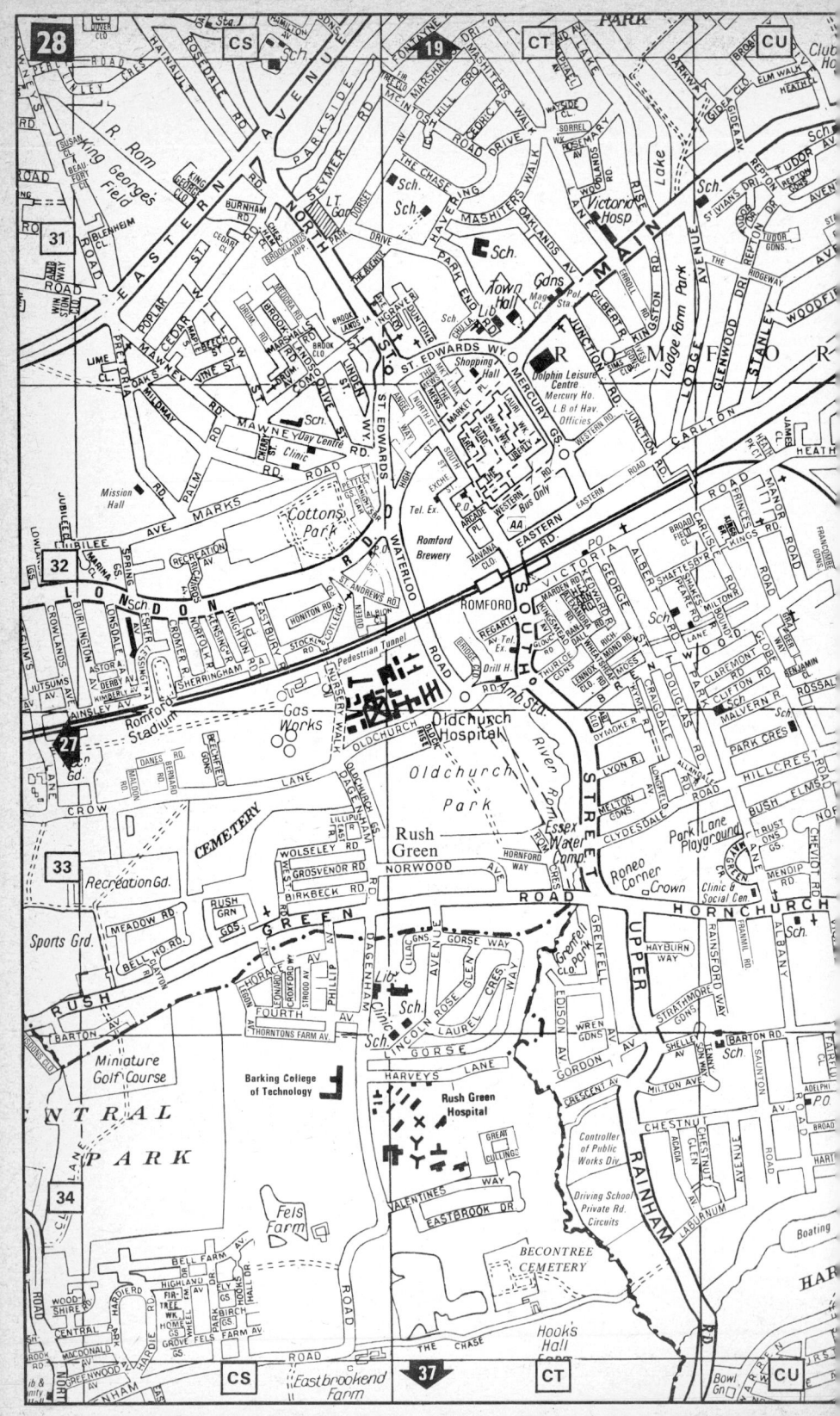

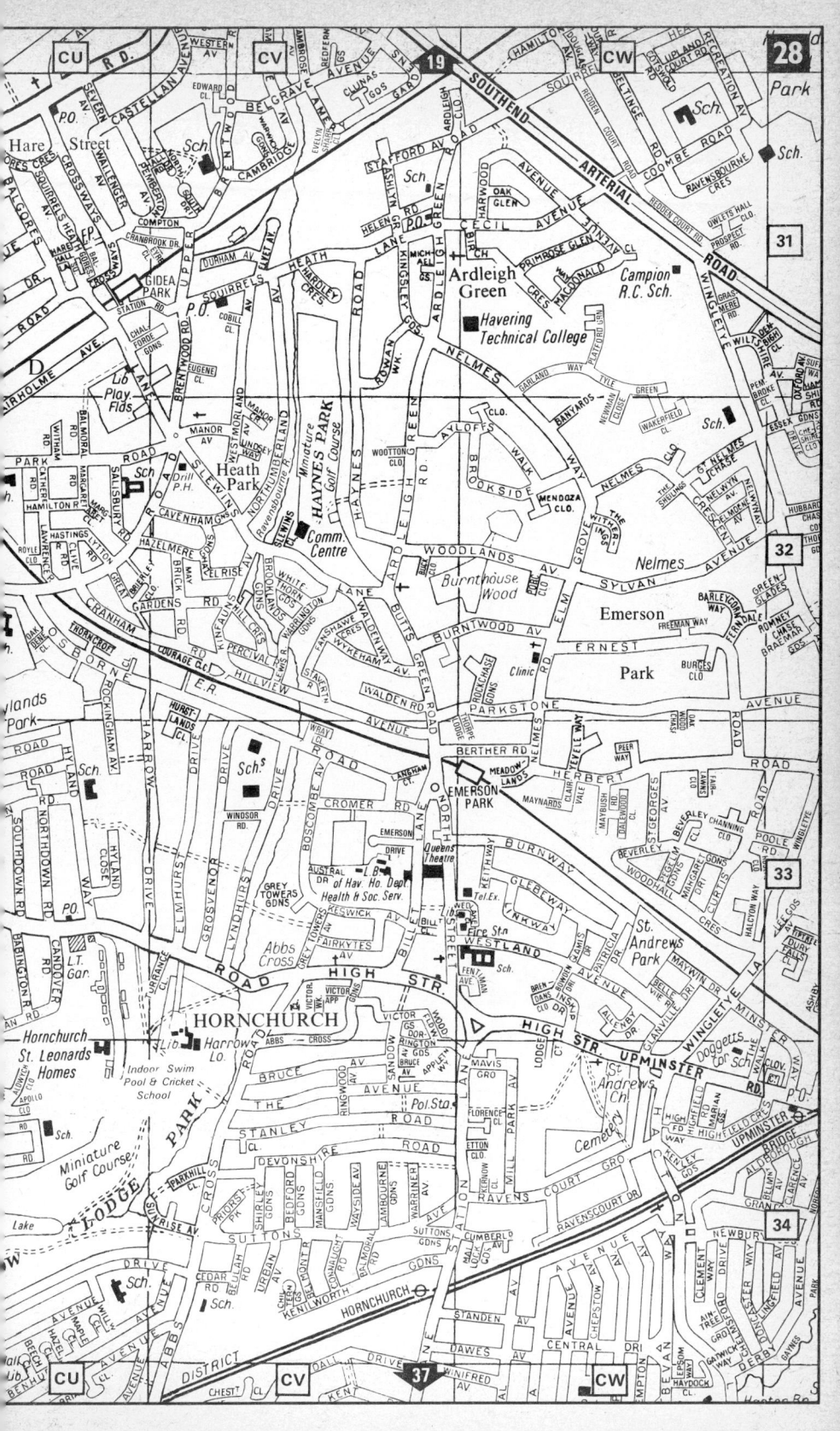

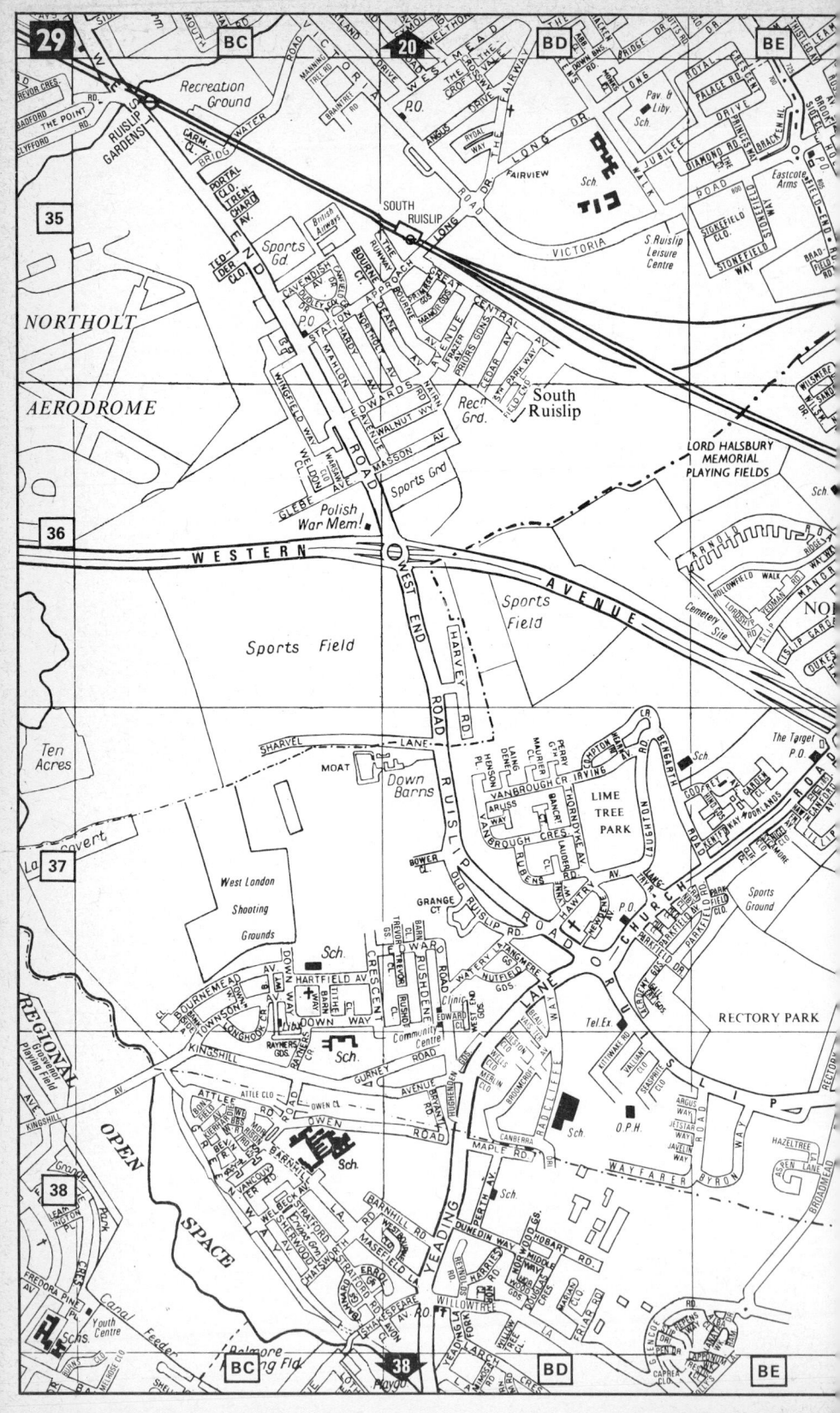

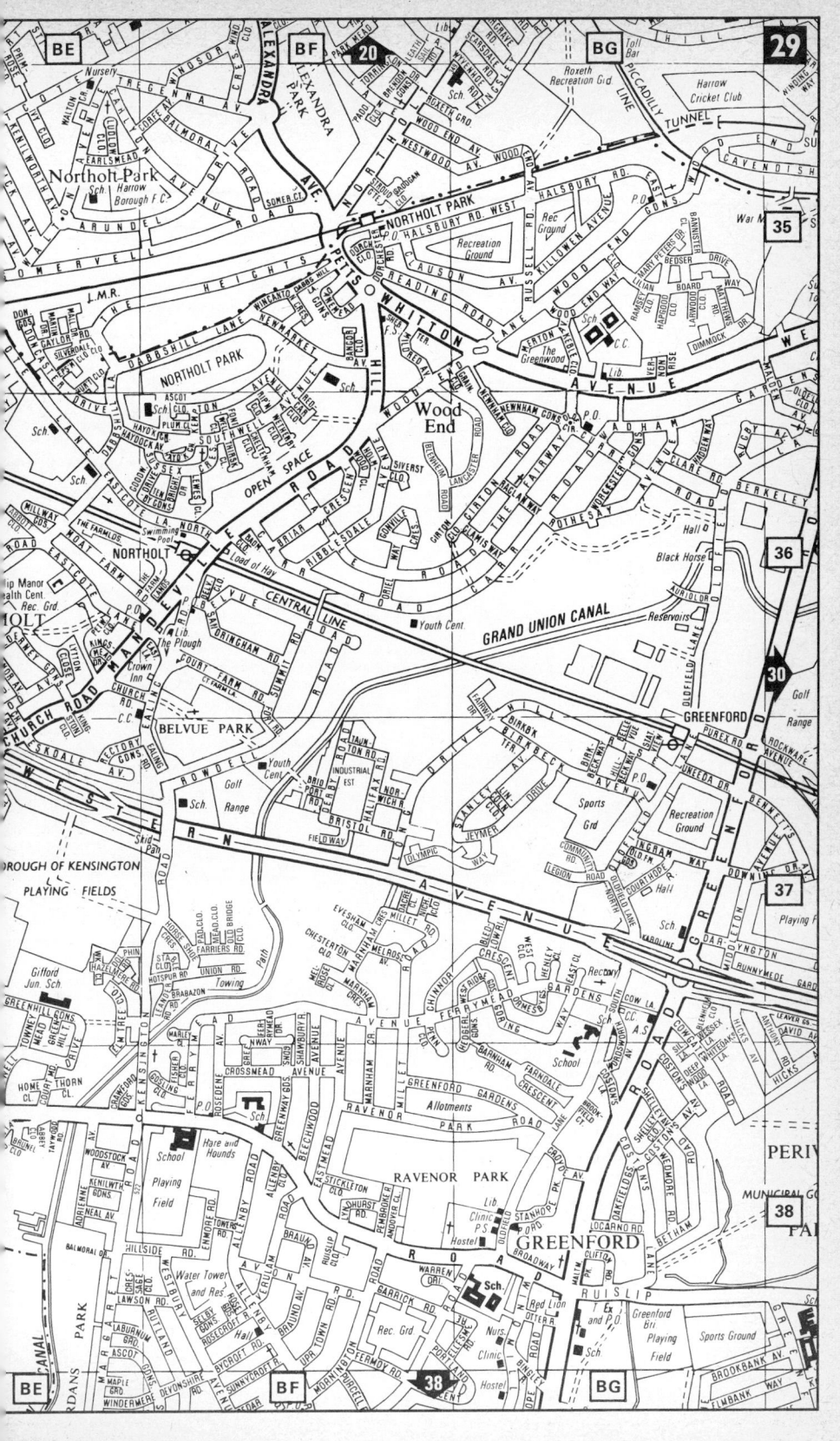

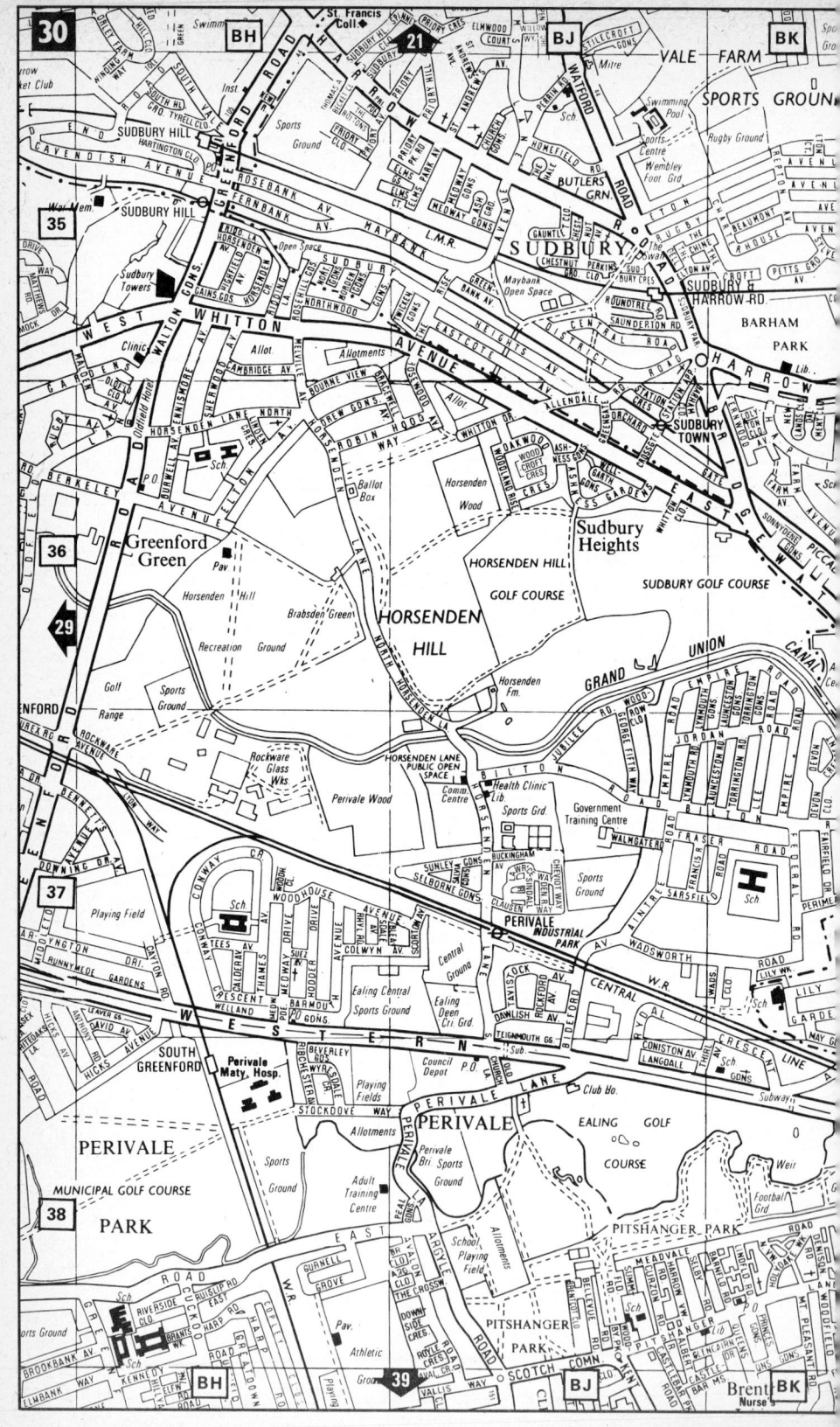

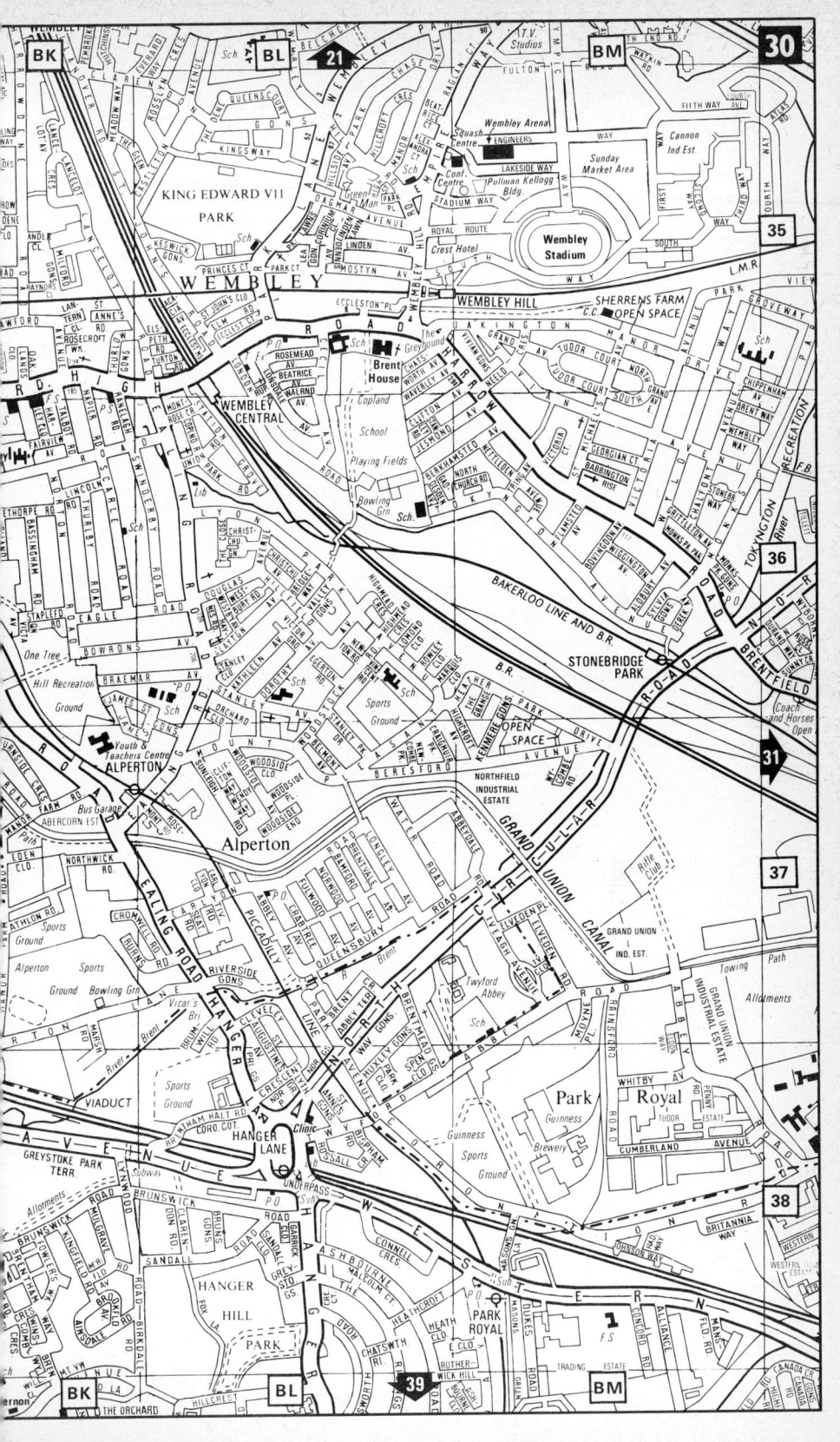

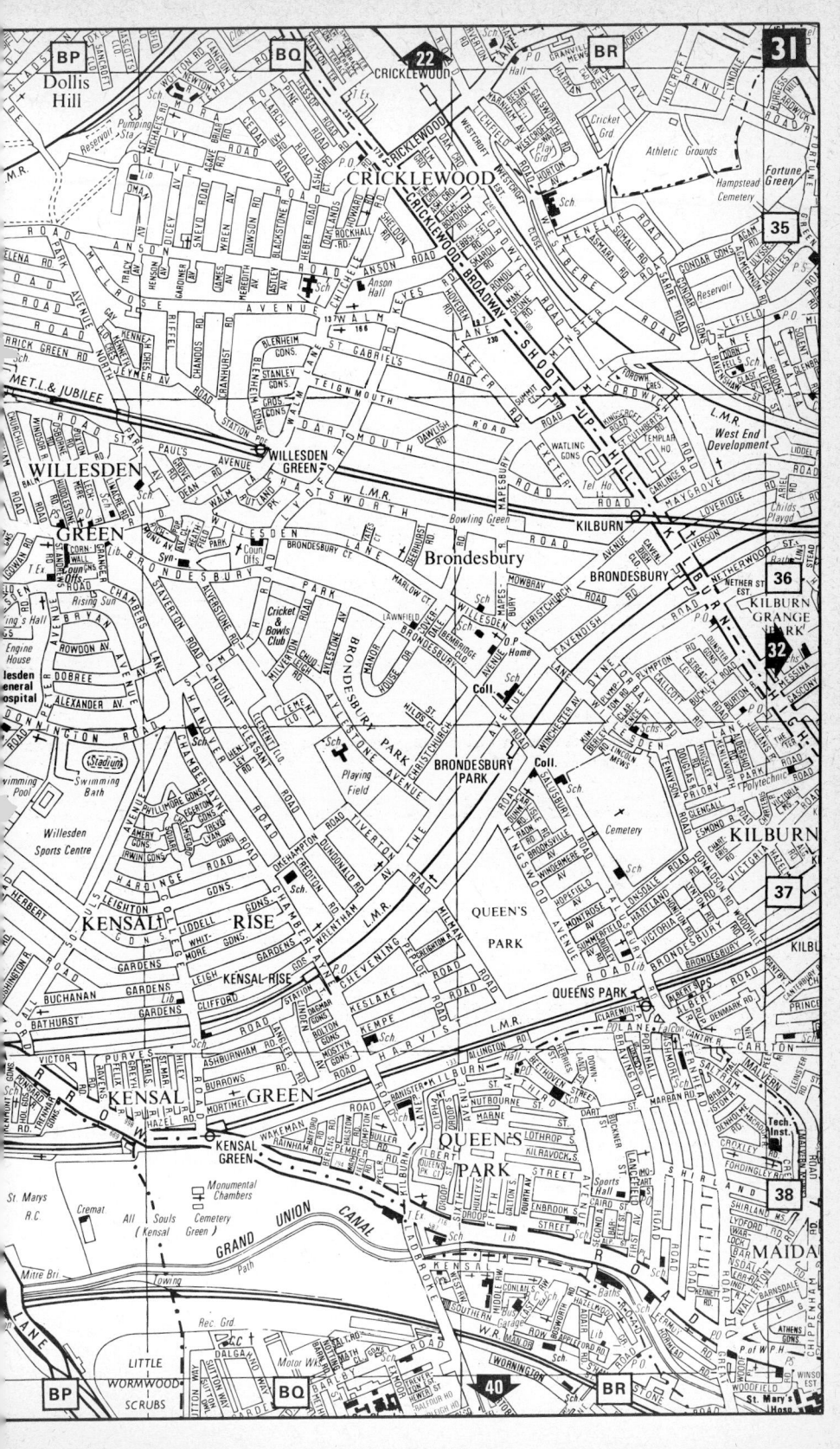

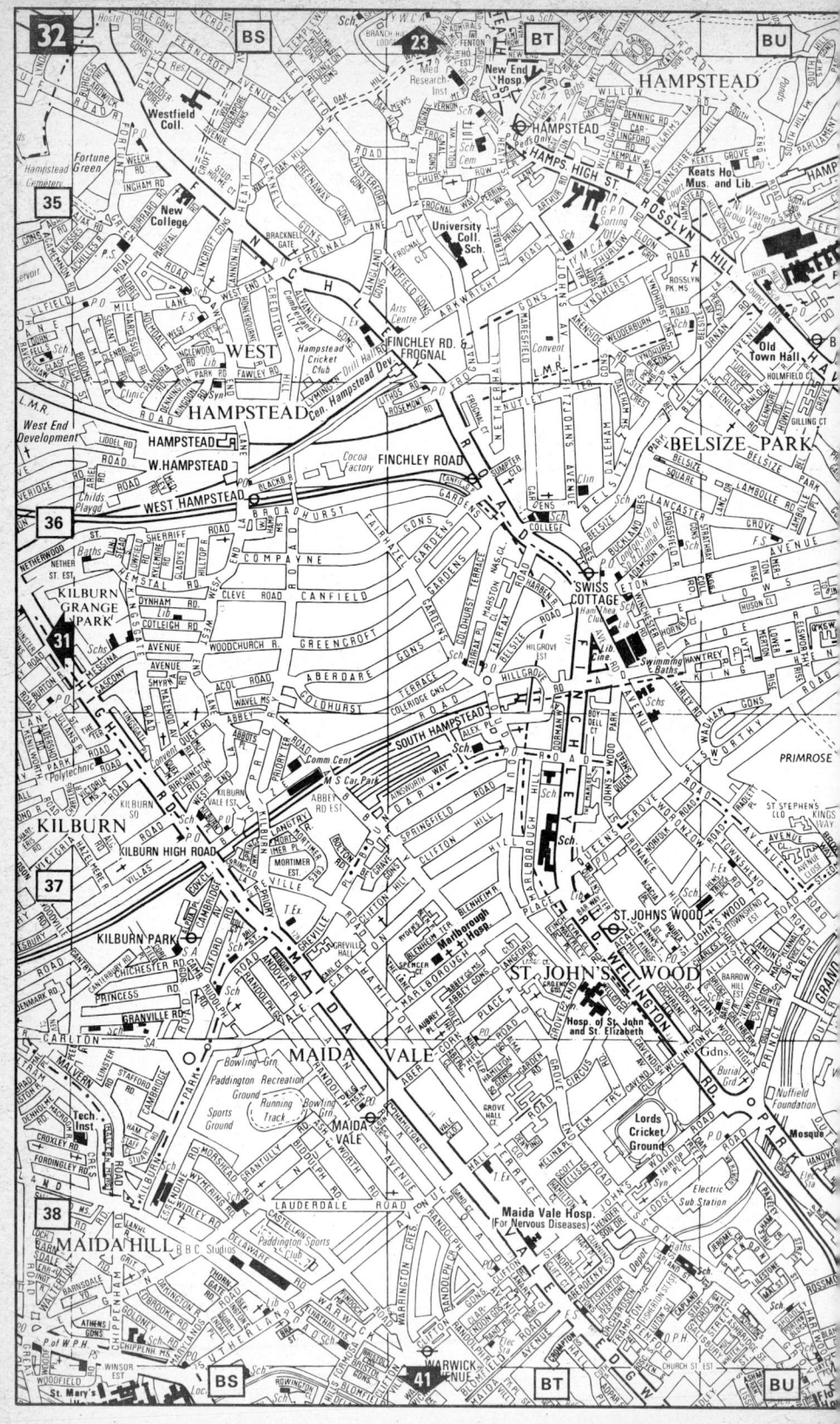

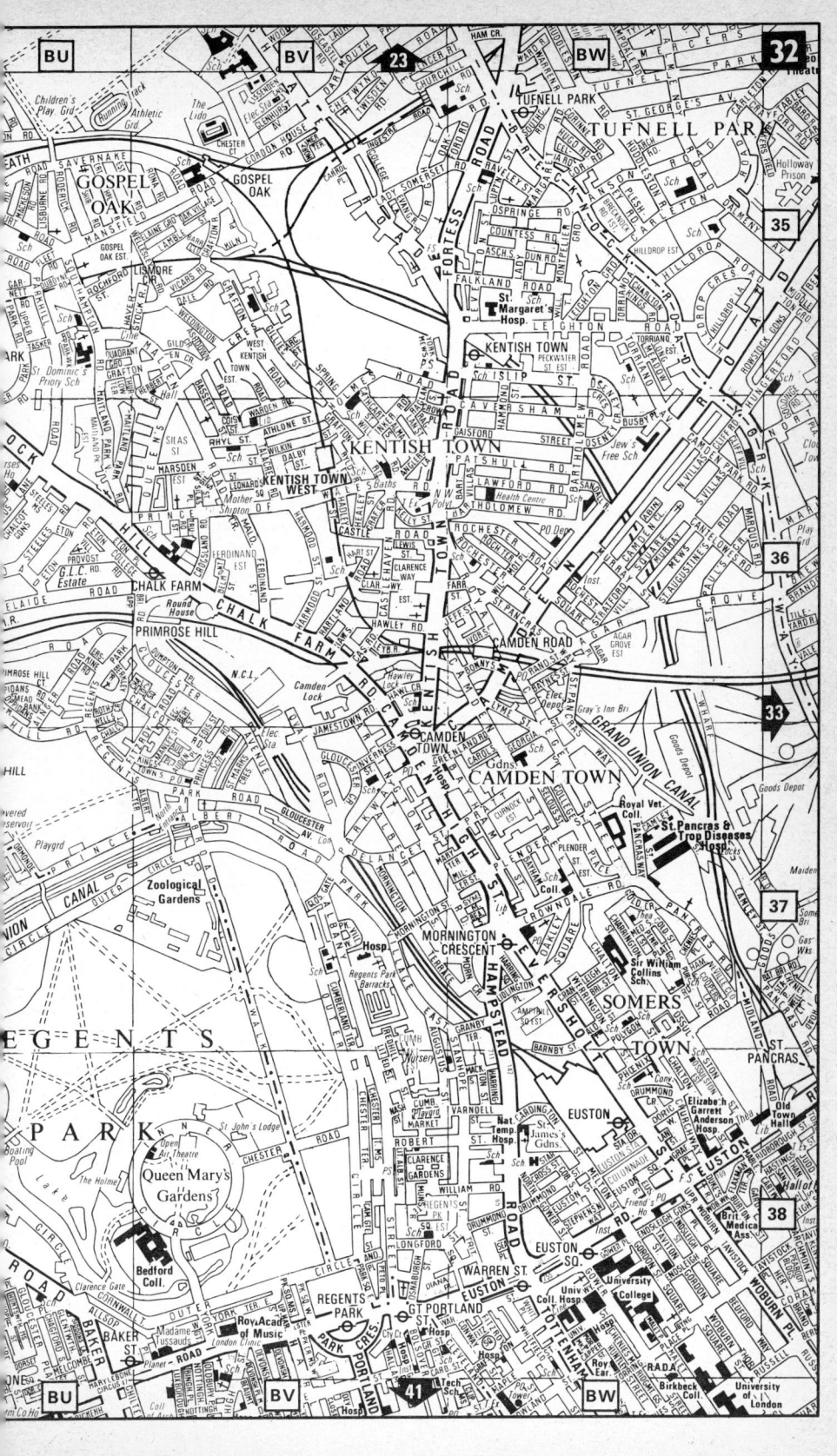

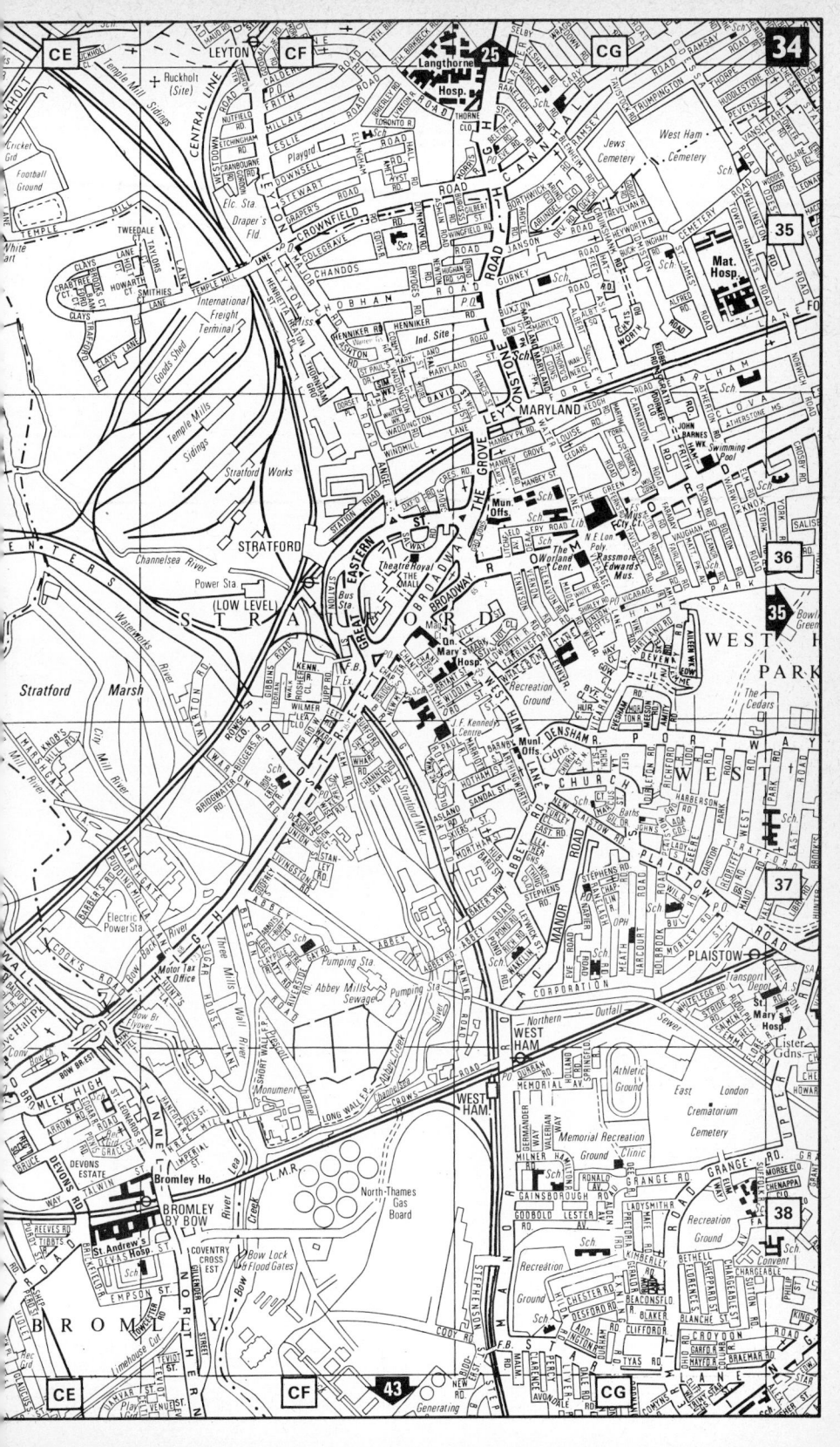

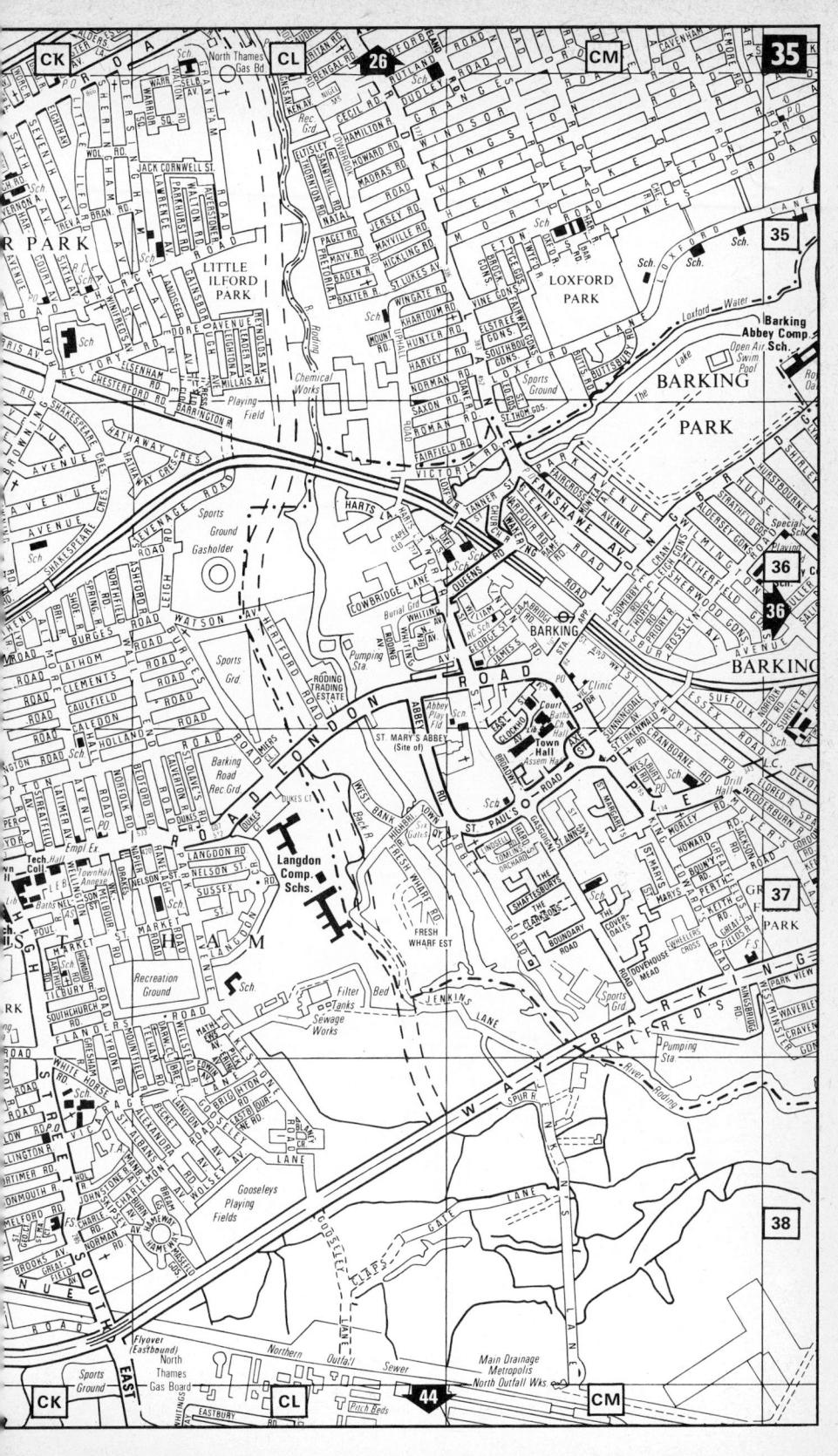

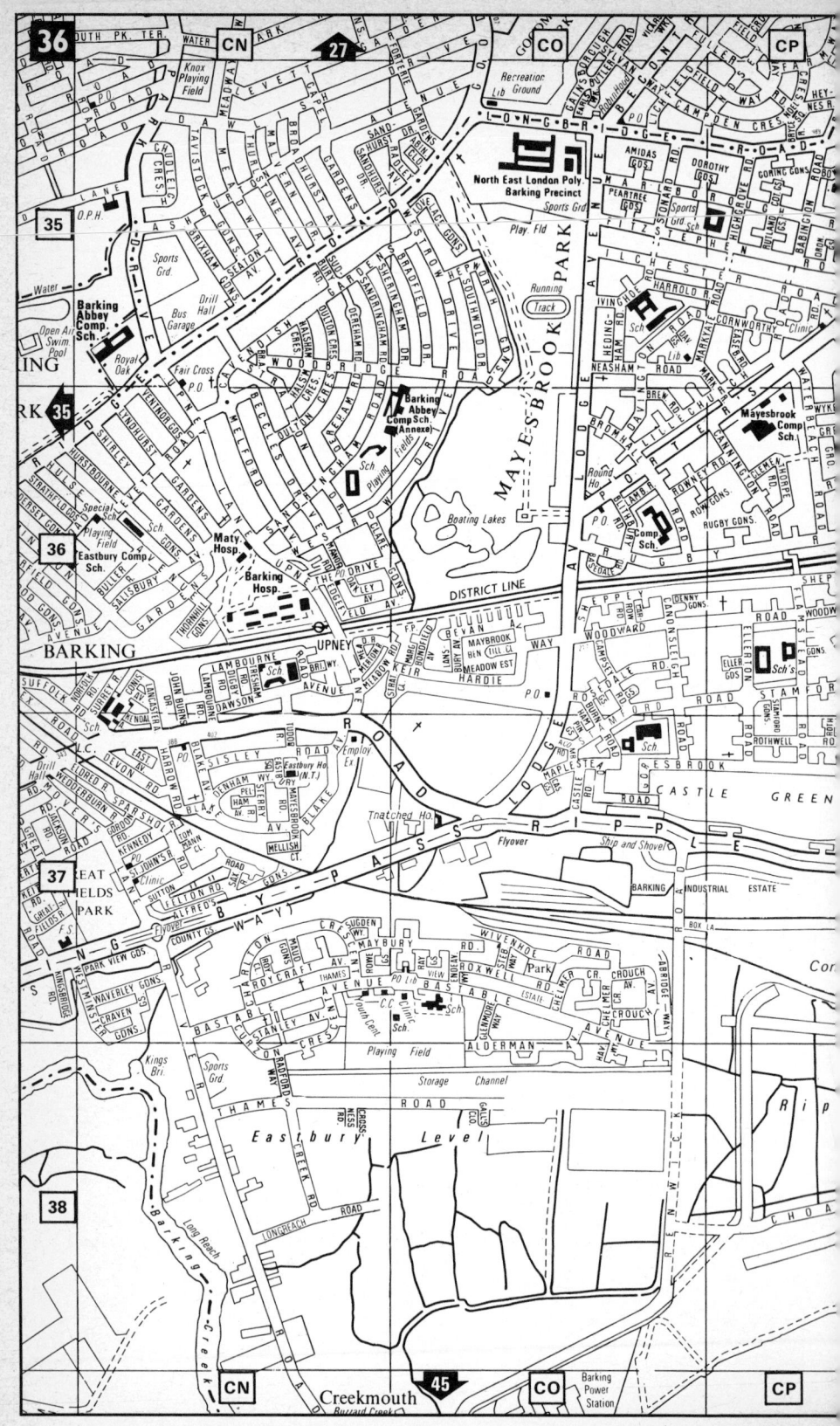

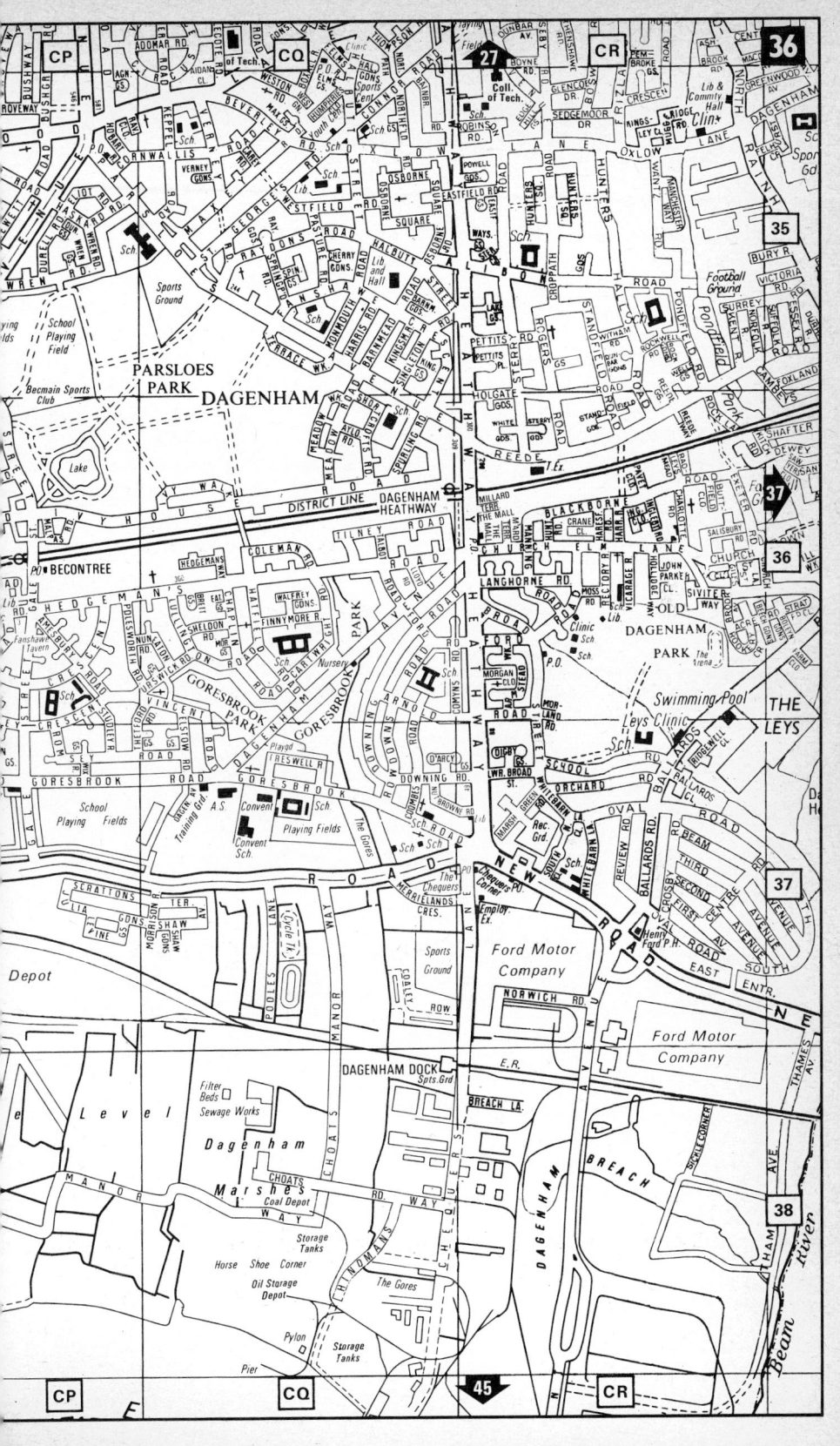

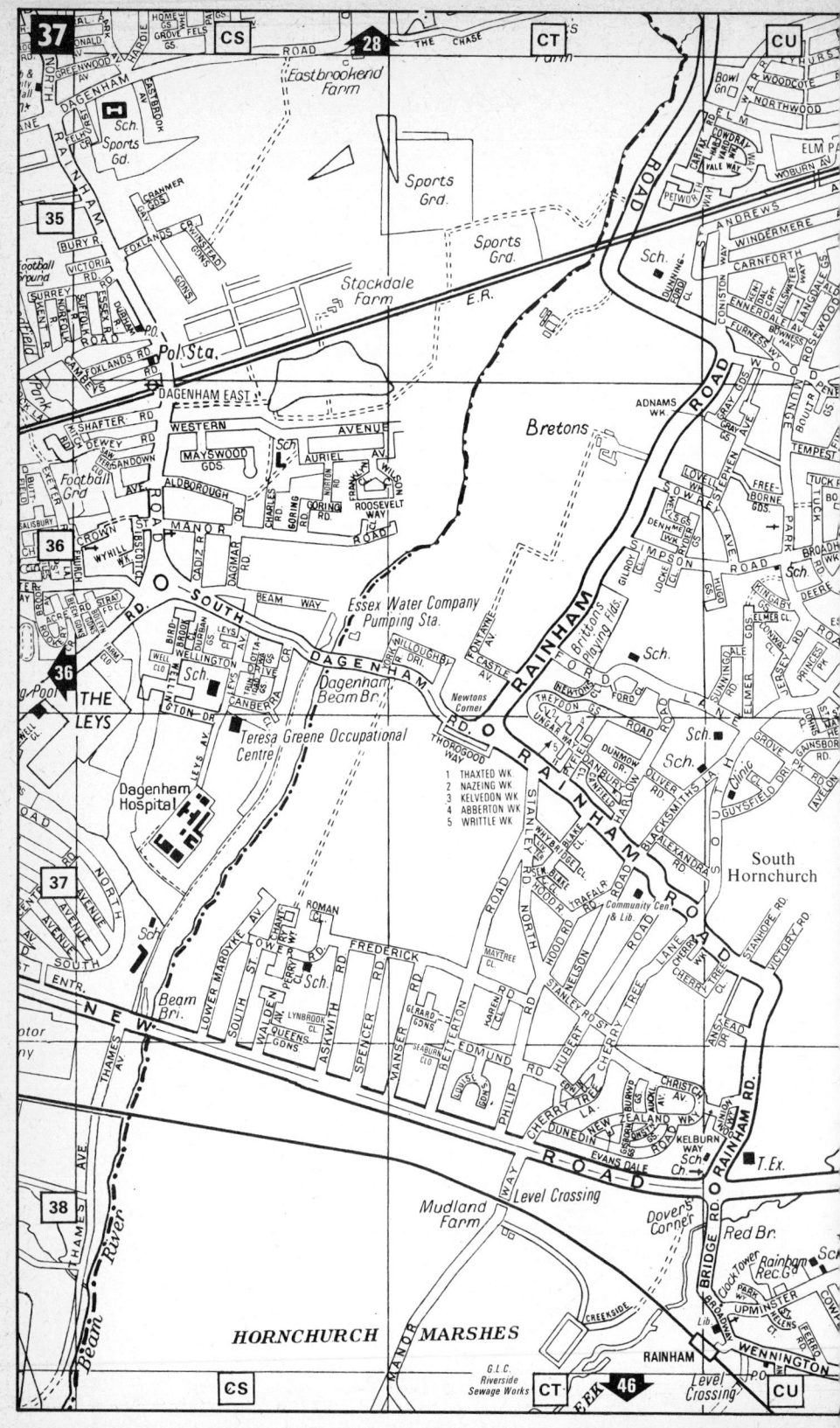

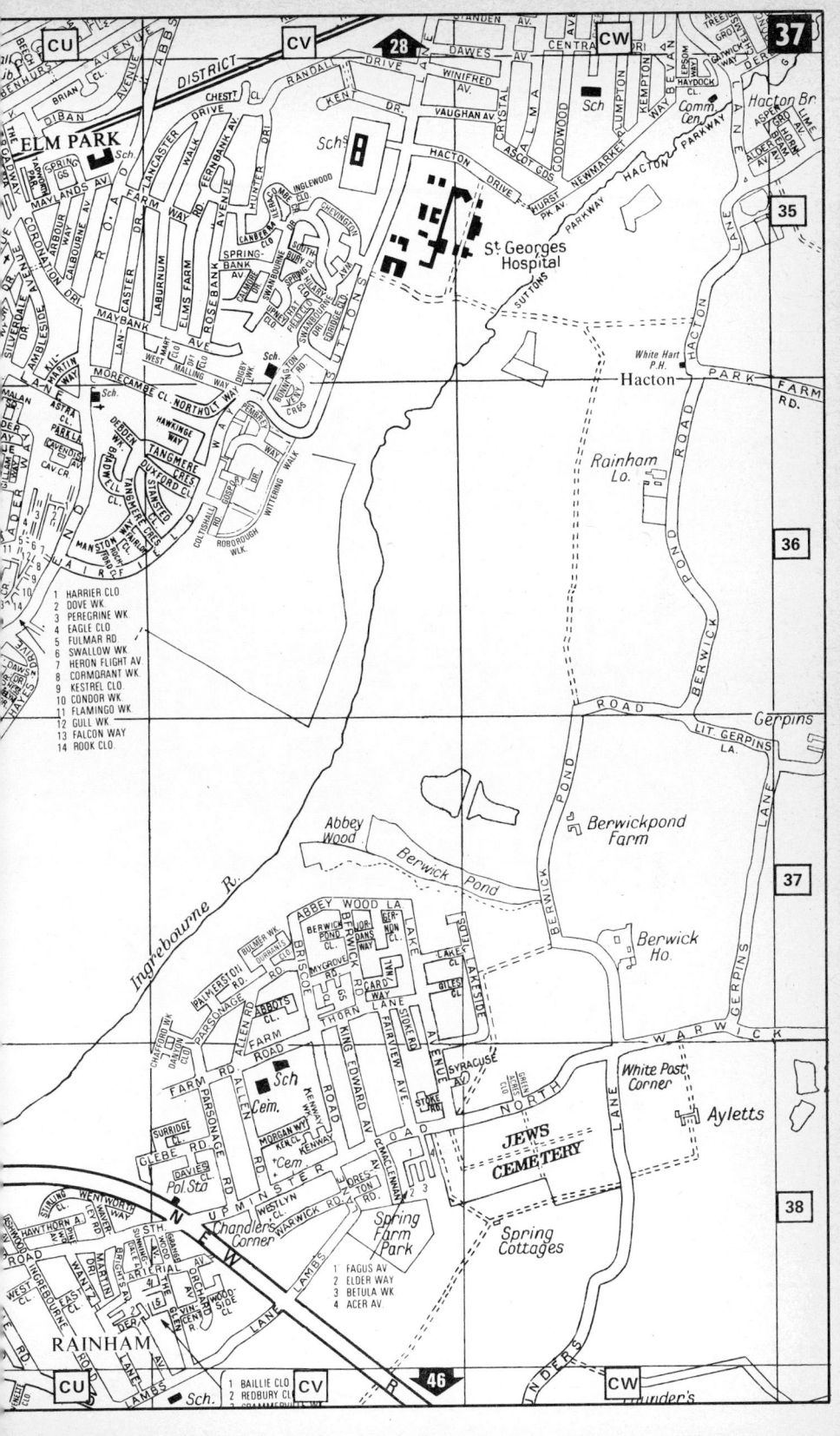

ELM PARK

St Georges Hospital

35

Hacton

White Hart P.H.

Rainham Lo.

36

1 HARRIER CLO.
2 DOVE WK.
3 PEREGRINE WK.
4 EAGLE CLO.
5 FULMAR RD.
6 SWALLOW WK.
7 HERON FLIGHT AV.
8 CORMORANT WK.
9 KESTREL CLO.
10 CONDOR WK.
11 FLAMINGO WK.
12 GULL WK.
13 FALCON WAY
14 ROOK CLO.

Gerpins
LIT GERPINS LA.

Abbey Wood

Berwick Pond

Berwickpond Farm

37

Berwick Ho.

ABBEY WOOD LA.

WARWICK

White Post Corner

Ayletts

Sch
Cem.

Pol. Sta.

JEWS CEMETERY

Chandlers Corner

Spring Farm Park

Spring Cottages

38

RAINHAM

1 FAGUS AV
2 ELDER WAY
3 BETULA WK
4 ACER AV

1 BAILLIE CLO
2 REDBURY CL

Thunder's

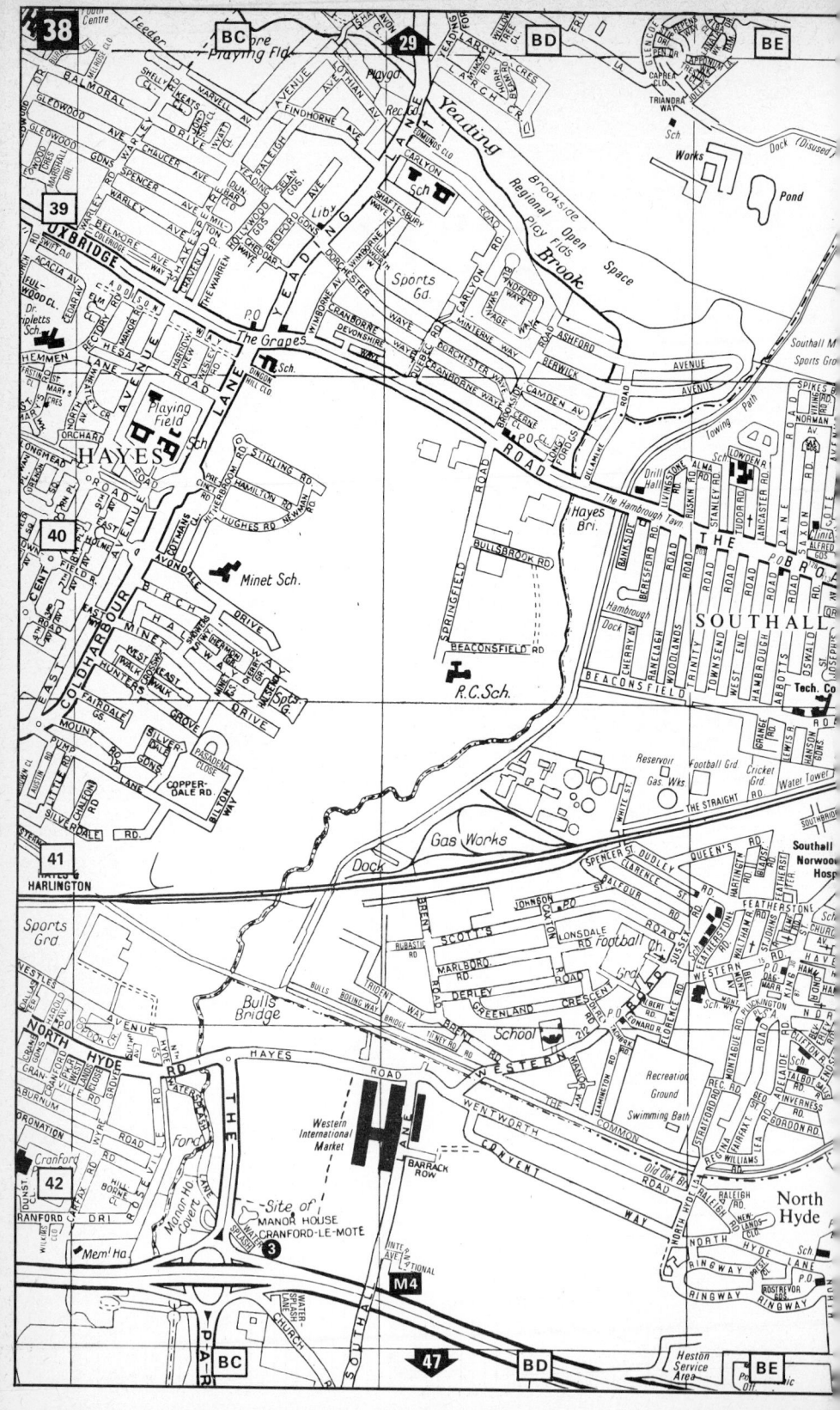

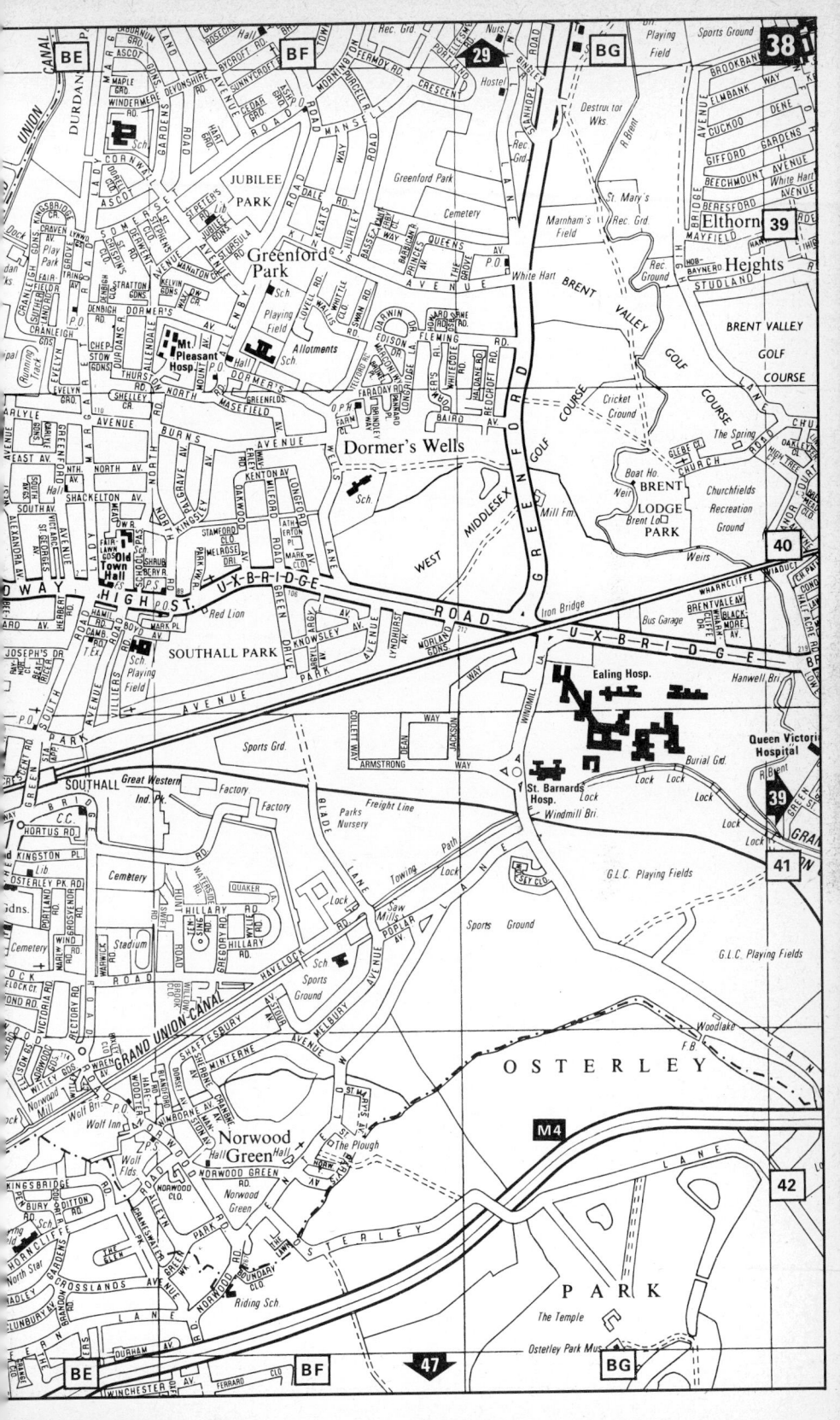

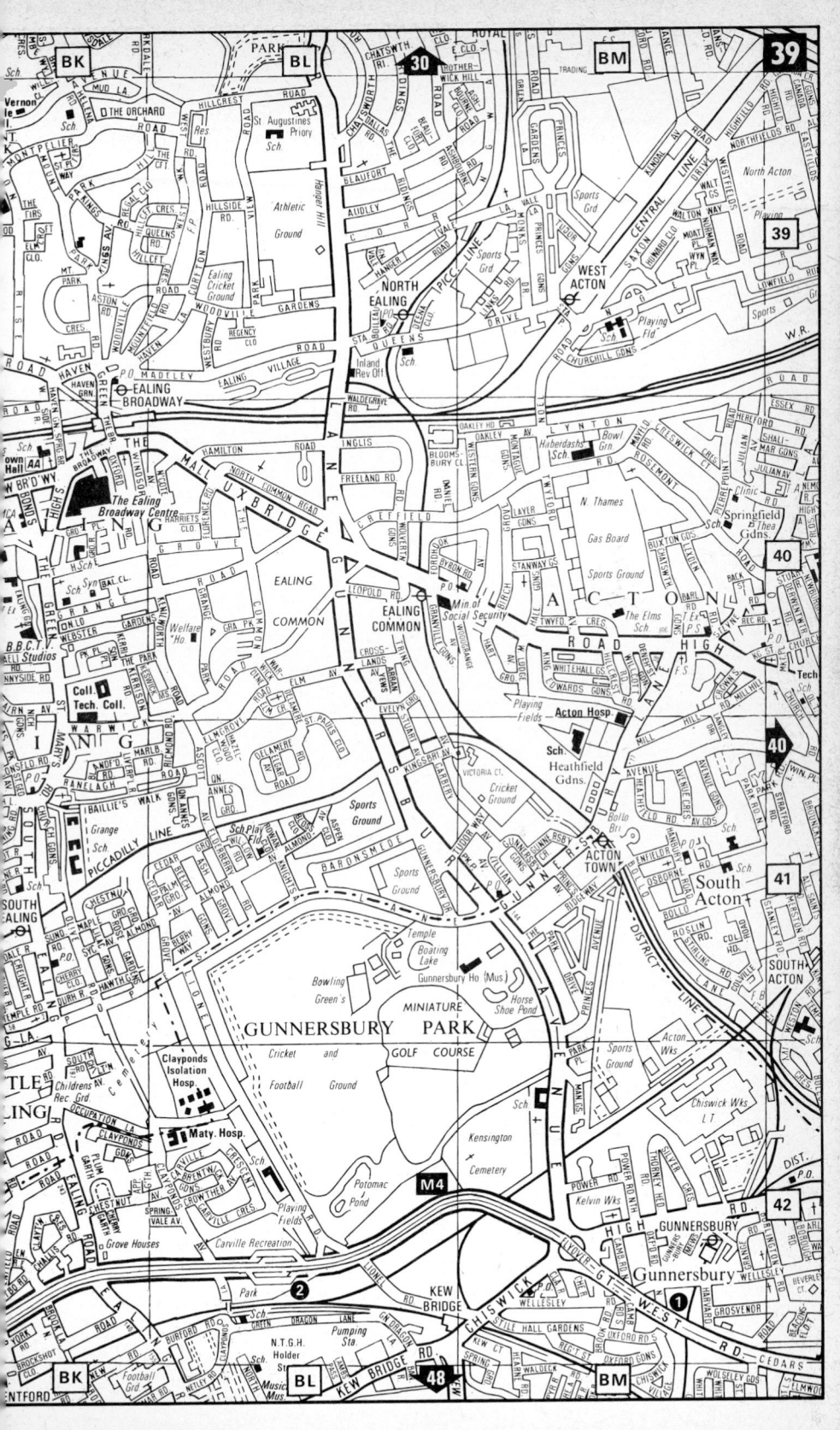

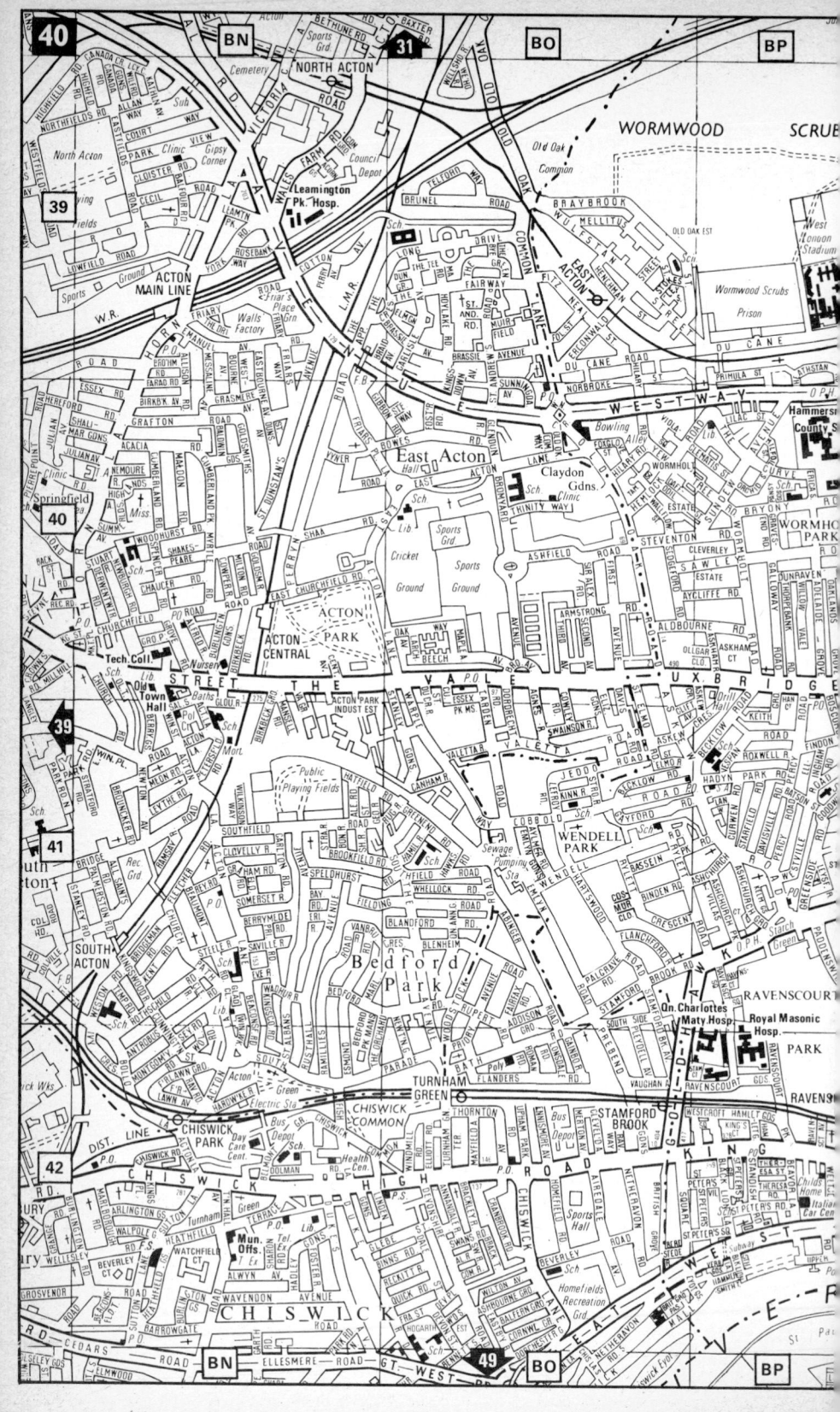

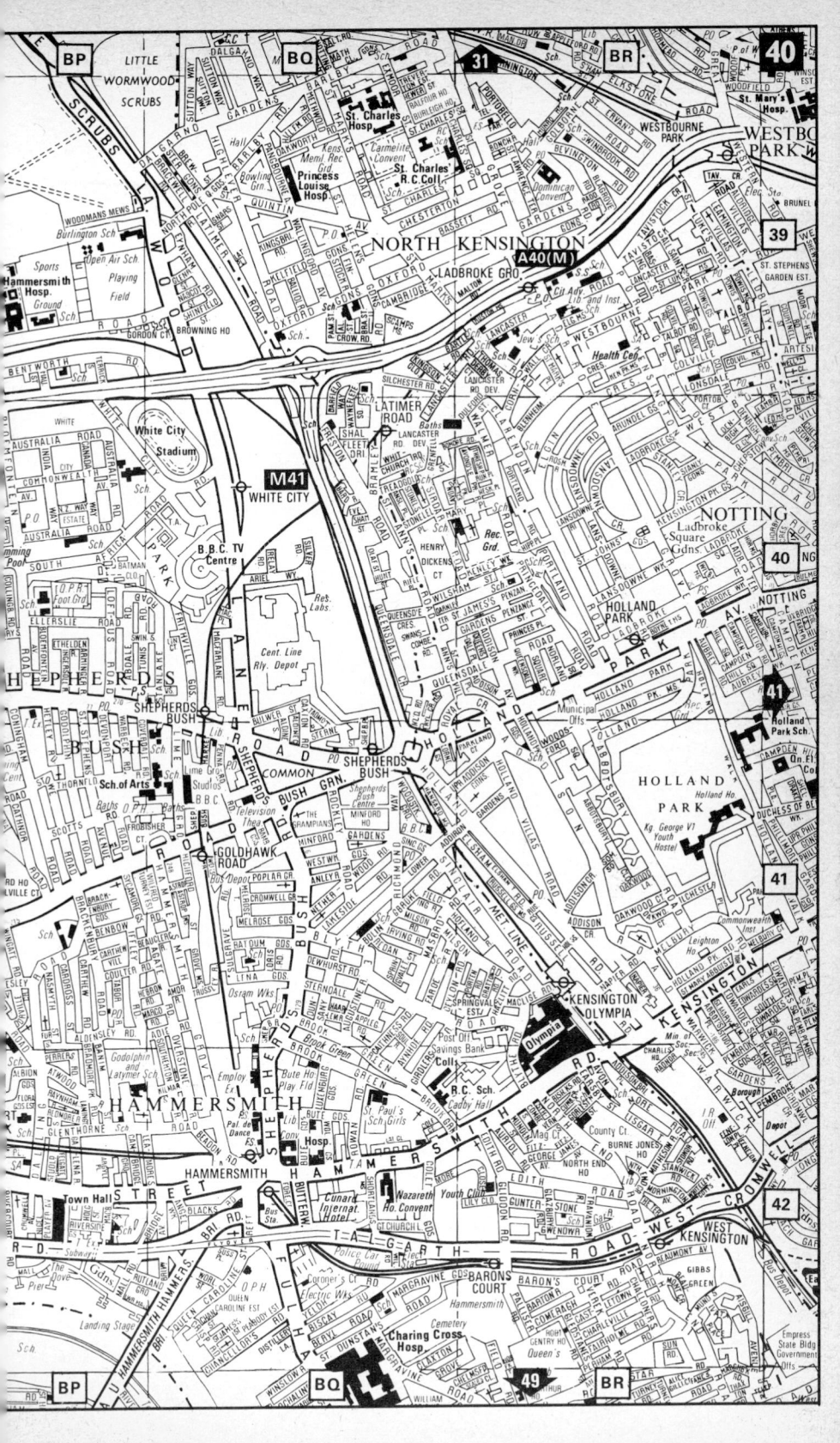

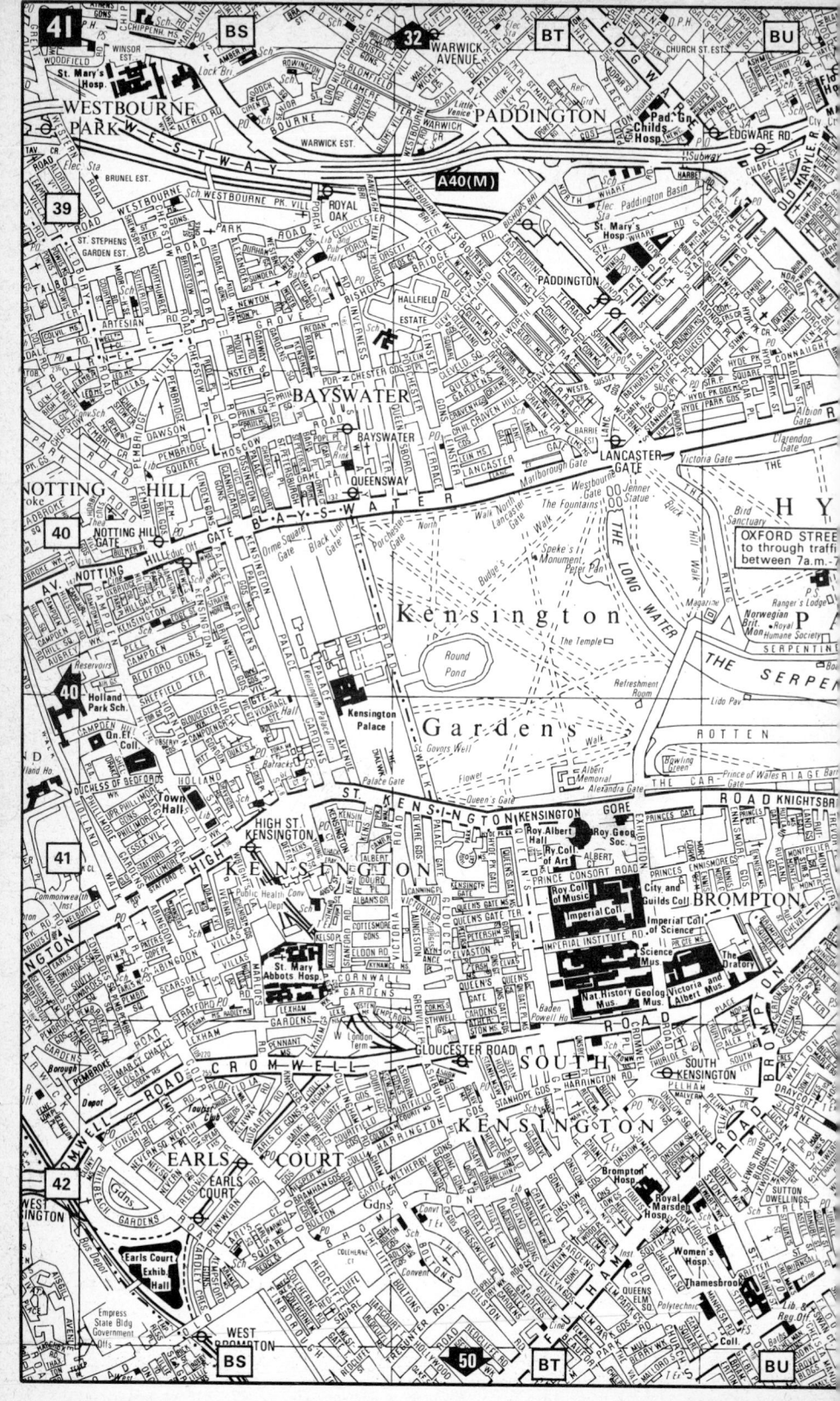

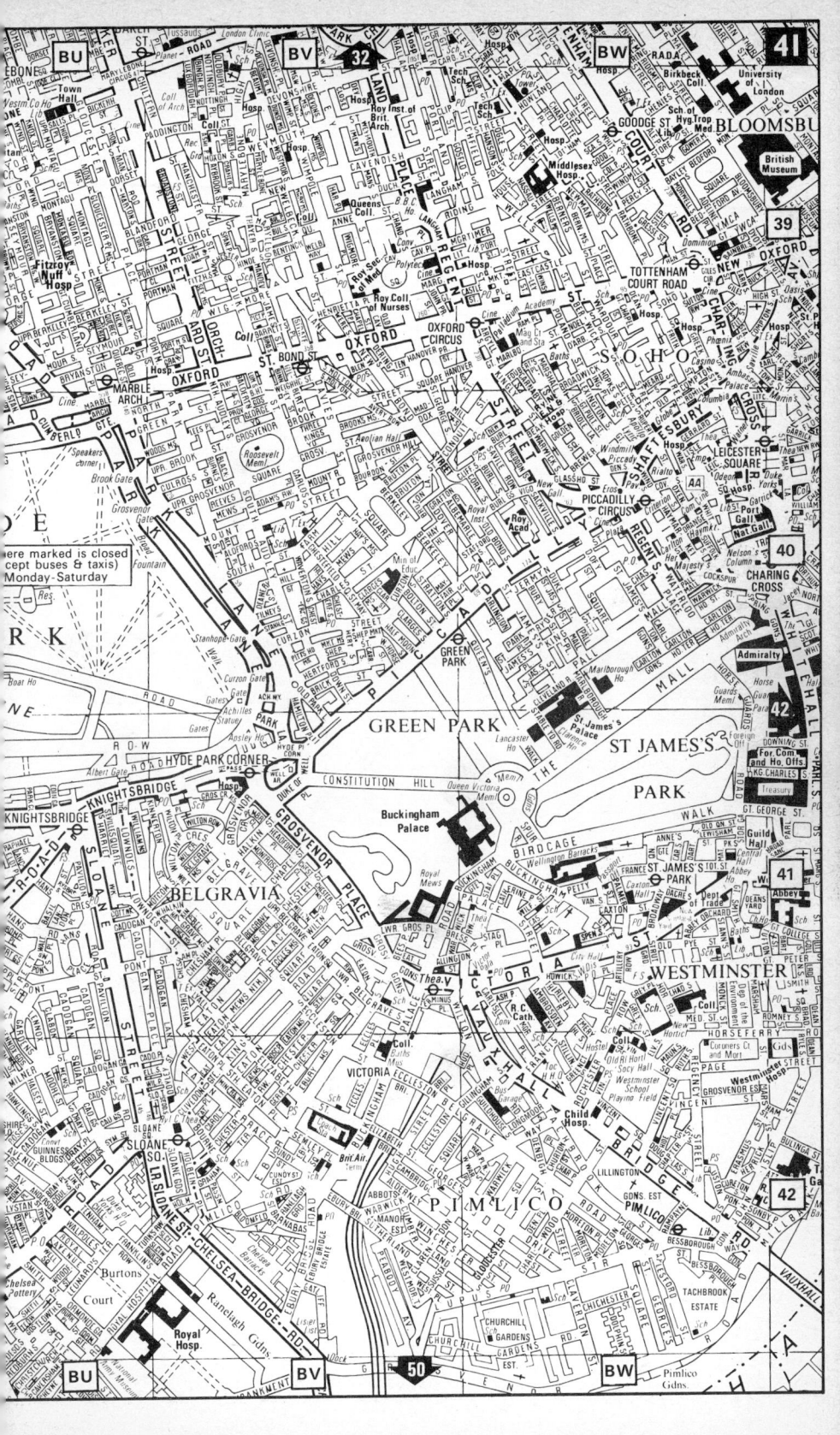

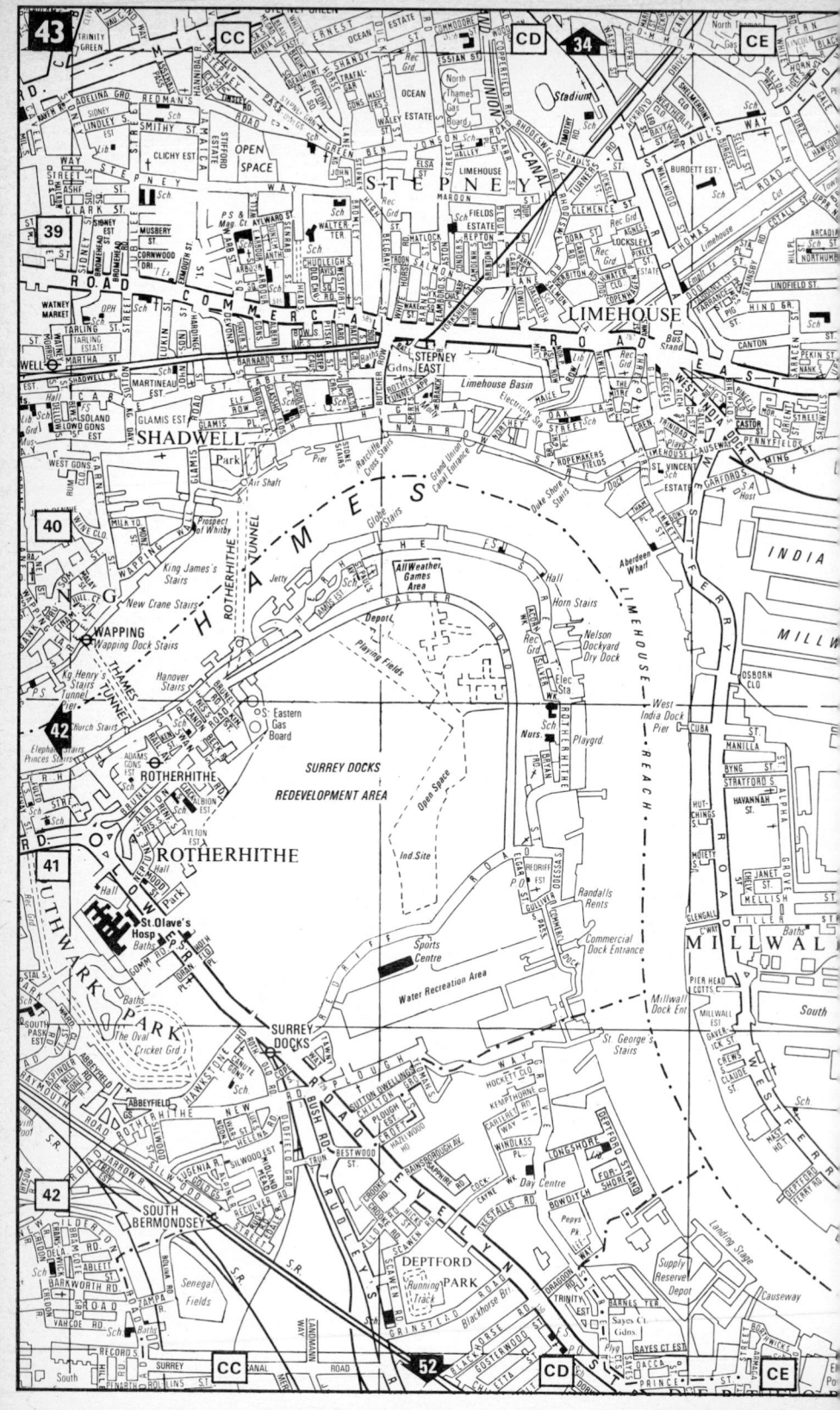

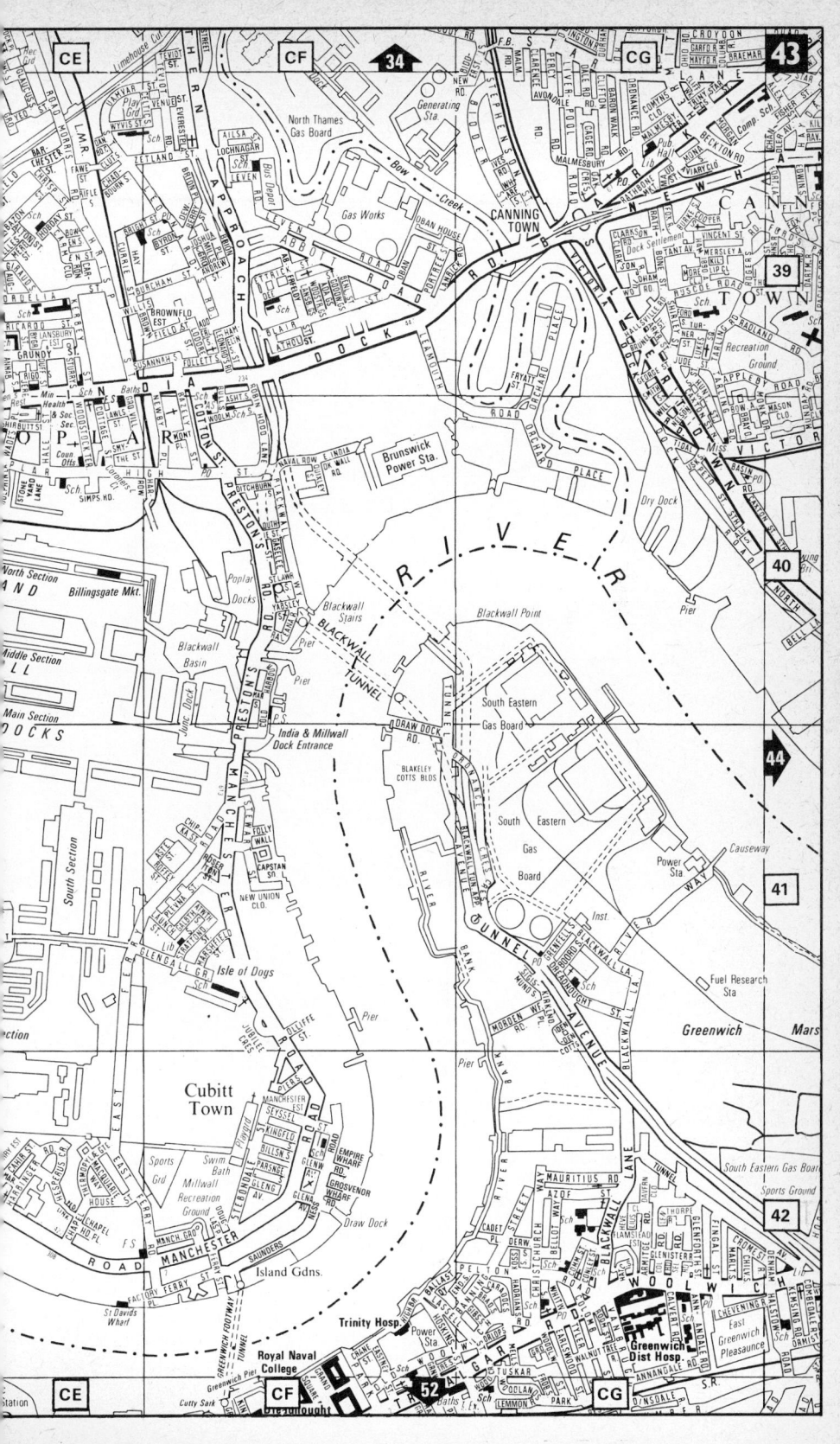

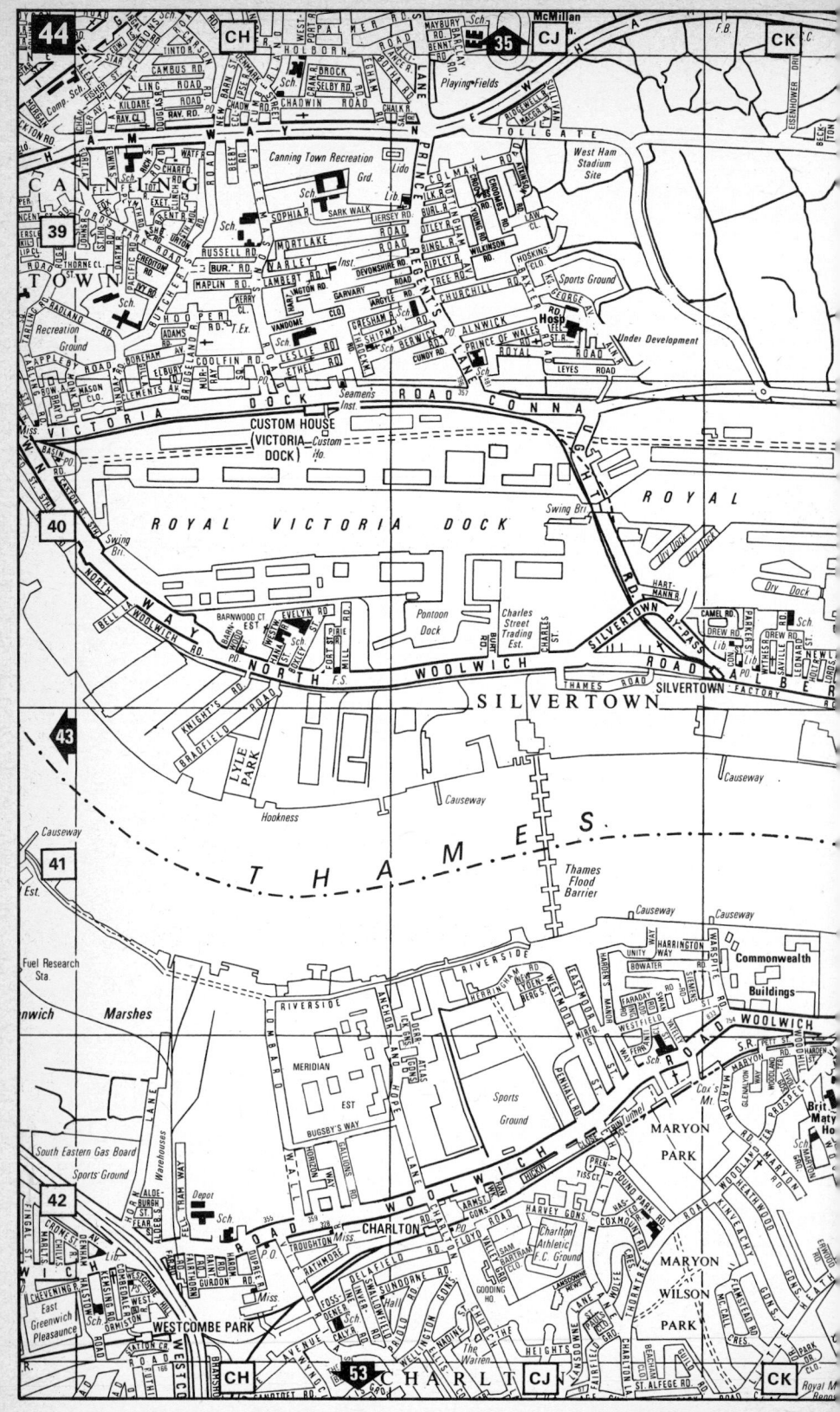

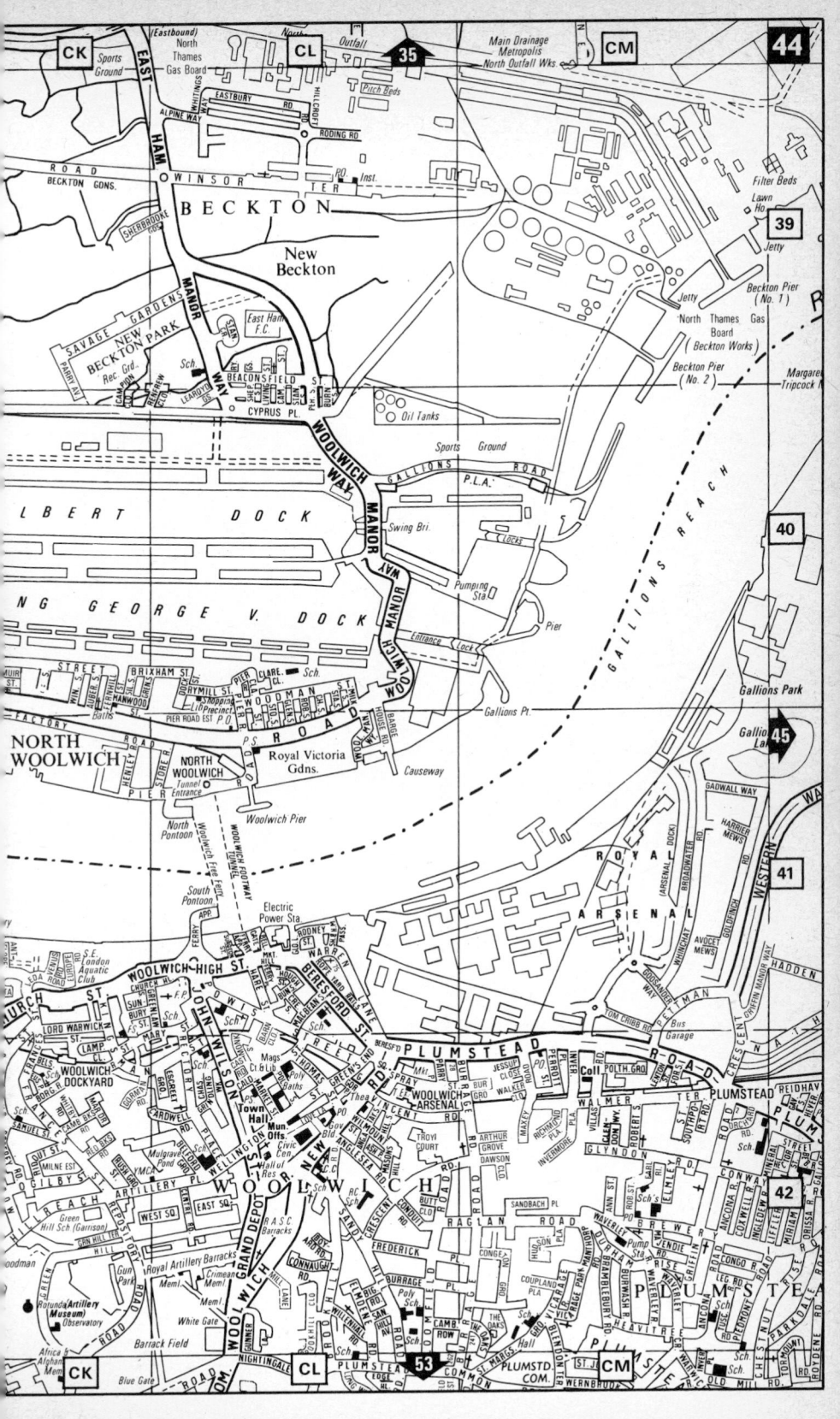

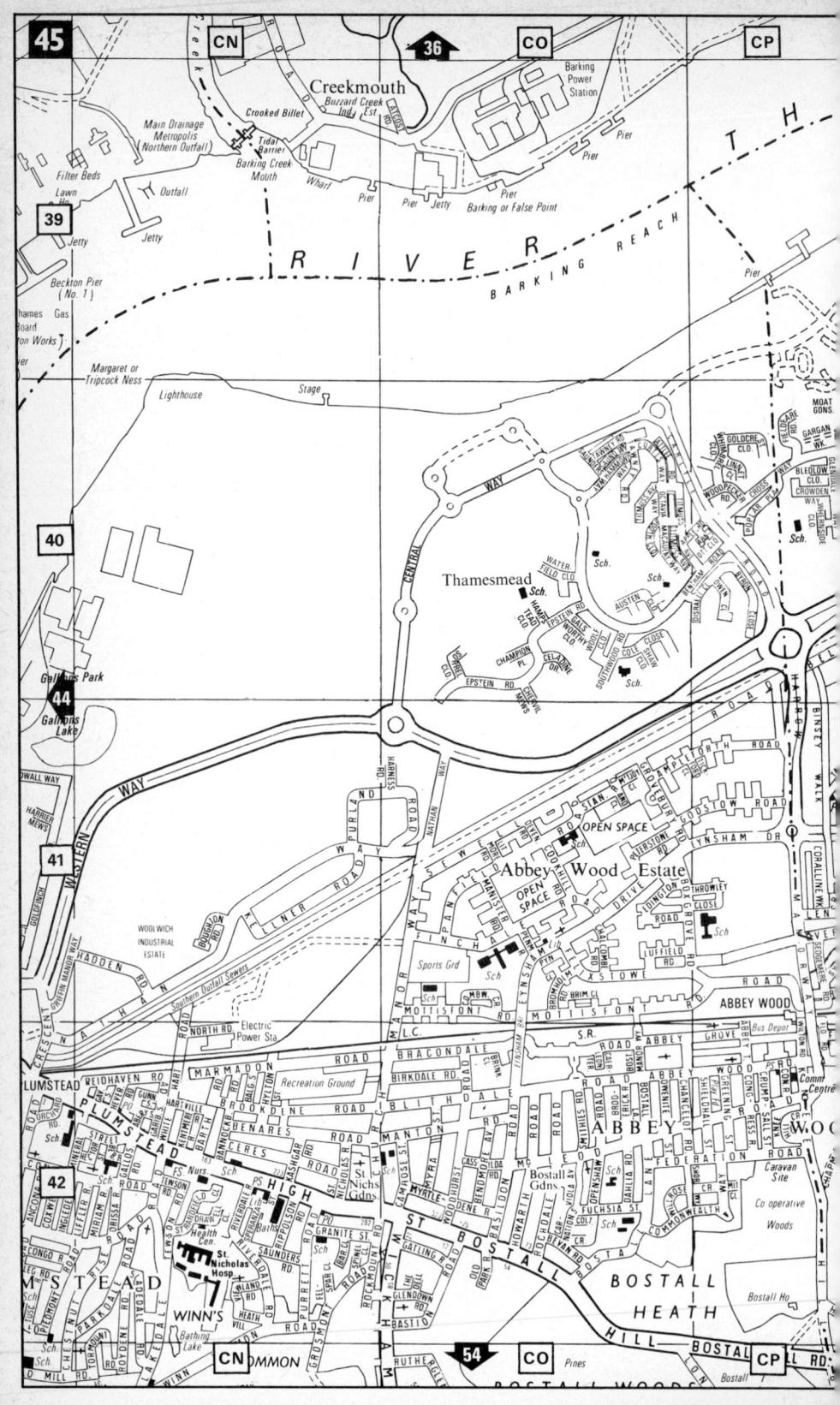

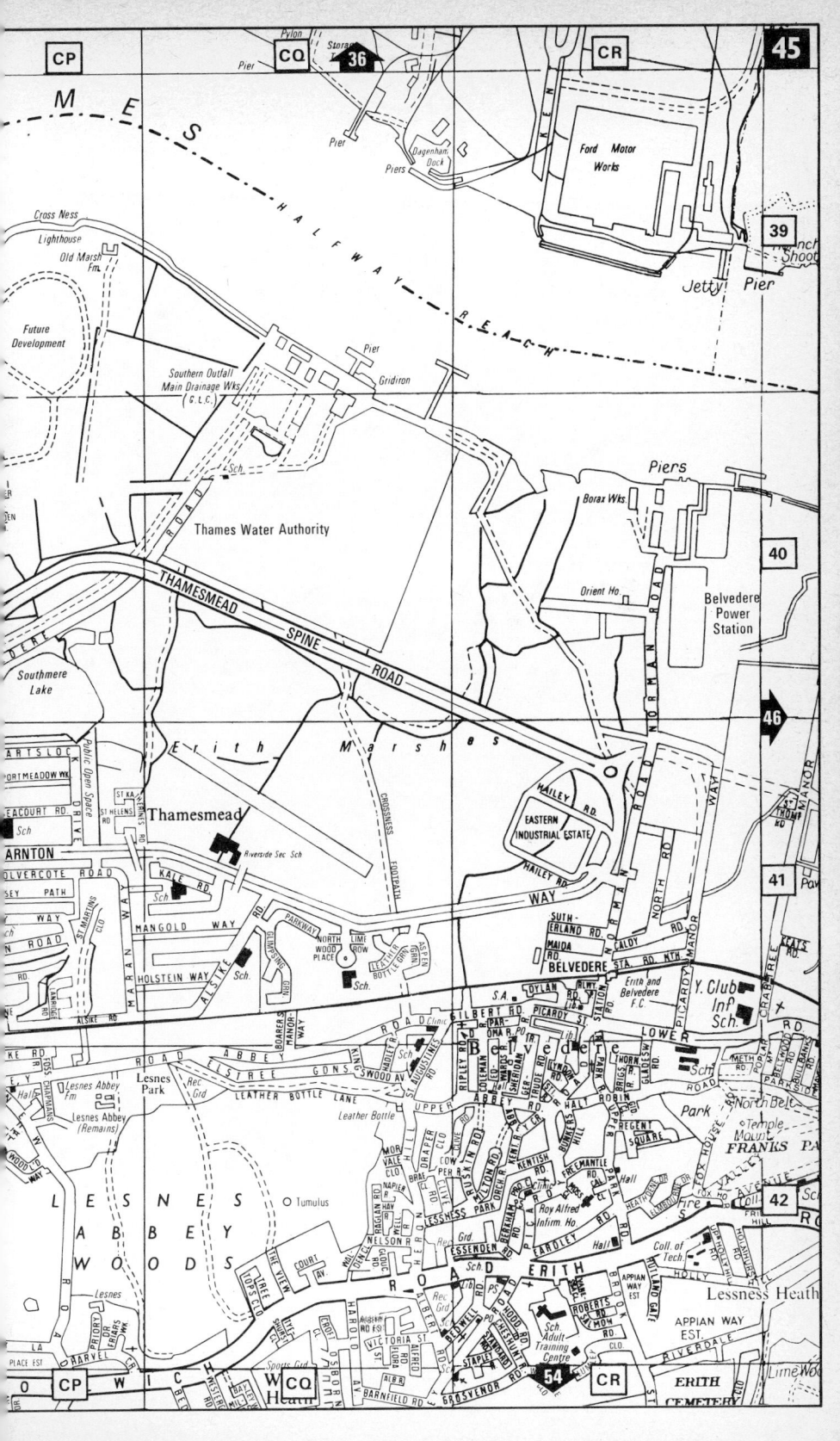

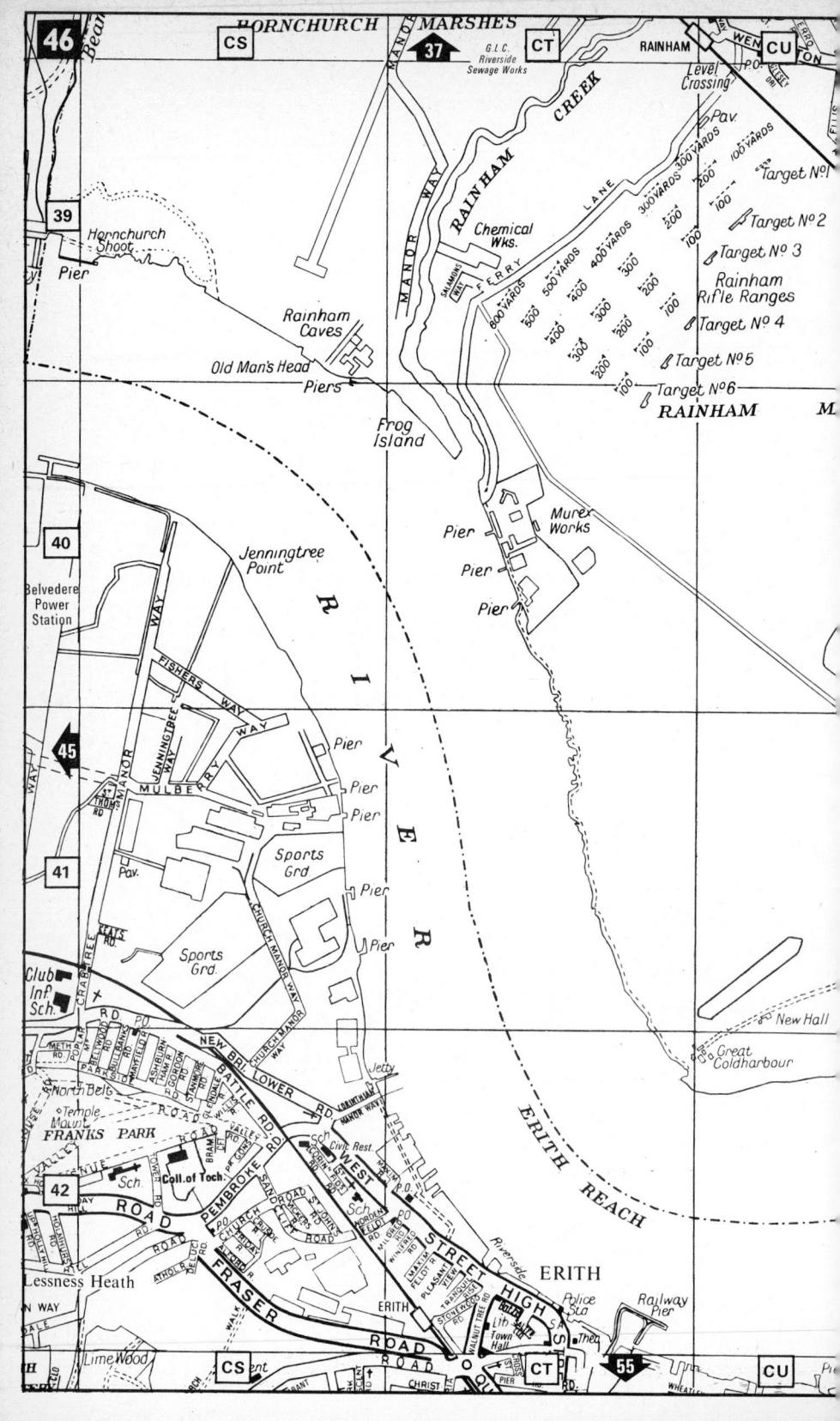

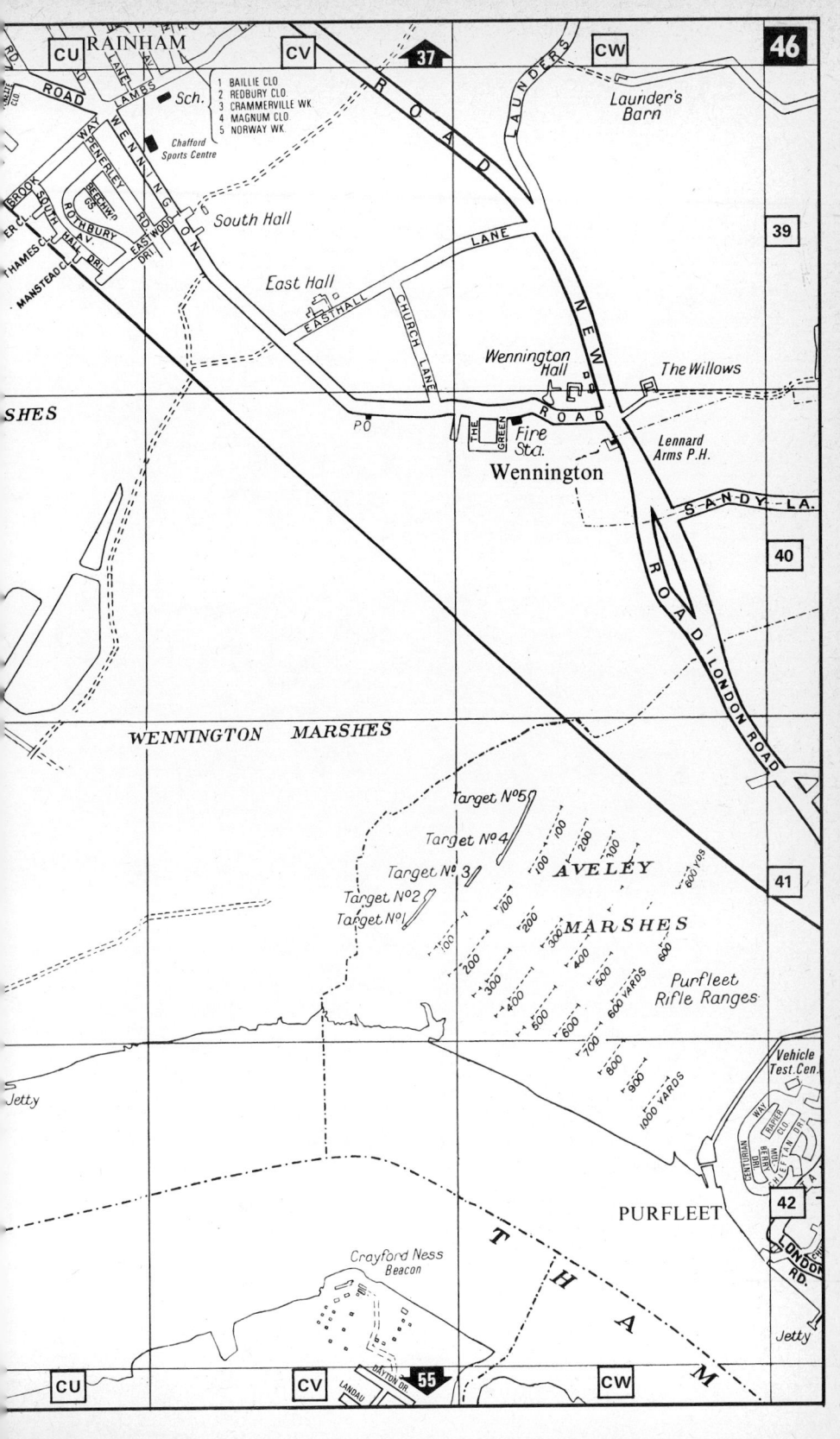

RAINHAM

CU

CV

37

CW

39

40

41

42

1 BAILLIE CLO
2 REDBURY CLO
3 CRAMMERVILLE WK.
4 MAGNUM CLO
5 NORWAY WK.

Sch.

Chafford
Sports Centre

Launder's
Barn

South Hall

East Hall

Wennington
Hall

The Willows

PO

THE GREEN

Fire
Sta.

Lennard
Arms P.H.

Wennington

SANDY LA.

WENNINGTON MARSHES

Target No 5

Target No 4

Target No 3

Target No 2

Target No 1

100

200

100

600 yds

AVELEY

MARSHES

100

200

300

400

500

600

100

200

300

400

500

600

700

800

900

1000 YARDS

600 YARDS

Purfleet
Rifle Ranges

Vehicle
Test.Cen.

Jetty

PURFLEET

LONDON RD.

Crayford Ness
Beacon

THAMES

Jetty

CU

CV

55

CW

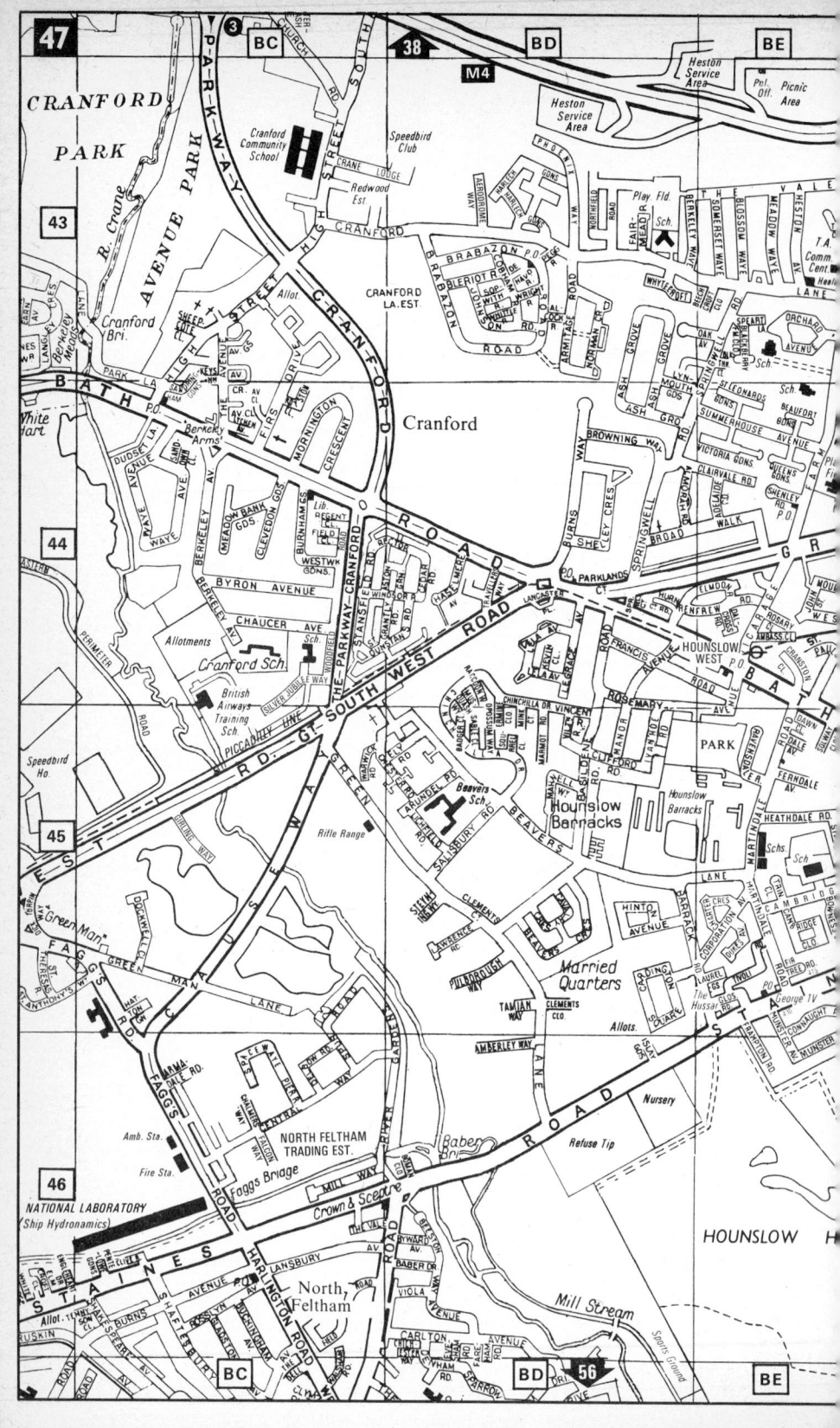

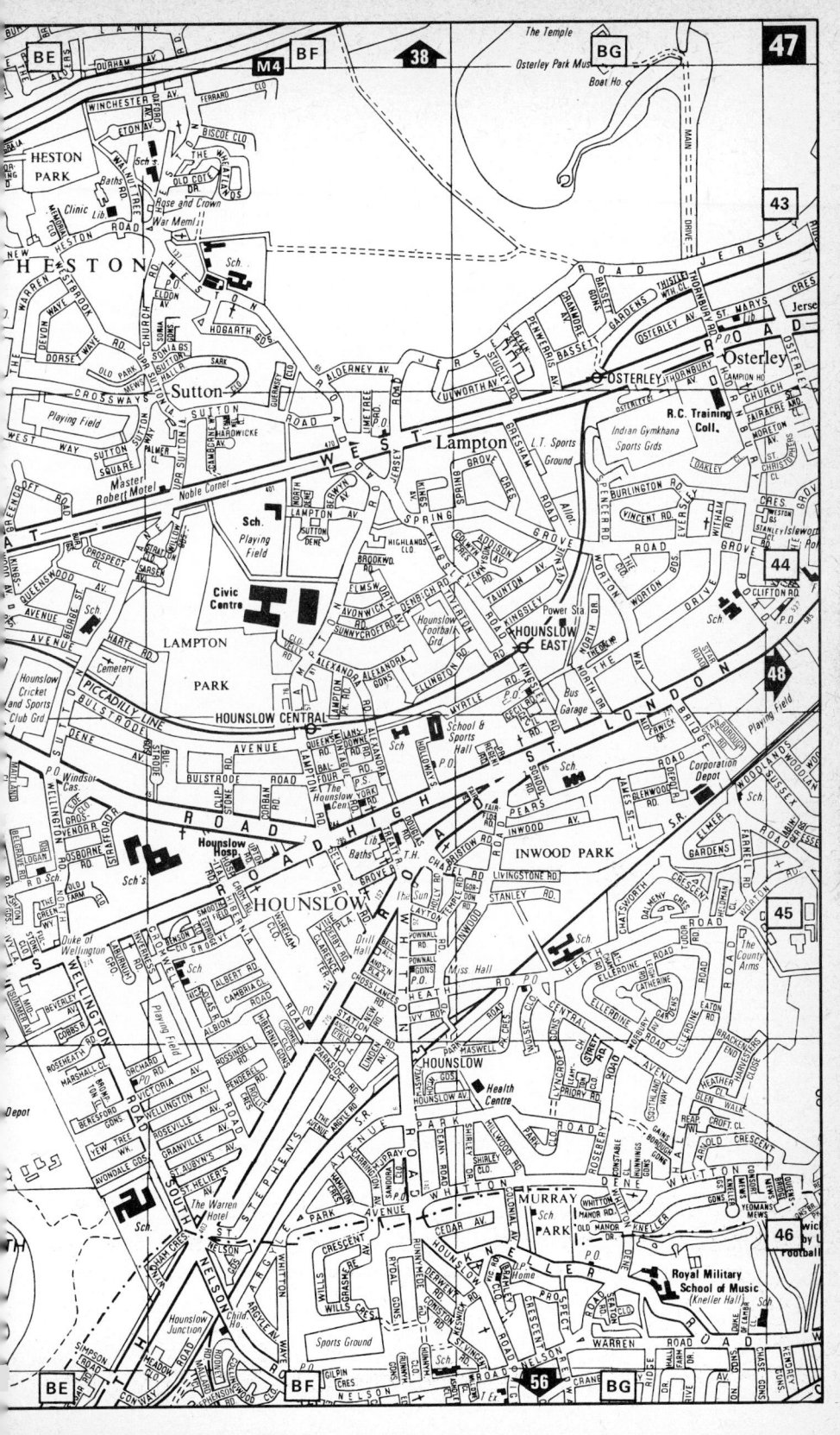

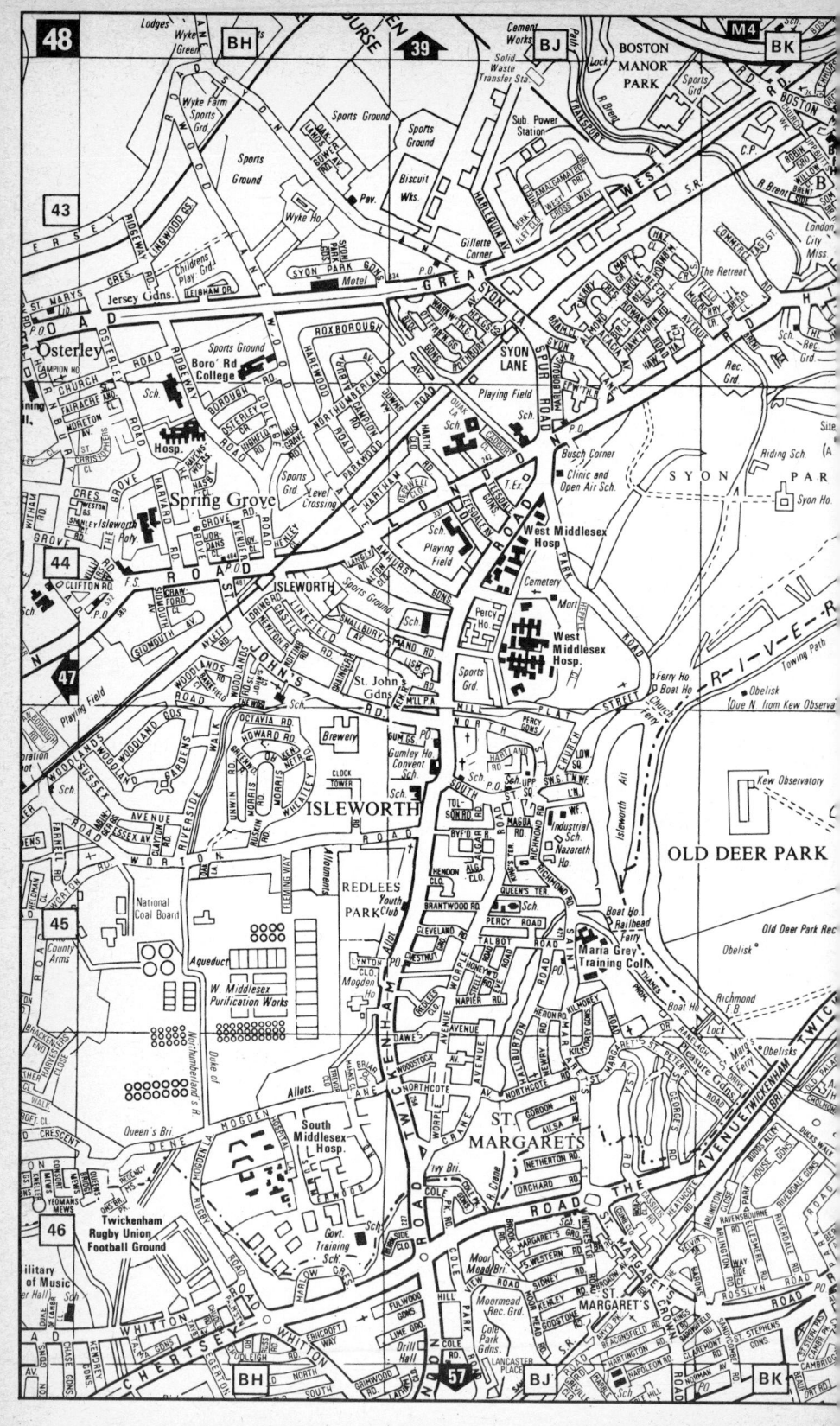

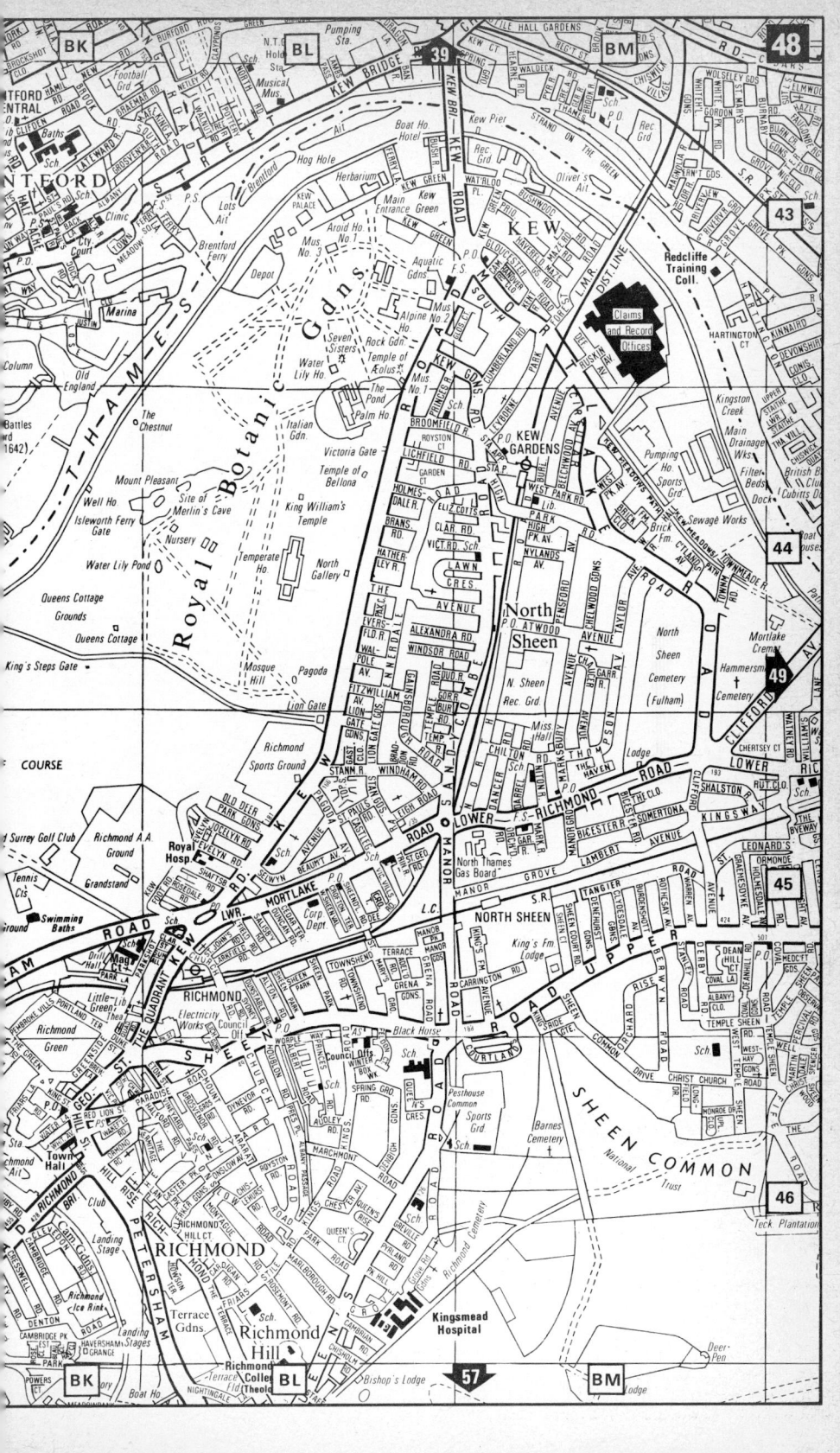

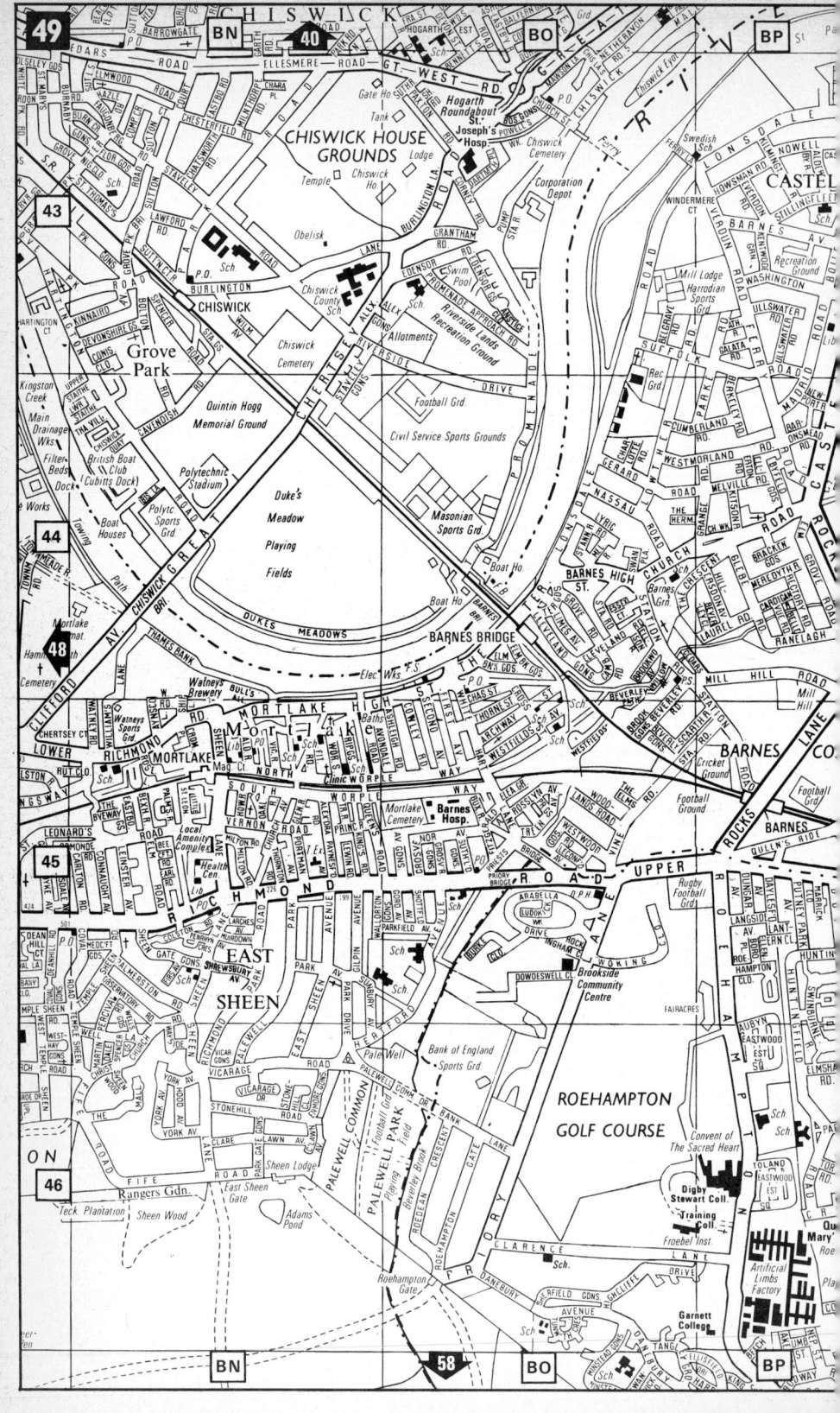

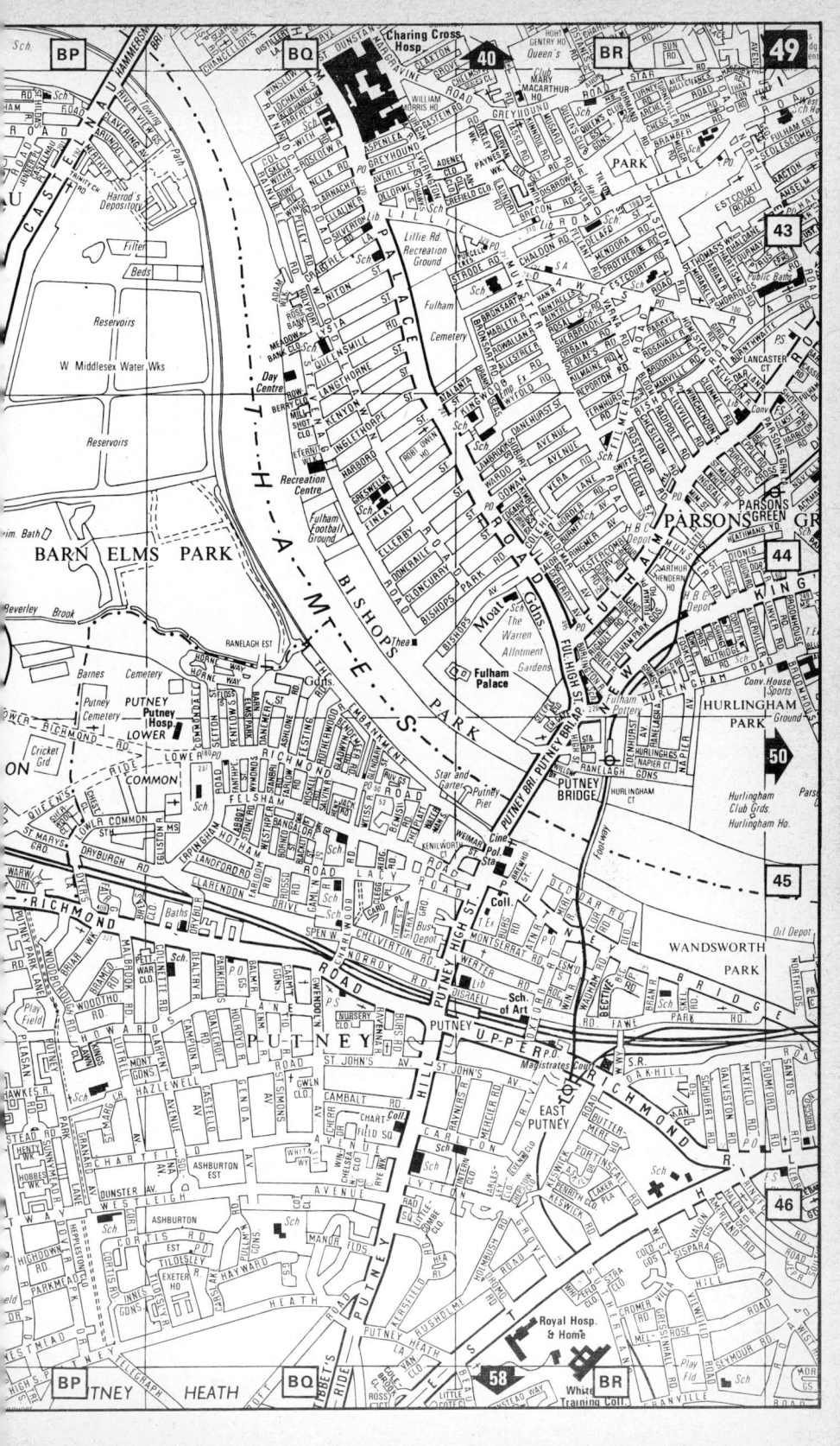

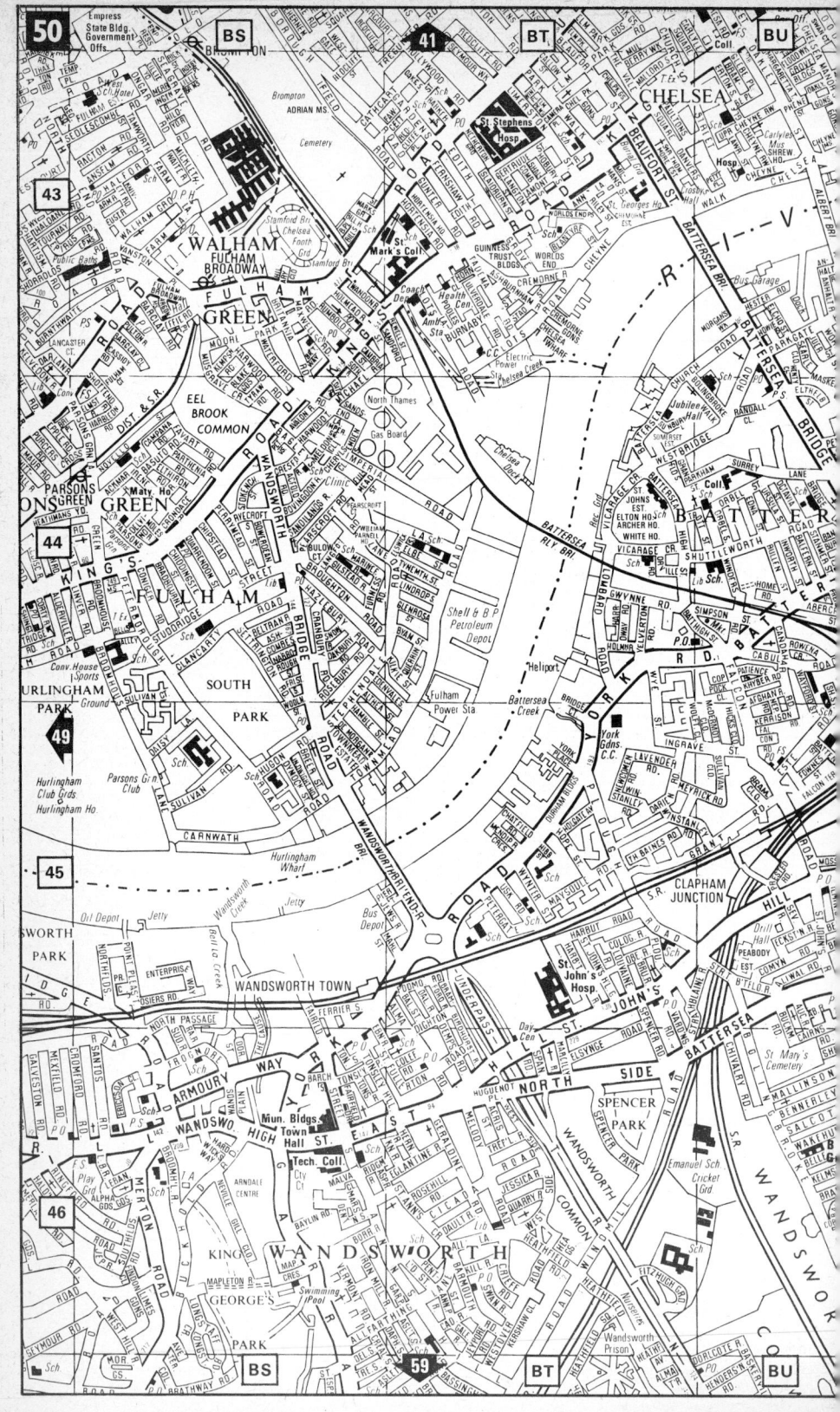

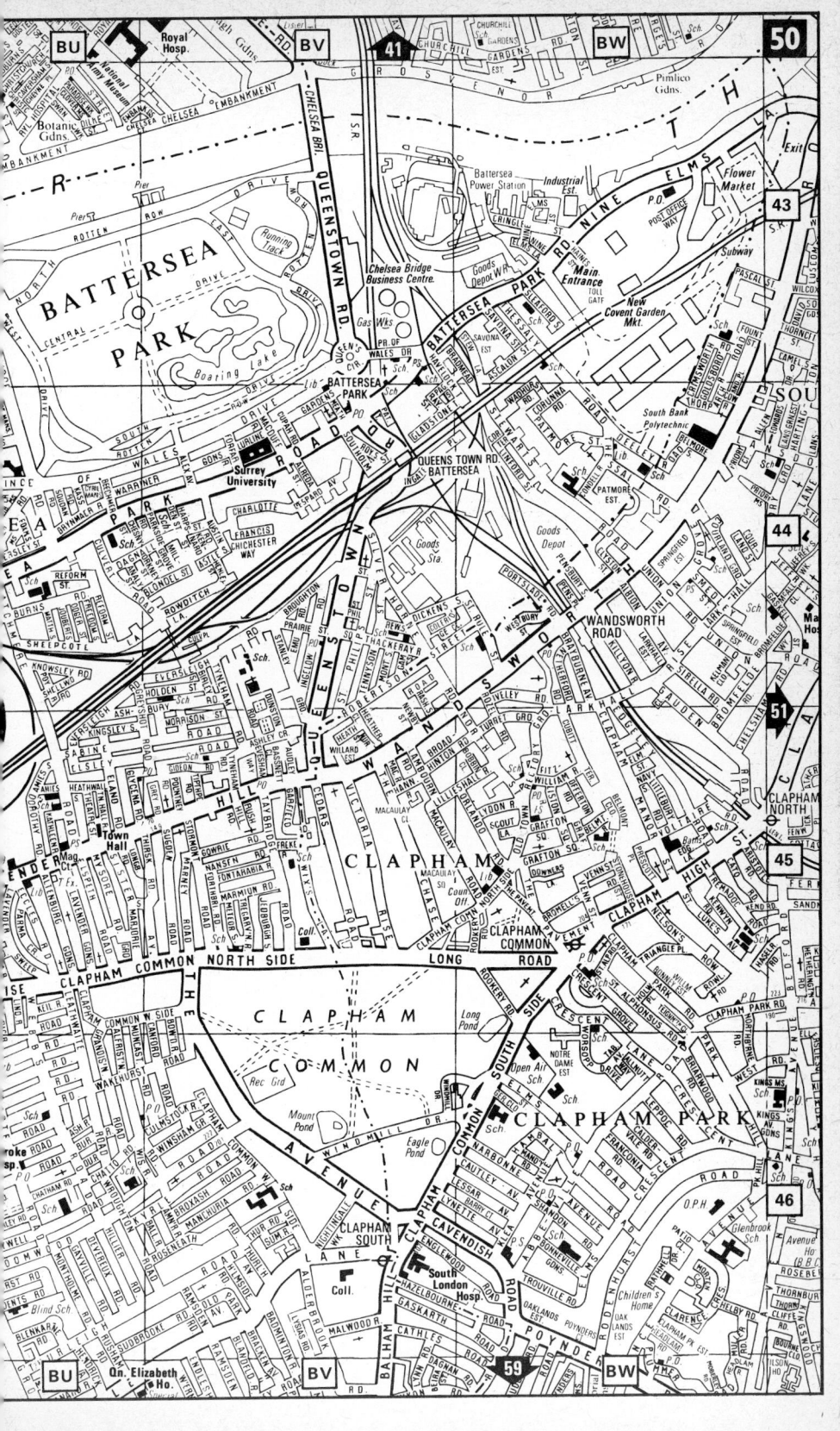

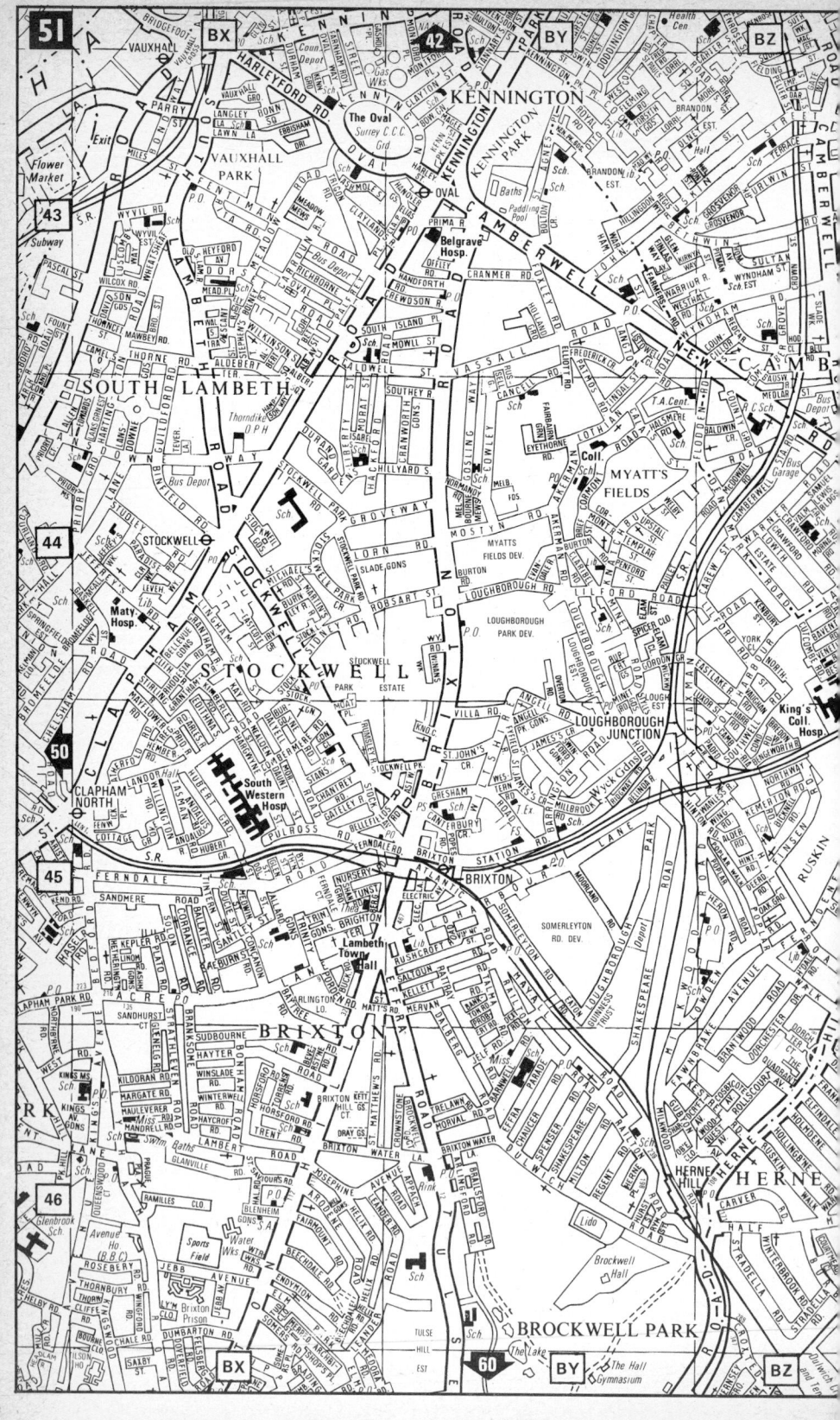

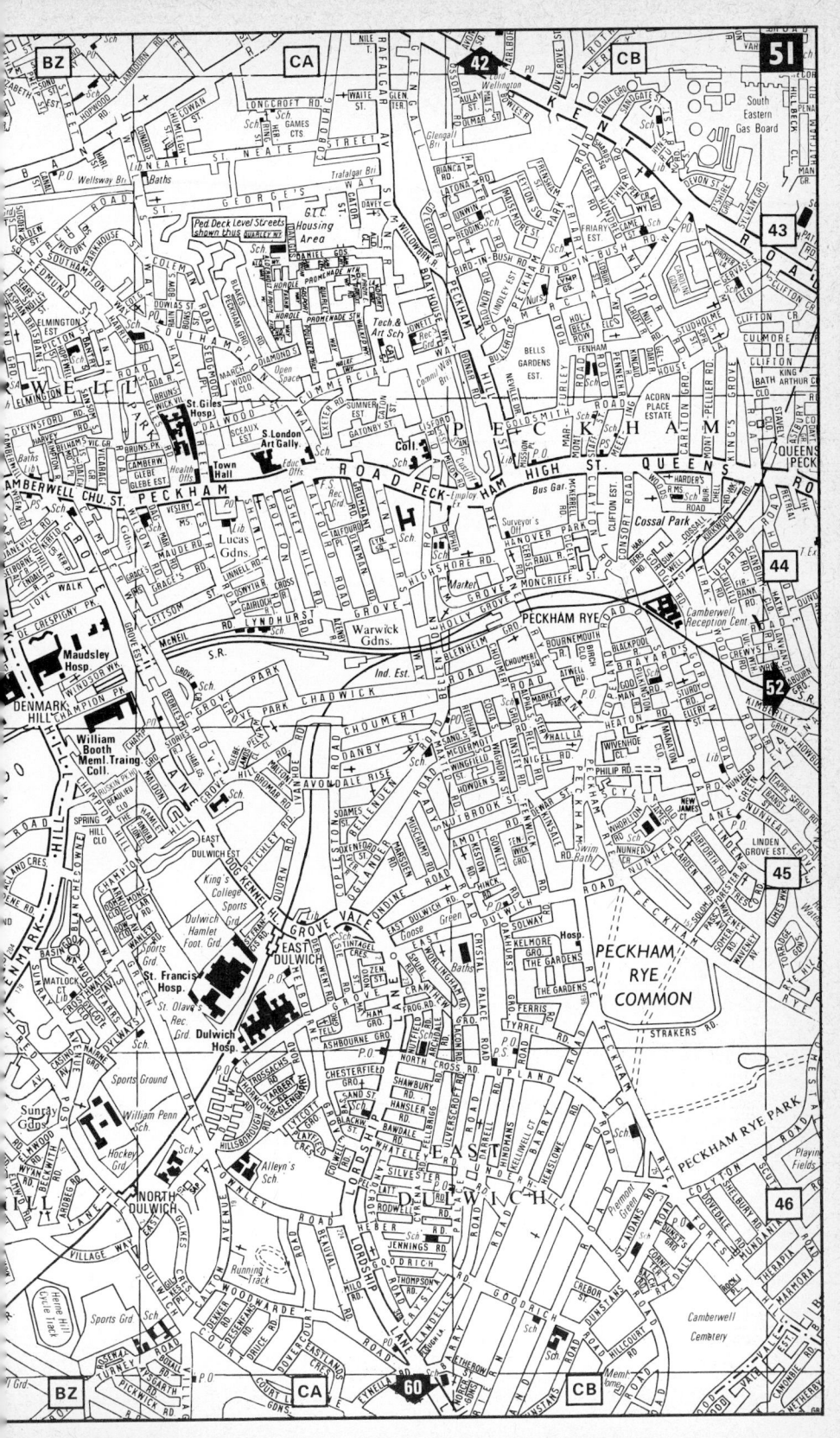

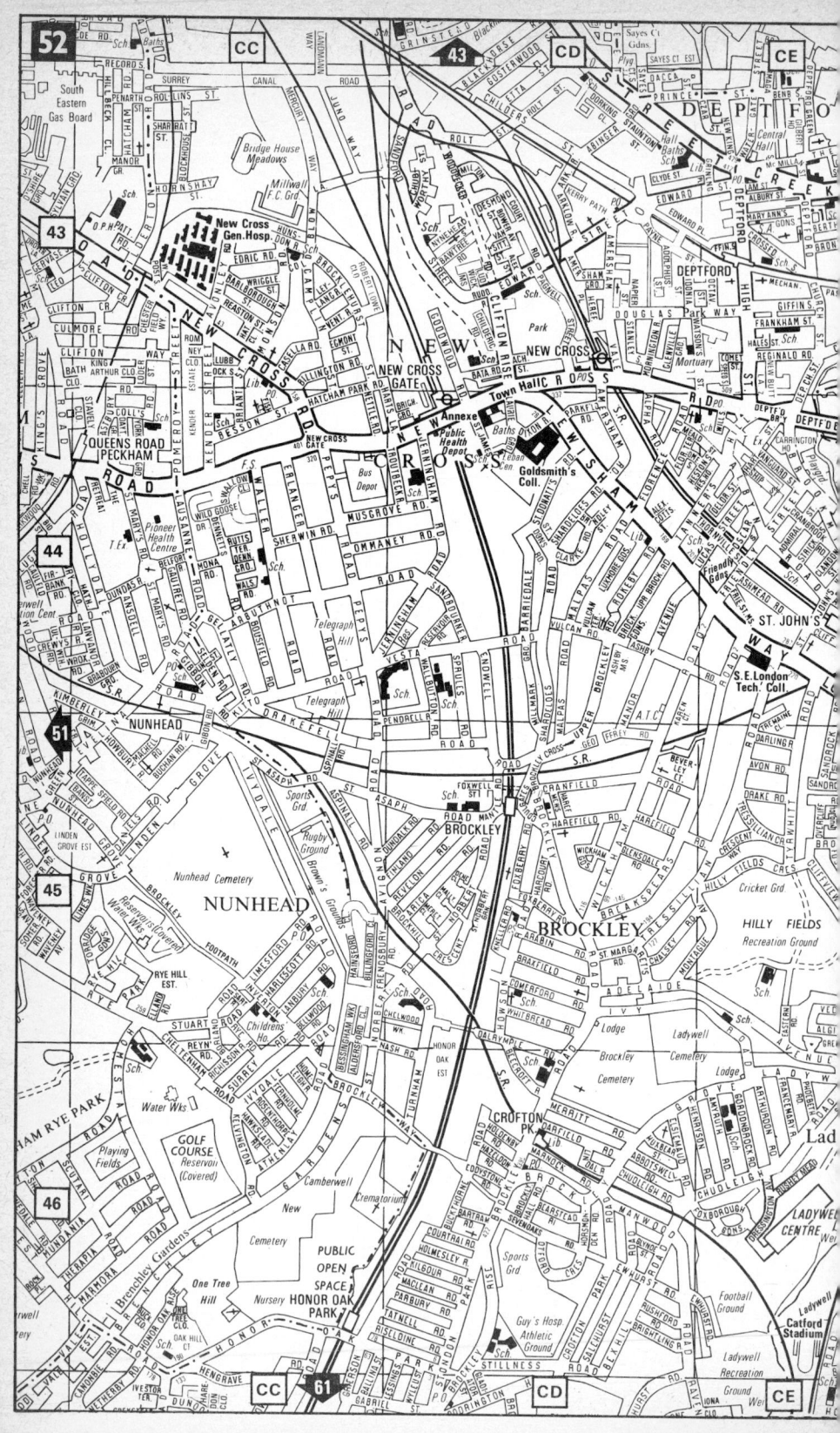

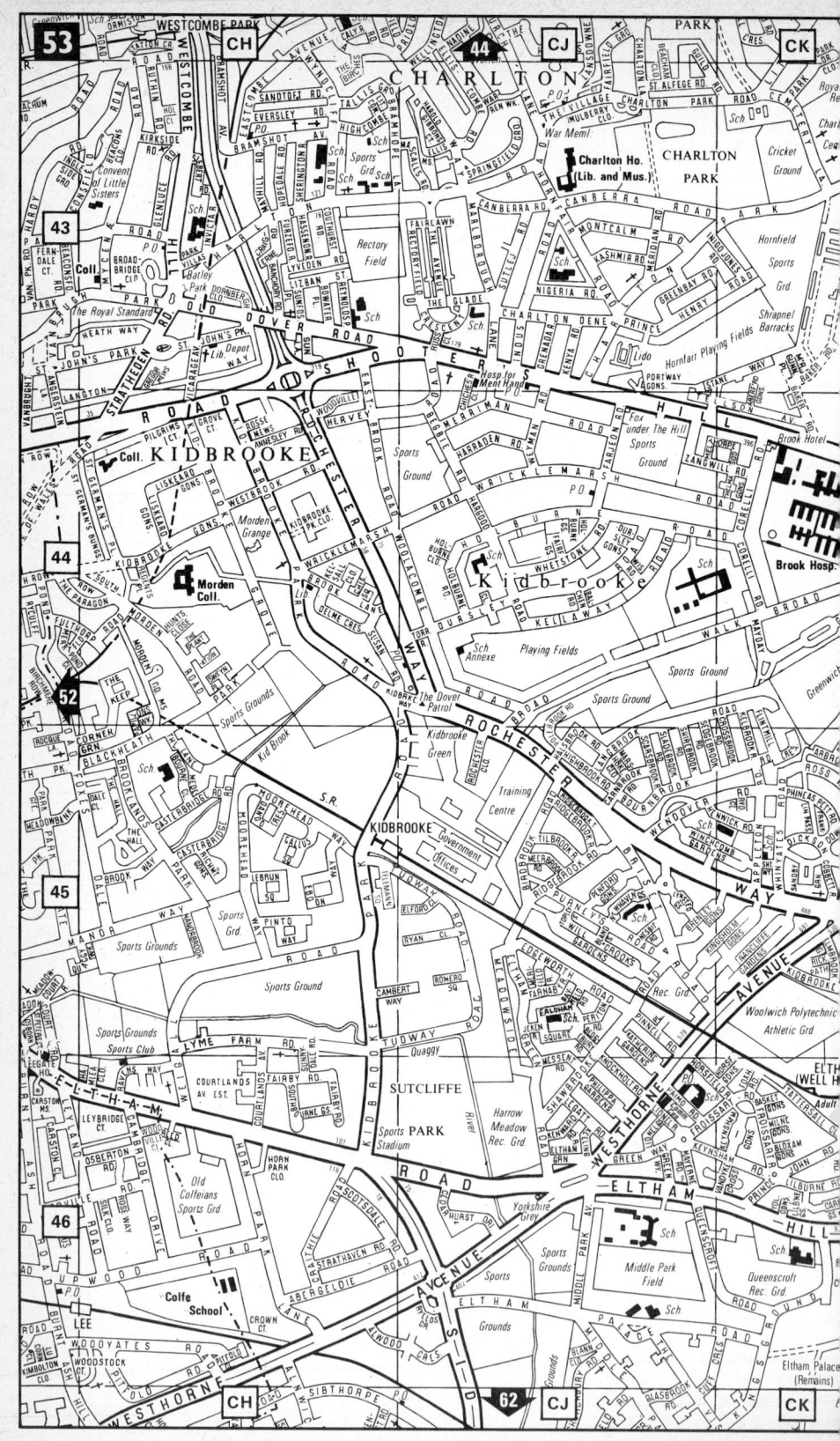

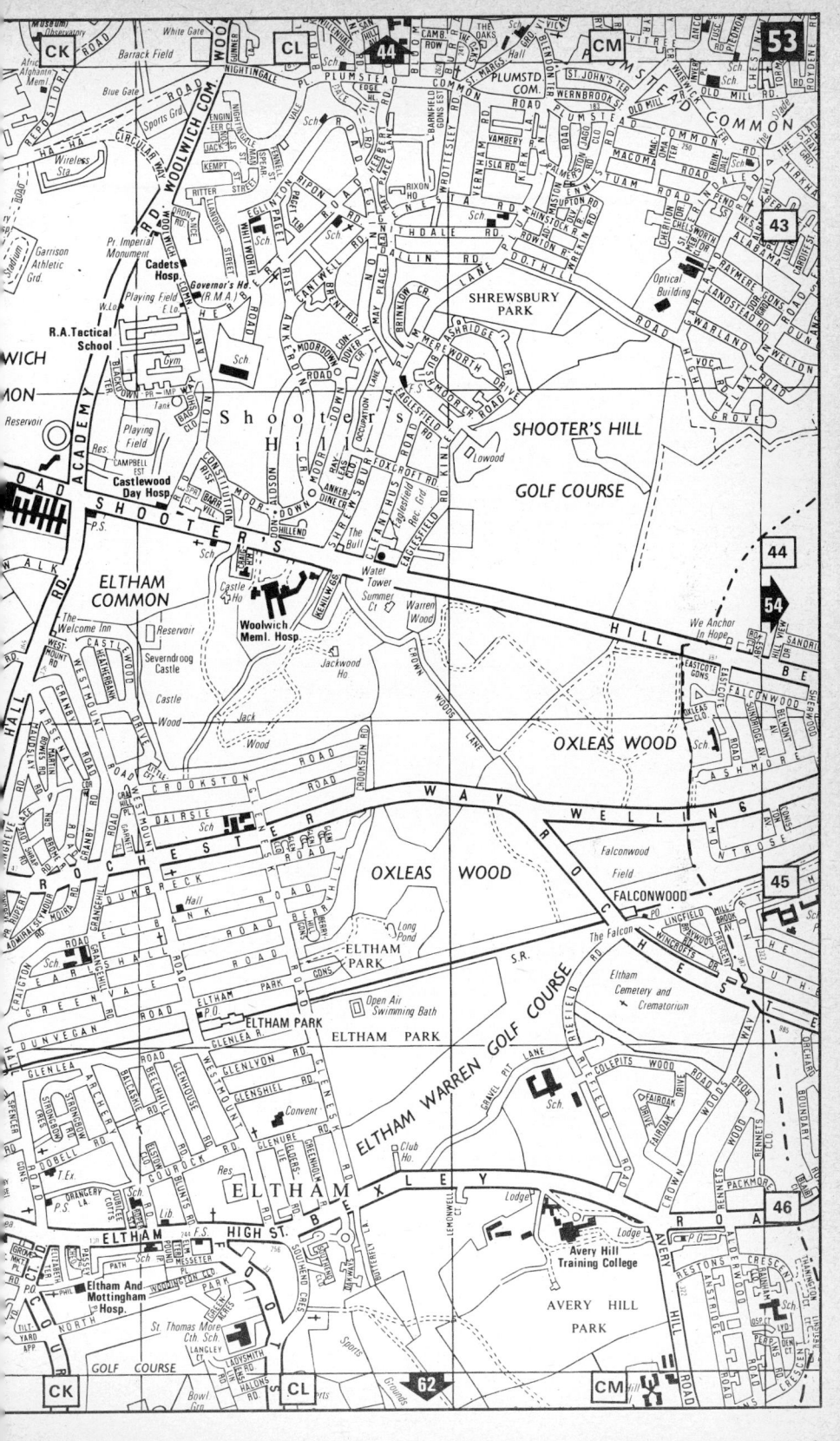

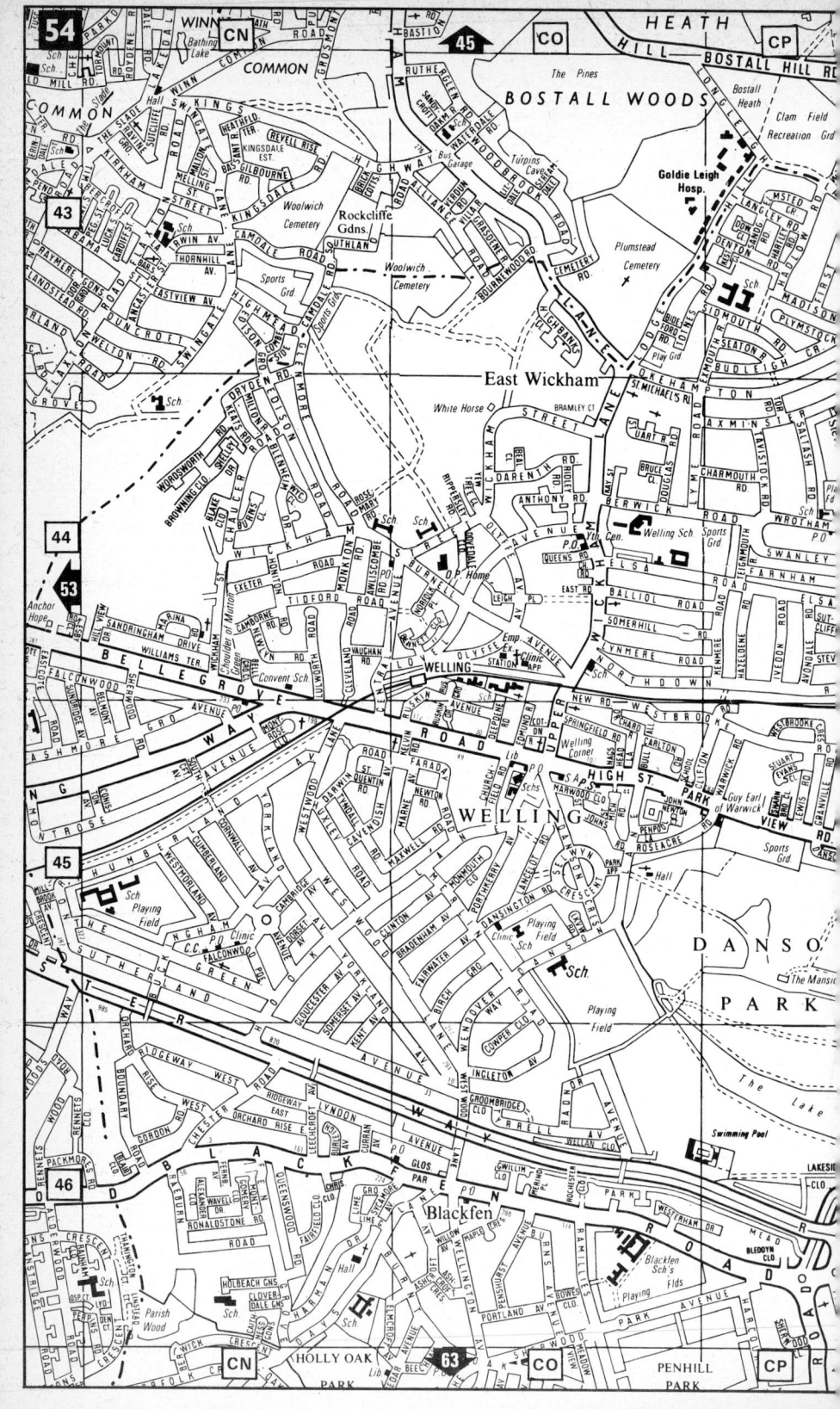

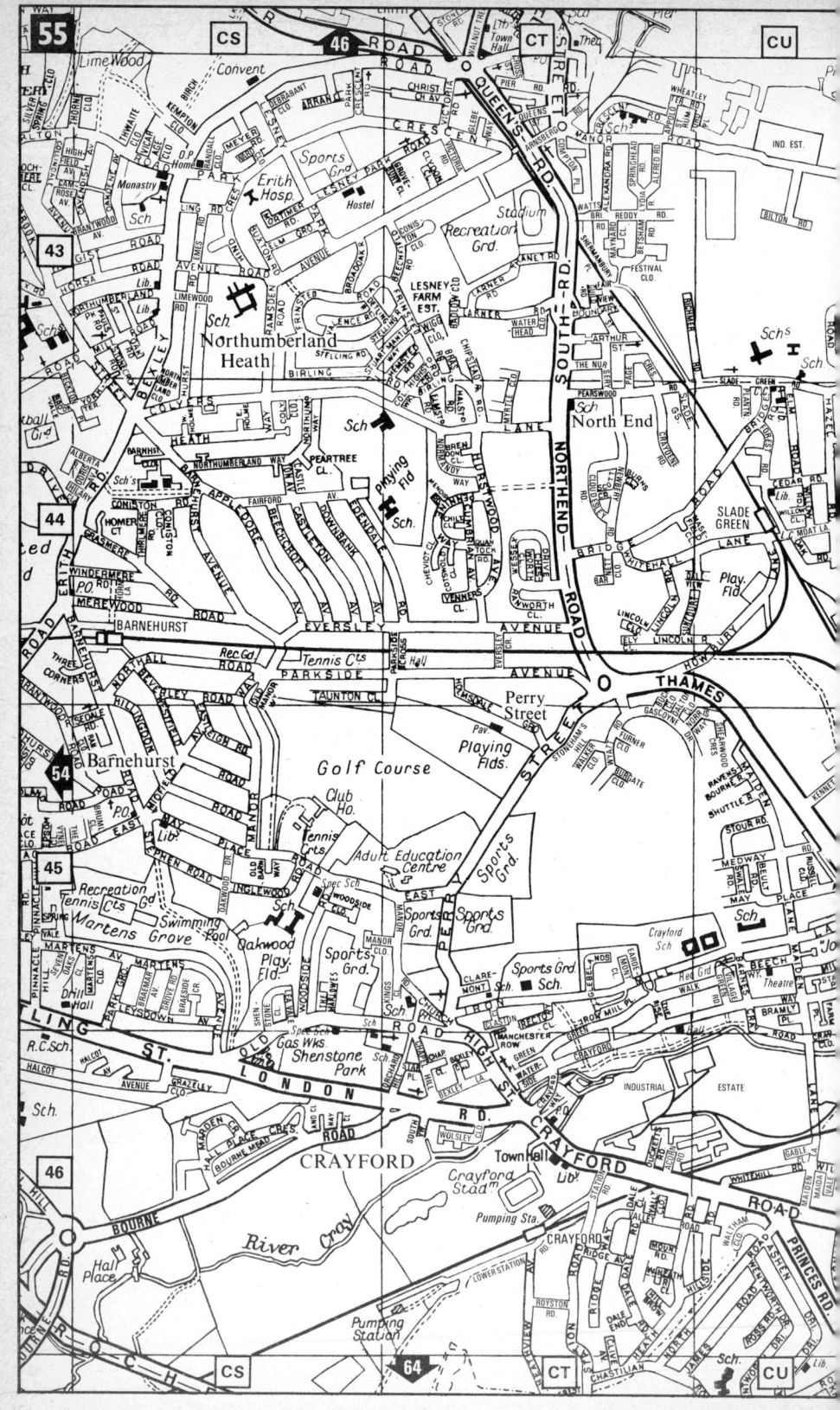

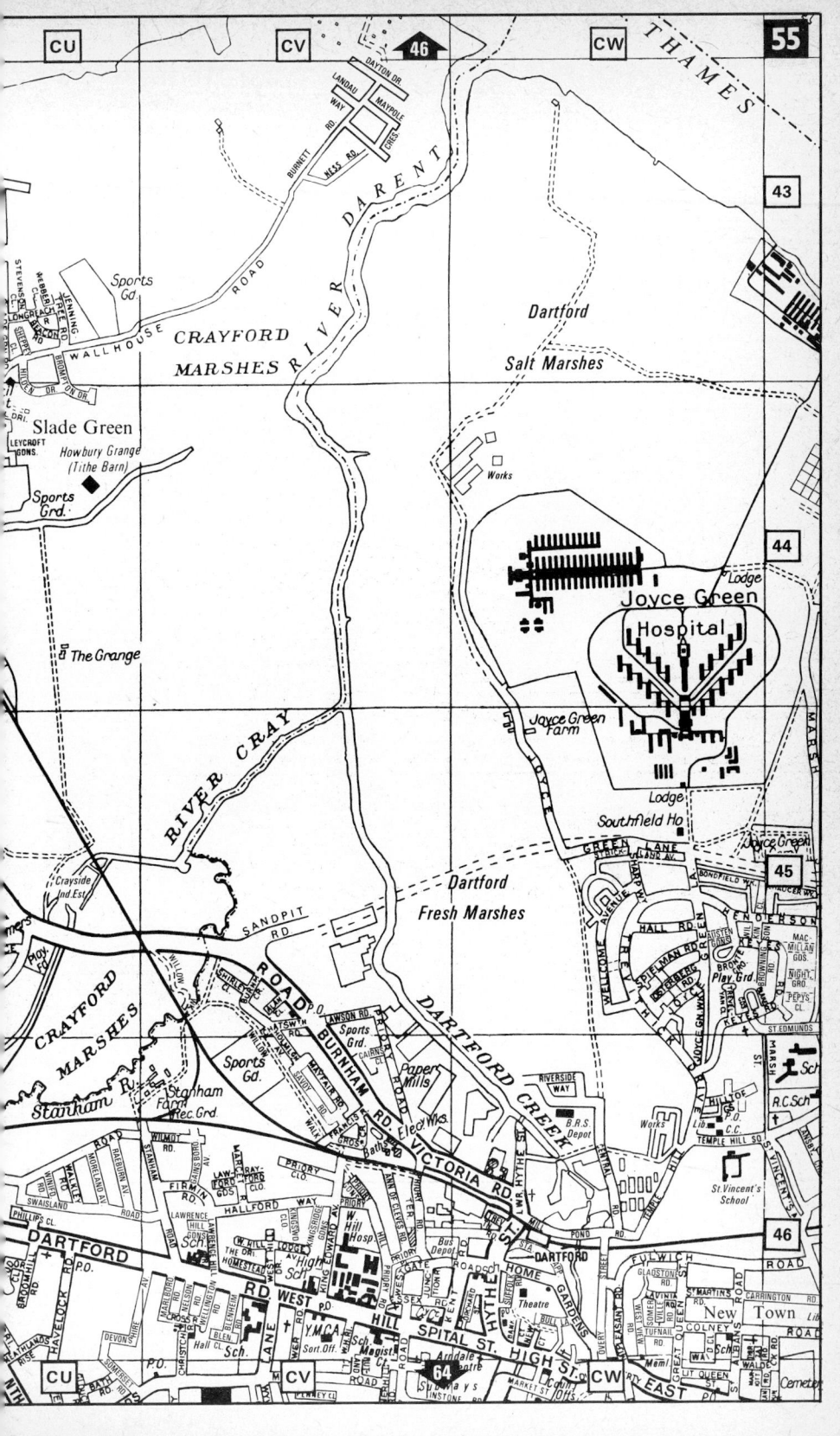

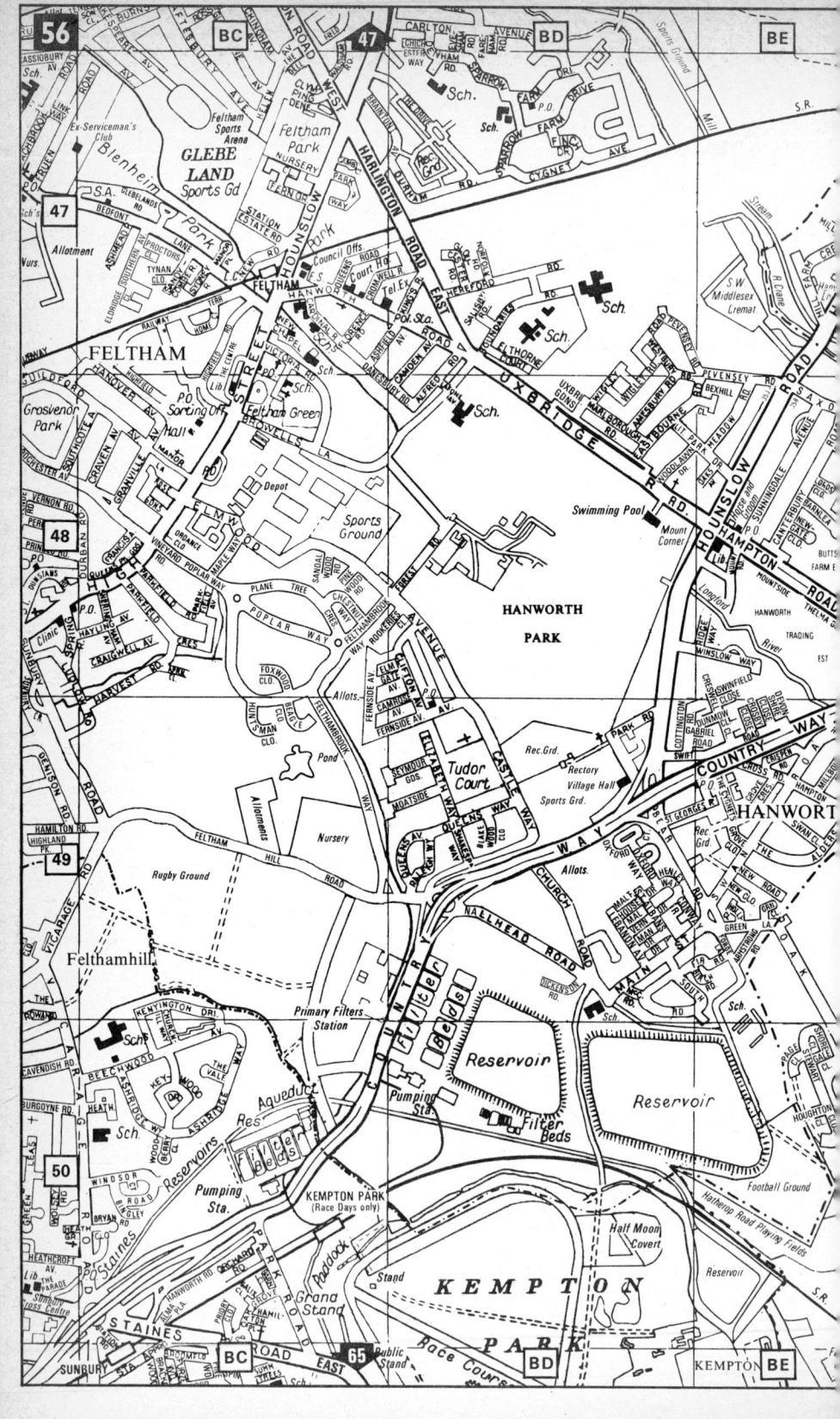

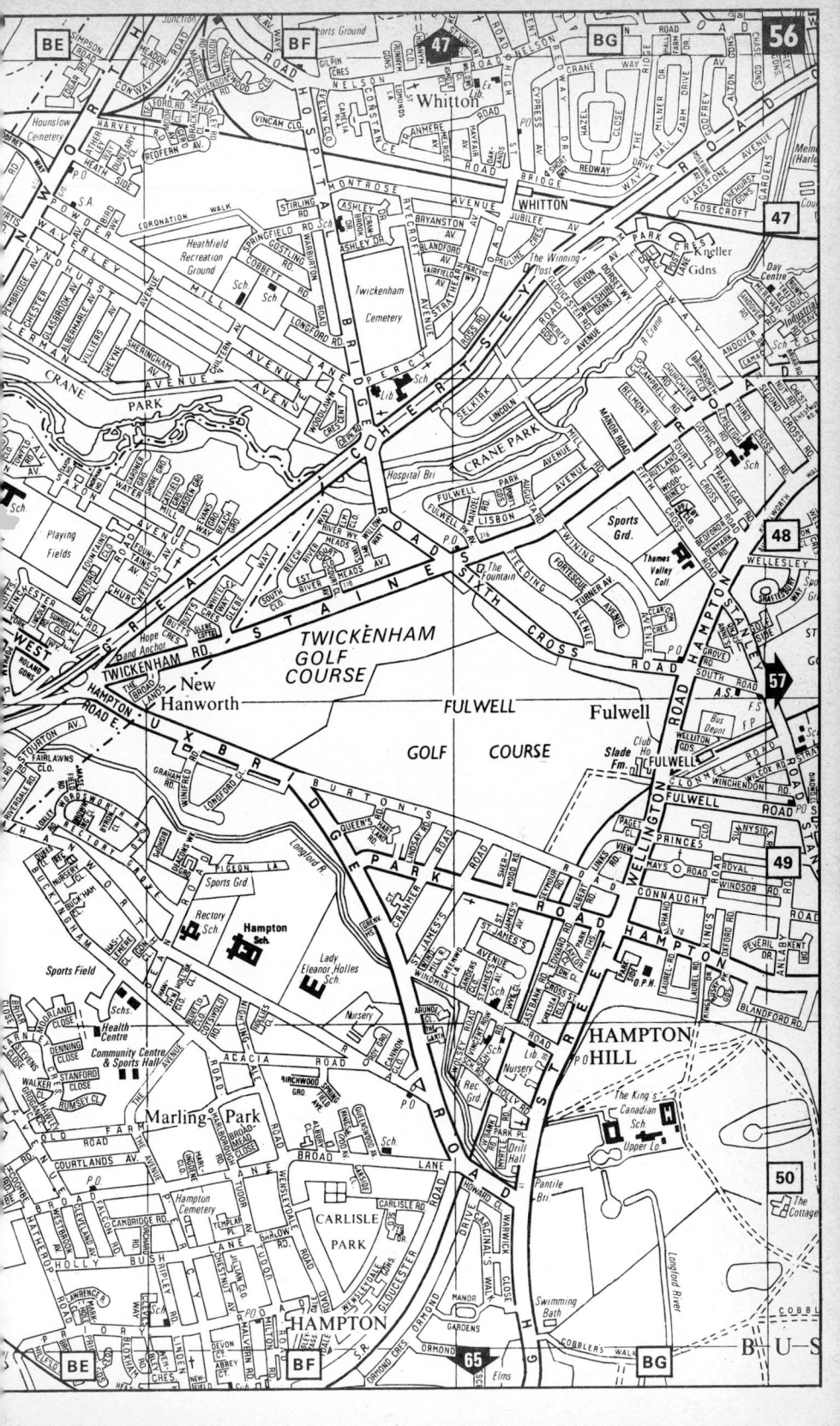

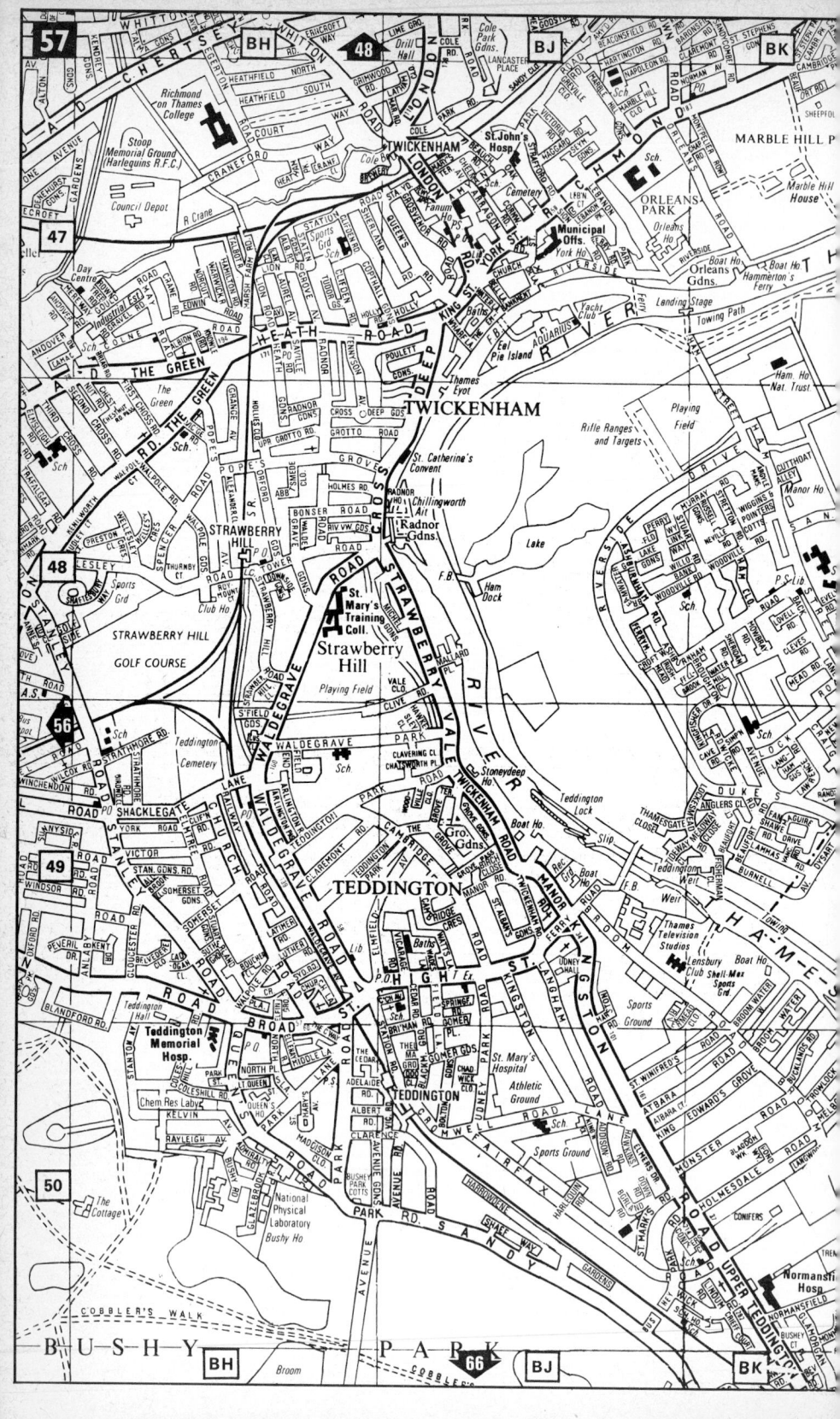

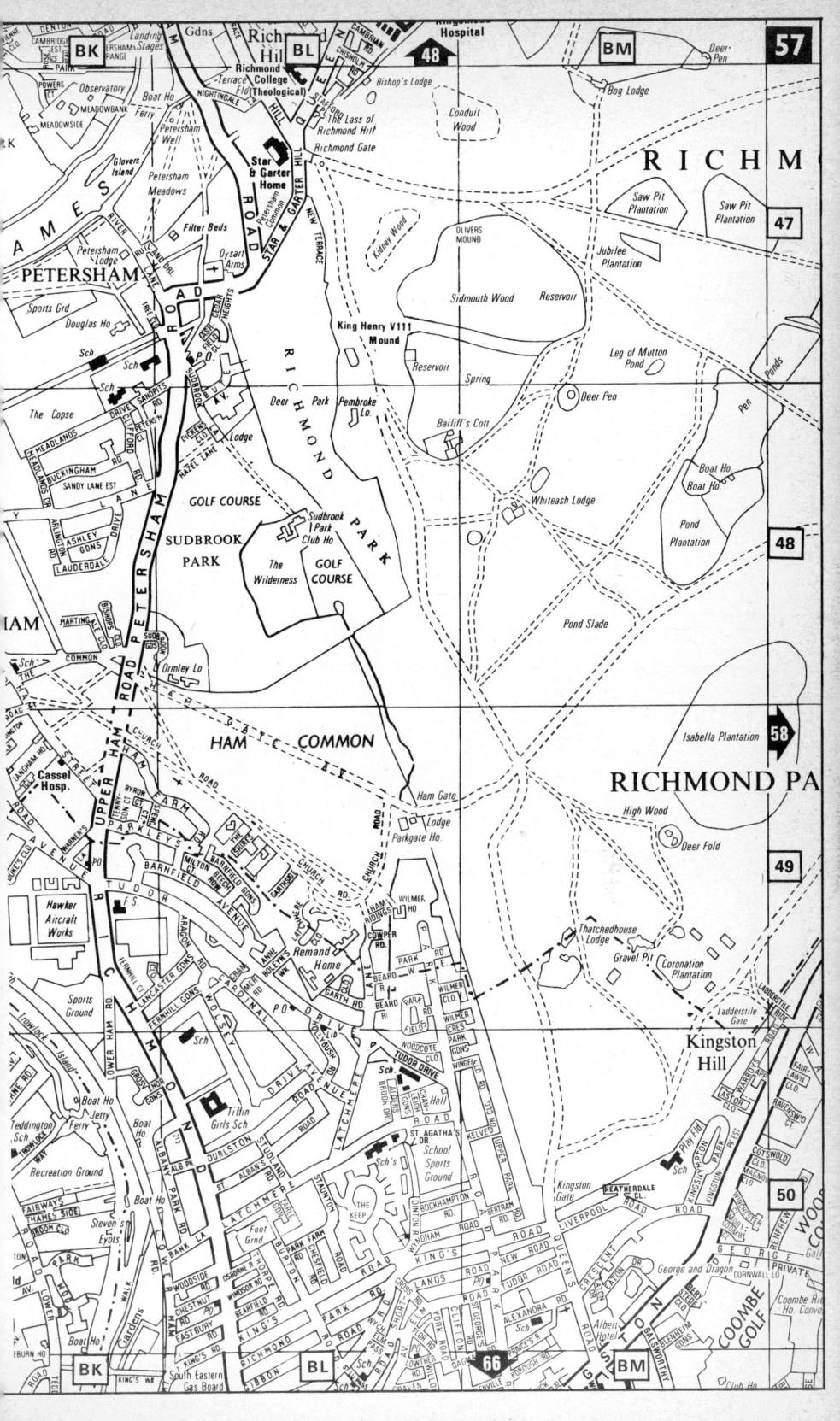

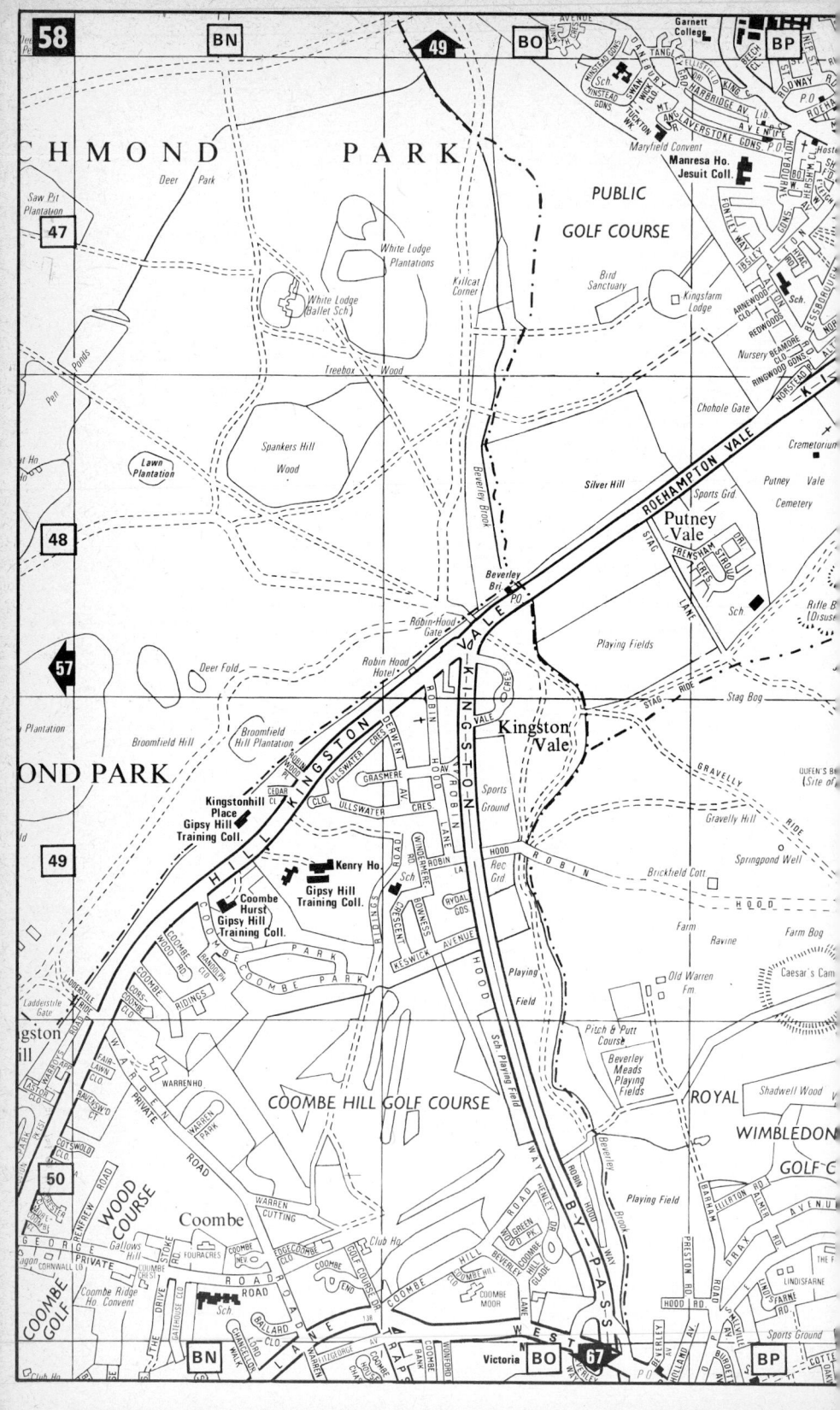

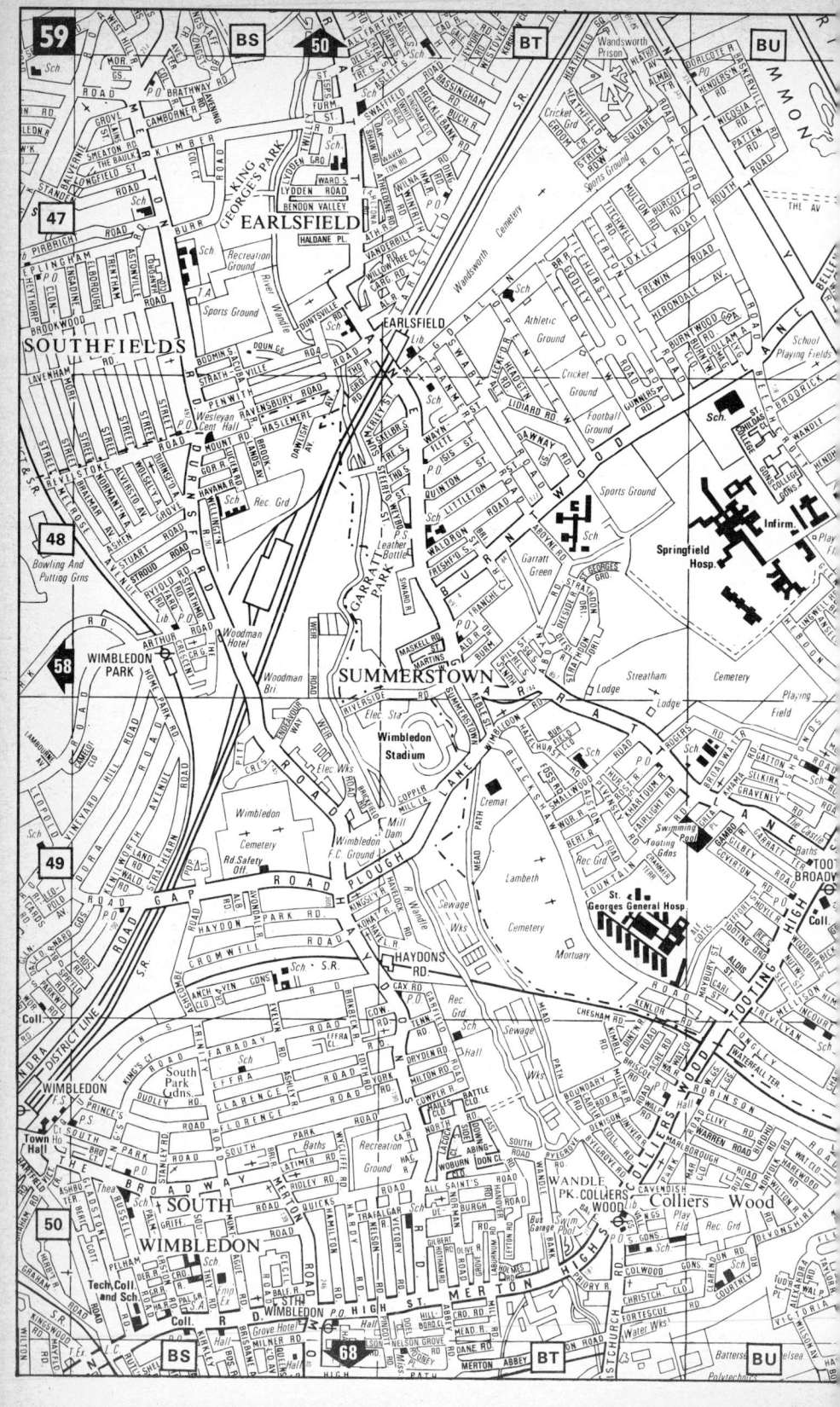

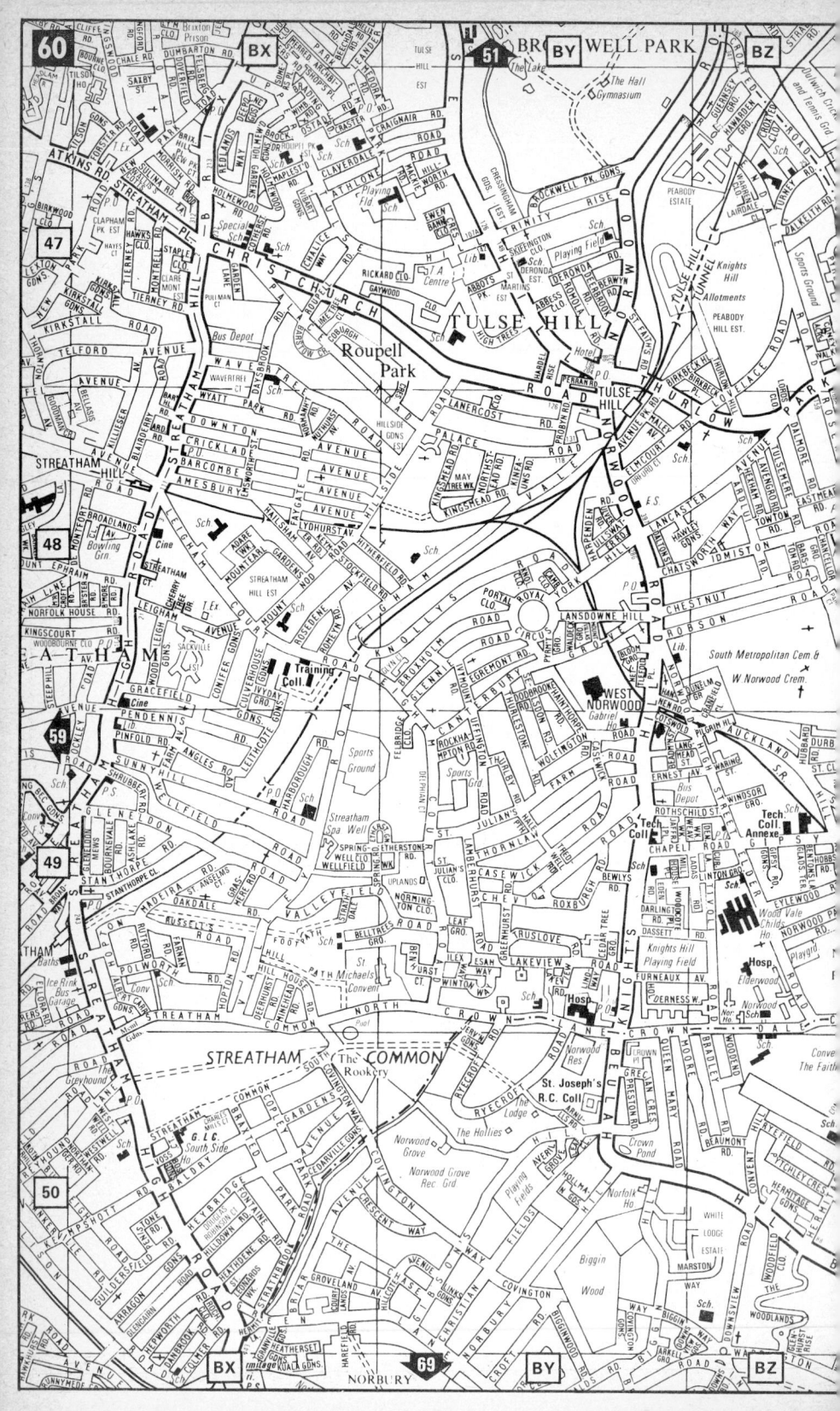

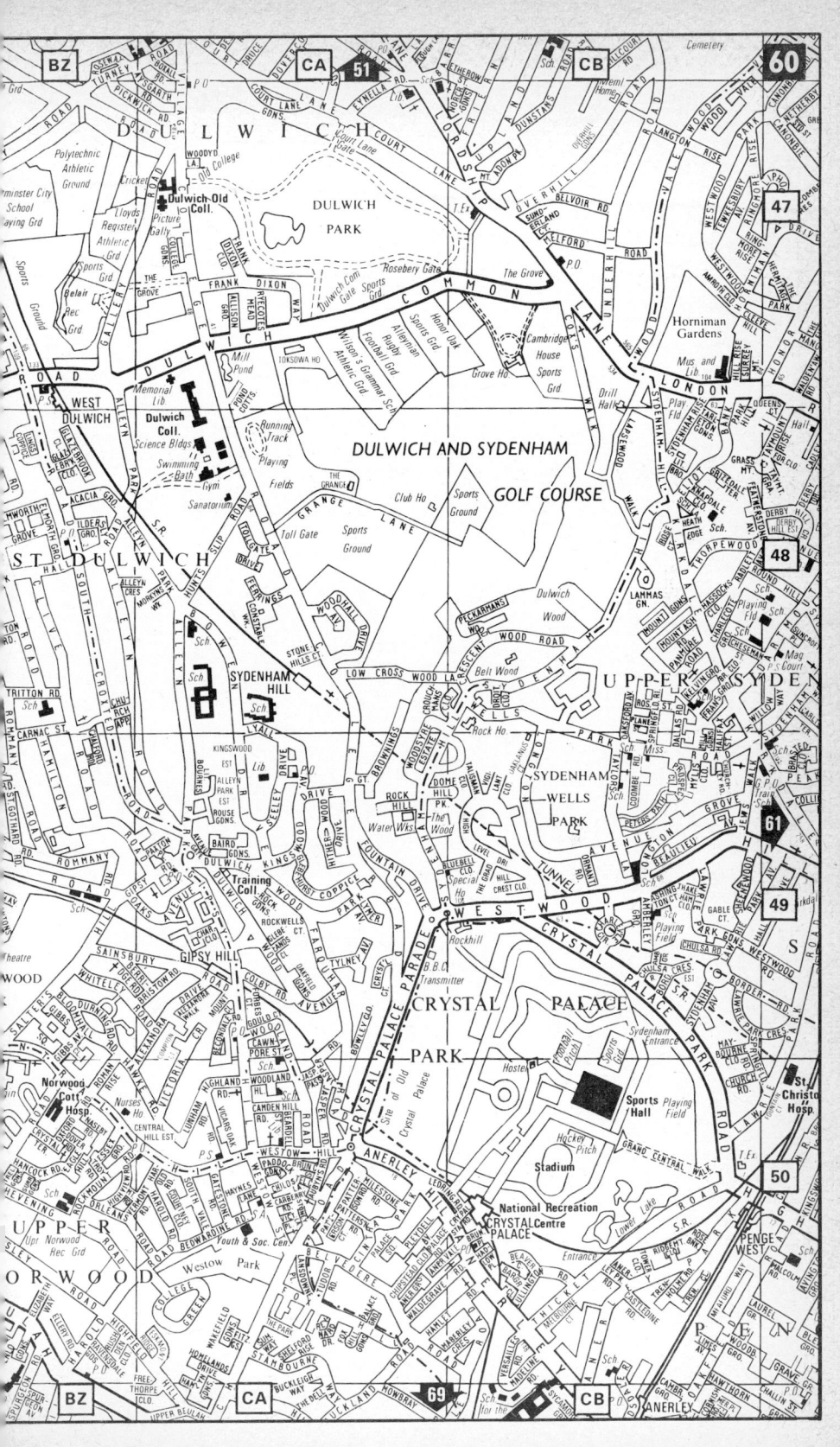

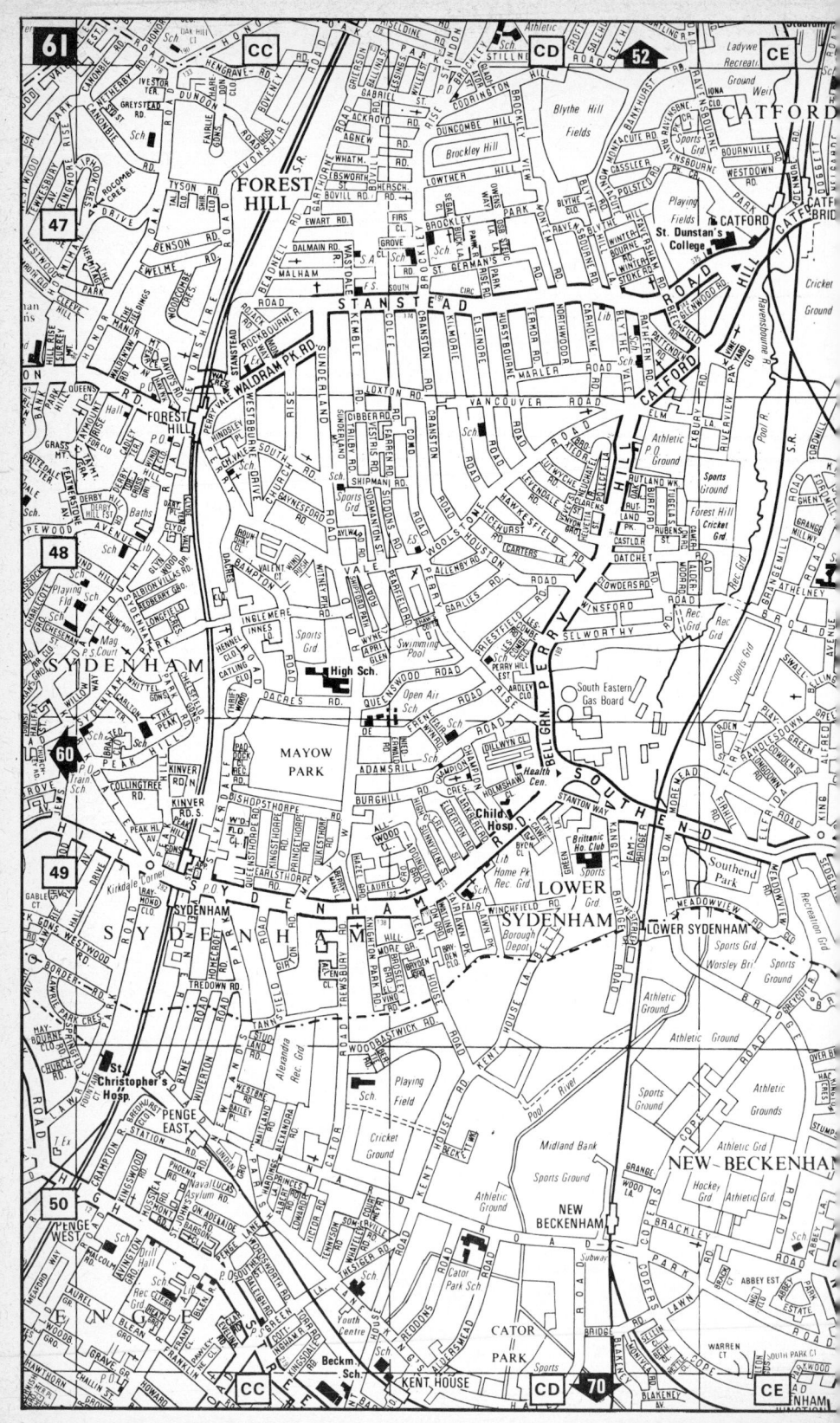

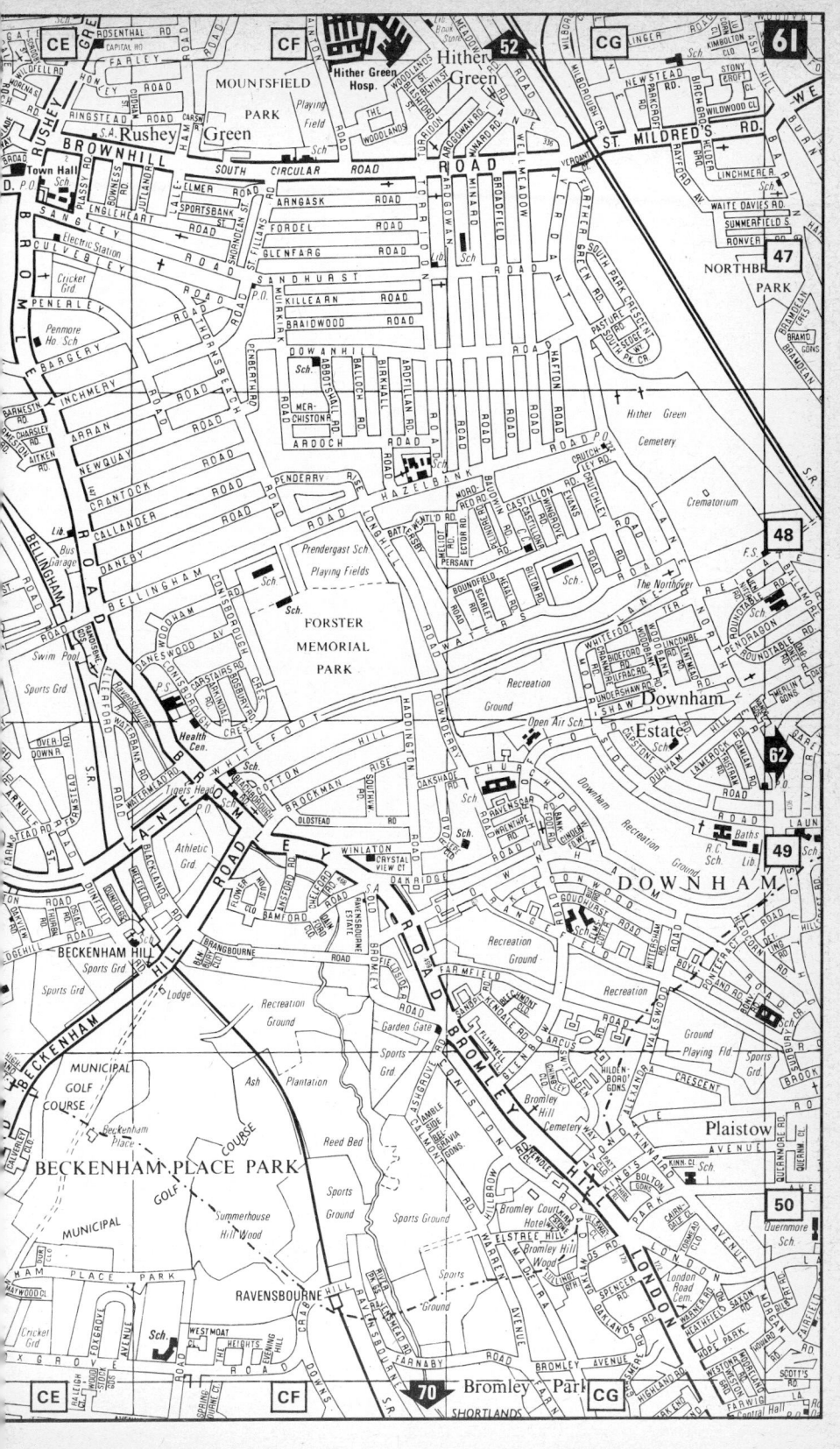

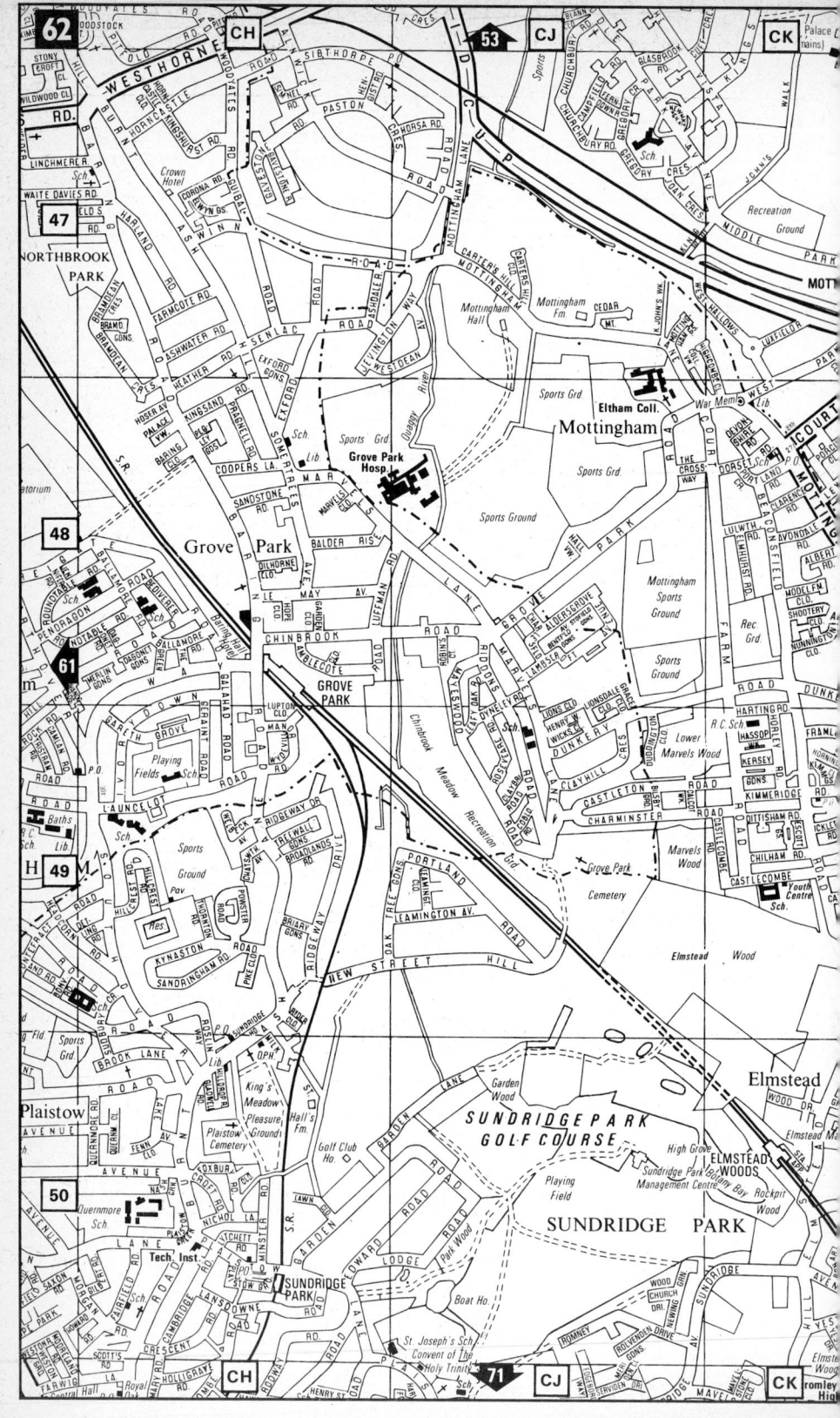

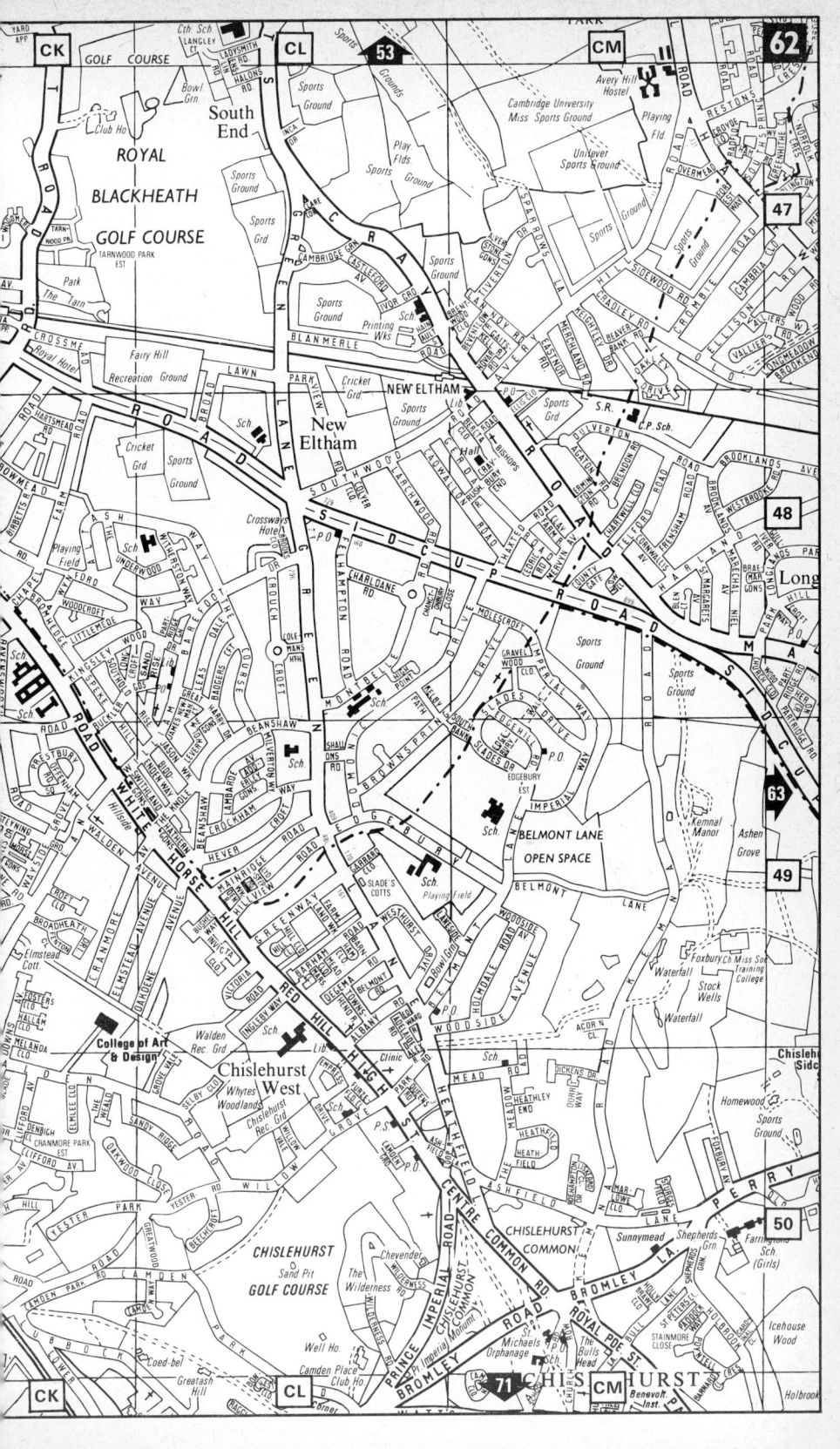

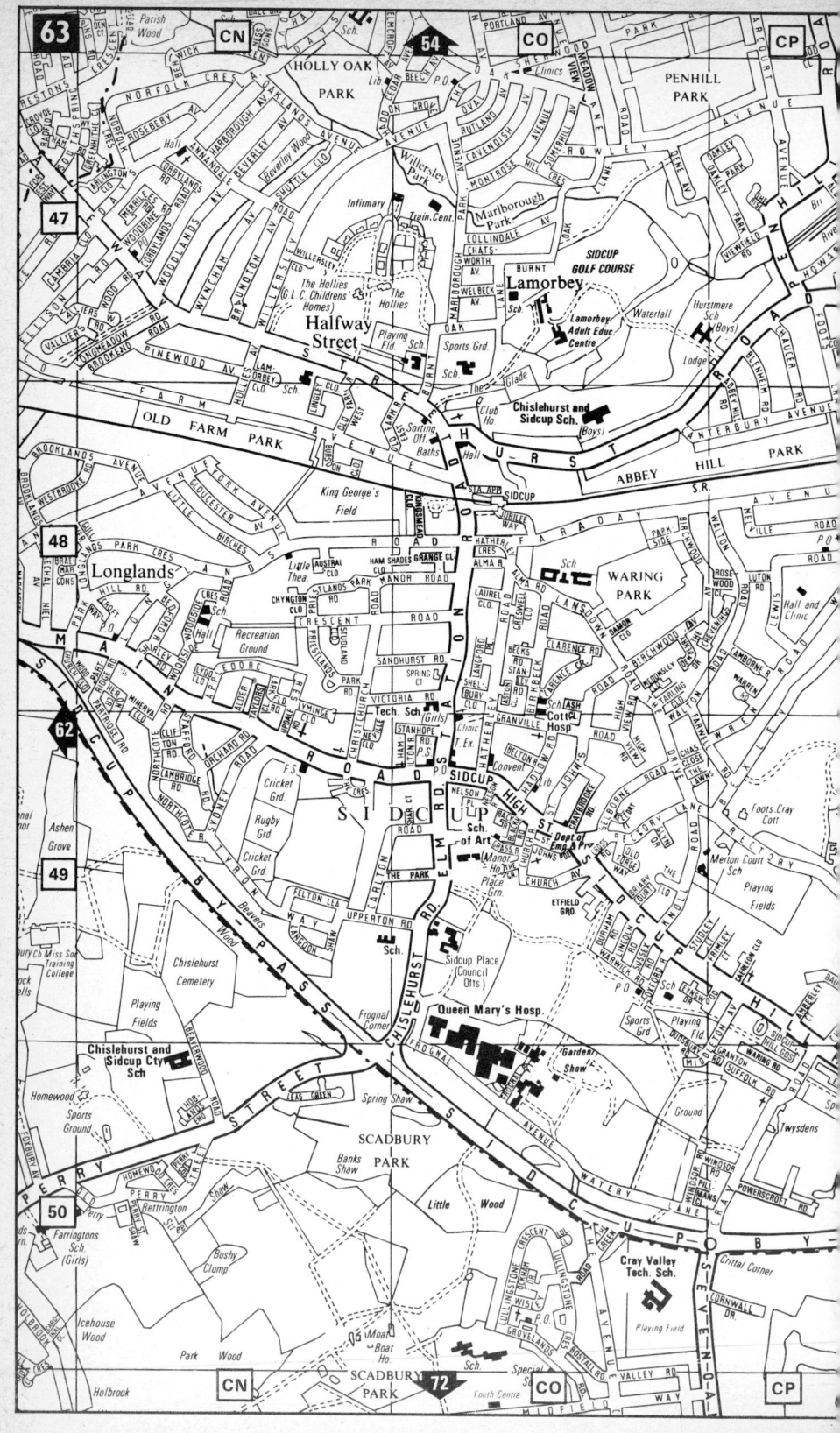

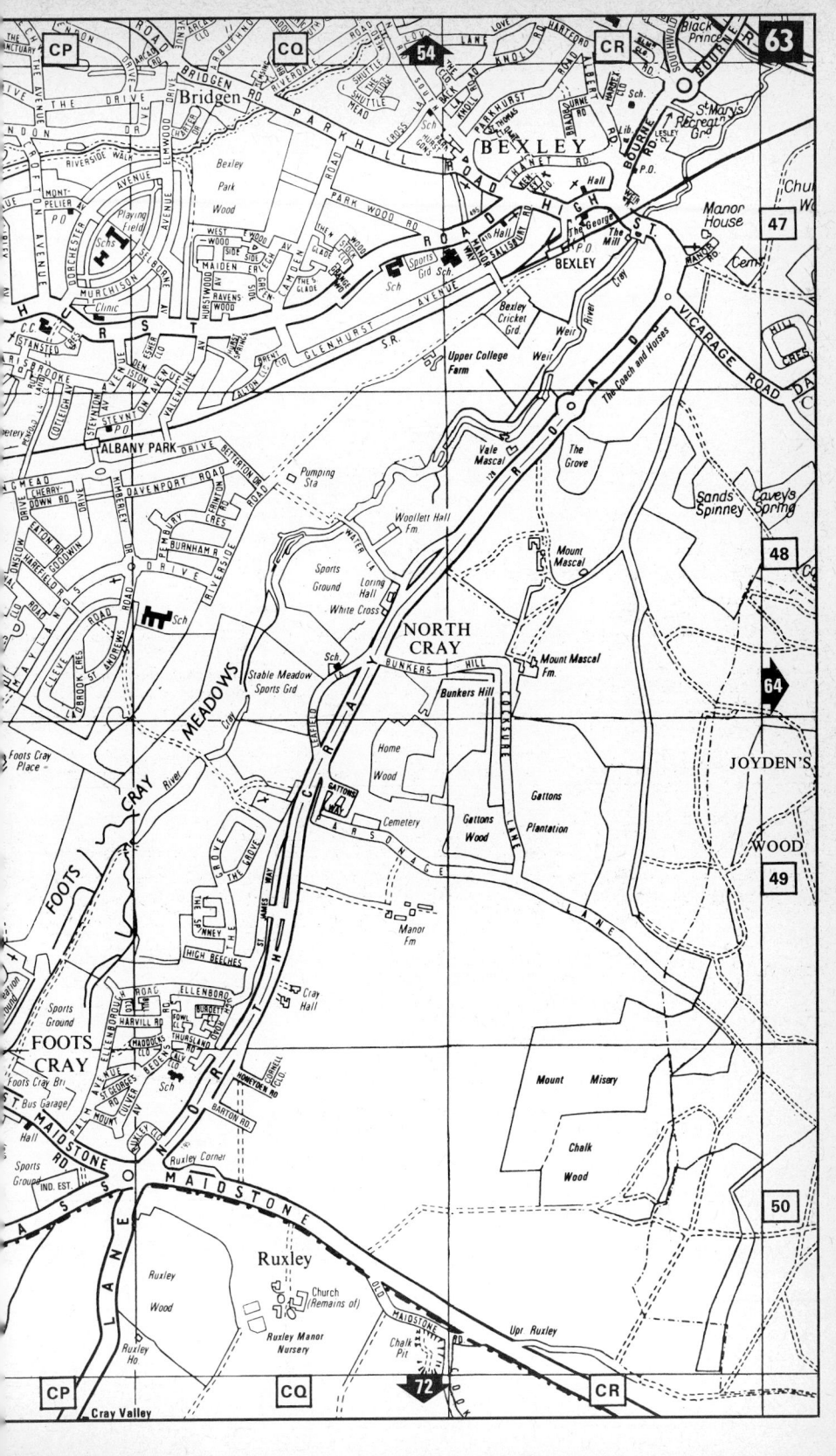

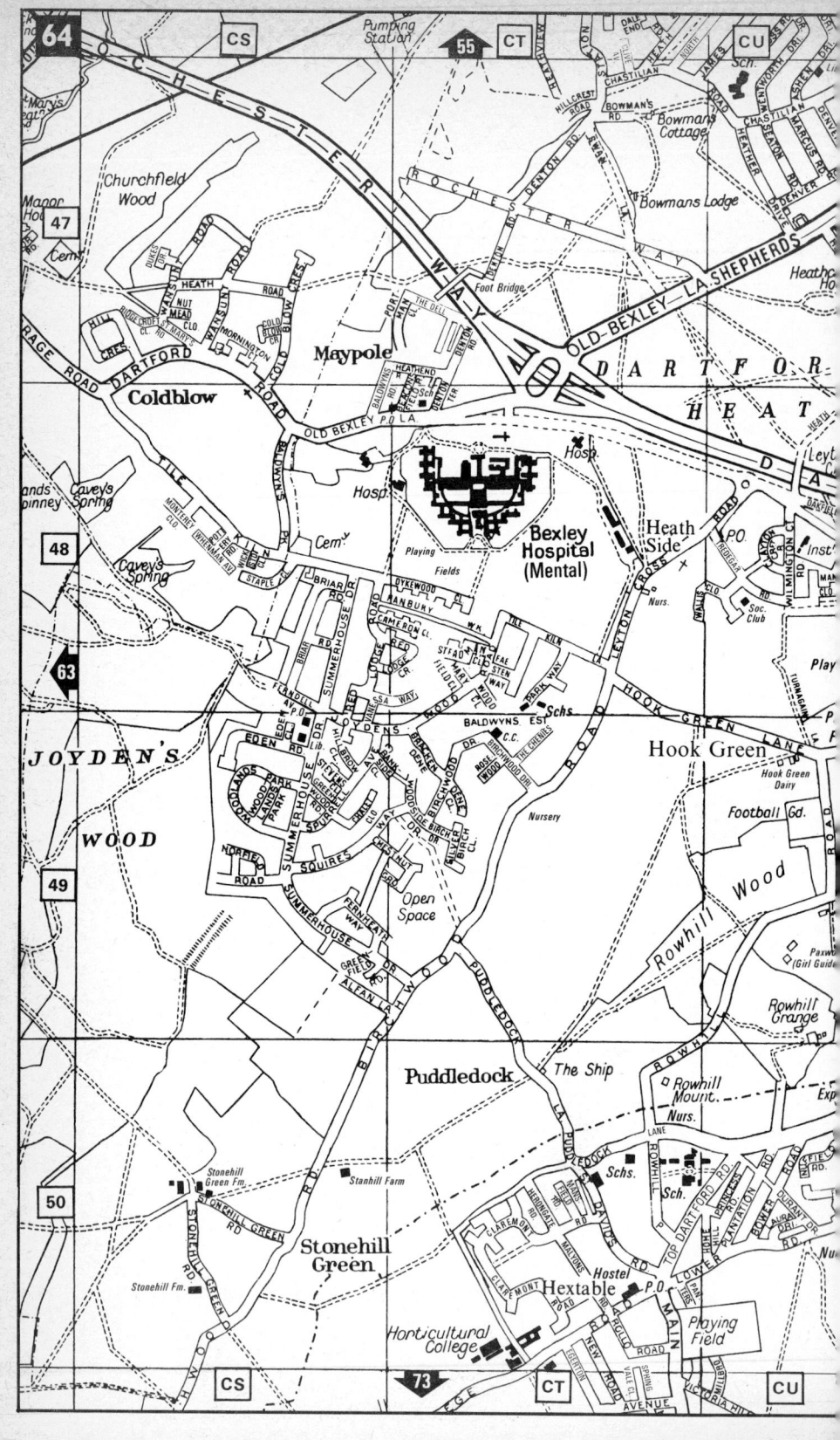

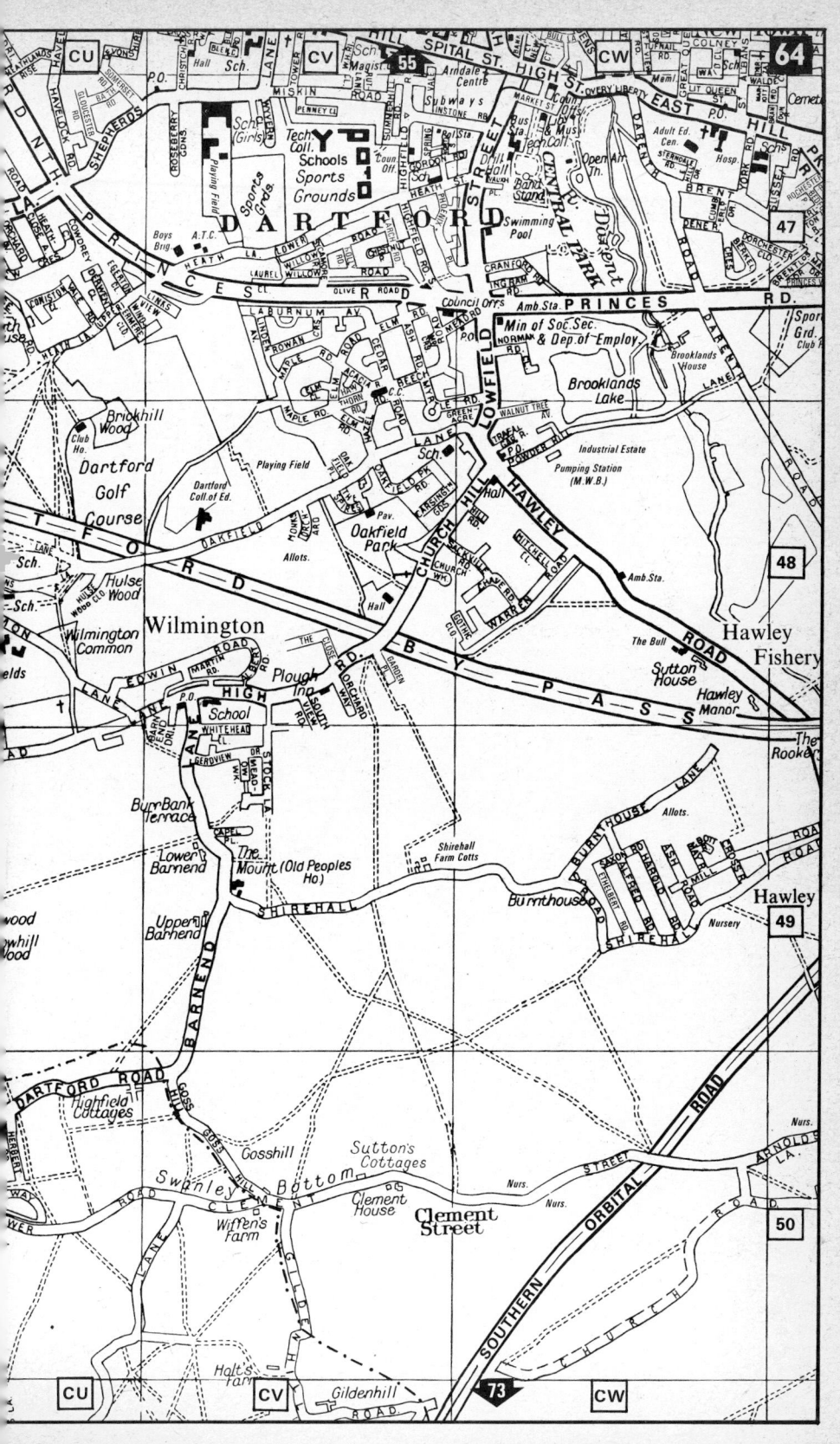

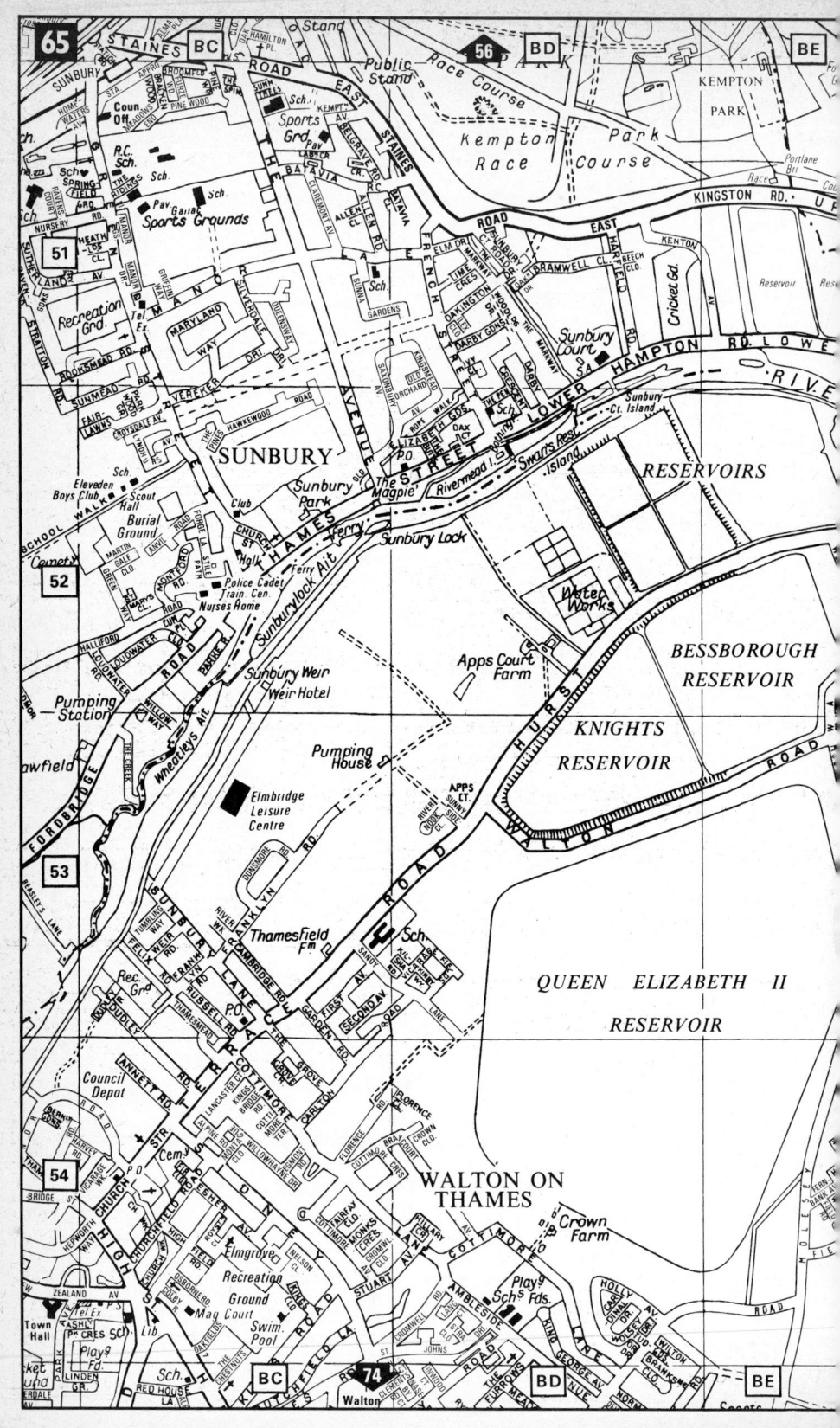

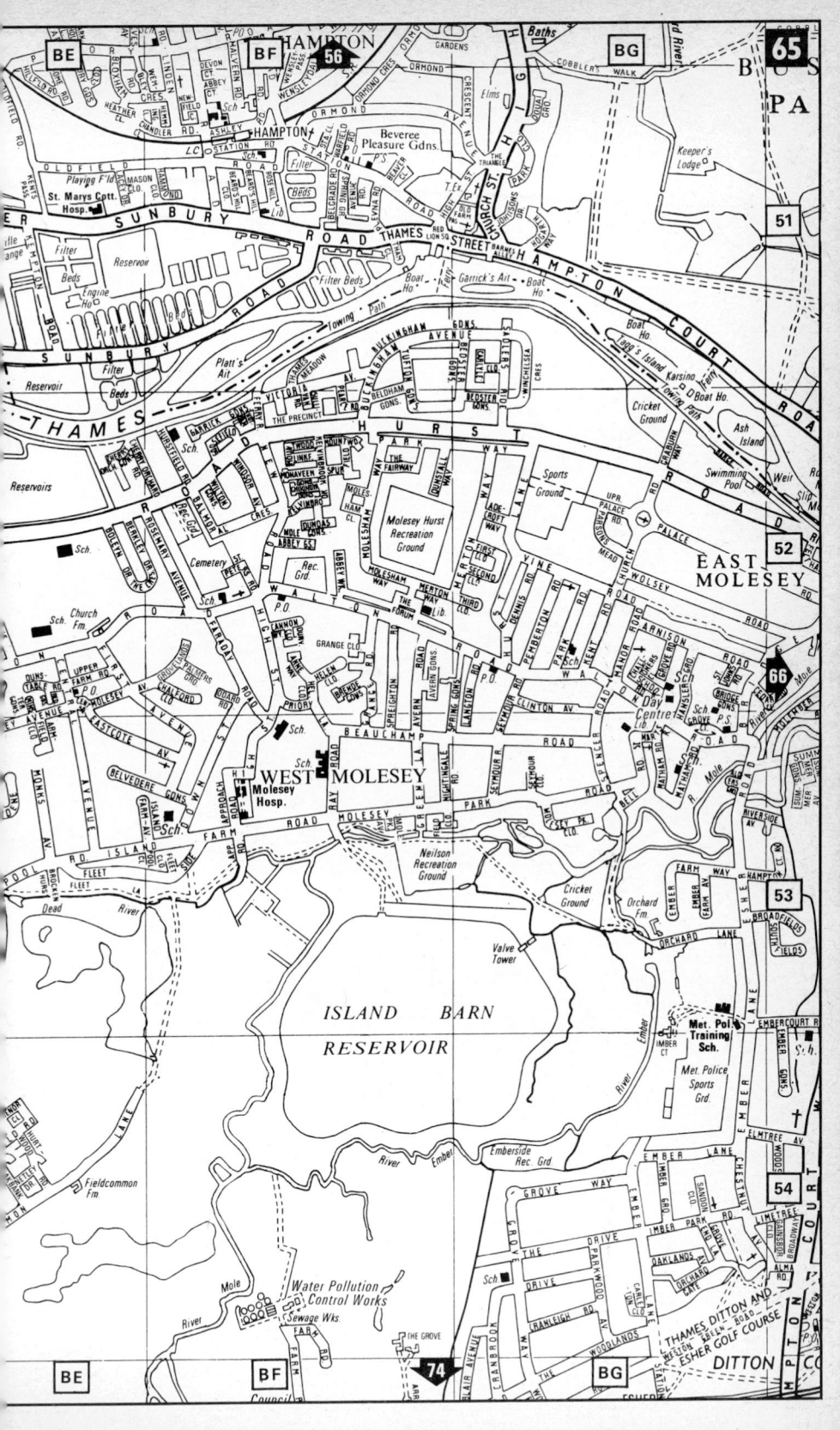

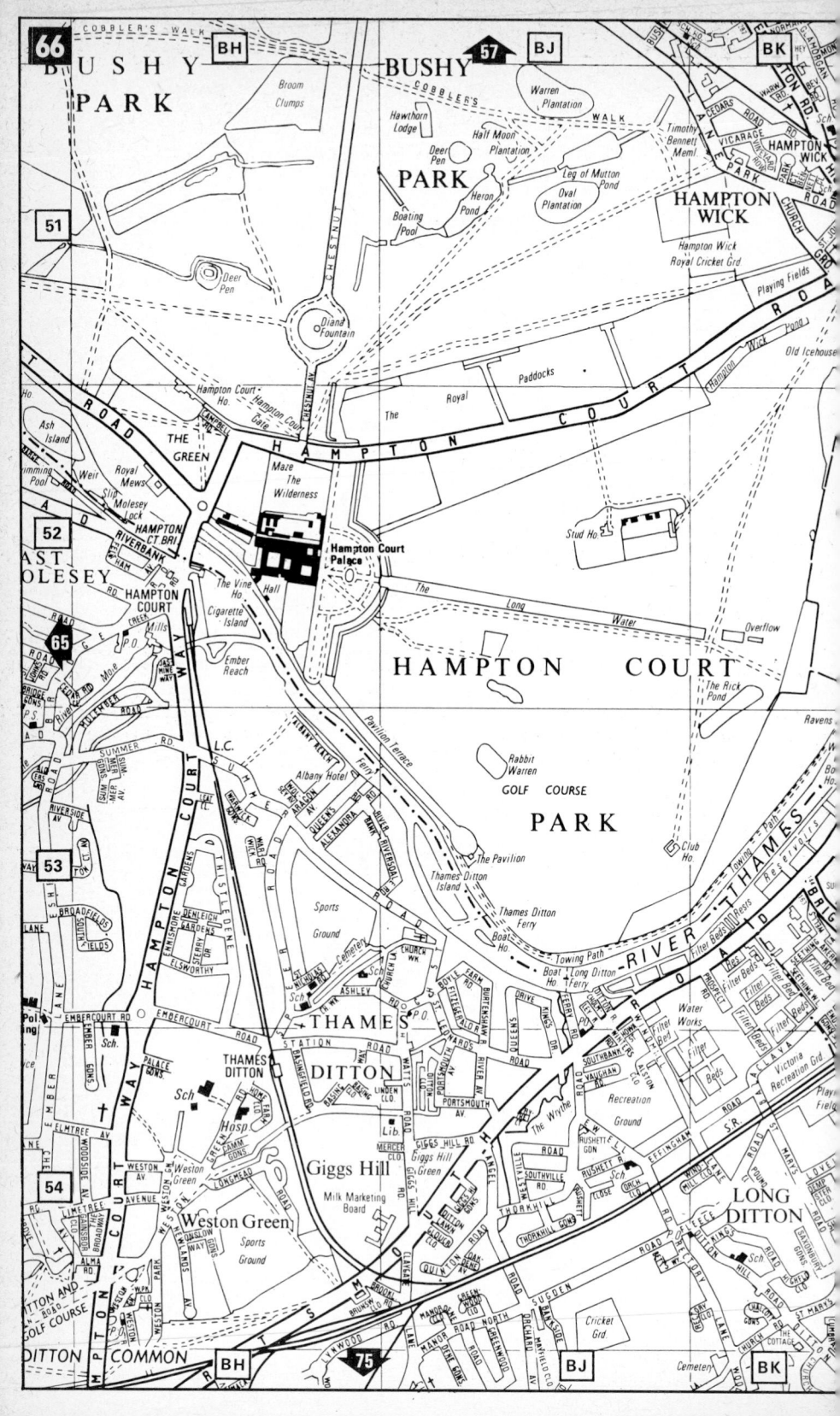

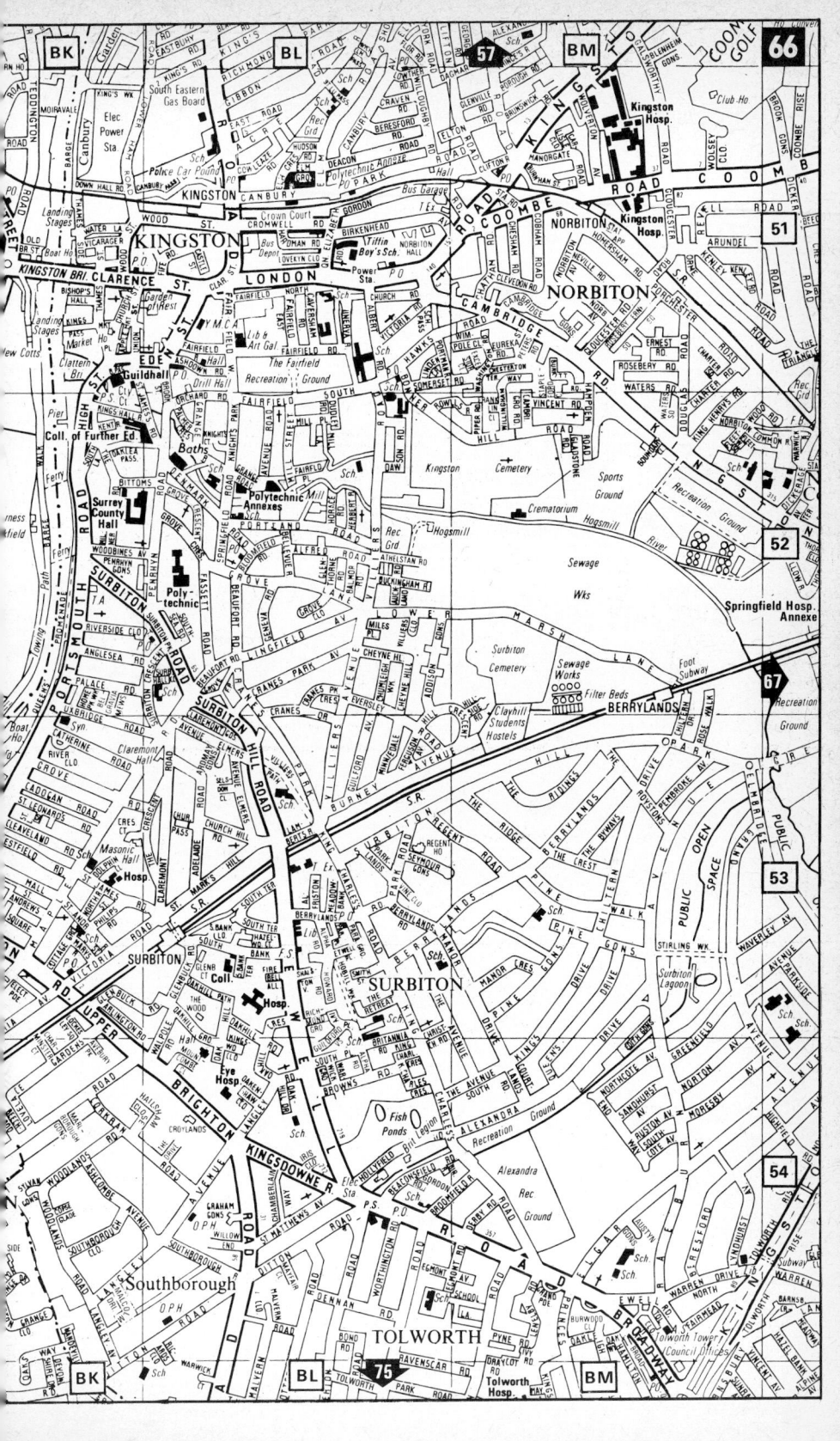

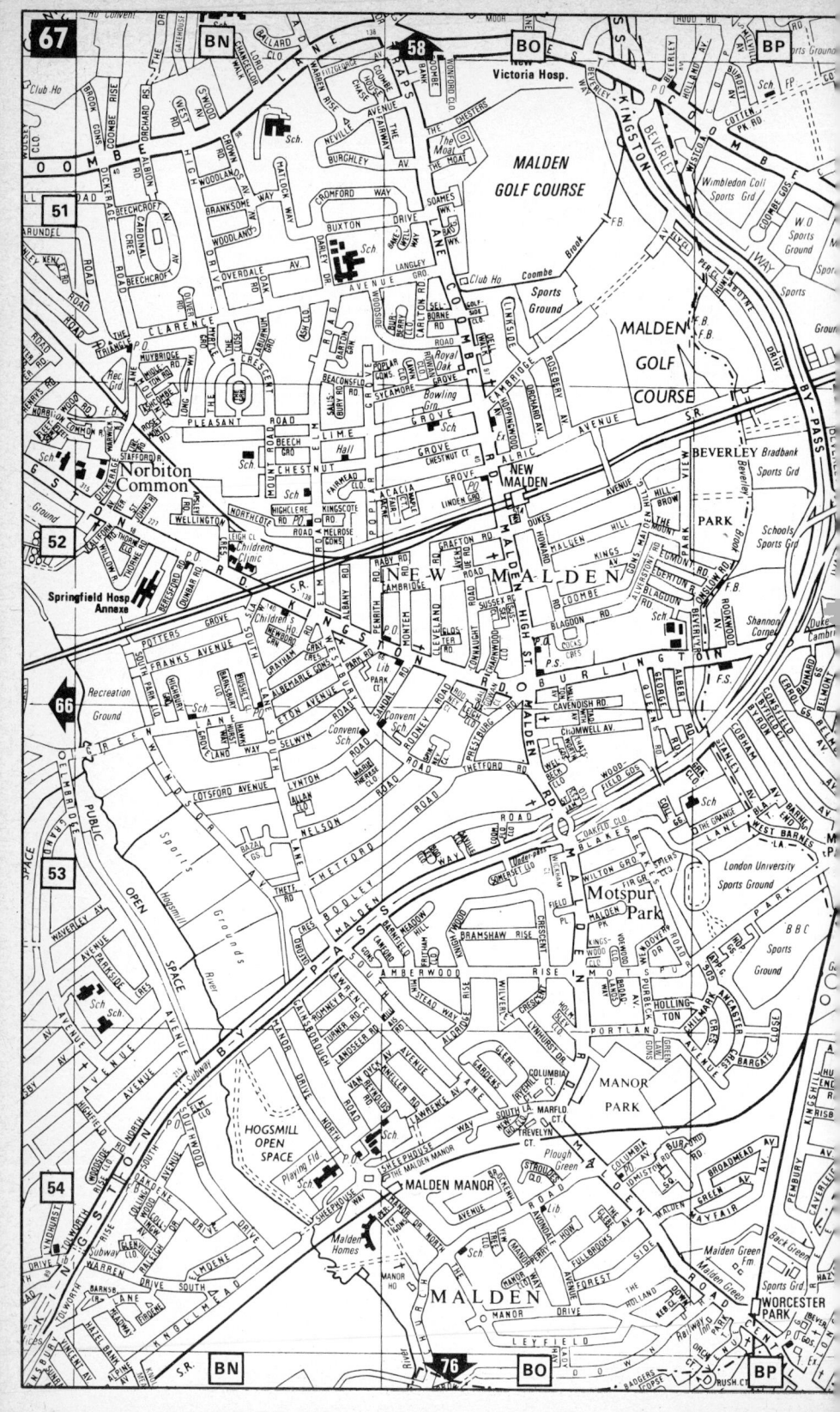

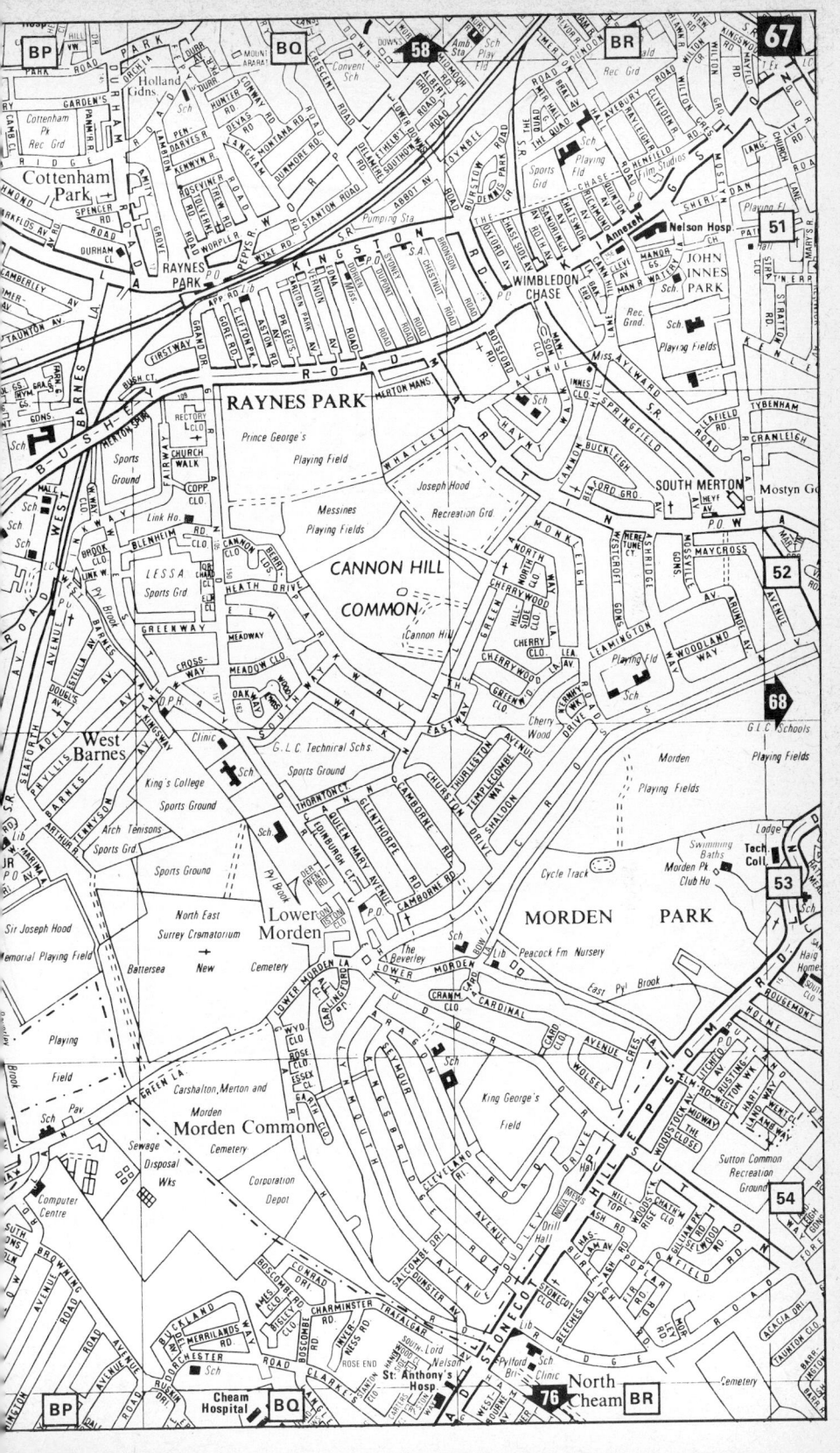

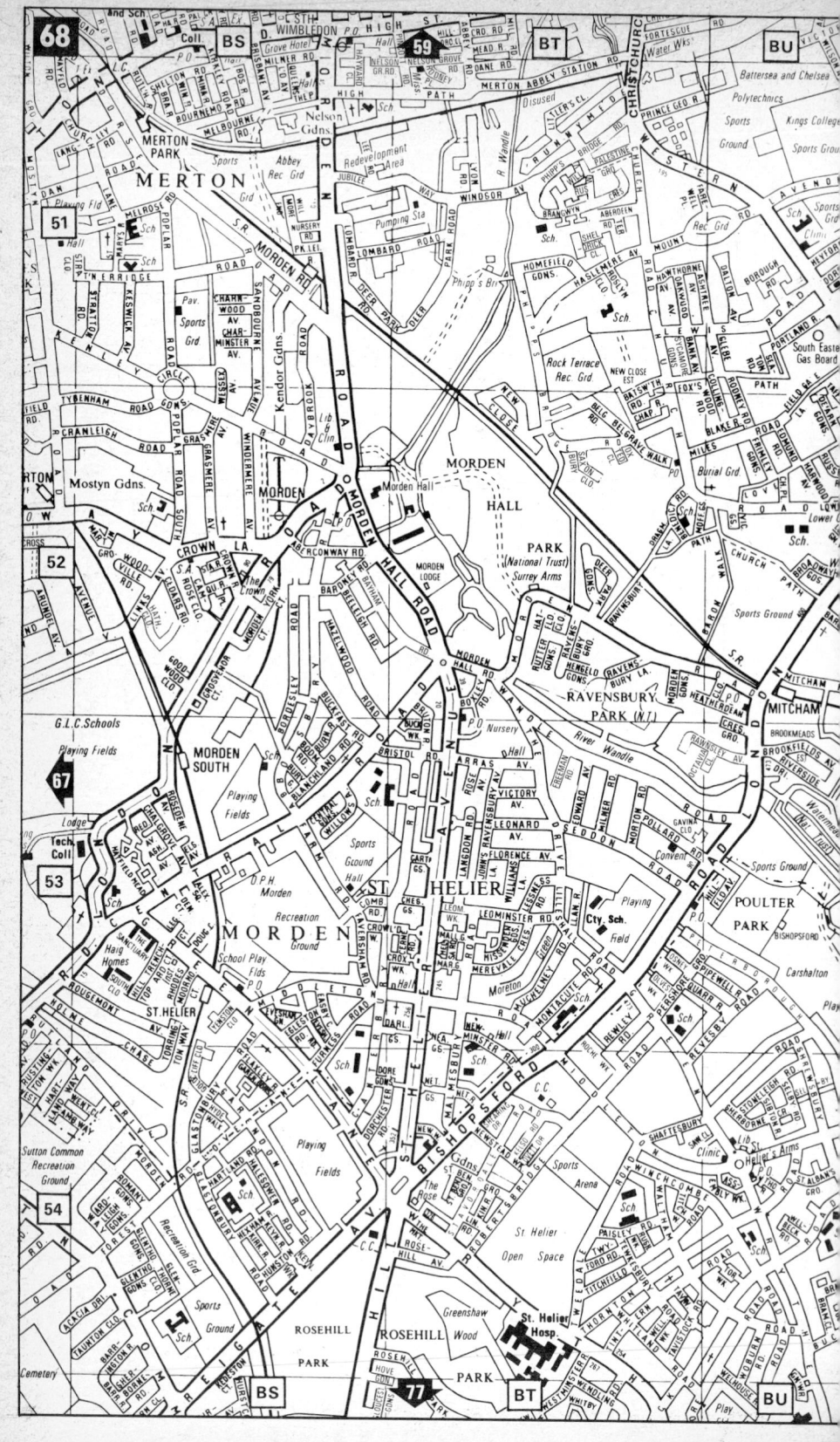

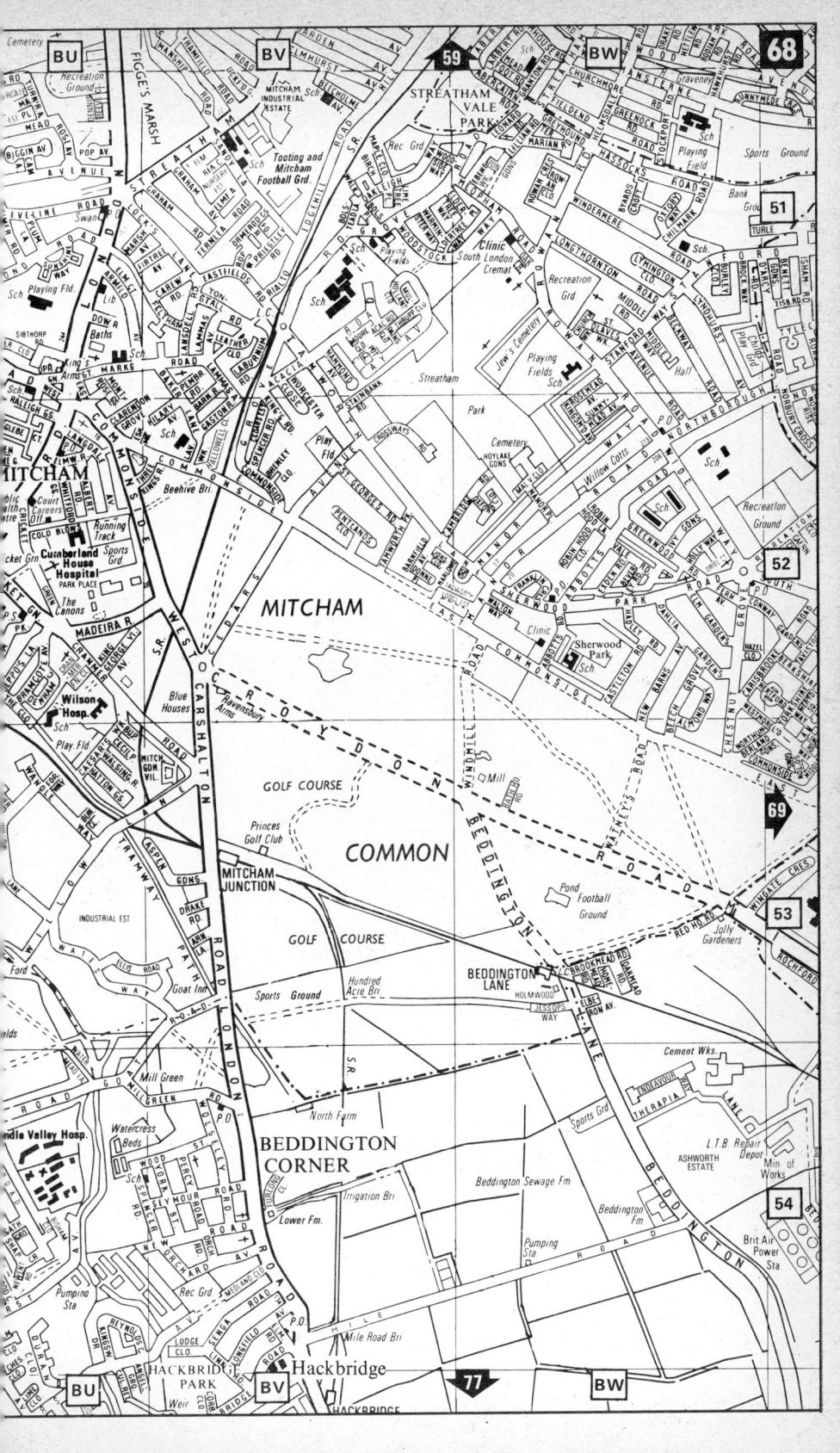

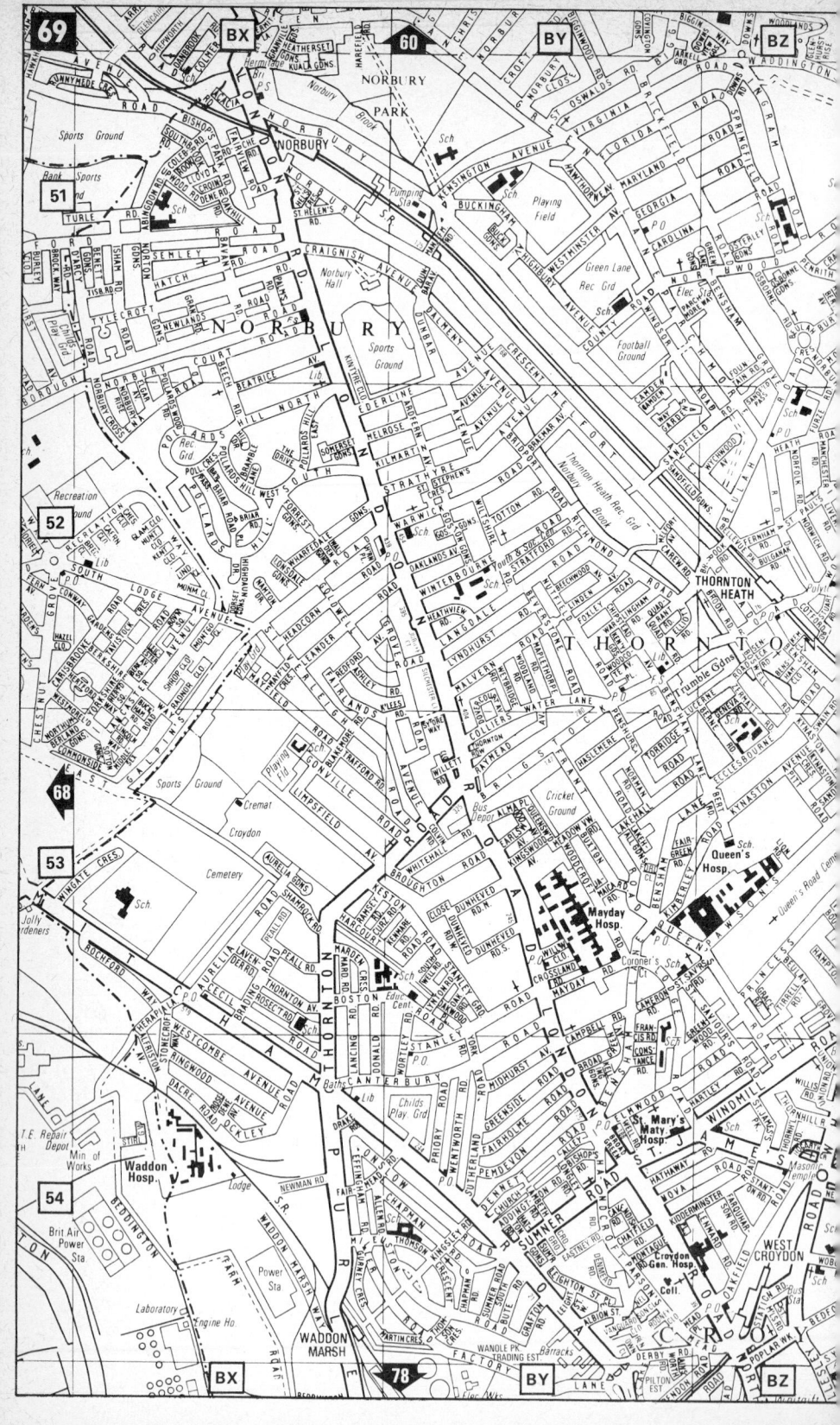

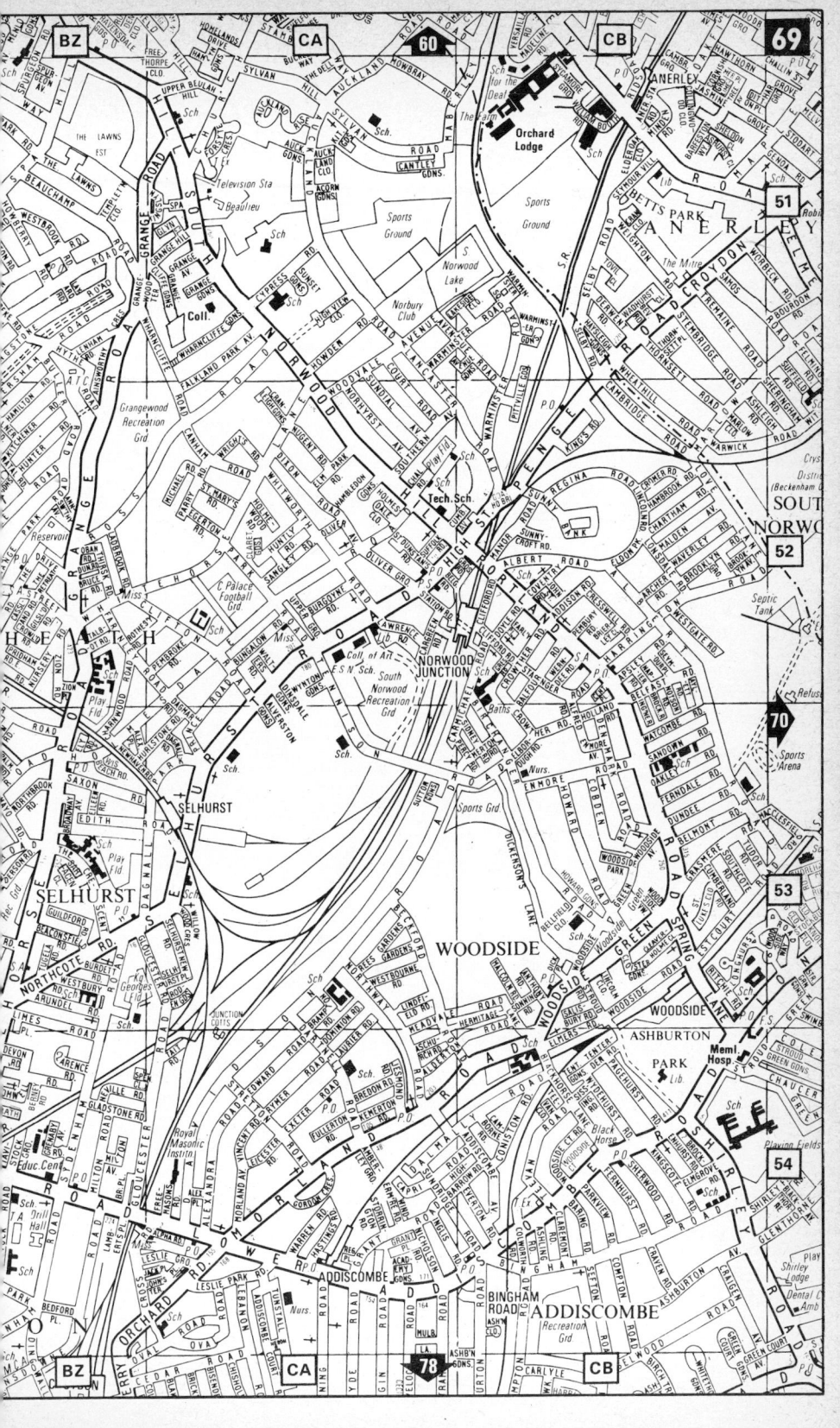

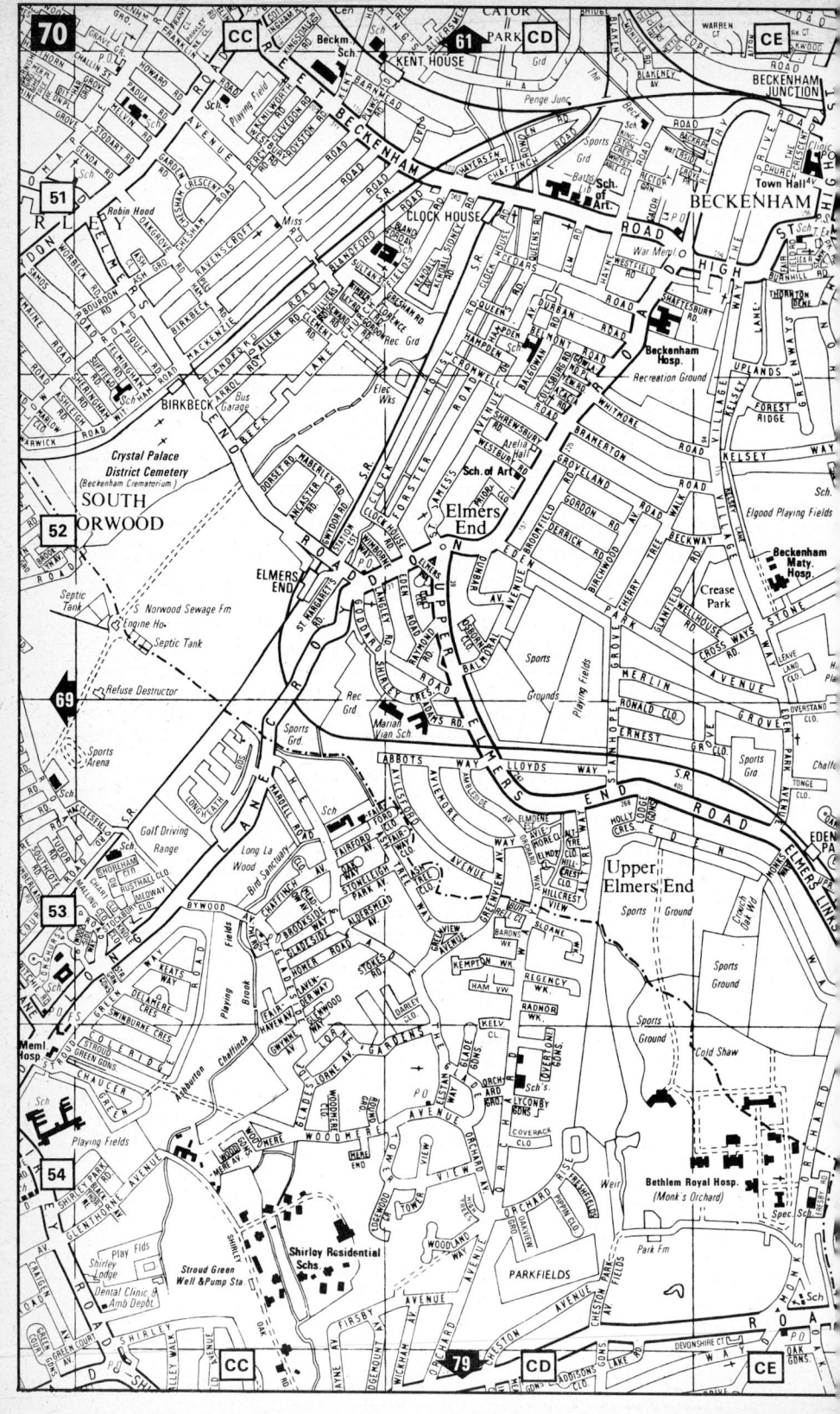

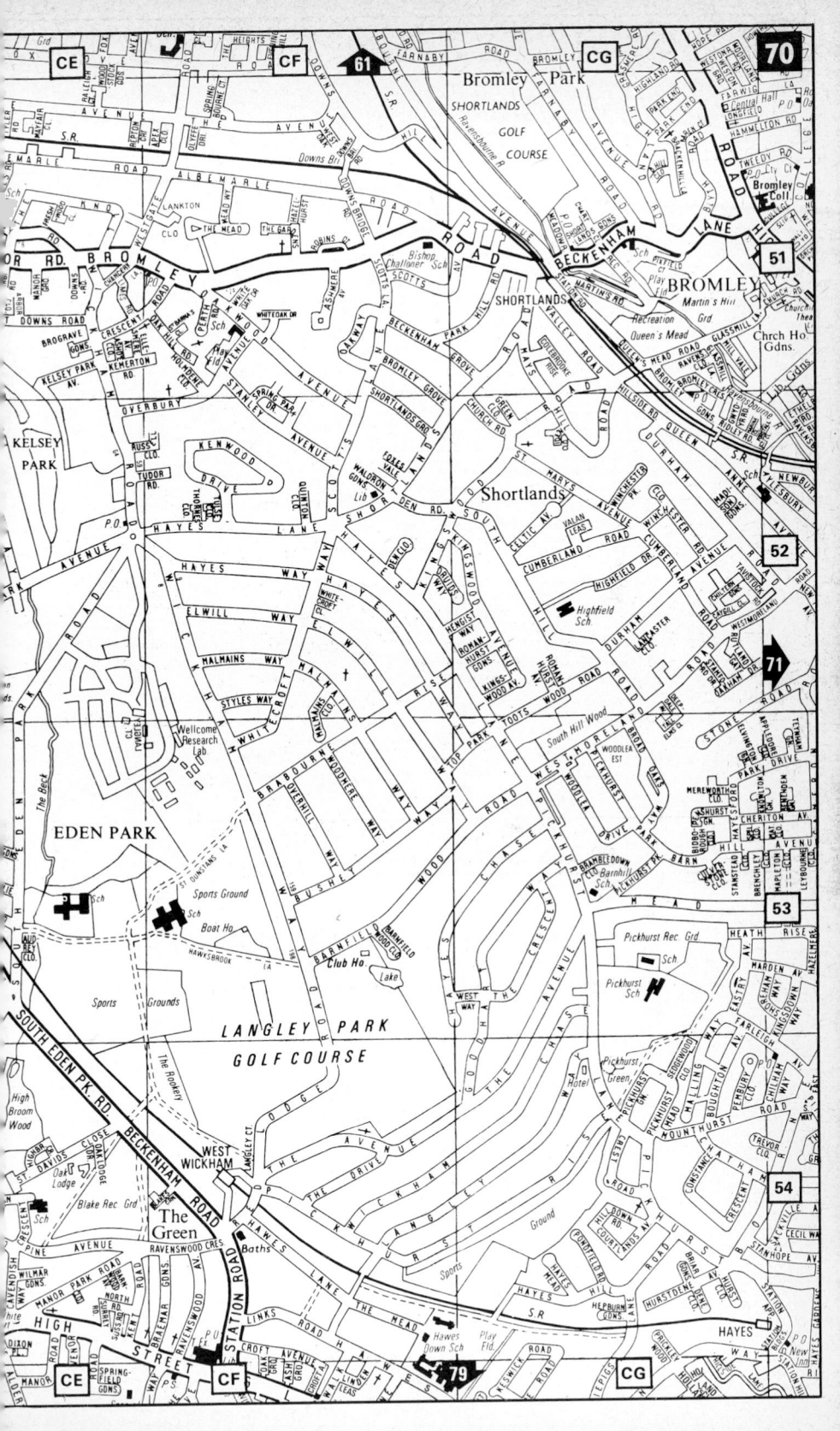

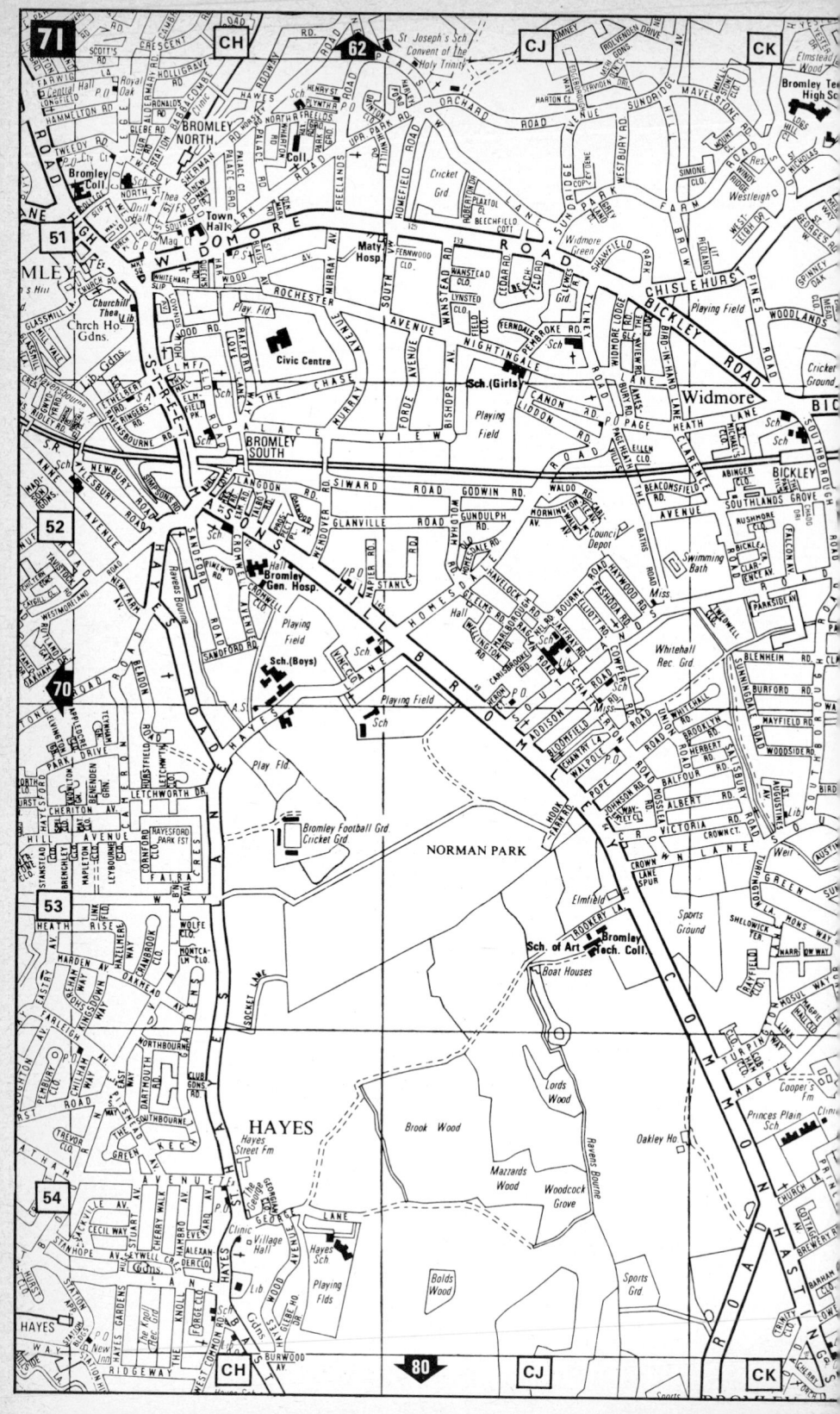

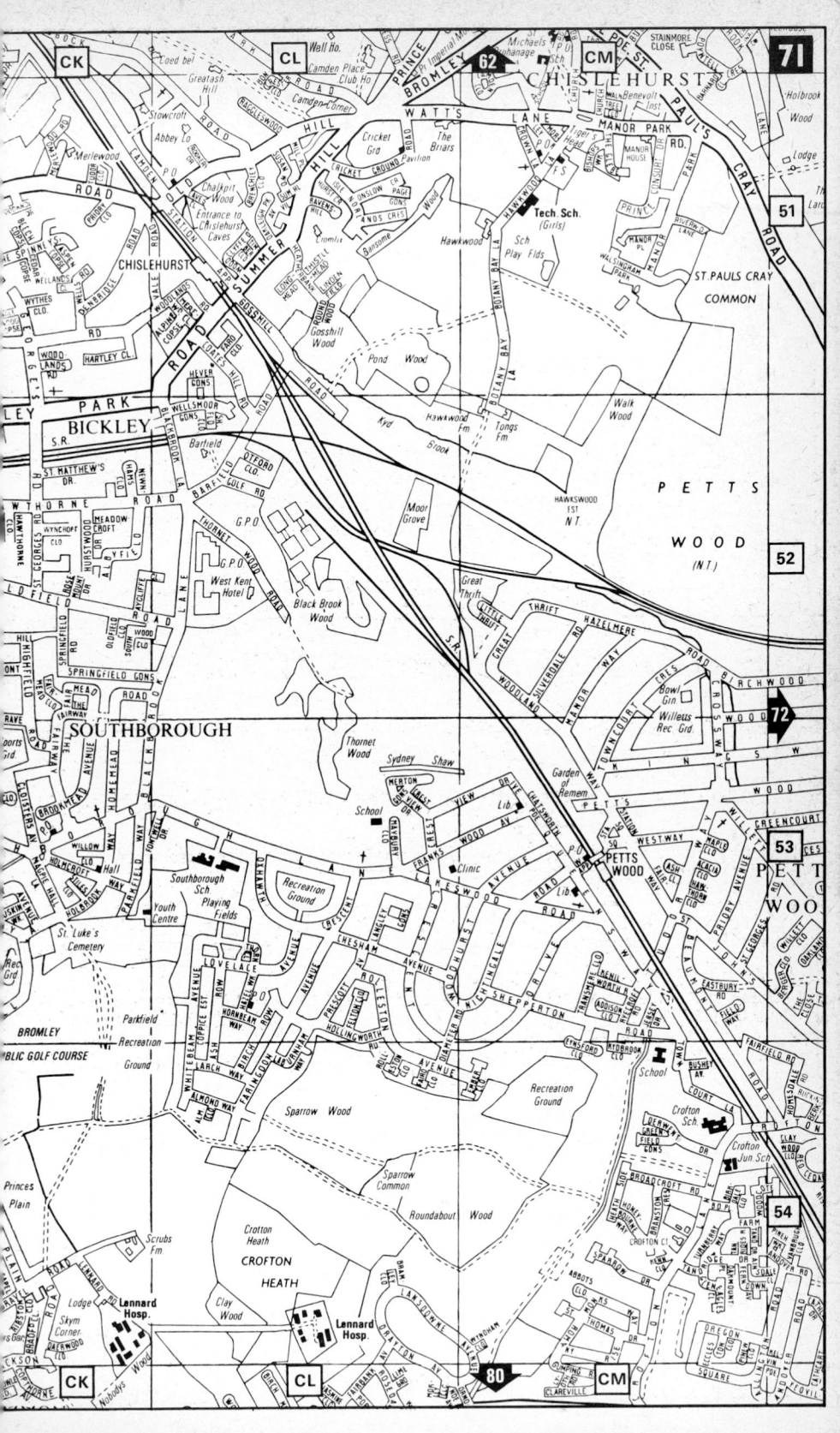

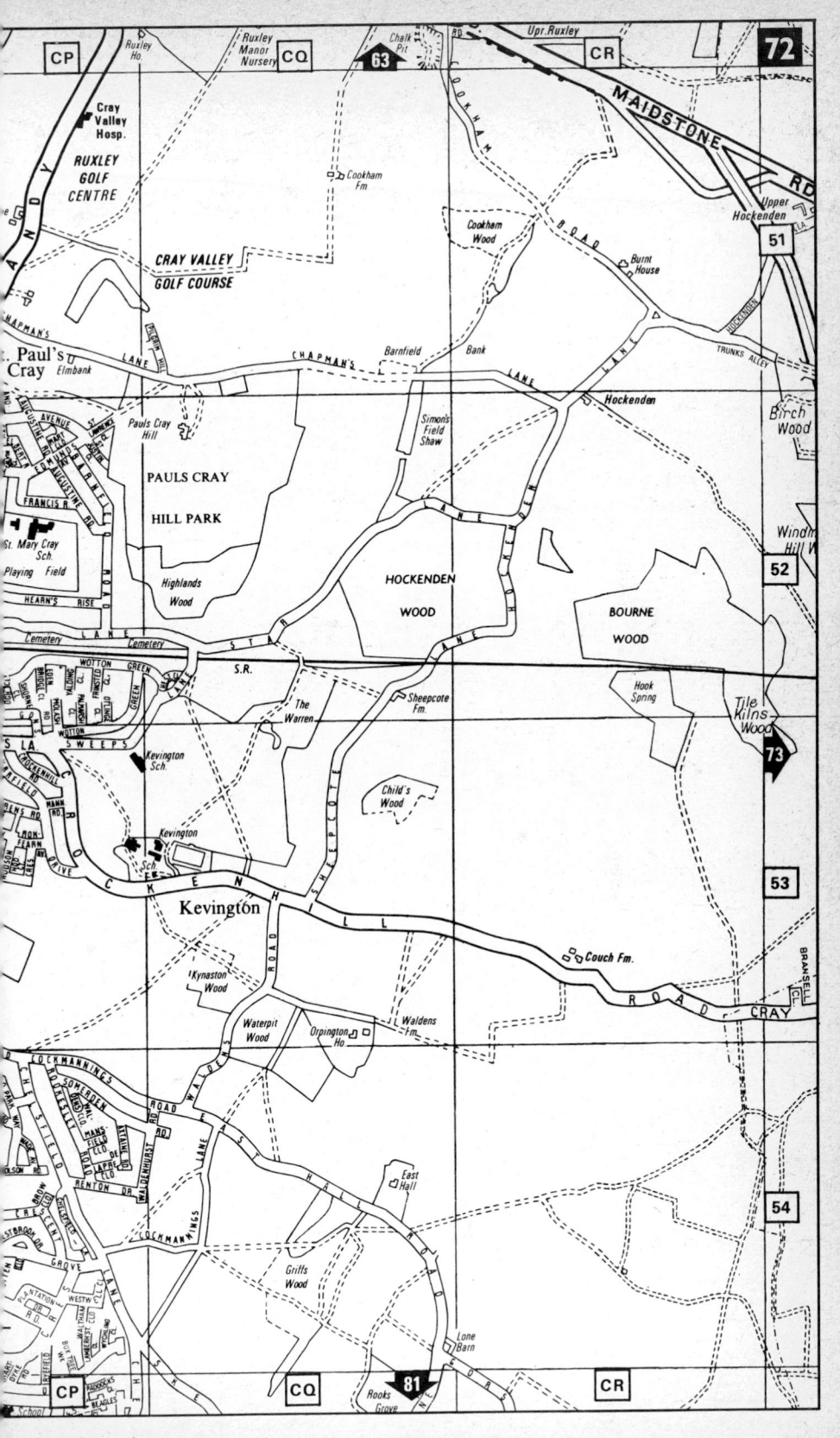

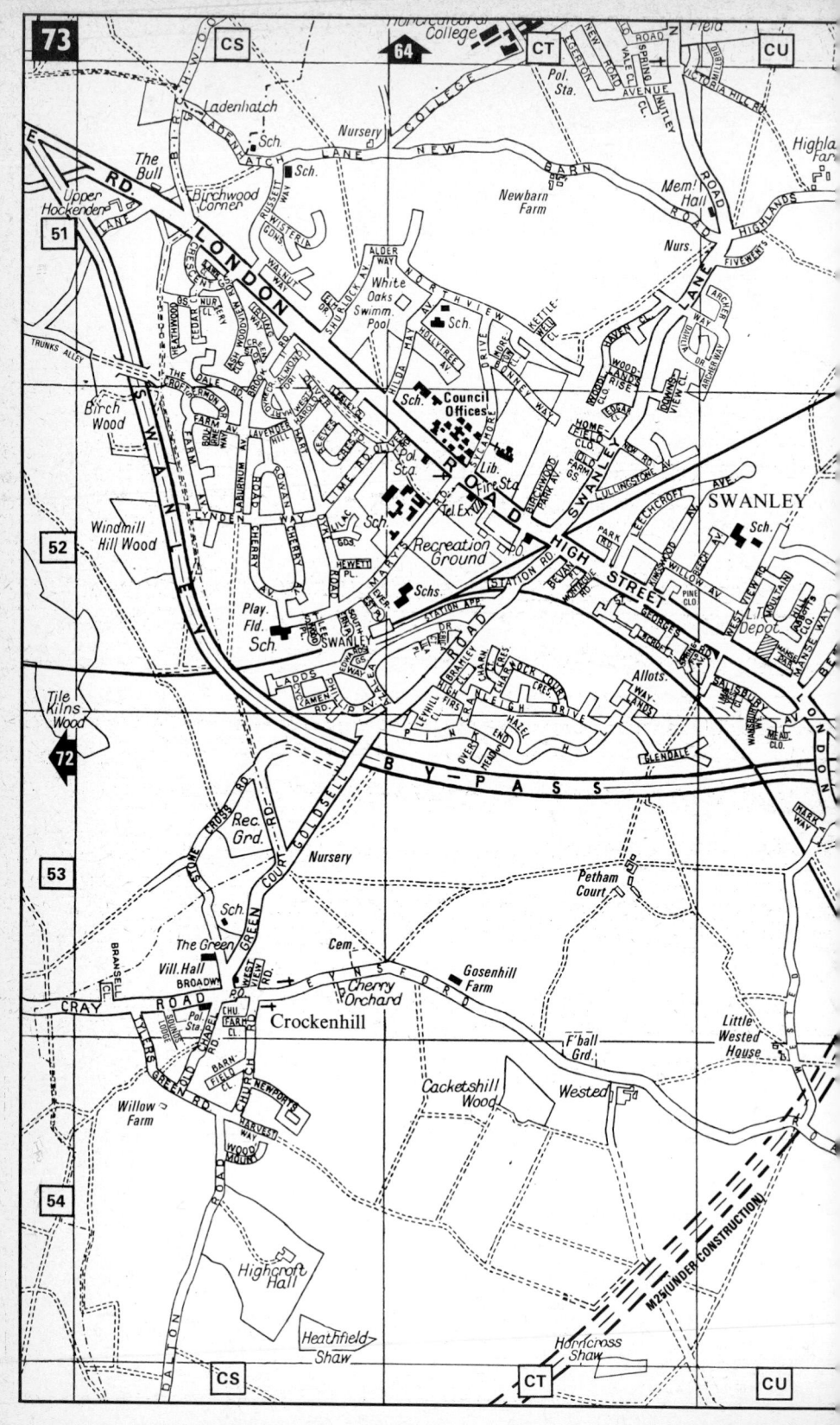

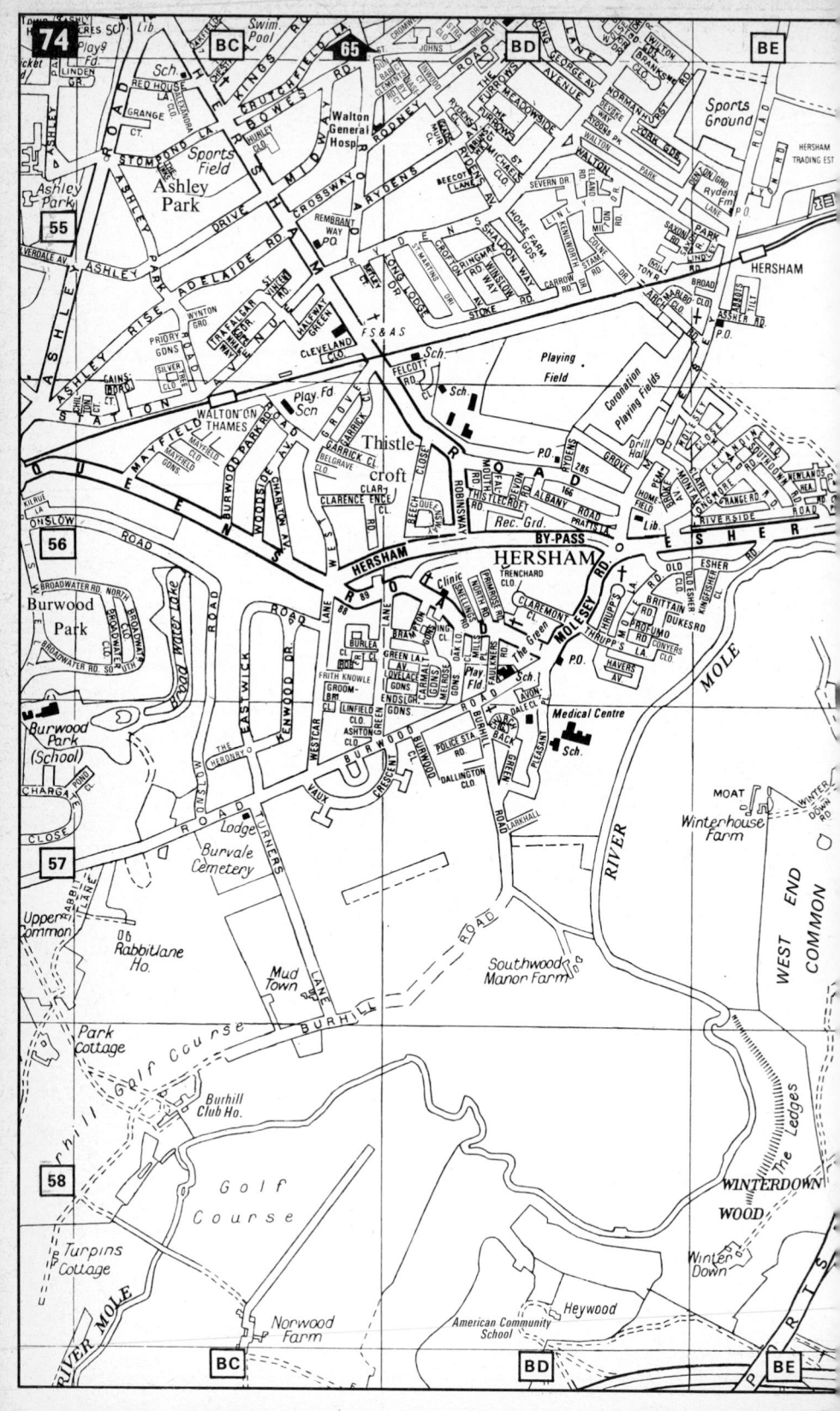

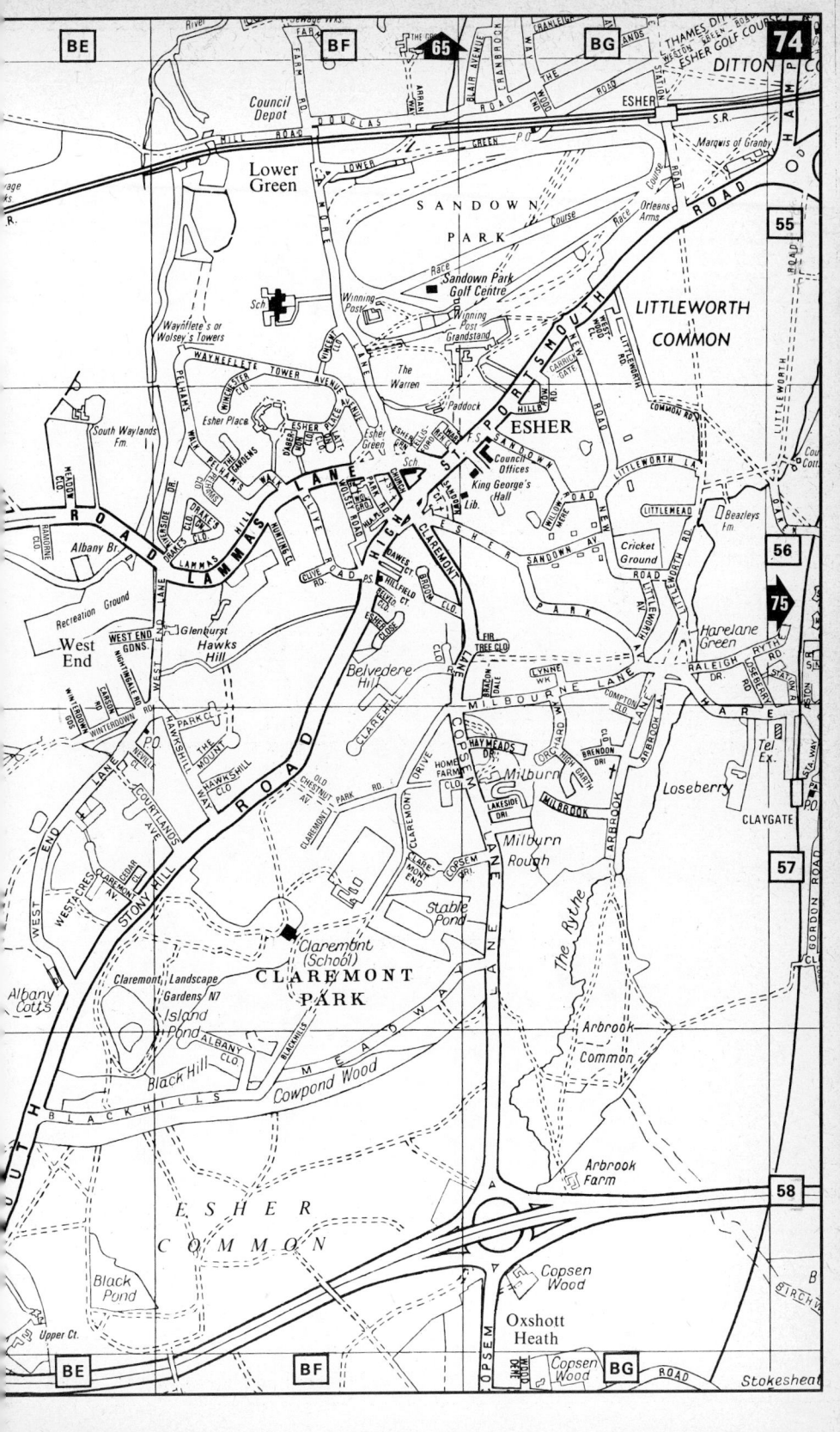

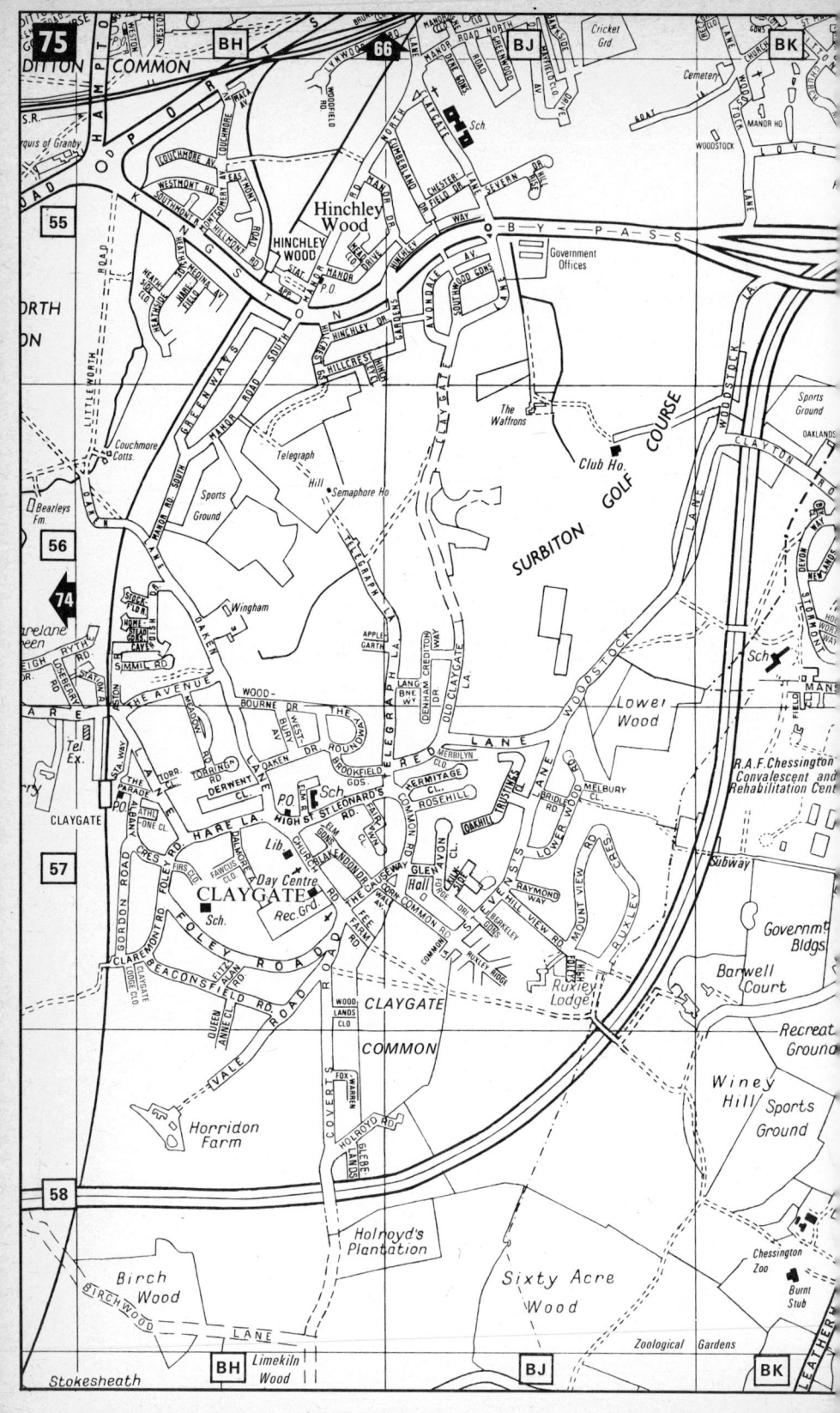

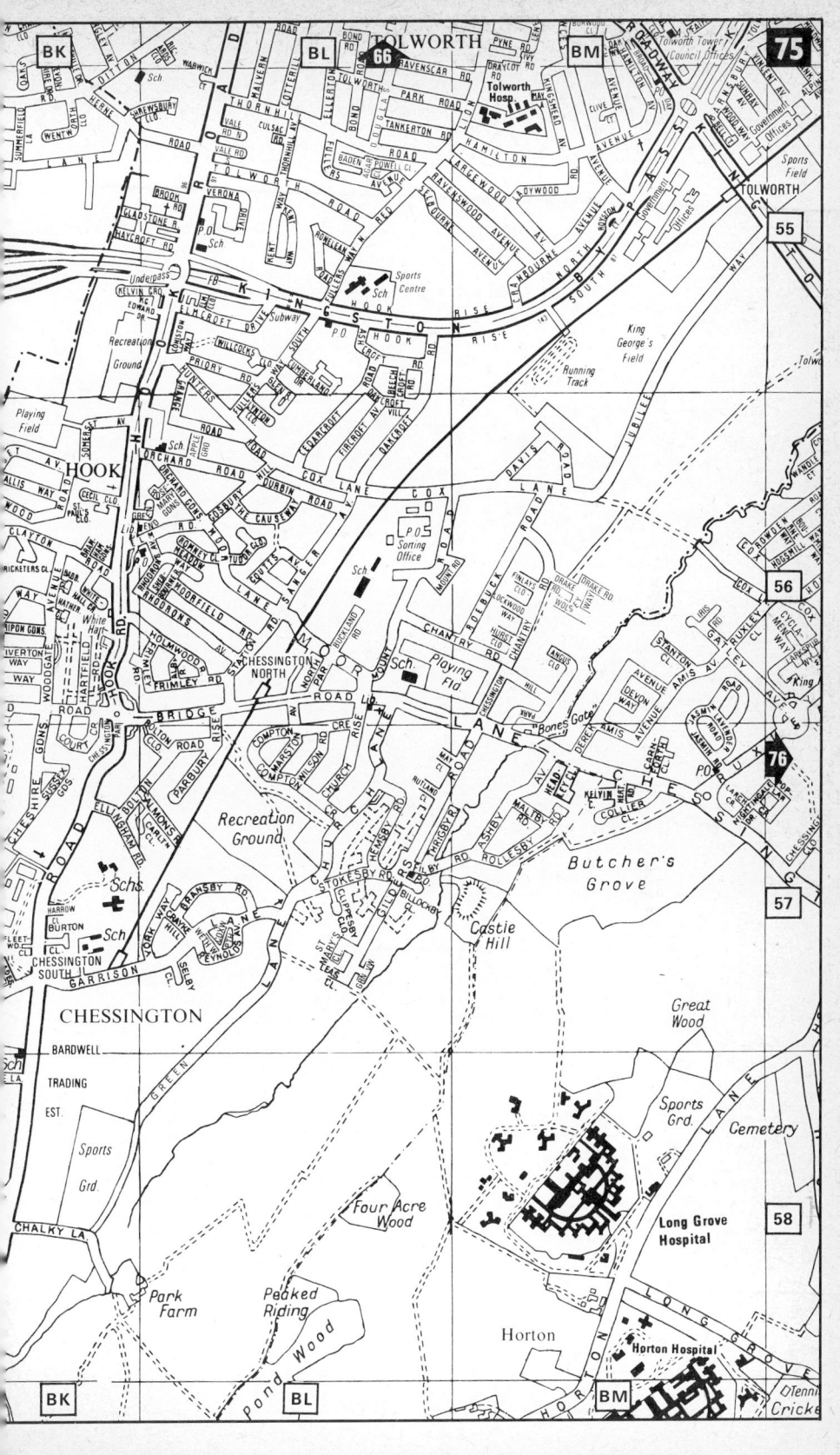

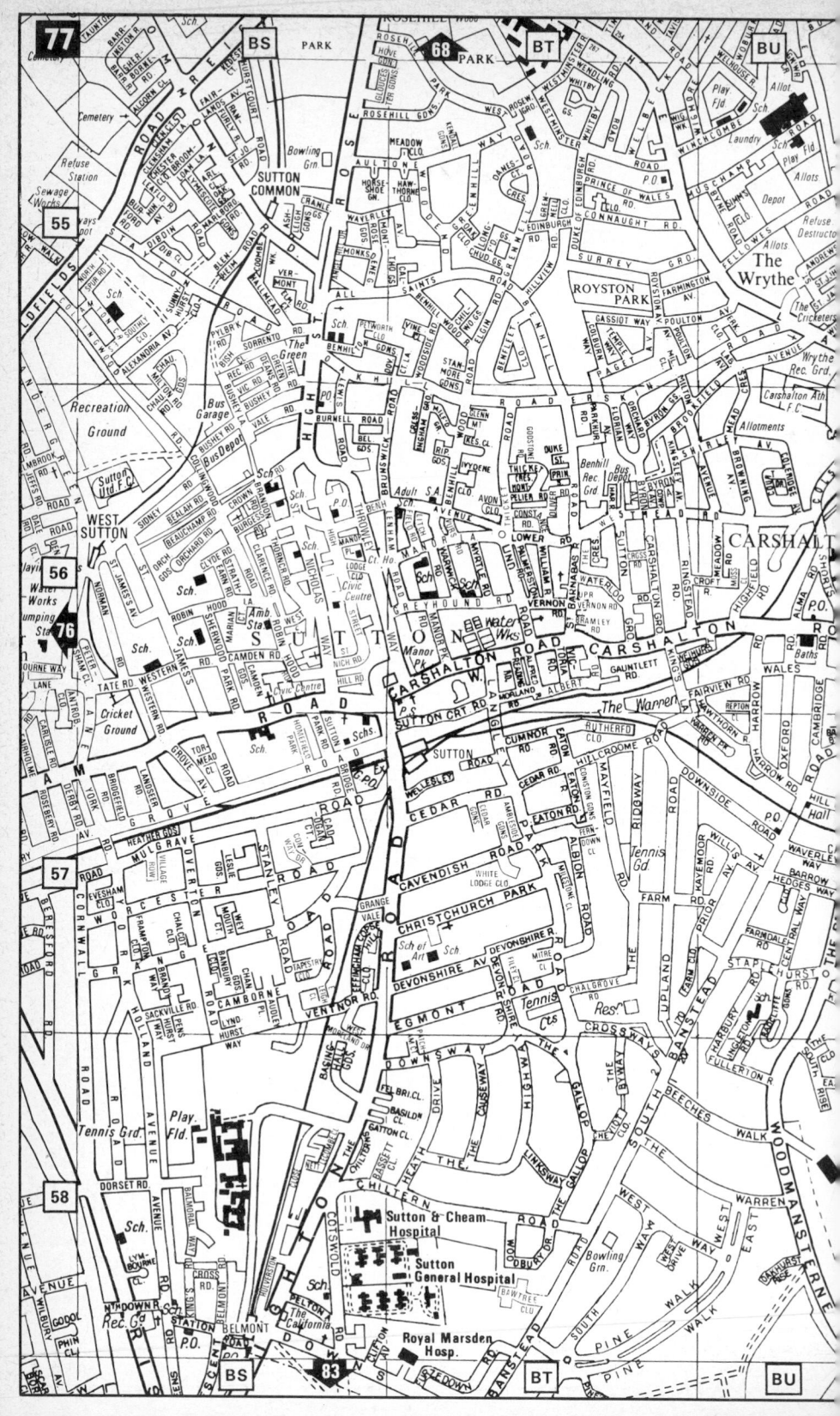

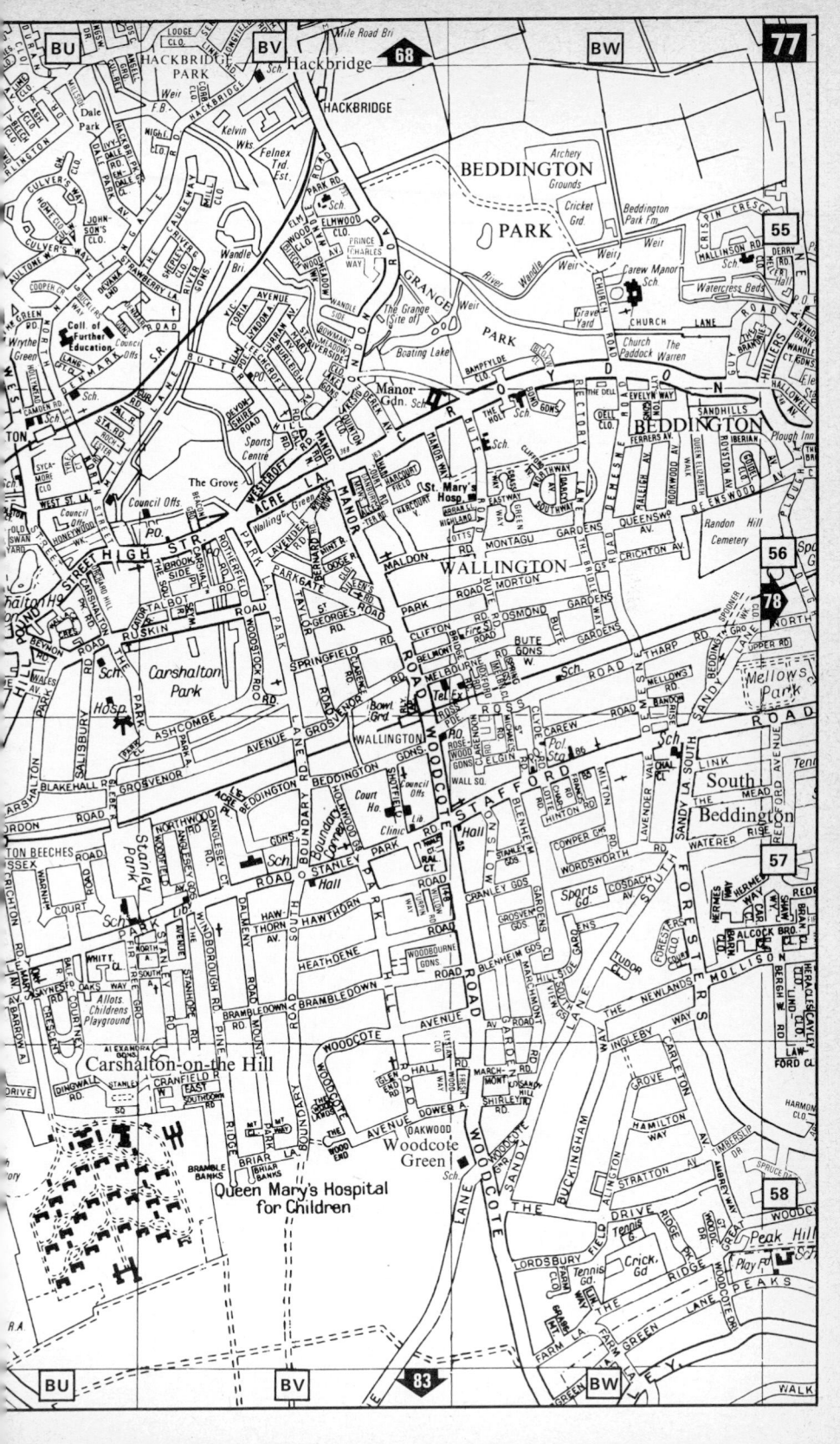

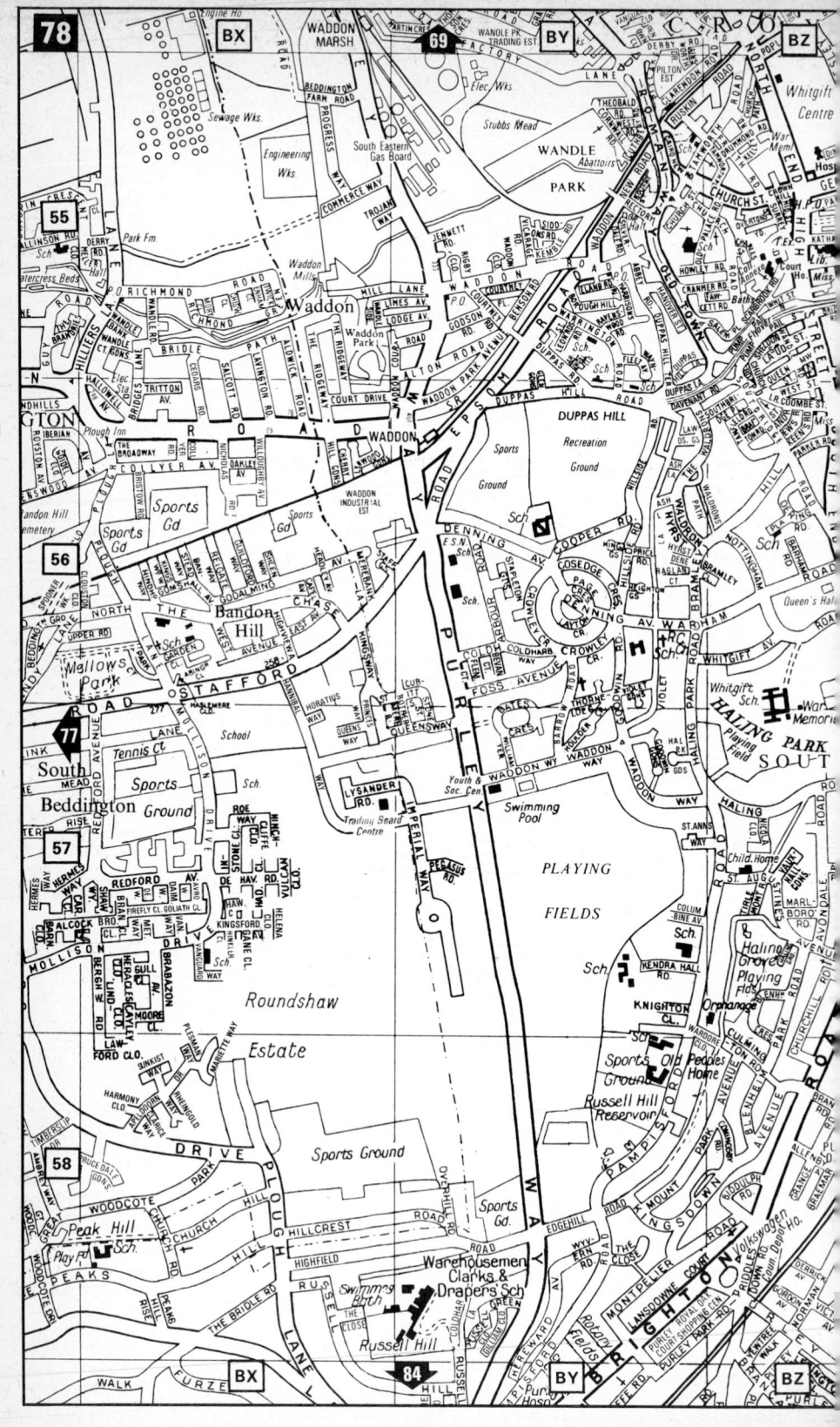

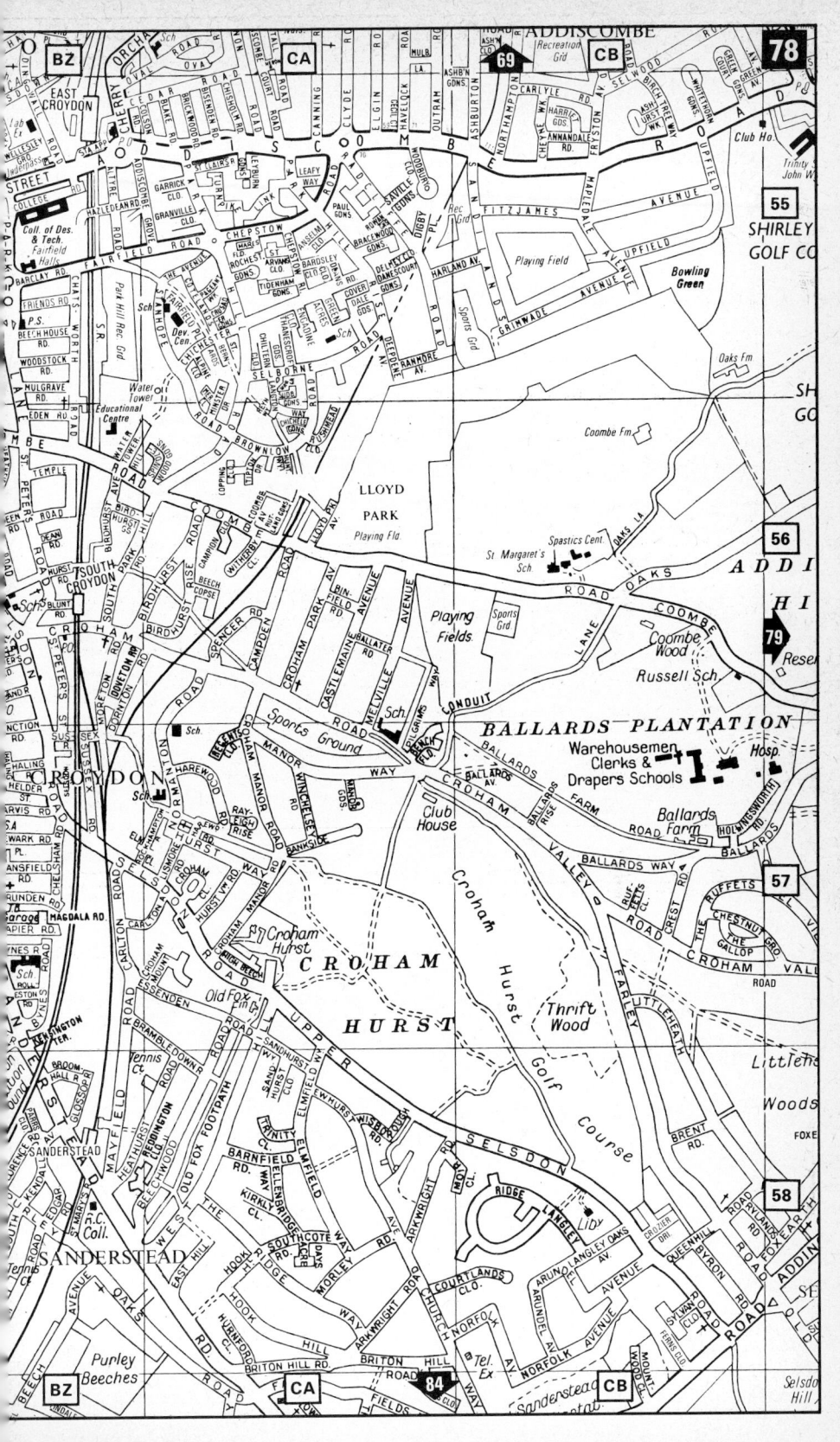

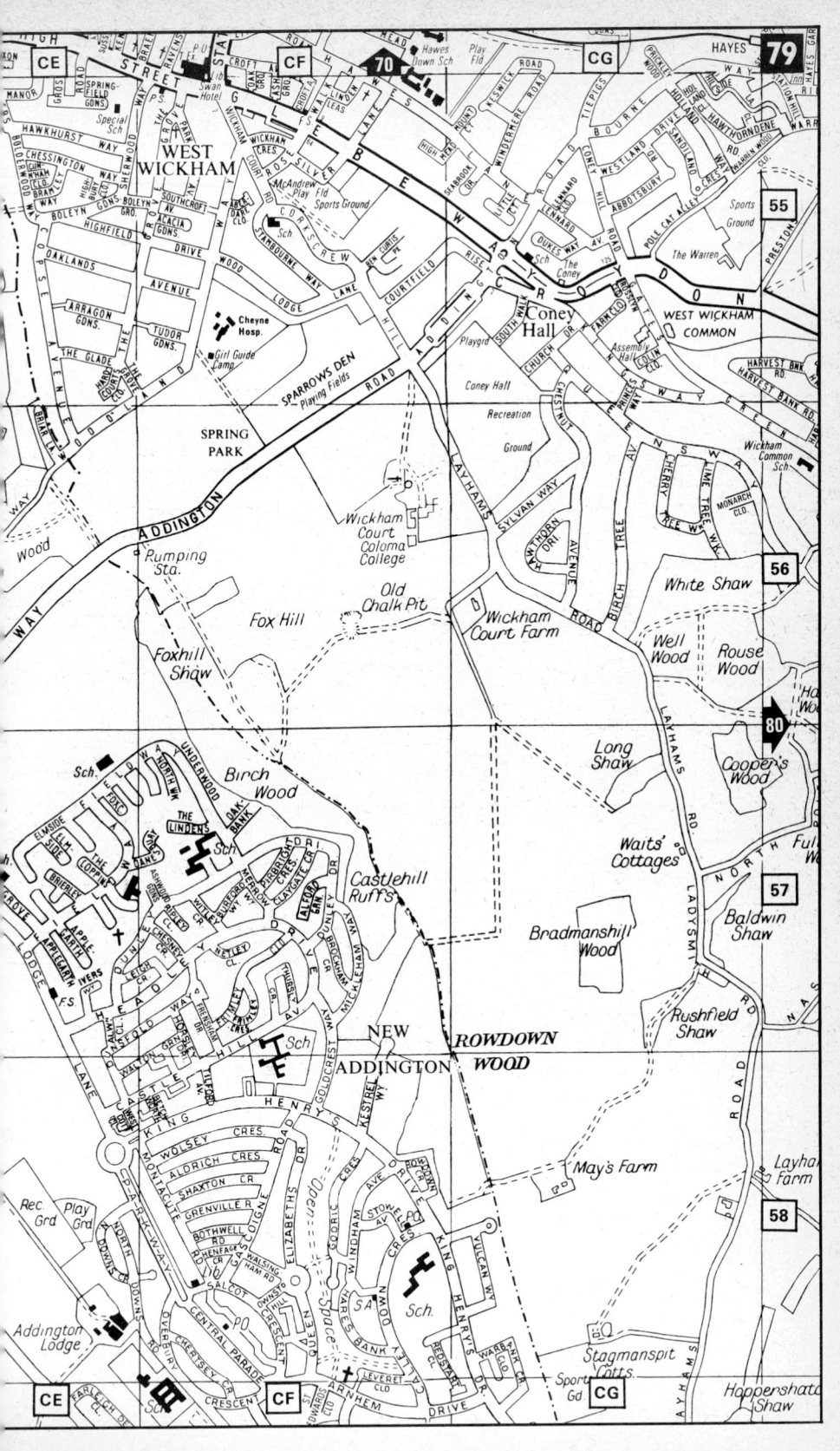

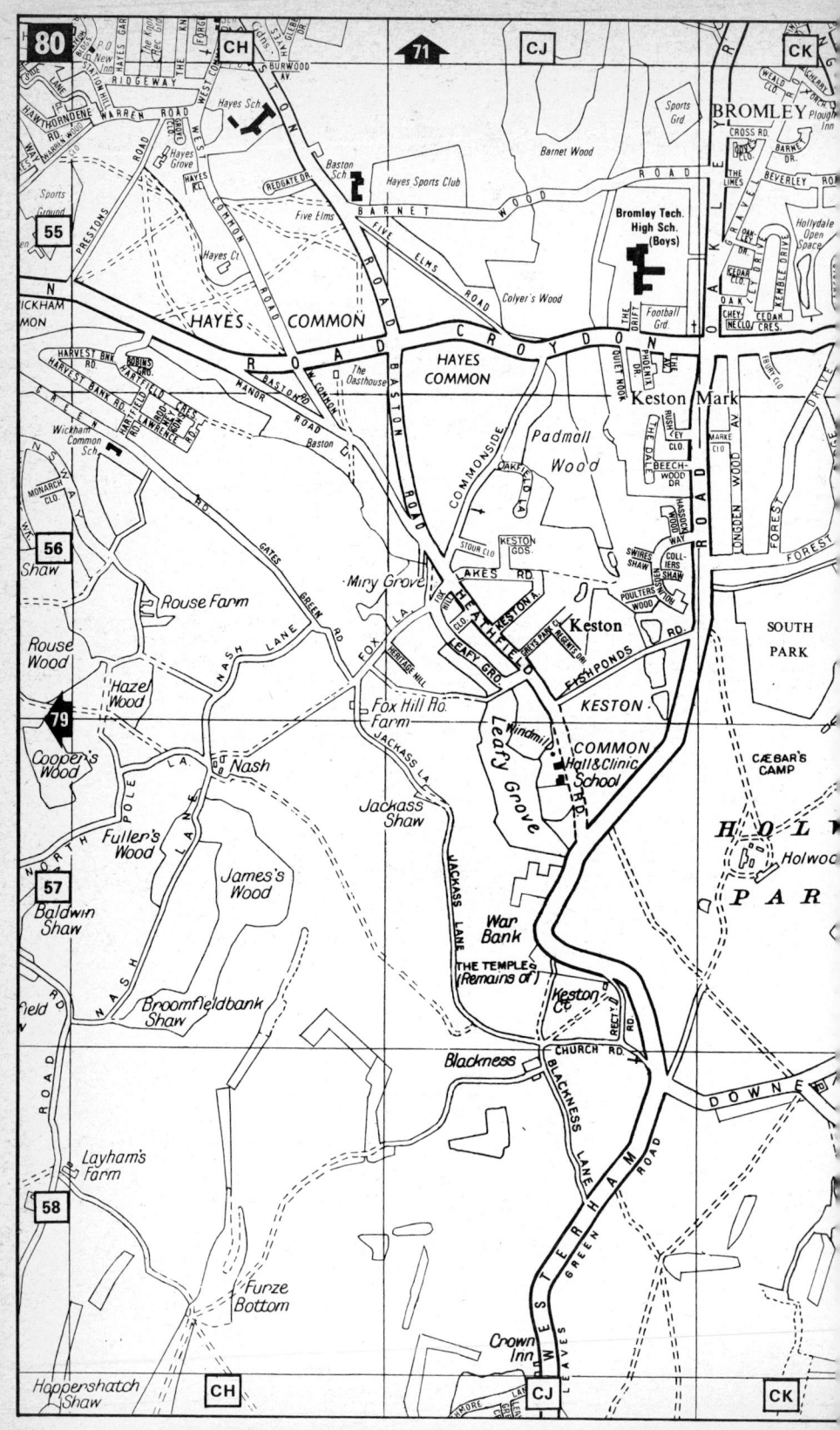

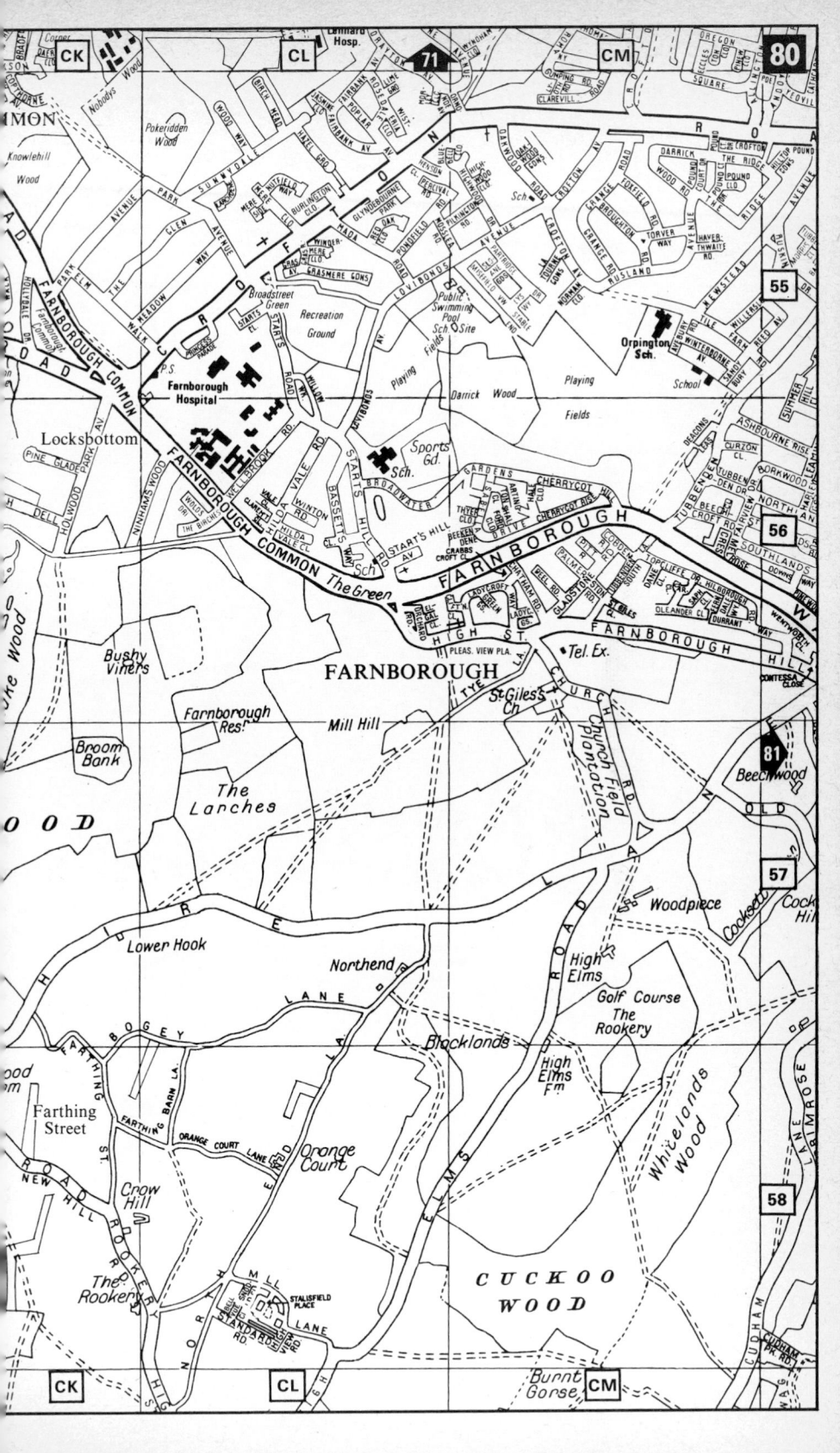

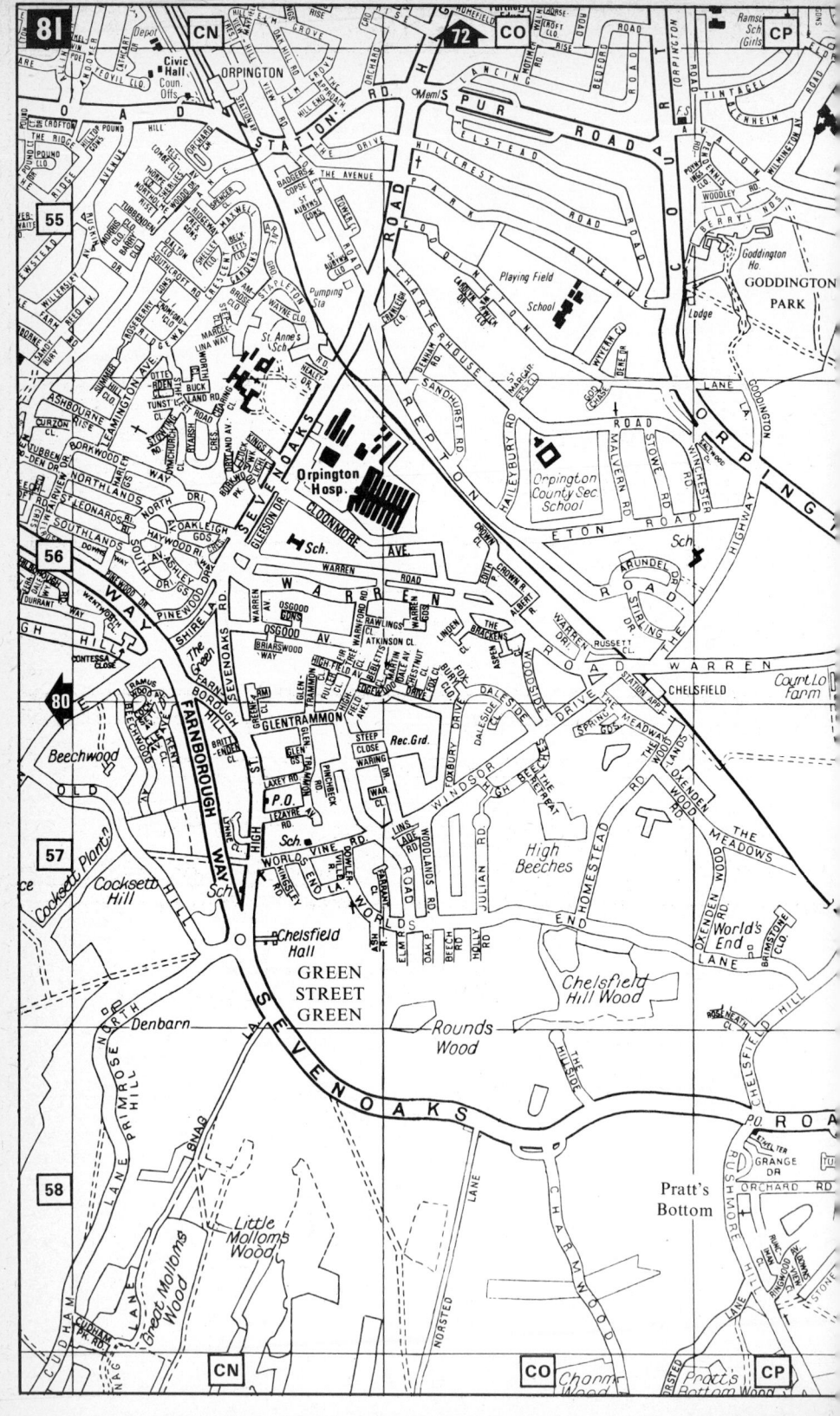

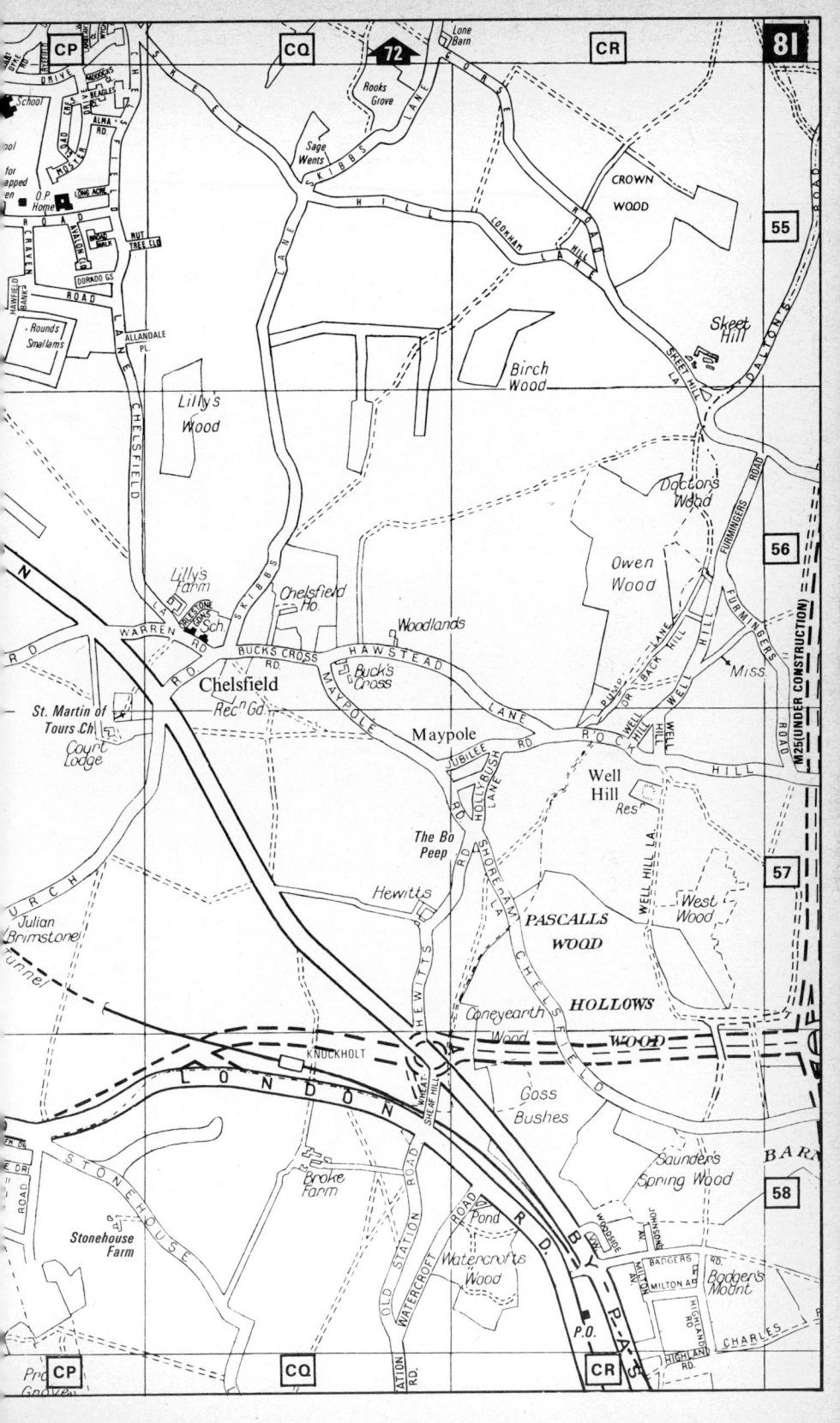

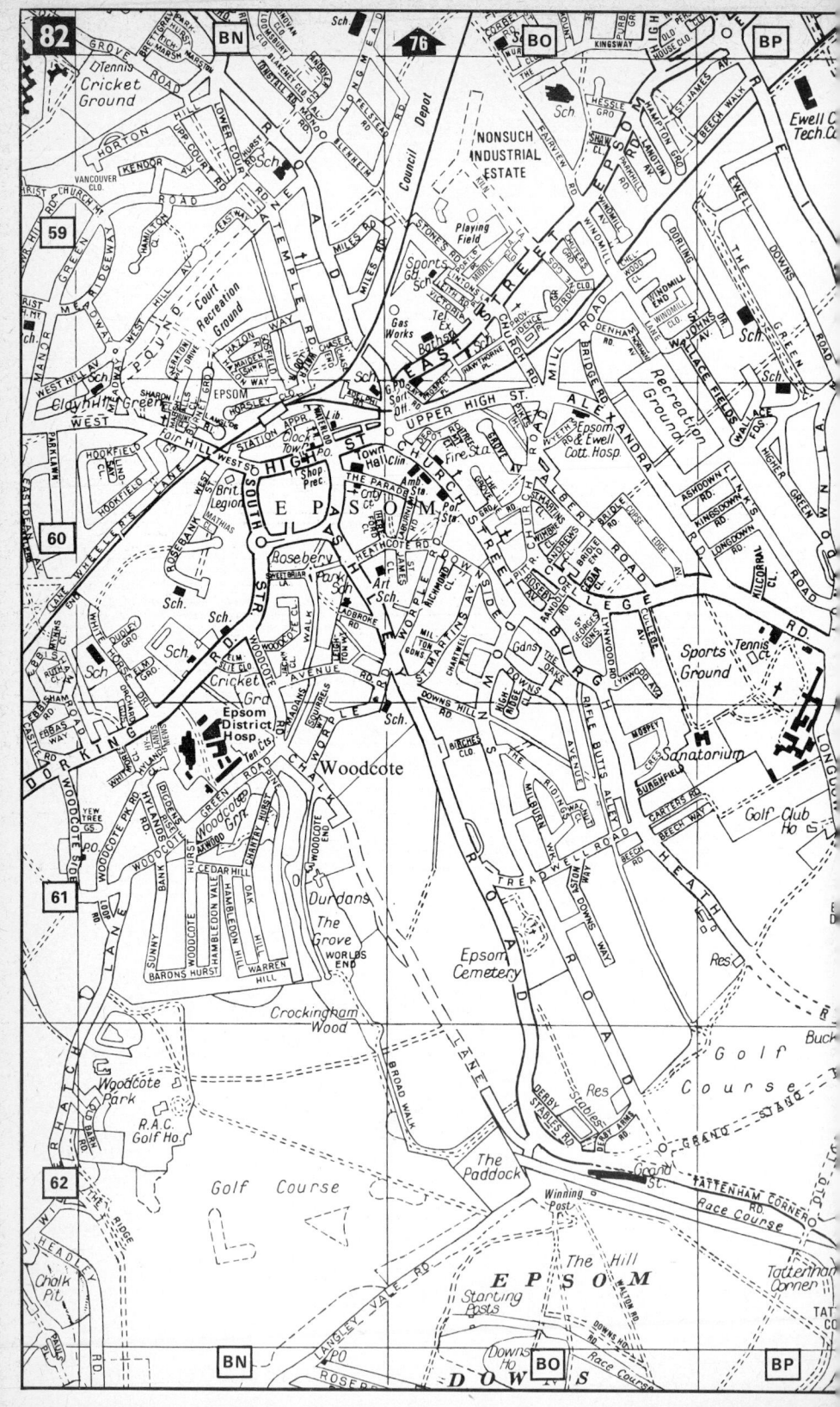

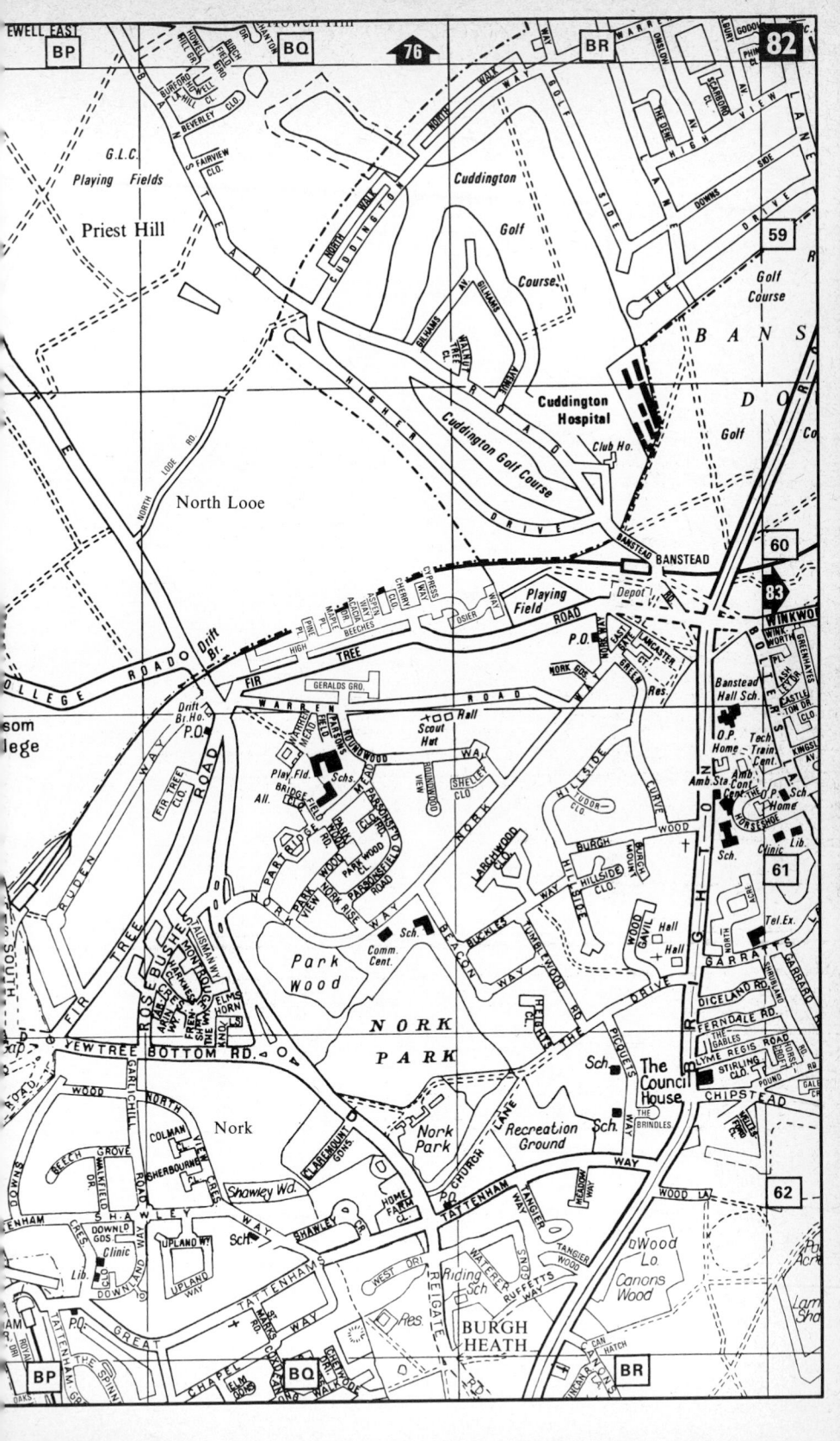

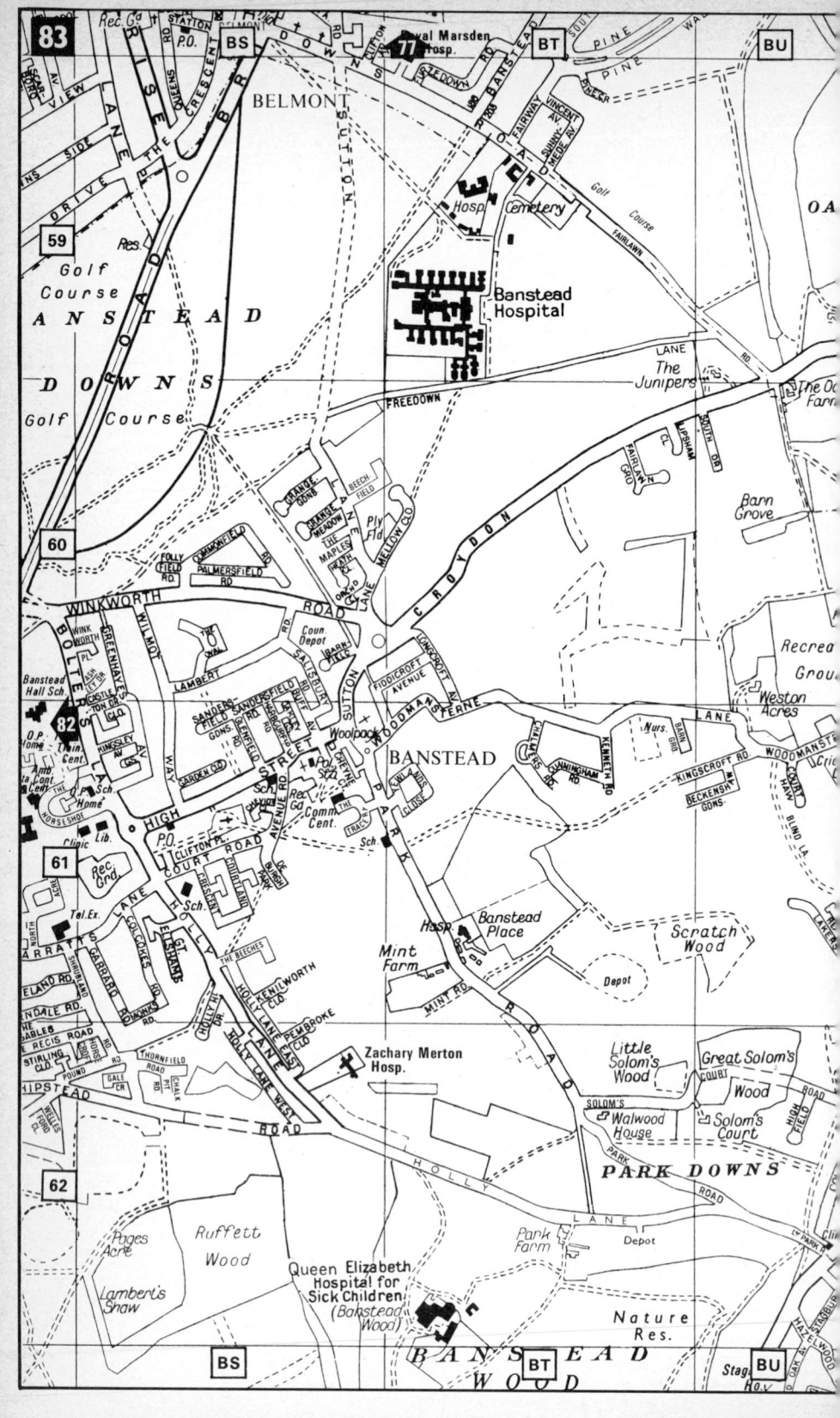

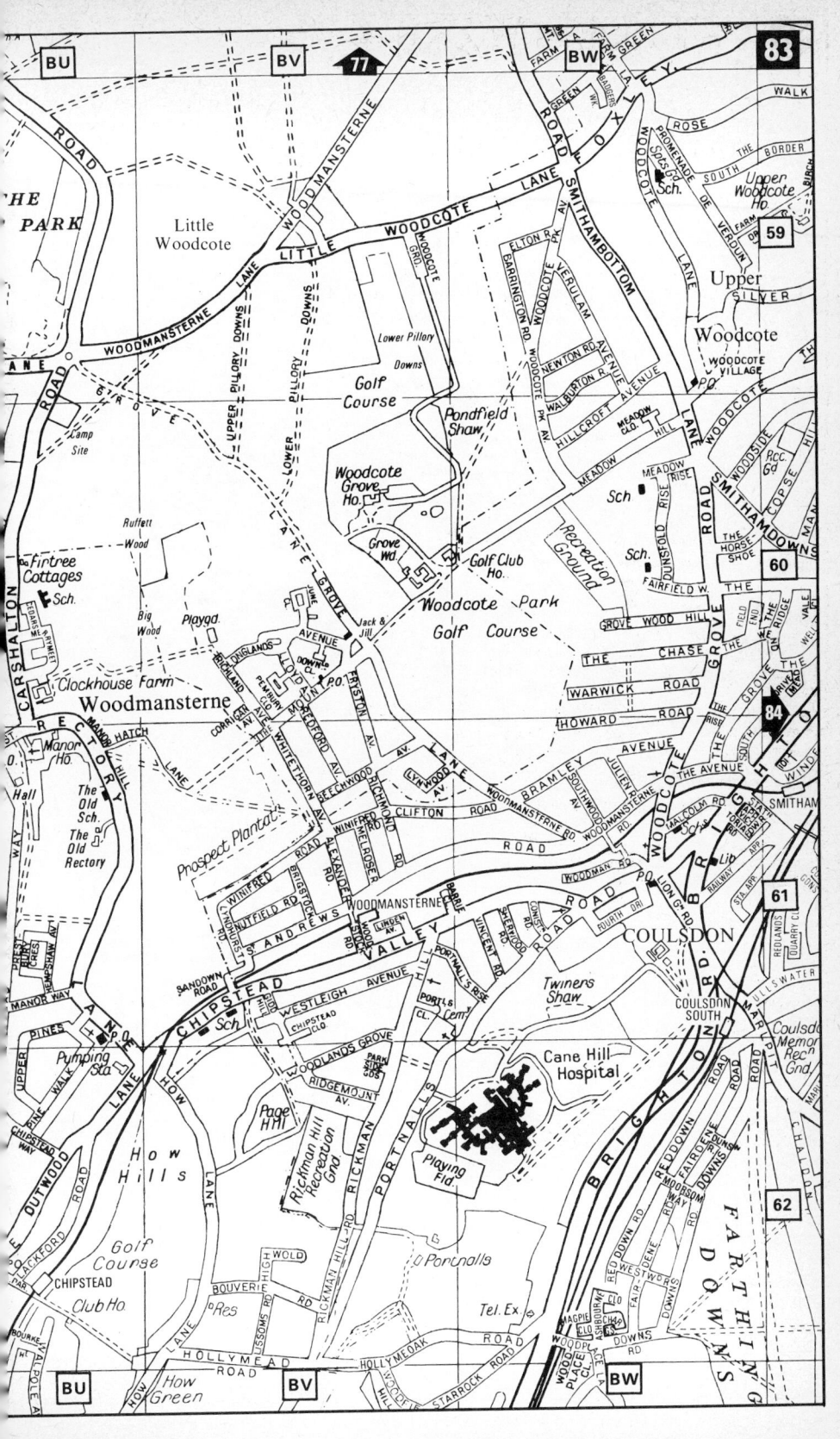

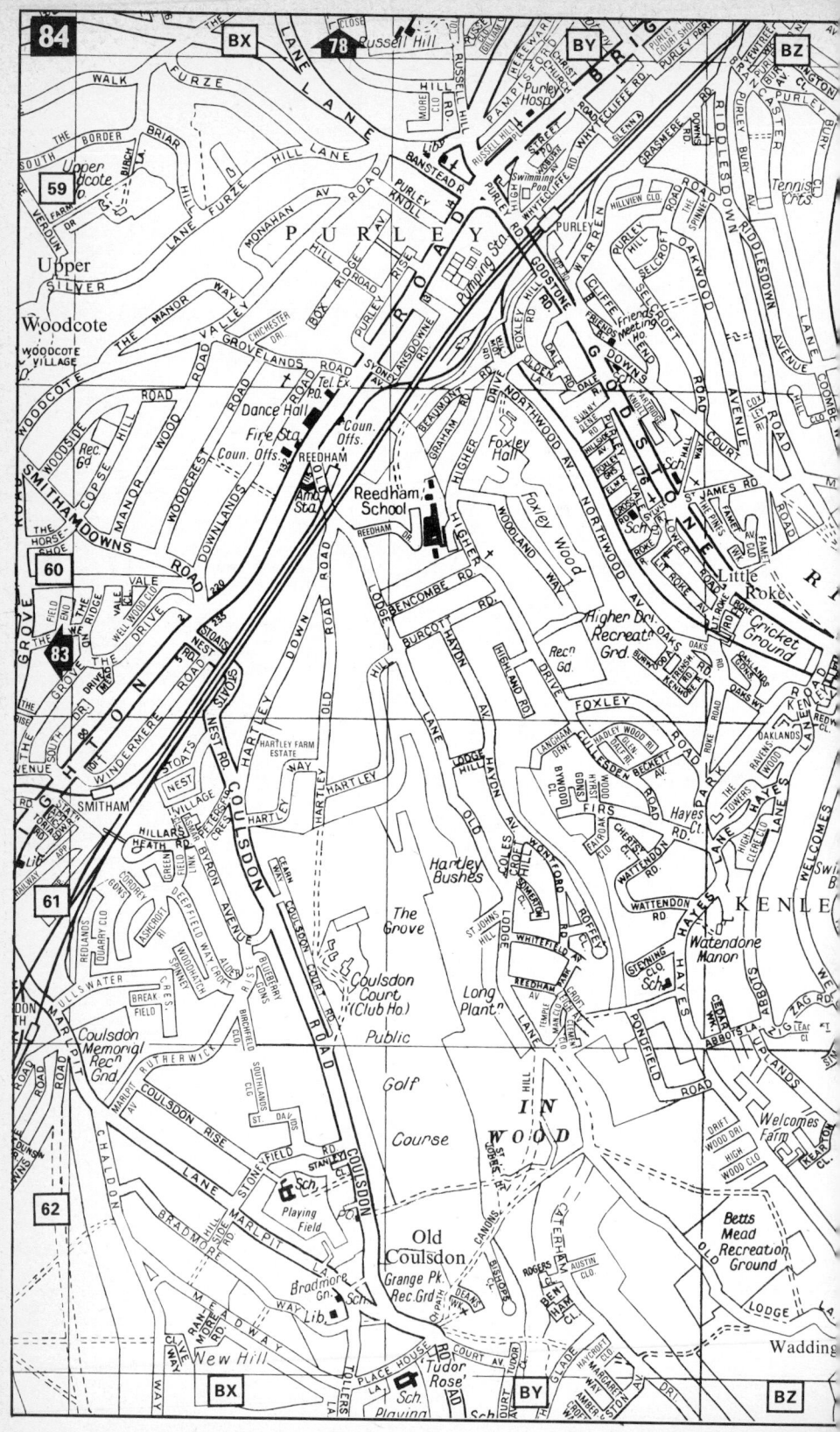

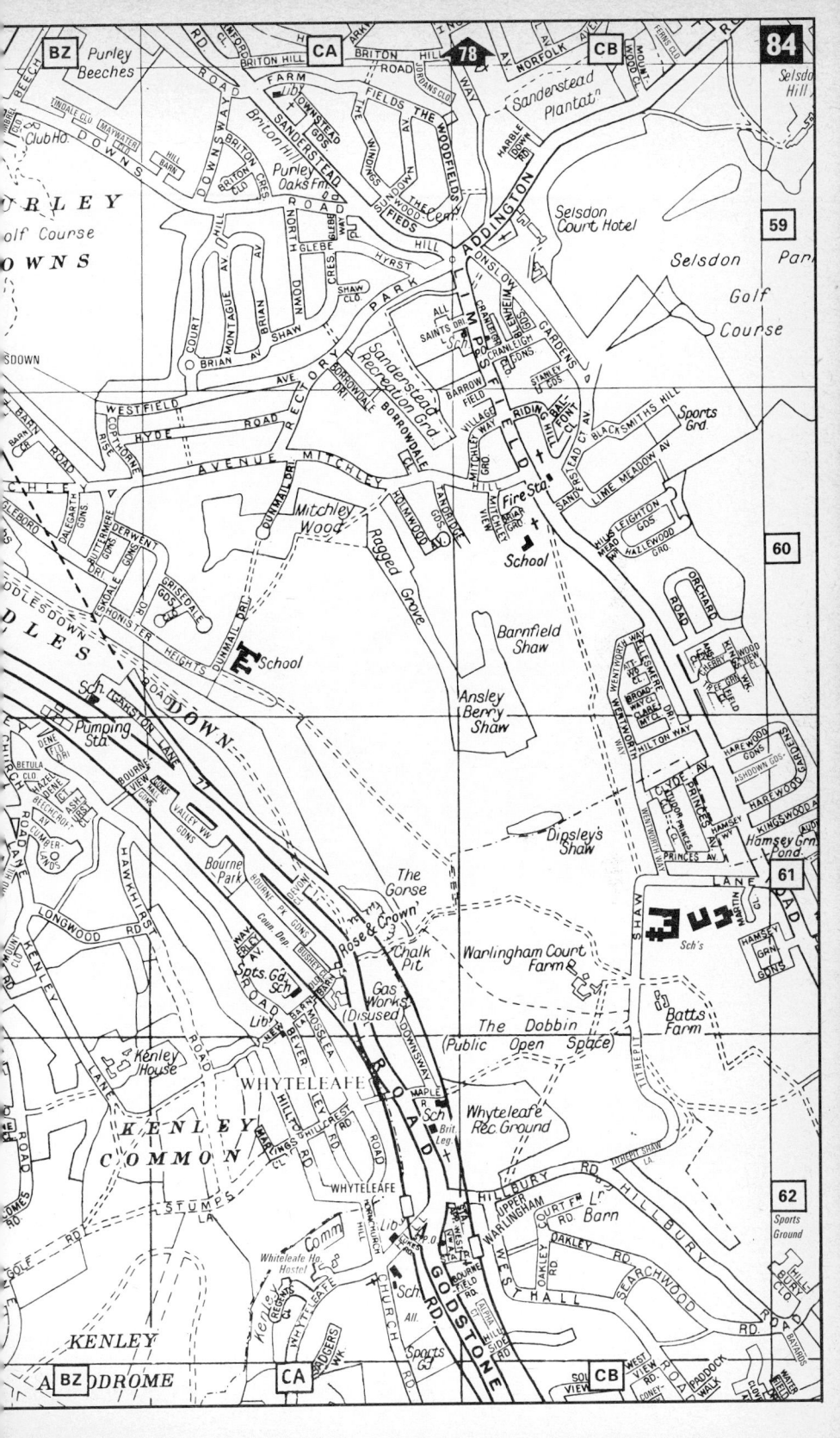

LONDON

Index to Streets

Indexing System

The street name and postal district or locality of an entry is followed by a grid reference and number of the map on which the name will be found e.g. Abbey Rd.SW19 will be found in square **BT50** on map **59** and Norfolk Crescent,Sidcup in square **CN47** on map **63**

The index contains some names for which there is insufficient space on the map.The adjoining thoroughfare to such roads is shown in italics e.g. Agar Place, NW1 is off *Agar Grove* the latter being found in square **BW36** on map **32**.

A strict alphabetical order is followed in which Avenue,Close,Gardens etc, although abbreviated,are read as part of the preceeding name. For example Andrews Rd. comes before Andrew St.and Abbey Orchard St. before Abbey Rd.

Abbreviations of District Names

Ashf.	Ashford	Grnf.	Greenford	Sid.	Sidcup
Ash.	Ashtead	Hamptn.	Hampton	Sthl.	Southall
Bans.	Banstead	Har.	Harrow	Sth.Croy.	South Croydon
Bark.	Barking	Hav.	Havering-atte-Bower	S.Dnth.	South Darenth
Barn.	Barnet	Horn.	Hornchurch	S.Ock.	South Ockendon
Beck.	Beckenham	Hort.K.	Horton Kirby	Stai.	Staines
Belv.	Belvedere	Houns.	Hounslow	Stan.	Stanmore
Bex.	Bexley	Ilf.	Ilford	Sun.	Sunbury-on-Thames
Bexh.	Bexleyheath	Islw.	Isleworth	Surb.	Surbiton
Borwd.	Boreham Wood	Ken.	Kenley	Sutt.	Sutton
Brent.	Brentford	Kes.	Keston	S.at H.	Sutton at Hone
Brom.	Bromley	Kings.on T.	Kingston on Thames	Swan.	Swanley
Buck.H.	Buckhurst Hill	Leath.	Leatherhead	Tad.	Tadworth
Bush.	Bushey	Loug.	Loughton	Tedd.	Teddington
Cars.	Carshalton	Lthd.	Leatherhead	Th.Hth.	Thornton Heath
Cat.	Caterham	Mitch.	Mitch.	Twick.	Twickenham
Chess.	Chessington	Mord.	Morden	Upmin.	Upminster
Chig.	Chigwell	N.Mal.	New Malden	Uxb.	Uxbridge
Chis.	Chislehurst	Nthlt.	Northolt	Wall.	Wallington
Cob.	Cobham	Nthwd.	Northwood	Wal.Abb.	Waltham Abbey
Couls.	Coulsdon	Orp.	Orpington	Wal.Cr.	Waltham Cross
Croy.	Croydon	Pnr.	Pinner	Walt.	Walton-on-Thames
Dag.	Dagenham	Pot.B.	Potters Bar	Warl.	Warlingham
Dart.	Dartford	Pur.	Purley	Wat.	Watford
E.Mol.	East Molesey	Rad.	Radlett	Well.	Welling
Edg.	Edgware	Rain.	Rainham	Wem.	Wembley
Enf.	Enfield	Rich.	Richmond	West Dr.	West Drayton
Epp.	Epping	Rick.	Rickmansworth	Wey.	Weybridge
Eyns.	Eynsford	Rom.	Romford	Whyt.	Whyteleafe
Farn.	Farningham	Ruis.	Ruislip	Wdf.Grn.	Woodford Green
Felt.	Feltham	Sev.	Sevenoaks	W.Wick.	West Wickham
Gds.	Godstone	Shep.	Shepperton	Wor.Pk.	Worcester Park

General Abbreviations

All.	Alley	Fld.	Field	Pk.	Park
App.	Approach	Gdns.	Gardens	Pass.	Passage
Arc.	Arcade	Gth.	Garth	Peritr.	Perimiter
Av.	Avenue	Gte.	Gate	Pl.	Place
Bk.	Back	Gra.	Grange	Pr.	Prince,Princess
Bnk.	Bank	Gt.	Great	Prom.	Promenade
Boul.	Boulevard	Grn.	Green	Qn.	Queen
Br.	Bridge	Gro.	Grove	Ri.	Rise
Bldgs.	Buildings	Hth.	Heath	Rd.	Road
Chyd.	Churchyard	Hr.	Higher	S.Sth.	South
Circ.	Circle	Hl.	Hill	Sq.	Square
Cir.	Circus	Ho.	House	Sta.	Station
Clo.	Close	Kg.	King	St.	Street
Cor.	Corner	La.	Lane	Ter.	Terrace
Cotts.	Cottages	Lit.	Little	Trd.	Trading
Ct.	Court	Lo.	Lodge	Upr.	Upper
Cres.	Crescent	Lwr.	Lower	Vall.	Valley
Cft.	Croft	Mans.	Mansion	Vw.	View
Dr.	Drive	Mkt.	Market	Vill.	Villas
Dws.	Dwellings	Ms.	Mews	Wk.	Walk
E.	East	Mt.	Mount	W.	West
Embk.	Embankment	N.Nth.	North	Wf.	Wharf
Esp.	Esplanade	Orch.	Orchard	Wd.	Wood
Est.	Estate	Pde.	Parade	Yd.	Yard

Abberton Wk. Rain. CT37 37
Abbess Clo. SW2 BY47 60
Abbeville Rd. N8 BW31 23
Abbeville Rd. SW4 BW46 50
Abbey Av. Wem. BL37 30
Abbey Clo. Nthlt. BE38 29
Abbey Clo. Pnr. BD31 20
Abbey Ct. Hmptn. BF51 65
Abbey Cres. Belv. CR42 45
Abbeydale Rd. Wem. BL37 30
Abbey Est. Beck. CE50 61
Abbeyfield Gdns. SE16 CC42 43
Abbeyfield Rd. SE16 CC42 43
Abbey Gdns. NW8 BT37 32
Abbey Gdns. Ms. NW8 BT37 32
Abbey Gro. SE2 CO42 45
Abbey Hill Rd. Sid. CP47 63
Abbey La. E15 CF37 34
Abbey La. Beck. CE50 61
Abbey Orchard St. SW1 BW41 41
Abbey Park Est. Beck. CE50 61
Abbey Rd. E15 CF37 34
Abbey Rd. NW6 BS37 32
Abbey Rd. NW8 BT37 32
Abbey Rd. NW10 BM37 30
Abbey Rd. SE2 CP42 45
Abbey Rd. SW19 BT50 59
Abbey Rd. Bark. CL36 35
Abbey Rd. Belv. CP42 45
Abbey Rd. Bexh. CQ45 54
Abbey Rd. Croy. BY55 78
Abbey Rd. Enf. CA25 8
Abbey Rd. Est. NW8 BS37 32
Abbey Rd. Ilf. CM32 26
Abbey Rd. Sth. Croy. CC58 79
Abbey St. E13 CH38 35
Abbey St. SE1 CA41 42
Abbey Ter. NW10 BL37 30
Abbey Ter. SE2 CP42 45
Abbey Vw. NW7 BO27 13
Abbey Wk. E. Mol. BF52 65
Abbey Wd. La. Rain. CV37 37
Abbey Wood Rd. SE2 CO42 45
Abbot Clo. Hamptn. BE50 56
Abbotsbury Clo. E15 CF37 34
Abbotsbury Clo. W14 BR41 40
Abbotsbury Gdns. Pnr. BD33 20
Abbotsbury Rd. W14 BR41 40
Abbotsbury Rd. Brom. CG55 79
Abbotsbury Rd. Mord. BS53 68
Abbots Clo. Orp. CM54 71
Abbots Clo. Rain. CV37 37
Abbots Clo. Ruis. BD34 20
Abbots Cres. Enf. BY23 8
Abbotsford Av. N15 BZ31 24
Abbotsford Gdns. Wdf. Grn. CH30 17
Abbotsford Rd. Ilf. CO34 27
Abbots Gdns. N2 BT31 23
Abbots Grn. Sth. Croy. CC57 79
Abbotshall Av. N14 BW27 14
Abbotshall Rd. SE6 CF47 61
Abbots La. SE1 CA40 42
Abbots La. Ken. BZ61 84
Abbotsleigh Cl. Sutt. BS57 77
Camborne Rd.
Abbotsleigh Rd. SW16 BW49 59
Abbots Manor Est. SW1 BV42 41
Abbots Pk. SW2 BY47 60
Abbots Pl. NW6 BS37 32
Abbot's Rd. E6 CJ37 35
Abbots Rd. Edg. BN29 13
Abbots Tilt Walt. BE55 74
Abbotstone Rd. SW15 BQ45 49
Abbot St. E8 CA36 33
Abbots Way Beck. CD53 70
Abbotswell Rd. SE4 CD46 52
Abbotswood Gdns. Ilf. CK31 26
Abbotswood Rd. SW16 BW48 59
Abbott Av. SW20 BQ51 67
Abbott Clo. Nthlt. BE36 29
Abbott Rd. E14 CF39 43
Abbotts Clo. Rom. CR31 27
Abbotts Clo. Swan. CU52 73
Abbotts Cres. E4 CF28 16
Abbotts Dr. Wem. BJ34 21
Abbottsmede Clo. Twick. BH48 57
Abbott's Park Rd. E10 CF33 25
Abbott's Rd. Barn. BS24 7
Abbotts Rd. Mitch. BW52 68
Abbotts Rd. Sthl. BE40 38
Abbotts Rd. Sutt. BR56 76
Abbott's Wk. Bexh. CP43 54
Abbscross Clo. Horn. CV33 28
Abbscross Gdns. Horn. CV33 28
Abbs Cross Rd. Horn. CV34 28
Abchurch La. EC4 BZ40 42
Abchurch Yd. EC4 BZ40 42
Abchurch La.
Abdale Rd. W12 BP40 40
Aberavon Rd. E3 CD38 34
Abercairn Rd. SW16 BW50 59
Aberconway Rd. Mord. BS52 68
Abercorn Clo. NW7 BR29 13
Abercorn Clo. NW8 BT38 32
Abercorn Cres. Har. BF33 20
Abercorn Est. Wem. BK37 30
Abercorn Gdns. Har. BK33 21
Abercorn Gdns. Rom. CO32 27
Abercorn Ms. NW8 BT37 32
Abercorn Pl. NW8 BT37 32
Abercorn Rd. NW7 BR29 13
Abercorn Rd. Stan. BK29 12
Abercrombie St. SW11 BU44 50
Aberdale Rd. Enf. CB24 8
Aberdare Clo. W. Wick. CF55 79
Aberdare Gdns. NW6 BS36 32
Aberdare Gdns. NW7 BQ29 13
Aberdeen La. N5 BY35 33
Aberdeen Pk. N5 BY35 33
Aberdeen Pl. NW8 BT38 32
Aberdeen Rd. N5 BZ35 33
Aberdeen Rd. N18 CB28 15
Aberdeen Rd. SW19 BT51 68
Aberdeen Rd. Croy. BZ56 78
Aberdeen Rd. Har. BH30 12
Aberdeen Ter. SE3 CF44 52
Aberdeen Ter. SW19 BT51 68
Aberdour Rd. Ilf. CO34 27
Aberdour St. SE1 CA42 42
Aberfeldy St. E14 CF39 43
Aberford Gdns. SE18 CK44 53
Aberford Rd. Borwd. BM23 5
Aberfoyle Rd. SW16 BW50 59
Abergeldie Rd. SE12 CH46 53
Abernethy Rd. SE13 CG45 52
Abersham Rd. E8 CA35 33
Abery St. SE18 CN42 45
Abingdon Clo. NW1 BW36 32
Camden Sq.
Abingdon Clo. SW19 BT50 59
Abingdon Rd. N3 BT30 14
Abingdon Rd. SW16 BX51 69
Abingdon Rd. W8 BS41 41
Abingdon St. SW1 BX41 42
Abingdon Vill. W8 BS41 41
Abinger Av. Sutt. BQ58 76
Abinger Clo. Brom. CK52 71
Abinger Clo. Ilf. CO35 36
Abinger Clo. Wall. BX56 78
Abinger Gdns. Islw. BH45 48
Abinger Gro. SE8 CD43 52
Abinger Rd. W4 BO41 40
Ablemarle Way EC1 BY38 33
Clerkenwell Rd.
Ablett St. SE16 CC42 43
Abourne St. W9 BS38 32
Amberley Rd.
Aboyne Dr. SW20 BP51 67
Aboyne Rd. NW10 BN34 22
Aboyne Rd. SW17 BT48 59
Abridge Gdns. Rom. CR29 18
Abridge Rd. Chig. CM26 10
Abridge Way, Bark. CO37 36
Abyssinia Clo. SW11 BU45 50
Cairns Rd.
Acacia Av. N17 BZ29 15
Acacia Av. Brent. BJ43 48
Acacia Av. Horn. CT34 28
Acacia Av. Ruis. BC33 20
Acacia Av. Wem. BL35 30
Acacia Clo. Orp. CM53 71
Acacia Clo. Stan. BH29 12
Acacia Dr. Bans. BQ60 82
Acacia Dr. Sutt. BR54 67
Acacia Gdns. W. Wick. CF55 79
Acacia Gro. SE21 BZ48 60
Acacia Gro. N. Mal. BN52 67
Acacia Pl. NW8 BT37 32
Acacia Rd. E11 CG34 25
Acacia Rd. E17 CD32 25
Acacia Rd. N22 BY30 15
Acacia Rd. NW8 BT37 32
Acacia Rd. SW16 BX51 69
Acacia Rd. W3 BN40 40
Acacia Rd. Beck. CD52 70
Acacia Rd. Dart. CV47 64
Acacia Rd. Enf. BZ23 8
Acacia Rd. Hamptn. BF50 56
Acacia Rd. Mitch. BV51 68
Academy Gdns. Croy. CA54 69
Academy Gdns. Nthlt. BD37 29
Academy Rd. SE18 CK44 53
Acanthus Rd. SW11 BV45 50
Gideon Rd.
Accommodation Rd. NW11 BR33 22
Accorn Rd. DA1 CT46 55
Acer Av. Rain. CV38 37
Acfold Rd. SW6 BS44 50
Achilles Rd. NW6 BS35 32
Achilles St. SE14 CD43 52
Achilles Way W1 BV40 41
Acklington Dr. NW9 BO30 13
Ackmar Rd. SW6 BS44 50
Ackroyd Dr. E3 CD39 43
Ackroydon Est. SW19 BQ47 58
Ackroyd Rd. SE23 CC47 61
Acland Cres. SE5 BZ45 51
Acland Rd. NW2 BP36 31
Linacre Rd.
Acol Cres. Ruis. BC35 29
Acol Rd. NW6 BS36 32
Acomb Rd. Dag. CO37 36
Aconbury Rd. Dag. CO37 36
Stamford Rd.
Acorn Clo. E4 CE28 16
Lawns, The
Acorn Clo. Enf. BY23 8
Acorn Clo. Enp. CH23 10
Drapers Rd.
Acorn Gdns. SE19 CA51 69
Acorn Gdns. W3 BN39 40
Acorn Pl. Est. SE15 CB44 51
Acorns, The, Chig. CN28 18
Acorn Wk. SE16 CD40 43
Acre La. SW2 BX45 51
Acre La. Cars. BV56 77
Acre Path Nthlt. BE36 29
Acre Rd. SW19 BT50 59
Acre Rd. Dag. CR36 36
Acre Rd. Kings. On T. BL51 66
Acris St. SW18 BT46 50
Acton Est. E8 CA37 33
Acton La. NW10 BN38 31
Acton la. W3 BO41 40
Acton La. W4 BN42 40
Acton Park Ind. Est. W3 BN40 40
Acton St. WC1 BX38 33
Acuba Rd. SW18 BS47 59
Ada Gdns. E14 CF39 43
Ada Gdns. E15 CG37 34
Adair Rd. W10 BR38 31
Adam And Eve Ct. W1 BW39 41
Eastcastle St.
Adam And Eve Ms. W8 BS41 41
Adam Pl. N16 CA34 24
High St.
Adams Clo. Wem. BM34 21
Adams Gdns. Est. SE16 CC41 43
Adamson Rd. E16 CH39 44
Adamson Rd. NW3 BT36 32
Adamsrill Clo. Enf. BZ25 8
Adamsrill Rd. SE26 CC49 61
Adams Rd. N17 CA30 15
Adams Rd. Beck. CD52 70
Adam's Row W1 BV40 41
Adams Sq. Bexh. CP45 54
Adam St. W1 BV39 41
Adam St. WC2 BX40 42
Adam Wk. SW6 BQ43 49
Ada Pl. E2 CB37 33
Adare Wk. SW16 BX48 60
Ada Rd. SE5 CA43 51
Ada Rd. Wem. BK34 21
Ada St. E8 CB37 33
Adcock Wk. Orp. CN56 81
Adderley Gdns. SE9 CL49 62
Adderley Gro. SW11 BV46 50
Culmstock Rd.
Adderley Rd. Har. BH30 12
Adderley St. E14 CF39 43
Addington Dr. N12 BT29 14
Addington Gro. SE26 CD49 61
Addington Rd. E3 CE38 34
Addington Rd. E16 CG38 34
Addington Rd. N4 BY32 24
Addington Rd. Croy. BY54 69
Addington Rd. Sth. Croy. CB59 84
Addington Rd. W. Wick. CF56 79
Addington Sq. SE5 BZ43 51
Addington St. SE1 BX41 42
Addington Village Rd. Croy. CD57 79
Addis Clo. Enf. CC23 9
Addiscombe Av. Croy. CB54 69
Addiscombe Clo. Har. BK32 21
Addiscombe Ct. Rd. Croy. CA54 69
Addiscombe Gro. Croy. BZ55 78
Addiscombe Rd. Croy. BZ55 78
Addiscombe Rd. Wat. BC24 4
Addison Av. N14 BV25 7
Addison Av. W11 BR40 40
Addison Av. Houns. BG44 47
Addison Bridge Pl. W14 BR42 40
Addison Clo. Orp. CM53 71
Addison Cres. W14 BR41 40
Addison Gdns. W14 BQ41 40
Addison Gdns. Surb. BL52 66
Addison Gro. W4 BO41 40
Addison Pl. W11 BR40 40
Addison Rd. E11 CH32 26
Addison Rd. E17 CE32 25
Addison Rd. SE25 CB52 69
Addison Rd. W14 BR41 40
Addison Rd. Brom. CJ53 71
Addison Rd. Enf. CC23 9
Addison Rd. Ilf. CM30 17
Addison Rd. Tedd. BJ50 57
Addisons Clo. Croy. CD55 79
Addison Way NW11 BR31 22
Addison Way Hayes BC39 38
Addle Hill EC4 BY39 42
Adecroft Way E. Mol. BG52 65
Adela Av. N. Mal. BP53 67
Adelaide Av. SE4 CD45 52
Adelaide Clo. Stan. BJ28 12
Adelaide Cotts. W7 BH40 39
Adelaide Gdns. Rom. CQ32 27
Adelaide Gro. W12 BP40 40
Adelaide Rd. E10 CF34 25
Adelaide Rd. NW3 BT36 32
Adelaide Rd. W13 BJ40 39
Adelaide Rd. Chis. CL49 62
Adelaide Rd. Houns. BE44 47
Adelaide Rd. Ilf. CL34 26
Adelaide Rd. Rich. BL45 48
Adelaide Rd. Sthl. BE42 38
Adelaide Rd. Surb. BL53 66
Adelaide Rd. Tedd. BH50 57
Adelaide Rd. Walt. BC55 74
Adelaide St. WC2 BX40 42
William Iv St.
Adela St. W10 BR38 31
Kensal Rd.
Adelina Gro. E1 CC39 43
Adeline Pl. WC1 BW39 41
Adelphi Cres. Horn. CU34 28
Adelphi Rd. Epsom BN60 82
Adelphi Ter. WC2 BX40 42
Adam St.
Adeney Clo. W6 BQ43 49
Aden Gro. N16 BZ35 33
Aden Lo. N16 BZ35 33
Adenmore Rd. SE6 CE47 61
Aden Rd. Enf. CD24 9
Aden Rd. Ilf. CL33 26
Aden Ter. N5 BZ35 33
Adie Rd. W6 BQ41 40
Adine Rd. E13 CH38 35
Adler St. E1 CB39 42
Adley Rd. E5 CD35 34
Admaston Rd. SE18 CM43 53
Admiral Seymour Rd. SE9 CK45 53
Admiral St. SE8 CE44 52
Admirals Wk. NW3 BT34 23
Admiralty Rd. Tedd. BH50 57
Adnams Wk. Rain. CT36 37
Adolf St. SE6 CE49 61
Adolphus Rd. N4 BY34 24
Adolphus St. SE8 CD43 52
Adomar Rd. Dag. CQ34 27
Adpar St. W2 BT39 41
Adrian Av. NW2 BP33 22
Adrian Ms. SW10 BS43 50
Adrienne Av. Sthl. BE38 29
Adys Rd. SE15 CA45 51
Aerodrome Rd. NW9 BO30 13
Aerodrome Way Houns. BD43 47
Aeroville, NW9 BO30 13
Affleck St. N1 BX37 33
Afghan Rd. SW11 BU44 50
Agamemnon Rd. NW6 BR35 31
Agar Gro. NW1 BW36 32
Agar Gro. Est. NW1 BW36 32
Agar Pl. NW1 BW36 32
Agar St. WC2 BX40 42
Chandos Pl.
Agate Rd. W6 BQ41 40
Agatha Clo. E1 CB40 42
Agaton Rd. SE9 CM48 62
Agave Rd. NW2 BQ35 31
Agdon St. EC1 BY38 33
Agincourt Rd. NW3 BU35 32
Agister Rd. Chig. CO28 18
Agnes Av. Ilf. CL34 26
Agnes Gdns. Dag. CP35 36
Agnes Rd. W3 BO40 40
Agnes St. E14 CD39 43
Agnew Rd. SE23 CC47 61
Agra Pl. E1 CB39 42
Agricola Pl. Enf. CA25 8
Main Av.
Aidan Clo. Rom. CQ34 27
Aileen Wk. E15 CG36 34
Ailsa Av. Twick. BJ46 48
Ailsa Rd. Twick. BJ46 48
Ailsa St. E14 CF39 43
Ainger Rd. NW3 BU36 32
Ainsdale Clo. Orp. CM54 71
Ainsdale Cres. Pnr. BF31 20
Ainsdale Rd. W5 BK38 30
Ainsdale Rd. Wat. BD27 11
Ainsley Av. Rom. CR32 27
Ainsley St. E2 CB38 33
Ainslie Wk. SW12 BV47 59
Ainslie Wood Cres. E4 CE28 16
Ainslie Wood Gdns. E4 CE28 16
Ainslie Wood Rd. E4 CE28 16
Ainsworth Clo. NW2 BP34 22
Ainsworth Rd. Croy. BY54 69
Ainsworth Way NW8 BT37 32
Aintree Av. E6 CK37 35
Aintree Cres. Ilf. CM30 17
Aintree Est. SW6 BR43 49
Aintree Gro. Upmin. CW34 28
Aintree Rd. Grnf. BJ37 30
Aintree St. SW6 BR43 49
Airdrie Clo. N1 BX36 33
Carnoustie Dr.
Airedale Av. W4 BO42 40
Airedale Rd. SW12 BU47 59
Airedale Rd. W5 BK41 39
Airfield Way. Horn. CU36 37
Airlie Gdns. W8 BS40 41
Airlie Gdns. Ilf. CL33 26
Air St. W1 BW40 41
Glasshouse St.
Airthrie Rd. Ilf. CO34 27
Aisgill Av. SW5 BR42 40
Aislibie Rd. SE12 CG45 52

Aitken Rd. SE6 CE48 61
Aitken Rd. Barn. BQ25 6
Ajax Av. NW9 BO31 22
Ajax Rd. NW6 BS35 32
Akehurst St. SW15 BP46 49
Akenside Rd. NW3 BT35 32
Akerman Rd. SW9 BY44 51
Akerman Rd. Surb. BK53 66
Alabama St. SE18 CM43 53
Alacross Rd. W5 BK41 39
Alan Clo. Dart. CV45 55
Alandale Dr. Pnr. BC30 11
Alan Dri. Barn. BR25 6
Alan Gdns. Rom. CR33 27
Alan Rd. SW19 BR49 58
Alanthus Clo. SE12 CG46 52
Alaska St. SE1 BY40 42
Cornwall Rd.
Albacore Cres. SE13 CE46 52
Alba Clo. Hayes BE38 29
Alba Gdns. NW11 BR32 22
Alban Cres. Borwd. BM23 5
Albans La. NW11 BS33 23
West Hth. Dr.
Albany Clo. SW14 BM45 48
Albany Clo. Bex. CP47 63
Albany Clo. Bush. BG25 4
Albany Clo. Esher BE57 74
Albany Cotts. Esher BE57 74
Albany Ct. NW9 BN30 13
Albany Cres. Edg. BM29 12
Albany Cres. Esher BH57 75
Albany Mans. SW11 BU43 50
Albany Ms. SE17 BZ43 51
Albany Rd.
Albany Pde. Brent BL43 48
Albany Rd.
Albany Pk. Av. Enf. CC23 9
Albany Pk. Rd. Kings. On T. BK50 57
Albany Pass. Rich. BL46 48
Albany Pl. N7 BY35 33
Albany Pl. Brent. BK43 48
Albany Rd.
Albany Reach Surb. BH53 66
Albany Rd. E10 CE33 25
Albany Rd. E12 CJ35 35
Albany Rd. E17 CD32 25
Albany Rd. N4 BY32 24
Albany Rd. N18 CC28 16
Albany Rd. SE5 BZ43 51
Albany Rd. SW19 BS49 59
Albany Rd. W13 BJ39 39
Albany Rd. Belv. CQ43 54
Albany Rd. Bex. CP47 63
Albany Rd. Brent. BK43 48
Albany Rd. Chis. CL49 62
Albany Rd. Horn. CU33 28
Albany Rd. N. Mal. BN52 67
Albany Rd. Rich. BL46 48
Albert Rd.
Albany Rd. Rom. CQ32 27
Albany Rd. Walt. BD56 74
Albany St. NW1 BV37 32
Albany Ter. NW1 BV38 32
Euston Rd.
Albany, The, W1 BW40 41
Vigo St.
Albany, The, Wdf. Grn. CG28 16
Albany Vw. Buck. H. CH26 10
Alba Pl. W11 BR39 40
Portobello Rd.
Albatross Gdns. Sth. Croy. CC58 79
Albatross St. SE18 CM43 53
Albemarle App. Ilf. CL32 26
Albemarle Gdns. Ilf. CL32 26
Albemarle Gdns. N. Mal. BN52 67
Albemarle Pk. Stan. BK28 12
Albemarle Rd. Barn. BU26 7
Albemarle Rd. Beck. CE51 70
Albemarle St. W1 BV40 41
Albemarle Way EC1 BY38 33
Clerkenwell Rd.
Albermarle Av. Twick. BE47 56
Alberon Gdns. NW11 BR31 22
Alberta Av. Sutt. BR56 76
Alberta Est. SE17 BY42 42
Alberta Rd. Enf. CA25 8
Alberta Rd. Erith CR44 54
Alberta St. SE17 BY42 42
Albert Av. E4 CE28 16
Albert Av. SW8 BX43 51
Albert Av. Ilf. CM34 26
High Rd.
Albert Br. SW3 BU43 50
Albert Br. SW11 BU43 50
Albert Br. Rd. SW11 BU43 50
Albert Carr Gdns. SW16 BX49 60
Albert Clo. N22 BW30 14
Albert Ct. SW7 BT41 41
Albert Ct. SW19 BR47 58
Albert Cres. E4 CE28 16
Albert Dr. SW19 BR48 58
Albert Embankment, SE1 BX42 42
Albert Gdns. E1 CC39 43
Albert Gdns. NW6 BR37 31
Albert Gate SW1 BU41 41
Rotten Row
Albert Gro. SW20 BQ51 67
Albert Hall Ms. SW7 BT41 41
Prince Consort Rd.
Albert Mans. SW11 BU44 50
Albert Pl. N3 BS30 14
Popes Dr.
Albert Pl. W8 BS41 41
Albert Rd. E10 CF34 25
Albert Rd. E17 CE32 25
Albert Rd. E18 CH31 26
Albert Rd. N4 BX33 24
Albert Rd. N15 CA32 24
Albert Rd. N22 BW30 14
Albert Rd. NW4 BQ31 22
Albert Rd. NW6 BR37 31
Albert Rd. NW7 BO28 13
Albert Rd. SE9 CK48 62
Albert Rd. SE20 CC50 61
Albert Rd. SE25 CB52 69
Albert Rd. W5 BJ38 30
Albert Rd. Barn. BT24 7
Albert Rd. Belv. CQ42 45
Albert Rd. Bex. CR47 63
Albert Rd. Brom. CJ53 71
Albert Rd. Buck. H. CJ27 17
Albert Rd. Dag. CQ33 27
Albert Rd. Dart. CV48 64
Albert Rd. Epsom BO60 82
Albert Rd. Har. BG31 20
Albert Rd. Hmptn. BG49 56
Albert Rd. Houns. BF45 47
Albert Rd. Ilf. CL34 26
Albert Rd. Kings. On T. BL51 66
Albert Rd. Mitch. BU52 68
Albert Rd. N. Mal. BO52 67
Albert Rd. Orp. CO53 72
Albert Rd. Orp. CO56 81
Albert Rd. Rich. BL46 48
Albert Rd. Rom. CT32 28
Albert Rd. Sthl. BD41 38
Albert Rd. Sutt. BT56 77
Albert Rd. Tedd. BH50 57
Albert Rd. Twick. BH47 57
Albert Rd. Est. Belv. CQ42 45
Albert Rd. N. Wat. BC24 4
Albert Rd. S. Wat. BC24 4
Albert Sq. E15 CG35 34
Albert Sq. SW8 BX43 51
Albert St. N12 BT28 14
Lodge La.
Albert St. NW1 BV37 32
Albert St. Wat. BD24 4
Queen's Rd.
Albert Ter. NW1 BV37 32
Albert Ter. NW10 BO37 31
Albion Av. N10 BV30 14
Albion Av. SW8 BW44 50
Albion Bldgs EC1 BZ39 42
Bartholomews Clo.
Albion Clo. W2 BU40 41
Albion St.
Albion Clo. Rom. CS32 28
Albion Dr. E8 CA36 33
Albion Est. SE16 CC41 43
Albion Gdns. W6 BP42 40
Albion Gdns. Dag. CR35 36
Albion Gte. W2 BT40 41
Albion Gro. N16 BZ35 33
Albion Hill SE13 CE44 52
Albion Hill Loug. CJ25 10
Albion Ms. N1 BX36 33
Albion Ms. W2 BU40 41
Albion Pk. Loug. CJ25 10
Albion Pl. EC1 BY39 42
Albion Pl. SE25 CB52 69
High St.
Albion Rd. E17 CF31 25
Albion Rd. N16 BZ34 24
Albion Rd. N17 CA30 15
Reform Row
Albion Rd. Bexh. CR45 54
Albion Rd. Houns. BF45 47
Albion Rd. Kings. On T. BN51 67
Albion Rd. Sutt. BT57 77
Albion Rd. Twick. BH47 57
Albion Sq. E8 CA36 33
Albion St. E15 CF37 34
Albion St. SE16 CC41 43
Albion St. W2 BU39 41
Albion St. Croy. BY54 69
Albion Ter. E8 CA36 33
Albion Villas Rd. SE26 CC48 61
Albion Way, SE13 CF45 52
Albion Way, Wem. BM34 21
Albuhera Clo. Enf. BY23 8
Albury Av. Islw. BH43 48
Albury Av. Sutt. BQ58 76
Albury Clo. Hamptn. BF50 56
Albury Dr. Pnr. BD29 11
Albury Dr. Pnr. BD30 11
Albury Rd. Bexh. CQ44 54
Albury Rd. Chess. BL56 75
Albury St. SE8 CE43 52
Albyfield Brom. CK52 71
Albyn Rd. SE8 CE44 52
Alcester Cres. E5 CB34 24
Alcester Rd. Wall. BV56 77
Alcock Clo. Wall. BW57 77
Alcock Rd. Houns. BD43 47
Alconbury Rd. E5 CB34 24
Alcorn Clo. Sutt. BS55 77
Aldam Pl. N16 CA34 24
High St.
Aldborough Rd. Dag. CS36 37
Aldborough Rd. Ilf. CN33 27
Aldborough Rd. Upmin. CW34 28
Aldborough Rd. N. Ilf. CN32 27
Aldborough Rd. S. Ilf. CN33 27
Aldbourne Rd. W12 BO40 40
Aldbridge St. SE17 CA42 42
Aldbury Av., Wem. BM36 30
Aldbury Ms. N9 BZ26 8
Aldbury St. SE8 CE43 52
Aldebert Ter. SW8 BX43 51
Aldeburgh Clo. E5 CB34 24
Southwold Rd.
Aldeburgh Pl. Wdf. Grn. CH28 17
Aldeburgh St. SE10 CH42 44
Alden Av. E15 CG38 34
Aldenham Rd. Borwd. BJ24 5
Aldenham Rd. Wat. BD25 4
Aldenham St. NW1 BW37 32
Aldensley Rd. W6 BP41 40
Alder Av. Upmin. CW35 37
Alderbrook Rd. SW12 BV46 50
Alderbury Rd. SW13 BP43 49
Alder Clo. SE15 CA43 51
Aldercroft Couls. BX61 84
Alder Gro. NW2 BP34 22
Alderholt Way, SE15 CA43 51
Bedenham Way
Alderman Av. Bark. CO38 36
Aldermanbury, EC2 BZ39 42
Aldermanbury Sq. EC2 BZ39 42
Alderman's Hill N13 BX28 15
Aldermans Wk. EC2 CA39 42
Bishopgate
Aldermary Rd. Brom. CH51 71
Aldermas St. W10 BQ39 40
Alderminster Rd. SE1 CB42 42
Aldermoor Rd. SE6 CD48 61
Alderney Av. Houns. BF43 47
Alderney Gdns. Nthlt. BE36 29
Alderney Rd. E1 CC38 34
Alderney Rd. Erith CU43 55
Alderney St. SW1 BV42 41
Alder Rd. SW14 BN45 49
Alder Rd. Sid. CN48 63
Alders Av. Wdf. Grn. CG29 16
Aldersbrook Av. Enf. CA23 8
Aldersbrook Dr. Kings. On T. -BL50 57
Aldersbrook La. E12 CK34 26
Aldersbrook Rd. E11 CH34 26
Aldersbrook Rd. E12 CH34 26
Alders Clo. Edg. BN28 13
Aldersey Gdns. Bark. CM36 35
Aldersford Clo. SE4 CC46 52
Aldersgate St. EC1 BZ39 42
Alders Gro. E. Mol. BG53 65
Aldersgrove Av. SE9 CJ48 62
Aldershot Rd. NW6 BR37 31
Aldershot Ter. SE18 CK43 53
Imperial Way
Aldersmead Av. Croy. CC53 70
Aldersmead Rd. Beck. CD50 61
Alderson St. W10 BR38 31
Kensal Rd.
Alders Rd. Edg. BO28 13
Alders, The, N21 BY25 8
Alders, The, Felt. BE49 56
Alders, The, Houns. BE42 38
Alders, The, W. Wick. CE54 70
Alderton Clo. SE5 BZ45 51
Alderton Clo. Loug. CL24 10
Alderton Cres. NW4 BP32 22
Alderton Hall La. Loug. CL24 10
Alderton Hill Loug. CK25 10
Alderton Rise, Loug. CL24 10
Alderton Rd. SE24 BZ45 51
Alderton Rd. Croy. CA54 69
Alderton Way NW4 BP32 22
Alderton Cres.
Alderton Way Loug. CK25 10
Alderville Rd. SW6 BR44 49
Alder Way Swan. CS51 73
Alderwick Dr. Houns. BG45 47
Alderwood Rd. SE9 CM46 53
Aldford St. W1 BV40 41
Aldgate Av. E1 CA39 42
Aldgate High St. EC3 CA39 42
Aldham Hall, E11 CH32 26
Aldine Ct. W12 BQ41 40
Aldine Cres. W12 BQ41 40
Aldine St. W12 BQ41 40
Aldingham Gdns. Horn. CU35 37
Aldington Rd. SE7 CJ41 44
Aldis Ms. SW17 BU49 59
Aldis St.
Aldis St. SW17 BU49 59
Aldred Rd. NW6 BS35 32
Aldren Rd. SW17 BT48 59
Aldrich Cres. Croy. CF58 79
Aldriche Way, E4 CF29 16
Aldridge Av. Edg. BM27 12
Aldridge Av. Ruis. BD34 20
Aldridge Av. Stan. BL30 12
Aldridge Rise N. Mal. BO54 67
Aldridge Rd. Vill. W11 BR39 40
Aldridge Wk. N14 BX26 8
Aldrington Rd. SW16 BW49 59
Aldsworth Clo. W9 BS38 32
Amberley Est.
Aldwick Clo. SE9 CM48 62
Aldwick Rd. Croy. BX55 78
Aldworth Gr. SE13 CF46 52
Aldworth Rd. E15 CG36 34
Aldwych, WC2 BX40 42
Aldwych Av. Ilf. CM31 26
Aldwych Clo. Horn. CU34 28
Alers Rd. Bexh. CP46 54
Alexander Av. NW10 BP36 31
Alexander Clo. Brom. CH54 71
Alexander Clo. Sid. CN46 54
Alexander Clo. Twick. BH48 57
Alexander Ct. N14 BW25 7
Alexander Ms. W2 BS39 41
Alexander St.
Alexander Pl. SW7 BU42 41
Alexander Rd. N19 BX34 24
Alexander Rd. W13 BJ40 39
Alexander Rd. Bexh. CP44 54
Alexander Rd. Chis. CL49 62
Alexander Rd. Couls. BV61 83
Alexander Sq. SW3 BU41 41
Alexander St. W2 BS39 41
Alexandra Av. N22 BW30 14
Alexandra Av. SW11 BV44 50
Alexandra Av. W4 BN43 49
Alexandra Av. Har. BE33 20
Alexandra Av. Sthl. BE40 38
Alexandra Av., Sutt. BS55 77
Alexandra Clo. Har. BF34 20
Alexandra Clo. Walt. BC55 74
Alexandra Cotts. SE14 CD44 52
Alexandra Ct. Wem. BL35 30
Alexandra Cres. Brom. CG50 61
Alexandra Dr. SE19 CA49 60
Alexandra Dr. Surb. BM54 66
Alexandra Gdns. N10 BV31 23
Alexandra Gdns. W4 BO43 49
Alexandra Gdns. Cars. BU58 77
Alexandra Gdns. Houns. BF44 47
Alexandra Gro. N4 BY33 24
Alexandra Gro. N12 BS29 14
Alexandra Ms. NW8 BS37 32
Alexandra Pk. Rd. N10 BV30 14
Alexandra Pk. Rd. N22 BV30 14
Alexandra Pl. NW8 BT37 32
Alexandra Pl. SE25 BZ53 69
Alexandra Pl. Croy. CA54 69
Alexandra Rd. E6 CL38 35
Alexandra Rd. E10 CF34 25
Alexandra Rd. E17 CD32 25
Alexandra Rd. E18 CH31 26
Alexandra Rd. N8 BY31 24
Alexandra Rd. N9 CB26 8
King Edwards Rd.
Alexandra Rd. N10 BV30 14
Alexandra Rd. N15 BZ32 24
Alexandra Rd. NW4 BQ31 22
Alexandra Rd. NW8 BS37 32
Alexandra Rd. NW8 BT37 32
Alexandra Rd. SE26 CC50 61
Alexandra Rd. SW14 BN45 49
Alexandra Rd. SW19 BR50 58
Alexandra Rd. W4 BN41 40
Alexandra Rd. Borwd. BN23 6
Alexandra Rd. Brent. BK43 48
Alexandra Rd. Croy. CA54 69
Alexandra Rd. Enf. CC24 9
Alexandra Rd. Epsom. BO60 82
Alexandra Rd. Erith CT43 55
Alexandra Rd. Houns. BF44 47
Alexandra Rd. Kings. On T. BM50 57
Alexandra Rd. Mitch. BU50 59
Alexandra Rd. Rain. CT37 37
Alexandra Rd. Rich. BL44 48
Alexandra Rd. Rom. CP32 27
Alexandra Rd. Rom. CQ32 27
Alexandra Rd. Rom. CT32 28
Alexandra Rd. Surb. BH53 66
Alexandra Rd. Twick. BK46 48
Alexandra Rd. Wat. BC23 4
Alexandra Sq. Mord. BS53 68
Alexandra St. E16 CH39 44
Alexandra St. SE14 CD43 52
Bowerman Av.
Alexandra Wk. SE19 CA49 60
Alexandria Rd. W13 BJ40 39
Alexis St. SE16 CB42 42
Alfan La. Dart. CS49 64
Alford Grn. Croy. CF57 79
Alford Pl. N1 BZ37 33
Shepherders Wk.
Alford Rd. SW8 BW44 50
Union Gro.
Alford Rd. Erith. CS42 46
Alfoxton Av. N15 BY31 24
Alfreda St. SW11 BV44 50
Alfred Cotts. SW17 BU49 59

Anchor Dr. RM13 CU38 37
Wentworth Way
Anchor St. SE16 CB42 42
Anchor Yd. EC1 BZ38 33
Old St.
Ancill Clo. W6 BR43 49
Ancona Rd. NW10 BP37 31
Ancona Rd. SE18 CM42 44
Andalus Rd. SW9 BX45 51
Ander Clo. Wem. BK35 30
Anderson Rd. E9 CC36 34
Digby Rd.
Anderson Rd. Ilf. CJ31 26
Anderson's Pl. Houns. BF45 47
Anderson St. SW3 BU42 41
Anderson St. W10 BR38 31
Kensal Rd.
Andoe Rd. SW11 BU45 50
Andover Clo. Epsom BN58 76
Andover Clo. Grnf. BF38 29
Andover Pl. NW6 BS37 32
Andover Rd. EC1 BX34 24
Andover Rd. Orp. CN54 72
Andover Rd. Twick. BG47 56
Andover St. N7 BX34 24
Andre St. E8 CB35 33
Andrew Borde St. WC2 BW39 41
Charing Cross Rd.
Andrew Clo. Bex. CS46 55
Bourne Rd.
Andrew Pl. SW8 BW44 50
Andrews Clo. Buck. H. CJ27 17
Andrews Clo. Epsom BO60 82
Andrews Clo. Orp. CP51 72
Andrew's Crosse WC2 CM39 44
Bell Yd.
Andrew's Rd. E8 CB37 33
Andrew St. E14 CF39 43
Andwell Clo. SE2 CO41 45
Anerley Gro. SE19 CA50 60
Anerley Hill SE19 CA50 60
Anerley Pk. SE20 CB50 60
Anerley Pk. Rd. SE20 CB50 60
Anerley Rd. SE19 CB50 60
Anerley Rd. SE20 CB50 60
Anerley Sta. Rd. SE20 CB51 69
Anerley Val. SE19 CA50 60
Angel Clo. N18 CA28 15
Angel Ct. EC2 BZ39 42
Throgmorton St.
Angel Ct. SW17 BU49 59
Angelfield Houns. BF45 47
Angel Hill Sutt. BS55 77
Angel Hill Dr. Sutt. BS55 77
Angel La. E15 CF36 34
Angell Pk. Gdns. SW9 BY45 51
Angel Ms. N1 BY37 33
Angel Pass. EC3 BZ40 42
Wharfside
Angel Pl. EC4 BZ40 42
Angel Pl. SE1 BZ41 42
Borough High St.
Angel Rd. N18 CB28 15
Angel Rd. Har. BH32 2
Angel Rd. Surb. BJ54 66
Angel St. EC1 BZ39 42
Angel Wk. W6 BQ42 40
Angel Way Rom. CT31 28
Angerstein La. SE3 CG43 52
Anglers Clo. Rich. BK49 57
Anglers La. NW5 BV36 32
Anglesea Av. SE18 CL42 44
Anglesea Rd. SE18 CL42 44
Anglesea Rd. BK52 66
Kings-on-t.
Anglesea Rd. Orp. CO53 72
Anglesea St. E1 CB38 33
Anglesey Ct. Rd. Cars. BV57 77
Aglesey Dr. Rain. CU38 37
Anglesey Gdns. Cars. BV57 77
Anglesey Rd. Enf. CB24 8
Anglesey Rd. Wat. BD28 11
Anglesmede Cres. Pnr. BF31 20
Anglesmede Wk. Pnr. BF31 20
Anglesmede Way Pnr. BF31 20
Angles Rd. SW16 BX49 60
Anglo Rd. E3 CD37 34
Usher Rd.
Angus Clo. Chess. BM56 75
Angus Dr. Ruis. BD35 29
Angus Gdns. NW9 BN30 13
Angus Rd. E13 CJ38 35
Angus St. SE14 CD43 52
Clifton Rise
Anhalt Rd. SW11 BU43 50
Ankerdine Cres. SE18 CL44 53
Anlaby Rd. Tedd. BH49 57
Anley Rd. W14 BQ41 40
Anmersh Gro. Stan. BK30 12
Annabel Clo. E14 CE39 43
Annandale Rd. SE10 CG42 43
Annandale Rd. W4 BO42 40
Annandale Rd. Croy. CB55 78
Annandale Rd. Sid. CN47 63
Annan Way Rom. CS30 19
Anne Boleyn's Wk. BL49 57
Kings. On T.
Anne Boleyn's Wk. Sutt. BQ57 76
Anne Of Cleves Rd. CV46 55
Dart.
Annersley Walk N19 BW34 23
Girdlestone Est.
Annesley Av. NW9 BN31 22
Annesley Clo. NW10 BO34 22
Annesley Dr. Croy. CD55 79
Annesley Rd. N19 BW34 23
Annesley Rd. SE3 CH44 53
Anne St. E13 CH38 35
Annette Clo. Har. BH30 12
Spencer Rd.
Annette Rd. N7 BX34 24
Annett Rd. Walt. BC54 65
Annetts Gro. N1 BZ36 33
Essex Rd.
Anne Way E. Mol. BF52 65
Anne Way, Ilf. CM29 17
Annie Besant Clo. E3 CD37 34
Anning St. EC2 CA38 33
New Inn Yd.
Annington Rd. N2 BU31 23
Annis Rd. E9 CD36 34
Ann La. SW10 BT43 50
Ann's Clo. SW1 BV41 41
Kinnerton St.
Ann St. SE18 CM42 44
Ann's Vill. W11 BQ41 40
Annsworthy Av. Th. Hth. BZ52 69
Annsworthy Cres. SE25 BZ52 69
Ansdell Rd. SE15 CC44 52
Ansdell St. W8 BS41 41
St Alban's Gro.
Ansdell Ter. W8 BS41 41
St. Alban's Gro.
Ansell Gro. Cars. BU54 68
Ansell Rd. SW17 BU48 59
Anselm Clo. Croy. CA55 78
Anselm Rd. SW6 BS43 50
Anselm Rd. Pnr. BE29 11
Ansford Rd. Brom. CF49 61
Ansleigh Pl. W11 BQ40 40
Ansley Clo. Sth. Croy. CB60 84
Anson Clo. Rom. CR30 18
Anson Rd. N7 BW35 32
Anson Rd. NW2 BQ35 31
Anson Ter. Nthlt. BF35 29
Wood End La.
Anstead Dr. Rain. CU37 37
Anstey Rd. SE15 CB45 51
Anstey Wk. N15 BY31 24
Anstridge Rd. SE9 CM46 53
Antelope Rd. SE18 CK41 44
Anthony Clo. NW7 BO28 13
Anthony Clo. Wat. BD26 4
Anthony Rd. SE25 CB53 69
Anthony Rd. Borwd. BL23 5
Anthony Rd. Grnf. BH37 30
Anthony Rd. Well. CO44 54
Anthony St. E1 CB39 42
Antill Rd. E3 CD38 34
Antill Rd. N15 CB31 24
Antill Ter. E1 CC39 43
Antlers Hill E4 CE24 9
Antoney's Clo. Pnr. BD30 11
Anton St. E8 CB35 33
Antrim Gro. NW3 BU36 32
Antrim Mans. NW3 BU36 32
Antrim Rd. NW3 BU36 32
Antrobus Clo. Sutt. BR56 76
Antrobus Rd. W4 BN42 40
Anvil Rd. Sun. BC52 65
Anworth Clo. Wdf. Grn. CH29 17
Apeldoorn Wall. BX58 78
Aperfield Rd. Erith CT43 55
Apex Clo. Beck. CF51 70
Apollo Av. Brom. CH51 71
Hawes Rd.
Apollo Av. Nthwd. BC28 11
Apollo Clo. Horn. CU34 28
Apollo Pl. SW10 BT43 50
Riley St.
Apollo Pl. SW10 BT44 50
Riley St.
Apothecary St. EC4 BY39 42
New Bridge St.
Appach Rd. SW2 BY46 51
Apper St. WC1 BW38 32
Appian Way Est. Erith. CR42 45
Appleby Clo. E4 CF29 16
Appleby Clo. N15 BZ32 24
Cornwall Rd.
Appleby Clo. Twick. BG48 56
Appleby Dr. Rom. CV28 19
Appleby Grn. Rom. CV28 19
Appleby Rd. E8 CB36 33
Appleby Rd. E16 CG39 43
Appleby St. E2 CA37 33
Appledore Av. Bexh. CS44 55
Appledore Av. Ruis. BC34 20
Appledore Clo. SW17 BU48 59
Appledore Clo. Brom. CG53 70
Appledore Clo. Edg. BM30 12
Appledore Clo. Rom. CV30 19
Appledore Cres. Sid. CN48 63
Appleford Rd. W10 BR38 31
Apple Garth, Brent. BK42 39
Applegarth Croy. CE57 79
Applegarth Dr. Ilf. CN31 27
Applegarth Esher BJ56 75
Applegarth Rd. SE28 CO40 45
Applegarth Rd. W14 BQ41 40
Apple Gro. Chess. BL56 75
Apple Gro. Enf. CA24 8
Apple Mkt. Kings-on-t. BK51 66
Appleton Gdns. N. Mal. BP53 67
Appleton Rd. SE9 CK45 53
Appleton Rd. Loug. CL24 10
Appleton Way, Horn. CV34 28
Apple Tree Yd. SW1 BW40 41
Duke Of York St.
Applewood Clo. NW2 BP34 22
Appold St. EC2 CA39 42
Appold St. Erith CT43 55
Approach Clo. N16 CA35 33
Cowper Rd.
Approach Rd. E2 CC37 34
Approach Rd. SW20 BQ51 67
Approach Rd. Barn. BT24 7
Approach Rd. Barn. BV24 7
Approach Rd. E. Mol. BF53 65
Approach, The, W3 BN39 40
Approach, The, Enf. CB23 8
Approach, The, Orp. CN55 81
Apps Ct. Walt. BD53 65
Aprey Gdns. NW4 BQ31 22
April Clo. Felt. BC48 56
April Clo. W13 BW40 39
April Glen SE23 CC48 61
Mayow Rd.
Apsley Clo. Har. BG32 20
Apsley Rd. E17 CD32 25
Apsley Rd. SE25 CB52 69
Apsley Rd. N. Mal. BN52 67
Apsley Sq. W1 BV40 41
Sth. Audley St.
Aquila Clo. Nthwd. BC28 11
Aquila St. NW8 BT37 32
Aquinas St. SE1 BY40 42
Arabella Dr. SW15 BO45 49
Arabia Clo. E4 CF26 9
Mornington Rd.
Arabin Rd. SE4 CD45 52
Aragon Av. Epsom BP58 76
Aragon Av. Surb. BH53 66
Aragon Dr. Ilf. CM29 17
Aragon Dr. Ruis. BD33 20
Aragon Rd. Kings. On T. BL49 57
Aragon Rd. Mord. BQ53 67
Arandora Cres. Rom. CO33 27
Arand St. W11 BQ40 40
Arbery Rd. E3 CD38 34
Arbor Clo. Beck. CE51 70
Arbor Ct. N16 BZ34 24
Arbor Rd. E4 CF27 16
Arbour Rd. Enf. CC24 9
Arbour Sq. E1 CC39 43
Arbour Way, Horn. CU35 37
Arbroath Grn. Wat. BC27 11
Arbroath Rd. SE9 CK45 53
Arbrook La. Esher BG57 74
Arbury Ter. SE26 CB48 60
Arbuthnot La. Bex. CQ47 63
Arbuthnot Rd. SE14 CC44 52
Arbutus St. E8 CA37 33
Arcade Pl. Rom. CT32 28
Arcade The, E17 CE31 25
Hoe St.
Arcade The, EC2 CA39 42
Liverpool St.
Arcadia Av. N3 BS30 14
Arcadian Av. Bex. CQ46 54
Arcadian Clo. Bex. CQ46 54
Arcadian Gdns. N22 BX29 15
Arcadian Rd. Bex. CQ46 54
Arcadia St. E14 CE39 43
Archbishop's Pl. SW2 BX47 60
Archdale Rd. SE22 CA46 51
Archel Rd. W14 BR43 49
Archer Ho. SW11 BT44 50
Archer Rd. SE25 CB52 69
Archer Rd. Orp. CO53 72
Archer St. W1 BW40 41
Rupert St.
Archers Walk SE15 CA44 51
Sumner Estate
Archer Way, Swan. CU51 73
Archery Clo. W2 BU39 41
Archery Clo. Har. BH31 21
Archery Rd. SE9 CK46 53
Arches, The, Har. BG34 20
Archibald Rd. E3 CE38 34
Archibald Rd. N7 BW35 32
Archibald Rd. Rom. CW30 19
Archibald St. E3 CE38 34
Arch Rd. Walt. BD55 74
Arch St. SE1 BZ41 42
Arch Way, Rom. CU29 19
Archway Mall N19 BW34 23
Archway Rd. N6 BU32 23
Archway Rd. N19 BU32 23
Archway St. SW13 BO45 49
Arcola St. E8 CA35 33
Arctic St. NW5 BV35 32
Arcus Rd. Brom. CG49 61
Ardbeg Rd. SE24 BZ46 51
Arden Clo. Bush. BH26 5
Arden Clo. Har. BG34 20
Arden Cres. Dag. CP36 36
Arden Est. N1 CA38 33
Arden Rd. N3 BR31 22
Arden Rd. W13 BJ40 39
Ardern BC31 20
Ardfern Av. SW16 BY52 69
Ardfillan Rd. SE6 CF47 61
Ardgowan Rd. SE6 CG47 61
Ardilaun Rd. N5 BZ35 33
Ardleigh Clo. Horn. CV31 28
Ardleigh Gdns. Sutt. BS54 68
Ardleigh Grn. Rd. Horn. CV32 28
Ardleigh Mews, Ilf. CL34 26
Benegal Rd.
Ardleigh Rd. E17 CD30 16
Ardleigh Rd. N1 CA36 33
Ardley Clo. NW10 BO34 22
Ardley Clo. SE23 CD48 61
Ardlui Rd. SE27 BZ48 60
Ardmay Gdns. Surb. BL53 66
Ardmere Rd. SE13 CF46 52
Ardmore La. Buck. H. CH26 10
Ardoch Rd. SE6 CF48 61
Ardrossan Gdns. Wor. Pk. BP55 76
Ardshiel Clo. SW15 BQ45 49
Bemish Rd
Ardwell Av. Ilf. CM32 26
Ardwell Rd. SW2 BX48 60
Ardwick Rd. NW2 BS35 32
Argall Av. E10 CC33 25
Argon Ms. SW6 BS43 50
Argus Way, W3 BM41 39
Argus Way, Nthlt. BE38 29
Argyle Av. Houns. BF46 47
Argyle Clo. W13 BJ38 30
Argyle Est. SW19 BQ48 58
Argyle Mans. Rich. BK48 57
Argyle Pass. N17 CB30 15
Argyle Rd.
Argyle Pl. W6 BP42 40
Argyle Rd. E1 CC38 34
Argyle Rd. E15 CG35 34
Argyle Rd. E16 CH39 44
Argyle Rd. N12 BS28 14
Argyle Rd. N17 CB30 15
Argyle Rd. N18 CB28 15
Argyle Rd. W13 BJ38 30
Argyle Rd. W13 BJ39 39
Argyle Rd. Barn. BQ24 6
Argyle Rd. Har. BF32 20
Argyle Rd. Houns. BF46 47
Argyle Rd. Ilf. CL34 26
Argyle Sq. WC1 BX38 33
Argyle St. WC1 BX38 33
Argyll Av. Sthl. BF40 38
Argyll Gdns. Edg. BM30 12
Argyll Rd. W8 BS41 41
Argyll St. W1 BW39 41
Arica Rd. SE4 CD45 52
Ariel Rd. NW6 BS36 32
Ariel Way W12 BQ40 40
Aristotle Rd. SW4 BW45 50
Arkell Gro. SE19 BY50 60
Arkindale Rd. SE6 CF48 61
Arkley Cres. E17 CD32 25
Arkley Dr. Barn. BP24 6
Arkley La. Barn. BO23 6
Arkley Rd. Barn. BP24 6
Arkley Vw. Barn. BP24 6
Arklow Rd. SE14 CD43 52
Arkwright Rd. NW3 BT35 32
Arkwright Rd. Sth. Croy. CA58 78
Arkwright St. E16 CG39 43
Arlesford Rd. SW9 BX45 51
Arlesley Clo. SW15 BR46 49
Arlingford Rd. SW2 BY46 51
Arlington N12 BS28 14
Arlington Av. N1 BZ37 33
Arlington Clo. Sid. CN47 63
Arlington Clo. Sutt. BS55 77
Arlington Clo. Twick. BK46 48
Arlington Dr. Cars. BU55 77
Arlington Gdns. W4 BN42 40
Arlington Gdns. Ilf. CL33 26
Arlington Gdns. Rom. CW30 19
Arlington Lo. SW2 BX45 51
Arlington Pass. Tedd. BH49 57
Arlington Rd. N14 BV27 14
Arlington Rd. NW1 BV37 32
Arlington Rd. W13 BJ39 39
Arlington Rd. Rich. BK48 57
Arlington Rd. Surb. BK53 66
Arlington Rd. Tedd. BH49 57
Arlington Rd. Wdf. Grn. CH30 17
Arlington Sq. N1 BZ37 33
Arlington St. SW1 BW40 41
Arlington Way EC1 BY38 33
Arliss Way Nthlt. BD37 29
Arlow Rd. N21 BY26 8
Armada Ct. SE8 CE43 52
Watergate St.
Armadale Cl. N15 CB31 24
Armadale Rd. SW6 BS43 50
Armadale Rd. Felt. BC46 47
Armada St. SE8 CE43 52
Armfield Av. Mitch. BU51 68

Armfield Clo. E. Mol. BE53 65
Armfield Rd. Enf. BZ23 8
Arminger Rd. W12 BP40 40
Armitage Rd. NW11 BR33 22
Armitage Rd. SE10 CG42 43
Armitage Rd. Houns. BD43 47
Armoury Way SW18 BS46 50
Armstead Wk. Dag. CR36 36
Armstrong Av. Wdf. Grn. CG29 16
Armstrong Clo. Ruis. BC32 20
Armstrong Cres. Barn. BT24 7
Armstrong Gdns. SE7 CJ42 44
Armstrong Rd. W3 BO40 40
Armstrong Rd. Felt. BE49 56
Armstrong Way, Grnf. BF41 38
Arnal Cres. SW18 BR47 58
Arndale Est. SW18 BS46 50
Arne Gro. Orp. CN55 81
Arne Wk. SE3 CG45 52
Arneways Av. Rom. CP31 27
Arneway St. SW1 BW41 41
Horseferry Rd.
Arnewood Clo. SW15 BP47 58
Arneys La. Mitch. BV53 68
Arngask Rd. SE6 CF47 61
Arnhem Dr. Croy. CF58 79
Arnhem Way SE22 CA46 51
Dulwich Gro.
Arnison Rd. E. Mol. BG52 65
Arnold Circus, E2 CA38 33
Arnold Clo. Har. BL32 21
Arnold Cres. Islw. BG46 47
Arnold Est. SE1 CA41 42
Arnold Gdns. N13 BY28 15
Arnold Rd. E3 CE38 34
Arnold Rd. N15 CA31 24
Arnold Rd. SW17 BU50 59
Arnold Rd. Dag. CQ36 36
Arnold Rd. Nthlt. BE36 29
Arnold's La. S. At. H. CW50 64
Arnos Gro. N14 BW28 14
Arnos Rd. N11 BW28 14
Arnott Rd. SE28 CP40 45
Arnould Av. SE5 BZ45 51
Arnsberg Rd. Erith. CT43 55
Arnside Gdns. Wem. BK33 21
Arnside Rd. Bexh. CR44 64
Arnside St. SE17 BZ43 51
Arnulf St. SE6 CF49 61
Arnulls Rd. SW16 BY50 60
Arodene Rd. SW2 BX46 51
Arondel Gr. N16 CA35 33
Arragon Gdns. SW16 BX50 60
Arragon Gdns. W. Wick. CE55 79
Arragon Rd. E6 CJ37 35
Arragon Rd. Twick. BJ47 57
Arran Clo. Erith CS43 55
Arran Clo. Wall. BW56 77
Arran Dr. E12 CJ33 26
Arran Dr. Stan. BJ28 12
Arran Rd. SE6 CE48 61
Arran Wk. N1 BZ36 33
Marquess Est.
Arran Way, Esher BF55 74
Arran Yews W5 BL40 39
Arras Av. Mord. BT53 68
Arrol Rd. Beck. CC52 70
Arrow Rd. E3 CE38 34
Arrowscout Wk. Nthlt. BE38 29
Wayfarer Rd.
Arrowsmith Clo. Chig. CN28 18
Arrowsmith Path. Chig. CN28 18
Arrowsmith Rd. Chig. CN28 18
Arrowsmith Rd. Loug. CK24 10
Arsenal Rd. SE9 CK44 53
Arterberry Rd. SW20 BQ50 58
Arterial Av. Rain. CU38 37
Artesian Clo. Horn. CT33 28
Artesian Rd. W2 BS39 41
Arthingworth St. E15 CG37 34
Arthur Ct. W2 BS39 41
Queensway
Arthurdon Rd. SE4 CE46 52
Arthur Grn. SE18 CM42 44
Arthur Gro. SE18 CM42 44
Arthur Henderson Ho. SW6 BR44 49
Arthur Ms. N7 BX36 33
Arthur Rd. E6 CK37 35
Arthur Rd. N7 BX34 24
Arthur Rd. N9 CA27 15
Arthur Rd. SW19 BS48 59
Arthur Rd. Kings. On T. BM50 57
Arthur Rd. N. Mal. BP53 67
Arthur Rd. Rom. CP33 27
Arthur St. EC4 BZ40 42
Arthur St. Bush. BD24 4
Arthur St. Erith. CT43 55
Artichoke Hl. E1 CB40 42
Pennington St.
Artichoke Pl. SE5 BZ44 51
Camberwell Church St.
Artillery Clo. Ilf. CM32 26
Artillery La. E1 CA39 42
Artillery Pl. SE18 CK42 44
Artillery Row, SW1 BW41 41
Artillery Yd. EC2 CA38 33
Worship St.

Artington Clo. Orp. CM56 80
Artizan St. E1 CA39 42
Harrow Pl.
Arty Pass. E1 CA39 42
Sandy's Row
Arundel Av. Epsom BP58 76
Arundel Av. Mord. BR52 67
Arundel Clo. E15 CG35 34
Arundel Clo. Bex. CQ46 54
Arundel Clo. Croy. BY55 78
Arundel Clo. Hamptn. BF49 56
Arundel Dr. Borwd. BN24 6
Arundel Dr. Har. BE35 29
Arundel Dr. Orp. CO56 81
Arundel Dr. Wdf. Grn. CH29 17
Arundel Gdns. N21 BY26 8
Arundel Gdns. W11 BR40 40
Arundel Gdn. Ilf. CO34 27
Arundell Gdns. Edg. BN29 13
Cressingham Rd.
Arundel Pl. N1 BY36 33
Arundel Rd. Barn. BU24 7
Arundel Rd. Croy. BZ53 69
Arundel Rd. Houns. BD45 47
Arundel Rd. Kings. On T. BM51 66
Arundel Rd. Rom. CW30 19
Arundel Rd. Sutt. BR57 76
Arundel Sq. N7 BY36 33
Arundel St. WC2 BX40 42
Arundel Ter. SW13 BP43 49
Arvon Rd. N5 BY35 33
Ascalon St. SW8 BW43 50
Ascension Rd. Rom. CS29 19
Ascham Dr. E4 CE29 16
Ascham End E17 CD30 16
Ascham St. NW5 BW35 32
Aschurch Rd. Croy. CA54 69
Ascot Av. W5 BL41 39
Ascot Clo. Borwd. BM25 5
Ascot Clo. Ilf. CN29 18
Ascot Clo. Nthlt. BF36 29
Ascot Gdns. Sthl. BE39 38
Ascot Rd. E6 CK38 35
Ascot Rd. N15 BZ32 24
Ascot Rd. N18 CB28 15
Ascot Rd. SW17 BV50 59
Ascot Rd. Orp. CN52 72
Ashbourne Av. E18 CH31 26
Ashbourne Av. N20 BU27 14
Ashbourne Av. NW11 BR32 22
Ashbourne Av. Bexh. CQ43 54
Ashbourne Av. Har. BG34 20
Ashbourne Clo. N12 BS28 14
Ashbourne Clo. W5 BM39 39
Ashbourne Clo. Couls. BW58 77
Ashbourne Clo. Couls. BW62 83
Ashbourne Ct. E5 CC35 34
Clapton Park Est.
Ashbourne Gro. NW7 BN28 13
Ashbourne Gro. SE22 CA45 51
Ashbourne Gro. W4 BO42 40
Ashbourne Rise, Orp. CM56 80
Ashbourne Rd. W5 BL38 39
Ashbourne Rd. Mitch. BV50 59
Ashbourne Rd. Rom. CV28 19
Ashbourne Ter. SW19 BR50 58
Ashbridge Rd. E11 CG33 25
Ashbridge St. NW8 BU38 32
Ashbrook Rd. N19 BW33 23
Ashbrook Rd. Dag. CR34 27
Ashburn Gdns. SW7 BT42 41
Asburnham Av. Har. BH32 21
Ashburnham Clo. N2 BT31 23
Ashburnham Clo. Wat. BC27 11
Ashburnham Dr.
Ashburnham Dr. Wat. BC27 11
Ashburnham Gdns. Har. BH32 21
Ashburnham Gro. SE10 CE43 52
Ashburnham Pl. SE10 CE43 52
Ashburnham Retreat SE10 CE43 52
Ashburnham Rd. NW10 BQ38 31
Ashburnham Rd. SW10 BT43 50
Ashburnham Rd. Belv. CE43 52
Ashburnham Rd. Rich. BJ48 57
Ashburn Ms. SW7 BT42 41
Ashburn Pl. SW7 BT42 41
Ashburton Av. Croy. CB54 69
Ashburton Av. Ilf. CN35 36
Ashburton Clo. Croy. CB54 69
Ashburton Ct. Pnr. BD30 11
Ashburton Est. SW15 BQ46 49
Ashburton Gdns. Croy. CB55 78
Ashburton Gro. N7 BY35 33
Ashburton Rd. E16 CH39 44
Ashburton Rd. Croy. CB55 78
Ashburton Rd. Ruis. BC34 20
Ashburton Ter. E13 CH37 35
Grasmere Rd.
Ashbury Gdns. Rom. CP32 27
Ashbury Rd. SW11 BU45 50
Ashby Av. Chess. BM57 75
Ashby Clo. Horn. CW33 28
Ashby Gro. N1 BZ36 33
Ashby Ms. SE4 CD44 52
Ashby Rd. N15 CB32 24
Ashby Rd. SE4 CD44 52
Ashby St. EC1 BY38 33

Ashchurch Clo. SE20 CB51 69
Ashchurch Ct. W12 BP41 40
Ashchurch Gro. W12 BP41 40
Ashchurch Pk. Vill. W12 BP41 40
Ashchurch Rd. W12 BP41 40
Ashchurch Ter. W12 BP41 40
Ash Clo. SE20 CC51 70
Ash Clo. Cars. BU55 77
Ash Clo. N. Mal. BN51 67
Ash Clo. Orp. CM53 71
Ash Clo. Rom. CR29 18
Ash Clo. Sid. CO48 63
Ash Clo. Stan. BJ29 12
Ash Clo. Swan. CS51 73
Ashcombe Av. Surb. BK54 66
Ashcombe Gdns. Edg. BM28 12
Ashcombe Pk. NW2 BO34 22
Ashcombe Rd. SW19 BS49 59
Ashcombe Rd. Cars. BV57 77
Ashcombe Sq. N. Mal. BN52 67
Ashcombe St. SW6 BS44 50
Ash Ct. Epsom BN56 76
Ashcroft Pnr. BF29 11
Ashcroft Av. Sid. CO46 54
Ashcroft Cres. Sid. CO46 54
Ashcroft Gdns. SW2 BY47 60
Trinity Rise
Ashcroft Rise, Couls. BX57 78
Ashcroft Rise, Couls. BX61 84
Ashcroft Rd. E3 CD38 34
Ashcroft Rd. Chess. BL55 75
Ashdale Clo. Twick. BF47 56
Ashdale Gro. Stan. BH29 12
Ashdale Rd. SE12 CH47 62
Ashdale Way, Twick. BF47 56
Ashdale Clo.
Ashdene Pnr. BD31 20
Ashdon Clo. Wdf. Grn. CH29 17
Ashdon Rd. NW10 BO37 31
Ashdown Cres. NW5 BV35 32
Ashdown Dr. Borwd. BL23 5
Ashdowne Clo. Beck. CE51 70
Ashdown Gdns. Sth. Croy. CB61 84
Ashdown Rd. Enf. CC24 9
Ashdown Rd. Epsom BO60 82
Ashdown Rd. Kings-on-t. BL51 66
Ashdown Rd. Wat. BD28 11
Woodhall La.
Ashdown Wk. Rom. CR30 18
Ashenden Rd. E5 CC35 34
Ashen Dr. Dart. CU46 55
Ashen Gro. SW19 BS48 59
Ashentree Ct. EC4 BY39 42
Whitefriars St.
Ashen Vale, Sth. Croy. CC58 79
Ashfield Av. Bush. BF26 4
Ashfield Av. Felt. BC47 56
Ashfield La. Chis. BL47 57
Ashfield Pde. N14 BW26 7
Ashfield Rd. N4 BZ32 24
Ashfield Rd. N14 BW27 14
Ashfield Rd. W3 BO40 40
Ashfields Loug. CK23 10
Ashfield St. E1 CB39 42
Ashford Av. N8 BX31 24
Ashford Av. Hayes BO39 38
Ashford Clo. E17 CD32 25
Ashford Cres. Enf. CC23 9
Ashford Grn. Wat. BD28 11
Ashford Rd. E6 CL36 35
Ashford Rd. E18 CH30 17
Ashford Rd. NW2 BQ35 31
Tenterden Rd.
Ashford Rd. NW2 BQ35 31
Ashford St. N1 CA38 33
Ash Grn. Loughton CK23 10
Ash Gro. E8 CB37 33
Ash Gro. N13 BZ27 15
Ash Gro. NW2 BQ35 31
Ash Gro. SE20 CC51 70
Ash Gro. W5 BL41 39
Ash Gro. Enf. CA26 8
Ash Gro. Houns. BD44 47
Ash Gro. Sthl. BF39 38
Ash Gro. Wem. BJ35 30
Ash Gro. W. Wick. CF55 79
Ashgrove Rd. Brom. CF50 61
Ashgrove Rd. Ilf. CN33 27
Ash Hill Clo. Bush. BF26 4
Ash Hill Dr. Pnr. BD31 20
Ashington Ct. SE26 CB49 60
Ashington Rd. SW6 BR44 49
Ashlake Rd. SW16 BX49 60
Ashland Pl. W1 BV39 41
Ashland Rd. E15 CF37 34
Ashlar Pl. SE18 CL42 44
Ashleigh Gdns. Sutt. BS55 77
Ashleigh Rd. SE20 CB52 69
Ashleigh Rd. SW14 BO45 49
Ashley Av. Ilf. CL30 17
Ashley Av. Mord. BS53 68
Ashley Clo. NW4 BO30 13
Ashley Clo. Pnr. BC30 11
Ashley Cres. N22 BY30 15
Ashley Cres. SW11 BV45 50

Ashley Dr. Bans. BS60 83
Ashley Dr. Borwd. BN25 6
Ashley Dr. Twick. BF47 56
Ashley Dr. Walt. BC55 74
Ashley Gdns. N13 BZ28 15
Ashley Gdns. Orp. CN56 81
Ashley Gdns. Rich. BK48 57
Ashley Gdns. Wem. BL34 21
Ashley Gro. Loug. CK24 10
Ashley La. NW4 BO30 13
Ashley La. Croy. BY56 78
Ashley Pk. Rd. Walt. BC55 74
Ashley Pl. SW1 BW41 41
Ashley Rd. E4 CE29 16
Ashley Rd. E7 CJ36 35
Ashley Rd. N17 CB31 24
Ashley Rd. N19 BX33 24
Ashley Rd. SW19 BS50 59
Ashley Rd. Enf. CC23 9
Ashley Rd. Epsom BN60 82
Ashley Rd. Hamptn. BF51 65
Ashley Rd. Rich. BL45 48
Jocelyn Rd.
Ashley Rd. Surb. BH53 66
Ashley Rd. Th. Hth. BX52 69
Ashley Wk. NW7 BQ29 13
Ashling Rd. Croy. CB54 69
Ashlin Rd. E15 CF35 34
Ashlone Rd. SW15 BO45 49
Ashlyn Clo. Bush. BE24 4
Ashlyn Gro. Horn. CV31 28
Ashmead N14 BW25 7
Ashmead Rd. SE8 CE44 52
Ashmead Rd. Felt. BC47 56
Ashmere Av. Beck. CF51 70
Ashmere Clo. Sutt. BO56 76
Ashmere Gdns. SW2 BX45 51
Ashmill St. NW1 BU39 41
Ashmole Pl. SW8 BX43 51
Ashmole St. SW8 BX43 51
Ashmore Gro. Well. CM45 53
Ashmore Rd. W9 BR38 31
Ashmount Rd. N15 CA32 24
Ashmount Rd. N19 BW33 23
Ashmour Gdns. Rom. CS30 19
Ashness Gdns. Grnf. BJ36 30
Ashness Rd. SW11 BU46 50
Ashridge Clo. Har. BK32 21
Ashridge Cres. SE18 CL43 53
Ashridge Dr. Wat. BC28 11
Ashridge Gdns. N13 BW28 14
Ashridge Gdns. Pnr. BE31 20
Ashridge Way, Mord. BR52 67
Ashridge Way, Sun. BC50 56
Ash Rd. E15 CG35 34
Ash Rd. Croy. CE55 79
Ash Rd. Dart. CV47 64
Ash Rd. Dart. CW49 64
Ash Rd. Orp. CN57 81
Ash Rd. Sutt. BR54 67
Ash Row, Brom. CL54 71
Ashstead Rd. E5 CB33 24
Ashton Clo. Sutt. BR56 76
Gandergreen La.
Ashton Clo. Walt. BC57 74
Ashton Gdns. Houns. BE45 47
Ashton Gdns. Rom. CO32 27
Ashton Rd. E15 CF35 34
Ashton Rd. Rom. CV29 19
Ashton St. E14 CF40 43
Ashtree Av. Mitch. BF51 68
Ash Tree Clo. Croy. CD53 70
Ash Tree Dell NW9 BN32 22
Ash Tree Way, Croy. CD53 70
Ashurst Clo. Ken. BZ61 84
Ashurst Dr. Ilf. CL32 26
Ashurst Rd. N12 BU28 14
Ashurst Rd. Barn. BU25 7
Ashurst Wk. Croy. CB55 78
Ashvale Gdns. Rom. CS28 19
Ashvale Rd. SW17 BU49 59
Ashville Rd. E11 CF33 25
Ashwater Rd. SE12 CH47 62
Ashwin St. E8 CA37 33
Ashwood Av. Rain. CU38 37
Ashwood Gdns. Croy. CF57 79
Ashwood Rd. E4 CE29 16
Ashworth Rd. W9 BS38 32
Askern Clo. Bexh. CP45 54
Aske St. N1 CA38 33
Pitfield St.
Askew Bldgs. W12 BP41 40
Askew Rd.
Askew Cres. W12 BO41 40
Askew Rd. W12 BO40 40
Askham Ct. W12 BP40 40
Askham Rd. W12 BP40 40
Askill Dr. SW15 BR46 49
Askwith Rd. Rain. CS38 37
Aslett St. SW18 BT47 59
Asmara Rd. NW2 BR35 31
Asmar Clo. Couls. BX61 84
Asmuns Hl. NW11 BS32 23
Asmuns Rd. NW11 BR32 22
Aspen Clo. W5 BL41 39
Aspen Clo. Orp. CO56 81
Aspen Copse, Brom. CK51 71

Aspen Dr. Wem. BJ34 21
Aspen Gdns. W6 BP42 40
Bridge Av.
Aspen Gdns. Mitch. BV53 68
Aspen Grn. Belv. CQ41 45
Aspen Gro. Upmin. CW35 37
Aspen La. Nthlt. BE38 29
Aspenlea Rd. W6 BQ43 49
Aspen Way, Bans. BQ60 82
Aspinall Rd. SE4 CC45 52
Aspinden Rd. SE16 CB42 42
Aspley Rd. SW18 BS46 50
Asplins Rd. N17 CB30 15
Assam St. E1 CB39 42
Assembly Ms. E1 CC39 43
Assembly Pass. E1 CC39 43
Assembly Wk. Cars. BU54 68
Assher Rd. Walt. BE55 74
Ass House La. Har. BF28 11
Astall Clo. Har. BH30 12
Astbury Rd. SE15 CC44 52
Astell St. SW3 BU42 41
Aster Pl. SE1 BZ41 42
Aste St. E14 CF41 43
Astey's Row N1 BY36 33
Asthall Gdns. Ilf. CM31 26
Astle St. SW11 BV44 50
Astley Av. NW2 BQ35 31
Astley St. SE1 CA42 42
Aston Av. Har. BK33 21
Aston Clo. Sid. CO48 63
Aston Grn. Houns. BC44 47
Aston Mews, Rom. CP33 27
Reynolds Av.
Aston Rd. SW20 BQ51 67
Approach Rd.
Aston Rd. W5 BK39 39
Aston Rd. Esher BH57 75
Aston St. E14 CD39 43
Astonville St. SW18 BS47 59
Aston Way, Epsom BO61 82
Astor Av. Rom. CS32 28
Astor Clo. King. On T. BM50 57
Astoria Wk. SW9 BY45 51
Astra Clo. Horn. CU36 37
Astrop Ms. W6 BQ41 40
Astrop Ter. W6 BQ41 40
Astwood Ms. SW7 BS42 41
Asylum Rd. SE15 CB43 51
Atalanta St. SW6 BQ44 49
Atbara Ct. Tedd. BJ50 57
Atbara Rd. Tedd. BJ50 57
Atcham Rd. Houns. BG45 47
Atcost Rd. Bark. CO39 45
Atheldene Rd. SW18 BS47 59
Athelney St. SE6 CE48 61
Athelstan Clo. Rom. CW30 19
Athelstane Gro. E3 CD37 34
Athelstane Rd. N4 BY34 24
Athelstan Rd. Kings-on-T. BL52 66
Villiers Rd.
Athelstan Rd. Rom. CW30 19
Athelstone Rd. Har. BG30 11
Athena Clo. Har. BG34 20
Athenaeum Rd. N20 BT26 7
Athenlay Rd. SE15 CC46 52
Athens Gdns. W9 BS38 32
Atherden Rd. E5 CC35 34
Atherfold Rd. SW9 BX45 51
Atherley Way, Houns. BE47 56
Atherstone Ms. E7 CG36 34
Atherstone Ms. SW7 BT42 41
Atherton Dr. SW19 BQ49 58
Atherton Heights, Wem. BK36 30
Bridgewater Rd.
Atherton Pl. Har. BG31 20
Atherton Pl. Sthl. BF40 38
Atherton Rd. E7 CG35 34
Atherton Rd. SW13 BP43 49
Atherton Rd. Ilf. CK30 17
Atherton St. SW11 BU44 50
Athlone Clo. Esher BH57 75
Albany Cres.
Athlone Gdns. W10 BR39 40
Wheatstone Rd.
Athlone Rd. SW2 BX47 60
Athlone St. NW5 BV36 32
Athlon Rd. Wem. BK37 30
Athol Clo. Pnr. BC30 11
Athole Gdns. Enf. CA25 8
Athol Gdns. Pnr. BC30 11
Atholl Rd. Ilf. CO33 27
Athol Rd. Erith CS42 46
Athol St. E14 CF39 43
Atkinson Clo. Orp. CN56 81
Atkinson Rd. E16 CJ39 44
Atkins Rd. E10 CE32 25
Atkins Rd. SW12 BW47 59
Atlantic Rd. SW9 BY45 51
Atlas Gdns. SE7 CJ42 44
Atlas Mews N7 BX36 33
Atlas Rd. E13 CH37 35
Atlas Rd. NW10 BO38 31
Atlas Rd. Wem. BN35 31
Atley Rd. E3 CE37 34
Atney Rd. SW15 BR45 49
Atria Rd. Nthwd. BC28 11
Atterbury Rd. N4 BY32 24
Wightman Rd.
Atterbury St. SW1 BX42 42
Attewood Av. NW10 BO34 22
Attewood Rd. Nthlt. BE36 29
Arnold Rd.
Attfield Clo. N20 BT27 14
Attle Clo. Hayes BC38 29
Attlee Dri. Dart. CW46 55
Attlee Rd. SE28 CO40 45
Attlee Rd. Hayes BC38 29
Attlee Ter. E17 CE31 25
Attneave St. WC1 BY38 33
Attwood Clo. Sth. Croy. CB60 84
Atwater Clo. SW2 BY47 60
Atwell Rd. SE15 CB44 51
Atwood Av. Rich. BM44 48
Atwood Rd. W6 BP42 40
Atwood's Alley, Rich. BM44 48
Kew Gardens Rd.
Atworth St. E14 CF41 43
Auberon St. E16 CK40 44
Aubert Ct. N5 BY35 33
Aubert Pk. N5 BY35 33
Aubrey Rd. E17 CE31 25
Aubrey Rd. N8 BX32 24
Aubrey Rd. W8 BR40 40
Aubrey Wk. W8 BR40 40
Aubrietia Clo. Rom. CW30 19
Sunflower Way
Aubyn Hl. SE27 BZ49 60
Aubyn Sq. SW15 BP46 49
Auckland Av. Rain. CT38 37
Auckland Clo. SE19 CA51 69
Auckland Gdns. SE19 CA51 69
Auckland Hl. SE27 BZ49 60
Auckland Ri. SE19 CA51 69
Auckland Rd. E10 CE34 25
Auckland Rd. SE19 CA51 69
Auckland Rd. SW11 BU45 50
Auckland Rd. Ilf. CL33 26
Auckland Rd. Kings. On T. BL52 66
Auckland St. SE11 BX42 42
Kennington La.
Auden Pl. NW1 BV37 32
Audleigh Pl. IG7 CL29 17
Audley SW11 BV45 50
Audley Clo. Borwd. BM24 5
Audley Ct. E18 CG31 25
Audley Ct. Pnr. BD30 11
Audley Ct. Twick. BG48 56
Audley Gdns. Ilf. CN34 27
Audley Gdns. Loug. CM23 10
Audley Pl. Sutt. BS57 77
Audley Rd. NW4 BP32 22
Audley Rd. W5 BL39 39
Audley Rd. Enf. BY23 8
Audley Rd. Rich. BL46 48
Audrey Clo. Beck. CE53 70
Audrey Gdns. Wem. BJ34 21
Audrey Rd. Ilf. CL34 26
Audrey St. E2 CB37 33
Augurs La. E13 CH38 35
Augusta Rd. Twick. BG48 56
Augusta St. E14 CE39 43
Augustine Rd. W14 BQ41 40
Augustine Rd. Har. BG30 11
Augustine Rd. Orp. CP52 72
Augustus Clo. Brent. BK43 48
Augustus Rd. SW19 BQ47 58
Augustus St. NW1 BV37 32
Aulay St. SE1 CB43 51
Aultone Way, Cars. BU55 77
Aultone Way, Sutt. BS55 77
Aulton St. SE11 BY42 42
Aurelia Gdns. Croy. BX53 69
Aurelia Rd. Croy. BX53 69
Auriel Av. Dag. CS36 37
Auriga Mews N1 BZ35 33
Auriol Dr. Grnf. BG36 29
Auriol Rd. W14 BR42 40
Austell Gdns. NW7 BO27 13
Austen Clo. SE28 CO40 45
Austen Clo. Loug. CM24 10
Austen Gdns. Dart. CW45 55
Austen Rd. Har. BF34 20
Austin Av. Brom. CK53 71
Austin Clo. Couls. BY62 84
Austin Friars, EC2 BZ39 42
Austin Friars Pass. EC2 BZ39 42
Austin Friars
Austin Rd. SW11 BV44 50
Austin Rd. Orp. CO53 72
Austin St. E2 CA38 33
Austral Clo. Sid. CN48 63
Longlands Rd.
Austral Dr. Horn. CV33 28
Australia Rd. W12 BP40 40
Austral St. SE11 BY42 42
Austyn Gdns. Surb. BM54 66
Autumn Clo. Enf. CB23 8
Autumn St. E3 CE37 34
Avalon Clo. Enf. BY23 8
Avalon Clo. Orp. CP55 81
Avalon Cres. W13 BJ39 39
Avalon Rd. SW6 BS44 50
Avalon Rd. W13 BJ38 30
Avalon Rd. Orp. CO55 81
Avalon Ter. Rich. BL45 48
Avard Gdns. Orp. CM56 80
Cherrycot Rise
Avarn Rd. SW17 BU50 59
Avcliffe Clo. Kings. On T. BM51 66
Avebury Est. E2 CB38 33
Avebury Pk. Surb. BK54 66
Avebury Rd. E11 CF33 25
Avebury Rd. SW19 BR51 67
Avebury Rd. Orp. CM55 80
Aveley Rd. Rom. CS31 28
Aveline St. SE11 BX42 42
Aveling Park Rd. E17 CE30 16
Avelon Rd. Rain. CU37 37
Avelon Rd. Rom. CS29 19
Ave Maria La. EC4 BY39 42
Ludgate Hill
Avenell Rd. N5 BY34 24
Avening Rd. SW18 BS47 59
Avening Ter. SW18 BS46 50
Avenons Rd. E13 CH38 35
Avenue Clo. N14 BW25 7
Avenue Clo. NW8 BU37 32
Avenue Clo. Houns. BC44 47
Avenue Clo. Rom. CW29 19
Avenue Cres. W3 BM41 39
Avenue Cres. Houns. BC43 47
Avenue Elmers. Surb. BL53 66
Avenue Gdns. SE25 CA51 69
Avenue Gdns. SW14 BO45 49
Avenue Gdns. W3 BM41 39
Avenue Gdns. Houns. BC43 47
Avenue Gdns. Tedd. BH50 57
Avenue Gte. SE21 CA49 60
Avenue Mews. N10 BV31 23
Queen's Av.
Avenue Ms. NW6 BS35 32
Finchley Rd.
Avenue Pk. Rd. SE27 BY48 60
Avenue Rise, Bush. BF25 4
Avenue Rd. E7 CH35 35
Avenue Rd. N6 BW33 23
Avenue Rd. N12 BT28 14
Avenue Rd. N14 BV26 7
Avenue Rd. N15 BZ32 24
Avenue Rd. NW3 BT36 32
Avenue Rd. NW10 BO37 31
Avenue Rd. SE20 CC51 70
Avenue Rd. SE25 CA51 69
Avenue Rd. SW16 BW51 68
Avenue Rd. SW20 BP51 67
Avenue Rd. W3 BM41 39
Avenue Rd. Bans. BS61 83
Avenue Rd. Beck. CC51 70
Avenue Rd. Belv. CR42 45
Avenue Rd. Bexh. CQ45 54
Avenue Rd. Brent. BK42 39
Avenue Rd. Epsom BN60 82
Avenue Rd. Erith CS43 55
Avenue Rd. Hmptn. BF51 65
Avenue Rd. Islw. BH44 48
Avenue Rd. Kings. On T. BL52 66
Avenue Rd. N. Mal. BO52 67
Avenue Rd. Pnr. BE31 20
Avenue Rd. Rom. CP33 27
Avenue Rd. Rom. CW29 19
Avenue Rd. Sthl. BE40 38
Avenue Rd. Sutt. BS58 77
Avenue Rd. Tedd. BJ50 57
Avenue Rd. Wall. BW57 77
Avenue Rd. Wdf. Grn. CJ29 17
Avenue S. The, Surb. BL54 66
Avenue Ter. N. Mal. BN52 67
Avenue Ter. Wat. BE25 4
Avenue, The, E4 CF29 16
Avenue, The, E11 CH32 26
Avenue, The, EC3 CA40 42
Avenue, The, N1 BV28 14
Avenue, The, N3 BS30 14
Avenue, The, N8 BY31 24
Avenue, The, N10 BW30 14
Avenue, The, N11 BV28 14
Avenue, The, N17 BZ31 24
Avenue, The, SE7 CJ43 53
Avenue, The, SE10 CF43 52
Avenue, The, SE19 CA49 60
Avenue, The, SW4 BV46 50
Avenue, The, SW17 BV48 59
Avenue, The, SW18 BU47 59
Avenue, The, W4 BO41 40
Avenue, The, W13 BJ40 39
Avenue, The, Barn. BR24 6
Avenue, The, Beck. CE51 70
Avenue, The, Bex. CP47 63
Avenue, The, Brom. CJ52 71
Avenue, The, Bush. BE24 4
Avenue, The, Cars. BV57 77
Avenue, The, Couls. BW61 83
Avenue, The, Croy. CA55 78
Avenue, The, Epsom BP57 76
Avenue, The, Esher BH57 75
Avenue, The, Hamptn. BE50 56
Avenue, The, Har. BH30 12
Avenue, The, Har. BH31 21
Marlborough Rd.
Avenue, The, Horn. CV34 28
Avenue, The, Houns. BC44 47
Avenue, The, Houns. BF46 47
Avenue, The, Kes. CJ55 80
Avenue, The, Loug. CJ25 10
Avenue, The, Orp. CN55 81
Avenue, The, Orp. CO50 63
Avenue, The, Pnr. BE29 11
Avenue, The, Pnr. BE32 20
Avenue, The, Rich. BL44 48
Avenue, The, Rom. CS31 28
Avenue, The, Sun. BC51 65
Avenue, The, Surb. BL53 66
Avenue, The, Sutt. BR58 76
Avenue, The, Twick. BJ46 48
Avenue, The, Wat. BC23 4
Avenue, The, Wem. BL33 21
Avenue, The, W. Wick. CF54 70
Avenue, The, Wor. Pk. BO55 76
Averil Gro. SW16 BY50 60
Averill St. W6 BQ43 49
Avern Gdns. KT8 BF52 65
Avern Rd. E. Mol. BF53 65
Avery Gdns. Ilf. CK32 26
Avery Hl. Rd. SE9 CM46 53
Avery Row. W1 BV40 41
Aviary Clo. E16 CG39 43
Aviary St. E16 CG39 43
Lawrence St.
Aviemore Clo. Beck. CD53 70
Aviemore Way, Beck. CD53 70
Avignon Rd. SE4 CC45 52
Avington Gro. SE20 CC50 61
Avington Way SE15 CA43 51
Avior Dr. Nthwd. BC28 11
Avis Sq. E1 CC39 43
Avoca Rd. SW17 BV49 59
Avocet Mews SE28 CM41 44
Avon Clo. Hayes BC38 29
Avon Clo. Sutt. BT56 77
Avon Clo. Wor. Pk. BP55 76
Avondale Av. N12 BS28 14
Avondale Av. NW2 BO34 22
Avondale Av. Barn. BU26 7
Avondale Av. Esher BJ55 75
Avondale Av. Wor. Pk. BO54 67
Avondale Clo. Loug. CK26 10
Avondale Dr.
Avondale Clo. Walt. BD56 74
Avondale Ct. E16 CG39 43
Avondale Rd.
Avondale Ct. E18 CH30 17
Avondale Cres. Enf. CD24 9
Avondale Cres. Ilf. CJ32 26
Avondale Dr. Hayes BC40 38
Avondale Dr. Loug. CK26 10
Avondale Gdns. Houns. BE46 47
Avondale Pk. W11 BR40 40
Avondale Pk. Gdns. W11 BQ40 40
Avondale Rise, SE15 CA45 51
Avondale Rd. E16 CG39 43
Avondale Rd. E17 CE33 25
Avondale Rd. N3 BT30 14
Avondale Rd. N13 BY27 15
Avondale Rd. N15 BY32 24
Avondale Rd. SE9 CK48 62
Avondale Rd. SW14 BN45 49
Avondale Rd. SW19 BS49 59
Avondale Rd. Brom. CG50 61
Avondale Rd. Har. BH31 21
Avondale Rd. Hayes BC40 38
Avondale Rd. Sth. Croy. BZ57 78
Avondale Rd. Well. CP44 54
Avondale Sq. SE1 CB42 42
Avonley Rd. SE14 CC43 52
Avon Ms. Pnr. BE29 11
Avonmore Pl. W14 BR42 40
Avonmore Rd.
Avonmore Rd. W14 BR42 40
Avon Path, Sth. Croy. BZ57 78
Avon Pl. SE1 BZ41 42
Avon Rd. E17 CF31 25
Avon Rd. SE4 CE45 52
Avon Rd. Grnf. BF38 29
Avon St. SE1 BZ41 42
Avon Way E18 CH31 26
Avonwick Rd. Houns. BF44 47
Avril Way, E4 CF28 16
Avro Way, Wall. BX57 78
Awlfield Av. N17 BZ30 15
Awliscombe Rd. Well. CN44 54
Axe St. Bark. CM37 35
Axholme Av. Edg. BM30 12
Axminster Cres. Well. CP44 54
Axminster Rd. N7 BX34 24
Axtaine Rd. Orp. CP54 72
Axwood, Epsom BN61 82
Aybrook St. W1 BV39 41
Aycliff Clo. Brom. CK52 71
Aycliffe, Kings. On T. BM51 66
Cambridge Rd.
Aycliffe Rd. W12 BP40 40
Aycliffe Rd. Borwd. BL23 5
Aylesbury Rd. SE17 BZ42 42
Aylesbury Rd. Brom. CH52 71
Aylesbury St. EC1 BY38 33
Aylesbury St. NW10 BN34 22
Aylesford Av. Beck. CD53 70
Aylesford St. SW1 BW42 41
Aylesham Rd. Orp. CN54 72
Ayles Rd. Hayes BC38 29

Aylestone Av. NW6 BQ36 31
Aylett Rd. SE25 CB52 69
Aylett Rd. Islw. BH44 48
Ayley Croft, Enf. CB25 8
Cambridge Rd.
Aylmer Clo. Stan. BJ28 12
Aylmer Dr. Stan. BJ28 12
Aylmer Rd. E11 CG33 25
Aylmer Rd. N2 BU32 23
Aylmer Rd. W12 BO41 40
Aylmer Rd. Dag. CQ34 27
Ayloffe Rd. Dag. CQ36 36
Ayloffs Clo. Horn. CW32 28
Ayloffs Wk. Horn. CV32 28
Aylsham La. Rom. CV28 19
Aylton Est. SE16 CC41 43
Aylward Rd. SE23 CC48 61
Aylward Rd. SW20 BR51 67
Aylwards Rise, Stan. BJ28 12
Aylward St. E1 CC39 43
Aylwin Est. SE1 CA41 42
Ayr 'hoe Rd. W14 BQ42 40
Aynno St. Wat. BC25 4
Aynscombe Angle, Orp. CO54 72
Aynscombe La. SW14 BN45 49
Aynscombe Path, SW14 BN44 49
Ayr Ct. W3 BM39 39
Monks Dr.
Ayres Clo. E13 CH38 35
Ayres Cres. NW10 BN36 31
Ayres St. SE1 BZ41 42
Ayr Grn. Rom. CS29 19
Ayrsome Rd. N16 CA34 24
Ayr Way, Rom. CT30 19
Aysgarth Rd. SE21 BZ46 51
Azalea Clo. W7 BH40 39
Azalea Dr. Swan. CS52 73
Azalea Wk. Pnr. BC32 20
Azenby Rd. SE15 CA44 51
Azof St. SE10 CG42 43
Baalbec Rd. N5 BY35 33
Babbacombe Clo. Chess. BK56 75
Babbacombe Gdns. Ilf. CK31 26
Babbacombe Rd. Brom. CH51 71
Babbington Ri. Wem. BM36 30
Baber Dr. Felt. BD46 47
Babington Rise, Wem. BM36 30
Babington Rd. NW4 BP31 22
Babington Rd. SW16 BW49 59
Babington Rd. Dag. CP35 36
Babington Rd. Horn. CU33 28
Babmaes St. SW1 BW40 41
Jermyn St.
Baccallay St. E3 CD39 43
Bacchus Wk. N1 CA37 33
Baches St. N1 BZ38 33
Back Church La. E1 CB39 42
Back Grn. Walt. BD57 74
Back Hill, EC1 BY38 33
Back La. N8 BX37 33
New Rd.
Back La. NW3 BT35 32
Flask Wk.
Back La. Bark. CM37 35
Broadway
Back La. Bex. CR47 63
Back La. Brent. BK43 48
Back La. Edg. BN30 13
Back La. Rich. BK48 57
Back La. Rom. CP33 27
Station Rd.
Back La. Wat. BH23 5
Back River Pk. Beck. CD51 70
Back Rd. Sid. CO49 63
Back Row, SW17 BU49 59
Totterdown St.
Back St. W3 BM40 39
Bermondsey St.
Bacon Gro. SE1 CA41 42
Bacon La. N9 BM31 21
Bacon La. Edg. BM30 12
Bacon Link, Rom. CR29 18
Bacons La. N6 BV33 23
Bacon St. E1 CA38 33
Bacon St. E2 CA38 33
Bacton St. E2 CC38 34
Digby St.
Baddow Clo. Wdf. Grn. CJ29 17
Baden Pl. SE1 BZ41 42
Baden Powell Clo. Surb. BL55 75
Baden Rd. N8 BW31 23
Baden Rd. Ilf. CL35 35
Bader Way, Rain. CU36 37
Badger Clo. Houns. BD45 47
Badgers Clo. Enf. BY24 8
Badgers Copse BR6 CN55 81
Badgers Copse, Wor. Pk. BO55 76
Badgers Croft N20 BR26 6
Badgers Croft SE9 CL48 62
Badgers Hole, Croy. CC56 79
Badgers Rd. Sev. CR58 81
Badgers Wk. N. Mal. BO51 67
Badgers Wk. Pur. BW59 83
Badlis Rd. E17 BW31 23
Badmaes St. SW1 BW40 41
Badminton Clo. Har. BH31 21
Badminton Clo. Nthlt. BF36 29
Badminton Rd. SW12 BV46 59
Badney House, SE5 CA44 51
Glebe Rd.
Badsworth Rd. SE5 BZ44 51
Badwell Clo. Horn. CU36 37
Bagley Rd. SW6 BS44 50
Bagleys Spring, Rom. CQ31 27
Bagshot Clo. SE18 CL44 53
Bagshot Rd. Enf. CA26 8
Bagshot St. SE17 CA42 42
Baildon St. SE8 CD43 52
Bailey Gdns. Rom. CO32 27
Bailey Pl. SE26 CC50 61
Baillie Clo. Rain. CV39 46
Baillie's Wk. W5 BK41 39
Bainbridge Rd. Dag. CQ35 36
Bainbridge St. WC1 BW39 41
Baird Av. Sthl. BF40 38
Baird Clo. NW9 BN32 22
Baird Clo. Bush. BF25 4
Baird Gdns. SE21 CA49 60
Baird Rd. Enf. CB24 8
Bairstow Clo. Borwd. BK23 5
Baizdon Rd. SE3 CG44 52
Baker La. Mitch. BV51 68
Baker Rd. NW10 BN37 31
Baker Rd. SE18 CK43 53
Bakers Alley, SE1 CA40 42
Abbots La.
Baker's Av. E17 CE32 25
Bakers End, SW20 BR51 67
Bakers Field, N19 BX35 33
Baker's Hill, E5 CB33 24
Bakers Hill, Barn. BT23 7
Bakers La. N6 BU32 23
North Hill
Bakers La. W5 BK40 39
Grove, The
Baker's Ms. W1 BV39 41
Adam St.
Baker's Row, E15 CG37 34
Baker's Row, EC1 BY38 33
Baker St. NW1 BU38 32
Baker St. W1 BU38 32
Baker St. Enf. BZ24 8
Bakewell Ct. E5 CC35 34
Clapton Park Est.
Bakewell Way, N. Mal. BO51 67
Balaams La. N14 BW27 14
Balaam St. E13 CH38 35
Balaclava Rd. SE1 CA42 42
Balaclava Rd. Surb. BK54 66
Balben Rd. E9 CC36 34
Balcaskie Rd. SE9 CK46 53
Balchen Rd. SE3 CJ44 53
Balchier Rd. SE22 CB46 51
Balcombe St. NW1 BU38 32
Balcon Way, Borwd. BN23 6
Wilcox Clo.
Balcorne St. E9 CC36 34
Balder Ri. SE12 CH48 62
Balderton St. W1 BV39 41
Baldock St. E3 CE37 34
Baldock Way, Browd. BL23 5
Baldry Gdns. SW16 BX50 60
Baldwin Cres. SE5 BZ44 51
Baldwin's Gdns. EC1 BY39 42
Baldwin's Hill, Loug. CK23 10
Baldwins Pond, Loug. CK23 10
Baldwin St. EC1 BZ38 33
Baldwin Ter. N1 BZ37 33
Baldwyn Gdns. W3 BN40 40
Baldwyn's Est. Dart. CT48 64
Baldwyn's Pk. Bex. CS48 64
Baldwyn's Rd. Bex. CS48 64
Balfern Gro. W4 BO42 40
Balfern St. SW11 BU44 50
Balfe St. N1 BX37 33
Balfont Clo. Sth. Croy. CB60 84
Balfour App. Ilf. CL34 26
Balfour Rd.
Balfour Av. W7 BH40 39
Balfour Av. N20 BU27 14
Balfour House, W10 BQ39 40
Balfour Ms. W1 BV39 41
Aldford St.
Balfour Pl. W1 BV40 41
Aldford St.
Balfour Pl. N5 BZ35 33
Balfour Rd. SE25 CB52 69
Balfour Rd. SW19 BS50 59
Balfour Rd. W3 BN39 40
Balfour Rd. W13 BJ41 39
Balfour Rd. Brom. CJ53 71
Balfour Rd. Cars. BU57 77
Balfour Rd. Har. BG32 20
Balfour Rd. Houns. BF45 47
Balfour Rd. Ilf. CL34 26
Balfour Rd. Sthl. BD41 38
Balfour St. SE17 BZ42 42
Balgonie Rd. E4 CF26 9
Balgores Cres. Rom. CU31 28
Balgores La. Rom. CU31 28
Balgores Sq. Rom. CU31 28
Balgowan Rd. N. Mal. BO52 67
Balgowan Rd. Beck. CD52 70
Balgowan St. SE18 CN42 45
Balgrove Rd. W10 BR39 40
Balham Gro. SW12 BV47 59
Balham High Rd. SW12 BV48 59
Balham High Rd. SW17 BV48 59
Balham Hill, SW12 BV47 59
Balham New Rd. SW12 BV47 59
Balham Pk. Rd. SW12 BU47 59
Balham Rd. N9 CB27 15
Balham Station Rd. SW12 BV47 59
Ballamore Rd. Brom. CH48 62
Ballance Rd. E9 CC36 34
Ballantine St. SW18 BT45 50
Ballard Clo. King. On T. BN50 58
Ballards Av. Sth. Croy. CB57 78
Ballards Clo. Dag. CR37 36
Ballards Fm. Rd.
  Sth. Croy. CB57 78
Ballard's La. N3 BS30 14
Ballard's La. N12 BS30 14
Ballards Rise, Sth. Croy. CB57* 78
Ballards Rd. NW2 BP34 22
Ballards Rd. Dag. CR37 36
Ballards Way, Croy. CB57 78
Ballards Way, Sth. Croy. CB57 78
Ballast Quay, SE10 CF42 43
Ballater Clo. Wat. BD28 11
Ballater Rd. SW2 BX45 51
Ballater Rd. Sth. Croy. CA56 78
Ballina St. SE23 CC47 61
Ballingdon Rd. SW11 BV46 50
Balliol Av. E4 CG28 16
Balliol Rd. N17 BZ30 15
Balliol Rd. W10 BQ39 40
Balliol Rd. Well. CO44 54
Ball La. N14 BW27 14
Balloch Rd. SE6 CF47 61
Ballogie Av. NW10 BO35 31
Ballow Close SE5 BZ43 51
Elmington Estate
Balls Pond Rd. N1 BZ36 33
Balmain Clo. W5 BK40 39
Balmer Rd. E3 CD37 34
Balmes Rd. N1 BZ37 33
Balmoral Av. Beck. CD52 70
Balmoral Av. N11 BV28 14
Balmoral Clo. SW15 BO46 49
Westleigh Ave.
Balmoral Cres. E. Mol. BF52 65
Balmoral Dr. Borwd. BN25 6
Balmoral Dr. Hayes BC39 38
Balmoral Dr. Sthl. BE38 29
Balmoral Gdns. W13 BJ41 39
Balmoral Gdns. Ilf. CN33 27
Balmoral Gro. N7 BX36 33
Brewery Rd.
Balmoral Rd. E7 CJ35 35
Balmoral Rd. E10 CE34 25
Balmoral Rd. Har. BF35 29
Balmoral Rd. Horn. CV34 28
Balmoral Rd. King. On T. BL52 66
Balmoral Rd. Rom. CU32 28
Balmoral Rd. Wor. Pk. BP55 76
Balmoral Way, Sutt. BS58 77
Balmore Cres. Barn. BV25 7
Balmore Gdns. SW15 BQ45 49
Balnacraig Av. NW10 BO35 31
Baltic Clo. SW19 BT50 59
Baltic Gdns. SW15 BQ45 49
Baltic St. EC1 BZ38 33
Baltimore Place, DA16 CN44 54
Balvernie Gro. SW18 BR47 58
Bamborough Gdns. W12 BQ41 40
Bamford Av. Wem. BL37 30
Bamford Ct. E15 CE35 34
Bamford Rd. Bark. CA36 33
Bamford Rd. Brom. CF49 61
Bamford Way, Rom. CR28 18
Bampfylde Clo. Wall. BW55 77
Bampton Rd. SE23 CC48 61
Bampton Rd. Rom. CW30 19
Banbury Clo. Sutt. BS57 77
Banbury Ct. WC2 BX40 42
Long Acre
Banbury Rd. E9 CC36 34
Banbury Rd. SW11 BU44 50
Banbury St. Wat. BC25 4
Banbury Wk. Nthlt. BF37 29
Leander Rd.
Banchory Rd. SE3 CH43 53
Bancroft Av. N2 BU32 23
Bancroft Av. Buck. H. CH27 17
Bancroft Ct. Nthlt. BD37 29
Bancroft Gdns. Har. BG30 11
Bancroft Gdns. Orp. CN54 72
Bancroft Rd. E1 CC38 34
Bancroft Rd. Har. BG30 11
Bandon Rise, Wall. BW56 77
Bangalore St. SW15 BQ45 49
Bangor Clo. Nthlt. BF35 29
Bangor Rd. Brent. BL43 48
Banim St. W6 BP41 40
Banister Rd. W10 BQ38 31
Bank Av. Mitch. BT51 68
Bank Ct. Dart. CW46 55
Bankfoot Rd. Brom. CG49 61
Bankhurst Rd. SE6 CD47 61
Bank La. SW15 BO46 49
Bank La. Kings. On T. BL50 57
Bankside, SE1 BY40 42
Bankside, Enf. BY23 8
Bankside, Sth. Croy. CA57 78
Bankside, Sthl. BD40 38
Bankside Av. Nthlt. BC37 29
Bankside Clo. Bex. CS49 64
Bankside Clo. Cars. BU57 77
Banks La. Bexh. CQ45 54
Bankside Dr. Surb. BJ54 66
Bankton Rd. SW2 BY45 51
Bankwell Rd. SE13 CG45 52
Banner St. EC1 BZ38 33
Banning St. SE10 CG42 43
Bannister Clo. SW2 BY47 60
Bannister Clo. Har. BG35 29
Bannister Ho. E9 CC35 34
Wells Clo.
Bannock Burn Rd. SE18 CN42 45
Banstead Gdns. N9 CA27 15
Banstead Rd. Cars. BT58 77
Banstead Rd. Epsom BN59 82
Banstead Rd. Pur. BY59 84
Banstead Rd. S. Sutt. BT59 83
Banstead St. SE15 CC45 52
Banstead Way, Wall. BX56 78
Banstock Rd. Edg. BM29 12
Banton Clo. Enf. CB23 8
Banyard Rd. SE16 CB41 42
Banyards, Horn. CW32 28
Bapchild Pl. Orp. CP52 72
Okemore Gdns.
Baptist Gdns. NW5 BV36 32
Queen's Cres.
Barbara St. N7 BX36 33
Barbauld Rd. N16 CA34 24
Barber Clo. N21 BY26 8
Barber's All. E13 CH38 35
Barber's Rd. E15 CE37 34
Barbican Grnf. BF39 38
Barb Ms. W6 BQ41 40
Shepherds Bush Rd.
Barbon Clo. WC1 BX39 42
Boswell St.
Barbot Clo. N9 CB27 15
Barchard St. SW18 BS46 50
Barchester Clo. Har. BG30 11
Barchester St. E14 CE39 43
Barclay Clo. SW6 BS43 50
Barclay Oval, Wdf. Grn. CH28 17
Barclay Rd. E11 CG33 25
Barclay Rd. E13 CJ38 35
Barclay Rd. E17 CE32 25
Barclay Rd. N18 BZ29 15
Barclay Rd. SW6 BS43 50
Barclay Rd. Croy. BZ55 78
Barcombe Av. SW2 BX48 60
Barden St. SE18 CN43 54
Barden Av. Rom. CP31 27
Bardfield Trd. Est. Chess. BK58 75
Bardney Rd. Mord. BS52 68
Bardolph Av. Croy. CD58 79
Bardolph Rd. N7 BX35 33
Bardolph Rd. Rich. BL45 48
St. George's Rd.
Bard Rd. W10 BQ40 40
Bardsey Wk. N1 BZ36 33
Marquess Est.
Bardsley Clo. Croy. CA55 78
Bardsley La. SE10 CF43 52
Bardwell St. N7 BX35 33
Barfett St. W10 BR38 31
Barfield Av. N20 BU27 14
Barfield Rd. E11 CG33 25
Barfield Rd. Brom. CL52 71
Barfields, Loug. CL24 10
Barfields Clo. Loug. CL24 10
Barfields Path, Loug. CL24 10
Barford Clo. NW4 BP30 13
Barford St. N1 BY37 33
Barford Rd. SE15 CB45 51
Bargate Clo. SE18 CN42 45
Bargate Clo. N. Mal. BP54 67
Barge House Rd. E16 CL41 44
Barge House St. SE1 BY40 42
Barge Rd. E. Mol. BG52 65
Bargery Rd. SE6 CE47 61
Barge Wk. Kings. On T. BK51 66
Barge Wk. Kings. On T. BK52 66
Bargrove Clo. SE20 CB50 60
Bargrove Cres. SE6 CD48 61
Barham Av. Borwd. BL24 5
Barham Clo. Brom. CK54 71
Barham Clo. Chis. CL49 62
Barham Clo. Rom. CR30 18
Barham Rd. SW20 BP50 58
Barham Rd. Chis. CL49 62
Barham Rd. Epsom BN58 76
Barham Rd. Sth. Croy. BZ56 78
Baring Clo. SE12 CH48 62
Baring Rd. SE12 CH47 62
Baring Rd. Barn. BT24 7
Baring Rd. Croy. CB54 69
Baring St. N1 BZ37 33
Barker St. SW10 BT43 50
Barker Wk. SW16 BW48 59

Barkham Rd. N17 BZ29 15
Bark Hart Rd. Orp. CO54 72
Barking By-pass, Bark. CM38 35
Barking Industrial Est. CO37 36
  Bark.
Barking Rd. E6 CJ37 35
Barking Rd. E13 CG39 43
Barking Rd. E16 CG39 43
Barkis Way, SE16 CB42 42
  Egan Way
Bark Pl. W2 BS40 41
Barkston Gdns. SW5 BS42 41
Barkway St. N16 BZ34 24
  Kings Crescent Est.
Barkworth Rd. SE16 CB42 42
Barlborough St. SE14 CC43 52
Barlby Gdns. W10 BQ38 31
Barlby Rd. W10 BQ39 40
Barley Clo. Bush. BF25 4
Barleycorn Way, Horn. CW32 28
Barley La. Ilf. CO33 27
Barley Mow Pass, W4 BN42 40
Barlow Pl. W1 BV40 41
  Burton La.
Barlow Rd. W3 BM40 39
Barlow Rd. Hamptn. BF50 56
Barlow St. SE17 BZ42 42
Barmeston Rd. SE6 CE48 61
Barmor Clo. Pnr. BF30 11
Barmouth Av. Grnf. BH37 30
Barmouth Rd. SW18 BT46 50
Barmouth Rd. Croy. CC55 79
Barnabas Rd. E9 CC35 34
Barnaby Way, Chig. CL27 17
Barnard Clo. SE18 CL42 44
Barnard Clo. Chis. CM51 71
Barnard Clo. Sun. BC50 56
Barnard Clo. Wall. BW57 77
Barnard Gdns. Hayes BC38 29
Barnard Gdns. N. Mal. BP52 67
Barnard Gro. E15 CG36 34
Barnard Hill, N10 BV30 14
Barnard Ms. SW11 BU45 50
  Barnard Rd.
Barnardo Dr. Ilf. CM31 26
  Ashurst Dr.
Barnardo St. E1 CC39 43
Barnard Rd. SW11 BU45 50
Barnard Rd. Enf. CB23 8
Barnards Inn, EC4 BY39 42
  Fetter La.
Barnard's Pl. Pur. BY59 84
  Pampisford Rd.
Barnby St. E15 CG37 34
Barnby St. NW1 BW37 32
Barn Clo. Nthlt. BD37 29
Barn Cres. Pur. BZ60 84
Barncroft Clo. Loug. CL25 10
Barncroft Rd. Loug. CL25 10
Barnehurst Av. Erith CS44 55
Barnehurst Clo. Erith CS44 55
Barnehurst Rd. Bexh. CR44 54
Barn Elms Pk. SW15 BQ44 49
Barnes Ct. E16 CJ39 44
  Ridgewell Rd.
Barnend Dr. Dart. CV49 64
Barnend La. Dart. CV49 64
Barnes Alley, Hmptn. BG51 65
Barnes Av. SW13 BP43 49
Barnes Br. SW13 BO44 49
Barnes Br. W4 BO44 49
Barnes Ct. E12 CJ35 35
Barnes Ct. N22 BX29 15
Barnes Ct. Wdf. Grn. CJ28 17
  Durham Ave.
Barnes Cray Rd. Dart. CU45 55
Barnesdale Cres. Orp. CO53 72
Barnes End, N. Mal. BP53 67
Barnes High St. SW13 BO44 49
Barnes Pikle W5 BJ40 39
  Mattock La.
Barnes Rd. Ilf. CM35 35
Barnes St. E14 CD39 43
Barnes Ter. SE8 CD42 43
Barnet By-pass N2 BT32 23
Barnet By-pass, Barn. BO23 6
Barnet Dr. Brom. CK55 80
Barnet Gate La. Barn. BO25 6
Barnet Hill, Barn. BR24 6
Barnet La. N20 BR26 6
Barnet La. Borwd. BK25 5
Barnet La. Barn. BN25 6
Barnett Clo. DA8 CT44 55
Barnet Way NW7 BN27 13
Barnfield, Bans. BS60 83
Barnfield N. Mal. BN53 67
Barnfield Av. Croy. CC55 79
Barnfield Av. Kings. On T. BK49 57
Barnfield Av. Mitch. BV52 68
Barnfield Clo. Swan. CS54 73
Barnfield Gdns. BL49 57
  Kings. On T.
Barnfield Gdns. Est. SE18 CL43 53
Barnfield Rd. SE18 CL43 53
Barnfield Rd. W5 BK38 30
Barnfield Rd. Belv. CQ43 54

Barnfield Rd. Edg. BN30 13
Barnfield Rd. Orp. CP52 72
Barnfield Rd. Sth. Croy. CA58 78
Barnfield Wd. Clo. Beck. CF53 70
Barnfield Wd. Rd. Beck. CF53 70
Barn Gro. Bans. BT61 83
Barnham Rd. Grnf. BG38 29
Barnham St. SE1 CA41 42
Barnhill, Pnr. BD32 20
Barn Hill, Wem. BM33 21
Barn Hill Av. Brom. CG53 70
Barnhill La. Hayes BC38 29
Barnhill Rd. Hayes BC38 29
Barnhill Rd. Wem. BN34 22
Barnhurst Path, Wat. BD28 11
Barnlea Clo. Felt. BE48 56
Barnmead Gdns. Dag. CQ35 36
Barnmead Rd. Beck. CC51 70
Barnmead Rd. Dag. CQ35 36
Barn Rise, Wem. BM34 21
Barn Rd. Mitch. BV52 68
Barnsbury Clo. N. Mal. BN52 67
Barnsbury Cres. Surb. BN54 67
Barnsbury Est. N1 BX37 33
Bransbury Gro. N7 BX36 33
  Roman Way
Barnsbury La. Surb. BM55 75
Barnsbury Ms. N1 BY36 33
  Brooksby St.
Barnsbury Pk. N1 BY36 33
Barnsbury Rd. N1 BY37 33
Barnsbury Sq. N1 BY36 33
Barnsbury St. N1 BY36 33
Barnsbury Ter. N1 BY36 33
Barnsdale Yard W9 BR38 31
Barnsdale Rd. W9 BR38 31
Barnsley Rd. Rom. CW29 19
Barnsley St. E1 CB38 33
Barnstaple La. SE13 CF45 52
  Lewisham High St.
Barnstaple Path, Rom. CV28 19
Barnstaple Rd. Rom. CV28 19
Barnstaple Rd. Ruis. BD34 20
Barn St. N16 CA34 24
  Church St.
Barn Way, Wem. BM33 21
Barnwell Rd. SW2 BY46 51
Barnwood Av. W. Wick. CE54 70
Barnwood Clo. W9 BS38 32
  Amberley Rd.
Barnwood Ct. E16 CH40 44
Barnwood Ct. Est. E16 CH40 44
Baroness Rd. E2 CA38 33
Baronet Gro. N17 CB30 15
Baronet Rd. N17 CB30 15
Baron Gdns. Ilf. CM31 26
Baron Gro. Mitch. BU52 68
Baron Rd. Dag. CP33 27
Baron's Court Rd. W14 BR42 40
Baronsfield Rd. Twick. BJ46 48
Barons Gate, Barn. BU25 7
Barons Hurst, Epsom BN61 82
Barons Keep W14 BR42 40
  Gliddon Rd.
Barons Mead, Har. BH31 21
Baronsmead Rd. SW13 BP44 49
Baronsmede W5 BL41 39
Barons Pl. N2 BU31 23
Barons Pl. SE1 BY41 42
Barons, The, Twick. BJ46 48
Baron St. N1 BY37 33
Barons Wk. Croy. CC53 70
Baron Wk. E16 CG39 43
Baron Wk. Mitch. BU52 68
Barque Mews SE8 CE43 52
  Watergate St.
Barque St. E14 CF42 43
Barrack Rd. Houns. BD45 47
Barrack Row, Sthl. BD42 38
Barratt Av. N22 BX30 15
Barrat Way, Har. BG31 20
Barrenger Rd. N10 BU30 14
Barrets Green Rd. NW10 BN38 31
Barrett Rd. E17 CF31 25
Barretts Gro. N16 CA35 33
Barrett St. W1 BV39 41
Barrhill Rd. SW2 BX48 60
Barrie Clo. Couls. BV61 83
Barriedale SE14 CD44 52
Barrie Est. W2 BT40 41
Barrington Clo. Loug. CM24 10
Barrington Ct. N10 BV30 14
Barrington Grn. Loug. CM24 10
Barrington Rd. E12 CL36 35
Barrington Rd. N8 BW32 23
Barrington Rd. SW9 BY45 51
Barrington Rd. Bexh. CP44 54
Barrington Rd. Loug. CM24 10
Barrington Rd. Pur. BW59 83
Barrington Rd. Sutt. BS54 68
Barrington Vills. SE18 CL44 53
Barrosa Dr. Hamptn. BF51 65
Barrosa Dr. Hamptn. BF51 65
  Oldfield Rd.
Barrow Av. Cars. BU57 77

Barrow Clo. N21 BY27 15
Barrowdene Clo. Pnr. BE30 11
  Paines La.
Barrowell Grn. N21 BY27 15
Barrowfield Clo. N9 CB27 15
Barrowfield La. N9 CB27 15
Barrowfield, Sth. Croy. CB59 84
Barrowgate Rd. W4 BN40 40
Barrow Hedges Clo. BU57 77
  Cars.
Barrow Hedges Way, BU57 77
  Cars.
Barrow Hill NW8 BU37 32
Barrow Hill Clo. Wor. Pk. BO55 76
Barrow Hill Est. NW8 BU37 32
Barrow Hill Rd. NW8 BU37 32
Barrow Point Av. Pnr. BE30 11
Barrow Point La. Pnr. BE30 11
Barrow Rd. SW16 BW50 59
Barrow Rd. Croy. BY57 78
Barrs Rd. NW10 BN36 31
Barry Av. N15 CD32 25
  Craven Pk. Rd.
Barry Av. Bexh. CQ43 54
Barry Clo. Orp. CN55 81
Barry Ct. SW4 BW46 50
Barry Rd. SE22 CA46 51
Barson Clo. SE20 CC50 61
Barston Rd. SE27 BZ48 60
Barstow Cres. SW2 BX47 60
Barter St. WC1 BX39 42
Bartholomew Clo. EC1 BY39 42
Bartholomew Clo. SW18 BT45 50
Bartholomew La. EC2 BZ39 42
  Theadneedle St.
Bartholomew Sq. EC1 BZ38 33
Bartholomew Vill. NW5 BW36 32
Bartle Rd. SE18 CN42 45
Bartle Av. E6 CK37 35
Bartle Rd. W11 BR39 40
Bartlett Ct. EC4 BY39 42
  New Fetter La.
Bartlett Sth. Croy. BZ56 78
Bartlett Rd. SW2 BX46 51
Barton Gdns. Rom. CS30 19
Barton Av. Rom. CR33 27
Barton Clo. E9 CC35 34
Barton Clo. Bexh. CQ46 54
Barton Clo. Chig. CM27 17
Barton Meadows, Ilf. CL31 26
  Brandville Gdns.
Barton Rd. W14 BR42 40
Barton Rd. Horn. CU33 28
Barton Rd. Sid. CQ50 63
Bartons, The, Borwd. BK25 5
Barton St. SW1 BX41 42
Barton Way NW8 BT37 32
Barton Way, Borwd. BM23 5
Bartram Rd. SE4 CD46 52
Bartrip St. E9 CD36 34
Barwick Rd. E7 CH35 35
Basden Gro. Felt. BF48 56
Basedale Rd. Dag. CO36 36
Bashley Rd. NW10 BN38 31
Basil Av. E6 CK38 35
Basildene Rd. Houns. BD45 47
Basildon Av. Ilf. CL30 17
Basildon Clo. Sutt. BT58 77
Basildon Rd. SE2 CO42 45
Basildon Rd. Bexh. CQ44 54
Basil St. SW3 BU41 41
Basing Clo. Surb. BH54 66
Basing Dr. Bex. CQ46 54
Basing Hl. N3 BS31 23
Basing Hl. NW11 BR33 22
Basing Hl. Wem. BL34 21
Basing Ho. Yard N1 CA38 33
  Kingsland Rd.
Basing Pl. N1 CA38 33
  Kingsland Rd.
Basing Rd. Bans. BR60 82
Basing St. W11 BR39 40
Basing Way N3 BS31 23
Basing Way, Surb. BH54 66
Basire St. N1 BZ37 33
Baskerville Rd. SW18 BU47 59
Basket Gdns. SE9 CK46 53
Baslow Clo. Har. BG30 11
Baslow Wk. E5 CC35 34
  Clapton Park Est.
Bassant Clo. SW11 BV45 50
Bassett Rd. SW11 BV45 50
Bassano St. SE22 CA46 51
Bassant Rd. SE18 CN43 54
Bassein Pk. Rd. W12 BO41 40
Bassett Clo. Sutt. BS58 77
Bassett Gdns. Islw. BG43 47

Bassett Rd. W10 BR39 40
Bassetts Clo. Orp. CL56 80
Bassett St. NW5 BV35 32
Bassetts Way, Orp. CL56 80
Bassett Way, Grnf. BF39 38
Bassingham Rd. SW18 BT47 59
Bassingham Rd. Wem. BK36 30
Basswood Clo. SE15 CB45 51
  Linden Gro.
Bastable Av. Bark. CN37 36
Bastion Rd. SE2 CO42 45
Baston Manor Rd. Brom. CH55 80
Baston Rd. Brom. CH54 71
Bastwick St. EC1 BY38 33
Basuto Rd. SW6 BS44 50
Batavia Clo. Sun. BC51 65
Batavia Rd. SE14 CD43 52
Batavia Rd. Sun. BC51 65
Batchelor St. N1 BY37 33
Batchwood Grn. Orp. CO52 72
Bateman Rd. E4 CE29 16
Bateman's Row EC2 CA38 33
Bateman St. W1 BW39 41
  Dean St.
Bates Cres. Croy. BY56 78
Bateson St. SE18 CN42 45
  Gunning St.
Bate St. E14 CD40 43
Bath Clo. SE15 CB43 51
Bathgate Rd. SW19 BO48 58
Bath Pass. Kings. On T. BK51 66
Bath Pl. EC2 CA38 33
  Old St.
Bath Pl. Barn. BR24 6
Bath Rd. E7 CJ36 35
Bath Rd. E15 CG36 34
Bath Rd. N9 CB27 15
Bath Rd. W4 BO42 40
Bath Rd. Dart. CU47 64
Bath Rd. Hayes BC43 47
Bath Rd. Houns. BC44 47
Bath Rd. Rom. CO32 27
Bath Rd. Mitch. BT50 59
Baths Rd. Brom. CJ52 71
Bath St. EC1 BZ38 33
Bath Ter. SE1 BZ41 42
Bathurst Av. SW19 BS51 68
  Brisbane Av.
Bathurst Gdns. NW10 BP37 31
Bathurst Ms. W2 BT40 41
Bathurst Rd. Ilf. CL33 26
Bathurst St. W2 BT40 41
Bathway SE18 CL42 44
  Market St.
Batley Rd. N16 CD34 25
  Stoke Newington High St.
Batley Rd. Enf. BZ23 8
Batman Clo. W12 BP40 40
Batoum Gdns. W6 BQ41 40
Batson St. W12 BP41 40
Batsworth Rd. Mitch. BT52 68
Batten St. SW11 BU45 50
Battenberg Rd. SE6 CF48 61
Battersby Rd. SW3 BT43 50
Battersea Br. Rd. SW11 BU43 50
Battersea Church Rd. SW11 BT44 50
  Dagnall St.
Battersea High St. SW11 BT44 50
Battersea Park Est. SW11 BV44 50
  Dagnall St.
Battersea Pk. Rd. SW8 BU44 50
Battersea Pk. Rd. SW11 BU44 50
Battersea Ri. SW11 BT46 50
Battishill St. N1 BY36 33
  Waterloo Ter.
Battle Bridge La. SE1 CA40 42
  Tooley St.
Battle Bridge Rd. NW1 BX37 33
Battle Clo. SW19 BT50 59
Battledean Rd. N5 BY35 33
Battle Rd. Erith CS42 46
Battle St. E1 CB39 42
Baudwin Rd. SE6 CG48 61
Baugh Rd. Sid. CP49 63
Baulk, The SW18 BS47 59
Bavant Rd. SW16 BX51 69
Bavaria Rd. N19 BX34 24
Bavent Rd. SE5 BZ44 51
Bavonne Rd. W6 BR44 49
Bawdale Rd. SE22 CA46 51
Bawdsey Av. Ilf. CN31 27
Bawtree Clo. Sutt. BT58 77
Bawtree Rd. SE14 CD43 52
Bawtree Rd. N20 BU27 14
Baxendale N20 BT27 14
Baxendale St. E2 CB38 33
Baxter Clo. E16 CJ39 44
Baxter Rd. N1 BZ36 33
Baxter Rd. N17 CB31 24
Baxter Rd. N18 CB28 15
Baxter Rd. Ilf. CL35 35
Bayfield Rd. SE9 CJ45 53
Bayford Rd. NW10 BQ38 31
Bayford St. E8 CC36 33
Bayham Pl. NW1 BW37 32
Bayham Rd. W4 BN41 40

Bayham Rd. W13 BJ40 39
Bayham Rd. Mord. BS52 68
Bayham Rd. NW1 BW37 32
Bayley St. WC1 BW39 41
Bayley Walk SE2 CQ43 54
Baylin Rd. SW18 BS46 50
Baylis Rd. SE1 BY41 42
Baynes Clo. Enf. CB23 8
Baynes St. NW1 BW36 32
Bayston Rd. N16 CA34 24
Bayswater Rd. W2 BS40 41
Baythorne St. E3 CD39 43
Baytree Rd. SW2 BX45 51
Baywood Sq. Chig. CO28 18
Bazalgette Gdns. N. Mal. BN53 67
Bazely St. E14 CF39 43
Bazile Rd. N21 BY25 8
Beacham Clo. SE7 CJ43 53
Beachborough Rd. Brom. CF49 61
Beachcroft Rd. E11 CG34 25
Beachcroft Way N19 BX33 24
Hornsey Rise
Beach Gro. Felt. * BF48 56
Beachy Rd. E3 CE36 34
Beacon Gro. Cars. BV56 77
Beacon Hl. N7 BX35 33
Beacon Rd. SE13 CF46 52
Beacon Rd. Erith. CU43 55
Beaconsfield Clo. N11 BV28 14
Beaconsfield Clo. SE3 CH43 53
Beaconsfield Pl. Epsom BO59 82
Beaconsfield Rd. E10 CF34 25
Beaconsfield Rd. E16 CG38 34
Beaconsfield Rd. E17 CD32 25
Beaconsfield Rd. E18 CG30 16
Beaconsfield Rd. N9 CB27 15
Beaconsfield Rd. N11 BV27 14
Beaconsfield Rd. N15 CA31 24
Beaconsfield Rd. NW10 BO36 31
Beaconsfield Rd. SE3 CG43 52
Beaconsfield Rd. SE9 CK48 62
Beaconsfield Rd. W4 BN41 40
Beaconsfield Rd. W5 BK41 39
Beaconsfield Rd. Bex. CT48 64
Beaconsfield Rd. Brom. CJ52 71
Beaconsfield Rd. Croy. BZ53 69
Beaconsfield Rd. Esher BH57 75
Beaconsfield Rd. Hayes BD40 38
Beaconsfield Rd. N. Mal. BN51 67
Beaconsfield Rd. Sthl. BD40 38
Beaconsfield Rd. Surb. BL54 66
Beaconsfield Rd. Twick. BJ46 48
Beaconsfield St. E6 CL39 44
Beaconsfield Ter. W4 BN42 40
Beaconsfield Ter. W4 BR41 40
Maclise Rd.
Beaconsfield Ter. Rom. CQ32 27
Beaconsfield Wk. SW6 BR44 49
Parsons Grn. La.
Beacons, The, Loug. CL23 10
Beacontree Av. E17 CF30 16
Beacontree Rd. E11 CG33 25
Beacon Way, Bans. BQ61 82
Beadlow Clo. Cars. BT53 68
Olveston Wk.
Beadman St. SE27 BY49 60
Beadnell Rd. SE23 CC47 61
Beadon Rd. W6 BQ42 40
Beadon Rd. Brom. CH52 71
Beaford Gro. SW20 BR52 67
Beagle Clo. Felt. BE49 56
Beagles Clo. Orp. CP55 81
Beak St. W1 BW40 41
Bealah Rd. Sutt. BS56 77
Beal Clo. Well. CO44 54
Beale Clo. N13 BY28 15
Beale Pl. E3 CD37 34
Beale Rd. E3 CD37 34
Beale St. E13 CH37 35
Beale St. W. E13 CH37 35
Beale St.
Beal Rd. Ilf. CL34 26
Beam Av. Dag. CR37 36
Beaminster Gdns. Ilf. CL31 26
Beamish Dr. Bush. BG26 4
Beamish Rd. N9 CB26 8
Beamish Rd., Orp. CP54 72
Beamore Rd. SW15 BP47 58
Beam Way, Dag. CS36 37
Beanacre Clo. E9 CD36 34
Bean Rd. Bexh. CP45 54
Beanshaw SE9 CL49 62
Beansland Gro. Rom. CQ30 18
Bear All. EC4 BY39 42
Farringdon St.
Beardell St. SE19 CA50 60
Beardow Gro. N14 BW25 7
Beard Rd. Kings. On T. BL49 57
Beard's Hill, Hampton BF51 65
Beard's Hill Clo. BF51 65
Hampton
Beardsley Way W3 BN41 40
Birkbeck Gro.
Bearfield Rd. Kings. On T. BL50 57
Bear Gdns. SE1 BZ40 42
Bearing Clo. Chig. CO28 18
Bearing Way, Chig. CO28 18
Bear La. SE1 BY40 42

Bear Rd. Felt. BD49 56
Bearstead Ri. SE4 CD46 52
Bear St. WC2 BW40 41
Cranbourn St.
Beasley's Ait La. Sun. BC53 65
Beatrice Av. SW16 BX52 69
Beatrice Av. Wem. BL35 30
Beatrice Clo. E13 CH38 35
Chargeable La.
Beatrice Clo. Pnr. BC31 20
Beatrice Ct. Wem. BL35 30
Beatrice Rd. E17 CE32 25
Beatrice Rd. N4 BY33 24
Beatrice Rd. N9 CC26 9
Beatrice Rd. SE1 CB42 42
Beatrice Rd. Rich. BL46 48
Beatrice Rd. Sthl. BE40 38
Beatrice St. E13 CH38 35
Beattock Rise N10 BV31 23
Beatty Rd. N16 CA35 33
Beatty Rd. Stan. BK29 12
Beatty St. NW1 BW37 32
Bettyville Gdns. Ilf. CL31 26
Beauchamp Pl. SW3 BU41 41
Beauchamp Rd. E7 CH36 35
Beauchamp Rd. SE19 BZ51 69
Beauchamp Rd. SW11 BU45 50
Beauchamp Rd. E. Mol. BF53 65
Beauchamp Rd. Sutt. BS56 77
Beauchamp Rd. Twick. BJ47 57
Beauchamp St. EC1 BY39 42
Leather La.
Beauclere Rd. W6 BP41 40
Beaufort Av. Har. BJ31 21
Beaufort Clo. W5 BL39 39
Beaufort Clo. Rom. CR31 27
Beaufort Ct. Rich. BK49 57
Beaufort Dr. NW11 BS31 23
Beaufort Gdns. NW4 BO32 22
Beaufort Gdns. SW3 BU41 41
Beaufort Gdns. SW16 BX50 60
Green La.
Beaufort Gdns. Houns. BE44 47
Beaufort Gdns. Ilf. CL33 26
Beaufort Pk. NW11 BS31 23
Beaufort Rd. W5 BL39 39
Beaufort Rd. Kings. On T. BL52 66
Beaufort Rd. Rich. BK49 57
Beaufort Rd. Twick. BK47 57
Beaufort St. SW3 BT43 50
Beaufort Way, Epsom BP57 76
Beaufoy Rd. N17 CA29 15
Beaufoy Rd. SW11 BV44 50
Beaufoy Wk. SE11 BX42 42
Beaulieu Av. SE26 CB49 60
Beaulieu Clo. SE5 CA45 51
Beaulieu Clo. NW9 BO31 22
Beaulieu Clo. Mitch. BV51 68
Beaulieu Clo. Twick. BK46 48
Beaulieu Dr. Pnr. BD32 20
Beaulieu Gdns. N21 BZ26 8
Beauly Way, Rom. CT30 19
Beaumanor Gdns. SE9 CL49 62
Beanshaw
Beaumaris Dr. Wdf. Grn. CJ29 17
Beaumont Av. W14 BR42 40
Beaumont Av. Har. BF32 20
Beaumont Av. Rich. BL45 48
Beaumont Av. Wem. BK35 30
Beaumont Clo. Rom. CV30 19
Beaumont Cres. W14 BR42 40
Beaumont Cres. Rain. CU36 37
Beaumont Gro. E1 CC38 34
Beaumont Ms. W1 BV39 41
Marylebone High St.
Beaumont Pl. W1 BW38 32
Beaumont Ri. N19 BW33 23
Beaumont Rd. E10 CE33 25
Beaumont Rd. E13 CH38 35
Beaumont Rd. SE19 BZ50 60
Beaumont Rd. W4 BN41 40
Beaumont Rd. Orp. CM53 71
Beaumont Rd. Pur. BY60 84
Beaumont Sq. E1 CC39 43
Beaumont St. W1 BV39 41
Beauvais Terr. Nthlt. BD38 29
Beauval Rd. SE22 CA46 51
Beaverbank Rd. SE9 CM47 62
Beaver Clo. SE20 CB50 60
Beavercote Wk. DA18 CQ43 54
Beaver Clo. Hamptn. BF51 65
Beaver Gro. Nthlt. BD38 29
Jetstar Way
Beavers Cres. Houns. BD45 47
Beavers La. Houns. BD45 47
Beaverwood Rd. Sid. CN49 63
Beavor La. W6 BP42 40
Bebbington Rd. SE18 CN42 45
Beblets Clo. Orp. CN56 81
Beccles Dr. Bark. CN36 36
Beccles St. E14 CD39 43
Bec Clo. Ruis. BD34 29
Beckenham Gdns. N9 CA27 15
Beckenham Gro. Brom. CF51 70

Beckenham Hill Rd. SE6 CE50 61
Beckenham Hill Rd. Beck. CE50 6
Beckenham La. Brom. CG51 70
Beckenham Pl. Pk. Beck. CE50 61
Beckenham Rd. Beck. CC51 70
Beckenham Rd. W. Wick. CE54 70
Beckenshaw Gdns. Bans. BT61 83
Beckers Est. The N16 CB34 24
Becket Av. E6 CL38 35
Becket Fold, Har. BH32 21
Becket Rd. N18 CC28 16
Becket St. SE1 BZ41 42
Beckett Av. Ken. BY61 84
Becketts Clo. Orp. CN55 81
Harlington Rd. West
Becketts Clo. Felt. CD50 61
Beckett Wk. Beck. CD50 61
Beckford Rd. Croy. CA53 69
Beck La. Beck. CC52 70
Becklow Gdns. W12 BP41 40
Becklow Rd.
Becklow Rd. W12 BO41 40
Beck Rd. E8 CB37 33
Becks Rd. Sid. CO48 63
Beckton Gdns. E6 CK39 44
Beckton Pl. Erith. CR44 54
Beckton Rd. E6 CG39 43
Beckton Rd. E16 CG39 43
Beckton Rd. E16 CG39 43
Beckway, Beck. CD52 70
Beckway Rd. SW16 BW51 68
Beckway St. SE17 BZ42 42
Beckwith Rd. SE24 BZ46 51
Beclands Rd. SW17 BV50 59
Becmead Av. SW16 BW49 59
Becmead Av. Har. BJ32 21
Becondale Rd. SE19 CA49 60
Beaconsfield Rd. W11 BV28 14
Becontree Av. Dag. CO35 36
Becontree Av. Dag. CO34 27
Bective Pl. SW15 BR45 49
Bective Rd. E7 CH35 35
Bective Rd. SW15 BR45 49
Bedale Rd. Enf. BZ23 8
Bedale St. SE1 BZ40 42
Beddington Fm. Rd. Croy. BX54 69
Beddington Gdns. Wall. BV57 77
Beddington Grn. Orp. CN51 72
Beddington Gro. Wall. BW56 77
Beddington La. Croy. BW53 68
Beddington Path, Orp. CN51 72
Beddington Rd. Ilf. CN33 27
Beddington Rd. Orp. CN51 72
Bede Clo. Pnr. BD30 11
Bedenham Way SE15 CA43 51
Hordle Prom N.
Bedens Rd. Sid. CQ50 63
Bede Rd. E3 CD39 43
Bede Rd. Rom. CP32 27
Bedfont Clo. Mitch. BU51 68
St. Marks Rd.
Bedfont La. Felt. BC47 56
Bedford Av. WC1 BW39 41
Bedford Av. Barn. BR25 6
Bedford Av. Hayes BC39 38
Bedfordbury WC2 BX40 42
Chandos Pl.
Bedford Clo. N10 BV29 23
Bedford Ct. WC2 BX40 42
Bedford St.
Bedford Gdns. W8 BS40 41
Bedford Gdns. Horn. CV34 28
Bedford Hill SW12 BV47 59
Bedford Hill SW16 BV47 59
Bedford Pk. Croy. BZ54 69
Bedford Pk. Mans. W4 BN42 40
Bedford Pl. W1 BW39 41
Charlotte St.
Bedford Pl. WC1 BX39 42
Bedford Pl. Croy. BZ54 69
Bedford Rd. E6 CL37 35
Bedford Rd. E17 CE31 25
Bedford Rd. E18 CH30 17
Bedford Rd. N2 BU31 23
Bedford Rd. N8 BW32 23
Bedford Rd. N9 CB26 8
Bedford Rd. N15 CA31 24
Bedford Rd. N22 BX30 15
Bedford Rd. NW7 BO27 13
Bedford Rd. SW4 BX45 51
Bedford Rd. W4 BN41 40
Bedford Rd. W13 BJ40 39
Bedford Rd. Brent. BL42 39
Clayponds La.
Bedford Rd. Enf. CC23 9
Bedford Rd. Har. BG32 20
Bedford Rd. Ilf. CL34 26
Bedford Rd. Orp. CO55 81
Bedford Rd. Sid. CN48 63
Bedford Rd. Twick. BG48 56
Bedford Rd. Wor. Pk. BQ55 76
Bedford Row WC1 BX39 42
Bedford Sq. WC1 BW39 41
Bedford St. WC2 BX40 42
Bedford St. Wat. BC23 4
Bedford Way WC1 BW38 32

Bedgebury Gdns. SW19 BR47 58
Bedgebury Rd. SE9 CJ45 53
Bedivere Rd. Brom. CH48 62
Bedlam Gdns. E. Mol. BF52 65
Bedonwell Rd. Belv. CQ43 54
Bedser Dr. Har. BG35 29
Bedster Gdns. E. Mol. BG51 65
Bedwardine Rd. SE19 CA50 60
Bedwell Rd. N17 CA30 15
Bedwell Rd. Belv. CQ42 45
Bedwin Way SE16 CB42 42
Catlin St.
Beeby Rd. E16 CH39 44
Beech Av. N20 BU26 7
Beech Av. W3 BO40 40
Beech Av. Brent. BJ43 48
Beech Av. Buck. H. CH27 17
Beech Av. Ruis. BC33 20
Beech Av. Sid. CO47 63
Beech Av. Sth. Croy. BZ59 84
Beech Av. Swan. CT52 73
Beech Clo. N9 CB25 8
Beech Clo. SW15 BP47 58
Beech Clo. SW19 BO50 58
Beech Clo. Cars. BU55 77
Beech Clo. Horn. CU34 28
Beech Clo. Walt. BD56 74
Beech Copse, Brom. CK51 71
Beech Copse, Sth. Croy. CA56 78
Beech Ct. SE9 CK46 53
Beech Ct. Tedd. BK50 57
Broom Water
Beechcroft, Chis. CL50 62
Beechcroft Av. NW11 BR33 22
Beechcroft Av. Bexh. CS44 55
Beechcroft Av. Har. BF33 20
Beechcroft Av. Ken. BZ61 84
Beechcroft Av. N. Mal. BE40 38
Beechcroft Clo. SW16 BX49 60
Beechcroft Clo. Houns. BE43 47
Beechcroft Gdns. Wem. BL34 21
Beechcroft Rd. E18 CH30 17
Beechcroft Rd. SW14 BN45 49
Beechcroft Rd. SW17 BU48 59
Beechcroft Rd. Bush. BE25 4
Beechcroft Rd. Chess. BL55 75
Beechcroft Rd. Orp. CM56 80
Beechdale N21 BX27 15
Beechdale Rd. SW2 BX46 51
Beech Dell, Orp. CK56 80
Beech Dr. N2 BU30 14
Beech Dr. Borwd. BL23 5
Beechen Clo. Pnr. BE31 20
Beechen Gro. Wat. BC24 4
Beechenlea La. Swan. CU52 73
Beechen Pl. SE23 CC48 61
Beeches Av. The, Cars. BU57 77
Beeches Rd. SW17 BU48 59
Beeches Rd. Sutt. BR54 67
Beeches The, Bans. BS61 83
Beeches Wk. Cars. BT58 77
Beechfield, Bans. BS60 83
Beechfield Cott. Brom. CJ51 71
Beechfield Gdns. Rom. BZ32 24
Beechfield Rd. N4 BZ32 24
Beechfield Rd. SE6 CD47 61
Beechfield Rd. Brom. CJ51 71
Beechfield Rd. Erith CT43 55
Beech Gdns. W5 BL41 39
Beech Gdns. Dag. CR36 36
Beech Gro. Epsom BP62 82
Beech Gro. Ilf. CN29 18
Beech Gro. Mitch. BW53 68
Beech Gro. N. Mal. BN52 67
Beech Hall Cres. E4 CF29 16
Beech Hall Rd. E4 CF29 16
Beech Hill Av. Barn. BT23 7
Beech Hill Rd. SE9
Beech House Rd. Croy. BZ55 78
Beech La. Buck. H. CH27 17
Beechlawns N2 BT28 14
Beechmont Rd. Brom. CG49 61
Beechmore Gdns. Sutt. BQ55 76
Beechmore Rd. SW11 BU44 50
Beechmount Av. W7 BG39 38
Beecholme Av. Mitch. BV51 68
Beecholme Est. E5 CB34 24
Beech Rd. N11 BX29 15
Beech Rd. SW16 BX51 69
Beech Rd. Dart. CV47 64
Beech Rd. Epsom BO61 82
Beech Rd. Orp. CO57 81
Beech Row, Kings. On T. BL49 57
Beech St. EC1 BZ39 42
Beech St. Rom. CS31 27
Beech Tree Clo. Stan. BJ28 12
Beech Tree Clo. Stan. BK28 12
Marsh La.
Beech Tree Glade E4 CG26 9
Beech Tree Pl. Sutt. BS56 77
West St.
Beech Wk. NW7 BN29 13
Beech Wk. Dart. CU45 55
Beech Way, Epsom BP59 82
Beech Way NW10 BN36 31

Beechway, Bex. CP46 54
Beech Way, Epsom BO61 82
Beech Way, Twick. BF48 56
Beechwood Av. N3 BF31 22
Beechwood Av. Couls. BV61 83
Beechwood Av. Grnf. BF38 29
Beechwood Av. Har. BF34 20
Beechwood Av. Orp. CN56 81
Beechwood Av. Rich. BM44 48
Beechwood Av. Sun. BC50 56
Beechwood Av. Th. Hth. BY52 69
Beechwood Clo. NW7 BO28 13
Beechwood Clo. Surb. BK54 66
Beechwood Cres. Bexh. CP45 54
Beechwood Dr. Kes. CJ56 80
Beechwood Dr. Wdf. Grn. CG28 16
Beechwood Gdns. W5 BL37 30
 St. Anne's Gdns.
Beechwood Gdns. Har. BF34 20
Beechwood Gdns. Ilf. CK32 26
Beechwood Gdns. Rain. CU39 46
Beechwood Pk. E18 CH31 26
Beechwood Rd. E8 CA36 33
Beechwood Rd. N8 BW31 23
Beechwood Rd. Sth. Croy. BZ58 78
Beechwood Ter. E4 CF29 16
 Larks Hall Rd.
Beechworth Clo. NW3 BS34 23
Beecot La. Walt. BD55 74
Beecroft Rd. SE4 CD46 52
Beehive Ct. Ilf. CK32 26
Beehive La. Ilf. CK32 26
Beeken Dene, Orp. LM56 80
Beeleigh Rd. Mord. BS52 68
Beeston Clo. Wat. BD28 11
Beeston Pl. SW1 BV41 41
Beeston Rd. Barn. BT25 7
 Berkeley Cres.
Beeston Way, Felt. BD46 47
Beethoven St. W10 BR38 31
Beeton Clo. Pnr. BF29 11
Begbie Rd. SE3 CJ44 53
Beggar's Hill, Epsom BO57 76
Begonia Wk. W12 BO39 40
 Du Cane Rd.
Beira St. SW12 BV47 59
Belcroft Clo. Brom. CG50 61
Beldham Gdns. E. Mdl. BG51 65
Belfairs Dr. Rom. CP33 27
Belfairs Grn. Wat. BD28 11
 Heysham Dr.
Belfast Rd. N16 CA34 24
Belfast Rd. SE25 CB52 69
Belfield Rd. Epsom BN57 76
Belfont Walk N7 BX35 33
 Warlters Rd.
Belford Gro. SE18 CL42 44
Belfort Rd. SE15 CC44 52
Belgrade Rd. N16 CA35 33
Belgrade Rd. Hamptn. CV31 65
Belgrave Av. Rom. CV31 28
Belgrave Clo. N14 BW25 7
Belgrave Clo. W3 BM41 39
 Avenue Rd.
Belgrave Clo. Walt. BC56 74
Belgrave Cres. Sun. BC51 65
Belgrave Gdns. N14 BW24 7
Belgrave Gdns. NW8 BS37 32
Belgrave Gdns. Stan. BK28 12
Belgrave Ms. N. SW1 BV41 41
Belgrave Ms. N. SW1 BV41 41
Belgrave Ms. S. SW1 BV41 41
Belgrave Pl. SW1 BV41 41
Belgrave Rd. E10 CF33 25
Belgrave Rd. E11 CH34 26
Belgrave Rd. E13 CJ38 35
Belgrave Rd. E17 CE32 25
Belgrave Rd. SE25 CB52 69
Belgrave Rd. SW1 BV42 41
Belgrave Rd. SW13 BO43 49
Belgrave Rd. Houns. BE45 47
Belgrave Rd. Ilf. CK33 26
Belgrave Rd. Mitch. BT52 68
Belgrave Rd. Sun. BC51 65
Belgrave Sq. SW1 BV41 41
Belgrave St. E1 CD39 43
Belgrave St. SW1 BV41 41
Belgrave Ter. Wdf. Grn. CH27 17
Belgrave Wk. Mitch. BT52 68
Belgravia Gdns. Brom. CF50 61
Belgravia Mews, Kings. On T. BK52 66
Belgrove St. WC1 BX38 33
Belham St. SE5 BZ44 51
Belhaven St. E3 CD38 34
Belinda Rd. SW9 BY45 51
Belitha Vill. N1 BX36 33
Bell Alley, Houns. BF45 47
Bellamy Clo. SW5 BR42 40
 Aisgill Av.
Bellamy Dr. Wat. BC23 4
Bellamy Dr. Stan. BJ30 12
Bellamy Rd. E4 CE29 16
Bellamy St. SW12 BV47 59
Bellasis Av. SW2 BX48 60
Bell Av. Rom. CU30 19
Bell Clo. Pnr. BD30 11

Bell Ct. Surb. BM55 75
Bell Dr. SW18 BR47 58
Bellefield Rd. Orp. CO53 72
Bellefields Rd. SW9 BX45 51
Belle Grove Clo. Well. CN44 54
Bellegrove Rd. Well. CN44 54
Bellenden Rd. SE15 CA45 51
Belleville Rd. SW11 BU46 50
Bellevue Gdns. SW9 BX44 51
Bellevue La. Bush. BG26 4
Bellevue Pk. Th. Hth. BZ52 69
Bellevue Rd. E17 CF30 16
Bellevue Rd. N11 BV28 14
Bellevue Rd. NW4 BQ31 22
Bellevue Rd. SW13 BP44 49
Bellevue Rd. SW17 BU47 59
Bellevue Rd. W13 BJ38 30
Belle Vue Rd. Bexh. CQ46 54
Belle Vue Rd. Horn. CW33 28
Belle Vue Rd., Kings. On T. BL52 66
Belle Vue Rd. Rom. CS29 19
Belle Vue Ter. NW4 BQ31 22
Bellew St. SW17 BT48 59
Bellfield, Croy. CD57 79
Bellfield Av. Har. BG29 11
Bellflower Path, Rom. CV29 19
Bell Fm. Av. Dag. CS34 28
Bell Gdns. Orp. CP53 72
Bell Grn. SE26 CD49 61
Bell Grn. La. SE26 CD49 61
Bell Ho. Rd. Rom. CS33 28
Bellingham Grn. SE6 CE48 61
Bellingham Rd. SE6 CE48 61
Bell La. E1 CA39 42
Bell La. E16 CH40 44
Bell La. NW4 BQ31 22
Bell La. Twick. BJ47 57
Bellot Way SE10 CG42 43
Bellring Clo. Belv. CR43 54
Bell Rd. E. Mol. BG53 65
Bell Rd. Enf. BZ23 8
Bell Rd. Houns. BF45 47
Bells All. SW6 BS44 50
Bells Gdns. Est. SE15 CB43 51
Bells Hill, Barn. BQ25 6
Bellstaines Pleasaunce E4 CE27 16
Bell St. NW1 BU39 41
Bellvue Clo. Orp. CL58 80
Bell Water Gate SE18 CL41 44
Bell Wharf La. EC4 BZ40 42
 Upr. Thames St.
Bell Yd. WC2 BY39 42
Bellwood Rd. SE15 CC45 52
Belmont Av. N9 CB26 8
Belmont Av. N13 BX28 15
Belmont Av. N17 BZ31 24
Belmont Av. Barn. BU25 7
Belmont Av. N. Mal. BP52 67
Belmont Av. N. Mal. BP53 67
Belmont Av. Sthl. BE41 38
Belmont Av. Well. CN44 54
Belmont Av. Wem. BL37 30
Belmont Circle, Har. BJ30 12
Belmont Clo. N20 BS26 7
Belmont Clo. SW4 BW45 50
Belmont Clo. Barn. BU24 7
Belmont Clo. Wdf. Grn. CH28 17
Belmont Gro. SE13 CF45 52
Belmont Gro. W4 BN42 40
Belmont Hall Ct. SE13 CF45 52
Belmont Hill SE13 CF45 52
Belmont La. Chis. CL49 62
Belmont La. Stan. BK29 12
Belmont Pk. SE13 CF45 52
Belmont Park Rd. E10 CE32 25
Belmont Rise, Sutt. BR57 76
Belmont Rd. N15 BZ31 24
Belmont Rd. N17 BZ31 24
Belmont Rd. NW11 BR32 22
Belmont Rd. SW4 BW45 50
Belmont Rd. Beck. CD51 70
Belmont Rd. Bexh. CR43 54
Belmont Rd. Bush. BE24 4
Belmont Rd. Chis. CL49 62
Belmont Rd. Har. BH31 21
Belmont Rd. Horn. CV34 29
Belmont Rd. Ilf. CM34 26
Belmont Rd. SE25 CB53 69
Belmont Rd. Sutt. BS58 77
Belmont Rd. Twick. BG48 56
Belmont Rd. Wall. BV56 77
Belmont St. NW1 BV36 32
Belmore, Borwd. BL25 5
Belmore Av. Hayes BC39 38
Belmore St. SW8 BW44 50
Belper Ct. E5 CC35 34
 Clapton Park Est.
Belper St. N1 CB36 33
 Lofting Rd.
Belsham St. E9 CC36 34
Belsize Av. N13 BX29 15
Belsize Av. NW3 BT36 32
Belsize Av. W13 BJ41 39
Belsize Cres. NW3 BT35 32
Belsize Gdns. Sutt. BS56 77

Belsize Gro. NW3 BU36 32
Belsize La. NW3 BT36 32
Belsize Ms. NW3 BT35 32
 Belsize La.
Belsize Pk. NW3 BT36 32
Belsize Pk. Gdns. NW3 BT36 32
Belsize Pk. Ms. NW3 BT35 32
 Belsize La.
Belsize Pl. NW3 BT35 32
 Belsize La.
Belsize Rd. NW6 BS37 32
Belsize Rd. NW6 BT36 32
Belsize Rd. Har. BG29 11
Belsize Sq. NW3 BT36 32
Belsize Ter. NW3 BT36 32
Belson Rd. SE18 CK42 44
Beltane Dr. SW19 BQ48 58
Beltinge Rd. Rom. CW31 28
Belton Rd. E7 CH36 35
Belton Rd. E11 CG35 34
Belton Rd. N17 CA31 24
Belton Rd. NW2 BP36 31
Belton Rd. Sid. CO49 63
Belton Way E3 CE39 43
Beltran Rd. SW6 BS44 50
Belvedere NW9 BO30 13
Belvedere Av. SW19 BR49 58
Belvedere Av. Ilf. CL30 17
Belvedere Bldgs. SE1 BY41 42
Belvedere Clo. Esher BF56 74
Belvedere Clo. Tedd. BH49 57
Belvedere Ct. N2 BT32 23
Belvedere Ct. SW15 BQ45 49
 Upr. Richmond Rd.
Belvedere Dr. SW19 BR49 58
Belvedere Gdns. E. Mol. BE53 65
Belvedere Gro. SW19 BR49 58
Belvedere Pl. SE1 BY41 42
 Borough Rd.
Belvedere Rd. E10 CD33 25
Belvedere Rd. SE1 BX41 42
Belvedere Rd. SE2 CP40 45
Belvedere Rd. SE19 CA50 60
Belvedere Rd. Bexh. CQ45 54
Belvedere Sq. SW19 BR49 58
Belvoir Rd. SE22 CB47 60
Belvue Rd. Nthlt. BF36 29
Bembridge Clo. NW6 BQ36 31
Bemerton St. N1 BX37 33
Bemish Rd. SW15 BQ45 49
Bempton Dr. Ruis. BC34 20
Bemsted Rd. E17 CD31 25
Benares Rd. SE18 CN42 45
Benbow Rd. W6 BP41 40
Benbow St. SE8 CE43 52
Bench Fld. Sth. Croy. CA57 78
Benchley Gdns. SE23 CC46 52
Bench, The, Rich. BK48 57
 Back La.
Bencombe Rd. Pur. BX60 84
Bencroft Rd. SW16 BW50 59
Bendall Ms. NW1 BU39 41
Bendemeer Rd. SW15 BQ45 49
Bendish Rd. E6 CK36 35
Bendmore Av. SE2 CO42 45
Bendon Vall. SW18 BS47 59
Bendysh Rd. Bush. BE24 4
Benedict Rd. SW9 BX45 51
Benedict Rd. Mitch. BT52 68
Benedict Way N2 BT31 23
Beneden Grn. Brom. CH53 71
Benett Gdns. SW16 BX51 69
Benfleet Clo. Sutt. BS55 77
Bengal Rd. Ilf. CL35 35
Bengarth Dr. Har. BG30 11
Bengarth Rd. Nthlt. BD37 29
Bengeworth Rd. SE5 BZ45 51
Bengeworth Rd. Har. BJ34 21
Benhale Clo. Stan. BJ28 12
Benham Clo. SW11 BT45 50
 Wayland Rd.
Benham Clo. Couls. BY62 84
Benham Rd. W7 BH39 39
Benhill Av. Sutt. BS56 77
 Throwley Way
Benhill Rd. SE5 BZ43 51
Benhill Rd. Sutt. BT55 77
Benhill Rd. Sutt. BT55 77
Benhilton Gdns. Sutt. BS55 77
Benhurst Av. Horn. CU35 37
Benhurst Clo. Sth. Croy. CC58 79
Benhurst Ct. SW16 BY49 60
Benhurst La. SW16 BY49 60
Benin St. SE13 CF47 61
Benjafield Clo. N18 CB28 15
 Craig Park Rd.
Benjamin Clo. Rom. CV32 28
Benjamin St. EC1 BY39 42
Ben Jonson Rd. E1 CC39 43
Benledi St. E14 CF39 43
Bennerley Rd. SW11 BU46 50

Bennet St. SW1 BW40 41
 Arlington St.
Bennett Clo. Kings. On T. BK51 66
Bennett Clo. Well. CO44 54
Bennett Gro. SE13 CE44 52
Bennett Pk. SE3 CG45 52
Bennett Rd. E13 CJ38 35
Bennett Rd. Rom. CQ32 27
Bennett Rd. Croy. CD55 79
Bennett's Av. Grnf. BG37 29
Bennett's Av. Croy. CD55 79
Bennetts Castle La. Dag. CP35 36
Bennetts Clo. N17 CB29 15
Bennetts Copse, Chis. CK50 62
Bennett St. W4 BO43 49
Bennetts Way, Croy. CD55 79
Bennetts Yd. SW1 BX41 42
 Marsham St.
Benningholme Rd. Edg. BO29 13
Bennington Av. Bark. CL36 35
Bennington Rd. N17 CA30 15
Bennington Rd. Wdf. Grn. CG29 16
 Forest Dr.
Benns Alley, Hamptn. BF51 65
 Thames St.
Benn St. E9 CD36 34
Benrek Clo. Ilf. CM30 17
Bensham Clo. Th. Hth. BZ52 69
Bensham Gro. Th. Hth. BZ51 69
Bensham La. Croy. BY54 69
Bensham Manor Rd. Th. Hth. BZ52 69
Benskin Rd. Wat. BC25 4
Benson Av. E6 CJ37 35
Benson Rd. SE23 CC47 61
Benson Rd. Croy. BY55 78
Bentfield Gdns. SE9 CJ48 62
Benthal Rd. N16 CB36 24
Bentham Rd. E9 CC36 34
Bentham Rd. SE28 CO40 45
Ben Tillet Clo. Bark. CO36 36
Ben Tillet Clo. W1 BV39 41
 Marylebone La.
Bentinck St. W1 BV39 41
Bentley Dr. Ilf. CM32 26
Bentley House SE5 CA44 51
 Glebe Est.
Bentley Rd. N1 CA36 33
 Tottenham Rd.
Bentley Way, Stan. BJ28 12
Bentley Way, Wdf. Grn. CH27 17
Benton Clo. Houns. BF45 47
Benton Rd. E16 CJ40 44
 Oriental Rd.
Benton Rd. Wat. BD28 11
 Staines Rd.
Benton Rd. E. Ilf. CM33 26
Benton Rd. W. Ilf. CM33 26
Bentons La. SE27 BZ49 60
Bentons Ri. SE27 BZ49 60
Bentry Clo. Dag. CQ34 27
Bentry Rd. Dag. CQ34 27
Bentworth Rd. W12 BP39 40
Benwell Rd. N7 BY35 33
Benworth St. E3 CD38 34
Benyon Rd. N1 BZ37 33
Berber Rd. SW11 BU46 50
 Ashness Rd.
Berens Rd. NW10 BQ38 31
Berens Rd. Orp. CP53 72
Berens Way, Chis. CN52 72
Beresford Av. N20 BU27 14
Beresford Av. W7 BG39 38
Beresford Av. Surb. BM54 66
Beresford Av. Twick. BK46 48
Beresford Av. Wem. BL37 30
Beresford Dr. Wdf. Grn. CJ28 17
Beresford Gdns. Enf. CA24 8
Beresford Gdns. Houns. BE46 47
Beresford Gdns. Rom. CO32 27
Beresford Rd. E4 CG26 9
Beresford Rd. E17 CE30 16
Beresford Rd. N2 BU31 23
Beresford Rd. N5 BZ35 33
Beresford Rd. N8 BY32 24
Beresford Rd. Kings. On T. BL51 66
Beresford Rd. N. Mal. BN52 67
Beresford Rd. Sthl. BD40 38
Beresford Rd. Sutt. BR58 76
Beresford St. SE18 CL41 44
Beresford Ter. N5 BZ35 33
Berestead Rd. W6 BO43 49
Berestede Rd. W6 BO42 40
Berger Clo. Orp. CN53 72
Berger Rd. E9 CC36 34
Bergholt Av. Ilf. CK32 26
Bergholt Cres. N16 CA33 24
Berkeley Av. Grnf. BG36 29
Berkeley Av. Houns. BC44 47
Berkeley Av. Ilf. CL30 17
Berkeley Av. Rom. CS29 19
Berkeley Clo. Borwd. BM25 5
Berkeley Clo. Brent. BJ43 48
Berkeley Clo. Ruis. BC34 20

Berkeley Ct. EC1 BY39 42
Briset St.
Berkeley Ct. N14 BW25 7
Berkeley Cres. Barn. BT25 7
Berkeley Cres. Dart. CW47 64
Berkeley Gdns. N21 BZ26 8
Berkeley Gdns. W8 BS40 41
Brunswick Gdns.
Berkeley Gdns. Esher BJ57 75
Berkeley Ms. W1 BU39 41
Berkeley Pl. SW19 BQ50 58
Berkeley Rd. E12 CK35 35
Berkeley Rd. N8 BW32 23
Berkeley Rd. N15 BZ32 24
Berkeley Rd. NW9 BM31 21
Berkeley Rd. SW13 BP44 49
Berkeley Rd. Bexh. CP44 54
Berkeley Sq. W1 BV40 41
Berkeley St. W1 BU39 41
Berkeley Wk. N7 BX34 24
Andover Est.
Berkeley Waye. Houns. BD43 47
Berkhampstead Rd. Belv. CR42 45
Berkhampstead Av. Wem. BL36 30
Berkley Dr. E. Mol. BE52 65
Berkley Gro. NW1 BU36 32
Berkley Rd.
Berkley Rd. NW1 BU36 32
Berkley Rd. NW13 BP44 49
Berkshire Gdns. N13 BX29 15
Berkshire Gdns. N18 CB28 15
Berkshire Rd. E9 CD36 34
Berskire Sq. Mitch. BX52 69
Berkshire Way, Mitch. BX52 69
Berman's Way NW10 BO35 31
Bermondsey Sq. SE1 CA41 42
Abbey St.
Bermondsey St. SE1 CA40 42
Bermondsey Wall E. SE16 CB41 42
Bermondsey Wall W. SE16 CB41 42
Mill St.
Bernard Av. W13 BJ41 39
Bernard Gdns. SW19 BR49 58
Bernard Rd. N15 CA32 24
Bernard Rd. Rom. CS33 28
Bernard Rd. Wall. BV56 77
Bernard St. WC1 BX38 33
Bernays Clo. Stan. BK29 12
Bernays Gro. SW9 BX45 51
Bernell Dr. Croy. CD55 79
Berner Est. E1 CB39 42
Berne Rd. Th. Hth. BZ52 69
Berners Ms. W1 BW39 41
Berners Pl. W1 BW39 41
Berners Rd. N1 BY37 33
Berners Rd. N22 BY30 15
Berners St. W1 BW39 41
Berney Rd. Croy. BZ54 69
Bernville Way, Harrow BL32 21
Kenton Rd.
Bernwell Rd. E4 CG27 16
Berota Rd. SE9 CM48 62
Berridge Est. Edg. BL29 12
Berridge Grn. Edg. BM29 12
Berridge Rd. SE19 BZ49 60
Berriman Rd. N7 BX34 24
Berrin Way N1 BY37 33
Berriton Rd. Har. BE33 20
Berry Clo. N21 BY26 8
Berry Clo. NW10 BO36 31
Berry Grn. La. Wat. BE23 4
Berry Gro. La. Wat. BF23 4
Berryhill SE9 CL45 53
Berry Hill, Stan. BK28 12
Berryhill Gdns. SE9 CL45 53
Berry Ho. Rd. SW11 BU44 50
Dagnall St.
Berrylands SW20 BQ52 67
Berrylands, Orp. CP55 81
Berrylands, Surb. BL53 66
Berrylands Rd. Surb. BL53 66
Berryman Clo. Dag. CP34 27
Berrymans La. SE26 CB49 60
Berrymead Gdns. W3 BN41 40
Berrymede Rd. W4 BN41 40
Berry Rd. SE17 BY42 41
Berrystede Clo.
Kings. On T. BM50 57
Berry St. EC1 BY38 33
Dallington St.
Berry Way W5 BL41 39
Bertal Rd. SW17 BT49 59
Berther Rd. Horn. CV33 28
Berthon St. SE8 CE43 52
Bertie Rd. NW10 BP36 31
Bertie Rd. SE26 CC50 61
Bertram Cott. SW19 BS50 59
Bertram Rd. NW4 BP32 22
Bertram Rd. Enf. CA24 8
Bertram Rd. BM50 57
Kings. On T.
Bertram St. N19 BV34 23
Bertram Way, Enf. CA24 8
Bertrand St. SE13 CE45 52
Bert Rd. Th. Hth. BZ53 69
Berwick Av. Hayes BD39 38
Berwick Cres. Sid. CN47 63

Berwick Pond Clo. Rain. CV37 37
Berwick Pond Rd. Rain. CW37 37
Berwick Rd. E16 CJ39 44
Berwick Rd. E16 CD31 25
Berwick Rd. N22 BY30 15
Berwick Rd. Rain. CV37 37
Berwick Rd. Well. CO44 54
Berwick St. W1 BW39 41
Berwyn Av. Houns. BF44 47
Berwyn Rd. SE24 BY47 60
Berwyn Rd. Rich. BM45 48
Beryl Rd. W6 BQ42 40
Besant Ct. N1 BZ35 33
Besant Walk N7 BX34 24
Andover Est.
Besant Way NW10 BN35 31
Besley St. SW16 BW50 59
Bessborough Gdns. SW1 BW42 41
Vauxhall Bridge Rd.
Bessborough Pl. SW1 BW42 41
Bessborough Rd. SW15 BP47 58
Bessborough Rd. Har. BG33 20
Bessborough St. SW1 BW42 41
Bessein Pk. Rd. W12 BO41 40
Bessemer Rd. SE5 BZ44 51
Bessingby Rd. Ruis. BC34 20
Bessingham Wk. SE4 CC46 52
Besson St. SE14 CC44 52
Bess St. E2 CC38 34
Roman Rd.
Bestwood St. SE8 CC42 43
Betchworth Rd. Ilf. CN34 27
Betchworth Way, Croy. CF58 79
Betham Rd. Grnf. BG38 29
Bethecar Rd. Har. BH32 21
Bethell Av. E16 CG38 34
Bethell Av. Ilf. CL33 26
Bethel Rd. Well. CP45 54
Bethesden Clo. Beck. CD50 61
Bethnal Green Est. E2 CC38 34
Bethnal Grn. Rd. E1 CA38 33
Bethnal Grn. Rd. E2 CA38 33
Bethune Av. N11 BU28 14
Bethune Rd. N16 BZ33 24
Bethune Rd. NW10 BN38 31
Bethwin Rd. SE5 BY43 51
Betley Ct. Walt. BC55 74
Betony Rd. Rom. CV29 19
Betoyne Av. E4 CG28 16
Betsham Rd. Erith. CT43 55
Betstyle Rd. N11 BV28 14
Betterton Dr. Sid. CQ48 63
Betterton Rd. Rain. CT38 37
Betterton St. WC2 BX39 42
Bette St. E1 CB40 42
Betton Pl. E2 CA37 33
Bettons Pk. E15 CG37 34
Bettridge Rd. SW6 BR44 49
Betts St. E1 CB40 42
Betts Way, Surb. BJ54 66
Betula Clo. Ken. BZ61 84
Betula Wk. Rain. CV38 37
Beulah Av. Th. Hth. BZ51 69
Beulah Clo. Edg. BM27 12
Beulah Cres. Th. Hth. BZ51 69
Beulah Gro. Croy. BZ53 69
Beulah Hill SE19 BY50 60
Beulah Path E17 CE32 25
Addison Rd.
Beulah Rd. E11 CG34 25
Beulah Rd. E17 CE32 25
Beulah Rd. SW19 BR50 58
Beulah Rd. Horn. CV34 28
Beulah Rd. Th. Hth. BZ52 69
Beult Rd. Dart. CU45 55
Bevan Av. Bark. CO36 36
Bevan Ct. Croy. BY56 78
Bevan Est. Barn. BT24 7
Bevan Pl. Swan. CT52 73
Bevan Rd. SE2 CO42 45
Bevan Rd. Barn. BU24 7
Bevan St. N1 BZ37 33
Bevan Way, Horn. CW35 37
Bevenden St. N1 BZ38 33
Beverley Av. SW20 BO51 67
Beverley Av. Houns. BE45 47
Beverley Av. Sid. CN47 63
Beverley Clo. N21 BZ26 8
Beverley Clo. SW13 BP44 49
Beverley Clo. Chess. BK56 75
Beverley Clo. Enf. CA24 8
Beverley Clo. Horn. CW33 28
Beverley Ct. N14 BW26 7
Beverley Ct. SE4 CD45 52
Beverley Ct. W4 BN42 40
Beverley Ct. Epsom BQ59 82
Beverley Cres. Wdf. Grn. CH30 17
Beverley Dr. Edg. BM31 21
Beverley Gdns. NW11 BR33 22
Beverley Gdns. SW13 BO45 49
Beverley Gdns. Grnf. BH38 39
Beverley Gdns. Horn. CW33 28
Beverley Gdns. Stan. BJ30 12
Beverley Gdns. Wem. BL33 21
Beverley Gdns. Wor. Pk. BP54 67
Beverley La. Kings. On T. BO50 58
Beverley Path SW13 BO44 49

Beverley Rd. E4 CF29 16
Beverley Rd. E6 CJ38 35
Beverley Rd. SE20 CB51 69
Beverley Rd. SW13 BO45 49
Beverley Rd. W4 BO42 40
Beverley Rd. Bexh. CS44 55
Beverley Rd. Brom. CK55 80
Beverley Rd. Dag. CQ35 36
Beverley Rd. Mitch. BW52 68
Beverley Rd. N. Mal. BP52 67
Beverley Rd. Ruis. BC34 20
Beverley Rd. Sthl. BE42 38
Beverley Rd. Whyt. CA61 84
Beverley Rd. Wor. Pk. BQ55 76
Beverley Rd. Kings. On T. BK51 66
Beverley Way SW20 BO51 67
Beverley Way, N. Mal. BP52 67
Beversbrook Rd. N19 BW34 23
Beverstone Rd. SW2 BX46 51
Beverstone Rd. Th. Hth. BY52 69
Bevill Allen Clo. SW17 BU49 59
Bevil St. SE8 CE43 52
Frankham St.
Bevin Ct. WC1 BX38 33
Bevington Rd. W10 BR39 40
Bevington Rd. Beck. CE51 70
Bevington St. SE16 CB41 42
Bevin Rd. Hayes BC38 29
Bevis Marks EC3 CA39 42
Bewcastle Gdns. Enf. BX24 8
Bewdley St. N1 BY36 33
Bewick St. SW8 BV44 50
Bewley St. E1 CC40 43
Bewlys Rd. SE27 BY49 60
Bexhill Clo. Felt. BE48 56
Bexhill Rd. N11 BW28 14
Bexhill Rd. SE4 CD46 52
Bexhill Rd. SW14 BO45 49
Bexley Clo. Dart. CT46 55
Bexley Gdns. N9 BZ27 15
Bexley La. Dart. CT46 55
Bexley La. Sid. CP49 63
Bexley Rd. SE9 CL46 53
Bexley Rd. Erith CS43 55
Beynon Rd. Cars. BU56 77
Bianca House N1 CA37 33
Purcell St.
Bianca Rd. SE15 CA43 51
Bibsworth Rd. N3 BR30 13
Bicester Rd. Rich. BM45 48
Bickenhall St. W1 BU39 41
Bickersteth Rd. SW17 BU50 59
Bickerton Rd. N19 BW34 23
Bickley Cres. Brom. CK52 71
Bickley Pk. Rd. Brom. CK52 71
Bickley Rd. E10 CE33 25
Bickley Rd. Brom. CJ51 71
Bickley St. SW17 BU49 59
Bicknell Rd. SE5 BZ45 51
Bicknoller Clo. Sutt. BS58 77
Cotswold Rd.
Bicknoller Rd. Enf. CA23 8
Bicknor Rd. Orp. CN54 72
Bidborough Clo. Brom. CG53 70
Bidborough St. WC1 BW38 32
Biddenden Way SE9 CL49 62
Bidder St. E16 CG38 34
Biddestone Rd. N7 BX35 33
Biddulph Rd. W9 BS38 32
Biddulph Rd. Sth. Croy. BZ58 78
Bideford Av. Grnf. BJ38 30
Bideford Clo. Edg. BM30 12
Bideford Clo. Felt. BE48 56
Bideford Rd. Rom. CV30 19
Bideford Rd. Brom. CG48 61
Bideford Rd. Ruis. BC34 20
Bideford Rd. Well. CO43 54
Bidwell Gdns. N11 BW29 14
Bidwell St. SE15 CB44 51
Bifron St. Bark. CM37 35
Biggerstaff Rd. E15 CF37 34
Biggerstaff St. N4 BY34 24
Biggin Av. Mitch. BU51 68
Biggin Hl. SE19 BY51 69
Biggin Way SE19 BY50 60
Bigginwood Rd. SW16 BY50 60
Big Hill E5 CB33 24
Bigland St. E1 CB39 42
Bignell Rd. SE18 CL42 44
Bignold Rd. E7 CH35 35
Bigwood Ct. NW11 BS32 23
Bigwood Rd.
Bigwood Rd. NW11 BS32 23
Billet Clo. Horn. CV33 28
Billet Clo. Rom. CP31 27
Billet Rd.
Billet La. Horn. CV33 28
Billet Rd. E17 CC30 16
Billet Rd. Rom. CO31 27
Billing Pl. SW10 BS43 50
Billing Rd. SW10 BS43 50
Billingsgate St. SE10 CF43 52
Billing St. SW10 BS43 50
Billington Pl. W3 BM40 39
Steyne Rd.
Billington Rd. SE14 CC43 52

Billiter Sq. EC3 CA39 42
Fenchurch Av.
Billiter St. EC3 CA39 42
Billockby Clo. Chess. BL57 75
Bilson St. E14 CF42 43
Bilsby Gro. SE9 CJ49 62
Bilton Rd. Erith. CU43 55
Bilton Rd. Grnf. BJ37 30
Bilton Way, Enf. CD23 9
Bilton Way, Hayes BC41 38
Bina Gdns. SW5 BT42 41
Bincote Rd. Enf. BX24 8
Binden Rd. W12 BO41 40
Bindon Grn. Mord. BS52 68
Bayham Rd.
Binfield Rd. SW4 BX44 51
Binfield Rd. Sth. Croy. CA56 78
Bingfield St. N1 BX37 33
Bingham Pl. W1 BV38 32
Bingham Rd. Croy. CB54 69
Bingham St. N1 BZ36 33
Bingley Rd. E16 CJ39 44
Bingley Rd. Grnf. BG38 29
Bingley Rd. Sun. BV39 41
Binney St. W1 BV39 41
Binns Rd. W4 BO42 40
Binsey Wk. SE2 CP41 45
Binyon Cres. Stan. BH28 12
Birbeck Rd. Sid. CO48 63
Birbetts Rd. SE9 CK48 62
Bircham Path SE4 CD45 52
Frendsbury Rd.
Birchanger Rd. SE25 CB53 69
Birch Av. N13 BZ27 15
Birch Clo. SE15 CB44 51
Birch Clo. Brent. BJ43 48
Birch Clo. Buck. H. CJ27 17
Birch Clo. Rom. CR31 27
Birch Clo. Tedd. BJ49 57
Birch Cres. Horn. CW31 28
Birchdale Gdns. Rom. CP33 27
Birchdale Rd. E7 CJ35 35
Birchen Clo. NW9 BN34 22
Birchen Gro. NW9 BN34 22
Birches Clo. Epsom BO61 82
Birches Clo. Mitch. BU51 68
Birches Clo. Pnr. BE32 20
Birches, The N21 BX25 8
Birches, The SE7 CH43 53
Birches, The, Bush. BG25 4
Birches, The, Orp. CL56 80
Birchfield Clo. Couls. BX61 84
Birchfield Gro. Epsom BQ58 76
Birchfield St. E14 CE40 43
Birch Gdns. Dag. CS34 28
Birch Gro. SE12 CG47 61
Birch Gro. W3 BM40 39
Birch Gro. Well. CO45 54
Birch Hill, Croy. CC56 79
Birchington Clo. Bexh. CR44 54
Birchington Rd. N8 BW32 23
Birchington Rd. NW6 BS37 32
Birchington Rd. Surb. BL54 66
Birchin La. EC3 CA39 42
Birchlands Av. SW12 BU47 59
Birch La. Pur. BX59 84
Birch Mead, Orp. CL55 80
Birchmead Av. Pnr. BD31 20
Birchmere Row. SE3 CG44 52
Birchmore Wk. N5 BY34 24
Birch Park, Har. BG29 11
Birch Rd. Felt. BE49 56
Birch Rd. Rom. CR31 27
Birch Row, Brom. BL54 71
Birch Tree Av. W. Wick. CG56 79
Birch Tree Way, Croy. CG55 78
Birchville Ct. Bush. BH26 5
Birch Wk. N5 BZ34 24
Birch Wk. Borwd. BM23 5
Birch Wk. Erith CS43 55
Birch Wk. Mitch. BV51 68
Birchway, Hayes BC40 38
Birchwood Av. N10 BV31 23
Birchwood Av. Beck. CD52 70
Birchwood Av. Sid. CO48 63
Birchwood Av. Wall. BV55 77
Birchwood Ct. N13 BY28 15
North Circular Rd.
Birchwood Ct. Edg. BN30 13
Birchwood Dr. Dart. CT49 64
Birchwood Gro. Hamptn. BF50 56
Birchwood Pk. Av. Swan. CT52 73
Birchwood Rd. SW17 BV49 59
Birchwood Rd. Dart. CS50 64
Birchwood Rd. Orp. CM52 71
Birchwood Rd. Swan. CS51 73
Birdbrook Clo. Dag. CS36 37
Birdbrook Rd. SE3 CJ45 53
Birdcage Wk. SW1 BW41 41
Bird Fm. Av. Rom. CR29 18
Birdham Clo. Brom. CK53 71
Birdhurst Av. Sth. Croy. BZ56 78
Birdhurst Gdns. Sth. Croy. BZ56 78
Birdhurst Rise, Sth. Croy. CA56 78
Birdhurst Rd. SW18 BT45 50
Birdhurst Rd. SW19 BU50 59
Birdhurst Rd. Sth. Croy. CA56 78
Bird-in-bush Rd. SE15 CB43 51

Bird-in-hand La. Brom. CJ51 71
Bird-in-hand Pass. SE23 CC48 61
Dartmouth Rd.
Birdlington Rd. Wat. BD27 11
Bird St. W1 BV39 41
Bird Wk. Twick. BE47 56
Birdwood Clo. Tedd. BH49 57
Birdwood Clo. Sth. Croy. CC58 79
Birkbeck Av. W3 BN40 40
Birkbeck Av. Grnf. BF37 29
Birkbeck Gdns. Wdf. Grn. CH27 17
Birkbeck Gro. W3 BN41 40
Birkbeck Hl. SE21 BJ47 60
Birkbeck Pl. SE21 BZ47 60
Birkbeck Rd. E8 CA35 33
Birkbeck Rd. N8 BX31 24
Birkbeck Rd. N12 BT28 14
Birkbeck Rd. N17 CA30 15
Birkbeck Rd. NW7 BO28 13
Birkbeck Rd. SW19 BS49 59
Birkbeck Rd. W3 BN40 40
Birkbeck Rd. W5 BK42 39
Birkbeck Rd. Beck. CC51 70
Birkbeck Rd. Enf. BZ23 8
Birkbeck Rd. Ilf. CM32 26
Birkbeck Rd. Rom. CS33 28
Birkbeck St. E2 CB38 33
Birkbeck Ter. Grnf. BG37 29
Birkbeck Way. Grnf. BG37 29
Birkdale Av. Pnr. BF31 20
Birkdale Av. Rom. CW29 19
Birkdale Clo. Orp. CM54 71
Birkdale Gdns. Wat. BD27 11
Birkdale Pl. Orp. CM54 71
Birkdale Rd. SE2 CO42 45
Birkenhead Av.
Kings. On T. BL51 66
Birkenhead St. WC1 BX38 33
St. Chads St.
Birkhall Rd. SE6 CF47 61
Birkwood Clo. SW12 BW47 59
Birley Rd. N20 BT27 14
Birley St. SW11 BV44 50
Birling Rd. Erith CS43 55
Birnam Rd. N4 BX34 24
Birstall Grn. Wat. BD27 11
Birstall Rd. N15 CA32 24
Biscay Rd. W6 BQ42 40
Biscoe Clo. Houns. BF43 47
Biscoe Way SE13 CF45 52
Bisenden Rd. Croy. CA55 78
Bisham Clo. Cars. BU54 68
Bisham Gdns. N6 BV33 23
Bishop Craven Rd. Enf. BY23 8
Bishop Ken Rd. Har. BH30 12
Bishop King's Rd. W14 BR42 40
Bishop Rd. N14 BV26 7
Bishop's Av. E13 CH37 35
Bishops Av. SW6 BQ44 49
Bishops Av. Borwd. BL25 5
Bishops Av. Brom. CJ52 71
Bishops Av. Rom. CP32 27
Bishop's Av. The N2 BT33 23
Bishop's Bri. W2 BT39 41
Bishop's Bridge Rd. W2 BS39 41
Bishops Clo. E17 CE31 25
Bishops Clo. Barn. BQ25 6
Bishops Clo. Couls. BY62 84
Bishop's Clo. Enf. CB23 8
Bishops Clo. Rich. BK48 57
Bishops Clo. Sutt. BS55 77
Bishops Ct. EC4 BY39 42
Old Bailey
Bishop's Ct. WC2 BY39 42
Chancery La.
Bishopsford, Cars. BU53 68
Bishopsford Rd. Mord. BT54 68
Bishopsgate EC2 CA39 42
Bishopsgate Churchyard EC2
Bishop's Gro. N2 BU32 23
Bishops Gro. Hamptn. BE49 56
Bishop's Hall,
Kings. On T. BK51 66
Bishops Pk. Rd. SW6 BQ44 49
Bishop's Park Rd. SW16 BX51 69
Bishops Rd. N6 BV32 23
Bishops Rd. SW6 BR44 49
Bishops Rd. W7 BH41 39
Bishop's Rd. Croy. BY54 69
Bishop's Ter. SE11 BY42 42
Bishopsthorpe Rd. SE26 CC49 61
Bishop St. N1 BZ37 33
Bishops Wk. Chis. CM51 71
Bishops Wk. Croy. CC56 79
Bishop's Way E2 CC37 34
Bishops Way NW10 BO36 31
Bishopswood Rd. N6 BU33 23
Bisley Clo. Wor. Pk. BQ54 67
Bispham Rd. NW10 BL38 30
Bisson Rd. E15 CF37 34
Bisterne Av. E17 CF31 25
Bittacy Clo. NW7 BQ29 13
Bittacy Hl. NW7 BQ29 13
Bittacy Pk. Av. NW7 BQ29 13
Bittacy Ri. NW7 BQ29 13
Bittacy Rd. NW7 BQ29 13
Bittern St. SE1 BZ41 42

Bittoms, The,
Kings. On T. BK52 66
Bixley Clo. Sthl. BE41 38
Blackall St. EC2 CA38 33
Blackberry Farm Clo. BE43 47
Houns.
Blackbird Hill NW9 BN34 22
Blackborne Rd. Dag. CR36 36
Black Boy La. N15 BZ32 24
Blackbrook La. Brom. CK53 71
Blackburn Rd. NW6 BS36 32
Blackburn's Ms. W1 BV40 41
Blackbush Av. Rom. CP32 27
Black Bush Clo. Sutt. BS57 77
Blackdown Ter. SE18 CK43 53
Fairlawn
Blackett St. SW15 BQ45 49
Blackfen Rd. Sid. CN46 54
Blackfield Path SW15 BP47 58
Roehampton High St.
Blackford Clo. Pur. BY59 84
Pampisford Rd.
Blackford Rd. Wat. BD28 11
Blackfords Path SW15 BP47 58
Roehampton High St.
Blackfriars Br. EC4 BY40 42
Blackfriars Br. SE1 BY40 42
Blackfriars La. EC4 BY40 42
Blackfriars Rd. SE1 BY40 42
Blackheath Av. SE10 CG43 52
Blackheath Gro. SE3 CG44 52
Blackheath Pk. SE3 CG45 52
Blackheath Ri. SE13 CF44 52
Blackheath Rd. SE10 CE44 52
Blackheath Vale SE3 CG44 52
Blackheath Vill. SE3 CG44 52
Black Hill Rd. Esher BE58 74
Blackhorse La. Croy. CC54 70
Blackhorse La. E17 CC31 25
Blackhorse Rd. E17 CC31 25
Blackhorse Rd. SE8 CD43 52
Blackhorse Rd. Sid. CO49 63
Black Horse Yd. E1 CA39 42
Middlesex St.
Blacklands SE6 CF49 61
Blacklands Ter. SW3 BU42 41
Black Lion La. W6 BP42 40
Black Lion Yd. E1 CB39 42
Old Montague St.
Blackmore Av. Sthl. BG40 38
Blackmore Rd. Buck. H. CK26 10
Blackmore's Gro. Tedd. BJ50 57
Blacknell Clo. Har. BG29 11
Blackness La. Kes. CJ57 80
Black Path E10 CC33 25
Black Prince Rd. SE1 BX42 42
Black Prince Rd. SE11 BX42 42
Black Raven Alley EC4 BZ40 42
Swan Wf.
Blackshaw Pl. N1 CD36 34
Hertford Rd.
Blackshaw Rd. SW17 BT49 59
Blacksmith's Hill, CB60 84
Sth. Croy.
Blacksmiths La. Orp. CP53 72
Blacksmith's La. Rain. CT37 37
Blacks Rd. W6 BQ42 40
Blackstock Rd. N4 BY34 24
Blackstock Rd. N5 BY34 24
Blackstone Rd. NW2 BQ35 31
Blackthorn Ct. Houns. BE43 47
Blackthorne Av. Croy. CC54 70
Blackthorne Dr. E4 CF28 16
Blackthorne Gro. Bexh. CQ45 54
Blackthorn Rd. E3 CE38 34
Blackthorn Rd. Ilf. CM32 26
Blackwall La. SE10 CG41 43
Blackwall Tunnel App. E14 CF39 43
Blackwall Tunnel App. SE10
CG41 43
Blackwall Tunnel N. App.
E3 CE37 34
Blackwall Way E14 CF39 43
Blackwater St. SE22 CA46 51
Blackwater Clo. E5 CC35 33
Blackwater Clo. Rain. CT37 37
Blackwell Dr. Wat. BD25 4
Blackwell Gdns. Edg. BM27 12
Blackwell Sq SW9 BY43 51
Brixton Rd.
Bladindon Dr. Bex. CP47 63
Bladgen's Clo. N14 BW27 14
Bladgen's La. N14 BW27 14
Blagdon Rd. SE13 CE46 52
Blagdon Rd. N. Mal. BO52 67
Blagdon Wk. Tedd. BK50 57
Blagrove Rd. W10 BR39 40
Blair Av. NW9 BO33 22
Blair Av. Esher BG55 74
Blair Clo. Sid. CN46 54
Boundary Rd.
Blairderry Rd. SW2 BX48 60
Blairhead Dr. Wat. BC27 11
Blair St. E14 CF39 43
Blake Av. Bark. CO37 36
Blake Clo. Rain. CT37 37
Blake Clo. Well. CN44 54

Blake Gdns. SW6 BS44 50
Blake Gdns. Dart. CW45 55
Blake Hall Cres. E11 CH33 26
Blake Hall Rd. E11 CH33 26
Blakeley Bldgs. SE10 CG41 43
Blakeley Cotts. SE10 CG41 43
Blakemore Rd. SW16 BX48 60
Blakemore Rd. Th. Hth. BX53 69
Blakendon Dr. Esher BH57 75
Blakeney Av. Beck. CD51 70
Blakeney Clo. N20 BT26 7
Blakeney Clo. Epsom BN59 82
Blakeney Rd. Beck. CD50 61
Blakenham Rd. SW17 BU49 59
Blaker Ct. SE7 CJ43 53
Fairlawn
Blake Rd. E16 CG38 34
Blake Rd. N11 BW29 14
Blake Rd. Croy. CA55 78
Blake Rd. Mitch. BU52 68
Blakes Av. N. Mal. BO53 67
Blakes Grn. W. Wick. CF54 70
Blakes La. N. Mal. BO53 67
Blakesware Gdns. N9 BZ26 8
Blake's Rd. SE15 CA43 51
Blakes Ter. N. Mal. BP53 67
Blakewood Clo. Felt. BD49 56
Blanchard Pl. E8 CB36 33
Blanchard St. E8 CB36 33
Blanchard Rd.
Blanchedowne SE5 BZ45 51
Blanch Clo. SE15 CC43 52
Clifton Way
Blanche St. E16 CG38 34
Blanchland Rd. Mord. BS53 68
Bland Av. SE15 CC43 52
King Arthur St.
Blandfield Rd. SW12 BV46 50
Blandford Av. Beck. CD51 70
Blandford Av. Twick. BF47 56
Blandford Clo. N2 BT32 23
Blandford Clo. Croy. BX55 78
Wandle Rd.
Blandford Clo. Rom. CR31 27
Blandford Cres. E4 CF26 9
Blandford Rd. Tedd. BG49 56
Blandford Sq. NW1 BU39 41
Blandford St. W1 BV39 41
Blandford Waye, Hayes BD39 38
Blandon St. SE9 CJ45 53
Blaney Cres. E6 CL38 35
Blanmerle Rd. SE9 CL47 62
Blann Cl. SE9 CJ46 53
Middle Pk. Av.
Blantyre St. SW10 BT43 50
Blashford St. SE13 CF47 61
Blawith Rd. Har. BH31 21
Blaydon Clo. N17 CB29 15
Bleak Hill Lane SE18 CN43 53
Bean Gro. SE20 CC50 61
Bleasdale Av. Grnf. BH37 30
Blechynden St. W10 BQ40 40
Bramley Rd.
Bleddyn Clo. Sid. CP46 54
Bledlow Clo. SE28 CP40 45
Bledlow Ri. Grnf. BG37 29
Blegborough Rd. SW16 BW49 59
Blendon Dr. Bex. CP46 54
Blendon Path, Brom. CG50 61
Hope Park
Blendon Rd. Bex. CP46 54
Blendon Ter. SE18 CM42 44
Blendworth Way SE15 CA43 51
Hordle Prom. N.
Blenheim Av. Ilf. CL32 26
Blenheim Clo. N21 BZ26 8
Blenheim Clo. SW20 BQ52 67
Blenheim Clo. Dart. CV46 55
Blenheim Clo. Grnf. BG37 29
Blenheim Clo. Rom. CS31 28
Blenheim Clo. Wall. BW57 77
Blenheim Clo. W11 BR40 40
Blenheim Ct. EC1 BX34 24
Blenheim Ct. Sid. CM48 62
Blenheim Cres. W11 BR40 40
Blenheim Cres. Sth. Croy. BZ57 78
Blenheim Dr. Well. CM44 54
Blenheim Gdns. NW2 BQ35 31
Blenheim Gdns. SW2 BX46 51
Blenheim Gdns. CB59 84
Sth. Croy.
Blenheim Gdns. Wall. BW57 77
Blenheim Gdns. Wem. BL34 21
Blenheim Gdns. BM50 57
Kings. on T.
Blenheim Gro. SE15 CA44 51
Blenheim Pk Rd. BZ58 78
Blenheim Pass. NW8 BT37 32
Loudoun Rd.
Blenheim Rd. E6 CJ38 35
Blenheim Rd. E15 CG35 34
Blenheim Rd. E17 CC31 25

Blenheim Rd. NW8 BT37 32
Blenheim Rd. SE20 CC50 61
Blenheim Rd. SW20 BQ52 67
Blenheim Rd. W4 BO41 40
Blenheim Rd. Barn. BQ24 6
Blenheim Rd. Brom. CK52 71
Blenheim Rd. Dart. CV46 55
Blenheim Rd. Epsom BN59 82
Blenheim Rd. Har. BF32 20
Blenheim Rd. Nthlt. BF36 29
Blenheim Rd. Orp. CP55 81
Blenheim Rd. Sid. CP47 63
Blenheim Rd. Sutt. BS55 77
Blenheim St. W1 BV39 41
Blenheim Ter. NW8 BT37 32
Blenkarne Rd. SW11 BU46 50
Bleriot Rd. Houns. BD43 47
Blessbury Rd. Edg. BM30 12
Blessington Clo. SE13 CF45 52
Blessington Rd. SE13 CF45 52
Bletchley Ct. N1 BZ37 33
Bletchley St. N1 BZ37 33
Blewett St. SE17 BZ42 42
Sandford Row
Blincoe Cl. SW19 BR48 58
Queensmere Rd.
Blind La. Bans. BU61 83
Blind La. Loug. CG23 9
Bliss Cres. SE13 CE44 52
Blissett St. SE10 CF44 52
Blithbury Rd. Dag. CO36 36
Blithdale Rd. SE2 CO42 45
Blockhouse St. SE15 CC43 52
Blockley Rd. Wem. BJ34 21
Bloemfontein Av. W12 BP40 40
Bloemfontein Rd. W12 BP40 40
Blomfield Rd. W9 BS39 41
Blomfield St. EC2 BZ39 42
Blomfield Vill. W2 BS39 41
Blomville Rd. Dag. CQ34 27
Blondel St. SW11 BV44 50
Blondin Av. W5 BK42 39
Blondin St. E3 CE37 34
Bloomburg St. SW1 BW42 41
Vincent Sq.
Bloomfield Cres. Ilf. CL32 26
Bloomfield Rd. N6 BV32 23
Bloomfield Rd. SE18 CL43 53
Bloomfield Rd. BL52 66
Kings. On T.
Bloomfield Ter. SW1 BV42 41
Bloom Gro. SE27 BY48 60
Bloomhall Rd. SE19 BZ49 60
Bloom Pk. Rd. SW6 BR43 49
Bloomsbury Clo. W5 BL40 39
Bloomsbury Clo. Epsom BN58 76
Bloomsbury Ct. Pnr. BE31 20
Bloomsbury Pl. WC1 BX39 42
Southampton Row
Bloomsbury Sq. WC1 BX39 42
Bloomsbury Way WC1 BX39 42
Bloxam Clo. W5 BL41 39
Blossom Av. Enf. BZ23 8
Blossom St. E1 CA38 33
Blossom Waye, Houns. BE43 47
Blount St. E14 CD39 43
Bloxam Gdns. SE9 CK46 53
Bloxhall Rd. E10 CD33 25
Bloxham Cres. Hamptn. BE50 56
Bloxworth Clo. Wall. BW55 77
Blucher Rd. SE5 BZ43 51
Blue Anchor All. Rich. BL45 48
Kew Rd.
Blue Anchor La. SE16 CB42 42
Blue Anchor Yd. E1 CA40 42
Derby St.
Bluebell Clo. SE26 CA49 60
Bluebell Clo. Orp. CL55 80
Blueberry Gdns. Couls. BX57 78
Blueberry Gdns. Couls. BX61 84
Blue Cedars, Bans. BR60 82
Bluefield Clo. Hamptn. BF49 56
Bluehouse Rd. E4 CG27 16
Blundell Rd. Edg. BN30 13
Blundell St. N7 BX36 33
Blunden Clo. Dag. CP34 27
Blunt Rd. Sth. Croy. BZ56 78
Blunts Rd. SE9 CL46 53
Blurton Rd. E5 CC35 34
Blyth Clo. SE6 CD47 61
Blythe Hl. SE6 CD47 61
Blythe Hl. Orp. CN51 72
Blythe Hl. La. SE6 CD47 61
Blythe Rd. W14 BR41 40
Blythe Rd. Brom. CG51 70
Blythe St. E2 CB38 33
Blythe Val. SE6 CD47 61
Blyth Rd. E15 CF36 34
Blyth Rd. E17 CD33 25
Blyth Rd. W14 BQ42 40
Blythswood Rd. Ilf. CO33 27
Blythwood Rd. N4 BX33 24
Blythwood Rd. Pnr. BD30 11
Boadicea St. N1 BX37 33
Boar Clo. Chig. CO28 18
Hart Cres.
Boardman Av. E4 CF25 9
Boarica St. N1 BX37 33

Boat House Wk. SE15 CA43 51
Bobbin Clo. SW4 BW45 50
Bobs La. Rom. CT30 19
Bockhampton Rd. BL50 57
Kings. On T.
Bocking St. E8 CB37 33
Boddicott Clo. SW19 BR48 58
Queensmere Rd.
Boddy's Bridge SE1 BY40 42
Bodiam Clo. Enf. BZ23 8
Bodiam Rd. SW16 BW50 59
Bodicott Clo. SW19 BR48 58
Bodley Clo. N. Mal. BO53 67
Bodley Rd. N. Mal. BN53 67
Bodley St. SE17 BZ42 42
Wansey St.
Bodmin Rd. Mord. BS53 68
Bodmin St. SW18 BS47 59
Bodnant Gdns. SW20 BP52 67
Bodney Rd. E8 CB35 33
Boeing Way, Sthl. BC41 38
Bogey La. Orp. CK57 80
Bognor Gdns. Wat. BD28 11
Bognor Rd. Well. CP44 54
Bognor St. SW8 BW44 50
Bohun Gro. Barn. BU25 7
Boileau Rd. SW13 BP43 49
Boileau Rd. W5 BL39 39
Bolden St. SE8 CE44 52
Bolderwood Way, CE55 79
W. Wick. CE55 79
Boldmere Rd. Pnr. BD33 20
Boleyn Av. Enf. CB23 8
Boleyn Av. Epsom BP58 76
Boleyn Dr. E. Mol. BE52 65
Boleyn Dr. Ruis. BD34 20
Boleyn Gdns. Dag. CS36 37
Boleyn Gdns. W. Wick. CE55 79
Boleyn Gro. W. Wick. CE55 79
Boleyn Rd. E6 CJ37 35
Boleyn Rd. E7 CH36 35
Boleyn Rd. N16 CA35 33
Boleyn Way, Ilf. CM29 17
Bolina Rd. SE16 CC42 43
Bolingbroke Gro. SW11 BU46 50
Bolingbroke Rd. W14 BQ41 40
Bolingbroke Wk. SW11 BU44 50
Bollo Bridge Rd. W3 BM41 39
Bollo La. W3 BM41 39
Bollo La. W4 BM41 39
Bolmer Wk. Rain. CV37 37
Bolney St. SW8 BX43 51
Bolsover St. W1 BV38 32
Bolstead La. Mitch. BV51 68
Bolstead Pl. Mitch. BV51 68
Bolster Gro. N22 BW29 14
Bolt Ct. EC4 BY39 42
Gough Sq.
Bolter's La. Bans. BR60 82
Boltimore Clo. NW4 BQ31 22
Bolton Clo. Chess. BL57 75
Bolton Cres. SE5 BY43 51
Bolton Gdns. NW10 BQ37 31
Bolton Gdns. SW5 BS42 41
Bolton Gdns. Brom. CG50 61
Bolton Gdns. Tedd. BJ50 57
Bolton Gdns. Ms. SW10 BT42 41
Bolton Rd. E15 CG36 34
Bolton Rd. N18 CA28 15
Bolton Rd. NW8 BS37 32
Bolton Rd. NW10 BO37 31
Bolton Rd. W4 BN43 49
Bolton Rd. Chess. BK57 75
Bolton Rd. Har. BG31 20
Boltons, The SW10 BT42 41
Boltons, The, Wem. BH35 30
Bolton St. W1 BV40 41
Bolton Wk. N7 BX34 24
Andover Est.
Bolwell St. SE11 BX42 42
Boman Clo. Felt. BD46 47
Bombay St. SE16 CB42 42
Bomore Rd. W11 BQ40 40
Bomore St. W11 BQ40 40
Bonar Pl. Chis. CK50 62
Bonar Rd. SE15 CB43 51
Bonchester Clo. Chis. CL51 71
Bonchurch Clo. Sutt. BS57 77
Ventnor Rd.
Bonchurch Rd. W10 BR39 40
Bonchurch Rd. W13 BJ40 39
Bond Ct. EC3 BZ39 42
Wallbrook
Bondfield Wk. Dart. CW45 55
Bond Gdns. Wall. BW56 77
Bond Rd. Mitch. BU51 68
Bond Rd. Surb. BL55 75
Bond St. E15 CG35 34
Bond St. W4 BO42 40
Chiswick Common Rd.
Bond St. W5 BK40 39
Bondway SW8 BX43 51
Boneta Rd. SE18 CK41 44
Bonfield Av. Hayes BC38 29
Bonfield Rd. SE13 CF45 52
Bonham Gdns. Dag. CP34 27
Bonham Rd. SW2 BX46 51
Bonham Rd. Dag. CP34 27
Bonheur Rd. W4 BN41 40
Bonhill St. EC2 BZ38 33
Boniface Gdns. Har. BF29 11
Boniface Wk. Har. BF29 11
Bonmarche Ter. Ms. SE19 CA49 60
Gipsy Rd.
Bonner Hill Rd. BL51 66
Kings. On T.
Bonner Rd. E2 CC37 34
Bonnersfield Clo. Har. BH32 21
Bonnersfield La. Har. BH32 21
Bonnersfield La. Har. BJ32 21
Bonner St. E2 CC37 34
Bonneville Gdns. SW4 BW46 50
Bonney Way, Swan. CT51 73
Bonnington Rd. Horn. CV36 37
Bonnington Sq. SW8 BX43 51
Bonny St. NW1 BW36 32
Bonser Rd. Twick. BH48 57
Bonsor St. SE5 CA43 51
Bonville Rd. Brom. CG49 61
Booker Rd. N18 CB28 15
Boones Rd. SE13 CG45 52
Boone St. SE13 CG45 52
Boord St. SE10 CG41 43
Boothby Rd. N19 BW34 23
Booth Clo. SE28 CO40 45
Booth Rd. NW9 BN30 13
Booth Rd. Croy. BY55 78
Bourne St.
Booths Pl. W1 BW39 41
Wells St.
Boot St. N1 CA38 33
Bordars Rd. W7 BH39 39
Bordars Wk. W7 BH39 39
Borden Av. Enf. BZ25 8
Border Cres. SE26 CB49 60
Border Gdns. Croy. CE56 79
Bordergate, Mitch. BU51 68
Bordergate Est. Mitch. BU51 68
Border Rd. SE26 CB49 60
Border's La. Loug. CL24 10
Bordesley Rd. Mord. BS53 68
Bordon Wk. SW15 BP47 58
Boreas Walk N1 BY37 33
Nelson Pl.
Boreham Av. E16 CH39 44
Boreham Clo. E11 CF33 25
Hainault Rd.
Boreham Holt, Borwd. BL24 5
Boreham Rd. N22 BZ30 15
Boreham St. E2 CA31 24
Roda St.
Borer's Pass. EC2 CA39 42
Borgard Rd. SE18 CK42 44
Borkwood Pk. Orp. CN56 81
Borkwood Way, Orp. CM56 80
Borland Rd. SE15 CC46 52
Borland Rd. Tedd. BJ50 57
Borneo St. SW15 BQ45 49
Borough High St. SE1 BZ41 42
Borough Hill, Croy. BY55 78
Borough Rd. SE1 BY41 42
Borough Rd. Islw. BH44 48
Borough Rd. Kings. On T. BM51 66
Borough Rd. Mitch. BU51 68
Borrodaile Rd. SW18 BS46 50
Borrowdale Av. Har. BJ30 12
Borrowdale Clo. Ilf. CK31 26
Borrowdale Clo. CA60 84
Sth. Croy.
Borrowdale Dri. CA59 84
Sth. Croy.
Borthwick Rd. E15 CG35 34
Borthwick Rd. NW9 BO32 22
Broadway, The
Borthwick St. SE8 CE42 43
Borwick Av. E17 CD31 25
Bosbury Rd. SE6 CF48 61
Boscastle Rd. NW5 BV34 23
Boscobel Pl. SW1 BV42 41
Boscobel St. NW8 BT38 32
Boscombe Av. E10 CF33 25
Boscombe Av. Horn. CV33 28
Boscombe Gdns. SW16 BX50 60
Boscombe Rd. SW17 BV50 59
Boscombe Rd. SW19 BS51 68
Boscombe Rd. W12 BP40 40
Boscombe Rd. Wor. Pk. BQ54 67
Bosgrove E4 CF26 9
Boss Rd. E3 CE36 34
Rothbury Rd.
Boss St. SE1 CA41 42
Bostall Hill SE2 CO42 45
Bostall Hill Rd. SE2 CP42 45
Bostall La. SE2 CO42 45
Bostall Manor Way SE2 CO42 45
Bostall Pk. Av. Bexh. CQ43 54
Bostal Rd. Orp. CO50 63
Bostal Row, Bexh. CQ45 54
Boston Gdns. Brent. BJ42 39
Boston Gdns. W4 BO43 49
Boston Manor Rd. Brent. BJ42 39
Boston Pk. Rd. Brent. BK43 48
Boston Pl. NW1 BU38 32
Boston Rd. E6 CK38 35
Boston Rd. E17 CE32 25
Boundary Rd.
Boston Rd. W7 BH40 39
Boston Rd. Croy. BX53 69
Boston Rd. Edg. BN29 13
Boston St. E2 CB37 33
Bostonthorpe Rd. W7 BH41 39
Boston Vale W7 BJ42 39
Boswell Ct. WC1 BX39 42
Boswell Rd. Th. Hth. BZ52 69
Boswell St. WC1 BX39 42
Bosworth Clo. E17 CD30 16
Bosworth Rd. N11 BW29 14
Bosworth Rd. W10 BR38 31
Bosworth Rd. Barn. BS24 7
Bosworth Rd. Dag. CR34 27
Botany Bay La. Chis. CM51 71
Boteley Clo. E4 CF27 16
Botha Rd. E13 CH39 44
Bothwell Clo. E16 CH39 44
Bothwell Rd. Croy. CF58 79
Bothwell St. W6 BQ43 49
Delorme St.
Botolph La. EC3 BZ40 42
Botolph Pass. E3 CE38 34
Botolph Rd.
Botolph Rd. E3 CE38 34
Botsford Rd. SW20 BR51 67
Bott Rd. Dart. CW49 64
Botts Ms. W2 BS39 41
Chepstow Rd.
Boucher Clo. Tedd. BH49 57
Bouchier Wk. Rain. CU36 37
Boughton Av. Brom. CG54 70
Boughton Rd. SE18 CN41 45
Boughton Rd. SE28 CN41 45
Boulcott St. E1 CC39 43
Boulevard, The, Pnr. BE31 20
Boulogne Rd. Croy. BZ53 69
Boulter Gdns. Horn. CU36 37
Boulton Rd. Dag. CQ34 27
Bounces La. N9 CB27 15
Bounces Rd. N9 CB26 8
Boundaries Rd. SW12 BU48 59
Boundaries Rd. Felt. BD47 56
Boundary Clo. CM35 35
Loxford La.
Boundary Clo. Ilf. CN35 36
Loxford La.
Boundary Clo. BM52 66
Kings. On T.
Boundary Clo. Sthl. BF42 38
Boundary La. E13 CJ38 35
Boundary La. SE17 BZ43 51
Boundary Rd. E13 CJ37 35
Boundary Rd. E17 CD33 25
Boundary Rd. N9 CC25 9
Boundary Rd. N22 BY31 24
Boundary Rd. NW8 BS37 32
Boundary Rd. SW19 BT50 59
Boundary Rd. Bark. CM37 35
Boundary Rd. Pnr. BD33 20
Boundary Rd. Rom. CU32 28
Boundary Rd. Sid. CN46 54
Boundary Rd. Wall. BV57 77
Boundary Rd. S. Wall. BV58 77
Boundary Row SE1 BY41 42
Boundary St. E2 CA38 33
Boundary St. Erith CT43 55
Boundary Way, Croy. CE56 79
Boundfield Rd. SE6 CG48 61
Bounds Green Ind. Est. BW29 14
N11
Bounds Green Rd. N11 BW29 14
Bounds Green Rd. N22 BW29 14
Bourchier St. W1 BW40 41
Wardour St.
Bourdon Pl. W1 BV40 41
Bourdon St.
Bourdon Rd. SE20 CC51 70
Bourdon St. W1 BV40 41
Bourke Clo. NW10 BO36 31
Bourke Clo. SW2 BX47 60
Bourke Hill, Couls. BU62 83
Bourlet Clo. W1 BW39 41
Wells St.
Bourn Av. N15 BZ31 24
Bourn Av. Barn. BT25 7
Bournbrook Rd. SE3 CJ45 53
Bourne Av. N14 BX27 15
Bourne Av. Ruis. BD35 29
Bournebridge Gdns. Ilf. CO29 18
Bourne Clo. SW2 BX46 51
Bourne Ct. Ruis. BC35 29
Bourne Est. EC1 BY39 42
Bournefield Rd. Whyt. CA62 84
Bourne Gdns. E4 CE28 16
Bournehall Av. Bush. BF25 4
Bournehall La. Bush. BF25 4
Bournehall Rd. Bush. BF25 4
Bourne Hill N13 BX27 15
Bourne Mead, Bex. CS46 55
Bournemead Av. Nthlt. BC37 29
Bournemead Way, Nthlt. BC37 29
Bournemouth Rd. SE15 CB44 51
Bournemouth Rd. SW19 BS51 68
Bourne Pk. Gdns. Ken. CA57 78
Bourne Pk. Gdns. Ken. CA61 84
Bourne Pl. W4 BN42 40
Dukes Av.
Bourne Rd. E7 CG34 25
Bourne Rd. N8 BX32 24
Bourne Rd. Bex. CR47 63
Bourne Rd. Brom. CJ52 71
Bourne Rd. Bush. BF25 4
Bourne St. SW1 BV42 41
Bourne St. Croy. BY55 78
Bourne Ter. W2 BS29 41
Bourne, The N14 BW26 7
Bourne Vale, Brom. CG54 70
Bournevale Rd. SW16 BX49 60
Bourne Vw. Grnf. BH36 30
Bourne Vw. Ken. BZ61 84
Bourne Way, Brom. CG55 79
Bourne Way, Epsom BN56 76
Bourne Way, Sutt. BR56 76
Bourne Way, Swan. CS52 73
Bournewood Rd. SE18 CO43 54
Bournewood Rd. Orp. CO54 72
Bournville Rd. SE6 CE47 61
Bournwell Clo. Barn. BU23 7
Bousfield Rd. SE14 CC44 52
Boutflower Rd. SW11 BU45 50
Bouverie Gdns. Har. BK33 21
Bouverie Pl. W2 BT39 41
Bouverie Rd. N16 CA33 24
Bouverie Rd. Couls. BV62 83
Bouverie Rd. Har. BG32 20
Bouverie St. EC4 BY39 42
Bovay Pl. N7 BX35 33
Holloway Rd.
Bovay St. N7 BX35 33
Boveney Rd. SE23 CC47 61
Bovill Rd. SE23 CC47 61
Bovingdon Av. Wem. BM36 30
Bovingdon Clo. N19 BW34 23
Bovingdon La. NW9 BO29 13
Bovingdon Rd. SW6 BS44 50
Bovingdon Sq. Mitch. BX52 69
Bowater Clo. NW9 BN32 22
Bowater Clo. NW9 BO32 22
Buck La.
Bowater Pl. SE3 CH43 53
Bowater Rd. SE18 CJ41 44
Bow Bridge Est. E3 CE38 34
Bow Common La. E3 CD38 34
Bowden Dr. Horn. CW33 28
Bowden Rd. E17 CE33 25
Bowden St. SE11 BY42 42
Bowditch SE8 CD42 43
Bowdon Rd. E17 CE33 25
Bowen Dr. SE21 CA48 60
Bowen Rd. Har. BG33 20
Bowen St. E14 CE39 43
Bowens Wood, Croy. CD58 79
Bower Av. SE10 CG44 52
Bower Clo. Nthlt. BD37 29
Bower Clo. Rom. CS29 19
Bowerdeans St. SW6 BS44 50
Bowerdean St. SW6 BS44 50
Bower Fm. Rd. Hay. CS27 19
Bowerman Av. SE14 CD43 52
Bower Rd. Swan. CU50 64
Bower St. E1 CC39 43
Bowes Clo. Sid. CO46 54
Bowes Rd. N11 BV28 14
Bowes Rd. N13 BV28 14
Bowes Rd. W3 BO40 40
Bowes Rd. Dag. CP35 36
Bowes Rd. Walt. BC55 74
Bowfell Rd. W6 BQ43 49
Bowford Av. Bexh. CQ44 54
Bowie Clo. SW4 BW47 59
Plummer Rd.
Bow La. EC4 BZ39 42
Bow La. N12 BT29 14
Bowland Rd. IG8 CJ28 17
Bowles Rd. SE1 CB43 51
Bowley Clo. SE19 CA49 60
Bowley St. E14 CD40 43
Bowling Green Clo. SW15 BP47 58
Bowling Green La. EC1 BY38 33
Bowling Green St. SE11 BY43 51
Bowling Green Wk. N1 CA38 33
Bowls Clo. Stan. BJ28 12
Bowls, The Chig. CN27 18
Bowman Av. E16 CG40 43
Bowmans Clo. W13 BJ40 39
Bowmans Meadow, Wall. BV55 77
Bowman's Rd. Dart. CT47 64
Bowmead SE9 CK48 62
Bowness Cres. SW15 BO49 58
Bowness Dr. Houns. BE45 47
Bowness Rd. SE6 CE47 61
Bowness Rd. Bexh. CR44 54
Bowness Way, Horn. CU35 37
Bowood Rd. SW11 BV45 50
Bowood Rd. Enf. CC23 9
Bowring Grn. Wat. BD28 11
Bow Rd. E3 CD38 34
Bowrons Av. Wem. BK36 30
Bow St. E15 CG35 34
Bow St. WC2 BX39 42
Boxall Rd. SE21 CA46 51
Boxall Rd. Dag. CQ35 36
Boxgrove Rd. SE2 CO41 45
Box La. Bark. CO37 36
Boxley Rd. Mord. BT52 68

Boxley St. E16 CH40 44
Boxmoor Rd. Har. BJ31 21
Boxmoor Rd. Rom. CS28 19
Box Ridge Av. Pur. BX59 84
Boxted Clo. Buck. H. CK26 10
Boxtree La. Har. BG30 11
Boxtree Rd. Har. BG29 11
Box Tree Wk. Orp. CP54 72
Boyard Rd. SE18 CL42 44
Boyce Clo. Borwd. BL23 5
Boyce St. SE1 BX40 42
Mepham St.
Boyce Way, E13 CH38 35
Boycroft Av. NW9 BN32 22
Boyd Av. Sthl. BE40 38
Boydell Ct. NW8 BT36 32
Boyd Rd. SW19 BT50 59
Boyd St. E1 CB39 42
Boyfield St. SE1 BY41 42
Boyland Rd. Brom. CG49 61
Boyle Av. NW4 BO31 22
Boyle Av. Stan. BJ29 12
Boyle Frn. Rd. Surb. BJ53 66
Boyle St. W1 BW40 41
Savile Row
Boyne Rd. SE13 CF45 52
Boyne Rd. Dag. CR34 27
Boyne Ter. Ms. W11 BR40 40
Boyseland Ct. Edg. BN27 13
Boyson Rd. SE17 BZ43 51
Boythorn Way SE16 CB42 42
Bonamy Estate East, The
Boyton Clo. N8 BX31 24
Boyton Rd. N8 BX31 24
Brabant Ct. EC3 BZ40 42
Philpot La.
Brabant Rd. N22 BX30 15
Brabazon Av. Wall. BX57 78
Brabazon Rd. Houns. BD43 47
Brabazon Rd. Nthlt. BF37 29
Brabazon St. E14 CE39 43
Brabourne Cres. Bexh. CQ43 54
Brabourne Ri. Beck. CF53 70
Brabourn Rd. SE15 CC44 52
Bracewell Av. Grnf. BH35 30
Bracewell Rd. W10 BQ39 40
Bracewood Gdns. Croy. CA55 78
Bracey St. N4 BX33 24
Bracken Av. SE12 BV46 50
Bracken Av. Croy. CE55 79
Bracken Bri. Dr. Ruis. BD34 20
Brackenbury Gdns. W6 BP41 40
Brackenbury Rd. N3 BT31 23
Brackenbury Rd. W6 BP41 40
Bracken Clo. Houns. BF47 56
Brackendale N21 BX26 8
Brackendale Clo. Brent. BF44 48
Brackendale Ct. Beck. CE50 61
Bracken Dene, Dart. CT49 64
Bracken Dr. Chig. CL29 17
Bracken End, Islw. BG45 47
Brackenfield Ho. N16 CB34 24
Downs Rd.
Bracken Gdns. SW13 BP44 49
Bracken Hill, Ruis. BE35 29
Bracken Hill Clo. Brom. CG51 70
Bracken Hill La. Brom. CG51 70
Brackens, The, Enf. CA26 8
Brackens, The, Orp. CO56 81
Bracken, The E4 CF26 9
Hortus Rd.
Brackenwood, Sun. BC51 65
Brackley Rd. W4 BO42 40
Brackley Rd. Beck. CD50 61
Brackley Sq. Wdf. Grn. CJ29 17
Brackley St. EC1 BZ38 33
Viscount St.
Brackley Ter. W4 BO42 40
Bracklyn Ct. N1 BZ37 33
Bracknell Gdns. NW3 BS35 32
Bracknell Gte. NW3 BS35 32
Brackwell Clo. N22 BY30 15
Bracondale, Esher BG57 74
Bracondale Rd. SE2 CO42 45
Bradbourne Rd. Bex. CR47 63
Bradbourne St. SW6 BS44 50
Bradbury Rd. Sthl. BE42 38
Blandford Rd.
Bradbury St. N16 CA35 33
Braddon Rd. Rich. BL45 48
Braddyll St. SE10 CG42 43
Bradenham Av. Well. CO45 54
Bradenham Rd. Har. BJ31 21
Braden St. W9 BS38 32
Bradfield Dr. Bark. CO35 36
Bradfield Rd. E16 CH41 44
Bradfield Rd. Ruis. BE35 29
Bradfields Av. Edg. BM28 12
Bradford Clo. SE26 CB49 60
Wells Park Rd.
Bradford Rd. Brom. CK54 71
Bradford Dr. Epsom BO57 76
Bradford Rd. W3 BO41 40
Warple Way
Bradford Rd. Ilf. CM33 26
Bradgate Rd. SE6 CE46 52
Brading Cres. E11 CH34 26
Brading Rd. SW2 BX47 60

Brading Rd. Croy. BX53 69
Bradiston Rd. W9 BR38 31
Bradlaugh St. N1 CA37 33
Bradley Clo. N7 BX36 33
Sutterton St.
Bradley Gdns. W13 BJ39 39
Bradley Rd. N22 BX30 15
Station Rd.
Bradley Rd. SE19 BZ50 60
Bradmead SW8 BV43 50
Bradmore La. W6
Beadon Rd.
Bradmore Pk. Rd. W6 BP41 40
Bradmore Way, Couls. BX62 84
Bradshaw Rd. Wat. BD23 4
Bradstock Rd. E9 CC36 34
Bradstock Rd. Epsom BP57 76
Bradstock Rd. Est. E9 CC36 34
Brad St. SE1 BY40 42
Bradwell Clo. E18 CG31 25
Bradwell Rd. Buck. H. CK26 10
Brady Av. Loug. CM23 10
Brady St. E1 CB38 33
Brae Clo. Belv. CQ42 45
Brae Ct. Kings. On T. BM51 66
Wolverton Av.
Braefoot Ct. SW15 BQ46 49
Putney Hill
Braemar Av. N22 BX30 15
Braemar Av. NW10 BN34 22
Braemar Av. SW19 BS48 59
Braemar Av. Bexh. CS45 55
Braemar Av. Sth. Croy. BZ58 78
Braemar Av. Th. Hth. BY52 69
Braemar Av. Wem. BK36 30
Braemar Gdns. NW9 BN30 13
Braemar Gdns. Horn. CW32 28
Braemar Gdns. Sid. CM48 62
Braemar Gdns. W. Wick. CF54 70
Braemar Rd. E13 CG38 34
Braemar Rd. N15 CA32 24
Braemar Rd. Brent. BK43 48
Braemar Rd. Wor. Pk. BP55 76
Braeside, Beck. CE49 61
Braeside Av. SW19 BR51 67
Braeside Clo. Pnr. BF29 11
Braeside Cres. Bexh. CS45 55
Braeside Rd. SW16 BW50 59
Braes St. N1 BY36 33
Brafton Rd. Croy. BZ56 78
Braganza St. SE17 BY42 42
Braham St. E1 CA39 42
Braid Av. W3 BN39 40
Braid Clo. Felt. BE48 56
Braidwood Rd. SE6 CF47 61
Braidwood St. SE1 CA40 42
Brailsford Rd. SW2 BY46 51
Brainton Av. Felt. BC47 56
Braintree Av. Ilf. CK32 26
Braintree Rd. Dag. CR34 27
Braintree Rd. Ruis. BE35 29
Braintree St. E2 CC38 34
Braithwaite Av. Rom. CR33 27
Braithwaite Gdns. Stan. BK30 12
Bramah Rd. SW9 BY44 51
Bramall Clo. E15 CG35 34
Bramber Rd. N12 BU28 14
Bramber Rd. W14 BR43 49
Bramblebury Rd. SE18 CM42 44
Bramble Clo. Croy. CE56 79
Bramble Cft. Erith CS42 46
Brambledown Clo.
W. Wick. CG53 70
Brambledown Rd. Cars. BV57 77
Brambledown Rd.
Sth. Croy. BZ57 78
Bramble Clo. Stan. BK30 12
Bramble Clo. W12 BO40 40
Wallflower St.
Bramble La. SW16 BX52 69
Brambles Clo. Brent. BJ43 48
Brambles, The, Chig. CM29 17
Bramblewood Clo. Cars. BU54 68
Bramblings, The E4 CF28 16
Bramcote Av. Mitch. BU52 68
Bramcote Gro. SE16 CC42 43
Bramcote Rd. SW15 BP45 49
Bramdean Cres. SE12 CH47 62
Bramdean Gdns. SE12 CH47 62
Bramerton Rd. Beck. CD52 70
Bramerton St. SW3 BU43 50
Kings Crescent Est.
Bramfield Rd. SW11 BU46 50
Bramford Ct. N14 BW27 14
Bramford Rd. SW18 BT45 50
Bramham Gdns. SW5 BS42 41
Bramham Gdns. Chess. BK56 75
Bramhope La. SE7 CH43 53
Bramlands Clo. SW11 BU45 50
Bramley Av. Couls. BW61 83
Bramley Clo. E17 CD30 16
Bramley Clo. N14 BV25 7
Bramley Clo. Orp. CL54 71
Bramley Clo. Sth. Croy. BY56 78
Bramley Clo. Swan. CT52 73
Bramley Clo. Twick. BG46 47

Bramley Ct. Well. CO44 54
Bramley Cres. Ilf. CL32 26
Bramley Gdns. Wat. BD28 11
Bramley Hill, Sth. Croy. BY56 78
Bramley Pl. Dart. CU45 55
Bramley Rd. N14 BV25 7
Bramley Rd. W5 BK41 39
Bramley Rd. W10 BQ40 40
Bramley Rd. W12 BQ40 40
Bramley Rd. Sutt. BD58 76
Bramley Rd. Sutt. BT56 77
Bramley St. W10 BQ39 40
Bramley Way, W. Wick. CE55 79
Brampton Clo. E5 CB34 24
Comberton Rd.
Brampton Gdns. N15 BZ32 24
Brampton Rd.
Brampton Gdns. Walt. BD56 74
Brampton Gro. NW4 BP31 22
Brampton Gro. Har. BJ31 21
Brampton Gro. Wem. BL33 21
Brampton La. NW4 BO31 22
Brampton Gro.
Brampton Pk. Rd. N8 BY31 24
High Rd.
Brampton Rd. E6 CJ38 35
Brampton Rd. N15 BZ32 24
Brampton Rd. NW9 BM31 21
Brampton Rd. SE2 CP43 54
Brampton Rd. Bexh. CP43 54
Brampton Rd. Croy. CA53 69
Brampton Rd. Wat. BC27 11
Bramsham Gdns. Wat. BD28 11
Bramshaw Rise, N. Mal. BO53 67
Bramshaw Rd. E9 CC36 34
Bramshill Clo. Chig. CN28 18
Tine Rd.
Bramshill Gdns. NW5 BV34 23
Bramshill Rd. NW10 BO37 31
Bramshot Av. SE7 CH43 53
Bramshot Way, Wat. BC27 11
Bramston Rd. NW10 BP37 31
Bramwell Clo. Sun. BD51 65
Brancaster La. Pur. BZ58 78
Brancaster Rd. E12 CK35 35
Brancaster Rd. SW16 BX48 60
Brancaster Rd. Ilf. CM32 26
Brancepeth Gdns. CH27 17
Buck. H.
Branch Hl. NW3 BT34 23
Branch Hl. Lo. NW3 BS34 23
Branch Pl. N1 BZ37 33
Branch Rd. E14 CD40 43
Branch Rd. Ilf. CO28 18
Brancker Clo. Wall. BX57 78
Brandlehow Rd. SW15 BR45 49
Brandon Est. SE17 BY43 51
Brandon Gro. Ilf. CL34 26
Brandon Rd. E17 CF31 25
Brandon Rd. N7 BX36 33
Brandon Rd. Sthl. BE42 38
Brandon St. SE17 BZ42 42
Brandram Rd. SE13 CG45 52
Brandreth Rd. SW17 BV48 59
Brandries, The, Wall. BW55 77
Brand St. SE10 CF43 52
Brandville Gdns. Ilf. CL31 26
Brandy Way, Sutt. BS57 77
Brangbourne Rd. Brom. CF49 61
Brangton Rd. SE11 BX42 42
Loughborough St.
Brangwyn Cres. SW19 BT51 68
Branksea St. SW6 BR43 49
Branksome Av. N18 CA29 15
Branksome Clo. Walt. BD54 65
Branksome Rd. SW2 BX45 51
Branksome Rd. SW19 BS51 68
Branksome Way, Har. BL32 21
Branksome Way, N. Mal. BN51 67
Bransby Rd. Chess. BL57 75
Branscombe Gdns. N21 BY26 8
Branscombe St. SE13 CE45 52
Bransell Clo. Swan. CS53 73
Bransgrove Rd. Edg. BL30 12
Branston Cres. Orp. CM54 71
Branstone Rd. Rich. BL44 48
Branstone St. W10 BQ38 31
Brants Wk. W7 BH38 30
Brantwood Av. Erith CS44 55
Brantwood Clo. E17 CE31 25
Brantwood Gdns. Enf. BX24 8
Brantwood Gdns. Ilf. CK31 26
Brantwood Rd. N17 CA29 15
Brantwood Rd. SE24 BZ46 51
Brantwood Rd. Bexh. CR44 54
Brantwood Rd. Islw. BJ45 48
Brantwood Rd. BZ58 78
Sth. Croy.
Brassey Sq. SW11 BV45 50
Ashbury Rd.
Brassie Av. W3 BO39 40
Brasted Clo. SE26 CC49 61
Brasted Clo. Bexh. CP46 54
Brasted Rd. Erith CT43 55
Brathway Rd. SW18 BS47 59

Bratley St. E1 CB38 33
Weaver St.
Bratten Ct. Croy. BZ53 69
Arundel Rd.
Braund Av. Grnf. BF38 29
Braundton Av. Sid. CN47 63
Bravington Rd. W9 BR38 31
Brawell Ms. N18 CB28 15
Lyndhurst Rd.
Braxfield Rd. SE4 CD45 52
Braxted Pk. SW16 BX50 60
Brayard's Rd. SE15 CB44 51
Braybrooke Gdns. SE19 CA50 60
Fox Hill
Braybrooke St. W12 BO39 40
Brayburne Av. SW4 BW44 50
Braycourt Av. Walt. BC54 65
Braydon Rd. N16 CA33 24
Brayfield Ter. N1 BY36 33
Lofting Rd.
Bray Pass. E16 CG40 43
Bray Dr.
Bray Pl. SW3 BU42 41
Bray Rd. E16 CG40 43
Bray Rd. NW7 BQ29 13
Brayton Gdns. Enf. BW24 7
Braywood Rd. SE9 CM45 53
Breach La. Dag. CR38 36
Bread St. EC4 BZ39 42
Breakfield, Couls. CD45 52
Breakspears Rd. SE4 CD45 52
Breakspears Rd. Croy. CO51 72
Bream Gdns. E6 CL38 35
Breamore Ct. Ilf. CO34 27
Breamore Rd. Ilf. CN34 27
Breams Bldgs. EC4 BY39 42
Bream St. E3 CE36 34
Breamwater Gdns. Rich. BJ48 57
Breasley Clo. SW15 BP45 49
Brechin Pl. SW7 BT42 41
Brecknock Rd. N7 BW35 32
Brecknock Rd. N19 BW35 32
Brecknock Rd. Est. N7 BW35 32
Brecon Clo. Mitch. BX52 69
Brecon Rd. W6 BR43 49
Brecon Rd. Enf. CC24 9
Brede Clo. E6 CL38 35
Bredgar Rd. N19 BW34 23
Bredhurst Clo. SE20 CC50 61
Bredon Rd. SE5 BZ45 51
Bredon Rd. Croy. CA54 69
Bredune, Ken. BZ61 84
Church Rd.
Breer St. SW6 BS45 50
Breezers St. E1 CB40 42
Pennington St.
Brember Rd. Har. BG57 74
Bremner Rd. SW7 BT41 41
Queen's Gte.
Brenchley Clo. Brom. CG53 70
Brenchley Clo. Chis. CL51 71
Brenchley Gdns. SE23 CC46 52
Brenchley Rd. Orp. CN51 72
Brendans Clo. Horn. CW33 28
Brenda Rd. SW17 BU48 59
Brende Gdns. E. Mol. BF52 65
Brendon Av. NW10 BO35 31
Brendon Clo. Erith CT44 55
Brendon Dr. Esher BG57 74
Brendon Gdns. Har. BF35 29
Brendon Gdns. Ilf. CN32 27
Brendon Rd. SE9 CM48 62
Brendon Rd. Dag. CQ33 27
Brendon Row St. W1 BU39 41
Brendon Way, Enf. CA26 8
Brenley Gdns. SE9 CJ45 53
Brent Clo. Bex. CQ47 63
Brentcot Clo. W13 BJ38 30
Brent Ct. NW11 BQ33 22
Highfield Av.
Brent Cross NW10 BL37 30
Brentfield NW10 BM36 30
Brentfield Clo. NW10 BM36 31
Brentfield Gdns. NW11 BQ33 22
Brentfield Rd. NW10 BM36 31
Brent Grn. NW4 BO32 22
Brentham Halt. Rd. W5 BL38 30
Brentham Way W5 BK38 30
Brenthouse St. E9 CC36 34
Brenthurst Rd. NW10 BO36 31
Brentlands Dr. Dart. CW47 64
Brent La. Dart. CW47 64
Brent Lea, Brent. BK43 48
Brentmead Clo. W7 BH40 39
Brentmead Gdns. NW10 BL37 30
Brentmead Pl. NW11 BQ32 22
Brenton St. E14 CD39 43
Brent Pk. Rd. NW4 BP33 22
Brent Pl. Barn. BR25 6
Brent Pl. E16 CH39 44
Brent Rd. SE18 CL43 53
Brent Rd. Brent. BK43 48
Brent Rd. Sth. Croy. CB58 78

Brent Rd. Sthl. BD41 38
Brent Side, Brent. BK43 48
Brentside Clo. W13 BJ38 30
Brent St. NW4 BQ31 22
Brent Ter. NW2 BQ33 22
Brentvale Av. Sthl. BG40 38
Brentvale Av. Wem. BL37 30
Brent Vw. Rd. NW4 BP32 22
Brent Way N3 BS29 14
Brent Way, Brent. BK43 48
Brent Way, Wem. BM36 30
Brentwick Gdns. Brent. BL42 39
Brentwood Av. Islw. BJ45 48
Brentwood Clo. SE9 CM47 62
Brentwood Rd. Rom. CT32 28
Brereton Rd. N17 CA29 15
Bressenden Pl. SW1 BV41 41
Bressey Gro. E18 CG30 16
Brett Clo. Nthlt. BD38 29
Broomcroft Av.
Brett Cres. NW10 BN36 31
Brettell St. SE17 BZ42 42
Merrow St.
Brettenham Av. E17 CE30 16
Brettenham Rd. E17 CD30 16
Brettenham Rd. N18 CB28 15
Brett Gdns. Dag. CQ36 36
Brettgrave, Epsom BN58 76
Brett Rd. E8 CB36 33
Brett Rd. NW10 BO37 31
Brett Rd. Barn. BQ25 6
Brewer Pl. SE18 CL42 44
Charles Grindling Wk.
Brewers La. Rich. BK46 48
Brewer St. W1 BW40 41
Brewery La. Twick. BJ47 57
London Rd.
Brewery Rd. N7 BX36 33
Brewery Rd. SE18 CM42 44
Brewery Rd. Brom. CK54 71
Brewery Sq. Twick. BH47 57
Brewhouse La. E1 CB40 42
Wapping La.
Brewhouse Rd. SE18 CK42 44
Red Bks. Rd.
Brewhouse St. SW15 BR45 49
Brewood Rd. Dag. CO36 36
Brewster Gdns. W10 BQ39 40
Brewster Rd. E10 CE33 25
Brian Av. Sth. Croy. CA59 84
Brian Clo. Horn. CU35 37
Brian Ct. N10 BV30 14
Briane Rd. Epsom BN58 76
Brian Rd. Rom. CP32 27
Briants Clo. Pnr. BE30 11
Briant St. SE4 CC44 52
Briar Av. SW16 BX50 60
Briarbank Rd. W13 BJ39 39
Briar Banks, Cars. BV58 77
Briar Clo. N13 BZ27 15
Briar Clo. Buck. H. CJ27 17
Briar Clo. Hamptn. BE49 56
Briar Clo. Islw. BH46 48
Briars Clo. N2 BS31 23
Briar Cres. Nthlt. BF36 29
Briardale Gdns. NW3 BS34 23
Briarfield Av. N3 BS30 14
Briar Gdns. Brom. CG54 70
Briar Gro. Sth. Croy. CB60 84
Briar Hill, Pur. BX59 84
Briar La. Cars. BV58 77
Briar La. Croy. CE56 79
Briar Pass. SW16 BX52 69
Briar Pl. SW16 BX52 69
Briar Rd. NW2 BQ35 31
Briar Rd. SW16 BX52 69
Briar Rd. Bex. CS48 64
Briar Rd. Har. BK32 21
Briar Rd. Rom. CV29 19
Briar Rd. Twick. BH47 57
Briars Clo N17 CB29 15
Briars Clo. N17 CB29 15
Park La.
Briars Wk. Rom. CW30 19
Briar Wk. SW15 BP45 49
Briar Wk. Edg. BN29 13
Briarswood Way, Orp. CN56 81
Briarwood Clo. NW9 BN32 22
Briarwood Dr. Nthwd. BC30 11
Briarwood Rd. SW4 BW46 50
Briarwood Rd. Epsom BP57 76
Briary Clo. NW3 BT36 32
Chalcotts Est.
Briary Ct. Sid. CO49 63
Briary Gdns. Brom. CH49 62
Briary La. N9 CA27 15
Brickbarn Clo. SW10 BT43 50
Edith Gro.
Brick Ct. EC4 BY39 42
Middle Temple La.
Brick Farm Clo. Rich. BM44 48
Brickfield Clo. Brent. BK43 48
Brickfield Cotts. SE18 CN43 54
Brickfield La. Barn. BO25 6
Brickfield Rd. E3 CE38 34
Brickfield Rd. SW19 BS49 59
Brickfield Rd. Th. Hth. BY51 69
Brickfields, Har. BG34 20
Brick La. E1 CA38 33
Brick La. E2 CA38 33
Brick La. Enf. CB23 8
Brick St. W1 BV40 41
Brickwood Clo. SE26 CB48 60
Kirkdale
Brickwood Rd. Croy. CA55 78
Bride La. EC4 BY39 42
Bride St. N7 BX36 33
Bridewell Pl. EC4 BY39 42
Tudor Rd.
Bridford Ms. W1 BV39 41
Devonshire St.
Bridge App. NW1 BV36 32
Bridge Av. W6 BQ42 40
Bridge Av. W7 BG39 38
Bridge Clo. Enf. CB23 8
Bridge Clo. Rom. CT32 28
Bridge Cotts. Surb. BJ54 66
Portsmouth Rd.
Bridge Ct. E10 CD33 25
Bridge Ct. SW11 BT45 50
Bridge End E17 CF30 16
Bridgefield Clo. Bans. BQ61 82
Bridgefield Rd. Sutt. BS57 77
Bridgefoot SE1 BX42 42
Bridge Gdns. E. Mol. BG52 65
Bridge Gate N21 BZ26 8
Bridgeland Rd. E16 CH40 44
Bridge La. NW11 BR31 22
Bridge La. SW11 BU44 50
Bridgeman Rd. N1 BX36 33
Bridgeman Rd. W4 BN41 40
Bridgeman St. NW8 BU37 32
Bridgend Rd. SW18 BT45 50
Bridgenhall Rd. Enf. CA23 8
Bridgen Rd. Bex. CQ47 63
Bridge Path, Wat. BC23 4
Bridge Pl. SW1 BV42 41
Bridge Pl. SW6 BR45 49
Bridge Pl. Croy. BZ54 69
Bridge Pl. Wat. BD25 4
Bridge Rd. E6 CK36 35
Bridge Rd. E15 CF36 34
Bridge Rd. E17 CD33 25
Bridge Rd. N9 CB27 15
Bridge Rd. N22 BX30 15
Bridge Rd. NW10 BO36 31
Bridge Rd. Beck. CD50 61
Bridge Rd. Bexh. CQ45 54
Bridge Rd. Chess. BL56 75
Bridge Rd. Croy. BZ55 78
Bridge Rd. E. Mol. BG53 65
Bridge Rd. Epsom BO59 82
Bridge Rd. Erith CT44 55
Bridge Rd. Houns. BG44 47
Bridge Rd. Orp. CO53 72
Bridge Rd. Rain. CU38 37
Bridge Rd. Sthl. BE41 38
Bridge Rd. Sutt. BS57 77
Bridge Rd. Twick. BJ46 48
Bridge Rd. Wall. BV56 77
Bridge Rd. Wem. BM34 21
Bridge Row, Croy. BZ54 69
Cross Rd.
Bridges La. Croy. BX56 78
Bridges Pl. SW6 BR44 49
Bridges Rd. SW19 BS50 59
Bridges Rd. Stan. BH28 12
Bridges St. SW1 BX41 42
Bridge St. Pnr. BE31 20
Bridge St. Rich. BK46 48
Bridge Ter. E15 CF36 34
Bridge, The, Har. BH31 21
Bridge Vw. W6 BQ42 40
Bridgewater Clo. Chis. CN52 72
Bridgewater Gdns. Edg. BL30 12
Bridewater Rd. Wem. BK36 30
Bridgewater St. EC1 BZ39 42
Viscount St.
Bridge Way N11 BW27 14
Bridge Way NW11 BR32 22
Bridge Way, Bark. CN36 36
Bridge Way, Twick. BG47 56
Bridgeway, Wem. BL36 30
Bridgeway St. NW1 BW37 32
Bridgewood Clo. SE20 CB50 60
Castledine Rd.
Bridgewood Rd. SW16 BW50 59
Bridgewood Rd. Wor. Pk. BP55 76
Bridgford St. SW18 BT48 59
Bridgman Rd. N1 BX36 33
Bridgman Rd. Tedd. BJ50 57
Bridgwater Clo. Rom. CV28 19
Bridgwater Rd. E15 CF37 34
Bridgwater Rd. Rom. CV28 19
Bridgwater Rd. Ruis. BC35 29
Bridgwater Wk. Rom. CV28 19
Bridgwater Way, Bush. BF26 4
Bridle Clo. Epsom BN56 76
Bridle Clo. Sun. BC52 65
Bridle End, Epsom BO60 82
Bridle La. W1 BW40 41
Bridle Path, Croy. BX55 78
Bridle Path Rd. Barn. BT23 7
Bridle Path, The, Wdf Grn. CG29 16
Bridle Rd. Croy. CE55 79
Bridle Rd. Epsom BO60 82
Bridle Rd. Esher BJ57 75
Bridle Rd. Pnr. BC32 20
Bridle, The, Rd. Pur. BX58 78
Bridle, The, Way, Croy. CD58 79
Bridle, The, Way Wall. BW56 77
Bridle Way, Croy. CE56 79
Bridport Av. Rom. CR32 27
Bridport Pl. N1 BZ37 33
Bridport Rd. N18 CA28 15
Bridport Rd. Grnf. BR37 29
Bridport Rd. Th. Hth. BY52 69
Bristow Pl. W2 BS39 41
Bridwell Pl. EC4 BY39 42
Tudor St.
Brief St. SE5 BY44 51
Brierley, Croy. CE57 79
Brierley Av. N9 CC26 9
Brierley Clo. SE25 CB52 69
Brierley Clo. Horn. CU32 28
Brierley Rd. E11 CF35 34
Brierley Rd. SW12 BW48 59
Brierley St. E2 CC38 34
Royston St.
Brigadier Av. Enf. BZ23 8
Brigadier Hill, Enf. BZ23 8
Brightfield Rd. SE12 CG46 52
Brightling Rd. SE4 CD46 52
Brightlingsea Pl. E14 CD40 43•
Brightman Rd. SW18 BT47 59
Brighton Av. E17 CD32 25
Brighton Dr. Nthlt. BF36 29
Brighton Gro. SE14 CD44 52
Brighton Rd. E6 CL38 35
Brighton Rd. N2 BT30 14
Brighton Rd. N16 CA35 33
Brighton Rd. Couls. BW62 83
Brighton Rd. Pur. BX60 84
Brighton Rd. Surb. BK53 66
Brighton Rd. Tad. BR62 82
Brighton Ter. SW9 BX45 51
Brights Av. Rain. CU38 37
Brightside Rd. SE13 CF46 52
Brightside, The, Enf. CC23 9
Bright St. E14 CE39 43
Brightwell Cres. SW17 BU49 59
Brightwell Rd. Wat. BC25 4
Brig Mews SE8 CE43 52
Watergate St.
Brigstock Rd. Belv. CR42 45
Brigstock Rd. Couls. BV61 83
Brigstock Rd. Th. Hth. BY53 69
Brig St. E14 CF42 43
Brim Hill N2 BT31 23
Brimpsfield Clo. SE2 CO41 45
Brimsdown Av. Enf. CD23 9
Brimstone Clo. Orp. CP57 81
Brindles, The, Bans. BR62 82
Brindley St. SE14 CD44 52
Brindley Way, Sthl. BF40 38
Brindwood Rd. E4 CD27 16
Brinkburn Clo. SE2 CO42 45
Brinkburn Clo. Edg. BM30 12
Brinkburn Gdns. Edg. BM31 21
Brinkley Rd. Wor. Pk. BP55 76
Brinklow Cres. SE18 CL43 53
Brinkworth Rd. Ilf. CK31 26
Brinkworth Way E9 CD36 34
Brinsdale Rd. NW4 BQ31 22
Brinsley Rd. Har. BG30 11
Brinsworth Clo. Twick. BG47 56
Brion Pl. E14 CF39 43
Brisbane Av. SW19 BS50 59
Brisbane Rd. E10 CE34 25
Brisbane Rd. W13 BJ40 39
Brisbane Rd. Ilf. CL33 26
Brisbane St. SE5 BZ43 57
Briscoe Rd. SW19 BT50 59
Briscoe Rd. Rain. CV37 37
Briset Rd. SE9 CJ45 53
Briset St. EC1 BY38 33
Briset Way N7 BX34 24
Bristol Gdns. SW15 BP47 58
Bristol Gdns. W9 BS38 32
Bristol Park Rd. E17 CD31 25
Hervey Park Rd.
Bristol Rd. E7 CJ36 35
Bristol Rd. E15 CG36 34
Bristol Rd. Grnf. BF37 29
Bristol Rd. Mord. BS53 68
Bristow Rd. SE19 CA49 60
Bristow Rd. Bexh. CQ44 54
Bristow Rd. Croy. BX56 78
Bristow Rd. Houns. BF45 47
Britannia Clo. Grnf. BE38 29
Ruislip Rd.
Britannia Rd. N12 BT27 14
Britannia Rd. SW6 BS43 50
Britannia Rd. Ilf. CL34 26
Britannia Rd. Surb. BL54 66
Britannia Row N1 BY37 33
Britannia St. WC1 BX38 33
Britannia Wk. N1 BZ38 33
Britannia Way NW10 BM38 30
Britannia Way SW6 BS43 50
Britannia Rd.
British Gro. W4 BO42 40
British Gro. Pass. W6 BO42 40
British Legion Rd. E4 CG27 16
British St. E3 CD38 34
Briton Clo. Sth. Croy. CA59 84
Briton Cres. Sth. Croy. CA59 84
Briton Hill Rd. Sth. Croy. CA58 78
Britten Clo. NW11 BS33 23
Brittain Rd. Dag. CQ34 27
Brittain Rd. Walt. BD56 74
Brittany St. SE11 BX42 42
Brittenden Clo. Orp. CN57 81
Britten St. SW3 BU42 41
Britton St. EC1 BY38 33
Brixham Cres. Ruis. BC33 20
Brixham Gdns. Ilf. CN35 36
Brixham Rd. E16 CH39 44
Brixham Rd. Well. CP44 54
Brixham St. E16 CK40 44
Brixton Est. Edg. BM30 12
Brixton Hill SW2 BX47 60
Brixton Hill Ct. SW2 BX46 51
Brixton Hill Pl. SW2 BX47 60
Brixton Oval SW2 BY45 51
Rushcroft Rd.
Brixton Rd. SW9 BY45 51
Brixton Rd. Wat. BC23 4
Brixton Sta. Rd. SW9 BY45 51
Brixton Water La. SW2 BX46 51
Broadbent St. W1 BV40 41
Bourdon St.
Broadbridge Clo. SE3 CH43 53
Broad Clo. Walt. BE55 74
Broadcoombe, Sth. Croy. CC57 79
Broad Ct. WC2 BX39 42
Broadcroft Av. Stan. BK30 12
Broadcroft Rd. Orp. CM54 71
Broadfield Clo. NW2 BQ34 22
Broadfield Clo. Rom. CT32 28
Broadfield Ct. Bush. BH27 12
Broadfield La. Wat. BC26 4
Broadfield Rd. SE6 CG47 61
Broadfields, E. Mol. BG53 65
Broadfields, Har. BF30 11
Broadfields Av. N21 BY26 8
Broadfields Av. Edg. BM28 12
Broadfield Sq. Enf. CB23 8
Broadfield Way, Buck. H. CJ28 17
Broadgates Av. Barn. BS23 7
Broadgates Rd. SW18 BT47 59
Ellerton Rd.
Broad Grn. Croy. BY54 69
Broad Grn. Av. Croy. BY54 69
Broadhead Strand NW9 BO30 13
Broadheath Dr. Brom. CK49 62
Broadhinton Rd. SW4 BV45 50
Broadhurst Av. Edg. BM28 12
Broadhurst Av. Ilf. CN35 36
Broadhurst Gdns. NW6 BS36 32
Broadhurst Gdns. Chig. CM28 17
Broadhurst Gdns. Ruis. BD34 20
Broadhurst Wk. Rain. CU36 37
Broadlands Av. SW16 BX48 60
Broadlands Av. Enf. CB24 8
Broadlands Clo. N6 BV33 23
Broadlands Clo. Enf. CB24 8
Broadlands Rd. N6 BU33 23
Broadlands Rd. Brom. CH49 62
Broadlands, The, Felt. BE48 56
Broadlands Way, N. Mal. BO53 67
Broad La. N8 BX32 24
Enfield Rd.
Broad La. N15 CA31 24
Broad La. Dart. CU49 64
Broad La. Hamptn. BE50 56
Broad Lawn SE9 CL48 62
Broadlawns Clo. Har. BH30 12
Broadley St. NW8 BT39 41
Broadley Ter. NW1 BU38 32
Broadmead SE6 CE48 61
Broadmead Av. Wor. Pk. BP54 67
Broadmead Clo. Hamptn. BF50 56
Broadmeacd Clo. Pnr. BE29 11
Broadmead Rd. Nthlt. BE38 29
Broadmead Rd. Wdf. Grn. CH29 17
Broad Oak, Wdf. Grn. CH28 17
Broadoak Rd. Erith CS43 55
Broad Oaks, Surb. BM54 66
Broad Oaks, Surb. BM55 75
Kingston By Pass
Broad Oaks Way, Brom. CG53 70
Broad St. Av. EC2 CA39 42
Old Broad St.
Broad Sanctuary SW1 BX41 42
Broadstone Pl. W1 BV39 41
Broadstone Rd. Horn. CU34 28
Broad St. E15 CF36 34
Broad St. Dag. CR36 36
Broad St. Tedd. BH49 57
Broad Strood, Loug. CL23 10
Broad, The, Walk W8 BS40 41
Broad Vw. NW9 BM32 21
Broadview Rd. SW16 BW50 59
Broad Wk. E18 CG31 25
Broad Wk. N21 BX27 15
Broad Wk. SE3 CJ44 53
Broad Wk. Epsom BO62 82
Broad Wk. Har. BF31 20
Broad Wk. Houns. BD44 47
Broad Wk. Orp. CP55 81

Broad Wk. La. NW11 BR33 22
Broadwall SE1 BY40 42
Broadwater Clo. Walt. BC56 74
Broadwater Gdns. Orp. CL56 80
Broadwater Rd. N17 CA30 15
Broadwater Rd. SW17 BU49 59
Broadwater Rd. N. Walt. BC56 74
Broadwater Rd. S. Walt. BC56 74
Broadway E13 CH37 35
Broadway E15 CF36 34
Broadway N16 CA33 24
Broad Way N20 BT27 14
Broadway SW1 BW41 41
Broadway SW16 BW49 59
Broadway W6 BQ42 40
Hammersmith Rd.
Broadway W7 BH40 39
Broadway W13 BJ40 39
Broadway, Bark. CM37 35
Broadway, Bexh. CQ45 64
Broadway, Edg. BM30 12
Broadway, Epsom BP56 76
Broadway, Grnf. BG38 29
Broadway, Rain. CU38 37
Broadway, Rom. CU30 19
Broadway, Surb. BM54 66
Broadway, Swan. CS53 73
Broadway Av. Croy. BZ53 69
Broadway Av. Twick. BJ46 48
Broadway Clo. Sth. Croy. CB60 84
Broadway Clo. Wdf. Grn. CH29 17
Broadway Ct. SW19 BS50 59
Broadway Gdns. Mitch. BU52 68
Broadway Ho. Brom. CF49 61
Broadway Mkt. E8 CB37 33
Broadway Mews N21 BY26 8
Compton Rd.
Broadway, The E4 CF29 16
Broadway, The N8 BX32 24
Broadway, The N9 CB27 15
Fore St.
Broadway, The NW7 BO28 13
Broadway, The NW9 BP33 22
Broadway, The SW14 BO44 49
Terrace, The
Broadway, The SW19 BR50 58
Broadway, The W3 BM41 39
Gunnersbury La.
Broadway, The W5 BK40 39
Broadway, The, Croy. BX56 78
Broadway, The, Dag. CQ34 27
Broadway, The, Har. BH30 12
Broadway, The, Horn. CU35 37
Broadway, The, Loug. CM24 10
Broadway, The N9 CB27 15
Broadway, The, Pnr. BE29 11
Broadway, The, Stan. BK28 12
Broadway, The, Sthl. BE40 38
Broadway, The, Surb. BH54 66
Broadway, The, Sutt. BR57 76
Broadway, The, Wat. BD24 4
Broadway, The, Wdf. Grn. CH29 17
Snakes La.
Broadwick St. W1 BW39 41
Broad Yd. EC1 BY38 33
Brocas Clo. NW3 BT36 32
Chalcotts Est.
Brockdene Dr. Nthwd. BC29 11
Brockdish Av. Bark. CN35 36
Brockell Hurst, E. Mol. BE53 65
Brockenhurst Av. BO54 67
Wor. Pk.
Brockenhurst Gdns. NW7 BO28 13
Brockenhurst Gdns. Ilf. CM35 35
Brockenhurst Rd. Croy. CB54 69
Brockenhurst Way SW16 BW51 68
Brocket Clo. Chig. CN28 18
Brocket Way. Chig. CN28 18
Brockham Clo. SW19 BR49 58
Brockham Cres. Croy. CF57 79
Brockham Dr. Ilf. CM32 26
Brockham St. SE1 BZ41 42
Brockhill Cres. SE4 CD45 52
Brockhurst Clo. Stan. BH29 12
Brocklebank Rd. SW18 BT47 59
Brocklebank St. SE14 CC43 52
Brocklesby Rd. SE25 CB52 69
Brockley Av. Stan. BL27 12
Brockley Av. N. Stan. BL27 12
Brockley Clo. Stan. BL28 12
Brockley Cres. Rom. CS29 19
Brockley Cross SE4 CD45 52
Brockley Footpath SE15 CC45 52
Brockley Gdns. SE4 CD44 52
Brockley Gro. SE4 CD46 52
Brockley Hall Rd. SE4 CD46 52
Brockley Hill, Stan. BK26 5
Brockley Pk. SE23 CD47 61
Brockley Ri. SE23 CD47 61
Brockley Rd. SE4 CD45 52
Brockley Side, Stan. BL28 12
Brockley Ter. SE17 CA42 42
Alvey St.
Brockley Vw. SE23 CD47 61
Brockley Way SE4 CC46 52
Brockman Ri. Brom. CF49 61
Brock Pl. E3 CE38 34
Brock Rd. E13 CH39 44

Brocks Dr. Sutt. BR55 76
Brockshot Clo. Brent. BK43 48
Brockwell Clo. Orp. CN53 72
Brockwell Ct. SW2 BY46 51
Brockwell Pk. Gdns. SE24 BY47 60
Broderick Rd. SE2 CO42 45
Brodewater Rd. Borwd. BM23 5
Brodia Rd. N16 CA34 24
Brodie Rd. E4 CF26 9
Brodie St. SE1 CA42 42
Brodlove La. E1 CC40 43
Brodrick Rd. SW17 BU48 59
Brograve Gdns. Beck. CE51 70
Brograve Rd. W17 CB31 24
Broke Fm. Dr. Orp. CP58 81
Broken Warf EC4 BZ40 42
Brokesley St. E3 CD38 34
Bromar Rd. SE5 CA45 51
Bromboro' Grn. Wat. BD28 11
Bromefield, Stan. BK30 12
Bromehead Rd. E1 CC39 43
Bromehead St. E1 CC39 43
Brome Rd. SE9 CK45 53
Bromell's Rd. SW4 BW45 50
Bromfelde Rd. SW4 BW45 50
Bromfelde Way SW4 BX44 51
Bromfield E17 CD33 25
Bromfield St. N1 BY37 33
Bromhall Rd. Dag. CO36 36
Bromhedge SE9 CK48 62
Bromholm Rd. SE2 CO41 45
Bromley Av. Brom. CG50 61
Bromley Common, Brom. CJ52 71
Bromley Cres. Brom. CG51 70
Bromley Gdns. Brom. CG51 70
Bromley Gro. Brom. CF51 70
Bromley Hall Rd. E14 CF39 43
Lochnager St.
Bromley Hall Rd. E14 CF39 43
Bromley High St. E3 CE38 34
Bromley Hill, Brom. CG49 61
Bromley La. Chis. CM50 62
Bromley Rd. E10 CE32 25
Bromley Rd. E17 CE31 25
Bromley Rd. N17 CA30 15
Bromley Rd. N18 BZ28 15
Bromley Rd. SE6 CE47 61
Bromley Rd. Beck. CE51 70
Bromley Rd. Brom. CE47 61
Bromley Rd. Chis. CL51 71
Bromley St. E1 CC39 43
Brompton Clo. Houns. BE46 47
Brompton Dr. Erith. CU43 55
Brompton Gro. N2 BU31 23
Brompton Pk. Rd. N8 BY31 24
High Rd.
Brompton Pl. SW3 BU41 41
Brompton Rd. SW1 BU42 41
Brompton Rd. SW3 BU42 41
Brompton Rd. SW7 BU42 41
Brompton Sq. SW3 BU41 41
Bromwich Av. N6 BV34 23
Bromyard Av. W3 BO40 40
Brondesbury Ct. NW2 BQ36 31
Brondesbury Pk. NW2 BP36 31
Brondesbury Pk. NW6 BP36 31
Brondesbury Rd. NW6 BR37 31
Brondesbury Vill. NW6 BR37 31
Lansdowne Rd.
Bronsart Rd. SW6 BR43 49
Bronson Rd. SW20 BQ51 67
Bronte Gro. Dart. CE43 52
Bronze St. SE8 CE36 36
Brook Av. Dag. CR36 36
Brook Av. Edg. BM29 12
Brook Av. Wem. BL34 21
Brookbank Av. W7 BG39 38
Brookbank Rd. SE13 CE45 52
Brook Clo. SW20 BP52 67
Brook Clo. Rom. CT30 19
Brook Ct. Edg. BM28 12
Brook Cres. E4 CE28 16
Brook Cres. N9 CB28 15
Brookdale N11 BW28 14
Brookdale Rd. E17 CD31 25
Brookdale Rd. SE6 CE46 52
Brookdale Rd. Bex. CQ47 63
Brookdene Av. Wat. BC26 4
Brookdene Rd. SE18 CN42 45
Brook Dr. SE11 BY41 42
Brook Dr. Har. BG31 20
Brooke Av. Har. BG34 20
Brooke Clo. Bush. BG26 4
Brookehowse Rd. SE6 CE48 61
Brookend Rd. Sid. CN47 63
Brooke Rd. E5 CA34 24
Brooke Rd. E17 CF31 25
Brooke Rd. N16 CA34 24
Brooke's Ct. EC1 BY39 42
Brooke St. EC1 BY39 42
Brookfield N6 BV34 23
Brookfield Av. E17 CF31 25
Brookfield Av. NW7 BP29 13
Brookfield Av. W5 BK38 30
Brookfield Av. Sutt. BT56 77
Brookfield Clo. NW7 BP29 13
Brookfield Ct. Grnf. BG38 29

Brookfield Ct. Har. BK32 21
Brookfield Cres. NW7 BP29 13
Brookfield Cres. Har. BK32 21
Brookfield Gdns. Esher BH57 75
Brookfield Pk. NW5 BV34 23
Brookfield Path, Wdf. Grn. CG29 16
Oak Hill
Brookfield Rd. E9 CD36 34
Brookfield Rd. N9 CB27 15
Brookfield Rd. W4 BN41 40
Brookfields Enf. CC24 9
Brookfields Av. Mitch. BU53 68
Brook Gdns. E4 CE28 16
Brook Gdns. SW13 BO45 49
Beverley Rd.
Brook Gdns. Kings. On T. BN51 67
Brook Grn. W6 BQ41 40
Brookhill Close SE18 CL42 44
Brookhill Clo. Barn. BU25 7
Brookhill Rd. SE18 CL43 53
Brookhill Rd. Barn. BT25 7
Brook Ho. Gdns. E4 CG28 16
Brookhowse Rd. SE6 CE48 61
Brookland Clo. NW11 BS31 23
Brookland Rise
Brookland Garth NW11 BS31 23
Brookland Hl. NW11 BS31 23
Brookland Ri. NW11 BS31 23
Brooklands App. Rom. CS31 28
Brooklands Av. SW19 BS48 59
Brooklands Av. Sid. CM48 62
Brooklands Clo. Rom. CS31 28
Brooklands Dr. Wem. BH37 30
Brooklands Gdns. Horn. CV34 28
Brooklands Gdns. Rom. CS31 28
Brooklands La. Rom. CS31 28
Brooklands Pk. SE3 CH45 53
Brooklands Rd. Rom. CS31 28
Brooklands Rd. Rom. BH54 66
Brooklands St. SW8 BW44 50
Brook La. SE3 CH44 53
Brook La. Bex. CP46 54
Brook La. Brom. CH50 62
Brook La. N. Brent. BK42 39
Brooklea Clo. NW9 BO30 13
Brooklyn Av. SE25 CB52 69
Brooklyn Av. Loug. CK24 10
Brooklyn Ct. Loug. CK24 10
Brooklyn Gro. SE25 CB52 69
Brooklyn Rd. SE25 CB52 69
Brooklyn Rd. Brom. CJ53 71
Brookmead, Epsom BO57 76
Brookmead Av. Brom. CK53 71
Brookmead Clo. Orp. CO54 72
Brook Meadow N12 BS28 14
Brookmead Rd. Croy. BW53 68
Brookmeads Est. Mitch. BU53 68
Brookmead Way. Orp. CO53 72
Brook Ms. NW2 BT40 41
Brook Ms. N. W2 BT40 41
Brookmill Rd. SE8 CE44 52
Brook Parade, Chig. CL27 17
Brook Path, Loug. CK24 10
Brook Pl. Barn. BS25 7
Brook Ri. Chig. CL27 17
Brook Rd. E7 CH35 35
Brook Rd. N8 BX31 24
Brook Rd. N22 BX31 24
Brook Rd. NW2 BQ34 22
Brook Rd. W4 BM43 48
Brook Rd. Borwd. BM23 5
Brook Rd. Buck. H. CH27 17
Brook Rd. Ilf. CN32 27
Brook Rd. Loug. CK24 10
Brook Rd. Rom. CT30 19
Brook Rd. Surb. BL55 75
Brook Rd. Swan. CS52 73
Brook Rd. Th. Hth. BZ52 69
Brook Rd. Twick. BJ46 48
Brook Rd. S. Brent. BK43 48
Brooks Av. E6 CK38 35
Brooksbank St. E9 CC36 34
Brooksby's Ms. N1 BY36 33
Brooksby St.
Brooksby's Wk. E9 CC35 34
Brooks Clo. SE9 CL48 62
Brooks Ct. E15 CE35 34
Brookscroft, Croy. CD58 79
Linton Glade
Brookscroft Rd. E17 CE30 16
Brookshill, Har. BG28 11
Brookshill Av. Stan. BG28 11
Brookshill Dr. Har. BG28 11
Brookside N21 BX25 8
Brookside, Barn. BU25 7
Brookside, Cars. BV56 77
Brookside, Horn. CW32 28
Brookside, Ilf. CM29 17
Brookside, Orp. CN54 72
Brookside, Wat. BC26 4
Brookside Clo. Barn. BQ25 6
Brookside Clo. Har. BE35 29
Brookside Cres. Wor. Pk. BP54 67
Green La.
Brookside Rd. N9 CB28 15
Brookside Rd. N19 BW34 23
Brookside Rd. NW11 BR32 22

Brookside Rd. Hayes BD40 38
Brookside, S. Barn. BV26 7
Brookside Wk. N3 BR29 13
Brookside Way, Croy. CC53 70
Brook's Ms. W1 BV40 41
Brook's Rd. E13 CH37 35
Brook St. N17 CA30 15
High Rd.
Brook St. W1 BV40 41
Brook St. W2 BT40 41
Brook St. Belv. CR42 45
Brook St. Kings. On T. BL51 66
Brooksville Av. NW6 BR37 31
Brooks Way, Orp. CP51 72
Richfield Rd.
Brooks Way, Rain. CU39 46
Brookwood Av. SW13 BO44 49
Brookwood Rd. SW18 BR47 58
Brookwood Rd. Houns. BF44 47
Broom Av. Orp. CO51 72
Broom Clo. Brom. CK53 71
Broom Clo. Esher BF56 74
Broom Clo. Tedd. BK50 57
Broom Ct. Rich. BM44 48
Lichfield Rd.
Broomcroft Av. Nthlt. BD38 29
Broome Rd. Hamptn. BE50 56
Broome Way SE5 BZ43 51
Broomfield E17 CS33 28
Alexandra Rd.
Broomfield, Sun. BC51 65
Broomfield Av. N13 BX28 15
Broomfield Av. Loug. CK25 10
Broomfield La. N13 BX28 15
Broomfield Pl. W13 BJ40 39
Mattock La.
Broomfield Rd. N13 BX28 15
Broomfield Rd. W13 BJ40 39
Broomfield Rd. Beck. CD52 70
Broomfield Rd. Bexh. CR46 64
Broomfield Rd. Brom. CJ53 71
Broomfield Rd. Rich. BL44 48
Broomfield Rd. Surb. BL54 66
Broomfield Rd. Tedd. BK50 57
Melbourne Rd.
Broomfield St. E14 CE33 43
Broom Gdns. Croy. CE55 79
Broomgrove Gdns. Edg. BM30 12
Broomgrove Rd. SW9 BX44 51
Stockwell Rd.
Broomhall Rd. Sth. Croy. BZ58 78
Broom Hill Ct. Wdf. Grn. CH29 17
Broomhill Rise, Bexh. CR46 54
Broomhill Rd. SW18 BS46 50
Broomhill Rd. Dart. CU46 55
Broomhill Rd. Ilf. CO34 27
Broomhill Rd. Orp. CO54 72
Broomhill Rd. Wdf. Grn. CH29 17
Broomhouse Gdns. E4 CG28 16
Broomhouse La. SW6 BS44 50
Abbotts Cres.
Broomhouse La. SW6 BS44 50
Broomhouse Rd. SW6 BS44 50
Broomloan La. Sutt. BS55 77
Broom Lock, Tedd. BK50 57
Broom Water
Broom Mead, Bexh. CR46 54
Broom Pk. Tedd. BK50 57
Broom Rd. Croy. CE55 79
Broom Rd. Tedd. BJ49 57
Broomsleigh St. NW6 BR35 31
Broom Water, Tedd. BK50 57
Broom Water W. Tedd. BK49 57
Broomwood Rd. SW11 BU46 50
Broomwood Rd. Orp. CO51 72
Broseley Gdns. Rom. CW28 19
Broseley Gro. SE26 CD49 61
Broseley Rd. Rom. CW28 19
Brott St. E1 CE38 34
Mantus Rd.
Brougham Rd. E8 CB37 33
Brougham Rd. W3 BN39 40
Broughinge Rd. Borwd. BM23 5
Brough St. SW8 BX43 51
Broughton Av. N3 BR31 22
Broughton Av. Rich. BK48 57
Broughton Ct. W13 BJ40 39
Broughton Rd.
Broughton Gdns. N6 BW32 23
Broughton Rd. SW6 BS44 50
Broughton Rd. W13 BJ40 39
Broughton Rd. Orp. CM55 80
Broughton Rd. Th. Hth. BY53 69
Broughton Rd. Th. SW8 BV44 50
Brouncker Rd. W3 BN41 40
Brow Clo. Orp. CP54 72
Brow Cres. Orp. CP54 72
Browells La. Felt. CB48 56
Brown Clo. Wall. BX57 78
Browne Clo. Rom. CR28 18

Brownfield Est. E14 CF39 43
Brownfield St. E14 CF39 43
Brown Hart Gdns. W1 BV40 41
Brownhill Rd. SE6 CE47 61
Browning Av. W7 BH39 39
Browning Av. Sutt. BU56 77
Browning Av. Wor. Pk. BP54 67
Browning Clo. W9 BT38 32
Randolph Av.
Browning Clo. Hamptn. BE49 56
Browning Clo. Well. CN44 54
Bronwing Est. SE17 BZ42 42
Browning Ho. W12 BQ39 40
Browning Ms. W1 BV39 41
New Cavendish St.
Browning Rd. E11 CG33 25
Browning Rd. E12 CK36 35
Browning Rd. Enf. BZ23 8
Browning St. SE17 BZ42 42
Browning Way, Houns. BD44 47
Brownlea Gdns. Ilf. CO34 27
Brownlow Ms. WC1 BX38 33
Brownlow Rd. E7 CH35 35
Brownlow Rd. E8 CA37 33
Brownlow Rd. N3 BS29 14
Brownlow Rd. N11 BW29 14
Brownlow Rd. NW10 BO36 31
Brownlow Rd. Borwd. BM24 5
Brownlow Rd. Croy. CA56 78
Brownlow St. WC1 BX39 42
Brownspring Dr. SE9 CL49 62
Browns Rd. E17 CE31 25
Brown's Rd. Surb. BL54 66
Brown St. W1 BU39 41
Brownswell Rd. N2 BT30 14
Brownswood Rd. N4 BY34 24
Broxash Rd. SW11 BV46 50
Broxbourne Av. E18 CH31 26
Broxbourne Rd. E7 CH34 26
Broxbourn Rd. Orp. CN54 72
Broxhill Rd. Hav. CT27 19
Broxholm Rd. SE27 BY48 60
Broxted Rd. SE6 CD48 61
Broxwood Way NW3 BU37 32
Bruce Av. Horn. CV34 28
Bruce Castle Rd. N17 CA30 15
Bruce Clo. Well. CO44 54
Bruce Dr. Sth. Croy. CC58 79
Bruce Gdns. N20 BU27 14
Bruce Gro. N17 CA30 15
Bruce Gro. Orp. CO54 72
Bruce Hall Mews SW17 BV49 59
Bruce Rd. E3 CE38 34
Bruce Rd. NW10 BN36 31
Bruce Rd. SE25 BZ52 69
Bruce Rd. Barn. BR24 6
Bruce Rd. Har. BH30 12
Bruce Rd. Mitch. BV50 59
Brudenell Rd. SW17 BU48 59
Bruffs Meadow, Nthlt. BE36 29
Arnold Rd.
Brumfield Rd. Epsom BN56 76
Brummel Clo. Bexh. CR45 54
Brumwill Rd. W5 BL37 30
Brunel Clo. SE19 CA50 60
Brunel Clo. Nthlt. BE38 29
Brunel Estate W2 BS39 41
Brunel Pl. Houns. BF39 38
Brunel Rd. SE16 CC41 43
Brunel Rd. W3 BO39 40
Brunel Rd. Wdf. Grn. CK28 17
Brunel Rd. E16 CD39 43
Brunel Wk. N15 CA31 24
Brunel Wk. Houns. BF47 56
Mallard Clo.
Brune St. E1 CA39 42
Brunner Clo. NW11 BS32 23
Brunner Rd. E17 CD32 25
Brunner Rd. W5 BK38 30
Brunswick Av. N11 BV27 14
Brunswick Cent. WC1 BX38 33
Brunswick Clo. EC1 BY38 33
Brunswick Clo. Bexh. CP45 54
Brunswick Rd.
Brunswick Clo. Pnr. BE32 20
Brunswick Clo. Surb. BH54 66
Brunswick Ct. SE1 CA41 42
Brunswick Ct. Walt. BD55 74
Brunswick Cres. N11 BV27 14
Brunswick Gdns. W5 BL38 30
Brunswick Gdns. W8 BS40 41
Brunswick Gdns. Ilf. CM29 17
Brunswick Gro. N11 BV27 14
Brunswick Ms. W1 BU39 41
Gt. Cumberland Pl.
Brunswick Pk. SE5 BZ44 51
Brunswick Pk. Gdns. N11 BV27 14
Brunswick Pk. Rd. N11 BV27 14
Brunswick Pl. N1 BZ38 33
Brunswick Pl. SE19 CB50 60
Brunswick Rd. E10 CF33 25
Brunswick Rd. E14 CF39 43
Brunswick Rd. N15 CA31 24
West Green Rd.
Brunswick Rd. W5 BK38 30
Brunswick Rd. Bexh. CP45 54

Brunswick Rd. BM51 66
Kings. On T.
Brunswick Rd. Sutt. BS56 77
Brunswick Sq. N17 CA29 15
Brunswick Sq. WC1 BX38 33
Brunswick St. E17 CF32 25
Brunswick Vil. SE5 CA44 51
Brunswick Way N11 BV28 14
Brunton Pl. E14 CD39 43
Brushfield St. E1 CA39 42
Brussels Rd. SW11 BT45 50
Bruton Clo. Chis. CK50 62
Bruton La. W1 BV40 41
Bruton Pl. W1 BV40 41
Bruton Rd. Mord. BT52 68
Bruton St. W1 BV40 41
Bruton Way W13 BJ39 39
Bryan Av. NW10 BP36 31
Bryan Clo. Sun. BC50 64
Bryan Rd. SE16 CD41 43
Bryanston Av. Twick. BF47 56
Bryanston Clo. Sthl. BE42 38
Blandford St.
Bryanstone Ms. W. W1 BU39 41
Bryanston Pl. W1 BU39 41
Bryanstone Rd. N8 BW32 23
Bryanstone Sq. W1 BU39 41
Bryanstone St. W1 BU39 41
Bryant Av. Rom. CV30 19
Bryant Clo. Barn. BR25 6
Bryant Clo. Nthlt. BD38 29
Bryant St. E15 CF36 34
Bryantwood Rd. N7 BY35 33
Brycedale Cres. N14 BW28 14
Bryce Rd. Dag. CP35 36
Bryden Clo. SE26 CD49 61
Bryden Gro. SE26 CD49 61
Brydges Rd. E15 CF35 34
Brydon Wk. N1 BX37 33
Havelock St.
Bryett Rd. N7 BX34 24
Tollington Way
Brynmaer Rd. SW11 BU44 50
Brynmawr Rd. Enf. CA24 8
Bryony Rd. W12 BP40 40
Buccleuch Cotts. E5 CB33 24
Buccleuch Ter. E5 CB33 24
Buchanan Gdns. NW10 BP37 31
Bucharest Rd. SW18 BT47 59
Buckbean Path, Rom. CV29 19
Buckbean Pth. Rom. CV29 19
Buck Clo. Horn. CV32 28
Buckden Clo. SE12 CG46 52
Buckenham St. SE1 BZ41 42
Old Kent Rd.
Buckfast Ct. BJ 40
Romsey Rd.
Buckfast Rd. Mord. BS52 68
Buckfast St. E2 CB38 33
Buckhold Rd. SW18 BS46 50
Buckhurst Av. Cars. BU54 68
Buckhurst St. E1 CB38 33
Buckhurst Way, Buck. H. CJ28 17
Buckingham Av. N20 BT26 7
Buckingham Av. E. Mol. BF52 65
Buckingham Av. Felt. BC46 47
Buckingham Av. Grnf. BJ37 30
Buckingham Av. Th. Hth. BY51 69
Buckingham Av. Well. CN45 54
Buckingham Clo. Enf. CA23 8
Buckingham Clo. Hamptn. BE49 56
Buckingham Clo. Orp. CN54 72
Buckingham Ct. NW4 BP30 13
Buckingham Gdns. E. Mol. BF51 65
Buckingham Gdns. Edg. BL29 12
Buckingham Gdns. BY51 69
Th. Hth.
Buckingham Gate SW1 BW41 41
Buckingham La. SE23 CD47 61
Buckingham Ms. NW6 BS35 32
West End La.
Buckingham Ms. NW10 BO37 31
Buckingham Rd.
Buckingham Palace Rd. BV42 41
SW1
Buckingham Pl. SW1 BW41 41
Palace St.
Buckingham Rd. E10 CE34 25
Buckingham Rd. E11 CJ32 26
Buckingham Rd. E15 CG35 34
Buckingham Rd. E18 CG30 16
Buckingham Rd. N1 CA36 33
Buckingham Rd. N22 BX30 15
Buckingham Rd. NW10 BO37 31
Buckingham Rd. Borwd. BN24 6
Buckingham Rd. Edg. BG32 20
Buckingham Rd. Hamptn. BE49 56
Buckingham Rd. Har. BG32 20
Buckingham Rd. Ilf. CM34 26
Buckingham Rd. BL52 66
Kings. On T.
Buckingham Rd. Mitch. BX53 69
Buckingham Rd. Rich. BK48 57

Buckingham St. WC2 BX40 42
Watergate Wk.
Buckingham Ter. Sthl. BF41 38
Havelock Rd.
Buckingham Way, Wall. BW58 77
Buckland Cres. NW3 BT36 32
Buckland Rd. E10 CF34 25
Buckland Rd. Chess. BL56 75
Buckland Rd. Orp. CN56 81
Buckland Rd. Sutt. BQ58 76
Bucklands Rd. Tedd. BK50 57
Buckland Ri. Pnr. BD30 11
Buckland St. N1 BZ37 33
Buckland Wk. Mord. BT53 68
Buckland Way, Wor. Pk. BQ54 67
Buck La. NW9 BN32 22
Buckleigh Av. SW20 BR52 67
Buckleigh Rd. SW16 BW50 59
Buckleigh Way SE19 CA50 60
Buckler Gdns. SE9 CK49 62
Bucklers All. SW6 BR43 49
Haldane Rd.
Bucklersbury EC4 BZ39 42
Walbrook
Buckle St. E1 CA39 42
Buckles Way, Bans. BR61 82
Buckley Clo. DA1 CT45 55
Buckley Rd. NW6 BR36 31
Buckley St. SE1 BX40 42
Mepham St.
Buckmaster Rd. SW11 BU45 50
Bucknall St. WC2 BW39 41
Buckner Rd. SW2 BX45 51
Buckner St. W10 BR38 31
Bucknills Clo. BN61 82
Bucknills Clo. Epsom BN60 82
Buckrell Rd. E4 CF27 16
Bucks Av. Wat. BE26 4
Bucks Cross Rd. Orp. CO56 81
Buckshaw Rd. N18 CB29 15
Buckstone Clo. SE23 CC46 52
Buckstone Rd. N18 CB29 15
Buck St. NW1 BV36 32
Camden High St.
Buckthorne Rd. SE4 CD46 52
Buck Wk. E17 CF31 25
Buddings Circ. Wem. BN34 22
Budd's All. Twick. BK46 48
Arlington Clo.
Budge Row EC4 BZ40 42
Cannon St.
Budleigh Cres. Well. CP43 54
Budoch Dr. Ilf. CO34 27
Budock Rd. SW6 BR44 49
Bugsby's Way SE7 CH42 44
Bulganak Rd. Th. Hth. BZ52 69
Bulinga St. SW1 BX42 42
John Islip St.
Bulingford Clo. SE4 CD45 52
Frensbury Rd.
Bullace La. Dart. CW46 55
Bullace Row SE5 BZ44 51
Camberwell Rd.
Bull All. Well. CO45 54
Bullard's Pl. E2 CC38 34
Bullbanks Rd. Belv. CS42 46
Bullen St. SW11 BU44 50
Buller Rd. N17 CB30 15
Buller Rd. NW10 BQ38 31
Buller Rd. Bark. CN36 36
Buller Rd. Th. Hth. BZ51 69
Bullers Clo. Sid. CQ49 63
Bullers Rd. N22 BY30 15
Bullerswood Dr. Chis. CK50 62
Bullescroft Rd. Edg. BM27 12
Bullfinch Rd. Sth. Croy. CC58 79
Bullhead Rd. Borwd. BN23 6
Bull Inn Ct. WC2 BX40 42
Strand
Bullivant St. E14 CF39 43
Bull La. N18 CA28 15
Bull La. Chis. CM50 62
Bull La. Dag. CR34 27
Bull's Alley SW14 BN44 49
Bullsbrook Rd. Hayes BD40 38
Bulls Gdns. SW3 BU42 41
Walton St.
Bulls Head Pass. EC3 CA39 42
Gracechurch St.
Bull Yd. N15 CA31 24
Stamford Hill High Rd.
Bulmer Gdns. Har. BK33 21
Bulmer Ms. W11 BS40 41
Kensington Pk. Rd.
Bulmer Rd. W11 BS40 41
Bulmer Wk. Rain. CV37 37
Bulstrode Av. Houns. BE44 47
Bulstrode Gdns. Houns. BF45 47
Bulstrode Pl. W1 BV39 41
Bulstrode Rd. Houns. BF45 47
Bulstrode St. W1 BV40 41
Bulwer Ct. Rd. E11 CF33 25
Bulwer Gdns. Barn. BT24 7
Bulwer Rd. E11 CF33 25
Bulwer Rd. N18 CA28 15
Bulwer Rd. Barn. BS24 7

Bulwer St. W12 BQ40 40
Bunces La. Wdf. Grn. CG29 16
Bungalow Rd. SE25 CA52 69
Bungalows, The SW16 BV50 59
Bungalows, The, Bush. BE24 4
Bunhill Row EC1 BZ38 33
Bunkers Hill NW11 BT33 23
Bunkers Hill, Belv. CR42 45
Bunkers Hill, Sid. CQ48 63
Bunns La. NW7 BO29 13
Bunsen St. E3 CD37 34
Buntingbridge Rd. Ilf. CM32 26
Bunting Clo. Mitch. BU53 68
Bunton St. SE18 CL41 44
Bunyan Rd. E17 BW31 23
Burbage Clo. SE21 BZ46 51
Burbage Rd. SE24 BZ46 51
Burberry Clo. N. Mal. BO51 67
Burcham St. E14 CE39 43
Burchabro Rd. SE2 CP43 54
Burchell Ct. Bush. BG26 4
Burchell Rd. E10 CE33 25
Burchell Rd. SE15 CB44 51
Burchett Way, Rom. CQ32 27
Burchwall Clo. Rom. CS29 19
Burcote Rd. SW18 BT47 59
Burcott Rd. Pur. BY60 84
Burdenshott Av. Rich. BM45 48
Burden Way E11 CH34 26
Burdett Av. SW20 BP51 67
Burdett Clo. Sid. CQ49 63
Burdett Est. E14 CE39 43
Burdett Rd. E3 CD38 34
Burdett Rd. E14 CD38 34
Burdett Rd. Croy. BZ53 69
Burdett Rd. Rich. BL44 48
Burdett St. SE1 BY41 42
Pearman St.
Burdon La. Sutt. BR57 76
Burdon Pk. Sutt. BR58 76
Burfield Clo. SW17 BT49 59
Burford Clo. Dag. CP34 27
Bennetts Castle La.
Burford Clo. Ilf. CM31 26
Burford Clo. Rom. CP34 27
Burford Gdns. N13 BX27 15
Burford La. Epsom BQ59 82
Burford Rd. E6 CK38 35
Burford Rd. E15 CF36 34
Burford Rd. SE6 CD48 61
Burford Rd. Brent. BL42 39
Burford Rd. Brom. CK52 71
Burford Rd. Sutt. BS55 77
Burford Rd. Wor. Pk. BO54 67
Burford Way, Croy. CF57 79
Burgate Clo. DA1 CT45 55
Burgess Clo. Horn. CW32 28
Ernest Rd.
Burges Rd. E6 CK36 35
Burgess Av. NW9 BN32 22
Burgess Clo. Felt. BE49 56
Creswell Clo.
Burgess Cotts. Belv. CR41 45
Burgess Hi. NW2 BS35 32
Burgess Rd. E15 CG35 34
Burgess Rd. Sutt. BS56 77
Burgess St. E14 CE39 43
Burgh St. N1 BZ41 42
Burghfield, Epsom BO61 82
Burgh Heath Rd. Epsom BO60 82
Burghill Rd. SE26 CC49 61
Burghley Av. Borwd. BN25 6
Burghley Av. N. Mal. BN51 67
Burghley Gdns. Ilf. CO29 18
Burghley Rd. E11 CG33 25
Burghley Rd. N8 BY31 24
Burghley Rd. NW5 BV35 32
Burghley Rd. SW19 BQ49 58
Burgh Mt. Bans. BR61 82
Burgh Wd. Bans. BR61 82
Burgon St. EC4 BY39 42
Carter La.
Burgos Gro. SE10 CE44 52
Burgoyne Rd. N4 BY32 24
Burgoyne Rd. SE25 CA52 69
Burgoyne Rd. SW9 BX45 51
Burgundy Clo. SE1 CA42 42
Burham Clo. SE20 CC50 61
Blenheim Rd.
Burhill Gro. Pnr. BE30 11
Burhill Rd. Walt. BC58 74
Burke Clo. SW15 BO45 49
Burke St. E16 CG39 43
Burland Rd. SW11 BU46 50
Burland Rd. Rom. CS29 19
Burlea Clo. Walt. BC56 74
Burleigh Av. Sid. CN46 54
Burleigh Av. Wall. BU57 77
Burleigh Gdns. N14 BW26 7
Burleigh House W10 BQ39 40
Burleigh Pl. Mitch. BU53 68
Burleigh Rd. Enf. CA24 8
Burleigh Rd. Sutt. BR54 67
Burleigh St. WC2 BX40 42
Tavistock St.
Burleigh Way, Enf. BZ24 8

Burley Clo. E4 CE28 16
Burley Clo. SW16 BW51 68
Burley Rd. E16 CJ39 44
Burlington Arcade W1 BW40 41
Burlington Gdns.
Burlington Av. Rich. BM44 48
Burlington Av. Rom. CS32 28
Burlington Clo. BR6 CL55 80
Burlington Gdns. W1 BW40 41
Burlington Gdns. W3 BN40 40
Burlington Gdns. W4 BN42 40
Burlington Gdns. Rom. CQ33 27
Burlington La. W4 BN43 49
Burlington Ms. W3 BO40 40
Burlington Gdns.
Burlington Ms. E. W2 BS39 41
Shrewsbury Rd.
Burlington Ms. W. W2 BS39 41
Ledbury Rd.
Burlington Pl. Wdf. Grn. CH27 17
Burlington Ri. Barn. BU26 7
Burlington Rd. N10 BV31 23
Tetherdown
Burlington Rd. N17 CB30 15
Burlington Rd. SW6 BR44 49
Burlington Rd. W4 BN42 40
Burlington Rd. Enf. BZ23 8
Burlington Rd. Islw. BG44 47
Burlington Rd. N. Mal. BO52 67
Burlington Rd. Th. Hth. BZ51 69
Burma Ct. N5 BZ35 33
Green Lanes
Burman St. SE1 BY41 42
Burma Rd. N16 BZ35 33
Burmester Rd. SW17 BT48 59
Burnaby Cres. W4 BN43 49
Burnaby Gdns. W4 BM43 48
Burnaby St. E6 CL40 44
Burnaby St. SW10 BT43 50
Burnaston Ho. N16 CB34 24
Downs Est.
Burn Brae Clo. N3 BS29 14
Burnbury Rd. SW12 BW47 59
Burn Clo. Wat. BG24 4
Burncroft Av. Enf. CC23 9
Burne Jones Ho. W14 BR42 40
Burnell Av. Twick. BK49 57
Burnell Av. Well. CO44 54
Burnell Gdns. Stan. BK30 12
Burnell Rd. Sutt. BS56 77
Burnells Av. E6 CL38 35
Burness Clo. N7 BX36 33
Roman Way
Burnet Gro. Epsom BN60 82
Burnett Clo. E9 CC35 34
Churchill Wk.
Burnett Rd. Erith CV43 55
Burney Av. Surb. BL53 66
Burney Dr. Loug. CL23 10
Burney Dr. Loug. CM23 10
Burney St. SE10 CF43 52
Burnfoot Av. SW6 BR44 49
Burnham Clo. E11 CJ31 26
Burnham Ct. NW4 BQ31 22
Burnham Cres. Dart. CV45 55
Burnham Dr. Wor. Pk. BQ55 76
Burnham Gdns. Houns. BC44 47
Burnham Rd. E4 CD28 16
Burnham Rd. Dag. CO36 36
Burnham Rd. Dart. CV45 55
Burnham Rd. Mord. BS53 68
Burnham Rd. Rom. CS31 28
Burnham Rd. Sid. CQ48 63
Burnham St. E2 CC38 34
Burnham St. Kings. On T. BM51 66
Burnham Way W13 BJ41 39
Burnhill Rd. Beck. CE51 70
Burnley Rd. NW10 BO35 31
Burnley Rd. SW9 BX44 51
Burnsall St. SW3 BU42 41
Burns Av. Felt. BC46 47
Burns Av. Sid. CO46 54
Burns Av. Sthl. BF40 38
Burns Clo. Erith CT44 55
Burns Clo. Well. CO44 54
Burn Side N9 CC27 16
Burnside Clo. Twick. BJ46 48
Burnside Cres. Wem. BK37 30
Burnside Rd. E3 CD38 34
Burnside Rd. Dag. CP34 27
Burns Rd. NW10 BO37 31
Burns Rd. SW11 BU44 50
Burns Rd. W13 BJ41 39
Burns Rd. Har. BH31 21
Burns Rd. Wem. BL37 30
Burns Way, Houns. BD44 47
Burnt Ash Hl. SE12 CG46 52
Burnt Ash La. Brom. CH50 62
Burnt Ash Rd. SE12 CG46 52
Burnthouse La. Dart. CW49 64
Burnthwaite Rd. SW6 BR43 49
Burnt Oak, Edg. BM29 12
Burnt Oak Fields, Edg. BM30 12
East Rd.
Burnt Oak La. Sid. CN46 54
Burntwood Av. Horn. CV32 28
Burntwood Clo. SW18 BU47 59
Burntwood Gra. Rd. SW18 BT47 59
Burntwood La. SW17 BT48 59
Burnway, Horn. CW33 28
Burrage Gro. SE18 CM42 44
Burrage Pl. SE18 CL42 44
Burrage Rd. SE18 CL43 53
Burrard Rd. E16 CH39 44
Burrard Rd. NW6 BS35 32
Burr Clo. E1 CB40 42
Burr Clo. Bexh. CQ45 54
Burrell Clo. Croy. CD53 70
Burrell Clo. Edg. BM27 12
Burrell Row, Beck. CE51 70
High St.
Burrell St. SE1 BY40 42
Burrfield Dr. Orp. CP53 72
Burritt Rd. Kings. On T. BM51 66
Burroughs Gdns. NW4 BP31 22
Burroughs, The NW4 BP31 22
Burrow Clo. Chig. CN28 18
Burrow Grn. Chig. CN28 18
Burrow Rd. Chig. CN28 18
Burrows Ms. SE1 BY41 42
Burrows Rd. NW10 BQ38 31
Burr Rd. SW18 BS47 59
Bursdon Clo. Sid. CN48 63
Bursland Rd. Enf. CC24 9
Burslem Av. Ilf. CO29 18
Burslem St. E1 CB39 42
Burstead Gdns. Ilf. CO29 18
Burstock Rd. SW15 BR45 49
Burston Rd. SW15 BQ45 49
Burstow Rd. SW20 BR51 67
Burtenshaw Rd. Surb. BJ53 66
Burtley Clo. N4 BZ33 24
Burton Av. Wat. BC24 4
Burton Clo. NW7 BQ28 13
Burton Clo. Chess. BK57 75
Burton Ct. SW3 BU42 41
Burton Dr. Loug. CM24 10
Burton Gdns. Houns. BE44 47
Burton Gro. SE17 BZ42 42
Burtonhole La. NW7 BQ28 13
Burton Pl. WC1 BW38 32
Burton St.
Burton Rd. E18 CH31 26
Burton Rd. NW6 BR36 31
Burton Rd. SW9 BY44 51
Burton Rd. Kings. On T. BL50 57
Burton Rd. Loug. CM24 10
Burton's Rd. Hamptn. BF49 56
Burton St. WC1 BW38 32
Burtop Rd. SW17 BT48 59
Burt Rd. E16 CJ40 44
Burwash Rd. SE18 CM42 44
Burwell Av. Grnf. BH36 30
Burwell Clo. E1 CB39 42
Burwell Rd. E10 CD33 25
Burwood Av. Brom. CH55 80
Burwood Av. Ken. BY60 84
Burwood Av. Pnr. BC32 20
Burwood Clo. Surb. BM54 66
Burwood Clo. Walt. BD57 74
Burwood Gdns. Rain. CT38 37
Burwood Pk. Rd. Walt. BC56 74
Burwood Pl. W2 BU39 41
Bury Ct. EC3 CA39 42
Bury Gro. Mord. BS53 68
Bury Hall Vill. N9 CA26 8
Bury Pl. WC1 BX39 42
Bury Rd. E4 CG25 9
Bury Rd. N22 BY30 15
Bury Rd. Dag. CR35 36
Bury St. EC3 CA39 42
Bury St. N9 CA26 8
Bury St. SW1 BW40 41
Bury St. W. N9 BZ26 8
Bury Wk. SW3 BU42 41
Busby Ms. NW5 BW36 32
Torriano Av.
Busby Pl. NW5 BW36 32
Busby St. E2 CA38 33
Bushberry Rd. E9 CD36 34
Bush Clo. Ilf. CM32 26
Bush Cotts. SW18 BS46 50
Putney Br. Rd.
Bush Ct. N14 BW26 7
Bushell Gro. Bush. BG27 11
Gleed Av.
Bushell St. E1 CB40 42
Hermitage Wall
Bushell Way, Chis. CL49 62
Bush Elms Rd. Horn. CU33 28
Bushey Av. E18 CG31 25
Bushey Av. Orp. CM54 71
Bushey Clo. Ken. CA61 84
Bushey Ct. SW20 BP51 67
Bushey Ct. Kings. On T. BK50 57
Bushey Gro. Rd. Bush. BD24 4
Bushey Hall Dr. Bush. BE24 4
Bushey Hall Rd. Bush. BD24 4
Bushey Hill Rd. SE5 CA44 51
Bushey La. Sutt. BS56 77
Bushey Lees, Sid. CN46 54
Fen Grove
Bushey Pk. Cotts. Tedd. BH50 57
Bushey Pk. Rd. Tedd. BJ50 57
Bushey Rd. E13 CJ37 35
Bushey Rd. N15 CA32 24
Albert Rd.
Bushey Rd. SW20 BP52 67
Bushey Rd. Croy. CE55 79
Bushey Rd. Sutt. BS56 77
Bushey Way, Beck. CF53 70
Bushfield Clo. Edg. BM27 12
Bushfield Cres. Edg. BM27 12
Bushfields, Loug. CL25 10
Bush Gro. NW9 BN33 22
Bush Gro. Stan. BK29 12
Bushgrove Rd. Dag. CP35 36
Bush Hill N21 BZ26 8
Bush Hill Rd. N21 BZ25 8
Bush Hill Rd. Har. BL32 21
Bush La. EC4 BZ40 42
Bushmoor Cres. SE18 CL43 53
Bushnell Rd. SW17 BV48 59
Bush Rd. E11 CG33 25
Bush Rd. SE8 CC42 43
Bush Rd. Buck. H. CJ28 17
Bush Rd. Rich. BL43 48
Bushway, Dag. CP35 36
Bushwood E11 CG33 25
Bushwood Rd. Rich. BM43 48
Bushy Pk. Gdns. Tedd. BG49 56
Bushy Rd. E13 CJ37 35
Bushy Rd. Tedd. BH50 57
Busk St. E2 CB37 33
Yorkton St.
Butcher Row E14 CC40 43
Butcher's Rd. E16 CH39 44
Bute Av. Rich. BL48 57
Bute Ct. Wall. BV56 77
Bute Gdns. W6 BQ42 40
Bute Gdns. Wall. BW56 77
Bute Gdns. W. Wall. BW56 77
Bute Rd. Croy. BY54 69
Bute Rd. Ilf. CL31 26
Bute Rd. Wall. BW56 77
Bute St. SW7 BT42 41
Bute Wk. N1 BZ36 33
Marquess Est.
Buthfield St. W8 BS41 41
Butler Av. Har. BG33 20
Butler Rd. Dag. CO35 36
Butler Rd. Har. BG33 20
Butler St. E2 CC38 34
Digby St.
Buttercup Clo. Rom. CV30 19
Copperfields Way
Butterfields E17 CF32 25
Butterfly La. SE9 CL46 53
Butterfly La. Borwd. BJ24 5
Butter Hill, Wall. BV55 77
Buttermere Clo. Mord. BQ53 67
Lower Morden La.
Buttermere Gdns. Pur. BZ60 84
Buttermere Rd. SW15 BR46 49
Butterwick W6 BQ42 40
Buttesland St. N1 BZ38 33
Pitfield St.
Buttfield Clo. Dag. CR36 36
Buttmarsh Clo. SE18 CL42 44
Button St. Swan. CV53 73
Buttsbury Rd. Ilf. CM35 35
Butts Cotts. Felt. BE48 56
Butts Cres. Felt. BF48 56
Butts Farm Est. Felt. BE48 56
Butts Grn. Rd. Horn. CV32 28
Butts Rd. Brom. CG49 61
Butt's, The, Brent. BK43 48
Butts, The, Bush. BG26 4
Butts, The, Sun. BD52 65
Buxted Clo. E8 CA36 33
Richmond Rd.
Buxted Rd. N12 BU28 14
Buxton Clo. Wdf. Grn. CJ29 17
Buxton Cres. Sutt. BR56 76
Buxton Dr. E11 CG31 25
Buxton Dr. N. Mal. BN51 67
Buxton Gdns. W3 BM40 39
Buxton Path, Wat. BD27 11
Buxton Rd. E4 CF26 9
Buxton Rd. E6 CK38 35
Buxton Rd. E15 CG35 34
Buxton Rd. E17 CD31 25
Buxton Rd. N19 BW33 23
Buxton Rd. NW2 BP36 31
Buxton Rd. SW14 BO45 49
Buxton Rd. Erith CS43 55
Buxton Rd. Ilf. CN32 27
Buxton Rd. Th. Hth. BY53 69
Buxton St. E1 CA38 33
Byam St. SW6 BT44 50
Byards Cft. SW16 BW51 68
Bycroft Rd. Sthl. BF38 29
Bycroft St. SE20 CC50 61
Parish La.
Bycullah Av. Enf. BY24 8
Bycullah Rd. Enf. BY23 8
Byegrove Rd. SW19 BT50 59
Bye, The W3 BO39 40
Bye Ways, Twick. BF48 56
Byeway, The SW14 BN45 49
Byeway, The, Epsom BO56 76
Bye Way, The, Har. BH30 12
Byfeld Gdns. SW13 BP44 49
Byfield Ct. N. Mal. BP52 67
Byfield Rd. Islw. BJ45 48
Byford Clo. E15 CG36 34
Bygrove, Croy. CE57 79
Bygrove St. E14 CE39 43
Byland Clo. N21 BX26 8
Byland Clo. SE2 CO41 45
Finchale Rd.
Byne Rd. SE26 CC50 61
Byne Rd. Cars. BU55 77
Bynes Rd. Sth. Croy. BZ57 78
Byng Pl. WC1 BW38 32
Byng Rd. Barn. BQ24 6
Byng St. E14 CE41 43
Bynon Av. Bexh. CQ45 54
Byon Clo. SE26 CD49 61
Byrne Rd. SW12 BV47 59
Byron Av. E12 CK36 35
Byron Av. E18 CG31 25
Byron Av. NW9 BM31 21
Byron Av. Borwd. BM25 5
Byron Av. Couls. BX61 84
Byron Av. Houns. BC44 47
Bryon Av. N. Mal. BP52 67
Byron Av. Sutt. BT56 77
Byron Av. Wat. BD23 4
Byron Clo. Walt. BE54 65
Byron Clo. SE28 CP40 45
Byron Clo. Hamptn. BE49 56
Byron Ct. Enf. BY23 8
Byron Ct. Rich. BK49 57
Byron Dr. N2 BT32 23
Byron Gdns. Sutt. BT56 77
Byron Hill Rd. Har. BG33 20
Byron Rd. E10 CE33 25
Byron Rd. E17 CE31 25
Byron Rd. NW2 BP34 22
Byron Rd. NW7 BP28 13
Byron Rd. W5 BL40 39
Byron Rd. Har. BH30 12
Byron Rd. Har. BH32 21
Byron Rd. Sth. Croy. CB58 78
Byron Rd. Wem. BK34 21
Byron St. E14 CF39 43
Byron Way, Nthlt. BE38 29
Byron Way, Rom. CV30 19
Bysouth Clo. Ilf. CL30 17
By The Wol. Wat. BE27 11
Bythorn St. SW9 BX45 51
Byton Rd. SW17 BU50 59
Byward Av. Felt. BD46 47
Byward St. EC3 CA40 42
Bywater St. SW3 BU42 41
Byways, The, Surb. BM53 66
Byway, The, Sutt. BT58 77
Bywood Av. Croy. CC53 70
Bywood Clo. Ken. BY61 84
Byworth Wk. N19 BX33 24
Cabbell St. NW1 BU39 41
Cable St. E1 CB40 42
Cabul Rd. SW11 BU44 50
Cactus Wk. W12 BO39 40
Du Cane Rd.
Cadbury Clo. Islw. BJ44 48
Cadbury Rd. SE16 CB41 42
Caddington Clo. Barn. BU25 7
Caddington Rd. NW2 BR34 22
Cade Rd. SE10 CF44 52
Cader Rd. SW18 BT46 50
Cadet Pl. SE10 CG42 43
Cadiz Rd. Dag. CS36 37
Cadiz St. SE17 BZ42 42
Cadley Ter. SE23 CC48 61
Cadogan Clo. Har. BF35 29
Cadogan Clo. Tedd. BH49 57
Cadogan Ct. SW3 BU42 41
Draycott Av.
Cadogan Ct. Sutt. BS57 77
Cadogan Gdns. E18 CH31 26
Cadogan Gdns. N3 BS30 14
Cadogan Gdns. N21 BY25 8
Cadogan Gdns. SW3 BU42 41
Cadogan Gte. SW1 BU42 41
Cadogan La. SW1 BV41 41
Cadogan Pl. SW1 BU41 41
Cadogan Rd. Surb. BK53 66
Cadogan Sq. SW1 BU41 41
Cadogan St. SW3 BU42 41
Cadogan Ter. E9 CD36 34
Cadoxton Av. N15 CA32 24
Cadwallon Rd. SE9 CL48 62
Caedmon Rd. N7 BX35 33
Caerleon Clo. Sid. CP49 63
Caerleon Ter. SE2 CO42 45
Caernarvon Clo. Mitch. BX52 69
Caernarvon Dr. Ilf. CL30 17
Caesar St. E2 CA38 33
Caesar's Wk. Mitch. BU53 68
Cahir St. E14 CE42 43
Caird St. W10 BR38 31
Cairn Av. W5 BK40 39
Cairndale Clo. Brom. CG50 61
Cairnfield Av. NW2 BO34 22
Cairns Av. IG8 CK29 17
Cairns Clo. Dart. CV46 55
Cairns Rd. SW11 BU46 50
Cairn Way, Stan. BH29 12
Cairo Rd. E17 CE31 25

Canonbury Av. N1	BY36	33
Canonbury Rd.		
Canonbury Gro. N1	BZ36	33
Canonbury La. N1	BY36	33
Canonbury Pk. N. N1	BZ36	33
Canonbury Pk. S. N1	BZ36	33
Canonbury Pl. N1	BY36	33
Canonbury Rd. N1	BY36	33
Canonbury Rd. Enf.	CA23	8
Canonbury Sq. N1	BY36	33
Canonbury St. N1	BZ36	33
Canonbury Vill. N1	BY36	33
Canon Ct. Edg.	BL29	12
Canon Pk. Est. Stan.	BK28	12
Canon Rd. Brom.	CJ52	71
Canons Clo. N2	BT33	23
Canons Clo. Edg.	BL29	12
Canons Dr. Edg.	BL29	12
Canons Hatch, Tad.	BR62	82
Canons Hill, Couls.	BY62	84
Canons La. Tad.	BR62	82
Canonsleigh Rd. Dag.	CO36	36
Canons Pk. Par. Edg.	BL29	12
Canon St. N1	BZ37	33
Prebend St.		
Canon's Wk. Croy.	CC55	79
Canopus Way, Nthwd.	BC28	11
Canrobert St. E2	CB37	33
Canterbury Av. Ilf.	CK33	26
Canterbury Av. Sid.	CO48	63
Canterbury Clo. Beck.	CE51	70
Tyler Rd.		
Canterbury Clo. Grnf.	BF39	38
Canterbury Cres. SW9	BY45	51
Canterbury Gro. SE27	BY49	60
Canterbury Rd. E10	CF33	25
Canberbury Rd. NW6	BR37	31
Canterbury Rd. Borwd.	BM23	5
Canterbury Rd. Croy.	BX54	69
Canterbury Rd. Felt.	BE48	56
Canterbury Rd. Har.	BF32	20
Canterbury Rd. Mord.	BS54	68
Canterbury Rd. Wat.	BC23	4
Canterbury Ter. NW6	BS37	32
Cantley Gdns. SE19	CA51	69
Cantley Gdns. Ilf.	CM32	26
Cantley Rd. W7	BJ41	39
Canton St. E14	CE39	43
Cantrell Rd. E3	CD38	34
Cantwell Rd. SE18	CL43	53
Canute Gdns. SE16	CC42	43
Canvey St. SE1	BZ40	42
Zoar St.		
Cape Clo. Bark.	CL36	35
Harts La.		
Capel Av. Wall.	BX56	78
Capel Clo. N20	BT27	14
Capel Clo. Brom.	CK54	71
Capel Ct. EC2	BZ39	42
Bartholomew La.		
Capel Gdns. Ilf.	CN35	36
Capel Gdns. Pnr.	BE31	20
Capel Pl. Dart.	CV49	64
Capel Rd. E7	CH35	35
Capel Rd. Barn.	BU26	7
Capel Rd. Wat.	BD25	4
Capener's Clo. SW1	BV41	41
Kinnerton St.		
Caperne Rd. SW18	BT47	59
Cargill Rd.		
Cape Rd. N17	CB31	24
Capital Ho. SE6	CE46	52
Capland St. NW8	BT38	32
Caple Rd. NW10	BO37	31
Caprea Clo. Hayes	BD39	38
Capri Rd. Croy.	CA54	69
Capstan Ride, Enf.	BY23	8
Capstan Sq. E14	CF41	43
Capstone Rd. Brom.	CG49	61
Capthorne Av. Har.	BE33	20
Capthorne Ct. Har.	BE33	20
Capworth St. E10	CE33	25
Caradoc St. SE10	CG42	43
Caradon Way N15	BZ31	24
Caravelle Gdns. Nthlt.	BD38	29
Javelin Way		
Caravel Mews SE8	CE43	52
Watergate St.		
Carberry Rd. SE19	CA50	60
Carbery Av. W3	BL41	39
Carbis Rd. E14	CD39	43
Carburton St. W1	BV38	32
Carden Rd. SE15	CB45	51
Cardiff Rd. W7	BJ41	39
Cardiff Rd. Enf.	CB24	8
Cardiff Rd. Wat.	BC25	4
Cardiff St. SE18	CN43	54
Cardigan Gdns. Ilf.	CO34	27
Cardigan Pl. NW6	BS38	32
Cardigan Rd. E3	CD37	34
Cardigan Rd. SW13	BP44	49
Cardigan Rd. SW19	BT50	59
Cardigan Rd. Rich.	BL46	48
Cardigan St. SE11	BY42	42
Cardigan Wk. N1	BZ36	33
Clephane Rd.		
Cardinal Av. Borwd.	BM54	66
Cardinal Av. Kings. On T.	BL49	57
Cardinal Av. Mord.	BR53	67
Cardinal Bourne St. SE1	BZ41	42
Burge St.		
Cardinal Clo. Chis.	CM50	62
Cardinal Clo. Mord.	BR53	67
Cardinal Clo. Wor. Pk.	BP56	76
Cardinal Cres. N. Mal.	BN51	67
Cardinal Dr. Ilf.	CM29	17
Cardinal Dr. Walt.	BD54	65
Cardinale Way N19	BW34	23
Cardinal Pl. SW15	BQ45	49
Cardinal Rd. Felt.	BC47	56
Cardinal Rd. Ruis.	BD33	20
Cardinal's Wk. Hmptn.	BG50	56
Cardinal Way, Rain.	CV37	37
Cardington Sq. Houns.	BD45	47
Cardington St. NW1	BW38	32
Cardozo Rd. N7	BX35	33
Cardrew Av. N12	BT28	14
Cardrew Clo. N12	BT28	14
Cardross St. W6	BP41	40
Cardwell Rd. N7	BX35	33
Cardwell Rd. SE18	CL42	44
Carew Rd. N17	CB30	15
Carew Rd. W13	BK41	39
Carew Rd. Mitch.	BV51	68
Carew Rd. Th. Hth.	BY52	69
Carew Rd. Wall.	BW57	77
Carew St. SE5	BZ44	51
Carey Gdns. SW8	BW44	50
Thessaly Rd.		
Carey La. EC2	BZ39	42
Gutter La.		
Carey Pl. Wat.	BD24	4
Clifford St.		
Carey Rd. Dag.	CQ35	36
Carey St. WC2	BX39	42
Carfax Rd. Horn.	CT35	37
Carfax Sq. SW4	BW45	50
Clapham Pk. Rd.		
Carfree Clo. N1	BY36	33
Bewdley St.		
Cargill Rd. SW18	BS47	59
Cargreen Pl. SE25	CA52	69
Cargreen Rd.		
Cargreen Rd. SE25	CA52	69
Carholme Rd. SE23	CD47	61
Carisbrook Av. Wat.	BD23	4
Carisbrook Clo. Stan.	BK30	12
Carisbrooke Av. Bex.	CP47	63
Carisbrooke Clo. Enf.	CA23	8
Carisbrooke Gdns. SE15	CA43	51
Rosemary Rd.		
Carisbrooke Rd. E17	CD31	25
Carisbrooke Rd. Brom.	CJ52	71
Carisbrooke Rd. Mitch.	BW52	68
Carleton Av. Wall.	BW57	77
Carleton Clo. Esher	BG54	65
Carleton Rd. N7	BW35	32
Carlile Clo. E3	CD37	34
Carlingford Gdns. Mitch.	BV50	59
Carlingford Rd. N15	BY31	24
Carlingford Rd. NW3	BT35	32
Carlingford Rd. Mord.	BQ53	67
Carlisle Av. EC3	CA39	42
Carlisle Av. W3	BO39	40
Carlisle Clo. Kings. On T.	BM51	66
Carlisle Gdns. Har.	BK33	21
Carlisle Gdns. Ilf.	CK32	26
Carlisle La. SE1	BX41	42
Carlisle Pl. N11	BV28	14
Carlisle Pl. N12	BV28	14
Oakleigh Rd.		
Carlisle Pl. SW1	BW41	41
Carlisle Rd. E10	CE33	25
Carlisle Rd. N4	BY33	24
Scarborough Rd.		
Carlisle Rd. NW6	BR37	31
Carlisle Rd. NW9	BN31	22
Carlisle Rd. Hmptn.	BF50	56
Carlisle Rd. Rom.	CU32	28
Carlisle Rd. Sutt.	BR57	76
Carlisle St. W1	BW39	41
Soho Sq.		
Carlos Pl. W1	BV40	41
Carlow St. NW1	BW37	32
Carlton Av. N14	BW25	7
Carlton Av. Felt.	BD46	47
Carlton Av. Harrow	BJ32	21
Carlton Av. Sth. Croy.	BZ57	78
Carlton Av. E. Wem.	BK34	21
Carlton Av. W. Wem.	BJ34	21
Carlton Clo. Borwd.	BN24	6
Carlton Clo. Chess.	BL57	75
Carlton Clo. Edg.	BM28	12
Carlton Ct. NW6	BS37	32
Carlton Cres. Sutt.	BR56	76
Carlton Dr. SW15	BQ46	49
Carlton Dr. Ilf.	CM31	26
Carlton Gdns SW1	BW40	41
Carlton Gdns. W5	BK39	39
Carlton Gro. SE15	CB44	51
Carlton Hl. NW8	BS37	32
Carlton Ho. Ter. SW1	BW40	41
Carlton Pde. Orp.	CO54	72
Carlton Pk. Av. SW20	BQ51	67
Carlton Rd. E11	CG33	25
Carlton Rd. E12	CJ35	35
Carlton Rd. E17	CD30	16
Carlton Rd. N4	BY33	24
Carlton Rd. N11	BV28	14
Carlton Rd. N15	CA31	24
Carlton Rd. SW14	BN45	49
Carlton Rd. W4	BN41	40
Carlton Rd. W5	BK40	39
Carlton Rd. Dart.	CW47	64
Carlton Rd. Erith	CR43	54
Carlton Rd. N. Mal.	BO51	67
Carlton Rd. Rom.	CT32	28
Carlton Rd. Sid.	CN49	63
Carlton Rd. Sth. Croy.	BZ57	78
Carlton Rd. Sun.	BC50	56
Carlton Rd. Walt.	BC54	65
Carlton Rd. Well.	CO45	54
Carlton Sq. E1	CC38	34
Carlton St. SW1	BW40	41
Regent St.		
Carlton Ter. E7	CJ36	35
Carlton Ter. E11	CH32	26
Carlton Ter. N18	BZ27	15
Carlton Ter. SE26	CC48	61
Carlton Vale NW6	BS38	32
Carlwell St. SW17	BU49	59
Carlyle Av. Brom.	CJ52	71
Carlyle Av. Sthl.	BE40	38
Carlyle Clo. N2	BT32	23
Carlyle Clo. NW10	BN37	31
Carlyle Clo. E. Mol.	BG51	65
Carlyle Gdns. Sthl.	BE40	38
Carlyle Rd. E12	CK35	35
Carlyle Rd. SE28	CO40	45
Carlyle Rd. W5	BK42	39
Carlyle Rd. Croy.	CB55	78
Carlyle Sq. SW3	BT42	41
Carlyon Av. Har.	BE35	29
Carlyon Clo. Wem.	BL37	30
Carlyon Rd. Hayes	BD39	38
Carlyon Rd. Wem.	BL37	30
Carmalt Gdns. SW15	BQ45	49
Carmalt Gdns. Walt.	BD56	74
Carmelite Clo. Har.	BG30	11
Carmelite Rd. Har.	BG30	11
Carmelite St. EC4	BY40	42
Carmelite Wk. Har.	BG30	11
Carmelite Way, Har.	BG30	11
Carmen St. E14	CE39	43
Carmichael Rd. SE25	CA53	69
Carminia Rd. SW17	BV48	59
Carnaby St. W1	BW39	41
Carnac St. SE27	BZ48	60
Carnanton Rd. E17	CF30	16
Carnarvon Av. Enf.	CA24	8
Carnarvon Rd. E10	CF32	25
Carnarvon Rd. E15	CG36	34
Carnarvon Rd. E18	CG30	16
Carnarvon Rd. Barn.	BR24	6
Carnation St. SE2	CO42	45
Carnbrook Rd. SE3	CJ45	53
Carnecke Gdns. SE9	CK46	53
Carnegie Clo. Surb.	BL55	75
Carnegie Pl. SW19	BQ48	58
Carnegie St. N1	BX37	33
Carnfield Dr. NW9	BO29	13
Carnforth Clo. Epsom	BM57	75
Carnforth Gdns. Horn.	CU35	37
Carnforth Rd. SW16	BW50	59
Carnoustie Dr. N1	BX36	33
Carnwath Rd. SW6	BS45	50
Carolina Rd. Th. Hth.	BY51	69
Caroline Clo. Croy.	CA56	78
Browlow Rd.		
Caroline Ct. Stan.	BJ29	12
Chase, The		
Caroline Gdns. N1	CA38	33
Kingsland Rd.		
Caroline Gdns. SE15	CB43	51
Caroline Pl. W2	BS40	41
Caroline Pl. Wat.	BD25	4
Caroline Pl. Ms. W2	BS40	41
Orme La.		
Caroline Rd. SW19	BR50	58
Caroline St. E1	CC39	43
Caroline Ter. SW1	BV42	41
Carol St. NW1	BW37	32
Carolyn Dr. Orp.	CO55	81
Carpenders Av. Wat.	BE27	11
Carpenter Gdns. N21	BY27	15
Carpenters Ct. Twick.	BH48	57
Hampton Rd.		
Carpenters Pl. SW4	BW45	50
Carpenters Rd. E15	CE36	34
Carriage, The, Rd. SW7	BT41	41
Carrick Gdns. N17	BZ30	15
Flexmere Rd.		
Carrick Gate, Esher	BG55	74
New Road		
Carrick Mews SE8	CE43	52
Watergate St.		
Carrington Av. Borwd.	BM25	5
Carrington Av. Houns.	BF46	47
Carrington Clo. Borwd.	BN25	6
Carrington Ho. SE8	CE44	52
Carrington Rd. Dart.	CW46	55
Carrington Rd. Rich.	BM45	48
Carrington St. W1	BV40	41
Carroll Cld. E15	CG35	34
Carroll Clo. E15	CG35	34
Ash Rd.		
Carroll Hill Loug.	CK24	10
Carrol Pl. NW5	BV35	32
Carron Clo. E14	CE39	43
Carrotway Grnf.	BG38	29
Deepwood La.		
Carroun Rd. SW8	BX43	51
Carrow Rd. Dag.	CO36	36
Carrow Rd. KT12	BD55	74
Carr Rd. E17	CD30	16
Carr Rd. Nthlt.	BF36	29
Carrs La. N21	BZ25	8
Carr St. E14	CD39	43
Carshalton Gro. Sutt.	BT56	77
Carshalton Pk. Rd. Cars.	BU56	77
Carshalton Pk. Rd. Cars.	BU57	77
Carshalton Pl. Cars.	BV56	77
Carshalton Rd. Bans.	BU60	83
Carshalton Rd. Mitch.	BV52	68
Carshalton Rd. Sutt.	BT56	77
Carshalton Rd. W. Sutt.	BS56	77
Carsington Gdns. Dart.	CV48	64
Carslake Rd. SW15	BQ46	49
Carson Rd. E16	CH38	35
Carson Rd. SE21	BZ47	60
Carson Rd. Barn.	BU24	7
Carstairs Rd. SE6	CF48	61
Carston Clo. SE12	CG46	52
Carston Ms. SE12	CG46	52
Carswell Clo. Ilf.	CJ31	26
Carswell Rd. SE6	CF47	61
Carter Clo. Rom.	CR29	18
Carter Clo. Wall.	BW57	77
Carter Ct. EC4	BY39	42
Carter La.		
Carter Dr. Rom.	CR29	18
Carteret St. SW1	BW41	41
Carteret Way SE8	CD42	43
Carterhatch La. Enf.	CA23	8
Carterhatch Rd. Enf.	CC23	9
Carter La. EC4	BY39	42
Carter Pl. SE17	BZ42	42
Carter Rd. E13	CH37	35
Carter Rd. SW19	BT50	59
Carters Clo. Loug.	CL25	10
Carters Clo. Wor. Pk.	BQ55	76
Carter's Hl. SE9	CJ47	62
Carter's Hl. Clo. SE9	CJ47	62
Carters La. SE23	CD48	61
Carters Rd. Epsom.	BO61	82
Carter St. SE17	BZ43	51
Carters Yd. SW18	BS46	50
Wandsworth High St.		
Carthew Rd. W6	BP41	40
Carthew Vill. W6	BP41	40
Carting La. WC2	BX40	42
Cart La. E4	CF26	9
Station Rd.		
Cartmel Clo. N17	CB29	15
Cartmell Gdns. Mord.	BT53	68
Cartmel Rd. Bexh.	CR44	54
Carton St. W1	BU39	41
Cartwright Gdns. WC1	BX38	33
Cartwright Rd. Dag.	CQ36	36
Cartwright St. E1	CA40	42
Carver Rd. SE24	BZ46	51
Carville Cres. Brent.	BL42	39
Carville St. N7	BY34	24
Durham Rd.		
Cary Rd. E11	CG35	34
Carysfort Rd. N8	BW32	23
Carysfort Rd. N16	BZ34	24
Cascade Av. N10	BW31	23
Cascade Rd. Buck. H.	CJ27	17
Cascades Croy.	CD58	79
Casella Rd. SE14	CC43	52
Casewick Rd. SE27	BY49	60
Casimir Rd. E5	CC34	25
Casino Av. SE24	BZ46	51
Casket St. E2	CA38	33
Caslon Pl. E1	CB38	33
Cudworth St.		
Caslte Rd. Islw.	BH44	48
Caspian St. SE5	BZ43	51
Caspian Wk. E16	CJ39	44
King George Av.		
Casselden Rd. NW10	BN36	31
Cassidy Rd. SW6	BS43	50
Cassilda Rd. SE2	CO42	45
Cassilis Rd. Twick.	BJ46	48
Cassiobury Rd. E17	CC32	25
Cassio Rd. Wat.	BC24	4
Cassland Rd. E9	CC36	34
Cassland Rd. Th. Hth.	BZ52	69
Casslee Rd. SE6	CD47	61
Casson St. E1	CB39	42
Castalia St. E14	CF41	43
Castellain Rd. W9	BS38	32
Castellan Av. Rom.	CU31	28
Castello Av. SW15	BQ46	49
Castell Rd. Loug.	CM23	10
Castelnau SW13	BP44	49
Castelnau Est. SW13	BP43	49
Castelnau Pl. SW13	BP43	49

Chaldon Rd. SW6 BR43 49
Chaldon Way Couls. BX62 84
Chale Rd. SW2 BX46 51
Chalet Clo. Bex. CS49 64
Chale Walk Sutt. BS58 77
Hulverston Clo.
Chalfont Av. Wem. BM36 30
Chalfont Grn. N9 CA27 15
Chalfont Rd. N9 CA27 15
Chalfont Rd. SE25 CA52 69
Chalfont Rd. Hayes BC41 38
Chalfont Way W13 BJ41 39
Chalford Clo. E. Mol. BF52 65
Chalforde Gdns. Rom. CU31 28
Chalford Rd. SE21 BZ49 60
Chalford Wk. Wdf. Grn. CJ30 17
Chalgrove Av. Mord. BS53 68
Chalgrove Cres. Ilf. CK30 17
Chalgrove Gdns. N3 BR31 22
Chalgrove Rd. E9 CC36 34
Chalgrove Rd. N17 CB30 15
Chalgrove Rd. Sutt. BT57 77
Chalice Clo. Wall. BW57 77
Chalice Way SW2 BX47 60
Chalkenden Clo. SE20 CB50 60
Castledine Rd.
Chalk Farm Rd. NW1 BV36 32
Chalk Hill. Wat. BD25 4
Chalk Hill Rd. W6 BQ42 40
Shortlands Ms.
Chalkhill Rd. Wem. BM34 21
Chalklands, The, Wem. BN34 22
Chalk La. Barn. BU24 7
Chalk La. Epsom BN61 82
Chalk Pit Av. Orp. CP52 72
Chalk Pit Rd. Bans. BS62 83
Chalk Rd. E13 CH39 44
Chalkwell Pk. Av. Enf. CA24 8
Chalky La. Chess. BK58 75
Challice Way, SW2 BX47 60
Challin St. SE20 CC51 70
Challis Rd. Brent. BK42 39
Challoner Clo. N2 BT30 14
Challoner Cres. W14 BR42 40
Challoner St.
Challoners Clo. E. Mol. BG52 65
Challoner St. W14 BR42 40
Chalmers Rd. Bans. BT61 83
Chalmers St. SW8 BW44 50
Chalmers Ter. N16 CD34 25
Victorian Rd.
Chalmers Way Felt. BC46 47
Chalsey Rd. SE4 CD45 52
Chalterton Rd. N4 BY34 24
Chalton Dr. N2 BT32 23
Chalton St. NW1 BW37 32
Chamberlain Cotts. SE5 BZ44 51
Camberwell Gro.
Chamberlain Cres. W. Wick. CE54 70
Chamberlain La. Pnr. BC31 20
Chamberlain Rd. N2 BT30 14
Chamberlain Rd. N9 CB27 15
Chamberlain Rd. W13 BJ41 39
Midhurst Rd.
Chamberlain Sq. N1 BY36 33
Lofting Rd.
Chamberlain St. NW1 BU36 32
Regents Pk. Rd.
Chamberlain Way Pnr. BC31 20
Chamberlain Way Surb. BL54 66
Chamberlayne Rd. NW10 BQ37 31
Chambers La. NW10 BP36 31
Chambers Rd. N7 BX35 33
Chambers St. SE16 CB41 42
Chamber St. E1 CA40 42
Chambord St. E2 CA38 33
Champion Cres. SE26 CD49 61
Champion Gro. SE5 CA45 51
Champion Hill, SE5 BZ45 51
Champion Pk. SE5 BZ44 51
Champion Pl. SE28 CQ40 45
Champneys Clo. Sutt. BR57 76
Chancellor Gro. SE21 BZ48 60
Chancellor's Rd. W6 BQ42 40
Chancellor's St. W6 BQ42 40
Chancelot Rd. SE2 CO42 45
Chancel St. SE1 BY40 42
Dolben St.
Chancery La. WC2 BX39 42
Chancery La. Beck. CE51 70
Chance St. E1 CA38 33
Chance St. E2 CA38 33
Chanctonbury Clo. SE9 CL48 62
Chanctonbury Gdns. Sutt. BS57 77
Chanctonbury Way N12 BR28 13
Chandler Av. E16 CH39 44
Chandler Clo. Hamptn. BF51 65
Chandler Rd. Loug. CL23 10
Chandler St. E1 CB40 42
Chandos Av. E17 CE30 16
Chandos Av. N14 BW27 14
Chandos Av. N20 BT26 7
Chandos Av. W5 BK42 39
Chandos Clo. Buck. H. CH27 17
Chandos Cres. Edg. BL29 12
Chandos Pl. WC2 BX40 42
Chandos Rd. E15 CF35 34
Chandos Rd. N2 BT30 14
Chandos Rd. N17 CA30 15
Chandos Rd. NW2 BQ35 31
Chandos Rd. NW10 BO38 31
Chandos Rd. Borwd. BL23 5
Chandos Rd. Har. BG32 20
Chandos Rd. Pnr. BD33 20
Chandos St. W1 BV39 41
Chandos Way NW11 BS33 23
Chandrye Clo. SE9 CK46 53
Change Alley EC3 BZ39 42
Birchin La.
Channel Clo. Houns. BF44 47
Channelsea Rd. E15 CF37 34
Channing Clo. Horn. CW33 28
Chanton Dr. Sutt. BQ58 76
Chantrey Rd. SW9 BX45 51
Chantry Clo. Har. BL32 21
Chantry Clo. Sid. CQ49 63
Ellenborough La.
Chantry Hurst Epsom BN61 82
Chantry La. Brom. CJ53 71
Chantry Pl. Har. BF30 11
Chantry Rd. Chess. BL56 75
Chantry Rd. Har. BF30 11
Chantry St. N1 BY37 33
Chantry Way Rain. CS37 37
Chant Sq. E15 CF36 34
Chant St. E15 CF36 34
Chapel All. Brent. BL43 48
High St.
Chapel Clo. N2 BU31 23
Chapel Ct.
Chapel Clo. Dart. CT46 55
Chapel Ct. N2 BU31 23
Chapel Farm Rd. SE9 CK48 62
Chapel Hill Dart. CT46 55
Chapel Hill Pnr. BC32 20
Chapel Ho. Pl. E14 CE42 43
Chapel Ho. St. E14 CE42 43
Chapel La. Chig. CN27 18
Chapel La. Houns. BF45 47
Staines Rd.
Chapel la. Loug. CK24 10
Chapel La. Pnr. BD31 20
Chapel Mkt. N1 BY37 33
Chapel Pl. N17 CA29 15
Chapel Pl. W1 BV39 41
Chapel Rd. SE27 BY49 60
Chapel Rd. W13 BJ40 39
Chapel Rd. Bexh. CR45 54
Chapel Rd. Houns. BF45 47
Chapel Rd. Mitch. BT52 68
Chapel Rd. Twick. BJ47 57
Chapel Row. Ilf. CL34 26
Station Rd.
Chapel Side W2 BS40 41
Chapel Stones N17 CA30 15
Kings Rd.
Chapel St. E15 CF36 34
Chapel St. NW1 BU39 41
Chapel St. SW1 BV41 41
Chapel St. W2 BS40 41
Chapel St. Enf. BZ24 8
Chapel Vw. Sth. Croy. CC57 79
Chapel Wk. NW4 BP31 22
Chapel Way Brom. CH51 71
Holwood Rd.
Chapel Yd. N18 CB29 15
Fore St.
Chaplaincy Gdns. Horn. CW33 28
Allenby Dr.
Chaplin Rd. E15 CG37 34
Chaplin Rd. NW2 BP36 31
Chaplin Rd. Dag. CQ36 36
Chaplin Rd. Wem. BK36 30
Chaplin St. SE23 CC48 61
Chapman Cres. Har. BL32 21
Chapman Rd. E9 CD36 34
Chapman Rd. Croy. BY54 69
Chapman's La. SE2 CP42 45
Chapman's La. Belv. CP42 45
Chapman's La. Orp. CP51 72
Chapman St. E1 CB39 42
Chapples Clo. Loug. CM25 10
Chapter Rd. NW2 BP35 31
Chapter Rd. SE17 BY42 42
Chapter St. SW1 BW42 41
Charcroft Gdns. Enf. CC24 9
Chardin Rd. W4 BO42 40
Elliott Rd.
Chadmore Rd. N16 CB33 24
Charford Rd. E16 CH39 44
Chargeable La. E13 CG38 34
Chargeable St. E16 CG38 34
Charing Clo. Orp. CN56 81
Charing Cros Rd. WC2 BW39 41
Charlbert St. NW8 BU37 32
Charlbury Av. Stan. BK28 12
Charlbury Clo. Rom. CV29 19
Charlbury Cres. Rom. CV29 19
Charlbury Gdns. Ilf. CN34 27
Charlbury Gro. W5 BK39 39
Charldane Rd. SE9 CL48 62
Charlecote Gro. SE26 CB48 60
Charlecote Rd. Dag. CQ34 27
Charlemont Rd. E6 CK38 35
Charles Burton Ct. E10 CD35 34
Meeson St.
Charles Clo. Sid. CO49 63
Charles Cres. Har. BG33 20
Charlesfield Way, SE9 CJ48 62
Charles Grinding Wk. SE18 CL42 44
Charles Ho. W14 BR42 40
Charles St. SW1 BW40 41
Charles La. NW8 BU37 32
Charles Mills Ct. SW16 BX50 60
Charles Rd. E7 CJ36 35
Charles Rd. W13 BJ39 39
Charles Rd. Dag. CS36 37
Charles Rd. Rom. CP33 27
Charles Rd. Sev. CR58 81
Charles Sevright Dr. NW7 BQ28 13
Charles Sq. N1 BZ38 33
Charles St. E16 CJ40 44
Charles St. SW13 BO45 49
Charles St. W1 BV40 41
Charles St. W5 BK40 39
Lancaster Rd.
Charles St. Croy. BZ55 78
Charles St. Enf. CA25 8
Charles St. Houns. BE44 47
Charleston St. SE17 BZ42 42
Charlesworth St. N7 BX36 33
Charleville Circs. SE26 CB49 60
Charleville Rd. W14 BR42 40
Charlieville Rd. Erith CS43 55
Mill Rd.
Charlmont Rd. SW17 BU50 59
Charlock Way Wat. BC25 4
Charlotte Despard Av. SW11 BV44 50
Charlotte Gdns. Rom. CR29 18
Charlotte Ms. W1 BW39 41
Tottenham St.
Charlotte Pl. W1 BW39 41
Goodge St.
Charlotte Rd. EC2 CA38 33
Charlotte Rd. Dag. CR36 36
Charlotte Rd. SW13 BO44 49
Charlotte Rd. Wall. BW57 77
Charlotte Row SW4 BW45 50
North St.
Charlotte St. W1 BW39 41
Charlotte Ter. N1 BX37 33
Charlton Av. Walt. BC56 74
Charlton Church La. SE7 CJ42 44
Charlton Cres. Bark. CN37 36
Charlton Dene, SE7 CJ43 53
Charlton Gdns. Couls. BW62 83
Charlton Gdns. Couls. BW58 77
Charlton Kings Rd. NW5 BW35 32
Charlton La. SE7 CJ42 44
Charlton Pk. La. SE7 CJ43 53
Charlton Pk. Rd. SE7 CJ43 53
Charlton Pl. N1 BY37 33
Charlton Rd. Har. BK31 21
Charlton Rd. N9 CC26 9
Charlton Rd. NW10 BO37 31
Charlton Rd. SE3 CH43 53
Charlton Rd. SE7 CH43 53
Charlton Rd. Wem. BL33 21
Charlton Way, SE3 CG43 52
Charlwood Clo. Har. BH29 12
Charlwood Croy. CD58 79
Charlwood Pl. SW1 BW42 41
Charlwood Rd. SW15 BQ45 49
Charlwood St. SW1 BW42 41
Charlwood Ter. SW15 BQ45 49
Cardinal Pl.
Charmian Av. Stan. BK30 12
Charminster Av. SW19 BS51 68
Charminster Ct. Surb. BK54 66
Charminster Rd. SE9 CJ49 62
Charminster Rd. Wor. Pk. BQ54 67
Charmouth Rd. Well. CP44 54
Charmwood La. Orp. CO58 81
Charnock Ct. Cres. Swan. CT52 73
Charnock Rd. E5 CB34 24
Charnwood Av. SW19 BS51 68
Charnwood Clo. N. Mal. BO52 67
Charnwood Dr. E18 CH31 26
Charnwood Pl. N20 BT27 14
Charnwood Rd. SE25 BZ53 69
Charnwood St. E5 CB34 24
Charrington St. NW1 BW37 32
Charsley Rd. SE6 CE48 61
Chart Clo. Brom. CG51 70
Chart Clo. Croy. CC53 70
Chartecote Gro. SE26 CB48 60
Charter Av. Ilf. CM33 26
Charter Cres. Houns. BE45 47
Charter Dr. Bex. CQ47 63
Charterhouse Av. Wem. BK35 30
Charterhouse Bldgs. EC1 BY38 33
Goswell Rd.
Charterhouse Rd. Orp. CO55 81
Charterhouse Sq. EC1 BY39 42
Charterhouse St. EC1 BY39 42
Charteris Rd. N4 BY33 24
Charteris Rd. NW6 BR37 31
Charteris Rd. Wdf. Grn. CH29 17
Charter Rd. Kings. On T. BM52 66
Charters Clo. SE19 CA49 60
Charter Sq. Kings. On T. BM51 66
Charter, The, Rd. Wdf. Grn. CG29 16
Charter Way, N3 BR31 22
Regents Pk. Rd.
Charter Way, N14 BW25 7
Chartfield Av. SW15 BP46 49
Chartfield Sq. SW15 BQ46 49
Chartham Gro. SE27 BY48 60
Royal Circus
Chartham Rd. SE25 CB52 69
Chartley Av. NW2 BO34 22
Chartley Av. Stan. BH29 12
Charton Clo. Belv. CQ43 54
Chart St. N1 BZ38 33
Chartwell Clo. SE9 CM48 62
Chartwell Pl. Epsom BO60 82
Chartwell Pl. Sutt. BR55 76
Charwood, SE27 BY49 60
Leigham Ct. Rd.
Chase Ct. Gdns. Enf. BZ24 8
Chase Cross Rd. Rom. CS29 19
Chase End, Epsom BN59 82
Chasefield Rd. SW17 BU49 59
Chase Gdns. E4 CE28 16
Chase Gdns. Twick. BG46 47
Chase Grn. Av. Enf. BY23 8
Chase Grn. Enf. BZ24 8
Chase Hill, Enf. BZ24 8
Chase La. Chig. CN27 18
Chase La. Ilf. CM32 26
Chaseley St. E14 CD39 43
Chase Ridings, Enf. BY23 8
Chase Rd. Epsom BN59 80
Chase Rd. E18 CG30 16
Chase Rd. N14 BW25 7
Chase Rd. NW10 BO39 40
Chase Rd. W3 BO39 40
Chase Side Av. Enf. BZ23 8
Chase Side Av. SW20 BR51 67
Chase Side Clo. Rom. CT29 19
Chase Side Cres. Enf. BZ23 8
Chase Side, Enf. BZ24 8
Chase Side, N14 BV25 7
Chase, The, Bexh. CR45 54
Chase, The, Brom. CH52 71
Chase, The, Chig. CM28 17
Chase, The, Couls. BW60 83
Chase, The, Edge. BM30 12
Chase, The, E12 CJ35 35
Chase, The, Pnr. BD32 20
Chase, The, Pnr. BE31 20
Chase, The, Rom. CQ32 27
Chase, The, Rom. CT31 28
Chase, The, Rom. CT34 28
Chase, The, Stan. BJ29 12
Chase, The, Sun. BC51 65
Staines Rd.
Chase, The, SW4 BV45 50
Chase, The, SW16 BX50 60
Chase, The, SW20 BR51 67
Chase, The, Trading Est. NW10 BN38 31
Chase, The, Wall. BX56 78
Chaseville Pk. Rd. N21 BX25 8
Chase Way, N14 BV27 14
Chasewood Av. Enf. BY23 8
Chastilian Rd. Dart. CT47 64
Chaston St. NW5 BV35 32
Herbert St.
Chatfield Pl. W5 BL39 39
Park View Rd.
Chatfield Rd. Croy. BY54 69
Chatfield Rd. SW11 BT45 50
Chatham Av. Brom. CG54 70
Chatham Av. N1 BZ38 33
Nile St.
Chatham Clo. NW11 BS32 23
Chatham Clo. Sutt. BR54 67
Chatham Pl. E9 CC36 34
Chatham Rd. E17 CD31 25
Chatham Rd. E18 CG30 16
Chatham Rd. Kings. On T. BM51 66
Chatham Rd. Orp. CM56 80
Chatham Rd. SW11 BU46 50
Chatham St. SE17 BZ42 42
Chatsfield, Epsom BP58 76
Chatsworth Av. Brom. CH49 62
Chatsworth Av. NW4 BQ30 13
Chatsworth Av. Sid. CO47 63
Chatsworth Av. SW20 BR51 67
Chatsworth Av. Wem. BL35 30
Chatsworth Clo. Borwd. BM24 5
Chatsworth Clo. NW4 BQ30 13
Chatsworth Ct. W8 BS42 41
Chatsworth Cres. Houns. BG45 47
Chatsworth Dr. Enf. CB26 8
Chatsworth Est. E5 CC35 34
Chatsworth Gdns. W3 BM40 39
Chatsworth Gdns. Har. BF33 20
Chatsworth Gdns. N. Mal. BO53 67
Chatsworth Pde. Orp. CM53 71
Chatsworth Pl. Tedd. BJ49 57
Chatsworth Ri. W5 BL38 30

Chatsworth Rd. Croy. BZ55 78
Chatsworth Rd. Dart. CV45 55
Chatsworth Rd. E5 CC34 25
Chatsworth Rd. E15 CG35 34
Chatsworth Rd. Hayes BC38 29
Chatsworth Rd. NW2 BQ36 31
Chatsworth Rd. Sutt. BQ56 76
Chatsworth Rd. W4 BN43 49
Chatsworth Rd. W5 BL39 39
Chatsworth Way, SE27 BY48 60
Chatteris Av. Rom. CV29 19
Chatterton Rd. Brom. CJ52 71
Chatto Rd. SW11 BU46 50
Chaucer Av. Hayes BC39 38
Chaucer Av. Houns. BC44 47
Chaucer Av. Rich. BM44 48
Chaucer Clo. N11 BW28 14
Chaucer Gdns. Sutt. BS55 77
Chaucer Grn. Croy. CC54 70
Chaucer Rd. E7 CH36 35
Chaucer Rd. E11 CH32 26
Chaucer Rd. E17 CF30 16
Chaucer Rd. Rom. CU29 19
Chaucer Rd. SE24 BY46 51
Chaucer Rd. Sid. CP47 63
Chaucer Rd. Sutt. BS56 77
Chaucer Rd. Well. CN44 54
Chaucer Rd. W3 BN40 40
Chaucer Wk. Dart. CW45 55
Chauncey Clo. N9 CB27 15
Chave Rd. Dart. CW48 64
Cheam Common Rd. BP55 76
Wor. Pk.
Cheam Park Way, Sutt. BQ57 76
Cheam Pl. SE5 BZ43 51
Cheam Rd. Epsom BP58 76
Cheam Rd. Sutt. BR57 76
Cheam St. SE15 CC45 52
Evelina Rd.
Cheapside, EC2 BZ39 42
Cheddar Waye, Hayes BC39 38
Cheddington Rd. N18 CA27 15
Chedworth Clo. E16 CG39 43
Hallsville Rd.
Chelford Rd. Brom. CF49 61
Chelmer Cres. Bark. CO37 36
Chelmer Rd. E9 CC35 34
Chelmsford Av. Rom. CS29 19
Chelmsford Clo. W6 BR43 49
Chelmsford Dr. Upmin. CW34 28
Chelmsford Gdns. Ilf. CK33 26
Chelmsford Rd. E11 CF33 25
Chelmsford Rd. E17 CE32 25
Chelmsford Rd. E18 CG30 16
Chelmsford Rd. N14 BW26 7
Chelmsford Sq. NW10 BQ37 31
Chelsea Br. BV43 50
Chelsea Br. Rd. SW1 BV42 41
Chelsea Cloisters SW3 BU42 41
Makins St.
Chelsea Clo. Edg. BM30 12
Chelsea Clo. Hamptn. BG49 56
Chelsea Clo. NW10 BR37 31
Chelsea Ct. SW3 BU43 50
Chelsea Embankment, BU43 50
SW3
Chelsea Manor Est. SW3 BU43 50
Alpha Pl.
Chelsea Manor Gdns. BU42 41
SW3
Chelsea Manor St. SW3 BU42 41
Chelsea Pk. Gdns. SW3 BT43 50
Chelsea Sq. SW3 BT42 41
Chelsea Wk. SW3 BU43 50
Chelsea Wharf, SW10 BT43 50
Chelsfield Av. N9 CC26 9
Mottingham Rd.
Chelsfield Gdns. SE26 CC48 61
Chelsfield Hill, Orp. CP58 81
Chelsfield La. Orp. CP54 72
Chelsfield La. Sev. CR57 81
Chelsfield Rd. Orp. CP53 72
Chelsham Clo. Sutt. BR58 76
Chelsham Rd. SW4 BW45 50
Chelsham Rd. S. Croy. BZ57 78
Chelston Rd. Ruis. BC33 20
Chelsworth Clo. Rom. CW30 19
Chelsworth Dr. Rom. CW30 19
Chelsworth Dr. SE18 CM43 53
Cheltenham Av. Twick. BJ47 57
Cheltenham Clo. Nthlt. BF36 29
Cheltenham Gdns. E6 CK37 35
Cheltenham Gdns. Loug. CK25 10
Cheltenham Pl. Har. BL31 21
Cheltenham Pl. W3 BM40 39
Cheltenham Rd. Orp. CO55 81
Cheltenham Rd. SE15 CC45 52
Cheltenham Ter. SW3 BU42 41
Chelval St. E14 CE41 43
Chelverton Rd. SW15 BQ45 49
Chelwood Clo. Epsom BO59 82
Chelwood Gdns. Pass. BM44 48
Rich.
Pensford Av.
Chelwood Gdns. Rich. BM44 48
Chelwood Wk. SE4 CD45 52
Chenappa Clo. E13 CG38 34
Chenduit Way, Stan. BH28 12
Cheney Rd. NW1 BX37 33
Cheneys Rd. E11 CG34 25
Cheney St. Pnr. BD32 20
Chenies Ms. WC1 BW38 32
Chenies Pl. NW1 BW37 32
Chenies St. WC1 BW39 41
Chenies, The, Dart. CT49 64
Chenies, The, Orp. CN53 72
Cheniston Gdns. W8 BS41 41
Chepstow Av. Horn. CW34 28
Chepstow Clo. SW15 BR46 49
Chepstow Cres. Ilf. CN32 27
Chepstow Cres. W11 BS40 41
Chepstow Gdns. Sthl. BE39 38
Chepstow Pl. W2 BS40 41
Chepstow Ri. Croy. CA55 78
Chepstow Rd. Croy. CA55 78
Chepstow Rd. W2 BS39 41
Chepstow Rd. W7 BJ41 39
Chepstow Vill. W11 BR40 40
Chepstow Way, SE15 CA44 51
Sumner Estate
Chequers Gdns. N13 BY28 15
Chequers La. Dag. CQ38 36
Chequers Pde. SE9 CK46 53
Chequers Rd. Loug. CL25 10
Chequers Rd. Rom. CW27 19
Chequer St. EC1 BZ38 33
Chequers Way N13 BY28 15
Cherbury Ct. N1 BZ37 33
Cherbury St. N1 BZ37 33
Cherimoya Gdns. E. Mol. BF52 65
Cherington Rd. W7 BH40 39
Cheriton Av. Brom. CG53 70
Cheriton Av. Ilf. CK30 17
Cheriton Clo. W5 BK59 39
Queens Wk.
Cheriton Ct. KT12 BD54 65
Stratton Clo.
Cheriton Dr. SE18 CM43 53
Cheriton Sq. SW17 BV48 59
Cherry Av. Sthl. BD40 38
Cherry Av. Swan. CS52 73
Cherry Clo. Bans. BQ60 82
Cherry Clo. Cars. BU55 77
Cherry Clo. Mord. BR52 67
Cherry Clo. W5 BK41 39
Cherrycot Hill, Orp. CM56 80
Cherrycot Rise, Orp. CM56 80
Cherry Cres. Brent. BJ43 48
Cherrycroft Gdns. Pnr. BE29 11
Westfield Pk.
Cherrydown Av. E4 CD27 16
Cherrydown Clo. E4 CD27 16
Cherrydown Rd. Sid. CP48 63
Cherrydown Wk. Rom. CR30 18
Cherry Gdns. Dag. CQ35 36
Cherry Garden St. SE16 CB41 42
Cherry Garth, Brent. BK42 39
Cherry Gro. Hayes BC40 38
Cherry Hill, Barn. BS25 7
Cherry Hill Gdns. Croy. BX56 78
Cherry Orchard Clo. Orp. CP53 72
Cherry Orchard Gdns. BE52 65
E. Mol.
Cherry Orchard La. Brom. CK55 80
Cherry Orchard Rd. Croy. BZ55 78
Cherry Orchard Rd. BE52 65
E. Mol.
Cherry St. Rom. CS32 28
Cherry Tree Clo. Rain. CT37 37
Cherry Tree Ct. Couls. BX62 84
Coulsdon Rd.
Cherry Tree Dr. SW16 BX48 60
Cherry Tree Gdns. Croy.. BZ54 69
Oval Rd.
Cherry Tree Gdns. N2 BU31 23
Cherry Tree Grn. CB60 84
S. Croy.
Cherry Tree La. Rain. CT38 37
Cherry Tree Rise, Buck. H. CJ28 17
Cherry Tree Wk. Beck. CD52 70
Cherry Tree Wk. W. Wick. CG56 79
Cherry Tree Way, Stan. BJ28 12
Cherry Wk. Brom. CH54 71
Cherry Wk. Rain. CT37 37
Cherry Way, Epsom BN57 76
Cherrywood Clo. BM50 57
Kings. On T.
Alexandra Rd.
Cherrywood Ct. Tedd. BJ49 57
Elmfield Av.
Cherrywood Dr. SW15 BQ46 49
Cherrywood La. Mord. BR52 67
Cherry Wood Way W3 BM39 39
Vale La.
Cherston Gdns. Loug. CL24 10
Cherston Rd. Loug. CL24 10
Chertsey Clo. Ken. BY61 84
Chertsey Ct. SW14 BM45 48
Chertsey Cres. Croy. CF58 79
Chertsey Dr. Sutt. BR55 76
Chertsey Rd. E11 CF34 25
Chertsey Rd. Ilf. CM35 35
Chertsey Rd. Twick. BJ46 48
Chertsey Rd. W4 BM42 39
Chertsey St. SW17 BV49 59
Chervil Ms. SE28 CQ40 45
Cherwell Ct. Epsom BN56 76
Cheseman St. SE26 CB48 60
Chesfield Rd. Kings. On T. BL50 57
Chesham Av. Orp. CL53 71
Chesham Clo. Rom. CS31 28
Chesham Cres. SE20 CC51 70
Chesham Ms. SW1 BV41 41
Chesham Pl. SW1 BV41 41
Chesham Rd. Kings. On T. BM51 66
Chesham Rd. SE20 CC51 70
Chesham Rd. SW19 BT49 59
Chesham St. NW10 BN34 22
Chesham St. SW1 BV41 41
Chesham Ter. W13 BJ41 39
Cheshire Clo. Mitch. BX52 69
Cheshire Ct. EC4 BY39 42
Fleet St.
Cheshire Gdns. Chess. BK57 75
Cheshire Rd. N22 BX29 15
Chesholm Rd. N16 CA34 24
Cheshunt Rd. E7 CH36 35
Cheshunt Rd. Belv. CR42 45
Chesilton Rd. SW6 BR44 49
Chesley Gdns. E6 CJ37 35
Chesney Cres. Croy. CF57 79
Chesnut Clo. Cars. BU54 68
Chesnut Gro. Mitch. BW53 68
Chesny St. SW11 BU44 50
Chess Ct. E1 CA39 42
Old Castle St.
Chessington Av. Bexh. CQ43 54
Chessington Av. N3 BR31 22
Chessington Clo. Epsom BN57 76
Chessington Ct. Pnr. BE31 20
Chessington Hill Pk. BM57 75
Chess.
Chessington Pde. Chess. BK56 75
Chessington Rd. Epsom BM57 75
Chessington Way, CE55 79
W. Wick.
Chesson Rd. W14 BR43 49
Chesswood Way, Pnr. BD30 11
Chester Av. Rich. BL46 48
Chester Av. Twick. BE47 56
Chester Clo. Loug. CM23 10
Chester Clo. Sutt. BS55 77
Chester Clo. SW1 BV41 41
Chester Clo. SW15 BP45 49
Chester Cotts. SW1 BV42 41
Bourne St.
Chester Ct. NW5 BV35 32
Chester Ct. W3 BM39 39
Monks Dr.
Chester Dr. Har. BE32 20
Chesterfield Clo. Orp. CQ52 72
Chesterfield Dr. Esher BJ55 75
Chesterfield Gdns. W1 BV40 41
Chesterfield Gro. SE22 CA46 51
Chesterfield Hill, W1 BV40 41
Chesterfield Rd. Barn. BQ25 6
Chesterfield Rd. Epsom BN57 76
Chesterfield Rd. E10 CF32 25
Chesterfield Rd. N3 BS29 14
Chesterfield Rd. N4 BY32 24
Chesterfield Rd. W4 BN43 49
Chesterfield St. W1 BV40 41
Chesterfield Wk. SE10 CF44 52
Chesterfield Way, SE15 CC43 52
Chesterford Gdns. NW3 BS35 32
Chesterford Rd. E12 CK35 35
Chester Gdns. Enf. CB25 8
Chester Gdns. Mord. BT53 68
Chester Gdns. W13 BJ39 39
Chester Gate NW1 BV38 32
Outer Circle
Chester Grn. E2 CB38 33
Dunbridge St.
Chester Grn. Loug. CM23 10
Chester Gro. SE18 CM42 44
Chester Ms. SW1 BV41 41
Chester Path, Loug. CM23 10
Chester Pl. NW1 BV38 32
Chester Ter. Ms.
Chester Rd. Borwd. BN24 6
Chester Rd. Chig. CL27 17
Chester Rd. E7 CJ36 35
Chester Rd. E11 CH32 26
Chester Rd. E16 CG38 34
Chester Rd. E17 CC32 25
Chester Rd. Houns. BC45 47
Chester Rd. Ilf. CN33 27
Chester Rd. Loug. CL23 10
Chester Rd. N9 CB26 8
Chester Rd. N17 BZ31 24
Chester Rd. N19 BV34 23
Chester Rd. NW1 BV38 32
Chester Rd. Sid. CN46 54
Chester Rd. SW19 BQ50 58
Chester Rd. Wat. BC25 4
Chester Row, SW1 BV42 41
Chester Sq. SW1 BV42 41
Chesters, The, N. Mal. BO51 67
Chester St. SW1 BV41 41
Chester Ter. NW1 BV38 32
Chester Ter. Ms. NW1 BV38 32
Chesterton Clo. Grnf. BF37 29
Chesterton Rd. E13 CH38 35
Chesterton Rd. W10 BQ39 40
Chesterton SW18 BS46 50
Chesterton Ter. E13 CH38 35
Chesterton Ter. BM51 66
Kings. On T.
Chester Way, SE11 BY42 42
Chesthunte Rd. N17 BZ30 15
Chestney St. SW11 BV44 50
Battersea Pk. Rd.
Chestnut Av. Brent. BK42 39
Chestnut Av. Buck. H. CJ27 17
Chestnut Av. Edg. BL29 12
Chestnut Av. Epsom BO56 76
Chestnut Av. Esher BG54 65
Chestnut Av. E7 CH35 35
Chestnut Av. Hamptn. BF50 56
Chestnut Av. Horn. CT34 28
Chestnut Av. N8 BX32 24
Chestnut Av. N. E17 CF31 25
Chestnut Av. S. E17 CF32 25
Chestnut Av. SW14 BO45 49
Thornton Rd.
Chestnut Av. Tedd. BH52 66
Chestnut Av. Wem. BJ35 30
Chestnut Av. W. Wick. CG55 79
Chestnut Clo. Buck. H. CJ27 17
Chestnut Clo. Horn. CV35 37
Chestnut Clo. N14 BW25 7
Chestnut Clo. Orp. CO56 81
Chestnut Ct. N. Mal. BO52 67
Chestnut Dr. Bexh. CP45 54
Chestnut Dr. E11 CH32 26
Chestnut Dr. Har. BH29 12
Chestnut Dr. Pnr. BD32 20
Chestnut Gdns. Sutt. BS56 77
Elm Gro.
Chestnut Glen, Horn. CU34 28
Chestnut Gro. Barn. BU25 7
Chestnut Gro. Dart. CS49 64
Chestnut Gro. Ilf. CN29 18
Chestnut Gro. Islw. BJ45 48
Chestnut Gro. N. Mal. BN52 67
Chestnut Gro. S. Croy. CB57 78
Chestnut Gro. SW12 BV47 59
Chestnut Gro. W5 BK41 39
Chestnut Gro. Wem. BJ35 30
Chestnut La. N20 BQ26 6
Chestnut Ri. SE18 CM43 53
Chestnut Ri. Bush. BF26 4
Chestnut Rd. Dart. CV47 64
Chestnut Rd. Kings. On T. BL50 57
Chestnut Rd. N17 CA31 24
Chestnut Rd. SE27 BY48 60
Chestnut Rd. SW20 BQ51 67
Chestnut Rd. Pass. Twick. BH48 57
Chestnut Rd. Twick. BH48 57
Chestnuts, The, Walt. BC55 74
Chestnut Wk. Wdf. Grn. CH28 17
Chestnut Way, Felt. BC48 56
Cheston Av. Croy. CD54 70
Chesworth Clo. Erith CT44 55
Chettle Clo. SE1 BZ41 42
Spurgeon St.
Chetwode Dr. Epsom BQ62 82
Chetwode Rd. SW17 BU48 59
Chetwynd Av. Barn. BU26 7
Chetwynd Rd. NW5 BV35 32
Cheval Pl. SW7 BU41 41
Cheveley Clo. Rom. CW30 19
Chevening Rd. NW6 BQ37 31
Chevening Rd. SE10 CG42 43
Chevening Rd. SE19 BZ50 60
Chevenings, The, Sid. CO48 63
Cheverton Rd. N19 BW33 23
Chevet St. E9 CD35 34
Chevington Way, Horn. CV35 37
Cheviot Clo. Bans. BS61 83
Cheviot Clo. Bexh. CT44 55
Cheviot Clo. Bush. BG25 4
Cheviot Clo. Enf. BZ23 8
Cheviot Clo. Sutt. BT58 77
Cheviot Gdns. NW2 BQ34 22
Cheviot Gte. NW2 BR34 22
Cheviot Rd. Horn. CU33 28
Cheviot Rd. SE27 BY49 60
Cheviot Way, Grnf. BJ37 30
Cheviot Way, Ilf. CN32 27
Chewton Rd. E17 CD31 25
Cheyham Way, Sutt. BR58 76
Cheyne Av. E18 CG31 25
Cheyne Av. Twick. BE47 56
Cheyne Clo. Brom. CK55 80
Cheyne Clo. NW4 BQ32 22
Cheyne Ct. Bans. BS61 83
Cheyne Ct. SW3 BU43 50
Flood St.
Cheyne Gdns. SW3 BU43 50
Cheyne Hill, Surb. BL52 66
Cheyne Ms. SW3 BU43 50
Cheyne Path W7 BH39 39
Copley Clo.
Cheyne Pl. SW3 BT43 50
Cheyne Row SW3 BU43 50
Cheyne Wk. Croy. CB55 78
Cheyne Wk. N21 BY25 8
Cheyne Wk. NW4 BQ32 22
Cheyne Wk. SW3 BT43 50
Cheyne Wk. SW10 BT43 50
Cheyney Row E17 CD30 16

Church Rd. Pur. BX58 78
Church Rd. Rich. BK49 57
Church Rd. Rich. BL45 48
Church Rd. Rom. CW30 19
Church Rd. SE19 CA51 69
Church Rd. SE26 CB50 60
Church Rd. Shortlands CC52 70
Church Rd. Sid. CG49 63
Church Rd. Stan. BJ28 12
Church Rd. Surb. BK54 66
Church Rd. S. At H CW50 64
Church Rd. Sutt. BR57 76
Church Rd. Swan. CV51 73
Church Rd. SW13 BO44 49
Church Rd. SW19 BR49 58
Church Rd. SW19 BT51 68
Church Rd. Tedd. BH49 57
Church Rd. Wall. BW55 77
Church Rd. Wat. BC23 4
Church Rd. Whyt. CA62 84
Church Rd. Wimb. BK49 58
Church Rd. Wor. Pk. BO55 76
Church Rd. W3 BN40 40
Church Rd. W7 BG40 38
Church Row, Chis. CM51 71
Church Row NW3 BT35 32
Church Row SW6 BS43 50
Britannia Rd.
Church St. Croy. BY55 78
Church St. Dag. CR36 36
Church St. E15 CG37 34
Church St. E16 CL40 44
Church St. Enf. BZ24 8
Church St. Epsom BO60 82
Church St. Esher BF56 74
Church St. Est. NW8 BT38 32
Church St. Ewell BP58 76
Church St. Hmptn. BG51 65
Church St. Islw. BJ45 48
Church St. Kings. On T. BK51 66
Clarence St.
Church St. N. E15 CG37 34
Church St. N9 BZ26 8
Church St. NW8 BT39 41
Church St. Pl. E15 CG37 34
Church St. Sun. BC52 65
Church St. Sutt. BS56 77
Church St. Twick. BJ47 57
Church St. Walt. BC54 65
Church St. Wat. BC24 4
Church St. W2 BT39 41
Church St. W4 BO43 49
Church Stretton Rd. BG46 47
Hours.
Church Ter. N1 CA36 33
Mortimer Rd.
Church Ter. NW4 BP31 22
Church Ter. Rich. BK46 48
Wakefield Rd.
Church Ter. SE13 CG45 52
Church Ter. SW8 BW44 50
Union Gro.
Churchdale Ct. W4 BM42 39
Harvard Rd.
Church Vale N2 BU31 23
Church Vale SE23 CC48 61
Churchview Rd. Twick. BG47 56
Church Wk. N16 BZ35 33
Church Wk. Brent. BK43 48
Church Wk. Dart. CV48 64
Church Wk. NW2 BR34 22
Church Wk. NW4 BQ31 22
Church Wk. NW9 BN34 22
Church Wk. Rich. BK46 48
Church Wk. Surb. BH53 66
Church Wk. SW13 BP44 49
Church Wk. SW16 BW51 68
Church Wk. SW20 BQ52 67
Church Wk. Walt. BC54 65
Church Way, Barn. BU24 7
Church Way, Cars. BU56 77
Church Way, Edg. BM29 21
Churchway N20 BT27 14
Churchway NW1 BW38 32
Church Way, S. Croy. CA58 78
Churchwell Path E9 CC34 34
Churchyard Pass. SE5 BZ44 51
Camberwell Gro.
Church Yard Row SE11 BY42 42
Churston Av. E13 CH37 35
Churston Dr. Mord. BQ53 67
Churston Gdns. N11 BW29 14
Churton Pl. SW1 BW42 41
Churton St.
Churton St. SW1 BW42 41
Chuters Gro. Epsom BO59 82
Chyngton Clo. Sid. CN48 63
Priestlands Pk. Rd.
Cicada Rd. SW18 BT46 50
Cicely Rd. SE15 CB44 51
Cinderford Way, Brom. CG49 61
Cinnamon St. E1 CB40 42
Cintra Pk. SE19 CA50 60
Cinzen St. N7 BY35 33
Circle, The, NW2 BO34 22
Circle, The, NW7 BN29 13

Circuits, The, Pnr. BD31 20
Circular Rd. N17 CA31 24
Circular Rd. SE1 BZ41 42
Circular Way SE18 CK43 53
Falmouth Rd.
Circus Pl. EC2 BZ39 42
London Wall
Circus Rd. NW8 BT38 32
Circus St. SE10 CF43 52
Cirencester St. W2 BS39 41
Cissbury Ring, N. N12 BR28 13
Cissbury Ring, S. N12 BR28 13
Cissbury Rd. N15 BZ32 24
City Garden Row N1 BY37 33
City Rd. EC1 BY37 33
Minories
Civic Way, Ilf. CM31 26
Clabon Ms. SW1 BU41 41
Clack St. SE16 CC41 43
Clacton Path SE4 CD45 52
Frendsbury Rd.
Clacton Rd. E6 CJ38 35
Clacton Rd. E17 CD32 25
Claigmar Gdns. N3 BS30 14
Clancarty Rd. SW6 BS44 50
Clandon Av. N3 BS31 23
Clandon Clo. Epsom BO57 76
Clandon Clo. W3 BM41 39
Avenue Rd.
Clandon Rd. Ilf. CN34 27
Clandon St. SE8 CE44 52
Clanfield Way SE15 CA43 51
Hordle Prom. W.
Clanricarde Gdns. W2 BS40 41
Clapgate Rd. Bush. BF25 4
Clapham Com. N. Side BU45 50
SW4
Clapham Com. S. Side BV46 50
SW4
Clapham Com. W. Side BU45 50
SW4
Clapham High St. SW4 BW45 50
Clapham Pk. Est. E5 CC35 34
Clapham Pk. Est. SW2 BX47 60
Clapham Pk. Rd. SW4 BW46 50
Clapham Pk. Rd. SW4 BW45 50
Clapham Rd. SW9 BX45 51
Claps Gate La. Bark. CL38 35
Clapton Common E5 CA33 24
Clapton Pass. E5 CC35 34
Lwr. Clapton Rd.
Clapton Sq. E5 CC35 34
Clapton Ter. N16 CB33 24
Clapton Way E5 CB35 33
Clara Pl. SE18 CL42 44
Monk St.
Clare Clo. Borwd. BL25 5
Clare Corner SE9 CL47 62
Clarecroft Way W14 BQ41 40
Claredale St. E2 CB37 33
Clare Gdns. Bark. CN36 36
Clare Gdns. E7 CH35 35
Clare Gdns. Stan. BK28 12
Clare Gdns. W11 BR39 40
Westbourne Pk. Rd.
Clare Hall Pl. SE16 CC42 43
Litlington St.
Clarehill Clo. Esher BF56 74
Clarehill Rd. Esher BF57 74
Clare La. N1 BZ36 33
Clare Lawn Av. SW14 BN46 49
Clare Mkt. WC2 BX39 42
Portugal St.
Claremont Av. Esher BE57 74
Claremont Av. Har. BL32 21
Claremont Av. N. Mal. BP53 67
Claremont Av. Sun. BC51 65
Claremont Av. Walt. BD56 74
Claremont Clo. E16 CL40 44
Claremont Clo. N1 BY37 33
Claremont Clo. Orp. CL56 80
Claremont Clo. S. Croy. CB60 84
Claremont Clo. Walt. BD56 74
Claremont Cres. Dart. CT45 55
Claremont Dr. Esher BF58 74
Claremont End, Esher BF57 74
Claremont Est. SW2 BX47 60
Claremont Gdns. Ilf. CN34 27
Claremont Gdns. Surb. BL52 66
Claremont Gro. Wdf. Grn. CJ29 17
Claremont La. Esher BF56 74
Claremont Pk. N3 BR30 13
Claremont Park Rd. Esher BF57 74
Claremont Rd. Brom. CK52 71
Claremont Rd. Croy. CB54 69
Claremont Rd. Esher BF57 75
Claremont Rd. E7 CH35 35
Claremont Rd. E11 CF34 25
Claremont Rd. E17 CD30 16
Claremont Rd. Har. BH30 12

Claremont Rd. Horn. CU32 28
Claremont Rd. N6 BV31 23
Claremont Rd. NW2 BQ33 22
Claremont Rd. Surb. BL53 66
Claremont Rd. Swan. CT50 64
Claremont Rd. Tedd. BH49 57
Claremont Rd. Twick. BJ46 48
Claremont Rd. W9 BR37 31
Claremont Rd. W13 BJ39 39
Claremont St. N1 BY37 33
Claremont St. E16 CL40 44
Claremont St. N18 CB29 15
Claremont St. SE10 CE43 52
Claremont Way NW2 BQ33 22
Claremount Gdns. Epsom BO62 82
Clarence Av. Brom. CK52 71
Clarence Av. Ilf. CL32 26
Clarence Av. N. Mal. BN51 67
Clarence Av. SW4 BW47 59
Clarence Clo. Bush. BH26 5
Clarence Clo. Walt. BC56 74
Clarence Cres. Sid. CO48 63
Clarence Cres. SW4 BW46 50
Clarence Gdns. NW1 BV38 32
Clarence La. SW15 BO46 49
Clarence Ms. E5 CB35 33
Clarence Pl.
Clarence Rd. Bexh. CQ45 54
Clarence Rd. Brom. CJ52 71
Clarence Rd. Croy. BZ54 69
Clarence Rd. Enf. CB25 8
Clarence Rd. E5 CB35 33
Clarence Rd. E5 CB35 33
Clarence Rd. E12 CJ35 35
Clarence Rd. E16 CG38 34
Clarence Rd. E17 CC30 16
Clarence Rd. N15 BZ32 24
Clarence Rd. N22 BX29 15
Clarence Rd. NW6 BR38 31
Clarence Rd. Rich. BL44 48
Clarence Rd. SE9 CK48 62
Clarence Rd. Sid. CO48 63
Clarence Rd. Sutt. BS56 77
Clarence Rd. SW19 BS50 59
Clarence Rd. Tedd. BH50 57
Clarence Rd. Walt. BV56 77
Clarence Rd. Walt. BC56 74
Clarence Rd. W4 BM42 39
Clarence St. Kings. On T. BK51 66
Clarence St. Rich. BL45 48
Clarence St. Sthl. BD41 38
Clarence Ter. Houns. BF45 47
Clarence Ter. NW1 BU38 32
Cornwall Ter.
Clarence Wk. SW4 BX44 51
Jeffrey's Rd.
Clarence Way Est. NW1 BV36 32
Clarence Way NW1 BV36 32
Clarendon Av. SE5 BZ43 51
Councillor St.
Clarendon Clo. Orp. CO52 72
Clarendon Clo. W2 BU40 41
Clarendon Cr.
Clarendon Ct. NW11 BR31 22
Finchley Rd.
Clarendon Cres. Twick. BG48 56
Clarendon Cross W11 BR40 40
Portland Rd.
Clarendon Dr. SW15 BQ45 49
Clarendon Gdns. Ilf. CK33 26
Clarendon Gdns. NW4 BP31 22
Clarendon Gdns. Wem. BK35 30
Clarendon Gdns. W9 BT32 38
Clarendon Grn. Orp. CO52 72
Clarendon Gro. Mitch. BU52 68
Clarendon Gro. NW1 BW38 32
Phoenix Rd.
Clarendon Gro. Orp. CO52 72
Clarendon Ms. W2 BU40 41
Clarendon Pl.
Clarendon Path, Orp. CO52 72
Clarendon Ri. SE13 CF45 52
Clarendon Rd. Borwd. BM24 5
Clarendon Rd. Croy. BY55 78
Clarendon Rd. E11 CF33 25
Clarendon Rd. E17 CE32 25
Clarendon Rd. E18 CH31 26
Clarendon Rd. Har. BH32 21
Clarendon Rd. N15 BY31 24
Clarendon Rd. N18 CB29 15
Clarendon Rd. N22 BX30 15
Clarendon Rd. SW19 BU50 59
Clarendon Rd. Wall. BW57 77
Clarendon Rd. Wat. BC24 4
Clarendon Rd. W5 BL38 30
Clarendon Rd. W11 BR40 40
Clarendon St. SW1 BV42 41
Clarendon Ter. W9 BT38 32
Maida Vale
Clarendon Way, Chis. CN52 72
Clarendon Way N21 BZ25 8
Clarendon Way, Orp. CN52 72
Clarens St. SE6 CD48 61
Clare Rd. E11 CF32 25
Clare Rd. Grnf. BG36 29
Clare Rd. Houns. BE45 47
Clare Rd. NW10 BP36 31

Clare St. E2 CB37 33
Claret Gdns. SE25 CA52 69
Clareville Gro. SW7 BT42 41
Clareville Rd. Orp. CM55 80
Clareville St. SW7 BT42 41
Gloucester Rd.
Clare Way, Bexh. CQ44 54
Clarewood Mans. SW9 BY45 51
Coldharbour La.
Clarges Ms. W1 BV40 41
Clarges St. W1 BV40 41
Claribel Rd. SW9 BY44 51
Clarice Way, Wall. BX58 78
Claridge Rd. Dag. CP33 27
Clarina Rd. SE20 CC50 61
Clarissa Rd. Rom. CP33 27
Clarissa St. E8 CA37 33
Clark Clo. Erith CU44 55
Forest Rd.
Clarke Rd. SE14 CD44 52
Clarke's Av. Wor. Pk. BO54 67
Clarks Mead, Bush. BG26 4
Clarkson Rd. E16 CG39 43
Clarksons, The, Bark. CM37 35
Clarkson St. E2 CB38 33
Clarks Rd. Ilf. CM34 26
Clark St. E1 CB39 42
Claston Clo. DA1 CT45 55
Clatre Ct. N12 BT28 14
Claude Rd. E10 CF34 25
Claude Rd. E13 CH37 35
Claude Rd. SE15 CB44 51
Godman Rd.
Claude St. E14 CE42 43
Claudia Pl. SW19 BR47 58
Augustus Rd.
Claughton Rd. E13 CJ37 35
Clausen Way, Grnf. BJ37 30
Clauson Av. Nthlt. BF35 29
Claverdale Rd. SW2 BX47 60
Clavering Av. SW13 BP43 49
Clavering Clo. Twick. BJ49 57
Clavering Rd. E12 CJ33 26
Claverley Gro. N3 BS30 14
Claverley Vill. N3 BS30 14
Claverley Gro.
Claverton St. SW1 BW42 41
Clave St. E1 CC40 43
Claxton Gro. W6 BQ42 40
Claybridge Rd. SE12 CJ49 62
Claybury Broadway, Ilf. CK31 26
Claybury, Bush. BF25 4
Claybury Rd. Wdf. Grn. CK29 17
Clay Farm Rd. SE9 CM48 62
Claygate Clo. Horn. CT35 37
Carfax Rd.
Claygate Cres. Croy. CF57 79
Claygate La. Esher BJ54 66
Claygate Lodge Clo. Esher BH57 75
Claygate Rd. W13 BJ41 39
Clayhall Av. Ilf. CK31 26
Clay Hill, Bush. BF25 4
Clayhill Cres. SE9 CJ49 62
Claylands Pl. SW8 BY43 51
Claylands Rd. SW8 BX43 51
Clay La. Bush. BH26 5
Clay La. Edg. BM27 12
Claypole Rd. E15 CF37 34
Claypoints Av. Brent. BL35 39
Claypoints Gdns. W5 BK42 39
Claypoints La. Brent. BL42 39
Clay Ride, Loug. CK25 10
Clay Side, Chig. CM28 17
Clays Lane Clo. E15 CE35 34
Clays La. E15 CE35 34
Clay's La. Loug. CL23 10
Clay St. W1 BU39 41
Dorset St.
Clayton Av. Wem. BL36 30
Clayton Cres. Brent. BK42 39
Clayton Croft, Dart. CU48 64
Clayton Mead NW9 BO29 13
Clayton Rd. Chess. BK56 75
Clayton Rd. Epsom BO60 82
Clayton Rd. Islw. BH45 48
Clayton Rd. Rom. CS33 28
Clayton Rd. SE15 CB44 51
Clayton St. SE11 BY43 51
Claywood Clo. Orp. CN54 72
Cleanthus Rd. SE18 CL44 53
Clearbrook Way E1 CC39 43
W. Arbour St.
Clearwell Dr. W9 BS38 31
Amberley Est.
Cleave Av. Orp. CN57 81
Cleaveland Rd. Surb. BK53 66
Cleaverholme Clo. SE25 CB53 69
Cleaver Sq. SE11 BY42 42
Cleeve Hill SE23 CB47 60
Clegg Pl. SW15 BQ45 49
Clegg St. E1 CB40 42
Prusom St.
Clegg St. E13 CH37 35
Cleland Path, Loug. CL23 10
Clematis Clo. Rom. CV29 19
Clematis St. W12 BP40 40
Clem Attlee Ct. SW6 BR43 49
Clem Av. SW4 BW45 50

Clemence St. E14 CD39 43
Clement Clo. NW6 BQ36 31
Clement Clo. Pur. BY57 78
Clement Clo. Pur. BY61 84
Clementhorpe Rd. Dag. CP36 36
Clementina Rd. E10 CD33 25
Clement Rd. Beck. CC51 70
Clement Rd. SW19 BR49 58
Clements Av. E16 CH40 44
Clements Ct. Houns. BD45 47
Clements Inn WC2 BX39 42
Strand
Clement's La. EC4 BZ40 42
Clements La. Ilf. CL34 26
Clement's Pl. Brent. BK42 39
Challis Rd.
Clements Rd. E6 CK36 35
Clements Rd. Ilf. CL34 26
Clement's Rd. SE16 CB41 42
Clements Rd. Walt. BC55 74
Clement St. Swan. CV50 64
Clement Way, Upmin. CW34 28
Clendon Way SE18 CM42 44
Polthorne Gro.
Clennam St. SE1 BZ41 42
Southwark Bridge Rd.
Clensham Ct. Sutt. BS55 77
Clensham La. Sutt. BS55 77
Clenston Ms. W1 BU39 41
Seymour Pl.
Clephane Rd. N1 BZ36 33
Cleremont Rd. E9 CC37 34
Clere Pl. EC2 BZ38 33
Clere St.
Clere St. EC2 BZ38 33
Clerkenwell Clo. EC1 BY38 33
Clerkenwell Grn. EC1 BY38 33
Clerkenwell Rd. EC1 BY38 33
Clerks Piece, Loug. CK24 10
Clevedon Clo. N16 CA34 24
Smalley Rd.
Clevedon Gdns. Houns. BC44 47
Clevedon Pass. N16 CA34 24
High St.
Clevedon Rd. Kings. On T. BM51 66
Clevedon Rd. SE20 CC51 70
Clevedon Rd. Twick. BK46 48
Clevedon St. N16 CA34 24
Sanford La.
Cleveland Av. Hamptn. BE50 56
Cleveland Av. SW20 BR51 67
Cleveland Av. W4 BO42 40
Cleveland Clo. Walt. BC55 74
Cleveland Ct. W13 BJ39 39
Kent Av.
Cleveland Cres. Borwd. BN25 6
Cleveland Est. E1 CC38 34
Cleveland Gdns. N4 BZ32 24
Cleveland Gdns. NW2 BQ34 22
Cleveland Gdns. SW13 BO44 49
Cleveland Gdns. W2 BT39 41
Cleveland Gdns. W13 BJ39 39
Argyle Rd.
Cleveland Gdns. Wor. Pk. BO55 76
Cleveland Gro. E1 CC38 34
Cleveland Way
Cleveland Ms. WC1 BW39 41
Maple St.
Cleveland Pk. Av. E17 CD31 25
Cleveland Pk. Cres. E17 CD31 25
Cleveland Ri. Mord. BQ54 67
Cleveland Rd. E18 CH31 26
Cleveland Rd. Ilf. CL34 26
Cleveland Rd. Islw. BJ45 48
Cleveland Rd. N. Mal. BO52 67
Cleveland Rd. N1 BZ36 33
Cleveland Rd. N9 CB26 8
Cleveland Rd. SW13 BO44 49
Cleveland Rd. W4 BN41 40
Cleveland Rd. W13 BJ39 39
Cleveland Rd. Well. CG44 52
Cleveland Rd. Wor. Pk. BO55 76
Cleveland Row SW1 BW40 41
Cleveland Sq. W2 BT39 41
Cleveland St. W1 BW38 32
Cleveland Ter. W2 BT39 41
Cleveland Way E1 CC38 34
Cleveley Cres. W5 BL37 30
Cleveleys Rd. E5 CB34 24
Cleverley Est. W12 BP40 40
Cleve Rd. NW6 BS36 32
Cleve Rd. Sid. CP48 63
Cleves Av. Epsom BP58 76
Cleves Rd. E6 CJ37 35
Cleves Rd. Rich. BK48 57
Cleves Wk. Ilf. CM29 17
Cleves Way, Hamptn. BE50 56
Cleves Way, Ruis. BD33 20
Clewer Cres. Har. BG30 11
Clichy Est. E1 CC39 43
Clifden Rd. E5 CC35 34
Clifden Rd. Brent. BK43 48
Clifden Rd. Twick. BH47 57
Cliffe End, Pur. BY59 84
Cliffe Rd. S. Croy. BZ56 78
Clifford Av. Chis. CK50 62
Clifford Av. Ilf. CL30 17
Clifford Av. SW14 BM45 48
Clifford Av. Wall. BW56 77
Clifford Clo. Nthlt. BE37 29
Clifford Ct. NW2 BO35 31
Clifford Gdns. NW10 BQ37 31
Clifford Gro. SE20 CC50 61
Clifford Rd. Barn. BS24 7
Clifford Rd. E16 CG38 34
Clifford Rd. E17 CF30 16
Clifford Rd. Houns. BD45 47
Clifford Rd. N9 CC25 9
Clifford Rd. Rich. BK48 57
Clifford Rd. SE25 CB52 69
Clifford Rd. Wem. BK36 30
Clifford's Inn EC4 BY39 42
Fleet St.
Clifford St. W1 BW40 41
Clifford St. Wat. BD24 4
Clifford Way NW10 BO35 31
Cliff Rd. NW1 BW36 32
Cliff Ter. SE8 CE44 52
Cliffview Rd. SE13 CE45 52
Cliff Vill. NW1 BW36 32
Cliff Wk. E16 CG39 43
Clifton Av. E17 CC31 25
Clifton Av. Felt. BD48 56
Clifton Av. N3 BR30 13
Clifton Av. Stan. BJ30 12
Clifton Av. Sutt. BS58 77
Clifton Av. Wem. BL36 30
Clifton Av. W12 BO41 40
Clifton Clo. Orp. CL56 80
Clifton Ct. NW8 BT38 32
Clifton Cres. SE15 CB43 51
Clifton Est. SE15 CB44 51
Clifton Gdns. Enf. BX24 8
Clifton Gdns. N15 CA32 24
Clifton Gdns. NW11 BR31 22
Clifton Gdns. W4 BN42 40
Dolman Rd.
Clifton Gdns. W9 BT38 32
Clifton Hill NW8 BS37 32
Clifton Pk. Av. SW20 BQ51 67
Approach Rd.
Clifton Pl. Bans. BS61 83
Clifton Pl. W2 BT40 41
Clifton Ri. SE14 CD43 52
Clifton Rd. E16 CG39 43
Clifton Rd. Couls. BV61 83
Clifton Rd. E7 CJ36 35
Clifton Rd. E16 CG39 43
Clifton Rd. Grnf. BG38 29
Clifton Rd. Har. BL32 21
Clifton Rd. Horn. CU32 28
Clifton Rd. Ilf. CM32 26
Clifton Rd. Islw. BG44 47
Clifton Rd. Kings. On T. BL50 57
Clifton Rd. Loug. CK24 10
Clifton Rd. N3 BT30 14
Clifton Rd. N8 BW32 23
Clifton Rd. N22 BW30 14
Clifton Rd. NW10 BP37 31
Clifton Rd. Sid. CN49 63
Clifton Rd. Sthl. BE42 38
Clifton Rd. SE25 BZ52 69
Clifton Rd. SW19 BQ50 58
Clifton Rd. Tedd. BH49 57
Clifton Rd. Wall. BV56 77
Clifton Rd. Wat. BC25 4
Clifton Rd. Well. CO45 54
Clifton Rd. W9 BT38 32
Clifton St. E8 CB36 33
Graham Rd.
Clifton St. EC2 CA39 42
Clifton Ter. N4 BY34 24
Clifton Vills. W9 BS39 41
Clifton Way SE15 CB43 51
Clifton Way, Wem. BL37 30
Climpsing Grn. Belv. CQ41 45
Clinch Ct. E16 CH39 44
Cline Rd. N11 BW29 14
Clink St. SE1 BZ40 42
Clinton Av. E. Mol. BG52 65
Clinton Av. Well. CN45 54
Clinton Cres. Ilf. CN29 18
Clinton Rd. E3 CD38 34
Clinton Rd. E7 CH35 35
Clinton Rd. N15 BZ31 24
Clipper Way SE13 CF45 52
Limes Gro.
Clippesby Clo. Chess. BL57 75
Clipstone Ms. BW39 41
Clipstone Rd. Houns. BF45 47
Clipstone St. W1 BV39 41
Clissold Clo. N2 BU31 23
Clissold Ct. N16 BZ34 24
Clissold Cres. N16 BZ35 33
Clissold Est. N16 BZ34 24
Clissold Rd. N16 BZ34 24
Clitheroe Av. Har. BF33 20
Clitheroe Gdns. Wat. BD27 11
Clitheroe Rd. Rom. CS28 19
Clitheroe Rd. SW9 BX44 51
Clitherow Av. W7 BJ41 39
Clitherow Rd. Brent. BJ42 39
Clitterhouse Cres. NW2 BQ33 22
Clitterhouse Rd. NW2 BQ33 22
Clive Av. Dart. CT46 55
Clive Av. N18 CB29 15
Claremont St.
Clive Ct. Surb. BM55 75
Cliveden Clo. N12 BT28 14
Cliveden Rd. SW19 BR51 67
Clivedon Pl. SW1 BV42 41
Clivedon Rd. E4 CG28 16
Clive Pass. SE21 BZ48 60
Clive Rd.
Clive Rd. Belv. CQ42 45
Clive Rd. Enf. CB24 8
Clive Rd. Esher BF56 74
Clive Rd. Felt. BC46 47
Clive Rd. Rom. CU32 28
Clive Rd. SE21 BZ48 60
Clive Rd. SW19 BU50 59
Clive Rd. Twick. BH49 57
Clivesdale Dr. Hayes BC40 38
Avondale Dr.
Clive Way, Couls. BX62 84
Clive Way, Enf. CB24 8
Clive Way, Wat. BD23 4
Cloak La. EC4 BZ40 42
Clockhouse Av. Bark. CM37 35
Clockhouse La. Rom. CR29 18
Clock House Rd. Beck. CD51 70
Clock Pl. SE17 BY42 42
Hampton St.
Clock Tower Ms. N1 BZ37 33
Arlington Av.
Clock Tower Rd. Islw. BH45 48
Cloister Gdns. Edg. BO28 13
Cloister Gdns. SE25 CB53 69
Cloister Rd. NW2 BR34 22
Cloister Rd. W3 BN39 40
Cloisters Av. Brom. CK53 71
Clonard Way, Pnr. BF29 11
Clonbrock Rd. N16 CA35 33
Cloncurry St. SW6 BQ44 49
Clonmell Rd. N17 BZ31 24
Clonmel Rd. SW6 BR43 49
Clonmel Rd. Tedd. BG49 56
Clonmore St. SW18 BR47 58
Clonnel Clo. Har. BG34 20
Cloonmore Av. Orp. CN56 81
Clorane Gdns. NW3 BS34 23
Close, The, Barn. BU25 7
Close, The, Beck. CD52 70
Close, The, Bex. CR47 63
Close, The, Bush. BF25 4
Close, The, Cars. BU58 77
Close, The, Dart. CV48 64
Close, The E4 CF29 16
Close, The, Har. BG30 11
Close, The, Islw. BG44 47
Close, The, Mitch. BU52 68
Close, The, N. Mal. BN51 67
Close, The N10 BV30 14
Close, The N14 BW27 14
Close, The N20 BR27 13
Close, The, Orp. CN53 72
Close, The, Pnr. BD33 20
Close, The, Pnr. BE33 20
Close, The, Pur. BX58 78
Close, The, Pur. BY58 78
Close, The, Rich. BM45 48
Close, The, Rom. CQ32 27
Close, The, Sid. CO49 63
Close, The, Sutt. BR54 67
Close, The, Wem. BL36 30
Close, The, Wem. BN34 22
Cloth Fair EC1 BY39 42
Clothier St. E1 CA39 42
Cutler St.
Cloudberry Rd. Rom. CV29 19
Cloudesdale Rd. SW17 BV48 59
Cloudesley Pl. N1 BY37 33
Cloudesley Rd. Bexh. CQ44 54
Cloudesley Rd. Erith CT44 55
Cloudesley Rd. N1 BY37 33
Cloudesley Sq. N1 BY37 33
Cloudesley St. N1 BY37 33
Clouston Clo. Wall. BW56 77
Clova Rd. E7 CG36 34
Clovelly Av. NW9 BO31 22
Clovelly Ct. Horn. CW34 28
Clovelly Gdns. Enf. CA26 8
Clovelly Gdns. Rom. CR30 18
Clovelly Rd. Bexh. CQ43 54
Clovelly Rd. Houns. BF44 47
Clovelly Rd. N8 BW31 23
Clovelly Rd. W4 BN41 40
Clovelly Rd. W5 BK41 39
Clovelly Way E1 CC39 43
Jamaica St.
Clovelly Way, Har. BE34 20
Clovelly Way, Orp. CN53 72
Clover Clo. E11 CF34 25
Cloverdale Gdns. Sid. CN46 54
Clover Ms. SW3 BU43 50
Clove St. E13 CH38 35
Clowders Rd. SE6 CD48 61
Cloysters Gn. E1 CB40 42
St. Anthonys Clo.
Cloyster Wood, Edg. BK29 12
Club Gdns. Rd. Brom. CH54 71
Club Row E1 CA38 33
Club Row E2 CA38 33
Clunas Gdns. Rom. CV31 28
Clunbury Av. Sthl. BE42 38
Cluny Ms. SW5 BS42 41
Cluny Pl. SE1 CA41 42
Clutton St. E14 CE39 43
Clydach Rd. Enf. CA24 8
Clyde Av. S. Croy. CB61 84
Clyde Cir. N15 CA31 24
Clyde Pl. E10 CE33 25
Clyde Pl. SE23 CC48 61
Clyde Rd. Croy. CA55 78
Clyde Rd. N15 BZ31 24
Clyde Rd. N22 BW30 14
Clyde Rd. Sutt. BS56 77
Clyde Rd. Wall. BW56 77
Clydesdale Av. Stan. BK31 21
Clydesdale Clo. Borwd. BN25 6
Clydesdale Gdns. Rich. BM45 48
Clydesdale Rd. Enf. CC24 9
Clydesdale Rd. Horn. CT33 28
Clydesdale Rd. W11 BR39 40
Clyde St. SE8 CD43 52
Clyde Ter. SE23 CC48 61
Clyde Vale SE23 CC48 61
Clyde Way, Rom. CT30 19
Clydon Clo. Erith CT43 55
Clymping Dene, Felt. BC47 56
Clyn Clo. SE25 CA51 69
Clyston St. SW8 BW44 50
Coach & Horses Yd. W1 BW40 41
Old Burlington St.
Coach House La. SW19 BQ49 58
Coalecroft Rd. SW15 BQ45 49
Coaley Row, Dag. CQ37 36
Coal Wharf Rd. W12 BQ42 40
Coalwith Rd. W6 BQ44 49
Coates Hill Rd. Brom. CL51 71
Coates Rd. Borwd. BK26 5
Coate St. E2 CB37 33
Cobbett Rd. SE9 CK45 53
Cobbett Rd. Twick. BF47 56
Cobbetts Av. Ilf. CJ32 26
Cobbler's Wk. Tedd. BG50 56
Cobbold Est. NW10 BO36 31
Cobbold Rd. E11 CG34 25
Cobbold Rd. NW10 BO36 31
Cobbold Rd. W12 BO41 40
Cobbs Rd. Houns. BE45 47
Cobb St. E1 CA39 42
Leyden St.
Cobden Rd. E11 CG34 25
Cobden Rd. Orp. CM56 80
Cobden Rd. SE25 CB53 69
Cobham Av. N. Mal. BP53 67
Cobham Clo. Brom. CK54 71
Cobham Clo. SW11 BU46 50
Cobham Rd. Bark. CM37 35
St. Margaret's Rd.
Cobham Rd. E17 CF30 16
Cobham Rd. Houns. BC43 47
Cobham Rd. Ilf. CN34 27
Cobham Rd. Kings. On T. BM51 66
Cobham Rd. N22 BY31 24
Cobill Clo. Horn. CV31 28
Cobland Rd. SE12 CJ49 62
Coborn Rd. E3 CD37 34
Coborn St. E3 CD38 34
Cobourg Rd. SE5 CA43 51
Cobourg St. NW1 BW38 32
Coburgh Cres. SW2 BX47 60
Coburg Rd. N22 BX31 24
Cochrane Clo. NW8 BT37 32
Cochrane Ms. NW8 BT37 32
Cochrane Rd. SW19 BR50 58
Cochrane St. NW8 BT37 32
Cockcayne Wk. SE8 CD42 43
Cock Hill E1 CA39 42
New St.
Cock La. EC1 BY39 42
Cockmannings La. Orp. CP54 72
Cockmannings Rd. Orp. CP54 72
Cockpit Yd. WC1 BX39 42
Northington St.
Cocks Cres. N. Mal. BO52 67
Cocksett Av. Orp. CN57 81
Cockspur St. SW1 BW40 41
Cocksure La. Sid. CR48 63
Cock Yd. SE5 BZ44 51
Denmark Hill
Code St. E1 CA38 33
Codicote Ter. N4 BZ34 24
Codling Way, Wem. BK35 30
Codrington Hl. SE23 CD47 61
Codrington Ms. W11 BR39 40
Blenheim Cres.
Cody Clo. Har. BK31 21
Cody Clo. Wall. BW57 77
Alcock Clo.
Cody Rd. E16 CF38 34
Cofers Circ. Wem. BM34 21
Cogan Av. E17 CD30 16
Coin St. SE1 BY40 42
Coisy St. NW5 BV36 32
Coity Rd. NW5 BV36 32
Cokers La. SE21 BZ47 60
Perifield
Coke St. E1 CB39 42

Colas Ms. NW6 BS37 32
Birchington Rd.
Colbeck Ms. SW7 BS42 41
Colbeck Rd. Har. BG33 20
Colberg Pl. N16 CA33 24
Colborne Way, Wor. Pk. BQ55 76
Colburn Av. Pnr. BE29 11
Colburn Way, Sutt. BT55 77
Colby Rd. SE19 CA49 60
Colby Rd. Walt. BC54 65
Colchester Av. E12 CK34 26
Colchester Dr. Pnr. BD32 20
Colchester Rd. E10 CF33 25
Colchester Rd. E17 CE32 25
Colchester Rd. Edg. BN30 13
Colchester Rd. Nthwd. BC30 11
Colchester Rd. Rom. CV30 19
Colcokes Rd. Bans. BS61 83
Coldbath Sq. EC1 BY38 33
Topham St.
Coldbath St. SE13 CE44 52
Cold Blow Cres. Bexh. CS47 64
Cold Blow La. SE14 CC43 52
Cold Blows, Mitch. BU52 68
Coldershaw Rd. W13 BJ40 39
Coldfall Av. N10 BV30 14
Coldhara La. Pur. BY54 69
Coldharbour La. Bush. BF25 4
Coldharbour La. Hayes BC41 38
Coldharbour La. Pur. CH58 80
Coldharbour La. SE5 BZ45 51
Coldharbour La. SW9 BY45 51
Coldharbour Rd. Croy. BY56 78
Coldharbour Way, Croy. BY56 78
Coldstream Gdns. SW18 BR46 49
Colebeck Ms. N1 BY36 33
Colebert Av. E1 CC38 34
Colebrook Av. W13 BJ39 39
Colebrook Clo. SW15 BQ47 58
Colebrooke Dr. E11 CH33 26
Colebrooke Ri. Brom. CG51 70
Colebrooke Row N1 BY37 33
Colebrook Gdns. Loug. CM23 10
Colebrook La. Loug. CL23 10
Colebrook Path, Loug. CL23 10
Colebrook Rd. E17 CD31 25
Colebrook Rd. SW16 BX51 69
Colebrook Way N11 BV28 14
High Rd.
Coleby Path SE5 BZ43 51
Elmington Est.
Cole Clo. SE28 CO40 45
Cole Ct. W6 BQ43 49
Coledale Dr. Stan. BK30 12
Coleford Clo. Loug. CL23 10
Coleford Rd. SW18 BT46 50
Colegrave Rd. E15 CF35 34
Colegrove Rd. SE15 CA43 51
Coleherne Ct. SW5 BS42 41
Coleherne Ms. SW10 BS42 41
Coleherne Rd. SW10 BS42 41
Colehill Gdns. SW6 BR44 49
Fulham Palace Rd.
Colehill La. SW6 BR44 49
Coleman Clo. E1 CC40 43
Garnet St.
Coleman Ct. E1 CC40 43
Garnet St.
Coleman Ct. SW18 BS47 59
Coleman Flds. N1 BZ37 33
Coleman Rd. Belv. CR42 45
Coleman Rd. Dag. CQ36 36
Coleman Rd. SE5 CA43 51
Coleman Hth. SE9 CL48 62
Coleman St. EC2 BZ39 42
Colenso Rd. E5 CC35 34
Colenso Rd. Ilf. CN33 27
Cole Pk. Gdns. Twick. BJ46 48
Cole Pk. Rd. Twick. BJ46 48
Colepits Wood Rd. SE9 CM46 53
Coleraine Rd. N8 BY31 24
Coleraine Rd. SE3 CG43 52
Coleridge Av. E12 CK36 35
Coleridge Clo. SW8 BV44 50
Coleridge Gdns. NW6 BT36 32
Coleridge La. N8 BX32 24
Coleridge Rd.
Coleridge Rd. Croy. CC54 70
Coleridge Rd. E17 CD31 25
Coleridge Rd. N4 BY34 24
Coleridge Rd. N8 BW32 23
Coleridge Rd. N12 BT28 14
Coleridge Rd. Rom. CU29 19
Coleridge Wk. NW11 BS31 23
Coleridge Way, Hayes BC39 38
Coleridge Way, Orp. CO53 72
Cole Rd. Twick. BJ46 48
Cole Rd. Wat. BC23 4
Colesburg Rd. Beck. CD52 70
Coles Cres. Har. BF34 20
Colescroft Hill, Pur. BY61 84
Coles Grn. Bush. BG26 4
Coles Grn. Ct. NW2 BP34 22
Coles Grn. Loug. CL23 10
Coles Grn. Rd. NW2 BP33 22
Coleshill Rd. Tedd. BH50 57
Colestown St. SW11 BU44 50
Cole St. SE1 BZ41 42
Cole St. SW11 BU44 50
Colet Ct. W6 BQ42 40
Colet Gdns. W14 BQ42 40
Coley St. WC1 BX38 33
Colfe Rd. SE23 CD47 61
Colina Ms. N8 BY31 24
Park Rd.
Colina Rd. N15 BY32 24
Colin Clo. Croy. CD55 79
Colin Clo. NW9 BO31 22
Colin Clo. W. Wick. CG55 79
Colin Cres. NW9 BO31 22
Colindale Av. Erith CR43 54
Colindale Av. NW9 BN31 22
Colindeep Gdns. NW4 BP31 22
Colindeep La. NW9 BN31 22
Colin Dr. NW9 BO32 22
Colinette Rd. SW15 BQ45 49
Colin Gdns. NW9 BO32 22
Colin Pk. Rd. NW9 BO31 22
Colin Rd. NW10 BP36 31
Colinton Rd. Ilf. CO34 27
Coliston Rd. SW18 BS47 59
Collamore Av. SW18 BU47 59
Collapit Clo. Har. BF32 20
Collard Av. Loug. CM23 10
Collard Grn. Loug. CM23 10
College App. SE10 CF43 52
College Av. Epsom BO60 82
College Av. Har. BH30 12
College Clo. E9 CC35 34
Churchill Wk.
College Clo. Har. BH29 12
College Clo. N18 CA28 15
College Clo. Twick. BG47 56
College Cres. NW3 BT36 32
College Cross N1 BY36 33
College Dr. Ruis. BC33 20
College Gdns. E4 CE26 9
College Gdns. Enf. BZ23 8
College Gdns. Ilf. CK32 26
College Gdns. N. Mal. BO53 67
College Gdns. N18 CA28 15
College Gdns. SE21 CA47 60
College Gdns. SW17 BU48 59
College Grn. SE19 CA50 60
College Hill EC4 BZ40 42
College Hill Rd. Har. BH29 12
College La. NW5 BV35 32
College Park Clo. SE13 CF45 52
College Pl. E17 CG31 25
College Pl. NW1 BW37 32
College Rd. Brom. CH51 71
College Rd. Croy. BZ55 78
College Rd. E17 CF32 25
College Rd. Enf. BZ23 8
College Rd. Epsom BO60 82
College Rd. Har. BH30 12
College Rd. Har. BH32 21
College Rd. Islw. BH44 48
College Rd. N17 CA29 15
College Rd. N21 BY27 15
College Rd. NW10 BQ37 31
College Rd. SE19 CA47 60
College Rd. SE21 CA47 60
College Rd. Swan. CT51 73
College Rd. SW19 BT50 59
College Rd. W13 BJ39 39
College Rd. Wem. BK33 21
College Slip, Brom. CH51 71
College St. EC4 BZ40 42
College Ter. E3 CD38 34
College Ter. N3 BR30 13
Hendon La.
College Vw. SE9 CJ47 62
Collent St. E9 CC36 34
Collerston Rd. SE10 CG42 43
Colless Rd. N15 CA32 24
Collett Rd. SE16 CB41 42
Collett Way, Grnf. BF41 38
Collier Clo. Epsom BM57 75
Collier Dr. Edg. BM30 12
Collier Row La. Rom. CR29 18
Collier Row Rd. Rom. CQ30 18
Collier St. N1 BX37 33
Colliers Shaw, Kes. CJ56 80
Colliers Water La. Th. Hth. BY53 69
Collindale Av. Sid. CO47 63
Collingbourne Rd. W12 BP40 40
Collingham Gdns. SW5 BS42 41
Collingham Pl. SW5 BS42 41
Collingham Rd. SW5 BS42 41
Collingtree Rd. SE26 CC49 61
Collingwood Av. N10 BV31 23
Collingwood Av. Surb. BN54 67
Collingwood Clo. SE20 CB51 69
Collingwood Clo. Twick. BF46 47
Collingwood Est. E1 CB38 33
Collingwood Pl. SE18 CL41 44
Rodney St.
Collingwood Rd. E17 CE32 25
Barnadas Rd.
Collingwood Rd. Mitch. BU51 68
Collingwood Rd. N15 CA31 24
Collingwood Rd. Sutt. BR55 76
Collingwood St. E1 CB38 33
Collins Av. Stan. BL30 12
Collins Dr. Ruis. BD34 20
Collinson St. SE1 BZ41 42
Collins Rd. N5 BZ35 33
Collins St. SE3 CG44 52
Collinwood Av. Enf. CC24 9
Collinwood Gdns. Ilf. CK32 26
Collis Alley, Twick. BH47 57
Albion Rd.
Coll's Rd. SE15 CC44 52
Collyer Av. Croy. BX56 78
Collyer Pl. SE15 CA44 51
Peckham High St.
Collyer Rd. Croy. BX56 78
Colman Clo. Epsom BP62 82
Colman Rd. E16 CJ39 44
Colmer Pl. Har. BG29 11
Colmer Rd. SW16 BX51 69
Colmore Rd. Enf. CC24 9
Colnbrook St. SE1 BY41 42
Colne Av. Wat. BC25 4
Colne Ct. Epsom BN56 76
Colne Dr. KT12 BD55 74
Colne Dr. Rom. CW29 19
Colne Rd. E5 CD35 34
Colne Rd. N21 BZ26 8
Colne Rd. Twick. BH47 57
Colne St. E13 CH38 35
Colney Hatch La. N10 BU29 14
Colney Hatch La. N11 BU29 14
Colney Rd. Dart. CW46 55
Cologne Rd. SW11 BT45 50
Colombo Rd. Ilf. CM33 26
Colombo St. SE1 BY40 42
Colomb St. SE10 CG42 43
Colonels Wk. Enf. BY24 8
Colonial Av. Twick. BG46 47
Colonial Way, Wat. BD23 4
Colonnade WC1 BX38 33
Herbrand St.
Colridge Av. Sutt. BU56 77
Colson Gdns. Loug. CL24 10
Colson Path, Loug. CL24 10
Colson Rd. Croy. CA55 78
Colson Rd. Loug. CL24 10
Colson Way SW16 BW49 59
Colsterworth Rd. N15 CA31 24
Colston Av. Cars. BU56 77
Colston Ct. Cars. BU56 77
Colston Rd. E7 CJ36 35
Colston Rd. SW14 BN45 49
Coltishall Rd. Horn. CV36 37
Coltness Cres. SE2 CO42 45
Colton Gdns. N17 BZ31 24
Coltsfoot Path, Rom. CV29 19
Columbia Av. Edg. BM30 12
Columbia Av. Wor. Pk. BO54 67
Columbia Ct. N. Mal. BO54 67
Columbia Rd. E2 CA38 33
Columbia Rd. E13 CG38 34
Columbia Sq. SW14 BN45 49
Upr. Richmond Rd.
Columbine Av. S. Croy. BY57 78
Columbine Way, Rom. CW30 19
Columbine Way SE13 CF44 52
Colvestone Cres. E8 CA35 33
Colville Est. N1 BZ37 33
Colville Gdns. W11 BR39 40
Colville Ho. W11 BR39 40
Colville Ms. W11 BR39 40
Colville Pl. W1 BW39 41
Colville Rd. E11 CF34 25
Colville Rd. E17 CD30 16
Colville Rd. N9 CB26 8
Colville Rd. W3 BM41 39
Colville Rd. W11 BR39 40
Colville Sq. Ms. W11 BR39 40
Portobello Rd.
Colville Sq. W11 BR39 40
Colville Ter. W11 BR39 40
Colvin Gdns. E4 CF27 16
Colvin Gdns. E11 CH31 26
Colvin Gdns. Ilf. CM30 17
Colvin Rd. E6 CK36 35
Colvin Rd. Th. Hth. BY53 69
Colvin St. W6 BQ42 40
Glenthorne Rd.
Colwell Rd. SE22 CA46 51
Colwick Clo. N6 BW33 23
Colwith Rd. W6 BQ43 49
Colwood Gdns. SW19 BT50 59
Colworth Gro. SE17 BZ42 42
Browning St.
Colworth Rd. Croy. CB54 69
Colworth Rd. E11 CG32 25
Colwyn Av. Grnf. BH37 30
Colwyn Cres. Houns. BG44 47
Colwyn Rd. NW2 BP34 22
Colyer Clo. SE9 CL48 62
Colyers Clo. Erith CS44 55
Colyers La. Erith CS44 55
Colyers Wk. Erith CT44 55
Colyton Clo. Well. CP44 54
Colyton Clo. Wem. BK36 30
Colyton Rd. SE22 CB46 51
Colyton Way N18 CB28 15
Combe Av. SE3 CG43 52
Combedale Rd. SE10 CH42 44
Combe Lea, Brom. CK52 71
Combemartin Rd. SW18 BR47 58
Combe Ms. SE3 CG43 52
Comber Clo. NW2 BP34 22
Comber Gro. SE5 BZ43 51
Combermere Rd. Mord. BS53 68
Combermere Rd. SW9 BX45 51
Combe Rd. N16 CD35 34
Trumans Rd.
Comberton Rd. E5 CB34 24
Comberton Wk. E12 CL35 35
Landseer Av.
Combeside SE18 CN43 54
Combwell Cres. SE2 CO41 45
Comely Bank Rd. E17 CE32 25
Comeragh Ms. W14 BR42 40
Comeragh Rd.
Comeragh Rd. W14 BR42 40
Comerford Rd. SE4 CD45 52
Comet Ho. Pl. SE8 CE43 52
Watson's St.
Comet St. SE8 CE43 52
Commerce Rd. Brent. BK43 48
Commerce Rd. N22 BX30 15
Commerce Way, Croy. BX55 78
Commercial Dock Pass. SE16 CD41 43
Commercial Rd. E1 CB39 42
Commercial Rd. N18 CA29 15
Commercial St. E1 CA38 33
Commercial Way SE15 CA43 51
Commerell Pl. SE10 CG42 43
Blackwall La.
Commerell St. SE10 CG42 43
Commodore St. E1 CD38 34
Commondale SW15 BQ45 49
Commonfield Rd. Bans. BS60 83
Common La. Dart. CU48 64
Common La. Esher BJ57 75
Common La. Wat. BH23 5
Common Rd. Esher BJ57 75
Common Rd. Stan. BG28 11
Common Rd. SW13 BP45 49
Commonside, E. Mitch. BV52 68
Commonside, Kes. CJ56 80
Commonside, Mitch. BW53 68
Commonside, W. Mitch. BU52 68
Common, The, Rich. BK48 57
Common, The, Stan. BH27 12
Common, The, Sthl. BD42 38
Common, The W5 BL40 39
Common Way, Esher BG58 74
Commonwealth Av. W12 BP40 40
Commonwealth Rd. N17 CB29 15
Commonwealth Way SE2 CO42 45
Community Rd. E15 CF35 34
Community Rd. Grnf. BG37 29
Como Rd. SE23 CD48 61
Como St. Rom. CS31 28
Compass Hill, Rich. BL46 48
Petersham Rd.
Compayne Rd. NW6 BS36 32
Compton Av. E6 CJ37 35
Compton Av. N1 BY36 33
Compton Av. N6 BU33 23
Compton Av. Rom. CU31 28
Compton Clo. Esher BG56 74
Compton Clo. NW1 BV38 32
Robert St.
Compton Clo. W13 BH39 39
Compton Ct. SE19 CA49 60
Compton Cres. Chess. BL57 75
Compton Cres. Nthlt. BD37 29
Compton Cres. N17 BZ29 15
Compton Cres. W4 BN43 49
Compton Pass. EC1 BY38 33
Compton St.
Compton Pl. Erith CT43 55
Compton Pl. Wat. BE28 11
Compton Ri. Pnr. BE32 20
Compton Rd. Croy. CB54 69
Compton Rd. N1 BY36 33
Compton Rd. N21 BY26 8
Compton Rd. NW10 BQ38 31
Compton Rd. SW19 BR50 58
Compton Sq. N1 BY36 33
Canonbury Rd.
Compton St. E13 CH37 35
Compton St. EC1 BY38 33
Compton Ter. N1 BY36 33
Comus Pl. SE17 CA42 42
Comus Rd. N19 BW34 23
Comyn Rd. SW11 BU45 50
Comyns Clo. E16 CG39 43
Comyns Rd. Dag. CR36 36
Comyns, The, Bush. BG26 4
Conaways Clo. Epsom BP58 76
Concanon Rd. SW2 BX45 51
Concert Hall App. SE1 BX40 42
Concord Clo. Grnf. BE38 29
Ruislip Rd.
Concord Rd. Enf. CB25 8
Concord Rd. W3 BM38 30
Concourse, The, N9 CB27 15
Condell Rd. SW8 BW44 50
Conder St. E14 CD39 43
Conderton Rd. SE5 BZ45 51
Condor Path, Nthlt. BF37 29
Leander Rd.
Condor Wk. Rain. CU36 37

Condover Cres. SE18 CL43 53
Condray St. SW11 BT44 50
Conduct Rd. Barn. BR25 6
Conduit Ct. WC2 BX40 42
Long Acre
Conduit La. Enf. CC25 9
Conduit La. N18 BZ28 15
Hermitage La.
Conduit La. S. Croy. CA56 78
Conduit Ms. W2 BT39 41
Conduit Pl. W2 BT39 41
London St.
Conduit Rd. SE18 CL42 44
Conduit St. W1 BV40 41
Conduit Way, NW10 BM36 30
Conewood Pl. N5 BY34 24
Conewood St.
Conewood St. N5 BY34 24
Coney Acre, SE21 BZ47 60
Coneybury Clo. Whyt. CB62 84
Coneygove Path, Nthlt. BE36 29
Arnold Rd.
Coney Hill Rd. W. Wick. CG55 79
Conference Rd. SE2 CP42 45
Congeton Gro. SE18 CM42 44
Congo Rd. SE18 CM42 44
Congress Rd. SE2 CP42 45
Congreve Rd. SE9 CK45 53
Congreve St. SE17 CA42 42
Conical Cor. Enf. BZ23 8
Conifer Av. Rom. CR28 18
Conifer Gdns. Enf. BZ26 8
Conifer Gdns. SW16 BX48 60
Conifers Clo. Tedd. BJ50 57
Coniger Rd. SW6 BS44 50
Coningham Ms. W12 BP40 40
Percy Rd.
Coningham Rd. W12 BP40 40
Coningsby Cott. W5 BK41 39
Coningsby Rd.
Coningsby Gdns. E4 CE29 16
Coningsby Rd. N4 BY33 24
Coningsby Rd. S. Croy. BZ58 78
Coningsby Rd. W5 BK41 39
Conington Rd. SE13 CE44 52
Conisbee Ct. N14 BW25 7
Conisborough Cres. SE6 CF48 61
Coniscliffe Rd. N13 BZ27 15
Coniston Av. Bark. CN36 36
Coniston Av. Grnf. BJ38 30
Coniston Av. Well. CN45 54
Coniston Clo. Bark. CN36 36
Coniston Av.
Coniston Clo. Bexh. CS44 55
Coniston Clo. Dart. CU47 64
Coniston Clo. Erith. CT43 55
Coniston Clo. Mord. BQ53 67
Coniston Clo. N20 BT27 14
Coniston Clo. W4 BN43 49
Conistone Way N7 BX36 33
Sutterton St.
Coniston Gdns. Ilf. CK31 26
Coniston Gdns. N9 CC26 9
Coniston Gdns. NW9 BN32 22
Coniston Gdns. Pnr. BC31 20
Coniston Gdns. Sutt. BT57 77
Coniston Gdns. Wem. BK33 21
Coniston Rd. Bexh. CS44 55
Coniston Rd. Brom. CF50 61
Coniston Rd. Couls. BW61 83
Coniston Rd. Croy. CB54 69
Coniston Rd. N10 BV30 14
Coniston Rd. N17 CB29 15
Coniston Rd. Twick. BF46 47
Coniston Wk. E9 CC35 34
Churchill Wk.
Coniston Way, Chess. BL55 75
Coniston Way, Horn. CU35 37
Conlan St. W10 BR38 31
Conley Rd. NW10 BO36 31
Conley St. SE10 CG42 43
Connaught App. E16 CJ39 44
Baxter Rd.
Connaught Av. Barn. BU26 7
Connaught Av. Enf. CA23 8
Connaught Av. E4 CF26 9
Connaught Av. Houns. BE45 47
Connaught Av. Loug. CJ24 10
Connaught Av. SW14 BN45 49
Connaught Clo. E10 CD34 25
Connaught Clo. Enf. CA23 8
Connaught Clo. Sutt. BT55 77
Connaught Clo. W2 BU39 41
Connaught St.
Connaught Dr. NW11 BS31 23
Connaught Gdns. N10 BW32 23
Connaught Gdns. N13 BY28 15
Connaught Hill, Loug. CJ24 10
Connaught La. Ilf. CM34 26
Connaught Ms. W2 BU39 41
Connaught Pl.
Connaught Pl. W2 BU39 41
Connaught Rd. E4 CG26 9
Springfield Rd.
Connaught Rd. Barn. BQ25 6
Connaught Rd. E11 CF33 25
Connaught Rd. E16 CJ40 44
Connaught Rd. E17 CE32 25
Connaught Rd. Har. BH30 12
Connaught Rd. Horn. CV34 28
Connaught Rd. Ilf. CM34 26
Connaught Rd. N. Mal. BO52 67
Connaught Rd. N4 BY33 24
Connaught Rd. NW10 BO37 31
Connaught Rd. Rich. BL46 48
Albert Rd.
Connaught Rd. SE18 CL42 44
Connaught Rd. Sutt. BT55 77
Connaught Rd. Tedd. BG49 56
Connaught Rd. W13 BJ40 39
Connaught Sq. W2 BU39 41
Connaught St. W2 BU39 41
Connaught Way N13 BY28 15
Connell Cres. W5 BL38 30
Connington Cres. E4 CF27 16
Connor Rd. Dag. CQ35 36
Connors All. W6 BR43 49
Bayonne Rd.
Connor St. E9 CC37 34
Lauriston Rd.
Conolly Rd. W7 BH40 39
Conquest St. SE1 BY41 42
Conrad Dr. Wor. Pk. BQ54 67
Conrad St. E9 CC36 34
Consfield Av. N. Mal. BP52 67
Consort Ms. Islw. BG46 47
Consort Rd. SE15 CB44 51
Cons St. SE1 BY41 42
Windmill Wk.
Constable Clo. Islw. BG46 47
Constable Clo. NW11 BS32 23
Constable Cres. W15 CB32 24
Constable Gdns. Edg. BM30 12
Constable Wk. SE21 CA48 60
Constance Cres. Brom. CG54 70
Constance Rd. Croy. BY54 69
Constance Rd. Enf. CA25 8
Constance Rd. Sutt. BT56 77
Constance Rd. Twick. BF47 56
Constance St. E16 CK40 44
Constantine Rd. NW3 BU35 32
Constitution Hill, SW1 BV41 41
Constitution Ri. SE18 CL44 53
Content St. SE17 BZ42 42
Contessa Clo. Orp. CM56 80
Convair Wk. Nthlt. BD37 29
Convent Est. SE19 BZ50 60
Convent Gdns. W5 BK42 39
Convent Hl. SE19 BZ50 60
Convent Way, Sthl. BD42 38
Conway Clo. Rain. CU36 37
Conway Clo. Stan. BJ29 12
Conway Cres. Grnf. BH37 30
Conway Cres. Rom. CP33 27
Conway Dr. Sutt. BS57 77
Conway Gdns. Mitch. BW52 68
Conway Gdns. Wem. BK33 21
Conway Gro. W3 BN39 40
Conway Rd. Felt. BD49 56
Conway Rd. Houns. BE47 56
Conway Rd. N14 BX27 15
Conway Rd. N15 BY32 24
Conway Rd. NW2 BQ34 22
Conway Rd. SE18 CM42 44
Conway Rd. SW20 BQ51 67
Conway St. E13 CH38 35
Philip St.
Conway St. W1 BW38 32
Conybeare NW3 BU36 32
Quickswood
Conyers Clo. Wdf. Grn. CG29 16
Conyers Rd. E3 CD37 34
Conyers Rd. SW16 BW49 59
Conyers Way, Loug. CL24 10
Cooden Clo. Brom. CH50 62
Plaistow La.
Cookes La. Sutt. BR57 76
Church Rd.
Cookham Hill, Orp. CR55 81
Cookham Rd. Sid. CR50 63
Cookhill Rd. SE2 CO41 45
Cookes Mead, Bush. BF25 4
Cook's Rd. E15 CE37 34
Cooks Rd. SE17 BY43 51
Coolfin Rd. E16 CH39 44
Coolgardie Av. Chig. CL27 17
Coolgardie Av. E4 CF28 16
Coolhurst Rd. N8 BW32 23
Cool Oak La. NW9 BO33 22
Cool Oak Rd. NW4 BP33 22
Coomassie Rd. W9 BR38 31
Bravington Rd.
Coombe Av. Croy. CA56 78
Coombe Bank, Kings. On T. BO51 67
Coombe Clo. Edg. BL30 12
Coombe Clo. Houns. BF45 47
Coombe Cor. N21 BY26 8
Coombe Cres. Hamptn. BE50 56
Coombe Crest, Kings. On T. BN50 58
Coombe Dr. Ruis. BC34 20
Coombe End, Kings. On T. BN50 58
Coombefield Clo. N. Mal. BO53 67
Coombe Gdns. N. Mal. BO52 67
Coombe Gdns. SW20 BP51 67
Coombe Hill Clo. Kings. On T. BO50 58
Coombe Hill Glade, Kings. On T. BO50 58
Coombe Hill Rd. Kings. On T. BO50 58
Coombe Ho. Chase, N. Mal. BN51 67
Coombehurst Clo. Barn. BU23 7
Coombe La. Croy. CB56 78
Coombe La. Kings. On T. BM51 66
Coombe Moor, Kings. On T. BO50 58
Coombe Nevile, Kings. On T. BN50 58
Coombe Pk. Kings. On T. BN49 58
Coombe Ridings, Kings. On T. BN49 58
Coombe Ri. Kings. On T. BN51 67
Coombe Rd. Bush. BG26 4
Coombe Rd. Croy. BZ56 78
Coombe Rd. Hamptn. BE50 56
Coombe Rd. Kings. On T. BM51 66
Coombe Rd. N. Mal. BO51 67
Coombe Rd. NW10 BN34 22
Coombe Rd. Rom. CW31 28
Coombe Rd. SE26 CB49 60
Coombe Rd. W4 BO42 40
Coombe Rd. W13 BJ41 39
Northcroft Rd.
Coombe Springs, Kings. On T. BN51 67
Coombes Rd. Dag. CQ37 36
Coombe Wk. Sutt. BS55 77
Coombe Wd. Hill, Pur. BZ59 84
Coombe W. Lane, Kings. On T. BM51 66
Coombewood Dr. Rom. CR32 27
Coombe Wood Rd. Kings. On T. BN49 58
Coomb's St. N1 BY37 33
Remington St.
Coomer Rd. SW6 BR43 49
Cooper Av. E17 CD30 16
Cooper Cres. Cars. BU55 77
Cooper La. NW1 BW37 32
Cooper Rd. Croy. BY56 78
Cooper Rd. NW10 BP35 31
Coopersale Clo. Wdf. Grn. CJ29 17
Coopersale Rd. E9 CC35 34
Coopers Clo. Chig. CO27 18
Coopers La. E10 CE33 25
Coopers La. SE12 CH48 62
Coopers Rd. NW4 BQ32 22
Renters Av.
Cooper's Rd. SE1 CA42 42
Cooper's Row, EC3 CA40 42
Cooper St. E16 CG39 43
Coote Gdns. Dag. CQ34 27
Coote Rd. Bexh. CQ44 54
Coote Rd. Dag. CQ34 27
Copeland Rd. E17 CE32 25
Copeland Rd. SE15 CB44 51
Copelia Rd. SE3 CG45 52
Copenhagen Pl. E14 CD39 43
Copenhagen St. N1 BX37 33
Copenhagen Way, Walt. BC55 74
Copen Rd. Dag. CQ33 27
Cope Pl. W8 BS41 41
Copers Cope Rd. Beck. CD50 61
Cope St. SE16 CC42 43
Copeswood Rd. Wat. BC23 4
Copford Clo. Wdf. Grn. CK29 17
Green Wk.
Copland Av. Wem. BK35 30
Copland Clo. Wem. BK35 30
Copland Rd. Wem. BL36 30
Copleston Rd. SE15 CA45 51
Copley Clo. W7 BH39 39
Copley Dene, Brom. CJ51 71
Copley Pk. SW16 BX50 60
Copley Rd. Stan. BK28 12
Copnor Way, SE15 CA43 51
Hordle Prom. W.
Copperas St. SE8 CE43 52
Copper Beech Clo. Ilf. CL30 17
Copperbeech Clo. NW3 BT35 32
Akenside Rd.
Copperdale Rd. Hayes BC41 38
Copperfield, Chig. CM28 17
Copperfield App. Chig. CM29 17
Copperfield App. Ilf. CM29 17
Copperfield Ms. N18 CA28 15
Copperfield Rd. E3 CD38 34
Copperfield St. SE1 BY41 42
Copperfields Way, Rom. CV30 19
Copperfield Way, Chis. CM50 62
Coppermill La. E17 CC32 25
Copper Mill La. SW17 BT49 59
Coppetts Clo. N12 BU29 14
Coppetts Rd. N10 BU29 14
Coppice Clo. SW20 BQ52 67
Coppice Dr. SW15 BP46 49
Coppice Est. Brom. CL53 71
Coppice Path, Chig. CO28 18
Coppice, The, Enf. BY24 8
Coppice, The, Wat. BD25 4
Coppice Wk. N20 BS27 14
Coppice Way E18 CG31 25
Coppifs Gro. N11 BV28 14
Copping Clo. Croy. CA56 78
Coppins, The, Croy. CE57 79
Coppins, The, Har. BH29 12
Coppock Clo. SW11 BU44 50
Copse Av. W. Wick. CE55 79
Copse Edge Av. Epsom. BO60 82
Copse Glade, Surb. BK54 66
Copse Hill, Pur. BX60 84
Copse Hill, Sutt. BS57 77
Copse Hill, SW20 BP51 67
Copsem Dr. Esher BG57 74
Copsem La. Esher BG57 74
Copse, The, E4 CG26 9
Copse Vw. S. Croy. CC58 79
Copthall Av. EC2 BZ39 42
Copthall Bldgs. EC2 BZ39 42
Telegraph St.
Copthall Clo. EC2 BZ39 42
Copthall Ct. EC2 BZ39 42
Throgmorton St.
Copthall Dr. NW7 BP29 13
Copthall Gdns. NW7 BP29 13
Copthall Gdns. Twick. BH47 57
Copthorne Av. Brom. CK55 80
Copthorne Av. Ilf. CL29 17
Copthorne Av. SW12 BW47 59
Copthorne Clo. Ilf. CL29 17
Copthorne Ri. S. Croy. BZ60 84
Copthorpe Ct. NW8 BT38 32
Maida Vale
Coptic St. WC1 BX39 42
Coralline Wk. SE2 CP41 45
Coral St. SE1 BY41 42
Coram St. WC1 BX38 33
Coran Clo. N9 CC26 9
Corban Rd. Houns. BF45 47
Corbar Clo. Barn. BT23 7
Corbet Clo. Wall. BV55 77
Corbet Pl. E1 CA39 42
Jerome St.
Corbet Rd. Epsom BO58 76
Corbett Rd. E11 CJ32 26
Corbett Rd. E17 CF31 25
Corbett Rd. N22 BX29 15
Trinity Rd.
Corbetts La. SE16 CC42 43
Rotherhithe New Rd.
Corbetts Pass. SE16 CC42 43
Rotherhithe New Rd.
Corbett St. SW8 BX43 51
Corbicum, E11 CG33 25
Corbiere Ct. SW19 BQ50 58
Thornton Rd.
Corbins La. Har. BF34 20
Corby Cres. Enf. BX24 8
Corbylands Rd. Sid. CN47 63
Corbyn St. N4 BX33 24
Corby Rd. NW10 BN37 31
Corby Way E3 CE38 34
Knapp Rd.
Cordelia St. E14 CE39 43
Cordelia St. N7 BX36 33
Cordova Rd. E3 CD38 34
Cordrey Gdns. Couls. BX61 84
Cord Way, E14 CE41 43
Mellish St.
Cordwell Rd. SE13 CF46 52
Corelli Rd. SE3 CK44 53
Corfe Av. Har. BF35 29
Corfield St. E2 CB38 33
Corfton Rd. W5 BL39 39
Corinium Clo. Wem. BL35 30
Corinne Rd. N19 BW35 32
Corinthian Manorway, Erith CS42 46
Corinthian Rd. Erith CS42 46
Corinth Rd. N7 BX35 33
Corker Wk. N7 BX34 24
Corker Way EC1 BX34 24
Corkran Rd. Surb. BK54 66
Corkscrew Hill, W. Wick. CF55 79
Cork St. W1 BW40 41
Corlett St. NW1 BU39 41
Bell St.
Cormont Rd. SE5 BY44 51
Cormorant Wk. Rain. CU36 37
Cornbury Rd. Edg. BK29 12
Cornell Clo. Sid. CQ50 63
Cornell Way, Rom. CR28 18
Corner Mead NW9 BO30 13
Corner St. E16 CH39 44
Beckton Rd.
Cornery Rd. W4 BO43 49
Cornfield Rd. Bush. BF24 4
Cornflower Ter. SE22 CB46 51
Cornflower Way, Rom. CW30 19
Cornford Clo. Brom. CH53 71

Street	Grid	Page
Cornford Gro. SW12	BV48	59
Cornhill, EC3	BZ39	42
Cornish Gro. SE20	CB51	69
Cornmill Dr. Orp.	CN54	72
Cornmill La. SE13	CF45	52
Cornshaw Rd. Dag.	CP33	27
Cornthwaite Rd. E5	CC34	25
Cornwall Av. E2	CC38	34
Cornwall Av. Esher.	BJ57	75
Cornwall Av. N3	BS29	14
Cornwall Av. N22	BX30	15
Cornwall Av. Sthl.	BE39	38
Cornwall AV. Well.	CN45	54
Cornwall Clo. Bark.	CN36	36
Cornwall Cres. W11	BR39	40
Cornwall Dr. Orp.	CP50	63
Cornwall Gdns. NW10	BP36	31
Cornwall Gdns. SW7	BS41	41
Cornwall Gdns. Wk. SW7	BS41	41
Cornwall Gro. W4	BO42	40
Cornwallis Av. N9	CB27	15
Cornwallis Av. SE9	CM48	62
Conrwallis Gro. N9	CB27	15
Cornwallis Rd. Dag.	CP35	36
Cornwallis Rd. E17	CC31	25
Cornwallis Rd. N9	CB27	15
Cornwallis Rd. N19	BX34	24
Cornwallis Wk. SE9	CK45	53
Cornwall Lo. Kings. On T.	BM50	57
Cornwall Ms. St. SW7	BT41	41
Cornwall Rd. Croy.	BY55	78
Cornwall Rd. Har.	BG32	20
Cornwall Rd. N4	BY33	24
Cornwall Rd. N15	BZ32	24
Cornwall Rd. N18	CB28	15
Fairfield Rd.		
Cornwall Rd. Pnr.	BE29	11
Cornwall Rd. Ruis.	BC34	20
Cornwall Rd. SE1	BY40	42
Cornwall Rd. Sutt.	BS58	77
Cornwall Rd. Twick.	BJ47	57
Cornwall Ter. NW1	BU38	32
Cornwood Clo. N2	BT32	23
Cornworthy Rd. Dag.	CP35	36
Corol Clo. Rom.	CP31	27
Corona Rd. SE12	CH47	62
Coronation Clo. Bex.	CP46	54
Coronation Clo. Ilf.	CM31	26
Coronation Cotts. W5	BL38	30
Coronation Dr. Horn.	CU35	37
Coronation Rd. E13	CJ38	35
Coronation Rd. NW10	BL38	30
Coronation Wk. Twick.	BF47	56
Coroner's Ct. W6	BQ42	40
Coronet St. N1	CA38	33
Corporation Av. Houns.	BE45	47
Corporation Row, EC1	BY38	33
Corporation St. E15	CG37	34
Corporation St. N7	BX35	33
Corrall Rd. N7	BX36	33
Lough Rd.		
Corrance Rd. SW2	BX45	51
Corri Av. N14	BW28	14
Corrigan Av. Couls.	BV61	83
Corringham Ct. NW11	BS33	23
Corringham Rd. NW11	BS33	23
Corringham Rd. Wem.	BM34	21
Corrington Ms. W11	BR40	40
Blenheim Cres.		
Corringway NW11	BS33	23
Corringway W5	BL39	39
Corscoombe Clo. Kings. On T.	BN49	58
Corsehill St. SW16	BW50	59
Corseley Way, E9	CD39	34
Trowbridge Est.		
Corsham St. N1	BZ38	33
Corsica St. N5	BY36	33
Cortayne Rd. SW6	BR44	49
Cortis Rd. SW15	BP46	49
Cortis Ter. SW15	BP46	49
Corunna Rd. SW8	BW44	50
Corunna Ter. SW8	BW44	50
Corwall Ct. SW17	BU49	59
Broadwater Rd.		
Corwood Dr. E1	CC39	43
Cosbycote Av. SE24	BZ46	51
Cosdach Av. Wall.	BW57	77
Cosedge Cres. Croy.	BY56	78
Cosmo Pl. WC1	BX39	42
Southampton Row		
Cosmur Clo. W12	BO41	40
Cossall St. SE15	CB44	51
Cossall Wk. SE15	CB44	51
Cosser St. SE1	BY41	42
Costa St. SE15	CB44	51
Coston's Av. Grnf.	BG38	29
Coston's La. Grnf.	BG38	29
Coston Wk. SE4	CD45	52
Frendsbury Rd.		
Cosway St. NW1	BU39	41
Cotall St. E14	CE39	43
Coteford Clo. Pnr.	BC32	20
Coteford St. SW17	BU49	59
Cotelands, Croy.	CA55	78
Cotesbach Rd. E5	CC34	25
Cotesmere Gdns. Dag.	CP35	36
Cotford Rd. Th. Hth.	BZ52	69
Cotham St. SE17	BZ42	42
Cotherstone, Epsom	BN58	76
Cotherstone Rd. SW2	BX47	60
Cotleigh Av. Bex.	CP48	63
Cotleigh Rd. NW6	BS36	32
Cotleigh Rd. Rom.	CS32	28
Cotman Clo. NW11	BT32	23
Cotman Clo. SW15	BQ46	49
Cotmandene Cres. Orp.	CO51	72
Cotman Gdns. Edg.	BM30	12
Cotmans Clo. Hayes	BC40	38
Coton Rd. Well.	CO45	54
Cotsford Av. N. Mal.	BN53	67
Cotswold Av. Bush.	BG25	4
Cotswold Clo. Bexh.	CT44	55
Cotswold Clo. Kings. On T.	BM50	57
Cotswold Gdns. E6	CJ38	35
Cotswold Gdns. Ilf.	CM33	26
Cotswold Gdns. NW2	BQ34	22
Cotswold Rd. Hamptn.	BF49	56
Cotswold Rd. Rom.	CW30	19
Cotswold Rd. Sutt.	BS58	77
Cotswold St. SE27	BY49	60
Cotswold Way, Enf.	BX24	8
Cottage Av. Brom.	CK54	71
Cottage Field Clo. Sid.	CP47	63
Cottage Gro. SE5	BZ43	51
Cottage Gro. Surb.	BK53	66
Cottage Gro. SW9	BX45	51
Cottage Pl. SW3	BU41	41
Cottage Rd. Epsom	BN57	76
Cottage St. E14	CE40	43
Cottage, The, Surb.	BK54	66
Cottage Wk. N16	CA34	24
Smalley Rd.		
Cottage Wk. SE15	CA44	51
Sumner Estate		
Cottage Wk. SW1	BU41	41
Cottenham Dr. SW20	BP50	58
Cottenham Pde. SW20	BP51	67
Durham Rd.		
Cottenham Pk. Rd. SW20	BP51	67
Cottenham Pl. SW20	BP50	58
Cottenham Rd. E17	CD31	25
Cotterill Rd. Surb.	BL55	75
Cottesbrook St. SE14	CD43	52
Nynehead St.		
Cottesmore Av. Ilf.	CL30	17
Cottesmore Gdns. W8	BS41	41
Cottimore Av. Walt.	BC54	65
Cottimore Cres. Walt.	BC54	65
Cottimore La. Walt.	BC54	65
Cottimore Ter. Walt.	BC54	65
Cottingham Chase, Ruis.	BC34	20
Cottingham Rd. SE20	CC50	61
Cottington Rd. Felt.	BD49	56
Cottington St. SE11	BY42	42
Cotton Av. W3	BN39	40
Cotton Hl. Brom.	CF49	61
Cottons App. Rom.	CS32	28
Pettley Gdns.		
Cotton's Gdns. E2	CA38	33
Hackney Rd.		
Cotton St. E14	CF40	43
Couchmore Av. Esher	BH55	75
Couchmore Av. Ilf.	CK30	17
Coulgate St. SE4	CD45	52
Coulsdon Ct. Rd. Couls.	BX61	84
Coulsdon Ri. Couls.	BX62	84
Coulsdon Rd. Couls.	BX61	84
Coulson St. SW3	BU42	41
Coulter Rd. W6	BP41	40
Councillor St. SE5	BZ43	51
Countess Rd. NW5	BW35	32
Countisbury Av. Enf.	CA26	8
Country Way, Felt.	BC50	56
County Gdns. Bark.	CN37	36
County Gate, Barn.	BS25	7
County Gate, SE9	CM48	62
County Gro. SE5	BZ44	51
County Rd. Th. Hth.	BY51	69
County St. SE1	BZ41	42
Coupland Pl. SE18	CM42	44
Courage Clo. Horn.	CV32	28
Courcy Rd. N8	BY31	24
Courland Gro. SW8	BW44	50
Courland St. SW8	BW44	50
Course, The, SE9	CL48	62
Courtauld Rd. N19	BW34	23
Court Av. Belv.	CQ42	45
Court Clo. Av. Twick.	BF48	56
Court Clo. Har.	BL31	21
Court Clo. Twick.	BF48	56
Court Clo. Wall.	BW57	77
Court Cres. Chess.	BK57	75
Court Downs Rd. Beck.	BE51	70
Court Dr. Croy.	BX56	78
Court Dr. Stan.	BL28	12
Court Dr. Sutt.	BU56	77
Courtenay Av. Har.	BG29	11
Courtenay Av. N6	BU33	23
Courtenay Gdns. Har.	BG30	11
Courtenay Pl. E17	CD32	25
Courtenay Rd. E11	CG34	25
Courtenay Rd. E17	CC31	25
Courtenay Rd. Wor. Pk.	BQ55	76
Courtenay Sq. SE11	BY42	42
Courtenay St.		
Courtenay St. SE11	BY42	42
Court Farm La. Nthlt.	BF36	29
Court Farm Rd. SE9	CK48	62
Court Farm Rd. Nthlt.	BF36	29
Courtfield Av. Har.	BH32	21
Courtfield Cres. Har.	BH32	21
Courtfield Gdns. SW5	BS42	41
Courtfield Gdns. W13	BJ39	39
Courftield Ms. SW7	BT42	41
Courtfield Ri. W. Wick.	CF55	79
Courtfield Rd. SW7	BT42	41
Court Fm. Av. Epsom.	BN56	76
Court Fm. Rd. Warl.	CB62	84
Court Haw, Bans.	BU61	83
Court Hill, Couls.	BU62	83
Courthill Rd. SE13	CF45	52
Court Hill, S. Croy.	CA59	84
Courthope Rd. Grnf.	BG37	29
Courthope Rd. NW3	BU35	32
Mansfield Rd.		
Courthope Rd. SW19	BR49	58
Courthope Vill. SW19	BR50	58
Court House Gdns. N3	BS29	14
Court House Rd. N12	BS29	14
Courtland Av. E4	CG27	16
Courtland Av. Ilf.	CK34	26
Courtland Cres. Bans.	BS61	83
Courtland Dr. Chig.	CL27	17
Courtlands, Rich.	BM46	48
Courtlands Av. Brom.	CG54	70
Courtlands Av. Esher	BE57	74
Courtlands Av. Est. SE12	CH46	53
Courtlands Av. Hamptn.	BE50	56
Courtlands Av. Rich.	BM44	48
Courtlands Av. NW7	BN27	13
Courtlands Av. SE12	CH46	53
Courtlands Av. SW16	BX50	60
Courtlands Clo. S. Croy.	CA58	78
Courtlands Dr. Epsom	BO57	76
Courtlands Rd. Surb.	BM54	66
Court La. Gdns. SE21	CA47	60
Court La. SE21	CA46	51
Courtleet Dr. Erith	CR44	54
Alberta Rd.		
Courtleigh Gdns. NW11	BR31	22
Courtman Rd. N17	BZ29	15
Courtmead Clo. SE24	BZ46	51
Court Mead, Nthlt.	BE38	29
Courtnell St. W2	BS39	41
Courtney Clo. SE19	CA50	60
Courtney Cres. Cars.	BU57	77
Courtney Pl. Croy.	BY55	78
Courtney Rd. Croy.	BY55	78
Courtney Rd. N7	BY35	33
Courtney Rd. SE20	CC50	61
Courtney Rd. SW19	BU50	59
Court Pde. Wem.	BJ34	21
Courtrai Rd. SE23	CD46	52
Court Rd. Bans.	BS61	83
Court Rd. Orp.	CO55	81
Court Rd. SE9	CK48	62
Court Rd. SE25	CA51	69
Court Rd. Sthl.	BE42	38
Court Side N8	BW32	23
Court Side SE26	CC48	61
Round Hill		
Court St. Brom.	CH51	71
South St.		
Court St. E1	CB39	42
Durward St.		
Court, The, Ruis.	BE35	29
Court Way, Ilf.	CM31	26
Court Way, NW9	BO31	22
Court Way, Rom.	CW30	19
Courtway, The, Wat.	BE27	11
Court Way, Twick.	BH47	57
Courtway, Wdf. Grn.	CJ28	17
Court Way W3	BN39	40
Court Yard, SE9	CK46	53
Courtyard, The, N1	BY36	33
Barnsbury Ter.		
Cousin La. EC4	BZ40	42
Couthurst Rd. SE3	CH43	53
Coutts Av. Chess.	BL56	75
Coval Gdns. SW14	BM45	48
Coval La. SW14	BM45	48
Coval Pass. SW14	BN45	49
Upper Richmond Rd.		
Coval Rd. SW14	BN45	49
Covent Gdn. WC2	BX40	42
Coventry Clo. NW6	BS37	32
Coventry Cross Est. E3	CF38	34
Coventry Rd. E1	CB38	33
Coventry Rd. Ilf.	CL34	26
Coventry Rd. SE25	CB52	69
Coventry St. W1	BW40	41
Coverack Clo. N14	BV25	7
Coverdale Clo. Stan.	BJ28	12
Coverdale Gdns. Croy.	CA55	78
Coverdale Rd. NW2	BQ36	31
Coverdale Rd. W12	BP41	40
Coverdales, The, Bark.	CM37	35
Coverley Clo. E1	CB39	42
Coverton Rd. SW17	BU49	59
Covert Rd. Chig.	CN28	18
Coverts Rd. Leath.	BH58	75
Covert, The, Orp.	CN53	72
Covert Way, Barn.	BT23	7
Covington Gdns. SW16	BY50	60
Covington Way, SW16	BX50	60
Cowan St. SE5	CA43	51
Cowbridge La. Bark.	CL36	35
Cowbridge Rd. Har.	BL31	21
Cowcross St. EC1	BY39	42
Cowdenbeath Path N1	BX37	33
Twyford St.		
Cowden Rd. Orp.	CN54	72
Cowden St. SE6	CE49	61
Cowdray Way, Horn.	CU35	37
Cowdrey Clo. Enf.	CA23	8
Cowdrey Ct. Dart.	CU47	64
Cowdrey Rd. SW19	BS49	59
Cowdry Rd. E9	CD36	34
Cowen Av. Har.	BG34	20
Cowgate Rd. Grnf.	BG37	29
Cowick Rd. SW17	BU49	59
Cowings Mead, Nthlt.	BE36	29
Arnold Rd.		
Cowland Av. Enf.	CC24	9
Cow La. Grnf.	BG37	29
Cowleaze Rd. Kings. On T.	BL51	66
Cowley Clo. S. Croy.	CC58	79
Cowley Cres. Walt.	BD56	74
Cowley Hill, Borwd.	BM23	5
Cowley La. E11	CG34	25
Cathall Rd.		
Cowley Rd. E11	CH32	26
Cowley Rd. Ilf.	CK33	26
Cowley Rd. Rom.	CU29	19
Cowley Rd. SW9	BY44	51
Cowley Rd. SW14	BO45	49
Cowley Rd. W3	BO40	40
Cowley St. SW1	BX41	42
Little College St.		
Cowling Clo. W11	BQ40	40
Wilsham St.		
Cowper Av. E6	CK36	35
Cowper Av. Sutt.	BT56	77
Cowper Clo. Well.	CO46	54
Cowper Gdns. N14	BV25	7
Cowper Gdns. Wall.	BW57	77
Cowper Rd. Belv.	CQ42	45
Cowper Rd. Brom.	CJ52	71
Cowper Rd. Kings On T.	BL49	57
Cowper Rd. N14	BV26	7
Cowper Rd. N16	CA35	33
Cowper Rd. N18	CB28	15
Cowper Rd. Rain.	CU38	37
Cowper Rd. SW19	BT50	59
Cowper Rd. W3	BN40	40
Cowper Rd. W7	BH40	39
Cowper St. EC2	BZ38	33
Cowper Ter. W10	BQ39	40
St. Quintin Av.		
Cowslip Rd. E18	CH31	26
Cowthorpe Rd. SW8	BW44	50
Coxdean, Epsom	BQ62	82
Cox La. Chess.	BL56	75
Cox La. Epsom	BM56	75
Coxley Ri. Pur.	BZ56	78
Coxley Ri. Pur.	BZ60	84
Coxmount Rd. SE7	CJ42	44
Coxon Clo. Har.	BF30	11
Cox's Ct. EC1	BZ39	42
Coxson Pl. SE1	CA41	42
Cox's Wk. SE21	CB47	60
Coxwell Rd. SE18	CM42	44
Coxwold Path, Chess.	BL57	75
Crabbs Croft Clo. Orp.	CL56	80
Crab Hill, Beck.	CF50	61
Crabtree Av. Rom.	CP31	27
Crabtree Av. Wem.	BL37	30
Crabtree Clo. Bush.	BF25	4
Crabtree Ct. E15	CE35	34
Crabtree La. SW6	BQ43	49
Crabtree Manorway, Belv.	CS41	46
Crabtree Walk SE15	CA44	51
Sumner Estate		
Crace St. NW1	BW38	32
Drummond Cres.		
Craddock Rd. Enf.	CA24	8
Craddock St. NW3	BV36	32
Prince Of Wales Rd.		
Cradley Rd. SE9	CM47	62
Craigdale Rd. Horn.	CT32	28
Craigen Av. Croy.	CB54	69
Craig Gdns. E18	CG30	16
Craigholm, SE18	CL44	53
Craigmuir Pk. Wem.	BL37	30
Craignair Rd. SW2	BX47	60
Craignish Av. SW16	BX51	69
Craig Park Rd. N18	CB28	15
Craig Rd. Rich.	BK49	57
Craig's Ct. SW1	BX40	42
Whitehall		
Craigton Rd. SE9	CK45	53
Craigwell Av. Felt.	BC48	56
Craigwell Clo. Stan.	BK28	12
Craigwell Dr. Stan.	BK28	12
Crail Row, SE17	BZ42	42
Darwin St.		
Craithie Rd. SE12	CH46	53
Cramer St. W1	BV39	41
Crammerville Wk. Rain.	CV39	46

Crampton Rd. SE20 CC50 61
Crampton St. SE17 BZ42 42
Cranberry Clo. Nthlt. BE37 29
Cranberry St. E1 CB38 33
Cranborne Av. Sthl. BF42 38
Cranborne Waye, Hayes BC39 38
Cranbourne Av. E11 CH31 26
Cranbourne Av. Surb. BM55 75
Cranbourne Ct. E18 CH31 26
Cranbourne Dr. Pnr. BD32 20
Cranbourne Gdns. Ilf. CM31 26
Cranbourne Gdns. NW11 BR32 22
Cranbourne Rd. E12 CK35 35
Cranbourne Rd. E15 CF35 34
Cranbourne Rd. N10 BV30 14
Cranbourn Pass. SE16 CB41 42
Wilson Gro.
Cranbourn St. WC2 BW40 41
Long Acre
Cranbrook Clo. Brom. CH53 71
Cranbrook Dr. Esher BG55 74
Cranbrook Dr. Rom. CU31 28
Cranbrook Dr. Twick. BF47 56
Cranbrook Est. E2 CC37 34
Cranbrook Pk. N22 BX30 15
Cranbrook Ri. Ilf. CK32 26
Cranbrook Rd. Barn. BT25 7
Cranbrook Rd. Bexh. CO44 54
Cranbrook Rd. Houns. BE45 47
Cranbrook Rd. Ilf. CL32 26
Cranbrook Rd. SE8 CE44 52
Cranbrook Rd. W4 BP50 58
Cranbrook Rd. Th. Hth. BZ51 69
Cranbrook Rd. W4 BQ42 40
Cranbrook St. E2 CC37 34
Gathorne St.
Cranbrook Ter. E2 CC37 34
Roman Rd.
Cranbury Rd. SW6 BS44 50
Crane Av. W3 BO40 40
Cumberland Pk.
Crane Av. Islw. BJ46 48
Crane Clo. Dag. CR36 36
Crane Ct. Epsom BN56 76
Craneford Clo. Twick. BH47 57
Craneford Way, Twick. BH47 57
Crane Gro. N7 BY36 33
Furlong Rd.
Crane Lo. Houns. BC43 47
Crane Pk. Rd. Twick. BF48 56
Crane Rd. Twick. BH47 57
Cranes Dr. Surb. BL52 66
Cranes Pk. Av. Surb. BL52 66
Cranes Pk. Cres Surb. BL52 66
Cranes Pk. Surb. BL52 66
Crane St. SE10 CF42 43
Craneswater Pk. Sthl. BE42 38
Creanes Way, Borwd. BN25 6
Crane Way, Twick. BG47 56
Cranfield Clo. SE27 BZ49 60
Cranfield Rd. E., Cars. BV58 77
Cranfield Rd. SE4 CD45 52
Cranfield Rd. W., Cars. BV58 77
Cranfield Vills. SE27 BZ48 60
Norwood High St.
Cranford Av. N13 BX28 15
Cranford Clo. SW20 BP50 58
Cranford Cotts. E1 CC40 43
Cranford St.
Cranford Dr. Hayes BC42 38
Cranford La. Est. Hous. BC43 47
Cranford La. Houns. BC43 47
Cranford Pk. Rd. Hayes BC42 38
Cranford Rd. Dart. CW47 64
Cranford St. E1 CC40 43
Cranham Rd. Horn. CU32 28
Cranhurst Rd. NW2 BQ35 31
Cranleigh Clo. SE20 CR46 54
Cranleigh Clo. Orp. CN55 81
Cranleigh Clo. SE20 CB51 69
Cranleigh Clo. S. Croy. BS59 84
Cranleigh Dr. Swan. CT53 73
Cranleigh Gdns. Bans. CM36 35
Cranleigh Gdns. Har. BL32 21
Cranleigh Gdns. Kings. On T. BL50 57
Cranleigh Gdns. Loug. CK25 10
Cranleigh Gdns. N21 BY24 8
Cranleigh Gdns. SE25 CA52 69
Cranleigh Gdns. Sth. BS59 84
Croy.
Cranleigh Gdns. Sthl. BE39 38
Cranleigh Gdns. Sutt. BS55 77
Cranleigh Rd. Esher BG54 65
Cranleigh Rd. N15 BZ32 24
Cranleigh Rd. N19 BR52 67
Cranleigh St. NW1 BW37 32
Cranley Dr. Ilf. CM33 26
Cranley Dr. Ruis. BC34 20
Cranley Gdns. N10 BV31 23
Cranley Gdns. N13 BX27 15
Cranley Gdns. SW7 BT42 41
Cranley Gdns. Wall. BW57 77
Cranley Ms. Ilf. CM32 26
Cranley Rd.
Cranley Pl. SW7 BT42 41
Cranley Pl. SW7 BT42 41
Cranley Rd. E13 CH39 44
Cranley Rd. Ilf. CM33 26

Cranmer Av. W13 BJ41 39
Cranmer Clo. Mord. BO53 67
Cranmer Clo. Ruis. BD33 20
Cranmer Clo. Stan. BK29 12
Cranmer Ct. SW3 BU42 41
Cranmer Ct. Hamptn. BF49 56
Cranmer Farn. Clo. BU52 68
Mitch.
Cranmer Gdns. Dag. CS35 37
Cranmer Rd. Croy. BY55 78
Cranmer Rd. Edg. BM27 12
Cranmer Rc. E7 CH35 35
Cranmer Rd. Hamptn. BF49 56
Cranmer Rd. BL49 57
Kings. On T.
Cranmer Rd. Mitch. BU52 68
Cranmer Rd. SW9 BY43 51
Cranmer Ter. SW17 BT49 59
Cranmer Ter. SW17 BU49 59
Tooting Gro.
Cranmore Av. Chis. CK49 62
Cranmore Av. Islw. BG43 47
Cranmore Pk. Est. Chis. CK50 62
Cranmore Rd. Brom. CG48 61
Cranmore Way N10 BW31 23
Cranston Clo. Houns. BE44 47
Cranston Est. N1 BZ37 33
Cranston Gdns. E4 CE28 16
Cranston Rd. SE23 CD47 61
Cranswick Rd. SE16 CB42 42
Crantock Rd. SE6 CE48 61
Cranwich Av. N21 BZ26 8
Cranwich Rd. N16 BZ33 24
Cranwood St. EC1 BZ38 33
Cranworth Cres. E4 CF26 9
Cranworth Gdns. SW9 BY44 51
Craster Rd. SW2 BX47 60
Craven Av. Felt. BC48 56
Craven Av. W5 BK40 39
Craven Av. Sthl. BE39 38
Craven Av. W5 BK40 39
Craven Clo. Hayes BC39 38
Craven Gdns. Bark. CN37 36
Craven Gdns. Ilf. CM30 17
Craven Gdns. Rom. CR28 18
Craven Gdns. SW19 BS49 59
Craven Hill Gdns. W2 BT40 41
Craven Hill Ms. W2 BT40 41
Craven Hill W2 BT40 41
Craven Ms. SW11 BV46 50
Craven Pk. NW10 BN37 31
Craven Pk. Rd. N15 CA32 24
Craven Pk. Rd. NW10 BO37 31
Craven Pl. WC2 BX40 42
Craven St.
Craven Rd. Croy. CB54 69
Craven Rd. Kings. On T. BL51 66
Craven Rd. NW10 BO37 31
Craven Rd. Orp. CP55 81
Craven Rd. W2 BT40 41
Craven Rd. W5 BK40 39
Craven St. WC2 BX40 42
Craven Ter. W2 BT40 41
Craven Wk. N16 CB33 24
Crawford Av. Wem. BK35 30
Crawford Clo. Islw. BH44 48
Crawford Est. SE5 BZ44 51
Crawford Gdns. N13 BY27 15
Crawford Gdns. Nthlt. BE38 29
Crawford Pl. W1 BU39 41
Crawford Rd. SE5 BZ44 51
Crawford St. W1 BU39 41
Crawley Rd. E10 CE33 25
Crawley Rd. Enf. CA26 8
Crawley Rd. N22 BZ30 15
Crawshay Rd. SW9 BY44 51
Crawthew Gro. SE22 CA45 51
Cray Av. Orp. CO54 72
Craybrooke Rd. Sid. CO49 63
Craybury End SE9 CM48 62
Cray Clo. Dart. CU45 55
Craydene Rd. Erith CT44 55
Crayford High St. Dart. CT45 55
Crayford Rd. Dart. CT46 55
Crayford Rd. N7 BW35 32
Crayford Way, Dart. CT46 55
Crayhill St. DA1 CU45 55
Norris Way
Crayke Hill, Chess. BL57 75
Craylands, Orp. CP52 72
Cray Rd. Belv. CR43 54
Cray Rd. Sid. CP50 63
Cray Rd. Swan. CR53 72
Cray Valley Rd. Orp. CO53 72
Crealock Gro. Wdf. Grn. CG28 16
Crealock St. SW18 BS46 50
Creasy St. SE1 CA41 42
Webb St.
Crebor St. SE22 CB46 51
Creden Hall Dr. Brom. CK54 71
Lwr. Gravel Rd.
Crediton Hl. NW6 BS35 32
Crediton Rd. E16 CH39 44
Crediton Rd. NW10 BQ37 31
Crediton Way, Esher BJ56 75
Credon Rd. E13 CJ37 35
Credon Rd. SE16 CB42 42

Creechurch La. EC3 CA39 42
Creed La. EC4 BY39 42
Creek Br. SE10 CE43 52
Creek Rd. Bark. CN38 36
Creek Rd. E. Mol. BH52 56
Creek Rd. SE8 CE43 52
Creek Rd. SE10 CE43 52
Creekside, SE8 CE43 52
Creekside, Rain. CT38 37
Creek, The, Sun. BC53 65
Creeland Gro. SE6 CD47 61
Catford Hill
Cree Way, Rom. CT29 19
Creffield Rd. W3 BL40 39
Creffield Rd. W5 BL40 39
Crefield Clo. W6 BQ43 49
Creighton Av. E6 CJ37 35
Creighton Av. N2 BU31 23
Creighton Av. N10 BU31 23
Creighton Rd. N17 CA29 15
Creighton Rd. NW6 BQ37 31
Creighton Rd. W5 BK41 39
Cremer St. E2 CA37 33
Cremorne Est. SW10 BT43 50
Cremorne Gdn. Epsom BN58 76
Cremorne Gdns. SW10 BT43 50
Cremorne Rd. SW10 BT43 50
Crescent Av. Horn. CT34 28
Crescent Ct. Surb. BK53 66
Crescent Dr. Orp. CL53 71
Crescent Gdns. Ruis. BC33 20
Crescent Gdns. Swan. CS51 73
Crescent Gdns. SW19 BS48 59
Crescent Gro. Mitch. BU53 68
Crescent Gro. SW4 BW45 50
Crescent La. SW4 BW45 50
Crescent Pl. SW3 BU42 41
Crescent Ri. N22 BW29 14
Crescent Rd. Barn. BT24 7
Crescent Rd. Beck. CE51 70
Crescent Rd. Brom. CH50 62
Crescent Rd. Dag. CR34 27
Crescent Rd. Enf. BY24 8
Crescent Rd. Erith CT43 55
Crescent Rd. E4 CG26 9
Crescent Rd. E6 CJ37 35
Crescent Rd. E10 CE34 25
Crescent Rd. E13 CH37 35
Crescent Rd. E18 CH30 17
Crescent Rd. Kings. BM50 57
On T.
Crescent Rd. N3 BR30 13
Crescent Rd. N8 BW32 23
Crescent Rd. N9 CB26 8
Crescent Rd. N15 CH31 26
Crescent Rd. N22 BW30 14
Crescent Rd. Sid. CN48 63
Crescent Rd. Sthl. BE41 38
Crescent Rd. SW20 BO50 58
Crescent Row, EC1 BZ38 33
Baltic St.
Crescent, The, Barn. BS23 7
Crescent, The, Beck. CE51 70
Crescent, The, Bex. CP47 63
Crescent, The, Croy. BZ53 69
Crescent, The, E17 CD32 25
Crescent, The, E. Mol. BE52 65
Crescent, The, Har. BG33 20
Crescent, The, Ilf. CL32 26
Crescent, The, Loug. CJ25 10
Crescent, The, N. Mal. BN52 67
Crescent, The, N11 BU28 14
Crescent, The, NW2 BP34 22
Crescent, The, Sid. CN49 63
Crescent, The, Surb. BL53 66
Crescent, The, Sutt. BS59 83
Crescent, The, Sutt. BT56 77
Crescent, The, SW13 BO44 49
Crescent, The, SW19 BS48 59
Crescent, The, W3 BN39 40
Crescent, The, Wat. BD24 4
Crescent, The, Wem. BJ34 21
Crescent, The, W. Wick. CG53 70
Crescent Vw. Loug. CJ25 10
Crescent Way N12 BU29 14
Crescent Way, Orp. CN56 81
Crescent Way SW16 BX50 60
Crescent W. Barn. BS23 7
Crescent, The, SW13 BO44 49
Crescent, The, W3 BN39 40
Cresfield Rd. SW6 BS44 50
Cresford Rd. SW6 BS44 50
Crespigny Rd. NW4 BP32 22
Cressage Av. W4 BM43 48
Cressage Clo. Sthl. BE38 29
Cresset Rd. E9 CC36 34
Cresset St. SW4 BW45 50
Clapham Manor St.
Cressida Rd. N19 BW33 23
Cressingham Gdns. Est. BY47 60
SE24
Cressingham Gro. Sutt. BT56 77
Cressingham Rd. Edg. BN29 13
Cressingham Rd. SE13 CF45 52
Cresswell Gdns. Houns. BE45 47
Cresswell Gdns. SW5 BT42 41

Cresswell Pk. SE3 CG45 52
Cresswell Pl. SW10 BT42 41
Cresswell Rd. SE25 CB52 69
Cresswell Rd. Twick. BK46 48
Cresswell Way N21 BY26 8
Cressy Ct. E1 CC39 43
Cressy Pl. E1 CC39 43
Cressy Rd. NW3 BU35 32
Cresta Ct. W5 BL38 30
Crestbrook Av. N13 BY27 15
Crestbrook Pl. N13 BY27 15
Crest Dr. Enf. CC23 9
Crestfield St. WC1 BX38 33
St. Chads St.
Creston Way, Wor. Pk. BQ55 76
Crest Rd. Brom. CG54 70
Crest Rd. NW2 BO34 22
Crest Rd. Sth. Croy. CB57 78
Crest St. N1 BX36 33
Huntingdon St.
Crest, The, N13 BY28 15
Crest, The, N14 BX32 22
Crest, The, NV4 BS23 7
Crest, The, Surb. BM53 66
Crest Vw. Pnr. BD31 20
Crestview Dr. Orp. CL53 71
Crestway, SW15 BP46 49
Creswell Clo. Felt. BE49 56
Creswick Ct. W3 BM40 39
Creswick Rd. W3 BM40 39
Creswick Wk. NW11 BR31 22
Creton St. SE18 CL41 44
Crevington Way, Horn. CV35 37
Crewdson Rd. SW9 BY43 51
Crewe Pl. NW10 BO38 31
Crews St. E14 CE42 43
Crewys Rd. NW2 BR34 22
Crewys Rd. SE15 CB44 51
Crichton Av. Wall. BW56 77
Crichton Rd. Cars. BU57 77
Cricketers Arms Rd. Enf. BZ23 8
Cricketers Clo. Chess. BK56 75
Cricketfield Rd. E5 CB35 33
Cricket Grn. Mitch. BU52 68
Cricket Ground Rd. Chis. CL51 71
Cricklade Av. Rom. CV29 19
Cricklade Av. SW2 BX48 60
Cricklewood Broadway BQ35 31
NW2
Cricklewood La. NW2 BQ35 31
Cridland St. E15 CG37 34
Church St.
Crieff Ct. Tedd. BK50 57
Crieff Rd. SW18 BT46 50
Criffel Av. SW2 BW48 59
Crighton Gdns. Rom. CR33 27
Crimscott St. SE1 CA41 42
Crimsworth Rd. SW8 BW44 50
Crinan St. N1 BX37 33
Cringle St. SW8 BW43 50
Crispen Rd. Felt. BD49 56
Crispin Clo. Croy. BX55 78
Crispin Cres. Croy. BW55 77
Crispin Rd. Edg. BN29 13
Crispin St. E1 CA39 42
Brushfield St.
Crisp Rd. W6 BQ42 40
Cristowe Rd. SW6 BR44 49
Criterion Ms. N19 BW34 23
St. John's Vill.
Crockenhill La. Dart. CV54 73
Crockenhill Rd. Orp. CP53 72
Crockenhill Rd. Swan. CQ53 72
Crockerton Rd. SW17 BU48 59
Crockham Way, SE9 CL49 62
Crockham Barn. BR25 6
Crodon Gro. Croy. BY54 69
Croft Clo. NW7 BO27 13
Croft Clo. Belv. CQ42 45
Croft Clo. Chis. CK49 62
Croft Av. W. Wick. CF54 70
Croftdown Rd. NW5 BV34 23
Crofters Clo. Islw. BG46 47
Crofters Mead, Croy. CD58 79
Croft Gdns. W7 BJ41 39
Croftleigh Av. Pur. BY57 78
Croftleigh Av. Pur. BY61 84
Croft Lodge Clo. Wdf. CH29 17
Grn.
Crofton Av. Bex. CP47 63
Crofton Av. Orp. CM55 80
Crofton Av. Walt. BD55 74
Crofton Ct. Orp. CM54 71
Crofton La. Orp. CM55 80
Crofton Pk. Rd. SE4 CD46 52
Crofton Pound Hill, Orp. CM55 80
Crofton Rd. E13 CH38 35
Crofton Rd. Orp. CL55 80
Crofton Rd. SE5 CA44 51
Crofton Ter. Rich. BL45 48
Crofton Way, Enf. BY23 8
Croft Rd. N17 CA29 15
Durban Rd.
Croft Rd. Brom. CH50 62
Croft Rd. Enf. CD23 9
Croft Rd. Sutt. BT56 77
Croft Rd. SW16 BY51 69
Croft Rd. SW19 BT50 59
Crofts Rd. Har. BJ32 21

Cumberland Ter. Ms. NW1   BV37 32
Albany St.
Cumberland Ter. NW1   BV37 32
Cumberlow Av. SE25   CA52 69
Cumberton Rd. N17   BZ30 15
Cumbrae Gdns. Surb.   BK54 66
Cumbrian Av. Bexh.   CT44 55
Cumbrian Gdns. NW2   BQ34 22
Cumming St. N1   BX37 33
Cummings Hall La. Rom.   CV27 19
Cumming St. N1   BX37 33
Cumnor Gdns. Epsom   BP57 76
Cumnor Rd. Sutt.   BT57 77
Bury St.
Cunard Pl. EC3   CA39 42
Cunard St. SE5   CA43 51
Cundy Rd. E16   CH39 44
Cundy St. Est. SW1   BV42 41
Cundy St. SW1   BV42 41
Cunliffe Rd. Wor. Pk.   BO56 76
Cunliffe Rd. SW16   BW50 59
Cunningham Clo. W. Wick.   CE55 79
Cunningham Pk. Har.   BG32 20
Cunningham Pl. NW8   BT38 32
Cunningham Rd. Bans.   BT61 83
Cunningham Rd. N15   CB31 24
Cunnington St. W4   BN41 40
Cupar Rd. SW11   BV44 50
Cupola Clo. BR1   CH49 62
Powster Rd.
Cureton St. SW1   BW42 41
Curlew Clo. SE28   CP40 45
Curlew Clo. S. Croy.   CC58 79
Curlew St. SE1   CA41 42
Curnick's La. SE27   BZ49 60
Curnock Est. NW1   BW37 32
Curran Av. Sid.   CN46 54
Curran Av. Wall.   BV55 77
Currey Rd. Grnf.   BG36 29
Curricle St. W3   BO40 40
Currie Hill Clo. SW19   BR49 58
Curry Ri. NW7   BQ29 13
Cursitor St. EC4   BY39 42
Curtain Rd. EC2   CA38 33
Curthwaite Gdns. Enf.   BW24 7
Curtismill Clo. Orp.   CO52 72
Curtismill Way, Orp.   CO52 72
Curtis Rd. Epsom   BN56 76
Curtis Rd. Horn.   CW33 28
Curtis Rd. Houns.   BE47 56
Curtis St. SE1   CA42 42
Curtis Way SE28   CO40 45
Curvan Clo. Epsom   BO58 76
Curve, The W12   BP40 40
Curwen Av. E7   CH35 35
Woodford Rd.
Curwen Rd. W12   BP41 40
Curzon Av. Enf.   CC25 9
Curzon Av. Stan.   BJ30 12
Curzon Clo. Orp.   CM56 80
Curzon Cres. Bark.   CN37 36
Curzon Ho. W5   BJ38 30
Castlebar Pk. Rd.
Curzon Pl. Pnr.   BD32 20
Curzon Pl. W1   BV40 41
Curzon St.
Curzon Rd. N10   BV30 14
Curzon Rd. Th. Hth.   BX53 69
Curzon Rd. W5   BJ38 30
Curzon St. W1   BV40 41
Cusack Clo. Twick.   BH49 57
Custom Ho. Wf. EC3   CA40 42
Cutcombe Rd. SE5   BZ44 51
Cuthbert Rd. Cfoy.   BY55 78
Cuthbert Rd. E17   CF31 25
Cuthbert Rd. N18   CB28 15
Fairfield Rd.
Cuthbert St. W2   BT39 41
Cuthill Rd SE5   BZ44 51
Cutler's Ter. N1   CA36 33
Balls Pond Rd.
Cutler St. E1   CA39 42
Cut, The SE1   BY41 42
Cutthoat All. Rich.   BK48 57
Cuxton Clo. Bexh.   CO46 54
Cyclamen Rd. Swan.   CS52 73
Cyclamen Way, Epsom   BN56 76
Cygnet Av. Felt.   BD47 56
Cygnets, The, Felt.   BE49 56
Cygnet St. E1   CA38 33
Sclater St.
Cynthia St. N1   BX37 33
Cyntra Pl. E8   CB36 33
Mare St.
Cypress Av. Twick.   BG47 56
Cypress Gro. Ilf.   CN29 18
Cypress Pl. W1   BW38 32
Maple St.
Cypress Rd. SE25   CA51 69
Cypress Rd. Har.   BG30 11
Cypress Way, Bans.   BQ60 82
Cypress Av. N3   BR30 13
Cypress Gdns. N3   BR30 13
Cypress Pl. E2   CC37 34
Cyprus St.

Cyprus Pl. E6   CL40 44
Cyprus Rd. N3   BR30 13
Cyprus Rd. N9   CA27 15
Cyprus St. E2   CC37 34
Cyprus St. EC1   BY38 33
Cyrena Rd. SE22   CA46 51
Cyril Mans. SW11   BU44 50
Cyril Rd. Bexh.   CQ44 54
Cyril Rd. Orp.   CQ54 72
Czar St. SE8   CD43 52
Dabbshill La. Nthlt.   BE35 29
D'abernon Clo. Esher   BF56 74
Dabin Cres. SE10   CF44 52
Lindsell St.
Dacca St. SE8   CD43 52
Dace Rd. E3   CE36 34
Dacre Av. Ilf.   CL30 17
Dacre Clo. Chig.   CM28 17
Dacre Clo. Grnf.   BF37 29
Dacre Gdns. Borwd.   BN25 6
Dacre Gdns. Chig.   CM28 17
Dacre Pk. SE13   CG45 52
Dacre Pl. SE13   CG45 52
Dacre Rd. Croy.   BX54 69
Dacre Rd. E11   CG33 25
Dacre Rd. E13   CH37 35
Dacres Clo. SE23   CC48 61
Dacres Rd. SE23   CC48 61
Dacre St. SW1   BW41 41
Daerwood Clo. Brom.   CK54 71
Daffodil St. W12   BO40 40
Dafforne Rd. SW17   BU48 59
Dagenham Av. Dag.   CQ37 36
Dagenham Rd. Dag.   CR35 36
Dagenham Rd. E10   CD33 25
Dagenham Rd. Rain.   CS36 37
Dagenham Rd. Rom.   CS33 28
Dagger La. Borwd.   BJ25 5
Dagmar Av. Wem.   BL35 30
Dagmar Gdns. NW10   BQ37 31
Dagmar Rd. Dag.   CS36 37
Dagmar Rd. Kings. On T.   BL51 66
Dagmar Rd. N4   BY33 24
Dagmar Rd. N15   BZ32 24
Cornwall Rd.
Dagmar Rd. N22   BW30 14
Dagmar Rd. SE5   CA44 51
Dagmar Rd. SE25   CA52 69
Dagmar Rd. Sthl.   BE41 38
Dagmar Ter. N1   BY37 33
Dagnall Pk. SE25   BZ53 69
Dagnall Rd. SE25   CA53 69
Dagnall St. SW11   BU44 50
Dagnam Pk. Dr. Rom.   CW28 19
Dagnan Rd. SW12   BV47 59
Dagonet Gdns. Brom.   CH48 62
Dagonet Rd. Brom.   CH48 62
Dahlia Dr. Swan.   CT51 73
Dahlia Gdns. Mitch.   BW52 68
Dahlia Rd. SE2   CO42 45
Dahomey Rd. SW16   BW50 59
Daimler Cotts. SE15   CA43 51
Cronin St.
Daimler Way, Wall.   BX57 78
Daines Clo. E12   CK34 26
Dainford Clo. Brom.   CF49 61
Dainton Clo. Brom.   CH51 71
Dainty Way E9   CD36 34
Dairsie Rd. SE9   CL45 53
Dairy Wk. SW19   BR49 58
Daisy Dormer Ct. SW9   BX45 51
Trinity Gdns.
Daisy La. SW6   BS45 50
Daisy Rd. E18   CH30 17
Dakota Gdns. Nthlt.   BD38 29
Dalberg Rd. SW2   BY45 51
Dalberg Way, Belv.   CP41 45
Dalby Rd. SW18   BT45 50
Dalby St. NW5   BU36 32
Dalcross Rd. Houns.   BE44 47
Dale Av. Edg.   BL30 12
Dale Av. Houns.   BE45 47
Dalebury Rd. SW17   BU48 59
Dale Clo. Barn.   BS25 7
Dale Clo. Dart.   CT46 55
Dale Clo. Pnr.   BC30 11
Dale Clo. SE3   CH45 53
Dale End, Dart.   CT46 55
Dale Gdns. Wdf. Grn.   CH28 17
Dale Gro. N12   BT28 14
Dale Grn. Rd. N11   BV27 14
Dale Pk. Av. Cars.   BU55 77
Dale Pk. Rd. SE19   BZ51 69
Dale Rd. Borwd.   BN25 6
Dale Rd. Dart.   CT46 55
Dale Rd. E16   CG38 34
Dale Rd. Grnf.   BF39 38
Dale Rd. NW5   BV35 32
Dale Rd. Pur.   BY59 84
Dale Rd. SE18   CL43 53
Dale Rd. Sutt.   BR56 76
Dale Rd. Swan.   CS51 73
Dale Rd. Walt.   BC54 65
Daleside Clo. Orp.   CO57 81

Daleside Gdns. Chig.   CM27 17
Daleside, Orp.   CO56 81
Daleside Rd. Epsom   BN57 76
Daleside Rd. SW16   BV49 59
Dale, The, Kes.   CJ56 80
Daleview, Erith   CT44 55
Dale View Av. E4   CF27 16
Dale View Gdns. E4   CF27 16
Dalewood Clo. Horn.   CW33 28
Dalewood Gdns. Wor. Pk.   BP55 76
Dale Wood Rd. Orp.   CN54 72
Daley St. E9   CC36 34
Dalgarno Gdns. W10   BQ39 40
Dalgarno Way W10   BQ38 31
Dalgleish St. E14   CD39 43
Daling Way E3   CD37 34
Dalkeith Gro. Stan.   BK28 12
Dalkeith Rd. Ilf.   CM34 26
Dalkeith Rd. SE21   BZ47 60
Dallas Rd. NW4   BP33 22
Dallas Rd. SE26   CB49 60
Dallas Rd. Sutt.   BR57 76
Dallas Rd. W5   BL39 39
Dallinger Rd. SE12   CG46 52
Dalling Rd. W6   BP42 40
Dallington Clo. Walt.   BD57 74
Dallington St. EC1   BY38 33
Dallin Rd. Bexh.   CP45 54
Dallin Rd. SE18   CL43 53
Dalmain Rd. SE23   CC47 61
Dalmally Rd. Croy.   CA54 69
Dalmeny Av. N7   BW35 32
Dalmeny Av. SW16   BY51 69
Dalmeny Clo. Wem.   BK36 30
Dalmeny Cres. Houns.   BG45 47
Dalmeny Rd. Barn.   BT25 7
Dalmeny Rd. Bexh.   CR44 54
Dalmeny Rd. Cars.   BV57 77
Dalmeny Rd. N7   BW34 23
Dalmeny Rd. Wor. Pk.   BP55 76
Dalmore Av. Esher   BH57 75
Dalmore Rd. SE21   BZ48 60
Dalrymple Rd. SE4   CD45 52
Dalston Gdns. Stan.   BL30 12
Dalston La. E8   CA36 33
Dalton Av. Mitch.   BU51 68
Dalton Clo. DA1   CT45 55
Dalton Clo. Orp.   CN55 81
Dalton Rd. Har.   BG30 11
Dalton St. SE27   BZ48 60
Dalton's Rd. Orp.   CR55 81
Dalton's Rd. Swan.   CS54 73
Dalwood St. SE5   CA44 51
Dalyell Rd. SW9   BX45 51
Damer Ter. SW10   BT43 50
Ashburnham Rd.
Dames Rd. E7   CH34 26
Dame St. N1   BZ37 33
Damien St. E1   CB39 42
Damon Clo. Sid.   CO48 63
Danbrook Rd. SW16   BX51 69
Danbury Clo. Rom.   CP31 27
Danbury Ms. Wall.   BV56 77
Danbury Rd. Loug.   CK26 10
Danbury Rd. Rain.   CT37 37
Danbury St. N1   BY37 33
Danbury Way, Wdf. Grn.   CJ29 17
Danby St. SE15   CA45 51
Dancer Rd. Rich.   BM45 48
Dancer Rd. SW6   BR44 49
Fulham Rd.
Dando Cres. SE3   CH45 53
Danebury Av. SW15   BO46 49
Daneby Rd. SE6   CE48 61
Dane Clo. Bex.   CR47 63
Dane Clo. Orp.   CM56 80
Dane Ct. N2   BT33 23
Danecroft Rd. SE24   BZ46 51
Danehill Wk. Sid.   CO48 63
Hatherley Rd.
Danehurst Gdns. Ilf.   CK32 26
Danehurst St. SW6   BR44 49
Daneland, Barn.   BU25 7
Danemead Gdns. Nthlt.   BF35 29
Danemere St. SW15   BQ45 49
Dane Rd. Ilf.   CM35 35
Dane Rd. N18   CC28 16
Dane Rd. Sthl.   BE40 38
Dane Rd. SW19   BT51 68
Dane Rd. W13   BK40 39
Danesbury, Croy.   CF57 79
Danesbury Rd. Felt.   BC47 56
Danescourt Cres. Sutt.   BT55 77
Danescroft Gdns. Croy.   CA55 78
Danescroft Av. NW4   BQ32 22
Danescroft Gdns. NW4   BQ32 22
Danesdale Rd. E9   CD36 34
Daneshurst St. SW6   BR44 49
Danes Rd. Rom.   CS33 28
Dane St. WC1   BX39 42
Red Lion Sq.
Daneswood Av. SE6   CF48 61
Danethorpe Rd. Wem.   BK36 30
Danetree Rd. Epsom   BN57 76
Danette Gdns. Dag.   CR34 27

Daneville Rd. SE5   BZ44 51
Dangan Rd. E11   CH32 26
Daniel Bolt Clo. E14   CE39 43
Daniel Gdns. SE15   CA43 51
Daniel Pl. NW4   BP32 22
Daniel Rd. W5   BL40 39
Daniels Rd. SE15   CC45 52
Dansington Rd. Well.   CO45 54
Danson Cres. Well.   CP45 54
Danson La. Well.   CO45 54
Danson Mead, Well.   CP45 54
Danson Rd. Bex.   CP46 54
Dante Rd. SE11   BY42 42
Danube St. SW3   BU42 41
Danvers Rd. N8   BW31 23
Danvers St. SW3   BT43 50
Danyon Clo. Rain.   CV38 37
Da Palma St. SW6   BS43 50
Racton Rd.
Daphne Gdns. E4   CF27 16
Gunners Gro.
Daphne St. SW18   BT46 50
Daplyn St. E1   CB39 42
D'arblay St. W1   BW39 41
Darby Cres. Sun.   BD51 65
Darby Gdns. Sun.   BD51 65
Darcy Av. Wall.   BW56 77
Darcy Clo. N20   BT27 14
D'arcy Dr. Har.   BK31 21
D'arcy Gdns. Dag.   CQ37 36
D'arcy Gdns. Har.   BK31 21
D'arcy Rd. Sutt.   BQ56 76
Darcy Rd. SW16   BW51 68
Dare Gdns. Dag.   CQ34 27
Darenth Rd. Dart.   CW47 64
Darenth Rd. N16   CA33 24
Darenth Rd. Well.   CO44 54
Darfield Rd. SE4   CD46 52
Darfield Way W10   BQ40 40
Darfur St. SW15   BQ45 49
Darien Rd. SW11   BT45 50
Darlan Rd. SW6   BS43 50
Darlaston Rd. SW19   BQ50 58
Darley Clo. Croy.   CD53 70
Darley Dr. N. Mal.   BN51 67
Darley Gdns. Mord.   BT53 68
Darley Rd. N9   CA26 8
Darley Rd. SW11   BU46 50
Darling Rd. SE4   CE45 52
Darling Row E1   CB38 33
Collingwood St.
Darlington Gdns. Rom.   CV28 19
Darlington Path, Rom.   CV28 19
Darnley Rd. E9   BY49 60
Darnley Rd. Wdf. Grn.   CH30 17
Darnley Ter. W11   BQ40 40
Darrell Rd. SE22   CB46 51
Darrel Rd. Rich.   BM45 48
Darren Clo. N4   BX33 24
Darrick Wd. Rd. Orp.   CM55 80
Darrington Rd. Borwd.   BL23 5
Dartfields Rd. Rom.   CV29 19
Dartford Av. N9   CC25 9
Dartford By-pass, Dart.   CU48 64
Dartford La. Bex.   CS48 64
Dartford Rd. Bex.   CS47 64
Dartford Rd. Dart.   CU46 55
Dartford Rd. Farn.   CW53 73
Dartford St. SE17   BZ43 51
Dartmouth Gro. SE10   CF44 52
Dartmouth Hl. SE10   CF44 52
Dartmouth Pk. Av. NW5   BV34 23
Dartmouth Pk. Hill N19   BV33 23
Dartmouth Pk. Hill NW5   BV33 23
Dartmouth Pk. Rd. NW5   BV35 32
Dartmouth Pl. SE23   CC48 61
Dartmouth Pl. W4   BO43 49
Dartmouth Rd. Brom.   CH54 71
Dartmouth Rd. E16   CH39 44
Dartmouth Rd. NW2   BQ36 31
Dartmouth Rd. NW4   BP32 22
Dartmouth Rd. Ruis.   BC34 20
Dartmouth Rd. SE23   CB48 60
Dartmouth Rd. SE26   CB48 60
Dartmouth Row SE10   CF44 52
Dartmouth St. SW1   BW41 41
Dartnell Rd. Croy.   CA54 69
Dart St. W10   BR38 31
Darville Rd. N16   CA34 24
Darwell Clo. E6   CL37 35
Darwin Clo. Orp.   CM56 80
Darwin Dr. Sthl.   BF39 38
Darwin Gdns. Wat.   BD28 11
Barnhurst Path
Darwin Pl. SE17   BZ42 42
Darwin St.
Darwin Rd. N22   BY30 15
Darwin Rd. W5   BK42 39
Darwin Rd. Well.   CN45 54
Darwin St. SE17   BZ42 42
Daryngton Dr. Grnf.   BG37 29
Dashwood Clo. Bexh.   CR46 54
Dashwood Rd. N8   BX32 24
Dassett Rd. SE27   BY49 60
Datchelor Pl. SE5   BZ44 51
Datchet Rd. SE6   CD48 61

Deptford High St. SE8 CE43 52
Deptford Strand SE8 CD42 43
De Quincey Rd. N17 BZ30 15
Derby Arms Rd. Epsom BO62 82
Derby Av. N12 BT28 14
Derby Av. Har. BG30 11
Derby Av. Rom. CS32 28
Derby Av. Upmin. CW35 37
Derby Ct. E5 CC35 34
Clapton Park Est.
Derby Gate SW1 BX41 42
Derby Gro. Croy. BY54 69
Derby Hl. SE23 CC48 61
Derby Hl. Cres. SE23 CC48 61
Derby Hl. Est. SE23 CC48 61
Derby Rd. E7 CJ36 35
Derby Rd. E9 CC37 34
Derby Rd. E18 CG30 16
Derby Rd. N15 BY31 24
Derby Rd. N18 CB28 15
Derby Rd. SW14 BM45 48
Derby Rd. SW19 BS50 59
Derby Rd. Croy. BY54 69
Derby Rd. Enf. CB25 8
Derby Rd. Grnf. BF37 29
Derby Rd. Houns. BF45 47
Derby Rd. Surb. BM54 66
Derby Rd. Sutt. BR57 76
Derby Rd. Wat. BD24 4
Derbyshire St. E2 CC38 33
Derby Stables Rd. Epsom BO62 82
Derby St. W1 BV40 41
Curzon St.
Dereham Pl. EC2 CA38 33
Dereham Rd. Bark. CN35 36
Derek Av. Epsom BM57 75
Derek Av. Wall. BV56 77
Derek Av. Wem. BM36 30
Deri Av. Rain. CU38 37
Dericote Rd. E8 CB37 33
Croston St.
Dering Pl. Croy. BZ56 78
Dering Rd. Croy. BZ56 78
Dering St. W1 BV39 41
Derinton Rd. SW17 BU49 59
Derley Rd. Sthl. BD41 38
Dermody Gdns. SE13 CF46 52
Dermody Rd. SE13 CF46 52
Derns Wk. Couls. BY58 78
Deronda Est. SE2 BY47 60
Deronda Rd. SE24 BY47 60
Derrick Av. Sth. Croy. BZ58 78
Derrick Gdns. SE7 CJ41 44
Derrick Rd. Beck. CD52 70
Derry Downs, Orp. CP53 72
Derry Rd. Croy. BX55 78
Derry St. W8 BS41 41
Dersingham Av. E12 CK35 35
Dersingham Rd. NW2 BR34 22
Derwent Av. N18 BZ28 15
Derwent Av. NW7 BN29 13
Derwent Av. NW9 BO32 22
Derwent Av. SW15 BN40 58
Derwent Av. Barn. BU26 7
Derwent Av. Pnr. BE29 11
Derwent Clo. Dart. CU47 64
Derwent Clo. Esher BH57 75
Derwent Clo. N20 BT27 14
Derwent Cres. Bexh. CR44 54
Derwent Cres. Stan. BK30 12
Derwent Dr. Orp. CM54 71
Derwent Dr. Pur. BZ60 84
Derwent Gdns. Ilf. CK31 26
Derwent Gdns. Wem. BK33 21
Derwent Ri. NW9 BO32 22
Derwent Rd. N13 BX28 15
Derwent Rd. SE20 CB51 69
Derwent Rd. SE22 CA45 51
Derwent Rd. W5 BK41 39
Derwent Rd. Mord. BQ53 67
Derwent Rd. Sthl. BE39 38
Derwent Rd. Twick. BF46 47
Derwent St. SE10 CG42 43
Derwent Wk. Wall. BV57 77
Woodbourne Gdns.
Derwentwater Rd. W3 BN40 40
Derwent Way, Horn. CU35 37
Desborough St. W2 BS39 41
Cirencester St.
Desenfans Rd. SE21 CA46 51
Desford Rd. E16 CG38 34
Desmond St. SE14 CD43 52
Despard Av. SW11 BV44 50
Despard Rd. N19 BW33 23
Detling Clo. Horn. CV35 37
Detling Rd. Brom. CH49 62
Detling Rd. Erith CS43 55
Devana End, Cars. BU55 77
Devas Rd. SW20 BQ51 67
Devas St. E3 CE38 34
Devenay Rd. E15 CG36 34
Devenish Rd. SE2 CO41 45
Deventer Cres. SE22 CA46 51
Dulwich Gro.
De Vere Gdns. W8 BT41 41
De Vere Gdns. Ilf. CK34 26
Deverell St. SE1 BZ41 42

Devereux Ct. EC4 BY40 42
Fountain Ct.
Devereux Rd. SW11 BU46 50
Deveron Way, Rom. CT30 19
Devoke Way, Walt. BD55 74
Devon Av. Twick. BG47 56
Devon Clo. N17 CA31 24
Devon Clo. Buck. H. CH27 17
Devon Clo. Grnf. BK37 30
Devon Clo. Ken. CA61 84
Devon Ct. W3 BM39 39
Links Rd.
Devon Ct. Hmptn. BF50 56
Devon Cres. Grnf. BK37 30
Devoncroft Gdns. Twick. BJ47 57
Oak La.
Devon Gdns. N4 BY32 24
Devonia Gdns. N18 BZ29 15
Devonia Rd. N1 BY37 33
Devonport Gdns. Ilf. CK32 26
Devonport Mews W12 BP41 40
Devonport Pass. E1 CC39 43
Devonport Rd. W12 BP40 40
Devonport St. E1 CC39 43
Devon Rise N2 BT31 23
Devon Rd. Bark. CN37 36
Devon Rd. Sutt. BR58 76
Devon Rd. Walt. BD56 74
Devon Rd. Wat. BD23 4
Devons Est. E3 CE38 34
Devonshire Av. Dart. CU46 55
Devonshire Av. Sutt. BT57 77
Devonshire Clo. E15 CG35 34
Devonshire Clo. N13 BY27 15
Devonshire Clo. W1 BV39 41
Devonshire Ct. Croy. CD54 70
Devonshire Ct. Rich. BL44 48
Holmesdale Rd.
Devonshire Cres. NW7 BQ29 13
Devonshire Dr. SE10 CE43 52
Devonshire Gdns. N17 BZ29 15
Devonshire Rd.
Devonshire Gdns. N21 BZ26 8
Devonshire Gdns. W4 BN43 49
Devonshire Gro. SE15 CB43 51
Devonshire Hill La. N17 BY29 15
Devonshire Ms. S. W1 BV39 41
Devonshire Ms. W. W1 BV38 32
Devonshire Pl. W1 BV38 32
Devonshire Pl. W4 BO42 40
Devonshire Pl. Ms. W1 BV38 32
Devonshire Rd. E15 CG35 34
Devonshire Rd. E16 CH39 44
Devonshire Rd. E17 CE32 25
Devonshire Rd. N9 CC26 9
Devonshire Rd. N13 BX28 15
Devonshire Rd. NW7 BQ29 13
Devonshire Rd. SE9 CK48 62
Devonshire Rd. SE23 CC47 61
Devonshire Rd. SW19 BU50 59
Devonshire Rd. W4 BO42 40
Devonshire Rd. W5 BK41 39
Devonshire Rd. Bexh. CQ45 54
Devonshire Rd. Croy. BZ54 69
Devonshire Rd. Felt. BE48 56
Devonshire Rd. Har. BG32 20
Devonshire Rd. Horn. CV34 28
Devonshire Rd. Ilf. CM33 26
Devonshire Rd. Orp. CO54 72
Devonshire Rd. Pnr. BD32 20
Devonshire Rd. Pnr. BE30 11
Devonshire Rd. Sthl. BF39 38
Devonshire Rd. Sutt. BT57 77
Devonshire Rd. Wall. BV56 77
Devonshire Row EC2 CA39 42
Devonshire Sq. EC2 CA39 42
Devonshire Sq. Brom. CH52 71
Masons Hill
Devonshire St. W1 BV40 41
Devonshire St. W4 BO42 40
Devonshire Ter. W2 BT39 41
Devonshire Way, Croy. CD55 79
Devonshire Waye, Hayes BC39 38
Devons Rd. E3 CE38 34
Devon St. SE15 CB44 51
Devon Way, Chess. BK56 75
Devon Way, Epsom BM56 75
Devon Waye, Houns. BE43 47
De Walden St. W1 BV39 41
Marylebone St.
Dewar St. SE15 CB45 51
Dewberry St. E14 CF39 43
Dewey Rd. N1 BY37 33
Dewey Rd. Dag. CS36 37
Dewey St. SW17 BU49 59
Dewhurst Rd. W14 BQ41 40
Dewport St. W6 BR43 49
Field Rd.
Dewsbury Clo. Pnr. BE32 20
Dewsbury Clo. Rom. CV29 19
Dewsbury Ct. W4 BN42 40
Chiswick Rd.
Dewsbury Gdns. Rom. CV29 19
Dewsbury Gdns. Wor. Pk. BP55 76
Dewsbury Rd. NW10 BP35 31
Dewsbury Rd. Rom. CV29 19

Dewsbury Ter. NW1 BV36 32
Camden High St.
Dexter Rd. SE24 BY45 51
Dexter Rd. Barn. BQ25 6
Deyncourt Rd. N17 BZ30 15
Deynecourt Gdns. E11 CJ31 26
D'eynsford Rd. SE5 BZ44 51
Dial, The, Wk. W8 BS41 41
Diameter Rd. Orp. CL54 71
Diamond Clo. E7 CH35 35
Stracey Rd.
Diamond Rd. Ruis. BE35 29
Diamond St. SE15 CA43 51
Diamond Ter. SE10 CF43 52
Diana Pl. NW1 BV38 32
Diana Rd. E17 CD31 25
Dianthus Clo. SE2 CO42 45
Carnation St.
Diban Av. Horn. CU35 37
Dibden Rd. SE23 CA48 61
Dibden's Cotts. SE27 BY49 60
Crown La.
Dibden St. N1 BZ37 33
Dibdin Clo. Sutt. BS55 77
Dibdin Ho. NW6 BS38 32
Dibdin Rd. NW6 BS37 32
Dibdin Rd. Sutt. BS55 77
Diceland Rd. Bans. BR61 82
Dicey Av. NW2 BQ35 31
Dickens Av. N3 BT30 14
Dickens Clo. Rich. BL48 57
Dickens Dr. Chis. CM50 62
Dickens Est. SE1 CB41 42
Dickens La. N18 CA28 15
Dickenson Rd. N8 BX33 24
Dickenson Rd. Felt. BD49 56
Dickenson's. La. SE25 CB53 69
Dickenson's Pl. SE25 CB53 69
Dickens Rise, Chig. CL27 17
Dickens Rd. E6 CJ37 35
Dickens Sq. SE1 BZ41 42
Dickens St. SW8 BV44 50
Dickerage La. N. Mal. BN52 67
Dickerage Rd. BN51 67
Kings. On T.
Dickson Rd. SE9 CK45 53
Dickson's Fold, Pnr. BD31 20
Didcot St. SW11 BT45 50
Digby Cres. N4 BZ34 24
Digby Est. E2 CC38 34
Digby Gdns. Dag. CR37 36
Digby Pl. Croy. CA55 78
Digby Rd. E9 CC35 34
Digby Rd. Bark. CN36 36
Digby St. E2 CC38 34
Digby Wk. Horn. CV35 37
Digdens Rise, Epsom BN61 82
Dighton Rd. SW18 BT46 50
Dignum St. N1 BY37 33
Digswell St. N7 BY36 33
Holloway Rd.
Dilhorne Clo. SE12 CH48 62
Dilke St. SW3 BU43 50
Dillan Pl. N7 BX34 24
Dillwyn Clo. SE26 CD49 61
Dilston Clo. Nthlt. BD38 29
Dilston Gro. SE16 CC42 43
Abbeyfield Rd.
Dilton Gdns. SW15 BP47 58
Dimes Pl. W3 BN42 40
King St.
Dimmock Dr. Har. BG35 29
Dimsdale Dr. NW9 BN33 22
Dimsdale Dr. Enf. CB25 8
Dimsdale Wk. E13 CG37 34
Dingle Gdns. E14 CE40 43
Dingley La. SW16 BW48 59
Dingley Pl. EC1 BZ38 33
Dingley Rd.
Dingley Rd. EC1 BZ38 33
Dingon Hill Clo. Hayes BE39 38
Dingwall Av. Croy. BZ55 78
Dingwall Gdns. NW11 BS32 23
Dingwall Pl. Croy. BZ55 78
Dingwall Rd. SW18 BT47 59
Dingwall Rd. Cars. BU58 77
Dingwall Rd. Croy. BZ54 69
Dinmont Est. E2 CB37 33
Dinmont St. E2 CB37 33
Dinsdale Gdns. SE25 CA52 69
Dinsdale Rd. SE3 CG43 52
Dinsmore Rd. SW12 BV47 59
Dinton Rd. SW19 BT50 59
Dinton Rd. Kings. On T. BL50 57
Dirdene Clo. Epsom BO59 82
Dirdene Gdns. Epsom BO59 82
Dirdene Gro. Epsom BO59 82
Dirleton Rd. E15 CG37 34
Disbrowe Rd. W6 BR43 49
Dishforth La. NW9 BO30 13
Disney Pl. SE1 BZ41 42
Disney St.
Disney St. SE1 BZ41 42
Dison Clo. Enf. CC23 9
Disraeli Clo. SE28 CP40 45

Disraeli Gdns. SW15 BR45 49
Fawe Pk. Rd.
Disraeli Rd. E7 CH36 35
Disraeli Rd. NW10 BN37 31
Disraeli Rd. SW15 BQ45 49
Disraeli Rd. W5 BK40 39
Diss St. E2 CA38 33
Distaff La. EC4 BZ39 42
Cannon St.
Distillery Wk. Brent. BL43 48
Pottery Rd.
District Rd. Wem. BJ35 30
Ditch Alley SE10 CE44 52
Ditchburn St. E14 CF40 43
Dittisham Rd. SE9 CK49 62
Ditton Clo. Surb. BJ54 66
Ditton Gra. Clo. Surb. BK54 66
Ditton Gdr. Dr. Surb. BK54 66
Ditton Hill Rd. Surb. BK54 66
Ditton Lawn, Surb. BJ54 66
Ditton Pl. SE20 CB51 69
Ditton Reach KT7 BJ53 66
Ditton Rd. Bexh. CP46 54
Ditton Rd. Sthl. BE42 38
Ditton Rd. Surb. BK55 75
Divis Way SW15 BP46 49
Dover Pk. Dr.
Dixon Pl. W. Wick. CE54 70
Dixon Rd. SE14 CD44 52
Dixon Rd. SE25 CA52 69
Dixons Alley SE16 CB41 42
West La.
Dobbin Clo. Har. BJ30 12
Dobell Rd. SE9 CK46 53
Dobtree Rd. NW10 BP36 31
Dobson Clo. NW6 BT36 32
Belsize Rd.
Dockhead SE1 CA41 42
Jamaica Rd.
Dockland St. E16 CL40 44
Dockley Rd. SE16 CB41 42
Dockley Rd. E16 CG40 43
Dock Rd. Brent. BK43 48
Dock St. E1 CB40 42
Dockwell Clo. Felt. BC45 47
Doctors Clo. SE26 CC49 61
Docwras Bldgs. N1 BZ36 33
Dodbrooke Rd. SE27 BY48 60
Doddington Gro. SE17 BY43 51
Doddington Pl. SE17 BY43 51
Kennington Pk. Pl.
Dodson St. SE1 BY41 42
Dod St. E14 CD39 43
Doel Clo. SW19 BT50 59
Doggets Clo. Barn. BU25 7
Doggett Rd. SE6 CE47 61
Dog Kennel Hill SE22 CA45 51
Dog La. NW10 BO35 31
Doherty Rd. E13 CH38 35
Dolben St. SE1 BY40 42
Dolby Rd. SW6 BR44 49
Ewald Rd.
Dole St. NW7 BQ29 13
Dolland St. SE11 BX42 42
Dollis Av. N3 BR30 13
Dollis Brook Wk. Barn. BQ25 6
Dollis Cres. Ruis. BD33 20
Dollis Hill Av. NW2 BP34 22
Dollis Hill La. NW2 BO35 31
Dollis Ms. N3 BS30 14
Dollis Pk.
Dollis Rd. N3 BR30 13
Dollis Valley Way, Barn. BR25 6
Dolman Rd. W4 BN42 40
Dolman St. SW4 BX45 51
Dolphin Clo. Surb. BK53 66
Dolphin Clo. E14 CE40 43
Dolphin La. E14 CE40 43
Dolphin Rd. Nthlt. BE37 29
Dolphin Sq. SW1 BW42 41
Dombey St. WC1 BX39 42
Dome Hill Pk. SE26 CA49 60
Domett Clo. SE5 BZ45 51
Dominic Dr. SE9 CL49 62
Dominion Dr. Rom. CR29 18
Dominion Rd. Croy. CA54 69
Dominion Rd. Sthl. BE41 38
Feathurst Ter.
Dominion St. EC2 BZ39 42
Dominion Way, Rain. CU38 37
Domville Clo. N20 BT27 14
Domville Gro. SE5 CA42 42
Donald Dr. Rom. CP32 27
Donald Rd. E13 CH37 35
Donald Rd. Croy. BX54 69
Donaldson Rd. NW6 BR37 31
Donaldson Rd. SE18 CL44 53
Doncaster Dr. Nthlt. BE35 29
Doncaster Gdns. N4 BY32 24
Stanhope Gdns.
Doncaster Gdns. Nthlt. BE35 29
Doncaster Rd. N9 CB26 8
Doncaster Way, Upmin. CW34 28
Donegal St. N1 BX37 33
Doneraile St. SW6 BQ44 49
Dongala Rd. E13 CH38 35
Dongala Rd. N17 CA31 24
Donington Av. Ilf. CM32 26
Donkey La. Enf. CB23 8

Donnefield Av. Edg. BL29 12
Donne Pl. SW3 BU42 41
Donne Pl. Mitch. BV52 68
Donne Rd. Dag. CP34 27
Donnington Rd. NW10 BP36 31
Donnington Rd. Har. BK32 21
Donnington Rd. Wor. Pk. BP55 76
Donnybrook Rd. SW16 BW50 59
Donovan Av. N10 BV30 14
Donovan Clo. Epsom BN58 76
Don Way, Rom. CT29 19
Doone Clo. Tedd. BJ50 57
Doon St. SE1 BY40 42
Dorado Gs. Orp. CP55 81
Doral Way, Cars. BU56 77
Carshalton Pk. Rd.
Doran Gro. SE18 CM43 53
Doran Mans. N2 BU32 23
Doran Wk. E15 CF36 34
Dora Rd. SW19 BS49 59
Dora St. E14 CD39 43
Dorchester Av. N13 BZ28 15
Dorchester Av. Bex. CP47 63
Dorchester Av. Har. BF32 20
Dorchester Clo. Dart. CW47 64
Dorchester Clo. Nthlt. BF35 29
Dorchester Ct. N14 BV26 7
Dorchester Ct. SE24 BZ46 51
Dorcester Dr. SE24 BZ46 51
Dorchester Gdns. E4 CE28 16
Dorchester Gdns. NW11 BS31 23
Gloucester Dr.
Dorchester Gdns. Mord. BS54 68
Dorchester Gro. W4 BO43 49
Dorchester Rd. Mord. BS54 68
Dorchester Rd. Nhtlt. BF35 29
Dorchester Rd. Wor. Pk. BQ54 67
Dorchester Way, Har. BL32 21
Dorchester Waye, Hayes BC39 38
Dorcis Av. Bexh. CQ44 54
Dordrecht Rd. W3 BO40 40
Dore Av. E12 CL35 35
Doreen Av. NW9 BN33 22
Dorell Clo. Sthl. BE39 38
Dorian Rd. Horn. CU33 28
Doria Rd. SW6 BR44 49
Doric Way NW1 BW38 32
Dorien Rd. SW20 BQ51 67
Dorincourt SW15 BN49 58
Dorinium Clo. Wem. BL35 30
Lea Gdns.
Doris Av. Erith. CR44 54
Doris Rd. E7 CH36 35
Doris St. SE11 BY42 42
Tracey St.
Dorking Clo. SE8 CD43 52
Dorking Clo. Wor. Pk. BQ55 76
Dorking Rise, Rom. CV28 19
Dorking Rd. Rom. CV28 19
Dorking Wk. Rom. CV28 19
Dorlcote Rd. SW18 BT47 59
Dorling Dr. Epsom BO59 82
Dorman Way NW8 BT37 32
Dormay St. SW18 BS46 50
Dormer Clo. E15 CG36 34
Dormer Clo. Barn. BQ25 6
Dormer's Av. Sthl. BE39 38
Dormer's Rise, Sthl. BF40 38
Dormer's Wells La. Sthl. BF39 38
Dornbergh Clo. SE3 CH43 53
Dornbery Rd. SE3 CH43 53
Banchory Rd.
Dorncliffe Rd. SW6 BR44 49
Dorney Rise, Orp. CN53 72
Dornfell St. NW6 BR35 31
Dornton Rd. SW12 BV48 59
Dornton Rd. Sth. Croy. BZ57 78
Dorothy Av. Wem. BL36 30
Dorothy Evans Clo. Bexh. CR45 54
Dorothy Gdns. Dag. CO35 36
Dorothy Rd. SW11 BU45 50
Dorran Pl. N9 CB27 15
Herthford Rd.
Dorrington Gdns. Horn. CV33 28
Dorrington St. EC1 BY39 42
Dorrit Mews N18 CA28 15
Dorrit Way, Chis. CM50 62
Dors Clo. NW9 BN33 22
Dorset Av. Rom. CS31 28
Dorset Av. Sthl. BF42 38
Dorset Av. Well. CN45 54
Dorset Bldgs. EC4 BY39 42
Dorset Rise
Dorset Clo. NW1 BU39 41
Dorset Dr. Edg. BL29 12
Dorset Est. E2 CA38 33
Dorset Gdns. Mitch. BX52 69
Dorset Ms. SW1 BV41 41
Wilton St.
Dorset Pl. E15 CF36 34
Dorset Pl. SW1 BW42 41
Rampayne St.
Dorset Rise EC4 BY39 42
Dorset Rd. E7 CJ36 35
Dorset Rd. N15 BZ31 24
Dorset Rd. N22 BX30 15
Dorset Rd. SE9 CK48 62
Dorset Rd. SW8 BX43 51
Dorset Rd. SW19 BS51 68
Dorset Rd. W5 BK41 39
Dorset Rd. Beck. CC52 70
Dorset Rd. Har. BG32 20
Dorset Rd. Mitch. BU51 68
Dorset Rd. Sutt. BS58 77
Dorset Sq. NW1 BU38 32
Dorset St. W1 BU39 41
Dorset Way, Twick. BG47 56
Dorset Waye, Houns. BE43 47
Dorville Cres. W6 BP41 40
Dorville Rd. SE12 CG46 52
Dothill Rd. SE18 CM43 53
Douai Gro. Hamptn. BG51 65
Doubleday Rd. Loug. CM24 10
Doughty Ms. WC1 BX38 33
Roger St.
Doughty St. WC1 BX38 33
Douglas Av. E17 CD30 16
Douglas Av. N. Mal. BP52 67
Douglas Av. Rom. CW30 19
Douglas Av. Wem. BL36 30
Douglas Clo. Stan. BJ28 12
Douglas Cres. Hayes BD38 29
Douglas Dr. Croy. CE55 79
Douglas Est. N1 BZ36 33
Douglas Pl. E14 CF42 43
Douglas Rd. E4 CG26 9
Douglas Rd. E16 CH39 44
Douglas Rd. N1 BZ36 33
Douglas Rd. N22 BY30 15
Douglas Rd. NW6 BR37 31
Douglas Rd. Esher BF55 74
Douglas Rd. Horn. CT32 28
Douglas Rd. Houns. BF45 47
Douglas Rd. Ilf. CO33 27
Douglas Rd. Kings. On T. BM52 66
Douglas Rd. Surb. BL55 75
Douglas Rd. Well. CO44 54
Douglas Robinson Ct. SW16 BX50 60
Douglas Sq. Mord. BS53 68
Douglas St. SW1 BW42 41
Douglas Way SE8 CD43 52
Dounesforth Gdns. SW18 BS47 59
Douro Pl. W8 BS41 41
Douro St. E3 CE37 34
Dovecote Av. N22 BY31 24
Dovedale Av. Har. BK32 21
Dovedale Av. Ilf. CL30 17
Dovedale Clo. Well. CO44 54
Dovedale Rise, Mitch. BU50 59
Dovedale Rd. SE22 CB46 51
Dove Ho. Gdns. E4 CE27 16
Dovehouse Mead, Bark. CM37 35
Dove House St. SW3 BT42 41
Dove Ms. SW5 BT42 41
Dove Pk. Pnr. BF29 11
Dover Clo. Rom. CS30 19
Dovercourt Av. Th. Hth. BY53 69
Dovercourt Gdns. Stan. BL28 12
Dovercourt La. Sutt. BT55 77
Dovercourt Rd. SE22 CA46 51
Doverfield Rd. SW2 BX46 51
Dover Ho Rd. SW15 BP45 49
Doveridge Gdns. N13 BY28 15
Dove Rd. N1 BZ36 33
Dove Row E2 CB37 33
Dover Pk. Dr. SW15 BP46 49
Dover Rd. E12 CJ34 26
Dover Rd. N9 CC27 16
Dover Rd. SE19 BZ50 60
Dover Rd. Rom. CQ32 27
Dover St. W1 BV40 41
Dover Yd. W1 BW40 41
Berkeley St.
Dove's Clo. Brom. CK55 80
Dove St. E1 CC38 34
Doveton Rd. Sth. Croy. BZ56 78
Dove Wk. Rain. CU36 37
Dowanhill Rd. SE6 CF47 61
Dowdeswell Clo. SW15 BO45 49
Dowding Pl. Stan. BJ29 12
Dower Av. Wall. BV58 77
Dowgate Hill EC4 BZ40 42
Dowlas St. SE5 CA43 51
Dowlerville Rd. Orp. CN57 81
Dowlings Par. Wem. BK37 30
Bridgewater Rd.
Downage NW4 BQ30 13
Downalong, Bush. BG26 4
Downbank Av. Bexh. CS44 55
Downbarns Rd. Ruis. BD34 20
Downderry Rd. Brom. CF48 61
Downe Clo. Well. CP43 54
Down End SE18 CL43 53
Moordown
Downe Rd. Kes. CK58 80
Downe Rd. Mitch. BU51 68
Downers La. SW4 BW45 50
Downes Ct. N21 BY26 8
Downe Ter. Rich. BL46 48
Richmond Hill
Downfield, Wor. Pk. BO54 67
Downfield Clo. W9 BS38 32
Amberley Est.
Down Hall Rd. Kings. On T. BK51 66
Downham Clo. Rom. CR29 18
Downham Rd. N1 BZ36 33
Downham Way, Brom. CF49 61
Downhills Av. N17 BZ31 24
Downhills Park Rd. N17 BZ31 24
Downhills Way N17 BZ31 24
Downhills Way N22 BZ30 15
Downhurst Av. NW7 BN28 13
Downing Clo. Har. BG31 20
Downing Dr. Grnf. BG37 29
Downing Rd. Dag. CQ37 36
Downing St. SW1 BW41 41
Downland Clo. Couls. BV60 83
Downland Clo. Epsom BP62 82
Downland Clo. N20 BT26 7
Downland Gdns. Epsom BP62 82
Downlands Rd. Pur. BX60 84
Downland St. W10 BR38 31
Downland Way, Epsom BP62 82
Downleys Clo. SE9 CK48 62
Downman Rd. SE9 CK45 53
Down Pl. W6 BP42 40
Bridge Av.
Down Rd. Tedd. BJ50 57
Downs Av. Chis. CK49 62
Downs Av. Epsom BO60 82
Downs Av. Pnr. BE32 20
Downs Bri. Rd. Beck. CF51 70
Downs Ct. SW20 BQ50 58
Downs Ct. Rd. Pur. BY59 84
Downsell Rd. E15 CF35 34
Downs Est. E5 CB34 24
Downsfield Rd. E17 CD32 25
Downshall Av. Ilf. CN32 27
Downs Hill, Beck. CF50 61
Downs Hill Rd. Epsom BO60 82
Downshire Hill NW3 BT35 32
Downside Clo. SW19 BT50 59
Downside Cres. NW3 BU35 32
Downside Cres. W13 BJ38 30
Downside, Epsom BO60 82
Downside Gdns. Twick. BH48 57
Downside Rd. Sutt. BT57 77
Downside Wk. Nthlt. BD38 29
Invicta Gro.
Downs Park Rd. E5 CA35 33
Downs Park Rd. E8 CA35 33
Downs Rd. E5 CB35 33
Downs Rd. Beck. CE51 70
Downs Rd. Couls. BW62 83
Downs Rd. Enf. CA24 8
Downs Rd. Epsom BO60 82
Downs Rd. Pur. BY59 84
Downs Rd. Sutt. BS58 77
Downs Rd. Th. Hth. BZ51 69
Downs Side, Sutt. BR59 82
Downs, The SW20 BQ50 58
Down St. W1 BV41 41
Down St. E. Mol. BF53 65
Downs Vw. Islw. BH44 48
Downsview Clo. Orp. CP58 81
Downsview Clo. Swan. CT52 73
Downsview Gdns. SE19 BY50 60
Downsview Rd. SE19 BZ50 60
Downs Way, Epsom BO61 82
Downs Way, Orp. CN56 81
Downsway, Sth. Croy. CA59 84
Downsway, Sutt. BS58 77
Downsway, Whyt. CA61 84
Downs Wood, Epsom BP62 82
Downton Av. SW2 BX48 60
Downway N12 BU29 14
Downway, Nthlt. BC37 29
Down Way Clo. Nthlt. BC37 29
Dowsett Rd. N17 CA30 15
Dowson Clo. SE5 BZ45 51
Doyce St. SE1 BZ41 42
Southwark Bridge Rd.
Doyle Gdns. NW10 BP37 31
Doyle Rd. SE25 CB52 69
Doyley St. SW1 BV42 41
Draco St. SE17 BZ43 51
Dragmire La. Mitch. BT52 68
Dragoon Rd. SE8 CD42 43
Dragor Rd. NW10 BN38 31
Drake Ct. Har. BE34 20
Drakefell Rd. SE4 CC44 52
Drakefell Rd. SE14 CC44 52
Drakefield Rd. SW17 BV48 59
Drakely Ct. N5 BY35 33
Highbury Hill
Drake Rd. SE4 CE45 52
Drake Rd. Chess. BM56 75
Drake Rd. Croy. BX54 69
Drake Rd. Har. BE34 20
Drake Rd. Mitch. BV53 68
Drakes Clo. Esher BF56 74
Drakes Grn. Esher BF56 74
Drake St. WC1 BX39 42
Theobalds Rd.
Drake St. Enf. BZ23 8
Drakes Wk. E6 CK37 35
Drakewood Rd. SW16 BW50 59
Draper Clo. Belv. CQ42 45
Draper's Rd. E15 CF35 34
Drapers Rd. N17 CA31 24
Drapers Rd. Enf. BY23 8
Draw Dock Rd. SE10 CF40 43
Drawell Clo. SE18 CN42 45
Drax Av. SW20 BP50 58
Draxmont App. SW19 BR50 58
Draycot Rd. E11 CH32 26
Draycot Rd. Surb. BM55 75
Draycott Av. SW3 BU42 41
Draycott Av. Har. BJ32 21
Draycott Clo. Har. BJ32 21
Draycott Pl. SW3 BU42 41
Draycott Ter. SW3 BU42 41
Dray Gdns. SW2 BX46 51
Drayson Ms. W8 BS41 41
Drayton Av. W13 BJ40 39
Drayton Av. Loug. CK26 10
Drayton Av. Orp. CL54 71
Drayton Br. Rd. W7 BH40 39
Drayton Br. Rd. W13 BH40 39
Drayton Gdns. N21 BY26 8
Drayton Gdns. SW10 BT42 41
Drayton Gdns. W13 BJ40 39
Drayton Gn. Rd. W13 BJ40 39
Drayton Grn. W13 BJ40 39
Drayton Gro. W13 BJ40 39
Drayton Pk. N5 BY34 24
Drayton Rd. E11 CF33 25
Drayton Rd. N17 CA30 15
Drayton Rd. NW10 BO37 31
Drayton Rd. W13 BJ39 39
Drayton Rd. Borwd. BM24 5
Drayton Rd. Croy. BY55 78
Ruskin Rd.
Drayton Waye, Harrow BJ32 21
Dreadnought St. SE10 CG41 43
Dresden Rd. N19 BW33 23
Dressington Av. SE13 CE46 52
Drew Av. NW7 BQ29 13
Drew Av. NW7 BR29 13
Drew Gdns. Grnf. BH36 30
Drew Rd. E16 CJ40 44
Drewstead Rd. SW16 BW48 59
Driffield Rd. E3 CD37 34
Drift, The, Kes. CJ55 80
Driftway, The, Mitch. BV51 68
Streatham Rd.
Driftwood Dr. Ken. BZ58 78
Driftwood Dr. Ken. BZ62 84
Drill Hall Rd. Epsom BP57 76
Drinkwater Est. Felt. BC46 47
Drinkwater Rd. Har. BF34 20
Dr. Johnsons Av. SW17 BV48 59
Drive Mead, Couls. BX60 84
Drive, The E4 CF26 9
Drive, The E18 CH31 26
Drive, The N3 BS29 14
Drive, The N6 BU32 23
Drive, The N7 BX36 33
Drive, The N11 BW29 14
Blake Rd.
Drive, The NW11 BR33 22
Drive, The SW6 BR44 49
Drive, The SW16 BX52 69
Drive, The SW20 BQ50 58
Drive, The W3 BN39 40
Drive, The, Bans. BR62 82
Drive, The, Bark. CN30 18
Drive, The, Barn. BR24 6
Drive, The, Barn. BT25 7
Drive, The, Beck. CE51 70
Drive, The, Bex. CP47 63
Drive, The, Buck. H. CJ26 10
Drive, The, Chis. CN51 72
Drive, The, Chis. CN52 72
Drive, The, Couls. BX60 84
Drive, The, Edg. BM28 12
Drive, The, Enf. BZ23 8
Drive, The, Epsom BO57 76
Drive, The, Erith CR43 54
Drive, The, Esher BG54 65
Drive, The E17 CE31 25
Drive, The, Felt. BC47 56
Drive, The, Har. BF32 20
Drive, The, Houns. BG44 47
Drive, The, Ilf. CK32 26
Drive, The, Kings. On T. BN51 67
Drive, The, Loug. CK24 10
Drive, The, Mord. BT53 68
Drive, The, Orp. CN55 81
Drive, The, Rom. CS29 19
Drive, The, Rom. CW30 19
Drive, The, Sid. CO48 63
Drive, The, Surb. BL54 66
Drive, The, Sutt. BR59 82
Drive, The, Th. Hth. BZ52 69
Drive, The, Wall. BW58 77
Drive, The, Wem. BN34 22
Drive, The, W. Wick. CF54 70
Droitwich Clo. SE26 CB48 60
Dromey Gdns. Har. BH29 12
Dromore Rd. SW15 BR46 49
Dronfield Gdns. Dag. CP35 36
Drood Yd. E1 CB40 42
Pennington St.
Droop St. W10 BQ38 31
Drover La. SE15 CB43 51
Drover's Rd. Sth. Croy. BZ56 78
Druce Rd. SE21 CA46 51
Druid St. SE1 CA41 42
Druids Way, Brom. CF52 70

Drumaline Ridge, Wor. Pk.	BO55	76
Drummond Av. Rom.	CS31	28
Drummond Cres. NW1	BW38	32
Drummond Dr. Stan.	BH29	12
Drummond Rd. E11	CH32	26
Drummond Rd. SE16	CB41	42
Drummond Rd. Croy.	BZ55	78
Drummond Rd. Rom.	CS31	28
Drummonds', The, Buck. H.	CH27	17
Drummond St. NW1	BW38	32
Drury La. WC2	BX39	42
Drury Rd. Har.	BG33	20
Dryad St. SW15	BQ45	49
Dryburgh Gdns. NW9	BM31	21
Dryburgh Rd. SW15	BP45	49
Dryden Av. W7	BH39	39
Dryden Clo. Ilf.	CN29	18
Dryden Clo. Orp.	CO54	72
Dryden House SE5	CA44	51
Glebe Est.		
Dryden Rd. SW19	BT50	59
Dryden Rd. Enf.	CA25	8
Dryden Rd. Har.	BH30	12
Dryden Rd. Well.	CN44	54
Dryfield Clo. NW10	BN36	31
Dryfield Rd. Edg.	BN29	13
Dryfield Wk. SE8	CE43	52
Czar St.		
Dryhill Rd. Belv.	CQ43	54
Dryland Av. Orp.	CN56	81
Drylands Rd. N8	BX32	24
Drysdale Av. E4	CE26	9
Drysdale Pl. N1	CA38	33
Drysdale St.		
Drysdale St. N1	CA38	33
Duboyne Rd. NW5	BU35	32
Du Cane Clo. W12	BQ39	40
Du Cane Rd.		
Du Cane Ct. SW17	BV47	59
Du Cane Rd. W12	BP39	40
Ducat St. E2	CA38	33
Duchess Ms. W1	BV39	41
Duchess St.		
Duchess of Bedford's Wk. W8	BS41	41
Duchess St. W1	BV39	41
Duchy St. SE1	BY40	42
Ducie St. SW4	BX45	51
Duckett Rd. N4	BY32	24
Ducketts Rd. Dart.	CT46	55
Duckett St. E1	CC38	34
Ducking Stool Ct. Rom.	CT31	28
Market Link		
Ducklees La. Enf.	CD24	9
Ducksfoot La. EC4	BZ40	42
Upr. Thames St.		
Duck's Wk. Twick.	BK46	48
Du Cross Dr. Stan.	BK29	12
Ducross Rd. W3	BO40	40
Dudden Hill La. NW10	BO35	31
Duddington Clo. SE9	CJ49	62
Dudley Av. Har.	BK31	21
Dudley Ct. NW11	BR31	22
Dudley Dr. Mord.	BR54	67
Dudley Dr. Ruis.	BC35	29
Dudley Gdns. W13	BJ41	39
Dudley Gdns. Har.	BG33	20
Dudley Gdns. Rom.	CV29	19
Dudley Gro. Epsom	BN60	82
Dudley Rd. E17	CE31	25
Dudley Rd. N3	BS30	14
Dudley Rd. NW6	BR37	31
Dudley Rd. SW19	BS50	59
Dudley Rd. Har.	BG34	20
Dudley Rd. Ilf.	CL35	35
Dudley Rd. Kings. On T.	BL52	66
Dudley Rd. Rich.	BL44	48
Dudley Rd. Rom.	CV29	19
Dudley Rd. Sthl.	BD41	38
Dudley Rd. Walt.	BC53	65
Dudlington Rd. E5	CC34	25
Dudmaston Ms. SW3	BT42	41
Dudsbury Rd. Dart.	CU46	55
Dudsbury Rd. Sid.	CO49	63
Dudset La. Houns.	BC44	47
Dufferin St. EC1	BZ38	33
Duffield Clo. Har.	BH32	21
Duffield Rd. SW11	BU45	50
Batten St.		
Duff St. E14	CE39	43
Dufours Pl. W1	BW39	41
Broadwick St.		
Duke Gdns. Ilf.	CM31	26
Duke Rd.		
Duke Humphrey Rd. SE3	CG44	52
Duke Of Cambridge Clo. Twick.	BG46	47
Duke Of Edinburgh Rd. Sutt.	BT55	77
Duke Of Wellington Pl. SW1	BV41	41
Duke Of York St. SW1	BW40	41
Duke Rd. W4	BN42	40
Duke Rd. Ilf.	CM31	26
Duke's Av. N3	BS30	14
Duke's Av. N10	BV31	23
Dukes Av. N10	BW31	23
Duke's Av. W4	BN42	40
Dukes Av. Edg.	BL29	12
Dukes Av. Har.	BE32	20
Duke's Av. Har.	BH31	21
Dukes Av. Houns.	BE45	47
Dukes Av. N. Mal.	BO52	67
Dukes Av. Nthlt.	BE36	29
Duke's Av. Twick.	BK49	57
Dukes Clo. Hamptn.	BE49	56
Duke's Clo. Kings. On T.	BK49	57
Dukes Ct. E6	CL37	35
Dukes La. W8	BS41	41
Dukes Meadows W4	BN44	49
Dukes Mews N10	BV31	23
Dukes Av.		
Duke's Ms. W1	BV39	41
Duke St.		
Dukes Orchard DA5	CS47	64
Dukes Pass. E17	CF31	25
Marlowe Rd.		
Duke's Pl. EC3	CA39	42
Dukes Rd. E6	CL37	35
Dukes Rd. W3	BM38	30
Duke's Rd. WC1	BW38	32
Dukes Rd. Walt.	BD56	74
Dukesthorpe Rd. SE26	CC49	61
Duke St. Hill SE1	BZ40	42
Duke St. Ms. NW8	BU38	32
Lisson Gro.		
Duke St. SW1	BW40	41
Duke St. W1	BV39	41
Duke St. Rich.	BK46	48
Duke St. Sutt.	BT56	77
Duke St. Wat.	BD24	4
Dukes Way, W. Wick.	CG55	79
Dulas St. N4	BX33	24
Everleigh St.		
Dulford St. W11	BR40	40
Dulka Rd. SW11	BU46	50
Dulverton Rd. SE9	CM48	62
Dulverton Rd. Rom.	CV29	19
Dulverton Rd. Ruis.	BC33	20
Dulverton Rd. Sth. Croy.	CC58	79
Dulwich Common SE21	CA47	60
Dulwich Rd. SE24	BY46	51
Dulwich Village SE21	CA46	51
Dulwich Wood Av. SE19	CA49	60
Dulwich Wood Pk. SE19	CA49	60
Dumbarton Rd. SW2	BX46	51
Dumbreck Rd. SE9	CK45	53
Dumfries Clo. Wat.	BC27	11
Dumont Rd. N16	CA34	24
Dumpton Pl. NW1	BV36	32
Dunbar Av. SW16	BY51	69
Dunbar Av. Beck.	CD52	70
Dunbar Av. Dag.	CR34	27
Dunbar Clo. Hayes	BC39	38
Dunbar Ct. KT12	BC54	65
Dunbar Gdns. Dag.	CR35	36
Dunbar Pl. SE27	BZ48	60
Dunbar Rd. E7	CH36	35
Dunbar Rd. N22	BY30	15
Dunbar Rd. N. Mal.	BN52	67
Dunbar St. SE27	BY48	60
Dunblane Rd. SE9	CK45	53
Dunboyne Rd. NW3	BU35	32
Dunbridge St. E2	CB38	33
Duncan Clo. Barn.	BT24	7
Duncan Gro. W3	BO39	40
Duncannon St. WC2	BX40	42
Adelaide St.		
Duncan Rd. E8	CB37	33
Duncan Rd. Rich.	BL45	48
Duncan Rd. Tad.	BR62	82
Duncan St. N1	BY37	33
Duncan Ter. N1	BY37	33
Duncan Way, Bush.	BE23	4
Duncombe Hl. SE23	CD47	61
Duncombe Rd. N19	BW33	23
Duncrievie Rd. SE13	CF46	52
Duncroft SE18	CN43	54
Dundalk Rd. SE4	CD45	52
Dundas Gdns. E. Mol.	BF52	65
Dundas Rd. SE15	CC44	52
Dundee Rd. E13	CH37	35
Dundee Rd. SE25	CB53	69
Dundee St. E1	CB40	42
Green Bank		
Dundela Gdns. Epsom	BP56	76
Dundonald Rd. NW10	BQ37	31
Dundonald Rd. SW19	BR51	67
Dundon Gdns. SE23	CC47	61
Dunedin Rd. E10	CE34	25
Dunedin Rd. Ilf.	CM33	26
Dunedin Rd. Rain.	CT38	37
Dunedin Way, Hayes	BD38	29
Dunelm Gro. SE27	BY48	60
Dunelm St. E1	CC39	43
Duneved Rd. S. Th. Hth.	BY53	69
Dunfield Gdns. SE6	CE49	61
Dunfield Rd. SE6	CE49	61
Dunford Rd. N7	BX35	33
Dungarvan Av. SW15	BP45	49
Dunheved Clo. Th. Hth.	BY53	69
Dunheved Rd. N. Th. Hth.	BY53	69
Dunheved Rd. W. Th. Hth.	BY53	69
Dunholme Grn. N9	CA27	15
Dunholme La. N9	CA27	15
Dunholme Rd.		
Dunholme Rd. N9	CA27	15
Dunkeld Rd. SE25	BZ52	69
Dunkeld Rd. Dag.	CO34	27
Dunkery Rd. SE9	CJ49	62
Dunlace Rd. E5	CC35	34
Dunleary Clo. Houns.	BE47	56
Dunley Dr. Croy.	CE57	79
Dunloe Av. N17	BZ31	24
Dunloe St. E2	CA37	33
Dunlop Pl. SE16	CA41	42
Dunmail Dr. Pur.	CA60	84
Dunmore Rd. NW6	BR37	31
Dunmore Rd. SW20	BQ51	67
Dunmow Clo. Felt.	BE49	56
Dunmow Clo. Loug.	CK25	10
Dunmow Clo. Rom.	CP32	27
Dunmow Dr. Rain.	CT37	37
Dunmow Rd. E15	CF35	34
Dunningford Clo. Horn.	CT35	37
Dunn St. E8	CA35	33
Dunnymans Rd. Bans.	BR60	82
Basing Rd.		
Dunollie Pl. NW5	BW35	32
Dunollie Rd.		
Dunollie Rd. NW5	BW35	32
Dunoon Rd. SE23	CC47	61
Dunraven Dr. Enf.	BY23	8
Dunraven Rd. W12	BP40	40
Dunraven St. W1	BU40	41
Green St.		
Dunsany Rd. W14	BQ41	40
Dunsbury Clo. Sutt.	BS58	77
Nettlecombe Clo.		
Dunsfold Rise. Couls.	BW60	83
Dunsford Way Croy.	CE58	79
Dunsford Cres. SW18	BS47	59
Merton Rd.		
Dunsford Way SW15	BP46	49
Dover Pk. Dr.		
Dunsmore Rd. Walt.	BC53	65
Dunsmore Way Bush.	BG25	4
Dunsmure Rd. N16	CA33	24
Dunspring La. Ilf.	CL30	17
Dunstable Clo. Rom.	CV29	19
Dunstable Ms. W1	BV39	41
Dunstable Rd. E. Mol.	BE52	65
Dunstable Rd. Rich.	BL45	48
Dunstable Rd. Rom.	CV29	19
Dunstall Rd. SW20	BP50	58
Dunstall Way E. Mol.	BF52	65
Dunstan Rd. NW11	BR33	22
Dunstan Rd. Couls.	BW62	83
Dunstan's Gro. SE22	CB46	51
Dunstans Rd. SE22	CB47	60
Dunstar Clo. Barn.	BQ24	6
Dunster Av. SW15	BP46	49
Dunster Av. Mord.	BQ54	67
Dunster Clo. Rom.	CS30	19
Dunster Ct. EC3	CA40	42
Mincing La.		
Dunster Dr. NW9	BN33	22
Dunster Gdns. NW6	BR36	31
Dunster Way Har.	BE34	20
Dunston Rd. E8	CA37	33
Dunston Rd. SW8	BV44	50
Dunston St. E8	CA37	33
Dunton Rd. E10	CE33	25
Dunton Rd. SE1	CA42	42
Dunton Rd. Rom.	CT31	28
Duntshill Rd. SW18	BS47	59
Dunvegan Clo. E. Mol.	BF52	65
Dunvegan Rd. SE9	CK45	53
Dunwich Rd. Bexh.	CQ44	54
Dupont Rd. SW20	BQ51	67
Dupont St. E14	CD39	43
Duppas Av. Croy.	BY56	78
Violet La.		
Duppas Hill La. Croy.	BY55	78
Duppas Hill Rd. Croy.	BY56	78
Duppas Hill Ter. Croy.	BY55	78
Duppash La. Croy.	BY55	78
Duppas Rd. Croy.	BY55	78
Dupree Rd. SE7	CH42	44
Duraden Clo. Beck.	CE50	61
Durand Clo. Cars.	BU54	68
Durand Gdns. SW9	BX44	51
Durand Way, NW10	BN36	31
Durant Dr. Swan.	CU50	64
Durants Pk. Av. Enf.	CC24	9
Durants Rd. Enf.	CC24	9
Durant St. E2	CB37	33
Durban Gdns. Dag.	CS36	37
Durban Ho. E7	CJ36	35
Durban Rd. E15	CG38	34
Durban Rd. E17	CD30	16
Durban Rd. N17	CA29	15
Durban Rd. SE27	BZ49	60
Durban Rd. Beck.	CD51	70
Durban Rd. Felt.	BC48	56
Durban Rd. Ilf.	CN33	27
Durban Rd. E. Wat.	BC24	4
Durban Rd. W. Wat.	BC24	4
Durbin Rd. Chess.	BL56	75
Durdans Rd. Sthl.	BE39	38
Durell Gdns. Dag.	CP35	36
Durell Rd. Dag.	CP35	36
Durford Cres. SW15	BP47	58
Durham Av. Brom.	CG52	70
Durham Av. Houns.	BE42	38
Durham Av. Rom.	CV31	28
Durham Av. Wdf. Grn.	CJ28	17
Durham Bldgs. SW11	BT45	50
Durham Clo. SW20	BP51	67
Durham Hl. Brom.	CG49	61
Durham Pl. SW3	BU42	41
Durham Rise, SE18	CM42	44
Durham Rd. E12	CJ35	35
Durham Rd. E16	CG38	34
Durham Rd. N2	BU31	23
Durham Rd. N7	BX34	24
Durham Rd. N9	CB27	15
Durham Rd. SW20	BP51	67
Durham Rd. W5	BK41	39
Durham Rd. Borwd.	BN24	6
Durham Rd. Brom.	CG52	70
Durham Rd. Dag.	CS35	37
Durham Rd. Felt.	BD47	56
Durham Rd. Har.	BF32	20
Durham Rd. Sid.	CO49	63
Durham Row, E1	CC39	43
Stepney High St.		
Durham St. SE11	BX42	42
Durham Ter. W2	BS39	41
Durley Av. Pnr.	BE33	20
Durley Rd. N16	CA33	24
Durlston Rd. E5	CB34	24
Durlston Rd. Kings. On T.	BL50	57
Durnell Way, Loug.	CL24	10
Durnford St. N15	CA32	24
Durnford St. SE10	CF43	52
Greenwich Church St.		
Durning Rd. SE19	BZ49	60
Durnsford Av. SW19	BS48	59
Durnsford Rd. N11	BW30	14
Durnsford Rd. SW19	BS48	59
Duro Pl. W8	BS42	41
Durrants Clo. Rain.	CV37	37
Durrant Way, Orp.	CM56	80
Durrell Rd. SW6	BR44	49
Durrington Av. SW20	BQ50	58
Durrington Pk. Rd. SW20	BQ51	67
Durringon Rd. E5	CD35	34
Dursley Clo. SE3	CJ44	53
Dursley Gdns. SE3	CJ44	53
Dursley Rd. SE3	CJ44	53
Durward St. E1	CB39	42
Dury Rd. Barn.	BR23	6
Duthie St. E14	CF40	43
Dutton St. SE10	CF44	52
Duxford Clo. Horn.	CV36	37
Dyers Bldgs. EC1	BY39	42
Holborn		
Dyers Hall Rd. E11	CF34	25
Dyer's La. SW15	BP45	49
Dyers Way, Rom.	CU29	19
Dyke Dr. Orp.	CP54	72
Dykewood Clo. Bex.	CT48	64
Dylan Rd. Belv.	CR41	45
Dylways, SE5	BZ45	51
Dymchurch Clo. Ilf.	CL30	17
Cymchurch Clo. Orp.	CN56	81
Dymes Path SW19	BR48	58
Queensmere Rd.		
Dymock St. SW6	BS45	50
Dymoke Rd. Horn.	CT33	28
Dyneley Rd. SE12	CJ49	62
Dyne Rd. NW6	BR36	31
Dynevor Rd. N16	CA34	24
Dynevor Rd. Rich.	BL46	48
Dynham Rd. NW6	BS36	32
Dyott St. WC1	BW39	41
Dysart Av. Kings. On T.	BK49	57
Dysart St. EC2	BZ38	33
Dyson Rd. E11	CG32	25
Dyson Rd. E15	CG36	34
Dyson's Rd. N18	CB29	15
Eade Rd. N4	BY33	24
Eagans Clo. N2	BT31	23
Eagle Av. Rom.	CQ32	27
Eagle Clo. Enf.	CC24	9
Eagle Clo. Rain.	CU36	37
Eagle Ct. EC1	BY39	42
Albion Pl.		
Eagle Hill, SE19	BZ50	60
Eagle La. E11	CH31	26
Eagle Pl. E1	CC38	34
Mile End Rd.		
Eagle Pl. SW1	BW40	41
Jerymyn St.		
Eagle Rd. Wem.	BK36	30
Eaglesfield Rd. SE18	CL44	53
Eagles, The, W6	BV34	23
Eagle St. WC1	BX39	42
High Holborn		
Eagle Ter. Wdf. Grn.	CH29	17
Eaglet Pl. E1	CC38	34
Mile End Rd.		
Eagle Wharf Rd. N1	BZ37	33
Ealdham Sq. SE9	CJ45	53
Ealing Gn. W5	BK40	39
Ealing Pk. Gdns. W5	BK42	39

Ealing Pk. Ms. W5 BK41 39
Ealing Rd.
Ealing Rd. Brent. BK42 39
Ealing Rd. Nthlt. BF37 29
Ealing Rd. Wem. BL36 30
Ealing Village, W5 BL39 39
Eamont St. NW8 BU37 32
Eardemont Clo. Dart. CT45 55
Eardley Cres. SW5 BS42 41
Eardley Rd. SW16 BU49 59
Eardley Rd. Belv. CR42 45
Earl Cotts. SE1 CA42 42
Earldom Rd. SW15 BQ45 49
Earle Gdn. Kings. On T. BL50 57
Earlesmead Rd. NW10 BQ38 31
Earlham Gro. E7 CG35 34
Earlham Gro. N22 BX29 15
Earlham St. WC2 BW39 41
Earl Rise, SE18 CM42 44
Earl Rd. SE1 CA42 42
Earl Rd. SW14 BN45 49
Earls Ct. Gdns. SW5 BS42 41
Earl's Ct. Rd. SW5 BS41 41
Earl's Ct. Rd. W8 BS41 41
Earl's Sq. SW5 BS42 41
Earls Cres. Har. BH31 21
Earlsferry Clo. N1 BX36 33
Carnoustie Dr.
Earlsferry Way N1 BX36 33
Earlsferry Clo.
Earlsfield Rd. SW18 BT47 59
Earlshall Rd. SE9 CK45 53
Ealrsmead Har. BE35 29
Earlsmead Rd. N15 CA32 24
Earlsmead Rd. NW10 BQ39 40
Earls Path. Loug. CJ23 10
Earls Ter. W8 BR41 40
Earlsthorpe Rd. SE26 CC49 61
Earlstoke St. EC1 BY38 33
Earlstone Gro. E9 CB37 33
Victoria Pk. Rd.
Earl St. EC2 BZ39 42
Earl St. Wat. BD24 4
Earl's Wk. W8 BS41 41
Earls Wk. Dag. CO35 36
Earlswood Av. Th. Hth. BY53 69
Earlswood Gdns. Ilf. CL31 26
Earlswood St. SE10 CG42 43
Early Ms. NW1 BV37 32
Arlington Rd.
Earnshaw St. WC2 BW39 41
Earsby St. W14 BR42 40
Easby Cres. Mord. BS53 68
Easebourne Rd. Dag. CP35 36
Easedale Dr. Horn. CU35 37
Easleys Ms. W1 BV39 41
Wigmore St.
East Acton La. W3 BO40 40
East Arbour Sq. E1 CC39 43
East Av. E12 CK36 35
East Av. E17 CE31 25
East Av. Hayes BC40 38
East Av. Sthl. BE40 38
East Av. Wall. BX56 78
Eastbank Rd. Hamptn. BG49 56
East Barnet Rd. Barn. BT24 7
Eastbourne Av. W3 BN39 40
Eastbourne Gdns. SW14 BN45 49
Eastbourne Ms. W2 BT39 41
Eastbourne Rd. E6 CL38 35
Eastbourne Rd. E15 CG37 34
Eastbourne Rd. N15 CA32 24
Eastbourne Rd. SW17 BV50 59
Eastbourne Rd. W4 BN43 49
Eastbourne Rd. Brent. BK42 39
Eastbourne Rd. Felt. BD48 56
Eastbourne Ter. W2 BT39 41
Eastbrook Av. N9 CC26 9
Eastbrook Av. Dag. CS35 37
Eastbrook Dr. Rom. CT34 28
Eastbrook Rd. SE3 CH43 53
Eastbury Av. Bark. CN37 36
Eastbury Av. Enf. CA23 8
Eastbury Gro. W4 BO43 49
Eastbury Rd.
Eastbury Rd. E6 CL39 44
Eastbury Rd. Kings. On T. BL50 57
Eastbury Rd. Orp. CM53 71
Eastbury Rd. Rom. CS32 28
Eastbury Rd. Wat. BC26 4
Eastbury Sq. Bark. CN37 36
Eastbury Ter. E1 CC38 34
Eastcastle St. W1 BW39 41
Eastcheap, EC3 CA40 42
East Churchfield Rd. W3 BN40 40
East Clo. W5 BM38 30
East Clo. Barn. BV24 7
East Clo. Grnf. BG37 29
East Clo. Rain. CU38 37
Eastcombe Av. SE7 CH43 53
Eastcote Orp. CN54 72
Eastcote Av. E. Mol. BE53 65
Eastcote Av. Grnf. BJ35 30
Eastcote Av. Har. BF34 20
Eastcote Gdns. Well. CM44 53

Eastcote High Rd. Pnr. BC32 20
Eastcote La. Har. BE35 29
Eastcote La. Nthlt. BE35 29
Eastcote La. N. Nthlt. BE36 29
Eastcote Rd. Har. BG34 20
Eastcote Rd. Pnr. BD32 20
Eastcote Rd. Well. CM44 53
Eastcote St. SW9 BX44 51
Eastcote Vw. Pnr. BD31 20
East Ct. Wem. BK34 21
East Cres. N11 BU28 14
East Cres. Enf. CA25 8
Eastcroft Rd., Epsom BO57 76
East Cross Route E9 CD36 34
Eastdene Dr. Rom. CV28 19
East Dr. SW11 BV43 50
East Dr. Cars. BU58 77
East Dr. Orp. CO53 72
East Dulwich Est. SE22 CA45 51
East Dulwich Gro. SE22 CA46 51
East Dulwich Rd. SE22 CA45 51
East End Rd. N2 BS30 14
East End Rd. N3 BS30 14
East End Way Pnr. BE31 20
East Entrance Dag. CR37 36
Eastern Av. E11 CH32 26
Eastern Av. Ilf. CH32 26
Eastern Av. Pnr. BD33 20
Eastern Av. Rom. CH32 26
Eastern Av. E. Rom. CS30 19
Eastern Av. W. Ilf. CN32 27
Eastern Industrial Est. CR41 45
Belv.
Eastern Rd. E13 CH37 35
Eastern Rd. E17 CF32 25
Eastern Rd. N2 BU31 23
Eastern Rd. N22 BX30 15
Eastern Rd. SE4 CE45 52
Eastern Rd. Rom. CT32 28
Easternville Gdns. Ilf. CM32 26
East Ferry Rd. E14 CE42 43
Eastfield Av. Wat. BD23 4
Eastfield Gdns. Dag. CR35 36
Eastfield Rd. E17 CE31 25
Eastfield Rd. N8 BX31 24
Eastfield Rd. Dag. CQ35 36
Eastfields Pnr. BD32 20
Eastfields Rd. W3 BN39 40
Eastfields Rd. Mitch. BV51 68
Eastfield St. E14 CD39 43
East Gdns. SW17 BU50 59
Eastgate Barn. BR60 82
East Glade Pnr. BE31 20
Easthall La. Rain. CV39 46
East Hall Rd. Orp. CQ54 72
East Ham Manor Way E6 CK38 35
East Ham Manor Way E16 CK38 35
East Harding St. EC4 BY39 42
East Heath Rd. NW3 BT34 23
East Hill, SW18 BS46 50
East Hill Dart. CW47 64
East Hill, Wem. BM34 21
East Hill Dr. Dart. CW47 64
Eastholm NW11 BS31 23
East Holme Erith CS44 55
East Holme Hayes BC40 38
E. India Dock Rd. E14 CF39 43
E. India Dock Wall Rd. E14 CF40 43
Eastlake Rd. SE5 BZ44 51
Eastlands Cres. SE21 CA46 51
East La. SE16 CB41 42
Chambers St.
East La. Kings. on T. BK52 66
High St.
East La. Wem. BJ34 21
Eastleigh Av. Har. BF34 20
E. Leigh Clo. Sutt. BS57 77
Eastleigh Clo. Bexh. CS44 55
Eastleigh Wk. SW15 BP47 58
Eastman Rd. W3 BN40 40
Eastman St. E1
Vale, The
Eastmead Ruis. BD34 20
East Mead Rd. Brom. CK51 71
Eastmearn Rd. SE21 BZ48 60
Eastminster, E1 CA40 42
Royal Mint St.
Eastmont Rd. Esher BH55 75
Eastmoor Pl. SE7 CJ41 44
Eastmoor St. SE7 CJ41 44
East Mount St. CB39 42
Eastney Rd. Croy. BY54 69
Eastney St. SE10 CF42 43
Eastnor Rd. SE9 CM47 62
Easton Gdns. Borwd. BN24 6
Easton St. WC1 BY38 33
E. Park Clo. Rom. CQ32 27
East Pl. SE27 BZ49 60
Dunkirk St.
East Rd. E11 CH32 26
East Rd. E15 CH37 35
East Rd. N1 BZ38 33
East Rd. SW19 BT50 59

East Rd. Barn. BV26 7
East Rd. Belv. CQ41 45
East Rd. Edg. BM30 12
East Rd. Kings. on T. BL51 66
East Rd. Rom. CQ31 27
East Rd. Rom. CS33 28
East Rd. Well. CO44 54
East Row W10 BR38 31
Eastry Av. Brom. CG53 70
Eastry Rd. Erith CR43 54
East Sheen Av. SW14 BN46 49
Eastside Rd. NW11 BR31 22
East Smithfield, E1 CA40 42
East Sq. SE18 CL42 44
East St. EC2 BZ39 42
Blomfield St.
East St. SE17 BZ42 42
East St. Bexh. CR45 54
East St. Brent. BK43 48
East St. Epsom BO59 82
East St. Ilf. CO47 63
East Tenter St. E1 CA39 42
East Towers Pnr. BD32 20
East Vw. E4 CF28 16
East Vw. NW3 BT34 23
East Vw. Barn. BR23 6
Eastview Av. SE18 CN43 54
Eastville Av. NW11 BR32 22
East Wk. Barn. BV26 7
East Wk. Hayes BC40 38
Eastway, E9 CD36 34
East Way, E11 CH32 26
East Way Brom. CH54 71
East Way Croy. CD55 79
Eastway Epsom BN59 82
East Way Hayes BC40 38
Eastway Mord. BQ53 67
Eastway Ruis. BC33 20
Eastway Wall. BW56 77
Eastwell Clo. Beck. CD51 70
King's Hall Rd.
Eastwick Rd. Walt. BC57 74
Eastwood Clo. E18 CH30 17
Eastwood Dr. Rain. CU39 46
Eastwood St. SW15 BP46 49
Eastwood Rd. E18 CH30 17
Eastwood Rd. N10 BV30 14
Eastwood Rd. Ilf. CO33 27
East Wood Side, Bex. CO47 63
Eastwood St. SW16 BW50 59
Eatington Rd. E10 CF32 25
Eaton Clo. SW1 BV42 41
Eaton Clo. Stan. BJ28 12
Eaton Dr. Kings. On T. BM50 57
Eaton Dr. Rom. CR29 18
Eaton Gte. SW1 BV42 41
Eaton La. SW1 BV41 41
Eaton Ms. N. SW1 BV41 41
Eaton Ms. S. SW1 BV42 41
Eaton Ms. W. SW1 BV42 41
Eaton Pk. Rd. N13 BY27 15
Eaton Pl. SW1 BV41 41
Eaton Rise W5 BK39 39
Eaton Rd. E11 CJ32 26
Eaton Rd. NW4 BQ32 22
Eaton Rd. SW9 BY45 51
Eaton Rd. Enf. CA24 8
Eaton Rd. Houns. BG45 47
Eaton Rd. Sid. CP48 63
Eaton Rd. Sutt. BT57 77
Eaton Row SW1 BV41 41
Eatons Mead E4 CE27 16
Eaton Sq. SW1 BV42 41
Eaton Ter. SW1 BV42 41
Eaton Ter. Ms. SW1 BV42 41
Eaton Ter.
Eatonville Rd. SW17 BU48 59
Eatonville Vills. SW17 BU48 59
Eatonville Rd.
Eaton Walk SE15 CA44 51
Sumner Estate
Ebbisham Dr. SW8 BX43 51
Ebbisham Rd. Epsom BN60 82
Ebbisham Rd. Wor. Pk. BO55 76
Ebbsfleet Rd. NW2 BR35 31
Ebdon Way, SE3 CH45 53
Ebenezer Rd. NW2 BS34 23
Ebenezer St. N1 BZ38 33
Ebenezer Wk. SW16 BW51 68
Ebner St. SW18 BS46 50
Ebor St. E1 CA38 33
Ebrington Rd. Har. BK32 21
Ebsworth St. SE23 CC47 61
Ebury Br. N7 BX34 24
Ebury Br. Est. SW1 BV42 41
Ebury Br. Rd. SW1 BV42 41
Ebury Clo. Kes. CK55 80
Ebury Ms. SW1 BV42 41
Ebury Rd. Wat. BD24 4
Ebury Sq. SW1 BV42 41
Ebury St. SW1 BV42 41
Ecclesbourne Clo. N13 BY28 15
Ecclesbourne Gdns. N13 BY28 15
Ecclesbourne Rd. N1 BZ36 33
Ecclesbourne Rd. Th. BZ53 69
Hth.

Eccles Clo. N13 BY28 15
Ecclesbourne Gdns.
Eccles Rd. SW11 BU45 50
Eccleston Br. SW1 BV42 41
Eccleston Clo. Barn. BU24 7
Carson Rd.
Eccleston Clo. Orp. CM54 71
Eccleston Cres. Rom. CO33 27
Ecclestone Ct. Wem. BL35 30
Ecclestone Ms. Wem. BL35 30
Eccleston Ms. SW1 BV41 41
Eccleston Pl. SW1 BV41 41
Eccleston Pl. Wem. BL35 30
Eccleston Rd. W13 BJ40 39
Eccleston Sq. SW1 BV42 41
Eccleston Sq. Ms. SW1 BV42 41
Warwick Pl. N.
Eccleston St. SW1 BV41 41
Echo Heights E4 CE26 9
Eckersley St. E1 CA38 33
Eckford St. N1 BY37 33
Eckstein Rd. SW11 BU45 50
Eclipse Rd. E13 CH39 44
Ector Rd. SE6 CG48 61
Edale Rd. SE16 CC42 43
Edbrooke Rd. W9 BS38 32
Eddington St. N4 BY33 24
Everleigh St.
Eddiscombe Rd. SW6 BR44 49
Eddy Clo. Rom. CR32 27
Eddystone Rd. SE4 CD46 52
Ede Clo. Houns. BE45 47
Edenbridge Clo. Orp. CP52 72
Edenbridge Rd. E9 CC36 34
Edenbridge Rd. Enf. CA25 8
Eden Clo. Bex. CS49 64
Eden Clo. Wem. BK37 30
Eden Court, W5 BL39 39
Station Rd.
Edencourt Rd. SW16 BV50 59
Edendale W3 BM40 39
Julian Av.
Edendale Rd. Bexh. CS44 55
Edenfield Gdns. Wor. BO55 76
Pk.
Eden Grn. E17 CE32 25
Eden Gro. N7 BX35 33
Edenhall Rd. Rom. CV28 19
Edenhurst Av. SW6 BR45 49
Eden Pk. Av. Beck. CD52 70
Eden Pk. Av. Beck. CE53 70
Eden Rd. E17 CE32 25
Eden Rd. Beck. CD52 70
Eden Rd. Bex. CS49 64
Eden Rd. Croy. BZ56 78
Edensor Gdns. W4 BO43 49
Edensor Rd. W4 BO43 49
Eden St. Kings. On T. BL51 66
Edenvale Rd. Mitch. BV50 59
Edenvale St. SW6 BS44 50
Eden Way Beck. CD53 70
Ederline Av. SW16 BX52 69
Edgar Rd. E3 CE38 34
Edgar Rd. Houns. BE47 56
Edgar Rd. Rom. CP33 27
Edgarton Rd. Wat. BC27 11
Edgbury, Chis. CL49 62
Edgebury Est. Chis. CM49 62
Edgebury Wk. Chis. CM49 62
Edgcombe Clo. BN50 58
Kings. On T.
Harvey Rd.
Edgecoombe Sth. Croy. CC57 79
Edgecot Gro. N15 CA32 24
Edgefield Av. Bark. CN36 36
Edge Hill, SE18 CL43 53
Edge Hill, SW19 BQ50 58
Edge Hill Av. N3 BS31 23
Edge Hill Ct. KT12 BD64 65
Rodney Rd.
Edgehill Ct. SW19 BQ50 58
Edgehill Gdns. Dag. CR35 36
Edgehill Rd. W13 BJ39 39
Edgehill Rd. Chis. CM48 62
Edgehill Rd. Mitch. BV51 68
Edgehill Rd. Pur. BY58 78
Edgeley La. SW4 BW45 50
Edgeley Rd. SW4 BW45 50
Edgel St. SW18 BS45 50
Ferrier St.
Edgepoint Clo. SE27 BY49 60
Knight's Hill
Edge St. W8 BS40 41
Edgewood Dr. Orp. CN56 81
Edgewood Grn. Croy. CC54 70
Edgeworth Av. NW4 BP32 22
Edgeworth Clo. NW4 BP32 22
Edgeworth Clo. Whyt. CB62 84
Edgeworth Cres. NW4 BP32 22
Edgeworth Rd. Barn. BU24 7
Edgeworth Rd. SE9 CJ45 53

Edgington Rd. SW16	BW50	59
Edgwarebury Gdns. Edg.	BM28	12
Edgwarebury La. Borwd.	BL26	5
Edgwarebury La. Edg.	BM28	12
Edgware Ct. Edg.	BM29	12
Edgware Rd. NW2	BP33	22
Edgware Rd. NW9	BN30	13
Edgware Rd. W2	BT38	32
Edgware Way Edg.	BK26	5
Edinburgh Ct. SW20	BQ53	67
Edinburgh Gte. SW1	BU41	41
Edinbugh Rd. E13	CH37	35
Edinburgh Rd. E17	CD32	25
Edinbrugh Rd. N18	CB28	15
Edinburgh Rd. W7	BH41	39
Edinburgh Rd. Sutt.	BT55	77
Edington Rd. SE2	CO41	45
Edington Rd. Enf.	CC23	9
Edison Av. Horn.	CT33	28
Edison Clo. Horn.	CT33	28
Edison Dr. Sthl.	BF39	38
Edison Gro. SE18	CN43	54
Edison Rd. N8	BW32	23
Edison Rd. Brom.	CH51	71
Church Rd.		
Edison Rd. Well.	CN44	54
Edis St. NW1	BV37	32
Edith Dr. N11	BW29	14
Edith Gdns. Surb.	BM54	66
Edith Gro. SW10	BT43	50
Edith Ms. SW6	BS44	50
Edith Row		
Edithna St. SW9	BX45	51
Edith Rd. E6	CJ36	35
Edith Rd. SE25	BZ53	69
Edith Rd. SW19	BS50	59
Edith Rd. W14	BR42	40
Edith Rd. Orp.	CO56	81
Edith Rd. Rom.	CP33	27
Edith Row SW6	BS44	50
Edith Ter. SW10	BT43	50
Edmondscote W13	BJ39	39
Cleveland Rd.		
Edmonton Grn. N9	CB27	15
Edmund Rd. Mitch.	BU52	68
Edmund Rd. Orp.	CP53	72
Edmund Rd. Rain.	CT38	37
Edmund Rd. Well.	CO45	54
Edmunds Av. Orp.	CP52	72
Edmunds Clo. Hayes	BD39	38
Edmund St. SE5	BZ43	51
Edmunds Wk. N2	BU31	23
Edna Rd. SW20	BQ51	67
Edna St. SW11	BU44	50
Edric Rd. SE14	CC43	52
Edrick Rd. Edg.	BN29	13
Edrick Wk. Edg.	BN29	13
Edridge Clo. Bush.	BG25	4
Edridge Clo. Horn.	CV35	37
Edridge Rd. Croy.	BZ55	78
Edulf Rd. Borwd.	BM23	5
Edward Av. E4	CE29	16
Edward Av. Mord.	BT53	68
Edward Clo. N9	CA26	8
Edward Clo. Barn.	BT25	7
Edward Clo. Hamptn.	BG49	56
Edward Clo. Nthlt.	BD37	29
Edward Clo. Rom.	CV31	28
Edward Ct. E16	CH39	44
Alexandra St.		
Edwardes Sq. W8	BR41	40
Edward Pl. SE8	CD43	52
Edward Rd. E17	CC31	25
Edward Rd. SE20	CC50	61
Edward Rd. Barn.	BT24	7
Edward Rd. Belv.	CR42	45
Edward Rd. Brom.	CH50	62
Edward Rd. Chis.	CL49	62
Edward Rd. Couls.	BW61	83
Edward Rd. Croy.	CA54	69
Edward Rd. Har.	BG31	20
Edward Rd. Hmptn.	BG49	56
Edward Rd. Nthlt.	BD37	29
Edward Rd. Rom.	CQ32	27
Edwards Av. Ruis.	BC36	29
Edwards Clo. Wor. Pk.	BQ55	76
Edwards Gdns. Swan.	CS52	73
Edward's La. N16	BZ34	24
Edward St. E16	CH38	35
Edward St. SE8	CD43	52
Edward St. SE14	CD43	52
Edwina Gdns. Ilf.	CK32	26
Edwin Av. E6	CL38	35
Edwin Clo. Bexh.	CQ43	54
Edwin Clo. Rain.	CT38	37
Edwin Rd. Dart.	CU48	64
Edwin Rd. Edg.	BN29	13
Edwin Rd. Twick.	BH47	57
Edwin St. E1	CC38	34
Edwin St. E16	CG39	43
Edw. Temme Av. E15	CG36	34
Edwyn Clo. Barn.	BQ25	6
Effie Pl. SW6	BS43	50
Effie Rd. SW6	BS43	50
Effingham Clo. Sutt.	BS57	77
Effingham Rd. N8	BY32	24
Effingham Rd. SE12	CG46	52
Effingham Rd. Croy.	BX54	69

Effingham Rd. Surb.	BJ54	66
Effort St. SW17	BU49	59
Effra Clo. SW19	BS50	59
Effra Pde. SW2	BY46	51
Effra Rd. SW2	BY45	51
Effra Rd. SW19	BS50	59
Egan Way SE16	CB42	42
Bonamy Estate East, The		
Egbert St. NW1	BV36	32
Egerton Av. Swan.	CT50	64
Egerton Clo. Dart.	CU47	64
Egerton Clo. Pnr.	BC31	20
Egerton Ct. E11	CF33	25
Edgerton Cres. SW3	BU42	41
Egerton Dr. SE10	CE44	52
Egerton Gdns. NW4	BP31	22
Egerton Gdns. NW10	BQ37	31
Egerton Gdns. SW3	BU41	41
Egerton Gdns. W13	BJ39	39
Egerton Gdns. Ilf.	CN34	27
Egerton Gdns. Ms. SW3	BU41	41
Egerton Ms. SW3	BU41	41
Egertn Pl. SW3	BU41	41
Egerton Rd. N16	CA33	24
Egerton Rd. SE25	CA52	69
Egerton Rd. N. Mal.	BO52	67
Egerton Rd. Twick.	BH46	48
Egerton Rd. Wem.	BL36	30
Egerton Ter. SW3	BU41	41
Egham Clo. Sutt.	BR55	76
Egham Cres. Sutt.	BQ55	76
Egham Rd. E13	CH39	44
Eglantine Rd. SW18	BT46	50
Egleston Rd. Mord.	BS53	68
Eglington Ct. SE17	BZ43	51
Carter St.		
Eglington Rd. E4	CF26	9
Eglinton Hill, SE18	CL43	53
Eglinton Rd. SE18	CL43	53
Egliston Ms. SW15	BQ45	49
Egliston Rd. SW15	BQ45	49
Egmont Av. Surb.	BL54	66
Egmont Rd. N. Mal.	BO52	67
Egmont Rd. Surb.	BL54	66
Egmont Rd. Sutt.	BT57	77
Egmont Rd. Walt.	BC54	65
Egmont St. SE14	CC43	52
Egremont Rd. SE27	BY48	60
Eider St. SE17	BZ42	42
Rodney Rd.		
Eighth Av. E12	CK35	35
Eighth Av. Hayes	BC40	38
Eileen Rd. SE25	BZ53	69
Eisenhower Dr. E6	CK39	44
Elaine Gro. NW5	BV35	32
Elam Clo. SE5	BY44	51
Elam St. SE5	BY44	51
Eland Rd. SW11	BU45	50
Eland Rd. Croy.	BY55	78
Elba Pl. SE17	BZ42	42
Rodney Rd.		
Elberon Av. Croy.	BW53	68
Elbe St. SW6	BT44	50
Elborough Rd. SE25	CB53	69
Elborough St. SW18	BS47	59
Elbury Dr. E16	CH39	44
Elcho St. SW11	BU43	50
Elcom St. W10	BR39	40
Kensal Rd.		
Elcot Av. SE15	CB43	51
Elder Av. N8	BX32	24
Elderberry Rd. W5	BL41	39
Elder Ct. Bush.	BH27	12
Elderfield Rd. E5	CC35	34
Elderfield Wk. E11	CH32	26
Elder Oak Clo. SE20	CB51	69
Elder Rd. SE27	BZ49	60
Elders Ct. Bush.	BH27	12
Elderslie Clo. Beck.	CE53	70
Elderslie Rd. SE9	CL46	53
Elder St. E1	CA39	42
Elderton Rd. SE26	CD49	61
Eldertree Pl. Mitch.	BV51	68
Eldertree Way, Mitch.	BV51	68
Elder Way, Rain.	CV38	37
Eldon Av. Borwd.	BM23	5
Eldon Av. Croy.	CC55	79
Eldon Av. Houns.	BF43	47
Eldon Gro. NW3	BT35	32
Eldon Pk. SE25	CB52	69
Eldon Rd. E17	CD32	25
Eldon Rd. N9	CC27	16
Eldon Rd. N22	BY30	15
Eldon Rd. NW3	BT35	32
Eldon R. W8	BS41	41
Eldon Way NW10	BM37	30
Eldred Dr. Orp.	CP54	72
Eldred Rd. Bark.	CM37	35
Eldridge Clo. Felt.	BC47	56
Eldridge Clo. Horn.	CV35	37
Eleanor Av. Epsom	BN58	76
Eleanor Cres. NW7	BQ28	13
Eleanor Gdns. Dag.	CQ34	27
Eleanor Gro. SW13	BO45	49
Eleanor Rd. E8	CB36	33
Eleanor Rd. E15	CG36	34
Eleanor Rd. N11	BX29	150

Eleanor St. E3	CE38	34
Electric Av. SW9	BY45	51
Electric La. SW9	BY45	51
Electric Pde. Surb.	BK53	66
Elephant & Castle SE1	BY42	42
Elephant La. SE16	CC41	43
Rotherhithe St.		
Elephant Rd. SE17	BZ42	42
Elers Rd. W13	BK41	39
Eleven Acre Rise Loug.	CK24	10
Eley Rd. N18	CC28	16
Eleys Est. N18	CC28	16
Elfindale Rd. SE24	BZ46	51
Elfin Gro. Tedd.	BH49	57
Elfin Rd. SE5	BY43	51
Warrior Rd.		
Elford Clo. SE3	CJ45	53
Elfort Rd. N5	BY34	24
Elfrida Cres. SE6	CE49	61
Elfrida Rd. Wat.	BD25	4
Elf Row, E1	CC40	43
Elfwine Rd. W7	BH39	39
Elgal Clo. Orp.	CL56	80
Elgar Av. SW16	BX52	69
Elgar Av. W5	BL41	39
Elgar Av. Surb.	BM54	66
Elgar Clo. Borwd.	BK26	5
Elgar St. SE16	CD41	43
Elgin Av. W9	BS38	32
Elgin Av. Har.	BJ30	12
Elgin Cres. W11	BM40	39
Elgin Ms. W11	BR39	40
Elgin Ms. N. W9	BS38	32
Elgin Ms. S. W9	BS38	32
Elgin Rd. N22	BW30	14
Elgin Rd. Croy.	CA55	78
Elgin Rd. Ilf.	CN33	27
Elgin Rd. Sutt.	BT55	77
Elgin Rd. Wall.	BW57	77
Elgin Ter. W11	BR40	40
Elia Mews N1	BY37	33
Elias Pl. SW8	BY43	51
Elia St. N1	BY37	33
Elibank Rd. SE9	CK45	53
Elim Est. SE1	CA41	42
Elim Way, E13	CG38	34
Eliot Bank, SE23	CB48	60
Eliot Cotts. SE3	CG44	52
Eliot Dr. Har.	BF34	20
Eliot Hill, SE13	CF44	52
Eliot Pk. SE13	CF44	52
Eliot Pl. SE3	CG44	52
Eliot Rd. Dag.	CP35	36
Eliot Vale, SE3	CF44	52
Elizabeth Av. N1	BZ37	33
Elizabeth Av. Enf.	BY24	8
Elizabeth Av. Ilf.	CM34	26
Elizabeth Br. SW1	BV42	41
Elizabeth Cl. Barn.	BQ24	6
Elizabeth Clo. E14	CE39	43
Grundy St.		
Elizabeth Clo. W9	BT38	32
Randolph Av.		
Elizabeth Clo. Rom.	CR30	18
Elizabeth Clo. Rom.	CR30	18
Elizabeth Clyde Cl. N15	CA31	24
Lawrence Rd.		
Elizabeth Clyde Clo. N15	CA31	24
Lawrence Road		
Elizabeth Cotts. Rich.	BL44	48
Elizabeth Est. SE17	BZ43	51
Elizabeth Gdns. Stan.	BK29	12
Elizabeth Gdns. W3	BN40	40
Elizabeth Gdns. Sun.	BC52	65
Elizabeth Ms. NW3	BU36	32
Elizabeth Pl. N15	CA31	24
Elizabeth Ride N9	CB26	8
Elizabeth Rd. E6	CJ37	35
Elizabeth Rd. N15	CA32	24
Elizabeth St. SW1	BV42	41
Elizabeth Ter. SE9	CK46	53
Elizabeth Way Felt.	BD49	56
Elizabeth Way Orp.	CP53	72
Elizabeth Way SE19	BZ50	60
Elkingtn Rd. E13	CH38	35
Elkins, The Rom.	CT30	19
Elkstone Rd. W10	BR39	40
Ellaline Rd. W6	BQ43	49
Ellanby Cres. N18	CB28	15
Elland Rd. KT12	BD55	74
Elland Rd. SE15	CC45	52
Ella Rd. N8	BX33	24
Ellement Clo. Pnr.	BD32	20
Ellenborough Pl. SW15	BP45	49
Ellenborough Rd. N22	BY30	15
Ellenborough Rd. Sid.	CP50	63
Ellenbridge Way Sth. Croy.	CA58	78
Ellen Clo. Brom.	CJ52	71
Ellen Ct. N9	CC27	16
Densworth Gro.		
Ellen St. E1	CB39	42
Elleray Rd. Tedd.	BH50	57
Ellerby St. SW6	BQ44	49
Ellerdale Clo. NW3	BT35	32
Ellerdale Rd. NW3	BT35	32
Ellerdale St. SE13	CE45	52
Ellerdine Rd. Houns.	BG45	47

Ellerker Gdns. Rich.	BL46	48
Ellerman Av. Twick.	BE47	56
Ellerslie Rd. W12	BP40	40
Ellers Rd. W13	BK41	39
Ellerton Gdns. Dag.	CP36	36
Ellerton Rd. SW13	BP44	49
Ellerton Rd. SW18	BT47	59
Ellerton Rd. SW20	BP50	58
Ellerton Rd. Dag.	CP36	36
Ellerton Rd. Surb.	BL55	75
Ellery Rd. SE19	BZ50	60
Ellery St. SE15	CB44	51
Ellesborough Clo. Way	BD28	11
Ellesmere Av. NW7	BN27	13
Ellesmere Av. Beck.	CE51	70
Ellesmere Clo. E11	CG32	25
Ellesmere Ct. W4	BN42	40
Great W. Rd.		
Ellesmere Dri. Sth. Croy.	CB60	84
Ellesmere Gdns. Ilf.	CK32	26
Elesmere Gro. Barn.	BR25	6
Elesmere Rd. E3	CD37	34
Ellesmere Rd. NW10	BP35	31
Ellesmere Rd. W4	BN43	49
Ellesmer Rd. Grnf.	BG38	29
Ellesmere Rd. Twick.	BK46	48
Ellesmere St. E14	CE39	43
Ellias St. E1	CD39	43
Ellingfort Rd. E8	CB36	33
Ellingham Rd. E15	CF35	34
Ellingham Rd. W12	BP41	40
Ellingham Rd. Chess.	BK57	75
Ellington Rd. N10	BV31	23
Ellington Rd. Houns.	BF44	47
Ellington St. N7	BY36	33
Elliot Clo. E15	CG36	34
Elliot Gdns. Rom.	CU30	19
Elliot Rd. NW4	BP32	22
Elliot Rd. Stan.	BJ29	12
Elliott Clo. Wem.	BM34	21
Elliott Rd. SW9	BY43	51
Elliott Rd. W4	BO42	40
Elliott Rd. Brom.	CJ52	71
Elliott Rd. Th. Hth.	BY52	69
Eliotts Ct. EC4	BY39	42
Old Bailey		
Elliotts Pl. N1	BY37	33
St. Peters St.		
Elliotts Row, SE11	BY42	42
Ellis Av. Rain.	CU39	46
Ellis Clo. SE9	CM48	62
Elliscombe Rd. SE7	CJ42	44
Ellis David Rd. Croy.	BY55	78
Ellisfield Dr. SW15	BP47	58
Ellis Ms. SE7	CJ43	53
Ellison Rd. SW13	BO44	49
Ellison Rd. SW16	BW50	59
Ellison Rd. Sid.	CM47	62
Ellis Rd. Mitch.	BU53	68
Ellis St. SW1	BU42	41
Ellmore Clo. Rom.	CU30	19
Ellora Rd. SW16	BW49	59
Ellsworth St. E2	CB38	33
Elmar Rd. N15	BZ31	24
Elm Av. W5	BL40	39
Elm Av. Ruis.	BC33	20
Elm AV. Wat.	BE26	4
Elm Bank N14	BX26	8
Elmbank Av. Barn.	BQ24	6
Elm Bank Gdns. SW13	BO44	49
Elmbank Way W7	BG39	38
Elmbourne Dr. DA17	CR42	45
Elmbourne Rd. SW17	BV48	59
Elmbridge Av. Surb.	BM53	66
Elmbridge Clo. Ruis.	BC32	20
Elmbridge Dr. Ruis.	BC32	20
Elmbridge Rd. Ilf.	CO29	18
Elmbridge Wk. E8	CB36	33
Wilman Gro.		
Elmbrook Gdns. SE9	CK45	53
Elmbrook Rd. Sutt.	BR56	76
Elm Clo. E11	CH32	26
Elm Clo. NW4	BQ32	22
Elm Clo. SW20	BQ52	67
Elm Clo. Buck. H.	CJ27	17
Elm Clo. Cars.	BU54	68
Elm Clo. Dart.	CV47	64
Elm Clo. Har.	BF32	20
Elm Clo. Hayes	BC39	38
Elm Clo. Rom.	CR30	18
Elm Clo. Sth. Croy.	BZ57	78
Elm Clo. Surb.	BN54	67
Elm Clo. Twick.	BF48	56
Elm Ct. Mitch.	BU51	68
Elmcourt Rd. SE27	BY48	60
Elm Cres. W5	BL40	39
Elm Cres. Kings. On T.	BL51	66
Elmcroft Av. E11	CH32	26
Elmcroft Av. N9	CB25	8
Elmcroft Av. NW11	BR33	22
Elmcroft Av. Sid.	CN46	54
Elmcroft Clo. E11	CH31	26
Elmcroft Clo. W5	BK39	39
Elmcroft Clo. Chess.	BL55	75
Elmcroft Clo. NW11	BQ33	22
Elmcroft Cres. Har.	BF31	20
Elmcroft Dr. Chess.	BL55	75
Elmcroft Gdns. NW9	BM32	21

Elmcroft Rd. Orp. CO54 72
Elmcroft St. E5 CC35 34
Elmdale Rd. N13 BX28 15
Elmdene Surb. BN54 67
Elmdene Av. Horn. CW32 28
Elmdene Clo. Beck. CD53 70
Elmdene Est. Beck. CD53 70
Elmdene Rd. SE18 CL42 44
Elmdon Rd. Houns. BE44 47
Elm Dr. Har. BF32 20
Elm Dr. Sun. BD51 65
Elm Dr. Swan. CS51 73
Elmer AY. Hav. CT27 19
Elmer Clo. Enf. BX24 8
Elmer Clo. Rain. CU36 37
Elmer Gdns. Edg. BM29 12
Elmer Gdns. Islw. BG45 47
Elmer Gdns. Rain. CU36 37
Elmer Rd. SE6 CF47 61
Elmers Dr. Tedd. BJ50 57
Elmers End Rd. SE20 CC51 70
Elmers End Rd. Beck. CC51 70
Elmerside Rd. Beck. CD52 70
Elmers Rd. SE25 CB54 69
Elmfield Av. N8 BX32 24
Elmfield Av. Mitch. BV51 68
Elmfield Av. Tedd. BH49 57
Elmfield Pk. Brom. CH52 71
Elmfield Rd. E4 CF27 16
Elmfield Rd. E17 CC32 25
Elmfield Rd. N2 BT31 23
Elmfield Rd. SW17 BV48 59
Elmfield Rd. Brom. CH51 71
Elmfield Rd. Sthl. BE41 38
Elmfield Way Sth. Croy. CA58 78
Elm Gdns. N2 BT31 23
Elm Gdns. Esher BH57 75
Elm Gdns. Mitch. BW52 68
Elmgate Av. Felt. BC48 56
Elmgate Gdns. Edg. BN28 13
Elm Grn. W3 BO39 40
Elm Gro. N18 BX32 24
Elm Gro. NW2 BQ35 31
Elm Gro. SE15 CA44 51
Elm Gro. SW19 BR50 58
Elm Gro. Epsom BN60 82
Elm Gro. Erith. CS43 55
Elm Gro. Har. BF33 20
Elm Gro. Horn. CW32 28
Elm Gro. Kings. On T. BL51 66
Elm Gro. Orp. CN54 72
Elm Gro. Sutt. BS56 77
High St.
Elm Gro. Wdf. Grn. CG28 16
Elmgrove Cres. Har. BH32 21
Elmgrove Gdns. Har. BJ32 21
Elmgrove Pde. Wall. BV55 77
Elm Grove Rd. SW13 BP44 49
Elmgrove Rd. W5 BL41 39
Elmgrove Rd. Croy. CB54 69
Elmgrove Rd. Har. BH32 21
Elmhall Gdns. E11 CH32 26
Elmhurst Gdns. E11 CH32 26
Elmhurst Belv. CQ43 54
Elmhurst Av. N2 BT31 23
Elmhurst Av. Mitch. BV50 59
Elmhurst Dr. E18 CH30 17
Elmhurst Dr. Horn. CV33 28
Elmhurst Gdns. E18 CH30 17
Elmhurst Rd. N17 CA30 15
Elmhurst Rd. SE9 CK48 62
Elmhurst St. SW4 BW45 50
Elmhurst Way Loug. CK26 10
Elmington Clo. Bex. CR46 54
Elmington Est. SE5 BZ43 51
Elmington Rd. SE5 BZ44 51
Elmira St. SE13 CE45 52
Elm La. SE6 CD48 61
Elmlee Clo. Chis. CK50 62
Elmley St. SE18 CM42 44
Elm Ms. W2 BT41 41
Elm Nursery Est. Mitch. BV51 68
Elmore Rd. E11 CF34 25
Elmores Loug. CL24 10
Elmore St. N1 BZ36 33
Elm Park, SW2 BX46 51
Elm Park, Stan. BJ28 12
Elm Pk. Av. N13 CA32 24
Elm Pk. Av. Horn. CU35 37
Elm Pk. Ct. Pnr. BD31 20
Elm Pk. Gdns. NW4 BQ32 22
Elm Pk. Gdns. SW10 BT42 41
Elm Pk. Gdns. W3 BN39 40
Noel Rd.
Elm Pk. Gdns. Sth. Croy. CC58 79
Elm Pk. La. SW3 BT42 41
Elm Pk. Mans. SW10 BT43 50
Elm Pk. Pde. W3 BN39 40
Noel Rd.
Elm Pk. Rd. E10 CD33 25
Elm Pk. Rd. N3 BR29 13
Elm Pk. Rd. N21 BZ26 8
Elm Pk. Rd. SE25 CA52 69
Elm Pk. Rd. SW3 BT43 50
Elm Pk. Rd. Pnr. BD30 11
Elm Pl. SW7 BT42 41
Elm Rd. E7 CG36 34
Elm Rd. E11 CF34 25
Elm Rd. E17 CF32 25
Elm Rd. N22 BY30 15
Elm Rd. SW14 BO45 49
Elm Rd. Barn. BR24 6
Elm Rd. Beck. CD51 70
Elm Rd. Chess. BK56 75
Elm Rd. Dart. CV47 64
Elm Rd. Epsom BO57 76
Elm Rd. Erith CU44 55
Elm Rd. Esher BH57 75
Elm Rd. Houns. BE44 47
Elm Rd. Kings. On T. BL51 66
Elm Rd. N. Mal. BN52 67
Elm Rd. Orp. CO57 81
Elm Rd. Pur. BY60 84
Elm Rd. Rom. CR30 18
Elm Rd. Sid. CO49 63
Elm Rd. Th. Hth. BZ52 69
Elm Rd. Wall. BV54 68
Elm Rd. Wem. BL35 30
Elm Rd. W. Sutt. BR54 67
Elm Row NW3 BT34 23
Elms Av. N10 BV31 23
Elms Av. NW4 BQ32 22
Elmscott Gdns. N21 BZ25 8
Elmscott Rd. Brom. CG49 61
Elms Ct. Wem. BJ35 30
Elms Cres. SW4 BW46 50
Elmsdale Rd. E17 CD31 25
Elms Fm. Rd. Horn. CV35 37
Elms Gdns. Dag. CQ35 36
Elms Gdns. Wem. BJ35 30
Elmshaw Rd. SW15 BP46 49
Elmshorn, Epsom BQ61 82
Elmshurst Cres. N2 BT31 23
Elmshurst Est. N2 BT31 23
Elmshurst Rd. E7 CH36 35
Elmside, Croy. CE57 79
Elmside Rd. Wem. BM34 21
Elms La. Wem. BJ34 21
Elmsleigh Av. Har. BJ31 21
Elmsleigh Ct. Sutt. BS55 77
Elmsleigh Rd. Twick. BG48 56
Elmslie Clo. Epsom BN60 82
Elms Ms. W2 BT40 41
Elms Pk. Av. Wem. BJ35 30
Elms Rd. SW4 BW46 50
Elms Rd. Har. BH29 12
Elmstead Av. Chis. CK49 62
Elmstead Av. Wem. BL33 21
Elmstead Clo. N20 BS27 14
Elmstead Clo. Epsom BO56 76
Elmstead Gdns. Wor. Pk. BP55 76
Elmstead Glade, Chis. CK50 62
Elmstead La. Chis. CK50 62
Elmstead Rd. Erith CT44 55
Elmstead Rd. Ilf. CN34 27
Elmstead Cres. Well. CP43 54
Elms, The SW13 BO45 49
Elms, The, Slou. BO53 67
Elmstone Rd. SW6 BS44 50
Elm St. WC1 BX38 33
Elmsworth Av. Houns. BF44 47
Elm Ter. NW3 BS34 23
Finchley Rd.
Elm Ter. SE9 CL46 53
Elm Ter. Har. BG29 11
Elmtree Av. Esher BG54 65
Elm Tree Clo. Nthlt. BE37 29
Elm Tree Gdns. Nthlt. BE37 29
Elm Tree Clo.
Elm Tree Rd. NW8 BT38 32
Elmtree Rd. Tedd. BH49 57
Elm Wk. NW3 BS34 23
West Heath Rd.
Elm Wk. SW20 BQ52 67
Elm Wk. Orp. CK55 80
Elm Wk. Rom. CU31 28
Elm Way NW3 BS33 23
West Heath Rd.
Elm Way NW10 BO35 31
Elm Way, Epsom BN56 76
Elm Way, Wor. Pk. BQ55 76
Elmwood Av. N13 BX28 15
Elmwood Av. Borwd. BM24 5
Elmwood Av. Felt. BC48 56
Elmwood Av. Har. BJ32 21
Elmwood Clo. Wall. BV55 77
Elmwood Ct. Wem. BJ34 21
Elmwood Cres. NW9 BN31 22
Elmwood Dr. Bex. CQ47 63
Elmwood Dr. Epsom BP57 76
Elmwood Gdns. W7 BH39 39
Elmwood Rd. SE24 BZ46 51
Elmwood Rd. W4 BN43 49
Elmwood Rd. Croy. BY54 69
Elmwood Rd. Mitch. BU52 68
Elmworth Gro. SE21 BZ48 60
Elnathan Ms. W9 BS38 32
Elphinstone Rd. E17 CD30 16
Elphinstone St. N5 BY34 24
Elrick Clo. DA8 CT43 55
Queens St.
Elrington Rd. E8 CB36 33
Elsa Rd. Well. CO44 54
Elsa St. E14 CD39 43
Elsdale St. E9 CC36 34
Elsden Ms. E2 CC37 34
Old Ford Rd.
Elsden Rd. N17 CA30 15
Elsenham Rd. E12 CK35 35
Elsenham St. SW18 BR47 58
Elsham Rd. E11 CG34 25
Elsham Rd. W14 BR41 40
Elsham Ter. W14 BR41 40
Elsiedene Rd. N21 BZ26 8
Elsiemaud Rd. SE4 CD46 52
Elsie Rd. SE22 CA45 51
Elsinore Rd. SE23 CD47 61
Elsley Rd. SW11 BU45 50
Elspeth Rd. SW11 BU45 50
Elspeth Rd. Wem. BL35 30
Elsrick Av. Mord. BS53 68
Elstan Way, Croy. CD54 70
Elsted St. SE17 BZ42 42
Elsthorpe Rd. Rom. CR29 18
Elstow Clo. SE9 CL46 53
Elstow Clo. Ruis. BD33 20
Elstow Gdns. Dag. CQ37 36
Elstow Rd. Dag. CQ36 36
Elstree Gdns. N9 CB26 8
Elstree Gdns. Belv. CQ42 45
Elstree Gdns. Ilf. CM35 35
Elstree Hill N. Borwd. BK25 5
Elstree Hill S. Borwd. BK26 5
Elstree Hl. Brom. CG50 61
Elstree Rd. Borwd. BJ25 5
Elstree Rd. Bush. BG26 4
Elstree Rd. Stan. BK27 12
Elstree Rd. Wat. BH23 5
Elstree Way, Borwd. BM23 5
Elswick Rd. SE13 CE44 52
Elswick St. SW6 BT44 50
Elsworth Rd. NW3 BU36 32
Elsworthy, E. Mol. BH53 66
Elsworthy Ri. NW3 BU36 32
Elsynge Rd. SW18 BT46 50
Eltham Grn. SE9 CJ46 53
Eltham Grn. Rd. SE9 CJ45 53
Eltham High St. SE9 CK46 53
Eltham Hill SE9 CJ46 53
Eltham Palace Rd. SE9 CJ46 53
Eltham Park Gdns. SE9 CL45 53
Eltham Rd. SE9 CG46 52
Eltham Rd. SE12 CG46 52
Eltham St. SE17 BZ42 42
Elthiron Rd. SW6 BS44 50
Elthorpe Av. W7 BH41 39
Elthorne Ct. Felt. BD47 56
Elthorne Pk. Rd. W7 BH41 39
Elthorne Rd. N19 BW34 23
Elthorne Way NW9 BN32 22
Elthruda Rd. SE13 CF46 52
Eltisley Rd. Ilf. CL35 35
Elton Av. Barn. BR25 6
Elton Av. Grnf. BH36 30
Elton Av. Wem. BJ35 30
Elton Clo. Kings. On T. BK50 57
Normansfield Av.
Elton Clo. Tedd. BK50 57
Elton Pl. N16 CA35 33
Elton Rd. Kings. On T. BL51 66
Elton Rd. Pur. BW59 83
Elton St. N16 CA35 33
Matthias Rd.
Elton Way, Wat. BF23 4
Eltringham St. SW11 BT45 50
Petergate
Elvaston Ms. SW7 BT41 41
Elvaston Pl. SW7 BT41 41
Elvaston Terr. SW7 BT42 41
Elveden Pl. NW10 BM37 30
Elveden Rd. NW10 BM37 30
Elvendon Rd. N13 BX29 15
Elver Gdns. E2 CB38 33
Avebury Est.
Elverson Rd. SE8 CE44 52
Elverton St. SW1 BW42 41
Elvet Av. Rom. CV31 28
Elvington Dr. NW9 BO30 13
Elvington Grn. Brom. CG53 70
Elvino Rd. SE26 CC49 61
Elwill Way, Beck. CF52 70
Elwin St. E2 CA38 33
Elwood St. N5 BY34 24
Elwyn Gdns. SE12 CH47 62
Ely Clo. Erith CT44 55
Ely Clo. N. Mal. BO51 67
Ely Gdns. Borwd. BN25 6
Ely Gdns. Dag. CS34 28
Elyne Rd. N4 BY32 24
Ely Pl. Wdf. Grn. CL29 17
Ely Pl. EC1 BY39 42
Ely Rd. E10 CF32 25
Ely Rd. Croy. BZ53 69
Ely Rd. Houns. BD45 47
Elysian Av. Orp. CN53 72
Elysium Pl. SW6 BR44 49
Elysium St.
Elysium St. SW6 BR44 49
Fulham Park Gdns.
Elystan Clo. Wall. BV57 77
Elystan Pl. SW3 BU42 41
Elystan St. SW3 BU42 41
Elystan Wk. N1 BY37 33
Emanuel Av. W3 BN39 40
Embankment Gdns. SW3 BU43 50
Embankment Pl. WC2 BX40 42
Villiers St.
Embankment, The SW15 BQ44 49
Embankment, The, Twick. BJ47 57
Emba St. SE16 CB41 42
Ember Clo. Orp. CM54 71
Embercourt Rd. Surb. BG53 65
Ember Fm. Av. E. Mol. BG53 65
Ember Fm. Way, E. Mol. BG53 65
Ember Gdns. Surb. BH53 66
Ember La. Esher BG54 65
Embleton Rd. SE13 CE45 52
Embleton Rd. Wat. BC27 11
Embleton Wk. Hamptn. BE49 56
Moorland Clo.
Embley Point N16 CB34 24
Downs Est.
Embry Clo. Stan. BJ28 12
Embry Dr. Stan. BJ29 12
Embry Way, Stan. BJ28 12
Emden St. SW6 BS44 50
Emerald Gdns. Dag. CR33 27
Emerald St. WC1 BX39 42
Emerson Dr. Horn. CV33 28
Emerson Gdns. Har. BL33 21
Emerson Rd. Ilf. CL33 26
Emerson St. SE1 BZ40 42
Emery Hill St. SW1 BW41 41
Emery St. SE1 BY41 42
Morley St.
Emes Rd. Erith CS43 55
Emily Pl. N7 BY35 33
Queensland Rd.
Emily St. E16 CG39 43
Jude St.
Emlyn Gdns. W12 BO41 40
Emlyn Rd. W12 BO41 40
Emmanuel Rd. SW12 BW47 59
Emma Rd. E13 CG37 34
Emma St. E2 CB37 33
Emma's Way IG7 CL28 17
Emma Ter. E11 CG34 25
Montague Rd.
Emmett St. E14 CD40 43
Emmot Clo. NW11 BT32 23
Emmott Av. Ilf. CM32 26
Emmott St. E1 CD38 34
Emperors Gte. SW7 BS41 41
Empire Av. N18 BZ28 15
Empire Ct. Wem. BM34 21
Empire Pde. N18 BZ29 15
Empire Rd. Grnf. BJ37 30
Empire Way, Wem. BL35 30
Empire Wharf Rd. E14 CF42 43
Empire Yd. N7 BX34 24
Empress Av. E4 CE29 16
Empress Av. E12 CJ34 26
Empress Av. Ilf. CK34 26
Empress Av. Wdf. Grn. CG29 16
Empress Dr. Chis. CL50 62
Empress Pl. SW6 BS42 41
Empress St. SE17 BZ43 51
Empson St. E3 CF38 34
Emsworth Clo. N9 CC26 9
Emsworth Rd. Ilf. CL30 17
Emsworth St. SW2 BX48 60
Emu Rd. SW8 BV44 50
Ena Rd. SW16 BX52 69
Enbrook St. W10 BR38 31
Endale Clo. Cars. BU55 77
Endeavour Way SW19 BS49 59
Endeavour Way, Bark. CO37 36
Endeavour Way, Croy. BW54 68
Endell St. WC2 BX39 42
Endensor Gdns. Est. W9 BS40 41
Enderby St. SE10 CG42 43
Enderley Clo. Har. BH30 12
Enderley Rd. Har. BH30 12
Endersby Rd. Barn. BQ25 6
Endersleigh Gdns. NW4 BP31 22
Endlebury Rd. E4 CE27 16
Endlesham Rd. SW12 BV47 59
Endsleigh Clo. Sth. Croy. CC58 79
Endsleigh Gdns. WC1 BW38 32
Endsleigh Gdns. Ilf. CK34 26
Endsleigh Gdns. Surb. BK53 66
Endsleigh Gdns. Walt. BC56 74
Endsleigh Pl. WC1 BW38 32
Endsleigh Rd. Sthl. BE42 38
Endsleigh St. WC1 BW38 32
End Way, Surb. BM54 66
Endwell Rd. SE4 CD44 52
Endymion Rd. N4 BY33 24
Endymion Rd. SW2 BX46 51
Enfield Rd. N1 CA36 33
Enfield Rd. N8 BX32 24
Enfield Rd. W3 BM41 39
Enfield Rd. Brent. BK42 39
Enfield Rd. Enf. BW24 7
Enfield Rd. E. Brent. BK42 39
Enford St. W1 BU39 41
Engadine Clo. Croy. CA55 78
Engadine St. SW18 BR47 58
Engate St. SE13 CF45 52
Engel Pk. NW7 BQ29 13
Engineer Clo. SE18 CL43 53

Engineers Dr. Bush. BE24 4
Engineers Way, Wem. BM35 30
Englands La. NW3 BU36 32
Englands La. Loug. CL23 10
Englefield Clo. Enf. BY23 8
Englefields Clo. Orp. CN53 72
Englefield Cres. Orp. CN52 72
Englefield Path, Orp. CN52 72
Englefield Rd. N1 BZ36 33
Engleheart Dr. Felt. BC46 47
Engleheart Rd. SE6 CE47 61
Englewood Rd. SW12 BV46 50
English Grounds SE1 CA40 42
English St. E3 CD38 34
Enid St. SE16 CA41 42
Enkel St. N7 BX34 24
Enmore Av. SE25 CB53 69
Enmore Gdns. SW14 BN46 49
Enmore Rd. SE25 CB53 69
Enmore Rd. SW15 BQ45 49
Enmore Rd. Sthl. BF38 29
Ennerdale Av. Horn. CU35 37
Ennerdale Av. Stan. BK31 21
Ennerdale Dr. NW9 BN32 22
Ennerdale Gdns. Wem. BK33 21
Ennerdale Rd. Bexh. CR44 54
Ennerdale Rd. Rich. BL44 48
Ennersdale Rd. SE13 CF46 52
Ennismore Av. W4 BO42 40
Ennismore Av. Grnf. BH36 30
Ennismore Gdns. SW7 BU41 41
Ennismore Gdns. E. Mol. BH53 66
Ennismore Gdns. Ms. BU41 41
SW7
Ennismore Ms. SW7 BU41 41
Ennismore St. SW7 BU41 41
Ennismore Gdns.
Ennis Rd. N4 BY33 24
Ennis Rd. SE18 CM43 53
Ennor Ct. Wor. Pk. BQ56 76
Ensign Dr. N13 BZ27 15
Ensign St. E1 CB40 42
Enslin Rd. SE9 CL46 53
Ensor Ms. SW7 BT42 41
Cranley Gdns.
Enterprise Way SW18 BS45 50
Epirus Ms. SW6 BR43 49
Epirus Rd. SW6 BR43 49
Epping Clo. Rom. CR31 27
Epping Glade E4 CF25 9
Epping New Rd. Buck. H. CH27 17
Epping New Rd. Loug. CH25 10
Epping Pl. N7 BY36 33
Liverpool Rd.
Epping Way E4 CE25 9
Epple Rd. SW6 BR44 49
Epsom Clo. Bexh. CR45 54
Epsom Clo. Nthlt. BE35 29
Epsom Rd. E10 CF32 25
Epsom Rd. Croy. BY56 78
Epsom Rd. Epsom BO59 82
Epsom Rd. Ilf. CN32 27
Epsom Rd. Mord. BR54 67
Epsom Way, Horn. CW35 37
Epstein Rd. SE28 CO40 45
Epworth Pl. EC2 BZ38 33
Epworth St.
Epworth Rd. Islw. BJ43 48
Epworth St. EC2 BZ38 33
Erasmus St. SW1 BW42 41
Erconwald St. W12 BO39 40
Eresby Rd. Beck. CE54 70
Eresey Rd. Croy. CD54 70
Erica Ct. Swan. CT52 73
Erica Gdns. Croy. CE55 79
Erica St. W12 BP40 40
Eric Clo. E7 CH35 35
Ericsson Clo. SW18 BS46 50
Eric Rd. E7 CH35 35
Eric Rd. Rom. CP33 27
Eric St. E3 CD38 34
Eridge Rd. W4 BN41 40
Erin Clo. Brom. CG50 61
Elstree Hill
Erindale SE18 CM43 53
Erith Cres. Rom. CS30 19
Erith Rd. Belv. CR42 45
Erith Rd. Bexh. CR45 54
Erith Rd. Erith CR44 54
Erlanger Rd. SE14 CC44 52
Erlesmere Gdns. W13 BJ41 39
Ermine Clo. Houns. BD45 47
Ermine Rd. N15 CA32 24
Ermine Rd. SE13 CE45 52
Ermine Side, Enf. CB25 8
Ermington Rd. SE9 CM48 62
Ernald Av. E6 CK37 35
Erncroft Way, Twick. BH46 48
Ernest Av. SE27 BY49 60
Ernest Gdns. W4 BM43 48
Ernest Gro. Beck. CD53 70
Ernest Gro. Clo. Beck. CE53 70
Ernest Rd. Horn. CW32 28
Ernest Rd. Kings. On T. BM51 66
Ernest Sq. Kings. On T. BM51 66
Ernest St. E1 CC38 34
Ernle Rd. SW20 BP50 58

Ernshaw Pl. SW15 BR46 49
Carlton Dr.
Erpingham Rd. SW15 BQ45 49
Erridge Rd. SW19 BS51 68
Errington Rd. W9 BR38 31
Errol Gdns. Hayes BC38 29
Errol Gdns. N. Mal. BP52 67
Erroll Rd. Rom. CT31 28
Errol St. EC1 BZ38 33
Erskine Clo. Sutt. BU55 77
Erskine Cres. N15 CB31 24
Erskine Hl. NW11 BS32 23
Erskine Rd. E17 CD31 25
Erskine Rd. NW3 BU36 32
Erskine Rd. Sutt. BT56 77
Erwood Rd. SE7 CK42 44
Esam Way SW16 BY49 60
Escott Gdns. SE9 CK49 62
Escot Way, Barn. BQ25 6
Escreet Gro. SE18 CL42 44
Esher Av. Rom. CS32 28
Esher Av. Sutt. BQ55 76
Esher Av. Walt. BC54 65
Esher Clo. Bex. CQ47 63
Esher Cres. Esher BF56 74
Esher Grn. Esher BF56 74
Esher Ms. Mitch. BU52 68
Esher Pk. Av. Esher BF56 74
Esher Pl. Av. Esher BF56 74
Esher Rd. E. Mol. BG53 65
Esher Rd. Ilf. CN34 27
Esher Rd. Walt. BE56 74
Eskdale Av. Nthlt. BE37 29
Eskdale Gdns. Pur. BZ60 84
Eskdale Rd. Bexh. CR44 54
Eskmont Ridge SE19 CA50 60
Esk Rd. E13 CH38 35
Esk Way, Rom. CS29 19
Esmar Cres. NW4 BP33 22
Esmeralda Rd. SE1 CB42 42
Esmond Clo. Rain. CU36 37
Esmond Gdns. W5 BK41 39
St. Marys Rd.
Esmond Rd. NW6 BR37 31
Esmond Rd. W4 BN42 40
Esmond St. SW15 BR45 49
Esparto St. SW18 BS47 59
Essenden Rd. Belv. CR42 45
Essenden Rd. Sth. Croy. BZ57 78
Essendine Rd. W9 BS38 31
Essex Av. Islw. BH45 48
Essex Clo. E17 CC31 25
Essex Clo. Mord. BQ54 67
Essex Clo. Rom. CR31 27
Essex Clo. Ruis. BD33 20
Essex Ct. EC4 BY39 42
Middle Temple La.
Essex Gdns. SW13 BP44 49
Essex Gdns. N4 BY32 24
Rutland Gdns.
Essex Gro. SE19 BZ50 60
Essex Pk. N3 BS29 14
Essex Pk. Ms. W3 BO40 40
Essex Pl. W4 BO42 40
Chiswick High Rd.
Essex Rd. E4 CG26 9
Essex Rd. E10 CF32 25
Essex Rd. E12 CK35 35
Essex Rd. E17 CD32 25
Essex Rd. E18 CH30 17
Essex Rd. N1 BY37 33
Essex Rd. NW10 BO36 31
Essex Rd. W3 BN40 40
Essex Rd. W4 BN42 40
Belmont Gr.
Essex Rd. Bark. CM36 35
Essex Rd. Borwd. BM24 5
Essex Rd. Dag. CS35 37
Essex Rd. Dart. CV46 55
Essex Rd. Enf. BZ24 8
Essex Rd. Rom. CP33 27
Essex Rd. Rom. CR31 27
Essex Rd. Wat. BC23 4
Essex St. E7 CH35 35
Essex St. WC2 BY39 42
Essex Vill. W8 BS41 41
Essex Wharf E10 CC34 25
Essian St. E1 CD39 43
Essoldo Way, Stan. BL31 21
Estate Way E10 CE33 25
Estcourt Rd. SE25 CB53 69
Estcourt Rd. SW6 BR43 49
Estcourt Rd. Wat. BD24 4
Estella Av. N. Mal. BP52 67
Estelle Rd. NW3 BU35 32
Mansfield Rd.
Esterbrooke St. SW1 BW42 41
Este Rd. SW11 BU45 50
Esther Clo. N21 CH26 10
Esther Rd. E11 CG33 25
Estreham Rd. SW16 BW50 59
Estridge Clo. Houns. BF45 47
Staines Rd.
Eswyn Rd. SW17 BU49 59
Etchingham Ct. N12 BS29 14
Etchingham Pk. Rd. N3 BS29 14
Etchingham Rd. E15 CF35 34

Eternit Wk. SW6 BQ44 49
Etfield Gro. Sid. CO49 63
Ethelbert Clo. Brom. CH52 71
Ethelbert Gdns. Ilf. CK32 26
Ethelbert Rd. SW20 BQ51 67
Ethelbert Rd. Brom. CH52 71
Ethelbert Rd. Dart. CW49 64
Ethelbert Rd. Erith CS43 55
Hengist Rd.
Ethelbert St. Orp. CP52 72
Ethelburga Rd. Rom. CW30 19
Ethelburga St. SW11 BU44 50
Etheldene Av. N10 BW31 23
Ethelden Rd. W12 BP40 40
Ethel Rd. Ashf. CR44 49
Ethel Rd. E16 CH39 44
Ethel Ter. Orp. CP58 81
Etheridge Grn. Loug. CM24 10
Etheridge Rd. NW4 BQ33 22
Etheridge Rd. Loug. CL23 10
Etherley Rd. N15 BZ32 24
Etherow St. SE22 CB46 51
Etherstone Gn. SW16 BX49 60
Etherstone Rd. SW16 BY49 60
Ethnard Rd. SE15 CB43 51
Ethronvi Rd. Bexh. CQ45 54
Etloe Rd. E10 CE34 25
Eton Av. N12 BT29 14
Eton Av. NW3 BT36 32
Eton Av. Barn. BU25 7
Eton Av. Houns. BE43 47
Eton Av. N. Mal. BN53 67
Eton Av. Wem. BJ35 30
Eton College Rd. NW3 BU36 32
Eton Ct. Wem. BK35 30
Eton Gdns. NW3 BU36 32
Lambolle Pl.
Eton Gro. N19 BW34 23
Wedmore St.
Eton Gro. NW9 BM31 21
Eton Gro. SE13 CG45 52
Eton Ho. N5 BT44 50
Eton Pl. NW3 BU36 32
Eton Rd. NW3 BU36 32
Eton Rd. Ilf. CM35 35
Eton Rd. Orp. CO56 81
Eton St. Rich. BL46 48
Eton Vill. NW3 BU36 32
Etta St. SE8 CD43 52
Etton Clo. Horn. CW34 28
Ettrick St. E14 CF39 43
Ettringham St. SW11 BT45 50
Petergate
Etwell Pl. Surb. BL53 66
Eugene Clo. Rom. CV31 28
Eugnia Rd. SE16 CC42 43
Eureka Rd. Kings. On T. BM51 66
Europe Rd. SE18 CK41 44
Eustace Pl. SE18 CK42 44
Eustace Rd. E6 CK38 35
Eustace Rd. SW6 BS43 50
Eustace Rd. Rom. CP33 27
Euston Gro. NW1 BW38 32
Euston Rd. N1 BW38 32
Euston Rd. NW1 BW38 32
Euston Rd. Croy. BX54 69
Euston Sq. NW1 BW38 32
Euston Sta. Colon. NW1 BW38 32
Euston St. NW1 BW38 32
Evandale Rd. SW9 BY44 51
Evangelist Rd. NW5 BV35 32
Evans Dale. Rain. CT38 37
Evans Gro. Felt. BE48 56
Evans Rd. SE6 CG48 61
Evanston Av. E4 CF29 16
Evanston Gdns. Ilf. CK32 26
Eva Rd. Rom. CP33 27
Evelina Clo. SE20 CC50 61
Evelina Rd. SE15 CB45 51
Eveline Lowe Est. SE16 CB41 42
Eveline Rd. Mitch. BU51 68
Evelyn Av. NW9 BN31 22
Evelyn Clo. Twick. BF47 56
Evelyn Ct. N1 BZ37 33
Evelyn Dr. Pnr. BD29 11
Evelyn Gdns. SW7 BT42 41
Evelyn Gdns. Rich. BL45 48
Kew Rd.
Evelyn Gro. W5 BL40 39
Evelyn Gro. Sthl. BE39 38
Evelyn Rd. E16 CH40 44
Evelyn Rd. E17 CF31 25
Evelyn Rd. SW19 BS49 59
Evelyn Rd. W4 BN41 40
Evelyn Rd. Barn. BU24 7
Evelyn Rd. Rich. BK48 57
Evelyn Rd. Rich. BL45 48
Evelyn Sharp Clo. Rom. CV31 28
Evelyn St. SE8 CD42 43
Evelyn Ter. Rich. BL45 48
Evelyn Wk. N1 BZ37 33
Evelyn Way, Wall. BW56 77
Evening Hill, Beck. CF50 61
Evenlode Way, Dag. CR34 27
Evenwood Clo. SW15 BR46 49
Everard Av. Brom. CH54 71
Everard Way, Wem. BL34 21

Everatt Clo. SW18 BR46 49
Amerland Rd.
Everdon Rd. SW13 BP43 49
Everest Pl. E14 CF39 43
Everest Pl. Swan. CS52 73
Everest Rd. SE9 CK46 53
Everett Wk. DA18 CQ43 54
Osbourne Rd.
Everglade Strand NW9 BO30 13
Everilda St. N1 BX37 33
Evering Rd. E5 CA34 24
Evering Rd. N16 CA34 24
Everington Rd. N10 BU30 14
Everington St. W6 BQ43 49
Everitt Rd. NW10 BN38 31
Everleigh St. N4 BX33 24
Eve Rd. E11 CG35 34
Eve Rd. E15 CG37 34
Eve Rd. N17 CA31 24
Eve Rd. Islw. BJ45 48
Eversfield Gdns. NW7 BO29 13
Eversfield Rd. Rich. BL44 48
Eversholt St. NW1 BW37 32
Eversholt St. NW3 BU36 32
Maitland Park Vw.
Evershot Rd. N4 BX33 24
Eversleigh Rd. E6 CJ37 35
Eversleigh Rd. N3 BR29 13
Eversleigh Rd. SW11 BU45 50
Eversleigh Rd. Barn. BT25 7
Eversley Av. Bexh. CS44 55
Eversley Av. Wem. BM34 21
Eversley Clo. N21 BX25 8
Eversley Cres. N21 BX25 8
Eversley Cres. Islw. BG44 47
Eversley Cres. Ruis. BB34 20
Eversley Cross, Bexh. CT44 55
Eversley Dr. Croy. CE55 79
Eversley Mt. N21 BX25 8
Eversley Pk. SW19 BP49 58
Eversley Pk. Rd. N21 BX25 8
Eversley Rd. SE7 CH43 53
Eversley Rd. SE19 BZ50 60
Eversley Rd. Surb. BL52 66
Everthorpe Rd. SE15 CA45 51
Everton Bldgs. NW1 BW38 32
Stanhope St.
Everton Dr. Stan. BL31 21
Everton Rd. Croy. CB54 69
Evesham Av. E17 CE30 16
Evesham Clo. Grnf. BF37 29
Evesham Clo. Sutt. BS57 77
Evesham Grn. Mord. BS53 68
Evesham Rd. E15 CG37 34
Evesham Rd. N11 BW28 14
Evesham Rd. Felt. BD46 47
Evesham Rd. Mord. BS53 68
Evesham Rd. W11 BQ40 40
Evesham Way SW11 BV45 50
Evesham Way, Ilf. CL31 26
Evry Rd. Sid. CP50 63
Ewald Rd. SW6 BR44 49
Ewanrigs Ter. Wdf. Grn. CJ28 17
Ewart Gro. N22 BX30 15
Ewart Rd. SE23 CC47 61
Ewe Clo. N19 BX36 33
Ewell By-pass, Epsom BP58 76
Ewell Ct. Av. Epsom BO56 76
Ewell Downs Rd. Epsom BP59 82
Ewell Ho. Gro. Epsom BO58 76
Ewellhurst Rd. Ilf. CK30 17
Ewell Pk. Way, Epsom BP57 76
Ewell Rd. Surb. BJ54 66
Ewell Rd. Surb. BL53 66
Ewell Rd. Sutt. BQ57 76
Ewen Cres. SW2 BY47 60
Ewer St. SE1
Ewhurst Av. Sth. Croy. CA58 78
Ewhurst Clo. Sutt. BQ58 76
Ewhurst Rd. SE4 CD46 52
Ewing St. E3 CD38 34
Maidman St.
Exbury Rd. SE6 CD48 61
Excelsior Clo. Kings. On T. BM51 66
Washington St.
Exchange Bldgs. EC1 CA39 42
Cutler St.
Exchange Rd. Wat. BC24 4
Exchange Sq. EC1 BZ38 33
Dingley St.
Exeter Clo. E6 CK38 35
Exeter Gdns. Ilf. CK33 26
Exeter Ho. SW15 BQ46 49
Exeter Rd. E16 CH39 44
Exeter Rd. E17 CE32 25
Exeter Rd. N9 CC27 16
Exeter Rd. N14 BV26 7
Exeter Rd. NW2 BR35 31
Exeter Rd. SE15 CA44 51
Exeter Rd. Croy. CA54 69
Exeter Rd. Dag. CR36 36
Exeter Rd. Enf. CC24 9
Exeter Rd. Felt. BE48 56
Exeter Rd. Har. BE34 20
Exeter Rd. Well. CN44 54
Exeter St. WC2 BX40 42
Exford Gdns. SE12 CH47 62
Exford Rd. SE12 CH48 62

Exhibition Rd. SW7 BT41 41
Exmoor St. W10 BQ39 40
Exmouth Mkt. EC1 BY38 33
Exmouth Rd. E17 CD32 25
Exmouth Rd. Brom. CH52 71
Exmouth Rd. Ruis. BD34 20
Exmouth Rd. Well. CO43 54
Exmouth St. E1 CC39 43
Exning Rd. E16 CG38 34
Exon St. SE17 CA42 42
Exton Cres. NW10 BN36 31
Exton Gdns. Dag. CP35 36
Exton St. SE1 BY40 42
Eyethorne Rd. SW9 BY44 51
Eyhurst Av. Horn. CU34 28
Eyhurst Clo. NW2 BP34 22
Eylewood Rd. SE27 BZ49 60
Eynella Rd. SE22 CA47 60
Eynham Rd. W12 BQ39 40
Eynsford Clo. Orp. CM54 71
Eynsford Cres. Bex. CP47 63
Eynsford Rd. Farn. CW54 73
Eynsford Rd. Ilf. CN34 27
Eynsford Rd. Swan. CS53 73
Eynsham Br. SE2 CO42 45
Eynsham Dr. SE2 CO41 45
Eynswood Dr. Sid. CO49 63
Eyot Gdns. W6 BO42 40
Eyre Clo. NW8 BT37 32
Eyre Clo. Rom. CV31 28
Eyre St. Hill EC1 BY38 33
Eythorne Rd. SW9 BY44 51
Ezra St. E2 CA38 33
Faber Gdns. NW4 BP32 22
Fabian Rd. SW6 BR43 49
Fabian St. E6 CK38 35
Factory La. N17 CA30 15
Factory La. Croy. BY54 69
Factory Pl. E14 CE42 43
Factory Rd. E16 CK40 44
Factory Yd. W7 BH40 39
Faesten Way, Bex. CT48 64
Fagg's Rd. Felt. BC45 47
Fagus Av. Rain. CV38 37
Fairacre, Islw. BG44 47
Fairacre, N. Mal. BO52 67
Fairacres SW15 BP45 49
Fairacres, Brom. CH53 71
Fairacres, Croy. CD58 79
Fairbairn Grn. SW9 BY44 51
Fairbairn Rd. SW9 BY44 51
Fairbank Av. Orp. CL55 80
Fairbank Rd. N17 CA31 24
Fairbourne Rd. N17 CA31 24
Fairbridge Rd. N19 BW34 23
Fairbrook Clo. N13 BY28 15
Fairbrook Rd. N13 BY29 15
Fairburn Clo. Borwd. BM23 5
Fairby Rd. SE12 CH46 53
Fair Child St. EC2 CA38 33
Gt. Eastern St.
Fair Clo. Bush. BF26 4
Fairclough St. E1 CB39 43
Faircross Av. Bark. CM36 35
Faircross Av. Rom. CS29 19
Fairdale Gdns. SW15 BP45 49
Fairdale Gdns. Hayes BC40 38
Fairdene Rd. Couls. BW62 83
Fairfax Av. Epsom BP58 76
Fairfax Clo. Walt. BC54 65
Fairfax Gdns. SE3 CJ44 53
Fairfax Pl. NW6 BT36 32
Fairfax Rd. N8 BY31 24
Fairfax Rd. NW6 BT36 32
Fairfax Rd. W4 BO41 41
Fairfax Rd. Tedd. BJ50 57
Fairfax Ter. Sthl. BE42 38
Fairfield Av. NW4 BP32 22
Fairfield Av. Edg. BM29 12
Fairfield Av. Twick. BF47 56
Fairfield Av. Wat. BD27 11
Fairfield Clo. N12 BT28 14
Torrington Park
Fairfield Clo. Enf. CC24 9
Scotland Green Rd. Nth.
Fairfield Clo. Rom. CU33 28
Fairfield Clo. Sid. CN46 54
Fairfield Ct. NW10 BP37 31
Longstone Av.
Fairfield Ct. Wdf. Grn. CH29 17
Fairfield Rd.
Fairfield Cres. Edg. BM29 12
Fairfield Dr. SW18 BS46 50
Fairfield St.
Fairfield Dr. SW18 BS46 50
Fairfield Dr. Grnf. BK37 30
Fairfield Dr. Har. BG31 20
Fairfield E. Kings. On T. BL51 66
Fairfield Gdns. N8 BX32 24
Elder Av.
Fairfield Gro. SE7 CJ43 53
Fairfield N. Kings. On T. BL51 66
Fairfield P. Croy. CA55 78
Fairfield Path, Croy. CA55 78
Fairfield Pl. Kings. BL52 66
Fairfield Rd. E3 CE37 34
Fairfield Rd. E17 CD30 16
Fairfield Rd. N8 BX32 24
Fairfield Rd. N18 CB28 15
Fairfield Rd. W7 BJ41 39
Fairfield Rd. Beck. CE51 70
Fairfield Rd. Bexh. CQ44 54
Fairfield Rd. Brom. CH50 62
Fairfield Rd. Croy. BZ55 78
Fairfield Rd. Ilf. CL36 35
Fairfield Rd. Kings. On T. BL51 66
Fairfield Rd. Orp. CM54 71
Fairfield Rd. Sthl. BE39 38
Fairfield Rd. Wdf. Grn. CH29 17
Fairfields, Croy. CA56 78
Fairfields Clo. NW9 BN32 22
Fairfields Cres. NW9 BN31 22
Fairfield S. Kings. On T. BL52 66
Fairfields Rd. Houns. BG45 47
Fairfield St. SW18 BS45 50
Fairfield Way, Barn. BS25 7
Fairfield Way, Couls. BW60 83
Fairfield Way, Epsom BO56 76
Fairfield W. Kings. On T. BL51 66
Fairfoot Rd. E3 CE38 34
Fairford Av. Bexh. CS44 55
Fairford Av. Croy. CC53 70
Fairford Clo. Croy. CC53 70
Fairford Gdns. Wor. Pk. BO55 76
Fairgreen, Barn. BU24 7
Fairgreen E. Barn. BU24 7
Fairgreen Rd. Th. Hth. BY53 69
Fairhaven Av. Croy. CC53 70
Fairhaven Cres. Wat. BC27 11
Fairhazel Gdns. NW6 BS36 32
Fairholme Av. Rom. CU32 28
Fairholme Clo. N3 BR31 22
Fairholme Gdns. N3 BR31 22
Fairholme Rd. W14 BR42 40
Fairholme Rd. Croy. BY54 69
Fairholme Rd. Har. BH32 21
Fairholme Rd. Sutt. BR57 76
Fairholm Rd. Ilf. CK32 26
Fairholt Rd. N16 BZ33 24
Fairholt St. SW7 BU41 41
Montpelier Wk.
Fairkytes Av. Horn. CV33 28
Fairland Rd. E15 CG36 34
Fairlands Av. Buck. H. CH27 17
Fairlands Av. Sutt. BS55 77
Fairlands Av. Th. Hth. BX52 69
Fairlawn SE7 CJ43 53
Fairlawn, Twick. BK46 48
Fairlawn, Wdf. Grn. CK29 17
Vicarage Rd.
Fairlawn Av. N2 BU31 23
Fairlawn Av. W4 BN42 40
Fairlawn Av. Bexh. CP44 54
Fairlawn Clo. N14 BW25 7
Fairlawn Clo. Esher BH57 75
Fairlawn Clo. Kings. On T. BN50 58
Fairlawn Ct. SE7 CJ43 53
Fairlawn
Fairlawn Ct. W4 BN42 40
Cunnington St.
Fairlawn Dr. Wdf. Grn. CH29 17
Fairlawn Gdns. Sthl. BE40 38
Fairlawn Gro. W4 BN42 40
Fairlawn Gro. Bans. BT60 83
Fairlawn Pk. SE26 CD49 61
Fairlawn Rd. SW19 BR50 58
Fairlawn Rd. Sutt. BT59 83
Fairlawns SW15 BQ46 49
Putney Hill
Fairlawns, Pnr. BD30 11
Fairlawns, Sun. BC52 65
Fairlawns Clo. Felt. BE49 56
Fairlawns Clo. Horn. CW33 28
Herbert Rd.
Fairlea Pl. W5 BK38 30
Fairleigh Pl. N16 CA35 33
Fairleigh Rd.
Fairlie Gdns. SE23 CC47 61
Fairlight Av. E4 CF27 16
Fairlight Av. NW10 BO37 31
Fairlight Av. Wdf. Grn. CH29 17
Fairlight Clo. E4 CF27 16
Fairlight Clo. Wor. Pk. BQ56 76
Fairlight Rd. SW17 BT49 59
Fairlop Clo. Horn. CU36 37
Fairlop Gdns. Ilf. CM29 17
Fairlop Pl. NW8 BT38 32
Fairlop Rd. E11 CF33 25
Fairlop Rd. Ilf. CM30 17
Fairmead, Brom. CK52 71
Fairmead, Surb. BM54 66
Fairmead Clo. Brom. CK52 71
Fairmead Clo. N. Mal. BN52 67
Fairmead Cres. Edg. BN27 13
Fairmead Gdns. Ilf. CJ32 26
Fairmead Rd. N19 BW34 23
Fairmead Rd. Croy. BX54 69
Fairmead Rd. Houns. BD43 47
Fairmead Rd. Loug. CH25 10
Fairmead Side, Loug. CJ25 10
Fairmile Av. SW16 BW49 59
Fairmount Rd. SW2 BX46 51
Fairoak Clo. Ken. BY61 84
Fairoak Clo. Orp. CL54 71
Fairoak Dr. SE9 CM46 53
Fairoak Gdns. Rom. CT30 19
Fairseat Clo. Bush. BH27 12
Hive Rd.
Fair St. SE1 CA41 42
Fair St. Houns. BG45 47
Fairthorn Rd. SE7 CH42 44
Fairview Av. Rain. CV37 37
Fairview Av. Wem. BK36 30
Fairview Clo. E17 CD30 16
Fairview Clo. Chig. CN28 18
Fairview Clo. Epsom BQ59 82
Fairview Cres. Har. BF33 20
Fairview Dr. Chig. CN28 18
Fairview Dr. Orp. CM56 80
Fairview Gdns. Wdf. Grn. CH30 17
Fairview Pl. SW2 BX47 60
Holmewood Gdns.
Fairview Rd. N15 CA32 24
Fairview Rd. SW16 BX51 69
Fairview Rd. Chig. CN28 18
Fairview Rd. Enf. BX23 8
Fairview Rd. Epsom BO59 82
Fairview Rd. Erith CT43 55
Fairview Rd. Sutt. CT56 77
Fairview Way, Edg. BM28 12
Fairwall House SE5 CA44 51
Glebe Est.
Fairwater Av. Well. CO45 54
Fairway SW20 BQ52 67
Fairway, Bexh. CQ46 54
Fairway, Cars. BT59 83
Fairway, Orp. CM53 71
Fairway, Wdf. Grn. CJ28 17
Fairway Av. NW9 BM31 21
Fairway Av. Borwd. BM23 5
Fairway Clo. NW11 BT33 23
Fairway Clo. Croy. CD53 70
Fairway Clo. Epsom BN55 76
Fairway Dr. Grnf. BG37 29
Fairway Gdns. Ilf. CM35 35
Fairways, Stan. BL30 12
Fairways, Tedd. BK50 57
Fairway, The N13 BZ27 15
Fairway, The W3 BO39 40
Fairway, The, Barn. BS25 7
Fairway, The, Brom. CK53 71
Fairway, The, E. Mol. BF52 65
Fairway, The, N. Mal. BN51 67
Fairway, The, Nthlt. BG36 29
Fairway, The NW7 BN27 13
Fairway, The N14 BV25 7
Fairway, The, Ruis. BD35 29
Fairway, The, Wem. BJ34 21
Fairweather Clo. N15 CA31 24
Lawrence Rd.
Fairweather Rd. N16 CB32 24
Fairwood Ct. E11 CF33 25
Fairwyn Rd. SE26 CD49 61
Fairy Lawns Clo. Horn. CW33 28
Falcon Av. Brom. CK52 71
Falconberg Ms. W1 BW39 41
Sutton Row
Falcon Clo. SE1 BZ40 42
Falcon Cres. Enf. CC25 9
Falconer Rd. Bush. BE25 4
Falconer Rd. Ilf. CO28 18
Falconer Wk. N7 BX34 24
Andover Est.
Falcon Pl. N16 CA34 24
Church St.
Falcon Rd. SW11 BU44 50
Falcon Rd. SW11 BU45 50
Falcon Rd. Enf. CC25 9
Falcon Rd. Hamptn. BE50 56
Falcon St. E13 CG38 34
Falcon Ter. Est. SW11 BU45 50
Falcon Way, Felt. BC46 47
Falcon Way, Har. BL32 21
Falcon Way, Rain. CU37 37
Falconwood Av. Well. CM44 53
Falconwood Pde. Well. CN45 54
Falconwood Rd. Croy. CD58 79
Falcourt Clo. Sutt. BS56 77
Robin Hood La.
Falkholt St. SW7 BU42 41
Rutland St.
Falkirk Gdns. Wat. BD28 11
Falkirk St. N1 CA37 33
Falkland Av. N3 BS29 14
Falkland Av. N11 BV28 14
Falkland Pk. Av. SE25 CA52 69
Falkland Pl. NW5 BW35 32
Falkland Rd.
Falkland Rd. N8 BY31 24
Falkland Rd. NW5 BW35 32
Falkland Rd. Barn. BR23 6
Fallaize Av. Ilf. CL35 35
Riverdene Rd.
Falloden Way NW11 BS31 23
Fallon Trading Est. NW10 BO35 31
Fallow Clo. Chig. CN28 18
Fallow Ct. Av. N12 BT29 14
Fallowfield, Stan. BJ27 12
Fallowfield Ct. Stan. BJ27 12
Fallow Hurst Path N3 BT29 14
Park Cres.
Fallsbrook Rd. SW16 BV50 59
Falmer Rd. E17 CE31 25
Falmer Rd. N15 BZ32 24
Falmer Rd. Enf. CA24 8
Falmouth Av. E4 CF28 16
Falmouth Clo. N22 BX29 15
Falmouth Clo. SE12 CG46 52
Taunton Rd.
Falmouth Gdns. Ilf. CJ31 26
Falmouth Rd. SE1 BZ41 42
Falmouth Rd. Walt. BD56 74
Falmouth St. E15 CF35 34
Falstaff House N1 CA37 33
Purcell St.
Fambridge Rd. SE26 CD49 61
Fambridge Rd. Dag. CR33 27
Famet Av. Pur. BZ60 84
Famet Clo. Pur. BZ60 84
Famet Wk. Pur. BZ60 84
Fane St. W14 BR43 49
Fann St. EC1 BZ38 33
Fanshawe Av. Bark. CM36 35
Fanshawe Cres. Dag. CQ35 36
Fanshawe Cres. Horn. CV32 28
Fanshawe Rd. Rich. BK49 57
Fanshaw St. N1 CA38 33
Fanthorpe St. SW15 BQ45 49
Fara Clo. N7 BX36 33
Faraday Av. Sid. CO48 63
Faraday Rd. E15 CG36 34
Faraday Rd. SE7 CJ41 44
Faraday Rd. SW19 BS50 59
Faraday Rd. W3 BN40 40
Faraday Rd. W10 BR39 40
Faraday Rd. E. Mol. BF52 65
Faraday Rd. Sthl. BF39 38
Faraday Rd. Well. CO45 54
Faraday Way, Orp. CO52 72
Fareham Rd. Felt. BD46 47
Farewell Pl. Mitch. BT51 68
Farifield Clo. N12 BT28 14
Faringdon Av. Brom. CL54 71
Faringdon Av. Rom. CV30 19
Farjeon Rd. SE3 CJ44 53
Farleigh Av. Brom. CG53 70
Farleigh Dean Cres. Croy. CE58 79
Farleigh Pl. N16 CA35 33
Farleigh Rd. N16 CA35 33
Farley Dr. Ilf. BT33 23
Farley Pl. SE25 CB52 69
Farley Rd. SE6 CE46 52
Farley Rd. Sth. Croy. CB58 78
Farlington Pl. SW15 BP47 58
Farlow Rd. SW15 BQ45 49
Farlton Rd. SW18 BS47 59
Farman Gro. Nthlt. BD38 29
Wayfarer Rd.
Farm Av. NW2 BR34 22
Farm Av. SW16 BX49 60
Farm Av. Har. BE33 20
Farm Av. Orp. CM54 71
Farm Av. Swan. CS52 73
Farm Av. Wem. BK36 30
Farm Clo. Barn. BP25 6
Farm Clo. Buck. H. CJ27 17
Farm Clo. Dag. CS36 37
Farm Clo. Sthl. BF40 38
Farm Clo. Sutt. BT57 77
Farm Clo. Wall. BW58 77
Farm Clo. W. Wick. CG55 79
Farmcote Rd. SE12 CH47 62
Farm Ct. NW4 BP31 22
Farmdale Rd. SE10 CH42 44
Farmdale Rd. Cars. BU57 77
Farm Dr. Croy. CD55 79
Farm Dr. Pur. BW59 83
Farm End E4 CG25 9
Farmer Rd. E10 CE33 25
Farmers Rd. SE5 BY43 51
Farmer St. W8 BS40 41
Farmfield Clo. N12 BS28 14
Farmfield Rd. Brom. CF49 61
Farm Fields, Sth. Croy. CA59 84
Farmhouse Rd. SW16 BW50 59
Farmilo Rd. E17 CD33 25
Farmington Av. Sutt. BT55 77
Farmlands, Enf. CH23 10
Farmlands, Pnr. BC31 20
Farmlands, The, Nthlt. BE36 29
Farmland Wk. Chis. CL49 62
Farm La. N14 BV25 7
Farm La. SW6 BS43 50
Farm La. Croy. CD55 79
Farm La. Pur. BW58 77
Farmleigh Rd. N14 BW26 7
Farm Pl. W8 BS40 41
Farm Pl. Dart. CU45 55
Farm Rd. N21 BY26 8
Farm Rd. Edg. BM29 12
Farm Rd. Epsom BN59 82
Farm Rd. Esher BF54 65
Farm Rd. Mord. BS53 68
Farm Rd. Rain. CV38 37
Farm Rd. Sutt. BT57 77
Farmstead Rd. SE6 CE49 61
Farmstead Rd. Har. BG30 11
Farm St. W1 BV40 41
Farm, The SW19 BQ47 58
Princes Way
Farm Vale, Bex. CR46 54
Farm Wk. NW11 BR32 22

Farm Way, Buck. H.	CJ28	17
Farm Way, Bush.	BF24	4
Farmway, Dag.	BP34	27
Farm Way, Horn.	CU35	37
Farm Way, Wor. Pk.	BQ55	76
Farnaby Rd. SE9	CJ45	53
Farnaby Rd. Brom.	CF50	61
Farnan Av. E17	CE30	16
Farnan Rd. SW16	BX49	60
Farnborough Av. E17	CD31	25
Farnborough Av. Sth. Croy.	CC58	79
Farnborough Clo. Wem.	BM34	21
Farnborough Comm. Orp.	CK55	80
Farnborough Cres. Sth. Croy.	CD58	79
Farnborough Hill, Orp.	CM56	80
Farnborough Hill, Orp.	CN56	81
Farnborough Way SE15	CA43	51
Farnborough Way, Orp.	CL56	80
Farncombe St. SE16	CB41	42
Farndale Av. N13	BY27	15
Farndale Cres. Grnf.	BG38	29
Farnell Ms. SW5	BS42	41
Farnell Point N16	CB34	24
Downs Est.		
Farnell Rd. Islw.	BG45	47
Farnes Dr. Rom.	CV30	19
Farnham Clo. N20	BT26	7
Farnham Gdns. SW20	BP51	67
Farnham Pl. SE1	BY40	42
Farnham Rd. Ilf.	CN33	27
Farnham Rd. Rom.	CV28	19
Farnham Rd. Well.	CP44	54
Farnham Royal SE11	BX43	51
Farningham Rd. N17	CB29	15
Farnley Rd. E4	CG26	9
Farnley Rd. SE25	BZ52	69
Faro Clo. Brom.	CL51	71
Faroe Rd. W14	BQ41	40
Farorna Wk. Enf.	BY23	8
Farquhar Rd. SE19	CA49	60
Farquhar Rd. SW19	BS48	59
Farquharson Rd. Croy.	BZ54	69
Farraline Rd. Wat.	BC24	4
Farrance Est. E14	CE39	43
Farrance Rd. Rom.	CQ32	27
Farrance St. E14	CE39	43
Farrant Av. N22	BY30	15
Farrant Clo. Orp.	CN57	81
Farrant Way, Borwd.	BL23	5
Farr Av. Bark.	CO37	36
Maybury Rd.		
Farren Rd. SE23	CD48	61
Farrer Rd. N8	BW31	23
Farrer Rd. Har.	BL32	21
Farrier Clo. Sun.	BC52	65
Farriers Rd. Nthlt.	BF37	29
Farrier St. NW1	BW36	32
Farriers Way, Borwd.	BN25	6
Farringdon La. EC1	BY38	33
Farringdon St. EC4	BY39	42
Farringford Rd. E15	CG36	34
Farrington Av. Orp.	CO52	72
Farr Rd. Enf.	BZ23	8
Farthing Alley SE1	CB41	42
Wolseley St.		
Farthing Barn La. Orp.	CK58	80
Farthing Flds. E1	CB40	42
Raine St.		
Farthings Clo. E4	CG27	16
Farthings Clo. Pnr.	BC32	20
Farthing St. Orp.	CK57	80
Farwell Rd. Sid.	CO48	63
Farwig La. Brom.	CG51	70
Fashion St. E1	CA39	42
Fashoda Rd. Brom.	CJ52	71
Fassett Rd. E8	CB36	33
Fassett Rd. Kings. On T.	BL52	66
Fassett Sq. E8	CB36	33
Fauconberg Rd. W4	BN43	49
Faulkners Rd. Walt.	BD56	74
Fauna Clo. Rom.	CP32	27
Faunce St. SE17	BY42	42
Favart Rd. SW6	BS44	50
Faversham Av. E4	CG26	9
Faversham Av. Enf.	BZ25	8
Faversham Clo. Chig.	CO27	18
Faversham Rd. SE6	CD47	61
Faversham Rd. Beck.	CD51	70
Croydon Rd.		
Faversham Rd. Mord.	BS53	68
Fawcett Est. E5	CB33	24
Fawcett Rd. NW10	BO37	31
Fawcett Rd. Croy.	BZ55	78
Fawcett St. SW10	BT43	50
Fawcus Clo. Esher	BH57	75
Fawe Pk. Rd. SW15	BR45	49
Fawe St. E14	CE39	43
Fawley Rd. N17	CB31	24
Fawley Rd. NW6	BS35	32
Fawnbrake Av. SE24	BY46	51
Fawn Rd. E13	CJ37	35
Fawn Rd. Chig.	CN28	18
Fawood Av. NW10	BN36	31
Faygate Cres. Bexh.	CR46	54
Faygate Rd. SW2	BX48	60
Fayland Av. SW16	BW49	59
Fayland Est. SW16	BW49	59
Fearnley Cres. Hamptn.	BE50	56
Fearnley Rd. SE5	CA44	51
Lettsom St.		
Fearnley St. Wat.	BC24	4
Fearon St. SE10	CH42	44
Featherbed La. Croy.	CD57	79
Feathers Pl. SE10	CF43	52
Featherstone Av. SE23	CB48	60
Featherstone Ct. EC1	BZ38	33
Featherstone St.		
Featherstone Gdns. Borwd.	BN24	6
Featherstone Rd. NW7	BP29	13
Featherstone Rd. Sthl.	BE41	38
Featherstone St. EC1	BZ38	33
Featherstone Ter. Sthl.	BE41	38
Featley Rd. SW9	BY45	51
Angell Rd.		
Federal Rd. Grnf.	BK37	30
Federal Way, Wat.	BD23	4
Federation Rd. SE2	CO42	45
Fee Farm Rd. Esher	BH57	75
Felbridge Av. Stan.	BJ30	12
Felbridge Clo. SW16	BY49	60
Felbridge Clo. Sutt.	BS58	77
Felbrigge Rd. Ilf.	CN34	27
Felcott Clo. Walt.	BD55	74
Felcott Rd. Walt.	BD55	74
Felday Rd. SE13	CE46	52
Felden Clo. Pnr.	BE29	11
Felden St. SW6	BR44	49
Felgate Ms. W6	BP42	40
Felhampton Rd. SE9	CL48	62
Felhurst Cres. Dag.	CR35	36
Felix Av. N8	BX32	24
Felix Rd. W13	BJ40	39
Felix Rd. Walt.	BC53	65
Felixstowe Rd. N9	CB27	15
Felixstowe Rd. N17	CA31	24
Felixstowe Rd. NW10	BP38	31
Felixstowe Rd. SE2	CP41	45
Sedgemere Rd.		
Felix St. E2	CB37	33
Cambridge Cres.		
Fellbrigg Rd. SE22	CA46	51
Fellbrigg St. E1	CB38	33
Headlam St.		
Fellbrook, Rich.	BJ48	57
Fellowes Rd. Cars.	BU55	77
Fellows Ct. E2	CA37	33
Fellows Rd. NW3	BT36	32
Fell Rd. Croy.	BZ55	78
Felltram Way SE7	CH42	44
Felmingham Rd. SE20	CC51	70
Felnex Est. NW10	BN38	31
Felsberg Rd. SW2	BX46	51
Fels Clo. Dag.	CR34	27
Fels Fm. Av. Dag.	CS34	28
Felsham Rd. SW15	BQ45	49
Felspar Clo. SE18	CN42	45
Felstead Av. Ilf.	CL30	17
Felstead Rd. E11	CH33	26
Felstead Rd. Epsom	BN59	82
Felstead Rd. Loug.	CK26	10
Felstead Rd. Orp.	CO55	81
Felstead Rd. Rom.	CS29	19
Felstead St. E9	CD36	34
Felsted Rd. E16	CJ39	44
Feltham Av. E. Mol.	BH52	66
Felthambrook Way, Felt.	BC48	56
Feltham Hill Rd. Felt.	BC49	56
Feltham Rd. Mitch.	BV51	68
Felton Clo. Brom.	CL53	71
Felton Lea, Sid.	CN49	63
Felton Rd. W13	BK41	39
Camborne Av.		
Felton Rd. Bark.	CN37	36
Felton St. N1	BZ37	33
Fencepiece Rd. Chig.	CM28	17
Fenchurch Av. EC3	CA39	42
Fenchurch Bldgs. EC3	CA39	42
Fenchurch St. EC3	BZ40	42
Fen Ct. EC3	CA39	42
Fendall Rd. Epsom	BN56	76
Fendall St. SE1	CA41	42
Fendt Clo. E16	CG40	43
Bowman Av.		
Fendyke Rd. Belv.	CP42	45
Fenelon Pl. W14	BR42	40
Fen Gro. Sid.	CN46	54
Fenham Rd. SE15	CB43	51
Fenismead Av. Sutt.	BQ58	76
Fenman Ct. N17	CB30	15
Shelbourne Rd.		
Fenn Clo. Brom.	CH50	62
Fennell St. SE18	CL43	53
Fennels Mead, Epsom	BO58	76
Fenning St. SE1	CA41	42
St. Thomas St.		
Fenn St. E9	CC35	34
Fenstanton Av. N12	BT29	14
Fen St. E16	CG40	43
Huntingdon St.		
Fens Way, Swan.	CU50	64
Fentiman Av. Horn.	CW33	28
Fentiman Rd. SW8	BX43	51
Fenton Clo. Brom.	CK49	62
Fenton Ho. Est. NW3	BT34	23
Fenton Rd. N17	BZ29	15
Fenton's Av. E13	CH37	35
Fenwick Gro. SE15	CB45	51
Fenwick Pl. SW9	BX45	51
Fenwick Rd. SE15	CB45	51
Ferdinand Est. NW1	BV36	32
Ferdinand Pl. NW1	BV36	32
Ferdinand St.		
Ferdinand Pl. NW5	BV36	32
Ferdinand St. NW1	BV36	32
Ferguson Av. Rom.	CV30	19
Ferguson Av. Surb.	BL53	66
Ferguson Cres. Rom.	CV30	19
Fergus Rd. N5	BY35	33
Calabria Rd.		
Ferme Pk. Rd. N4	BX32	24
Ferme Pk. Rd. N8	BX32	24
Fermor Rd. SE23	CD47	61
Fermoy Rd. W9	BR38	31
Fermoy Rd. Grnf.	BF38	29
Fern Av. Mitch.	BW52	68
Fernbank Av. Horn.	CV35	37
Fernbank Av. Walt.	BE54	65
Fernbank Av. Wem.	BH35	30
Fernbrook Av. Sid.	CN46	54
Fernbrook Cres. SE13	CG46	52
Fernbrook Dr. Har.	BF33	20
Fernbrook Rd. SE13	CG46	52
Ferncliff Rd. E8	CB35	33
Ferncroft Av. N12	BU29	14
Ferncroft Av. NW3	BS34	23
Ferncroft Av. Ruis.	BD34	20
Ferndale, Brom.	CJ51	71
Ferndale, Horn.	CW32	28
Ferndale Av. E17	CF32	25
Ferndale Av. Houns.	BE45	47
Ferndale Ct. SE3	CG43	52
Ferndale Ct. SW9	BX45	51
Ferndale Rd. E7	CH36	35
Ferndale Rd. E11	CG34	25
Ferndale Rd. N15	CA32	24
Ferndale Rd. SE25	CB53	69
Ferndale Rd. SW4	BX45	51
Ferndale Rd. SW9	BX45	51
Ferndale Rd. Bans.	BR61	82
Ferndale Rd. Rom.	CS30	19
Ferndale St. E6	CL40	44
Ferndale Ter. Har.	BH31	21
Ferndale Way, Orp.	CM56	80
Ferndell Av. Bex.	CS48	64
Fern Dene W13	BJ39	39
Dene, The		
Ferndene Rd. SE24	BZ45	51
Fernden Way, Rom.	CR32	27
Ferndown Av. Orp.	CM54	71
Ferndown Clo. Pnr.	BE29	11
Ferndown Clo. Sutt.	BT57	77
Ferndown Rd. SE9	CJ47	62
Ferndown Rd. Nthwd.	BC30	11
Ferndown Rd. Wat.	BD27	11
Fern Dr. Felt.	BC47	56
Ferney Rd. Barn.	BV26	7
Fernhall Dr. Ilf.	CJ32	26
Fernham Rd. Th. Hth.	BZ52	69
Fernhead Rd. W9	BR38	31
Fernhead Yd. W9	BR39	40
Fernhead Rd.		
Fernheath Way, Dart.	CS49	64
Fernhill Ct. E17	CF30	16
Fernhill Ct. Kings. On T.	BK49	57
Fernhill Gdns. Kings. On T.	BK49	57
Fernhill St. E16	CK40	44
Fernholme Rd. SE15	CC46	52
Fernhurst Gdns. Edg.	BM29	12
Fernhurst Rd. SW6	BR44	49
Fernhurst Rd. Croy.	CB54	69
Fernie Clo. Chig.	CO28	18
Fern La. Houns.	BE42	38
Fernlea Rd. SW12	BV47	59
Fernlea Rd. Mitch.	BV51	68
Fernleigh Ct. Har.	BF30	11
Fernleigh Ct. Wem.	BL34	21
Fernleigh Rd. N21	BY27	15
Fernsbury St. WC1	BY38	33
Margery St.		
Ferns Clo. Sth. Croy.	CB58	78
Fernshaw Rd. SW10	BT43	50
Fernside NW3	BS34	23
Fernside NW4	BQ30	13
Fernside, Buck. H.	CH26	10
Fernside Av. NW7	BN27	13
Fernside Av. Felt.	BC49	56
Fernside IG9	CH26	10
Fernside Rd. SW12	BU47	59
Ferns Rd. E15	CG36	34
Fern St. E3	CE38	34
Fernthorpe Rd. SW16	BW50	59
Ferntower Rd. N5	BZ35	33
Fernways, Ilf.	CL35	35
Cecil Rd.		
Fernwood Av. SW16	BW49	59
Fernwood Av. Wem.	BK36	30
Fernwood Clo. Brom.	CJ51	71
Fernwood Cres. N20	BU27	14
Ferrant Clo. SE7	CJ41	44
Ferrard Clo. Houns.	BF43	47
Ferrers Av. Wall.	BW56	77
Ferrers Rd. SW16	BW49	59
Ferrestone Rd. N8	BX31	24
Glebe Rd.		
Ferriby Clo. N1	BY36	33
Bewdley St.		
Ferrier St. SW18	BS45	50
Ferring Clo. Har.	BG33	20
Ferrings SE21	CA48	60
Ferris Av. Croy.	CD55	79
Ferris Rd. SE22	CB45	51
Ferron Rd. E5	CB34	24
Ferro Rd. Rain.	CU38	37
Ferry Approach SE18	CL41	44
Ferryhills Clo. Wat.	BD27	11
Ferry La. N17	CB31	24
Ferry La. Brent.	BL43	48
Ferry La. Rain.	CT39	46
Ferry La. Rich.	BL43	48
Ferry La. SW13	BO43	49
Ferrymead Av. Grnf.	BF38	29
Ferrymead Dr. Grnf.	BF37	29
Ferrymead Gdns. Grnf.	BF37	29
Ferrymoor, Rich.	BJ48	57
Ferry Pl. SE18	CL41	44
Ferry Rd. SW13	BP43	49
Ferry Rd. E. Mol.	BF52	65
Ferry Rd. Surb.	BJ53	66
Ferry Rd. Tedd.	BJ49	57
Ferry Rd. Twick.	BJ47	57
Ferry Sq. Brent.	BK43	48
Ferry St. E14	CF42	43
Festing Rd. SW15	BQ45	49
Festival Clo. Bex.	CP47	63
Festival Clo. Erith	CT43	55
Fetter La. EC4	BY39	42
Ffinch St. SE8	CE43	52
Fiddicroft Av. Bans.	BS60	83
Fidler Pl. Bush.	BF26	4
Field Clo. E4	CE29	16
Field Clo. Brom.	CJ51	71
Field Clo. Buck. H.	CJ27	17
Field Clo. Chess.	BK57	75
Field Clo. E. Mol.	BF53	65
Field Clo. Houns.	BC44	47
Field Clo. Sth. Croy.	CB60	84
Fieldcommon La. Walt.	BE54	65
Field Ct. WC1	BX39	42
Field End, Barn.	BP24	6
Field End, Couls.	BW56	77
Field End, Couls.	BW60	83
Field End, Nthlt.	BD36	29
Arnold Rd.		
Field End, Ruis.	BD36	29
Field End, Twick.	BH49	57
Fieldend Rd. SW16	BW51	68
Field End Rd. Pnr.	BC32	20
Field End Rd. Ruis.	BD33	20
Fieldfare Rd. SE28	CP40	45
Field Gate La. Mitch.	BU52	68
Fieldgate St. E1	CB39	42
Fieldhouse Rd. SW12	BW47	59
Fielding Av. Twick.	BG48	56
Fielding Rd. W4	BN41	40
Fielding Rd. W14	BQ41	40
Fieldings, The SE23	CC47	61
Fielding St. SE17	BZ43	51
Fielding Ter. W5	BL40	39
Uxbridge Rd.		
Field La. Brent.	BK43	48
Field La. Tedd.	BJ49	57
Field Mead NW9	BO29	13
Field Pk. Cres. Rom.	CP32	27
Field Pl. EC1	BY37	33
St. John St.		
Field Pl. N. Mal.	BO53	67
Field Rd. E7	CG35	34
Field Rd. E17	CE31	25
Field Rd. N17	BZ31	24
Field Rd. NW10	BQ38	31
Field Rd. W6	BR42	40
Field Rd. Felt.	BC46	47
Field Rd. Wat.	BE25	4
Fieldsend Rd. Sutt.	BR56	76
Fields Est. E8	CB36	33
Fieldside Rd. Brom.	CF49	61
Field St. WC1	BX38	33
Fieldview SW18	BT47	59
Field Way, Grnf.	BF37	29
Field Way NW10	BN36	31
Fieldway, Croy.	CE57	79
Field Way, Dag.	CO35	36
Fieldway, Orp.	CM53	71
Fieldway Cres. N5	BY35	33
Fife Ct. W3	BM39	39
Links Rd.		
Fife Rd. E16	CH39	44
Fife Rd. N22	BY29	15
Fife Rd. SW14	BN46	49
Fife Rd. Kings. On T.	BL51	66
Fife Ter. N1	BX37	33
Wynford Rd.		
Fife Way, Brom.	CH51	71
White Hart Slip.		
Fifield Path SE23	CC48	61
Bampton Rd.		
Fifth Av. E12	CK35	35
Fifth Av. W10	BR38	31

Fifth Av. Enf.	CA25	8
Fifth Cross Rd. Twick.	BG48	56
Fifth Way, Wem.	BM35	30
Figg's Rd. Mitch.	BV50	59
Filby Rd. Chess.	BL57	75
Filey Av. N16	CD33	25
Filey Clo. Sutt.	BT57	77
Filey Waye, Ruis.	BC34	20
Fillebrook Av. Enf.	CA23	8
Fillebrook Rd. E11	CF33	25
Filmer Rd. SW6	BR44	49
Filston Rd. Belv.	CR42	45
Riverdale Rd.		
Filston Rd. Erith	CR42	45
Riverdale Rd.		
Filston Rd. Erith	CS42	46
Holly Hill Rd.		
Finborough Rd. SW10	BS42	41
Finborough Rd. SW17	BU50	59
Finchale Rd. SE2	CO41	45
Finch Av. SE27	BZ49	60
Finch Clo. NW10	BN36	31
Finchdean Way SE15	CA43	51
Finch Dr. Felt.	BD47	56
Finchingfield Av. Wdf. Grn.	CJ29	17
Finch La. EC3	BZ39	42
Finch La. Bush.	BE24	4
Finchley Ct. N3	BS29	14
Finchley La. NW4	BQ31	22
Finchley Pk. N12	BT28	14
Finchley Pl. NW8	BT37	32
Finchley Rd. NW2	BR31	22
Finchley Rd. NW3	BR31	22
Finchley Rd. NW8	BR31	22
Finchley Rd. NW8	BT37	32
Finchley Rd. NW11	BR31	22
Finchley Way N3	BS29	14
Finck St. SE1	BX41	42
Finden Rd. E7	CH35	35
Findhorne Av. Hayes	BC39	38
Findhorn St. E14	CF39	43
Aberfeldy St.		
Findon Clo. Har.	BF34	20
Findon Clo. SW18	BS46	50
Findon Gdns. Rain.	CU39	46
Findon Rd. N9	CB26	8
Findon Rd. W12	BP41	40
Fingal St. SE10	CG42	43
Finistock Rd. W10	BQ39	40
Finland Rd. SE4	CD45	52
Finlays Clo. Chess.	BM56	75
Finlay St. SW6	BQ44	49
Finnis St. E2	CB38	33
Finnymore Rd. Dag.	CQ36	36
Finsbury Av. EC2	BZ39	42
Finsbury Circus EC2	BZ39	42
Finsbury Cotts. N22	BX29	15
Finsbury Mkt. EC2	CA38	33
Finsbury Pk. Av. N4	BZ33	24
Finsbury Pk. Rd. N4	BY34	24
Finsbury Pavement EC2	BZ39	42
Finsbury Rd. N22	BX29	15
Finsbury Sq. EC2	BZ38	33
Finsbury St. EC2	BZ39	42
Finsen Rd. SE5	BZ45	51
Finstock Rd. W10	BQ40	40
Finucane Dr. Orp.	CP54	72
Finucane Gdns. Rain.	CU36	37
Finucane Rise, Bush.	BG27	11
Firbank Dr. Wat.	BE26	4
Firbank Rd. SE15	CB44	51
Firbank Rd. Rom.	CR28	18
Fir Clo. Walt.	BC54	65
Fircroft Av. Chess.	BL56	75
Fircroft Gdns. Har.	BH34	21
Fircroft Rd. SW17	BU48	59
Firdene, Surb.	BN54	67
Fire Bell Alley, Surb.	BL53	66
Firefly Clo. Wall.	BX57	78
Fir Gro. N. Mal.	BO53	67
Firhill Rd. SE6	CE49	61
Firmin Rd. Dart.	CV46	55
Fir Rd. Felt.	BD49	56
Fir Rd. Sutt.	BR54	67
Firs Av. N10	BV31	23
Firs Av. SW14	BN45	49
Firsby Av. Croy.	CC54	70
Firsby Rd. N16	CA33	24
Firs Clo. Esher	BH57	75
Firs Clo. N10	BV31	23
Firs Av.		
Firs Clo. SE23	CD47	61
Firscroft N13	BZ27	15
Firs Dr. Loug.	CK23	10
Firs Dr. Houns.	BC44	47
Firs Dr. Loug.	CK23	10
Firs La. N13	BZ27	15
Firs La. N21	BZ27	15
Firs Pk. Av. N21	BZ26	8
Firs Pk. Gdns. N21	BZ26	8
Firs Rd. Ken.	BY61	84
First Av. E12	CK35	35
First Av. E13	CH38	35
First Av. E17	CE32	25
First Av. N18	CC28	16
First Av. NW4	BQ31	22
First Av. SW14	BO45	49
First Av. W3	BO40	40
First Av. W10	BR38	31
First Av. Dag.	CR37	36
First Av. E. Mol.	BE52	65
First Av. Enf.	CA25	8
First Av. Epsom	BO58	76
First Av. Hayes	BC40	38
Glebe Rd.		
First Av. Rom.	CP32	27
First Av. Walt.	BC53	65
First Av. Well.	CP43	54
First Av. Wem.	BK34	21
First Clo. E. Mol.	BG52	65
First Cross Rd. Twick.	BH48	57
Firs, The, Borwd.	BL24	5
Firs, The DA5	CS47	64
Dartford Road		
Firs, The N20	BT26	7
Athenaeum Rd.		
Firs, The SW20	BP50	58
Firs, The W5	BK39	39
Firs, The, Bex.	CS47	64
Dartford Rd.		
Firs, The, Bex.	CS48	64
Dartford Rd.		
First St. SW3	BU42	41
First Way SW20	BQ51	67
First Way, Wem.	BM35	30
Firs Wk. Wdf. Grn.	CH28	17
Firswood Av. Epsom	BO56	76
Firth Gdns. SW6	BR44	49
Firtree Av. Mitch.	BV51	68
Fir Tree Clo. Epsom	BO56	76
Fir Tree Clo. Epsom	BQ61	82
Fir Tree Clo. Esher	BG56	74
Firtree Clo. Orp.	CN56	81
Fir Tree Clo. Rom.	CT31	28
Fir Tree Ct. Borwd.	BL24	5
Fir Tree Gdns. Croy.	CE56	79
Firtree Gro. Cars.	BU57	77
Fir Tree Rd. Epsom	BP61	82
Fir Tree Rd. Houns.	BE45	47
Firtree Wk. Dag.	CS34	28
Fir Tree Wk. Enf.	BZ24	8
Fir Wk. Sutt.	BQ57	76
Fisher Clo. Grnf.	BF38	29
Fisherman Clo. Rich.	BK49	57
Fisher Rd. Har.	BH30	12
Fishers Ct. SE14	CC44	52
Fisher's La. W4	BN42	40
Fisher St. E16	CH39	44
Fisher St. WC1	BX39	42
Fishers Way, Belv.	CS40	46
Fisherton St. Est. NW8	BT38	32
Fisherton St. NW8	BT38	32
Fishmongers Hall St. EC4	BZ40	42
Swan Wf.		
Fishponds Rd. SW17	BU49	59
Fishponds Rd. Kes.	CJ56	80
Fish St. Hill EC3	BZ40	42
Lwr. Thames St.		
Fitzalan Rd. N3	BR31	22
Fitzalan Rd. Esher	BH57	75
Fitzegerald Rd. Surb.	BJ53	66
Fitz George Av. W14	BR42	40
Fitzgeorge Av. N. Mal.	BN51	67
Fitzgerald Av. SW14	BO45	49
Fitzgerald Rd. E11	CH32	26
Fitzgerald Rd. SW14	BN45	49
Fitzhardinge St. W1	BV39	41
Fitzhugh Gro. SW18	BT46	50
Fitzilian Av. Rom.	CW30	19
Fitz James Av. W14	BR42	40
Fitzjames Av. Croy.	CB55	78
Fitzjohn Av. Barn.	BR25	6
Fitzjohn's Av. NW3	BT35	32
Fitzmaurice Pl. W1	BV40	41
Curzon St.		
Fitz Neal St. W12	BO39	40
Fitzroy Clo. Har.	BH30	12
Fitzroy Gdns. SE19	CA50	60
Fitzroy Ms. NW1	BW38	32
Cleveland St.		
Fitzroy Pk. N6	BU33	23
Fitzroy Rd. NW1	BU37	32
Fitzroy Sq. W1	BW38	32
Fitzroy St. W1	BW38	32
Fitzstephen Rd. Dag.	CO35	36
Fitzwarren Gdns. N19	BW33	23
Fitzwilliam Av. Rich.	BL44	48
Fitzwillian Rd. SW4	BW45	50
Fitz Wygram Clo. Hamptn.	BG49	56
Five Acre NW9	BO30	13
Five Elms Rd. Dag.	CQ34	27
Five Oaks La. Chig.	CQ29	18
Fivewents, Swan.	CU51	73
Fladbury Rd. N15	BZ32	24
Fladgate Rd. E11	CG32	25
Flag Wk. Pnr.	BC32	20
Flambard Rd. Har.	BJ32	21
Flamborough Rd. Ruis.	BC34	20
Flamborough St. E14	CD39	43
Flamingo Gdns. Nthlt.	BD38	29
Jetstar Way		
Flamingo Wk. Rain.	CU36	37
Flamstead Est. SE10	CG42	43
Flamstead Gdns. Dag.	CP36	36
Flamstead Rd. SE7	CK42	44
Flamstead Rd. Dag.	CP36	36
Flamsted Av. Wem.	BM36	30
Flanchford Rd. W12	BO41	40
Flanders Cres. SW17	BU50	59
Flanders Rd. E6	CK37	35
Flanders Rd. W4	BO42	40
Flanders Way E9	CC36	34
Flank St. E1	CB40	42
Dock St.		
Flask Wk. NW3	BT35	32
Flexley Rd. Mord.	BS54	68
Flaxman Rd. SE5	BY45	51
Flaxman Ter. WC1	BW38	32
Flaxton Rd. SE18	CM44	53
Flecker Clo. Stan.	BH28	12
Fleece Rd. Surb.	BK54	66
Fleece Wk. N19	BX36	33
Fleeming Clo. E17	CD30	16
Pennant Ter.		
Fleeming Rd. E17	CD30	16
Fleet Clo. E. Mol.	BF53	65
Fleet La. EC4	BY39	42
Fleet La. E. Mol.	BE53	65
Fleet Rd. NW3	BU35	32
Fleet Rd. Hill E1	CB38	33
Weaver St.		
Fleet Side E. Mol.	BE53	65
Fleet St. EC4	BY39	42
Fleetwood Clo. Chess.	BK57	75
Fleetwood Rd. NW10	BP35	31
Fleetwood Rd. N. Mal.	BM52	66
Fleetwood Sq. N. Mal.	BM52	66
Fleetwood St. N16	CA34	24
Fleetwood Way, Wat.	BD28	11
Fleming Ct. W2	BT39	41
St. Mary's Ter.		
Fleming Ct. Croy.	BY56	78
Fleming Mead, Mitch.	BU50	59
Fleming Rd. SE17	BY43	51
Fleming Rd. Sthl.	BF39	38
Flempton Rd. E10	CD33	25
Fletcher La. E10	CF33	25
Fletcher Path SE8	CE43	52
New Butt La.		
Fletcher Rd. W4	BN41	40
Fletcher Rd. Chig.	CN28	18
Fletching Rd. E5	CC34	25
Fletch St. E1	CB40	42
Cable St.		
Fletton Rd. N11	BX29	15
Fleur De Lis St. E1	CA38	33
Fleur Gates SW19	BQ47	58
Princes Way		
Flexmere Rd. N17	BZ30	15
Flimwell Clo. Brom.	CG49	61
Flintmill Cres. SE3	CK44	53
Flinton St. SE17	CA42	42
Flint St. SE17	BZ42	42
Flitcroft St. WC2	BW39	41
Flockton St. SE16	CB41	42
George Row		
Flodden Rd. SE5	BZ44	51
Flood St. SW3	BU42	41
Flood Wk. SW3	BU43	50
Flora Clo. E14	CE39	43
Flora Gdns. Rom.	CP32	27
Flora Gdns. Est. W6	BP42	40
Floral St. WC2	BX40	42
Flora St. Belv.	CQ42	45
Florence Av. Enf.	BZ24	8
Florence Av. Mord.	BT53	68
Florence Clo Horn.	CW34	28
Florence Ct. NW4	BP32	22
Vivian Av.		
Florence Dr. Enf.	BZ24	8
Florence Gdns. W4	BN43	49
Florence Rd. E6	CJ37	35
Florence Rd. E13	CG38	34
Florence Rd. N4	BX33	24
Florence Rd. SE2	CP42	45
Florence Rd. SE14	CD44	52
Florence Rd. SW19	BS50	59
Florence Rd. W4	BN41	40
Florence Rd. W5	BL40	39
Florence Rd. Beck.	CC51	70
Florence Rd. Brom.	CH51	71
Florence Rd. Felt.	BC47	56
Florence Rd. Kings. On T.	BL50	57
Florence Rd. Sth. Croy.	BZ58	78
Florence Rd. Sthl.	BD42	38
Florence Rd. Walt.	BC54	65
Florence St. E16	CG38	34
Florence St. N1	BY36	33
Florence St. NW4	BQ31	22
Florence Ter. SE14	CD44	52
Florian Av. Sutt.	BT56	77
Florian Rd. SW15	BR45	49
Florida Clo. Bush.	BG27	11
Florida Rd. Th. Hth.	BY51	69
Florida St. E2	CB38	33
Floriston Clo. Stan.	BJ30	12
Floriston Gdns. Stan.	BJ30	12
Florys Ct. SW19	BR47	58
Floss St. SW15	BQ44	49
Flower & Dean St. E1	CA39	42
Flower Ho. Clo. SE6	CF49	61
Flower La. NW7	BO28	13
Flower Ms. N19	BW34	23
St. Johns Way		
Flowers Mews SW17	BV48	59
Floyd Rd. SE7	CJ42	44
Fludyer St. SE13	CG45	52
Flyover, Croy.	BZ56	78
Folair Way SE16	CB42	42
Catlin St.		
Foley Rd. Esher	BH57	75
Foley St. W1	BW39	41
Folgate St. E1	CA39	42
Foliot St. W12	BO39	40
Folkestone Gdns. SE8	CD43	52
Trundley's Rd.		
Folkestone Rd. E6	CL37	35
Folkestone Rd. E17	CE31	25
Folkestone Rd. N18	CB28	15
Folkington Cor. N12	BR28	13
Folk La. NW9	BN30	13
Follett St. E14	CF39	43
Follyfield Rd. Bans.	BS60	83
Folly La. E17	CD30	16
Folly Ms. W11	BR39	40
Kensington Pk.		
Folly Wall E14	CF41	43
Fontaine Rd. SW16	BX50	60
Fontarabia Rd. SW11	BV45	50
Fontayne Av. Chig.	CM28	17
Fontayne Av. Rain.	CT36	37
Fontayne Av. Rom.	CT30	19
Fontenoy Rd. SW12	BV48	59
Fonteyne Gdns. Wdf. Grn.	CJ30	17
Fonthill Ms. N4	BX34	24
Lennox Rd.		
Fonthill Rd. N4	BX33	24
Fontley Way SW15	BP47	58
Fonts Hill N2	BT30	14
Fontwell Clo. Har.	BG29	11
Fontwell Clo. Nthlt.	BF36	29
Fontwell Dr. Brom.	CK53	71
Football La. Har.	BH33	21
Footbury Hill Rd. Orp.	CO54	72
Foots Cray High St. Sid.	CP50	63
Foots Cray La. Sid.	CP47	63
Foots Cray Rd. SE9	CL46	53
Forbes St. E1	CB39	42
Forburg Rd. N16	CB33	24
Ford Clo. Bush.	BG24	4
Ford Clo. Croy.	BY54	69
Bensham		
Ford Clo. Har.	BG33	20
Ford Clo. Rain.	CT36	37
Fordcroft Rd. Orp.	CO53	72
Forde Av. Brom.	CJ52	71
Fordel Rd. SE6	CF47	61
Ford End, Wdf. Grn.	CH29	17
Fordham Clo. Barn.	BT24	7
Fordham Rd. Barn.	BT24	7
Fordham St. E1	CB39	42
Fordhook Av. W5	BL40	39
Fordingley Rd. W9	BR38	31
Fordington Rd. N6	BU32	23
Ford La. Rain.	CT36	37
Fordmill Rd. SE6	CE48	61
Ford Rd. E3	CD37	34
Ford Rd. Dag.	CQ36	36
Ford's Gro. N21	BZ26	8
Ford's Pk. Rd. E16	CH39	44
Ford Sq. E1	CB39	42
Cavell St.		
Ford St. E3	CD37	34
Ford St. E16	CG39	43
Fordwich Clo. Orp.	CN54	72
Fordwych Cres. NW2	BR35	31
Fordwych Rd. NW2	BR35	31
Fordyce Rd. SE13	CF46	52
Fordyke Rd. Dag.	CQ34	27
Foreland St. SE18	CM42	44
Forel Ct. NW4	BR30	13
Foreman Ct. W6	BQ42	40
Foremark Clo. Chig.	CN28	18
Fore St. Av. EC2	BZ39	42
Foreshore SE8	CD42	43
Forest App. Wdf. Grn.	CG29	16
Forest Av. E4	CG26	9
Forest Av. Chig.	CL28	17
Forest Clo. E11	CG32	25
Forest Clo. Wdf. Grn.	CH28	17
Forest Ct. E4	CG26	9
Forest Ct. E11	CG31	25
Forestdale N14	BW28	14
Forest Dr. E12	CJ34	26
Forest Dr. Kes.	CK56	80
Forest Dr. Sun.	BC50	56
Forest Dr. Wdf. Grn.	CF29	16
Forest Dr. E. E11	CF33	25
Forest Dr. W. E11	CF33	25
Forest Edge, Buck. H.	CJ28	17
Forester Rd. SE15	CB45	51
Foresters Clo. Wall.	BW57	77
Foresters Cres. Bexh.	CR45	54
Foresters Dr. E17	CF31	25
Foresters Dr. Wall.	BW57	77
Forester St. E3	CD38	34
Forest Gdns. N17	CA30	15
Bruce Gro.		
Forest Gate NW9	BO32	22
Forest Glade E4	CG28	16

Forest Glade E11 CG32 25
Forest Gro. E8 CA36 33
Forest Rd.
Forest Hill Rd. SE22 CB46 51
Forest Hill Rd. SE23 CB46 51
Forestholme Clo. SE23 CC48 61
Taymount Rise
Forest La. E7 CG36 34
Forest La. E15 CG36 34
Forest La. Chig. CL28 17
Forest Mount Rd. Wdf. Grn. CF29 16
Fore St. EC2 BZ39 42
Fore St. N9 CA29 15
Fore St. N18 CA29 15
Forest Ridge, Beck. CE52 70
Forest Ridge, Kes. CK56 80
Forest Rise E17 CF31 25
Forest Rd. E7 CH35 35
Forest Rd. E8 CA36 33
Forest Rd. E11 CF33 25
Forest Rd. E17 CB31 24
Forest Rd. N9 CB26 8
Forest Rd. N17 CB31 24
Forest Rd. Erith CU44 55
Forest Rd. Felt. BD48 56
Forest Rd. Ilf. CM30 17
Forest Rd. Loug. CJ24 10
Forest Rd. Rich. BM43 48
Forest Rd. Rom. CR31 27
Forest Rd. Sutt. BS54 68
Forest Rd. Wdf. Grn. CH27 17
Forest Side E4 CG26 9
Forest Side E7 CH35 35
Capel Rd.
Forest Side, Buck. H. CJ26 10
Forest Side, Wor. Pk. BO54 67
Forest St. E7 CH35 35
Forest, The E11 CG31 25
Forest Vw. E4 CF25 9
Forest Vw. E11 CG33 25
Forest Vw. Av. E10 CF32 25
Forest Vw. Rd. E12 CK35 35
Forest Vw. Rd. E17 CF30 16
Forest Vw. Rd. Loug. CJ24 10
Forest Wk. Bush. BE23 4
Forest Way, Loug. CK24 10
Forest Way, Orp. CN53 72
Forest Way, Sid. CM47 62
Forest Way, Wdf. Grn. CH28 17
Forfar Rd. N22 BY30 15
Forfar Rd. SW11 BV44 50
Forge Clo. Brom. CH54 71
Forge Dr. Esher BJ57 75
Forge La. Felt. BE49 56
Forge La. Sun. BC52 65
Forge La. Sutt. BR57 76
Forlong Path, Nthlt. BE36 29
Ridgeway Wk.
Forman Pl. N16 CA35 33
Farleigh Rd.
Formby Av. Stan. BK31 21
Formosa St. W9 BS38 32
Formunt Clo. E16 CG39 43
Forres Ct. SE19 CA49 60
Forres Gdns. NW11 BS32 23
Forrester Path SE26 CC49 61
Queensthorpe Rd.
Forresters Cres. Bexh. CR45 54
Forrest Gdns. SW16 BX52 69
Forset St. W1 BU39 41
Forstall Clo. Brom. CG52 70
Forster Rd. E17 CD32 25
Forster Rd. N17 CA31 24
Forster Rd. SW2 BX47 60
Forster Rd. Beck. CD52 70
Forster Rd. Croy. BZ53 69
Forsters Clo. Rom. CQ32 27
Forston St. N1 BZ37 33
Forsyth Gdns. SE17 BY43 51
Forsyth Pl. Enf. CA25 8
Forter Clo. E18 CH30 17
Latchett Rd.
Forterie Gdns. Ilf. CO34 27
Fortescue Av. E8 CB36 33
Mentmore Ter.
Fortescue Av. Twick. BG48 56
Fortescue Rd. SW19 BT50 59
Fortescue Rd. Edg. BN30 13
Fortess Gro. NW5 BW35 32
Fortess Rd.
Fortess Rd. NW5 BV35 32
Fortess Wk. NW5 BV35 32
Fortess Rd.
Fortess Way NW5 BV35 32
Fortess Rd.
Forth Av. W10 BR39 40
Forthbridge Rd. SW11 BV45 50
Forth Way, Wem. BN35 31
Fortis Gn. N2 BU31 23
Fortis Gn. Av. N2 BU31 23
Fortis Gn. Rd. N10 BV31 23
Fortismere Av. N10 BV31 23
Fortnum Rd. N19 BW34 23
Fortnums Acre, Stan. BH29 12
Fort Pass. SE1 CA42 42
Fort Pass. SE18 CL42 44
Sandy Hill Rd.
Fort Rd. SE1 CA42 42
Fort Rd. Nthlt. BF36 29
Fortrose Gdns. SW2 BX47 60
New Park Rd.
Fortune Gate Rd. NW10 BO37 31
Fortune Gn. Rd. NW6 BS35 32
Fortune La. Borwd. BL25 5
Fortunes Mead, Nthlt. BE36 29
Arnold Rd.
Fortune St. EC1 BZ38 33
Forty Acre La. E16 CG39 43
Forty Av. Wem. BL34 21
Forty Clo. Wem. BL34 21
Forty La. Wem. BM34 21
Forum, The, Edg. BM29 12
Forval Way, Mitch. BU53 68
Fosbury Ms. W2 BS41 41
Inverness Terr.
Foscote Ms. W9 BS39 41
Amberley Rd.
Foscote Rd. NW4 BP32 22
Foskett Rd. SW6 BR44 49
Foss Av. Croy. BY56 78
Fossdene Rd. SE7 CH42 44
Fosse Way W13 BJ39 39
Fossil Rd. SE13 CE45 52
Fossington Rd. Belv. CP42 45
Fossway, Dag. CP34 27
Foster La. EC2 BZ39 42
Foster Rd. E13 CH38 35
Foster Rd. W3 BO40 40
Foster Rd. W4 BN42 40
Fosters Clo. Chis. CK49 62
Foster St. NW4 BQ31 22
Foster Wk. NW4 BQ31 22
Fothergill Clo. E13 CG37 34
Fotheringham Rd. Enf. CA24 8
Foubert's Pl. W1 BW39 41
Foulden Rd. N16 CA35 33
Foulis Ter. SW7 BT42 41
Foulser Rd. SW17 BU48 59
Foulsham Rd. Th. Hth. BZ52 69
Founders Gdns. SE19 BZ50 60
Hermitage Rd.
Fountain Ct. EC4 BY40 42
Fountain Dr. SE19 CA49 60
Fountain Rd. SW17 BT49 59
Fountain Rd. Mitch. BU51 68
Fountain Rd. Th. Hth. BZ51 69
Fountains Av. Felt. BE48 56
Fountains Clo. Felt. BE48 56
Fountains Cres. N14 BX26 8
Fountain St. E2 CA38 33
Columbia Rd.
Fountayne Rd. N15 CB31 24
Fountayne Rd. N16 CB34 24
Fount St. SW8 BW43 50
Fouracres, Enf. CD23 9
Fouracres, Kings. On T. BN50 58
Fourland Wk. Edg. BN29 13
Fournier St. E1 CA39 42
Four Seasons Cres. Sutt. BR55 76
Fourth Av. E12 CK35 35
Fourth Av. W10 BR38 31
Fourth Av. Enf. CA25 8
Fourth Av. Rom. CS33 28
Fourth Av. Wem. BM35 30
Fourth Cross Rd. Twick. BG48 56
Fourth Dr. Couls. BW57 77
Fourth Dr. Couls. BW61 83
Four Tubs, Bush. BG26 4
Fowell St. W11 BQ40 40
Fowney Av. Ilf. CJ32 26
Fowler Clo. SW11 BT45 50
Plough Rd.
Fowler Clo. Sid. CQ49 63
Fowler Rd. E7 CH35 35
Fowler Rd. Ilf. CO28 18
Fowler Rd. Mitch. BV51 68
Fowler's Wk. W5 BK38 30
Fownes St. SW11 BU45 50
Foxberry Rd. SE4 CD45 52
Foxborough Gdns. SE4 CE46 52
Foxbourne Rd. SW17 BV48 59
Fox Burrow Rd. Chig. CP28 18
Foxbury Av. Chis. CM50 62
Foxbury Clo. Brom. CH50 62
Foxbury Clo. Orp. CO56 81
Foxbury Dr. Orp. CO57 81
Foxbury Rd. Brom. CH50 62
Fox Clo. E1 CC38 34
Colebert Av.
Fox Clo. E16 CH39 44
Fox Clo. Orp. CO56 81
Fox Clo. Rom. CR28 18
Foxcombe, Croy. CE57 79
Foxcombe Rd. SW15 BP47 58
Alton Rd.
Fox Ct. EC1 BY39 42
Brooke St.
Foxcroft Rd. SE18 CL44 53
Foxearth Rd. Sth. Croy. CB58 78
Foxearth Spur, Sth. Croy. CC58 79
Foxes Dale SE3 CG45 52
Foxes Vale, Brom. CF52 70
Foxfield Rd. Orp. CM55 80
Foxglove St. W12 BO40 40
Foxgrove N14 BX27 15
Foxgrove Av. Beck. CE50 61
Foxgrove Path, Wat. BD28 11
Foxgrove Rd. Beck. CE50 61
Foxham Rd. N19 BW34 23
Fox Hill SE19 CA50 60
Fox Hill, Kes. CJ56 80
Fox Hl. Gdns. SE19 CA50 60
Foxhole Rd. SE9 CK46 53
Foxholt Gdns. NW10 BN36 31
Fox Ho. Rd. Belv. CR42 45
Foxlands Cres. Dag. CS35 37
Foxlands Rd. Dag. CS35 37
Fox La. N13 BX27 15
Fox La. W5 BL38 30
Fox La. Kes. CH56 80
Foxley Clo. Loug. CL24 10
Foxley Gdns. Pur. BY60 84
Foxley Hill Rd. Pur. BY59 84
Foxle La. Pur. BW59 83
Foxley Rd. SW9 BY43 51
Foxley Rd. Ken. BY60 84
Foxley Rd. Th. Hth. BY52 69
Foxleys Wat. BE27 11
Foxmore St. SW11 BU44 50
Fox Rd. E16 CG39 43
Fox's Path Mitch. BT51 68
Foxwarren Esher BH58 75
Foxwell St. SE4 CD45 52
Foxwood Clo. Felt. BC48 56
Foxwood Rd. SE3 CG45 52
Foyle Rd. N17 CB30 15
Foyle Rd. SE3 CG43 52
Framfield Clo. N12 BS27 14
Framfield Rd. N5 BY35 33
Framfield Rd. W7 BH39 39
Framfield Rd. Mitch. BV50 59
Framlingham Clo. E5 CB34 24
Southwold Rd.
Framlingham Cres. SE9 CK49 62
Frampton Clo. Sutt. BS57 77
Frampton Pk. Est. E9 CC36 34
Frampton Pk. Rd. E9 CC36 34
Frampton Rd. Houns. BE45 47
Frampton St. NW8 BT38 32
Francemary Rd. SE4 CE46 52
Frances Rd. E4 CE29 16
Frances St. SE18 CK42 44
Franche Ct. Rd. SW17 BT48 59
Francis Av. Bexh. CR44 54
Francis AV. Felt. BC48 56
Francis Av. Har. BJ32 21
Francis Av. Ilf. CM34 26
Franciscan Rd. SW17 BU49 59
Francis Chichester Way SW11 BV44 50
Francis Clo. Epsom BN56 76
Francis Gro. SW19 BR50 58
Francis Rd. E10 CF33 25
Francis Rd. N2 BU31 23
Lynmouth Rd.
Francis Rd. Croy. BY53 69
Francis Rd. Dart. CV46 55
Francis Rd. Grnf. BJ37 30
Francis Rd. Har. BJ32 21
Francis Rd. Houns. BD44 47
Francis Rd. Ilf. CM34 26
Francis Rd. Orp. CP52 72
Francis Rd. Pnr. BD32 20
Francis Rd. Wall. BW57 77
Francis Rd. Wat. BC24 4
Francis St. E15 CG36 34
Francis St. SW1 BW42 41
Francis St. Ilf. CM34 26
Francis Ter. N19 BW34 23
Francklyn Gdns. Edg. BM27 12
Francombe Gdns. Rom. CV32 28
Franconia Rd. SW4 BW46 50
Frank Bailey Wk. E12 CK35 35
Frank Dixon Clo. SE21 CA47 60
Frank Dixon Way SE21 CA47 60
Frankfurt Rd. SE24 BZ46 51
Frankham St. SE8 CE43 52
Frankland Clo. Wdf. Grn. CJ28 17
Frankland Rd. E4 CE28 16
Franklin Clo. N20 BT26 7
Franklin Clo. Kings. On T. BM52 66
Willingham Way
Franklin Cres. Mitch. BW52 68
Franklin Pass. SE9 CK45 53
Franklin Rd. NW10 BO36 31
Franklin Rd. SE20 CC50 61
Franklin Rd. Bexh. CQ44 54
Franklin Rd. Wat. BC23 4
Franklin Sq. SW5 BR42 40
Marchbank Rd.
Franklin's Row SW3 BU42 41
Franklin St. E3 CE38 34
Bromley High St.
Franklin St. N15 CA32 24
Franklyn Clo. Dag. CS36 37
Franklyn Gdns. Ilf. CM29 17
Franklyn Rd. NW10 BO36 31
Franklyn Rd. Walt. BC53 65
Franks Av. N. Mal. BN52 67
Frank St. E13 CH38 35
Franks Wd. Av. Orp. CL53 71
Frankton Rd. SE15 CB44 51
Franlaw Cres. N13 BZ28 15
Franmil Rd. Horn. CU33 28
Fransfield Gro. SE26 CB48 60
Frant Clo. SE20 CC50 61
Franthorne Way SE6 CE48 61
Randlesdown Rd.
Frant Rd. Th. Hth. BY53 69
Fraser Clo. Bex. CS47 64
Dartford Rd.
Fraser Clo. Bex. CS48 64
Dartford Rd.
Fraser Rd. E17 CE32 25
Fraser Rd. N9 CB27 15
Fraser Rd. Erith CS42 46
Fraser Rd. Grnf. BJ37 30
Fraser St. W4 BO42 40
Frating Cres. Wdf. Grn. CH29 17
Frazer Av. Ruis. BD35 29
Frazier St. SE1 BY41 42
Frean St. SE16 CB41 42
Frederica Rd. E4 CF26 9
Frederica St. N7 BX36 33
Caledonian Rd.
Frederick Clo. W2 BU39 41
Frederick Clo. Sutt. BR56 76
Frederick Cres. SW9 BY43 51
Frederick Cres. Enf. CC23 9
Frederick Pl. SE18 CL42 44
Frederick Rd. SE17 BY43 51
Frederick Rd. Rain. CS37 37
Frederick Rd. Sutt. BR56 76
Frederick's Pl. EC2 BZ39 42
Old Jewry
Frederick's Pl. N12 BT28 14
Frederick St. E17 CD32 25
Frederick Ter. E8 CA36 33
Haggerston Rd.
Freeborne Gdns. Rain. CU36 37
Freedom St. SW11 BU44 50
Freedown La. Sutt. BT60 83
Freegrove Rd. N7 BX35 33
Freeland Pk. NW4 BR30 13
Freeland Rd. W5 BL40 39
Freelands Av. Sth. Croy. CC58 79
Freelands Gro. Brom. CH51 71
Freelands Rd. Brom. CH51 71
Freeling St. N1 BX36 33
Freeman Rd. Mord. BT53 68
Freeman Way Horn. CW32 28
Freemantle Av. Enf. CC25 9
Freemantle Rd. Belv. CR42 45
Freemason's Rd. E16 CH39 44
Freemasons Rd. Croy. CA54 69
Freethorpe Clo. SE19 CA51 69
Freke Rd. SW11 BV45 50
Fremantle Rd. Ilf. CM30 17
Fremantle St. SE17 CA42 42
Fremont St. E9 CC37 34
French St. Sun. BD51 65
Frendsbury Rd. SE4 CC45 52
Frensham Dr. SW15 BO48 58
Frensham Dr. Croy. CF57 79
Frensham Rd. SE9 CM48 62
Frensham Rd. Ken. BY60 84
Frensham St. SE15 CB43 51
Frensham Way Epsom BQ61 82
Frere St. SW11 BU44 50
Freshfield Dr. N14 BV26 7
Freshfields Croy. CD54 70
Freshford St. SW18 BT48 59
Freshwater Clo. SW17 BV50 59
Freshwater Rd. SW17 BV50 59
Freshwater Rd. Dag. CP33 27
Freshwell Av. Rom. CP31 27
Fresh Wharf Est. Bark. CL37 35
Fresh Wharf Rd. Bark. CL37 35
Freshwood Clo. Beck. CE51 70
Freshwood Way Wall. BV57 77
Fresley Clo. N15 BZ31 24
Clinton Rd.
Freston Gdns. Barn. BU25 7
Freston Rd. W10 BQ40 40
Freta Rd. Bexh. CQ46 54
Frewin Rd. SW18 BT47 59
Friar Rd. Hayes BD38 29
Friar Rd. Orp. CO53 72
Friars Av. N20 BU27 14
Friars Clo. N2 BT31 23
Friars Clo. Nthlt. BD38 29
Broomcroft Av.
Friars Gdns. W3 BN40 40
St. Dunstan's Av.
Friars La. Rich. BK46 48
Friars Place La. W3 BN40 40
Friars Rd. E6 CJ37 35
Friars Stile Pl. Rich. BL46 48
Friars Stile Rd.
Friars Stile Rd. Rich. BL46 48
Friars, The, Chig. CN28 18
Friar St. EC4 BY39 42
Carter La.
Friars Wk. N14 BV26 7
Friars Wk. SE2 CP42 45
Friars Wk. Har. BH29 12
Friars Way, N14 BV26 7
Friars Way W3 BN39 40

Friars Way Bush.	BE23	4
Friars Wd. Croy.	CD58	79
Friary Clo. N12	BU28	14
Friary Est. SE15	CB43	51
Friary La. Wdf. Grn.	CH28	17
Friary Rd. N12	BT28	14
Friary Rd. SE15	CB43	51
Friary Rd. W3	BN39	40
Friary Way N12	BU28	14
Friday Hill E4	CG27	16
Friday Hill Belv.	CR42	45
Friday Hill West E4	CG27	16
Friday Rd. Erith	CS42	46
Friday Rd. Mitch.	BU50	59
Friday St. EC4	BZ39	42
Cannon St.		
Frideswide Pl. NW5	BW35	32
Islip St.		
Friendly St. Ms. SE8	CE44	52
Friendly St. SE8	CE44	52
Friendship Wk. Nthlt.	BD38	29
Wayfarer Rd.		
Friendshop Wk. Nthlt.	BD38	29
Javelin Way		
Friends Rd. Croy.	BZ55	78
Friends Rd. Pur.	BY59	84
Friend St. EC1	BY38	33
Friern Barnet La. N11	BT27	14
Friern Barnet La. N20	BT27	14
Friern Barnet Rd. N11	BU28	14
Friern Ct. N20	BT27	14
Friern Mt. Dr. N20	BT26	7
Friern Pk. N12	BT28	14
Friern Rd. SE22	CB47	60
Friern Watch Av. N12	BT28	14
Frigate Mews SE8	CE43	52
Watergate St.		
Frigo Ct. Epsom	BN59	82
Frimley Clo. SW19	BR48	58
Frimley Clo. Croy.	CF57	79
Frimley Ct. Sid.	CP49	63
Frimley Cres. Croy.	CF57	79
Frimley Gdns. Mitch.	BU52	68
Frimley Rd. Chess.	BL56	75
Frimley Rd. Ilf.	CN34	27
Frinstead Rd. Erith	CS43	55
Frinsted Clo. Orp.	CP52	72
Frinton Clo. Wat.	BC27	11
Frinton Dr. Wdf. Grn.	CF29	16
Frinton Mews Ilf.	CL32	26
Bramley Cres.		
Frinton Rd. E6	CJ38	35
Frinton Rd. N15	CA32	24
Frinton Rd. SW17	BV50	59
Frinton Rd. Rom.	CR29	18
Frinton Rd. Sid.	CQ48	63
Friston Path Chig.	CN28	18
Manford Way		
Friston St. SW6	BS44	50
Fritham Clo. N. Mal.	BO53	67
Frith Ct. NW7	BR29	13
Frith Gdns. SW6	BR45	49
Frith Knowle Walt.	BC57	74
Frith La. NW7	BR29	13
Frith Rd. E11	CF35	34
Frith Rd. Croy.	BZ55	78
Frith St. W1	BW39	41
Frithville Gdns. W12	BQ40	40
Frithwood Av. Nthwd.	BC29	11
Frizlands La. Dag.	CR34	27
Frobish Ct. SE23	CB48	60
Sydenham Rise		
Frobisher Clo. Pnr.	BD33	20
Frobisher Ct. W12	BP41	40
Frobisher Rd. N8	BY31	24
Frobisher St. SE10	CG43	52
Froghall La. Chig.	CM28	17
Frogley Rd. SE22	CA45	51
Frogmore SW18	BS46	50
Frogmore Clo. Sutt.	BQ55	76
Frogmore Gdns. Sutt.	BQ56	76
Frognal, NW3	BT35	32
Frognal Av. Sid.	CO50	63
Frognal Clo. NW3	BT35	32
Frognal Gdns. NW3	BT35	32
Frognal La. NW3	BS35	32
Frognall Av. Har.	BH31	21
Frognal Pl. Sid.	CO50	63
Frognal Rise NW3	BT34	23
Frognal Rd. NW3	BT36	32
Frognal Way NW3	BT35	32
Froissart Rd. SE9	CJ46	53
Frome St. N1	BZ37	33
Fromondes Rd. Sutt.	BR56	76
Frostic Pl. E1	CA39	42
Hopetown St.		
Frostic Wk. E1	CA39	42
Chicksand St.		
Froude St. SW8	BV44	50
Dickens St.		
Froude St.	BW44	50
Robertson St.		
Fryatt Rd. N17	BZ29	15
Fryatt St. E14	CG39	43
Fry Clo. Rom.	CR28	18
Fryent Clo. NW9	BM32	21
Fryent Cres. NW9	BO32	22
Fryent Fields NW9	BO32	22
Fryent Gro. NW9	BO32	22
Fryent Way NW9	BM32	21
Frying Pan Alley E1	CA39	42
Bell La.		
Fry Rd. E6	CJ36	35
Fry Rd. NW10	BO37	31
Fryston Av. Couls.	BV60	83
Fryston Av. Croy.	CB55	78
Fuchsia St. SE2	CO42	45
Fulbeck Dr. NW9	BO30	13
Fulbeck Way Har.	BG30	11
Fulbourne Est. E1	CB38	33
Fulbourne Rd. E17	CF30	16
Fulbrook Ms. N19	BW35	32
Tufnell Pk.		
Fulbrook Rd. N19	BW35	32
Fulford Gro. Wat.	BC27	11
Fulford Rd. Epsom	BN57	76
Fulford St. SE16	CB41	42
Fulham Broadway SW6	BS43	50
Fulham Ct. SW6	BS44	50
Fulham Est. SW6	BR43	49
Fulham High St. SW6	BR44	49
Fulham Palace Rd. SW6	BQ42	40
Fulham Palace Rd. W6	BQ42	40
Fulham Pk. Gdns. SW6	BR44	49
Fulham Pk. Rd. SW6	BR44	49
Fulham Rd. SW3	BR44	49
Fulham Rd. SW6	BR44	49
Fulham Rd. SW8	BR44	49
Fulham Rd. SW10	BR44	49
Fullbrooks Av. Wor. Pk.	BO54	67
Fuller Clo. Orp.	CN56	81
Fuller Rd. Dag.	CO34	27
Fullers Av. Surb.	BL55	75
Fullers Clo. Rom.	CS29	19
Fullers La. Rom.	CS29	19
Fuller's Rd. E18	CG30	16
Fullers Rd. Wdf. Grn.	CG30	16
Fuller St. E2	CB38	33
Fuller St. NW4	BQ31	22
Fullers Way N. Surb.	BL55	75
Fullers Way S. Chess.	BL56	75
Fullers Wd. Croy.	CE56	79
Fullerton Rd. SW18	BT46	50
Fullerton Rd. Cars.	BU58	77
Fullerton Rd. Croy.	CA54	69
Fullwell Av. Ilf.	CK30	17
Fulmar Rd. Rain.	CU36	37
Fulmead St. SW6	BS44	50
Fulmer Way W13	BJ41	39
Fulready Rd. E10	CF32	25
Fulstone Clo. Houns.	BE45	47
Fulthorp Rd. SE3	CG44	52
Fulton Ms. W2	BT40	41
Porchester Ter.		
Fulton Rd. Wem.	BM34	21
Fulton St. E16	CG39	43
George St.		
Fulwell Pk. Av. Twick.	BF48	56
Fulwell Rd. Tedd.	BG49	56
Fulwich Rd. Dart.	CW46	55
Fulwood Av. Wem.	BL37	30
Fulwood Gdns. Twick.	BH46	48
Fulwood Pl. WC1	BX39	42
Fulwood Wk. SW19	BR47	58
Furber St. W6	BP41	40
Furham Flds. Pnr.	BF29	11
Furley Pl. N1	BY36	33
Islington Pk. St.		
Furley Rd. SE15	CB44	51
Furlong Clo. Mitch.	BV54	68
Furlong Rd. N7	BY36	33
Furmage St. SW18	BS47	59
Furminger Rd. Orp.	CR56	81
Furneaux Av. SE27	BY49	60
Furner Clo. DA155	CT45	55
Furness Rd. NW10	BP37	31
Furness Rd. SW6	BS44	50
Furness Rd. Har.	BF33	20
Furness Rd. Mord.	BS54	68
Furness Way, Horn.	CU35	37
Furnival St. EC4	BY39	42
Furrow La. E9	CC35	34
Furrows, The, Walt.	BD55	74
Fursby Av. N3	BS29	14
Further Acre NW9	BO30	13
Further Grn. Rd. SE6	CG47	61
Furzedown Dr. La. SW17	BV49	59
Furzedown Rd. SW17	BV49	59
Furzedown Rd. Sutt.	BT59	83
Furze Farm Clo. Rom.	CQ30	18
Furzefield Clo. Chis.	CL50	62
Furzefield Rd. SE3	CH43	53
Furze Hill Pur.	BX59	84
Furzehill Rd. Borwd.	BM24	5
Furze La. Pur.	BX59	84
Furze Rd. Th. Hth.	BZ52	69
Furze St. E3	CE39	43
Furze Wd. Sun.	BC51	65
Fyfield Rd. E17	CF31	25
Fyfield Rd. SW9	BY45	51
Fyfield Rd. Enf.	CA24	8
Fyfield Rd. Rain.	CT37	37
Fyfield Rd. Wdf. Grn.	CJ29	17
Fyness St. SW1	BW42	41
Regency St.		
Gable Clo. Dart.	CU46	55
Gable Clo. Pnr.	BF29	11
Gable Ct. SE26	CB49	60
Gables Av. Borwd.	BL24	5
Gables, The, Bans.	BR62	82
Gabriel Clo. Felt.	BE49	56
Gabriel Clo. Rom.	CS29	19
Gabriel St. SE23	CC47	61
Gaddesden Av. Wem.	BL36	30
Gadesden Rd. Epsom	BN57	76
Gadwall Way SE28	CM41	44
Gage Rd. E16	CG39	43
Gage St. WC1	BX39	42
Boswell St.		
Gainford St. N1	BY37	33
Gainsboro' Av. Dart.	CV46	55
Gainsboro' Ct. Walt.	BC55	74
Gainsboro' Rd. Rain.	CU37	37
Gainsborough Av. E12	CL35	35
Gainsborough Clo. KT7	BH54	66
Gainsborough Gdns. NW3	BT34	23
Gainsborough Gdns. NW11	BR33	22
Bainsborough Gdns. Edg.	BL30	12
Gainsborough Gdns. Grnf.	BH35	30
Gainsborough Gdns. Islw.	BG46	47
Gainsborough Mews SE26	CB48	60
Panmure Road		
Gainsborough Rd. E11	CG33	25
Gainsborough Rd. E15	CG38	34
Gainsborough Rd. N12	BS28	14
Gainsborough Rd. W4	BO42	40
Gainsborough Rd. Dag.	CO35	36
Gainsborough Rd. Epsom	BN58	76
Gainsborough Rd. N. Mal.	BN53	67
Gainsborough Rd. Rich.	BL44	48
Gainsborough Rd. Wdf. Grn.	CK29	17
Gainsford Rd. E17	CD31	25
Gainsford St. SE1	CA41	42
Gairloch Rd. SE5	CA44	51
Gaisford St. NW5	BW36	32
Gaitskell Rd. SE9	CM47	62
Galahad Rd. Brom.	CH48	62
Galata Rd. SW13	BP43	49
Galatia Sq. SE15	CB45	51
Scylla Rd.		
Galbraith St. E14	CF41	43
Galeboroguh Av. Wdf. Grn.	CF29	16
Gale Clo. Hamptn.	BE50	56
Gale Clo. Mitch.	BT52	68
Gale Cres. Bans.	BS62	83
Galena Rd. W6	BP42	40
Galen Pl. WC1	BX39	42
Bury Pl.		
Galesbury Rd. SW18	BT46	50
Gale's Gdns. E2	CB38	33
Bethnal Green Rd.		
Gale St. E3	CE39	43
Gale St. Dag.	CP37	36
Galesway Wdf. Grn.	CK29	17
Gallants Farm Rd. Barn.	BU26	7
Gallery Gdns. Nthlt.	BD37	29
Gallery Rd. SE21	BZ47	60
Galley La. Barn.	BP23	6
Galleywall Rd. SE16	CB42	42
Galleywood Cres. Rom.	CS29	19
Galliard Ct. N9	CC25	9
Galliard Rd. N9	CB26	8
Gallia Rd. N5	BY35	33
Calabria Rd.		
Gallions Clo. Bark.	CO38	36
Gallions Rd. E16	CL40	44
Gallions Rd. SE7	CH42	44
Gallop, The, Sth. Croy.	CB57	78
Gallop, The, Sutt.	BT58	77
Gallosson Rd. SE18	CN42	45
Galloway Rd. W12	BP40	40
Gallus Sq. SE3	CH45	53
Galpin's Rd. Th. Hth.	BX53	69
Galsworth Av. Rom.	CO33	27
Galsworthy Clo. SE28	CQ40	45
Galsworthy Cres. SE3	CJ44	53
Merriman Rd.		
Galsworthy Rd. NW2	BR35	31
Galsworthy Rd. Kings. On T.	BM50	57
Galton St. W10	BR38	31
Galva Clo. Barn.	BV24	7
Galveston Rd. SW15	BR46	49
Gambetta St. SW8	BV44	50
Gambia St. SE1	BY40	42
Gambole Rd. SW17	BU49	59
Games Rd. Barn.	BU24	7
Gamlen Rd. SW15	BQ45	49
Gandergreen La. Sutt.	BR55	76
Gane Clo. Wall.	BX57	78
Ganton St. W1	BW40	41
Kingly St.		
Gantshill Cres. Ilf.	CL32	26
Gantshill Cross Ilf.	CL32	26
Gap Rd. SW19	BS49	59
Garage Rd. NW4	BP32	22
Garage Rd. W3	BM39	39
Garbrand Walk Epsom	BO58	76
Garbutt Pl. W1	BV39	41
Garden Av. Bexh.	CR45	54
Garden Av. Mitch.	BV50	59
Garden City Edg.	BM29	12
Garden Clo. E4	CE28	16
Garden Clo. SE12	CH48	62
Garden Clo. SW9	BX45	51
Garden Clo. SW15	BP47	58
Bristol Gdns.		
Garden Clo. Bans.	BS61	83
Garden Clo. Hamptn.	BE49	56
Garden Clo. Nthlt.	BE37	29
Garden Clo. Wall.	BX56	78
Garden Cotts. Epsom	BN59	82
Garden Cotts. Orp.	CP51	72
Garden Court SE15	CA44	51
Sumner Estate		
Garden Ct. Rich.	BL44	48
Gardener Gro. Felt.	BE48	56
Gardeners Rd. E3	CC37	34
Gardenia Rd. Enf.	CA25	8
Garden La. SW2	BX47	60
Garden La. Brom.	CH50	62
Garden Pl. Dart.	CV48	64
Garden Pl. Sid.	CP50	63
Garden Rd. NW8	BT38	32
Garden Rd. SE20	CC51	70
Garden Rd. Brom.	CH50	62
Garden Rd. Rich.	BM45	48
Garden Rd. Walt.	BC53	65
Garden Row SE1	BY41	42
Gardens, The, N8	BX31	24
Rectory Gdns.		
Gardens, The, N16	CA33	24
Gardens, The, SE22	CB45	51
Gardens, The, Beck.	CF51	70
Gardens, The, Esher	BF56	74
Gardens, The, Har.	BG32	20
Gardens, The, Pnr.	BE32	20
Garden Wk. EC2	CA38	33
Rivington St.		
Garden Wk. Beck.	CD51	70
Garden Way NW10	BN36	31
Gardiner Av. NW2	BQ35	31
Gardner Clo. E11	CH32	26
Gardner Rd. E13	CH38	35
Gardnor Rd. NW3	BT35	32
Flask Wk.		
Garendon Gdns. Mord.	BS54	68
Garendon Rd. Mord.	BS54	68
Gareth Clo. Wor. Pk.	BQ55	76
Burnham Dr.		
Gareth Gro. Brom.	CH49	62
Garfield Rd. E4	CF26	9
Garfield Rd. E13	CG38	34
Garfeld Rd. SW11	BV45	50
Garfield Rd. SW19	BT49	59
Garfield Rd. Twick.	BJ47	57
York St.		
Garfield Rd. Enf.	CC24	9
Garford St. E14	CE40	43
Garganey Wk. SE28	CP40	45
Garibaldi St. SE18	CN42	45
Garland Rd. SE18	CM43	53
Garland Rd. Stan.	BL30	12
Garland Way, Horn.	CW31	28
Garlichill Rd. Epsom	BP62	82
Garlick Hill EC4	BZ40	42
Queen Victoria St.		
Garlies Rd. SE23	CD48	61
Garlinge Rd. NW2	BR36	31
Garman Rd. N17	CB29	15
Garnault Ms. EC1	BY38	33
Hardwick St.		
Garnautt Pl. EC1	BY38	33
Myddelton St.		
Garner Rd. E17	CF30	16
Garner St. E3	CB37	33
Coate St.		
Garnet Rd. NW10	BO36	31
Garnet Rd. Th. Hth.	BZ52	69
Garnet St. E1	CC40	43
Garnett Clo. SE9	CK45	53
Garnett Rd. NW3	BU35	32
Garnett Way E17	CD30	16
Mcentee Ave.		
Garnham Clo. N16	CA34	24
Smalley Rd.		
Garnies Clo. SE15	CA43	51
Garrad's Rd. SW16	BW48	59
Garrard Clo. Bexh.	CR45	54
Garrard Clo. Chis.	CL49	62
Garrard Rd. Bans.	BS61	83
Garratt Clo. Croy.	BX56	78
Croydon Rd.		
Garratt La. SW17	BS46	50
Garratt La. SW18	BS46	50
Garratt Rd. Edg.	BM29	12
Garratts La. Bans.	BR61	82
Garratts Rd. Bush.	BG26	4
Garratt Ter. SW17	BU49	59
Garrett St. EC1	BZ38	33
Garrick Av. NW11	BR32	22
Garrick Clo. SW18	BT45	50
Garrick Clo. W5	BL38	30
Garrick Clo. Croy.	CA55	78
Garrick Clo. Rich.	BK46	48
Green, The		

Golborne Rd. W10 BR39 40
Golda Clo. Barn. BQ25 6
Goldbeaters Gro. Edg. BO29 13
Goldchiff Clo. Mord. BS54 68
Goldcrest Clo. SE28 CP40 45
Goldcrest Way, Bush. BG26 4
Golcrest Way, Croy. CF58 79
Golden Ct. Rich. BK46 48
George St.
Golden La. EC1 BZ38 33
Golden La. Est. EC1 BZ38 33
Golden Manor W7 BH40 39
Golden Sq. W1 BW40 41
Golders Clo. Edg. BM28 12
Golders Gdns. NW11 BR33 22
Golders Grn. Cres. NW11 BR33 22
Golders Grn. Rd. NW11 BR32 22
Golders Manor Dr. NW4 BQ32 22
Golders Pk. Clo. NW11 BS33 23
Golders Ri. NW4 BQ32 22
Golders Way NW11 BR33 22
Goldfinch Rd. SE28 CM41 44
Goldfinch Rd. Sth. Croy. CC58 79
Goldhawk Rd. W6 BO42 40
Goldhawk Rd. W12 BO42 40
Goldhawk Rd. W12 BP41 40
Gold Hill Edg. BN29 13
Goldhurst Ter. NW6 BS36 32
Goldingham Av. Loug. CM23 10
Goldings Hill Loug. CK23 10
Goldings Ri. Loug. CL23 10
Goldings Rd. Loug. CL23 10
Golding St. E1 CB39 42
Goldington Cres. NW1 BW37 32
Goldington St. NW1 BW37 32
Gold La. Edg. BN29 13
Goldman Clo. E2 CB38 33
Goldney Rd. W9 BS38 32
Goldsboro' Rd. SW8 BW44 50
Goldsborough Ces. E4 CE27 16
Goldsdown Clo. Enf. CD23 9
Goldsdown Rd. Enf. CC23 9
Goldsell Rd. Swan. CS53 73
Goldsmid St. SE18 CN42 45
Sladedale Rd.
Goldsmith Av. E12 CK36 35
Goldsmith Av. NW9 BO32 22
Golsmith Av. Rom. CR33 27
Goldsmith Clo. W3 BN40 40
East Acton La.
Goldsmith Clo. Har. BF33 20
Goldsmith La. NW9 BM31 21
Goldsmith Rd. E10 CE33 25
Goldsmth Rd. E17 CC30 16
Goldsmith Rd. N11 BU28 14
Goldsmith Rd. SE15 CB44 51
Goldsmith Rd. W3 BN40 40
Goldsmiths Av. W3 BN40 40
Goldsmith's Row E2 CB37 33
Goldsmiths Sq. E2 CB37 33
Goldsmith St. EC2 BZ39 42
Gutter La.
Goldsworthy Gdns. SE16 CC42 43
Goldwell Rd. Th. Hth. BX52 69
Golf Clo. Bush. BD23 4
Golf Clo. Stan. BK29 12
Golf Course Dr. BN50 58
Kings. On T.
Golfe Rd. Ilf. CM34 26
Golford Pl. NW1 BU38 32
Lisson Gro.
Golf Rd. W5 BL39 39
Boileau Rd.
Golf Rd. Brom. CL52 71
Golf Rd. Ken. BZ62 84
Golf Side, Sutt. BR59 82
Golf Side Twick. BG48 56
Golfside Clo. N. Mal. BO51 67
Goliath Clo. Wall. BX57 78
Gollogly Terr. SE7 CJ42 44
Nadine St.
Gomer Gdns. Tedd. BJ50 57
Gomer Pl. Tedd. BJ49 57
Gomm Rd. SE16 CC41 43
Gomshall Av. Wall. BX56 78
Gomshall Gdns. Ken. BZ61 84
Gomshall Rd. Sutt. BQ58 76
Gonson Pl. SE8 CE43 52
Gonson St. SE8 CE43 52
Gonston Cl. SW19 BR48 58
Queensmere Rd.
Gonville Cres. Nthlt. BF36 29
Gonville Rd. Th. Hth. BX53 69
Gonville St. SW6 BR45 49
Putney Br. App.
Goodall Rd. E11 CF35 34
Gooden Ct. Har. BH34 21
Goodenough Rd. SW19 BR50 58
Goodge Pl. W1 BW39 41
Goodge St.
Goodge St. W1 BW39 41
Goodhall St. NW10 BO38 31
Goodhart Way W. Wick. CG54 70
Goodinge Clo. N7 BX35 33
North Rd.
Goodinge Rd. N7 BX36 33
Goodman Cres. SW2 BW48 59
Goodman Rd. E10 CF33 25
Goodmans Fields E1 CB39 42
Goodmans Stile Rd. E1 CA39 42
Alie St.
Goodmayes Av. Ilf. CO33 27
Goodmayes La. Ilf. CO35 36
Goodmayes Rd. Ilf. CO33 27
Goodmead Rd. Orp. CO54 72
Goodrich Rd. SE22 CA46 51
Goodson Rd. NW10 BO36 31
Goodway Gdns. E14 CF39 43
Goodwin Dr. Sid. CP48 63
Goodwin Gdns. Croy. BY57 78
Goodwin Rd. N9 CC26 9
Goodwin Rd. W12 BP41 40
Goodwin Rd. Croy. BY57 78
Goodwin's Ct. WC2 BX40 42
St. Martins La.
Goodwin St. N4 BY34 24
Fonthill Rd.
Goodwood Av. Horn. CW35 37
Goodwood Clo. Mord. BS52 68
Goodwood Dr. Nthlt. BF36 29
Goodwood Rd. SE14 CD43 52
Goodwyn Av. NW7 BO28 13
Goodwyn's Vale N10 BV30 14
Goodyear Pl. SE5 BZ43 51
Addington Sq.
Goodyers Gdns. NW4 BQ32 22
Brent Green
Goosander Way SE28 CM41 44
Gooseacre La. Har. BK32 21
Goosefield E. Mol. BF52 65
Gooseley La. E6 CL38 35
Goose Yd. EC1 BY37 33
St. John St.
Gooshays Dr. Rom. CW28 19
Gooshays Gdns. Rom. CW29 19
Gophir La. EC4 BZ40 42
Bush La.
Gopsall St. N1 BZ37 33
Gorden Gdns. Edg. BM30 12
Gordon Av. E4 CG29 16
Gordon Av. SW14 BO45 49
Gordon Av. Horn. CT34 28
Gordon Av. Stan. BH29 12
Gordon Av. Stan. BJ29 12
Gordon Av. Sth. Croy. BZ58 78
Gordon Av. Twick. BJ46 48
Gordonbrock Rd. SE4 CE46 52
Gordon Clo. E17 CE32 25
Lennox Rd.
Gordon Clo. N19 BW33 23
Highgate Hill
Gordon Ct. W12 BQ39 40
Gordon Cres. Croy. CA54 69
Gordon Cres. Hayes BC41 38
Gordondale Rd. SW19 BS48 59
Gordon Gro. SE5 BY44 51
Gordon Hill Enf. BZ23 8
Gordon House Rd. NW5 BV35 32
Gordon Pl. W8 BS41 41
Gordon Rd. E4 CG26 9
Gordon Rd. E11 CH32 26
Gordon Rd. E12 CL34 26
Gordon Rd. E15 CF35 34
Gordon Rd. E17 CD32 25
Gordon Rd. E18 CH30 17
Gordon Rd. N3 BR29 13
Gordon Rd. N9 CB27 15
Gordon Rd. N11 BW29 14
Gordon Rd. NW6 BS38 32
Gordon Rd. SE15 CB44 51
Gordon Rd. W4 BM43 48
Gordon Rd. W5 BK40 39
Gordon Rd. W13 BJ40 39
Gordon Rd. Bark. CN37 36
Gordon Rd. Beck. CC51 70
Gordon Rd. Beck. CD52 70
Gordon Rd. Belv. CS42 46
Gordon Rd. Cars. BU57 77
Gordon Rd. Dart. CV47 64
Gordon Rd. Enf. BZ23 8
Gordon Rd. Esher BH57 75
Gordon Rd. Har. BH31 21
Gordon Rd. Houns. BG45 47
Gordon Rd. Ilf. CM34 26
Gordon Rd. Kings. On T. BL51 66
Gordon Rd. Rich. BL44 48
Gordon Rd. Rom. CQ32 27
Gordon Rd. Sid. CN46 54
Gordon Rd. Sthl. BE42 38
Gordon Rd. Surb. BL54 66
Gordon Sq. WC1 BW38 32
Gordon St. E13 CH38 35
Gordon St. WC1 BW38 32
Gordon St. Twick. BJ46 48
Gordon Way Barn. BR24 6
Gore Ct. NW9 BM32 21
Gorefield Pl. NW6 BS37 32
Gore Rd. E9 CC37 34
Gore Rd. SW20 BQ51 67
Gorsebrook Rd. Dag. CO37 36
Gore St. SW7 BT41 41
Gorham Pl. W11 BR40 40
Mary Pl.
Goring Clo. Rom. CS30 19
Goring Gdns. Dag. CP35 36
Goring Rd. N11 BX29 15
Goring Rd. Dag. CS36 37
Goring Rd. N. Dag. CS36 37
Goring St. EC3 CA39 42
Houndsditch
Goring Way Grnf. BG37 29
Gorle Station St. W14 BR43 49
Gorleston Rd. N15 BZ32 24
Gorleston St. W14 BR42 40
Gorman Rd. SE18 CK42 44
Gorringe Pk. Av. Mitch. BU50 59
Gorse Rise SW17 BV49 59
Gorse Rd. Croy. CE56 79
Gorse Rd. Orp. CQ54 72
Gorse Way Rom. CT33 28
Gorst Rd. XW10 BN38 31
Gorst Rd. SW11 BU46 50
Gorsuch St. E2 CA38 33
Gorsberton Rd. SW12 BU47 59
Gosbury Hill, Chess. BL56 75
Gosfield Rd. Dag. CR33 27
Gosfield Rd. Dag. CR34 27
Gosfield Rd. Epsom BN59 82
Gosfield St. W1 BW39 41
Gosford Gdns. Ilf. CK32 26
Gosforth La. Wat. BC27 11
Goslett Yd. WC2 BW39 41
Charing Cross Rd.
Gosling Clo. Grnf. BF38 29
Gosling Way SW9 BY44 51
Gospatrick Rd. N17 BZ29 15
Gospel Oak Est. NW5 BU35 32
Gosport Dr. Horn. CV36 37
Gosport Rd. E17 CD32 25
Gosport Way SE15 CA43 51
Gossage Rd. SE18 CM42 44
Ancona Rd.
Gossett St. E2 CA38 33
Goss Hill, Swan. CV50 64
Gosshill Rd. Brom. CL51 71
Gosterwood St. SE8 CD43 52
Gostling Rd. Twick. BF47 56
Goston Gdns. Th. Hth. BY52 69
Goswell Rd. EC1 BY37 33
Gothic Clo. Dart. CW48 64
Gothic Rd. Twick. BG48 56
Goudhurst Rd. Brom. CG49 61
Gough Rd. E15 CG35 34
Gough Rd. Enf. CB23 8
Gough Sq. EC4 BY39 42
Gough St. WC1 BX38 33
Gould Ct. SE19 CA49 60
Gould Rd. Twick. BH47 57
Goulston St. E1 CA39 42
Goulton Rd. E5 CB35 33
Gourley St. N15 CA32 24
Gourock Rd. SE9 CL46 53
Govan St. E2 CB37 33
Whiston Rd.
Gowan Av. SW6 BR44 49
Gowan Rd. NW10 BP36 31
Gower Clo. E15 CG36 34
Gower Ms. WC1 BW39 41
Gower Pl. WC1 BW38 32
Gower Rd. E7 CH36 35
Gower Rd. Islw. BH43 48
Gower St. WC1 BW38 32
Gowers Wk. E1 CB39 42
Gowland Pl. Beck. CD51 70
Gowlett Rd. SE15 CB45 51
Gowrie Rd. SW11 BV45 50
Graburn Way, E. Mol. BG52 65
Grace Av. Bexh. CQ44 54
Gracechurch St. EC3 BZ40 42
Grace Clo. SE9 CS49 62
Gracedale Rd. SW16 BV49 59
Gracefield Gdns. SW16 BX48 60
Grace Path. SE26 CC49 61
Windfield Clo.
Grace Rd. Croy. BZ53 69
Grace's Alley E1 CB40 42
Ensign St.
Grace's Ms. SE5 BZ44 51
Grace's Rd. SE5 CA44 51
Grace St. E3 CE38 34
Gradient, The, SE26 CB49 60
Graeme Rd. Enf. BZ23 8
Graemesdyke Av. SW14 BM45 48
Grafton Clo. W13 BJ39 39
Grafton Clo. Wor. Pk. BO55 76
Grafton Cres. NW1 BV36 32
Grafton Gdns. N4 CB32 24
Rutland Gdns.
Grafton Gdns. Dag. CQ34 27
Grafton Ms. W1 BW38 32
Grafton Way
Grafton Park Rd. BO55 76
Wor. Pk.
Grafton Pl. NW1 BW38 32
Grafton Rd. NW5 BV35 32
Grafton Rd. W3 BN40 40
Grafton Rd. Croy. BY54 69
Grafton Rd. Dag. CQ34 27
Grafton Rd. Enf. BX24 8
Grafton Rd. Har. BG32 20
Grafton Rd. N. Mal. BO52 67
Grafton Rd. Wor. Pk. BN55 76
Grafton Sq. SW4 BW45 50
Grafton St. NW3 BS34 23
Hermitage La.
Grafton St. W1 BV40 41
Grafton Ter. NW2 BS34 23
Hermitage La.
Grafton Ter. NW5 BU35 32
Grafton Way NW1 BW38 32
Grafton Way WC1 BW38 32
Grafton Yd. NW5 BV36 32
Prince Of Wales Rd.
Graham Av. W13 BJ41 39
Graham Av. Mitch. BV51 68
Graham Clo. Croy. CE55 79
Grahame Park Est. NW9 BO30 13
Grahame Park Way NW9 BO30 13
Graham Gdns. Surb. BL54 66
Graham Rd. E8 CB36 33
Graham Rd. E13 CH38 35
Graham Rd. N15 BY31 24
Graham Rd. NW4 BP32 22
Graham Rd. SW19 BR50 58
Graham Rd. W4 BN41 40
Graham Rd. Bexh. CQ45 54
Graham Rd. Hamptn. BF49 56
Graham Rd. Har. BG31 20
Graham Rd. Mitch. BV51 68
Graham Rd. Pur. BY60 84
Graham St. N1 BY37 33
Graham Ter. SW1 BV42 41
Grainger Clo. Ntlt. BG35 29
Grainger Rd. N22 BZ30 15
Grainger Rd. Islw. BH44 48
Grampian Gdns. NW2 BR33 22
Grampians, The, W14 BQ41 40
Granada St. SW17 BU49 59
Granard Av. SW15 BP46 49
Granard Rd. SW12 BU47 59
Granby Rd. SE9 CK44 53
Granby's Bldgs. SE11 BX42 42
Salamanca St.
Granby St. E2 CA38 33
Granby Ter. NW1 BW37 32
Grand Av. EC1 BY39 42
Charterhouse St.
Grand Av. N10 BV31 23
Grand Av. Surb. BM53 66
Grand Av. Wem. BM35 30
Grand Av. E. Wem. BM35 30
Grand Central Wk. SE19 CB50 60
Grand Depot Rd. SE18 CL42 44
Grand Dr. SW20 BQ52 67
Granden Rd. SW16 BX51 69
Grandfield Av. Wat. BC23 4
Grandison Rd. SW11 BU45 50
Grandison Rd. Wor. Pk. BQ55 76
Grand Pde. Surb. BM54 66
Grand Pde. Wem. BM34 21
Grand Sq. SE10 CF42 43
Grand Stand Rd. Epsom BO62 82
Grand, The, Pde. NW3 BT35 32
Finchley Rd.
Grand Wk. E1 CD38 34
Solebay St.
Granfield St. SW11 BT44 50
Grange Av. N12 BT28 14
Grange Av. N20 BR26 6
Grange Av. SE25 CA51 69
Grange Av. Barn. BU26 7
Grange Av. Stan. BJ30 12
Grange Av. Twick. BH48 57
Grange Av. Wdf. Grn. CH29 17
Grangecliffe Gdns. SE25 CA51 69
Grange Clo. E. Mol. BF52 65
Grange Rd.
Grange Clo. Edg. BN28 13
Grange Clo. Houns. BE42 38
Grange Clo. N. Mal. BP53 67
Grange Clo. Sid. CO48 63
Grange Clo. Wdf. Grn. CH29 17
Grange Ct. WC2 BX39 42
Grange Ct. Chig. CM27 17
Grange Ct. Loug. CJ25 10
Grange Ct. Nthlt. BD37 29
Grange Ct. Walt. BC55 74
Grange Ct. Rd. N16 CA33 24
Grange Cres. Chig. CM28 17
Grange Dr. Chis. CK50 62
Grange Dr. Orp. CP58 81
Grange Est. N2 BT30 14
Grange Farm Clo. Har. BG34 20
Grange Gdns. N14 BW26 7
Grange Gdns. SE25 CA51 69
Grange Gdns. Bans. BS60 83
Grange Gdns. Pnr. BE31 20
Grange Gro. N1 BY36 33
Grange Hill SE25 CA51 69
Grange Hill Edg. BN28 13
Grangehill Pl. SE9 CK45 53
Grangehill Rd. SE9 CK45 53
Grange La. SE21 CA48 60
Grange Meadow Bans. BS60 83
Grangemill Rd. SE6 CE48 61
Grangemill Way SE6 CE48 61
Grange Pk. W5 BL40 39
Grange Pk. Av. N21 BY25 8
Grange Pk. Rd. E10 CE33 25
Grange Pk. Rd. Th. Hth. BZ52 69
Grange Rd. E10 CE33 25

Grange Rd. E13	CG38	34
Grange Rd. E17	CD32	25
Grange Rd. N6	BU32	23
Grange Rd. N17	CB29	15
Grange Rd. NW10	BP36	31
Grange Rd. SE1	CA41	42
Grange Rd. SW13	BP44	49
Grange Rd. W4	BM42	39
Grange Rd. W5	BK40	39
Grange Rd. Borwd.	BL25	5
Grange Rd. Bush.	BE25	4
Grange Rd. Chess.	BL55	75
Grange Rd. E. Mol.	BF53	65
Grange Rd. Edg.	BN29	13
Grange Rd. Har.	BG34	20
Grange Rd. Har.	BJ47	57
Grange Rd. Ilf.	CL35	35
Grange Rd. Kings. On T.	BL52	66
Grange Rd. Orp.	CM55	80
Grange Rd. Rom.	CU29	19
Grange Rd. Sth. Croy.	BZ58	78
Grange Rd. Sthl.	BE41	38
Grange Rd. Sutt.	BS57	77
Grange Rd. Th. Hth.	BZ52	69
Grange Rd. Walt.	BE56	74
Granger Way, Rom.	CU32	28
Grange St. N1	BZ37	33
Grange, The, N20	BT26	7
Grange, The, NW3	BS34	23
Grange, The, SE1	CA41	42
Grange, The, SW19	BQ49	58
Grange, The, Croy.	CD55	79
Grange, The, N. Mal.	BP53	67
Grange, The, Wem.	BM36	30
Grange, The, Wor. Pk.	BN55	76
Grange, The Dr. N21	BY25	8
Grange Vale, Sutt.	BS57	77
Grange View Rd. N20	BT26	7
Grange Wk. SE1	CA41	42
Grange Way N12	BS28	14
Grangeway NW6	BS36	32
Messina Av.		
Grange Way, Erith	CU43	55
Grangeway Wdf. Grn.	CJ28	17
Grangeway Gdns. Ilf.	CK32	26
Grangeway, The, N21	BY25	8
Grangewood Bex.	CQ47	63
Hurst Rd.		
Grangewood Av. Rain.	CV38	37
Grangewood Clo. Pnr.	BC32	20
Grangewood La. Beck.	CD50	61
Grangewood St. E6	CJ37	35
Grangewood Ter. SE25	BZ51	69
Grange Yd. SE1	CA41	42
Granham Gdns. N9	CA27	15
Granite St. SE18	CN42	45
Granleigh Rd. E11	CG34	25
Gransden Av. E8	CB36	33
London La.		
Gransden Rd. W12	BO41	40
Wendell Rd.		
Grantbridge St. N1	BY37	33
Grant Clo. N14	BW26	7
Grantham Clo. Edg.	BL27	12
Grantham Gdns. Rom.	CQ32	27
Grantham Grn. Borwd.	BN25	6
Grantham Pl. W1	BV40	41
Old Park La.		
Grantham Rd. E12	CL34	26
Grantham Rd. SW9	BX44	51
Grantham Rd. W4	BO43	49
Grantley Rd. Houns.	BC44	47
Grantley St. E1	CC38	34
Grantock Rd. E17	CF30	16
Granton Av. Upmin.	CW34	28
Granton Rd. SW16	BW51	68
Granton Rd. Ilf.	CO33	27
Granton Rd. Sid.	CP50	63
Grant Pl. Croy.	CA54	69
Grant Rd. SW11	BT45	50
Grant Rd. Croy.	CA54	69
Grant Rd. Har.	BH31	21
Grants Clo. NW7	BQ29	13
Grant St. E13	CH38	35
Grantully Rd. W9	BS38	32
Granville Av. Felt.	BC48	56
Granville Av. Houns.	BF46	47
Granville Clo. Croy.	CA55	78
Granville Gdns. SW16	BX51	69
Granville Gdns. W5	BL40	39
Granville Gro. SE13	CF45	52
Granville Ms. NW2	BR34	22
Granville Pk. SE13	CF45	52
Granville Pl. N12	BT29	14
High Rd. Nth. Finchley		
Granville Pl. W1	BV39	41
Granville Rd. E17	CE32	25
Granville Rd. E18	CH30	17
Granville Rd. N4	BX32	24
Granville Rd. N9	CC27	16
Granville Rd. N12	BS29	14
Granville Rd N13	BX29	15
Granville Rd. N22	BY30	15
Granville Rd. NW2	BR34	22
Granville Rd. NW6	BS37	32
Granville Rd. SW18	BR47	58
Granville Rd. SW19	BS50	59
Granville Rd. Barn.	BQ24	6
Granville Rd. Ilf.	CL33	26
Granville Rd. Sid.	CO49	63
Granville Rd. Wat.	BD24	4
Granville Rd. Well.	CP45	54
Granville Sq. WC1	BX38	33
Granville St. WC1	BX38	33
Wharton St.		
Grape St. WC2	BX39	42
High Holborn		
Grasdene Rd. SE18	CO43	54
Grasmere Av. SW15	BN49	58
Grasmere Av. SW19	BS51	68
Grasmere Av. W3	BN40	40
Grasmere Av. Houns.	BF46	47
Grasmere Av. Orp.	CL55	80
Grasmere Av. Wem.	BK33	21
Grasmere Clo. Loug.	CK23	10
Grasmere Ct. N22	BX29	15
Palmerston Rd.		
Grasmere Gdns. Har.	BJ30	12
Grasmere Gdns. Ilf.	CK32	26
Grasmere Gdns. Orp.	CL55	80
Grasmere Rd. E13	CH37	35
Grasmere Rd. N10	BV30	14
Grasmere Rd. N17	CB29	15
Grasmere Rd. SE25	CB53	69
Grasmere Rd. SW16	BX49	60
Grasmere Rd. Bexh.	CS44	55
Grasmere Rd. Brom.	CG51	70
Grasmere Rd. Horn.	CW31	28
Grasmere Rd. Orp.	CL55	80
Grasmere Rd. Pur.	BY59	84
Grassington Rd. Sid.	CO49	63
Grass Mt. SE23	CB48	60
Grassmount Pur.	BW58	77
Grass Pk. N3	BR30	13
Grassway Wall.	BW56	77
Grasvenor Av. Barn.	BS25	7
Gratley Way SE15	CA43	51
Hordle Prom. N.		
Gratton Rd W14	BR41	40
Gratton Ter. NW2	BQ34	22
Gratton Way W1	BW38	32
Gravel Clo. Chig.	CO27	18
Graveley Av. Borwd.	BN24	6
Gravel Hill N3	BR30	13
Broadway, The		
Gravel Hill Bexh.	CR46	54
Gravel Hill Loug.	CH23	10
Gravel Hll Sth. Croy.	CC57	79
Gravel Hill Clo. Bexh.	CR46	54
Gravel La. E1	CA39	42
Gravel La. Chig.	CO27	18
Gravelly Ride SW19	BP49	58
Gravel Pit La. SE9	CM46	53
Gravel Rd. Brom.	CK55	80
Gravel Rd. Twick.	BH47	57
Gravel Wood Clo. Chis.	CM48	62
Graveney Gro. SE20	CC50	61
Graveney Rd. SW17	BU49	59
Gravesend Rd. W12	BP40	40
Gray Av. Dag.	CQ33	27
Graydon St. SE18	CL43	53
Gray Gdns. Rain.	CU36	37
Grayham Cres. N. Mal.	BN52	67
Grayham Rd. N. Mal.	BN52	67
Grayling Rd. N16	BZ34	24
Grayling Sq. E2	CB38	33
Avebury Est.		
Grayscroft Rd. SW16	BW50	59
Grays Farm Rd. Orp.	CO51	72
Gray's Inn Rd. WC1	BX38	33
Gray's Inn Sq. WC1	BY39	42
Gray St. SE1	BY41	42
Grayswood Gdns. SW20	BP51	67
Graywood Ct. N12	BT29	14
Grazebrook Rd. N16	BZ34	24
Grazeley Clo. Bexh.	CS46	55
Gt. Acre Ct. SW4	BW45	50
Clapham Pk. Rd.		
Gt. Brownings SE21	CA49	60
Gt. Bushey Dr. N20	BS26	7
Gt. Cambridge Rd. N9	BZ27	15
Gt. Cambridge Rd. N17	BZ30	15
Gt. Cambridge Rd N18	BZ28	15
Gt. Cambridge Rd. Enf.	CB24	8
Gt. Cambridge Rd. Wal. Cr.	CB25	8
Gt. Castle St. W1	BV39	41
Gt. Central Av. Ruis.	BD35	29
Gt. Central St. NW1	BU39	41
Melcombe Sq.		
Gt. Chapel St. W1	BW39	41
Gt. Chertsey Rd. W4	BN44	49
Gt. Chertsey Rd. Felt.	BE48	56
Gt. Church La. W6	BQ42	40
Gt. College St. SW1	BX41	42
Gt. Cross Av. SE10	CG43	52
Gt. Cullings Rom.	CT34	28
Gt. Cumberland Ms. W1	BU39	41
Seymour Pl.		
Gt. Cumberland Pl. W1	BU39	41
Gt. Dover St. SE1	BZ41	42
Greatdown Rd. W7	BH38	30
Gt. Eastern Rd. E15	CF36	34
Gt. Eastern St. EC2	CA38	33
Gt. Ellshams Bans.	BS61	83
Gt. Elms Rd. Brom.	CJ52	71
Great Field NW9	BO30	13
Greatfield Av. E6	CK38	35
Greatfield Clo. N19	BW34	23
Warrender Rd.		
Greatfield Clo. SE4	CE45	52
Greatfields Rd. Bark.	CM37	35
Great Field Strand NW9	BO30	13
Gt. Gardens Rd. Horn.	CU32	28
Gt. George St. SW1	BW41	41
Gt. Grove Bush.	BF24	4
Gt. Guildford St. SE1	BZ40	42
Greatham Rd. Bush.	BD24	4
Gt. Harry Dr. SE9	CL48	62
Gt. James St. WC1	BX39	42
Gt. Marlborough St. W1	BW39	41
Gt. Nelmes Chase Horn.	CW32	28
Gt. Newport St. WC2	BW40	41
Upr. St. Martin's La.		
Great North Rd. N20	BT26	7
Gt. North Rd. Barn.	BR24	6
Gt. North Way NW4	BP30	13
Great Oaks Chig.	CM28	17
Greatorex St. E1	CB39	42
Gt. Ormond St. WC1	BX39	42
Gt. Owl Rd. Chig.	CL27	17
Gt. Percy St. WC1	BX38	33
Gt. Peter St. SW1	BX41	42
Gt. Portland St. W1	BV38	32
Gt. Pulteney St. W1	BW40	41
Gt. Queen St. WC2	BX39	42
Gt. Queen St. Dart.	CW47	64
Gt. Russell St. WC1	BW39	41
Gt. St. Helen's EC3	CA39	42
St. Mary Axe		
Gt. St. Thomas Apostle EC4	BZ40	42
Queen St.		
Great Saplings SE21	CA46	51
Gt. Scotland Yd. SW1	BX40	42
Gt. Smith St. SW1	BW41	41
Gt. Spilmans SE22	CA46	51
Gt. Suffolk St. SE1	BY40	42
Gt. Sutton St. EC1	BY38	33
Gt. Swan Alley EC2	BZ39	42
Gt. Tattenams Epsom	BP62	82
Gt. Thrift Orp.	CM52	71
Gt. Tichfield St. W1	BV38	32
Gt. Tower St. EC3	CA40	42
Gt. Trinity La. EC4	BZ40	42
Queen Victoria St.		
Gt. Turnstile WC1	BX39	42
High Holborn		
Gt. Western Rd. W9	BR38	31
Gt. Western Rd. W11	BR38	31
Gt. West Rd. W4	BM42	39
Gt. West Rd. Brent.	BJ43	48
Gt. West Rd. Houns.	BE44	47
Gt. Winchester St. EC2	BZ39	42
Gt. Windmill St. W1	BW40	41
Greatwood Chis.	CL50	62
Gt. Woodcote Dr. Pur.	BW58	77
Gt. Woodcote Pk. Pur.	BW58	77
Greaves Pl. SW17	BU49	59
Grecian Cres. SE19	BY50	60
Greek Ct. W1	BW39	41
Old Compton St.		
Greek St. W1	BW39	41
Greenacre Dart.	CV48	64
Greenacres Bush.	BG27	11
Greenacres SE9	CL46	53
Green Acres, Croy.	CA55	78
Greenacres Clo. Rain.	CW38	37
Greenacres Dr. Stan.	BJ29	12
Greenacre Wk. N14	BX27	15
Cannon Hill		
Green Arbour Ct. EC4	BY39	42
Old Bailey		
Green Av. NW7	BN28	13
Green Av. W13	BJ41	39
Greenaway Gdns. NW3	BS35	32
Green Bank E1	CB40	42
Green Bank, N12	BS28	14
Greenbank Av. Wem.	BJ35	30
Green Bank Clo. Rom.	CV27	19
Greenbank Cres. NW4	BR31	22
Green Banks Walt.	BE54	65
Greenbay Rd. SE7	CJ43	53
Greenberry St. NW8	BU37	32
Greenbrook Av. Barn.	BT23	7
Green Clo. NW9	BN32	22
Green Clo. NW11	BT33	23
Green Clo. Brom.	CG52	70
Green Clo. Cars.	BU55	77
Green Clo. Felt.	BE49	56
Greencoat Pl. SW1	BW42	41
Greencoat Row SW1	BW41	41
Francis St.		
Green Ct. Edg.	BM29	12
Green Ct. Av. Croy.	CB55	78
Greencourt Av. Edg.	BM30	12
Green Ct. Gdns. Croy.	CB54	69
Greencourt Rd. Orp.	CM53	71
Green Ct. Rd. Swan.	CS53	73
Greencroft Av. Ruis.	BD34	20
Greencroft Gdns. NW6	BS36	32
Greencroft Gdns. Enf.	CA24	8
Greencroft Rd. Houns.	BE44	47
Green Curve Bans.	BR60	82
Green Dale NW7	BO28	13
Green Dale SE22	BZ45	51
Green Dale Clo. SE22	CA46	51
Green Dale		
Green Dragon La. N21	BX25	8
Green Dragon La. Brent.	BL42	39
Green Dragon Yd. E1	CB39	42
Old Montague St.		
Green Dr. Sthl.	BF40	38
Green End N21	BY27	15
Green End. Chess.	BK56	75
Greenend Rd. W4	BO41	40
Greenfarm Clo. Orp.	CN57	81
Greenfield Av. Surb.	BM54	66
Greenfield Av. Wat.	BE27	11
Greenfield Gdns. NW2	BR34	22
Greenfield Gdns. Dag.	CP37	36
Greenfield Gdns. Orp.	CM54	71
Greenfield Link Couls.	BX57	78
Greenfield Rd. E1	CB39	42
Greenfield Rd. N15	CA32	24
Greenfield Rd. Dag.	CP37	36
Greenfield Rd. Dart.	CS49	64
Greenfields Loug.	CL24	10
Greenfields Sthl.	BF40	38
Greenfields Clo. Loug.	CL24	10
Greenfield Way, Har.	BF31	20
Greenford Av. W7	BH38	30
Greenford Av. Sthl.	BE40	38
Greenford Gdns. Grnf.	BF38	29
Greenford Rd. Grnf.	BG37	29
Greenford Rd. Har.	BH35	30
Greenford Rd. Sthl.	BG40	38
Greenford Rd. Sutt.	BS56	77
Thorncroft Rd.		
Green Gdns. Orp.	CM56	80
Greengate Grnf.	BJ36	30
Greengate St. E13	CH37	35
Greenglades Horn.	CW32	28
Greenhalgh Wk. N2	BT31	23
Greenham Rd. N10	BV30	14
Greenhayes Av. Bans.	BS60	83
Greenhayes Gdns. Bans.	BS61	83
Greenheys Dr. E18	CG31	25
Green Hill SE18	CK42	44
Green Hill Buck. H.	CJ26	10
Greenhill Sutt.	BT55	77
Greenhill Wem.	BM34	21
Greenhill Cres. Har.	BH32	21
Greenhill Gdns. Nthlt.	BE37	29
Greenhill Gro. E12	CK35	35
Greenhill Pk. NW10	BO37	31
Greenhill Pk. Barn.	BS25	7
Greenhill Rd. NW10	BO37	31
Greenhill Rd. Har.	BH32	21
Greenhill's Rents. EC1	BY39	42
Cowcross St.		
Greenhills Ter. N1	BZ36	33
Baxter Rd.		
Green Hill Ter. SE18	CK42	44
Green Hill Ter. SE18	CK42	44
Greenhill Ter. Nthlt.	BE37	29
Greenhill Way, Wem.	BM34	21
Greenhithe Clo. Sid.	CN47	63
Greenholm Rd. SE9	CL46	53
Green Hundred Rd. SE15	CB43	51
Greenhurst Rd. SE27	BY49	60
Greening St. SE2	CP42	45
Greenland Cres. Sthl.	BD41	38
Greenland Rd. NW1	BT37	32
Greenland Rd. Barn.	BQ25	6
Green La. E4	CF24	9
Green La. NW4	BQ31	22
Green La. SE9	CL47	62
Green La. SE20	CC50	61
Green La. SW16	BX50	60
Green La. W7	BH41	39
Green La. Chess.	BK58	75
Green La. Chis.	CL49	62
Green La. Dag.	CN34	27
Green La. Dag.	CP34	27
Green La. E. Mol.	BF53	65
Green La. Edg.	BL28	12
Green La. Felt.	BE49	56
Green La. Har.	BH34	21
Green La. Houns.	BC45	47
Green La. Ilf.	CM34	26
Green La. Mord.	BQ52	67
Green La. Mord.	BS53	68
Green La. N. Mal.	BN53	67
Green La. Pur.	BW59	83
Green La. Stan.	BJ28	12
Green La. Th. Hth.	BX50	60
Green La. Walt.	BC57	74
Green La. Wat.	BD26	4
Green La. Wor. Pk.	BP54	67
Green La. Av. Walt.	BD56	74
Green La. Gdns. Th. Hth.	BZ51	69
Green Lanes N4	BY31	24
Green Lanes N8	BY31	24
Green Lanes N16	BY31	24
Green Lanes, Epsom	BO58	76
Greenlaw Gdns. N. Mal.	BO54	67
Green Lawns, Ruis.	BD33	20
Greenlaw St. SE18	CK41	44
Greenleafe Dr. Ilf.	CL31	26

Grove Pk. E11 CH32 26
Grove Pk. NW9 BN31 22
Grove Pk. SE5 CA44 51
Grove Pk. Av. E4 CE29 16
Grove Pk. Gdns. W4 BM43 48
Grove Pk. Rd. N15 CA31 24
Grove Pk. Rd. SE9 CJ48 62
Grove Pk. Rd. W4 BM43 48
Grove Pk. Rd. Rain. CU37 37
Grove Pk. Ter. W4 BM43 48
Grove Pass. E2 CB37 33
Grove Pass. Tedd. BJ49 57
Grove Pl. NW3 BT34 23
Christchurch Hill
Grove Pl. W3 BN40 40
Grove Pl. W5 BK40 39
Grove Pl. Croy. BY54 69
Grove Pl. Wat. BF23 4
Grove Rd. E3 CD37 34
Grove Rd. E4 CE28 16
Grove Rd. E9 CC37 34
Grove Rd. E11 CG33 25
Grove Rd. E17 CE32 25
Grove Rd. N11 BV28 14
Grove Rd. N12 BT28 14
Grove Rd. N15 CA32 24
Grove Rd. NW2 BQ36 31
Grove Rd. SW13 BO44 49
Grove Rd. SW19 BT50 59
Grove Rd. W3 BN40 40
Grove Rd. W5 BK40 39
Grove Rd. Barn. BU24 7
Grove Rd. Belv. CQ43 54
Grove Rd. Bexh. CS45 55
Grove Rd. Borwd. BL23 5
Grove Rd. Brent. BK42 39
Grove Rd. E. Mol. BG52 65
Grove Rd. Edg. BM29 12
Grove Rd. Epsom BO60 82
Grove Rd. Houns. BF45 47
Grove Rd. Islw. BH44 48
Grove Rd. Mitch. BV51 68
Grove Rd. Pnr. BE32 20
Grove Rd. Rich. BL46 48
Grove Rd. Rom. CO33 27
Grove Rd. Surb. BK53 66
Grove Rd. Sutt. BS57 77
Grove Rd. Twick. BG48 56
Grover Rd. Wat. BD26 4
Groveside Rd. E4 CG27 16
Grove St. N18 CA28 15
Grove St. SE8 CD42 43
Grove Ter. Tedd. BJ49 57
Grove, The E15 CG36 34
Grove, The N3 BS30 14
Grove, The N4 BX33 24
Grove, The N6 BV33 23
Grove, The N8 BW32 23
Grove, The N13 BY28 15
Grove, The NW5 BV35 32
Lissenden Gdns.
Grove, The NW9 BN32 22
Grove, The NW11 BR33 22
Grove, The SE21 CA47 60
Grove, The SW16 BW49 59
Grove, The W5 BK40 39
Grove, The, Bexh. CP45 54
Grove, The, Couls. BW61 83
Grove, The, Edg. BM28 12
Grove, The, Enf. BY23 8
Grove, The, Epsom BO58 76
Grove, The, Epsom BO60 82
Grove, The, Esher BF54 65
Grove, The, Grnf. BG39 38
Grove, The, Islw. BH44 48
Grove, The, Sid. CQ49 63
Grove, The, Tedd. BJ49 57
Grove, The, Walt. BC53 65
Grove, The, W. Wick. CE55 79
Grove Val. SE22 CA45 51
Grove Val. Chis. CL50 62
Grove Vill. E14 CE40 43
Groveway SW9 BX44 51
Groveway, Dag. CP35 36
Grove Way, Esher BG54 65
Groveway, Wem. BM35 30
Grove Wd. Hill, Couls. BW60 83
Grovewood Rd. Rich. BM44 48
Sandycombe Rd.
Grummant Rd. SE15 CA44 51
Grundy St. E14 CE39 43
Gruneisen Rd. N3 BS29 14
Gubbins La. Rom. CW29 19
Gubyon Av. SE24 BY46 51
Guerin St. E3 CD38 34
Guernsey Clo. Houns. BF43 47
Guernsey Gro. SE24 BZ47 60
Guernsey Rd. E11 CF33 25
Guibal Rd. SE12 CH47 62
Guildersfield Rd. SW16 BX50 60
Guildford Av. Surb. BL53 66
Guildford Gro. SE10 CE44 52
Guildford Rd. E17 CF30 16
Guildford Rd. SW8 BX44 51
Guildford Rd. Croy. BZ53 69
Guildford Rd. Ilf. CN34 27
Guildford Rd. Rom. CV29 19
Guildford Way, Wall. BX56 78
Guildhall Bldgs. EC2 BZ39 42
Basinghall St.
Guildhall Yd. EC2 BZ39 42
Gresham St.
Guildhouse St. SW1 BW42 41
Guildown Av. N12 BS28 14
Holden Rd.
Guild Rd. SE7 CJ42 44
Guild Rd. Erith CT43 55
Guilds Way E17 CD30 16
Guilford Pl. WC1 BX38 33
Guilford St. WC1 BX38 33
Guilsborough Clo. NW10 BO36 31
Guinness Bldgs. SE1 CA42 42
Guinness Bldgs. SE17 BZ42 42
Guinness Bldgs. SW3 BU42 41
Guinness Bldgs. W6 BQ42 40
Fulham Palace Rd.
Guinness Sq. SE1 CA42 42
Page's Walk
Guinness Trust SE24 BY45 51
Guinness Trust Bldgs. SW10 CN44 54
Guinness Trust Dws. N16 CA33 24
Guion Rd. SW6 BR44 49
Gulland Clo. Bush. BG24 4
Gulland Wk. N1 BZ36 33
Gull Clo. Wall. BX57 78
Gulliver Clo. Nthlt. BE37 29
Gulliver Rd. Sid. CN48 63
Gulliver St. SE16 CD41 43
Gull Wk. Rain. CU37 37
Gumleigh Rd. W5 BK42 39
Gumley Gdns. Islw. BJ45 48
Gumping Rd. Orp. CM55 80
Gundulph Rd. Brom. CJ52 71
Gunmaker's La. E3 CD37 34
Gunner La. SE18 CL42 44
Gunnersbury Av. W3 BM41 39
Gunnersbury Av. W5 BL40 39
Gunnersbury Ct. W3 BM41 39
Bollo La.
Gunnersbury Cres. W3 BM41 39
Gunnersbury Dr. W5 BL41 39
Gunnersbury Gdns. W3 BM41 39
Gunnersbury La. W3 BM41 39
Gunnersbury Ms. Brent. BM42 39
Gunners Gro. E4 CF27 16
Gunners Rd. SW18 BT48 59
Gunning St. SE18 CN42 45
Gunstor Rd. N16 CA35 33
Gunter Gro. SW10 BT43 50
Gunter Gro. Edg. BN30 13
Gunterstone Rd. W14 BR42 40
Gunthorpe St. E1 CA39 42
Wentworth St.
Gunton Rd. E5 CB34 24
Gunton Rd. SW17 BV50 59
Gurdon Rd. SE7 CH42 44
Gurnell Gro. W13 BH38 30
Gurney Clo. E15 CG35 34
Gurney Rd.
Gurney Cres. Croy. BX54 69
Gurney Dr. N2 BT31 23
Gurney Rd. E15 CG35 34
Gurney Rd. Cars. BU56 77
Gurney Rd. Nthlt. BC38 29
Gurnsey Clo. Houns. BF44 47
Sutton Rd.
Gutherie St. SW3 BU42 41
Cale St.
Gutter La. EC2 BZ39 42
Guyatt Gdns. Mitch. BV51 68
Ormerod Gdns.
Guy Rd. Wall. BW55 77
Guyscliff Rd. SE13 CF46 52
Guysfield Clo. Rain. CU37 37
Guysfield Dr. Rain. CU37 37
Guy St. SE1 BZ41 42
Gwendolen Av. SW15 BQ45 49
Gwendolen Clo. SW15 BQ46 49
Gwendoline Av. E13 CH37 35
Gwendwr Rd. W14 BR42 40
Gwillim Clo. Sid. CO46 54
Gwydor Rd. Beck. CC52 70
Gwydyr Rd. Brom. CG52 70
Gwynne Av. Croy. CC54 70
Gwynne Park IG8 CK29 17
Gwynne Pl. WC1 BX38 33
King's Cross Rd.
Gwynne Rd. SW11 BT44 50
Gylcote Clo. SE5 BZ45 51
Gyles Pk. Stan. BK29 12
Gyllyngdune Gdns. Ilf. CN34 27
Haarlem Rd. W14 BQ41 40
Haberdasher St. W1 BZ38 33
Habgood Rd. Loug. CK24 10
Hackbridge Pk. Gdns. Cars. BU55 77
Hackbridge Rd. Wall. BV55 77
Hackford Rd. SW9 BX44 51
Hackforth Clo. Barn. BP25 6
Hackington Cres. Beck. CE50 61
Hackney Clo. Borwd. BN25 6
Hackney Gro. E8 CB36 33
Hackney Rd. E2 CA38 33
Hacombe Rd. SW19 BT50 59
Hacton Dr. Horn. CV35 37
Hacton La. Horn. CW34 28
Hacton Parkway, Upmin. CW35 37
Hadden Way SE28 CN41 45
Hadden Way, Grnf. BG36 29
Haddington Rd. Brom. CF48 61
Haddon Clo. Borwd. BM23 5
Haddon Clo. Enf. CB25 8
Haddon Clo. N. Mal. BO53 67
Cromwell Av.
Haddon Gro. Sid. CN47 63
Haddon Rd. Orp. CP53 72
Haddon Rd. Sutt. BS56 77
Haddo St. SE10 CE43 52
Haden Ct. N4 BY34 24
Hadleigh Clo. E1 CC38 34
Martus Rd.
Hadleigh Rd. N9 CB26 8
Hadleigh St. E2 CC38 34
Hadley Clo. Borwd. BL25 5
Hadley Clo. N21 BY25 8
Hadley Common, Barn. BS23 7
Hadley Gdns. W4 BN42 40
Hadley Gdns. Sthl. BE42 38
Hadley Grn. Rd. Barn. BR23 6
Hadley Grn. W. Barn. BR23 6
Hadley Gro. Barn. BR23 6
Hadley Highstone, Barn. BR23 6
Hadley Ridge, Barn. BR24 6
Hadley Rd. Barn. BS24 7
Hadley Rd. Barn. BX23 8
Hadley Rd. Belv. CQ42 45
Hadley Rd. Mitch. BW52 68
Hadley St. NW1 BV36 32
Hadley Way N21 BY25 8
Hadley Wood Rise, Ken. BY61 84
Hadlow Pl. SE19 CB50 60
Hadlow Rd. Sid. CO49 63
Hadlow Rd. Well. CP43 54
Hadrian Clo. Wall. BX57 78
Hadrian Est. E2 CB37 33
Hadrians Ride, Enf. CA25 8
Hadrian St. SE10 CG42 43
Haydn Pk. Rd. W12 BP41 40
Hafer Rd. SW11 BU45 50
Hafton Rd. SE6 CG47 61
Hagden La. Wat. BC25 4
Haggard Rd. Twick. BJ47 57
Haggerston Est. E8 CA37 33
Haggerston Rd. E8 CA36 33
Hague St. E2 CB38 33
Derbyshire St.
Ha-ha Rd. SE18 CK43 53
Haig Rd. Stan. BK28 12
Haig Rd. E. E13 CJ38 35
Haig Rd. W. E13 CJ38 35
Haigville Gdns. Ilf. CL31 26
Hailes Clo. SW19 BT50 59
Haileybury Av. Enf. CA25 8
Haileybury Rd. Orp. CO56 81
Hailey Rd. Belv. CR41 45
Hailsham Av. SW2 BX48 60
Hailsham Clo. Rom. CV28 19
Hailsham Clo. Surb. BK54 66
Hailsham Gdns. Rom. CV28 19
Hailsham Rd. SW17 BV50 59
Hailsham Rd. Rom. CV28 19
Hailsham Ter. N18 BZ28 15
Haimo Rd. SE9 CJ46 53
Hainault Ct. E17 CF31 25
Hainault Gore, Rom. CQ32 27
Hainault Gro. Chig. CM28 17
Hainault Rd. E11 CF33 25
Hainault Rd. Chig. CL27 17
Hainault Rd. Rom. CO29 18
Hainault Rd. Rom. CQ32 27
Hainault Rd. Rom. CS30 19
Hainault St. SE9 CL47 62
Hainault St. Ilf. CM34 26
Haines St. SW8 BW43 50
Hainsford Clo. SE4 CC45 52
Hainthorpe Rd. SE27 BY48 60
Halbred Ms. E5 CB34 24
Knightland Rd.
Halbutt Gdns. Dag. CQ34 27
Halbutt St. Dag. CQ34 27
Halcomb St. N1 CA37 33
Orsman Rd.
Halcot Av. Bexh. CR46 54
Halcrow St. E1 CB39 42
Halcyon Way, Horn. CW33 28
Haldane Clo. N10 BV29 14
Haldane Clo. N11 BV29 14
Hampden Rd.
Haldane Pl. SW18 BS47 59
Haldane Rd. E6 CJ38 35
Haldane Rd. SW6 BR43 49
Haldane Rd. Sthl. BG40 38
Haldan Rd. E4 CF29 16
Haldon Clo. Chig. CN28 18
Arrowsmith Rd.
Haldon Rd. SW18 BR46 49
Hale Clo. E4 CF27 16
Hale Clo. Edg. BN28 13
Hale Clo. Orp. CM56 80
Hale Ct. Edg. BN28 13
Hale Dr. NW7 BN29 13
Hale End, Rom. CU29 19
Hale End Rd. E4 CF29 16
Hale End Rd. E17 CF29 16
Hale End Rd. Wdf. Grn. CF29 16
Halefield Rd. N17 CB30 15
Hale Gdns. N17 CB31 24
High Cross Rd.
Hale Gdns. W3 BM40 39
Hale Grove Gdns. NW7 BN28 13
Hale La. NW7 BN28 13
Hale La. Edg. BM28 12
Hale Path SE27 BY49 60
Hale Rd. E6 CK38 35
Hale Rd. N17 CB31 24
Halesowen Rd. Mord. BS54 68
Hales St. SE8 CE43 52
Hale St. E14 CE40 43
Halesworth Clo. E5 CB34 24
Southwold Rd.
Halesworth Clo. Rom. CW29 19
Halesworth Rd. SE13 CE45 52
Halesworth Rd. Rom. CW29 19
Hale, The E4 CF29 16
Hale, The N17 CB31 24
Hale, The, Wem. BJ35 30
Hale Wk. W7 BH39 39
Haley Rd. NW4 BQ32 22
Half Acre, Brent. BK43 48
Half Acre Rd. W7 BH40 39
Half End Clo. Ruis. BC32 20
Halfield Est. W2 BT40 41
Half Moon Cres. N1 BX37 33
Half Moon La. SE24 BZ46 51
Half Moon Pass. E1 CA39 42
Braham St.
Half Moon St. W1 BV40 41
Halford Rd. E10 CF32 25
Halford Rd. SW6 BS43 50
Halford Rd. Rich. BL46 48
Halfway Grn. Walt. BC55 74
Halfway St. Sid. CM47 62
Haliburton Rd. Twick. BJ46 48
Halidon Clo. E9 CC35 34
Churchill Wk.
Halifax Rd. Enf. BZ23 8
Halifax Rd. Grnf. BF37 29
Halifax St. SE26 CB49 60
Haling Gro. Sth. Croy. BZ57 78
Haling Pk. Gdns. Sth. Croy. BY57 78
Haling Rd. Sth. Croy. BZ57 78
Halkin Arc. SW1 BV41 41
Motcomb St.
Halkin Ms. SW1 BV41 41
Motcomb St.
Halkin Pl. SW1 BV41 41
Halkin St. SW1 BV41 41
Hallam Clo. Chis. CK49 62
Hallam Gdns. Pnr. BE29 11
Hallam Ms. W1 BV39 41
Hallam St.
Hallam Rd. N15 BY31 24
Hallam St. W1 BV38 32
Hall Clo. W5 BK39 39
Regal Clo.
Hall Dr. SE26 CB49 60
Hall Dr. W7 BH39 39
Halley Pl. E14 CD39 43
Halley Rd. E7 CJ36 35
Halley Rd. E12 CJ36 35
Halley St. E14 CD39 43
Hall Farm Clo. Stan. BJ27 12
Hall Farm Dr. Twick. BG47 56
Hallfield Est. W2 BT39 41
Hallford Way, Dart. CV46 55
Hall Gdns. E4 CD28 16
Hall Gate NW8 BT38 32
Hall Rd.
Halliford St. N1 BZ36 33
Halliwell Rd. SW2 BX46 51
Halliwick Rd. N10 BV30 14
Hall La. E4 CD28 16
Hall La. NW4 BP30 13
Hallmead, Sutt. BS55 77
Hallowell Av. Croy. BX56 78
Hallowell Clo. Mitch. BV52 68
Hallowes Cres. Wat. BC27 11
Hayling Rd.
Hall Pl. W2 BT38 32
Hall Pl. Cres. Bex. CS46 55
Hall Rd. E6 CK37 35
Hall Rd. E15 CF35 34
Hall Rd. NW8 BT38 32
Hall Rd. Dart. CW45 55
Hall Rd. Islw. BG46 47
Hall Rd. Rom. CP32 27
Hall Rd. Rom. CV31 28
Hall Rd. Wall. BV58 77
Hall St. EC1 BY38 33
Hall St. N12 BT28 14
Hallsville Rd. E16 CG39 43
Hallswelle Rd. NW11 BR32 22
Hall, The SE3 CH45 53
Hall View SE9 CJ49 62
Hall Way, Pur. BY60 84
Halons Rd. SE9 CL47 62
Halpin Pl. SE17 BZ42 42
Halsbrook Rd. SE3 CJ45 53
Halsbury Clo. Stan. BJ28 12
Halsbury Rd. E. Nthlt. BG25 29

Halsbury Rd. W. Nthlt. BF35 29
Halsbury St. W12 BP40 40
Halsend, Hayes BC41 38
Halsey St. SW3 BU42 41
Halsham Cres. Bark. CN35 36
Halsmere Rd. SE5 BY44 51
Halstead Gdns. N21 BZ26 8
Halstead Rd. E11 CH32 26
Halstead Rd. N21 BZ26 8
Halstead Rd. Enf. CA24 8
Halstead Rd. Erith CT44 55
Halston Clo. SW11 BU46 50
Northcote Rd.
Halstow Rd. NW10 BQ38 31
Halstow Rd. SE10 CH42 44
Halsway, Hayes BC40 38
Halton Rd. N1 BY36 33
Halt Robin La. Belv. CR42 45
Halt Robin Rd.
Halt Robin Rd. Belv. CR42 45
Hambalt Rd. SW4 BW46 50
Hambledon Gnds. SE25 CA52 69
Hambledon Hill, Epsom BN61 82
Hambledon Rd. SW18 BR47 58
Hambledon Rd. Sid. CM47 62
Hambledon Vale, Epsom BN61 82
Hamble St. SW6 BS45 50
Hamble Wk. Nthlt. BF37 29
Leander Rd.
Hambro Av. Brom. CH54 71
Hambrook Rd. SE25 CB52 69
Hambro Rd. SW16 BW50 59
Hambrough Rd. SW16 BW50 59
Hambrough Rd. Sthl. BE40 38
Ham Clo. Rich. BK48 57
Hamden Cres. Dag. CR34 27
Hamelin St. E14 CF39 43
Hameway E6 CL38 35
Ham Farm Rd. Rich. BK49 57
Hamfrith Rd. E15 CG36 34
Ham Gate Av. Rich. BK48 57
Hamilton Av. N9 CB26 8
Hamilton Av. Ilf. CL31 26
Hamilton Av. Rom. CS30 19
Hamilton Av. Surb. BM55 75
Hamilton Av. Sutt. BR54 67
Hamilton Clo. N17 CA31 24
Hamilton Clo. NW8 BT38 32
Hamilton Clo. Barn. BU24 7
Hamilton Clo. Epsom BN59 82
Hamilton Clo. Stan. BH27 12
Hamilton Ct. W5 BL40 39
Hamilton Rd.
Hamilton Ct. W9 BT38 32
Hamilton Cres. N13 BY28 15
Hamilton Cres. Har. BE34 20
Hamilton Cres. Houns. BF46 47
Hamilton Dr. Rom. CW30 19
Hamilton Gdns. NW8 BT38 32
Hamilton La. N5 BY35 33
Hamilton Pk. Way N5 BY35 33
Hamilton Pl. SE20 CB51 69
Hamilton Pl. W1 BV40 41
Hamilton Pl. Sun. BC50 56
Hamilton Rd. E15 CG38 34
Hamilton Rd. E17 CD30 16
Hamilton Rd. N2 BT31 23
Hamilton Rd. N9 CB26 8
Hamilton Rd. NW10 BP35 31
Hamilton Rd. NW11 BQ33 22
Hamilton Rd. SE27 BZ49 60
Hamilton Rd. SW19 BS50 59
Hamilton Rd. W4 BO41 40
Hamilton Rd. W5 BL40 39
Hamilton Rd. Barn. BU24 7
Hamilton Rd. Bexh. CQ44 54
Hamilton Rd. Brent. BK43 48
Hamilton Rd. Har. BH32 21
Hamilton Rd. Hayes BC40 38
Hamilton Rd. Ilf. CL35 35
Hamilton Rd. Rom. CU32 28
Hamilton Rd. Sid. CO49 63
Hamilton Rd. Sthl. BE40 38
Hamilton Rd. Th. Hth. BZ52 69
Hamilton Rd. Twick. BH47 57
Hamilton Rd. Wat. BC27 11
Hamilton Sq. SE1 BZ41 42
Kipling St.
Hamilton St. SE8 CE43 52
Deptford High St.
Hamilton St. Wat. BD25 4
Hamilton Ter. NW8 BT38 32
Hamilton Way N3 BS29 14
Hamilton Way, Wall. BW58 77
Hamish St. SE11 BX42 42
Lambeth Wk.
Hamlea Clo. SE12 CH46 53
Hamlet Clo. Rom. CR29 18
Hamlet Clo. Wdf. Grn. CH29 17
Hamlet Gdns. W6 BP42 40
Hamlet Rd. SE19 CA50 60
Hamlet Rd. Rom. CQ29 18
Hamlets Way E3 CD38 34
Hamlet, The SE5 BZ45 51
Hamlin Cres. Pnr. BD32 20
Hamlyn Gdns. SE19 CA50 60
Hammelton Rd. Brom. CG51 70
Hammers La. NW7 BP28 13
Hammersley Av. E16 CG39 43
Hammersmith Br. Rd. W6 BQ42 40
Hammersmith Gro. W6 BQ41 40
Hammersmith Rd. W6 BQ42 40
Hammersmith Rd. W14 BR42 40
Hammersmith Ter. W6 BP42 40
Hammett St. EC3 CA40 42
Minories
Hammond Av. Mitch. BV51 68
Hammond Clo. Barn. BR25 6
Hammond Clo. Hamptn. BF51 65
Hammond Rd. Enf. CB23 8
Hammond Rd. Sthl. BE41 38
Hammond St. NW5 BW36 32
Hammond Way SE28 CO40 45
Hamonde Clo. Edg. BM27 12
Hamond Sq. N1 CA37 33
Hoxton St.
Ham Park Rd. E7 CH36 35
Ham Park Rd. E15 CG36 34
Hampden Av. Beck. CD51 70
Hampden Clo. NW1 BW37 32
Hampden Ct. N10 BV29 14
Hampden Gurney St. W1 BU39 41
Seymour Pl.
Hampden La. N17 CA30 15
Hampden Rd. N8 BY31 24
Hampden Rd. N10 BV29 14
Hampden Rd. N17 CB30 15
Hampden Rd. Beck. CD51 70
Hampden Rd. Har. BG30 11
Hampden Rd. Kings. On T. BM51 66
Hampden Rd. Rom. CR29 18
Hampden Sq. N14 BV26 7
Hampden Way N14 BV26 7
Hamper Mill La. Wat. BC26 4
Hampshire Clo. N18 CB28 15
Berkshire Gdns.
Hampshire Rd. N22 BX29 15
Hampshire St. NW5 BW36 32
Torriano Av.
Hampson Way SW8 BX44 51
Hampstead Clo. SE28 CQ40 45
Hampstead Gdns. NW3 BU35 32
Rosslyn Hill
Hampstead Gdns. NW11 BS32 23
Hampstead Gro. NW3 BT34 23
Hampstead High St. NW3 BT35 32
Hampstead Hill Gdns. NW3 BT35 32
Hampstead La. N6 BU33 23
Hampstead La. NW3 BT33 23
Hampstead Rd. NW1 BW37 32
Hampstead Sq. NW3 BT34 23
Hampstead Way NW11 BS32 23
Hampton Clo. NW6 BS38 32
Hampton Clo. SW20 BQ50 58
Hampton Ct. Av. E. Mol. BG53 65
Hampton Ct. Rd. E. Mol. BH51 66
Hampton Ct. Rd. E. Mol. BH52 66
Hampton Ct. Rd. Kings. On T. BH52 66
Hampton Ct. Way, E. Mol. BH53 66
Hampton Ct. Way, Surb. BH55 75
Hampton Gro. NW3 BU35 32
Pond St.
Hampton Gro. Epsom BO59 82
Hampton La. Felt. BE49 56
Hampton Mead, Loug. CL24 10
Hampton Pl. NW6 BS39 41
Hampton Rise, Har. BL32 21
Hampton Rd. E4 CD28 16
Hampton Rd. E7 CH35 35
Hampton Rd. E11 CF33 25
Hampton Rd. NW6 BS38 32
Hampton Rd. Croy. BZ53 69
Hampton Rd. Ilf. CL35 35
Hampton Rd. Tedd. BG49 56
Hampton Rd. Twick. BG48 56
Hampton Rd. Wor. Pk. BP55 76
Hampton Rd. E. Felt. BE48 56
Hampton Rd. W. Felt, BE48 56
Hampton St. SE17 BY42 42
Ham Ridings, Rich. BL49 57
Hamsey Grn. Gdns. Warl. CB61 84
Hamsey Way, Sth. Croy. CB61 84
Ham Shades Clo. Sid. CO48 63
Ham St. Rich. BK47 57
Ham, The, Brent. BK43 48
Ham Vw. Croy. CD53 70
Hanameel St. E16 CH40 44
Hanbury Rd. W3 BM41 39
Hanbury Rd. W17 CB30 15
Hanbury St. E1 CB39 42
Hanbury Wk. Bex. CS48 64
Hancock Rd. E3 CF38 34
Hancock Rd. SE19 BZ50 60
Handa Wk. N1 BZ36 33
Hand Ct. WC1 BX39 42
Sandland St.
Handcroft Rd. Croy. BY54 69
Handel Clo. Edg. BL29 12
Handel St. WC1 BX38 33
Handel Way, Edg. BM29 12
Handen Rd. SE12 CG46 52
Handforth Rd. SW9 BY43 51
Handley Rd. E9 CC37 34
Victoria Pk. Rd.
Handside Clo. Wor. Pk. BQ54 67
Handsworth Av. E4 CF29 16
Handsworth Clo. Wat. BC27 11
Handsworth Rd. N17 BZ31 24
Handtrough Way, Bark. CL37 35
Fresh Wharf Rd.
Handford Clo. SW18 BS47 59
Handford Row SW19 BQ50 58
Hanger Ct. W5 BL38 30
Heathcroft
Hanger Grn. W5 BM38 30
Western Av.
Hanger La. W5 BL37 30
Hanger Ruding, Wat. BE27 11
Hanger Vale La. W5 BL39 39
Hankey Pl. SE1 BZ41 42
Hankins La. NW7 BO27 13
Hanley Rd. N4 BX33 24
Hanmer Walk N7 BX34 24
Andover Est.
Hannards Wy, Ilf. CO28 18
Hannell Rd. SW6 BR43 49
Hannen Rd. SE27 BY48 60
Hannibal Rd. E1 CC39 43
Hannibal Way, Croy. BX56 78
Hannington Rd. SW4 BV45 50
Hanover Av. Felt. BC47 56
Hanover Clo. Rich. BM43 48
Cambridge Rd.
Hanover Clo. Sutt. BR56 76
Hanover Ct. W12 BP40 40
Hanover Gdns. SE11 BY43 51
Hanover Gdns. Ilf. CM29 17
Hanover Gate NW1 BU38 32
Hanover Pk. SE15 CB44 51
Hanover Pl. WC2 BX40 42
Long Acre
Hanover Rd. N15 CA31 24
Hanover Rd. NW10 BQ36 31
Hanover Rd. SW19 BT50 59
Hanover Sq. W1 BV39 41
Hanover St. Croy. BY55 78
Hanover St. W1 BV39 41
Hanover Ter. NW1 BU38 32
Hanover Ter. Ms. NW1 BU38 32
Hanover Way, Bexh. CP45 54
Hansard Ms. W14 BQ41 40
Hansart Way, Enf. BY23 8
Hans Cres. SW1 BU41 41
Hanshades Clo. Sid. CO48 63
Hanshaw Dr. Edg. BN30 13
Hansler Gro. E. Mol. BG53 65
Hansler Rd. SE22 CA46 51
Hansol Rd. Bexh. CQ46 54
Hanson Clo. SW12 BV47 59
Hanson Clo. Loug. CM23 10
Hanson Dr. Loug. CM23 10
Hanson Gdns. Sthl. BE41 38
Hanson Grn. Loug. CM23 10
Hanson St. W1 BW39 41
Hans Pl. SW1 BU41 41
Hans Rd. SW3 BU41 41
Hans St. SW1 BU41 41
Pavilion Rd.
Hanway Pl. W1 BW39 41
Hanway St.
Hanway Rd. W7 BF39 38
Hanway St. W1 BW39 41
Hanworth Clo. Felt. BE49 56
Hanworth Rd. Felt. BC47 56
Hanworth Rd. Hamptn. BE49 56
Hanworth Rd. Houns. BE47 56
Hanworth Rd. Sun. BC50 56
Hanworth Trd. Est. Felt. BE48 56
Hapgood Clo. Har. BG35 29
Harad's Pl. E1 CB40 42
Ensign St.
Harben Rd. NW6 BT36 32
Harberson Rd. E15 CG37 34
Harberson Rd. SW12 BV47 59
Harberton Rd. N19 BW33 23
Harbet Rd. E4 CC28 16
Harbet Rd. W2 BT39 41
Harbex Clo. Bex. CR47 63
Harbinger Rd. E14 CE42 43
Harbledown Pl. Orp. CP52 72
Okemore Gdns.
Harbledown Rd. SW6 BS44 50
Harbledown Rd. Sth. Croy. CB59 84
Harbord St. SW6 BQ44 49
Harborough Av. Sid. CN47 63
Harborough Rd. SW16 BX49 60
Harbourer Clo. Ilf. CO28 18
Harbourer Rd. Ilf. CO29 18
Harbourfield Rd. Bans. BS61 83
Harbour Rd. SE5 BZ45 51
Harbridge Av. SW15 BO47 58
Harbury Rd. Cars. BU58 77
Harbut Rd. SW11 BT45 50
Harcombe Rd. N16 CA34 24
Harcourt Av. E12 CK35 35
Harcourt Av. Edg. BN27 13
Harcourt Av. Sid. CP46 54
Harcourt Av. Wall. BV56 77
Harcourt Clo. Islw. BJ45 48
Silvern St.
Harcourt Fld. Wall. BV56 77
Harcourt Rd. E15 CG37 34
Harcourt Rd. N22 BW30 14
Harcourt Rd. SE4 CD45 52
Harcourt Rd. SW19 BS50 59
Harcourt Rd. Bexh. CQ45 54
Harcourt Rd. Bush. BF25 4
Harcourt Rd. Th. Hth. BX53 69
Harcourt Rd. Wall. BV56 77
Harcourt St. W1 BU39 41
Harcourt Ter. SW10 BS42 41
Hardcourts Clo. W. Wick. CE55 79
Hardel Rise SW2 BY47 60
Harden's Manor Way SE7 CJ41 44
Harden St. SE18 CK42 44
Harders Rd. SE15 CB44 51
Hardie Clo. NW10 BN35 31
Hardie Rd. Dag. CS34 28
Hardinge Rd. N18 CA29 15
Hardinge Rd. NW10 BP37 31
Hardinge St. E1 CC39 43
Harding Rd. Bexh. CQ44 54
Hardings La. SE20 CC50 61
Hardley Cres. Horn. CV31 28
Hardman Rd. SE7 CH42 44
Hardman Rd. Kings. On T. BL51 66
Hardwick Clo. Stan. BK28 12
Marsh La.
Hardwicke Av. Houns. BF44 47
Hardwicke Rd. N13 BX29 15
Hardwicke Rd. W4 BN42 40
Hardwicke Rd. Rich. BK49 57
Hardwicke St. Bark. CM37 35
Hardwick Gn. W13 BJ39 39
Templewood
Hardwick St. EC1 BY38 33
Hardwicks Way SW18 BS46 50
Hardwidge St. SE1 CA41 42
Snows Fields
Hardy Av. Ruis. BC35 29
Hardy Clo. Pnr. BD33 20
Hardy Pass. N22 BY30 15
Cranbrook Pk.
Hardy Rd. SE3 CG43 52
Hardy Rd. SW19 BS50 59
Hardy Way, Enf. BY23 8
Harebell Way, Rom. CV29 19
Hare & Billet Rd. SE3 CF44 52
Harecourt Rd. N1 BZ36 33
Haredale Rd. SE24 BZ45 51
Haredon Clo. SE23 CC47 61
Harefield, Esher BH55 75
Harefield Av. Sutt. BQ58 76
Harefield Gn. NW7 BQ29 13
Harefield Ms. SE4 CD45 52
Harefield Rd. N8 BW32 23
Harefield Rd. SE4 CD45 52
Harefield Rd. SW16 BX50 60
Harefield Rd. Sid. CP48 63
Hare Hall La. Rom. CU31 28
Hare La. Esher BG57 74
Hare La. Esher BH57 75
Hares Bank, Croy. CF58 79
Haresfield Rd. Dag. CR36 36
Hare St. SE18 CL41 44
Harewood SW15 BN49 58
Harewood Av. NW1 BU38 32
Harewood Av. Nthlt. BE36 29
Harewood Clo. Nthlt. BE36 29
Harewood Dr. Ilf. CK30 17
Harewood Gdns. Sth. Croy. CB61 84
Harewood Rd. SW19 BU50 59
Harewood Rd. Islw. BH43 48
Harewood Rd. Sth. Croy. CA57 78
Harewood Rd. Wat. BC27 11
Harewood Ter. Sthl. BE42 38
Harew Pl. W1 BV39 41
Hanover Sq.
Harfield Dr. Sun. BD51 65
Harfield Gdns. SE5 CA45 51
Harfield Rd. Sun. BD51 65
Harford Clo. E4 CE26 9
Harford Rd. E4 CE26 9
Harford St. E1 CD38 34
Harford Wk. N2 BT31 23
Hargood Rd. SE3 CJ44 53
Hargrave Pk. N19 BW34 23
Hargrave Pl. N7 BW35 32
Brecknock Rd.
Hargrave Rd. N19 BW34 23
Hargwyne St. SW9 BX45 51
Haringey Pk. N8 BX32 24
Haringey Pass. N4 BY31 24
Haringey Rd. N8 BX31 24
Harkett Clo. Har. BH30 12
Church La.
Harkness Clo. Epsom BQ61 82
Harkness Clo. Rom. CW28 19
Harland Av. Croy. CA55 78
Harland Av. Sid. CM48 62
Harland Rd. SE12 CH47 62
Harlech Gdns. Houns. BD43 47
Harlech Rd. N14 BX27 15
Harlequin Av. Brent. BJ43 48
Harlequin Rd. Tedd. BJ50 57

Havering Dr. Rom. CT31 28
Havering Gdns. Rom. CP32 27
Havering Pl. Hav. CS27 19
Havering Rd. Rom. CS30 19
Havering St. E1 CC39 43
Havering Way, Bark. CO38 36
Havers Av. Walt. BD56 74
Haversham Clo. Twick. BK46 48
Haversham Gra. Twick. BK46 48
Haverstock Hl. NW3 BU35 32
Haverstock Rd. NW5 BU35 32
Haverstock St. N1 BY37 33
Haverthwaite Rd. Orp. CM55 80
Havil St. SE5 CA43 51
Hawarden Gro. SE24 BZ47 60
Hawarden Rd. E17 CC31 25
Hawbridge Rd. E11 CF33 25
Hawes La. W. Wick. CF54 70
Hawes Rd. N18 CB29 15
Hawes Rd. Brom. CH51 71
Hawes St. N1 BY36 33
Hawfield Bk. Orp. CP55 81
Hawgood St. E3 CE39 43
Hawkdene E4 CE25 9
Hawker Clo. Wall. BX57 78
Hawke Rd. SE19 BZ50 60
Hawkesbury Rd. SW15 BP46 49
Hawkesfield Rd. SE23 CD48 61
Hawkesley Clo. Twick. BJ49 57
Hawkes Mews SE10 CF43 52
Luton Pl.
Hawkes Rd. Mitch. BU51 68
Hawkesworth Rd. Brom. CH52 71
Hawkewood Rd. Sun. BC52 65
Hawkfield Ct. Islw. BH44 48
Hawkhirst Rd. Ken. BZ61 84
Hawkhurst Gdns. Rom. CS29 19
Hawkhurst Rd. SW16 BW51 68
Hawkhurst Way, N. Mal. BN53 67
Hawkhurst Way, W. Wick. CE55 79
Hawkinge Wk. Orp. CO52 72
Robin Way
Hawkinge Way, Horn. CV36 37
Hawkins Clo. Har. BG33 20
Hawkins Rd. Tedd. BJ50 57
Hawk Park Rd. N22 BY31 24
Hawkridge Clo. Rom. CP32 27
Hawks Brook La. Beck. CF53 70
Hawkshaw Clo. SW2 BX47 60
Hawkshead Clo. Brom. CG50 61
Coniston Rd.
Hawkshead Rd. NW10 BO36 31
Hawkshead Rd. W4 BO41 40
Hawkshill Clo. Esher BF57 74
Hawkshill Way, Esher BF57 74
Hawkslade Rd. SE15 CC46 52
Hawksley Rd. N16 BZ34 24
Hawks Mews SE10 CF43 52
Luton Pl.
Hawksmoor St. W6 BQ43 49
Hawksmouth E4 CF26 9
Hawks Rd. Kings. On T. BL51 66
Hawkstone Rd. SE16 CC42 43
Hawkswood Est. Chis. CM52 71
Hawkswood Cres. E4 CE25 9
Hawkwood La. Chis. CM51 71
Hawkwood Mt. E5 CB33 24
Hawlands Dr. Pnr. BE33 20
Hawley Clo. Hamptn. BE50 56
Hawley Cres. NW1 BV36 32
Hawley Gdns. SE27 BY48 60
Hawley Rd. NW1 BV36 32
Hawley Rd. Dart. CW48 64
Hawley St. NW1 BV36 32
Hawstead La. Orp. CQ56 81
Hawstead Ranch, Buck. H. CH26 10
Hawstead Rd. SE6 CE46 52
Hawthorn Av. N13 BX28 15
Hawthorn Av. Cars. BV57 77
Hawthorn Av. Rain. CU38 37
Hawthorn Clo. Hamptn. BR49 56
Hawthorn Clo. Orp. CM53 71
Hawthorndene Clo. Brom. CG55 79
Hawthorndene Rd. Brom. CG55 79
Hawthorn Dr. Har. BE32 20
Hawthorn Dr. W. Wick. CG56 79
Hawthorne Av. Har. BJ32 21
Hawthorne Av. Mitch. BT51 68
Hawthorne Av. Ruis. BC32 20
Hawthorne Av. Th. Hth. BY51 69
Hawthorne Clo. Brom. CK52 71
Hawthorne Clo. N1 CA36 33
Hawthorne Clo. Sutt. BT55 77
Hawthorne Cres. Sth. Croy. CC58 79
Hawthorne Gro. NW9 BN33 22
Hawthorne Pl. Epsom BO59 82
Hawthorne Rd. Brom. CK52 71
Hawthorn Farm Av. Nthlt. BE37 29
Hawthorn Gdns. W5 BK41 39
Hawthorn Hatch, Brent. BJ43 48
Hawthorn Pl. Hayes BC40 38
Hawthorn Rd. E17 CE31 25
Hawthorn Rd. N8 BW31 23
Hawthorn Rd. N18 CA29 15
Hawthorn Rd. NW10 BP36 31
Hawthorn Rd. Bexh. CQ46 54
Hawthorn Rd. Brent. BJ43 48
Hawthorn Rd. Buck. H. CJ28 17
Hawthorn Rd. Dart. CV47 64
Hawthorn Rd. Sutt. BT56 77
Hawthorn Rd. Wall. BV57 77
Hawthorns, Wdf. Grn. CG27 16
Hawthorns, The, Loug. CL24 10
Hawthorn Way N9 CA27 15
Hawtrey Dr. Ruis. BC33 20
Hawtrey Rd. NW3 BU36 32
Hawtry Av. Nthlt. BD37 29
Haxtead Rd. Brom. CH51 71
Hayburn Way, Horn. CT33 28
Hay Clo. E15 CG36 34
Haycroft Clo. Couls. BY62 84
Haycroft Gdns. NW10 BP37 31
Haycroft Rd. SW2 BX46 51
Haycroft Rd. Surb. BK55 75
Hay Currie St. E14 CE39 43
Hayday Rd. E16 CH39 44
Haydens Clo. Orp. CP54 72
Hayden Way, Rom. CS30 19
Haydn Av. Pur. BY60 84
Haydock Av. Nthlt. BE36 29
Haydock Clo. Horn. CW35 37
Haydock Grn. Nthlt. BF36 29
Haydon Clo. NW9 BN31 22
Haydon Dr. Pnr. BC31 20
Haydon Rd. Dag. CP34 27
Haydon Rd. Wat. BE25 4
Haydons Pk. Rd. SW19 BS49 59
Haydon Sq. E1 CA39 42
Haydons Rd. SW19 BS49 59
Haydon St. EC3 CA40 42
Hayes Chase, W. Wick. CF53 70
Hayes Clo. Brom. CH55 80
Hayes Ct. SW2 BX47 60
Hayes Cres. NW11 BR32 22
Hayes Cres. Sutt. BQ56 76
Hayes Dr. Rain. CU36 37
Hayesford Pk. Dr. Brom. CG53 70
Hayesford Pk. Est. Brom. CH53 71
Hayes Gdns. Brom. CH55 80
Hayes Hill Rd. Brom. CG54 70
Hayes La. Beck. CF52 70
Hayes La. Brom. CH54 71
Hayes La. Ken. BY61 84
Hayes Mead, Brom. CG54 70
Hayes Pl. NW1 BU40 41
Hayes Rd. Brom. CH52 71
Hayes Rd. Sthl. BC42 38
Hayes St. Brom. CH54 71
Hayes Way, Beck. CF52 70
Hayes Way Av. Brom. CH54 71
Hayfield Clo. Bush. BF24 4
Hayfield Pass. E1 CC38 34
Hayfield Rd. Orp. CQ53 72
Haygarth Pl. SW19 BQ49 58
Hay Hill W1 BV40 41
Hayland Clo. NW9 BN31 22
Hay La. NW9 BN31 22
Hayles St. SE11 BY42 42
Hayling Av. Felt. BC48 56
Hayling Rd. Wat. BC27 11
Haymarket SW1 BW40 41
Haymeads Dr. Esher BG57 74
Haymer Gdns. Wor. Pk. BP55 76
Haymerle Rd. SE15 CB43 51
Hayne Rd. Beck. CD51 70
Haynes Clo. N17 CB29 15
Haynes Clo. SE3 CG45 52
Haynes La. SE19 CA50 60
Haynes Rd. Horn. CV32 28
Haynes Rd. Wem. BL36 30
Hayne St. EC1 BY39 42
Haynt Wk. SW20 BR52 67
Hays La. SE1 CA40 42
Haysleigh Gdns. SE20 CB51 69
Hays Ms. W1 BV40 41
Hays Wk. Sutt. BQ58 76
Hayter Rd. SW2 BX46 51
Hayward Clo. Bex. CS46 55
Bourne Rd.
Hayward Clo. SW19 BS51 68
Hayward Gdns. SW15 BQ46 49
Hayward Rd. N20 BT27 14
Haywards Pl. EC1 BY38 33
Sekforde St.
Haywood Rise, Orp. CN56 81
Haywood Rd. Brom. CJ52 71
Haywoods Clo. Pnr. BD30 11
Hazel Bnk. Surb. BN54 67
Hazelbank Rd. SE6 CF48 61
Hazelbourne Rd. SW12 BV46 50
Hazelbrouck Gdns. Ilf. CM29 17
Hazelbury Grn. N9 CA27 15
Hazelbury La. N9 CA27 15
Hazel Gro. N13 BZ27 15
Hazel Clo. Brent. BJ43 48
Hazel Clo. Horn. CU34 28
Hazel Clo. Mitch. BW52 68
Hazel Clo. SE15 CB44 51
Atwell Rd.
Hazel Clo. Twick. BG47 56
Hazel Cft. Pnr. BF29 11
Hazeldean Rd. NW10 BN36 31
Hazeldean Rd. Croy. BZ55 78
Hazeldene Ct. Ken. BZ61 84
Hazeldene Dr. Pnr. BD31 20
Hazeldene Rd. Ilf. CO34 27
Hazeldene Rd. Well. CP44 54
Hazeldon Rd. SE4 CD46 52
Hazel Dr. Erith CU44 55
Hazeleigh Gdns. Wdf. Grn. CK28 17
Hazel End, Swan. CT53 73
Hazel Gdns. Edg. BM28 12
Hazel Gro. SE26 CC49 61
Hazel Gro. Enf. CB25 8
Dimsdale Dr.
Hazel Gro. Orp. CL55 80
Hazel Gro. Rom. CQ31 27
Hazel Gro. Wem. BL37 30
Carlyon Rd.
Hazelhurst, Beck. CF51 70
Hazelhurst Rd. SW17 BT49 59
Hazel La. Rich. BL48 57
Hazell Cres. Rom. CR29 18
Hazellville Rd. N19 BW33 23
Hazel Mead, Barn. BP25 6
Hazel Mead, Epsom BP58 76
Hazelmere Clo. Grnf. BE37 29
Hazelmere Rd.
Hazelmere Dr. Grnf. BE37 29
Hazelmere Rd.
Hazelmere Gdns. Horn. CU32 28
Hazelmere Rd. Nthlt. BE37 29
Hazelmere Rd. NW6 BS37 32
Hazelmere Rd. Orp. CM52 71
Hazelmere Wk. Grnf. BD37 29
Hazelmere Rd.
Hazelmere Way, Brom. CH53 71
Hazel Ri. Horn. CV32 28
Hazel Rd. NW10 BQ38 31
Hazel Rd. Dart. CV47 64
Hazel Rd. Erith CU44 55
Hazeltree La. Nthlt. BE38 29
Hazel Wk. Brom. CL53 71
Hazel Way E4 CD29 16
Hazelwood Clo. W5 BL41 39
Hazelwood Ct. Surb. BL53 66
Hazelwood Cres. N13 BY28 15
Hazelwood Cres. W10 BR38 31
Hazelwood Dr. Pnr. BC30 11
Hazelwood Ho. SE8 CD42 43
Hazelwood La. N13 BY28 15
Hazelwood La. Couls. BU62 83
Hazelwood Rd. E17 CD32 25
Hazelwood Rd. Enf. CA25 8
Hazelwood Rd. Mord. BS52 68
Hazlebury Rd. SW6 BS44 50
Hazledene Rd. W4 BN43 49
Hazlemere Gdns. Wor. Pk. BP54 67
Hazlemere Rd. Ilf. CN34 27
Hazlemere Rd. SW15 BP46 49
Hazlewood, Loug. CJ25 10
Hazlewood Cres. W10 BR39 40
Hazlewood Gro. Sth. Croy. CB60 84
Hazlitt Rd. W14 BR41 40
Hazon Way, Epsom BN59 82
Headcorn Pl. Th. Hth. BX52 69
Headcorn Rd. N17 CA29 15
Tenterden Rd.
Headcorn Rd. Brom. CG49 61
Headcorn Rd. Th. Hth. BX52 69
Headfort Pl. SW1 BV41 41
Heading St. NW4 BQ31 22
Headington Rd. SW18 BT47 59
Headlam Rd. SW4 BW46 50
Headlam St. E1 CB38 33
Headley App. Ilf. CL32 26
Headley Av. Wall. BX56 78
Headley Clo. Epsom BM57 75
Headley Dr. Croy. CE57 79
Headley Dr. Ilf. CL32 26
Headley St. SE15 CB44 51
Gordon Rd.
Head Mews W10 BS39 41
Needham Rd.
Head Mews W10 BS39 41
Headstone Dr. Har. BG31 20
Headstone Gdns. Har. BG31 20
Headstone La. Har. BF31 20
Headstone Rd. Har. BG32 20
Head St. E1 CC39 43
Headway, The, Epsom BO58 76
Headworth Rd. SW17 BT48 59
Heald St. SE8 CE44 52
Heston St.
Healey Dr. Orp. CN56 81
Healey St. NW1 BV36 32
Heanor Ct. E5 CC35 34
Clapton Park Est.
Hearne Rd. W4 BM42 39
Hearn Ri. Nthlt. BD37 29
Hearn Rd. Rom. CT32 28
Hearn's Bldgs. SE17 BZ42 42
Elsted St.
Hearns Clo. Orp. CP52 72
Hearns Ri. Orp. CP52 72
Hearns Rd. Orp. CP52 72
Hearn St. EC2 CA38 33
Curtain Rd.
Hearnville Rd. SW12 BV47 59
Heatham Pk. Twick. BH47 57
Heath Av. Bexh. CP43 54
Heathbourne Rd. Bush. BH26 5
Heathbrow NW3 BT34 23
North End Way
Heath Clo. NW11 BS33 23
Heath Clo. W5 BL38 30
Heath Clo. Bans. BS60 83
Heath Clo. Orp. CP54 72
Heath Clo. Rom. CU31 28
Heathclose Av. Dart. CU47 64
Heathclose Rd. Dart. CU47 64
Heathcote Av. Ilf. CK30 17
Heathcote Gro. E4 CF27 16
Heathcote Rd. Epsom BN60 82
Heathcote Rd. Twick. BJ46 48
Heathcote St. WC1 BX38 33
Heath Ct. W5 BL38 30
Heath Croft NW11 BS33 23
Heathcroft W5 BL38 30
Heathdale Rd. Houns. BE45 47
Heathdene Dr. DA17 CR42 45
Heathdene Rd. SW16 BX50 60
Heathdene Rd. Wall. BV57 77
Heath Dr. NW3 BS35 32
Heath Dr. SW20 BQ52 67
Heath Dr. Rom. CU30 19
Heath Dr. Sutt. BT58 77
Heath Edge SE26 CB48 60
Heathend Rd. Bex. CT47 64
Heather Av. Rom. CS30 19
Heatherbank SE9 CK44 53
Heatherbank, Chis. CL51 71
Heather Clo. Hamptn. BE51 65
Heather Clo. Islw. BG46 47
Heather Clo. Rom. CS30 19
Heather Clo. SW8 BV44 50
Heatherdale Clo. Kings. On T. BM50 57
Heatherdean Clo. Mitch. BT52 68
Heather Dr. Rom. CS30 19
Heather Dr. Dart. CU47 64
Heather Gdns. NW11 BR32 22
Heather Gdns. Rom. CS30 19
Heather Gdns. Sutt. BS57 77
Heather Glen. Rom. CS30 19
Heatherlands, Sun. BC50 56
Heatherley Pl. E5 CB34 24
Heatherley St.
Heatherley St. E5 CB34 24
Heatherly Dr. Ilf. CK31 26
Heather Park Dr. Wem. BM36 30
Heather Pl. Esher BF56 74
Heather Ri. Bush. BD23 4
Heather Rd. NW2 BO34 22
Heather Rd. SE12 CH48 62
Heatherset Gdns. SW16 BX50 60
Heatherside Rd. Epsom BN57 76
Heatherside Rd. Sid. CP48 63
Bexley La.
Heatherton Ter. N3 BS30 14
Squires La.
Heather Wk. Edg. BM28 12
Heather Wk. Houns. BF47 56
Stephenson Rd.
Heather Way, Rom. CS30 19
Heather Way, Stan. BH29 12
Heather Way, Sth. Croy. CC58 79
Heatherwood Clo. E12 CJ34 26
Heathfield E4 CF27 16
Heathfield SW17 BU47 59
Burntwood Gra. Rd.
Heathfield, Chis. CM50 62
Heathfield Av. SW18 BT47 59
Heathfield Clo. Kes. CJ56 80
Heathfield Ct. W4 BN42 40
Heathfield Ter.
Heathfield Gdns. NW11 BQ32 22
Heathfield Gdns. SW18 BT46 50
Heathfield Gdns. W4 BN42 40
Heathfield La. Chis. CL50 62
Heathfield North Twick. BH47 57
Heathfield Pk. NW2 BQ36 31
Heathfield Rd. SW18 BT46 50
Heathfield Rd. W3 BM41 39
Heathfield Rd. Bexh. CQ45 54
Heathfield Rd. Brom. CG50 61
Heathfield Rd. Bush. BE24 4
Heathfield Rd. Croy. BZ56 78
Heathfield Rd. Kes. CJ56 80
Heathfield Rd. Walt. BE56 74
Heathfield South Twick. BH47 57
Heathfield Sq. SW18 BT47 59
Heathfield St. W11 BR40 40
Portland Rd.
Heathfield Ter. SE18 CN43 54
Heathfield Ter. W4 BN42 40
Heathfield Vale, Sth. Croy. CC58 79
Heath Gdns. Twick. BH47 57
Heathgate NW11 BS32 23
Heath Gro. SE20 CC50 61
Heath Hurst Rd. NW3 BT35 32
Keats Gro.
Heathland Rd. N16 CA33 24
Heathlands NW3 BT34 23
Heathlands Clo. Sun. BC51 65
Heathlands Ri. Dart. CU46 55
Heath La. SE3 CF44 52

Heston St. SE8 CD44 52
Hetherington Rd. SW4 BX45 51
Hetley Gdns. SE19 CA50 60
Fox Hill
Hetley Rd. W12 BP40 40
Hetton St. W6 BQ42 40
Glenthorne Rd.
Hevelius Clo. SE10 CG42 43
Hever Croft SE9 CL49 62
Hever Gdns. Brom. CL51 71
Heverham Rd. SE18 CN42 45
Heversham Rd. Bexh. CR44 54
Hewer St. W10 BQ39 40
Hewett Clo. Stan. BJ28 12
Stanmore Hill
Hewett Pl. Swan. CS52 73
Hewett Rd. Dag. CP35 36
Hewish Rd. N18 CA28 15
Hewitt Av. N22 BY30 15
Hewitt Rd. N8 BY32 24
Hewitts Rd. Orp. CQ57 81
Hewlett Rd. E3 CD37 34
Hexagon, The N6 BU33 23
Hexal Rd. SE6 CG48 61
Hexham Gdns. Islw. BJ43 48
Hexham Rd. SE27 BZ48 60
Hexham Rd. Barn. BS24 7
Hexham Rd. Mord. BS54 68
Heybourne Rd. N17 CB29 15
Heybridge Av. SW16 BX50 60
Heybridge Dr. Ilf. CM31 26
Heybridge Way E10 CD33 25
Heyford Av. SW8 BX43 51
Heyford Av. SW20 BR52 67
Heyford Rd. Mitch. BU51 68
Heygate St. SE17 BZ42 42
Heynes Rd. Dag. CP35 36
Heysham Dr. Wat. BD28 11
Heysham Rd. N15 BZ32 24
Heythorp St. SW18 BR47 58
Heyworth Rd. E5 CB35 33
Heyworth Rd. E15 CG35 34
Hibbert Rd. E17 CD33 25
Hibbert Rd. Har. BH30 12
Hibbert St. SW11 BT45 50
Hibernia Gdns. Houns. BF45 47
Hibernia Rd. Houns. BF45 47
Hichisson Rd. SE15 CC46 52
Hickin Clo. SE7 CJ42 44
Hickling Rd. Ilf. CL35 35
Hickman Av. E4 CF28 16
Hickman Rd. Rom. CP33 27
Hicks Av. Grnf. BG37 29
Hicks Clo. SW11 BU45 50
Hicks St. SE8 CD42 43
Hide Pl. SW1 BW42 41
Hide Rd. Har. BG31 20
Higgs Row SW15 BQ45 49
Felsham Rd.
Higham Hill Rd. E17 CD30 16
Higham Pl. E17 CD31 25
Higham Rd. E17 CD30 16
Higham Rd. N17 BZ31 24
Higham Rd. Wdf. Grn. CH29 17
Higham Station Av. E4 CE29 16
Highams, The, Park. Wdf. Grn. CG28 16
Highbanks Clo. Well. CO43 54
High Banks Rd. Pnr. BF29 11
Highbarrow Rd. Croy. CA54 69
High Beeches, Bans. BQ60 82
High Beeches, Orp. CO57 81
High Beeches, Sid. CQ49 63
High Beech Rd. Loug. CK24 10
High Beech, Sth. Croy. CA57 78
High Bridge SE10 CF42 43
Highbridge Rd. Bark. CL37 35
Highbrook Rd. SE3 CJ45 53
Highbroom Cres. W. Wick. CE54 70
Highbury Av. Th. Hth. BY51 69
Highbury Clo. W. Wick. CE55 79
Highbury Cres. N5 BY35 33
Highbury Gdns. Ilf. CN34 27
Highbury Gra. N5 BY35 33
Highbury Gro. N5 BY36 33
Highbury Gro. N. Mal. BN52 67
Highbury Hill N5 BY34 24
Highbury Mews N5 BY35 33
Ronalds Rd.
Highbury New Pk. N5 BZ34 24
Highbury Pk. N5 BY35 33
Highbury Pl. N5 BY36 33
Highbury Quadrant N5 BY34 24
Highbury Quadrant Est. N5 BZ34 24
Highbury Rd. SW19 BR49 58
Highbury Station Rd. N1 BY36 33
Highbury Ter. N5 BY35 33
Highbury Ter. Ms. N5 BY35 33
Ronalds Rd.
High Clere Clo. Ken. BZ57 78
Highclere Clo. Ken. BZ61 84
Highclere Rd. N. Mal. BN52 67
Highclere St. SE26 CD49 61
Highcliffe Dr. SW15 BO46 49
Highcliffe Gdns. Ilf. CK32 26
Highcombe SE7 CH43 53
Highcombe Clo. SE9 CJ47 62
High Ct. Wdf. Grn. CH29 17
Higham Rd.
Highcroft NW9 BO32 22
Highcroft Av. Wem. BL36 30
Highcroft Gdns. NW11 BR32 22
Highcroft Rd. N19 BX33 24
High Cross Rd. N17 CA31 24
Highdaun Dr. SW16 BX52 69
Highdown, Wor. Pk. BO55 76
Highdown Rd. SW15 BP46 49
High Dr. N. Mal. BN51 67
High Elms, Chig. CN28 18
High Elms, Wdf. Grn. CH28 17
Higher Dr. Bans. BQ59 82
Higher Dr. Pur. BY60 84
Higher Grn. Epsom BP60 82
High Field, Bans. BU62 83
Highfield, Rom. CS29 19
Highfield Av. NW9 BN32 22
Highfield Av. NW11 BQ33 22
Highfield Av. Erith CR43 54
Highfield Av. Grnf. BH35 30
Highfield Av. Orp. CN56 81
Highfield Av. Pnr. BE32 20
Highfield Av. Wem. BL34 21
Highfield Clo. NW9 BN32 22
Highfield Clo. Rom. CS29 19
Highfield Clo. Surb. BK54 66
Highfield Ct. N14 BW25 7
Highfield Cres. Horn. CW34 28
Highfield Dr. Brom. CG52 70
Highfield Dr. Epsom BO57 76
Highfield Dr. W. Wick. CE55 79
Highfield Gdns. NW11 BR32 22
Highfield Hill SE19 BZ50 60
Highfield Rd. N21 BY26 8
Highfield Rd. NW11 BR32 22
Highfield Rd. W3 BM39 39
Highfield Rd. Bexh. CQ46 54
Highfield Rd. Brom. CK52 71
Highfield Rd. Bush. BE25 4
Highfield Rd. Chis. CN52 72
Highfield Rd. Dart. CV47 64
Highfield Rd. Felt. BC47 56
Highfield Rd. Horn. CW34 28
Highfield Rd. Islw. BH44 48
Highfield Rd. Pur. BX58 78
Highfield Rd. Rom. CS29 19
Highfield Rd. Surb. BN54 67
Highfield Rd. Sutt. BU56 77
Highfield Rd. Walt. BC54 65
Highfield Rd. Wdf. Grn. CK29 17
Highfield Rd. S. Dart. CV47 64
Highfield Way, Horn. CW34 28
High Firs, Swan. CT52 73
High Foleys, Esher BJ57 75
High Gables, Loug. CJ25 10
High Garth, Esher BG57 74
Orchard Way
Highgate Av. N6 BV32 23
Highgate High St. N6 BV33 23
Highgate Hill N19 BV33 23
Highgate La. N6 BV33 23
Highgate Rd. NW5 BV34 23
Highgate West Hill N6 BV33 23
High Grove SE18 CM43 53
Highgrove Rd. Dag. CP35 36
Highgrove Way, Ruis. BC33 20
High Hill Est. E5 CB33 24
High Holborn WC1 BX39 42
Highland Av. W7 BH39 39
Highland Av. Dag. CS34 28
Highland Av. Loug. CK25 10
Highland Cotts. Wall. BV56 77
Highland Croft, Beck. CE50 61
Highland Dr. Bush. BG26 4
Highland Park, Felt. BC49 56
Highland Rd. SE19 CA50 60
Highland Rd. Bexh. CR46 54
Highland Rd. Brom. CG51 70
Highland Rd. Nthwd. BC31 20
Highland Rd. Pur. BY60 84
Highlands, Wat. BD27 11
Highlands Av. W3 BN40 40
Highlands Clo. Houns. BF44 47
Highlands Gdns. Ilf. CK33 26
Highlands Heath SW15 BQ47 58
Bristol Gdns.
Highlands Hill, Swan. CU51 73
Highlands Rd. Barn. BS25 7
Highlands Rd. Orp. CO54 72
Highlands, The, Edg. BM30 12
Highlands, Wat. BD26 4
High La. W7 BG39 38
Highlea Clo. NW9 BO30 13
High Level Dr. SE26 CB49 60
High Lever Rd. W10 BQ39 40
Highmead SE18 CN43 54
Highmead, Chig. CL27 17
High Mead, Har. BH32 21
High Mead, W. Wick. CF55 79
Highmead Cres. Wem. BL36 30
Highmeadow Clo. Pnr. BC31 20
High Meadow Cres. NW9 BN32 22
High Meadows, Chig. CM28 17
Highmore Rd. SE3 CG43 52
High Pk. Av. Rich. BM44 48
High Pk. Rd. Rich. BM44 48
High Path SW19 BS51 68
High Point SE9 CL48 62
High Ridge N10 BV30 14
High Ridge Clo. Epsom BO60 82
Highridge Pl. Enf. BY23 8
The Ridgeway
High Rd. E11 CG34 25
High Rd. E18 CG29 16
High Rd. N11 BW28 14
High Rd. N15 CA32 24
High Rd. N17 CA30 15
High Rd. N22 BX29 15
High Rd. NW10 BO36 31
High Rd. NW10 BP36 31
High Rd. Buck. H. CH27 17
High Rd. Bush. BG26 4
High Rd. E. Finchley N2 BT30 14
High Rd. Har. Weald BH29 12
High Rd. Ilf. CL34 26
High Rd. Leytonstone E11 CG35 34
High Rd. Leytonstone E15 CG35 34
High Rd. Loug. CJ25 10
High Rd. N. Finchley N12 BT28 14
High Rd. Rom. CO33 27
High Rd. Wdf. Grn. CG29 16
High Rd. Wem. BK35 30
High Rd. Barn. BT26 7
High Rd. Willesden Grn. NW10 BO36 31
High Rd. Wilmington CV48 64
High St. Ms. SW19 BR49 58
Courthope Rd.
High St. Southgate N14 BW26 7
Highshore Rd. SE15 CA44 51
High Silver, Loug. CJ24 10
Highstead Cres. Erith CT43 55
Highstone E11 CG32 25
Highstone Av. E11 CH32 26
High St. E11 CH32 26
High St. E13 CH37 35
High St. E15 CF37 34
High St. E17 CD32 25
High St. N8 BX31 24
High St. NW7 BP28 13
High St. NW10 BO37 31
High St. SE20 CB50 60
High St. SW6 BR45 49
High St. SW19 BQ49 58
High St. W3 BM40 39
High St. W5 BK40 39
High St. Bans. BS61 83
High St. Barkingside CM31 26
High St. Barn. BR24 6
High St. Beck. CE51 70
High St. Bex. CR47 63
High St. Brent. BK43 48
High St. Brom. CG51 70
High St. Bush. BE25 4
High St. Cars. BU56 77
High St. Cheam. BR57 76
High St. Chis. CL50 62
High St. Claygate BH57 75
High St. Colliers Wood SW19 BT50 59
High St. Cranford BC43 47
High St. Croy. BZ55 78
High St. Dart. CW46 55
High St. Edg. BM29 12
High St. Elstree BK25 5
High St. E. Mol. BR52 65
High St. Epsom BN60 82
High St. Erith CT42 46
High St. Esher BF56 74
High St. Ewell BO58 76
High St. Farn. CW53 73
High St. Farnborough CL56 80
High St. Felt. BC48 56
High St. Green St. Green CN57 81
High St. Har. BH33 21
High St. Hmptn. BF51 65
High St. Horn. CV33 28
High St. Houns. BF45 47
High St. Kings. On T. BK52 66
High St. Merton SW19 BS50 59
High St. Orp. CO55 81
High St. Pnr. BE31 20
High St. Ponders End CC25 9
High St. Pur. BY59 84
High St. Rom. CT32 28
High St. St. Mary Cray CP53 72
High St. S. Norwood SE25 CA52 69
High St. Sid. CP50 63
High St. Sthl. BE40 38
High St. Sutt. BS56 77
High St. Swan. CT52 73
High St. Tedd. BJ49 57
High St. Thames Ditton BJ54 66
High St. Th. Hth. BZ52 69
High St. Walt. BC54 65
High St. Wat. BD24 4
High St. Wealdstone BH30 12
High St. Well. CO45 54
High St. Wem. BL35 30
High St. Whitton BG47 56
High St. W. Wick. CE54 70
High St. N. E6 CK36 35
High St. N. E12 CK35 35
High Timber St. EC4 BZ40 42
Broken Wharf
Hightor Clo. Brom. CH50 62
High Tree Ct. W13 BH40 39
High Trees, Barn. BU25 7
High Trees SW2 BY47 60
High Trees, Croy. CD54 70
Highview, Pnr. BD31 20
High Vw. Sutt. BR59 82
Highview Av. Edg. BN28 13
Highview Av. Wall. BX56 78
High Vw. Clo. SE19 CA51 69
High Vw. Clo. Loug. CJ25 10
Highview Gdns. N3 BR31 22
High View Gdns. N11 BW28 14
Highview Gdns. Edg. BN28 13
Highview Pk. Bans. BS61 83
High View Rd. E18 CG31 25
High View Rd. SE19 BZ50 60
Highview Rd. W13 BJ39 39
High Vw. Rd. Orp. CL58 80
High View Rd. Sid. CO49 63
Highway, The E1 CB40 42
Highway, The, Orp. CO56 81
Highway, The, Stan. BH30 12
Highway, The, Sutt. BT58 77
Highwold, Couls. BV62 83
Highwood Av. N12 BT28 14
Highwood Av. Bush. BE23 4
Highwood Clo. Ken. BZ58 78
Highwood Clo. Ken. BZ62 84
Highwood Clo. Orp. CM55 80
Highwood Dr. Orp. CM55 80
Highwood Gdns. Ilf. CK32 26
Highwood Gro. NW7 BN28 13
Highwood Hill NW7 BO27 13
Highwood La. Loug. CL25 10
Highwood Rd. N19 BX34 24
High Worple, Har. BE33 20
Highworth Rd. N11 BW29 14
Hilary Av. Mitch. BV52 68
Hilary Clo. SW6 BS43 50
Hilary Clo. Bexh. CR44 54
Hilary Clo. Horn. CV35 37
Hilary Rd. W12 BO39 40
Hilbert Rd. Sutt. BQ55 76
Hilborough Rd. Orp. CM56 80
Hilda May Av. Swan. CT52 73
Hilda Rd. E6 CJ36 35
Hilda Rd. E16 CG38 34
Hilda Vale Clo. Orp. CL56 80
Hilda Vale Rd. Orp. CL56 80
Hildenborough Gdns. Brom. CG50 61
Hilden Dr. Erith CU43 55
Hildreth St. SW12 BV47 59
Hildyard Rd. SW6 BS43 50
Hiley Rd. NW10 BQ38 31
Hilfield La. Wat. BF23 4
Hilfield La. S. Bush. BH25 5
Hilgrove Est. NW6 BT36 32
Hiliard Rd. Nthwd. BC30 11
Hiliary Gdns. Stan. BK30 12
Hillars Hth. Rd. Couls. BX61 84
Hillary Cres. Walt. BD54 65
Hillary Ri. Barn. BSI 4 7
Hillary Rd. Sthl. BF41 38
Hill Barn. Pur. CA59 84
Hillbeck Clo. SE15 CB43 51
Hillbeck Way, Grnf. BG37 29
Hillborne Clo. Hayes BC42 38
Hillborough Clo. SW19 BT50 59
Hillbrook Rd. SW17 BU48 59
Hill Brow, Brom. CJ51 71
Hill Brow, Dart. CT46 55
Hillbrow, N. Mal. BO52 67
Hillbrow Clo. Bex. CS49 64
Hillbrow Rd. Brom. CG50 61
Hillbrow Rd. Esher BG56 74
Hillbury Av. Har. BJ32 21
Hillbury Rd. SW17 BV48 59
Hillbury Rd. Warl. CB62 84
Hill Clo. NW2 BP34 22
Hill Clo. NW11 BS32 23
Hill Clo. Barn. BQ25 6
Hill Clo. Chis. CL49 62
Hill Clo. Har. BH34 21
Hill Clo. Pur. BZ60 84
Hill Clo. Rom. CV28 19
Hill Clo. Stan. BJ28 12
Hillcote Av. SW16 BY50 60
Hill Ct. SW15 BQ46 49
Putney Hill
Hill Ct. W5 BL38 30
Ridings, The
Hillcourt Av. N12 BS29 14
Hillcourt Est. N16 BZ33 24
Hillcourt Rd. SE22 CB46 51
Hill Cres. N20 BS27 14
Hill Cres. Bex. CS47 64
Hill Cres. Har. BJ32 21
Hill Cres. Horn. CV32 28
Hill Cres. Surb. BL53 66
Hill Cres. Wor. Pk. BQ55 76
Hillcrest N6 BV33 23
Hillcrest N21 BY26 8
Hill Crest, Sid. CO47 63

Hillcrest Av. NW11 BR32 22
Hillcrest Av. Edg. BM28 12
Hillcrest Av. Pnr. BD31 20
Hill Crest Gdns. Beck. CD53 70
Hill Crest Gdns. N3 BR31 22
Hillcrest Gdns. NW2 BP34 22
Hillcrest Gdns. Esher BH55 75
Hillcrest Gdns. Ruis. BD34 20
Hillcrest Rd. E17 CF30 16
Hillcrest Rd. E18 CG30 16
Hillcrest Rd. SE26 CB49 60
Hillcrest Rd. W3 BM40 39
Hillcrest Rd. W5 BL39 39
Hillcrest Rd. Brom. CH49 62
Hillcrest Rd. Dart. CT47 64
Hillcrest Rd. Horn. CU33 28
Hillcrest Rd. Loug. CJ25 10
Hillcrest Rd. Orp. CO55 81
Hillcrest Rd. Pur. BX58 78
Hillcrest Rd. Whyt. CB62 84
Hillcrest Vw. Beck. CD53 70
Hillcroft, Loug. CL23 10
Hill Croft Av. Pnr. BE32 20
Hillcroft Av. Pur. BW60 83
Hillcroft Cres. W5 BK39 39
Hillcroft Cres. Ruis. BD34 20
Hillcroft Cres. Wat. BC26 4
Hillcroft Cres. Wem. BL35 30
Hillcroft Rd. E6 CL39 44
Hillcroome Rd. Sutt. BT57 77
Hillcross Av. Mord. BQ53 67
Hilldale Rd. Sutt. BR56 76
Hilldene Av. Rom. CV29 19
Hilldown Rd. SW16 BX50 60
Hilldown Rd. Brom. CG54 70
Hill Dr. NW9 BN33 22
Hill Dr. SW16 BX52 69
Hilldrop Cres. N7 BW35 32
Hilldrop Est. N7 BW35 32
Hilldrop La. N7 BW35 32
Hilldrop Rd. N7 BW35 32
Hilldrop Rd. Brom. CH50 62
Hillend SE18 CL44 53
Hillerdon Av. Edg. BL28 12
Hillersdon Av. SW13 BP44 49
Hill Farn Rd. W10 BQ39 40
Hillfield Av. N8 BX32 24
Hillfield Av. NW9 BO32 22
Hillfield Av. Mitch. BU53 68
Hillfield Av. Wem. BL36 30
Hillfield Clo. Har. BG31 20
Hillfield Ct. NW3 BU35 32
Belsize Av.
Hillfield Ct. Esher BF56 74
Hillfield Pk. N10 BV31 23
Hillfield Pk. N21 BY27 15
Hillfield Pk. Ms. N10 BV31 23
Hillfield Pk.
Hillfield Rd. NW6 BR35 31
Hillfield Rd. Hamptn. BE50 56
Hillfoot Av. Rom. CS30 19
Hillfoot Rd. Rom. CS30 19
Hillgate Pl. W8 BS40 41
Hillgate St. W8 BS40 41
Hill Gro. Rom. CT31 28
Hillgrove Rd. NW6 BT36 32
Hill Ho. Av. Stan. BH29 12
Hill Ho. Clo. N21 BY26 8
Hill House Rd. SW16 BX49 60
Hilliards Ct. E1 CC40 43
Hillier Clo. Barn. BS25 7
Hillier Rd. SW11 BU46 50
Hilliers La. Croy. BX55 78
Hillingdon Rd. Bexh. CS44 55
Hillingdon St. SE17 BY43 51
Hillingdon St. SW9 BY43 51
Hillington Gdns. CJ30 17
Wdf. Grn.
Hillman St. E8 CB36 33
Hillmarton Rd. N7 BX35 33
Hillmont Rd. Esher BH55 75
Hillmore Gro. SE26 CC49 61
Hill Path SW16 BX49 60
Hill Place St. E14 CE39 43
Hillreach SE18 CK42 44
Hill Rise N9 CB25 8
Hill Rise NW11 BS31 23
Hill Rise SE23 CB47 60
Hill Rise, Esher BJ55 75
Hill Rise, Grnf. BG36 29
Hill Rise, Rich. BK46 48
Hillrise Rd. N19 BX33 24
Hillrise Rd. Rom. CS29 19
Hill Rd. N10 BU30 14
Hill Rd. NW8 BT38 32
Hill Rd. Cars. BU57 77
Hill Rd. Dart. CW48 64
Hill Rd. Har. BJ32 21
Hill Rd. Mitch. BV51 68
Hill Rd. Pnr. BE32 20
Hill Rd. Pur. BX59 84
Hill Rd. Sutt. BS56 77
Hill Rd. Wem. BJ34 21
Hillsborough Grn. Wat. BC27 11
Ashburnham Dr.
Hillsborough Rd. SE22 CA46 51
Hillside NW9 BN31 22

Hillside SW19 BQ50 58
Hillside, Bans. BR61 82
Hillside, Barn. BT25 7
Hillside, Farn. CW54 73
Hill Side, Surb. BK54 66
Hillside Av. N11 BU29 14
Hillside Av. Borwd. BM24 5
Hillside Av. Pur. BY60 84
Hillside Av. Wdf. Grn. CJ28 17
Hillside Av. Wem. BL35 30
Hillside Clo. Bans. BR61 82
Hillside Clo. Mord. BR52 67
Hillside Clo. Wdf. Grn. CJ28 17
Hillside Ct. NW4 BQ30 13
Hill Side Cres. Har. BG33 20
Hill Side Cres. Nthwd. BC30 11
Hillside Cres. Wat. BD25 4
Hillside Dr. Edg. BM28 12
Hillside Est. N15 CA32 24
Hillside Gdns. E17 CF31 25
Hillside Gdns. N6 BV32 23
Hillside Gdns. N11 BW29 14
Hillside Gdns. Barn. BR25 6
Hillside Gdns. Edg. BL28 12
Hillside Gdns. Har. BL33 21
Hillside Gdns. Nthwd. BC29 11
Hillside Gdns. Wall. BW57 77
Hillside Gdns. Est. SW2 BY48 60
Hillside Gro. N14 BW26 7
Hillside Gro. NW7 BP29 13
Hillside La. Brom. CG55 79
Hillside Ri. Nthwd. BC29 11
Hillside Rd. N15 CA33 24
Hillside Rd. SW2 BX48 60
Hillside Rd. W5 BL39 39
Hillside Rd. Bush. BE25 4
Hillside Rd. Couls. BX62 84
Hillside Rd. Croy. BY56 78
Hillside Rd. Dart. CT46 55
Hillside Rd. Epsom BQ58 76
Hillside Rd. Nthwd. BC29 11
Hillside Rd. Sthl. BE38 29
Hillside Rd. Surb. BM52 66
Hillside Rd. Sutt. BR57 76
Hillside Rd. Whyt. CB62 84
Hillside, The, Orp. CO58 81
Hillsleigh Rd. W8 BR40 40
Hillsmead Way, Sth. Croy. CB60 84
Hills Pl. W1 BW39 41
Ramillies Pl.
Hills Rd. Buck. H. CH26 10
Hillstowe St. E5 CC34 25
Hill St. W1 BV40 41
Hill St. Rich. BK46 48
Hill Top NW11 BS31 23
Hill Top, Loug. CL23 10
Hilltop, Mord. BS53 68
Hilltop, Sutt. BR54 67
Hill Top Clo. Loug. CL24 10
Hilltop Gdns. NW4 BP30 13
Hilltop Gdns. Dart. CW46 55
Hilltop Gdns. Orp. CN55 81
Hilltop Rd. NW6 BS36 32
Hilltop Rd. Whyt. CA62 84
Hill Top View, Chig. CK29 17
Hilltop Way, Stan. BJ27 12
Hillview SW20 BP50 58
Hillview Av. Har. BL32 21
Hillview Av. Horn. CV32 28
Hillview Clo. Pnr. BE29 11
Hillview Clo. Pur. BY59 84
Hill View, The, Wat. BD27 11
Hill View Cres. Ilf. CK32 26
Hill View Cres. Orp. CN54 72
Hill View Dr. Well. CN44 54
Hillview Gdns. NW4 BQ31 22
Hillview Gdns. NW9 BN32 22
Hill View Gdns. Har. BF31 20
Hill View Rd. NW7 BQ28 13
Hill Vw. Rd. Chis. CL49 62
Hill Vw. Rd. Esher BJ57 75
Hill Vw. Rd. Orp. CN54 72
Hillview Rd. Sutt. BT55 77
Hill View Rd. Twick. BJ46 48
Hillway N6 BV34 23
Hill Way NW9 BK33 21
Hillworth Rd. SW2 BY47 60
Hillyard Rd. W7 BH36 38
Hillyard St. SW9 BX44 51
Hillyfields E17 CD30 16
Hillyfields, Loug. CL23 10
Hilly Fields Cres. SE4 CE45 52
Hillyfields Est. Loug. CL23 10
Hilsea St. E5 CC35 34
Hilton Av. N12 BT28 14
Hilton Way, Sth. Croy. CB61 84
Hilversum Cres. SE22 CA46 51
Dulwich Gro.
Himley Rd. SW17 BU49 59
Hinchcliffe Clo. Wall. BX57 78
Hinchley Clo. Esher BH55 75
Hinchley Dr. Esher BH55 75
Hinchley Way, Esher BJ55 75
Hinckley Rd. SE15 CB45 51
Hind Clo. Chig. CN28 18
Hind Ct. EC4 BY39 42
Gough Sq.

Hind Cres. Erith CS43 55
Hindehead Gdns. Nthlt. BE37 29
Hindes Rd. Har. BG32 20
Hinde St. W1 BV39 41
Hind Gro. E14 CE39 43
Hindhead Clo. N16 CA33 24
Hindhead Grn. Wat. BD28 11
Hindhead Way, Wall. BX56 78
Hindman's Rd. SE22 CB46 51
Hindmans Way, Dag. CQ38 36
Hindmarsh Clo. E1 CB40 42
Hindrey Rd. Est. N16 CB35 33
Pembury Est.
Hindsley Pl. SE23 CC48 61
Hinkler Clo. Wall. BX57 78
Hinkler Rd. Har. BK31 21
Hinksey Path SE2 CP41 45
Hinstock Rd. SE18 CM43 53
Hinton Av. Houns. BD45 47
Hinton Clo. SE9 CK47 62
Hinton Rd. N18 CA28 15
Hinton Rd. SE24 BZ45 51
Hinton Rd. Wall. BW57 77
Hippodrome Pl. W11 BR40 40
Hitcham Rd. E17 CD33 25
Hitchins Clo. Rom. CV28 19
Hitherbroom Rd. Hayes BE40 38
Hitherfield Rd. SW16 BX48 60
Hitherfield Rd. Dag. CQ34 27
Hither Grn. La. SE13 CF46 52
Hitherwell Rd. Har. BG30 11
Hitherwood Clo. Horn. CV35 37
Swanbourne Dr.
Hither Wood Dr. SE19 CA49 60
Hive Clo. Bush. BG27 11
Hive Rd. Bush. BG27 11
Hoadly Rd. SW16 BW48 59
Hobart Clo. N20 BU27 14
Hobart Gdns. Th. Hth. BZ52 69
Hobart Pl. SW1 BV41 41
Belgrave St.
Hobart Pl. TW10 BL46 48
Chisholm Rd.
Hobart Rd. Dag. CP35 36
Hobart Rd. Hayes BD38 29
Hobart Rd. Ilf. CM30 17
Hobart Rd. Wor. Pk. BP55 76
Hobbayne Rd. W7 BG39 38
Hobbes Wk. SW15 BP46 58
Hobbs Gn. N2 BT31 23
Hobbs Rd. SE27 BZ49 60
Hobday St. E14 CE39 43
Hobill Wk. Surb. BL53 66
Hoblands End, Chis. CN50 63
Hobury St. SW10 BT43 50
Hockenden La. Swan. CR52 72
Hocker St. E2 CA38 33
Arnold Circus
Hockett Clo. SE8 CD42 43
Hockley Av. E6 CK37 35
Hockley Dr. Rom. CU30 19
Hocroft Rd. NW2 BR35 31
Hodder Dr. Grnf. BH37 30
Hoddesdon Rd. Belv. CR42 45
Hodford Rd. NW11 BR33 22
Hodister Clo. BZ43 51
Hodnet Gro. SE16 CC42 43
Suffolk Gro.
Hodsoll Ct. Orp. CP53 72
Hodson Cres. Orp. CP53 72
Hoe St. E17 CE31 25
Hoe, The, Wat. BD27 11
Hofland Rd. W14 BR41 40
Hogarth Clo. W3 BM40 39
Hillcrest Rd.
Hogarth Cres. Croy. BZ54 69
Hogarth Est. W4 BO42 40
Hogarth Gdns. Houns. BF43 47
Hogarth H. NW11 BR31 22
Hogarth Rd. SW5 BS42 41
Hogarth Rd. Edg. BM30 12
Hogarth Way, Hamptn. BG51 65
Hoggsmill Way, Epsom BN56 76
Hog Hill Rd. Rom. CQ29 18
Holbeach Gdns. Sid. CN46 54
Holbeach Rd. SE6 CE47 61
Holbeck Row SE15 CB43 51
Holbein Ms. SW1 BV42 41
Holbein Pl. SW1 BV42 41
Holberton Gdns. NW10 BP38 31
Holborn EC1 BY39 42
Holborn Circus EC1 BY39 42
Holborn Rd. E13 CH38 35
Holborn Viaduct EC1 BY39 42
Holbrook Clo. Enf. CA23 8
Holbrook Clo. N19 BW34 23
Dartmouth Park Hill
Holbrooke Ct. N7 BX35 33
Holbrook La. Chis. CM50 62
Holbrook Rd. E15 CG37 34
Holbrook Way, Brom. CK53 71
Holburne Clo. SE3 CJ44 53
Holburne Gdns. SE3 CJ44 53
Holburne Rd. SE3 CJ44 53
Holcombe Dale NW7 BP27 13
Holcombe Hill NW7 BP27 13
Holcombe Rd. N17 CA31 24

Holcombe Rd. Ilf. CL33 26
Holcombe Rd. W6 BP42 40
Holcroft Rd. E9 CC36 34
Holden Av. N12 BS28 14
Holden Av. NW9 BN33 22
Holdenhurst Av. N12 BT29 14
Holden Rd. N12 BS28 14
Holden St. SW11 BV44 50
Holdernesse Rd. SW17 BU48 59
Holderness Way SE27 BY49 60
Holder's Hl. Cir. NW4 BR29 13
Holder's Hl. Cres. NW4 BQ30 13
Holder's Hl. Dr. NW4 BQ31 22
Holder's Hl. Gdns. NW4 BR30 13
Holder's Hl. Rd. NW4 BQ30 13
Holdgate St. SE7 CJ41 44
Westmoor St.
Hole La. NW7 BR28 13
Holford Pl. WC1 BY38 33
Holford Rd. NW3 BT34 23
Hampstead Sq.
Holgate Av. SW11 BT45 50
Holgate Ct. Edg. BL28 12
Holgate Gdns. Dag. CR35 36
Holgate Rd. Dag. CR35 36
Holland Av. SW20 BO51 67
Holland Av. Sutt. BS57 77
Holland Clo. Barn. BT25 7
Holland Clo. Barn. BT26 7
Holland Clo. Brom. CG55 79
Holland Clo. Stan. BJ28 12
Holland Gdns. W14 BR41 40
Holland Gro. SW9 BY43 51
Holland La. W14 BR41 40
Holland Pk. W11 BR40 40
Holland Pk. Av. W11 BR40 40
Holland Pk. Av. Ilf. CN32 27
Holland Pk. Gdns. W14 BR41 40
Holland Pk. Ms. W11 BR40 40
Holland Pk. Rd. W14 BR41 40
Holland Rd. E6 CK37 35
Holland Rd. E15 CG38 34
Holland Rd. NW10 BP37 31
Holland Rd. SE25 CB53 69
Holland Rd. W14 BQ41 40
Holland Rd. Wem. BK36 30
Hollands, The, Wor. Pk. BO54 67
Holland St. SE1 BY40 42
Holland St. W8 BS41 41
Holland Villas Rd. W14 BR41 40
Holland Wk. N19 BW33 23
Holland Wk. W8 BS41 41
Holland Wk. Stan. BJ28 12
Holland Way, Brom. CG55 79
Hollar Rd. N16 CA34 24
Stoke Newington High St.
Hollen St. W1 BW39 41
Wardour St.
Holles St. W1 BV39 41
Cavendish Sq.
Hollickwood Av. N12 BU29 14
Hollidge Way, Dag. CR36 36
Hollies Av. Sid. CN48 63
Hollies Clo. Hamptn. BF49 56
Hollies Clo. Twick. BH48 57
Hollies End NW7 BP28 13
Hollies Rd. W5 BK42 39
Hollies St. W1 BV39 41
Cavendish Sq.
Hollies, The E11 CH32 26
New Wanstead
Hollies Way SW12 BV47 59
Bracken Av.
Holligrave Rd. Brom. CH51 71
Hollingbourne Av. Bexh. CQ44 54
Hollingbourne Gdns. W13 BJ39 39
Hollingbourne Rd. SE24 BZ46 51
Hollingsworth Rd. Croy. CB57 78
Hollington Cres. N. Mal. BO53 67
Hollington Rd. E6 CK38 35
Hollington Rd. N17 CB30 15
Sherringham Av.
Hollingworth Rd. Brom. CL53 71
Hollit St. N7 BY35 33
Holloway Gdns. SW16 BY50 60
Holloway Rd. E6 CK38 35
Holloway Rd. E11 CF34 25
Holloway Rd. N7 BX34 24
Holloway Rd. N19 BW34 23
Holloway Rd. SW11 BT45 50
Holloway St. Houns. BF45 47
Hollowfield Wk. Nthlt. BE36 29
Hollow, The IG8 CG28 17
Holly Av. Stan. BL30 12
Holly Av. Walt. BD54 65
Hollybank Clo. Hamptn. BF49 56
Hollybrane Clo. Chis. CM50 62
Hollybush Clo. E11 CH32 26
Woodford Rd.
Hollybush Clo. Har. BH30 12
Hollybush Gdns. E2 CB38 33
Hollybush Hill E11 CG32 25
Holly Bush Hill NW3 BT35 32
Hollybush La. Hamptn. BE50 56
Hollybush Pl. E2 CB38 33
Bethnal Green Rd.

Hollybush Rd. Kings. On T. BL49 57
Hollybush St. E13 CH37 35
Holly Bush Vale NW3 BT35 32
Heath St.
Holly Clo. NW10 BO36 31
Holly Clo. Buck. H. CJ27 17
Holly Clo. Felt. BE49 56
Holly Cres. Beck. CD53 70
Holly Cres. Wdf. Grn. CF29 16
Hollycroft Av. Wem. BL34 21
Hollydale Dr. Brom. CK55 80
Hollydale Rd. SE15 CC44 52
Hollydown Way E11 CF34 25
Holly Dr. E4 CE26 9
Hollyfield Av. N11 BU28 14
Hollyfield Rd. Surb. BL54 66
Hollygrove, Bush. BG26 4
Holly Gro. NW9 BN33 22
Holly Gro. SE15 CA44 51
Hollyhedge Ter. SE13 CF46 52
Holly Hill N21 BX25 8
Holly Hill NW3 BT35 32
Holly Hill Dr. Bans. BS61 83
Holly Hill Rd. Belv. CR41 45
Holly La. Bans. BS61 83
Holly La. E. Bans. BS61 83
Holly La. W. Bans. BS62 83
Hollylodge Gdns. N6 BV33 23
Hollymead Rd. Couls. BV62 83
Hollymeoak Rd. Couls. BV62 83
Holly Mews SW10 BT42 41
Drayton Gdns.
Hollymoor La. Epsom BN58 76
Hollymount Clo. SE10 CF44 52
Holly Mt. NW3 BT35 32
Holly Bush Hill
Hollymount Rd. SE10 CF44 52
Blackheath Hill
Hollyoak Rd. SE11 BY42 42
Holly Park N3 BR31 22
Holly Park N4 BX33 24
Holly Pk. Rd. N11 BV28 14
Holly Pk. Rd. W7 BH40 39
Holly Rd. E11 CG33 25
Holly Rd. Dart. CV47 64
Holly Rd. Hamptn. BG50 56
Holly Rd. Houns. BF45 47
Holly Rd. Orp. CO57 81
Holly Rd. Twick. BH47 57
Holly St. E1 CB39 42
Holly St. E8 CA36 33
Holly Ter. N20 BT27 14
Hollytree Av. Swan. CT51 73
Hollytree Clo SW19 BQ47 58
Holly Wk. NW3 BT35 32
Holly Wk. Enf. BZ24 8
Holly Way, Mitch. BW52 68
Hollywood Gdns. Hayes BC39 38
Hollywood Rd. E4 CD28 16
Hollywood Rd. SW10 BT43 50
Hollywoods, Croy. CD58 79
Hollywood Way, Wdf. Grn. CF29 16
Holman Ct. Epsom BP58 76
Holman Hunt Ho. W14 BR42 40
Field Rd.
Holman Rd. SW11 BT44 50
Holman Rd. Epsom BN56 76
Holmbridge Gdns. Enf. CC24 9
Holmbrook Dr. NW4 BQ32 22
Holmbury Ct. SW17 BU48 59
Holmbury Gro. Croy. CD57 79
Holmbury Vw. E5 CB33 24
Holmbush Rd. SW15 BR46 49
Holmcroft Way, Brom. CK53 71
Holmdale Clo. Borwd. BL23 5
Holmdale Gdns. NW4 BQ32 22
Holmdale Lo. Ct. W3 BM40 39
Whitehall Gdns.
Holmdale Rd. NW6 BS35 32
Holmdale Rd. Chis. CM49 62
Holmdale Ter. N15 CA32 24
Holmdene Av. NW7 BP29 13
Holmdene Av. SE24 BZ46 51
Holmdene Av. Har. BF31 20
Holmdene Clo. Beck. CF51 70
Holmdene Rd. N15 CA33 24
Holmead Rd. SW6 BS43 50
Holmebury Clo. Bush. BH27 12
Holme Chase, Mord. BR53 67
Holmecote Gdns. N1 BZ35 33
Holme Lacey Rd. SE12 CG46 52
Holme Park, Borwd. BL23 5
Holme Rd. E6 CK37 35
Holme Rd. Horn. CW33 28
Holmes Av. E17 CD31 25
Holmes Av. NW7 BR28 13
Homesdale Av. SW14 BM45 48
Holmesdale Clo. SE25 CA52 69
Holmesdale Rd. N6 BV32 23
Holmesdale Rd. Bexh. CP44 54
Holmesdale Rd. Croy. BZ53 69
Holmesdale Rd. Rich. BL44 48
Holmesdale Rd. Tedd. BK50 57
Holmesley Rd. SE23 CD46 52
Holmes Rd. NW5 BV36 32
Holmes Rd. SW19 BT50 59
Holmes Rd. Twick. BH48 57
Holmes Way, Stan. BH29 12
Holmewood Gdns. SW2 BX47 60
Holmewood Rd. SE25 CA52 69
Holmewood Rd. SW2 BX47 60
Holmfield Av. NW4 BQ32 22
Holmfield Ct. NW3 BU35 32
Holmhurst Rd. Belv. CR42 45
Holmleigh Av. Dart. CV45 55
Holmleigh Rd. N16 CA33 24
Holmoak Clo. SW18 BR46 49
West Hill
Holmsdale Gro. Bexh. CT44 55
Holmsdale Rd. N11 BV28 14
Holmshaw Clo. SE26 CD49 61
Holmside Rise, Wat. BC27 11
Holmside Rd. SW12 BV46 50
Homsley Clo. N. Mal. BO53 67
Holms St. E2 CA37 33
Holmstall Av. Edg. BN30 13
Holmsworth Ct. Har. BG32 20
Holmwood, Croy. BW53 68
Holmwood, Kings. On T. BN49 58
Holmwood Av. Sth. Croy. CA60 84
Holmwood Clo. Har. BG31 20
Holmwood Clo. Nthlt. BF36 29
Holmwood Clo. Sutt. BQ58 76
Holmwood Gdns. N3 BS30 14
Holmwood Gdns. Wall. BV57 77
Holmwood Gro. NW7 BN28 13
Holmwood Rd. Chess. BL56 75
Holmwood Rd. Ilf. CN34 27
Holmwood Rd. Sutt. BQ58 76
Holne Chase N2 BT32 23
Holness Rd. E15 CG36 34
Holroyd Rd. SW15 BQ45 49
Holroyd Rd. Leath. BH58 75
Holstein Way, Belv. CP41 45
Holsworthy Way, Chess. BK56 75
Holt Clo. Borwd. BL24 5
Holt Clo. Chig. CN28 18
Holt Clo. N10 BV31 23
Holt Ct. E15 CE35 34
Holton St. E1 CC38 34
Holt Rd. E16 CK40 44
Holt Rd. Wem. BJ34 21
Holt, The, Ilf. CM29 17
Holt, The, Wall. BW56 77
Holt Way, Chig. CN28 18
Holtwhites Av. Enf. BZ23 8
Holtwhite's Hill, Enf. BY23 8
Holwell Pl. Pnr. BE31 20
Holwood Clo. Walt. BD55 74
Holwood Pk. Av. Orp. CK56 80
Holwood Pl. SW4 BW45 50
Holwood Rd. Brom. CH51 71
Holybourne Av. SW15 BP47 58
Holyoake Wk. N2 BT31 23
Holyoake Wk. W5 BK38 30
Holyport Rd. SW6 BQ43 49
Holyrood Av. Har. BE35 29
Holyrood Gdns. Edg. BM31 21
Holyrood Rd. Barn. BT25 7
Holywell La. EC2 CA38 33
Holywell Rd. Wat. BC25 4
Holywell Row EC2 CA38 33
Scrutton St.
Home Clo. Cars. BU55 77
Home Clo. Nthlt. BE38 29
Homecroft Gdns. Loug. CL24 10
Homecroft Rd. N22 BY30 15
Homecroft Rd. SE26 CC49 61
Home Farm Clo. Esher BF57 74
Homefarm Rd. W7 BH39 39
Homefield, Walt. BD56 74
Homefield Av. Ilf. CN32 27
Homefield Clo. NW10 BN36 31
Homefield Clo. Swan. CT52 73
Homefield Gdns. Mitch. BT51 68
Homefield Gdns. N2 BT31 23
Homefield Park, Sutt. BS57 77
Homefield Rise, Orp. CO54 72
Homefield Rd. SW19 BQ50 58
Homefield Rd. W4 BO42 40
Homefield Rd. Brom. CJ51 71
Homefield Rd. Bush. BF25 4
Homefield Rd. Edg. BN29 13
Homefield Rd. Walt. BE54 65
Homefield Rd. Wem. BJ35 30
Homefield St. N1 CA37 33
Regan Way
Home Fm. Clo. Epsom BQ62 82
Home Fm. Clo. Surb. BH54 66
Home Fm. Gdns. Walt. BD55 74
Home Gdns. Dag. CS34 28
Home Gdns. Dart. CW46 55
Home Hill, Swan. CT50 64
Homelands Dr. SE19 CA50 60
Home Lea, Orp. CN56 81
Osgood Av.
Homeleigh Rd. SE15 CC46 52
Home Mead, Stan. BK30 12
Homemead Rd. Brom. CK53 71
Homemead Rd. Croy. BW53 68
Home Orchard, Dart. CW46 55
Home Pk. Rd. SW19 BR49 58
Home Pk. Wk. Kings. On T. BK52 66
Homer Ct. Bexh. CS44 55
Home Rd. SW11 BU44 50
Homer Rd. E9 CD36 34
Homer Rd. Croy. CC53 70
Homer Row W1 BU39 41
Homersham Rd. Kings. On T. BM51 66
Homer St. W1 BU39 41
Homerton Gro. E9 CC35 34
Homerton High St. E9 CC35 34
Homerton Row E9 CC35 34
Homerton Ter. E9 CC36 34
Homesdale Clo. E11 CH32 26
New Wanstead
Homesdale Rd. Brom. CJ52 71
Homesdale Rd. Orp. CN54 72
Homesdale Rd. NW11 BS32 23
Homestall Rd. SE22 CC46 52
Homestead Gdns. Esher BH56 75
Homestead Paddock N14 BV25 7
Homestead Pk. NW2 BO34 22
Homestead Rd. SW6 BR43 49
Homestead Rd. Dag. CQ34 27
Homestead Rd. Orp. CO57 81
Homestead, The, Dart. CV46 55
Homeswood Cres. Chis. CN50 63
Homland Dr. Sutt. BS58 77
Hulverston Clo.
Honduras St. EC1 BZ38 33
Baltic St.
Hone Par. SE11 BX42 42
Lambeth Wk.
Honeybourne Rd. NW6 BS35 32
Honeybourne Way, Orp. CM54 71
Honeybrook Rd. SW12 BW47 59
Honeycroft, Loug. CL24 10
Honey La. EC2 BZ39 42
Cheapside
Honeypot Clo. Har. BL31 21
Honeysett Rd. N17 CA30 15
Reform Row
Honeysuckle Clo. Rom. CV29 19
Cloudberry Rd.
Honeysuckle La. N22 BZ30 15
Crawley Rd.
Honeywell Rd. SW11 BU46 50
Honeywood Rd. NW10 BO37 31
Honeywood Rd. Islw. BJ45 48
Honeywood Wk. Cars. BU56 77
Honister Clo. Stan. BJ30 12
Honister Gdns. Stan. BJ29 12
Honister Heights, Pur. BZ60 84
Honister Pl. Stan. BJ30 12
Honiton Rd. NW6 BR37 31
Honiton Rd. Rom. CS32 28
Honiton Rd. Well. CN44 54
Honley Rd. SE6 CE47 61
Honor Oak Est. SE4 CD45 52
Honor Oak Ri. SE23 CC46 52
Honor Oak Rd. SE23 CC47 61
Hood Av. N14 BV25 7
Hood Av. N14 BW26 7
Hood Av. SW14 BN46 49
Hood Av. Orp. CO53 72
Hood Clo. Croy. BY54 69
Hoodcote Gdns. N21 BY26 8
Hood Rd. SW20 BO50 58
Hood Rd. Rain. CT37 37
Hood Wk. Rom. CR30 18
Hookers Rd. E17 CC31 25
Hookfarm Rd. Brom. CJ53 71
Hookfield, Epsom BN60 82
Hook Grn. La. Dart. CT48 64
Hook Hill, Sth. Croy. CA58 78
Hooking Grn. Har. BF32 20
Hook La. Well. CN46 54
Hook Rise N. Surb. BL55 75
Hook Rise S. Surb. BL55 75
Hook Rd. Chess. BK56 75
Hook Rd. Epsom BN57 76
Hooks Hall Dr. Dag. CS34 28
Hook, The, Barn. BT25 7
Hook Wk. Edg. BN29 13
Hooper Rd. E16 CH39 44
Hoopers Ct. SW3 BU41 41
Basil St.
Hooper's Ms. W3 BM40 39
Churchfield Rd.
Hooper St. E1 CB39 42
Hoop La. NW11 BR33 22
Hope Clo. IG8 CJ29 17
West Gro.
Hope Clo. SE12 CH48 62
Hopedale Rd. SE7 CH43 53
Hopefield Av. NW6 BR37 31
Hope Pk. Brom. CG50 61
Hope St. SW11 BT45 50
Hopewell St. SE5 BZ43 51
Hop Gdns. WC2 BX40 42
Bedfordbury
Hopgood St. W12 BQ40 40
Macfarlane Rd.
Hopkins St. W1 BW39 41
Hopland Rd. W14 BR42 40
Hopper's Rd. N13 BY27 15
Hopper's Rd. N21 BY27 15
Hoppett Rd. E4 CG27 16
Hopping La. N1 BY36 33
St. Mary's Gro.
Hoppingwood Av. N. Mal. BO52 67
Hopton Gdns. N. Mal. BP53 67
Hopton Rd. SW16 BX49 60
Hopton St. SE1 BY40 42
Hopwood Rd. SE17 BZ43 51
Hopwood Wk. E8 CB36 33
Wilman Gro.
Horace Av. Rom. CS33 28
Horace Rd. E7 CH35 35
Horace Rd. Ilf. CM31 26
Horace Rd. Kings. On T. BL52 66
Horatio St. E2 CA37 33
Horatius Way, Croy. BX56 78
Horbury Cres. W11 BS40 41
Horbury Ms. W11 BR40 40
Horder Rd. SW6 BR44 49
Hordle Promenade E. SE15 CA43 51
Hordle Promenade N. SE15 CA43 51
Hordle Promenade S. SE15 CA43 51
Hordle Promenade W. SE15 CA43 51
Horizon Way SE7 CH42 44
Horley Clo. Bexh. CR46 54
Horley Rd. SE9 CK49 62
Hormead Rd. W9 BR38 31
Hornbeam Clo. Borwd. BM23 5
Hornbeam Cres. Brent. BJ43 48
Hornbeam Gro. E4 CG27 16
Hornbeam La. DA7 CS44 55
Hornbeam La. E4 CG25 9
Hornbeam Rd. Buck. H. CJ27 17
Hornbeam Rd. Hayes BD39 38
Hornbeam Way, Brom. CL53 71
Hornbuckle Clo. Har. BG34 20
Hornby Clo. NW3 BT36 32
Horncastle Clo. SE12 CH47 62
Horncastle Rd. SE12 CH47 62
Hornchurch Hill, Whyt. CA62 84
Hornchurch Rd. Horn. CT33 28
Horndean Clo. SW15 BP47 58
Bessborough Rd.
Horndean Clo. Rom. CS30 19
Horndon Grn. Rom. CS30 19
Horndon Rd. Rom. CS30 19
Hornets, The, Wat. BC24 4
Horne Way SW15 BQ44 49
Hornfair Rd. SE7 CJ43 53
Hornford Way, Rom. CT33 28
Horniman Dr. SE23 CB47 60
Horning Clo. SE9 CK49 62
Horn La. SE10 CH42 44
Horn La. W3 CB40 40
Horn La. Wdf. Grn. CH29 17
Horn Pk. Clo. SE12 CH46 53
Horn Pk. La. SE12 CH46 53
Hornsey La. N6 BX33 24
Hornsey La. Est. N19 BW33 23
Hornsey La. Gdns. N6 BW33 23
Hornsey Pk. Rd. N8 BX31 24
Hornsey Rise N19 BW33 23
Hornsey Rise Gdns. N19 BW33 23
Hornsey Rd. N7 BX34 24
Hornsey Rd. N19 BX33 24
Hornsey St. N7 BX35 33
Hornshay Pl. SE15 CC43 52
Hornshay St. SE15 CC43 52
Horns Rd. Ilf. CM32 26
Hornton Pl. W8 BS41 41
Hornton St.
Hornton St. W8 BS41 41
Horsa Rd. SE12 CJ47 62
Horsa Rd. Erith CR43 54
Horsbury Cres. W11 BS40 41
Horsecroft, Bans. BS62 83
Horsecroft Clo. Orp. CO54 72
Horsecroft Rd. Edg. BN29 13
Horse Fair, Kings. On T. BL51 66
Wood St.
Horseferry Pl. SE10 CF43 52
Horseferry Rd. SW1 BW41 41
Horse Guards Av. SW1 BX40 42
Horse Guards Rd. SW1 BW40 41
Horsell Rd. N5 BY35 33
Horsell Rd. Orp. CO51 72
Horselydown La. SE1 CA41 42
Horsemonden Clo. Orp. CN54 72
Horsendon Av. Grnf. BH35 30
Horsendon Cres. Grnf. BH35 30
Horsendon La. N. Grnf. BH36 30
Horsendon La. S. Grnf. BJ37 30
Horseshoe Alley SE1 BZ40 42
Bankside
Horseshoe Clo. NW2 BP34 22
Horse Shoe Cres. Nthlt. BF37 29
Horseshoe Grn. Sutt. BS55 77
Horseshoe La. Enf. BZ24 8
Chase Side
Horseshoe La. N20 BQ26 6
Horseshoe, The, Bans. BR61 82
Horseshoe, The, Couls. BW60 83
Horsfield Gdns. SE9 CK46 53
Horsfield Rd. SE9 CJ46 53
Horsford Rd. SW2 BX46 51

Ilkeston Ct. E5	CC35	34
Clapton Park Est.		
Ilkley Rd. E16	CJ39	44
Ilkley Rd. Wat.	BD28	11
Illingworth Way, Enf.	CA25	8
Ilmington Rd. Har.	BK32	21
Ilminster Gdns. SW11	BU45	50
Imber Clo. N14	BW26	7
Imber Ct. E. Mol.	BG54	65
Imber Gro. Esher	BG54	65
Imber Pk. Rd. Esher	BG54	65
Imperial Av. N16	CA35	33
Imperial Clo. Har.	BF32	20
Imperial Dr. Har.	BF33	20
Imperial Institute Rd. SW7	BT41	41
Imperial Mews E6	CJ37	35
Imperial Rd. N22	BX30	15
Imperial Rd. SW6	BS44	50
Imperial Sq. SW6	BS44	50
Imperial St. E3	CF38	34
Imperial Way SE18	CK43	53
Imperial Way, Chis.	CM48	62
Imperial Way, Croy.	BY57	78
Imperial Way, Har.	BL32	21
Imperial Way, Wat.	BD23	4
Inca Dr. SE9	CL47	62
Inchmery Rd. SE6	CE48	61
Independents Rd. SE3	CG45	52
Inderwick Rd. N8	BX32	24
India St. EC3	CA39	42
Jewry St.		
India Way W12	BP40	40
Indus Rd. SE7	CJ43	53
Industrial Est. Grnf.	BF37	29
Industrial Est. Mitch.	BU53	68
Ingate Pl. SW8	BV44	50
Ingatestone Rd. E12	CJ33	26
Ingatestone Rd. SE25	CB52	69
Ingatestone Rd. Wdf. Grn.	CH29	17
Ingelow Rd. SW8	BV44	50
Ingersoll Rd. W12	BP40	40
Ingestre Pl. W1	BW39	41
Ingestre Rd. E7	CH35	35
Ingham Clo. Sth. Croy.	CC58	79
Ingham Rd. NW6	BS35	32
Ingham Rd. Sth. Croy.	CC58	79
Inglebert St. EC1	BY38	33
Ingelboro' Dr. Pur.	BZ60	84
Ingleby Clo. Dag.	CR36	36
Ingleby Dr. Har.	BG34	20
Ingleby Gdns. Chig.	CO27	18
Lambourne Rd.		
Ingleby Rd. N7	BX34	24
Tollington Way		
Ingleby Rd. Dag.	CR36	36
Ingleby Rd. Ilf.	CL33	26
Ingleby Rd. N7	BX34	24
Ingleby Way, Chis.	CL49	62
Ingleby Way, Wall.	BW57	77
Ingle Clo. Pnr.	BE31	20
Ingledew Rd. SE18	CM42	44
Inglehurst Gdns. Ilf.	CK32	26
Inglemere Rd. SE23	CC48	61
Inglemere Rd. Mitch.	BU50	59
Inglesham Wk. E9	CD36	34
Trowbridge Est.		
Ingleside Clo. Beck.	CE50	61
Ingleside Gro. SE3	CG43	52
Inglethorpe St. SW6	BQ44	49
Ingleton Av. Well.	CO46	54
Ingleton Rd. N18	CB29	15
Ingleton Rd. Cars.	BU58	77
Ingleway N12	BT29	14
Inglewood Clo. Chig.	CN29	18
Inglewood Clo. Horn.	CV35	37
Inglewood Copse, Brom.	CK51	71
Inglewood Rd. NW6	BS35	32
Inglewood Rd. Bexh.	CS45	55
Inglis Rd. W5	BL40	39
Inglis Rd. Croy.	CA54	69
Ingram Av. NW11	BT33	23
Ingram Clo. Stan.	BK28	12
Ingram Ho. Kings. On T.	BK51	66
Park Rd.		
Ingram Rd. N2	BU31	23
Ingram Rd. Dart.	CW47	64
Ingram Rd. Th. Hth.	BZ51	69
Ingrams Clo. Walt.	BD56	74
Ingram Way, Grnf.	BG37	29
Ingrave Rd. Rom.	CS31	28
Ingrave St. SW11	BT45	50
Ingrebourne Rd. Rain.	CU38	37
Inigo Jones Rd. SE7	CJ43	53
Inkerman Rd. NW5	BV36	32
Inks Green E4	CE28	16
Inman Rd. NW10	BO37	31
Inman Rd. SW18	BT47	59
Inman's Row, Wdf. Grn.	CH28	17
Inner Cir. NW1	BV38	32
Inner Pk. Rd. SW19	BQ47	58
Inner Staithe W4	BN43	49
Upper Staithe		
Innes Clo. SW20	BR51	67
Innes Gdns. SW15	BP46	49
Innes Lo. SE23	CC48	61
Innes Yd. Croy.	BZ55	78
Whitgift St.		
Inniskilling Rd. E13	CJ37	35
Inskip Dr. Horn.	CW33	28
Inskip Rd. Dag.	CP33	27
Institute Pl. E8	CB35	33
Amhurst Rd.		
Instone Clo. Wall.	BX57	78
Instone Rd. Dart.	CV47	64
Instow Pl. N7	BY35	33
Queensland Rd.		
Insurance St. WC1	BY38	33
Margery St.		
International Av. Houns.	BD42	38
Inverarey Pl. SE18	CM43	53
Inver Clo. E5	CB34	24
Southwold Rd.		
Inverclyde Gdns. Rom.	CP31	27
Inveresk Gdns. Wor. Pk.	BO55	76
Inverforth Rd. N11	BV28	14
High Rd.		
Inverine Rd. SE7	CH42	44
Invermore Pl. SE18	CM42	44
Inverna Gdns. W8	BS42	41
Inverness Av. Enf.	CA23	8
Inverness Ct. W3	BM39	39
Lings Rd.		
Inverness Dr. Ilf.	CN29	18
Inverness Gdns. W8	BS40	41
Inverness Ms. W2	BS40	41
Inverness Ter.		
Inverness Pl. W2	BS40	41
Inverness Ter.		
Inverness Rd. N18	CB28	15
Inverness Rd. Houns.	BE45	47
Inverness Rd. Sthl.	BE42	38
Inverness Rd. Wor. Pk.	BQ54	67
Inverness St. NW1	BV37	32
Inverness Ter. W2	BS40	41
Inverton Rd. SE15	CC45	52
Invicta Clo. Chis.	CL49	62
Invicta Gro. Nthlt.	BE38	29
Invicta Rd. SE3	CH43	53
Inwood Av. Houns.	BG45	47
Inwood Clo. Croy.	CD55	79
Inwood Ct. Walt.	BD55	74
Inwood Rd. Houns.	BF45	47
Inworth St. SW11	BU44	50
Iona Clo. SE6	CE47	61
Ravensbourne Pk.		
Ipswich Rd. SW17	BV50	59
Ireland Yd. EC4	BY39	42
St. Andrew's Hill		
Irene Rd. SW6	BS44	50
Irene Rd. Orp.	CN54	72
Iris Av. Bex.	CQ46	54
Iris Clo. Surb.	BL54	66
Iris Ct. Pnr.	BD31	20
Nursery Rd.		
Iris Cres. Bexh.	CQ43	54
Iris Path, Rom.	CV29	19
Iris Rd. Epsom	BM56	75
Iris Way E4	CD29	16
Irkdale Av. Enf.	CA23	8
Irongate Wf. Rd. W2	BT39	41
Iron Mill La. Dart.	CT45	55
Iron Mill Pl. Dart.	CT45	55
Iron Mill Rd. SW18	BS46	50
Ironmonger La. EC2	BZ39	42
Ironmonger Row EC1	BZ38	33
Irons Way, Rom.	CS29	19
Irvine Av. Har.	BJ31	21
Irvine Clo. N20	BU27	14
Irvine Way, Orp.	CN54	72
Irving Av. Nthlt.	BD37	29
Irving Gro. SW9	BX44	51
Irving Rd. W14	BQ41	40
Irving St. WC2	BW40	41
Leicester Sq.		
Irving Way, Swan.	CS51	73
Irwin Av. SE18	CN43	54
Irwin Gdns. NW10	BP37	31
Isabella Dr. Orp.	CM56	80
Isabella Rd. E9	CC35	34
Isabel St. SW9	BX44	51
Isbell Gdns. Rom.	CT29	19
Isel Way SE22	CA46	51
Dulwich Gro.		
Isham Rd. SW16	BX51	69
Isis St. SW18	BT48	59
Island Fm. Av. E. Mol.	BF53	65
Island Fm. Rd. E. Mol.	BE53	65
Island Rd. Mitch.	BU50	59
Island Row E14	CD39	43
Isla Rd. SE18	CM43	53
Islay Gdns. Houns.	BD46	47
Islay Wk. N1	BZ36	33
Isledon Rd. N7	BX34	24
Islehurst Clo. Chis.	CL51	71
Summer Hill		
Islehurst Cr. Chis.	CL51	71
Islington High St. N1	BY37	33
Islington Pk. St. N1	BY36	33
Islip Gdns. Edg.	BN29	13
Islip Gdns. Nthlt.	BE36	29
Islip Manor Rd. Nthlt.	BE36	29
Islip St. NW5	BW35	32
Ismailia Rd. E7	CH36	35
Ivanhoe Dr. Har.	BJ31	21
Ivanhoe Rd. SE5	CA45	51
Ivanhoe Rd. Houns.	BD45	47
Ivatt Way N17	BZ31	24
Iveagh Av. NW10	BM37	30
Iveagh Clo. NW10	CM37	30
Ivedon Rd. Well.	CP44	54
Ive Farm Clo. E10	CE34	25
Iveley Rd. SW4	BW44	50
Ivere Dr. Barn.	BS25	7
Iverna Gdns. W8	BS41	41
Iverson Rd. NW6	BR36	31
Ivers Way, Croy.	CE57	79
Ives Gdns. Rom.	CT31	28
Ives Rd. E16	CG39	43
Ives St. SW3	BU42	41
Ivestor Ter. SE23	CC47	61
Ivimey St. E2	CB38	33
Pollard Row		
Ivinghoe Clo. Enf.	BZ23	8
Ivinghoe Rd. Bush.	BG26	4
Ivinghoe Rd. Dag.	CO35	36
Ivor Gro. SE9	CL47	62
Ivor Pl. NW1	BU38	32
Ivor St. NW1	BW36	32
Ivorydown, Brom.	CH49	62
Ivychurch Clo. SE20	CC50	61
Laurel Gro.		
Ivy Clo. Har.	BE35	29
Ivy Clo. Pnr.	BD33	20
Ivy Clo. Sun.	BD51	65
Ivy Cres. W4	BN42	40
Ivydale Rd. SE15	CC46	52
Ivydale Rd. Cars.	BU55	77
Ivyday Gro. SW16	BX48	60
Ivydene, E. Mol.	BE53	65
Ivydene Clo. Sutt.	BT56	77
Ivydene Rd. E8	CB36	33
Ivy Gdns. N8	BX32	24
Ivy Gdns. Mitch.	BW52	68
Ivyhouse Rd. Dag.	CP36	36
Ivy La. Houns.	BE45	47
Ivymount Rd. SE27	BY48	60
Ivy Pl. Surb.	BL53	66
Ivy Rd. E16	CH39	44
Ivy Rd. E17	CE32	25
Ivy Rd. N14	BW26	7
Ivy Rd. NW2	BQ35	31
Ivy Rd. SE4	CD45	52
Ivy Rd. Houns.	BF45	47
Ivy Rd. Surb.	BM54	66
Ivy St. N1	CA37	33
Ivy Wk. Dag.	CQ36	36
Ixworth Pl. SW3	BU42	41
Izane Rd. Bexh.	CQ45	54
Jackass La. Kes.	CH57	80
Jack Cornwell St. E12	CL35	35
Jacklin Grn. Wdf. Grn.	CH28	17
Jackman Ms. NW10	BO34	22
North Cicular Rd.		
Jackman St. E8	CB37	33
Jackson Clo. Epsom	BN60	82
Jackson Rd. N7	BX35	33
Jackson Rd. Bark.	CM37	35
Jackson Rd. Barn.	BT25	7
Jackson Rd. Brom.	CK55	80
Jackson's La. N6	BV33	23
Jackson's Ms. NW10	BO34	22
Jackson's Pl. Croy.	BZ54	69
Jackson St. SE18	CL43	53
Jackson Way, Grnf.	BF41	38
Jack Walker Ct. N5	BY35	33
Jacob St. SE1	CB41	42
Jacobs Well Ms. W1	BV39	41
George St.		
Jaffray Pl. SE27	BY49	60
Jaffray Rd. Brom.	CJ52	71
Jago Clo. SE18	CM43	53
Jago Wk. SE5	BZ43	51
Lomond Gro.		
Jamaica Rd. SE16	CA41	42
Jamaica Rd. Th. Hth.	BY53	69
Jamaica St. E1	CC39	43
James Av. NW2	BQ35	31
James Av. Dag.	CQ33	27
James Bedford Clo. Pnr.	BD30	11
James Clo. Bush.	BE25	4
James Clo. Rom.	CU32	28
James Ct. N1	BZ36	33
Morton Rd.		
James Gdns. N22	BY29	15
James Gdns. Wem.	BK36	30
James La. E10	CF33	25
James La. E11	CF33	25
James Newman Ct. SE9	CL49	62
Jameson St. W8	BS40	41
James Pass. N17	CA29	15
Church Rd.		
James Rd. Dart.	CT47	64
James's Cotts. Rich.	BM43	48
Kew Rd.		
James St. W1	BV39	41
James St. WC2	BX40	42
Long Acre		
James St. Bark.	CM36	35
James St. Enf.	CA25	8
James St. Houns.	BG45	47
James St. Wem.	BK36	30
Jamestown Rd. NW1	BV37	32
Janet St. E14	CE41	43
Janeway Pl. SE16	CB41	42
Janeway St.		
Janeway St. SE16	CB41	42
Janice Mews, Ilf.	CL34	26
Oakfield Rd.		
Jansen Wk. SW11	BT45	50
Wayland Rd.		
Janson Clo. E15	CG35	34
Janson Rd.		
Janson Rd. E15	CG35	34
Jansons Rd. N15	CA31	24
Japan Cres. N4	BX33	24
Japan Rd. Rom.	CP32	27
Jaqueline Clo. Nthlt.	BE37	29
Jarrow Rd. N15	CB31	24
Jarrow Rd. SE16	CC42	43
Jarrow Rd. Rom.	CP32	27
Jarvis Clo. Barn.	BQ25	6
Jarvis Rd. SE22	CA45	51
Jarvis Rd. Sth. Croy.	BZ57	78
Jasmine Clo. Orp.	CL55	80
Jasmine Gdns. Croy.	CE55	79
Jasmine Gro. SE20	CB51	69
Jasmine Way, E. Mol.	BH52	66
Jasmin Rd. Epsom	BM57	75
Jason Clo. E15	CG35	34
Jason Rd.		
Jason Wk. SE9	CL49	62
Jasper Pass. SE19	CA50	60
Jasper Rd. SE19	CA50	60
Javelin Way, Nthlt.	BD38	29
Jay Ms. SW7	BT41	41
Jays Bldgs. W1	BX37	33
Rodney St.		
Jebb Av. SW2	BX46	51
Jebb St. E3	CE37	34
Jedburgh Rd. E13	CJ38	35
Jedburgh St. SW11	BV45	50
Jeddo Rd. W12	BO41	40
Jefferson Clo. Ilf.	CL32	26
Jefferson Clo. W13	BJ41	39
Jeffreys Pl. NW1	BW36	32
Jeffreys St.		
Jeffrey's Rd. SW4	BX44	51
Jeffreys Rd. Enf.	CD24	9
Jeffrey's St. NW1	BW36	32
Jeffreys Way, Enf.	CD24	9
Jeffs Rd. Sutt.	BR56	76
Jefrey's Wk. SW4	BX44	51
Jeken Rd. SE9	CJ45	53
Jelf Rd. SW2	BY46	51
Jellicoe Gdns. Stan.	BJ29	12
Jellicoe Rd. N17	BZ29	15
Jenkins La. Bark.	CL37	35
Jenkins Rd. E13	CH38	35
Jenner Pl. SW13	BP43	49
Jenner Rd. N16	CA34	24
Jennett Rd. Croy.	BY55	78
Jennifer Rd. Brom.	CG48	61
Jennings Rd. SE22	CA46	51
Jennings Way, Barn.	BQ24	6
Jenningtree Rd. Erith	CU43	55
Jenningtree Way, Belv.	CS41	46
Jenny Path, Rom.	CV29	19
Jenson Way SE19	CA50	60
Fox Hill		
Jenton Av. Bexh.	CQ44	54
Jephson Rd. E7	CJ36	35
Jephson St. SE5	BZ44	51
Grove La.		
Jephtha Rd. SW18	BS46	50
Jeppo's La. Mitch.	BU52	68
Jerdan Pl. SW6	BS43	50
Fulham Broadway		
Jeremiah St. E14	CE39	43
Jeremys Grn. N18	CB28	15
Jermyn St. SW1	BW40	41
Jerningham Av. Ilf.	CL30	17
Jerningham Rd. SE14	CD44	52
Jerome Cres. NW8	BU38	32
Jerome Pl. SE17	BZ43	51
Hillingdon St.		
Jerome St. E1	CA38	33
Calvin St.		
Jerome St. E1	CA39	42
Jerrard St. SE13	CE45	52
Jersey Av. Stan.	BJ30	12
Jersey Dr. Orp.	CM53	71
Jersey Rd. E11	CF33	25
Jersey Rd. E16	CH39	44
Jersey Rd. SW17	BV50	59
Jersey Rd. W7	BJ41	39
Jersey Rd. Houns.	BF44	47
Jersey Rd. Ilf.	CL35	35
Jersey Rd. Rain.	CU36	37
Jersey St. E2	CB38	33
Bethnal Green Rd.		
Jerusalem Pl. EC1	BY38	33
Aylesbury St.		
Jerviston Gdns. SW16	BY50	60
Jesmond Av. Wem.	BL36	30
Jesmond Rd. Croy.	CA54	69
Jesmond Way, Stan.	BL28	12
Jessam Av. E5	CB33	24
Jessamine Rd. W7	BH40	39
Jessel Dr. Loug.	CM23	10
Jesse Rd. E10	CF33	25
Jessica Rd. SW18	BT46	50

Jessops Way, Croy. BW53 68
Jessup Clo. SE18 CM42 44
Jetstar Way, Nthlt. BE38 29
Jevington Way SE12 CH47 62
Jewel Rd. E17 CE31 25
Jewry St. EC3 CA39 42
Jews Row SW18 BS45 50
Jews Wk. SE26 CB49 60
Jeymer Av. NW2 BP35 31
Jeymer Dr. Grnf. BG37 29
Jeypore Rd. SW18 BT46 50
Jillan Clo. Hamptn. BF50 56
Joan Cres. SE9 CJ47 62
Joan Gdns. Dag. CQ34 27
Joan Rd. Dag. CO34 27
Joan St. SE1 BY40 42
Jocelyn Rd. Rich. BL45 48
Jockey's Fields WC1 BX39 42
Jodrell Rd. E3 CD37 34
Joel St. Nthwd. BC31 20
Johanna St. SE1 BY41 42
John Adam St. WC2 BX40 42
Villiers St.
John Aird Ct. W2 BT39 41
Howley Pl.
John Barnes Wk. E15 CG36 34
John Bradshaw Rd. N14 BW26 7
John Burns Dr. Bark. CN36 36
John Campbell Rd. N16 CA35 33
John Carpenter St. EC4 BY40 42
John Clynge Ct. SW15 BP45 49
Woodborough Rd.
John Dwight Ho. SW6 BS45 50
John Felton Rd. SE16 CB41 42
John Fisher St. E1 CB40 42
John Islip St. SW1 BW42 41
John Newton Ct. Well. CO45 54
John Parker Clo. Dag. CR36 36
John Penn St. SE13 CE44 52
John Perrin Pl. Har. BL33 21
Preston Hill
John Prince's St. W1 BV39 41
John Rennie Wk. E1 CB40 42
John Ruskin St. SE5 BY43 51
John's Av. NW4 BQ31 22
John's Gro. Rich. BL45 48
Kew Foot Rd.
John's La. Mord. BT53 68
John's Ms. WC1 BX38 33
Johnson Clo. Mitch. BV52 68
Johnson Rd. Brom. CJ53 71
Johnson Rd. Croy. BZ54 69
Cromwell Rd.
Johnson Rd. Houns. BD43 47
Johnsons Av. Sev. CR58 81
Johnson's Clo. Cars. BU55 77
Johnsons Dr. Hamptn. BG51 65
Johnson St. E1 CC40 43
Johnson St. Sthl. BD41 38
Johnson Way NW10 BM38 30
John Spencer Sq. N1 BY36 33
John's Pl. E1 CB39 42
Nelson St.
John's Ter. Croy. CA54 69
Johnstone Rd. E6 CK38 35
Johnstone Ter. NW2 BQ34 22
Johnstone Rd. Wdf. Grn. CH29 17
John St. E15 CG37 34
John St. SE25 CB52 69
John St. WC1 BX38 33
John St. Enf. CA25 8
John St. Houns. BE44 47
John Wilson St. SE18 CL41 44
John Woolley Clo. SE13 CF45 52
Joiner St. SE1 BZ40 42
Jolleys La. Har. BG33 20
Jolly's La. Hayes BD39 38
Jonathan St. SE11 BX42 42
Jones Rd. E13 CH38 35
Holborn Rd.
Jones St. W1 BV40 41
Bourdon St.
Jonson Clo. Hayes BC39 38
Joram Way SE16 CB42 42
Egan Way
Jordan Rd. Grnf. BJ37 30
Jordans Clo. Islw. BH44 48
Jordans Clo. Sth. Croy. CA59 84
Jordans Way, Rain. CV37 37
Josephine Av. SW2 BX46 51
Joseph St. E3 CD38 34
Joshua St. E14 CF39 43
Joubert St. SW11 BU44 50
Jowett St. SE15 CA43 51
Joyce Av. N18 CA28 15
Joyce Grn. La. Dart. CW45 55
Joyce Grn. Wk. Dart. CW46 55
Joydens Wd. Rd. Bex. CS48 64
Joydon Dr. Rom. CO32 27
Jubilee Av. E4 CF28 16
Jubilee Av. Rom. CR32 27
Jubilee Av. Twick. BG47 56
Jubilee Clo. NW9 BN32 22
Jubilee Clo. Pnr. BD30 11
Jubilee Clo. Rom. CR32 27
Jubilee Cotts. SE9 CK46 53
Jubilee Ct. Th. Hth. BY52 69
Jubilee Cres. E14 CF41 43
Jubilee Cres. N9 CB26 8
Jubilee Dr. Ruis. BD35 29
Jubilee Gdns. Sthl. BF39 38
Jubilee Pl. SW3 BU42 41
Jubilee Rd. Grnf. BJ37 30
Jubilee Rd. Orp. CQ57 81
Jubilee Rd. Sutt. BQ57 76
Jubilee St. E1 CC39 43
Jubilee Way, Chess. BM56 75
Jubilee Way, Sid. CO48 63
Jubilee Way SW19 BS51 68
Judd St. WC1 BX38 33
Jude St. E16 CG39 43
Judges Wk. NW3 BT34 23
Judith Av. Rom. CR29 18
Juer St. SW11 BU43 50
Julia Gdns. Bark. CP37 36
Julian Av. W3 BM40 39
Julian Clo. Barn. BS24 7
Julian Hill, Har. BH34 21
Julian Rd. Orp. CO57 81
Julian Tayler Path SE23 CB48 60
School La.
Julia St. NW5 BV35 32
Oak Village
Julien Rd. W5 BK42 39
Julien Rd. Couls. BW61 83
Juliet House N1 CA37 33
Purcell St.
Junction App. SE13 CF45 52
Junction Ms. W2 BU39 41
Sale Pl.
Junction Rd. E13 CH37 35
Junction Rd. N9 CB26 8
Junction Rd. N17 CB31 24
Junction Rd. N19 BW35 32
Junction Rd. W5 BK42 39
Junction Rd. Dart. CV46 55
Junction Rd. Har. BG32 20
Junction Rd. Rom. CT31 28
Junction Rd. Sth. Croy. BZ57 78
Junction Rd. E. Rom. CQ33 27
Kenneth Rd.
Junction Rd. W. Rom. CQ33 27
June Clo. Couls. BV60 83
Juniper Gdns. SW16 BW51 68
Lillian Rd.
Juniper Rd. Ilf. CL35 35
Riverdene Rd.
Juniper Wk. Rom. CW30 19
Juno Way SE14 CC43 52
Jupiter Way N7 BX36 33
Jupp Rd. W. E15 CF37 34
Jupp Rd. E15 CF36 34
Justice Wk. SW3 BU43 50
Lawrence St.
Justin Clo. Brent. BK43 48
Jute La. Enf. CD23 9
Jutland Rd. E13 CH38 35
Jutland Rd. SE6 CF47 61
Jutsums Av. Rom. CR32 27
Jutsums La. Rom. CR32 27
Juxon St. SE11 BX42 42
Kadona Clo. Pnr. BC32 20
Kale Rd. Belv. CQ41 45
Kambala Rd. SW11 BT44 50
Kangley Br. Rd. SE26 CD49 61
Karen Clo. Rain. CT37 37
Karen Ct. Brom. CG51 70
Karen Ter. E11 CG34 25
Montague Rd.
Karoline Gdns. Grnf. BG37 29
Kashgar Rd. SE18 CN42 45
Kashmir Rd. SE7 CJ43 53
Kassala Rd. SW11 BU44 50
Kate St. SW12 BV47 59
Katharine St. Croy. BZ55 78
Katherine Gdns. SE9 CJ45 53
Katherine Gdns. Ilf. CM29 17
Katherine Rd. E6 CJ36 35
Katherine Rd. E7 CJ35 35
Katherine Sq. W11 BQ40 40
Wilsham St.
Kathleen Av. W3 BN39 40
Kathleen Av. Wem. BL36 30
Kathleen Rd. SW11 BU45 50
Kayemoor Rd. Sutt. BT57 77
Kay Rd. SW9 BX44 51
Kay St. E2 CB37 33
Kay St. E15 CF36 34
New Mk. St.
Kay St. Well. CO44 54
Kean St. WC2 BX39 42
Kearton Clo. Ken. BZ62 84
Keats Av. Rom. CU30 19
Keats Clo. NW3 BU35 32
Keats Gro.
Keats Clo. Chig. CM29 17
Keats Clo. Hayes BC39 38
Keats Gro. NW3 BT35 32
Keats Pl. EC2 BZ39 42
Moor La.
Keats Rd. Belv. CS41 46
Keats Rd. Well. CN44 54
Keats Way, Croy. CC53 70
Keats Way, Grnf. BF39 38
Keble Clo. Nthlt. BG35 29
Keble Clo. Wor. Pk. BO54 67
Keble St. SW17 BT49 59
Kechill Gdns. Brom. CH54 71
Kedeston Ct. Sutt. BS55 77
Kedleston Ct. E5 CC35 34
Clapton Park Est.
Kedleston Dr. Orp. CN53 72
Kedleston Wk. E2 CB38 33
Keedonwood Rd. Brom. CG49 61
Keeley Rd. Croy. BZ55 78
Keeley St. WC2 BX39 42
Keeling Rd. SE9 CJ46 53
Keen's Rd. Croy. BZ56 78
Keen's Yd. N1 BY36 33
St. Paul's Rd.
Keep, The, Kings. On T. BL50 57
Keep, The SE3 CH44 53
Keeton's Rd. SE16 CB41 42
Keevil Dr. SW19 BQ47 58
Keighley Clo. N7 BX35 33
Penn Rd.
Keighley Rd. Rom. CW29 19
Keightley Dr. SE9 CK47 62
Keilder Clo. Chig. CN29 18
Keildon Rd. SW11 BU45 50
Keir Hardie Est. N16 CB33 24
Keir Hardie Ho. W6 BQ43 49
Keir Hardie Way, Bark. CO36 36
Keir, The SW19 BQ49 58
Keith Gro. W12 BP41 40
Keith Rd. E17 CD30 16
Keith Rd. Bark. CM37 35
Keithway, Horn. CW33 28
Kelbrook Rd. SE3 CK44 53
Kelburn Way, Rain. CT38 37
Kelby Path SE9 CL48 62
Kelceda Clo. NW2 BP34 22
Kelfield Gdns. W10 BQ39 40
Kelland Clo. N8 BW32 23
Kelland Rd. E13 CH38 35
Kellaway Rd. SE3 CJ44 53
Kellerton Rd. SE13 CG46 52
Kellett Rd. SW2 BY45 51
Kelling Gdns. Croy. BY54 69
Kellino St. SW17 BU49 59
Kelliwell Ct. SE22 CB46 51
Kellner Rd. SE28 CN41 45
Kellway Pl. W14 BR42 40
Kelly Rd. NW7 BR29 13
Kelly St. NW1 BV36 32
Kelly Way, Rom. CQ32 27
Kelman Clo. SW4 BW44 50
Kelmore Gro. SE22 CB45 51
Kelmscott Clo. E17 CD30 16
Kelmscott Clo. Wat. BC25 4
Kelmscott Cres. Wat. BC25 4
Kelmscott Gdns. W12 BP41 40
Kelmscott Rd. SW11 BU46 50
Kelross Rd. N5 BY35 33
Kelsall Clo. SE3 CH44 53
Kelsey La. Beck. CE52 70
Kelsey Pk. Av. Beck. CE51 70
Kelsey Pk. Rd. Beck. CE51 70
Kelsey Rd. Orp. CO51 72
Kelsey St. E2 CB38 33
Kelsey Way, Beck. CE52 70
Kelshall Ct. N16 BZ34 24
Kings Crescent Est.
Kelsie Way, Ilf. CN29 18
Kelso Pl. W8 BS41 41
Kelso Rd. Cars. BT54 68
Kelston Rd. Ilf. CL30 17
Kelvedon Clo. Kings. On T. BM50 57
Kelvedon Rd. SW6 BR43 49
Kelvedon Wk. Rain. CT37 37
Kelvedon Way, Wdf. Grn. CK29 17
Kelvin Av. N13 BX29 15
Kelvin Av. Tedd. BH50 57
Kelvinbrook, E. Mol. BF52 65
Kelvin Clo. Epsom BM57 75
Kelvin Cres. Har. BH29 12
Kelvin Dr. Twick. BJ46 48
Kelvin Gdns. Sthl. BF39 38
Kelvin Gro. SE26 CB48 60
Kelvin Gro. Chess. BK55 75
Kelvington Clo. Croy. CD53 70
Kelvington Clo. Orp. CO52 72
Kelvington Rd. SE15 CC46 52
Kelvin Pde. Orp. CN54 72
Kelvin Rd. N5 BZ35 33
Kelvin Rd. Well. CO45 54
Kelway Pl. W14 BR43 49
Kemble Dr. Brom. CK55 80
Kemble Rd. N17 CB30 15
Kemble Rd. SE23 CC47 61
Kemble Rd. Croy. BY55 78
Kemble St. WC2 BX39 42
Kemerton Rd. SE5 BZ45 51
Kemerton Rd. Beck. CE51 70
Kemerton Rd. Croy. CA54 69
Kemeys St. E9 CD35 34
Kemnal Rd. Chis. CM50 62
Kempe Rd. NW6 BQ37 31
Kempis Way SE22 CA46 51
Dulwich Gro.
Kemp Pl. Bush. BF25 4
Kemplay Rd. NW3 BT35 32
Kemple Rd. NW6 BQ37 31
Kemp Rd. Dag. CP33 27
Kempsford Gdns. SW5 BS42 41
Kempshott Rd. SW16 BW50 59
Kempson Rd. SW6 BS43 50
Kempthorne Rd. SE8 CD42 43
Kempton Av. Horn. CW35 37
Kempton Av. Nthlt. BF36 29
Kempton Av. Sun. BC51 65
Kempton Clo. Erith CS43 55
Kempton Rd. E6 CK37 35
Kempton Rd. Hamptn. BE51 65
Kempton Wk. Croy. CD53 70
Kempt St. SE18 CL43 53
Kemsing Clo. Bex. CQ47 63
Kemsing Clo. Brom. CG55 79
Kemsing Clo. Th. Hth. BZ52 69
Kemsing Rd. SE10 CH42 44
Kenbrick Ms. SW7 BT43 50
Reece Ms.
Kenbury St. SE5 BZ44 51
Kendal Av. N18 BZ28 15
Kendal Av. W3 BM39 39
Kendal Av. Bark. CN37 36
Kendal Clo. Wdf. Grn. CG27 16
Kendal Croft, Horn. CU35 37
Kendale Rd. Brom. CG49 61
Kendal Gdns. N18 BZ28 15
Kendall Av. Beck. CD51 70
Kendall Av. Sth. Croy. BZ58 78
Kendall Av. S. Sth. Croy. BZ58 78
Kendall Gdns. Sutt. BT55 77
Kendall Rd. Beck. CD51 70
Kendall Rd. Islw. BJ44 48
Kendal Rd. NW10 BP35 31
Kendal St. W2 BU39 41
Kender Est. SE14 CC44 52
Kender St. SE14 CC44 52
Kendoa Rd. SW4 BW45 50
Kendon Clo. E11 CH32 26
Avenue, The
Kendor Av. Epsom BN59 82
Kendra Hall Rd. Sth. Croy. BY57 78
Kendrey Gdns. Twick. BH46 48
Kendrick Ms. SW7 BT42 41
Brompton Rd.
Kenelm Clo. Har. BJ34 21
Kenerne Dr. Barn. BR25 6
Kenilford Rd. SW12 BV47 59
Kenilworth Av. E17 CE30 16
Kenilworth Av. SW19 BS49 59
Kenilworth Av. Har. BE35 29
Kenilworth Clo. Bans. BS61 83
Kenilworth Clo. Borwd. BN24 6
Kenilworth Ct. SW15 BQ45 49
Kenilworth Ct. Twick. BG48 56
Kenilworth Ct. Wat. BC23 4
Kenilworth Cres. Enf. CA23 8
Kenilworth Dr. Borwd. BN24 6
Kenilworth Dr. KT12 BD55 74
Kenilworth Gdns. SE18 CL44 53
Kenilworth Gdns. Horn. CV34 28
Kenilworth Gdns. Ilf. CN34 27
Kenilworth Gdns. Loug. CK25 10
Kenilworth Gdns. Sthl. BE38 29
Kenilworth Gdns. Wat. BD28 11
Kenilworth Rd. E3 CD37 34
Kenilworth Rd. NW6 BR37 31
Kenilworth Rd. SE20 CC51 70
Kenilworth Rd. W5 BL40 39
Kenilworth Rd. Edg. BN27 13
Kenilworth Rd. Epsom BP56 76
Kenilworth Rd. Orp. CM53 71
Kenley Av. NW9 BO30 13
Kenley Clo. Bex. CR47 63
Kenley Clo. Chis. CN52 72
Kenley Gdns. Horn. CW34 28
Kenley Gdns. Th. Hth. BY52 69
Kenley La. Ken. BZ61 84
Kenley Rd. SW19 BR51 67
Kenley Rd. Kings. On T. BM51 66
Kenley Rd. Twick. BJ46 48
Kenley St. W11 BR40 40
Kenley Wk. W11 BR40 40
Kenley Wk. Sutt. BQ56 76
Kenlor St. SW17 BT49 59
Kenmare Dr. Mitch. BU50 59
Kenmare Gdns. N13 BZ28 15
Kenmare Rd. Th. Hth. BX53 69
Kenmere Rd. Well. CP44 54
Kenmont Gdns. NW10 BP38 31
Kenmore N2 BT33 23
Kenmore Av. Har. BJ32 21
Kenmore Gdns. NW10 BM37 30
Kenmore Gdns. Edg. BM30 12
Kenmore Rd. Har. BK31 21
Kenmore Rd. Ken. BY60 84
Kenmure Rd. E8 CB35 33
Kennard Rd. E15 CF36 34
Kennard Rd. N11 BU28 14
Kennard St. E16 CK40 44
Kennard St. SW11 BV44 50
Kennedy Av. Enf. CC25 9
Kennedy Clo. E13 CH37 35
Kennedy Clo. Orp. CM54 71
Kennedy Clo. Pnr. BE29 11
Kennedy Rd. W7 BH39 39
Kennedy Rd. Bark. CN37 36
Kenneth Av. Ilf. CL35 35

Kenneth Cres. NW2	BP35	31
Kenneth Gdns. Stan.	BJ29	12
Kenneth Rd. Bans.	BT61	83
Kenneth Rd. Rom.	CP33	27
Kennet Rd. W9	BR38	31
Kennet Rd. Dart.	CU45	55
Kennet Rd. Islw.	BH45	48
Kenninghall Rd. E5	CB34	24
Kenninghall Rd. N18	CC28	16
Kennings Est. SE11	BY42	42
Kenning St. SE16	CC41	43
Kennings Way SE11	BY42	42
Kenning Ter. N1	CA37	33
Branch Pl.		
Kennington Gro. SE11	BX43	51
Kennington La. SE11	BX42	42
Kennington Oval SE11	BX43	51
Kennington Park Est. SE11	BY43	51
Kennington Pk. Gdns. SE11	BY43	51
Kennington Pk. Pl. SE11	BY43	51
Kennington Park Rd. SE11	BY43	51
Kennington Rd. SE1	BY41	42
Kennington Rd. SE11	BY42	42
Kennyland Ct. NW4	BP32	22
Kenny Rd. NW7	BR29	13
Kenrick Pl. W1	BV39	41
Dorset St.		
Kensal Rd. W10	BR38	31
Kensington Av. E12	CK36	35
Kensington Av. Th. Hth.	BY51	69
Kensington Church St. W8	BS40	41
Kensington Church Wk. W8	BS41	41
Holland St.		
Kensington Ct. W8	BS41	41
Kensington Ct. Ms. W8	BS41	41
Kensington Ct. Pl.		
Kensington Ct. Pl. W8	BS41	41
Kensington Dr. Wdf. Grn.	CJ30	17
Kensington Gdns. Ilf.	CK33	26
Kensington Gdns. Sq. W2	BS39	41
Kensington Gte. W8	BT41	41
Kensington Gore SW7	BT41	41
Kensington High St. W8	BR41	40
Kensington High St. W14	BR41	40
Kensington Mall W8	BS40	41
Kensington Palace Gdns. W8	BS40	41
Kensington Pk. Gdns. W11	BR40	40
Kensington Pk. Ms. W11	BR39	40
Kensington Pk. Rd. W11	BR39	40
Kensington Pl. W8	BS40	41
Kensington Rd. SW7	BT41	41
Kensington Rd. W8	BT41	41
Kensington Rd. Nthlt.	BF38	29
Kensington Rd. Rom.	CS32	28
Kensington Sq. W8	BS41	41
Kensington Ter. Sth. Croy.	BZ57	78
Kent Av. W13	BJ39	39
Kent Av. Dag.	CR39	45
Kent Av. Well.	CN46	54
Kent Clo. Mitch.	BX52	69
Kent Clo. Orp.	CN57	81
Kent Dr. Barn.	BV24	7
Kent Dr. Horn.	CV35	37
Kent Dr. Tedd.	BH49	57
Kentford Way, Nthlt.	BE37	29
Kent Gdns. W13	BJ39	39
Kent Gdns. Ruis.	BC32	20
Kent Gate Way, Croy.	CD57	79
Kent House La. Beck.	CD50	61
Kent House Rd. Beck.	CD50	61
Kentish Rd. Belv.	CR42	45
Kentish Town Rd. NW1	BV36	32
Kentish Town Rd. NW5	BV36	32
Kentmere Rd. SE18	CN42	45
Kenton Av. Har.	BH33	21
Kenton Av. Sthl.	BF40	38
Kenton Av. Sun.	BD51	65
Kenton Ct. Har.	BJ32	21
Kenton Gdns. Har.	BK32	21
Kenton La. Har.	BH29	12
Kenton Park Av. Har.	BK31	21
Kenton Park Clo. Har.	BK32	21
Kenton Park Cres. Har.	BK31	21
Kenton Park Rd. Har.	BK31	21
Kenton Rd. E9	CC36	34
Kenton Rd. Har.	BH33	21
Kenton St. WC1	BX38	33
Kent Pass. NW1	BU38	32
Kent Rd. N15	CA32	24
Kent Rd. N21	BZ26	8
Kent Rd. W4	BN41	40
Kent Rd. Dag.	CR35	36
Kent Rd. Dart.	CV46	55
Kent Rd. E. Mol.	BG52	65
Kent Rd. Kings. On T.	BK52	66
Kent Rd. Orp.	CO53	72
Kent Rd. Rich.	BM43	48
Kent Rd. W. Wick.	CE54	70
Kents Pass, Hamptn.	BE51	65
Kent St. E2	CA37	33
Kent St. E13	CH38	35
Kent Ter. NW1	BU38	32
Kent Vw. Gdns. Ilf.	CN34	27
Kent Way SE15	CA44	51
Sumner Estate		
Kent Way, Surb.	BL55	75
Kentwode Grn. SW13	BP43	49
Kenver Av. N12	BT29	14
Kenward Rd. SE9	CJ46	53
Kenway, Rain.	CV38	37
Kenway, Rom.	CS30	19
Ken Way, Wem.	BN34	22
Kenway Clo. Rain.	CV38	37
Kenway Rd. SW5	BS42	41
Kenway Wk. Rain.	CV38	37
Kenwood Av. N14	BW25	7
Kenwood Dr. Beck.	CF52	70
Kenwood Dr. Walt.	BC57	74
Kenwood Gdns. E18	CH31	26
Kenwood Gdns. Ilf.	CL31	26
Kenwood Rd. N6	BU32	23
Kenwood Rd. N9	CB26	8
Kenworthy Rd. E9	CD35	34
Kenwyn Dr. NW2	BO34	32
Kenwyn Rd. SW4	BW45	50
Kenwyn Rd. SW20	BQ51	67
Kenya Rd. SE7	CJ43	53
Kenyington Dr. Sun.	BC49	56
Kenyington Pl. Har.	BK32	21
Kenyon St. SW6	BQ44	49
Keogh Rd. E15	CG36	34
Kepler Rd. SW4	BX45	51
Keppel Rd. E6	CK36	35
Keppel Rd. Dag.	CQ35	36
Keppel Row SE1	BZ40	42
Great Guildford St.		
Keppel St. WC1	BW39	41
Malet St.		
Kerbela St. E2	CA38	33
Kerbey St. E14	CE39	43
Kerfield Cres. SE5	BZ44	51
Kerfield Pl. SE5	BZ44	51
Kernick Clo. N7	BX36	33
Sutterton St.		
Kernow Clo. Horn.	CW34	28
Kerrison Rd. E15	CF37	34
Kerrison Rd. SW11	BU45	50
Kerrison Rd. W5	BK40	39
Kerry Av. Stan.	BK28	12
Kerry Clo. E16	CH39	44
Kerry Clo. Barn.	BQ24	6
Kerry Path SE14	CD43	52
Kersey Gdns. SE9	CK49	62
Kersey Gdns. Rom.	CW29	19
Kersfield Rd. SW15	BQ46	49
Kershaw Clo. SW18	BT46	59
Kershaw Rd. Dag.	CR34	27
Kersley Ms. SW11	BU44	50
Kersley Rd. N16	CA34	24
Kersley St. SW11	BU44	50
Kerswell Cres. N15	CA32	24
Keslake Rd. NW6	BQ37	31
Kessock Clo. N15	CB32	24
Kesteven Clo. Chig.	CN29	18
Keston Av. Kes.	CJ56	80
Keston Clo. N18	BZ27	15
Keston Clo. Well.	CP43	54
Keston Gdns. Kes.	CJ56	80
Keston Rd. N17	BZ31	24
Keston Rd. SE15	CB45	51
Keston Rd. Th. Hth.	BX53	69
Kestrel Ct. E17	CC30	16
Kestrel Clo. Ilf.	CP28	18
Kestrel Clo. Rain.	CU36	37
Kestrel Rd. SE24	BY46	51
Kestrel Way, Croy.	CF58	79
Keswick Av. SW15	BO49	58
Keswick Av. SW19	BS51	68
Keswick Av. Horn.	CV33	28
Keswick Clo. Sutt.	BT56	77
Keswick Gdns. Ilf.	CK31	26
Keswick Gdns. Wem.	BL35	30
Keswick Mews. W5	BK40	39
Keswick Rd. SW15	BR46	49
Keswick Rd. Bexh.	CR44	54
Keswick Rd. Orp.	CN54	72
Keswick Rd. Twick.	BF46	47
Keswick Rd. W. Wick.	CG55	79
Kettering Rd. Rom.	CW29	19
Kettering St. SW16	BV50	59
Kett Gdn. SW2	BX46	51
Kettlebaston Rd. E10	CD33	25
Kettlewell Clo. Swan.	CT51	73
Kevelioc Rd. N17	BZ30	15
Kevin Clo. Houns.	BD44	47
Kevington Dr. Chis.	CN52	72
Kew Bridge Rd. Brent.	BL43	48
Kew Ct. W4	BM42	39
Kew Cres. Sutt.	BR55	76
Kew Foot Rd. Rich.	BL45	48
Kew Gdns. Rd. Rich.	BL43	48
Kew Grn. Rich.	BL43	48
Kew Meadow Path, Rich.	BM44	48
Kew Palace, Rich.	BL43	48
Kew Rd. Rich.	BL43	48
Kew Rd. Rich.	BL45	48
Key Clo. E1	CC38	34
Cambridge Heath Rd.		
Keyes Rd. NW2	BQ35	31
Keyes Rd. Dart.	CW45	55
Keymer Rd. SW2	BX48	60
Keynes Clo. N2	BU31	23
Keynsham Av. Wdf. Grn.	CG28	16
Keynsham Gdns. SE9	CK46	53
Keynsham Rd. SE9	CJ46	53
Keynsham Rd. Mord.	BS54	68
Keynsham Wk. Mord.	BS54	68
Keysham Av. Houns.	BC43	47
Keystone Cres. N1	BX37	33
Caledonian Rd.		
Keywood Dr. Sun.	BC50	56
Keyworth St. SE1	BY41	42
Khama Rd. SW17	BU49	59
Khartoum Rd. E13	CH38	35
Khartoum Rd. SW17	BT49	59
Khartoum Rd. Ilf.	CL35	35
Khyber Rd. SW11	BU44	50
Kibworth St. SW8	BX43	51
Dorset Rd.		
Kidbrooke Gdns. SE3	CH44	53
Kidbrooke Gro. SE3	CH44	53
Kidbrooke La. SE9	CK45	53
Kidbrooke Pk. Clo. SE3	CH44	53
Kidbrooke Pk. Rd. SE3	CH44	53
Kidbrooke Way SE3	CH44	53
Kidington Way NW9	BN30	13
Kidderminster Pl. Croy.	BY54	69
Kidderminster Rd. Croy.	BY54	69
Kidderpore Av. NW3	BS35	32
Kidrow Way E9	CC37	34
Cleremont Rd.		
Kielder Clo. Ilf.	CN29	18
New North Rd.		
Kier Hardie Way, Hayes	BC38	29
Kilburn Bldgs. NW6	BS37	32
Kilburn High St.		
Kilburn Gate NW6	BS37	32
Kilburn High Rd. NW6	BR36	31
Kilburn La. W10	BQ38	31
Kilburn Pk. Rd. NW6	BS38	32
Kilburn Pl. NW6	BS37	32
Kilburn Priory NW6	BS37	32
Kilburn Sq. NW6	BS37	32
Kilburn Vale Est. NW6	BS37	32
Kilcorral Clo. Epsom	BP60	82
Kildare Clo. Ruis.	BD33	20
Kildare Gdns. W2	BS39	41
Kildare Rd. E16	CH39	44
Kildare Ter. W2	BS39	41
Kildonan Clo. Wat.	BC23	4
Kildoran Rd. SW2	BX46	51
Kildowan Rd. Ilf.	CO33	27
Kilgour Rd. SE23	CD46	52
Kilgowan Rd. Ilf.	CO33	27
Kilkie St. SW6	BT44	50
Killarney Rd. SW18	BT46	50
Killearn Rd. SE6	CF47	61
Killester Gdns. Wor. Pk.	BP56	76
Killick St. N1	BX37	33
Kilieser Av. SW2	BX48	60
Killip Clo. E16	CG39	43
Killowen Av. Nthlt.	BG35	29
Killowen Rd. E9	CC36	34
Killyon Rd. SW8	BW44	50
Kilmaine Rd. SW6	BR43	49
Kilmarnock Rd. Wat.	BD28	11
Woodhall La.		
Kilmarsh St. W6	BQ42	40
Kilmartin Av. SW16	BX52	69
Kilmartin Rd. Ilf.	CO34	27
Kilmartin Way, Horn.	CU35	37
Kilmeston Way SE15	CA43	51
Kilmington Rd. SW13	BP43	49
Kilmorey Gdns. Twick.	BJ46	48
Kilmorey Rd. Twick.	BJ45	48
Kilmorie Rd. SE23	CD47	61
Kiln La. Epsom	BO59	82
Kiln Pl. NW5	BV35	32
Kilnside, Esher	BJ57	75
Kilravock St. W10	BR38	31
Kilsby Wk. Dag.	CO36	36
Rugby Rd.		
Kilsha Rd. Walt.	BD53	65
Kimbell Gdns. SW6	BR44	49
Kimberley Av. E6	CK37	35
Kimberley Av. SE15	CB44	51
Kimberley Av. Ilf.	CM33	26
Kimberley Av. Rom.	CS32	28
Kimberley Dr. Sid.	CP48	63
Kimberley Gdns. Enf.	CA24	8
Kimberley Pl. Pur.	BY59	84
Brighton Rd.		
Kimberley Rd. E4	CG26	9
Kimberley Rd. E11	CF34	25
Kimberley Rd. E16	CG38	34
Kimberley Rd. E17	CD30	16
Kimberley Rd. N4	BY32	24
Kimberley Rd. N17	CB30	15
Kimberley Rd. N18	CB29	15
Kimberley Rd. NW6	BR37	31
Kimberley Rd. SW9	BX44	51
Kimberley Rd. Beck.	CC51	70
Kimberley Rd. Croy.	BY53	69
Kimberley Way E4	CG26	9
Kimber Rd. SW18	BS47	59
Kimble Cres. Bush.	BG26	4
Kimble Rd. SW19	BT50	59
Kimbolton Clo. SE12	CG46	52
Kimbolton Grn. Borwd.	BN24	6
Kimbolton Row SW3	BU42	41
Fulham Rd.		
Kimbur Rd. SW18	BS47	59
Kimmeridge Gdns. SE9	CK49	62
Kimmeridge Rd. SE9	CK49	62
Kimpton Rd. SE5	BZ44	51
Kimpton Rd. Sutt.	BR55	76
Kinburn St. SE16	CC41	43
Kincaid Rd. SE15	CB43	51
Kinch Gro. Har.	BL33	21
Kinder Clo. SE2	CP40	45
Kinder St. E1	CB39	42
Kinfauns Av. Horn.	CV32	28
Kinfauns Rd. SW2	BY48	60
Kinfauns Rd. Ilf.	CO33	27
Kingaby Gdns. Rain.	CU36	37
King Alfred Av. SE6	CE49	61
King Alfred Rd. Rom.	CW30	19
King Arthur Clo. SE15	CC43	52
King Charles Cres. Surb.	BL54	66
King Charles's Rd. Surb.	BL53	66
King Charles St. SW1	BW41	41
King Charles Wk. SW13	BR47	58
Princes Way		
King David La. E1	CC40	43
Kingdon Rd. NW6	BS36	32
King Edward Av. Dart.	CV46	55
King Edward Av. Rain.	CV37	37
King Edward Dr. Chess.	BK55	75
King Edward Rd. E10	CF33	25
King Edward Rd. E17	CD31	25
King Edward Rd. Barn.	BS24	7
King Edward Rd. Rom.	CT32	28
King Edward Rd. Wat.	BE25	4
King Edwards Gdns. W3	BM40	39
King Edward's Gro. Tedd.	BJ50	57
King Edward's Rd. E9	CB37	33
King Edward's Rd. N9	CB26	8
King Edward's Rd. Bark.	CM37	35
King Edward's Rd. Enf.	CC24	9
King Edward St. EC1	BZ39	42
King Edward Wk. SE1	BY41	42
Kingfield Rd. W5	BK38	30
Kingfield St. E14	CF42	43
Kingfisher Clo. Walt.	BE56	74
Kingfisher Dr. Rich.	BK49	57
Kingfisher Gdns. Sth. Croy.	CC58	79
Kingfisher Sq. SE8	CE43	52
Staunton St.		
King George Av. E16	CJ39	44
King George Av. Bush.	BF25	4
King George Av. Walt.	BD54	65
King George Clo. Rom.	CS31	28
King George St. SE10	CF43	52
King George VI Av. Mitch.	BU52	68
Kingham Clo. SW18	BT47	59
King Harold's Way, Bexh.	CP43	54
King Henry's Dr. Croy.	CF58	79
King Henry's Rd. NW3	BT36	32
King Henry's Rd. Kings. On T.	BM52	66
King Henry St. N16	CA35	33
King Henry's Wk. N1	CA36	33
King James St. SE1	BY41	42
King John Ct. EC2	CA38	33
New Inn Yd.		
King John's Ct. EC2	CA38	33
New Inn Yd.		
King John St. E1	CC39	43
King John's Wk. SE9	CJ47	62
Kinglake Est. SE17	CA42	42
Kinglake St. SE17	CA42	42
Kingly St. W1	BW39	41
King & Queen St. SE17	BZ42	42
Kingsand Rd. SE12	CH48	62
Kings Arms Ct. E1	CB39	42
Old Montague St.		
King's Arms Yd. EC2	BZ39	42
Kings Arms Yd. Rom.	CT32	28
Quad. Arcade		
Kings Arms Yd. SW18	BS46	50
Wandsworth High St.		
Kings Av. N10	BV31	23
King's Av. N21	BY26	8
King's Av. SW4	BW47	59
King's Av. SW12	BW47	59
Kings Av. W5	BK39	39
King's Av. Brom.	CG50	61
King's Av. Buck. H.	CJ27	17
King's Av. Cars.	BU57	77
King's Av. Grnf.	BF39	38
Kings Av. Houns.	BF44	47
Kings Av. N. Mal.	BO52	67
Kings Av. Rom.	CQ32	27
King's Av. Wdf. Grn.	CH29	17
King's Av. Gdns. SW4	BX46	51
King's Bench St. SE1	BY41	42
Kingsbridge Av. W3	BL41	39
Kingsbridge Cir. Rom.	CW29	19
Kingsbridge Clo. Rom.	CW29	19
Kingsbridge Cres. Sthl.	BE39	38
Kingsbridge Rd. W10	BQ39	40
Kingsbridge Rd. Bark.	CM37	35
Kingsbridge Rd. Mord.	BQ53	67
Kingsbridge Rd. Rom.	CW29	19

Kingsbridge Rd. Sthl. BE42 38
Kingsbridge Rd. Walt. BC54 65
Kingsbury Cir. NW9 BM32 21
Kingsbury Rd. N1 CA36 33
Kingsbury Rd. NW9 BM32 21
Kingsbury Ter. N1 CA36 33
Kingsclere Clo. SW15 BP47 58
Kingscliffe Gdns. SW19 BR47 58
Kings Clo. E10 CE33 25
Kings Clo. NW4 BQ31 22
King's Clo. Dart. CT45 55
King's Clo. Walt. BC54 65
Kings College Rd. NW3 BT36 32
Eton Av.
Kingscote Rd. W4 BN41 40
Kingscote Rd. Croy. CB54 69
Kingscote Rd. N. Mal. BN52 67
Kingscote St. EC4 BY40 42
Tudor St.
King's Ct. E13 CH37 35
King's Ct. SW19 BS50 59
Kings Ct. W5 BK39 39
Castlebar Pk.
King's Ct. W6 BP42 40
Kings Ct. Har. BF34 20
King's Ct. Wem. BM34 21
Kingscourt Rd. SW16 BW48 59
Kings Cres. Est. N16 BZ34 24
King's Cres. N4 BZ34 24
Kingscroft Rd. NW2 BR36 31
Kingscroft Rd. Bans. BT61 83
King's Cross Rd. WC1 BX38 33
Kingsdale Est. SE18 CN43 54
Kingsdale Rd. SE18 CN43 54
Kingsdale Rd. SE20 CC50 61
Kingsdown Av. W3 BO40 40
Kingsdown Av. W13 BJ41 39
Kingsdown Av. Sth. Croy. BY58 78
Kingsdown Clo. W11 BQ39 40
Kingsdowne Rd. Surb. BL54 66
Kingsdown Rd. E11 CG34 25
Kingsdown Rd. N19 BX34 24
Kingsdown Rd. Epsom BP60 82
Kingsdown Rd. Sutt. BR56 76
Kingsdown Way, Brom. CH53 71
Kings Dr. Edg. BL28 12
King's Dr. Surb. BJ53 66
King's Dr. Surb BM54 66
King's Dr. Tedd. BG49 56
King's Dr. Wem. BM34 21
King's Farm Av. Rich. BM45 48
Kingsfield Av. Har. BF31 20
Kingsfield Rd. Har. BG33 20
Kingsfield Rd. Wat. BD26 4
Kingsfield Ter. Dart. CV46 55
Kingsfield Ter. Har. BG33 20
Kingsford Av. Wall. BX57 78
Kings Gdns. Croy. CY56 78
Kings Gdns. Ilf. CM33 26
Kingsgate, Wem. BM34 21
Kingsgate Av. N3 BS31 23
Kingsgate Clo. Bexh. CQ44 54
Kingsgate Pl. NW6 BS36 32
Kingsgate Rd. NW6 BS36 32
Kings Grn. Loug. CK24 10
Kingsground SE9 CK47 62
King's Gro. SE15 CB44 51
Kings Gro. Rom. CU32 28
King's Hall Rd. Beck. CD50 61
Kings Head Ct. EC3 BZ40 42
Fish St. Hill
King's Head Hill E4 CE26 9
Kings Head Pass. SW4 BW45 50
Clapham Pk. Rd.
King's Head Yd. SE1 BZ40 42
Borough High St.
Kings Highway SE18 CN43 54
Kings Hill, Loug. CK23 10
Kingshill Av. Har. BJ31 21
Kingshill Av. Nthlt. BD38 29
Kingshill Av. Rom. CS29 19
Kingshill Av. Wor. Pk. BP54 67
Kingshill Dr. Har. BJ30 12
Kingshill Rd. E2 CA38 33
Kingshold Est. E9 CC37 34
Kingshold Rd. E9 CC36 34
Kingsholm Gdns. SE9 CJ45 53
Kingshurst Rd. SE12 CH47 62
Kingsland Est. E2 CA37 33
Kingsland Grn. N16 CA36 33
Kingsland High St. E8 CA36 33
Kingsland Rd. E8 CA38 33
Kingsland Rd. E13 CJ38 35
King's La. Sutt. BT56 77
Kings Lawn Clo. SW15 BP46 49
Kingsley Av. W13 BJ39 39
Kingsley Av. Bans. BS61 83
Kingsley Av. Borwd. BL23 5
Kingsley Av. Houns. BG44 47
Kingsley Av. Sthl. BF40 38
Kingsley Av. Sutt. BT56 77
Kingsley Clo. N2 BT32 23
Kingsley Clo. Dag. CR35 36
Kingsley Cres. SE19 CA51 69
Kingsley Dr. Wor. Pk. BO55 76
Kingsley Gdns. E4 CE28 16
Kingsley Gdns. Horn. CV31 28
Kingsley Ms. W8 BS41 41
Stanford Rd.
Kingsley Pl. N6 BV33 23
Kingsley Rd. E7 CH36 35
Kingsley Rd. E17 CF30 16
Kingsley Rd. Loug. CM24 10
Kingsley Rd. N13 BY28 15
Kingsley Rd. NW6 BR37 31
Kingsley Rd. SW19 BS49 59
Kingsley Rd. Croy. BY54 69
Kingsley Rd. Har. BG35 29
Kingsley Rd. Houns. BF44 47
Kingsley Rd. Ilf. CM30 17
Kingsley Rd. Orp. CN57 81
Kingsley Rd. Pnr. BE31 20
Kingsley St. SW11 BU45 50
Kingsley Way N2 BT32 23
Kingsley Wood Dr. SE9 CK48 62
Kings Lynn Clo. Rom. CV29 19
Kings Lynn Dr. Rom. CV29 19
Kings Lynn Path, Rom. CV29 19
Kings Lynn Dr.
Kingsman St. SE18 CK41 44
Kingsmead, Barn. BS24 7
Kingsmead Av. N9 CB26 8
Kingsmead Av. NW9 BN33 22
Kingsmead Av. Mitch. BW52 68
Kingsmead Av. Rom. CT32 28
Kingsmead Av. Sun. BD51 65
Kingsmead Av. Surb. BM55 75
Kingsmead Av. Wor. Pk. BP55 76
Kingsmead Clo. Epsom BN57 76
Kingsmead Clo. Sid. CO48 63
Kingsmead Dr. Nthlt. BE36 29
King's Mead Est. E9 CD35 34
Kingsmead Rd. SW2 BY48 60
King's Mead Way E9 CD35 34
Kingsmere Pk. NW9 BM33 21
Kingsmere Rd. SW19 BQ48 58
King's Ms. WC1 BX38 33
Kingsmill Gdns. Dag. CQ35 36
Kingsmill Rd. Dag. CQ35 36
Kingsmill Ter. NW8 BT37 32
Kingsnympton Pk. BM50 57
Kings. On T.
Kings Orchard SE9 CK46 53
Kingspark Ct. E18 CH31 26
Kingspark Ct. Ilf. CK33 26
Drive, The
Kings Pass. King. On T. BK51 66
Kings Pl. SE1 BZ41 42
Kings Pl. W4 BN42 40
Kings Pl. Buck. H. CJ27 17
Kings Sq. Est. EC1 BY38 33
Kings Ride Gate, Rich. BM45 48
Kingsridge Gdns. Dart. CV46 55
King's Rd. E4 CF26 9
King's Rd. E6 CJ37 35
King's Rd. E11 CG33 25
Kings Rd. N17 CA30 15
King's Rd. N18 CB28 15
King's Rd. N22 BX30 15
King's Rd. NW10 BP36 31
King's Rd. SE25 CB52 69
Kings Rd. SW1 BV42 41
King's Rd. SW3 BS43 50
King's Rd. SW6 BS44 50
King's Rd. SW10 BT43 50
King's Rd. SW14 BN45 49
King's Rd. SW19 BS50 59
Kings Rd. W5 BK39 39
Kings Rd. Bark. CM36 35
North St.
Kings Rd. Barn. BQ24 6
King's Rd. Felt. BD47 56
Kings Rd. Har. BE34 20
King's Rd. Kings. On T. BL51 66
King's Rd. Mitch. BV52 68
King's Rd. Orp. CN56 81
Kings Rd. Rich. BL46 48
Kings Rd. Rom. CU31 28
King's Rd. Surb. BK54 66
Kings Rd. Sutt. BS58 77
King's Rd. Tedd. BG49 56
Kings Rd. Twick. BJ46 48
Kings Rd. Walt. BC55 74
Kings Scholars Pass. BW41 41
SW1
Carlisle Pl.
Kings Ter. NW1 BW37 32
Plender St.
King's Ter. Islw. BJ45 48
Kingsthorpe Rd. SE26 CC49 61
Kingston Av. Sutt. BR55 76
Kingston Br. Kings. On T. BK51 66
Kingston By-pass, Esher BH55 75
Kingston Clo. Nthlt. BE36 29
Kingston Clo. Tedd. BJ50 57
Kingston Cres. Beck. CD51 70
Kingston Hall Rd. BK52 66
Kings. On T.
Kingston Hill Av. Rom. CQ30 18
Kingston Hl. Kings. On T. BM51 66
Kingston La. Tedd. BJ49 57
Kingston Pk. Est. BM50 57
Kings. On T.
Kingston Pl. Sthl. BE41 38
Kingston Rd. N9 CB27 15
Kingston Rd. SW15 BP48 58
Kingston Rd. SW19 BR51 67
Kingston Rd. SW20 BQ51 67
Kingston Rd. Barn. BT25 7
Kingston Rd. Epsom BO57 76
Kingston Rd. Ilf. CL35 35
Kingston Rd. N. Mal. BM52 66
Kingston Rd. Rom. CT31 28
Kingston Rd. Sun. BE51 65
Kingston Rd. Surb. BM55 75
Kingston Rd. Tedd. BJ49 57
Kingston Vale SW15 BN49 58
Kingstown St. NW1 BV37 32
King St. E13 CH38 35
King St. EC2 BZ39 42
King St. N2 BT31 23
King St. SW1 BW40 41
King St. W3 BM40 39
King St. W6 BO42 40
King St. WC2 BX40 42
King St. Rich. BK46 48
King St. Sthl. BE41 38
King St. Twick. BJ47 57
King St. Wat. BC24 4
King's Wk. Kings. On T. BK51 66
Kings Wk. Sth. Croy. CB60 84
Kingsway N12 BT29 14
Kingsway NW8 BU37 32
Kingsway SW14 BM45 48
Kingsway WC2 BX39 42
Kingsway, Croy. BX56 78
Kingsway, Enf. CB25 8
King's Way, Har. BH31 21
Kingsway, N. Mal. BQ52 67
Kingsway, Orp. CM53 71
Kingsway, Wdf. Grn. CJ28 17
Kingsway, Wem. BL35 30
Kingsway, W. Wick. CG55 79
Kingsway Av. Sth. Croy. CC58 79
Kingsway Cres. Har. BG31 20
Kingsway Ind. Est. N18 CC29 16
Kingsway Rd. Sutt. BR57 76
Kingsway, The, Epsom BO59 82
Kingswear Rd. NW5 BV34 23
Kingswear Rd. Ruis. BC34 20
Kingswood Av. NW6 BR37 31
Kingswood Av. Belv. CQ42 45
Kingswood Av. Houns. BE44 47
Kingswood Av. Sth. Croy. CB61 84
Kingswood Av. Swan. CT52 73
Kingswood Av. Th. Hth. BY53 69
Kingswood Clo. N20 BT26 7
Kingswood Clo. Dart. CV46 55
Kingswood Clo. N. Mal. BO53 67
Kingswood Clo. Orp. CM54 71
Woodcote Rd.
Kingswood Clo. Surb. BL54 66
Kingswood Ct. Rich. BL46 48
Marchmont Rd.
Kingswood Dr. SE19 CA49 60
Kingswood Dr. Cars. BU54 68
Kingswood Est. SE21 CA49 60
Kingswood Pk. N3 BR30 13
Kingswood Pl. SE13 CG45 52
Kingswood Rd. SE20 CC50 61
Kingswood Rd. SW2 BX46 51
Kingswood Rd. SW19 BR50 58
Kingswood Rd. W4 BN41 40
Kingswood Rd. Brom. CF52 70
Kingswood Rd. Ilf. CO33 27
Kingswood Way, Wall. BX56 78
Kingthorpe Rd. NW10 BN36 31
Kingthorpe Ter. NW10 BN36 31
King William St. EC4 BZ39 42
King William Wk. SE10 CF43 52
Kingwood Av. Brom. CG52 70
Kingwood Av. Hmptn. BF50 56
Kingwood Rd. SW6 BR44 49
Kinlet Clo. SE18 CM44 53
Kinlet Rd.
Kinlet Rd. SE18 CL44 53
Kinloch Dr. NW9 BO33 22
Kinloch St. N7 BX34 24
Kinloss Ct. N3 BR31 22
Kinloss Gdns.
Kinloss Gdns. N3 BR31 22
Kinloss Rd. Cars. BT54 68
Kinnaid Way IG8 CK29 17
Kinnaird Av. W4 BN43 49
Kinnaird Av. Brom. CG50 61
Kinnaird Clo. Brom. CG50 61
Kinnear Rd. W12 BO41 40
Kinnerton Pl. N. SW1 BU41 41
Kinnerton St.
Kinnerton Pl. S. SW1 BU41 41
Kinnerton St.
Kinnerton St. SW1 BV41 41
Kinnoul Rd. W6 BR43 49
Kinross Av. Wor. Pk. BP55 76
Kinross Clo. Har. BL32 21
Kinsale Rd. SE15 CB45 51
Kintore St. SE1 CA42 42
Kintore Way SE1 CA42 42
Grange Rd.
Kintyre Clo. SW16 BX52 69
Kinveachy Gdns. SE7 CK42 44
Kinver Rd. North SE26 CC49 61
Kinver Rd. South SE26 CC49 61
Kipling Estate SE1 BZ41 42
Kipling Pl. Stan. BH29 12
Kiplig Rd. Bexh. CQ44 54
Kipling St. SE1 BZ41 42
Kipling Ter. N9 BZ27 15
Kirby Clo. Epsom BO56 76
Kirby Clo. Ilf. CN29 18
Kirby Clo. Loug. CK26 10
Kirby Est. SE16 CB41 42
Kirby Gro. SE1 CA41 42
Kirby St. EC1 BY39 42
Kirby Way, Walt. BD53 65
Kircaldy Grn. Wat. BD27 11
Trevose Way
Kirchen Rd. W13 BJ40 39
Kirkdale SE26 CB48 60
Kirkdale Rd. E11 CG33 25
Kirkham St. SE18 CN43 54
Kirkland Av. Ilf. CL30 17
Kirkland Pl. SE10 CG41 43
Kirk La. SE18 CM43 53
Kirklees Rd. Th. Hth. BX52 69
Kirkley Rd. SW19 BS50 59
Kirkley Clo. Sth. Croy. CA58 78
Kirk Rd. E17 CD32 25
Kirkside Rd. SE3 CH43 53
Kirkstall Av. N17 BZ31 24
Kirkstall Gdns. SW2 BW47 59
Kirkstall Rd. SW2 BW47 59
Kirkstead Ct. E5 CC35 34
Clapton Park Est.
Kirksted Rd. Mord. BS54 68
Kirkstone Way, Brom. CG50 61
Kirkton Gdns. E2 CA38 33
Chambo Rd.
Kirkton Rd. N15 CA31 24
Kirkwall Pl. E2 CC38 34
Kirkwood Rd. SE15 CB44 51
Kirn Rd. W13 BJ40 39
Kirchen Rd.
Kirton Rd. E13 CJ37 35
Kirton Wk. Edg. BN29 13
Kirwyn Way SE5 BY43 51
Kitchener Rd. E7 CH36 35
Kitchener Rd. E17 CE30 16
Kitchener Rd. N2 BU31 23
Kitchener Rd. N17 BZ31 24
Kitchener Rd. Dag. CR36 36
Kitchener Rd. Th. Hth. BZ52 69
Kitkat Ter. E3 CE38 34
Kitley Gdns. SE19 CA51 69
Kitson Rd. SE5 BZ43 51
Kitson Rd. SW13 BP44 49
Kittiwake Clo. Sth. Croy. CD58 79
Kittiwake Rd. Nthlt. BD38 29
Kitto Rd. SE14 CC44 52
Kiver Rd. N19 BW34 23
Klea Av. SW4 BW46 50
Knapdale Clo. SE23 CB48 60
Knapmill Rd. SE6 CE48 61
Knapmill Way SE6 CE48 61
Knapp Clo. NW10 BO36 31
Knapp Rd. E3 CE38 34
Knaresborough Pl. SW5 BS42 41
Knaresborough Rd. SW5 BS43 50
Knatchbull Rd. NW10 BN37 31
Knatchbull Rd. SE5 BY44 51
Knebworth Av. E17 CE30 16
Knebworth Path, Borwd. BN24 6
Knebworth Rd. N16 CA35 33
Knee Hill SE2 CP42 45
Knee Hill Cres. SE2 CP42 45
Kneller Gdns. Islw. BG46 47
Kneller Rd. SE4 CD45 52
Kneller Rd. N. Mal. BN54 67
Kneller Rd. Twick. BG46 47
Knightland Rd. E5 CB34 24
Knighton Clo. Rom. CS32 28
Knighton Clo. Sth. Croy. BY57 78
Knighton Clo. Wdf. Grn. CH28 17
Knighton Dr. Wdf. Grn. CH28 17
Knighton La. Buck. H. CH27 17
Knighton Park Rd. SE26 CC49 61
Knighton Rd. E7 CH34 26
Knighton Rd. Rom. CS32 28
Knightrider St. EC4 BZ39 42
Godliman St.
Knightrider St. EC4 BZ40 42
Queen Victoria St.
Knights Av. W5 BL41 39
Knightsbridge Clo. SW1 BU41 41
Knightsbridge
Knightsbridge Gdns. Rom. CS32 28
Knightsbridge SW1 BU41 41
Knights Clo. E9 CC35 34
Churchill Wk.
Knights Ct. Kings. On T. BL52 66
Knight's Hill SE27 BY49 60
Knight's Hill Sq. SE27 BY49 60
Knight's Hill
Knights La. N9 CB27 15
Knights Pk. Kings. On T. BL52 66
Knight's Rd. E16 CH41 44
Knights Rd. Stan. BK28 12
Knights Way, Ilf. CM29 17
Knightswood Clo. Edg. BN27 13
Knightswood Cres. BO53 67
N. Mal.

Knivett Rd. SW6 BS43 50
Knob's Hill Rd. E15 CE37 34
Knockholt Rd. SE9 CJ46 53
Knole Clo. Croy. CC53 70
Knole Rd. Dart. CU47 64
Knole, The SE9 CL49 62
Knoll Dr. N14 BV26 7
Knollmead, Surb. BN54 67
Knoll Rise, Orp. CN54 72
Knoll Rd. SW18 BT46 50
Knoll Rd. Bex. CR47 63
Knoll Rd. Sid. CO49 63
Knolls Clo. Wor. Pk. BP55 76
Knolls, The, Epsom BQ61 82
Knoll, The W13 BK39 39
Knoll, The, Beck. CE51 70
Knoll, The, Brom. CH55 80
Knoll, The, Orp. CN54 72
Knollys Clo. SW16 BY48 60
Knollys Rd. SW16 BX48 60
Knottisford St. E2 CC38 34
Knott's Grn. Rd. E10 CE32 25
Knowle Av. Bexh. CQ43 54
Knowle Clo. SW9 BY45 51
Knowle Rd. Brom. CK55 80
Knowle Rd. Twick. BH47 57
Knowles Hl. Cres. SE13 CF46 52
Knowlton Grn. Brom. CG53 70
Knowl Way Borwd. BL24 5
Knowsley Av. Sthl. BF40 38
Knowsley Rd. SW11 BU44 50
Knox Rd. E7 CG36 34
Knox St. W1 BU39 41
Knoyle St. SE14 CD43 52
Chubworthy St.
Kohat Rd. SW19 BS49 59
Koh-i-noor Av. Bush. BF25 4
Kossuth St. SE10 CG42 43
Kramer Ms. SW5 BS42 41
Kreisel Wk. Rich. BL43 48
Kuala Gdns. SW16 BX51 69
Kuhn Way E7 CH35 35
Kydbrook Clo. Orp. CM54 71
Kylemore Rd. NW6 BS36 32
Kymberley Rd. Har. BH32 21
Kyme Rd. Rom. CT32 28
Kynance Clo. Rom. CV28 19
Kynance Gdns. Stan. BK30 12
Kynance Ms. SW7 BS41 41
Kynance Pl. SW7 BT41 41
Kynaston Av. N16 CA34 24
Dynevor Rd.
Kynaston Av. Th. Hth. BZ53 69
Kynaston Clo. Har. BG29 11
Kynaston Cres. Th. Hth. BZ53 69
Kynaston Rd. N16 CA34 24
Kynaston Rd. Brom. CH49 62
Kynaston Rd. Enf. BZ23 8
Kynaston Rd. Orp. CO54 72
Kynaston Rd. Th. Hth. BZ53 69
Kynaston Wood, Har. BG29 11
Kynnersley Clo. Cars. BU55 77
Kynock Rd. N18 CC28 16
Kyrle Rd. SW11 BU46 50
Kyverdale Rd. N16 CA33 24
Laburnham Ct. Stan. BK28 12
Laburnham Way, Brom. CL54 71
Laburnham Av. N9 CA27 15
Laburnham Av. N17 BZ29 15
Laburnham Av. Dart. CV47 64
Laburnham Av. Horn. CT34 28
Laburnham Av. Sutt. BU55 77
Laburnham Av. Swan. CS52 73
Laburnham Clo. E4 CD29 16
Maple Av.
Laburnham Cres. Sun. BC51 65
Laburnham Gdns. N21 BZ27 15
Laburnham Gro. N21 BZ27 15
Laburnham Gro. NW9 BN33 22
Laburnham Gro. Houns. BE45 47
Laburnham Gro. N. Mal. BN51 67
Laburnham Gro. Sthl. BE38 29
Laburnham Pl. SE9 CL46 53
Laburnham Rd. SW19 BT50 59
Laburnham Rd. Epsom BO60 82
Laburnham Rd. Hayes BC42 38
Laburnham Rd. Mitch. BV51 68
Laburnham St. E2 CA37 33
Laburnham Wk. Horn. CV35 37
Lacey Dr. Edg. BL28 12
Lacey Dr. Hamptn. BE51 65
Lackford Rd. Couls. BU62 83
Lackington St. EC2 BZ39 42
Lacock Clo. SW19 BZ50 59
Lacon Rd. SE22 CB45 51
Lacy Rd. SW15 BQ45 49
Ladas Rd. SE27 BZ49 60
Ladbroke Cres. W11 BR39 40
Ladbroke Gro.
Ladbroke Gdns. W11 BR40 40
Ladbroke Gro. W10 BQ38 31
Ladbroke Gro. W11 BR39 40
Ladbroke Rd. W11 BR40 40
Ladbroke Rd. Enf. CA25 8
Ladbroke Rd. Epsom BN60 82
Ladbroke Sq. W11 BR40 40
Ladbroke Sq. Gdns. W11 BR40 40
Ladbroke Ter. W11 BR40 40
Ladbroke Wk. W11 BR40 40
Ladbrook Clo. Pnr. BE32 20
Ladbrook Cres. Sid. CP48 63
Ladbrook Rd. SE25 BZ52 69
Ladderstile Ride, Kings. On T. BN49 58
Ladderswood Rd. N11 BW28 14
Ladds Way, Swan. CS52 73
Ladenhatch La. Swan. CS51 73
Ladworth Lo. Est. Mitch. BV52 68
Ladybower Ct. E5 CC35 34
Clapton Park Est.
Ladycroft Gdns. Orp. CM56 80
Ladycroft Rd. SE13 CE45 52
Ladycroft Wk. Stan. BK30 12
Ladycroft Way, Orp. CM56 80
Ladyfields, Loug. CL24 10
Ladyfields Clo. Loug. CL24 10
Lady Grove, Croy. CD58 79
Lady Hay, Wor. Pk. BO55 76
Lady Margaret Rd. N19 BW35 32
Lady Margaret Rd. NW5 BW35 32
Lady Margaret Rd. Sthl. BE40 38
Lady's Clo. Wat. BD26 4
Lady's Clo. Wat. BC24 4
Vicarage Rd.
Ladysmith Av. E6 CK37 35
Ladysmith Av. Ilf. CM33 26
Ladysmith Rd. E16 CG38 34
Ladysmith Rd. N17 CB30 15
Ladysmith Rd. N18 CB28 15
Ladysmith Rd. SE9 CL46 53
Ladysmith Rd. Enf. CA24 8
Ladysmith Rd. Har. BH30 12
Ladysmith Rd. Kes. CG57 79
Lady Somerset Rd. NW5 BV35 32
Ladywell Rd. SE13 CE46 52
Ladywell St. E15 CG37 34
Ladywood Av. Orp. CN53 72
Ladywood Rd. Surb. BM55 75
Lafone Av. Felt. BD48 56
Lafone St. SE1 CA41 42
Lagonda Av. Ilf. CN29 18
Lagoon Rd. Orp. CO53 72
Lahore Rd. Croy. BZ53 69
Sydenham Rd.
Laindon Av. E15 CG35 34
Leytonstone Rd.
Laing Dene, Nthlt. BD37 29
Laings Av. Mitch. BU51 68
Lainson St. SW18 BS47 59
Lairdale Clo. SE21 BZ46 51
Lairs Clo. N19 BX36 33
Laitwood Rd. SW12 BV47 59
Lake Av. Brom. CH50 62
Lake Av. Rain. CV37 37
Lake Clo. SW19 BR49 58
Lake Rd.
Lakedale Rd. SE18 CG43 52
Lakefield Rd. N22 BY30 15
Lakefields Clo. Rain. CV37 37
Lake Gdns. Dag. CR35 36
Lake Gdns. Rich. BJ48 57
Lake Gdns. Wall. BV55 77
Lakehall Gdns. Th. Hth. BY53 69
Lakehall Rd. Th. Hth. BY53 69
Lake House Rd. E11 CH34 26
Lakehurst Rd. Epsom BO56 76
Lakeland Clo. Chig. CO28 18
Lakeland Clo. Har. BG29 11
Lakenheath N14 BW25 7
Lake Rise, Rom. CT30 19
Lake Rd. SW19 BR49 58
Lake Rd. Croy. CD55 79
Lake Rd. Rom. CP31 27
Laker Pl. SW15 BR46 49
Lakers Rise, Bans. BU61 83
Lakeside W13 BK39 39
Edge Hill Rd.
Lakeside, Enf. BW24 7
Lakeside, Rain. CV37 37
Lakeside, Wall. BV56 77
Lakeside Av. Ilf. CJ31 26
Lakeside Clo. SE25 CA51 69
Lakeside Clo. Sid. CP46 54
Lakeside Ct. Borwd. BM25 5
Lakeside Cres. Barn. BU25 7
Lakeside Dr. Brom. CK55 80
Lakeside Dr. Esher BG57 74
Lakeside Rd. N13 BX28 15
Lakeside Rd. W14 BQ41 40
Lakeside Way, Wem. BM35 30
Lakes Rd. Kes. CJ56 80
Lakeswood Rd. Orp. CL53 71
Lake, The, Bush. BG26 4
Lake Vw, Edg. BL28 12
Lakeview Rd. SE27 BY49 60
Lake View Rd. Well. CO45 54
Lakis Clo. NW3 BT35 32
Flask Walk
Laleham Av. NW7 BN27 13
Laleham Rd. SE6 CF47 61
Lalor St. SW6 BR44 49
Lambarde Av. SE9 CL49 62
Lamberhurst Clo. Orp. CP54 72
Lamberhurst Rd. SE27 BY49 60
Lamberhurst Rd. Dag. CQ33 27
Lambert Av. Rich. BM45 48
Lambert Ct. Bush. BD24 4
Lamberton Ct. Borwd. BL23 5
Lambert Rd. E16 CH39 44
Lambert Rd. N12 BT28 14
Lambert Rd. SW2 BX46 51
Lambert Rd. Bans. BS60 83
Lamberts Pl. Croy. BZ54 69
Lamberts Rd. Surb. BL53 66
Lambert St. N1 BY36 33
Lambert Way N12 BT28 14
Lambeth High St. SE1 BX42 42
Lambeth Hill EC4 BZ40 42
Upper Thames St.
Lambeth Mews SE11 BX42 42
Lambeth Palace Rd. SE1 BX41 42
Lambeth Rd. SE1 BX41 42
Lambeth Rd. Croy. BY54 69
Lambeth Wk. SE11 BX42 42
Lamb La. E8 CB36 33
Lamble St. NW5 BV35 32
Lambley Rd. Dag. CO36 36
Lamb Mews N1 BY37 33
Colebrooke Row
Lambolle Pl. NW3 BU36 32
Lambolle Rd. NW3 BU36 32
Lambourn Clo. W7 BH41 39
Lambourne Av. SW19 BR49 58
Lambourne Cres. Chig. CO27 18
Lambourne Gdns. E4 CE27 16
Lambourne Gdns. Bark. CN36 36
Lambourne Rd.
Lambourne Gdns. Enf. CA23 8
Lambourne Gdns. Horn. CV34 28
Lambourne Rd. E11 CF33 25
Lambourne Rd. Bark. CN36 36
Lambourne Rd. Chig. CN28 18
Lambourne Rd. Ilf. CN34 27
Lambourn Rd. SE17 BZ43 51
Lambourn Rd. SW4 BV45 50
Lambrook Ter. SW6 BR44 49
Lambs Bldgs. EC1 BZ38 33
Errol St.
Lamb's Conduit Pass. WC1 BX39 42
Red Lion Sq.
Lamb's Conduit St. WC1 BX38 33
Lambscroft Av. SE9 CJ48 62
Lambs La. Rain. CU39 46
Lambs Meadow, Wdf. Grn. CJ30 17
Lambs Pass. EC1 BZ39 42
Lambs Pass. Brent. BL42 39
Lambs Ter. N9 BZ27 15
Lamb St. E1 CA39 42
Lambs Wk. Enf. BZ23 8
Lambton Pl. W11 BS40 41
Lambton Rd. N19 BX33 24
Lambton Rd. SW20 BQ51 67
Lamb Wk. SE1 CA41 42
Lamb Yd. Wat. BD24 4
Lamerock Rd. Brom. CG49 61
Lamerton Rd. Ilf. CL30 17
Lamerton St. SE8 CE43 52
Lamford Clo. N17 BZ29 15
Lamington St. W6 BP42 40
Lamlash St. SE11 BY42 42
Hayles St.
Lammas Av. Mitch. BV51 68
Lammas Gn. SE26 CB48 60
Lammas Hill, Esher BF56 74
Lammas La. Esher BF56 74
Lammas Pk. Gdns. W5 BK40 39
Lammas Pk. Rd. W5 BK40 39
Lammas Rd. E9 CC36 34
Lammas Rd. E10 CD33 25
Lammas Rd. Rich. BK49 57
Lammas Rd. Wat. BD25 4
Lammermoor Rd. SW12 BV47 59
Lamont Rd. SW10 BT43 50
Lamorbey Clo. Sid. CN47 63
Lamorna Clo. Orp. CO54 72
Larmorna Gro. Stan. BK30 12
Lampard Gro. N16 CA33 24
Lampern Sq. E2 CB38 33
Nelson Gdns.
Lampeter St. N1 BZ37 33
Lampeter Sq. W6 BR43 49
Humbolt Rd.
Lampmead Rd. SE12 CG46 52
Lamport Clo. SE18 CK42 44
Lampton Av. Houns. BF44 47
Lampton Ho. Clo. SW19 BQ49 58
Lampton Pk. Rd. Houns. BF44 47
Lampton Rd. Houns. BF44 47
Lanacre Av. NW9 BO30 13
Lanark Clo. W5 BK39 39
Lanark Pl. W9 BT38 32
Lanark Rd. W9 BS38 32
Lanata Wk. Hayes BE38 29
Lanbury Rd. SE15 CC45 52
Lancashire Rd. E17 CD30 16
Lancaster Av. E18 CH31 26
Lancaster Av. SE27 BY48 60
Lancaster Av. SW19 BQ49 58
Lancaster Av. Bark. CN36 36
Lancaster Av. Mitch. BX53 69
Lancaster Clo. Brom. CG52 70
Lancaster Clo. Kings. On T. BK49 57
Lancaster Cotts. Rich. BL46 48
Lancaster Pk.
Lancaster Ct. SW6 BR43 49
Lancaster Ct. W2 BT40 41
Lancaster Ct. Bans. BR60 82
Lancaster Ct. Walt. BC54 65
Lancaster Dr. NW3 BU36 32
Lancaster Dr. Horn. CU35 37
Lancaster Gdns. SW19 BR49 58
Lancaster Gdns. W2 BT40 41
Lancaster Gdns. W13 BJ40 39
Lancaster Gdns. Kings. On T. BK49 57
Lancaster Gte. W2 BT40 41
Lancaster Gds. NW3 BU36 32
Lambolle Pl.
Lancaster Gro. NW3 BT36 32
Lancaster Ms. W2 BT40 41
Lancaster Ms. Rich. BL46 48
Richmond Hill
Lancaster Pk. Rich. BL46 48
Lancaster Pl. SW19 BQ49 58
Lancaster Pl. Twick. BJ47 57
Lancaster Pl. WC2 BX40 42
Lancaster Pl. Houns. BD44 47
Lancaster Rd. E7 CH36 35
Lancaster Rd. E11 CG34 25
Lancaster Rd. E17 CC30 16
Lancaster Rd. N4 BY33 24
Lancaster Rd. N11 BW29 14
Lancaster Rd. N18 CA28 15
Lancaster Rd. NW10 BP35 31
Lancaster Rd. SE25 CA51 69
Lancaster Rd. SW19 BQ49 58
Lancaster Rd. W5 BK40 39
Lancaster Rd. W10 BQ41 40
Lancaster Rd. W11 BQ40 40
Lancaster Rd. Barn. BT24 7
Lancaster Rd. Enf. BZ23 8
Lancaster Rd. Har. BF32 20
Lancaster Rd. Nthlt. BG36 29
Lancaster Rd. Sthl. BE40 38
Lancaster St. SE1 BY41 42
Lancefield St. W10 BR38 31
Lancell St. N16 CA34 24
Lancelot Av. Wem. BK35 30
Lancelot Cres. Wem. BK35 30
Lancelot Gdns. Barn. BV26 7
Lancelot Pl. SW7 BU41 41
Lancelot Rd. Ilf. CN29 18
Lancelot Rd. Well. CO45 54
Lancelot Rd. Wem. BK35 30
Lance Rd. Har. BG33 20
Lanchester Rd. N6 BU32 23
Lancing Gdns. N9 CA26 8
Lancing Rd. W13 BJ40 39
Drayton Gn. Rd.
Lancing Rd. Croy. BX54 69
Lancing Rd. Ilf. CM32 26
Lancing Rd. Orp. CO55 81
Lancing Rd. Rom. CW29 19
Lancing St. NW1 BW38 32
Landau Way, Erith CV43 55
Landcroft Rd. SE22 CA46 51
Landells Rd. SE22 CA46 51
Landfield St. E5 CB34 24
Landford Rd. SW15 BQ45 49
Landgrove Rd. SW19 BS49 59
Landmann Way SE14 CC42 43
Landon Pl. SW1 BU41 41
Landor Ct. N16 CA35 33
Arundel Gro.
Landor Rd. SW9 BX45 51
Landor Wk. W12 BP41 40
Landport Way SE15 CA43 51
Landra Gdns. N21 BY25 8
Landridge Rd. SW6 BR44 49
Landrock Rd. N8 BX32 24
Landsbury Dr. Hayes BC38 29
Landscape Rd. Wdf. Grn. CH29 17
Landseer Av. E12 CL35 35
Landseer Clo. Edg. BM30 12
Landseer Rd. N19 BX34 24
Landseer Rd. Enf. CB25 8
Landseer Rd. N. Mal. BN54 67
Landseer Rd. Sutt. BS57 77
Lands End, Bush. BK25 5
Landstead Rd. SE18 CM43 53
Landway, The, Orp. CP52 72
Lane App. NW7 BR28 13
Lane Clo. NW2 BP34 22
Lane Ct. SW11 BU46 50
Thurleigh Rd.
Lane End, Bexh. CR45 54
Lane End, Epsom BN60 82
Lane Gdns. Bush. BH26 5
Lanercost Gdns. N14 BX26 8
Lanercost Rd. SW2 BY48 60
Laneside, Chis. CL49 62
Laneside, Edg. BN28 13
Laneside Av. Dag. CQ33 27
Lane, The NW8 BT37 32
Lane, The SE3 CH45 53
Casterbridge Rd.
La Tourne Gdns. Orp. CM55 80

Lane Way SW15 BP46 49
Sunnymead Rd.
Lanfranc Rd. E3 CD37 34
Lanfrey Pl. W14 BR42 40
North End Rd.
Langbourne Av. N6 BV34 23
Langbourne Way, Esher BJ56 75
Langbrook Rd. SE3 CJ45 53
Langcroft Clo. Cars. BU55 77
Langdale Av. Mitch. BU52 68
Langdale Cr. Bexh. CR43 54
Langdale Gdns. Grnf. BJ38 30
Langdale Gdns. Horn. CU35 37
Langdale Rd. SE10 CE43 52
Langdale Rd. Th. Hth. BY52 69
Langdale St. E1 CB39 42
Langdon Ct. NW10 BO37 31
Craven Park
Langdon Cres. E6 CL37 35
Langdon Dr. NW9 BN33 22
Langdon Pk. Rd. N6 BW33 23
Langdon Pl. SW14 BN45 49
Rosemary La.
Langdon Rd. E6 CL37 35
Langdon Rd. Brom. CH52 71
Langdon Rd. Mord. BT53 68
Langdon Shaw, Sid. CN49 63
Langford Clo. NW8 BT37 32
Langford Ct. NW8 BT37 32
Langford Cres. Barn. BU24 7
Langford Grn. Gdns. SE5 CA45 51
Langford Pl. NW8 BT37 32
Langford Pl. Sid. CO48 63
Langford Rd. SW6 BS44 50
Langford Rd. Barn. BU24 7
Langford Rd. Wdf. Grn. CJ29 17
Langfords, Buck. H. CJ27 17
Langham Clo. N15 BY31 24
Langham Rd.
Langham Ct. Horn. CV33 28
Langham Dene, Ken. BY57 78
Langham Dene, Ken. BY61 84
Langham Dr. Rom. CO32 27
Langham Gdns. N21 BY25 8
Langham Gdns. W13 BJ40 39
Garden Rd.
Langham Gdns. Edg. BN29 13
Langham Gdns. Rich. BK49 57
Langham Gdns. Wem. BK34 21
Langham Ho. Rich. BK49 57
Langham Pl. N15 BY31 24
Langham Pl. W1 BV39 41
Langham Rd. N15 BY31 24
Langham Rd. SW20 BQ51 67
Langham Rd. Edg. BN29 13
Langham Rd. Tedd. BJ49 57
Langham St. W1 BV39 41
Langhedge Clo. N18 CA29 15
Langholme, Bush. BG26 4
Sparrows Herne
Langhorne Rd. Dag. CR36 36
Langland Cres. E. Stan. BK30 12
Langland Cres. N. Stan. BK30 12
Langland Cres. S. Stan. BL31 21
Langland Cres. W. Stan. BK30 12
Langland Dr. Pnr. BE29 11
Langland Gdns. NW3 BS35 32
Langland Gdns. Croy. CD55 79
Langlands Rise, Epsom BN60 82
Langler Rd. NW10 BQ37 31
Langler Rd. NW10 BQ38 31
Langley Av. Ruis. BC34 20
Langley Av. Surb. BK54 66
Langley Av. Wor. Pk. BQ54 67
Langley Clo. Rom. CV29 19
Faringdon Av.
Langley Ct. SE9 CL46 53
Langley Ct. WC2 BX40 42
Long Acre
Langley Ct. Beck. CF53 70
Langley Ct. W. Wick. CF54 70
Langley Cres. E11 CH33 26
Langley Cres. Dag. CP36 36
Langley Cres. Edg. BN27 13
Langley Dr. E11 CH33 26
Langley Dr. W3 BM41 39
Langley Gdns. Dag. CP36 36
Langley Gdns. Orp. CL53 71
Langley Gro. N. Mal. BO51 67
Langley La. SW8 BX43 51
Langley Meadows, Loug. CM23 10
Langley Oaks Av. Sth. Croy. CB58 78
Langley Pk. NW7 BO29 13
Langley Pk. Rd. Sutt. BT56 77
Langley Rd. SW19 BR51 67
Langley Rd. Beck. CC52 70
Langley Rd. Sth. Croy. CC58 79
Langley Rd. Surb. BL54 66
Langley Rd. Well. CP43 54
Langley St. WC2 BX39 42
Langley Vale Rd. Epsom BN61 82
Langley Way, W. Wick. CF54 70
Langly Rd. Islw. BH44 48
Lang Mead SE27 BY49 60
Langmead Dr. Bush. BH26 5
Langport Ct. KT12 BD54 65
Langroyd Rd. SW17 BU48 59
Langside Av. SW15 BP45 49
Langside Cres. N14 BW27 14
Langston Rd. Loug. CM25 10
Lang St. E1 CC38 34
Langthorne Rd. E11 CF34 25
Langthorne St. SW6 BQ43 49
Langton Av. E6 CL38 35
Lanton Av. N20 BT26 7
Langton Av. Epsom BO59 82
Lanton Clo. WC1 BX38 33
Wren St.
Langton Ri. SE23 CB47 60
Langton Rd. NW2 BQ34 22
Langton Rd. SW9 BY43 51
Langton Rd. E. Mol. BG53 65
Langton Rd. Har. BG29 11
Langton St. SW10 BT43 50
Langton Way SE3 CB44 52
Langton Way, Croy. CA55 78
Langtry Rd. NW8 BT37 32
Langtry Rd. Nthlt. BD37 29
Langwood Chase, Tedd. BK50 57
Broom Rd.
Langwood Gdns. Wat. BC23 4
Lanhill Rd. W9 BS38 32
Lanier Rd. SE13 CF46 52
Lankaster Gdns. N2 BT30 14
Lankers Dr. Har. BE32 20
Lankton Clo. Beck. CF51 70
Lannoy Rd. SE9 CM47 62
Lanrick Rd. E14 CF39 43
Lanridge Rd. SE2 CP41 45
Lansbury Av. N18 BZ28 15
Lansbury Av. Bark. CO36 36
Lansbury Av. Felt. BC46 47
Lansbury Av. Rom. CQ32 27
Lansbury Clo. NW10 BN35 31
Lansbury Est. E14 CE39 43
Lansbury Gdns. E14 CF39 43
Lansbury Rd. Enf. CC23 9
Lansbury Way N18 CA28 15
Lansbury Av.
Lansdell Rd. Mitch. BV51 68
Lansdown Clo. KT12 BD54 65
St. Johns Dr.
Lansdowne Av. Orp. CL54 71
Lansdowne Av. Well. CP43 54
Lansdowne Clo. SW20 BQ50 58
Lansdowne Clo. Twick. BH47 57
Lansdowne Ct. Pur. BY58 78
Lansdowne Ct. Wor. Pk. BP55 76
Lansdowne Cres. W11 BR40 40
Lansdowne Dr. E8 CB36 33
Lansdowne Gdns. SW8 BX44 51
Lansdowne Gdns. W11 BR41 40
Lansdowne Grn. Est. SW8 BX44 51
Lansdowne Gro. NW10 BO35 31
Lansdowne Hill SE27 BY48 60
Lansdowne La. SE7 CJ43 53
Lansdowne Mews SE7 CJ42 44
Lansdowne Pl. SE1 BZ41 42
Lansdowne Pl. SE19 CA50 60
Lansdowne Rise W11 BR40 40
Lansdowne Rd. E4 CE27 16
Lansdowne Rd. E11 CG34 25
Lansdowne Rd. E17 CE32 25
Lansdowne Rd. E18 CH31 26
Lansdowne Rd. N3 BR29 13
Lansdowne Rd. N10 BW30 14
Lansdowne Rd. N17 CA30 15
Lansdowne Rd. SW20 BQ50 58
Lansdowne Rd. W11 BR40 40
Lansdowne Rd. Brom. CH50 62
Lansdowne Rd. Croy. BZ55 78
Lansdowne Rd. Epsom BN57 76
Lansdowne Rd. Har. BH33 21
Lansdowne Rd. Houns. BF45 47
Lansdowne Rd. Ilf. CN33 27
Lansdowne Rd. Pur. BX59 84
Lansdowne Rd. Stan. BK29 12
Lansdowne Row W1 BV40 41
Berkeley St.
Lansdowne Ter. WC1 BX38 33
Lansdowne Wk. W11 BR40 40
Lansdowne Way SW8 BW44 50
Lansdown Rd. E7 CJ36 35
Lansdown Rd. Sid. CO48 63
Lanseer Clo. SW19 BT51 68
Brangwyn Cres.
Lansfield Av. N18 CB28 15
Lanswell Est. SE1 BY41 42
Lantern Clo. SW15 BP45 49
Lant St. SE1 BZ41 42
Lanturn Clo. Wem. BK35 30
Lanvanor Rd. SE15 CC44 52
Lapponum Wk. Hayes BD39 38
Lapsewood Wk. SE23 CB48 60
Lapstone Gdns. Har. BK32 21
Lapwing Clo. Sth. Croy. CD58 79
Larbert Rd. SW16 BW50 59
Larby Pl. Epsom BO58 76
Larch Av. W3 BO40 40
Larch Clo. SW12 BV47 59
Larch Cres. Epsom BM57 75
Larch Cres. Hayes. BD38 29
Larchdene, Orp. CL55 80
Larches Av. SW14 BN45 49
Larches, The N13 BZ27 15
Larches, The, Wat. BE25 4
Larch Rd. NW2 BQ35 31
Larch Rd. Dart. CV47 64
Larch Tree Way, Croy. CE55 79
Larch Way, Brom. CL54 71
Larchwood Av. Rom. CR29 18
Larchwood Clo. Bans. BR61 82
Larchwood Clo. Rom. CS29 19
Larchwood Rd. SE9 CL48 62
Larcom St. SE17 BZ42 42
Larden Rd. W3 BO40 40
Lardo Av. SW6 BR45 49
Largewood Av. Surb. BL55 75
Larkbere Rd. SE26 CD49 61
Larken Dr. Bush. BG26 4
Larkfield Av. Har. BJ31 21
Larkfield Clo. Brom. CG55 79
Larkfield Rd. Rich. BL45 48
Larkfield Rd. Sid. CN48 63
Larkhall Est. SW8 BW44 50
Larkhall La. SW4 BW44 50
Larkhall Rise, SW4 BW44 50
Larkhall, Walt. BD57 74
Lark Row E3 CC37 34
Larksfield Gro. Enf. CB23 8
Larkshall Cres. E4 CF28 16
Larkshall Rd. E4 CF28 16
Larkspur Clo. N17 BZ29 15
Larkspur Clo. Orp. CP55 81
Berrylands
Larkspur Way, Epsom BN56 76
Larkswood Rise, Pnr. BD31 20
Larkswood Rd. E4 CE28 16
Larkway Clo. NW9 BN31 22
Larnach Rd. W6 BQ43 49
Larner Rd. Erith CT43 55
Larpent Av. SW15 BQ45 49
Larson Rd. Esher BE57 74
Larwood Clo. Har. BG35 29
Lascelles Av. Har. BG33 20
Lascelles Clo. E11 CF34 25
Lascott's Rd. N22 BX29 15
Lassa Rd. SE9 CK46 53
Lassell St. SE10 CF42 43
Latchet Rd. E18 CH30 17
Latchford Pl. Chig. CO28 18
Latchington Gdns. Wdf. Grn. CK29 17
Latchmere Clo. Rich. BL49 57
Latchmere Ho. Kings. On T. BL49 57
Latchmere La. Kings. On T. BL50 57
Latchmere Rd. SW11 BU44 50
Latchmere Rd. Kings. On T. BL50 57
Latchmere St. SW11 BU44 50
Burns Rd.
Lateward Rd. Brent. BK43 48
Latham Clo. Twick. BJ47 57
Latham Rd. Bexh. CR46 54
Latham Rd. Twick. BH47 57
Lathkill Clo. Enf. CA26 8
Lathom Rd. E6 CK36 35
Latimer Av. E6 CK37 35
Latimer Clo. N15 CA32 24
Latimer Clo. N15 CA32 24
St. Ann's Rd.
Latimer Clo. Pnr. BD30 11
Latimer Clo. Wor. Pk. BP56 76
Latimer Gdns. Pnr. BD30 11
Latimer Ms. W10 BQ39 40
Latimer Pl. W10 BQ39 40
Latimer Rd. E7 CH35 35
Latimer Rd. SW19 BS50 59
Latimer Rd. W10 BQ39 40
Latimer Rd. W11 BQ40 40
Latimer Rd. Barn. BS24 7
Latimer Rd. Tedd. BH49 57
Latona Rd. SE15 CA43 51
Latton Clo. Walt. BE54 65
Latton Clo. Esher BF56 74
Latymer Ct. W6 BQ42 40
Latymer Rd. N9 CA26 8
Latymer Way N9 BZ27 15
Lauder Clo. Nthlt. BD37 29
Lauderdale Dr. Rich. BK48 57
Lauderdale Rd. W9 BS38 32
Laud St. SE11 BX42 42
Laud St. Croy. BZ55 78
Laughedge La. N18 CA28 15
Laughton Rd. Nthlt. BD37 29
Launcelot Rd. Brom. CH49 62
Launceston Clo. Rom. CV30 19
Launceston Gdns. Grnf. BK37 30
Launceston Pl. W8 BT41 41
Launceston Rd. Grnf. BK37 30
Launch St. E14 CF41 43
Launder's La. Rain. CW39 46
Laundry Rd. E4 CF26 9
Station Rd.
Laundry Rd. W6 BR43 49
Lauradale Rd. N2 BU31 23
Laura Clo. E11 CJ32 26
Laura Dr. Swan. CU50 64
Laura Pl. E5 CC35 34
Laurel Av. Twick. BH47 57
Laurel Bank Rd. Enf. BZ23 8
Laurel Clo. Dart. CV47 64
Laurel Clo. Ilf. CM29 17
Laurel Clo. Sid. CO48 63
Laurel Cres. Croy. CE55 79
Laurel Cres. Rom. CT33 28
Laurel Dene, Tedd. BG49 56
Laurel Dr. N21 BY26 8
Laurel Gdns. E4 CE26 9
Laurel Gdns. NW7 BN27 13
Laurel Gdns. W7 BH40 39
Laurel Gdns. Houns. BE45 47
Laurel Gro. SE20 CB50 60
Laurel Gro. SE26 CC49 61
Laurel Rd. SW13 BP44 49
Laurel Rd. SW20 BP51 67
Laurel Rd. Tedd. BG49 56
Laurel St. E8 CA36 33
Laurel View N12 BS27 14
Laurel Way N20 BS27 14
Laurel Way E11 CG31 25
Laurence Mews W12 BP41 40
Askew Rd.
Laurence Pountney Hill EC4 BZ40 42
Cannon St.
Laurence Pountney La. EC4 BZ40 42
Laurie Gdns. W7 BH39 39
Laurie Gro. SE14 CD44 52
Laurie Rd. W7 BH39 39
Laurier Rd. NW5 BV34 23
Laurier Rd. Croy. CA54 69
Laurie Wk. Rom. CT31 28
Laurimel Clo. Stan. BJ29 12
September Way
Lauriston Rd. E9 CC36 34
Lauriston Rd. SW19 BQ50 58
Lausanne Rd. N8 BY31 24
Lausanne Rd. SE15 CC44 52
Lavell St. N16 BZ35 33
Lavender Av. NW9 BN33 22
Lavender Av. Mitch. BU51 68
Lavender Av. Wor. Pk. BQ55 76
Lavender Clo. Cars. BV56 77
Lavender Rd.
Lavender Clo. Rom. CV29 19
Lavender Gdns. SW11 BU45 50
Lavender Gro. E8 CA36 33
Lavender Gro. Mitch. BU51 68
Lavender Hill SW11 BU45 50
Lavender Hill, Enf. BY23 8
Lavender Hill, Swan. CS52 73
Lavender Rd. SW11 BT45 50
Lavender Rd. Cars. BV56 77
Lavender Rd. Croy. BX53 69
Lavender Rd. Enf. BZ23 8
Lavender Rd. Epsom BM57 75
Lavender Rd. Sutt. BT56 77
Parkhurst Rd.
Lavender St. E15 CG36 34
Lavender Sweep SW11 BU45 50
Lavender Vale, Wall. BW57 77
Lavender Wk. SW11 BU45 50
Lavender Hill
Lavender Wk. Mitch. BV52 68
Lavender Way, Croy. CC53 70
Lavengro Rd. SE27 BZ48 60
Lavenham Rd. SW18 BR48 58
Lavernock Rd. Bexh. CR44 54
Lavers Rd. N16 CA34 24
Laverstock Gdns. SW15 BO47 58
Lavidge Rd. SE9 CK48 62
Lavie Ms. W10 BR38 31
Portobello Rd.
Lavina Gro. N1 BX37 33
Wharfdale Rd.
Lavington Rd. W13 BJ40 39
Lavington Rd. Croy. BX55 78
Lavington St. SE1 BY40 42
Lavinia Rd. Dart. CW46 55
Lawdons Gdns. Croy. BY56 78
Lawford Clo. Wall. BX58 78
Lawford Gdns. Dart. CV46 55
Lawford Rd. NW5 BW36 32
Lawford Rd. W4 BN43 49
Lawless St. E14 CE40 43
Lawley Rd. N14 BV26 7
Lawley St. E5 CB35 33
Lawn Clo. N9 CA26 8
Lawn Clo. Brom. CH50 62
Lawn Clo. N. Mal. BO51 67
Lawn Clo. Swan. CS51 73
Lawn Cres. Rich. BL44 48
Lawn Farm Gro. Rom. CQ31 27
Lawnfield NW6 BQ36 31
Lawn Gdns. W7 BH40 39
Lawn La. SW8 BX43 51
Lawn Place SE15 CA44 51
Sumner Estate
Lawn Rd. NW3 BU35 32
Lawn Rd. Beck. CD50 61
Lawns Ct. Wem. BM34 21
Avenue, The
Lawnside SE3 CG45 52
Lawns, The E4 CE28 16
Lawns, The SE3 CG45 52
Lawns, The SE19 BZ51 69
Lawns, The, Pnr. BR29 11

Lingwood Rd. E5 CB33 24
Linhope St. NW1 BU38 32
Linkfield, Brom. CH53 71
Linkfield, E. Mol. BF52 65
Linkfield Rd. Islw. BH44 48
Link La. Wall. BW57 77
Linklea Clo. NW9 BO29 13
Link Rd. N11 BV28 14
Link Rd. Wall. BV54 68
Link Rd. Wat. BD23 4
Links Av. Mord. BS52 68
Links Av. Rom. CU30 19
Links Dr. N20 BS26 7
Links Dr. Borwd. BL24 5
Links Gdns. SW16 BY50 60
Linkside N12 BR29 13
Linkside Chig. CM28 17
Linkside N. Mal. BO51 67
Linkside Clo. Enf. BX24 8
Linkside Gdns. Enf. BX24 8
Links Rd. NW2 BO34 22
Links Rd. SW17 BU50 59
Links Rd. W3 BM39 39
Links Rd. Epsom BP60 82
Links Rd. Wdf. Grn. CH28 17
Links Rd. W. Wick. CF54 70
Links Side, Enf. BX24 8
Links, The, E17 CD31 25
Links, The, Walt. BC55 74
Link St. E9 CC35 34
Links View N2 BU31 23
Gt. North Rd.
Links Vw. N3 BR29 13
Links Vw. Dart. CU47 64
Links Vw. Clo. Stan. BJ29 12
Links Vw. Rd. Croy. CE55 79
Links View Rd. Hamptn. BG49 56
Links Way NW4 BQ30 13
Links Way Beck. CE53 70
Linksway, Sutt. BT58 77
Link, The, W3 BM40 39
Saxon Dr.
Link, The, Enf. CD23 9
Link, The, Pnr. BD33 20
Link, The, Wem. BK33 21
Nathans Rd.
Linkway N4 BZ33 24
Vale Rd.
Linkway SW20 BP52 67
Link Way, Brom. CK53 71
Link Way, Dag. CP34 27
Linkway Horn. CW33 28
Linkway Pnr. BD30 11
Linkway Rich. BJ48 57
Linkway, The, Barn. BS25 7
Linley Cres. Rom. CS31 28
Linley Dr. KT12 BD55 65
Linley Rd. N17 CA30 15
Linnell Clo. NW11 BS32 23
Linnell Dr. NW11 BS32 23
Linnell Rd. N18 CB28 15
Fairfield Rd.
Linnel Rd. SE5 CA44 51
Linnet Clo. SE28 CP40 45
Linnet Clo. Bush. BG26 4
Linnett Clo. E4 CF28 16
Linom Rd. SW4 BX45 51
Linscott Rd. E5 CC35 34
Linsey St. SE16 CB42 42
Linslade Rd. Orp. CO57 81
Linstead Ct. SE9 CN46 54
Linstead Rd. NW6 BS36 32
Linstead Way SW18 BR47 58
Linster Gro. Borwd. BN25 6
Lintaine Clo. W6 BR43 49
Moylan Rd.
Linthorpe Av. Wem. BK36 30
Linthorpe Rd. N16 CA33 24
Linthorpe Rd. Barn. BU24 7
Linton Av. Borwd. BL23 5
Linton Clo. Well. CO44 54
Anthony Rd.
Linton Ct. Rom. CT30 19
Rise Pk. Par.
Linton Glade Croy. CD58 79
Linton Gro. SE27 BZ49 60
Linton Rd. Bark. CM36 35
Lintons La. Epsom BO59 82
Linton St. N1 BZ37 33
Linver Rd. SW6 BS44 50
Linwood Way SE15 CA43 51
Linzee Rd. N8 BX31 24
Lion Av. Twick. BH47 57
Lionel Gdns. SE9 CJ46 53
Lionel Ms. W10 BR39 40
Telford Rd.
Lionel Rd. SE9 CJ46 53
Lionel Rd. Brent. BL41 39
Lion Gate Gdns. Rich. BL45 48
Lion Grn. Rd. Couls. BW61 83
Lion Rd. N9 CB27 15
Lion Rd. Bexh. CQ45 54
Lion Rd. Croy. BZ53 69
Lion Rd. Twick. BH47 57
Lions Clo. SE9 CJ49 62
Lionsdale Clo. SE9 CJ49 62
Lion Way, Brent. BK43 48
Lion Wf. Islw. BJ45 48
Liphook Clo. Horn. CT35 37
Petworth Way
Liphook Cres. SE23 CC47 61
Liphook Rd. Wat. BD28 11
Lippitts Hill, Loug. CG23 9
Lipsham Clo. Bans. BT60 83
Lipton St. E1 CC39 43
Lisbon Av. Twick. BG48 56
Lisburne Rd. NW3 BU35 32
Lisford St. SE15 CA44 51
Lisgar Ter. W14 BR42 40
Liskeard Clo. Chis. CM50 62
Liskeard Gdns. SE3 CH44 53
Lisle St. WC2 BW40 41
Lismore Cir. NW5 BV35 32
Lismore Clo. Islw. BJ44 48
Lismore Rd. N17 BZ31 24
Lismore Rd. Sth. Croy. CA57 78
Lismore Wk. N1 BZ36 33
Marquess Est.
Lissenden Gdns. NW5 BV35 32
Lissoms Rd. Couls. BV62 83
Lisson Green Est. NW8 BT39 41
Lisson Gro. NW1 BT38 32
Lisson St. NW1 BU39 41
Liss Way SE15 CA43 51
Hordle Prom. S.
Lister Ct. N16 CA34 24
Lister Gdns. N18 BZ28 15
Lister Ms. N7 BX35 33
Holloway Rd.
Lister Rd. E11 CG33 25
Liston Rd. N17 CB30 15
Liston Rd. SW4 BW45 50
Liston Way Wdf. Grn. CJ29 17
Listowell Clo. SW9 BY43 51
Listowel Rd. Dag. CR34 27
Listria Pk. N16 CA34 24
Litchfield Av. E15 CG36 34
Litchfield Av. Mord. BR54 67
Litchfield Gdns. NW10 BP36 31
Litchfield Rd. Sutt. BT56 77
Litchfield St. WC2 BW40 41
Litchfield Way NW11 BS32 23
Litford Rd. SE5 BY44 51
Lithos Rd. NW3 BS36 32
Litlington St. SE16 CB42 42
Little Acre Beck. CE52 70
Lit. Albany St. NW1 BV38 32
Lit. Argyll St. W1 BW39 41
Argyll St.
Lit. Aston Rd. Rom. CW30 19
Lit. Birches Sid. CN48 63
Little Boltons, The, SW10 BS42 41
Lit. Bournes SE21 CA49 60
Little Britain EC1 BY39 42
Little Brownings SE23 CB48 60
Littlebury Rd. SW4 BW45 50
Little Bury St. N9 BZ26 8
Lit. Bushey La. Bush. BF23 4
Little Cedars N12 BT28 14
Woodside Av.
Little Chester St. SW1 BV41 41
Wilton Ms.
Little College La. EC4 BZ40 42
College St.
Little College St. SW1 BX41 42
Littlecombe SE7 CH43 53
Littlecombe Clo. SW15 BQ46 49
Littlecote Clo. SW19 BR47 58
Beaumont Rd.
Littlecote Pl. Pnr. BE30 11
Little Ct. W. Wick. CG55 79
Little Croft SE9 CL45 53
Littledale SE2 CO43 54
Little Dimocks SW12 BV48 59
Little Dorrit Ct. SE1 BZ41 42
Lit. Ealing La. W5 BK42 39
Lit. Edward St. NW1 BV37 32
Little Ferry Rd. Twick. BJ47 57
Ferry Rd.
Littlefield Clo. N19 BW35 32
Junction Rd.
Littlefield Rd. Edg. BN29 13
Little Friday Hill E4 CG27 16
Lit. Friday Rd. E4 CG27 16
Little Gearies Ilf. CL31 26
Little George St. SW1 BX41 42
Gt. George St.
Lit. Gerpins La. Upmin. CW37 37
Lit. Green St. NW5 BV35 32
Collage La.
Little Gro. Barn. BU25 7
Lit. Grove Bush. BF24 4
Little Heath SE7 CK43 53
Little Heath Rom. CO31 27
Little Heath Rd. Bexh. CQ44 54
Littleheath Rd. Sth. Croy. CB58 78
Little Holt E11 CH32 26
Little Ilford La. E12 CK35 35
Little John Rd. W7 BH39 39
Littlejohn Rd. Orp. CO53 72
Little Marlborough St. W1 BW39 41
Gt. Marlborough St.
Lit. Martins Bush. BF25 4
Littlemead Esher BG56 74
Littlemede SE9 CK48 62
Littlemore Rd. SE2 CO41 45
Littlemore Rd. Ilf. CM34 26
Lit. Moss La. Pnr. BE30 11
Little Newport St. WC2 BW40 41
Charing Cross Rd.
Little New St. EC4 BY39 42
Lit. Orchard Clo. Pnr. BE30 11
Lit. Oxhey La. Wat. BE28 11
Little Pk. Dr. Felt. BD48 56
Little Pk. Gdns. Enf. BZ24 8
Litte Plucketts Way IG9 CJ26 10
Lit. Portland St. W1 BW39 41
Little Potters Bush. BG26 4
Lit. Queen's Rd. Tedd. BH50 57
Lit. Queen St. Dart. CW47 64
Lit. Queen St. Tedd. BH50 57
Little Redlands, Brom. CK51 71
Lit. Roke Av. Ken. BY60 84
Lit. Roke Rd. Ken. BZ60 84
Littler's Clo. SW19 BT51 68
Lit. Russell St. WC1 BX39 42
Little St. James's St. SW1 BW40 41
Little St. Leonard's SW14 BO45 49
Little Sanctuary SW1 BX41 42
Broad Sanctuary
Little Somerset St. E1 CA39 42
Littlestone Clo. Beck. CE50 61
Abbey La.
Little Strand NW9 BO30 13
Lit. Thrift Orp. CM52 71
Lit. Tichfield St. W1 BW39 41
Gt. Tichfield St.
Littleton Av. E4 CG26 9
Valance Av.
Littleton Cres. Har. BH34 21
Littleton Rd. Har. BH34 21
Littleton St. SW18 BT48 59
Lit. Trinity La. EC4 BZ40 42
Queen Victoria St.
Lit. Turnstile WC1 BX39 42
High Holborn
Littlewood Clo. W13 BJ41 39
Lit. Woodcote La. Cars. BV59 83
Littlewood Rd. SE13 CF46 52
Littleworth Av. Esher BG56 74
Littleworth Common Rd. Esher BG55 74
Littleworth La. Esher BG56 74
Littleworth Rd. Esher BG56 74
Liverpool Gro. SE17 BZ42 42
Liverpool Rd. E10 CF32 25
Liverpool Rd. E16 CG38 34
Liverpool Rd. N1 BY36 33
Liverpool Rd. N7 BY35 33
Liverpool Rd. W5 BK41 39
Liverpool Rd. Kings. On T. BM50 57
Liverpool Rd. Th. Hth. BZ52 69
Liverpool Rd. Wat. BC25 4
Liverpool St. EC2 CA39 42
Livesey Pl. SE15 CB43 51
Peckham Park Rd.
Livingstone Ct. E10 CF32 25
Livingstone Pl. E14 CF42 43
Ferry St.
Livingstone Rd. E15 CF37 34
Livingstone Rd. E17 CE32 25
Livingstone Rd. N13 BX29 15
Livingstone Rd. SW11 BT45 50
Winstanley Rd.
Livingstone Rd. Houns. BG45 47
Livingstone Rd. Sthl. BD40 38
Livingstone Rd. Th. Hth. BZ51 69
Livingstone St. E6 CL40 44
Livingstone Ter. Rain. CT37 37
Livingstone Wk. SW11 BT45 50
Plough Rd.
Livonia St. W1 BW39 41
Berwick St.
Lizard St. EC1 BZ38 33
Lizban St. SE3 CH43 53
Llanavor Rd. NW2 BR34 22
Llanelly La. NW2 BR34 22
Llanelly Rd. NW2 BR34 22
Crewys Rd.
Llanover Rd. Wem. BK34 21
Llanover St. SE18 CL43 53
Llanthony Rd. Mord. BT53 68
Llewellyn St. SE16 CB41 42
Chambers St.
Lloyd Av. SW16 BX51 69
Lloyd Av. Couls. BV60 83
Lloyd Baker St. WC1 BX38 33
Lloyd Ct. Pnr. BD32 20
Lloyd Pk. Av. Croy. CA56 78
Lloyd Rd. E6 CK37 35
Lloyd Rd. E17 CC31 25
Lloyd Rd. Dag. CQ36 36
Lloyd Rd. Wor. Pk. BQ55 76
Lloyd's Av. EC3 CA39 42
Lloyds Pl. SE3 CG44 52
Lloyd Sq. WC1 BY38 33
Lloyds Row EC1 BY38 33
Lloyd St. WC1 BY38 33
Lloyds Way Beck. CD53 70
Loampit Hill SE13 CE44 52
Loampit Vale SE13 CE45 52
Loates La. Wat. BD24 4
Local Board Rd. Wat. BD25 4
Locarno Rd. W3 BO41 40
High St.
Locarno Rd. Grnf. BG38 29
Lochaline St. W6 BQ43 49
Lochinvar St. SW12 BV47 59
Lochmere Clo. Erith CR43 54
Lochnagar St. E14 CF39 43
Lockaber Rd. SE13 CG45 52
Lock Chase SE3 CG45 52
Locke Clo. Rain. CT36 37
Lockesley Dr. Orp. CN53 72
Lockesley Sq. Surb. BK53 66
Lockesmead Rd. Rich. BK49 57
Locket Rd. Har. BH30 12
Lockfield Av. Enf. CD23 9
Lockhart Clo. N7 BX36 33
Mackenzie Rd.
Lockhart St. E3 CD38 34
Lockhurst Rd. E5 CC35 34
Lockmead Rd. N15 CB32 24
Lockmead Rd. SE13 CF45 52
Lock Rd. Rich. BK49 57
Lock's La. Mitch. BU51 68
Locksley Est. E14 CD39 43
Locksley St. E14 CD39 43
Lock St. W10 BQ40 40
Lockwood Rd. Ilf. CM34 26
Lockwood Sq. SE16 CB41 42
Southwark Pk. Rd.
Lockwood Wk. Rom. CT32 28
Liberty The
Lockwood Wy. E17 CC31 25
Lockwood Way, Chess. BM56 75
Locton Est. E3 CD37 34
Loddiges Rd. E9 CC36 34
Loder St. SE15 CC43 52
Lodge Av. SW14 BN45 49
South Worple Way
Lodge Av. Borwd. BL25 5
Lodge Av. Croy. BX55 78
Lodge Av. Dag. CO37 36
Lodge Av. Dart. CV46 55
Lodge Av. Har. BL31 21
Lodge Av. Rom. CT32 28
Lodge Clo. N18 BZ28 15
Lodge Clo. Chig. CO28 18
Lodge Clo. Edg. BL29 12
Lodge Clo. Orp. CO54 72
Lodge Clo. Sutt. BS56 77
Lodge Clo. Wall. BV54 68
Lodge Ct. Horn. CW34 28
Lodge Cres. Orp. CO54 72
Lodge Dr. N13 BY28 15
Lodge Gdns. Beck. CD53 70
Lodge Hill, Ilf. CK31 26
Lodge Hill, Pur. BY61 84
Lodge Hill, Well. CO43 54
Lodge La. N12 BT28 14
Lodge La. Bex. CP46 54
Lodge La. Croy. CE57 79
Lodge La. Est. Rom. CR29 18
Lodge Rd. NW4 BQ31 22
Lodge Rd. NW8 BT38 32
Lodge Rd. Brom. CH50 62
Lodge Rd. Croy. BY53 69
Lodge Rd. Sutt. BS56 77
Lodge Rd. Wall. BV56 77
Lodge Vill. Wdf. Grn. CG29 16
Lodore Gdns. NW9 BN32 22
Lodore St. E14 CF39 43
Loftie St. SE16 CB41 42
Chambers St.
Lofting Rd. N1 BX36 33
Loftus Rd. W12 BP40 40
Logan Clo. Enf. CC23 9
Logan Ms. W8 BS42 41
Logan Pl. W8 BS42 41
Logan Rd. N9 CB27 15
Logan Rd. Houns. BE45 47
Logan Rd. Wem. BK34 21
Logs Hill, Chis. CK51 71
Logs Hill Clo. Chis. CK51 71
Lolesworth St. E1 CA39 42
Wentworth St.
Lollard St. SE11 BX42 42
Loman St. SE1 BY41 42
Lomas St. E1 CB39 42
Lombard Av. N11 BW28 14
Lombard Av. Enf. CC23 9
Lombard Av. Ilf. CN33 27
Lombard La. EC4 BY39 42
Lombard Rd. SW11 BT44 50
Lombard Rd. SW19 BS51 68
Lombard St. EC3 BZ39 42
Lombard Wall. SE7 CH41 44
Lomond Clo. Wem. BL36 30
Lomond Gro. SE5 BZ43 51
Londale Clo. Pnr. BE29 11
Londesborough Rd. N16 CA35 33
London Av. Gro. SE26 CB49 60
London Br. St. SE1 BZ40 42
London Br. Wk. SE1 BZ40 42
Tooley St.
London Fields, E8 CB37 33

London Fields, E. Side E8 — CB37 33
London Fields, W. Side, E8 — CB36 33
London La. E8 — CB36 33
London La. Brom. — CG50 61
London Ms. W2 — CJ39 44
London St.
London Rd. E13 — CH37 35
London Rd. SE1 — BY41 42
London Rd. SE23 — CB47 60
London Rd. SW16 — BX51 69
London Rd. SW17 — BU53 68
London Rd. Bark. — CL37 35
London Rd. Brom. — CG50 61
London Rd. Bush. — BE25 4
London Rd. Dart. — CS46 55
London Rd. Enf. — BZ25 8
London Rd. Epsom — BP58 76
London Rd. Har. — BH34 21
London Rd. Houns. — BG45 47
London Rd. Kings. On T. — BL51 66
London Rd. Mitch. — BU53 68
London Rd. Mitch. — BV53 68
London Rd. Mord. — BS53 68
London Rd. Purfleet — CW42 46
London Rd. Rom. — CR32 27
London Rd. Sev. — CQ58 81
Londodn Rd. Stan. — BK28 12
London Rd. Swan. — CS51 73
London Rd. Swan. — CU52 73
London Rd. Th. Hth. & Croy. — BX51 69
London Rd. Twick. — BJ47 57
London Rd. Wall. — BV55 77
London Rd. Wem. — BL35 30
London Stile W4 — BM42 39
Wellesley Rd.
London St. EC3 — CA40 42
Fenchurch St.
London St. W2 — BT39 41
London Tilbury Rd. — CV38 37
Rain.
London Wall, EC2 — BZ39 42
Long Acre, WC2 — BX40 42
Long Acre Orp. — CP55 81
Longacre Pl. Cars. — BV57 77
Longacre Rd. E17 — CF30 16
Longbeach Rd. SW11 — BU45 50
Longberrys Rd. NW2 — BR34 22
Longbridge Rd. Bark. — CM36 35
Longbridge Way, SE13 — CF46 52
Longbury Dr. Orp. — CO52 72
Longcliffe Path Wat. — BC27 11
Gosforth La.
Longcroft SE9 — CK48 62
Long Cft. Wat. — BC26 4
Longcroft Av. Bans. — BT60 83
Longcroft Rise Loug. — CL25 10
Longcroft Rd. SE5 — CA43 51
Longcroft Rd. Edg. — BK29 12
Long Deacon Rd. E4 — CG26 9
Longden Wood Av. Kes. — CK56 80
Longdown La. N. Epsom — BP60 82
Longdown La. S. Epsom — BP61 82
Longdown Rd. SE6 — CE49 61
Longdown Rd. Epsom — BP60 82
Long Dr. W3 — BO39 40
Long Dr. Grnf. — BF37 29
Long Dr. Ruis. — BD35 29
Long Elmes Har. — BF30 11
Longfellow Rd. E17 — CE32 25
Longfellow Rd. Wor. Pk. — BP54 67
Longfield NW9 — BO29 13
Longfield, Brom. — CG51 70
Longfield Loug. — CJ25 10
Longfield Av. E17 — CD31 25
Longfield Av. NW7 — BP29 13
Longfield Av. W5 — BK40 39
Longfield Av. Horn. — CT33 28
Longfield Av. Wall. — BV54 68
Longfield Av. Wem. — BL33 21
Longfield Cres. SE26 — CC48 61
Longfield Dr. SW14 — BM46 48
Longfield Est. SE1 — CA42 42
Longfield Rd. W5 — BK40 39
Longfield St. SW18 — BS47 59
Longfield Wk. W5 — BK39 39
Longford Av. Sthl. — BF40 38
Longford Cl. N15 — CA32 24
Albert Rd.
Longford Clo. Hamptn. — BF49 56
Longford Ct. E5 — CC35 34
Clapton Park Est.
Longford Ct. Epsom — BN56 76
Longford Gdns. Hayes — BD40 38
Longford Gdns. Sutt. — BT55 77
Longford Rd. Twick. — BF47 56
Longford St. NW1 — BV38 32
Long Grn. Chig. — CN28 18
Long Gro. Rd. Epsom — BM58 75
Longhayes Av. Rom. — CP31 27
Longheath Gdns. Croy. — CC53 70
Longhill Rd. SE6 — CF48 61
Longhook Gdns. Nthlt. — BE37 29
Longhurst Rd. SE13 — CF46 52
Longhurst Rd. Croy. — CB53 69
Longland Dr. N20 — BS27 14

Longlands Av. Couls. — BV60 83
Longlands Ct. W11 — BR40 40
Westbourne Gro.
Longlands Pk. Cres. — CN48 63
Sid.
Longlands Rd. Sid. — CN48 63
Long La. EC1 — BY39 42
Long La. N2 — BT30 14
Long La. N3 — BS30 14
Long La. SE1 — BZ41 42
Long La. Bexh. — CP43 54
Long La. Croy. — CB53 69
Longleat Rd. Enf. — CA25 8
Longleigh House SE5 — CA44 51
Glebe Est.
Longleigh La. SE2 — CO43 54
Longley Av. Wem. — BL37 30
Longley Rd. SW17 — BU50 59
Longley Rd. Croy. — BY54 69
Longley Rd. Har. — BG32 20
Long Leys. E4 — CE29 16
Long Lodge Dr. Walt. — BD55 74
Long Mead NW9 — BO30 13
Longmead Chis. — CL51 71
Longmead Dr. Sid. — CP48 63
Long Meadow NW5 — BW35 32
Longmeadow Rd. Sid. — CN47 63
Longmead Rd. SW17 — BU49 59
Longmead Rd. Epsom — BN59 82
Longmead Rd. Surb. — BH54 66
Longmoor St. SW1 — BW42 41
Longmore Av. Barn. — BT25 7
Longmore Rd. Walt. — BE56 74
Longnor Rd. E1 — CC38 34
Long Pond Rd. SE3 — CG44 52
Longport Clo. Ilf. — CO29 18
Long Reach Rd. Bark. — CN38 36
Longreach Rd. Erith — CU43 55
Longridge La. Sthl. — BF40 38
Longridge Rd. SW5 — BS42 41
Longshaw Rd. E4 — CF27 16
Longshore SE8 — CD42 43
Longstaff Cres. SW18 — BS46 50
Longstaff Rd. SW18 — BS46 50
Longstone Av. NW10 — BO37 31
Longstone Rd. SW17 — BV49 59
Long St. E2 — CA38 33
Longthornton Rd. SW16 — BW51 68
Longton Gro. SE26 — CB49 60
Longtown Clo. Rom. — CV28 19
Longtown Rd. Rom. — CV28 19
Longview Way, Rom. — CS30 19
Longville Rd. SE11 — BY42 42
Long Wk. SE18 — CL43 53
Long Wk. SW13 — BO44 49
Terrace, The.
Long Wk. N. Mal. — BN52 67
Longwood Dr. SW15 — BP46 49
Longwood Gdns. Ilf. — CK31 26
Longwood Rd. Ken. — BZ61 84
Long Yd. WC1 — BX38 33
Loning, The, NW9 — BO31 22
Lonsdale Av. E6 — CJ38 35
Lonsdale Av. Rom. — CS32 28
Lonsdale Av. Wem. — BL35 30
Lonsdale Clo. E6 — CJ38 35
Lonsdale Clo. Edg. — BL29 12
Lonsdale Cres. Ilf. — CL32 26
Lonsdale Dr. Enf. — BW24 7
Lonsdale Dr. N. Enf. — BX25 8
Lonsdale Gdns. Th. Hth. — BX52 69
Lonsdale Mews. Rich. — BL44 48
Lonsdale Ms. Rich. — BM44 48
Elizabeth Cott.
Lonsdale Pl. N1 — BY36 33
Lonsdale Rd. E11 — CG33 25
Lonsdale Rd. NW6 — BR37 31
Lonsdale Rd. SE25 — CB52 69
Lonsdale Rd. SW13 — BO48 49
Lonsdale Rd. W4 — BO42 40
Lonsdale Rd. W11 — BR39 40
Lonsdale Rd. Bexh. — CQ44 54
Lonsdale Rd. Sthl. — BD41 38
Lonsdale Sq. N1 — BY36 33
Loobert Rd. N15 — CA31 24
Looe Gdns. Ilf. — CL31 26
Loop Rd. Chis. — CL50 62
Loop Rd. Epsom — BN61 82
Lopen Rd. N18 — CA28 15
Loraine Clo. Enf. — CC25 9
Loraine Rd. N7 — BX35 33
Loraine Rd. W4 — BM43 48
Lord Av. Ilf. — CK31 26
Lord Chancellor Wk. — BN51 67
Kings. On T.
Lorden Wk. E2 — CB38 33
Lord Gdns. Ilf. — CK31 26
Lord Hills Br. W2 — BS39 41
Lord Hills Rd. W2 — BS39 41
Lord Napier Pl. W6 — BP42 40
Upper Mall.
Lord North St. SW1 — BX41 42
Lord Roberts Mews SW6 — BS43 50
Waterford Rd.
Lordsbury Fld. Wall. — BW58 77
Lords Clo. SE21 — BZ47 60
Lords Clo. Felt. — BE48 56

Lordship La. N17 — BZ30 15
Lordship La. N22 — BY30 15
Lordship La. SE22 — CA46 51
Lordship Pk. N16 — BZ34 24
Lordship Pl. SW3 — BU43 50
Cheyne Row
Lordship Rd. N16 — BZ33 24
Lordship Rd. Nthlt. — BE36 29
Lordsmead Rd. N17 — CA30 15
Lord St. E16 — CK40 44
Lord St. Wat. — BD24 4
Lord Warwick St. SE18 — CK41 44
Lorenzo St. WC1 — BX38 33
Loretto Gdns. Har. — BL31 21
Lorian Av. N12 — BS28 14
Holden Rd.
Lorian Clo. N12 — BS28 14
Guildown Av.
Loring Rd. N20 — BU27 14
Loring Rd. Islw. — BH44 48
Loris Rd. W6 — BQ41 40
Lorne Av. Croy. — CC54 70
Lorne Gdns. E11 — CJ31 26
Lorne Gdns. W11 — BQ41 40
Lorne Gdns. W14 — BQ32 22
Lorne Gdns. Croy. — CC54 70
Lorne Rd. E7 — CH35 35
Lorne Rd. E17 — CE32 25
Lorne Rd. N4 — BX33 24
Lorne Rd. Har. — BH30 12
Lorne Rd. Rich. — BL46 48
Albert Rd.
Lorn Rd. SW9 — BX44 51
Lorraine Pk. Har. — BH29 12
Lorrimore Rd. SE17 — BY43 51
Lorrimore Sq. SE17 — BY43 51
Losberne Way SE16 — CB42 42
Bonamy Estate West, The
Loseberry Rd. Esher — BG56 74
Lothair Rd. W5 — BK41 39
Lothair Rd. N. N4 — BY32 24
Lothair Rd. S. N4 — BY33 24
Lothbury, EC2 — BZ39 42
Lothian Av. Hayes — BC39 38
Lothian Rd. SW9 — BY44 51
Lothrop St. W10 — BR38 31
Lots Rd. SW6 — BT44 50
Lots Rd. SW10 — BT43 50
Loubet St. SW17 — BU50 59
Loudoun Av. Ilf. — CL32 26
Loudoun Rd. NW8 — BT36 32
Loudwater Clo. Sun. — BC52 65
Loudwater Rd. Sun. — BC52 65
Loughborough Pk. SW9 — BY45 51
Loughborough Rd. SW9 — BY44 51
Loughborough St. SE11 — BX42 42
Loughton Way Buck. H. — CJ26 10
Louisa Gdns. E1 — CC38 34
Louisa St. E1 — CC38 34
Louise Gdns. Rain. — CT38 37
Louise Rd. E15 — CG36 34
Louisville Rd. SW17 — BV48 59
Lourdon Rd. Mews NW8 — BT37 32
Lourdon Rd.
Louvaine Rd. SW11 — BT45 50
Lovat Clo. NW2 — BO34 22
Lovat La. EC3 — CA40 42
Lovatt Clo. Edg. — BM29 12
Lovatt Dr. Ruis. — BC32 20
Loveday Rd. W13 — BJ40 39
Lovegrove St. SE1 — CB43 51
Lovekyn Clo. — BL51 66
Kings. On T.
Lovelace Av. Brom. — CL53 71
Lovelace Gdns. Bark. — CO35 36
Lovelace Gdns. Surb. — BK54 66
Lovelace Gdns. Walt. — BD56 74
Lovelace Grn. SE9 — CK45 53
Lovelace Rd. SE21 — BZ48 60
Lovelace Rd. Barn. — BU26 7
Lovelace Rd. Surb. — BK54 66
Love La. EC2 — BZ39 42
Love La. N17 — CA29 15
Love La. SE18 — CL42 44
Love La. SE25 — CB52 69
Love La. Bex. — CQ46 54
Love La. Brom. — CH51 71
Love La. Mitch. — BU52 68
Love La. Mord. — BS54 68
Love La. Pnr. — BE31 20
Love La. Surb. — BK55 75
Love La. Sutt. — BR57 76
Love La. Wdf. Grn. — CK29 17
Lovel Av. Well. — CO44 54
Lovelinch Clo. SE14 — CC43 51
Rollins St.
Lovell Rd. Rich. — BK48 57
Lovell Rd. Sthl. — BF39 38
Lovell Wk. Rain. — CT36 37
Loveridge Rd. NW6 — BR36 31
Lovers Wk. N3 — BS29 14
Lover's Wk. SE10 — CF43 52
Lover's Wk. Rom. — CS28 19
Lovett Dr. Cars. — BT54 68
Lovetts Pl. SW18 — BS45 50
York Rd.

Lovett Way NW10 — BN35 31
Love Wk. SE5 — BZ44 51
Lovibonds Av. Orp. — CL56 80
Lowbrook Rd. Ilf. — CL35 35
Low Cross Wood La. — CA48 60
SE21
Lowden Rd. N9 — CB26 8
Lowden Rd. SE24 — BY45 51
Lowden Rd. Sthl. — BE40 38
Lowe Av. E16 — CH39 44
Watford Rd.
Lowe Clo. Chig. — CO28 18
Lowell St. E14 — CD39 43
Lwr. Addiscombe Rd. — CA54 69
Croy.
Lwr. Barn. Rd. Pur. — BZ59 84
Lwr. Bedfords Rd. Rom. — CT29 19
Lwr. Belgrave St. SW1 — BV41 41
Lower Boston Rd. W7 — BH40 39
Lwr. Broad St. Dag. — CR37 36
Lwr. Camden, Chis. — CK50 62
Lwr. Clapton Rd. E5 — CB34 24
Lwr. Common S. SW15 — BP45 49
Lwr. Coombe St. Croy. — BZ56 78
Lwr. Court Rd. Epsom — BN59 82
Lwr. Croft Swan. — CT52 73
Lwr. Downs Rd. SW20 — BQ51 67
Lwr. Drayton Pl. Croy. — BY55 78
Lwr. Form Wor. Pk. — BO55 76
Lwr. George St. Rich. — BK46 48
George St.
Lwr. Gravel Rd. Brom. — CK54 71
Lwr. Green Rd. Esher — BF55 74
Lwr. Green W. Mitch. — BU52 68
Lwr. Grosvenor Pl. SW1 — BV41 41
Lower Hall La. E4 — CD28 16
Lwr. Hampton Rd. Sun. — BD52 65
Lwr. Ham Rd. — BK50 57
Kings. On T.
Lwr. Hythe St. Dart. — CW46 55
Lwr. James St. W1 — BW40 41
Brewer St.
Lwr. John St. W1 — BW40 41
Brewer St.
Lwr. Kenwood Av. Enf. — BW25 7
Lwr. Maidstone Rd. N11 — BW29 14
Lower Mall W6 — BP42 40
Lwr. Marsh, SE1 — BY41 42
Lwr. Marsh La. — BL52 66
Kings. On T.
Lower Merton Rd. NW3 — BU36 32
Lwr. Morden La. Mord. — BQ53 67
Lwr. Mortlake Rd. Rich. — BL45 48
Lower Paddock Rd. Wat. — BE25 4
Lower Pk. Rd. Bans. — BU62 83
Lower Pk. Rd. Belv. — CR41 45
Lower Pk. Rd. Loug. — CJ25 10
Lower Pk. Rd. N11 — BW28 14
Lwr. Pillory Downs. Cars. — BV60 83
Lwr. Queen's Rd. Buck. H. — CJ27 17
Lwr. Richmond Rd. — BM45 48
SW14
Lower Richmond Rd. Rich. — BM45 48
SW15
Lower Rd. E13 — CH38 35
Lower Rd. SE8 — CC42 43
Lower Rd. SE16 — CC41 43
Lower Rd. Belv. — CR41 45
Lower Rd. Erith — CS42 46
Lower Rd. Har. — BG34 20
Lower Rd. Ken. — BY60 84
Lower Rd. Loug. — CL23 10
Lower Rd. Orp. — CO54 72
Lower Rd. Rain. — CS37 37
Lower Rd. Sutt. — BT56 77
Lower Rd. Swan. — CT50 64
Lwr. Sloane St. SW1 — BV42 41
Lower Sq. Islw. — BJ45 48
Lower Staithe W4 — BN44 49
Lwr. Station Rd. — CT46 55
Crayford
Lower Strand NW9 — BO30 13
Lwr. Sunbury Rd. — BE51 65
Hamptn.
Lower Tail. Wat. — BE27 11
Lwr. Teddington Rd. — BK50 57
Kings. On T.
Lower Ter. NW3 — BT34 23
Lower Tub Bush. — BG26 4
Lwr. Vernon Rd. Sutt. — BT56 77
Lwr. Wood Rd. Esher — BJ57 75
Lowestoft Clo. E5 — CB34 24
Southwold Rd.
Lowestoft Rd. Wat. — BC23 4
Lowe, The, Chig. — CO28 18
Lowfield Rd. NW6 — BS36 32
Lowfield Rd. W3 — BM39 39
Lowfield St. Dart. — CW48 64
Low Hall Clo. E4 — CE26 9
Lowhall La. E17 — CD32 25
Lowick Rd. Har. — BH31 21
Lowlands Gdns. Rom. — CR32 27
Lowlands Rd. Har. — BH33 21
Lowlands Rd. Pnr. — BD33 20
Lowman Rd. N7 — BX35 33

Lowndes Clo. SW1	BV41	41
Lowndes Pl. SW1	BV41	41
Lowndes Sq. SW1	BU41	41
Lowndes St. SW1	BU41	41
Lownds Av. Brom.	CH51	71
Lowood St. E1	CB40	42
Lowshoe La. Rom.	CR30	18
Lowshoe Gro. Wat.	BE26	4
Lowther Dr. Enf.	BW24	7
Lowther Hl. SE23	CD47	61
Lowther Rd. E17	CD30	16
Lowther Rd. N7	BY35	33
Mackenzie Rd.		
Lowther Rd. SW13	BO44	49
Lowther Rd. Kings. On T.	BL51	66
Lowther Rd. Stan.	BL31	21
Lowth Rd. SE5	BZ44	51
Loxford Av. E6	CJ37	35
Loxford La. Ilf.	CM35	35
Loxford Rd. Bark.	BL36	35
Loxham Rd. E4	CE29	16
Loxham St. WC1	BX38	33
Argyle Wk.		
Loxley Clo. SE26	CC49	61
Trewsbury Rd.		
Loxley Rd. SW18	BT47	59
Loxley Rd. Hamptn.	BE49	56
Loxton Rd. SE23	CC47	61
Loxwood Rd. N17	CA31	24
Lubbock Rd. Chis.	CK50	62
Lubbock St. SE14	CC43	52
Lucan Pl. SW3	BU42	41
Lucan Rd. Barn.	BR24	6
Lucas Av. E13	CH37	35
Lucan Av. Har.	BF34	20
Lucas Ct. Har.	BF33	20
Lucas Rd. SE20	CC50	61
Lucas St. SE8	CD44	52
Lucerne Clo. N13	BX27	15
Lucerne Gro. E17	CF31	25
Lucerne Ms. W8	BS40	41
Kensington Mall		
Lucerne Rd. N5	BY35	33
Lucerne Rd. Orp.	CN54	72
Lucerne Rd. Th. Hth.	BY52	69
Lucerne Way, Rom.	CV29	19
Lucien Rd. SW17	BV49	59
Lucien Rd. SW19	BS48	59
Lucknow St. SE18	CN43	54
Lucorn Clo. SE12	CG46	52
Luctons Av. Buck. H.	CJ26	10
Lucy Cres. W3	BN39	40
Lucy Gdns. Dag.	CQ34	27
Luddesdon Rd. Erith	BR43	54
Ludford Clo. NW9	BO30	13
Ludgate Circus EC4	BY39	42
Ludgate Hill EC4	BY39	42
Ludgate Sq. EC4	BY39	42
Creed La.		
Ludlow Clo. Har.	BE35	29
Ludlow Mead, Wat.	BC27	11
Ludlow Rd. W5	BK38	30
Ludlow Wy. N2	BT31	23
Ludovick Wk. SW15	BO45	49
Ludwick Mews SE14	CD43	52
Luffield Rd. SE2	CO41	45
Luffman Rd. SE12	CH48	62
Lugard Rd. SE12	CH48	62
Lugard Rd. SE15	CB44	51
Luke St. EC2	CA38	33
Lukin Cres. E4	CF27	16
Lukin St. E1	CC39	43
Lullingstone Av. Swan.	CT52	73
Lullingstone Clo. Orp.	CO50	63
Lullingstone Cres. Orp.	CO50	63
Lullingstone Rd. Belv.	CQ43	54
Barnfield Rd.		
Lullington Garth N12	BR28	13
Lullington Garth, Brom.	CG50	61
Lullington Gth. Borwd.	BM25	5
Lullington Rd. SE20	CB50	60
Lullington Rd. Dag.	CQ36	36
Lulot St. N6	BV34	23
Lulworth Av. Houns.	BF44	47
Lulworth Av. Wem.	BK33	21
Lulworth Clo. Har.	BE34	20
Lulworth Dr. Pnr.	BD33	20
Lulworth Dr. Rom.	CR28	18
Lulworth Gdns. Har.	BE34	20
Lulworth Rd. SE9	CK48	62
Lulworth Rd. SE15	CB44	51
Lulworth Rd. Well.	CN44	54
Lulworth Waye, Hayes	BC39	38
Lumley Clo. Belv.	CR43	45
Lumley Gdns. Sutt.	BR56	76
Lumley Rd. Sutt.	BR57	76
Lumley St. W1	BV39	41
Luna Rd. Th. Hth.	BZ52	69
Lundin Wk. Wat.	BD28	11
Lundy St. W6	BR43	49
Field Rd.		
Lundy Wk. N1	BZ36	33
Marquess Est.		
Lunham Rd. SE19	CA50	60
Luntly Pl. E1	CA39	42
Chicksand St.		
Lupin Clo. SW2	BY48	60
Palace Rd.		
Lupton Clo. SE12	CH49	62
Lupton St. NW5	BW35	32
Lupus St. SW1	BV42	41
Lurgan Av. W6	BQ43	49
Lurline Gdns. SW11	BV44	50
Luscombe Way SW8	BX43	51
Lushes Rd. Loug.	CL25	10
Lushington Rd. NW10	BP37	31
Lushington Rd. SE6	CE49	61
Lutheran Pl. SW2	BX47	60
Upr. Tulse Hill		
Luther Clo. Edg.	BN27	13
Luther Rd. Tedd.	BH49	57
Luton Pl. SE10	CF43	52
Luton Rd. E17	CD31	25
Luton Rd. Sid.	CP48	63
Luton St. NW8	BT37	32
Lutrell Av. SW15	BP45	49
Lutwyche Rd. SE6	CD48	61
Luxborough La. Chig.	CK27	17
Luxborough Pl. W1	BV38	32
Luxborough St. W1	BV39	41
Luxemburg Gdns. W6	BQ42	40
Luxfield Rd. SE9	CK47	62
Luxford St. SE16	CC42	43
Luxmore Gdns. SE4	CD44	52
Luxmore St. SE4	CC44	52
Luxor St. SE5	BZ44	51
Lyall Av. SE21	CA48	60
Lyall Ms. E. SW1	BV41	41
Lyall St. SW1	BV41	41
Lyal Rd. E3	CD37	34
Lycett Pl. W12	BP41	40
Vespan Rd.		
Lychen Av. Houns.	BC44	47
Lyconby Gdns. Croy.	CD54	70
Lydd Clo. Sid.	CN48	63
Lydden Ct. SE9	CN46	54
Lydden Rd. SW18	BS47	59
Lyddon Gro. SW18	BS47	59
Lydd Rd. Bexh.	CQ43	54
Lydeard Rd. E6	CK36	35
Lydenburg St. SE7	CJ41	44
Lydford Rd. NW2	BQ36	31
Lydford Rd. W9	BR38	31
Lydhurst Av. SW2	BX48	60
Lydia Rd. Erith	CT43	55
Lydney Clo. SW15	BR48	58
Princes Way		
Lydon Rd. SW4	BW45	50
Lydstep Rd. Chis.	CL49	62
Lyford Rd. N15	BZ32	24
Lyford Rd. SW18	BT47	59
Lyford St. SE18	CK42	44
Lyham Clo. SW2	BX46	51
Lyham Rd. SW2	BX46	51
Lymborne Clo. Sutt.	BS58	77
Lyme Farm Rd. SE12	CH45	53
Lymer Av. SE19	CA49	60
Lyme Regis Rd. Bans.	BR62	82
Lyme Rd. Well.	CO44	54
Lymescote Gdns. Sutt.	BS55	77
Lyme St. NW1	BW36	32
Lyme Ter. NW1	BW36	32
Royal College St.		
Lyminge Clo. Sid.	CN48	63
Lyminge Gdns. SW18	BU47	59
Lymington Av. N22	BY30	15
Lymington Clo. SW16	BW51	68
Lymington Gdns. Epsom	BO56	76
Lymington Rd. NW6	BS36	32
Lymington Rd. Dag.	CP33	27
Lympstone Gdns. SE15	CB43	51
Lynbridge Gdns. N13	BY28	15
Lynbrook Clo. Rain.	CS37	37
Lyncroft Av. Pnr.	BE32	20
Lyncroft Gdns. NW6	BS35	32
Lyncroft Gdns. W13	BK41	39
Lyncroft Gdns. Epsom	BO58	76
Lyncroft Gdns. Houns.	BG46	47
Lyndale NW2	BR35	31
Lyndale Av. NW2	BR34	22
Lyndale Clo. SE3	CG43	52
Lynden Way, Swan.	BS52	73
Lyndhurst Av. N12	BU29	14
Lyndhurst Av. NW7	BO29	13
Lyndhurst Av. SW16	BW51	68
Lyndhurst Av. Pnr.	BC30	11
Lyndhurst Av. Sthl.	BF40	38
Lyndhurst Av. Sun.	BC52	65
Lyndhurst Av. Surb.	BM54	66
Lyndhurst Av. Twick.	BE47	56
Lyndhurst Clo. NW10	BN34	22
Lyndhurst Clo. Bexh.	CR45	54
Lyndhurst Clo. Croy.	CA55	78
Selborne Rd.		
Lyndhurst Ct. E18	CH30	17
Lyndhurst Dr. E10	CF33	25
Lyndhurst Dr. Horn.	CV33	28
Lyndhurst Dr. N. Mal.	BO53	67
Lyndhurst Gdns. N3	BR30	13
Lyndhurst Gdns. NW3	BT35	32
Lyndhurst Gdns. Bark.	CN36	36
Lyndhurst Gdns. Enf.	CA24	8
Lyndhurst Gdns. Ilf.	CM32	26
Lyndhurst Gdns. Pnr.	BC30	11
Lyndhurst Gro. SE15	CA44	51
Lyndhurst Rise, Chig.	CL28	17
Lyndhurst Rd. E4	CF29	16
Lyndhurst Rd. N18	CB28	15
Lyndhurst Rd. N22	BX29	15
Lyndhurst Rd. NW3	BT35	32
Lyndhurst Rd. Bexh.	CR45	54
Lyndhurst Rd. Couls.	BV61	83
Lyndhurst Rd. Grnf.	BF38	29
Lyndhurst Rd. Th. Hth.	BY52	69
Lyndhurst Sq. SE15	CA44	51
Lyndhurst Ter. NW3	BT35	32
Lyndhurst Way SE15	CA44	51
Lyndhurst Way, Sutt.	BS58	77
Lyndon Av. Pnr.	BE29	11
Lyndon Av. Sid.	CN46	54
Lyndon Av. Wall.	BV55	77
Lyndon Rd. Belv.	CR42	45
Lynette Av. SW4	BV46	50
Lynett Rd. Dag.	CP34	27
Lynford Clo. Edg.	BN29	13
Lynford Gdns. Edg.	BM27	12
Lynford Gdns. Ilf.	CN34	27
Lynford Ter. N9	CA26	8
Lynmere Rd. Well.	CO44	54
Lynmouth Av. Enf.	CA25	8
Lynmouth Av. Mord.	BQ53	67
Lynmouth Dr. Ruis.	BC34	20
Lynmouth Gdns. Grnf.	BJ37	30
Lynmouth Gdns. Houns.	BD43	47
Lynmouth Rise, Orp.	CO52	72
Lynmouth Rd. E17	CD32	25
Lynmouth Rd. N2	BU31	23
Lynmouth Rd. N16	CA33	24
Lynmouth Rd. Grnf.	BJ37	30
Lynn Clo. Har.	BG30	11
Lynne Clo. Orp.	CN57	81
Lynne Clo. Sth. Croy.	CC58	79
Lynne Wk. Esher	BG56	74
Lynne Way NW10	BO36	31
Lynne Way, Nthlt.	BD37	29
Lynn Rd. E11	CG34	25
Lynn Rd. SW12	BV47	59
Lynn Rd. Ilf.	CM33	26
Lynn St. Enf.	BZ23	8
Lyncross Clo. Rom.	CW30	19
Lynsted Clo. Bexh.	CR46	54
Lynsted Clo. BR1	CH51	71
Lynsted Gdns. SE9	CJ45	53
Lynton Av. N12	BT28	14
Lynton Av. NW9	BO31	22
Lynton Av. W13	BJ39	39
Lynton Av. Orp.	CO52	72
Lynton Av. Rom.	CR30	18
Lynton Clo. Chess.	BL56	75
Lynton Clo. Islw.	BH45	48
Lynton Cres. Ilf.	CL32	26
Lynton Est. SE1	CB42	42
Lynton Gdns. N11	BW29	14
Lynton Gdns. Enf.	CA26	8
Lynton Mead N20	BS27	14
Lynton Rd. E4	CE28	16
Lynton Rd. E11	CF35	34
Lynton Rd. N8	BW32	23
Lynton Rd. NW6	BR38	31
Lynton Rd. SE1	CA42	42
Lynton Rd. W3	BM40	39
Lynton Rd. Croy.	BY53	69
Lynton Rd. Har.	BE34	20
Lynton Rd. N. Mal.	BN53	67
Lynwood Av. Couls.	BV61	83
Lynwood Av. Epsom	BO60	82
Lynwood Clo. E18	CJ30	17
Lynwood Clo. Har.	BE34	20
Lynwood Clo. Rom.	CR29	18
Lynwood Dr. Rom.	CR29	18
Lynwood Dr. Wor. Pk.	BP55	76
Lynwood Gdns. Croy.	BX56	78
Lynwood Gdns. Sthl.	BE39	38
Lynwood Gro. N21	BY26	8
Lynwood Gro. Orp.	CN54	72
Lynwood Rd. SW17	BU49	59
Lynwood Rd. W5	BK38	30
Lynwood Rd. Epsom	BO60	82
Lynwood Rd. Surb.	BH55	75
Lyon Meade, Stan.	BK30	12
Lyon Park Av. Wem.	BL36	30
Lyon Rd. SW19	BT51	68
Lyon Rd. Har.	BH32	21
Lyon Rd. Rom.	CT33	28
Lyon Rd. Walt.	BE55	74
Lyonsdown Av. Barn.	BT25	7
Lyonsdown Rd. Barn.	BT25	7
Lyons Pl. NW8	BT38	32
Lyoth Rd. Orp.	CM55	80
Lyric Rd. SW13	BO44	49
Lysander Gro. N19	BW33	23
Lysander Rd. Croy.	BX57	78
Lysander Way, Orp.	CM55	80
Lysias Rd. SW12	BV46	50
Lysia St. SW6	BQ44	49
Lysons Wk. SW15	BP46	49
Swinburne Rd.		
Lytchett Rd. Brom.	CH50	62
Lytchet Way, Enf.	CC23	9
Lytcott Gro. SE22	CA46	51
Lytham Av. Wat.	BD28	11
Lytham Gro. W5	BL38	30
Lytham St. SE17	BZ42	42
Lyton Gdns. Wall.	BW56	77
Lyttelton Clo. NW3	BU36	32
Lyttelton Ct. N2	BT32	23
Lyttelton Rd. E10	CE34	25
Lyttelton Rd. N2	BT32	23
Lyttelton Rd. N8	BY31	24
Lytton Av. N13	BY27	15
Lytton Clo. Nthlt.	BE36	29
Lytton Clo. Loug.	CM24	10
Lytton Gro. SW15	BQ46	49
Lytton Rd. Barn.	BT24	7
Lytton Rd. E11	CG33	25
Lytton Rd. Pnr.	BE29	11
Lytton Rd. Rom.	CU32	28
Lytton Strachey Path SE28	CO40	45
Curtis Way		
Lyveden Rd. SE3	CH43	53
Lyvedon Rd. SW17	BU50	59
Mabel Rd. Swan.	CU50	64
Maberley Ct. SE19	CB50	60
Maberley Cres. SE19	CB50	60
Maberley Rd. Beck.	CC52	70
Maberley Rd. SE19	CA51	69
Mabledon Pl. WC1	BW38	32
Mablethorpe Rd. SW6	BR43	49
Mabley St. E9	CD35	34
Mabyn Rd. SE18	CN42	45
Macarthur Ter. SE7	CJ43	53
Charlton La.		
Macaulay Av. Esher	BH55	75
Macaulay Ct. SW4	BV45	50
Macaulay Rd. E6	CJ37	35
Macaulay Rd. SW4	BV45	50
Macaulay Sq. SW4	BV45	50
Macbean St. SE18	CL41	44
Macbeth House N1	CA37	33
Purcell St.		
Macbeth St. W6	BP42	40
Macclesfield Br. NW8	BU37	32
Macclesfield Rd. EC1	BZ38	33
Macclesfield Rd. SE25	CB53	69
Macclesfield St. W1	BW40	41
Gerrard St.		
Macdonald Av. Horn.	CW31	28
Macdonald Clo. Horn.	CW31	28
Macdonald Rd. Dag.	CR34	27
Macdonald Rd. E7	CH35	35
Macdonald Rd. E17	CF30	16
Macdonald Rd. N11	BU28	14
Macdonald Rd. N19	BW34	23
Macdonald Way, Horn.	CW31	28
Macduff Rd. SW11	BV44	50
Macfarlane Pl. W12	BQ40	40
Macfarlane Rd. W12	BQ40	40
Macgregor Rd. E16	CJ39	44
Machell Rd. SE15	CC45	52
Macintosh La. E9	CC35	34
High St.		
Mackay Rd. SW4	BV45	50
Mackennal St. NW8	BU37	32
Mackenzie Rd. Beck.	CC51	70
Mackenzie Rd. N7	BX36	33
Mackeson Rd. NW3	BU35	32
Mackie Rd. SW2	BY47	60
Macklin St. WC2	BX39	42
Mack's Rd. SE16	CB42	42
Mackworth St. NW1	BW38	32
Maclean Rd. SE23	CD46	52
Maclennan Av. Rain.	CV38	37
Macleod St. SE17	BZ42	42
Maclise Rd. W14	BR41	40
Macoma Rd. SE18	CM43	53
Macoma Ter. SE18	CM43	53
Macquarie Way E14	CE42	43
Macready Pl. N7	BX35	24
Macroom Rd. W9	BR38	31
Maculay Way SE28	CO40	45
Madans Wk. Epsom	BN61	82
Mada Rd. Orp.	CL55	80
Maddams St. E3	CE38	34
Maddison Clo. Tedd.	BH50	57
Maddocks Clo. Sid.	CQ49	63
Maddock Way SE17	BY43	51
Maddox St. W1	BV40	41
Madeira Av. Brom.	CG50	61
Madeira Gro. Wdf. Grn.	CJ29	17
Madeira Rd. E11	CF33	25
Madeira Rd. Mitch.	BU52	68
Madeira Rd. N13	BY28	15
Madeley Rd. W5	BK39	39
Madeline Rd. SE20	CB50	60
Madison Cres. Well.	CP43	54
Madison Gdns. Brom.	CG52	70
Madison Gdns. Well.	CP43	54
Madras Pl. N7	BY36	33
Madras Rd. Ilf.	CL35	35
Madrid Rd. SW13	BP44	49
Madron St. SE17	CA42	42
Mafeking Av. Brent.	BK43	48
Mafeking Av. E6	CK37	35
Mafeking Av. Ilf.	CM33	26
Mafeking Rd. E16	CG38	34
Mafeking Rd. Enf.	CA24	8
Mafeking Rd. N17	CB30	15
Magdala Rd. Islw.	BJ45	48
Magdala Rd. N19	BV34	23
Magdala Rd. S. Croy.	BZ57	78

Magdalene Gdns. E6 CL38 35
Homeway
Magdalene Pass. E1 CA40 42
Chamber St.
Magdalen Rd. SW18 BT47 59
Magdalen St. SE1 CA40 42
Magee St. SE11 BY43 51
Magnaville Rd. Bush. BH26 5
Magnolia Clo. Kings. On T. BM50 57
Magnolia Ct. Har. BL33 21
Magnolia Rd. W4 BM43 48
Magnolia Way, Epsom BN56 76
Magnum Clo. Rain. CV39 46
Magpie Alley EC4 BY39 42
Whitefriars St.
Magpie Clo. Couls. BW58 77
Magpie Clo. Couls. BW62 83
Magpie Hall Clo. Brom. CK53 71
Magpie Hall La. Brom. CK54 71
Magpie Hall Rd. Bush. BH27 12
Maguire Dr. Rich. BK49 57
Maguire St. SE1 CA41 42
Mahlon Av. Ruis. BC35 29
Maida Av. E4 CE26 9
Maida Av. W2 BT39 41
Maida Rd. Belv. CR41 45
Maida Rd. W2 BT40 41
Maida Vale Rd. Dart. CU46 55
Maida Vale W9 BS37 32
Maida Way E4 CE26 9
Maiden Erlegh Av. Bex. CQ47 63
Maiden La. Dart. CU45 55
Maiden La. WC2 BX40 42
Bedford St.
Maiden Rd. E15 CG36 34
Maidenshaw Rd. Epsom BN59 82
Maidenstone Hl. SE10 CF44 52
Maidman St. E3 CD38 34
Maid Of Honour Row, Rich. BK46 48
Green, The
Maidstone Av. Rom. CS30 19
Maidstone Bldgs. SE1 BZ40 42
Maidstone Rd. N11 BW29 14
Maidstone Rd. Sid. CP50 63
Maidstone St. E2 CB37 33
Mail Coach Yard N1 CA38 33
Kingsland Rd.
Main Av. Enf. CA25 8
Main Dr. Islw. BG43 48
Mainridge Rd. Chis. CL49 62
Main Rd. Farn. CW53 73
Main Rd. Orp. CP51 72
Main Rd. Rom. CT31 28
Main Rd. Sid. CM48 62
Main Rd. Swan. CT50 64
Main St. Felt. BD49 56
Maisemore St. SE15 CB43 51
Maitland Clo. Houns. BE45 47
Maitland Clo. SE10 CE43 52
Maitland Pk. Est. NW3 BU36 32
Maitland Pk. Rd. NW3 BU36 32
Maitland Pk. Vill. NW3 BU35 32
Maitland Rd. E15 CG36 34
Maitland Rd. SE26 CC50 61
Maize Row E14 CD40 43
Majendie Rd. SE18 CM42 44
Major Rd. E15 CF35 34
Makepeace Av. N6 BV34 23
Makins St. SW3 BU42 41
Malam Gdns. E14 CE40 43
Wade's Pl.
Malan Sq. Rain. CU36 37
Malay St. E1 CC40 43
Malbrook Rd. SW15 BP45 49
Malcolm Ct. Stan. BK28 12
Malcolm Ct. W5 BL38 30
Malcolm Cres. NW4 BP32 22
Malcolm Dr. Surb. BK54 66
Malcolm Pl. E2 CC38 34
Malcolm Rd. Couls. BW61 83
Malcolm Rd. E1 CC38 34
Malcolm Rd. SE20 CC50 61
Malcolm Rd. SE25 CB53 69
Malcolm Rd. SW19 BR50 58
Malcolm Way E11 CH31 26
Malcomb House N1 CA37 33
Purcell St.
Malden Av. Grnf. BH35 30
Malden Av. SE25 CB52 69
Malden Cres. NW5 BV36 32
Malden Grn. Av. Wor. Pk. BO54 67
Malden High St. N. Mal. BO52 67
Malden Hill Gdns. N. Mal. BO52 67
Malden Hill, N. Mal. BO52 67
Malden Manor, The, N. Mal. BO54 67
Malden Pk. N. Mal. BO53 67
Malden Pl. NW5 BV35 32
Grafton Ter.
Malden Rd. Borwd. BM24 5
Malden Rd. N. Mal. BO53 67
Malden Rd. NW5 BV35 32
Malden Rd. Sutt. BO56 76
Malden Rd. Wat. BC23 4
Malden Way, N. Mal. BN53 67
Maldon Clo. SE5 CA45 51
Maldon Ct. N. Mal. BP52 67
Maldon Rd. N9 CA27 15
Maldon Rd. Rom. CS33 28
Maldon Rd. W3 BN40 40
Maldon Rd. Wall. BV56 77
Maldon Wk. Wdf. Grn. CJ29 17
Malet Pl. WC1 BW38 32
Malet St. WC1 BW38 32
Maley Av. SE27 BY48 60
Malfont Rd. SE5 CA45 51
Malford Ct. E18 CH30 17
Malford Gro. E18 CG31 25
Malham Rd. SE23 CC47 61
Malins Clo. Barn. BP25 6
Mallard Clo. Barn. BT25 7
Mallard Clo. E9 CD36 34
Mallard Clo. Houns. BF47 56
Mallard Ct. NW9 BN33 22
Mallard Rd. S. Croy. CC58 79
Mallards Rd. Wdf. Grn. CH29 17
Mallard Wk. Sid. CP50 63
Cray Rd.
Mallard Way NW9 BN33 22
Mallet Av. SE13 CF46 52
Mallet Dr. Nthlt. BE35 29
Malling Clo. Croy. CC53 70
Malling Gdns. Mord. BT53 68
Malling Way, Brom. CG54 70
Mallinson Rd. Croy. BW55 77
Mallinson Rd. SW11 BU46 50
Mallord St. SW3 BT43 50
Mallory Clo. SE4 CD45 52
Mallory Gdns. Barn. BV26 7
Mallory St. W1 BU38 32
Mallow Mead NW7 BR29 13
Mallow St. EC1 BZ38 33
Mall Rd. W6 BP42 40
Mall, The, Brom. CH52 71
Mall, The, Dag. CR36 36
Mall, The E15 CF36 34
Mall, The, Har. BL33 21
Mall, The N14 BX27 15
Mall, The, Surb. BK53 66
Mall, The SW1 BW41 41
Mall, The SW14 BN46 49
Mall, The W5 BK40 39
Malmains Clo. Beck. CF53 70
Malmains Way, Beck. CF52 70
Malmesbury Clo. Pnr. BC31 20
Malmesbury Rd. E3 CD38 34
Malmesbury Rd. E16 CG39 43
Malmesbury Rd. E18 CG30 16
Malmesbury Rd. Mord. BT54 68
Malmesbury Ter. E16 CG39 43
Malpas Dr. Pnr. BD32 20
Malpas Rd. E8 CB36 33
Malpas Rd. Dag. CP36 36
Malpas Rd. SE4 CD44 52
Malta Rd. E10 CE33 25
Malta St. EC1 BY38 33
Maltby Rd. Chess. BM57 75
Maltby St. SE1 CA41 42
Malthouse Dr. Felt. BD49 56
Malt House Pass. SW13 BO44 49
Terrace, The
Malthus Path SE28 CP40 45
Owen Clo.
Maltings, The BR6 CN54 72
Maltman's Pk. Grnf. BG38 29
Malton Ms. W10 BR39 40
Cambridge Gdns.
Malton Rd. W10 BR39 40
Malton St. SE18 CN43 54
Maltravers St. WC2 BX40 42
Arundel St.
Malt St. SE1 CB43 51
Malva Clo. SW18 BS46 50
Malvern Av. Bexh. CQ43 54
Malvern Av. E4 CF29 16
Malvern Av. Har. BE34 20
Malvern Clo. Mitch. BW52 68
Malvern Clo. Surb. BL54 66
Malvern Ct. SW7 BT42 41
Malvern Dr. Felt. BD49 56
Malvern Dr. Ilf. CN35 36
Malvern Dr. Wdf. Grn. CJ28 17
Malvern Gdns. Har. BL31 21
Malvern Gdns. Loug. CK25 10
Malvern Gdns. NW2 BR34 22
Malvern Gdns. NW6 BR37 31
Canterbury Rd.
Malvern Gdns. W9 BR39 40
Canterbury Rd.
Malvern Ms. NW6 BS38 32
Malvern Pl. W9 BS38 32
Malvern Rd. E6 CK37 35
Malvern Rd. E8 CB36 33
Malvern Rd. E11 CG34 25
Malvern Rd. Hamptn. BF50 56
Malvern Rd. Horn. CU32 28
Malvern Rd. N8 BX31 24
Malvern Rd. N17 CB31 24
Malvern Rd. NW6 BS38 32
Malvern Rd. Orp. CO56 81
Malvern Rd. Surb. BL55 75
Malvern Rd. Th. Hth. BY52 69
Malvern Ter. N1 BY37 33
Malvern Ter. N9 CA26 8
Malvern Way W13 BJ39 39
Malverton Rd. E3 CE37 34
Malyons Rd. SE13 CE46 52
Malyons Rd. Swan. CT50 64
Malyons Ter. SE13 CE46 52
Manaton Clo. SE15 CB45 51
Manaton Clo. W. Borwd. BK25 5
Manaton Cres. Sthl. BF39 38
Manbey Gro. E15 CG36 34
Manbey Pk. Rd. E15 CG36 34
Manbey Rd. E15 CG36 34
Manbey St. E15 CG36 34
Manborough Av. E6 CK38 35
Manchester Dr. W10 BR38 31
Manchester Est. E14 CF42 43
Manchester Gro. E14 CF42 43
Manchester Rd. E14 CF41 43
Manchester Rd. N15 BZ32 24
Manchester Rd. Th. Hth. BZ52 69
Manchester Row, Dart. CT45 55
Manchester Sq. W1 BV39 41
Manchester St. W1 BV39 41
Manchester Way, Rom. CR35 36
Manchuria Rd. SW11 BV46 50
Manciple St. SE1 BZ41 42
Mandalay Rd. SW4 BW46 50
Mandarin St. E14 CE40 43
Salter St.
Mandeville Clo. SE3 CG43 52
Vanbrugh Pk.
Mandeville Dr. Surb. BK54 66
Mandeville Houses N1 BY37 33
Mandeville Pl. W1 BV39 41
Mandeville Rd. N14 BV27 14
Mandeville Rd. Islw. BJ44 48
Mandeville Rd. Nthlt. BE36 29
Mandeville St. E5 CD34 25
Mandon St. E14 CE40 43
Salter St.
Mandrake Rd. SW17 BU48 59
Mandrell Rd. SW2 BX46 51
Manette St. W1 BW39 41
Charing Cross Rd.
Manford Clo. Chig. CO28 18
Manford Cross, Chig. CO28 18
Manford Way, Chig. CN28 18
Manfred Ct. SW15 BR46 49
Manfred Rd.
Manfred Rd. SW15 BR46 49
Manger Rd. N19 BX36 33
Mangold Way, Belv. CP41 45
Manilla St. E14 CE41 43
Manister Rd. SE2 CO41 45
Manlays Yd. SW11 BU44 50
Manley Ct. N16 CA34 24
Stoke Newington High St.
Manley St. NW1 BV37 32
Manningford Clo. EC1 BY38 33
Manning Gdns. Har. BK33 21
Manning Rd. Dag. CR36 36
Manning Rd. E17 CC32 25
Manning Rd. Orp. CP53 72
Manningtree Clo. SW19 BR47 58
Manningtree Rd. Ruis. BC35 29
Manningtree St. E1 CB39 42
Mannin Rd. Rom. CO33 27
Mannock Dr. Loug. CM23 10
Mannock Rd. N22 BY31 24
Manns Clo. Islw. BH46 48
Manns Rd. Edg. BM29 12
Manoel Rd. Twick. BG48 56
Manor Alley W4 BO42 40
Devonshire Rd.
Manor Av. Horn. CV32 28
Manor Av. Houns. BD45 47
Manor Av. Nthlt. BE36 29
Manor Av. Pde . Chig. CM28 17
Grange Cres.
Manor Av. SE4 CD44 52
Manorbrook SE3 CH45 53
Manor Clo. Dag. CS36 37
Manor Clo. Dart. CS45 55
Manor Clo. Dart. CU48 64
Manor Clo. NW7 BN28 13
Manor Dr.
Manor Clo. NW9 BM32 21
Manor Clo. Wor. Pk. BO54 67
Manor Cotts. N2 BT30 14
Manor Ct. N2 BU32 23
Manor Ct. N14 BW27 14
Manor Ct. Rd. W7 BH40 39
Manor Ct. W3 BM42 39
Manor Gdns.
Manor Cres. Horn. CV32 28
Manor Cres. Surb. BM53 66
Manordene Clo. Surb. BJ54 66
Manor Dr. Epsom BO57 76
Manor Dr. Esher BH55 75
Manor Dr. Felt. BD49 56
Manor Dr. N14 BV26 7
Manor Dr. N20 BU28 14
Manor Dr. NW7 BN28 13
Manor Dr. N. N. Mal. BN54 67
Manor Dr. Sun. BC51 65
Manor Dr. Surb. BL53 66
Manor Dr. The, Wor. Pk. BO54 67
Manor Dr. Wem. BL35 30
Manor Farm. Dr. E4 CG27 16
Manor Farm Rd. SW16 BY51 69
Manor Farm Rd. Wem. BK37 30
Manorfield Clo. N19 BW35 32
Junction Rd.
Manor Flds. SW15 BQ46 49
Manor Gdns. Hamptn. BF50 56
Manor Gdns. N7 BX34 24
Manor Gdns. Rich. BL45 48
Manor Gdns. Ruis. BD35 29
Manor Gdns. S. Croy. CA57 78
Manor Gdns. Sun. BC51 65
Manor Gdns. SW20 BR51 67
Manor Gdns. W3 BM42 39
Manorgate Rd. Kings. On T. BM51 66
Manor Gro. Beck. CE51 70
Manor Gro. Rich. BM45 48
Manor Gro. SE15 CC43 52
Manor Hall Av. NW4 BQ30 13
Manor Hall, Bans. BU60 83
Manor Hall Dr. NW4 BQ30 13
Manor Hall Gdns. E10 CE33 25
Manor Ho. Chis. CM51 71
Manor Ho. Dr. NW6 BQ36 31
Manor Ho. Surb. BK55 75
Manor Ho. Wor. Pk. BN54 67
Manor La. Felt. BC48 56
Manor La. SE12 CG46 52
Manor La. SE13 CG45 52
Manor La. Sun. BC51 65
Manor La. Sutt. BS56 77
Manor La. Ter. SE13 CG45 52
Manor Mt. SE23 CC47 61
Manor Pk. Chis. CM51 71
Manor Pk. Clo. W. Wick. CE54 70
Manor Pk. Cres. Edg. BM29 12
Manor Pk. Dr. Har. BF31 20
Manor Pk. Gdns. Edg. BM28 12
Manor Pk. Rich. BL45 48
Manor Pk. Rd. Chis. CM51 71
Manor Pk. Rd. E12 CJ35 35
Manor Pk. Rd. N2 BT31 23
Manor Pk. Rd. Sutt. BT56 77
Manor Pk. Rd. W. Wick. CE54 70
Manor Pk. SE13 CF45 52
Manor Pl. Chis. CM51 71
Manor Pl. Felt. BC47 56
Manor Pl. Mitch. BW52 68
Manor Pl. SE17 BY42 42
Manor Pl. Sutt. BS56 77
Manor Rd. Bark. CN36 36
Manor Rd. Barn. BR25 6
Manor Rd. Beck. CE51 70
Manor Rd. Bex. CR47 63
Manor Rd. Chig. CK29 17
Manor Rd. Dag. CS36 37
Manor Rd. Dart. CT45 55
Manor Rd. E10 CE33 25
Manor Rd. E15 CG37 34
Manor Rd. E16 CG38 34
Manor Rd. E17 CD30 16
Manor Rd. E. Mol. BG52 65
Manor Rd. Enf. BZ23 8
Manor Rd. Erith CT43 55
Manor Rd. Har. BJ32 21
Manor Rd. Hayes BC39 38
Manor Rd. Loug. CH23 10
Manor Rd. Loug. CH25 10
Manor Rd. Mitch. BW52 68
Manor Rd. N. Esher BH55 75
Manor Rd. N. Wall. BV56 77
Manor Rd. N16 BZ34 24
Manor Rd. N17 CB30 15
Manor Rd. N22 BX29 15
Manor Rd. Rich. BL45 48
Manor Rd. Rom. CP27 18
Manor Rd. Rom. CP32 27
Manor Rd. Rom. CU32 28
Manor Rd. SE25 CB52 69
Manor Rd. Sid. CN48 63
Manor Rd. S. Esher BH56 75
Manor Rd. Sutt. BR57 76
Manor Rd. SW20 BR51 67
Manor Rd. Tedd. BJ49 57
Manor Rd. Twick. BG48 56
Manor Rd. W13 BJ40 39
Manor Rd. Wall. BV56 77
Manor Rd. Wat. BC23 4
Manor Rd. W. Wick. CE55 79
Manor Rd. Wdf. Grn. CK29 17
Manorside, Barn. BR24 6
Manor Sq. Dag. CP34 27
Manor Vale, Brent. BK42 39
Manor Vw. Beck. CE51 70
High St.
Manor Vw. N3 BS30 14
Manor Way, Bans. BU61 83
Manor Way, Beck. CE51 70
Manor Way, Bex. CR47 63
Manor Way, Bexh. CS45 55
Manor Way, Borwd. BN23 6
Manor Way, Brom. CK53 71
Manor Way E4 CF28 16
Manor Way, Enf. CA26 8
Manor Way, Har. BF31 20
Manor Way, Mitch. BW52 68
Manor Way NW9 BO31 22
Manor Way, Orp. CM53 71

Manor Way, Rain. CT39 46
Manor Way SE3 CG45 52
Manor Way, S. Croy. CA57 78
Manor Way, Sthl. BD42 38
Manor Way, The, Pur. BX59 84
Manor Way, Wall. BV56 77
Manorway, Wdf. Grn. CJ28 17
Manor Way, Wor. Pk. BO54 67
Manor Wd. Rd. Pur. BX60 84
Manresa Rd. SW3 BU42 41
Mansard Beeches SW17 BV49 59
Mansard Clo. Pnr. BD31 20
Mansell Gro. E17 CE30 16
Mansell Rd. Grnf. BF39 38
Mansell Rd. W3 BN41 40
Mansell St. E1 CA39 42
Mansel Rd. SW19 BR50 58
Manse Pde. Swan. CU52 73
Manse Rd. N16 CA34 24
Manser Rd. Rain. CT38 37
Manse Way, Swan. CU52 73
Mansfield Av. Barn. BU25 7
Mansfield Av. N15 BZ31 24
Mansfield Clo. N9 CB25 8
Mansfield Clo. Orp. CP54 72
Mansfield Gdns. Horn. CV34 28
Mansfield Hill E4 CE26 9
Mansfield Ms. W1 BV39 41
Duchess St.
Mansfield Pl. S. Croy. BZ57 78
Mansfield Rd. Chess. BK56 75
Mansfield Rd. E11 CH32 26
Mansfield Rd. E17 CD31 25
Mansfield Rd. W3 BM38 30
Mansfield Rd. Ilf. CL33 26
Mansfield Rd. NW3 BU35 32
Mansfield Rd. S. Croy. BZ57 78
Mansfield Rd. Swan. CT50 64
Mansfield St. W1 BV39 41
Mansford St. E2 CB37 33
Manship Rd. Mitch. BV50 59
Mansion House La. EC4 BZ39 42
St. Swithin's La.
Mansion House Pl. EC4 BZ39 42
Mansion Ho. St.
Mansion House St. EC2 BZ39 42
Cornhill
Manson Ms. SW7 BT42 41
Manson Pl. SW7 BT42 41
Mans St. SW10 BT44 50
Manstead Clo. Rain. CU39 46
Mansted Gdns. Rom. CP33 27
Manston Av. Sthl. BF42 38
Manstone Rd. NW2 BR35 31
Manston Way, Horn. CU36 37
Mantell St. N1 BY37 33
Manthorp Rd. SE18 CM42 44
Mantilla Rd. SW17 BV49 59
Mantle Rd. SE4 CD45 52
Manton Av. W7 BH41 39
Manton Rd. SE2 CO42 45
Mantua Rd. SW11 BT45 50
Mantus Clo. E1 CC38 34
Mantus Rd.
Mantus Rd. E1 CC38 34
Manus Way N20 BT27 14
Manville Gdns. SW17 BV48 59
Manville Rd. SW17 BX48 60
Manwood Rd. SE4 CD46 52
Manwood St. E16 CK40 44
Many Gates SW12 BV48 59
Mapesbury Rd. NW2 BR36 31
Mape St. E2 CB38 33
Maple Av. E4 CD28 16
Maple Av. Har. BF34 20
Maple Av. W3 BO40 40
Maple Clo. Buck. H. CJ27 17
Maple Clo. Bush. BE23 4
Maple Clo. Horn. CU34 28
Maple Clo. Mitch. BV51 68
Maple Clo. N16 CB32 24
Timberwharf Rd.
Maple Clo. Orp. CM53 71
Maple Clo. Ruis. BC32 20
Maple Clo. SW4 BW46 50
Clarence Av.
Maple Ct. N. Mal. BO52 67
Maple Cres. Sid. CO46 54
Mapledale Av. Croy. CB55 78
Mapledene, Chis. CM50 62
Mapledene Rd. E8 CA36 33
Maple Gdns. Edg. BO29 13
Maple Gro. Brent. BJ43 48
Maple Gro. NW9 BN33 22
Maple Gro. Sthl. BE39 38
Maple Gro. W5 BK41 39
Maple Ho. SE8 CD43 52
Idonia St.
Mapleleafe Gdns. Ilf. CL31 26
Maple Pl. Bans. BQ60 82
Maple Pl. E1 CB39 42
Maple Pl. W1 BW38 32
Maple St.
Maple Rd. Dart. CV47 64
Maple Rd. E11 CG32 25
Maple Rd. Hayes BD38 29
Maple Rd. SE20 CB51 69
Maple Rd. Surb. BK53 66
Maple Rd. Whyt. CA62 84
Maples Pl. E1 CB39 42
Raven Row
Maplestead Rd. Dag. CO37 36
Maplestead Rd. SW2 BX47 60
Maples, The, Bans. BS60 83
Maple St. Rom. CS31 28
Maple St. W1 BW39 41
Maplethorpe Rd. Th. Hth. BY52 69
Mapleton Clo. Brom. CH53 71
Mapleton Cres. SW18 BS46 50
Mapleton Rd. Enf. CB23 8
Mapleton Rd. SW18 BS46 50
Maple Way, Felt. BC48 56
Maplin Clo. N21 BX25 8
Maplin Rd. E16 CH39 44
Mapperley Dr. Wdf. Grn. CG29 16
Forest Drive
Maran Way, Belv. CP41 45
Marble Arch W1 BU40 41
Marble Clo. W3 BM40 39
Gunnersbury La.
Marble Hill. Clo. Twick. BJ47 57
Marble Hill. Gdns. Twick. BJ47 57
Marble Hill River Path, BK47 57
Twick.
Orleans Rd.
Marbrook Ct. SE12 CJ48 62
Marcellina Way BR6 CN55 81
Marcet Rd. Dart. CV46 55
Marchant Rd. E11 CF34 25
Marchbank Rd. SW5 BR43 49
Marchmont Rd. Rich. BL46 48
Marchmont Rd. Wall. BW57 77
Marchmont St. WC1 BX38 33
March Rd. Twick. BJ47 57
Marchwood Clo. SE5 CA43 51
Marchwood Cres. W5 BK39 39
Marcia Rd. SE1 CA42 42
Marcilly Rd. SW18 BT46 50
Marconi Way, Sthl. BF39 38
Marcon Pl. E8 CB35 33
Marco Rd. W6 BP41 40
Marcus Ct. E15 CG37 34
Marcus Rd. Dart. CU47 64
Marcus St. E15 CG37 34
Marcus St. SW18 BS46 50
Marcus Ter. SW18 BS46 50
Denton St.
Mardale Dr. NW9 BN32 22
Mardell Rd. Croy. CC53 70
Marden Av. Brom. CG53 70
Marden Clo. Chig. CO27 18
Marden Cres. Bex. CS46 55
Marden Cres. Croy. BX53 69
Marden Rd. Croy. BX53 69
Marden Rd. N17 CA30 15
Avenue, The
Marden Rd. Rom. CT32 28
Marden Sq. SE16 CB41 42
Drummond Rd.
Marder Rd. W13 BJ41 39
Mardyke St. SE17 BZ42 42
Townsend St.
Marechal Niel Av. Sid. CM48 62
Mares Field, Croy. CA55 78
Maresfield Gdns. NW3 BT35 32
Mare St. E8 CB37 33
Marfield Ct. N. Mal. BO54 67
Margaret Av. E4 CE25 9
Margaret Bondfield Av. CO36 36
Bark.
Margaret Bldgs. N16 CA33 24
Margaret Rd.
Margaret Clo. Rom. CU32 28
Margaret Ct. W1 BW39 41
Margaret St.
Margaret Dr. Horn. CW33 28
Margaret Rd. Barn. BT24 7
Margaret Rd. Bex. CP46 54
Margaret Rd. N16 CA33 24
Margaret Rd. Rom. CU32 28
Margaret St. W1 BV39 41
Margaretta Ter. SW3 BU43 50
Margaretting Rd. E12 CJ34 26
Margaret Way, Couls. BY62 84
Margaret Way, Ilf. CJ32 26
Margate Rd. SW2 BX46 51
Margeholes, Wat. BE27 11
Margery Pk. Rd. E7 CH36 35
Margery Rd. Dag. CP34 27
Margery St. WC1 BY38 33
Margin Dr. SW19 BQ49 58
Margravine Gdns. W6 BQ42 40
Margravine Rd. W6 BQ42 40
Marham Gdns. Mord. BT53 68
Marham Gdns. SW18 BU47 59
Marian Clo. Hayes BD38 29
Marian Ct. Sutt. BS56 77
Marian Gdns. Horn. CW34 28
Marian Pl. E2 CB37 33
Marian Rd. SW16 BW51 68
Marian Way NW10 BO36 31
Maria Ter. E1 CC38 34
Maricas Av. Har. BG29 11
Marie Therese Clo. BN53 67
N. Mal.
Mariette Way, Wall. BX58 78
Marigold St. SE16 CB41 42
Marina Av. N. Mal. BP53 67
Marina Clo. Brom. CG52 70
Marina Clo. Rom. CS32 28
Marina Dr. Dart. CW47 64
Marina Dr. Well. CN44 54
Marina Pl. SW8 BX44 51
Priory Gro.
Marina Way, Tedd. BK50 57
Fairways
Marine Av. Well. CO45 54
Marine Dr. SE18 CK42 44
Marinefield Rd. SW6 BS44 50
Mariner Gdns. Rich. BJ48 57
Ashburnham Rd.
Mariner Rd. E12 CL35 35
Dersingham Av.
Marine St. SE16 CB41 42
Enid St.
Marion Clo. Bush. BE23 4
Marion Clo. Ilf. CM29 17
Marion Cres. Orp. CO53 72
Marion Gro. Wdf. Grn. CG28 16
Marion Rd. NW7 BP28 13
Marion Rd. Th. Hth. BZ53 69
Mariott Rd. Dart. CW47 64
Marischal Rd. SE13 CF45 52
Maritime St. E3 CD38 34
Marius Rd. SW17 BV48 59
Marjorams Av. Loug. CL23 10
Marjorie Gro. SW11 BU45 50
Markab Rd. Nthwd. BC28 11
Mark Av. E4 CE25 9
Mark Clo. Bexh. CQ44 54
Mark Clo. Sthl. BF40 38
Marke Clo. Kes. CK56 80
Market Ct. W1 BW39 41
Market Pl.
Market Hill SE18 CL41 44
Market Link, Rom. CT31 28
Market Meadow Pl. Orp. CP52 72
Market Ms. W1 BV40 41
Market Pde. SE15 CB44 51
Market Pl. Brent. BK43 48
Market Pl. Dart. CW47 64
Market Pl. Enf. BZ24 8
Market Pl. Kings. On T. BK51 66
Market Pl. N2 BU31 23
Market Pl. NW11 BS31 23
Market Pl. Rom. CT32 28
Market Pl. Wat. BD24 4
Market Pl. W1 BW39 41
Market Pl. W3 BN40 40
Market Rd. Rich. BM45 48
Market Rd. N7 BX36 33
Market Row SW9 BY45 51
Atlantic Rd.
Market Sq. Brom. CH51 71
Market St. Dart. CW47 64
Market St. E6 CK37 35
Market St. SE18 CL42 44
Market St. Wat. BC24 4
Market, The, Cars. BT54 68
Markfield, Croy. CB58 78
Markfield Gdns. E4 CE26 9
Markfield Rd. N15 CB31 24
Mark Grn. Wat. BD28 11
Markham Sq. SW3 BU42 41
Markham St. SW3 BU42 41
Markhole Clo. Hamptn. BE50 56
Markhouse Av. E17 CD32 25
Markhouse Rd. E17 CD32 25
Mark La. EC3 CA40 42
Markmanor Av. E17 CD33 25
Mark Pl. Sthl. BF40 38
Mark Rd. N22 BY30 15
Marksbury Av. Rich. BM45 48
Marks Rd. Rom. CS32 28
Mark St. E1 CA39 42
Mark St. E15 CG36 34
Mark St. EC2 CA38 33
Markway, Sun. BD51 65
Mark Way, Swan. CU53 73
Markwell Clo. SE26 CB49 60
Longton Gro.
Markyate Rd. Dag. CO35 36
Marlands Rd. Ilf. CK31 26
Marlborough Av. Edg. BM27 12
Marlborough Av. N14 BW27 14
Marlborough Bldgs. SW3 BU42 41
Marlborough Clo. N20 BU27 14
Marlborough Clo. Orp. CN54 72
Marlborough Clo. SW19 BU50 59
Marlborough Clo. Walt. BD55 74
Marlborough Ct. W8 BS42 41
Marlborough Cres. W4 BN41 40
Marlborough Dr. Ilf. CK31 26
Marlborough Gdns. N20 BU27 14
Marlborough Gdns. Surb. BK54 66
Marlborough Gro. SE1 CB42 42
Marlborough Hill, Har. BG32 20
Marlborough Hill NW8 BT37 32
Marlborough Pk. Av. Sid. CO47 63
Marlborough Pl. NW8 BT37 32
Marlborough Rd. W4 BN42 40
Marlborough Rd. W5 BK41 39
Marlborough Rd. Bexh. CP45 54
Marlborough Rd. Brom. CJ52 71
Marlborough Rd. Dag. CO35 36
Marlborough Rd. Dart. CV46 55
Marlborough Rd. E4 CE29 16
Marlborough Rd. E7 CJ36 35
Marlborough Rd. E15 CG35 34
Borthwick Rd.
Marlborough Rd. E18 CH31 26
Marlborough Rd. Felt. BD48 56
Marlborough Rd. Hamptn. BF50 56
Marlborough Rd. Har. BH31 21
Marlborough Rd. Islw. BJ44 48
Marlborough Rd. N9 CA26 8
Marlborough Rd. N19 BW34 23
Marlborough Rd. N22 BX29 15
Marlborough Rd. Rich. BL46 48
Marlborough Rd. Rom. CR31 27
Marlborough Rd. SE7 CJ43 53
Marlborough Rd. S. Croy. BZ57 78
Marlborough Rd. Sthl. BD41 38
Marlborough Rd. Sutt. BS55 77
Marlborough Rd. SW1 BW40 41
Marlborough Rd. SW19 BT50 59
Marlborough Rd. Wat. BC24 4
Marlborough St. SW3 BU42 41
Marler Rd. SE23 CD47 61
Marlescroft, Loug. CL25 10
Marley Av. Bexh. CP43 54
Marley Clo. Grnf. BF37 29
Marlingdene Clo. Hamptn. BF50 56
Marlings Clo. Chis. CN52 72
Marlings Clo. Whyt. CA62 84
Marlings Pk. Av. Chis. CN52 72
Marloes Clo. Wem. BK35 30
Marloes Rd. W8 BS41 41
Marloes, The NW8 BT37 32
Marlow Clo. SE20 CB52 69
Marlow Ct. NW6 BQ36 31
Marlow Cres. Twick. BH46 48
Marlow Dr. Sutt. BQ55 76
Marlowe Clo. Chis. CM50 62
Marlowe Clo. Ilf. CM30 17
Marlowe Gdns. Rom. CV30 19
Shenstone Gdns.
Marlowe Rd. E17 CF31 25
Marlowe Sq. Mitch. BV52 68
Marlowes, The, Dart. CS45 55
Marlow Gdns. Rom. CV30 19
Marlow Rd. E6 CK38 35
Marlow Rd. SE20 CB52 69
Marlow Rd. Sthl. BE41 38
Marlpit Av. Couls. BX62 84
Marlpit La. Couls. BW61 83
Marl St. SW18 BT45 50
Marlton St. SE10 CG42 43
Marlyon Rd. Ilf. CO28 18
Marmadon Rd. SE18 CN42 45
Marmion App. E4 CE28 16
Marmion Clo.
Marmion Av. E4 CD28 16
Marmion Clo. E4 CE28 16
Marmion Rd. SW11 BV45 50
Marmont Rd. SE15 CB44 51
Marmora Rd. SE22 CC46 52
Marmot Rd. Houns. BD45 47
Marne Av. N11 BV28 14
Marne Av. Wells. CO45 54
Marnell Way, Houns. BD45 47
Marne St. W10 BR38 31
Marney Rd. SW11 BV45 50
Marnham Av. NW2 BR35 31
Marnham Cres. Grnf. BF37 29
Marnock Rd. SE4 CD46 52
Maroon St. E14 CD39 43
Marquess Gro. N1 BZ36 33
Marquess Rd. N1 BZ36 33
Marquis Clo. Wem. BL36 30
Marquis Rd. N4 BX33 24
Marquis Rd. N22 BX29 15
Marquis Rd. NW1 BW36 32
Marrick Clo. SW15 BP45 49
Marriot Rd. Barn. BQ24 6
Marriots Clo. BO32 22
Marriott Rd. E15 CG37 34
Marriott Rd. N4 BX33 24
Marriott Rd. N10 BU30 14
Marryat Pl. SW19 BR49 58
Marryat Rd. SW19 BQ49 58
Marsala Rd. SE13 CE45 52
Marsden Rd. N9 CB27 15
Marsden Rd. SE15 CA45 51
Marsden St. NW5 BV36 32
Marshall Clo. Houns. BE46 47
Marshall Clo. SW18 BT46 50
St. Ann's Cres.
Marshall Dr. Hayes BC39 38
Marshall Gdns. SE1 BY41 42
London Rd.
Marshall Path SE28 CO40 45
Titmuss Av.
Marshall Rd. N17 BZ30 15
Marshalls Clo. Epsom BN60 82
Marshalls Clo. Sutt. BS56 77
High St.
Marshalls Dr. Rom. CT31 28
Marshall's Gro. SE18 CK42 44
Marshalls Rd. Rom. CS31 28
Marshalls Rd. Sutt. BS56 77
Lewis Rd.

Merrion Av. Stan.	BK28	12
Merritt Rd. SE4	CD46	52
Merritts Bldgs. EC2	CA38	33
Worship St.		
Merrivale Av. Ilf.	CJ31	26
Merrivale N14	BW25	7
Merrow Rd. Sutt.	BQ58	76
Merrow St. SE17	BZ43	51
Merrow Way, Croy.	CF57	79
Merryfield Gdns. Stan.	BK28	12
Merryfield SE3	CG44	52
Merryhill Clo. E4	CE26	9
Merry Hill Mt. Bush.	BF26	4
Merry Hill Rd. Bush.	BE25	4
Merryhills Dr. Enf.	BW24	7
Merrymeet, Bans.	BU60	83
Mersey Rd. E17	CD31	25
Mersey Wk. Nthlt.	BF37	29
Leander Rd.		
Mersham Dr. NW9	BM32	21
Mersham Pl. SE20	CB51	69
Mersham Rd. Th. Hth.	BZ52	69
Merten Rd. Rom.	CQ33	27
Merthyr Ter. SW13	BP43	49
Merton Abbey Sta. Rd. SW19	BT51	68
Merton Av. W4	BO42	40
Merton Av. Nthlt.	BG35	29
Merton Gdns. Orp.	CL53	71
Merton Hall Gdns. SW20	BR51	67
Merton Hall Rd. SW19	BR50	58
Merton La. N6	BU34	23
Merton Mansions SW20	BQ51	67
Merton Pl. SE10	CF44	52
Merton Ri. NW3	BU36	32
Merton Rd. Bark.	CN36	36
Merton Rd. E17	CF32	25
Merton Rd. Har.	BG33	20
Merton Rd. Ilf.	CN33	27
Merton Rd. SE25	CB53	69
Merton Rd. SW18	BS46	50
Merton Rd. SW19	BS50	59
Merton Rd. Wat.	BC24	4
Merton Spur SW20	BP52	67
Bushey Rd.		
Merton Way, E. Mol.	BF52	65
Merttins Rd. SE15	CC46	52
Ivydale Rd.		
Mervan Rd. SW2	BY45	51
Mervyn Av. SE9	CM48	62
Mervyn Rd. W13	BJ41	39
Messaline Av. W3	BN39	40
Messent Rd. SE9	CJ46	53
Messeter Pl. SE9	CL46	53
Messina Av. NW6	BS36	32
Metcalf Wk. Felt.	BE49	56
Creswell Clo.		
Meteor St. SW11	BV45	50
Meteor Way, Wall.	BX57	78
Methley St. SE11	BY42	42
Methuan Rd. Edg.	BM29	12
Methuen Clo. Edg.	BM29	12
Methuen Pk. N10	BV30	14
Methuen Rd. Belv.	CR42	45
Methuen Rd. Bexh.	CQ45	54
Methwold Rd. W10	BQ39	40
Mews, The, Ilf.	CJ32	26
Mews, The N1	BZ37	33
St. Paul St.		
Mews, The, Rom.	CT32	28
Mews, The, Twick.	BJ46	48
Bridge Rd.		
Mexfield Rd. SW15	BR46	49
Meyer Rd. Erith	CS43	55
Meymott St. SE1	BY40	42
Meynell Cres. E9	CC36	34
Meynell Rd. E9	CC36	34
Meynell Rd. Rom.	CU29	19
Meyrick Rd. NW10	BP36	31
Meyrick Rd. SW11	BT45	50
Micawber St. N1	BZ38	33
Michael Gdns. Horn.	CV31	28
Michael Gaynor Clo. W13	BJ40	39
Michael Rd. E11	CG33	25
Michael Rd. SE25	CA52	69
Michael Rd. SW6	BS44	50
Michaels Clo. SE13	CG45	52
Michael's Rd. NW2	BQ35	31
Micheldever Rd. SE12	CG46	52
Michelham Gdns. Twick.	CB57	78
Michel's Row, Rich.	BL45	48
Kew Foot Rd.		
Michigan Av. E12	CK35	35
Michleham Down N12	BR28	13
Mickleham Clo. Orp.	CN51	72
Mickleham Rd.		
Mickleham Gdns. Sutt.	BR57	76
Mickleham Rd. Orp.	CN51	72
Mickleham Way, Croy.	CF57	79
Micklethwaite Rd. SW6	BS43	50
Middle Clo. Epsom	BO59	82
Middle Dene NW7	BN27	13
Boundary Rd.		
Middlefield Cres. Ilf.	CL32	26
Middle Fielde W13	BJ39	39
Templewood		
Middlefield NW8	BT37	32
Middlefields, Croy.	CD58	79
Middle Furlong, Bush.	BF24	4
Middleham Gdns. N18	CB29	15
Middleham Rd. N18	CB29	15
Middle La. Epsom	BO59	82
Middle La. Mews. N8	BX32	24
Middle La.		
Middle La. N8	BX32	24
Middle La. Tedd.	BH50	57
Middle Pk. Av SE9	CJ46	53
Middle Path Har.	BG33	20
Middle Rd. Barn.	BU25	7
Middle Rd. Har.	BG34	20
Middle Rd. SW16	BW51	68
Middle Row Pl. WC1	BY39	42
High Holborn		
Middle Row W10	BR38	31
Middlesborough Rd. N18	CB29	15
Middlesex Rd. Mitch.	BX53	69
Middlesex St. E1	CA39	42
Middlesex Wharf E5	CC34	25
Southwold Rd.		
Middle St. Croy.	BZ55	78
Middle St. EC1	BZ39	42
Bartholomew Clo.		
Middle Temple La. EC4	BY39	42
Middleton Av. E4	CD28	16
Middleton Av. Grnf.	BG37	29
Middleton Av. Sid.	CO50	63
Middleton Clo. E4	CD27	16
Middleton Dr. Pnr.	BC31	20
Middleton Gdns. Ilf.	CL32	26
Middleton Gro. N7	BX35	33
Middleton Rd. E8	CA36	33
Middleton Rd. Epsom	BN58	76
Eleanor Av.		
Middleton Rd. Mord.	BS53	68
Middleton Rd. N. Mal.	BN51	67
Middleton Rd. NW11	BS33	23
Middleton St. E2	CB38	33
Canrobert St.		
Middleton Way SE13	CF45	52
Middle Way, Hayes	BD38	29
Middleway NW11	BS32	23
Middle Way SW16	BW51	68
Middle Way, The, Har.	BH30	12
Midfield Av. Bexh.	CS45	55
Midfield Way, Orp.	CO51	72
Midford Pl. W1	BW38	32
Tottenham Court Rd.		
Midholm Clo. NW11	BS31	23
Midholm NW11	BS31	23
Midholm Rd. Croy.	CD55	79
Midholm, Wem.	BM33	21
Midhope St. WC1	BX38	33
Argyle Wk.		
Midhurst Av. Croy.	BY54	69
Midhurst Av. N10	BV31	23
Midhurst Clo. Horn.	CU35	37
Cowdray Way		
Midhurst Hill, Bexh.	CR46	54
Midhurst Rd. W13	BJ41	39
Midland Mead SE16	CC42	43
Midland Pl. E14	CF42	43
Ferry St.		
Midland Rd. E10	CF33	25
Midland Rd. NW1	BW37	32
Midland Ter. NW2	BQ34	22
Midland Ter. NW10	BO38	31
Midmoor Rd. SW12	BW47	59
Midmoor Rd. SW19	BQ50	58
Midstrath Rd. NW10	BO35	31
Ballogie Av.		
Midsummer Av. Houns.	BE45	47
Midway, Sutt.	BR54	67
Midway, Walt.	BC55	74
Midwood Clo. NW2	BP34	22
Miers Clo. E6	CL37	35
Mighell Av. Ilf.	CJ32	26
Milborne Gro. SW10	BT42	41
Gilston Rd.		
Milborne St. E9	CC36	34
Well St.		
Milborough Cres. SE12	CG46	52
Milbourne La. Esher	BG57	74
Milbourne Rd. Felt.	BE49	56
Milbrook, Esher	BG57	74
Milburn Wk. Epsom	BO61	82
Milcote St. SE1	BY41	42
Mildenhall Rd. E5	CB35	33
Mildmay Av. N1	BZ36	33
Mildmay Gro. N1	BZ35	33
Mildmay Pk. N1	BZ35	33
Mildmay Rd. N1	BZ35	33
Mildmay Rd. Rom.	CS32	28
Mildmay St. N1	BZ36	33
Mildred Av. Borwd.	BM24	5
Mildred Av. Nthlt.	BF35	29
Mildred Rd. Erith	CS42	46
Mile End Pl. E1	CC38	34
Mile End Rd. E1	CC38	34
Mile End Rd. E3	CD38	34
Mile End, The, E17	CC30	16
Mile Rd. Wall.	BV54	68
Miles La. EC4	BZ40	42
Arthur St.		
Milespit Hl. NW7	BP28	13
Miles Pl. NW1	BT39	41
Miles Pl. Surb.	BL52	66
Miles Rd. Epsom	BN59	82
Miles Rd. Mitch.	BT52	68
Miles St. SW8	BX43	51
Milestone Clo. Sutt.	BT57	77
Milestone Rd. SE19	CA50	60
Miles Way N20	BU27	14
Milfoil St. W12	BP40	40
Milford Clo. SE2	CQ43	54
Milford Gdns. Edg.	BM29	12
Milford Gdns. Wem.	BK35	30
Milford Gro. Sutt.	BT56	77
Milford La. WC2	BX39	42
Milford Rd. Sthl.	BF40	38
Milford Rd. W13	BJ40	39
Milford Way SE15	CA44	51
Sumner Estate		
Milk St. Brom.	CH50	62
Milk St. E16	CL40	44
Milk St. EC2	BZ39	42
Milkwell Gdns. Wdf. Grn.	CH29	17
Milkwell Yd. SE5	BZ44	51
Denmark Hill		
Milkwood Rd. SE24	BY46	51
Milk Yd. E1	CC40	43
Millais Av. E12	CL35	35
Millais Rd. E11	CF35	34
Millais Rd. Enf.	CA25	8
Millais Rd. N. Mal.	BO53	67
Millais Way, Epsom	BN56	76
Millard Ter. Dag. CR36		36
Millbank SW1	BW42	41
Millbrook Av. Well.	CM45	53
Millbrook Ct. SW15	BR46	49
Keswick Rd.		
Millbrook Gdns. Rom.	CQ32	27
Millbrook Gdns. Rom.	CT30	19
Millbrook Rd. N9	CB26	8
Millbrook Rd. St. Mary Cray	CP52	72
Millbrook Rd. SW9	BY45	51
Millbrooks Rd. Bush.	BE23	4
Millbourne Rd. Felt.	BE49	56
Millbro. Swan.	CU51	73
Mill Clo. Cars.	BV55	77
Mill Corner, Barn.	BR23	6
Miller Clo. Pnr.	BD30	11
Miller Rd. Croy.	BX54	69
Miller Rd. SW19	BT50	59
Millers Av. E8	CA35	33
Miller's Clo. Chig.	CO27	18
Millers Ct. W6	BO42	40
Millers Grn. Enf.	BY24	8
Miller's La. Chig.	CO27	18
Millers Ter. E8	CA35	33
Miller St. NW1	BW37	32
Millet Rd. Grnf.	BF38	39
Mill Farm Clo. Pnr.	BD30	11
Mill Farm Cres. Houns.	BE47	56
Millfield Av. E17	CD30	16
Millfield La. N6	BU34	23
Millfield Pl. N6	BV34	23
Millfield Rd. Edg.	BN30	13
Millfield Rd. Houns.	BE47	56
Millfields Clo. Orp.	CO52	72
Millfields Rd. E5	CC35	34
Mill Gdns. SE26	CB49	60
Millgreen Rd. Mitch.	BU54	68
Millgrove St. SW11	BV44	50
Mill Hill Cir. NW7	BO28	13
Mill Hill Gro. W3	BN40	40
Mill Hill NW3	BT34	23
Mill Hill Rd. SW13	BP44	49
Mill Hill Rd. W3	BM41	39
Mill Hill Ter. W3	BM40	39
Mill House Pl. SE27	BY49	60
Millias Gdns. Edg.	BM30	12
Millicent Rd. E10	CD33	25
Milling Rd. Edg.	BN29	13
Mill La. Cars.	BU56	77
Mill La. Croy.	BX55	78
Mill La. E4	CE24	9
Mill La. Epsom	BO58	76
Mill La. Kings. On T.	BL52	66
Mill La. NW6	BR35	31
Mill La. Orp.	CL58	80
Mill La. Orp.	CO52	72
Mill La. Rom.	CQ32	27
Mill La. SE18	CL42	44
Mill La. Wdf. Grn.	CG28	16
Millman Ms. WC1	BX38	33
Millman St. WC1	BX38	33
Millmark Gro. SE14	CD44	52
Millmarsh La. Enf.	CD23	9
Mill Mead Rd. N17	CB31	24
Millmead Way, Loug.	CK23	10
Mill Pk. Av. Horn.	CW34	28
Mill Pl. Chis.	CL51	71
Mill Pl. Dart.	CU45	55
Mill Pl. E14	CD39	43
Mill Pl. Kings. On T.	BL52	66
Mill Plat Av. Islw.	BJ44	48
Mill Plat, Islw.	BJ44	48
Millpond Est. SE16	CB41	42
Mill Pond Rd. Dart.	CW46	55
Mill Ridge Edg.	BL28	12
Mill Rd. Dart.	CW49	64
Mill Rd. E16	CH40	44
Mill Rd. Epsom	BO59	82
Mill Rd. Erith	CS43	55
Mill Rd. Esher	BF55	74
Mill Rd. Eyns.	CW54	73
Mill Rd. Ilf.	CL34	26
Mill Rd. SW19	BT50	59
Mill Rd. Twick.	BG48	56
Mill Row N1	CA37	33
Mill Row W4	BN42	40
Belmont Rd.		
Mills Ct. E11	CG34	25
Harrow Rd.		
Mills Gro. E14	CF39	43
Mills Gro. NW4	BQ31	22
Mill Shot Clo. SW6	BQ44	49
Millside, Cars.	BU55	77
Millson Clo. N20	BT27	14
Domville Clo.		
Mills Rd. Walt.	BD56	74
Millstream Rd. SE1	CA41	42
Mill St. Kings. On T.	BL52	66
Mill St. SE1	CA41	42
Mill St. W1	BV40	41
Mill Vale, Brom.	CG51	70
Mill Vw. Gdns. Croy.	CC55	79
Millwall Dock Rd. E14	CE41	43
Tiller St.		
Millwall Est. E14	CE41	43
Millward St. W10	BQ40	40
Charles Sq.		
Mill Way, Bush.	BE23	4
Mill Way, Felt.	BC46	47
Millway Gdns. Nthlt.	BE36	29
Millway NW7	BO28	13
Millwell Cres. Chig.	CM28	17
Millwood Rd. Houns.	BG46	47
Millwood Rd. Orp.	CP52	72
Millwood St. W10	BQ39	40
Chesterton Rd.		
Milman Clo. Pnr.	BD30	11
Milman Rd. NW6	BQ37	31
Milmans St. SW10	BT43	50
Milne Est. SE18	CK42	44
Milne Fld. Pnr.	BF29	11
Milne Gdns. SE9	CK46	53
Milner Ct. Bush.	BF26	4
Bridgewater Way		
Milner Dr. Twick.	BG47	56
Milner Pl. N1	BY37	33
Milner Rd. Dag.	CP34	27
Milner Rd. E15	CG38	34
Milner Rd. Kings. On T.	BK52	66
Milner Rd. Mord.	BT53	68
Milner Rd. SW19	BS51	68
Milner Rd. Th. Hth.	BZ52	69
Milner Sq. N1	BY36	33
Milner St. SW1	BU42	41
Milner St. SW3	BU42	41
Milnthorpe Rd. W4	BN43	49
Milo Rd. SE22	CA46	51
Milson Rd. W14	BQ41	40
Milton Av. Barn.	BR25	6
Milton Av. Croy.	BZ54	69
Milton Av. E6	CJ36	35
Milton Av. Horn.	CT34	28
Milton Av. N6	BW33	23
Milton Av. NW9	BM31	21
Milton Av. NW10	BO37	31
Milton Av. Sev.	CR58	81
Milton Av. Sutt.	BT55	77
Milton Clo. Hayes	BC39	38
Milton Clo. N2	BT32	23
Milton Clo. Pnr.	BE29	11
Milton Clo. Sutt.	BT55	77
Milton Ct. EC2	BZ39	42
Milton St.		
Milton Ct. Kings. On T.	BL49	57
Milton Ct. Rd. SE14	CD43	52
Milton Cres. Ilf.	CM33	26
Milton Dr. Borwd.	BM25	5
Milton Gdns. Epsom	BO60	82
Milton Gro. N11	BW28	14
Milton Gro. N16	BZ35	33
Milton Park N6	BW33	23
Milton Pl. N7	BY35	33
George's Rd.		
Milton Rd. Belv.	CR42	45
Milton Rd. Croy.	BZ54	69
Milton Rd. E17	CE31	25
Milton Rd. Hamptn.	BF50	56
Milton Rd. Har.	BH31	21
Milton Rd. KT12	BD55	74
Milton Rd. Mitch.	BV50	59
Milton Rd. N6	BW33	23
Milton Rd. N15	BY31	24
Milton Rd. NW7	BP28	13
Milton Rd. NW9	BP33	22
Milton Rd. Rom.	CU32	28
Milton Rd. SE24	BY46	51
Milton Rd. Sutt.	BS55	77
Milton Rd. SW14	BN45	49
Milton Rd. SW19	BT50	59
Milton Rd. W3	BN40	40
Milton Rd. W7	BH40	39
Milton Rd. Wall.	BW57	77
Milton Rd. Walt.	BD55	74
Milton Rd. Well.	CN44	54

Milton St. E13 CH37 35
Greenwood Rd.
Milton St. EC2 BZ39 42
Milverton Gdns. Ilf. CN34 27
Milverton Rd. NW6 BQ36 31
Milverton St. SE11 BY42 42
Milverton Way SE9 CL49 62
Milward St. E1 CB39 42
Mimosa Clo. Orp. CP55 81
Berrylands
Mimosa Clo. Rom. CV29 19
Mimosa Rd. Hayes BD39 38
Mimosa Rd. Rom. CV29 19
Mimosa St. SW6 BR44 49
Minard Rd. SE6 CG47 61
Mina Rd. SE17 CA42 42
Mina Rd. SW19 BS51 68
Minchenden Cres. N14 BW27 14
Mincing La. EC3 CA40 42
Minden Rd. SE20 CB51 69
Minehead Ct. Har. BF34 20
Minehead Rd. Har. BF34 20
Minehead Rd. SW16 BX49 60
Mineral St. SE18 CM42 44
Minerva Clo. Sid. CN49 63
Minerva Est. E2 CB37 33
Minerva Ms. SW1 BV42 41
Minerva Rd. E4 CE29 16
Minerva Rd. Kings. On T. BL51 66
Minerva Rd. NW10 BN38 31
Minerva St. E2 CB37 33
Minet Av. NW10 BO37 31
Minet Dr. Hayes BC40 38
Minet Gdns. Hayes BC40 38
Minet Gdns. NW10 BO37 31
Minet Rd. SW9 BY44 51
Minford Gdns. W14 BQ41 40
Minford Ho. W14 BQ41 40
Mingard Wk. EC1 BX34 24
Ming St. E14 CE40 43
Miniver Pl. Garlick Hill EC4 BZ40 42
Queen Victoria St.
Mink Ct. Houns. BD45 47
Minniedale, Surb. BL53 66
Minnow St. SE17 CA42 42
Minories EC3 CA39 42
Minshull St. SW8 BW44 50
Wandsworth Rd.
Minson Rd. E9 CC37 34
Minstead Gdns. SW15 BO46 49
Minstead Way, N. Mal. BO53 67
Minster Av. Sutt. BS55 77
Minster Dr. Croy. CA56 78
Minster Gdns. E. Mol. BE52 65
Minster Rd. Brom. CH50 62
Minster Rd. NW2 BR35 31
Minster Walk N8 BX31 24
Minster Way, Horn. CW33 28
Mintern Clo. N13 BY27 15
Minterne Av. Sthl. BF42 38
Minterne Way, Hayes BD39 38
Mintern Rd. Har. BL32 21
Mintern St. N1 BZ37 33
Mint Rd. Bans. BT61 83
Mint Rd. Wall. BV56 77
Mint St. SE1 BZ41 42
Mint Wk. Croy. BZ55 78
Mirabel Rd. SW6 BR43 49
Miranda House N1 CA37 33
Purcell St.
Miranda Rd. N19 BW33 23
Mirfield St. SE7 CJ41 44
Miriam Rd. SE18 CN42 45
Misefield View, Orp. CM55 80
Miskin Rd. Dart. CV47 64
Missenden Gdns. Mord. BT53 68
Mission Pl. SE15 CB44 51
Mission Sq. Brent. BL43 48
Pottery Rd.
Mitcham Gdn. Vill. Mitch. BV53 68
Mitcham La. SW16 BV50 59
Mitcham Pk. Mitch. BU52 68
Mitcham Rd. Croy. BW53 68
Mitcham Rd. E6 CK38 35
Mitcham Rd. Ilf. CN33 27
Mitcham Rd. SW17 BU49 59
Mitchell Clo. Dart. CW48 64
Mitchell Clo. SE2 CP42 45
Mitchell Rd. N13 BY28 15
Mitchell St. EC1 BZ38 33
Mitchell Way NW10 BN36 31
Mitchison Rd. N1 BZ36 33
Mitchley Av. Pur. BZ60 84
Mitchley Gro. S. Croy. CB60 84
Mitchley Hill, S. Croy. CA60 84
Mitchley Rd. N17 CB31 24
Mitchley Vw. S. Croy CB60 84
Mitford Rd. N19 BX34 24
Mitre Clo. Sutt. BT57 77
Mitre Ct. EC2 BZ39 42
Wood St.
Mitre Rd. SE1 BY41 42
Mitre St. EC3 CA39 42
Mitre, The E14 CD40 43
Moat Cres. N3 BS31 23
Basing Way
Moat Dr. E13 CJ37 35
Boundary Rd.
Moat Dr. Har. BG31 20
Moat Farm Rd. Nthlt. BE36 29
Moatfield Clo. Bush. BF25 4
Moatfield Rd. Bush. BF25 4
Moat Gdns. SE28 CP40 45
Moat La. Erith CU44 55
Moat Pl. SW9 BX45 51
Moat Pl. W3 BM39 39
Moat Side, Enf. CC24 9
Durantspk Av.
Moatside, Felt. BD49 56
Moat, The, N. Mal. BO51 67
Moberley Rd. SW4 BW47 59
Modbury Gdns. NW5 BV36 32
Queen's Cres.
Model Cotts. SW14 BN45 49
Upr. Richmond Rd.
Model Cotts. W13 BJ41 39
Glenfield Rd.
Model Farm Clo. SE9 CK48 62
Modena St. W10 BR39 40
Kensal Rd.
Modern Ct. EC4 BY39 42
Farringdon St.
Moelyn Mews, Har. BJ32 21
Moffat Ct. SW19 BS49 59
Gap Rd.
Moffat Gdns. Mitch. BT52 68
Moffat Rd. N13 BX29 15
Moffat Rd. SW17 BU49 59
Moffat Rd. Th. Hth. BZ51 69
Mogden La. Islw. BH46 48
Moiety St. E14 CE41 43
Moira Clo. N17 CA30 15
Moirant Gdns. Rom. CR28 18
Moira Rd. SE9 CK45 53
Moiravale, Kings. Qn T. BK51 66
Moir Clo. S. Croy. CB58 78
Moland Mead SE16 CC42 43
Molash Rd. Orp. CP52 72
Mole Abbey Gdns. E. Mol. BF52 65
Mole Ct. Epsom BN56 76
Molember Rd. E. Mol. BH53 66
Mole Rd. Walt. BD56 74
Molescroft SE9 CM48 62
Molesey Av. E. Mol. BE53 65
Molesey Clo. Walt. BE56 74
Molesey Dr. Sutt. BR55 76
Molesey Pk. Av. E. Mol. BF53 65
Molesey Pk. Clo. E. Mol. BG53 65
Molesey Pk. Rd. E. Mol. BF53 65
Molesey Rd. E. Mol. BE53 65
Molesey Rd. Walt. BD56 74
Molesford Rd. SW6 BS44 50
Molesham Clo. E. Mol. BF52 65
Molesham Way, E. Mol. BF52 65
Molesworth St. SE13 CF45 52
Mollison Av. Enf. CD25 9
Mollison Dr. Wall. BW57 77
Mollison Way, Edg. BL30 12
Molyneaux St. W1 BU39 41
Monahan Av. Pur. BX59 84
Monarch Clo. W. Wick. CG56 79
Monarch Ct. N2 BT32 23
Monarch Rd. Belv. CR42 45
Gertrude Rd.
Mona Rd. SE15 CC44 52
Monastery Gdns. Enf. BZ23 8
Mona St. E16 CG39 43
Monaveen Gdns. E. Mol. BF52 65
Monck St. SW1 BW41 41
Monclar Rd. SE5 BZ45 51
Moncrieff St. SE15 CB44 51
Monega Rd. E7 CJ36 35
Monega Rd. E12 CJ36 35
Mongers La. Epsom BO58 76
Monier Rd. E3 CE36 34
Monivea Rd. Beck. CD50 61
Monk Dr. E16 CG39 43
Monkfrith Av. N14 BV25 7
Monkfrith Clo. N14 BV26 7
Monkfrith Way N14 BV26 7
Monkham's Av. Wdf. Grn. CH28 17
Monkham's Dr. Wdf. Grn. CH28 17
Monkham's La. Wdf. Grn. CH28 17
Monkleigh Rd. Mord. BR52 67
Monks Av. Barn. BT25 7
Monks Av. E. Mol. BE53 65
Monks Clo. Enf. BZ23 8
Monks Clo. Ruis. BD35 29
Monks Clo. SE2 CP42 45
New Rd.
Monks Cres. Walt. BC54 65
Monksdene Gdns. Sutt. BS55 77
Monks Dr. W3 BM39 39
Monks Gro. Loug. CL25 10
Monksmead, Borwd. BN24 6
Monks Orchard, Dart. CV48 64
Monks Orchard Rd. Croy. CE54 70
Monks Pk. Gdns. Wem. BM36 30
Monks Pk. Pde. Wem. BM36 30
Monks Pk. Wem. BM36 30
Monks Rd. Bans. BS61 83
Monks Rd. Enf. BY23 8
Monk St. SE18 CL42 44
Monks Way, Beck. CE53 70
Monks Way, Orp. CM54 71
Monkswell Ct. N10 BV30 14
Monkswood Gdns. Borwd. BN24 6
Monkswood Gdns. Ilf. CL31 26
Monkton Rd. Well. CN44 54
Monkton St. SE11 BY42 42
Monkville Av. NW11 BR31 22
Monkwell Sq. EC2 BZ39 42
Monmouth Av. E18 CH31 26
Monmouth Av. Kings. On T. BK50 57
Monmouth Clo. Mitch. BX52 69
Monmouth Clo. Well. CO45 54
Monmouth Pl. W2 BS39 41
Monmouth Rd. Dag. CQ35 36
Monmouth Rd. E6 CK38 35
Monmouth Rd. N9 CB27 15
Monmouth Rd. W2 BS39 41
Monmouth Rd. Wat. BC24 4
Monmouth St. WC2 BX39 42
Monnery Rd. N19 BW34 23
Monnow Rd. SE1 CB42 42
Monoux Gro. E17 CD30 16
Monroe Cres. Enf. CB23 8
Monroe Dr. SW14 BM46 48
Monsal Ct. E5 CC35 34
Clapton Park Est.
Monsell Rd. N4 BY34 24
Monson Rd. NW10 BP37 31
Monson Rd. SE14 CC43 52
Mons Way, Brom. CK53 71
Montacute Rd. Bush. BH26 5
Montacute Rd. Croy. CF58 79
Montacute Rd. SE6 CD47 61
Montagu Cres. N18 CB28 15
Montague Av. SE4 CD45 52
Montague Av. S. Croy. CA59 84
Montague Av. W7 BH40 39
Montague Clo. SE1 BZ40 42
Montague Clo. Walt. BC54 65
Montague Ct. EC1 BZ39 42
Bartholomew Clo.
Montague Gdns. W3 BM40 39
Montague Ind. Est. N18 CC28 16
Montague Pl. E14 CF40 43
Montague Pl. WC1 BW39 41
Montague Rd. Croy. BY54 69
Montague Rd. E8 CB35 33
Montague Rd. E11 CG34 25
Montague Rd. Houns. BF45 47
Montague Rd. N8 BX32 24
Montague Rd. N15 CB31 24
Montague Rd. N18 CB28 15
Montague Rd. NW4 BP32 22
Montague Rd. Rich. BL46 48
Montague Rd. Sthl. BE42 38
Montague Rd. Swan. CT52 73
Montague Rd. SW19 BS50 59
Montague Rd. W7 BH40 39
Montague Rd. W13 BJ39 39
Montague Sq. SE15 CC43 52
Clifton Way
Montague St. WC1 BX39 42
Montague Way, Sthl. BE41 38
Montagu Gdns. N18 CB28 15
Montagu Gdns. Wall. BW56 77
Montagu Mans. W1 BU39 41
Montagu Ms. S. W1 BU39 41
George St.
Montagu Ms. W1 BU39 41
Montagu Pl. W1 BU39 41
Montagu Rd. N9 CC27 16
Montagu Rd. N18 CB28 15
Montagu Sq. W1 BU39 41
Montagu St. W1 BU39 41
Montalt Rd. Wdf. Grn. CG28 16
Montana Rd. SW17 BV49 59
Montana Rd. SW20 BQ51 67
Montbelle Rd. SE9 CL48 62
Montcalm Clo. Brom. CH53 71
Montcalm Clo. Hayes BC38 29
Ayles Rd.
Montcalm Rd. SE7 CJ43 53
Montclare St. E2 CA38 33
Monteagle Av. Bark. CM36 35
Monteagle Way N16 CB34 24
Downs Est.
Montecute Rd. Mord. BT53 68
Montefiore St. SW8 BV44 50
Monteith Rd. E3 CD37 34
Montem Rd. N. Mal. BO52 67
Montem Rd. SE23 CD47 61
Montem St. N4 BX33 24
Montenotte Rd. N8 BW32 23
Monterey Clo. Bex. CS48 64
Montesole Ct. Pnr. BD30 11
Montford Pl. SE11 BY43 51
Montford Pl. SW19 BQ47 58
Montford Rd. Sun. BC52 65
Montford St. E1 CB39 42
Montfort Gdns. Ilf. CM29 17
Montgolfier Wk. Nthlt. BD38 29
Montgomery Av. Esher BH55 75
Montgomery Clo. Mitch. BX52 69
Montgomery Clo. Sid. CN46 54
Montgomery Cres. Rom. CV28 19
Montgomery Rd. W4 BN42 40
Montgomery Rd. Edg. BL29 12
Montholme Rd. SW11 BU46 50
Montolieu Gdns. SW15 BP46 49
Montpelier Av. Bex. CP47 63
Montpelier Av. N3 BT30 14
Montpelier Av. W5 BK39 39
Montpelier Gdns. E6 CJ38 35
Montpelier Gdns. Rom. CP33 27
Montpelier Gro. NW5 BW35 32
Montpelier Pl. SW7 BU41 41
Montpelier Ri. NW11 BR33 22
Montpelier Ri. Wem. BK33 21
Montpelier Rd. Pur. BY58 78
Montpelier Rd. Sutt. BT56 77
Montpelier Rd. W5 BK39 39
Montpelier Row SE3 CG44 52
Montpelier Row, Twick. BK47 57
Montpelier Sq. SW7 BU41 41
Montpelier Ter.
Montpelier St. SW7 BU41 41
Montpelier Ter. SW7 BU41 41
Montpelier Vale SE3 CG44 52
Montpelier Wk. SW7 BU41 41
Montpelier Way NW11 BR33 22
Montpellier Rd. SE15 CB44 51
Montrave Rd. SE20 CC50 61
Montreal Pl. WC2 BX40 42
Aldwych
Montreal Rd. Ilf. CM33 26
Montrell Rd. SW2 BX47 60
Montrose Av. Edg. BN30 13
Montrose Av. NW6 BR37 31
Montrose Av. Rom. CV30 19
Montrose Av. Sid. CO47 63
Montrose Av. Twick. BF47 56
Montrose Av. Well. CM45 53
Montrose Clo. Well. CN45 54
Montrose Clo. Wdf. Grn. CH28 17
Montrose Ct. NW9 BN30 13
Montrose Ct. NW11 BR31 22
Addison Way
Montrose Cres. N12 BT29 14
Montrose Cres. Wem. BL36 30
Montrose Gdns. Mitch. BU51 68
Montrose Gdns. Sutt. BS55 77
Montrose Pl. SW1 BV41 41
Montrose Rd. Har. BH30 12
Montrose Way SE23 CC47 61
Montrouge Cres. Epsom BQ61 82
Montserrat Av. Wdf. Grn. CF29 16
Montserrat Rd. SW15 BR45 49
Monument St. EC3 BZ40 42
Monza St. E1 CC40 43
Moodkee St. SE16 CC41 43
Moody St. E1 CC38 34
Moon Ct. SE12 CH45 53
Lyme Farm Rd.
Moon La. Barn. BR24 6
Moon St. N1 BY37 33
Moorcroft Rd. SW16 BW48 59
Moorcroft Way, Pnr. BE32 20
Moordown SE18 CL43 53
Moore Clo. Mitch. BV51 68
Moore Clo. SW14 BN45 49
Little St. Leonard's
Moore Clo. Wall. BX57 78
Moorefield Rd. N17 CA30 15
Moorehead Way SE3 CH45 53
Mooreland Rd. Brom. CG50 61
Moore Pk. Rd. SW6 BS43 50
Moore Rd. SE19 BY50 60
Moore St. SW3 BU42 41
Moore Wk. E7 CH35 35
Moorey Clo. E15 CG37 34
Stephen's Rd.
Moorfield Av. W5 BK38 30
Moorfield Highbank EC2 BZ39 42
St. Alphages Gdns.
Moorfield Rd. Chess. BL56 75
Moorfield Rd. Enf. CC23 9
Moorfield Rd. Orp. CO54 72
Moorfields EC2 BZ39 42
Moorgate EC2 BZ39 42
Moorhouse Rd. Har. BK31 21
Moorhouse Rd. W2 BS39 41
Moorland Clo. Hamptn. BE49 56
Moorland Clo. Houns. BF47 56
Moorland Clo. Rom. CR29 18
Moorland Rd. Har. BL32 21
Moorland Rd. SW9 BY45 51
Moorlands Av. NW4 BP29 13
Moorlands, Nthlt. BE37 29
Moor La. Chess. BL56 75
Moor La. EC2 BZ39 42
Moormead Dr. Epsom BO56 76
Moor Mead Rd. Twick. BJ46 48
Moorside Rd. Brom. CG48 61
Moorsom Way, Couls. BW62 83
Moor St. W1 BW39 41
Old Compton St.
Moortown Rd. Wat. BD28 11
Moor Vw. Wat. BC26 4
Moot Clo. NW9 BM32 21
Morant Path E14 CE40 43
Pennyfields
Morant Pl. N22 BX30 15
Morant St. E14 CE40 43
Mora Rd. NW2 BQ35 31

Mora St. EC1	BZ38	33
Morat St. SW9	BX44	51
Moravian Pl. SW10	BT43	50
Milmans St.		
Moravian St. E2	CC38	34
Gawber St.		
Moray Clo. Rom.	CT29	19
Moray Mews N4	BX34	24
Durham Rd.		
Moray Rd. N4	BX34	24
Moray St. E2	CC38	34
Cyprus St.		
Moray Way, Rom.	CS29	19
Morcambe Gdns. Stan.	BK28	12
Morcorvo Clo. SW7	BT41	41
Mordaunt Gdns. Dag.	CQ36	36
Mordaunt Rd. NW10	BN37	31
Mordaunt St. SW9	BX45	51
Morden Ct. Mord.	BS52	68
Morden Gdns. Grnf.	BH35	30
Morden Gdns. Mitch.	BT52	68
Morden Hall Rd. Mord.	BS52	68
Morden Hl. Clo. SE13	CF44	52
Morden Hl. SE13	CF44	52
Morden La. SE13	CF44	52
Morden Lodge, Mord.	BT52	68
Morden Rd. Ms. SE3	CH44	53
Morden Rd. Mitch.	BT52	68
Morden Rd. Rom.	CQ33	27
Morden Rd. SE3	CH44	53
Morden Rd. SW19	BS50	59
Morden St. SE13	CE44	52
Morden Way, Sutt.	BS54	68
Morden Wharf Rd. SE10	CG41	43
Mordon Rd. Ilf.	CN33	27
Mordred Rd. SE6	CG48	61
Morecambe Clo. Horn.	CU35	37
Morecambe St. SE17	BZ42	42
Morecambe Ter. N18	BZ28	15
More Clo. E16	CG39	43
More Clo. Pur.	BY55	78
More Clo. Pur.	BY59	84
More Clo. W14	BQ42	40
Morecoombe Clo.	BM50	57
Kings. On T.		
Kingston Hill		
Moree Way N18	CB28	15
Moreland Av. Dart.	CU46	55
Moreland Gdns. Sthl.	BF40	38
Moreland St. EC1	BY38	33
Moreland Way E4	CE27	16
More La. Esher	BF55	74
Morella Rd. SW12	BU47	59
Moremead Rd. SE6	CD49	61
Morena St. SE6	CE47	61
Moresby Av. Surb.	BM54	66
Moresby Rd. E5	CB33	24
Moreton Av. Islw.	BG44	47
Moreton Clo. E5	CC34	25
Moreton Clo. N15	BZ32	24
Moreton Clo. NW7	BQ29	13
Moreton Clo. Swan.	CT51	73
Moreton Gdns. Wdf. Grn.	CK28	17
Moreton Pl. SW1	BW42	41
Moreton Rd. N15	BZ32	24
Moreton Rd. S. Croy.	BZ56	78
Moreton Rd. Wor. Pk.	BP55	76
Moreton St. SW1	BW42	41
Moreton Ter. SW1	BW42	41
Morford Clo. Ruis.	BC33	20
Morford Way, Ruis.	BC33	20
Morgan Av. E17	CF31	25
Morgan Clo. RM10	CR36	36
Morgan Rd. Brom.	CG50	61
Morgan Rd. N7	BY35	33
Morgan's La. SE1	CA40	42
Morgan St. E3	CD38	34
Morgan St. E16	CG39	43
Morgan's Wk. SW11	BU43	50
Morgan Way IG8	CK29	17
Morgan Way, Rain.	CV38	37
Morie St. SW18	BS45	50
Ferrier St.		
Morieux Rd. E10	CD33	25
Moring Rd. SW17	BV49	59
Morkyns Wk. SE21	CA48	60
Morland Av. Croy.	CA54	69
Morland Clo. NW11	BS33	23
Morland Rd. E17	CC32	25
Morland Rd. Croy.	CA54	69
Morland Rd. Dag.	CR36	36
Morland Rd. Ilf.	CL34	26
Morland Rd. Sutt.	BT56	77
Morley Av. E4	CF29	16
Morley Av. N18	CB28	15
Morley Av. N22	BY30	15
Morley Clo. Orp.	CL55	80
Morley Cres. E. Stan.	BK30	12
Morley Cres. Edg.	BN27	13
Morley Cres. Ruis.	BD34	20
Morley Cres. W. Stan.	BK30	12
Morley Ho. E5	CB34	24
Morley Rd. Bark.	CM37	35
Morley Rd. Chis.	CM51	71
Morley Rd. E10	CF33	25
Morley Rd. E15	CG37	34
Morley Rd. Rom.	CQ32	27
Morley Rd. SE13	CF45	52
Morley Rd. S. Croy.	CA58	78
Morley Rd. Sutt.	BR54	67
Morley Rd. Twick.	BK46	48
Morley St. SE1	BY41	42
Morna Rd. SE5	BZ44	51
Morning La. E9	CB36	33
Morningside Est. E9	CB36	34
Morningside Rd. Wor. Pk.	BP55	76
Mornington Av. Brom.	CJ52	71
Mornington Av. Ilf.	CL33	26
Mornington Av. W14	BR42	40
Mornington Clo.	BH28	12
Wdf. Grn.		
Mornington Ct. Bex.	CS47	64
Mornington Cres. Bex.	CS47	64
Mornington Cres. Houns.	BC44	47
Mornington Cres. NW1	BW37	32
Mornington Gro. E3	CE38	34
Mornington Mews SE5	BZ44	51
County Gro.		
Mornington Pl. NW1	BW37	32
Mornington Ter.		
Mornington Rd. E4	CF26	9
Mornington Rd. E11	CG33	25
Mornington Rd. Grnf.	BF39	38
Mornington Rd. Loug.	CM24	10
Mornington Rd. SE8	CD43	52
Mornington Rd.	CG28	16
Wdf. Grn.		
Mornington St. NW1	BV37	32
Mornington Ter. NW1	BV37	32
Mornington Wk. Rich.	BK49	57
Morocco St. SE1	CA41	42
Morpeth Gro. E9	CC37	34
Morpeth St. E2	CC38	34
Morpeth Ter. SW1	BW41	41
Morrab Gdns. Ilf.	CN34	27
Morres Rd. E1	CC39	43
Morris Av. E12	CK35	35
Morris Clo. Orp.	CN55	81
Morris Gdns. SW18	BS47	59
Morrish Rd. SW2	BX47	60
Morrison Av. N17	CA31	24
Morrison Rd. Bark.	CQ37	36
Morrison Rd. Hayes	BC38	29
Morrison St. SW11	BV45	50
Morris Pl. N19	BY34	24
Morris Rd. Dag.	CQ34	27
Morris Rd. E14	CE39	43
Morris Rd. E15	CG35	34
Morris Rd. Islw.	BH45	48
Morris Rd. Rom.	CU29	19
Morris St. E1	CB39	43
Morriston Clo. Wat.	BD28	11
Morse Clo. E13	CG38	34
Morshead Rd. W9	BS38	32
Morston Gdns. SE9	CK49	62
Morten Clo. SW4	BW46	50
Morteyne Rd. N17	BZ30	15
Mortgramit Sq. SE18	CL41	44
Hare St.		
Mortham St. E15	CG37	34
Mortimer Clo. SW16	BW48	59
Mortimer Cres. NW6	BS37	32
Mortimer Cres. Wor. Pk.	BN55	76
Mortimer Est. NW6	BS37	32
Mortimer Mkt. WC1	BW38	32
Mortimer Pl. NW6	BS37	32
Mortimer Rd. E6	CK38	35
Mortimer Rd. Erith	CS43	55
Mortimer Rd. Mitch.	BU51	68
Mortimer Rd. N1	CA36	33
Mortimer Rd. NW10	BQ38	31
Mortimer Rd. Orp.	CO55	81
Mortimer Rd. W13	BK39	39
Mortimer Sq. W11	BQ40	40
St. Ann's Rd.		
Mortimer St. W1	BW39	41
Mortimer Ter. NW5	BV35	32
Mortlake Clo. Croy.	BX55	78
Mortlake High St. SW14	BN45	49
Mortlake Rd. E16	CH39	44
Mortlake Rd. Ilf.	CM35	35
Mortlake Rd. Rich.	BM43	48
Morton Cres. N14	BW28	14
Morton Gdns. Wall.	BW56	77
Morton Rd. E15	CG36	34
Morton Rd. Mord.	BT53	68
Morton Rd. N1	BZ36	33
Morton Way N14	BW27	14
Morvale Clo. Belv.	CQ42	45
Morval Rd. SW2	BY46	51
Morven Rd. SW17	BU48	59
Morville St. E3	CE37	34
Morwell St. WC1	BW39	41
Moscow Pl. W2	BS40	41
Moscow Rd.		
Moscow Rd. W2	BS40	41
Mosedale St. SE5	BZ44	51
Moselle Av. N22	BY30	15
Moselle Clo. N8	BX31	24
Moselle Pl. N17	CA29	15
High Rd.		
Moselle St. N17	CA29	15
Mospey Cres. Epsom	BO61	82
Moss Borough Clo. N12	BS29	14
Mossbury Rd. SW11	BU45	50
Moss Clo. Pnr.	BE30	11
Moss Clo. Sutt.	BU56	77
Mossdown Clo. Belv.	CR42	45
Mossford Grn. Ilf.	CL31	26
Mossford La. Ilf.	CL30	17
Mossford St. E3	CD38	34
Mosshall Cres. N12	BS29	14
Mosshall Gro. N12	BS29	14
Moss La. Pnr.	BE30	11
Moss La. Rom.	CT32	28
Mosslea Rd. Brom.	CJ53	71
Mosslea Rd. Orp.	CL55	80
Mosslea Rd. SE20	CC50	61
Mosslea Rd. Whyt.	CA61	84
Mossop St. SW3	BU42	41
Moss Rd. Dag.	CR36	36
Mossville Gdns. Mord.	BR52	67
Mostyn Av. Wem.	BL35	30
Mostyn Gdns. NW10	BQ37	31
Mostyn Gro. E3	CD37	34
Mostyn Rd. Bush.	BG25	4
Mostyn Rd. Edg.	BN29	13
Mostyn Rd. E3	CD37	34
Mostyn Rd. SW9	BY44	51
Mostyn Rd. SW19	BR51	67
Mosul Way, Brom.	CK53	71
Mosyer Dr. Orp.	CP55	81
Motcomb St. SW1	BV41	41
Motor Race Track SE19	CB50	60
Motspur Pk. N. Mal.	BO53	67
Mottingham Gdns. SE9	CJ47	62
Mottingham La. SE9	CJ47	62
Mottingham La. SE12	CJ47	62
Mottingham Rd. N9	CC26	9
Mottingham Rd. SE9	CK48	62
Mottisfont Rd. SE2	CO41	45
Mottscroft Clo. Loug.	CL25	10
Mott St. E4	CF23	9
Moulins Rd. E9	CB36	33
Lauriston Rd.		
Moulins Rd. E9	CC36	34
Moultain Hill, Swan.	CU52	73
Moulton Av. Houns.	BE44	47
Moundfield Rd. N16	CB32	24
Mound, The SE9	CL48	62
William Barefoot Dr.		
Mount Adon Pk. SE22	CB47	60
Mount Angelus Rd. SW15	BO47	58
Mount Ararat Rd. Rich.	BL46	48
Mount Ararat SW20	BQ50	58
Mount Ash Rd. SE26	CB48	60
Mount Av. E4	CE27	16
Mount Av. Sthl.	BF39	38
Mount Av. W5	BK39	39
Mountbatten Clo. SE18	CN43	54
Mountbatten Clo. SE19	CA49	60
Mountbell Rd. Stan.	BJ30	12
Mount Clo. Barn.	BV24	7
Mount Clo. Brom.	CK51	71
Mount Clo. Cars.	BV58	77
Mount Clo. Ken.	BZ57	78
Mount Clo. Ken.	BZ61	84
Mount Clo. W5	BM39	39
Castlebar Rd.		
Mountcombe Clo. Surb.	BL54	66
Mount Ct. W. Wick.	CG55	79
Mount Culver Av. Sid.	CP50	63
Mount Culver Pde. Sid.	CP50	63
Maidstone Rd.		
Mount Dr. Bexh.	CQ46	54
Mount Dr. Har.	BE32	20
Mount Dr. Wem.	BN34	22
Mountearl Gdns. SW16	BX48	60
Mount Echo Av. E4	CE26	9
Mount Echo Dr. E4	CE26	9
Mount Ephraim La. SW16	BW48	59
Mount Ephraim Rd.	BW48	59
SW16		
Mount Est. The, E5	CB34	24
Mountfield Clo. Orp.	CP52	72
Mountfield Rd. E6	CK37	35
Mountfield Rd. N3	BR31	22
Mountfield Rd. W5	BK39	39
Mountfield Way, Orp.	CP52	72
Mountford St. E1	CB39	42
Adler St.		
Mountfort Cres. N1	BY36	33
Barns Sq.		
Mount Gdns. SE26	CB48	60
Mount Gro. Edg.	BN27	13
Mount Grove Rd. N5	BY34	24
Mounthurst Rd. Brom.	CG54	70
Mountjoy Clo. SE2	CO41	45
Mount Nod. Rd. SW16	BX48	60
Mount Pk. Av. Har.	BG34	20
Mount Pk. Av. S. Croy.	BY58	78
Mount Pk. Cars.	BV57	77
Mount Park Cres. W5	BK39	39
Mount Pk. Rd. Har.	BG34	20
Mount Pk. Rd. Pnr.	BC32	20
Mount Park Rd. W5	BK39	39
Mount Pleasant, Barn.	BU24	7
Mount Pleasant Cres. N4	BX33	24
Mount Pleasant, Epsom	BO58	76
Mount Pleasant Hill E5	CB34	24
Mount Pleasant La. E5	CB33	24
Mount Pleasant NW4	BP31	22
Church End		
Mount Pleasant Rd. Chig.	CM28	17
Mount Pleasant Rd. Dart.	CW46	55
Mount Pleasant Rd. E17	CD30	16
Mount Pleasant Rd. N17	CA30	15
Mount Pleasant Rd.	BN52	67
N. Mal.		
Mount Pleasant Rd. NW10	BQ36	31
Mount Pleasant Rd. Rom.	CS29	19
Mount Pleasant Rd. SE13	CE46	52
Mount Pleasant Rd. W5	BK38	30
Mount Pleasant, Ruis.	BD34	20
Mount Pleasant Vill. N4	BX33	24
Mount Pleasant Wk. Bex.	CS46	55
Mount Pleasant WC1	BY38	33
Mount Pleasant, Wem.	BL37	30
Mount Rd. Barn.	BU25	7
Mount Rd. Bexh.	CP46	54
Mount Rd. Chess.	BL56	75
Mount Rd. Dag.	CQ33	27
Mount Rd. Dart.	CT46	55
Mount Rd. Felt.	BE48	56
Mount Rd. Hayes	BC41	38
Mount Rd. Ilf.	CL35	35
Mount Rd. Mitch.	BT51	68
Mount Rd. N. Mal.	BN52	67
Mount Rd. NW2	BP34	22
Mount Rd. NW4	BP32	22
Mount Rd. SE19	BZ50	60
Mount Rd. SW19	BS48	59
Mount Row W1	BV40	41
Mountsfield Ct. SE13	CF46	52
Mountside, Felt.	BE48	56
Mount Side, Stan.	BH30	12
Mounts Pond Rd. SE3	CF44	52
Mount Sq. The, NW3	BT34	23
Heath St.		
Mount Stewart Av. Har.	BK33	21
Mount St. W1	BV40	41
Mount, The, Couls.	BV61	83
Mount, The, Epsom	BO58	76
Mount, The, Epsom	BP56	76
Mount, The, Esher	BF57	74
Mount, The, N. Mal.	BO52	67
Mount, The N20	BT27	14
Mount, The NW3	BT34	23
Heath St.		
Mount, The, Rom.	CV27	19
Mount, The, Wem.	BN34	22
Mount Vernon NW3	BT35	32
Mountview Ct. N8	BY31	24
Mount View NW7	BN27	13
Mount View Rd. E4	CF26	9
Mount View Rd. Esher	BJ57	75
Mount View Rd. N4	BX33	24
Mount View Rd. NW9	BN32	22
Mountview Rd. Orp.	CO54	72
Mount View W5	BK38	30
Mount Vill. SE27	BY48	60
Canterbury Gro.		
Mount Way, Cars.	BV58	77
Mountwood Clo. S.	CB58	78
Croy.		
Mountwood, E. Mol.	BF52	65
Mover's La. Bark.	CM37	35
Move St. E2	CB37	33
Mowatt Clo. N19	BW33	23
Mowbray Rd. Barn.	BT24	7
Mowbray Rd. Edg.	BM28	12
Mowbray Rd. NW6	BR36	31
Mowbray Rd. Rich.	BK48	57
Mowbray Rd. SE19	CA51	69
Mowbrays Clo. Rom.	CS30	19
Mowbrays Rd. Rom.	CS30	19
Mowbrey Gdns. Loug.	CM23	10
Mowlem St. E2	CB37	33
Mowll St. SW9	BY43	51
Moxon Clo. E13	CG37	34
Whitelegg Rd.		
Moxon St. Barn.	BR24	6
Moxon St. W1	BV39	41
Moyers Rd. E10	CF33	25
Moylan Rd. W6	BR43	49
Moyne Pl. NW10	BM37	30
Moyser Rd. SW16	BV49	59
Mozart Sq. SW1	BV42	41
Ebury Br. Rd.		
Mozart St. W10	BR38	31
Muchelney Rd. Mord.	BT53	68
Mud La. W5	BK39	39
Muggeridge Rd. Dag.	CR35	36
Muirdown Av. SW14	BN45	49
Muirfield Clo. Wat.	BD28	11
Muirfield Grn. Wat.	BC28	11
Muirfield Rd. Wat.	BC28	11
Muirfield W3	BO39	40
Muirkirk Rd. SE6	CF47	61
Muir Rd. E5	CB35	33
Muir St. E16	CK40	44
Mulbanton Ct. Chis.	CM49	62
Mulberry Clo. E4	CE27	16
Mulberry Clo. Nthlt.	BD37	29
Mulberry Clo. NW3	BT35	32
Hampstead High St.		
Mulberry Clo. NW4	BQ31	22
Mulberry Clo. SE7	CJ43	53
Mulberry Cres. Brent.	BJ43	48
Mulberry Dr. Grays	CW42	46
Mulberry Gdns. NW4	BQ31	22

Mulberry La. Croy.	CA54	69
Mulberry St. E1	CB39	42
Mulberry Wk. SW3	BT43	50
Mulberry Way, Belv.	CS41	46
Mulberry Way, Ilf.	CM31	26
Mulbery Rd. E18	CH30	17
Mulgrave Rd. Croy.	BZ55	78
Mulgrave Rd. Har.	BJ34	21
Mulgrave Rd. NW10	BO35	31
Mulgrave Rd. Sutt.	BR57	76
Mulgrave Rd. SW6	BR43	49
Mulgrave Rd. W5	BK38	30
Mulholland Clo. Mitch.	BV51	68
Mulkern Rd. N19	BW33	23
Muller Rd. SW4	BW46	50
Mullet Gdns. E2	CB38	33
St. Peters Clo.		
Mullins Path SW14	BN45	49
North Worple Way		
Mullion Clo. Har.	BF30	11
Mullion Wk. Wat.	BD28	11
Mull Wk. N1	BZ36	33
Marquess Est.		
Mulready St. NW8	BU38	32
Multi-way W3	BO41	40
Valetta Rd.		
Multon Rd. SW18	BT47	59
Mumford Ct. EC2	BZ39	42
Milk St.		
Muncaster Rd. SW11	BU45	50
Mundania Rd. SE22	CB46	51
Munday Rd. E16	CH40	44
Munden Pl. W14	BR42	40
Munden St.		
Munden St. W14	BR42	40
Mundford Rd. E5	CC34	25
Mundon Gdns. Ilf.	CM33	26
Mund St. W14	BR42	40
Mundy St. N1	CA38	33
Hoxton Sq.		
Mungo Pk. Clo. Bush.	BG27	11
Mungo Pk. Rd. Rain.	CU35	37
Mungo Pk. Way, Orp.	CP54	72
Munnings Gdns. Islw.	BG46	47
Munroe Rd. Bush.	BF24	4
Munro Gdns. Har.	BH29	12
Munro Ms. W10	BR39	40
Munster Av. Houns.	BE45	47
Munster Gdns. N13	BY28	15
Munster Rd. SW6	BR43	49
Munster Rd. Tedd.	BJ50	57
Munster Sq. NW1	BV38	32
Munton Rd. SE17	BZ42	42
Murchison Av. Bex.	CP47	63
Murchison Rd. E10	CF34	25
Murdock Cotts. E3	CD38	34
Clinton Rd.		
Murdock St. SE15	CB43	51
Murfett Clo. SW19	BR48	58
Muriel Av. Wat.	BD25	4
Muriel St. N1	BX37	33
Murillo Rd. SE13	CF45	52
Murphy St. SE1	BY41	42
Murray Av. Brom.	CH52	71
Murray Av. Houns.	BF46	47
Murray Clo. Pnr.	BD30	11
Murray Gro. N1	BZ37	33
Murray Ms. NW1	BW36	32
Murray Rd. Orp.	CO52	72
Murray Rd. Rich.	BJ48	57
Murray Rd. SW19	BQ50	58
Murray Rd. W5	BK42	39
Murray Sq. E16	CH39	44
Murray St. NW1	BW36	32
Murrays Yd. SE18	CL42	44
Powis St.		
Murray Ter. NW3	BT35	32
Flask Wk.		
Murton St. EC1	BZ38	33
Lever St.		
Murtwell Dr. Chig.	CM29	17
Musard Rd. W6	BR43	49
Musbery St. E1	CC39	43
Muscatel Pl. SE5	CA44	51
Dalwood St.		
Muschamp Rd. Cars.	BT55	77
Muschamp Rd. SE15	CA45	51
Muscovy St. EC3	CA40	42
Seething La.		
Museum St. WC1	BX39	42
Musgrave Clo. Barn.	BT23	7
Musgrave Cres. SW6	BS43	50
Musgrave Rd. Islw.	BH44	48
Musgrove Rd. SE14	CC44	52
Musjid Rd. SW11	BT44	50
Musquash Way, Houns.	BD44	47
Muston Rd. E5	CB34	24
Muswell Av. N10	BV30	14
Muswell Hill, Broadway N10	BV31	23
Muswell Hill Est. N10	BU30	14
Muswell Hill Pl. N10	BV31	23
Muswell Hill N10	BV31	23
Muswell Hill Rd. N6	BV32	23
Muswell Hill Rd. N10	BV32	23
Muswell Mews N10	BV31	23
Muswell Rd.		
Muswell Rd. N10	BV31	23
Mutrix Rd. NW6	BS37	32
Muybridge Rd. N. Mal.	BN51	67
Mycenae Rd. SE3	CH43	53
Mychurch Lane SE17	CA42	42
Myddelton Gdns. N21	BY26	8
Myddelton Pk. N20	BT27	14
Myddelton Pass. EC1	BY38	33
Myddelton Rd. N8	BX31	24
Myddelton Rd. N22	BX29	15
Myddelton Sq. EC1	BY38	33
Myddelton St. EC1	BY38	33
Mygrove Clo. Rain.	CV37	37
Mygrove Gdns. Rain.	CV37	37
Mygrove Rd. Rain.	CV37	37
Mylis Clo. SE26	CB49	60
Myra St. SE2	CO42	45
Myrdle St. E1	CB39	42
Myron Pl. SE13	CF45	52
Myrtle All. SE18	CL41	44
Hare St.		
Myrtle Av. Ruis.	BC33	20
Myrtle Clo. Barn.	BU26	7
Myrtle Clo. Erith	CT44	55
Myrtledene Rd. SE2	CO42	45
Myrtle Gdns. W7	BH40	39
Myrtle Gro. N. Mal.	BN51	67
Myrtle Rd. Croy.	CE55	79
Myrtle Rd. Dart.	CV47	64
Myrtle Rd. E6	CK37	35
Myrtle Rd. E17	CD32	25
Myrtle Rd. Hamptn.	BG50	56
Myrtle Rd. Houns.	BF44	47
Myrtle Rd. Ilf.	CL34	26
Myrtle Rd. N13	BZ27	15
Myrtle Rd. Rom.	CV29	19
Myrtle Rd. Sutt.	BT56	77
Myrtle Rd. W3	BN40	40
Mysore Rd. SW11	BU45	50
Myton Rd. SE21	BZ48	60
M1 Motorway, Edg.	BL26	5
M1 Motorway NW4	BN28	13
M1 Motorway NW7	BN28	13
M1 Motorway, Wat.	BF23	4
M4 Hayes	BC42	38
Nadine St. SE7	CJ42	44
Nagle Clo. E17	CF30	16
Nag's Head Est. E2	CA37	33
Nags Head La. Well.	CO45	54
Nag's Head Rd. Enf.	CC24	9
Nairne Gro. SE24	BZ45	51
Nairn Grn. Wat.	BC27	11
Nairn Rd. Ruis.	BD35	29
Nallhead Rd. Felt.	BD49	56
Nampton Dr. Th. Hth.	BX52	69
Nan-Clark's La. NW7	BO27	13
Nancy Downs, Wat.	BD26	4
Nankin St. E14	CE39	43
Nansen Rd. SW11	BV45	50
Nantes Clo. SW18	BT45	50
Nantes Pass. E1	CA39	42
Lamb St.		
Nant Rd. NW2	BR34	22
Nant St. E2	CB38	33
Paradise Row		
Napier Av. SW6	BR45	49
Napier Clo. W14	BR41	40
Napier Rd.		
Napier Ct. SW6	BR45	49
Napier Dr. Bush.	BE24	4
Napier Gro. N1	BZ37	33
Napier Pl. W14	BR41	40
Napier Rd. Belv.	CQ42	45
Napier Rd. Brom.	CH52	71
Napier Rd. Enf.	CC25	9
Napier Rd. E6	CL37	35
Napier Rd. E11	CG35	34
Napier Rd. E15	CG37	34
Napier Rd. Islw.	BJ45	48
Napier Rd. N17	CA31	24
Napier Rd. NW10	BP38	31
Napier Rd. SE25	CB52	69
Napier Rd. S. Croy.	BZ57	78
Napier Rd. W14	BR41	40
Napier Rd. Wem.	BK35	30
Napier St. SE8	CD43	52
Napier Ter. N1	BY36	33
Napoleon Rd. Twick.	BJ47	57
Narbonne Av. SW4	BW46	50
Narborough St. SW6	BS44	50
Narcissus Rd. NW6	BS35	32
Narford Rd. E5	CB34	24
Narrow St. E14	CD40	43
Narrow St. W3	BM40	39
Steyne Rd.		
Narrow Way, Brom.	CK53	71
Nasby Clo. Islw.	BH44	48
Nascot Pl. Wat.	BC23	4
Nascot Rd. Wat.	BC23	4
Nascot St. W12	BQ39	40
Nascot St. Wat.	BC23	4
Naseby Clo. NW6	BT36	32
Naseby Ct. KT12	BD55	74
Naseby Rd. Dag.	CR34	27
Naseby Rd. Ilf.	CK30	17
Naseby Rd. SE19	BZ50	60
Nash Grn. Brom.	CH50	62
Nash La. Kes.	CH57	80
Nash Rd. N9	CC27	16
Nash Rd. Rom.	CP31	27
Nash Rd. SE4	CD45	52
Nash St. NW1	BV38	32
Nasmyth St. W6	BP41	40
Nassau Path SE28	CP40	45
Disraeli Clo.		
Nassau Rd. SW13	BO44	49
Nassau St. W1	BW39	41
Nassington Rd. NW3	BU35	32
Natal Rd. Ilf.	CL35	35
Natal Rd. N11	BW29	14
Natal Rd. SW16	BW50	59
Natal Rd. Th. Hth.	BZ52	69
Nathans Rd. Wem.	BK33	21
Nathan Way SE28	CM42	44
Naval Row E14	CF40	43
Navarino Gro. E8	CB36	33
Navarino Rd. E8	CB36	33
Navarre Gdns. Rom.	CR28	18
Navarre Rd. E6	CK37	35
Navarre St. E2	CA38	33
Navestock Cres. Wdf. Grn.	CJ29	17
Navy St. SW4	BW45	50
Naylor Rd. N20	BT27	14
Naylor Rd. SE15	CB43	51
Nazeing Wk. Rain.	CT37	37
Neagle Clo. Borwd.	BN23	6
Neal Av. Sthl.	BE38	29
Nealden St. SW9	BX45	51
Neale Clo. N2	BT31	23
Neals Rd. Erith	CS43	55
Brook St.		
Neal St. WC2	BX39	42
Neal St. Wat.	BD25	4
Near Acre NW9	BO30	13
Neasden Clo. NW10	BO35	31
Neasden La. NW10	BN34	22
Neasham Rd. Dag.	CO35	36
Neate St. SE5	CA43	51
Neath Gdns. Mord.	BT53	68
Neathouse Pl. SW1	BW42	41
Vauxhall Bridge Rd.		
Neave Cres. Rom.	CV30	19
Nebraska St. SE1	BZ41	42
Neckinger Est. SE16	CA41	42
Neckinger St. SE1	CA41	42
Nectarine Way SE13	CE44	52
Needham Rd. W11	BS39	41
Artesian Rd.		
Needham Ter. NW2	BQ34	22
Neeld Cres. NW4	BP32	22
Neeld Cres. Wem.	BM35	30
Nelgarde Rd. SE6	CE47	61
Nella Rd. W6	BQ43	49
Nelmes Clo. Horn.	CW32	28
Nelmes Cres. Horn.	CW32	28
Nelmes Rd. Horn.	CW33	28
Nelmes Way, Horn.	CV31	28
Nelson Clo. Croy.	BY54	69
Nelson Clo. Rom.	CR30	18
Nelson Clo. Walt.	BC54	65
Nelson Gdns. E2	CB38	33
Bethnal Green Rd.		
Nelson Gdns. Houns.	BF46	47
Nelson Gro. Rd. SW19	BT51	68
Nelson Pl. N1	BY37	33
Nelson Pl. NW6	BS38	32
Percy Rd.		
Nelson Pl. Sid.	CO49	63
Nelson Pl. W3	BM40	39
Steyne Rd.		
Nelson Rd. Belv.	CQ42	45
Nelson Rd. Brom.	CJ52	71
Nelson Rd. Dart.	CV46	55
Nelson Rd. E4	CE29	16
Nelson Rd. E11	CH31	26
Nelson Rd. Enf.	CC25	9
Nelson Rd. Har.	BG33	20
Nelson Rd. Houns.	BF46	47
Nelson Rd. N8	BX32	24
Nelson Rd. N9	CB27	15
Nelson Rd. N15	CA31	24
Nelson Rd. N. Mal.	BN53	67
Nelson Rd. Rain.	CT37	37
Nelson Rd. SE10	CF43	52
Nelson Rd. Sid.	CO49	63
Nelson Rd. Stan.	BK29	12
Nelson Rd. SW19	BS50	59
Nelson Sq. SE1	BY41	42
Nelson's Row SW4	BW45	50
Nelson St. E1	CB39	42
Nelson St. E6	CK37	35
Nelson St. E16	CG40	43
Nelwyn Av. Horn.	CW32	28
Nemoure Rd. W3	BN40	40
Nepaul Rd. SW11	BU44	50
Nepean St. SW15	BP46	49
Neptune Rd. Har.	BG32	20
Neptune St. SE16	CC41	43
Nesbit Clo. SE3	CG45	52
Hurren Clo.		
Nesbit Rd. SE9	CJ45	53
Ness Rd. Erith	CV43	55
Ness St. SE16	CB41	42
Spa Rd.		
Nesta Rd. Wdf. Grn.	CG29	16
Neston St. SE16	CC41	43
Nestor Av. N21	BY25	8
Netheravon Rd. S. W4	BO42	40
Netheravon Rd. W4	BO42	40
Netheravon Rd. W7	BH40	39
Netherbury Rd. W5	BK41	39
Netherby Gdns. Enf.	BX24	8
Netherby Rd. SE23	CC47	61
Nether Clo. N3	BS29	14
Nethercourt Av. N3	BS29	14
Netherfield Gdns. Bark.	CM36	35
Netherfield Rd. N12	BS28	14
Netherfield St. W11	BR40	40
Portland Rd.		
Netherford Rd. SW4	BW44	50
Netherhall Gdns. NW3	BT36	32
Netherhall Way NW3	BT35	32
Netherhall Gdns.		
Netherland Rd. Barn.	BT25	7
Netherland Rd. SW17	BV48	59
Netherpark Dr. Rom.	CT30	19
Nether St. N3	BS30	14
Nether St. N12	BS29	14
Netherton Gro. SW10	BT43	50
Netherton Rd. Twick.	BJ46	48
Netherwood Rd. W14	BQ41	40
Netherwood St. Est. NW6	BR36	31
Netherwood St. NW6	BR36	31
Netley Clo. Croy.	CF57	79
Netley Clo. Sutt.	BQ56	76
Netley Dr. Walt.	BE54	65
Netley Gdns. Mord.	BT54	68
Netley Rd. Brent.	BL43	48
Netley Rd. E17	CD32	25
Netley Rd. Ilf.	CM32	26
Netley Rd. Mord.	BT54	68
Nettlecombe Clo. Sutt.	BS58	77
Nettleden Av. Wem.	BM36	30
Nettlefold Pl. SE27	BY48	60
Nettlestead Clo. Beck.	CD50	61
Nettleton Rd. SE14	CC44	52
Nettlewood Rd. SW16	BW50	59
Neuchatel Rd. SE6	CD48	61
Nevada St. SE10	CF43	52
Nevern Pl. SW5	BS42	41
Nevern Rd. SW5	BS42	41
Nevern Sq. SW5	BS42	41
Neville Av. N. Mal.	BN51	67
Neville Clo. E11	CG34	25
Neville Clo. Esher	BE57	74
Neville Clo. NW1	BW37	32
Neville Clo. NW6	BR37	31
Neville Clo. Sid.	CN49	63
Neville Clo. W3	BN41	40
Acton La.		
Neville Dr. N2	BT32	23
Neville Dr. SE15	CB44	51
Neville Gdns. Dag.	CP34	27
Neville Rd.		
Neville Gill Clo. SW18	BS46	50
Buckhold Rd.		
Neville Pl. NW6	BR37	31
Neville Rd. Croy.	BZ54	69
Neville Rd. Dag.	CP34	27
Neville Rd. E7	CH36	35
Neville Rd. Ilf.	CM30	17
Neville Rd. Kings. On T.	BM51	66
Neville Rd. NW6	BR37	31
Neville Rd. Rich.	BK48	57
Neville Rd. W5	BK38	30
Neville's Ct. NW2	BP34	22
Neville St. SW7	BT42	41
Neville Ter. SW7	BT42	41
Neville Wk. Cars.	BU54	68
Green Wrythe La.		
Nevill Gro. Wat.	BC23	4
Nevill Rd. N16	CA34	24
Nevill Way, Loug.	CK26	10
Nevin Dr. E4	CE26	9
Nevis Clo. Rom.	CT29	19
Nevis Rd. SW17	BV48	59
Newark Ct. KT12	BD54	65
Stratton Clo.		
Newark Cres. NW10	BN38	31
Newark Grn. Borwd.	BN24	6
Newark Rd. S. Croy.	BZ57	78
Newark St. E1	CB39	42
Newark Way NW4	BP31	22
New Barn La. Whyt.	CA61	84
New Barn Rd. Swan.	CT51	73
New Barns Av. Mitch.	BW52	68
New Barn St. E13	CH39	44
New Barns Way, Chig.	CL27	17
Newberry Est. N1	BZ36	33
Newbery Rd. Erith	CT44	55
Newbiggin Path, Wat.	BD28	11
Newbold Rd. SE15	CB44	51
Asylum Rd.		
Newbolt Av. Sutt.	BQ56	76
Newbolt Rd. Stan.	BH28	12
New Bond St. W1	BV39	41
Newborough Grn. N. Mal.	BN52	67
New Bowyer Pl. SE5	BZ43	51
New Brent St. NW4	BQ32	22
New Bridge, Erith	CS42	46
New Bridge St. EC4	BY39	42
New Broad St. EC2	BZ39	42
New Broadway W5	BK40	39
Newburgh Rd. W3	BN40	40

Newburgh St. W1 BW39 41
Foubert's Pl.
New Burlington Pl. W1 BW40 41
Savile Row
New Burlington St. W1 BW40 41
Newburn St. SE11 BX42 42
Newbury Clo. Nthlt. BE36 29
Newbury Clo. Rom. CV29 19
Newbury Ct. E11 CH31 26
Newbury Gdns. Epsom BO56 76
Newbury Gdns. Rom. CV29 19
Newbury Gdns. Upmin. CW34 28
Newbury Rd. Brom. CH52 71
Newbury Rd. E4 CF29 16
Newbury Rd. Ilf. CN32 27
Newbury Rd. Rom. CV28 19
Newbury St. EC1 BZ39 42
Newbury Wk. Rom. CV28 19
Newbury Way, Nthlt. BE36 29
New Butt La. SE8 CE43 52
Newby Clo. Enf. CA23 8
Newby Pl. E14 CF39 43
Newby St. SW8 BV44 50
Newcastle Av. Ilf. CO29 18
Newcastle Ct. EC4 BY39 42
Farringdon St.
Newcastle Pl. W2 BT39 41
Newcastle St. W8 BS41 41
New Cavendish St. W1 BV39 41
New Change EC4 BZ39 42
New Chapel Rd. Felt. BC47 56
New Church Rd. SE5 BZ43 51
New City Rd. E13 CJ38 35
New Clo. Est. Mitch. BT51 68
New Clo. Felt. BE49 56
New Clo. SW19 BT51 68
Newcombe Pk. NW7 BO28 13
Newcombe Pk. Wem. BL37 30
Newcombe St. W8 BS40 41
Newcomen Rd. E11 CG34 25
Newcomen Rd. SW11 BT45 50
Newcomen St. SE1 BZ41 42
New Compton St. WC2 BW39 41
New Ct. Dart. CW46 55
Newcourt St. NW8 BU37 32
New Coventry St. W1 BW40 41
Coventry St.
New Cross Gate SE14 CC44 52
New Cross Rd. SE14 CC43 52
New Dale Clo. N9 CB27 15
Balham Rd.
Newdene Av. Nthlt. BD37 29
Newell St. E14 CD39 43
New End NW3 BT35 32
New End Sq. NW3 BT35 32
Newent Clo. Cars. BU54 68
New Fetter La. EC4 BY39 42
Newfield Clo. Hamptn. BF51 65
Newfield Ri. NW2 BP34 22
New Fm. Av. Brom. CH52 71
New Forest La. Chig. CL29 17
Newgale Gdns. Edg. BL30 12
Newgate Clo. Felt. BE48 56
Newgate, Croy. BZ54 69
Newgate St. E4 CG27 16
Newgate St. EC1 BY39 42
New Goulston St. E1 CA39 42
Middlesex St.
New Hall Dr. Rom. CW30 19
Newhams Clo. Brom. CK52 71
Newhams Row SE1 CA41 42
Newham Way E16 CJ39 44
Newham Way, Har. BL31 21
Newhaven Gdns. SE9 CJ45 53
Newhaven Rd. SE25 BZ53 69
New Heston Rd. Houns. BE43 47
New Hill Rd. Orp. CK58 80
Newhouse Av. Rom. CP31 27
Newhouse Clo. N. Mal. BQ54 67
Newhouse Way, Mord. BT54 68
Newick Clo. Bex. CR46 54
Newick Rd. E5 CB34 24
Newing Grn. Brom. CJ50 62
Newington Barrow Way N7 BX34 24
Andover Est.
Newington Butts SE11 BY42 42
Newington Causeway SE1 BY41 42
Newington Grn. N16 BZ35 33
Newington Grn. Rd. N1 BZ35 33
Newington Way EC1 BX34 24
New Inn Broadway EC2 CA38 33
New Inn Yd.
New Inn Yd. EC2 CA38 33
New James Ct. SE15 CB45 51
New James St. SE15 CB45 51
Scylla Rd.
New Kent Rd. SE1 BZ41 42
New King's Rd. SW6 BR44 49
New King St. SE8 CE43 52
Newland Clo. Pnr. BE29 11
Newland Gdns. W13 BJ41 39
Newland Rd. N8 BX31 24
Newlands Av. Surb. BH54 66
Newlands Clo. Edg. BL27 12
Newlands Clo. Sthl. BE42 38
Newlands Clo. Walt. BE56 74
Newlands Clo. Wem. BK36 30
Newlands Ct. Wem. BM34 21
Newlands Dr. Enf. CB23 8
Newlands Est. SW17 BV49 59
Newlands Pk. SE26 CC50 61
Newlands Pl. Barn. BQ25 6
Newlands Rd. SW16 BX51 69
Newlands Rd. Wdf. Grn. CG27 16
Newlands, The, Wall. BW57 77
Newland St. E16 CK40 44
Newlands Way, Chess. BK56 75
Newlands Wd. Croy. CD58 79
Newling Est. E2 CA38 33
New London St. EC3 CA40 42
Hart St.
Newlyn Gdns. Har. BE33 20
Newlyn Rd. Barn. BR24 6
Newlyn Rd. N17 CA30 15
Newlyn Rd. NW2 BQ33 22
Newlyn Rd. Well. CN44 54
Newman Clo. Horn. CW32 28
Newman Pass. W1 BW39 41
Newman St.
Newman Rd. Brom. CH51 71
Newman Rd. Croy. BX54 69
Newman Rd. E13 CH38 35
Newman Rd. E17 CC32 25
Newman Rd. Hayes BC40 38
Newmans Clo. Loug. CL24 10
Newmans La. Loug. CL24 10
Newman's Row WC2 BV39 41
Gt. Turnstile
Newman St. W1 BW39 41
Newmans Way, Barn. BT23 7
Newmarket Av. Nthlt. BF35 29
Newmarket Grn. SE9 CJ47 62
Newmarket Way, Horn. CW35 37
New Martan St. E1 CA40 42
New Meadows Path, Rich. BM44 48
Townmead Rd.
Newminster Rd. Mord. BT53 68
New Mount St. E15 CF36 34
Newnham Av. Ruis. BD33 20
Newnham Clo. Loug. CJ25 10
Newnham Clo. Nthlt. BG35 29
Newnham Gdns. Nthlt. BG36 29
Newnham Rd. N22 BX30 15
Newnham Ter. SE1 BY41 42
New North Pl. EC2 CA38 33
Luke St.
New North Rd. Ilf. CM29 17
New North Rd. N1 BZ36 33
New North St. WC1 BX39 42
Newnton Clo. N4 BZ33 24
New Oak Rd. N2 BT30 14
New Orleans Walk N19 BW33 23
New Oxford St. WC1 BW39 41
New Park Av. N13 BY27 15
New Park Clo. Nthlt. BE36 29
Arnold Rd.
New Park Ct. SW2 BX47 60
New Park Rd. SW2 BW47 59
Newpiece, Loug. CL24 10
New Pl. Sq. SE16 CB41 42
Southwark Pk. Rd.
New Plaistow Rd. E15 CG37 34
Newport Av. E13 CH38 35
Palmer Rd.
Newport Ct. WC2 BW40 41
Charing Cross Rd.
Newport Mead, Wat. BD28 11
Newport Pl. WC2 BW40 41
Shaftesbury Av.
Newport Rd. E10 CF34 25
Newport Rd. E17 CD31 25
Newport Rd. SW13 BP44 49
Newports, Swan. CS54 73
Newport St. SE11 BX42 42
Newport St. WC2 BW41 41
Charing Cross Rd.
Newquay Cres. Har. BE34 20
Newquay Gdns. Wat. BC27 11
Newquay Rd. SE6 CE48 61
New Quebec St. W1 BU39 41
New River Ct. N5 BZ35 33
New River Cres. N13 BY28 15
New River Gdns. N22 BY29 15
New River Wk. N1 BZ36 33
New Rd. Borwd. BK25 5
High St.
New Rd. Brent. BK43 48
New Rd. Dag. CR37 36
New Rd. E. Mol. BF52 65
New Rd. Esher BG55 74
New Rd. E1 CB39 42
New Rd. E4 CE28 16
New Rd. E16 CG39 43
New Rd. Felt. BC47 56
New Rd. Felt. BE49 56
New Rd. Har. BH35 30
New Rd. Houns. BF45 47
New Rd. Ilf. CN34 27
New Rd. Kings. On T. BM50 57
New Rd. Leath. BH58 75
New Rd. Mord. BU54 68
New Rd. N8 BX32 24
New Rd. N9 CB27 15
New Rd. N17 CA30 15
New Rd. N22 BZ30 15
New Rd. NW7 BO26 6
New Rd. NW7 BR29 13
New Rd. Orp. CO54 72
New Rd. Rain. CS37 37
New Rd. Rich. BK49 57
New Rd. SE2 CP42 45
New Rd. Swan. CT50 64
New Rd. Swan. CT52 73
New Rd. Wat. BD24 4
New Rd. Wat. BH23 5
New Rd. Well. CO44 54
New Row WC2 BX40 42
Newry Rd. Twick. BJ46 48
Newsam Av. N15 BZ32 24
New Scotland Yd. SW1 BX41 42
Derby Gate
New Spring Gdns. Brent. BK43 48
Albany Rd.
New Sq. WC2 BX39 42
Newstead Av. Orp. CM55 80
Newstead Rd. SE12 CG47 61
Newstead Wk. Cars. BT54 68
Newstead Way SW19 BQ49 58
New St. EC2 CA39 42
New St. Hill, Brom. CH49 62
New St. Hill EC4 BY39 42
(Lit. New St.) •
New St. Sq. EC4 BY39 42
New St. Wat. BD24 4
Church St.
Newton Av. N10 BV30 14
Newton Av. W3 BN41 40
Newton Gro. N1 BZ37 33
Northport St.
Newton Gro. W4 BO41 40
Newton Rd. Chig. CO28 18
Newton Rd. E15 CF35 34
Newton Rd. Har. BH30 12
Newton Rd. Islw. BH44 48
Newton Rd. N15 CA32 24
Newton Rd. NW2 BQ35 31
Newton Rd. Pur. BW59 83
Newton Rd. SW19 BR50 58
Newton Rd. W2 BS39 41
Newton Rd. Well. CO45 54
Newton Rd. Wem. BL36 30
Newtons Clo. Rain. CT36 37
Newton St. WC2 BX39 42
Newton's Yd. SW18 BS46 50
Wandsworth High St.
Newton Way N18 BZ28 15
Newtown St. SW11 BV44 50
New Trinity Rd. N2 BT31 23
New Turnstile WC1 BX39 42
High Holborn
New Union Clo. E14 CF41 43
New Wanstead E11 CG32 25
New Way Rd. NW9 BO31 22
New Wharf Rd. N1 BX37 33
New Zealand Way, Rain. CT38 37
New Zealand Way W12 BP40 40
Niagara Av. W5 BK42 39
Nibthwaite Rd. Har. BH32 21
Nicholas Clo. Grnf. BF37 29
Nicholas Gdns. W5 BK40 39
Nicholas La. EC4 BZ40 42
Nicholas Rd. Borwd. BL25 5
Nicholas Rd. Croy. BX56 78
Nicholas Rd. Dag. CQ34 27
Nicholas Rd. E1 CC38 34
Nicholas Rd. Houns. BF45 47
Nicholay Rd. N19 BW33 23
Nichol Clo. N14 BW26 7
Nichol La. Brom. CH50 62
Nichols Grn. W5 BK39 39
Montpelier Rd.
Nicholson Dr. Bush. BG26 4
Nicholson Rd. Croy. CA54 69
Nicholson St. SE1 BY40 42
Nicola Clo. Har. BG30 11
Nicola Clo. S. Croy. BZ57 78
Nicol Clo. Twick. BJ46 48
Cassilis Rd.
Nicoll Rd. NW10 BO37 31
Nichollsfield Wk. N19 BX35 33
Nicoll Way, Borwd. BN24 6
Nicol Pl. NW4 BP32 22
Nicolson Rd. Orp. BP54 72
Nicosia Rd. SW18 BU47 59
Niederwald Rd. SE26 CD49 61
Nigel Clo. Nthlt. BE37 29
Nigel Ms. Ilf. CL35 35
Nigel Playfair Av. W6 BP42 40
Nigel Rd. E7 CJ35 35
Nigel Rd. SE15 CB45 51
Nigeria Rd. SE7 CJ43 53
Nightingale Av. E4 CG28 16
Nightingale Clo. Cars. BV55 77
Nightingale Clo. E4 CF28 16
Nightingale Clo. W4 BN43 49
Nightingale Cres. SW11 BU46 50
Blenkarne Rd.
Nightingale Dr. Epsom BM57 75
Nightingale Gro. SE13 CF46 52
Nightingale La. Brom. CJ51 71
Nightingale La. E11 CH32 26
Nightingale La. N6 BU33 23
Nightingale La. N8 BX31 24
Nightingale La. Rich. BL47 57
Nightingale La. SW4 BV46 50
Nightingale La. SW12 BV47 59
Nightingale Pl. SE18 CL43 53
Nightingale Rd. Bush. BF25 4
Nightingale Rd. Cars. BU55 77
Nightingale Rd. E5 CB34 24
Nightingale Rd. E. Mol. BF53 65
Nightingale Rd. Esher BE56 74
Nightingale Rd. Hamptn. BF49 56
Nightingale Rd. N9 CC26 9
Nightingale Rd. N22 BX30 15
Nightingale Rd. NW10 BO37 31
Nightingale Rd. Orp. CM53 71
Nightingale Rd. S. Croy. CC58 79
Nightingale Rd. W7 BH40 39
Nightingale Sq. SW12 BV47 59
Nightingale Vale SE18 CL43 53
Nightingale Wk. SW4 BV46 50
Nijmegen Way SE22 CA46 51
Dulwich Gro.
Nile Rd. E13 CJ37 35
Nile St. N1 BZ38 33
Nile Ter. SE15 CA42 42
Nimbus Rd. Epsom BN58 76
Nimmo Dr. Bush. BG26 4
Nimrod Clo. Grnf. BE38 29
Ruislip Rd.
Nimrod Rd. SW17 BV50 59
Nine Acres Clo. E12 CK35 35
Nine Elms La. SW8 BW43 50
Nine Elms St. SW8 BW43 50
Ninhams Wd. Orp. CK56 80
Ninth Av. Hayes BC40 38
Nisbet Ho. E9 CC35 34
Nithdale Rd. SE18 CL43 53
Niton Clo. Barn. BQ25 6
Niton Rd. Rich. BM45 48
Niton St. SW6 BQ43 49
Noak Hill Rd. Rom. CV28 19
Noble Rd. N18 CC28 16
Noble St. EC2 BZ39 42
Noel Park Rd. N22 BY30 15
Noel Rd. E6 CK38 35
Noel Rd. N1 BY37 33
Noel Rd. W3 BM40 39
Noel Sq. Dag. CP35 36
Noel St. W1 BW39 41
Nolan Way E5 CB35 33
Nolton Pl. Edg. BL30 12
Nonsuch Clo. Ilf. CL29 17
Nonsuch Court Av. Epsom BP58 76
Nonsuch Wk. Sutt. BQ58 76
Nora Gdns. NW4 BQ31 22
Nora Ter. Har. BH33 21
Norbiton Av. Kings. On T. BM51 66
Norbiton Common Rd. N. Mal. BM52 66
Nobiton Hall, Kings. On T. BL51 66
Norbiton Rd. E14 CD39 43
Norbreck Gdns. NW10 BL38 30
Norbreck Pde. W5 BL38 30
Norbroke St. W12 BO40 40
Norburn St. W10 BR39 40
Chesterton Rd.
Norbury Av. Houns. BG45 47
Norbury Av. SW16 BX51 69
Norbury Av. Th. Hth. BY51 69
Norbury Clo. SW16 BY51 69
Norbury Ct. E5 CC35 34
Clapton Park Est.
Norbury Ct. Rd. SW16 BX52 69
Norbury Cres. SW16 BX51 69
Norbury Cross SW16 BX52 69
Norbury Gdns. Rom. CP32 27
Norbury Gro. NW7 BO27 13
Norbury Hl. SW16 BY50 60
Norbury Ms. SW16 BX51 69
Norbury Cres.
Norbury Ri. SW16 BX52 69
Norbury Rd. E4 CE28 16
Norbury Rd. Th. Hth. BZ51 69
Norcombe Gdns. Har. BK32 21
Norcott Clo. Hayes BD38 29
Norcott Rd. N16 CB34 24
Norcroft Gdns. SE22 CB47 60
Norcutt Rd. Twick. BH47 57
Nordenfeldt Rd. Erith CS42 46
Norfield Rd. Bex. CS49 64
Norfolk Av. N13 BY29 15
Norfolk Av. N15 CA32 24
Norfolk Av. S. Croy. CA58 78
Norfolk Clo. Barn. BV24 7
Norfolk Clo. N2 BU31 23
Park Rd.
Norfolk Clo. N13 BY29 15
Norfolk Clo. Twick. BJ46 48
Norfolk Cres. Sid. CN47 63
Norfolk Cres. W2 BU39 41
Norfolk Est. E1 CC38 34
Norfolk Gdns. Bexh. CQ44 54
Norfolk Gdns. Borwd. BN24 6
Norfolk Ho. Rd. SW16 BW48 59
Norfolk Pl. W2 BT39 41
Norfolk Pl. Well. CO44 54
Norfolk Rd. Bark. CN36 36
Norfolk Rd. Barn. BS24 7

Norfolk Rd. Dag. CR35 36
Norfolk Rd. E6 CK37 35
Norfolk Rd. E17 CC30 16
Norfolk Rd. Enf. CB25 8
Norfolk Rd. Felt. BD47 56
Norfolk Rd. Har. BF32 20
Norfolk Rd. Ilf. CN33 27
Norfolk Rd. NW8 BT37 32
Norfolk Rd. NW10 BO36 31
Norfolk Rd. Rom. CS32 28
Norfolk Rd. SW19 BU50 59
Norfolk Rd. Th. Hth. BZ52 69
Norfolk Row SE11 BX42 42
Old Paradise St.
Norfolk Sq. W2 BT39 41
Norfolk St. E7 CH35 35
Norfolk Ter. W6 BR42 40
Norgrove St. SW12 BV47 59
Norhyrst Av. SE25 CA52 69
Nork Gdns. Bans. BR60 82
Nork Rise, Bans. BQ61 82
Nork Way, Bans. BQ61 82
Norland Pl. W11 BR40 40
Norland Rd. W11 BQ40 40
Norlands Cres. Chis. CL51 71
Norland Sq. W11 BR40 40
Norley Rd. SE13 CF45 52
Norley Vale SW15 BP47 58
Norlington Rd. E10 CF33 25
Norlington Rd. E11 CF33 25
Norman Av. Epsom BO59 82
Norman Av. Felt. BE48 56
Norman Av. N22 BY30 15
Norman Av. S. Croy. BZ58 78
Norman Av. Sthl. BE40 38
Norman Av. Twick. BJ47 57
Normanby Rd. NW10 BO35 31
Norman Clo. Dart. CW47 64
Norman Clo. N22 BZ30 15
Norman Av.
Norman Clo. Orp. CM55 80
Norman Clo. Rom. CR30 18
Norman Ct. N4 BY33 24
Norman Cres. Houns. BD43 47
Norman Cres. Pnr. BD30 11
Normand Ms. W14 BR43 49
Normand Rd. W14 BR43 49
Normandy Av. Barn. BR25 6
Normandy Rd. SW9 BY44 51
Normandy Ter. E16 CH39 44
Normandy Way, Erith CT44 55
Norman Gro. E3 CD37 34
Norman Ho. Felt. BE48 56
Normanhurst Dr. Twick. BJ46 48
St. Margaret's Rd.
Normanhurst Rd. Orp. CO51 72
Normanhurst Rd. SW2 BX48 60
Normanhurst Rd. Walt. BD55 74
Norman Rd. Belv. CR41 45
Norman Rd. Dart. CW47 64
Norman Rd. E6 CK38 35
Norman Rd. Horn. CU33 28
Norman Rd. Ilf. CL35 35
Norman Rd. N15 CA32 24
Norman Rd. SE10 CE43 52
Norman Rd. Sutt. BS56 77
Norman Rd. SW19 BT50 59
Norman Rd. Th. Hth. BY53 69
Norman's Bldgs. EC1 BZ38 33
Norman's Clo. NW10 BN36 31
Normansfield Av. Tedd. BK50 57
Normansfield Clo. Bush. BF26 4
Normansfield, Tedd. BK50 57
Normanshire Av. E4 CF28 16
Normanshire Dr. E4 CE28 16
Normansmead NW10 BN36 31
Norman St. EC1 BZ38 33
Normanton Av. SW19 BS48 59
Normanton Pk. E4 CG27 16
Normanton Rd. S. Croy. CA57 78
Normanton St. SE23 CC48 61
Norman Way N14 BX27 15
Norman Way W3 BM39 39
Normington Clo. SW16 BY49 60
Norrice Lea N2 BT32 23
Norris St. SW1 BW40 41
Haymarket
Norris Way DA1 CT45 55
Norroy Rd. SW15 BQ45 49
Norrys Clo. Barn. BU24 7
Norrys Rd. Barn. BU24 7
Norstead Pl. SW15 BP48 58
North Access Rd. E17 CC32 25
North Acre, Bans. BR61 82
North Acre NW9 BO30 13
North Acton Rd. NW10 BN37 31
Northallerton Way, Rom. CV28 19
Northall Rd. Bexh. CS44 55
Northampton Bldgs. EC1 BY38 33
Skinner St.
Northampton Gro. N1 BZ35 33
St. Pauls Pl.
Northampton Pk. N1 BZ36 33
Northampton Rd. Croy. CB55 78
Northampton Rd. EC1 BY38 33
Northampton Rd. Enf. CD25 9
Northampton Sq. EC1 BY38 33
Northampton St. N1 BZ36 33
Northanger Rd. SW16 BX50 60
North Audley St. W1 BV39 41
North Av. Cars. BU57 77
North Av. Har. BF32 20
North Av. Hayes BC40 38
North Av. N18 CB28 15
North Av. Rich. BM44 48
Sandycombe Rd.
North Av. Sthl. BE40 38
North Av. W13 BJ39 39
Northbank NW8 BU38 32
Northbank Rd. E17 CF30 16
North Birbeck Rd. E11 CF34 25
Northborough Rd. SW16 BW52 68
Northbourne, Brom. CH54 71
Northbourne Rd. SW4 BW45 50
Northbrook Rd. Barn. BR25 6
Northbrook Rd. Croy. BZ53 69
Northbrook Rd. Ilf. CL34 26
Northbrook Rd. N22 BX29 15
Northbrook Rd. SE13 CF46 52
Northburgh St. EC1 BY38 33
Northchurch Rd. N1 BZ36 33
North Church Rd. Wem. BM36 30
North Circular Rd. E4 CD28 16
North Circular Rd. E18 CG30 16
North Circular Rd. N3 BS31 23
North Circular Rd. N12 BU29 14
North Circular Rd. N13 BY28 15
North Circular Rd. NW2 BQ34 22
North Circular Rd. NW10 BL38 30
North Circular Rd. NW11 BQ32 22
Northcliffe Dr. N20 BR26 6
North Clo. Barn. BQ25 6
North Clo. Bexh. CP45 54
North Clo. Chig. CO28 18
North Clo. Dag. CR37 36
North Clo. Mord. BR52 67
North Common Rd. W5 BL40 39
Northcote Av. Islw. BJ46 48
Northcote Av. Sthl. BE40 38
Northcote Av. Surb. BM54 66
Northcote Av. W5 BL40 39
Northcote Rd. Croy. BZ53 69
Northcote Rd. E17 CD31 25
Northcote Rd. N. Mal. BN52 67
Northcote Rd. NW10 BO36 31
Northcote Rd. Sid. CN49 63
Northcote Rd. SW11 BU46 50
Northcote Rd. Twick. BJ46 48
Northcott Av. N22 BX30 15
North Countess Rd. E17 BX30 15
North Ct. W1 BW39 41
Chitty St.
North Cray Rd. Bex. CQ50 63
North Cray Rd. Sid. CQ50 63
North Cres. N3 BR30 13
North Cres. WC1 BW39 41
Store St.
Northcroft Rd. Epsom BN57 76
Northcroft Rd. W13 BJ41 39
North Cross Rd. Ilf. CM31 26
North Cross Rd. SE22 CA46 51
Northdene, Chig. CM28 17
Northdene Gdns. N15 CA32 24
North Dene, Houns. BF44 47
North Dene NW7 BN27 13
Northdown Gdns. Ilf. CN32 27
Northdown Rd. Horn. CU33 28
North Down Rd. Sutt. BS58 77
Northdown Rd. Well. CO44 54
Northdowns Cres. Croy. CE58 79
North Down, S. Croy. CA59 84
North Downs Rd. Croy. CE58 79
Northdown St. N1 BX37 33
North Dr. Houns. BG44 47
North Dr. Orp. CN56 81
North Dr. Rom. CV31 28
North Dr. SW11 BU43 50
North Dr. SW16 BW49 59
Northeast Pl. N1 BY37 33
Chapel Mkt.
North End Av. NW3 BT34 23
North End, Buck. H. CJ26 10
North End Cres. W14 BR42 40
North End, Croy. BZ54 69
North End Ho. W14 BR42 40
North End La. Orp. CL58 80
North End NW3 BT34 23
North End Pde. W14 BR42 40
North End Rd.
Northend Rd. Erith CT44 55
North End Rd. NW11 BS33 23
North End Rd. SW6 BR43 49
North End Rd. W14 BR42 40
North End Rd. Wem. BM34 21
North End Way NW3 BT34 23
North End W14 BR43 49
Northern Av. N9 CA27 15
Northernhay Wk. Mord. BR52 67
Northern Rd. E13 CH37 35
Northey Av. Sutt. BQ58 76
North Eyot Gdns. W6 BO42 40
Berestead Rd.
North Eyot St. W6 BP42 40
St. Peter's Sq.
Northey St. E14 CD40 43
North Feltham Trd. Est. BC46 47
Felt.
Northfield Av. Orp. CP53 72
Northfield Av. Pnr. BD31 20
Northfield Av. W5 BK41 39
Northfield Av. W13 BJ40 39
Northfield Cres. Sutt. BR56 76
Abbotts Rd.
Northfield Gdns. Dag. CQ35 36
Northfield Ind. Est. BM37 30
Wem.
Northfield Path, Dag. CQ34 27
Northfield Rd. Barn. BU24 7
Northfield Rd. Borwd. BM23 5
Northfield Rd. Dag. CQ35 36
Northfield Rd. E6 CK36 35
Northfield Rd. Enf. CB25 8
Northfield Rd. Houns. BD43 47
Northfield Rd. N16 CA33 24
Northfield Rd. W13 BJ41 39
Northfields Rd. W3 BM39 39
Northfields SW18 BS45 50
North Gdns. SW19 BT50 59
Northgate Path, Borwd. BL23 5
North Glade, The, Bex. CQ47 63
North Gower St. NW1 BW38 32
North Gro. N6 BV33 23
North Gro. N15 BZ32 24
North Harrow Est. Har. BF31 20
North Hill Av. N6 BU32 23
North Hill Dr. Rom. CV27 19
North Hill Grn. Rom. CV28 19
North Hill N6 BU32 23
North Hyde Gdns. Hayes BC42 38
North Hyde La. Houns. BE42 38
North Hyde La. Sthl. BD42 38
Northiam N12 BS28 14
Northiam St. E8 CB37 33
Northington St. WC1 BX38 33
Northlands Av. Orp. CM56 80
Northlands St. SE5 BZ44 51
North La. Tedd. BH50 57
North Lodge Clo. SW15 BQ46 49
North Lodge W5 BK40 39
North Looe Rd. Epsom BQ60 82
North Mall N9 CB27 15
North Ms. WC1 BX38 33
Northolm, Edg. BN28 13
Northolme Gdns. Edg. BM30 12
Northolme Rise, Orp. CN55 81
Northolme Rd. N5 BY35 33
Northolt Av. Ruis. CB35 29
Northolt Gdns. Grnf. BH35 30
Northolt Rd. Har. BF35 29
Northolt Way, Horn. CV36 37
Northover, Brom. CG48 61
North Pde. Chess. BL56 75
North Pk. SE9 CK46 53
North Pass. SW18 BS46 50
North Pl. Mitch. BU50 59
North Pl. Tedd. BH50 57
North Pole La. Kes. CG57 79
Northpole Rd. W10 BQ39 40
North Pole W10 BQ40 40
North Quay E1 CB40 42
North Ride W2 BU40 41
North Rd. Belv. CR41 45
North Rd. Brent. BL43 48
North Rd. Brom. CH51 71
North Rd. Dart. CT46 55
North Rd. Edg. BM30 12
North Rd. Hav. CT27 19
North Rd. Ilf. CN34 27
North Rd. N2 BT30 14
North Rd. N6 BV33 23
North Rd. N7 BX36 33
North Rd. N9 CB26 8
North Rd. Rich. BM45 48
North Rd. Rom. CQ32 27
North Rd. SE18 CN42 45
North Rd. Sthl. BF40 38
North Rd. Surb. BK53 66
North Rd. SW19 BT50 59
North Rd. Walt. BD56 74
North Rd. W. Wick. CE54 70
North Row W1 BU40 41
North Several SE3 CF44 52
North Side SW18 BT46 50
Northspur Rd. Sutt. BS55 77
North Sq. N9 CB27 15
North Sq. NW11 BS32 23
Northstead Rd. SW2 BY48 60
North St. Bark. CL36 35
North St. Bexh. CR45 54
North St. Brom. CH51 71
North St. Cars. BU55 77
North St. Dart. CV47 64
North St. E13 CH37 35
North St. Horn. CV33 28
North St. Islw. BJ45 48
North St. NW4 BQ32 22
Heriot Rd.
North St. Pass. E13 CH37 35
North St. Rom. CS31 28
North St. Rom. CT32 28
North St. SW4 BW45 50
North Tenter St. E1 CA39 42
North Ter. SW3 BU41 41
Northumberland Alley EC3 CA39 42
Northumberland Av. E12 CJ33 26
Northumberland Av. CV32 28
Horn.
Northumberland Av. BH44 48
Islw.
Northumberland Av. WC2 BX40 42
Northumberland Av. CM45 53
Well.
Northumberland Clo. CS43 55
Erith
Northumberland Gdns. BW53 68
Mitch.
Northumberland Gdns. N9 CA27 15
Northumberland Gro. N17 CB29 15
Northumberland Pk. Clo. CB29 15
N17
Northumberland Pk. CR43 54
Erith
Northumberland Pk.
Northumberland Pk. N17 CA29 15
Northumberland Pl. W2 BS39 41
Northumberland Rd. Barn. BT26 7
Northumberland Rd. E17 CE33 25
Northumberland Rd. Har. BE32 20
Northumberland Row, BH47 57
Twick.
Colne Rd.
Northumberland St. WC2 BX40 42
Northumberland Way CS44 55
Erith
Northumbria St. E14 CE39 43
North Verbena Gdns. W6 BP42 40
St. Peter's Sq.
North View Cres. Epsom BP62 82
North View Cres. NW10 BO35 31
North View Dr. Wdf. Grn. CJ30 17
North View Ilf. CO29 18
North View Pnr. BD33 20
North View Rd. N8 BW31 23
Northview, Swan. CT51 73
North View SW19 BP49 58
North View W5 BK38 30
North Vill. NW1 BW36 32
North Wk. Croy. CF57 79
North Wk. Sutt. BQ59 82
Northway Cir. NW7 BN28 13
Northway Cres. NW7 BN28 13
North Way, Mord. BR52 67
North Way N9 CC27 16
Northway N11 BW29 14
North Way NW9 BM31 21
Northway NW11 BS32 23
North Way, Pnr. BD31 20
Northway Rd. Croy. CA53 69
Northway Rd. SE5 BZ45 51
Northways NW3 BT36 32
College Cres.
Northway, Wall. BW56 77
North Western Av. Wat. BG24 4
Northwest Pl. N1 BY37 33
Chapel Mkt.
North Wharf. Rd. W2 BT39 41
Northwick Av. Har. BJ32 21
Northwick Circle, Har. BK32 21
Northwick Clo. NW8 CJ38 35
Northwick Ter.
Northwick Rd. Wat. BD28 11
Northwick Ter. NW8 BT38 32
Northwick Wk. Har. BH33 21
Northwold Dr. Pnr. BD30 11
Northwold Est. E5 CB34 24
Northwold Rd. E5 CB34 24
Northwold Rd. N16 CA34 24
North Wolf W13 BJ40 39
Northwood Av. Horn. CU35 37
Northwood Av. Pur. BY59 84
Northwood Gdns. Grnf. BH35 30
Northwood Gdns. Ilf. CL31 26
Northwood Gdns. N12 BT28 14
Northwood Hills Cir. BC30 11
Nthwd.
Northwood Hills, Nthwd. BC30 11
Northwood Pl. DA18 CQ41 45
Northwood Rd. Cars. BV57 77
Northwood Rd. N6 BV33 23
Northwood Rd. SE23 CD47 61
Northwood Rd. Th. Hth. BY51 69
North Woolwich Rd. E16 CH40 44
North Worple Way SW14 BN45 49
Norton Av. Surb. BM54 66
Norton Clo. E4 CE28 16
Norton Clo. Enf. CB23 8
Norton Folgate E1 CA39 42
Norton Gdns. SW16 BX51 69
Norton Rd. Dag. CS36 37
Norton Rd. E10 BX33 24
Dagenham Rd.
Norton Rd. Wem. BK36 30
Norval Rd. Wem. BJ34 21
Norway Pl. E14 CD39 43
Norway St. SE10 CE43 52
Norway Wk. Rain. CV39 46
Norwich Mews, Ilf. CO33 27
Ashgrove Rd.
Norwich Rd. Dag. CR37 36
Norwich Rd. E7 CH35 35
Norwich Rd. E8 CB37 33

Norwich Rd. Grnf.	CF37	29
Norwich Rd. Th. Hth.	BZ52	69
Norwich St. EC4	BY39	42
Norwich Wk. Edg.	BN29	13
Norwood Av. Rom.	CS33	28
Norwood Av. Wem.	BL37	30
Norwood Clo. Sthl.	BF42	38
Norwood Dr. Har.	BE32	20
Norwood Gdns. Hayes	BD38	29
Norwood Gdns. Sthl.	BE42	38
Norwood Grn. Rd. Sthl.	BF42	38
Norwood High St. SE27	BY48	60
Norwood Pk. Rd. SE27	BZ49	60
Norwood Rd. SE24	BY47	60
Norwood Rd. SE27	BY47	60
Norwood Rd. SE27	BY48	60
Norwood Rd. Sthl.	BE41	38
Norwood Ter. Sthl.	BF42	38
Notley St. SE5	BZ43	51
Notre Dame Est. SW4	BW46	50
Notson Rd. SE25	CB52	69
Nott Ct. WC2	BX39	42
Shorts Gdns.		
Notting Barn Rd. W10	BQ38	31
Nottingham Av. E16	CJ39	44
Nottingham Pl. W1	BV38	32
Nottingham Rd. E10	CF32	25
Nottingham Rd. Islw.	BH44	48
Nottingham Rd. S. Croy.	BZ56	78
Nottingham Rd. SW17	BU47	59
Nottingham St. W1	BV39	41
Nottingham Ter. NW1	BV38	32
Allsop Pl.		
Notting Hill Gte. W11	BS40	41
Nova Mews, Mord.	BR54	67
Nova Rd. Croy.	BY54	69
Novar Rd. SE9	CM47	62
Novello St. SW6	BS44	50
Nowell Rd. SW13	BP43	49
Nower Hill, Pnr.	BE31	20
Noyna Rd. SW17	BU48	59
Nuding Clo. SE13	CE45	52
Nuding Rd. SE13	CE45	52
Nufield Rd. Swan.	CU50	64
Nugent Rd. N19	BX33	24
Nugent Rd. SE25	CA52	69
Nugents Ct. Pnr.	BE30	11
Nugents Pk. Pnr.	BE30	11
Nugent Ter. NW8	BT38	32
Nuneaton Rd. Dag.	CP36	36
Nunhead Cres. SE15	CB45	51
Nunhead Grn. SE15	CB45	51
Nunhead Gro. SE15	CB45	51
Nunhead La. SE15	CB45	51
Nunnington Clo. SE9	CK48	62
Nunns Rd. Enf.	BZ23	8
Nupton Dr. Barn.	BQ25	6
Nursery Av. Bexh.	CQ45	54
Nursery Av. Croy.	CC55	79
Nursery Av. N3	BT30	14
Nursery Clo. Croy.	CC55	79
Nursery Clo. Enf.	CC23	9
Nursery Clo. Epsom	BO58	76
Nursery Clo. Felt.	BC47	56
Nursery Clo. Hamptn.	BE49	56
Nursery Clo. Orp.	CO54	72
Nursery Clo. Rom.	CP32	27
Nursery Clo. Swan.	CS51	73
Nursery Clo. SW15	BQ45	49
Nursery Clo. Wdf. Grn.	CH28	17
Nursery Gdns. Enf.	CC23	9
Nursery Gdns. Sun.	BC51	65
Nursery La. E7	CH36	35
Nursery Rd. Loug.	CH23	10
Nursery Rd. Loug.	CJ25	10
Nursery Rd. N2	BT30	14
Nursery Rd. N14	BW26	7
Nursery Rd. Pnr.	BD31	20
Nursery Rd. SW9	BX45	51
Nursery Rd. SW19	BR50	58
Nursery Rd. Th. Hth.	BZ52	69
Nursery Row SE17	BZ42	42
Nursery St. N17	CA29	15
Nursery St. SW4	BV45	50
Heath Rd.		
Nursery, The, Erith	CT43	55
Nursery Wk. NW4	BP31	22
Nursery Wk. Rom.	CS32	28
Nunstead Rd. Erith	CR43	54
Nutbourne St. W10	BR38	31
Nutbrook St. SE15	CA45	51
Nut Browne Rd. Dag.	CQ37	36
Nutcroft Rd. SE15	CB43	51
Nutfield Clo. N18	CB29	15
Fore St.		
Nutfield Gdns. Ilf.	CO34	27
Nutfield Gdns. Nthlt.	BD37	29
Nutfield Pl. Th. Hth.	BY52	69
Nutfield Rd. Couls.	BV61	83
Nutfield Rd. E15	CF35	34
Nutfield Rd. NW2	BP34	22
Nutfield Rd. SE22	CA45	51
Nutfield Rd. Th. Hth.	BY52	69
Nutfield Way, Orp.	CL55	80
Nutford Pl. W1	BU39	41
Nuthurst Av. SW2	BX48	60
Nutley Clo. Swan.	CT51	73
Nutley Ter. NW3	BT36	32
Nutmead Clo. Bex.	CS47	64
Nuttall St. N1	CA37	33
Nutter La. E11	CH32	26
Nutters La. E11	CJ32	26
Nutt Gro. Stan.	BK27	12
Nut Tree Clo. Orp.	CP55	81
Nutt St. SE15	CA43	51
Sumner Rd.		
Nutwell St. SW17	BU49	59
Nuxley Rd. Belv.	CQ43	54
Nyanza St. SE18	CM43	53
Nye Bevan Est. E5	CC34	25
Nylands Av. Rich.	BM44	48
Nymans Gdns. SW20	BP51	67
Nynehead St. SE14	CD43	52
Nyon Gro. SE6	CD48	61
Nyssa Clo. IG8	CK29	17
Nyton Clo. N19	BX33	24
Oak Av. Croy.	CE54	70
Oak Av. Hamptn.	BE49	56
Oak Av. Houns.	BE43	47
Oak Av. N8	BX31	24
Oak Av. N10	BV29	14
Oak Av. N17	BZ29	15
Oakbank Av. Walt.	BE54	65
Oakbank, Croy.	CF57	79
Oakbank Gro. SE24	BZ45	51
Oakbury Rd. SW6	BS44	50
Oak Clo. N14	BV26	7
Oak Clo. Sutt.	BT55	77
Oak Common W3	BO39	40
Common La.		
Oak Cottage Clo. SE6	CG47	61
Oak Ct. Barn.	BU25	7
Oak Cres. E16	CG39	43
Oakcroft Rd. Chess.	BL56	75
Oakcroft Rd. SE13	CF44	52
Oakcroft Vill. Chess.	BL56	75
Oakdale Av. Har.	BL32	21
Oakdale Av. Nthwd.	BC30	11
Oakdale Clo. Wat.	BD28	11
Oakdale Ct. E4	CF28	16
Oakdale N14	BV26	7
Oakdale Rd. E7	CH36	35
Oakdale Rd. E11	CF34	25
Oakdale Rd. E18	CH30	17
Oakdale Rd. Epsom	BN57	76
Oakdale Rd. N4	BZ32	24
Oakdale Rd. SW16	BX49	60
Oakdale Rd. Wat.	BD27	11
Oakdene Av. Chis.	CL49	62
Oakdene Av. Erith	CS43	55
Oakdene Av. Surb.	BJ54	66
Oakdene Clo. Horn.	CU32	28
Oakdene Clo. Pnr.	BE29	11
Oakdene Dr. Surb.	BN54	67
Oakdene Pk. N3	BR29	13
Oakdene Rd. Orp.	CN53	72
Oak Dene W13	BJ39	39
Dene, The		
Oakden St. SE11	BY42	42
Oaken Dr. Esher	BH57	75
Oaken La. Esher	BH56	75
Oakenshaw Clo. Surb.	BL54	66
Oakes Gro. E4	CG27	16
Oakeshott Av. N6	BV34	23
Oakey La. SE1	BY41	42
King Edward Wk.		
Oak Farm Clo. Borwd.	BN25	6
Oakfield Av. Har.	BJ31	21
Oakfield Clo. N. Mal.	BO53	67
Oakfield Ct. N8	BX33	24
Oakfield Ct. NW11	BQ33	22
Oakfield Gdns. Beck.	CE53	70
Oakfield Gdns. Cars.	BU54	68
Oakfield Gdns. Grnf.	BG38	29
Oakfield Gdns. N18	CA28	15
Oakfield Gdns. SE19	CA49	60
Oakfield La. Dart.	CU48	64
Oakfield La. Kes.	CJ56	80
Oakfield Pk. Rd. Dart.	CV48	64
Oakfield Pl. Dart.	CV48	64
Oakfield Rd. Croy.	BZ54	69
Oakfield Rd. E6	CK37	35
Oakfield Rd. E17	CD30	16
Oakfield Rd. Ilf.	CL34	26
Oakfield Rd. N3	BS30	14
Oakfield Rd. N4	BY32	24
Oakfield Rd. N14	BX27	15
Oakfield Rd. Orp.	CO54	72
Oakfield Rd. SE20	CB51	69
Oakfield Rd. SW19	BQ48	58
Oakfields Rd. NW11	BR32	22
Oakfield St. SW10	BT43	50
Oakfields, Walt.	BC54	65
Oakford Rd. NW5	BW35	32
Oak Gdns. Croy.	CE54	70
Oak Gdns. Edg.	BN30	13
Oak Glen. Horn.	CW31	28
Oak Gro. NW2	BQ35	31
Oakgrove Rd. SE20	CC51	70
Oak Gro. Ruis.	BC33	20
Oak Gro. Sun.	BC50	56
Oak Gro. W. Wick.	CF55	79
Oakhall Ct. E11	CH32	26
Cambridge Pk.		
Oakhall Rd. E11	CH32	26
Oakham Clo. SE6	CD48	61
Oakham Dr. Brom.	CG52	70
Oakhampton Clo. N12	BT28	14
Oakhampton Rd. NW7	BQ29	13
Oakhampton Rd. Rom.	CV29	19
Oakhampton Sq. Rom.	CV29	19
Oakhill Av. Pnr.	BE30	11
Oak Hill Clo. Wdf. Grn.	CF29	16
Oak Hill Ct. SE23	CC46	52
Oak Hill Ct. Wdf. Grn.	CF29	16
Oakhill Cres. Surb.	BL54	66
Oak Hill Cres. Wdf. Grn.	CF29	16
Oak Hill Dr. Surb.	BL54	66
Oak Hill, Epsom	BN61	82
Oakhill, Esher	BJ57	75
Oakhill Gdns. Wdf. Grn.	CG30	16
Oakhill Gro. Surb.	BL53	66
Oak Hill Ms. NW3	BT35	32
Oak Hill Pk. NW3	BS35	32
Oakhill Path, Surb.	BL53	66
Oakhill Pl. SW15	BS46	50
Oakhill Rd.		
Oak Hill Rd. Beck.	CF51	70
Oak Hill Rd. Orp.	CN54	72
Oakhill Rd. Surb.	BL53	66
Oakhill Rd. Sutt.	BS55	77
Oakhill Rd. SW15	BR46	49
Oakhill Rd. SW16	BX51	69
Oak Hill, Surb.	BL54	66
Oak Hill Way NW3	BS35	32
Oak Hill, Wdf. Grn.	CF29	16
Oakhouse Rd. Bexh.	CR46	54
Oakhurst Av. Barn.	BU26	7
Oakhurst Av. Bexh.	CQ44	54
Oakhurst Clo. E17	CG31	25
Oakhurst Gdns. Bexh.	CQ43	54
Oakhurst Gdns. E4	CG26	9
Oakhurst Gdns. E17	CG31	25
Oakhurst Gro. SE22	CB45	51
Oakhurst Ri. Cars.	BU58	77
Oakhurst Rd. Epsom	BN57	76
Oakhurst Rd. Kings. On T.	BL50	57
Oakington Av. Har.	BF33	20
Oakington Av. Wem.	BL34	21
Oakington Dr. Sun.	BD51	65
Oakington Manor Dr. Wem.	BM35	30
Oakington Rd. W9	BS38	32
Oakington Way N8	BX32	24
Oaklands Av. Esher	BG54	65
Oaklands Av. Islw.	BH43	48
Oaklands Av. N9	CB25	8
Oaklands Av. Rom.	CT31	28
Oaklands Av. Sid.	CN47	63
Oaklands Av. Th. Hth.	BY52	69
Oaklands Av. Wat.	BC26	4
Oaklands Av. W. Wick.	CE55	79
Oaklands, Chess.	BK56	75
Oaklands Clo. Bexh.	CQ46	54
Oaklands Clo. Orp.	CN53	72
Oaklands Clo. Wem.	BK35	30
Oaklands Ct. SE26	CB49	60
Oaklands Ct. Wat.	BC23	4
Oaklands Est. SW4	BW46	50
Oaklands Gdns. Ken.	BZ60	84
Oaklands Gro. W12	BP40	40
Oaklands, Ken.	BZ61	84
Oaklands La. Barn.	BP24	6
Oaklands N21	BX27	15
Oaklands Pk. Av. Ilf.	CM34	26
Oaklands Rd. Bexh.	CQ45	54
Oaklands Rd. Brom.	CG50	61
Oaklands Rd. N20	BR26	6
Oaklands Rd. NW2	BQ35	31
Oaklands Rd. SW14	BN45	49
Oaklands Rd. W7	BH41	39
Oaklands, Twick.	BG47	56
Oakland Way, Epsom	BN57	76
Oak La. E14	CD40	43
Oak La. Islw.	BH45	48
Oak La. N2	BT30	14
Oak La. N11	BW29	14
Oak La. Twick.	BJ47	57
Oak La. Wdf. Grn.	CG28	16
Oakleafe Gdns. Ilf.	CL31	26
Oaklea Pass. Kings. On T.	BK52	66
Oakleigh Av. Edg.	BM30	12
Oakleigh Av. N20	BT27	14
Oakleigh Av. Surb.	BM54	66
Oakleigh Clo. N20	BU27	14
Oakleigh Ct. Edg.	BN30	13
Oakleigh Cres. N20	BU27	14
Oakleigh Gdns. Edg.	BL28	12
Oakleigh Gdns. N20	BT26	7
Oakleigh Gdns. Orp.	CN56	81
Oakleigh Pk. Av. Chis.	CL51	71
Oakleigh Pk. N. N20	BT27	14
Oakleigh Pk. S. N20	BU26	7
Oakleigh Rd. N. N20	BT27	14
Oakleigh Rd. Pnr.	BE29	11
Oakleigh Rd. S. N11	BV27	14
Oakleigh Way, Mitch.	BV51	68
Oakleigh Way, Surb.	BM54	66
Oakley Av. Bark.	CN36	36
Oakley Av. Croy.	BX56	78
Oakley Av. W5	BM40	39
Oakley Clo. Islw.	BG44	47
Oakley Clo. W7	BH40	39
Oakley Cres. N1	BY37	33
City Rd.		
Oakley Dr. Brom.	CK55	80
Oakley Dr. Sid.	CM47	62
Oakley Gdns. Bans.	BS61	83
Oakley Gdns. N8	BX32	24
Oakley Gdns. SW3	BU43	50
Oakley Ho. W5	BM40	39
Oakley Pk. Sid.	CP47	63
Oakley Pl. SE1	CA42	42
Oakley Rd. Brom.	CK55	80
Oakley Rd. Har.	BH32	21
Oakley Rd. N1	BZ36	33
Oakley Rd. SE25	CB53	69
Oakley Rd. Warl.	CB62	84
Oakley Sq. NW1	BW37	32
Oakley St. SW3	BU42	41
Oakley Wk. W6	BR43	49
Oak Lodge Av. Chig.	CM28	17
Oak Lodge Clo. Stan.	BJ28	12
Oak Lodge Clo. Walt.	BD56	74
Oak Lodge Dr. W. Wick.	CE54	70
Oakmead Av. Brom.	CH53	71
Oakmeade, Pnr.	BF29	11
Oakmead Gdns. Edg.	BN28	13
Oakmead Rd. Croy.	BW53	68
Oakmead Rd. SW12	BV47	59
Oakmere Rd. SE2	CO43	54
Oakmoor Way, Chig.	CN28	18
Parkes Rd.		
Oakmount Pl. Orp.	CM54	71
Oakridge Rd. Brom.	CF49	61
Oak Ri. Buck. H.	CJ27	17
Oak Rd. Erith	CS43	55
Oak Rd. Erith	CU44	55
Oak Rd. N. Mal.	BN51	67
Oak Rd. Orp.	CO57	81
Oak Rd. Rom.	CW30	19
Oak Row, Mitch.	BW51	68
Oaks Av. Felt.	BE48	56
Oaks Av. Rom.	CS30	19
Oaks Av. SE19	BZ49	60
Oaks Av. Wor. Pk.	BP55	76
Oaksford Av. SE26	CB49	60
Oakshade Rd. Brom.	CF49	61
Oakshaw Rd. SW18	BS47	59
Oaks La. Croy.	CB56	78
Oaks La. Ilf.	CN31	27
Oaks Rd. Croy.	CB56	78
Oaks Rd. Ken.	BY60	84
Oaks, The, Epsom	BO60	82
Oaks, The SE18	CM42	44
Oaks, The, Wdf. Grn.	CG29	16
Oak St. Bakers La. W5	BK40	39
Grove, The		
Oak St. Rom.	CS31	28
Oaks Way, Cars.	BU57	77
Oaks Way, Ken.	BZ60	84
Oaks Way, Surb.	BK55	75
Oakthorpe Rd. N13	BY28	15
Oaktree Av. N13	BY27	15
Oak Tree Clo. Stan.	BK29	12
Oak Tree Clo. W13	BK39	39
Pinewood Gro.		
Oak Tree Ct. Borwd.	BK25	5
Oak Tree Dell NW9	BN32	22
Oak Tree Dr. N20	BS26	7
Oak Tree Gdns. Brom.	CH49	62
Oak Tree Pl. NW8	BT38	32
Oakview Gdns. N2	BT31	23
Oakview Gro. Croy.	CD54	70
Oakview Rd. SE6	CE49	61
Oak Village NW5	BV35	32
Oakway, Brom.	CF51	70
Oakway Clo. Bex.	CQ46	54
Oak Way, Croy.	CC53	70
Oakway La. Brom.	CF51	70
Oak Way N14	BV26	7
Oakways SE9	CL46	53
Oakway SW20	BQ52	67
Oak Way W3	BO40	40
Oakwood Av. Beck.	CF51	70
Oakwood Av. Borwd.	BM24	5
Oakwood Av. Brom.	CH52	71
Oakwood Av. Mitch.	BT51	68
Oakwood Av. N14	BW26	7
Oakwood Av. Pur.	BY59	84
Oakwood Av. Sthl.	BF40	38
Oak Wood Chase, Horn.	CW32	28
Oakwood Clo. Chis.	CK50	62
Oakwood Clo. N14	BW25	7
Oakwood Clo. Wdf. Grn.	CK29	17
Oakwood Ct. W14	BR41	40
Oakwood Cres. Grnf.	BJ36	30
Oakwood Cres. N21	BX25	8
Oakwood Dr. Bexh.	CS45	55
Oakwood Dr. Edg.	BN29	13
Oakwood Gdns. Ilf.	CN34	27
Oakwood Gdns. Orp.	CM55	80
Oakwood Gdns. Sutt.	BS55	77
Oakwood Hill, Loug.	CK25	10
Oakwood La. W14	BR41	40
Oakwood Pde. Enf.	CA25	8
Queen Anne's Pl.		
Oakwood Pk. Rd. N14	BW26	7
Oakwood Pl. Croy.	BY53	69
Oakwood Rd. Croy.	BY53	69
Oakwood Rd. NW11	BS31	23

Oakwood Rd. Orp. CM55 80
Oakwood Rd. Pnr. BC30 11
Oakwood Rd. SW20 BP51 67
Oakwood Vw. N14 BW25 7
Oakwood, Wall. BV58 77
Oakworth Rd. W10 BQ39 40
Oasthouse Way, Orp. CO52 72
Oates Rd. Rom. CR28 18
Oatfield Rd. Orp. CN54 72
Oatland Rise E17 CD30 16
Oatlands Rd. Enf. CC23 9
Oat La. EC2 BZ39 42
Oban House E14 CF39 43
Oban Rd. SE25 BZ52 69
Oban St. E14 CF39 43
Oberon House N1 CA37 33
Purcell St.
Oberstein Rd. SW11 BT45 50
Oborne Clo. SE24 BY46 51
Observatory Gdns. W8 BS41 41
Observatory Rd. SW14 BN45 49
Occupation La. SE18 CL44 53
Occupation La. W5 BK42 39
Occupation Rd. Belv. CQ41 45
Occupation Rd. SE17 BZ42 42
Manor Pl.
Occupation Rd. W7 BJ42 39
Occupation Rd. Wat. BC25 4
Ocean Est. E1 CC38 34
Ocean St. E1 CC39 43
Masters St.
Ockendon Rd. N1 BZ36 33
Ockham Clo. Orp. CO50 63
Ockley Ct. Sutt. BT56 77
Ockley Rd. Croy. BX54 69
Ockley Rd. SW16 BX49 60
Octavia Clo. Mitch. BT53 68
Octavia Rd. Islw. BH45 48
Octavia St. SW11 BU44 50
Octavia Way SE28 CO40 45
Octavius St. SE8 CE43 52
Odard Rd. E. Mol. BF52 65
Oddesey Rd. Borwd. BM23 5
Odessa Rd. E7 CG35 34
Odessa Rd. NW10 BP37 31
Odessa St. SE16 CD41 43
Odger St. SW11 BU44 50
Offenham Rd. SE9 CK49 62
Offerton Rd. SW4 BW45 50
Offham Slope N12 BR28 13
Offley Rd. SW9 BY43 51
Offord Clo. N17 CB29 15
Offord Rd. N1 BX36 33
Offord St. N1 BX36 33
Ogilby St. SE18 CK42 44
Oglander Rd. SE15 CA45 51
Ogle St. W1 BW39 41
Oglethorpe Rd. Dag. CQ34 27
Ohio Rd. E13 CG38 34
Oil Mill La. W6 BP42 40
Okeburn Rd. SW17 BV49 59
Okehampton Cres. Well. CO43 54
Okehampton Rd. NW10 BQ37 31
Okehampton Rd. Rom. CV29 19
Okehampton Sq. Rom.
Okemore Gdns. Orp. CP52 72
Olaf St. W11 BQ40 40
Old Bailey, EC4 BY39 42
Old Barn Clo. Sutt. BR57 76
Old Barn La. Whyt. CA61 84
Old Barn La. Epsom BN62 82
Old Barn Way, Bexh. CS45 55
Old Barrowfield, E15 CG37 34
Oldberry Rd. Edg. BN29 13
Old Bethnal Green Rd. E2 CB38 33
Old Bexley La. Bex. CS48 64
Old Bexley La. Dart. CT47 64
Old Bond St. W1 BW40 41
Oldborough Rd. Wem. BK34 21
Old Brewery Mews NW3 BT35 32
Hampstead High St.
Old Bridge Clo. Nthlt. BF37 29
Old Bridge St. Kings. On T. BK51 66
Thames St.
Old Broad St. EC2 BZ39 42
Old Bromley Rd. Brom. CF49 61
Old Brompton Rd. SW5 BS42 41
Old Brompton Rd. SW7 BT42 41
Old Burlington St. W1 BW40 41
Oldbury Pl. W1 BV38 32
Oldbury Rd. Enf. CB23 8
Old Castle St. E1 CA39 42
Old Cavendish St. W1 BV39 41
Old Change Ct. EC4 BZ39 42
Peters Hill
Old Chapel Rd. Swan. CS54 73
Old Chestnut Av. Esher BF57 74
Oldchurch Gdns. Rom. CS33 28
Old Church La. NW9 BN34 22
Old Church La. Stan. BJ28 12
Old Church La. W13 BJ38 30
Oldchurch Rise, Rom. CT33 28
Old Church Rd. E1 CC39 43
Old Church Rd. E4 CE28 16
Oldchurch Rd. Rom. CS33 28
Old Church St. SW3 BT42 41
Old Claygate La. Esher BJ57 75
Old Compton St. W1 BW40 41
Old Cote Dr. Houns. BF43 47
Old Court Pl. W8 BS41 41
Old Ct. W5 BL39 39
Hillcrest Rd.
Old Deer Park Gdns. Rich. BL45 48
Old Devonshire Rd. SW12 BV47 59
Old Dock Clo. Rich. BM43 48
Bushwood Rd.
Old Dover Rd. SE3 CH43 53
Olden La. Pur. BY59 84
Oldershaw Rd. N7 BX36 33
Old Esher Clo. Walt. BD56 74
Old Esher Rd. Walt. BD56 74
Old Farleigh Rd. S. Croy. CC58 79
Old Farm Av. N14 BW26 7
Old Farm Av. Sid. CM47 62
Old Farm Clo. Houns. BE45 47
Old Farm Gdns. Swan. CT52 73
Old Farm Pass. Hamptn. BG51 65
Old Farm Rd. E. Sid. CO48 63
Old Farm Rd. Hamptn. BG50 56
Old Farm Rd. N2 BT30 14
Old Farm Rd. W. Sid. CN48 63
Oldfield Cir. Nthlt. BG36 29
Oldfield Clo. Brom. CK52 71
Oldfield Clo. Grnf. BH35 30
Oldfield Clo. Stan. BJ28 12
Oldfield Farm Gdns. Grnf. BG37 29
Oldfield Gro. SE16 CC42 43
Oldfield La Grnf. BG38 29
Oldfield La. N. Grnf. BG37 29
Oldfield La. S. Grnf. BG38 29
Oldfield Rd. Bexh. CQ44 54
Oldfield Rd. Brom. CK52 71
Oldfield Rd. Hamptn. BE51 65
Oldfield Rd. N16 CA34 24
Oldfield Rd. NW10 BO36 31
Oldfield Rd. SW19 BR50 58
Oldfields Rd. Sutt. BR55 76
Old Fold La. Barn. BR23 6
Old Fold Vw. Barn. BQ24 6
Old Force Clo. Stan. BJ28 12
Old Ford Clo. Barn. BR23 6
Old Ford Rd. E2 CC37 34
Old Ford Rd. E3 CD37 34
Old Forge Way, Sid. CO49 63
Old Fox Footpath, S. Croy. CA58 78
Old Gloucester St. WC1 BX39 42
Old Hall Clo. Pnr. BE30 11
Old Hall Dr. Pnr. BE30 11
Oldhams Ter. W3 BN40 40
Old Hill, Chis. CL51 71
Old Hill, Orp. CM57 80
Oldhill Pl. N16 CB33 24
Oldhill St. N16 CB33 24
Old Homesdale Rd. Brom. CJ52 71
Old Ho. Clo. Epsom BO58 76
Old Ho. Clo. SW19 BR49 58
Old Jamaica Rd. SE16 CB41 42
Old James St. SE15 CB45 51
Old Jewry, EC2 BZ39 42
Old Kenton La. NW9 BM32 21
Old Kent Rd. SE1 BZ41 42
Old Kent Rd. SE15 CB43 51
Old Kent Rd. SE15 BZ41 42
Old Kingston Rd. Wor. Pk. BN55 76
Old Lambeth Palace Rd. SE1 BX41 42
Old Lodge La. Pur. BX60 84
Old Lodge Way, Stan. BJ28 12
Old London Rd. Epsom BP62 82
Old Maidstone Rd. Sid. CQ50 63
Old Malden La. Wor. Pk. BN55 76
Old Manor Dr. Islw. BG46 47
Old Manor Way, Bexh. CS44 55
Old Manor Yd. SW5 BS42 41
Earl's Ct. Rd.
Old Marylebone Rd. NW1 BU39 41
Old Mill Ct. E18 CJ31 26
Old Mill La. NW6 BP42 40
Upper Mall
Old Mill La. W6 BP42 40
Upper Mall
Old Mill Rd. SE18 CM43 53
Old Montague St. E1 CA39 42
Old Nichol St. E2 CA38 33
Old North St. WC1 BX39 42
Theobalds Rd.
Old Oak Av. Couls. BU62 83
Old Oak Common La. NW10 BO39 40
Old Oak Common La. W3 BO39 40
Old Oak Common Way, W3 BO40 40
Old Oak Est. W12 BO40 40
Old Oak La. NW10 BO38 31
Old Oak Rd. W3 BO40 40
Old Orchard, Sun. BD51 65
Old Palace La. Rich. BK46 48
Old Palace Rd. Croy. BY55 78
Old Palace Yd. Rich. BK46 48
Old Paradise St. SE11 BX42 42
Old Park Av. Enf. BZ24 8
Old Park Av. SW12 BV46 50
Old Park Gro. Enf. BZ24 8
Old Park La. W1 BV40 41
Old Park Mews, Houns. BE43 47
Old Park Ridings N21 BY25 8
Old Park Rd. Enf. BY24 8
Old Park Rd. N13 BX28 15
Old Park Rd. SE2 CO42 45
Old Park Rd. S. Enf. BY24 8
Old Park Vw. Enf. BY24 8
Old Perry St. Chis. CN50 63
Old Pye St. SW1 BW41 41
Old Quebec St. W1 BU40 41
Old Queen St. SW1 BW41 41
Old Rectory Gdns. Edg. BM29 12
Old Rectory Ho. SW19 BR49 58
Old Redding, Har. BF28 11
Oldridge Rd. SW12 BV47 59
Old Rd. Dart. CS46 55
Old Rd. Enf. CC23 9
Old Rd. SE13 CG45 52
Old Ruislip Rd. Nthlt. BD37 29
Old Schools La. Epsom BO58 76
Old South Lambeth Rd. SW8 BX43 51
Old Station Rd. Knockholt CQ58 81
Old Station Rd. Loug. CK25 10
Oldstead Rd. Brom. CF49 61
Old St. E13 CH37 35
Old St. EC1 BZ38 33
Old Swan Yard, Cars. BU56 77
Old Town, Croy. BY55 78
Old Town SW4 BW45 50
Old Woolwich Rd. SE10 CF43 52
Oleander Clo. Orp. CM56 80
Olinda Rd. N16 CA32 24
Oliphant St. W10 BQ38 31
Oliver Av. SE25 CA52 69
Oliver Clo. E10 CE34 25
Oliver Cres. Farn. CW54 73
Oliver Goldsmith Est. SE15 CB44 51
Goldsmith Rd.
Oliver Gro. SE25 CA52 69
Olive Rd. Dart CV47 64
Olive Rd. E13 CJ38 35
Olive Rd. NW2 BQ35 31
Olive Rd. SW19 BT50 59
Olive Rd. W5 BK41 39
Oliver Rd. E10 CE34 25
Oliver Rd. E17 CF32 25
Oliver Rd. N. Mal. BN51 67
Oliver Rd. Rain. CT37 37
Oliver Rd. Sutt. BT56 77
Oliver Rd. Swan. CS52 73
Olive St. Rom. CS31 28
Ollard's Gro. Loug. CJ24 10
Ollerton Grn. E3 CD37 34
Ollerton Rd. N11 BW28 14
Ollgar Clo. W12 BO40 40
Olliffe St. E14 CF41 43
Olmar St. SE1 CB43 51
Olney St. SE17 BY43 51
Olron Cres. Bexh. CP46 54
Olven Rd. SE18 CM43 53
Olveston Wk. Cars. BT53 68
Olyffe Av. Well. CO44 54
Olyffe Dr. Beck. CF51 70
Olympic Way, Grnf. BF37 29
Olympic Way, Wem. BM34 21
Olympus Sq. N16 CB34 24
Downs Est.
Oman Av. NW2 BP35 31
O'Meara St. SE1 BZ40 42
Omega Pl. N1 BX37 33
Caledonian Rd.
Omega St. SE14 CD44 52
Ommaney Rd. SE14 CC44 52
Ondine Rd. SE15 CA45 51
One Tree Clo. SE23 CC46 52
Ongar Clo. Rom. CP32 27
Ongar Rd. SW6 BS43 50
Ongar Way, Rain. CT37 37
Onra Rd. E17 CE33 25
Onslow Av. Rich. BL46 48
Onslow Av. Sutt. BR58 76
Onslow Clo. E4 CF27 16
Onslow Ct. Surb. BH54 66
Onslow Way
Onslow Cres. Chis. CL51 71
Onslow Cres. SW7 BT42 41
Old Brompton Rd.
Onslow Dr. Sid. CP48 63
Onslow Gdns. E18 CH31 26
Onslow Gdns. N10 BV32 23
Onslow Gdns. N21 BY25 8
Onslow Gdns. S. Croy. CB59 84
Onslow Gdns. Surb. BH54 66
Onslow Gdns. SW7 BT42 41
Onslow Gdns. Wall. BW57 77
Onslow Ms. W. SW7 BT42 41
Cranley Pl.
Onslow Rd. Croy. BX54 69
Onslow Rd. N. Mal. BP52 67
Onslow Rd. Rich. BL46 48
Onslow Sq. SW7 BT42 41
Onslow St. EC1 BY38 33
Clerkenwell Rd.
Onslow Way, Surb. BH54 66
Ontario St. SE1 BY41 42
On The Hill, Wat. BE27 11
Opal Mews, Ilf. CL34 26
Ley St.
Opal St. SE11 BY42 42
Openshaw Rd. SE2 CO42 45
Openview, SW18 BT47 59
Ophir Ter. SE15 CB44 51
Opossum Way, Houns. BD45 47
Oppidans Ms. NW3 BU36 32
Oppidans Rd. NW3 BU36 32
Orange Ct. E1 CB40 42
Hermitage Wall
Orange Hill Rd. Edg. BN29 13
Orange Pl. SE16 CC41 43
Orangery La. SE9 CC46 53
Orange St. WC2 BW40 41
Orange Tree Hill, Hav. CS28 19
Orantham Clo. Edg. BL27 12
Orbain Rd. SW6 BR43 49
Orbel St. SW11 BU44 50
Orb St. SE17 BZ42 42
Orchard Av. Belv. CQ43 54
Orchard Av. Croy. CD54 70
Orchard Av. Dart. CU47 64
Orchard Av. Houns. BE43 47
Orchard Av. Mitch. BV54 68
Orchard Av. N. Mal. BO51 67
Orchard Av. N3 BS31 23
Orchard Av. N14 BW26 7
Orchard Av. N20 BT27 14
Orchard Av. Rain. CV38 37
Orchard Av. Sthl. BE40 38
Orchard Av. Surb. BJ54 66
Orchard Clo. Bans. BS60 83
Orchard Clo. Bexh. CQ44 54
Orchard Clo. Borwd. BL24 5
Orchard Clo. Bush. BG26 4
Orchard Clo. Edg. BL29 12
Orchard Clo. Surb. BJ54 66
Orchard Clo. SW20 BQ52 67
Orchard Clo. Wem. BL36 30
Orchard Ct. Edg. BL28 12
Orchard Ct. Islw. BG44 47
Thornbury Av.
Orchard Ct. Wor. Pk. BP54 67
Orchard Cres. Edg. BN28 13
Orchard Cres. Enf. CA23 8
Orchard Dr. Edg. BL28 12
Orchard Dr. SE3 CF44 52
Orchard Est. SE13 CE44 52
Orchard Gdns. Chess. BL56 75
Orchard Gdns. Epsom BN60 82
Orchard Gdns. Sutt. BS56 77
Orchard Gate, Esher BG54 65
Orchard Gate, Grnf. BJ36 30
Orchard Gate, NW9 BO31 22
Orchard Grn. Orp. CN55 81
Orchard Gro. Croy. CD54 70
Orchard Gro. Edg. BM30 12
Orchard Gro. Har. BL32 21
Orchard Gro. Orp. CN55 81
Orchard Hill, Cars. BU56 77
Orchard Hill, Dart. CT46 55
Orchard Hill SE13 CE44 52
Coldbath St.
Orchard La. E. Mol. BG53 65
Orchard La. SW20 BP51 67
Orchard La. Wdf. Grn. CJ28 17
Orchardleigh Av. Enf. CC23 9
Orchardmede N21 BZ25 8
Orchard Pl. Brom. CH51 71
Orchard Pl. E14 CG40 43
Orchard Pl. N17 CA29 15
Orchard Rise, Croy. CD54 70
Orchard Rise, E. Sid. CN46 54
Orchard Rise Kings. On T. BN51 67
Orchard Rise, Rich. BM46 48
Orchard Rise, W. Sid. CN45 54
Orchard Rd. Barn. BR24 6
Orchard Rd. Belv. CR42 45
Orchard Rd. Brent. BK43 48
Orchard Rd. Brom. CJ51 71
Orchard Rd. Chess. BL56 75
Orchard Rd. Dag. CR37 36
Orchard Rd. Enf. CC25 9
Orchard Rd. Hamptn. BE50 56
Orchard Rd. Hayes BC40 38
Orchard Rd. Houns. BE46 47
Orchard Rd. Kings. On T. BL52 66
Orchard Rd. Mitch. BV54 68
Orchard Rd. N6 BV33 23
Orchard Rd. Orp. CL56 80
Orchard Rd. Orp. CP58 81
Orchard Rd. Rich. BM45 48
Orchard Rd. Rom. CR30 18
Orchard Rd. SE3 CG44 52
Orchard Rd. SE18 CM42 44
Orchard Rd. Sid. CN49 63
Orchard Rd. S. Croy. CB60 84
Orchard Rd. Sun. BC50 56
Orchard Rd. Sutt. BS56 77
Orchard Rd. Twick. BJ46 48

Park Rd. E10 CE33 25
Park Rd. E12 CH33 26
Park Rd. E15 CH37 35
Park Rd. E17 CD32 25
Park Rd. Felt. BD49 56
Park Rd. Hamptn. BF49 56
Park Rd. Houns. BF46 47
Park Rd. Ilf. CM34 26
Park Rd. Islw. BJ44 48
Park Rd. Ken. BY61 84
Park Rd. Kings. On T. BK51 66
Park Rd. Kings. On T. BL49 57
Park Rd. N. Mal. BN52 67
Park Rd. N. W3 BM41 39
Park Rd. N. W4 BN42 40
Park Rd. N2 BT31 23
Park Rd. N8 BW31 23
Park Rd. N11 BW29 14
Park Rd. N14 BW26 7
Park Rd. N15 BY31 24
Park Rd. N18 CA28 15
Park Rd. NW1 BU38 32
Park Rd. NW4 BP33 22
Park Rd. NW8 BU38 32
Park Rd. NW9 BN33 22
Park Rd. NW10 BN37 31
Park Rd. Orp. CP53 72
Park Rd. Rich. BL46 48
Park Rd. SE7 CK43 53
Park Rd. SE25 CA52 69
Park Rd. Sun. BC50 56
Park Rd. Surb. BL53 66
Park Rd. Sutt. BR57 76
Park Rd. Swan. CT52 73
Park Rd. SW19 BT50 59
Park Rd. SW20 BP50 58
Park Rd. Tedd. BH50 57
Park Rd. Twick. BK46 48
Park Rd. Wall. BV55 77
Park Rd. Wall. BV56 77
Park Rd. Wat. BC23 4
Park Rd. Wem. BL36 30
Park Rd. W. Kings. On T. BL49 57
Park Rd. W4 BN43 49
Park Rd. W7 BH40 39
Park Row SE10 CF42 43
Park Royal Rd. NW10 BN38 31
Park Royal Rd. W3 BN38 31
Parkshot Rd. Rich. BK45 48
Parkside Av. Bexh. CS44 55
Parkside Av. Brom. CK52 71
Parkside Av. Rom. CS31 28
Parkside Av. SW19 BQ49 58
Parkside, Buck. H. CH27 17
Parkside Cres. Surb. BN53 67
Parkside Cross, Bexh. CT44 55
Parkside Dr. Edg. BM27 12
Parkside Gdns. Barn. BU26 7
Parkside Gdns. Couls. BV62 83
Parkside Gdns. SW19 BQ49 58
Parkside, Hamptn. BG49 56
Parkside N3 BS30 14
Parkside NW2 BP34 22
Parkside NW7 BO29 13
Parkside Rd. Belv. CS42 46
Parkside Rd. Houns. BF46 47
Parkside Rd. SW11 BV44 50
Parkside SE3 CG43 52
Parkside, Sid. CO48 63
Parkside St. SW11 BV44 50
Battersea Pk. Rd.
Park Side, Sutt. BR57 76
Parkside SW19 BQ47 58
Parkside Ter. N18 BZ28 15
Parkside Way, Har. BF31 20
Park Sq. E. NW1 BV38 32
Park Sq. Esher BF56 74
Park Sq. Ms. NW1 BV38 32
Park Sq. W. NW1 BV38 32
Parkstead Rd. SW15 BP46 49
Parkstone Av. Horn. CW32 28
Parkstone Av. N18 CA29 15
Parkstone Rd. SE15 CB44 51
Rye La.
Park St. Croy. BZ55 78
Park St. N1 BZ37 33
Park St. SE1 BZ40 42
Park St. Tedd. BH50 57
Park St. W1 BV40 41
Park Ter. Wor. Pk. BP54 67
Park, The, Cars. BU56 77
Park, The N6 BV32 23
Park, The NW11 BS33 23
Park, The SE19 CA50 60
Park, The, Sid. CO49 63
Park, The W5 BK40 39
Parkthorne Clo. Har. BF32 20
Parkthorne Dr. Har. BF32 20
Parkthorne Rd. SW12 BW47 59
Parkvale Rd. SW6 BR44 49
Park Vw. Cres. N11 BV28 14
Park Vw. Est. E2 CC37 34
Park Vw. Gdns. Bark. CN37 36
Park Vw. Gdns. Ilf. CK31 26
Woodford Av.
Park Vw. Gdns. NW4 BQ32 22
Park Vw. N5 BZ35 33
Park Vw. N21 BX26 8
Park Vw. N. Mal. BO52 67
Park Vw. Pnr. BE30 11
Parkview Rd. Croy. CB54 69
Park Vw. Rd. N3 BS30 14
Park Vw. Rd. N17 CB31 24
Park Vw. Rd. NW10 BO35 31
Park Vw. Rd. Pnr. BC29 11
Parkview Rd. SE9 CL47 62
Park Vw. Rd. Sthl. BF40 38
Park Vw. Rd. Well. CO45 54
Park Vw. Wem. BM35 30
Park Vw. Rd. W5 BL39 39
Park Vw. W3 BN39 40
Park Vill. E. NW1 BV37 32
Park Vill. Rom. CP32 27
Park Vill. SE3 CH43 53
Park Vill. W. NW1 BV37 32
Parkville Rd. SW6 BR43 49
Park Vista SE10 CF43 52
Park Wk. Barn. BT24 7
Park Wk. SW10 BT43 50
Parkway, Belv. CQ41 45
Park Way, Bex. CT48 64
Parkway Cranford, The, Sthl. BC42 38
Parkway, Croy. CE58 79
Park Way, E. Mol. BF52 65
Park Way, Edg. BM30 12
Park Way, Enf. BY23 8
Park Way, Felt. BC47 56
Park Way, Ilf. CN34 27
Parkway N14 BX27 15
Parkway NW1 BV37 32
Park Way NW11 BR32 22
Park Way, Rain. CU38 37
Parkway, Rom. CT30 19
Park Way, Ruis. BC33 20
Parkway SW20 BQ52 67
Parkway, Wdf. Grn. CJ28 17
Park West Pl. W2 BU39 41
Park West W2 BU39 41
Parkwood Av. Esher BG54 65
Parkwood, Beck. CE50 61
Park Wood Clo. Bans. BQ61 82
Parkwood Gro. Sun. BC52 65
Parkwood N20 BU27 14
Park Wood Rd. Bans. BQ61 82
Park Wood Rd. Bex. CQ47 63
Parkwood Rd. Islw. BH44 48
Parkwood Rd. SW19 BR49 58
Park Wood Vw. Bans. BQ61 82
Parliament Hl. NW3 BU35 32
Parliament Sq. SW1 BX41 42
Parliament St. SW1 BX41 42
Parma Cres. SW11 BU45 50
Parmiter Pl. E2 CC37 34
Parmiter St.
Parmiter St. E2 CB37 33
Parnell Clo. Edg. BM27 12
Parnell Rd. E3 CD37 34
Parnham St. E14 CD39 43
Parolles Rd. N19 BW33 23
Paroma Rd. Belv. CR41 45
Parr Av. Epsom BP58 76
Parr Rd. E6 CJ37 35
Parr Rd. Stan. BK30 12
Parrs Clo. Pur. BZ58 78
Parrs Pl. Hamptn. BF50 56
Parry Av. E6 CK39 44
Parry Clo. Epsom BP57 76
Parry Pl. SE18 CL42 44
Parry Rd. SE25 CA52 69
Parry St. SW8 BX43 51
Parsifal Rd. NW6 BS35 32
Parsloes Av. Dag. CP35 36
Parsonage Gdns. Enf. BZ23 8
Parsonage La. Enf. BZ23 8
Parsonage La. Sid. CQ49 63
Parsonage Manorway, Belv. CR43 54
Parsonage Rd. Rain. CV37 37
Parsonage St. E14 CF42 43
Parsons Cres. Edg. BM27 12
Parsons Field, Bans. BQ61 82
Parsonsfield Clo. Bans. BQ61 82
Parsonsfield Rd. Bans. BQ61 82
Parsons Grn. La. SW6 BS44 50
Parsons Grn. SW6 BR44 49
Parsons Gro. Edg. BM27 12
Parson's Hill SE18 CL41 44
Powis St.
Parsons La. Dart. CU48 64
Parson's Mead, Croy. BY54 69
Parsons Mead, E. Mol. BG52 65
Parson's Rd. E13 CJ37 35
Parson St. NW4 BQ31 22
Pathenia Rd. SW6 BS44 50
Partingdale La. NW7 BQ28 13
Partington Clo. N19 BW34 23
Partridge Clo. Bush. BF26 4
Partridge Dr. Orp. CM55 80
Partridge Grn. SE9 CL48 62
Partridge Knoll, Pur. BY60 84
Partridge Mead, Bans. BQ61 82
Partridge Rd. N22 BX30 15
Partridge Rd. Sid. CN48 63
Partridge Way N22 BX30 15
Pasadena Clo. Hayes BC41 38
Pascal St. SW8 BW43 50
Pascoe Rd. SE13 CF46 52
Pasquier Rd. E17 CD31 25
Passage, The, Rich. BL45 48
Quadrant, The
Passey Pl. SE9 CK46 53
Passfield Path SE28 CO40 45
Booth Clo.
Passfields Dr. E14 CE39 43
St. Leonard's Rd.
Passfields W14 BR42 40
Passmore Gdns. N11 BW29 14
Passmore St. SW1 BV42 41
Pasteur Gdns. N18 BY28 15
Paston Cres. SE12 CH47 62
Pastor St. SE11 BY42 42
Pasture Clo. Bush. BG26 4
Pasture Clo. Wem. BJ34 21
Pasture Rd. Dag. CQ35 36
Pasture Rd. SE6 CG47 61
Pasture Rd. Wem. BJ34 21
Pastures, The N20 BR26 6
Patcham Ct. Sutt. BT57 77
Patcham Ter. SW8 BV44 50
Paternoster Row EC4 BY39 42
St. Pauls Churchyard
Paternoster Row, Hav. CV27 19
Paternoster Sq. EC4 BY39 42
Warwick La.
Pater St. W8 BS41 41
Pathfield Rd. SW16 BW50 59
Path, The SW19 BS51 68
Pathway, The, Wat. BD26 4
Patience Rd. SW11 BU44 50
Patio Clo. SW4 BW46 50
Patmore Est. SW8 BW44 50
Patmore St. SW8 BW44 50
Patmore Way, Rom. CR28 18
Patmos Rd. SW9 BY43 51
Paton Clo. E3 CE38 34
Paton St. EC1 BZ38 33
Patricia Ct. Well. CO43 54
Wickham La.
Patricia Dr. Horn. CW33 28
Patrick Connolly Gdns. E3 CF38 34
Talwin St.
Patrick Rd. E13 CJ38 35
Patriot Sq. E2 CB37 33
Patshull Rd. NW5 BW36 32
Patten All. Rich. BK46 48
Ormond Rd.
Patten Av. Rich. BK46 48
Ormond Rd.
Pattenden Rd. SE6 CD47 61
Patten Rd. SW18 BU47 59
Patterdale Clo. Brom. CG50 61
Patterdale Rd. SE15 CC43 52
Patterson Ct. SE19 CA50 60
Patterson Rd. SE19 CA50 60
Pattison Rd. NW3 BS34 23
Pattison Wk. SE18 CM42 44
Sandbach Pl.
Paul Clo. E15 CG37 34
Paul St.
Paulet Rd. SE5 BY44 51
Paul Gdns. Croy. CA55 78
Paulham Rd. Har. BK31 21
Paulin Dr. N21 BY26 8
Pauline Cres. Twick. BG47 56
Paulinus Clo. Orp. CP51 72
Paul St. E15 CF37 34
Paul St. EC2 BZ38 33
Paultons Sq. SW3 BT43 50
Paulton's St. SW3 BT43 50
Old Church St.
Pauntley St. N19 BW33 23
Paved Ct. Rich. BK46 48
King St.
Paveley St. W1 BU38 32
Pavement Mews, Rom. CP33 27
Pavement, The SW4 BW45 50
Pavet Clo. Dag. CR36 36
Pavilion Rd. E15 CG38 34
Springfield Rd.
Pavilion Rd. Ilf. CK33 26
Pavilion Rd. SW1 BU41 41
Pavilion Ter. Ilf. CN32 27
Pavilion Way, Ruis. BD34 20
Pawleyne Clo. SE20 CC50 61
Pawson's Rd. Croy. BZ53 69
Paxford Rd. Wem. BJ34 21
Paxton Ct. Rich. BL44 48
Paxton Pl. SE27 CA49 60
Paxton Rd. Brom. CH50 62
College Rd.
Paxton Rd. N17 CA29 15
Paxton Rd. W4 BO43 49
Paxton Ter. SW1 BV42 41
Churchill Gdns. Rd.
Payne Rd. E3 CE37 34
Old Ford Rd.
Paynesfield Av. SW14 BN45 49
Paynesfield Rd. Bush. BH26 5
Payne St. SE8 CD43 52
Paynes Wk. W6 BR43 49
Peabody Av. SW1 BV42 41
Peabody Bldgs. WC1 BX38 33
Peabody Clo. SE10 CE44 52
Devonshire Dr.
Peabody Est. EC1 BZ38 33
Peabody Est. N17 CA30 15
Peabody Est. SE1 BY40 42
Peabody Est. SE24 BY47 60
Peabody Est. SW11 BU45 50
Peabody Est. W6 BQ42 40
Peabody Hill Est. SE21 BZ47 60
Peabody Trust SW3 BU43 50
Chelsea Manor St.
Peace St. E1 CB38 33
Peaches Clo. Sutt. BR57 76
Peachum Rd. SE3 CG43 52
Peacock Gdns. S. Croy. CD58 79
Peacock St. SE17 BY42 42
Peacock Wk. N6 BV33 23
Chomeley Cres.
Peacock Yd. SE17 BY42 42
Iliffe St.
Peahen Ct. EC2 CA39 42
Bishopsgate
Peaketon Av. Ilf. CJ31 26
Peak Hill Av. SE26 CC49 61
Peak Hill Gdns. SE26 CC49 61
Peak Hill SE26 CC49 61
Peaks Hill, Pur. BW58 77
Peaks Hill Rise, Pur. BX58 78
Peak, The SE26 CC49 61
Peal Gdns. W13 BJ38 30
Peall Rd. Croy. BX53 69
Pearcefield Av. SE23 CC47 61
Pear Clo. NW9 BN31 22
Pearcroft Rd. E11 CF34 25
Peardon St. SW8 BV44 50
Silverthorne Rd.
Peareswood Gdns. Stan. BK30 12
Pearfield Rd. SE23 CD48 61
Pearle Rd. E. Mol. BF51 65
Pearl Rd. E17 CE31 25
Pearl St. E1 CB40 42
Penang St.
Pearman St. SE1 BY41 42
Pearscroft Ct. SW6 BS44 50
Pearscroft Rd. SW6 BS44 50
Pearson's Av. SE14 CD44 52
Tanner's Hill
Pearson St. E2 CA37 33
Pears Rd. Houns., BG45 47
Peaswood Rd. Erith CT43 55
Peartree Clo. Erith CS44 55
Peartree Clo. Mitch. BU51 68
Peartree Ct. EC1 BY38 33
Peartree Gdns. Dag. CO35 36
Peartree Gdns. Rom. CR30 18
Peartree Rd. Enf. CA24 8
Pear Tree St. EC1 BY38 33
Peary Pl. E2 CC38 34
Kirkwall Pl.
Pebworth Rd. Har. BJ34 21
Peckarmans Wood SE26 CB48 60
Peckett Sq. N5 BZ35 33
Peckford Yd. SW9 BY44 51
Peckford Pl.
Peckham Gro. SE15 CA43 51
Peckham High St. SE15 CA44 51
Peckham Hill St. SE15 CB43 51
Peckham Pk. Rd. SE15 CB43 51
Peckham Rd. SE5 CA44 51
Peckham Rd. SE15 CA44 51
Peckham Rye SE15 CB45 51
Peckham Rye SE22 CB45 51
Peckwater St. Est. NW5 BW35 32
Peckwater St. NW5 BW35 32
Islip St.
Pedlars Wk. N19 BX36 33
Pedley St. NE1 CA38 33
Pedro St. E5 CC34 25
Peek Cres. SW19 BQ49 58
Peel Dr. Ilf. CK31 26
Peel Gro. E2 CC37 34
Peel Prec. NW6 BS37 32
Peel Rd. E18 CG30 16
Peel Rd. Har. BH31 21
Peel Rd. NW6 BR37 31
Peel Rd. NW9 BP31 22
Peel Rd. Orp. CM56 80
Peel Rd. Wem. BK34 21
Peel St. W8 BS40 41
Peel Way, Rom. CW30 19
Peerless St. EC1 BZ38 33
Peer Way, Horn. CW33 28
Pegamoid Rd. N9 CC27 16
Pegasus Pl. SE11 BY43 51
Clayton St.
Pegasus Rd. Croy. BY57 78
Pegelm Gdns. Houns. CW33 28
Pegg Rd. Houns. BC43 47
Pegley Gdns. SE12 CH48 62
Pegmire La. Wat. BF23 4
Pegwell St. SE18 CN43 54
Pekin St. E14 CE49 43
Peldon Ct. Rich. BL45 48
Peldon Pass. Rich. BL45 48
Sheen Rd.
Pelham Av. Bark. CN37 36
Pelham Clo. SE5 CA44 51

Pelham Ct. SW3 — BU42 41
Fulham Rd.
Pelham Cres. SW7 — BU42 41
Pelham Pl. SW7 — BU42 41
Pelham Rd. Beck. — CC51 70
Pelham Rd. Bexh. — CR45 54
Pelham Rd. E18 — CH31 26
Pelham Rd. Ilf. — CM34 26
Pelham Rd. N15 — CA31 24
Pelham Rd. N22 — BY30 15
Pelham Rd. SW19 — BS50 59
Pelhams Clo. Esher — BF56 74
Pelham Rd. SW7 — BT42 41
Pelham's Wk. Esher — BF55 74
Pelier St. SE17 — BZ43 51
Pelinore Rd. SE6 — CG48 61
Pellant Rd. SW6 — BR43 49
Pellatt Gro. N22 — BY30 15
Pellatt Rd. SE22 — CA44 51
Pellatt Rd. Wem. — CA35 33
Pelling St. E14 — CE39 43
Pellipar Clo. N13 — BY27 15
Pellipar Clo. SE18 — CK42 44
*Hillreach ***
Pellipar Gdns. SE18 — CK42 44
Ogilby St.
Pellipar Rd. SE18 — CK42 44
Pell St. E1 — CB40 42
Pelly Rd. E13 — CH37 35
Pelter St. E2 — CA38 33
Diss St.
Pelton Av. Sutt. — BS58 77
Pelton Rd. SE10 — CG42 44
Pember Av. E17 — CD31 25
Pember Rd. NW10 — BO38 31
Pemberton Av. Rom. — CV31 28
Pemberton Gdns. N19 — BW34 23
Pemberton Gdns. Rom. — CQ32 27
Pemberton Rd. E. Mol. — BG52 65
Pemberton Rd. N4 — BY32 24
Pemberton Row EC4 — BY39 42
E. Harding St.
Pemberton Ter. N19 — BW34 23
Pembrey Way, Horn. — CV36 37
Pembridge Av. Twick. — BE47 56
Pembridge Cres. W11 — BS40 41
Pembridge Gdns. W2 — BS40 41
Pembridge Ms. W8 — BS43 50
Earls Ct. Rd.
Pembridge Ms. W11 — BS40 41
Pembridge Pl. W2 — BS40 41
Pembridge Pl. W8 — BS42 41
Pembridge Rd. W11 — BS40 41
Pembridge Sq. W2 — BS40 41
Pembridge Studios W8 — BS42 41
Pembridge Vill. W11 — BS40 41
Pembroke Av. Enf. — CB23 3
Pembroke Av. Har. — BJ31 21
Pembroke Av. Surb. — BM53 66
Pembroke Av. Walt. — BD56 74
Pembroke Clo. Bans. — BS63 83
Pembroke Clo. Horn. — CW31 28
Pembroke Clo. SW1 — BV41 41
Pembroke Gdns. Clo. W8 — BS41 41
Pembroke Gdns. Dag. — CR34 27
Pembroke Gdns. W8 — BR42 40
Pembroke Ms. W8 — BS41 41
Earl's Ct. Rd.
Pembroke Pl. Edg. — BM29 12
Pembroke Pl. Islw. — BH44 48
Clifton Rd.
Pembroke Pl. W8 — BS41 41
Pembroke Rd. W8 — BR42 40
Pembroke Rd. Brom. — CJ51 71
Pembroke Rd. Erith — CS42 46
Pembroke Rd. E17 — CE32 25
Pembroke Rd. Grnf. — BF38 29
Pembroke Rd. Ilf. — CN33 27
Pembroke Rd. Mitch. — BV51 68
Pembroke Rd. N8 — BX31 24
Pembroke Rd. N10 — BV29 14
Pembroke Rd. N13 — BZ27 15
Pembroke Rd. N15 — CA32 24
Pembroke Rd. SE25 — CA52 69
Pembroke Rd. Wem. — BK34 21
Pembroke Sq. W8 — BS41 41
Pembroke St. N1 — BX36 33
Gifford St.
Pembroke Studios W8 — BR41 40
Pembroke Vill. Rich. — BK45 48
Pembroke Vill. W8 — BS41 41
Pembury Av. Wor. Pk. — BP54 67
Pembury Clo. Brom. — CG54 70
Pembury Clo. Couls. — BV60 83
Pembury Clo. E5 — CB35 33
Pembury Cres. Sid. — CQ48 63
Pembury Pl. E5 — CB35 33
Pembury Rd. E5 — CB35 33
Pembury Rd.
Pembury Rd. Bexh. — CQ43 54
Pembury Rd. E5 — CB35 33
Pembury Rd. N17 — CA30 15
Pembury Rd. SE25 — CB52 69
Pemdevon Rd. Croy. — BY54 69
Pemell Clo. E1 — CC38 34
Colebert Av.
Penang St. E1 — CB40 42
Farthing Flds.

Penarth St. SE15 — CC43 52
Penberth Rd. SE6 — CF47 61
Penbury Rd. Sthl. — BE42 38
Pencombe Mews W11 — BS40 41
Ledbury Ms. W.
Pencraig Way SE15 — CB43 51
Penda Rd. Erith — CS43 55
Pendarves Rd. SW20 — BQ51 67
Pendennis Rd. N17 — BZ31 24
Pendennis Rd. Orp. — CP55 81
Pendennis Rd. SW16 — BX49 60
Penderel Rd. Houns. — BF46 47
Penderyn Way N7 — BW35 32
Carlton Rd.
Pendle Rd. SW16 — BV50 59
Pendlestone Rd. E17 — CE32 25
Pendragon Rd. Brom. — CG48 61
Pendrell Rd. SE4 — CD44 52
Pendrell St. SE18 — CM43 53
Penduca Dr. Hayes — BD38 29
Penerley Rd. Rain. — CU39 46
Penerley Rd. SE6 — CE47 61
Penfold La. Bex. — CP48 63
Penfold Pl. NW1 — BU39 41
Penfold Rd. N9 — CC26 9
Penfold St. NW1 — BT38 32
Penfold St. NW8 — BT38 32
Penford Gdns. SE5 — CJ45 53
Penford St. SE5 — BY44 51
Pengarth Rd. Bex. — CP46 54
Penge La. SE20 — CC50 61
Penge Rd. E13 — CJ37 35
Penge Rd. SE20 — CB51 69
Penge Rd. SE25 — CB52 69
Penhall Rd. SE7 — CJ42 44
Penhill Rd. Bex. — CP47 63
Penhurst Rd. Ilf. — CL29 17
Penhurst Rd. Th. Hth. — BY53 69
Penifather La. Grnf. — BG38 29
Whiteoaks La.
Penistone Rd. SW16 — BX50 60
Peniston Wk. Rom. — CV29 19
Penketh Dr. Har. — BG34 20
Penmon Rd. SE2 — CO41 45
Pennack Rd. SE15 — CA43 51
Sumner Rd.
Pennant Ms. W8 — BS42 41
Willowbrook Estate
Pennant Ter. E17 — CD30 16
Pennard Rd. W12 — BQ41 40
Pennards, The, Sun. — BD52 65
Penn Clo. Grnf. — BF37 29
Penn Clo. Har. — BK31 21
Penn Ct. Chis. — CL51 71
Penner Clo. SW19 — BR48 58
Queensmere Rd.
Pennethorne Clo. E9 — CC37 34
Victoria Pk. Rd.
Pennethorne Rd. SE15 — CB43 51
Pennethorpe Clo. E2 — CC37 34
Penn Gdns. Chis. — CL51 71
Penn Gdns. Rom. — CR29 18
Pennine Dr. NW2 — BQ34 22
Pennine La. NW2 — BQ34 22
Pennine Dr.
Pennine Way, Bexh. — CT44 55
Pennington Clo. Rom. — CR28 18
Pennington St. E1 — CB40 42
Penn La. Bex. — CP46 54
Penn Rd. N7 — BX35 33
Penn Rd. Wat. — BC23 4
Penn St. N1 — BZ37 33
Pennycroft, Croy. — CD58 79
Pennyfields E14 — CE40 43
Penny Rd. NW10 — BM38 30
Penpoll Rd. E8 — CB36 33
Penpool La. Well. — CO45 54
Penrhyn Av. E17 — CD30 16
Penrhyn Cres. E17 — CE30 16
Penrhyn Cres. SW14 — BN45 49
Penrhyn Gdns. — BK52 66
Kings. On T.
Penrhyn Gro. E17 — CE30 16
Penrhyn Rd. Kings. On T. — BL52 66
Penrith Clo. SW15 — BR46 49
Penrith Cres. Horn. — CU35 37
Penrith Rd. Ilf. — CN29 18
Penrith Rd. N15 — BZ32 24
Penrith Rd. N. Mal. — BN52 67
Penrith Rd. Th. Hth. — BZ51 69
Penrith St. SW16 — BW50 59
Penrose Av. Wat. — BE27 11
Penrose Gro. SE17 — BZ42 42
Penrose House SE17 — BZ42 42
Penrose St. SE17 — BZ42 42
Penryn St. NW1 — BW37 32
Marcia Rd.
Penry St. SE1 — CA42 42
Pensbury Pl. SW8 — BW44 50
Pensbury St. SW8 — BW44 50
Penscroft Gdns. Borwd. — BN24 6
Pensford Av. Rich. — BM44 48
Penshurst Av. Sid. — CO46 63
Penshurst Gdns. Edg. — BM28 12
Penshurst Grn. Brom. — CG53 70
Penshurst Rd. Bexh. — CQ44 54

Penshurst Rd. E9 — CC36 34
Penshurst Rd. N17 — CA29 15
Penshurst Way, Sutt. — BS57 77
Pensmead Ter. E4 — CF28 16
Mead Cres.
Penstock N8 — BX31 24
Pentelowe Gdns. Felt. — BC46 47
Pentire Rd. E17 — CF30 16
Pentland Rd. Bush. — BG25 4
Pentlands Clo. Mitch. — BV52 68
Pentland St. SW18 — BT46 50
Pentlow St. SW15 — BQ45 49
Pentlow Way, Buck. H. — CK26 10
Pentney Rd. E4 — CF26 9
Pretoria Rd.
Pentney Rd. SW12 — BW47 59
Pentney Rd. SW19 — BR51 67
Midmoor Rd.
Penton Gro. N1 — BY37 33
Penton Pl. SE17 — BY42 42
Penton Ri. WC1 — BX38 33
Penton St. N1 — BY37 33
Pentonville Rd. N1 — BX37 33
Pentridge St. SE15 — CA43 51
Pentyre Av. N18 — BZ28 15
Penwerris Av. Islw. — BG43 47
Penwith Rd. SW18 — BS48 59
Penwortham Rd. S. — BZ58 78
Croy.
Penwortham Rd. SW16 — BV50 59
Penylan Pl. Edg. — BM29 12
Penywern Rd. SW5 — BS42 41
Penzance Pl. W11 — BR40 40
Penzance St. W11 — BR40 40
Peony Gdns. W12 — BP40 40
Curve, The
Peploe Rd. NW6 — BQ37 31
Pepper Alley, Loug. — CG23 9
Pepper St. SE1 — BZ41 42
Pepys Cres. Barn. — BQ25 6
Pepys Rd. SE14 — CC44 52
Pepys Rd. SW20 — BQ50 58
Pepys St. EC3 — CA40 42
Perceval Av. NW3 — BU35 32
Perch St. E8 — CA35 33
Percival Ct. N17 — CA29 15
High Rd.
Percival Gdns. Rom. — CP32 27
Percival Rd. Enf. — CA24 8
Percival Rd. Horn. — CV32 28
Percival Rd. Orp. — CL55 80
Percival Rd. SW14 — BN46 49
Percival St. EC1 — BY38 33
Percival Rd. SE2 — CP41 45
Harrow Manorway
Percival Way, Epsom — BN56 76
Percy Circus WC1 — BX38 33
Percy Gdns. Enf. — CC25 9
Percy Gdns. Islw. — BJ45 48
Percy Gdns. Wor. Pk. — BN54 67
Percy Ms. W1 — BW39 41
Rathbone Pl.
Percy Rd. SE20 — CC51 70
Percy Rd. SE25 — CA53 69
Percy Rd. W12 — BP41 40
Percy Rd. Bexh. — CQ44 54
Percy Rd. E11 — CB33 25
Percy Rd. E16 — CG38 34
Percy Rd. Hamptn. — BF50 56
Percy Rd. Ilf. — CO33 27
Percy Rd. Islw. — BJ45 48
Percy Rd. Mitch. — BV54 68
Percy Rd. N12 — BT28 14
Percy Rd. N21 — BZ26 8
Percy Rd. NW6 — BS38 32
Percy Rd. Rom. — CR30 18
Percy Rd. Twick. — BF47 56
Percy St. W1 — BW39 41
Percy Way, Twick. — BG47 56
Peregrine Rd. Ilf. — CO28 18
Peregrine Wk. Rain. — CU36 37
Peregrine Way Rain. — BQ50 58
Perham Rd. W14 — BR42 40
Perifield SE21 — BZ47 60
Periton Rd. SE9 — CJ45 53
Perivale Gdns. W13 — BJ38 30
Bellevue Rd.
Perivale La. Grnf. — BJ38 30
Perivale La. W13 — BJ38 30
Perkins Rents SW1 — BW41 41
Old Pye St.
Perkins Rd. Ilf. — CO45 54
Perks Clo. SE3 — CG45 52
Hurren Clo.
Perpins Rd. SE9 — CM46 53
Perran Rd. SW2 — BY47 60
Perrers Rd. W6 — BP42 40
Perring Est. E3 — CE39 43
Perrin Rd. Wem. — BJ35 30
Perrins La. NW3 — BT35 32
Hampstead High St.
Perrins Wk. NW3 — BT35 32

Perrott St. SE18 — CM42 44
Perry Av. W3 — BN39 40
Perry Clo. Rain. — CS37 37
Perryfield Way NW9 — BN32 22
Broadway, The
Perryfield Way NW9 — BO32 22
Perryfield Way, Rich. — BJ48 57
Perry Gdns. N9 — BZ27 15
Perry Gth. Nthlt. — BD37 29
Perry Hall Clo. Orp. — CO54 72
Perry Hall Rd. Orp. — CN54 72
Perry Hl. SE23 — CD48 61
Perry Hl. SE6 — CD48 61
Perry How, Wor. Pk. — BO54 67
Perrymans Farm Rd. Ilf. — CM32 26
Perry Mead, Bush. — BF26 4
Perry Mead, Enf. — BY23 8
Perrymead St. SW6 — BS44 50
Perryn Rd. SE16 — CB41 42
Perryn Rd. W3 — BN40 40
Perry Rise SE23 — CD48 61
Perry St. Chis. — CM50 62
Perry St. Dart. — CT45 55
Perry St. Gdns. Chis. — CN50 63
Perry St. Shaw. Chis. — CN50 63
Perry Vale SE23 — CC48 61
Persant Rd. SE6 — CG48 61
Persfield Clo. Epsom — BO58 76
Pershore Clo. Ilf. — CL32 26
Pershore Gro. Cars. — BT53 68
Pert Clo. N10 — BV29 14
Perth Av. Hayes — BD38 29
Perth Av. NW9 — BN33 22
Perth Clo. SW20 — BP51 67
Perth Rd. Bark. — CM37 35
Perth Rd. Beck. — CF51 70
Perth Rd. E10 — CD33 25
Perth Rd. E13 — CH37 35
Perth Rd. Ilf. — CL32 26
Perth Rd. N4 — BY33 24
Perth Rd. N22 — BY30 15
Perth Ter. Ilf. — CM33 26
Perwell Av. Har. — BE33 20
Perwell Ct. Har. — BE33 20
Perys Rise BR6 — CN54 72
Pescot Av. NW10 — BP36 31
Peterborough Gdns. Ilf. — CK33 26
Peterborough Ms. SW6 — BS45 50
Peterborough Rd. Cars. — BT53 68
Peterborough Rd. E10 — CF32 22
Peterborough Rd. Har. — BH33 21
Peterborough Rd. SW6 — BS44 50
Peterborough Vill. SW6 — BS44 50
Petergate SW11 — BT45 50
Peters Clo. Stan. — BK29 12
Petersfield Av. Rom. — CW29 19
Petersfield Clo. N18 — BZ28 15
Petersfield Cres. Couls. — BX61 84
Petersfield Ri. SW15 — BP47 58
Petersfield Rd. W3 — BN41 40
Petersham Clo. Rich. — BK48 57
Petersham Clo. Sutt. — BS56 77
Petersham Dr. Orp. — CN51 72
Petersham Gdns. Orp. — CN51 72
Petersham Dr.
Petersham La. SW7 — BT42 41
Petersham Ms. SW7 — BT41 41
Petersham Pl. SW7 — BT41 41
Petersham Rd. Rich. — BK46 48
Petersham Way, Orp. — CO52 72
Peters Hill EC4 — BZ39 42
Peter's La. EC1 — BY39 42
Cowcross St.
Peters Path SE26 — CB49 60
Peterstone Rd. SE2 — CO41 45
Peterstow Clo. SW15 — BR48 58
Princes Way
Peter St. W1 — BW40 41
Petherton Rd. N5 — BZ35 33
Petley Rd. W6 — BQ43 49
Peto Pl. NW1 — BV38 32
Peto St. S. E16 — CG40 43
Petten Clo. Orp. — CP54 72
Petten Gro. Orp. — CP54 72
Pettits Boul. Rom. — CT30 19
Pettits Clo. Rom. — CT30 19
Pettits La. Rom. — CT30 19
Pettits La. N. Rom. — CS30 19
Pettits Pl. Dag. — CR35 36
Pettits Rd. Dag. — CR35 36
Pettley Gdns. Rom. — CS32 28
Pettman Cres. SE28 — CM41 44
Petts Gro. Av. Wem. — BK35 30
Petts Hill, Nthlt. — BF35 29
Pett St. SE18 — CK42 44
Pett's Wd. Rd. Orp. — CM53 71
Petty France SW1 — BW41 41
Petty La. Brom. — CH51 71
Church Rd.
Petworth Clo. Nthlt. — BE36 29
Petworth Clo. Sutt. — BS55 77
Petworth Gdns. SW20 — BP52 67
Petworth Rd. Bexh. — CR46 54
Petworth Rd. N12 — BU28 14
Petworth St. SW11 — BU44 50
Petworth Way, Horn. — CT35 37
Petyt Pl. SW3 — BU43 50

Petyward SW3	BU42	41
Pevensey Av. Enf.	BZ23	8
Pevensey Av. N11	BW28	14
Pevensey Clo. Houns.	BG43	47
Pevensey Rd. E7	CG35	34
Pevensey Rd. Felt.	BD47	56
Pevensey Rd. SW17	BT49	59
Peveril Dr. Tedd.	BG49	56
Pewsey Clo. E4	CE28	16
Peyton Pl. SE10	CF43	52
Phelp St. SE17	BZ42	42
Phene St. SW3	BU43	50
Philan Way, Rom.	CS29	19
Philbeach Gdns. SW5	BS42	41
Philchurch St. E1	CB39	42
Ellen St.		
Philip Gdns. Croy.	CD55	79
Philip La. N15	BZ31	24
Philipot Path SE9	CK46	53
Philippa Gdns. SE9	CJ46	53
Philip Rd. Rain.	CT38	37
Philip Rd. SE15	CB45	51
Philip St. E13	CH38	35
Philip Wk. SE15	CB45	51
Phillimore Clo. W8	BS41	41
Phillimore Gdns. NW10	BQ37	31
Phillimore Gdns. W8	BS41	41
Phillimore Pl. W8	BS41	41
Phillimore Wk. W8	BS41	41
Phillip Av. Rom.	CS33	28
Phillip Av. Swan.	CS52	73
Phillip Clo. Rom.	CS33	28
Phillip Av.		
Phillipp St. N1	CA37	33
Phillips Clo. Dart.	CU46	55
Phillips Way, Brom.	CH51	71
Lownds Av.		
Philpot La. EC3	CA40	42
Philpot St. E1	CB39	42
Phineas Pett Rd. SE9	CK45	53
Phipp's Br. Rd. SW19	BT51	68
Phipp's Ter. SW19	BT51	68
Phipp St. EC2	CA38	33
Phoebeth Rd. SE4	CE46	52
Phoenix Dr. Kes.	CJ55	80
Phoenix Lo. Mans. W6	BQ42	40
Brook Green		
Phoenix Pl. Dart.	CV47	64
Phoenix Rd. NW1	BW38	32
Phoenix Rd. SE20	CC50	61
Phoenix St. WC2	BW39	41
Stacey St.		
Phoenix Way, Houns.	BD43	47
Phyllis Av. N. Mal.	BP53	67
Phyllis Ct. NW3	BS34	23
Picardy Manorway, Belv.	CR41	45
Picardy Rd. Belv.	CR42	45
Picardy St. Bel.	CR41	45
Piccadilly Pl. W1	BW40	41
Piccadilly		
Piccadilly W1	BV40	41
Pickard St. EC1	BY38	33
Moreland St.		
Pickering Av. E6	CL37	35
Pickering Ms. W2	BS39	41
Bishop's Br. Rd.		
Picket Croft, Stan.	BK30	12
Pickets Clo. Bush.	BH26	5
Pickets St. SW12	BV47	59
Picketts Lock La. N9	CC27	16
Pickford Clo. Bexh.	CQ44	54
Pickford La. Bexh.	CQ44	54
Pickford Rd. Bexh.	CQ44	54
Pickhurst Grn. Brom.	CG54	70
Pickhurst La. Brom.	CG54	70
Pickhurst La. W. Wick.	CG53	70
Pickhurst Mead, Brom.	CG54	70
Pickhurst Pk. Brom.	CG53	70
Pickhurst Rise, W. Wick.	CF54	70
Pickle Herring St. SE1	CA40	42
Pickwick Mews N18	CA28	15
Pickwick Pl. Har.	BH33	21
Pickwick Rd. SE21	BZ47	60
Pickwick St. SE1	BZ41	42
Pickwick Way BR7	CM50	62
Dickens Dr.		
Pickwick Way, Chis.	CM50	62
Picquets Way, Bans.	BR61	82
Picton Pl. W1	BV39	41
Picton St. SE5	BZ43	51
Piedmont Rd. SE18	CM42	44
Pier Head Cotts. E14	CE41	43
Piermont Grn. SE22	CB46	51
Peckham Rye		
Piermont Rd. SE22	CB46	51
Upland Rd.		
Pierrepoint Rd. W3	BM40	39
Pier Rd. E16	CL40	44
Pier Rd. Erith	CT43	55
Pier Rd. Est. E16	CL41	44
Pier Rd. Felt.	BC46	47
Pier St. E14	CF42	43
Pier Ter. SW18	BS45	50
Pigeon La. Hamptn.	BF49	56
Piggot St. E14	CE39	43
Pike Clo. Brom.	CH49	62
Pike Gdns. SE1	BY40	42
Pikes End, Pnr.	BC31	20
Pikes Hill, Epsom	BO60	82
Pilgrimage St. SE1	BZ41	42
Pilgrim Hill, Orp.	CQ51	72
Pilgrim Hill SE27	BZ49	60
Pilgrims Ct. SE3	CH44	53
Pilgrim's La. NW3	BT35	32
Pilgrims Rise, Barn.	BU25	7
Pilgrim St. EC4	BY39	42
Pilgrims Way N19	BW34	23
Pilgrims Way, S. Croy.	CA57	78
Pilkington Rd. Orp.	CL55	80
Pilkington Rd. SE15	CB44	51
Pillmans Clo. Sid.	CO50	63
Piltdown Rd. Wat.	BD28	11
Pilton Rd. SW6	BS44	50
Pimlico Rd. SW1	BV42	41
Pimlico Wk. N1	CA38	33
Pimlico Wk. N1	CA37	33
Pimpernel Way, Rom.	CV29	19
Pinchbeck Rd. Orp.	CN57	81
Pinchin St. E1	CB40	42
Pincott Rd. Bexh.	CR46	54
Pincott Rd. SW19	BT50	59
Pindar St. EC2	CA39	42
Pindock Ms. W9	BS38	32
Pine Av. W. Wick.	CE54	70
Pine Clo. Swan.	CT52	73
High St.		
Pine Clo. Ken.	BZ62	84
Pine Clo. N14	BW26	7
Pine Clo. Stan.	BJ28	12
Pine Clo. Swan.	CT52	73
Pine Coombe, Croy.	CC56	79
Pine Cres. Cars.	BT59	83
Pinecroft, Pnr.	BF29	11
Pinefield Clo. E14	CD40	43
Pine Gdns. Ruis.	BC33	20
Pine Gdns. Surb.	BM53	66
Pineglade, Orp.	CK56	80
Pine Gro. Bush.	BE23	4
Pine Gro. N4	BX34	24
Pine Gro. N20	BR26	6
Pine Gro. SW19	BR49	58
Pine Hill, Epsom	BN61	82
Pinehurst Wk. Orp.	CN54	72
Pinelands Clo. SE3	CG43	52
St. John's Pk.		
Pine Pl. Bans.	BQ60	82
Pine Ridge, Cars.	BV57	77
Pine Rd. N11	BV27	14
Pine Rd. NW2	BQ35	31
Pines Rd. Brom.	CK51	71
Pines, The, Pur.	BY60	84
Pines, The, Pur.	BY56	78
Pines, The, Sun.	BC52	65
Pines, The, Wdf. Grn.	CH27	17
Pine St. EC1	BY38	33
Pine Wk. Bans.	BU62	83
Pine Wk. E. Cars.	BT59	83
Pine Wk. Surb.	BM53	66
Pine Wk. W. Cars.	BT58	77
Pinewood Av. Pnr.	BF29	11
Pinewood Av. Rain.	CU38	37
Pinewood Av. Sid.	CN47	63
Pinewood Clo. Croy.	CD55	79
Pinewood Clo. Orp.	CM54	71
Pinewood Dr. Orp.	CM56	80
Pinewood Gro. W13	BK39	39
Pinewood Rd. Brom.	CH52	71
Pinewood Rd. Felt.	BC48	56
Pinewood Rd. Hav.	CS28	19
Pinewood Rd. SE2	CP43	54
Pine Wd. Sun.	BC51	65
Pinfold Rd. Bush.	BE23	4
Pinfold Rd. SW16	BX49	60
Pinks Hill, Swan.	CT53	73
Pinley Gdns. Dag.	CO37	36
Pinnacle Hill, Bexh.	CR45	64
Pinnell Pl. SE9	CJ45	53
Pinnell Rd. SE9	CJ45	53
Pinner Ct. Pnr.	BF31	20
Pinner Cres. Pnr.	BE32	20
Pinner Grn. Pnr.	BD30	11
Pinner Hill, Pnr.	BD30	11
Pinner Hill Rd. Pnr.	BD30	11
Pinner Pk. Av. Har.	BF30	11
Pinner Pk. Gdns. Har.	BG30	11
Pinner Rd. Pnr.	BE31	20
Pinner Rd. Wat.	BD25	4
Pinner Vw. Har.	BG32	20
Pintail Rd. IG8	CH29	17
Pinto Clo. Borwd.	BN25	6
Pinto Way SE3	CH45	53
Piper Clo. N7	BX35	33
Piper Rd. Kings. On T.	BM52	66
Pipers Grn. La. Edg.	BL27	12
Pipers Grn. NW9	BN32	22
Pipewell Rd. Mitch.	BU53	68
Pippin Clo. Croy.	CD54	70
Piquet Rd. SE20	CC51	70
Pirbright Cres. Croy.	CF57	79
Pirbright Rd. SW18	BR47	58
Pirie Rd. E16	CH40	44
Pitcairn Clo. Rom.	CR31	27
Pitcairn Rd. Mitch.	BU50	59
Pitchford St. E15	CF36	34
Pitfield Est. N1	CA38	33
Pitfield St. N1	CA38	33
Pitfield Way, Enf.	CC23	9
Pitfield Way NW10	BN36	31
Pitfold Clo. SE12	CH46	53
Pitfold Rd. SE12	CH46	53
Pitlake, Croy.	BY55	78
Pitman St. SE5	BZ43	51
Pitsea Pl. E1	CC39	43
Pitsea St.		
Pitsea St. E1	CC39	43
Pitshanger La. W5	BJ38	30
Pitsmead Av. Brom.	CH54	71
Pitt Cres. SW19	BS49	59
Pitt Rd. Epsom	BO60	82
Pitt Rd. Orp.	CM56	80
Pitt Rd. Th. Hth.	BZ53	69
Pitts Head Ms. W1	BV40	41
Pitt St. SE15	CA44	51
Sumner Estate		
Pitt St. W8	BS41	41
Pittville Gdns. SE25	CB52	69
Pixfield Ct. Brom.	CG51	70
Pixley St. E14	CD39	43
Pixton Way, Croy.	CD58	79
Plaistow Grn. Brom.	CH50	62
Plaistow Gro. Brom.	CH50	62
Plaistow Gro. E15	CG37	34
Plaistow La. Brom.	CH50	62
Plaistow Park Rd. E13	CH37	35
Plaistow Rd. E15	CG37	34
Plane St. SE26	CB48	60
Plantagenet Clo. Wor. Pk.	BN56	76
Plantagenet Gdns. Rom.	CP33	27
Plantagenet Pl. Rom.	CP33	27
Broomfield Rd.		
Plantagenet Rd. Barn.	BT24	7
Plantation Dr. Orp.	CP54	72
Plantation Rd. Erith	CU43	55
Plantation Rd. Swan.	CU50	64
Plantation, The SE3	CH44	53
Plane Tree Cres. Felt.	BC48	56
Plashet Gro. E6	CJ36	35
Plashet Rd. E13	CH37	35
Plassy Rd. SE6	CE47	61
Platford Grn. Horn.	CW31	28
Plato Rd. SW2	BX45	51
Platts Av. Wat.	BC24	4
Platt's La. NW3	BS35	32
Platts Rd. Enf.	CC23	9
Platt St. NW1	BW37	32
Platt, The SW15	BQ45	49
Plawsfield Rd. SE20	CC51	70
Plaxtol Clo. Brom.	CJ51	71
Plaxtol Pl. SE10	CH42	44
Westcombe Hill		
Plaxtol Rd. Erith	CR43	54
Playfield Av. Rom.	CS30	19
Playfield Cres. SE22	CA46	51
Playfield Rd. Edg.	BN30	13
Playford Rd. N4	BX34	24
Playgreen Way SE6	CE48	61
Playhouse Yd. EC4	BY39	42
Pleany Clo. N15	BY31	24
Pleasance Rd. Orp.	CO51	72
Pleasance Rd. SW15	BP45	49
Pleasance, The SW15	BP45	49
Pleasance Rd.		
Pleasant Gro. Croy.	CD55	79
Pleasant Pl. Har.	BG33	20
Pleasant Pl. N1	BY36	33
Pleasant Pl. Walt.	BD57	74
Pleasant Rd. Kings. On T.	BN52	67
Pleasant Row NW1	BV37	32
Camden High St.		
Pleasant Vw. Erith	CT42	46
Pleasant Vw. Pl. Orp.	CM56	80
Pleasant Way, Wem.	BK37	30
Plender Pl. NW1	BW37	32
Plender St.		
Plender St. Est. NW1	BW37	32
Plender St. NW1	BW37	32
Pleshey Rd. N7	BW35	32
Plesman Way, Wall.	BX57	78
Plevna Rd. N9	CB27	15
Plevna Rd. Hamptn.	BF51	65
Plevna St. E6	CL40	44
Shepstone St.		
Plevna St. E14	CF41	43
Pleydell Av. SE19	CA50	60
Pleydell Av. W6	BO42	40
Pleydell Ct. EC4	BZ39	42
Fleet St.		
Plimsoll Rd. N4	BY34	24
Plough All. E1	CB40	42
Hermitage Wall		
Plough Est. SE8	CC42	43
Plough La. Clo. Wall.	BX56	78
Plough La. Pur.	BX58	78
Plough La. SE22	CA46	51
Plough La. SW17	BS49	59
Plough La. SW19	BS49	59
Plough La. Wall.	BX56	78
Ploughmans End, Islw.	BG46	47
Reapers Way		
Plough Pl. EC4	BY39	42
New Fetter La.		
Plough Rd. Brent.	BK43	48
Brent Way		
Plough Rd. Epsom	BN58	76
Plough Rd. SW11	BT45	50
Plough St. SW11	BT45	50
Plough Way SE16	CC42	43
Plough Yd. EC2	CA38	33
Hearn St.		
Pluckington Pl. Sthl.	BE41	38
Plumbers Row E1	CB39	42
Plumbridge St. SE10	CE44	52
Plum Garth, Brent.	BK42	39
Plum La. SE18	CL43	53
Plummer La. Mitch.	BU51	68
Plummer Rd. SW4	BW46	50
Plumpton Av. Horn.	CW35	37
Plumpton Clo. Nthlt.	BF36	29
Plumpton Way, Cars.	BU55	77
Plumstead Common Rd. SE18	CL43	53
Plumstead High St. SE18	CN42	45
Plumstead Rd. SE18	CL42	44
Plumtree Ct. EC4	BY39	42
Shoe La.		
Plumtree Mead, Loug.	CL24	10
Plymouth Rd. Brom.	CH51	71
Plymouth Rd. E16	CH39	44
Plympton Rd. NW6	BR36	31
Plympton St. NW8	BU38	32
Plymstock Rd. Well.	CP43	54
Pocklington Clo. NW9	BO30	13
Pocock St. SE1	BY41	42
Podmore Rd. SW18	BT45	50
Poet's Rd. N5	BZ35	33
Pointalls Clo. N3	BT30	14
Point Clo. SE10	CF44	52
Point Hill		
Point Hl. SE10	CF43	52
Point Pleasant SW18	BS45	50
Poiters Clo. Loug.	CK23	10
Poland St. W1	BW39	41
Polebrook Rd. SE3	CJ45	53
Pole Cat Alley, Brom.	CG55	79
Polecroft La. SE6	CD48	61
Pole Hill Rd. E4	CF26	9
Polesden Gdns. SW20	BP51	67
Polesworth Rd. Dag.	CP36	36
Police Sta. La. Bush.	BF26	4
School La.		
Police Sta. Rd. Walt.	BD57	74
Pollard Clo. Chig.	CO28	18
Pollard Clo. E16	CH40	44
Munday Rd.		
Pollard Clo. N7	BX35	33
Pollard Rd. Mord.	BT53	68
Pollard Rd. N20	BU27	14
Pollard Rd. NW9	BO33	22
Broadway, The		
Pollard Row E2	CB38	33
Pollards Cres. SW16	BX52	69
Pollards Hill E. SW16	BX52	69
Pollards Hill N. SW16	BX52	69
Pollards Hill S. SW16	BX52	69
Pollards Hill W. SW16	BX52	69
Pollards, Loug.	CJ25	10
Pollard St. E2	CB38	33
Pollards Wd. Rd. SW16	BX52	69
Pollen St. W1	BW39	41
Hanover St.		
Polsted Rd. SE6	CD47	61
Polthorne Gro. SE18	CM42	44
Poltimore Ms. W3	BN41	40
Mill Hill Gro.		
Polworth Rd. SW16	BX49	60
Polygon Rd. NW1	BW37	32
Polytechnic St. SE18	CL42	44
Pomeroy Sq. SE14	CC43	52
Pomeroy St.		
Pomeroy St. SE14	CC44	52
Pomfret Rd. SE5	BZ45	51
Flaxman Rd.		
Pond Clo. SE3	CG44	52
Pond Clo. Walt.	BC57	74
Pond Cotts. SE21	CA47	60
Ponder St. N7	BX36	33
Pondfield Rd. Brom.	CG54	70
Pondfield Rd. Dag.	CR35	36
Pondfield Rd. Ken.	BY61	84
Pondfield Rd. Orp.	CL55	80
Pond Hill Gdns. Sutt.	BR57	76
Pond Hill, Sutt.	BR57	76
Pond Ho SW3	BU42	41
Pond Pl. SW3	BU42	41
Pond Rd. E15	CG37	34
Pond Sq. N6	BV33	23
Pond St. NW3	BU35	32
Pond Way, Tedd.	BK50	57
Pondwood Rise, Orp.	CN54	72
Ponler St. E1	CB39	42
Ponsard Rd. NW10	BP38	31
Ponsford St. E9	CC36	34
Ponsonby Pl. SW1	BW42	41
Ponsonby Rd. SW15	BP47	58
Ponsonby Ter. SW1	BW42	41
Pontefract Rd. Brom.	CG49	61
Ponton Rd. SW8	BW43	50
Pont St. Ms. SW1	BU41	41
Pont St. SW1	BU41	41
Pontypool Wk. Rom.	CV29	19
Pool Clo. E. Mol.	BF53	65
Poole Ct. Rd. Houns.	BE44	47

Poole Rd. E9	CC36	34
Poole Rd. Epsom	BN57	76
Pooles Cotts. Rich.	BK48	57
Clifford Rd.		
Pooles La. Dag.	CQ37	36
Pooles La. SW10	BT43	50
Pooles Pk. N4	BY34	24
Poole St. N1	BZ37	33
Pool Rd. E. Mol.	BE53	65
Poolsford Rd. NW9	BO31	22
Pope Rd. Brom.	CJ53	71
Pope's Av. Twick.	BH48	57
Popes Dr. N3	BS30	14
Popes Gro. Croy.	CD55	79
Pope's Gro. Twick.	BH48	57
Popes La. W5	BK41	39
Popes Rd. SW9	BY45	51
Pope St. SE1	CA41	42
Popham Clo. Felt.	BE48	56
Popham Gdns. Rich.	BM45	48
Marksbury Av.		
Popham Rd. N1	BZ37	33
Popham St. N1	BY37	33
Poplar Av. Mitch.	BU51	68
Poplar Av. Orp.	CL55	80
Poplar Av. Sthl.	BF41	38
Poplar Bath St. E14	CE40	43
Lawless St.		
Poplar Clo. Pnr.	BD30	11
Poplar Ct. SW19	BS49	59
Poplar Cres. Epsom	BN57	76
Poplar Fm. Clo. Epsom	BN57	76
Poplar Gdns. N. Mal.	BN51	67
Poplar Gro. N. Mal.	BN52	67
Poplar Gro. W6	BQ41	40
Poplar Gro. Wem.	BN34	22
Poplar High St. E14	CE40	43
Poplar Mt. Belv.	CR42	45
Poplar Pl. SE28	CP40	45
Poplar Pl. W2	BS40	41
Poplar Rd. SE24	BZ45	51
Poplar Rd. S. SW19	BS52	68
Poplar Rd. Sutt.	BR54	67
Poplar Rd. SW19	BS51	68
Poplars Av. NW2	BQ36	31
High Rd.		
Poplars Rd. E17	CE32	25
Poplars, The, Borwd.	BM23	5
Grove Rd.		
Poplars, The N14	BV25	7
Poplar St. Rom.	CS31	28
Poplar Wk. Croy.	BZ54	69
Poplar Wk. SE24	BZ45	51
Poplar Way, Felt.	BC48	56
Poplar Way, Ilf.	CM31	26
Poppin's Ct. EC4	BY39	42
St. Bride St.		
Poppleton Rd. E11	CG32	25
Porchester Clo. Horn.	CW32	28
Porchester Gdns. W2	BS40	41
Porchester Pl. NW1	BW36	32
Agar Gro.		
Porchester Pl. W2	BU39	41
Porchester Rd. Kings. On T.	BM51	66
Porchester Rd. W2	BS39	41
Porchester Sq. W2	BS39	41
Porchester Ter. N. W2	BS40	41
Porchester Ter. W2	BT39	41
Porchfield Clo. Sutt.	BS58	77
Hulverston Clo.		
Porch Way N20	BU27	14
Porcupine Clo. SE9	CK48	62
Porden Rd. SW2	BX45	51
Porlock Av. Har.	BG33	20
Porlock Rd. Enf.	CA26	8
Porlock St. SE1	BZ41	42
Portal Clo. Ruis.	BC35	29
Portal Clo. SE27	BY48	60
Port Cres. E13	CH38	35
Portcullis Lo. Rd. Enf.	BZ24	8
Baker St.		
Portelet Rd. E1	CC38	34
Porten Rd. W14	BR41	40
Porter's Av. Dag.	CO36	36
Porter's Rd. W2	BT39	41
Porter St. W1	BU39	41
Baker St.		
Porteus Rd. W2	BT39	41
Porthcawe Rd. SE26	CD49	61
Porthester Mead, Beck.	CE50	61
Porthkerry Av. Well.	CO45	54
Portinscale Rd. SW15	BR46	49
Portland Av. N16	CA33	24
Portland Av. N. Mal.	BO54	67
Portland Av. Sid.	CO46	54
Portland Cres. SE9	CK48	62
Portland Cres. Grnf.	BF38	29
Portland Cres. Stan.	BK30	12
Portland Gdns. N4	BY32	24
Portland Gdns. Rom.	CP32	27
Portland Gro. SW8	BX44	51
Lansdowne Way		
Portland Pl. Epsom	BO59	82
Portland Pl. W1	BV38	32
Portland Rise Est. N4	BY33	24
Portland Rise N4	BY33	24
Portland Rd. Brom.	CJ49	62
Portland Rd. Kings. On T.	BL52	66
Portland Rd. Mitch.	BU51	68
Portland Rd. N15	CA31	24
Portland Rd. SE9	CK48	62
Portland Rd. SE25	CB52	69
Portland Rd. Sthl.	BE41	38
Portland Rd. W11	BR40	40
Portland St. SE17	BZ42	42
Portland Ter. Rich.	BK45	48
Portman Av. SW14	BN45	49
Portman Bldgs. NW1	BU38	32
Portman Clo. Bex.	CS47	64
Portman Clo. DA7	CP45	54
Glynde Rd.		
Portman Clo. W1	BV39	41
Portman Dr. Wdf. Grn.	CJ30	17
Portman Gdns. NW9	BN30	13
Portman Ms. S. W1	BV39	41
Portman Pl. E2	CC38	34
Portman Rd. Kings. On T.	BL51	66
Portman Sq. W1	BV39	41
Portman St. W1	BV39	41
Portmeadow Wk. SE2	CP41	45
Portmore Gdns. Rom.	CR28	18
Portnall Rd. W9	BR38	31
Portnall's Clo. Couls.	BV61	83
Portnall's Rise, Couls.	BV61	83
Portnalls Rd. Couls.	BV62	83
Portnoi Clo. Rom.	CS30	19
Portobello Ct. W11	BR40	40
Portobello Ms. W11	BS40	41
Portobello Rd.		
Portobello Rd. W10	BR39	40
Portobello Rd. W11	BR39	40
Portpool La. EC1	BY39	42
Portree Clo. N22	BX29	15
Nightingale Rd.		
Portree St. E14	CF39	43
Portsdown Av. NW11	BR32	22
Portsdown La. Edg.	BM28	12
Portsea Ms. W2	BU39	41
Kendal St.		
Portsea Pl. W2	BU39	41
Kendal St.		
Portslade Rd. SW8	BW44	50
Portsmouth Av. Surb.	BJ54	66
Portsmouth Bldgs. NW1	BU39	41
Portsmouth Rd. Esher	BE58	74
Portsmouth St. WC2	BX39	42
Portugal St.		
Portsoken St. E1	CA40	42
Portswood Pl. SW15	BO46	49
Danebury Av.		
Portswood SW15	BO46	58
Danebury Av.		
Portugal Gdns. Twick.	BG48	56
Portugal St. WC2	BX39	42
Portway Cres. Epsom	BP58	76
Portway E15	CG37	34
Portway, Epsom	BP58	76
Portway Gdns. SE18	CJ43	53
Postern Grn. Enf.	BY24	8
Post La. Twick.	BG47	56
Post Office Alley, Hamptn.	BF51	65
Thames St.		
Post Office App. E7	CH35	35
Post Office Way SW8	BW43	50
Postway Mews, Ilf.	CL34	26
Chadwick Rd.		
Potier St. SE1	BZ41	42
Potter Clo. Mitch.	BV51	68
Potter Heights Clo. Pnr.	BC29	11
Potterne Clo. SW19	BQ47	58
Castlecombe Dr.		
Potters Clo. Loug.	CK23	10
Potter's Fields SE1	CA40	42
Potters Gro. N. Mal.	BN52	67
Potter's La. Barn.	BS24	7
Potters La. Borwd.	BN23	6
Potter's La. SW16	BW50	59
Potter's Rd. Barn.	BS24	7
Potter St. Hill, Pnr.	BC29	11
Potter St. Pnr.	BC30	11
Pottery La. W11	BR40	40
Portland Rd.		
Pottery Rd. DA5	CS48	64
Pottery Rd. Brent.	BL43	48
Pottery St. SE16	CB41	42
Wilson Gro.		
Pott St. E2	CB38	33
Bethnal Green Rd.		
Poulett Gdns. Twick.	BJ47	57
Poulett Rd. E6	CK37	35
Poulner Way SE15	CA43	51
Poulters Wood, Kes.	CJ56	80
Poulton Av. Sutt.	BT55	77
Poultry EC2	BZ39	42
Cheapside		
Pound Clo. Surb.	BK54	66
Pound Ct. Dr. Orp.	CM55	80
Poundfield Rd. Loug.	CL25	10
Pound La. Epsom	BN59	82
Pound La. NW10	BP36	31
Pound Park Rd. SE7	CJ42	44
Pound Pl. SE9	CL46	53
Pound Rd. Bans.	BR62	82
Pound St. Cars.	BU56	77
Pountey Rd. SW11	BV45	50
Poverest Rd. Orp.	CN53	72
Powder Mill La. Dart.	CW48	64
Powder Mill La. Twick.	BE47	56
Powell Clo. Edg.	BL29	12
Powell Gdns. Dag.	CR35	36
Powell Rd. Buck. H.	CJ26	10
Powell Rd. E5	CB34	24
Powell's Wk. W4	BO43	49
Power Rd. W4	BM42	39
Powers Ct. Twick.	BK47	57
Cambridge Pk.		
Powerscroft Rd. E5	CC35	34
Powerscroft Rd. Sid.	CP50	63
Powis Gdns. NW11	BR33	22
Powis Gdns. W11	BR39	40
Powis Ms. W11	BR39	40
Powis Pl. WC1	BX38	33
Powis Rd. E3	CE38	34
Powis Sq. W11	BR39	40
Powis St. SE18	CL41	44
Powis Ter. W11	BR39	40
Powlett Pl. NW1	BV36	32
Harmood St.		
Pownall Gdns. Houns.	BF45	47
Pownall Rd. E8	CA37	33
Pownall Rd. Houns.	BF45	47
Powster Rd. Brom.	CH49	62
Powy's Clo. Bexh.	CP43	54
Powys La. N13	BX28	15
Powys La. N14	BX28	15
Poynders Ct. SW4	BW46	50
Poynders Gdns. SW4	BW47	59
Poynder's Rd. SW4	BW46	50
Plummer Rd.		
Poynings Clo. Orp.	CO55	81
Poynings Rd. N19	BW34	23
Poynings Way N12	BS28	14
Poynings Way, Rom.	CW30	19
Poyntell Cres. Chis.	CM50	62
Poynter Rd. Enf.	CA25	8
Poynton Rd. N17	CB30	15
Poyntz Rd. SW11	BU44	50
Poyser St. E2	CB37	33
Old Bethnal Grn. Rd.		
Praed Ms. W2	BT39	41
Norfolk Pl.		
Praed St. W2	BT39	41
Pragel St. E13	CJ37	35
Pragnell Rd. SE12	CH48	62
Prague Pl. SW2	BX46	51
Prah Rd. N4	BY34	24
Prairie St. SW8	BV44	50
Pratt Ms. NW1	BW37	32
Pratt St.		
Pratts La. Walt.	BD56	74
Pratt St. NW1	BW37	32
Pratt Wk. SE11	BX42	42
Prayle Gro. NW2	BQ33	22
Prebend Gdns. W4	BO42	40
Prebend Gdns. W6	BO42	40
Prebend St. N1	BZ37	33
Precinct Rd. Hayes	BC40	38
Precinct, The KT8	BF52	65
Premier Pl. SW15	BQ45	49
Putney High St.		
Prendergast Rd. SE3	CG45	52
Prentis Rd. SW16	BW49	59
Prentiss Ct. SE7	CJ42	44
Prescelly Pl. Edg.	BL30	12
Prescot St. E1	CA40	42
Prescott Av. Orp.	CL53	71
Prescott Grn. Loug.	CM24	10
Prescott Pl. SW4	BW45	50
Pressland St. W10	BR39	40
Kensal Rd.		
Press Rd. NW10	BN34	22
Prestage St. E14	CF40	43
Quixley St.		
Prestburg Rd. N. Mal.	BO53	67
Prestbury Cres. Bans.	BU61	83
Prestbury Rd. E7	CJ36	35
Prestbury Sq. SE9	CK49	62
Prested Rd. SW11	BU45	50
Preston Clo. SE1	CA42	42
Preston Clo. Twick.	BH48	57
Preston Ct. KT12	BD54	65
Lansdown Clo.		
Preston Dr. E11	CJ32	26
Preston Dr. Bexh.	CP44	54
Preston Dr. Epsom	BO57	76
Preston Gdns. Ilf.	CK32	26
Preston Hill, Har.	BL32	21
Preston Pk. N3	BR30	13
Preston Pl. NW2	BP36	31
Belton Rd.		
Preston Pl. NW6	BP36	31
Preston Pl. Rich.	BL46	48
Preston Rd. E4	CF29	16
Preston Rd. E11	CG32	25
Preston Rd. SE19	BY50	60
Preston Rd. SW20	BO50	58
Preston Rd. Har.	BL33	21
Preston Rd. Rom.	CV28	19
Preston's Rd. E14	CF40	43
Preston Waye, Har.	BL33	21
Prestwick Clo. Sthl.	BE42	38
Prestwick Rd. Wat.	BC28	11
Prestwood Av. Har.	BJ31	21
Prestwood Clo. Har.	BJ31	21
Prestwood Dr. Rom.	CS28	19
Prestwood St. N1	BZ37	33
Wenlock Rd.		
Pretoria Av. E17	CD31	25
Pretoria Clo. N17	CA29	15
Pretoria Cres. E4	CF26	9
Pretoria Rd. E4	CF26	9
Pretoria Rd. E11	CF33	25
Pretoria Rd. E16	CG38	34
Pretoria Rd. N17	CA29	15
Pretoria Rd. N18	CA29	15
Pretoria Rd. SW16	BV50	59
Pretoria Rd. Ilf.	CL35	35
Pretoria Rd. Rom.	CS31	28
Pretoria Rd. Wat.	BC24	4
Pretoria Rd. N. N18	CA29	15
Preyost Rd. N11	BV27	14
Price Clo. NW7	BR29	13
Price Clo. SW17	BU48	59
Price Rd. Croy.	BY56	78
Prices St. SE1	BY40	42
Pricklers Hill, Barn.	BS25	7
Prickley Wd. Brom.	CG54	70
Prideaux Pl. WC1	BX38	33
Prideaux Rd. SW9	BX45	51
Pridham Rd. E. Th. Hth.	BZ52	69
Priestfield Rd. SE23	CD48	61
Priestlands Pk. Rd. Sid.	CN48	63
Priestley Gdns. Rom.	CO32	27
Priestley Rd. Mitch.	BV51	68
Priestley Way E17	CC31	25
Priestley Way NW2	BP33	22
Priests Av. Rom.	CS30	19
Priests Ct. EC2	BZ39	42
Foster La.		
Prima Rd. SW9	BY43	51
Primrose Av. Enf.	BZ23	8
Primrose Av. Rom.	CO33	27
Primrose Clo. Har.	BE34	20
Primrose Ct. E15	CF36	34
Angel La.		
Primrose Gdns. NW3	BU36	32
Primrose Gdns. Bush.	BF26	4
Primrose Gdns. Ruis.	BD35	29
Primrose Glen, Horn.	CW31	28
Primrose Hill EC4	BY39	42
Primrose Hill Ct. NW3	BU36	32
Primrose Hill, Orp.	CN58	81
Primrose Hill Rd. NW3	BU36	32
Primrose Rd. E10	CE33	25
Primrose Rd. E18	CH30	17
Primrose Rd. Walt.	BD56	74
Primrose St. EC2	CA39	42
Primrose Way, Wem.	BK37	30
Primula St. W12	BP39	40
Prince Albert Rd. NW1	BU37	32
Prince Albert Rd. NW8	BU38	32
Prince Arthur Ms. NW3	BT35	32
Perrins La.		
Prince Arthur Rd. NW3	BT35	32
Prince Charles Dr. NW4	BQ33	22
Prince Charles Rd. SE3	CG44	52
Prince Charles Way, Wall.	BV55	77
Prince Consort Dr. Chis.	CM51	71
Prince Consort Rd. SW7	BT41	41
Princedale Rd. W11	BR40	40
Prince Edward Rd. E9	CD36	34
Prince George Av. N14	BW24	7
Prince George Rd. N16	CA35	33
Prince George Rd. SW19	BT51	68
Prince Georges Av. SW20	BQ51	67
Prince Henry Rd. SE7	CJ43	53
Prince Imperial Rd. Chis.	CL50	62
Prince Imperial Way SE18	CL43	53
Prince John Rd. SE9	CK46	53
Princelet St. E1	CA39	42
Prince Of Wales Clo. NW4	BP31	22
Prince Of Wales Dr. SW11	BU44	50
Prince Of Wales Rd. E16	CJ39	44
Prince Of Wales Rd. NW5	BV36	32
Prince Of Wales Rd. SE3	CG44	52
Prince Of Wales Rd. Sutt.	BT55	77
Prince Of Wales Ter. W4	BO42	40
Devonshire Rd.		
Prince Of Wales Ter. W8	BS41	41
Prince Regent Rd. Houns.	BG45	47
Prince Regent's La. E13	CH38	35
Prince Regent's La. E16	CJ39	44
Prince Rd. SE25	CA53	69
Prince Rupert Rd. SE9	CK45	53
Prince's Av. N10	BV31	23
Princes Av. N13	BY28	15
Prince's Av. N22	BW30	14
Princes Av. NW9	BM31	21
Princes Av. W3	BM41	39
Princes Av. Cars.	BU57	77
Princes Av. Grnf.	BF39	38
Princes Av. Orp.	CN53	72
Princes Av. S. Croy.	CB61	84
Princes Av. Surb.	BM54	66
Princes Av. Wdf. Grn.	CH28	17
Princes Clo. NW9	BM31	21
Princes Clo. Edg.	BM28	12
Princes Clo. Sid.	CP48	63
Princes Clo. S. Croy.	CB61	84
Princes Clo. Tedd.	BG49	56

Queensbury Rd. Wem. BL37 30
Queensbury Sta. Par. Edg. BL31 21
Queensbury St. N1 BZ36 33
Morton Rd.
Queen's Cir. SW11 BV43 50
Queens Clo. Dag. BM28 12
Queen's Club Gdns. W14 BR43 49
Queens Ct. NW11 BR32 22
Queens Ct. SE23 CC48 61
Queens Ct. W5 BK39 39
Queen's Wk.
Queen's Ct. Rich. BL46 48
Queens Ct. Wem. BL35 30
Queen's Cres. NW5 BV36 32
Queen's Cres. Rich. BL46 48
Queenscroft Rd. SE9 CJ46 53
Queensdale Cres. W11 BQ40 40
Queensdale Pl. W11 BR40 40
Queensdale Rd. W11 BQ40 40
Queensdale Wk. W11 BR40 40
Queensdown Rd. E5 CB35 33
Queens Dr. E10 CE33 25
Queen's Dr. N4 BY34 24
Queens Dr. W3 BL39 39
Queen's Dr. Surb. BJ54 66
Queen's Dr. Surb. BM54 66
Queen's Elm Sq. SW3 BT42 41
Queen's Gdns. NW4 BQ32 22
Queen's Gdns. W2 BT40 41
Queens Gdns. W5 BK38 30
Queens Gdns. Houns. BE43 47
Queens Gdns. Rain. CS37 37
Queen's Gte. SW7 BT41 41
Queen's Gte. Gdns. SW7 BT41 41
Queensgate Gdns. Chis. CM51 71
Prince Consort Dr.
Queen's Gte. Ms. SW7 BT41 41
Queen's Gte Pl. SW7 BT41 41
Queen's Gte Pl. Ms. SW7 BT41 41
Queen's Gte Ter. SW7 BT41 41
Queen's Gro. NW8 BT37 32
Queens Gro. Rd. E4 CF26 9
Queens Head St. N1 BY37 33
Raleigh St.
Queen's Ho. Tedd. BH50 57
Queensland Av. N18 BZ29 15
Queensland Av. SW19 BS51 68
Queensland Pl. N7 BY35 33
Queensland Rd.
Queensland Rd. N7 BY35 33
Queens Mans. NW4 BP32 22
Queen's Mkt. E13 CJ37 35
Queensmead NW8 BT37 32
Queensmead Av. Epsom BP58 76
Queens Mead Rd. Brom. CG51 70
Queensmere Clo. SW19 BQ48 58
Queensmere Rd. SW19 BQ48 58
Queen's Rd. SW19 BQ48 58
Queen's Ms. W2 BS40 41
Salem Rd.
Queensmill Rd. SW6 BQ43 49
Queens Pde. Clo. N12 BU28 14
Hollyfield Av.
Queens Pk. Ct. W10 BQ38 31
Queen's Pk. Gdns. Felt. BC48 56
Queens Pk. Rd. Rom. CW30 19
Queen's Pl. Mord. BS52 68
Queen's Pl. Wat. BD24 4
Queens Prom. BK52 66
Kings. On T.
Queen Sq. WC1 BX38 33
Queen's Ride SW13 BP45 49
Queen's Rise, Rich. BL46 48
Queen's Rd. E11 CF33 25
Queen's Rd. E13 CH37 35
Queen's Rd. E17 CD32 25
Queen's Rd. N3 BT30 14
Queens Rd. N9 CB27 15
Queen's Rd. N11 BX29 15
Queen's Rd. NW4 BQ32 22
Queen's Rd. SW14 BN45 49
Queen's Rd. SW19 BR50 58
Queen's Rd. Bark. CM36 35
Queens Rd. Barn. BQ24 6
Queens Rd. Beck. CD51 70
Queen's Rd. Brom. CH51 71
Queen's Rd. Buck. H. CH27 17
Queens Rd. Chis. CL50 62
Queen's Rd. Croy. BY53 69
Queen's Rd. Enf. CA24 8
Queen's Rd. Erith CT43 55
Queens Rd. Felt. BC47 56
Queen's Rd. Hamptn. BF49 56
Queen's Rd. Houns. BF45 47
Queens Rd. Ilf. CM34 26
Queen's Rd. Kings. On T. BM50 57
Queen's Rd. Loug. CK24 10
Queen's Rd. Mord. BS52 68
Queens Rd. N. Mal. BO52 67
Queen's Rd. Rich. BL47 57
Queen's Rd. Sthl. BE41 38
Queens Rd. Surb. BH53 66
Queen's Rd. Sutt. BS59 83
Queen's Rd. Tedd. BH50 57
Queen's Rd. Twick. BJ47 57
Queen's Rd. Wall. BV56 77
Queen's Rd. Wat. BD23 4
Queen's Rd. Well. CO44 54
Queens Rd. W5 BL39 39
Queen's Rd. W. E13 CH37 35
Queen's Row SE17 BZ43 51
Queens Sq. WC1 BX39 42
Gt. Ormond St.
Queen's Ter. E13 CH37 35
Queen's Ter. NW8 BT37 32
Queen's Ter. Islw. BJ45 48
Queensthorpe Rd. SE26 CC49 61
Queenstown Gdns. Rain. CT38 37
Queenstown Rd. SW8 BV45 50
Queen St. EC4 BZ40 42
Queen St. N17 CA29 15
Queen St. Bexh. CQ45 54
Queen St. Croy. BZ55 78
Queen St. Erith CT43 55
Queen St. Mayfair W1 BV40 41
Queen St. Rom. CS32 28
Queensville Rd. SW12 BW47 59
Queens Wk. E4 CF26 9
Green Wk.
Queen's Wk. NW9 BN34 22
Queen's Wk. SW1 BW40 41
Queens Wk. W5 BK38 30
Queen's Wk. Har. BH31 21
Queen's Wk. Ruis. BD34 20
Queensway NW4 BQ32 22
Queensway W2 BS40 41
Queensway, Croy. BX57 78
Queensway, Enf. CB24 8
Queen's Way, Felt. BD49 56
Queensway, Orp. CM53 71
Queensway, Sun. BC51 65
Queensway, Walt. BD56 74
Queensway, W. Wick. CG55 79
Queenswell Av. N20 BU28 14
Queenswood Av. E17 CF30 16
Queenswood Av. Hamptn. BF50 56
Queenswood Av. Houns. BE44 47
Queenswood Av. Th. Hth. BY53 69
Queenswood Av. Wall. BW56 77
Queenswood Ct. SW4 BX46 51
Queenswood Gdns. E11 CH33 26
Queenswood Pk. N3 BR30 13
Queen's Wood Rd. N10 BV32 23
Queenswood Rd. SE23 CC48 61
Queenswood Rd. Sid. CN46 54
Queen Victoria Av. Wem. BK36 30
Queen Victoria St. EC4 BY40 42
Quemerford Rd. N7 BX35 33
Quentin Pl. SE13 CG45 52
Quentin Rd. SE13 CG45 52
Quenton Pl. SE13 CG45 52
Quernmore Clo. Brom. CH50 62
Quernmore Rd. N4 BY32 24
Quernmore Rd. Brom. CH50 62
Querrin St. SW6 BT44 50
Quex Ms. NW6 BS37 32
Quex Rd. NW6 BS37 32
Quick Rd. W4 BO42 40
Quicks Rd. SW19 BS50 59
Quick St. N1 BY37 33
Quickswood NW3 BU36 32
Quiet Nook, Kes. CJ55 80
Quilp St. SE1 BZ41 42
Redcross Way
Quilter Gdns. Orp. CP54 72
Quilter Rd. Orp. CP54 72
Quilter St. E2 CA38 33
Quinta Dr. Barn. BP25 6
Quinton Av. SW20 BR51 67
Quinton Clo. Beck. CF52 70
Quinton Clo. Wall. BV56 77
Quinton Rd. Surb. BJ54 66
Quinton St. SW18 BT48 59
Quixley St. E14 CF40 43
Quorn Rd. SE22 CA45 51
Rabbit Row W8 BS40 41
Kensington Mall
Rabbits Rd. E12 CK35 35
Raby Rd. N. Mal. BN52 67
Raby St. E14 CD39 43
Raccoon Way, Houns. BD44 47
Rachel Point N16 CB34 24
Downs Est.
Rack Rd. W3 BM41 39
Racquet Ct. EC4 BY39 42
Fleet St.
Racton Rd. SW6 BS43 50
Radbourne Av. W5 BK42 39
Radbourne Clo. E5 CC35 34
Glyn Rd.
Radbourne Ct. E5 CC35 34
Clapton Park Est.
Radbourne Cres. E17 CF31 25
Radbourne Rd. SW12 BW47 59
Radcliffe Av. NW10 BP37 31
Radcliffe Av. Enf. BZ23 8
Radcliffe Gdns. Cars. BU57 77
Radcliffe Path SW8 BW44 50
St. Rule St.
Radcliffe Rd. N21 BY26 8
Radcliffe Rd. Croy. CA55 78
Radcliffe Rd. Har. BJ30 12
Radcliffe Sq. SW15 BQ46 49
Radcliffe Way, Nthlt. BD38 29
Radcot St. SE11 BY42 42
Raddington Rd. W10 BR39 40
Radfield Way, Sid. CM47 62
Radford Rd. SE13 CF46 52
Radipole Rd. SW6 BR44 49
Radland Rd. E16 CG39 43
Radlett Av. SE26 CB48 60
Radlett Pl. NW8 BU37 32
Radlett Rd. Wat. BD24 4
Radley Rd. Wat. BD24 4
Radley Av. Ilf. CO35 36
Radley Gdns. Har. BL31 21
Radley La. E18 CH30 17
Radley Ms. W8 BS41 41
Radley Rd. N17 CA30 15
Radleys Mead, Dag. CR36 36
Radlix Rd. E10 CE33 25
Radnor Av. Har. BH32 21
Radnor Av. Well. CO46 54
Radnor Clo. Mitch. BX52 69
Radnor Cres. Ilf. CK32 26
Radnor Gdns. Enf. CA23 8
Radnor Gdns. Twick. BH48 57
Radnor Ho. Twick. BJ48 57
Radnor Ms. W2 BT39 41
Radnor Pl. W2 BU39 41
Radnor Rd. NW6 BR37 31
Radnor Rd. SE15 CB43 51
Radnor Rd. Har. BG32 20
Radnor Rd. Twick. BH47 57
Radnor St. EC1 BZ38 33
Radnor Ter. SW8 BX43 51
South Lambeth Rd.
Radnor Ter. W14 BR42 40
Radnor Wk. SW3 BU42 41
Radnor Wk. Croy. CD53 70
Radnor Way NW10 BM38 30
Radstock Av. Har. BJ31 21
Radstock St. SW11 BU43 50
Parkgate Rd.
Raebarn Gdns. Barn. BP25 6
Raeburn Av. Dart. CU46 55
Raeburn Av. Surb. BM54 66
Raeburn Clo. NW11 BS32 23
Raeburn Clo. Kings. On T. BK50 57
Lower Teddington Rd.
Raeburn Ho. Kings. On T. BK50 57
Raeburn Rd. Edg. BM30 12
Raeburn Rd. Sid. CN46 54
Raeburn St. SW2 BX45 51
Rafford Way, Brom. CH51 71
Raft Rd. SW18 BS46 50
Raggleswood, Chis. CL51 71
Raglan Ct. S. Croy. BY56 78
Raglan Ct. Wem. BL35 30
Raglan Gdns. Wat. BC26 4
Raglan Rd. E17 CF32 25
Raglan Rd. SE18 CL42 44
Raglan Rd. Belv. CQ42 45
Raglan Rd. Brom. CJ52 71
Raglan Rd. Enf. CA26 8
Raglan St. NW5 BV36 32
Raglan Way, Nthlt. BG36 29
Ragley Clo. W3 BM41 39
Avenue Rd.
Raider Clo. Rom. CR30 18
Railey Ms. NW5 BW35 32
Leverton St.
Railshed Rd. Twick. BJ45 48
St. Margaret's Rd.
Railston Way, Wat. BD27 11
Railton Rd. SE24 BY45 51
Railway App. SE1 BZ40 42
Railway App. Cheam. BR57 76
Railway App. Couls. BW61 83
Railway App. Har. BH31 21
Railway App. Twick. BJ47 57
Railway App. Wall. BV56 77
Railway Arches SE8 CE43 52
Deptford High St.
Railway Av. SE16 CC41 43
Railway Cotts. Ilf. CN29 18
Railway Ms. W10 BR39 40
Ladbroke Gro.
Railway Pass. Tedd. BJ50 57
Clarence Rd.
Railway Pl. EC3 CA40 42
Fenchurch St.
Railway Pl. SW19 BR50 58
Hartfield Rd.
Railway Pl. Belv. CR41 45
Railway Rd. Tedd. BH49 57
Railway Side SW13 BO45 49
White Hart La.
Railway St. N1 BX37 33
Railway St. Rom. CP33 27
Railway Ter. SE13 CE46 52
Railway Ter. Felt. BC47 56
Rainborough Clo. NW10 BN36 31
Rainbow St. SE5 CA43 51
Raine St. E1 CB40 42
Rainham Clo. SE9 CN46 54
Rainham Clo. SW11 BU46 50
Rainham Rd. NW10 BQ38 31
Rainham Rd. Rain. CT36 37
Rainham Rd. N. Dag. CR34 27
Rainham Rd. S. Dag. CR35 36
Rainhill Way E3 CE38 34
Rainsborough Av. SE8 CD42 43
Rainsford Rd. NW10 BM38 30
Rainsford St. W2 BU39 41
Sale Pl.
Rainsford Way, Horn. CU33 28
Rainton Rd. SE7 CH42 44
Rainville Rd. W6 BQ43 49
Raisins Hill, Pnr. BC31 20
Raith Av. N14 BW27 14
Raleana Rd. E14 CF40 43
Raleigh Av. Hayes BC39 38
Raleigh Av. Wall. BW56 77
Raleigh Clo. N13 BZ27 15
Raleigh Clo. NW4 BQ32 22
Raleigh Clo. Pnr. BD33 20
Raleigh Ct. Beck. CE51 70
Raleigh Ct. Wall. BV57 77
Raleigh Dr. N20 BU27 14
Raleigh Dr. Esher BG56 74
Raleigh Dr. Surb. BN54 67
Raleigh Gdns. Mitch. BU51 68
Raleigh Rd. N8 BY31 24
Raleigh Rd. SE20 CC50 61
Raleigh Rd. Enf. BZ24 8
Raleigh Rd. Rich. BL45 48
Raleigh Rd. Sthl. BE42 38
Raleigh St. N1 BY37 33
Raleigh Way N14 BW26 7
Raleigh Way, Felt. BD49 56
Ralph St. SE1 BZ41 42
Ralston St. SW3 BU42 41
Tedworth Sq.
Rama Ct. Har. BH34 21
Rambler Clo. SW16 BW49 59
Ramblings, The E4 CF28 16
Ramilles Clo. SW2 BX46 51
Ramilles St. W1 BW39 41
Gt. Marlborough St.
Ramillies Pl. W1 BW39 41
Ramillies Rd. NW7 BO27 13
Ramillies Rd. W4 BN42 40
Ramillies Rd. Sid. CO46 54
Ramorne Clo. Walt. BE56 74
Rampart St. E1 CB39 42
Commercial Rd.
Rampayne St. SW1 BW42 41
Ram Pl. E9 CC36 34
Chatham Pl.
Rampton Clo. E4 CE27 16
Ramsay Gdns. Rom. CV30 19
Ramsay Rd. E7 CG35 34
Ramsay Rd. W3 BN41 40
Ramscroft Clo. N9 CA26 8
Ramsdale Rd. SW17 BV49 59
Ramsden Clo. Orp. CP54 72
Ramsden Rd.
Ramsden Dr. Rom. CR29 18
Ramsden Rd. N11 BU28 14
Ramsden Rd. SW12 BV46 50
Ramsden Rd. Erith CS43 55
Ramsden Rd. Orp. CO54 72
Ramsey Clo. Har. BG35 29
Ramsey Ct. N8 BW32 23
Ramsey Rd. NW9 BO32 22
Broadway, The
Ramsey Rd. Th. Hth. BX53 69
Ramsey St. E2 CB38 33
Ramsey Wk. N1 BZ36 33
Marquess Est.
Ramsey Way N14 BW26 7
Windsor Ct.
Ramsgate St. E8 CA36 33
Ramsgill App. Ilf. CN31 27
Ramsgill Dr. Ilf. CN32 27
Rams Gro. Rom. CQ31 27
Ramuls Dr. Hayes BE38 29
Ramus Wd. Av. Orp. CN56 81
Rancliffe Gdns. SE9 CK45 53
Rancliffe Rd. E6 CK38 35
Randall Av. NW2 BO34 22
Randall Clo. Erith CS43 55
Randall Clo. SW11 BU44 50
Randall Dr. Horn. CV35 37
Randall Pl. SE10 CF43 52
Randall Rd. SE11 BX42 42
Randall Rd. Rom. CT32 28
Randall Row SE11 BX42 42
Randall's Mkt. E14 CE39 43
Ricardo St.
Randell's Rd. N1 BX37 33
Randle Rd. Rich. BK49 57
Randlesdown Gdns. SE6 CE48 61
Randlesdown Rd. SE6 CE49 61
Randolph App. E16 CJ39 44
Baxter Rd.
Randolph Av. W9 BT38 32
Randolph Clo. Bexh. CS45 55
Randolph Clo. BN49 58
Kings. On T.
Randolph Cres. W9 BT38 32
Randolph Gdns. NW6 BS37 32
Randolph MS. W9 BT38 32
Randolph Rd. E17 CE32 25
Randolph Rd. W9 BT38 32
Randolph Rd. Epsom BO60 82
Randolph Rd. Sthl. BE41 38
Randolph St. NW1 BW36 32
Randon Clo. Har. BF30 11
Ranelagh Av. SW6 BR45 49
Ranelagh Av. SW13 BP44 49

Ranelagh Br. W2 BS40 41
Ranelagh Clo. Edg. BM28 12
Ranelagh Dr. Edg. BM28 12
Ranelagh Dr. Twick. BJ45 48
Ranelagh Est. SW15 BQ44 49
Ranelagh Gdns. E11 CJ32 26
Ranelagh Gdns. SW6 BR45 49
Ranelagh Gdns. W4 BN43 49
Grove Pk. Gdns.
Ranelagh Gdns. Ilf. CK33 26
Ranelagh Gro. SW1 BV42 41
Ranelagh Pl. N. Mal. BO53 67
Ranelagh Rd. E6 CL37 35
Ranelagh Rd. E11 CG35 34
Ranelagh Rd. E15 CG37 34
Ranelagh Rd. N17 CA31 24
Ranelagh Rd. N22 BX30 15
Ranelagh Rd. NW10 BO37 31
Ranelagh Rd. SW1 BW42 41
Lupus St.
Ranelagh Rd. W5 BK41 39
Ranelagh Rd. Sthl. BD40 38
Ranelagh Rd. Wem. BK35 30
Ranfurly Rd. Sutt. BS55 77
Rangefield Rd. Brom. CG49 61
Rangemoor Rd. N15 CA32 24
Ranger Rd. E4 CG26 9
Rangers Sq. SE10 CF44 52
Ragoon St. EC3 CA39 42
Northumberland Alley
Rankins Clo. NW9 BO31 22
Ranleigh Gdns. Bexh. CQ43 54
Ranmere St. SW12 BV47 59
Ranmoor Av. Croy. CA55 78
Ranmoor Clo. Har. BG31 20
Ranmoor Gdns. Har. BG31 20
Ranmore Path. Orp. CO52 72
Ranmore Rd. Couls. BX62 84
Ranmore Rd. Sutt. BQ58 76
Rannoch Rd. W6 BQ43 49
Rannock Dr. NW9 BN33 22
Ranskill Rd. Borwd. BM23 5
Ransom Walk SE7 CJ42 44
Ranston St. NW1 BU39 41
Ranulf Rd. NW2 BR35 31
Ranworth Clo. Erith CT44 55
Ranworth Rd. N9 CC27 16
Raphael Av. Rom. CT30 19
Raphael St. SW7 BU41 41
Rapier Clo. Grays CW42 46
Rashleigh St. SW8 BV44 50
Rasper Rd. N20 BT27 14
Rastell Av. SW2 BW48 59
Ratcliffe Cross St. E1 CC39 43
Ratcliffe La. E14 CC39 43
Ratcliff Orchard E1 CC40 43
Ratcliff Rd. E7 CJ35 35
Rathbone Mkt. E16 CG39 43
Rathbone Pl. W1 BW39 41
Rathbone Point N16 CB34 24
Downs Est.
Rathbone St. E16 CG39 43
Rathbone St. W1 BW39 41
Rathcoole Av. N8 BX31 24
Rathcoole Gdns. N8 BX32 24
Rathfern Rd. SE6 CD47 61
Rathgar Av. W13 BJ40 39
Rathgar Clo. N3 BR30 13
Rathlin Wk. N1 BZ36 33
Marquess Est.
Rathmell Dr. SW4 BW46 50
Rathmore Rd. SE7 CH42 44
Rattray Rd. SW2 BY45 51
Raul Rd. SE15 CB44 51
Raveley St. NW5 BW35 32
Ravenet St. SW11 BV44 50
Ravensfield Rd. SW17 BU48 59
Ravenhill Rd. E13 CJ37 35
Ravenna Rd. SW15 BQ45 49
Ravenor Pk. Rd. Grnf. BF38 29
Raven Rd. E18 CJ30 17
Raven Row E1 CB39 42
Ravensbourne Av. Brom. CF50 61
Ravensbourne Av. Ilf. CL30 17
Ravensbourne Cres. SE6 CD47 61
Ravensbourne Est. Brom. CF49 61
Ravensbourne Gdns. W13 BJ39 39
Ravensbourne Gdns. Ilf. CL30 17
Ravensbourne Pk. SE6 CD47 61
Ravensbourne Pk. Cres. SE6 CD47 61
Ravensbourne Rd. SE6 CD47 61
Ravensbourne Rd. Brom. CH52 71
Ravensbourne Rd. Dart. CU45 55
Ravensbourne Rd. Twick. BK46 48
Ravensbury Av. Mord. BT53 68
Ravensbury Gro. Mitch. BT52 68
Ravensbury La. Mitch. BT52 68
Ravensbury Path, Mitch. BT52 68
Ravensbury Rd. SW18 BS48 59
Ravensbury Rd. Orp. CN52 72
Ravenscar Rd. Brom. CG49 61
Ravenscar Rd. Surb. BL55 75
Ravens Clo. Brom. CG51 70
Ravens Clo. Enf. CA23 8
Ravenscourt Av. W6 BP42 40
Ravenscourt Clo. Horn. CW28 28
Ravenscourt Dr. Horn. CW34 28

Ravenscourt Gdns. W6 BP42 40
Ravenscourt Gro. Horn. CW34 28
Ravenscourt Pk. W6 BP42 40
Ravenscourt Pl. W6 BP42 40
King St.
Ravenscourt Rd. W6 BP41 40
Ravenscourt Rd. Orp. CO51 72
Ravenscourt Sq. W6 BP41 40
Ravenscraig Rd. N11 BW28 14
Ravenscroft Av. NW11 BR33 22
Ravenscroft Av. Wem. BL33 21
Ravenscroft Clo. E16 CH39 44
Ravenscroft Pk. Barn. BQ24 6
Ravenscroft Pk. Rd. Barn. BQ24 6
Ravenscroft Rd. E16 CG39 43
Ravenscroft Rd. W4 BN42 40
Ravenscroft Rd. Beck. CC51 70
Ravenscroft St. E2 CA37 33
Ravensdale Av. N12 BT28 14
Ravensdale Gdns. SE19 BZ50 60
Ravensdale Rd. N16 CA33 24
Ravensdale Rd. Houns. BE45 47
Ravensdon St. SE11 BY42 42
Ravensfield Clo. Dag. CP35 36
Ravensfield Gdns. Epsom BO56 76
Ravenshaw St. NW6 BR35 31
Ravenshill, Chis. CL51 71
Ravenshurst Av. NW4 BQ31 22
Ravenslea Rd. SW12 BU47 59
Ravensmead Rd. Brom. CF50 61
Ravensmede Way W4 BO42 40
Ravenstone Rd. N8 BY31 24
Ravenstone Rd. NW9 BO32 22
Broadway, The
Ravenstone St. SW12 BV47 59
Ravens Way SE12 CH46 53
Ravenswold, Ken. BZ61 84
Ravenswood, Bex. CQ47 63
Ravenswood, Ken. BZ57 78
Ravenswood Av. Surb. BL55 75
Ravenswood Av. CF54 70
W. Wick.
Ravenswood Clo. Croy. BY55 78
Ravenswood Rd.
Ravenswood Clo. Rom. CR28 18
Ravenswood Ct. BN50 58
Kings. on T.
Ravenswood Cres. Har. BE34 20
Ravenswood Cres. CF54 70
W. Wick.
Ravenswood Gdns. Islw. BH44 48
Ravenswood Pk. Nthwd. BC29 11
Ravenswood Rd. E17 CE31 25
Ravenswood Rd. SW12 BV47 59
Ravenswood Rd. Croy. BY55 78
Ravenswood Rd. NW10 BP38 31
Ravensworth Rd. SE9 CK48 62
Ravent Rd. SE11 BX42 42
Ravey St. EC2 CA38 33
Ravine Gro. SE18 CN43 53
Rawcester Clo. SW18 BR47 58
Rawchester Clo. SW18 BR47 58
Rawlings Clo. Orp. CN56 81
Rawlings Clo. SW13 BU42 41
Rawlins Clo. N3 BR30 13
Rawlins Clo. Sth. Croy. CD57 79
Rawlinson Ter. N17 CA31 24
Rawnsley Av. Mitch. BU53 68
Rawson St. SW11 BV44 50
Despard Av.
Rawsthorne Pl. EC1 BY38 33
Rawstorne St. EC1 BY38 33
Raydean Rd. Barn. BS25 7
Raydons Gdns. Dag. CQ35 36
Raydons Rd. Dag. CQ35 36
Raydon St. N19 BV34 23
Rayfield Clo. Brom. CK53 71
Rayford Av. SE12 CG47 51
Rayford Clo. Dart. CV46 55
Ray Gdns. Bark. CO37 36
Ray Gdns. Stan. BJ28 12
Rayleas Clo. SE18 CL44 53
Rayleigh Av. Tedd. BH50 57
Rayleigh Clo. N22 BZ30 15
Rayleigh Ct. Kings. on T. BM51 66
Cambridge Rd.
Rayleigh Rise, S. Croy. CA57 78
Rayleigh Rd. N13 BZ27 15
Rayleigh Rd. SW19 BR51 67
Rayleigh Rd. Wdf. Grn. CJ29 17
Ray Lodge Rd. Wdf. Grn. CJ29 17
Raymead NW4 BQ31 22
Raymead Av. Th. Hth. BY53 69
Raymead Clo. Loug. CM25 10
Raymere Gdns. SE18 CM43 53
Raymond Av. E18 CG31 25
Raymond Av. W13 BJ41 39
Raymond Bldgs. WC1 BX39 42
Raymond Clo. SE26 CC49 60
Raymond Gdns. Chig. CO27 18
Raymond Rd. E13 CJ36 35
Raymond Rd. SW19 BR50 58
Raymond Rd. Beck. CD52 70
Raymond Rd. Ilf. CM33 26
Raymond Way, Esher BJ57 75

Raymouth Rd. SE16 CB42 42
Rayne Ct. E18 CG31 25
Rayners Ct. Har. BF33 20
Rayners Cres. Nthlt. BC38 29
Rayners Gdns. Nthlt. BC38 29
Rayners La. Har. BF33 20
Rayners La. Pnr. BE32 20
Rayners Rd. SW15 BQ46 49
Rayner St. E9 CC36 34
Raynes Av. E11 CJ33 26
Raynham Av. N18 CB29 15
Raynham Rd. N18 CB28 15
Raynham Rd. W6 BP42 40
Raynham Ter. N18 CB28 15
Raynor Clo. Sthl. BE40 38
Raynors Clo. Wem. BK35 30
Raynton Clo. Har. BE33 20
Raynton Dr. Hayes BC38 29
Ray Rd. E. Mol. BF53 65
Rays Av. N18 CC28 16
Rays Rd. N18 CB28 15
Rays St. EC1 BY38 33
Ray Walk N7 BX34 24
Andover Rd.
Raywood St. SW8 BV44 50
Reading La. E8 CB36 33
Reading Rd. Nthlt. BF35 29
Reading Rd. Sutt. BT56 77
Reading Way NW7 BR28 13
Reapers Way, Islw. BG46 47
Reardon Path E1 CB40 42
Reardon St. E1 CB40 42
Reaston St. SE14 CC43 52
Reckitt Rd. W4 BO42 40
Recovery St. SW17 BU49 59
Recreation Av. Rom. CS32 28
Recreation Av. Rom. CW30 19
Recreation Rd. SE26 CC49 60
Recreation Rd. Brom. CG51 70
Recreation Rd. Sthl. BE42 38
Recreation Way, Mitch. BW52 68
Rector St. N1 BZ37 33
Rectory Clo. E4 CE27 16
Brindwood Rd.
Rectory Clo. N3 BR30 13
Rectory Clo. SW20 BQ52 67
Rectory Clo. Dart. CT45 55
Rectory Clo. Sid. CO49 63
Rectory Clo. Stan. BJ28 12
Rectory Cres. E11 CK35 35
Rectory Gdns. N8 BX31 24
Rectory Gdns. SW4 BW45 50
Rectory Gro.
Rectory Gdns. Nthlt. BE37 29
Rectory Grn. Beck. CD51 70
Rectory Gro. SW4 BW45 50
Rectory Gro. Croy. BY55 78
Rectory Gro. Hamptn. BE49 56
Rectory La. SW17 BV50 59
Rectory La. Bans. BU60 83
Rectory La. Edg. BM29 12
Rectory La. Loug. CL23 10
Rectory La. Sid. CO49 63
Rectory La. Stan. BJ28 12
Rectory La. Surb. BL54 66
Rectory La. Wall. BW55 77
Rectory Pk. S. Croy. CA60 84
Rectory Pk. Av. Nthlt. BE38 29
Rectory Pl. SE18 CL42 44
Rectory Rd. E12 CK35 35
Rectory Rd. E17 CE31 25
Rectory Rd. N16 CA34 24
Rectory Rd. SW13 BP44 49
Rectory Rd. W3 BM40 39
Rectory Rd. Beck. CE51 70
Rectory Rd. Dag. CR36 36
Rectory Rd. Hayes BC39 38
Rectory Rd. Houns. BC44 47
Rectory Rd. Kes. CJ57 80
Rectory Rd. Sthl. BE41 38
Rectory Rd. Sutt. BS55 77
Rectory Sq. E1 CC39 43
Reculver Mews N18 CB28 15
Lyndhurst Rd.
Reculver Rd. SE16 CC42 43
Red Anchor Clo. SW3 BU43 50
Redan Pl. W2 BS40 41
Redan St. W14 BQ41 40
Redan Ter. SE5 BZ44 51
Flaxman Rd.
Redberry Gro. SE26 CC48 61
Red Bks. Rd. SE18 CK42 44
Redbourne Av. N3 BS30 14
Redbridge Gdns. SE5 CA44 51
Redbridge La. E. Ilf. CJ32 26
Redbridge La. W. E11 CH32 26
Red Bull Yard EC4 BZ40 42
Upr. Thames St.
Redburn St. SW3 BU43 50
Redburn Ter. Enf. CC25 9
South St.
Redbury Clo. Rain. CV39 46

Redcar Clo. Nthlt. BF36 29
Redcar Rd. Rom. CW28 19
Redcar St. SE5 BZ43 51
Redcastle Clo. E1 CC40 43
Juniper St.
Red Cedars Rd. Orp. CN54 72
Redchurch St. E2 CA38 33
Redcliffe Clo. SW5 BS42 41
Redcliffe Gdns. SW10 BS42 41
Redcliffe Gdns. Ilf. CL33 26
Redcliffe Ms. SW10 BS42 41
Redcliffe Pl. SW10 BT43 50
Redcliffe Rd. SW10 BT42 41
Redcliffe Sq. SW10 BS42 41
Redcliffe St. SW10 BS43 50
Redclose Av. Mord. BS53 68
Redclyffe Rd. E6 CJ37 35
Redcroft Rd. Sthl. BG40 38
Redcross Pl. SE1 BZ41 42
Redcross Way
Redcross Way SE1 BZ41 42
Redden Ct. Rd. Horn. CW31 28
Redden Ct. Rd. Horn. CW31 28
Reddings Av. Bush. BF25 4
Reddings Clo. NW7 BO27 13
Reddings, The NW7 BO27 13
Reddington Clo. CA58 78
S. Croy.
Reddins Rd. SE15 CB43 51
Reddons Rd. Beck. CD50 61
Reddown Rd. Couls. BW62 83
Reddy Rd. Erith CT43 55
Rede Pl. W2 BS39 41
Redesdale Gdns. Islw. BJ43 48
Redesdale St. SW3 BU42 41
Redfern Av. Twick. BF47 56
Redfern Gdns. Rom. CV30 19
Redfern Rd. NW10 BO36 31
Redfern Rd. SE6 CF47 61
Brownhill Rd.
Redfield La. SW5 BS42 41
Redford Av. Couls. BV60 83
Redford Av. Th. Hth. BX52 69
Redford Av. Wall. BX57 78
Redgate Dr. Brom. CH55 80
Redgate Ter. SW15 BR46 49
Lytton Gro.
Redgrave Rd. SW15 BQ45 49
Red Hill, Chis. CL49 62
Redhill Dr. Edg. BN30 13
Redhill St. NW1 BV37 32
Redholm Vill. N16 BZ35 33
Winston Rd.
Red House La. Bexh. CP45 54
Red Ho. La. Walt. BC55 74
Red Ho. Rd. Croy. BW53 68
Redington Gdns. NW3 BS35 32
Redington Rd. NW3 BS35 32
Redlands, Couls. BX61 84
Redlands Gdns. E. Mol. BE52 65
Redlands Rd. Enf. CD23 9
Redlands Way SW2 BX47 60
Red La. Esher BJ57 75
Bonamy Estate West, The
Redleaf Clo. Belv. CR43 54
Redlees Clo. Islw. BJ45 48
Red Lion Clo. Orp. CP53 72
Red Lion Ct. EC4 BY39 42
Fleet St.
Red Lion Hl. N2 BT30 14
Red Lion La. SE18 CL44 53
Red Lion Rd. Surb. BL55 75
Red Lion Row SE17 BZ43 51
Red Lion Sq. WC1 BX39 42
Red Lion Sq. Hamptn. BF51 65
Red Lion St. WC1 BX39 42
Theobalds Rd.
Red Lion St. Rich. BK46 48
Red Lion Yd. Wat. BD24 4
Red Lo. Cres. Bex. CS48 64
Red Lo. Rd. Bex. CS48 64
Red Lo. Rd. W. Wick. CF54 70
Redman's Rd. E1 CC39 43
Redmead La. E1 CB40 42
Redmore Rd. W6 BP42 40
Red Oak Clo. Orp. CL55 80
Red Pl. W1 BV40 41
Park St.
Redpoll Way SE2 CP41 45
Maran Way
Redpoll Way, Belv. CP41 45
Maran Way
Red Post Hl. SE21 BZ45 51
Red Post Hl. SE24 BZ45 51
Redriffe Rd. E13 CG37 34
Redriff Est. SE16 CD41 43
Redriff Rd. SE16 CC41 43
Redriff Rd. Rom. CR30 18
Red Rd. Borwd. BL24 5
Redruth Clo. N22 BX29 15
Redruth Clo. Rom. CW28 19
Redruth Rd. Rom. CW28 19
Redruth Wk. Rom. CW28 19
Redstart Clo. Croy. CF58 79
Redston Rd. N8 BW31 23
Redvers Rd. N22 BY30 15

Ridings, The W5 BL39 39
Ridings, The, Epsom BO61 82
Ridings, The, Loug. CJ24 10
Ridings, The, Sun. BC51 65
Ridings, The, Surb. BM53 66
Riding, The NW11 BR33 22
Golders Green Rd.
Ridley Av. W13 BJ41 39
Ridley Clo. Brom. CG52 70
Ridley Clo. Rom. CU30 19
Ridley Rd. E7 CJ35 35
Ridley Rd. E8 CA35 33
Ridley Rd. NW10 BP37 31
Ridley Rd. SW19 BS50 59
Ridley Rd. Brom. CG52 70
Ridley Rd. Well. CO44 54
Ridley Several SE3 CH44 53
Blackheath Pk.
Ridout St. SE18 CK42 44
Ridsdale Rd. SE20 CB51 69
Riefield Rd. SE9 CM46 53
Riesco Dr. Croy. CC57 79
Riffel Rd. NW2 BQ35 31
Rifle Butts Alley, Epsom BO60 82
Rifle Ct. SE11 BY43 51
Kennington Pk. Rd.
Rifle Pl. W11 BQ40 40
Rifle St. E14 CE39 43
Rigault Rd. SW6 BR44 49
Rigby Clo. Croy. BY55 78
Rigby Mews, Ilf. CL34 26
Rigden St. E14 CE39 43
Rigeley Rd. NW10 BP38 31
Rigg App. E10 CC33 25
Riggindale Rd. SW16 BW49 59
Riley Rd. SE1 CA41 42
Riley St. SW10 BT43 50
Ringcroft Rd. N7 BY35 33
Madras Pl.
Ringers Rd. Brom. CH52 71
Ringford Rd. SW18 BR46 49
Ringmer Av. SW6 BR44 49
Ringmer Gdns. EC1 BX34 24
Ringmer Pk. N7 BX36 33
Ringmer Pl. N21 BZ25 8
Ringmore Ri. SE23 CB47 60
Ringmore Rd. Walt. BD55 74
Ring Rd. D. Hayes BC40 38
Ringshall Rd. Orp. CO52 72
Ringslade Rd. N22 BX30 15
Ringstead Rd. SE6 CE47 61
Ringstead Rd. Sutt. BT56 77
Ring, The W2 BT40 41
Ringway N11 BW29 14
Ringway, Sthl. BE42 38
Ringwell Clo. Enf. CB23 8
Bishops Clo.
Ringwold Clo. Beck. CD50 61
Aldersmead Rd.
Ringwood Av. N2 BU30 14
Ringwood Av. Croy. BX54 69
Ringwood Av. Horn. CV34 28
Ringwood Av. Orp. CP58 81
Ringwood Clo. Pnr. BD31 20
Ringwood Gdns. SW15 BS48 58
Ringwood Rd. E17 CD32 25
Ringwood Way N21 BY26 8
Ringwood Way, Hamptn. BF49 56
Ripley Clo. Croy. CF57 79
Ripley Gdns. SW14 BN45 49
Ripley Gdns. Sutt. BT56 77
Ripley Rd. E16 CJ39 44
Ripley Rd. Belv. CR42 45
Ripley Rd. Hamptn. BF50 56
Ripley Rd. Ilf. CN34 27
Riplington Ct. SW15 BP47 58
Ripon Clo. Nthlt. BF36 29
Ripon Gdns. Chess. BK56 75
Ripon Gdns. Ilf. CK32 26
Ripon Rd. N9 CB26 8
Ripon Rd. N17 BZ31 24
Ripon Rd. SE18 CL43 53
Ripon Way, Borwd. BN24 6
Rippersley Rd. Well. CO44 54
Ripple Rd. Bark. CM36 35
Ripplevale Gro. N1 BX36 33
Rippolson Rd. SE18 CN42 45
Risborough Dr. Wor. Pk. BP54 67
Risborough St. SE1 BY41 42
Risdon St. SE16 CC41 43
Risebridge Chase, Rom. CT29 19
Risebridge Rd. Rom. CT30 19
Risedale Rd. Bexh. CR45 54
Riseldine Rd. SE23 CD46 52
Rise Pk. Boul. Rom. CT30 19
Rise Pk. Pde. Rom. CT30 19
Rise, The E11 CH32 26
Rise, The N13 BY28 15
Rise, The NW7 BO29 13
Rise, The NW10 BN34 22
Rise, The, Bex. CP47 63
Rise, The, Borwd. BL25 5
Rise, The, Buck. H. CJ26 10
Rise, The, Couls. BW60 83
Rise, The, Dart. CT45 55
Rise, The, Edg. BM28 12
Rise, The, Epsom BO58 76
Rise, The, Grnf. BJ35 30
Rise, The, Sid. CP47 63
Rise, The, S. Croy. CC58 79
Risinghill St. N1 BX37 33
Risingholme Clo. Bush. BF26 4
Risingholme Clo. Har. BH30 12
Risingholme Rd. Har. BH30 12
Risings, The E17 CF31 25
Risley Av. N17 BZ30 15
Rita Rd. SW8 BX43 51
Ritches Rd. N15 BZ32 24
Ritchie Rd. Croy. CB53 69
Ritchie St. N1 BY37 33
Ritchings Av. E17 CD31 25
Ritherdon Rd. SW17 BV48 59
Ritson Rd. E8 CB36 33
Ritter St. SE18 CL43 53
Rivaz Pl. E9 CC36 34
Rivenhall Gdns. E18 CG31 25
River Av. N13 BY27 15
River Av. Surb. BJ54 66
River Bank N21 BZ26 8
River Bank SE10 CF41 43
Riverbank, E. Mol. BH52 66
River Bank, Surb. BH53 66
River Barge Clo. E14 CF41 43
Capstan Sq.
River Clo. E11 CJ32 26
River Clo. Rain. CU39 46
River Clo. Surb. BK53 66
Rivercourt Rd. W6 BP42 40
Riverdale Gdns. Twick. BK46 48
Riverdale Rd. SE18 CN42 45
Riverdale Rd. Bex. CQ47 63
Riverdale Rd. Erith CR42 45
Riverdale Rd. Felt. BE49 56
Riverdale Rd. Twick. BK46 48
Riverdene, Edg. BN27 13
Riverdene Rd. Ilf. CL34 26
River Front, Enf. BZ24 8
River Gdns. Cars. BV55 77
River Gdns. Felt. BD46 47
River Grove Pk. Beck. CD51 70
Riverhead Clo. E17 CC30 16
Riverholme Dr. Epsom BN58 76
River La. Rich. BK47 57
Rivermead Clo. Tedd. BJ49 57
River Mead Ct. SW6 BR45 49
Ranelagh Gdns.
River Meads Av. Twick. BF48 56
River Meads Est. Twick. BF48 56
River Nook Clo. Walt. BD53 65
River Pk. Gdns. Brom. CF50 61
River Park Rd. N22 BX30 15
River Pl. N1 BZ36 33
River Reach, Tedd. BK49 57
Broom Water
River Reach, Tedd. BK50 57
River Rd. Bark. CN37 36
River Rd. Buck. H. CK26 10
Riversdale Rd. N5 BY34 24
Riversdale Rd. Rom. CR29 18
Riversdale Rd. Surb. BJ53 66
Riversfield Rd. Enf. CA24 8
Riverside NW4 BP33 22
Riverside SE7 CH41 44
Riverside Av. E. Mol. BG53 65
Riverside Clo. W7 BH38 30
Riverside Clo. BK52 66
Kings. On T.
Riverside Clo. Wall. BV55 77
Riverside Dr. W4 BN43 49
Riverside Dr. Esher BF56 74
Riverside Dr. Mitch. BU53 68
Riverside Dr. Rich. BJ48 57
Riverside Gdns. W6 BP42 40
Riverside Gdns. Enf. BZ23 8
Riverside Gdns. Wem. BL37 30
Riverside Rd. E15 CF37 34
Riverside Rd. N15 CB32 24
Riverside Rd. SW17 BS49 59
Riverside Rd. Sid. CQ48 63
Riverside Rd. Walt. BE56 74
Riverside Rd. Wat. BC25 4
Riverside, Twick. BJ47 57
Riverside Wk. SE1 BX41 42
Riverside Wk. Bex. CP47 63
Riverside Wk. Islw. BH45 48
Riverside Way, Dart. CW46 55
River St. EC1 BY38 33
River Ter. W6 BQ42 40
Crisp Rd.
River Vw. Enf. BZ24 8
River Vw. Gdns. SW13 BP43 49
River Vw. Gdns. Twick. BH48 57
Riverview Gro. W4 BM43 48
Riverview Pk. SE6 CE48 61
Riverview Rd. W4 BM43 48
Riverview Rd. Epsom BN55 76
Riverview Rd. Epsom BN56 76
River Wk. Walt. BC53 65
Riverway N13 BY28 15
River Way SE10 CG41 43
River Way, Epsom BN56 76
River Way, Loug. CK25 10
River Way, Twick. BF48 56
Riverwood La. Chis. CM51 71
Rivington Av. Wdf. Grn. CJ30 17
Rivington Ct. NW10 BO36 31
Rivington Ct. NW10 BP37 31
Longstone Av.
Rivington Cres. NW9 BO29 13
Rivington St. EC2 CA38 33
Rivulet Rd. N17 BZ29 15
Rixon Ho. SE18 CL43 53
Rixsen Rd. E12 CK35 35
Roach Rd. E3 CE36 34
Roads Pl. N4 BW34 23
Hornsey Rd.
Roads Pl. N4 BX34 24
Roan ST. SE10 CE43 52
Robart House E11 CG33 25
Robb Rd. Stan. BJ29 12
Robert Adam St. W1 BV39 41
Roberta St. E2 CB38 33
Robert Clo. W9 BT38 32
Robert Clo. Chig. CN28 18
Robert Clo. Walt. BC56 74
Robert Gentry Ho. W14 BR42 40
Robert Lowe Clo. SE14 CC43 52
Robert Ms. NW1 BW38 32
Hampstead Rd.
Roberton Dr. Brom. CJ51 71
Robert Owen Ho. SW6 BQ44 49
Roberts Alley W5 BK41 39
Church Gdns.
Robertsbridge Rd. Cars. BT54 68
Roberts Clo. Rom. CU30 19
Robertson Rd. E15 CF37 34
Robertson St. SW8 BV45 50
Roberts Rd. E17 CE30 16
Roberts Rd. NW7 BR29 13
Roberts Rd. Belv. CR42 45
Roberts St. Wat. BD25 4
Robert St. E2 CB37 33
Old Bethnal Green Rd.
Robert St. E16 CL40 44
Robert St. NW1 BV38 32
Robert St. SE18 CM42 44
Robert St. WC2 BX40 42
Savoy Pl.
Robert St. Croy. BZ55 78
High St.
Robina Clo. Bexh. CP45 54
Brunswick Rd.
Robin Clo. Rom. CS29 19
Robin Gro. N6 BV34 23
Robin Gro. Brent. BK43 48
Robin Gro. Har. BL32 21
Robin Hill Dr. Chis. CK50 62
Wood Dr.
Robin Hood Clo. Mitch. BW52 68
Robin Hood Dr. Bush. BE23 4
Robin Hood Dr. Har. BH29 12
Robin Hood Grn. Orp. CO53 72
Robin Hood La. E14 CF39 43
Robin Hood La. SW15 BO48 58
Robin Hood La. Bexh. CQ46 54
Robin Hood La. Mitch. BW52 68
Robin Hood Rd. Sutt. BS56 77
Robin Hood Rd. SW19 BO49 58
Robin Hood Way SW15 BO49 58
Robin Hood Way SW20 BO49 58
Robin Hood Way, Grnf. BH36 30
Robin Hood Yd. EC1 BY39 42
Leather Lane
Robinia Clo. Chig. CN29 18
Robin's Ct. SE12 CJ48 62
Robins Ct. Beck. CF51 70
Robins Gro. Kes. CH55 80
Robinson Cres. Bush. BG27 11
Robinson Rd. E2 CC37 34
Robinson Rd. SW17 BU50 59
Robinson Rd. Dag. CR35 36
Robinsons Clo. W13 BJ39 39
Robin St. SW3 BU42 41
Flood St.
Robinsway, Walt. BD56 74
Robin Way, Orp. CO52 72
Roborough Wk. Horn. CV36 37
Robsart St. SW9 BX44 51
Robson Clo. Enf. BY23 8
Robson Rd. SE27 BY48 60
Roch Av. Edg. BL30 12
Rochdale Rd. E17 CE33 25
Rochdale Rd. SE2 CO42 45
Rochdale Way SE8 CD43 52
Idonia St.
Rochelle St. E2 CA38 33
Swanfield St.
Roche Rd. SW16 BX51 69
Rochester Av. E13 CJ37 35
Rochester Av. Brom. CH51 71
Rochester Clo. SE3 CJ45 53
Rochester Clo. SW16 BX50 60
Rochester Clo. Enf. CA23 8
Rochester Clo. Sid. CO46 54
Rochester Dr. Bex. CR46 54
Rochester Dr. Pnr. BD32 20
Rochester Gdns. Croy. CA55 78
Rochester Gdns. Ilf. CK33 26
Rochester Mews NW1 BW36 32
Rochester Rd.
Rochester Pl. NW1 BW36 32
Rochester Rd. NW1 BW36 32
Rochester Rd. Cars. BU56 77
Rochester Row SW1 BW42 41
Rochester Sq. NW1 BW36 32
Rochester St. SW1 BW42 41
Rochester Row
Rochester Ter. NW1 BW36 32
Rochester Way SE3 CH44 53
Rochester Way SE9 CH44 53
Rochester Way, Dart. CT47 64
Rochester Way, Sid. CN46 54
Roche Wk. Cars. BT53 68
Rochford Av. Loug. CM24 10
Rochford Av. Rom. CP32 27
Rochford Clo. E13 CJ37 35
Boleyn Rd.
Rochford Clo. Horn. CU36 37
Rochford Grn. Loug. CM24 10
Rochford St. NW5 BU35 32
Rochford Wk. E8 CB36 33
Wilman Gro.
Rochford Way, Croy. BX53 69
Rock Av. SW14 BN45 49
South Worple Way
Rockbourne Rd. SE23 CC47 61
Rockchase Gdns. Horn. CW32 28
Rockells Pl. SE22 CB46 51
Rockford Av. Grnf. BJ37 30
Rock Gro. SE16 CB42 42
Blue Anchor La.
Rockhall Rd. NW2 BQ35 31
Rockhampton Rd. SE27 BY49 60
Rockhampton Rd. BZ57 78
S. Croy.
Rock Hill SE26 CA49 60
Rock Hill, Orp. CR57 81
Rockingham Av. Horn. CU32 28
Rockingham Est. SE1 BZ41 42
Rockingham St. SE1 BZ41 42
Rocklands Dr. Stan. BJ30 12
Rocklands Rd. SW15 BR45 49
Rock La. Dag. CR35 36
Rockley Rd. W14 BQ41 40
Rockmount Rd. SE18 CN42 45
Rockmount Rd. SE19 BZ50 60
Rocks La. SW13 BP44 49
Rock St. N4 BY34 24
Rockware Av. Grnf. BH37 30
Rockways, Barn. BO25 6
Rockwell Gdns. Dag. CR35 36
Rockwell Rd. Dag. CR35 36
Rockwells Ct. SE19 CA49 60
Rockwells Gdns. SE19 CA49 60
Rockwood Gdns. CH27 17
Wdf. Grn.
Whitehall La.
Rockwood Pl. W12 BQ41 40
Shepherds Bush Grn.
Rocliffe St. N1 BY37 33
Roscombe Cres. SE23 CC47 61
Rocque La. SE3 CG45 52
Rodborough Rd. NW11 BS33 23
Rodd La. EC3 CA40 42
Roden Gdns. Croy. CA53 69
Rodenhurst Rd. SW4 BW46 50
Roden St. N7 BX34 24
Roden St. Ilf. CL34 26
Roden Way, Ilf. CL34 26
Roden St.
Roderick Rd. NW3 BU35 32
Roding Av. Bark. CL36 35
Roding Av. Wdf. Grn. CK29 17
Roding Ho. Wdf. Grn. CK29 17
Roding La. Buck. H. CJ27 17
Roding La. N. Wdf. Grn. CJ30 17
Roding La. S. Ilf. CJ31 26
Roding Rd. E5 CC35 34
Roding Rd. E6 CL39 44
Roding Rd. Loug. CK25 10
Rodings, The, Wdf. Grn. CJ29 17
Snakes La.
Roding St. E7 CH35 35
Oakhurst Rd.
Roding Trading Est. Bark. CL36 35
Roding Vw. Buck. H. CJ26 10
Roding Way, Rain. CV37 37
Briscoe Rd.
Rodmarton St. W1 BU39 41
Rodmell Slope N12 BR28 13
Rodmere St. SE10 CG42 43
Rodmill La. SW2 BX47 60
Rodney Clo. KT12 BD54 65
Rodney Rd.
Rodney Clo. Croy. BY54 69
Rodney Clo. N. Mal. BO52 67
Rodney Clo. Pnr. BE33 20
Rodney Gdns. Kes. CH56 80
Rodney Gdns. Pnr. BC32 20
Rodney Pl. E17 CD30 16
Rodney Pl. SE17 BZ42 42
Rodney Pl. SW19 BT51 68
Rodney Rd. E11 CH31 26
Rodney Rd. SE17 BZ42 42
Rodney Rd. Mitch. BU51 68
Rodney Rd. N. Mal. BO53 67
Rodney Rd. Twick. BF46 47
Rodney Rd. Walt. BC55 74
Rodney St. N1 BX37 33
Rodney St. SE18 CL41 44
Rodney Way, Rom. CR30 18

Rodsley Pl. SE15 CB44 51
Commercial Way
Rodsley St. SE1 CB43 51
Old Kent Rd.
Rodway Rd. SW15 BP47 58
Rodwell Clo. Ruis. BD33 20
Rodwell Rd. SE22 CA46 51
Roebourne Way E16 CL40 44
Pier Rd.
Roebuck Clo. N17 CA29 1V
High Rd.
Roebuck La. Buck. H. CJ26 10
Roebuck Rd. Chess. BM56 75
Roebuck Rd. Ilf. CO28 18
Roedean Av. Enf. CC23 9
Roedean Clo. Enf. CC23 9
Roedean Cres. SW15 BO46 49
Roe End NW9 BN31 22
Roe Grn. NW9 BN32 22
Roehampton Clo. SW15 BP45 49
Roehampton Dr. Chis. CM50 62
Roehampton Gate SW15 BO46 49
Roehampton High St. SW15 BP47 58
Roehampton La. SW15 BP45 49
Roehampton Vale SW15 BO48 58
Roe La. NW9 BM31 21
Roe Way, Wall. BX57 78
Roffey Clo. Pur. BY61 84
Roffey St. E14 CF41 43
Rogate Ho. N16 CB34 24
Downs Est.
Rogers Clo. Couls. BY62 84
Rogers Gdns. Dag. CR35 36
Rogers Rd. E16 CG39 43
Rogers Rd. SW17 BT49 59
Rogers Rd. Dag. CR35 36
Roger St. WC1 BX38 33
Rojack Rd. SE23 CC47 61
Rokesby Clo. Wem. BK35 30
Rokeby Gdns. Wdf. Grn. CH30 17
Rokeby Rd. SE4 CD44 52
Rokeby St. E15 CF37 34
Roke Clo. Ken. BZ60 84
Roke Lodge Rd. Ken. BY60 84
Roke Rd. Ken. BY61 84
Rokesby Clo. Well. CM44 53
Rokesly Av. N8 BX32 24
Roland Gdns. SW7 BT42 41
Roland Gdns. SW10 BT42 41
Roland Gdns. Felt. BE48 56
Roland Rd. E17 CF32 25
Roland Way SE17 BZ42 42
Roland Way SW7 BT42 41
Roland Way, Wor. Pk. BO55 76
Roles Gro. Rom. CP31 27
Rolfe Clo. Barn. BU24 7
Rolinsden Way, Kes. CJ56 80
Rollesby Rd. Chess. BM57 75
Rolleston Av. Orp. CL53 71
Rolleston Clo. Orp. CL54 71
Rolleston Rd. S. Croy. BZ57 78
Roll Gdns. Ilf. CL32 26
Rollins St. SE15 CC43 52
Rollit Cres. Houns. BF46 47
Rollit St. N7 BY35 33
Hornsey Rd.
Rollo Rd. Swan. CT50 64
Rolls Bldgs. EC4 BY39 42
Fetter La.
Rollscourt Av. SE24 BZ46 51
Rolls Park Av. E4 CE28 16
Rolls Park Rd. E4 CE28 16
Rolls Pass. EC4 BY39 42
Chancery La.
Rolls Rd. SE1 CA42 42
Rolt St. SE8 CD43 52
Rolvenden Gdns. Brom. CJ50 62
Roman Clo. Felt. BC46 47
Roman Clo. Rain. CS37 37
Romanhurst Av. Brom. CG52 70
Romanhurst Gdns. Brom. CG52 70
Roman Rise SE19 BZ50 60
Roman Rd. E2 CC38 34
Roman Rd. E3 CC38 34
Roman Rd. E6 CK38 35
Roman Rd. N10 BV29 14
Roman Rd. W4 BO42 40
Roman Rd. Ilf. CL36 35
Roman Way N7 BX36 33
Roman Way SE15 CC43 52
Clifton Way
Roman Way, Enf. CA25 8
Romany Gdns. E17 CD30 16
Mcentee Av.
Romany Gdns. Sutt. BS54 68
Romany Rise, Orp. CM54 71
Roma Rd. E17 CD31 25
Romberg Rd. SW17 BV48 59
Romborough Gdns. SE13 CF46 52
Romborough Way SE13 CE46 52
Romero Sq. SE9 CJ45 53
Romeyn Rd. SW16 BX48 60
Romford Cres. Rom. CT33 28
Romford Rd. E7 CG36 34
Romford Rd. E12 CG36 34
Romford Rd. E15 CG36 34
Romford Rd. Chig. CO27 18
Romford Rd. Rom. CP28 18
Romford St. E1 CB39 42
Romilly Dr. Wat. BE28 11
Romilly Rd. N4 BY34 24
Romilly St. W1 BW40 41
Rommany Rd. SE27 BZ48 60
Romney Chase, Horn. CW32 28
Romney Clo. N17 CB30 15
Romney Clo. NW11 BT33 23
Romney Clo. SE14 CC43 52
Romney Clo. Chess. BL56 75
Romney Clo. Har. BF33 20
Romney Dr. Brom. CJ50 62
Romney Dr. Har. BF33 20
Romney Gdns. Bexh. CQ44 54
Romney Rd. SE10 CF43 52
Romney Rd. N. Mal. BN53 67
Romney St. SW1 BW41 41
Romola Rd. SE24 BY47 60
Romsey Gdns. Dag. CP37 36
Romsey Rd. W13 BJ40 39
Romsey Rd. Dag. CP37 36
Ronald Av. E15 CG38 34
Ronald Clo. Beck. CD52 70
Ronalds Rd. N5 BY35 33
Ronald's Rd. Brom. CH51 71
Ronaldstone Rd. Sid. CN46 54
Rona Rd. NW3 BV35 32
Rona Wk. N1 BZ36 33
Marquess Est.
Rondu Rd. NW2 BR35 31
Ronelean Rd. Surb. BL55 75
Ronfearn Av. Orp. CP53 72
Ronver Rd. SE12 CG47 61
Rook Clo. Rain. CU37 37
Rookeries Clo. Felt. BC48 56
Rookery Clo. NW9 BO32 22
Rookery Cres. Dag. CR36 36
Rookery Dr. Chis. CL51 71
Rookery Gdns. Orp. CP53 72
Rookery La. Brom. CJ53 71
Rookery Rd. SW4 BW45 50
Rookery Rd. Orp. CK58 80
Rookery, The, Wat. BC26 4
Rookery Way NW9 BO32 22
Rookesley Rd. Orp. CP54 72
Rooke Way SE10 CG42 43
Glenister Rd.
Rookfield Av. N10 BW31 23
Rookfield Clo. N10 BW31 23
Cranmore Way
Rookley Clo. Sutt. BS58 77
Hulverston Clo.
Rookstone Rd. SW17 BU49 59
Rookwood Av. Loug. CM24 10
Rookwood Av. N. Mal. BP52 67
Rookwood Av. Wall. BW56 77
Rookwood Gdns. E4 CG27 16
Whitehall Rd.
Rookwood Gdns. Ilf. CO29 18
Rookwood Gdns. Loug. CM24 10
Rookwood Rd. N16 CA32 24
Roosevelt Way, Dag. CS36 37
Ropemakers Flds. E14 CD40 43
Ropemaker St. EC2 BZ39 42
Ropers Av. E4 CE28 16
Roper St. SE9 CK46 53
Roper Way, Mitch. BV51 68
Ropery St. E3 CD38 34
Rope Wk. Sun. BD52 65
Rope Wk. Gdns. E1 CB39 42
Commercial Rd.
Rope Yard Rails SE18 CL41 44
Ropley St. E2 CB37 33
Shipton St.
Rosa Alba Ms. N5 BZ35 33
Kelross Rd.
Rosaline Rd. SW6 BR43 49
Rosamond St. SE26 CB48 60
Rosary Clo. Houns. BE44 47
Rosary Gdns. SW7 BT42 41
Rosavale Rd. SW6 BR43 49
Roscoe St. EC1 BZ38 33
Roscoff Clo. Edg. BM30 12
East Rd.
Roseacre Clo. W13 BJ39 39
Roseacre Clo. Horn. CW33 28
Curtis Rd.
Roseacre Rd. Well. CO45 54
Rose Alley SE1 BZ40 42
Rose Av. E18 CH30 17
Rose Av. Mitch. BU51 68
Rose Av. Mord. BT53 68
Rose Bank SE20 CB50 60
Rose Bank SW6 BQ43 49
Rosebank Av. Horn. CV35 37
Rosebank Av. Wem. BH35 30
Rosebank, Epsom BN60 82
Rosebank Gdns. E3 CD37 34
St. Stephens Rd.
Rosebank Gdns. W3 BN39 40
York Rd.
Rosebank Gro. E17 CD31 25
Rosebank Rd. E3 CD37 34
Norman Gro.
Rosebank Rd. E17 CE32 25
Rosebank Rd. W7 BH41 39
Rosebank Vill. E17 CE31 25
High St.
Rosebank Way W3 BN39 40
Roseberry Av. EC1 BY38 33
Roseberry Clo. Mord. BQ53 67
Roseberry Gdns. Dart. CV47 64
Roseberry Gdns. Orp. CN55 81
Roseberry Pl. E8 CA36 33
Roseberry St. SE16 CB42 42
Rosebery Av. E12 CK36 35
Rosebery Av. EC1 BY38 33
Rosebery Av. N17 CB30 15
Rosebery Av. Epsom BO60 82
Rosebery Av. Har. BE35 29
Rosebery Av. N. Mal. BO51 67
Rosebery Av. Sid. CN47 63
Rosebery Av. Th. Hth. BZ51 69
Rosebery Gdns. N4 BY32 24
Rosebery Gdns. N8 BX32 24
Rosebery Gdns. W13 BJ39 39
Rosebery Rd. N9 CB27 15
Rosebery Rd. N10 BW30 14
Rosebery Rd. SW2 BX46 51
Rosebery Rd. Bush. BF26 4
Rosebery Rd. Houns. BG46 47
Rosebery Rd. Kings. On T. BM51 66
Rosebery Rd. Sutt. BR57 76
Rosebery Sq. Kings. On T. BM51 66
Rosebine Av. Twick. BG47 56
Rosebrook Villas E17 CD31 25
High St.
Rosebury Rd. SW6 BS44 50
Rosebushes, Epsom BP61 82
Rose Cott. W5 BK40 39
Western Rd.
Rose Ct. SE26 CB48 60
Rose Ct. Pnr. BD31 20
Nursery Rd.
Rosecourt Rd. Croy. BX53 69
Rosecroft Av. NW3 BS34 23
Rosecroft Clo. Orp. CP53 72
Rosecroft Gdns. NW2 BP34 22
Rosecroft Gdns. Twick. BG47 56
Rosecroft Rd. Sthl. BF38 29
Rosecroft Wk. Pnr. BD32 20
Rosecroft Wk. Wem. BK35 30
Rose & Crown Ct. EC2 BZ39 42
Foster La.
Rose & Crown La. W6 BQ42 40
Talgarth Rd.
Rose & Crown Yd. SW1 BW40 41
King St.
Rosedale, Orp. CL55 80
Rosedale Clo. SE2 CO41 45
Finchale Rd.
Rosedale Clo. Stan. BJ29 12
Rosedale Cotts. W7 BH41 39
Rosedale Ct. N5 BY35 33
Leigh Rd.
Rosedale Gdns. Dag. CO36 36
Rosedale Rd. E7 CJ35 35
Rosedale Rd. Dag. CO36 36
Rosedale Rd. Epsom BP56 76
Rosedale Rd. Rich. BL45 48
Rosedale Rd. Rom. CS30 19
Rosedene Av. SW16 BX48 60
Rosedene Av. Croy. BX54 69
Rosedene Av. Grnf. BF38 29
Rosedene Av. Mord. BS53 68
Rosedene Gdns. Ilf. CL31 26
Rosedene Ter. E10 CE34 25
Rosedew Rd. W6 BQ43 49
Rose End, Wor. Pk. BQ54 67
Rosefield Gdns. E14 CE40 43
Morant St.
Rose Garden Clo. Edg. BL29 12
Rose Gdns. W5 BK41 39
Rose Gdns. Felt. BC48 56
Rose Gdns. Sthl. BF38 29
Rose Gdns. Wat. BC25 4
Rose Glen NW9 BN31 22
Rose Glen, Rom. CT33 28
Rosehatch Av. Rom. CP31 27
Roseheath Rd. Houns. BE46 47
Rosehill, Esher BJ57 75
Rose Hill, Hamptn. BF51 65
Rose Hill, Sutt. BS55 77
Rosehill Av. Sutt. BT54 68
Rosehill Gdns. Grnf. BH35 30
Rosehill Gdns. Sutt. BS55 77
Rosehill Pk. W. Sutt. BS54 68
Rosehill Rd. SW18 BT46 50
Roseland Clo. N17 BZ29 15
Rose La. Rom. CP31 27
Rose Lawn, Bush. BG26 4
Roseleigh Av. N5 BY35 33
Roseleigh Clo. Twick. BK46 48
Rosemary Av. N3 BS30 14
Rosemary Av. N9 CB26 8
Rosemary Av. E. Mol. BF52 65
Rosemary Av. Enf. BZ23 8
Rosemary Av. Houns. BD44 47
Rosemary Av. Rom. CT31 28
Rosemary Dr. Ilf. CJ32 26
Rosemary Gdns. SW14 BN45 49
Rosemary La.
Rosemary Gdns. Chess. BL56 75
Rosemary Gdns. Dag. CQ33 27
Rosemary La. SW14 BN45 49
Rosemary Rd. Well. CN44 54
Rosemary St. N1 BZ37 33
Rose Mead BO33 22
Rosemead Av. Mitch. BW52 68
Rosemead Av. Wem. BL35 30
Rosemont Av. N12 BT29 14
Rosemont Cotts. Wem. BL37 30
Rosemont Ct. W3 BM40 39
Rosemont Rd.
Rosemont Rd. NW3 BT36 32
Rosemont Rd. W3 BM40 39
Rosemont Rd. N. Mal. BN52 67
Rosemont Rd. Rich. BL46 48
Rosemoor St. SW3 BU42 41
Rosemount Dr. Brom. CK52 71
Rosemount Rd. W13 BJ39 39
Rosenau Cres. SW11 BU44 50
Rosenau Rd. SW11 BU44 50
Rosendale Rd. SE21 BZ47 60
Rosendale Rd. SE24 BZ47 60
Rosendale St. E5 CB34 24
Roseneath Av. N21 BY26 8
Roseneath Clo. Orp. CP57 81
Roseneath Rd. SW11 BV46 50
Roseneath Wk. Enf. BZ24 8
Rosens Wk. Edg. BM27 12
Rosenthal Rd. SE6 CE46 52
Rosenthorpe Rd. SE15 CC46 52
Roserton St. E14 CF41 43
Rosert St. E14 CF41 43
Manchester Rd.
Rosery, The, Croy. CC53 70
Roses, The, Wdf. Grn. CG29 16
Bunces La.
Rose St. WC2 BX40 42
Floral St.
Rosevale Rd. SW6 BR44 49
Roseveare Rd. SE12 CJ49 62
Roseville Av. Houns. BF46 47
Roseville Rd. Hayes BC42 38
Rosevine Rd. SW20 BQ51 67
Rose Wk. Pur. BW59 83
Rose Wk. Surb. BM53 66
Rose Wk. W. Wick. CF55 79
Rose Way SE12 CH46 53
Roseway SE21 BZ46 51
Rosewood, Bex. CT49 64
Rosewood Av. Grnf. BJ35 30
Rosewood Av. Horn. CU35 37
Rosewood Clo. DA14 CP48 63
Rose Wood Gdns. Wall. BV57 77
Rosewood Gro. Sutt. BT55 77
Rosher Clo. E15 CF36 34
Rosina St. E9 CC35 34
Roskell Rd. SW15 BQ45 49
Roslin Rd. W3 BM41 39
Roslin Way, Brom. CH49 62
Roslyn Clo. Mitch. BT51 68
Roslyn Gdns. Rom. CT30 19
Roslyn Rd. N15 BZ32 24
Rosmead Rd. W11 BR40 40
Rosoman St. EC1 BY38 33
Rossall Clo. Horn. CU32 28
Rossall Cres. NW10 BL38 30
Ross Av. NW7 BR28 13
Ross Av. Dag. CQ34 27
Ross Clo. E4 CF27 16
Rossdale, Sutt. BU56 77
Rossdale Dr. N9 CC25 9
Rossdale Dr. NW9 BN33 22
Rossdale Rd. SW15 BQ45 49
Rosse Ms. SE3 CH44 53
Rossendale St. E5 CB34 24
Rossindale Rd. Houns. BF46 47
Rossington St. E5 CB34 24
Rossiter Rd. SW12 BV47 59
Rossland Clo. Bexh. CR46 54
Rosslyn Av. E4 CG27 16
Rosslyn Av. SW13 BO45 49
Rosslyn Av. Barn. BU25 7
Rosslyn Av. Dag. CQ33 27
Rosslyn Av. Felt. BC46 47
Rosslyn Av. Rom. CW30 19
Rosslyn Clo. W. Wick. CG55 79
Rosslyn Cres. Har. BH31 21
Rosslyn Cres. N. Har. BH31 21
Rosslyn Cres. S. Har. BH32 21
Rosslyn Cres. Wem. BL35 30
Rosslyn Hl. NW3 BT35 32
Rosslyn Pk. Ms. NW3 BT35 32
Rosslyn Rd. E17 CF31 25
Rosslyn Rd. Bark. CM36 35
Rosslyn Rd. Twick. BK46 48
Rosslyn Rd. Wat. BC24 4
Rossmore Rd. NW1 BU38 32
Ross Pde. Wall. BV57 77
Ross Rd. SE25 BZ52 69
Ross Rd. Dart. CU46 55
Ross Rd. Twick. BG47 56
Ross Rd. Wall. BW56 77
Ross Way SE9 CK45 53
Rostella Rd. SW17 BT49 59
Rostrevor Av. N15 CA32 24
Rostrevor Gdns. Sthl. BE42 38
Rostrevor Mews SW6 BR45 49

Rostrevor Rd. SW6 BR44 49
Rostrevor Rd. SW19 BS49 59
Rotary St. SE1 BY41 42
Rothbury Av. Rain. CU39 46
Rothbury Gdns. Islw. BJ43 48
Rothbury Rd. E9 CD36 34
Rothbury Wk. N17 CB29 15
Park La.
Rotherfield Rd. Cars. BV56 77
Rotherfield St. N1 BZ36 33
Rotherhill Av. SW16 BW50 59
Rotherhithe New Rd. SE16 CB42 42
Rothgerhithe Old Rd. SE16 CC42 43
Rotherhithe St. SE16 CB41 42
Rotherhithe Tunnel App. E14 CD40 43
Rothermere Rd. Croy. BX56 78
Rotherwick Hill W5 BL38 30
Rotherwick Rd. NW11 BS33 23
Rotherwood Rd. SW15 BQ45 49
Rothery St. N1 BY37 33
Gaskin St.
Rothersay Av. SW20 BR51 67
Rothesay Av. Grnf. BG36 29
Rothesay Av. Rich. BM45 48
Rothesay Rd. SE25 BZ52 69
Rothfield Rd. Cars. BV56 77
Rothsay Rd. E7 CJ36 35
Rothsay St. SE1 CA41 42
Rothschild Rd. W4 BN41 40
Rothschild St. SE27 BY49 60
Roth Wk. N7 BX34 24
Andover Est.
Rothwell Gdns. Dag. CP37 36
Rothwell Rd. Dag. CP37 36
Rothwell St. NW1 BU36 32
Rotten Row SE3 CG44 52
Rotten Row SW7 BU41 41
Rotten Row SW11 BU43 50
Rouel Rd. SE16 CB41 42
Rougemount Av. Mord. BS53 68
Roundacre Est. SW19 BQ48 58
Roundaway Rd. Ilf. CK30 17
Round Gro. Croy. CC54 70
Roundhay Clo. SE23 CC48 61
Roundhill Dr. Enf. BX24 8
Round Hl. SE26 CB48 60
Roundmead Av. Loug. CL24 10
Roundmead Clo. Loug. CL24 10
Roundtable Rd. Brom. CG48 61
Roundtree Rd. Wem. BJ36 30
Roundway, The N17 BZ29 15
Roundway, The, Esher BH57 75
Roundwood, Chis. CL51 71
Roundwood Rd. NW10 BO36 31
Round Wood Vw. Bans. BQ61 82
Roundwood Way, Bans. BQ61 82
Rounton Rd. E3 CE38 34
Roupell Rd. SW2 BX47 60
Roupell St. SE1 BY40 42
Rouse Gdns. SE21 CA49 60
Rous Rd. Buck. H. CK26 10
Routh Rd. SW18 BU47 59
Rover Av. Ilf. CN29 18
Rowallan Rd. SW6 BR43 49
Rowan Av. E4 CD29 16
Rowan Clo. N. Mal. BO51 67
Rowan Clo. SW16 BW51 68
Rowan Clo. W5 BL41 39
Rowan Ct. SE12 CG47 61
Rowan Cres. SW16 BW51 68
Rowan Cres. Dart. CV47 64
Rowan Gdns. W6 BQ42 40
Bute Gdns.
Rowan Gdns. Croy. CA55 78
Rowan Rd. SW16 BW51 68
Rowan Rd. W6 BQ42 40
Rowan Rd. Bexh. CQ45 54
Rowan Rd. Brent. BJ43 48
Rowan Rd. Swan. CS52 73
Rowans, The N13 BY27 15
Rowantree Clo. N21 BZ26 8
Rowantree Rd. N21 BZ26 8
Rowantree Rd. Enf. BY23 8
Rowan Wk. N2 BT32 23
Rowan Wk. Brom. CK55 80
Rowan Wk. Horn. CV31 28
Rowan Way, Rom. CP31 27
Rowben Clo. N20 BS26 7
Rowberry Clo. SW6 BQ44 49
Rowcross St. SE1 CA42 42
Rowdell Rd. Nthlt. BF37 29
Rowden Pk. Gdns. E4 CE29 16
Rowden Rd.
Rowden Rd. E4 CE29 16
Rowden Rd. Beck. CD51 70
Rowden Rd. Epsom BM56 75
Rowditch La. SW11 BV44 50
Rowdon Av. NW10 BP36 31
Rowdown Cres. Croy. CF58 79
Rowdowns Rd. Dag. CQ37 36
Rowe Gdns. Bark. CN37 36
Rowe La. E9 CC35 34
Urswick Rd.
Rowena Cres. SW11 BU44 50
Rowe Wk. Har. BF34 20
Rowfant Rd. SW17 BV47 59
Rowhill Rd. E5 CB35 33
Rowhill Rd. Dart. CT50 64
Row Hill St. NW3 BU35 32
Rowington Clo. W2 BS39 41
Rowland Av. Har. BK31 21
Rowland Ct. E13 CG38 34
Rowland Ct. E16 CG38 34
Beaconsfield Rd.
Rowland Cres. Chig. CN28 18
Fairview Dr.
Rowland Gro. SE26 CB48 60
Dallas Rd.
Rowland Hill Av. N17 BZ29 15
Rowland Hill St. NW3 BU35 32
Rowland Rd. SW4 BW45 50
Rowlands Av. Pnr. BF28 11
Rowlands Clo. NW4 BP29 13
Rowlands Rd. Dag. CQ34 27
Rowland Wk. Hav. CT27 19
Rowlatt Clo. Dart. CV49 64
Rowlatt Rd. Dart. CV49 64
Rowley Av. Sid. CO47 63
Rowley Clo. Wat. BE25 4
Rowley Clo. Wem. BL36 30
Rowley Gdns. N4 BZ33 24
Rowley Grn. Rd. Barn. BO25 6
Rowley La. Borwd. BN23 6
Rowley Rd. N15 BZ32 24
Rowley Ter. NW5 BV36 32
Rowlls Rd. Kings. On T. BL52 66
Rowney Gdns. Dag. CO36 36
Rowney Rd. Dag. CO36 36
Rowns Way, Loug. CK24 10
Rowntree Path SE28 CO40 45
Macauley Way
Rowntree Rd. Twick. BH47 57
Rowse Clo. E15 CF37 34
Rowsley Av. NW4 BQ31 22
Rowstock Gdns. N7 BW35 32
Rowton Rd. SE18 CM43 53
Roxborough Av. Har. BG33 20
Roxborough Av. Islw. BH43 48
Roxborough Pk. Har. BH33 21
Roxborough Rd. Har. BG32 20
Roxbourne Clo. Nthlt. BE36 29
Arnold Rd.
Roxburgh Rd. SE27 BY49 60
Roxby Pl. SW6 BS43 50
Roxeth Grn. Av. Har. BF34 20
Roxeth Gro. Har. BF35 29
Roxeth Hill, Har. BG34 20
Roxley Rd. SE13 CE46 52
Roxton Gdns. Croy. CE56 79
Roxwell Rd. W12 BP41 40
Roxwell Rd. Bark. CO37 36
Roxwell Way, Wdf. Grn. CJ29 17
Roxy Av. Rom. CP33 27
Royal Arc. SW1 BW40 41
Pall Mall
Royal Av. SW3 BU42 41
Royal Av. Wor. Pk. BO55 76
Royal Cir. SE27 BY48 60
Royal Clo. Wor. Pk. BO55 76
Royal College St. NW1 BW36 32
Royal Cres. W11 BQ40 40
Royal Cres. Ruis. BE35 29
Royal Cres. Ms. W11 BQ40 40
Royal Dr. Epsom BP62 82
Royal Exchange Bldgs. EC3 BZ39 42
Cornhill
Royal Hl. SE10 CF43 52
Royal Hospital Rd. SW3 BU43 50
Royal Mint St. E1 CA40 42
Royal Naval Pl. SE14 CD43 52
Hereford Pl.
Royal Oak Ct. N1 CA38 33
Royal Oak Rd. E8 CB36 33
Wilton Way
Royal Oak Rd. Bexh. CQ46 54
Royal Oak Shopping Cen. Pur. BY54 69
Royal Pde. SE3 CG44 52
Royal Pde. W5 BL38 30
Western Av.
Royal Pde. Chis. CM50 62
Royal Pl. SE10 CF43 52
Royal Rd. E16 CJ39 44
Royal Rd. SE17 BY43 51
Royal Rd. Sid. CP48 63
Royal Rd. Tedd. BG49 56
Royal St. SE1 BX41 42
Royal Victor Pl. E3 CC37 34
Old Ford Rd.
Roycraft Av. Bark. CN37 36
Roycraft Clo. Bark. CN37 36
Roycroft Clo. E18 CH30 17
Roydene Rd. SE18 CN43 54
Roydon Clo. SW11 BU44 50
Reform St.
Roydon Clo. Loug. CK26 10
Roydon St. SW11 BV44 50
Roy Gdns. Ilf. CN31 27
Roy Gro. Hamptn. BF50 56
Royle Clo. Rom. CU32 28
Royle Cres. W13 BJ38 30
Roymount Ct. Twick. BH48 57
Roysdon Rd. Rom. CW29 19
Royston Av. E4 CE28 16
Royston Av. Sutt. BT55 77
Royston Av. Wall. BW56 77
Royston Clo. Houns. BC44 47
Royston Clo. Walt. BC54 65
Royston Ct. Rich. BL44 48
Royston Ct. Surb. BM55 75
Royston Gdns. Ilf. CJ32 26
Royston Gro. Pnr. BE29 11
Royston Pk. Rd. Pnr. BE29 11
Royston Rd. SE20 CC51 70
Royston Rd. Dart. CT46 55
Royston Rd. Rich. BL46 48
Roystons, The, Surb. BM53 66
Royston St. E2 CC38 34
Rozel Rd. SW4 BW44 50
Rubastic Rd. Sthl. BC41 38
Rubens Rd. Nthlt. BD37 29
Rubens St. SE6 CD48 61
Ruberoid Rd. Enf. CD24 9
Ruby Rd. E17 CE31 25
Ruby St. SE15 CB43 51
Ruckholt Clo. E10 CE34 25
Ruckholt Rd. E10 CE35 34
Rucklidge Av. NW10 BO37 31
Rudall Clo. NW3 BT35 32
Wiloughby Rd.
Rudall Cres. NW3 BT35 32
Willoughby Rd.
Ruddigore Rd. SE14 CD43 52
Ruden Way, Epsom BP61 82
Rudford Rd. SE16 CC42 43
Rudland Rd. Bexh. CR45 54
Rudloe Rd. SW12 BW47 59
Rudolph Rd. E13 CG37 34
Rudolph Rd. NW6 BS37 32
Rudolph Rd. Bush. BF25 4
Rudyard Gro. NW7 BN29 13
Ruffetts Clo. S. Croy. CB57 78
Ruffetts, The, S. Croy. CB57 78
Ruffetts Way, Tad. BR62 82
Rufford Clo. Har. BJ32 21
Rufford St. N1 BX37 33
Rufus Clo. Ruis. BE34 20
Rufus St. EC1 CA38 33
Old St.
Rugby Av. N9 CA26 8
Rugby Av. Grnf. BG36 29
Rugby Av. Wem. BJ35 30
Rugby Clo. Har. BH31 21
Rugby Gdns. Dag. CP36 36
Rugby La. Sutt. BQ58 76
Rugby Rd. NW9 BM31 21
Rugby Rd. W4 BO41 40
Rugby Rd. Dag. CO36 36
Rugby Rd. Twick. BH46 48
Rugby St. WC1 BX38 33
Rugg St. E14 CE40 43
Ruislip Clo. Grnf. BF38 29
Ruislip Rd. Grnf. BE38 29
Ruislip Rd. Nthlt. BD37 29
Ruislip Rd. E. W7 BG38 29
Ruislip Rd. E. W13 BG38 29
Ruislip Rd. E. Grnf. BG38 29
Ruislip St. SW17 BU49 59
Rumbold Rd. SW6 BS43 50
Rum Clo. E1 CC40 43
Rumford Pl. Erith CT43 55
Rumsey Clo. Hamptn. BE50 56
Rumsley Rd. SW9 BX45 51
Runbury Circle NW9 BN34 22
Runciman Clo. Orp. CP58 81
Runcorn Pl. W11 BR40 40
Rundell Cres. NW4 BP32 22
Runham St. SE17 BZ42 42
Runnelfield, Har. BH34 21
Running Horse Yd. Brent. BL43 48
Pottery Rd.
Runn Way E7 CH35 35
Runnymede SW19 BT51 68
Runnymede Clo. Twick. BF46 47
Runnymede Cres. SW16 BW51 68
Runnymede Gdns. Grnf. BG37 29
Runnymede Gdns. Twick. BF46 47
Runnymede Rd. Twick. BF46 47
Runton St. N19 BW33 23
Runway, The, Ruis. BD35 29
Rupack St. SE16 CC41 43
St. Mary Church St.
Rupert Av. Wem. BL35 30
London Rd.
Rupert Gdns. SW9 BY44 51
Rupert Rd. N19 BW34 23
Rupert Rd. NW6 BR37 31
Rupert Rd. W4 BO41 40
Rupert St. W1 BW40 41
Rural Way SW16 BV50 59
Ruscoe Rd. E16 CG39 43
Rusham Rd. SW12 BU46 50
Rushbrook Cres. E17 CD30 16
Rushbrook Rd. SE9 CM48 62
Rushbrook Rd. E4 CE29 16
Rushcroft Rd. SW2 BY45 51
Rushden Clo. SE19 BZ50 60
Rushdene SE2 CP41 45
Rushdene Av. Barn. BU26 7
Rushdene Clo. Nthlt. BD37 29
Rushdene Cres. Nthlt. BD37 29
Rushdene Rd. Pnr. BD32 20
Rushdene Gdns. NW7 BQ29 13
Rushdene Gdns. Ilf. CL30 17
Rushen Wk. Cars. BT54 68
Rushet Rd. Orp. CO51 72
Rushett Clo. Surb. BJ54 66
Rushett Gdns. Surb. BJ54 66
Rushett Rd. Surb. BJ54 66
Rushey Clo. N. Mal. BN52 67
Rushey Grn. SE6 CE47 61
Rushey Hill, Enf. BX24 8
Rushey Mead SE13 CE46 52
Rushford Rd. SE4 CD46 52
Rush Grn. Gdns. Rom. CS33 28
Rush Grn. Rd. Rom. CR33 27
Rushgrove Av. NW9 BO32 22
Rushgrove St. SE18 CK42 44
Rush Hill Rd. SW11 BV45 50
Rushley Clo. Kes. CJ56 80
Rushmead, Rich. BJ48 57
Rushmead Clo. Croy. CA56 78
Rushmead Clo. Edg. BM27 12
Rushmoor Clo. Pnr. BC31 20
Rushmore Clo. Brom. CK52 71
Rushmore Ct. Wor. Pk. BP55 76
Rushmore Cres. E5 CC35 34
Rushmore Hill, Orp. CP58 81
Rushmore Rd. E5 CC35 34
Rusholme Rd. SW15 BQ46 49
Rusholme Rd. Dag. CR34 27
Rushout Av. Har. BJ32 21
Rushton St. N1 BZ37 33
Rushworth Gdns. NW4 BP31 22
Rushworth St. SE1 BY41 42
Ruskin Av. E12 CK36 35
Ruskin Av. Rich. BM43 48
Ruskin Av. Well. CO45 54
Ruskin Clo. NW11 BS32 23
Ruskin Ct. N21 BX26 8
Ruskin Dr. Orp. CN55 81
Ruskin Dr. Well. CO45 54
Ruskin Dr. Wor. Pk. BP55 76
Ruskin Dr. Wor. Pk. BQ54 67
Ruskin Gdns. W5 BK38 30
Ruskin Gdns. Har. BL32 21
Ruskin Gdns. Rom. CU30 19
Ruskin Gro. Well. CO44 54
Ruskin Pk. Ho. SE5 BZ45 51
Ruskin Rd. N17 CA30 15
Ruskin Rd. Belv. CR42 45
Ruskin Rd. Cars. BU56 77
Ruskin Rd. Croy. BY55 78
Ruskin Rd. Islw. BH45 48
Ruskin Rd. Sthl. BE40 38
Ruskin Wk. N9 CB27 15
Ruskin Wk. SE24 BZ46 51
Ruskin Wk. Brom. CK53 71
Ruskin Way SW19 BT51 68
Brangwyn Cres.
Rusland Av. Orp. CM55 80
Rusland Rd. Har. BH31 21
Rusper Clo. Stan. BK28 12
Rusper Ct. SW9 BX44 51
Clapham Rd.
Rusper Rd. N22 BY30 15
Rusper Rd. Dag. CP36 36
Russel Clo. SE7 CJ43 53
Russell Av. N22 BY30 15
Russell Clo. NW10 BN36 31
Russell Clo. Beck. CE52 70
Russell Clo. Bexh. CR45 54
Russell Clo. Dart. CU45 55
Russell Clo. Ruis. BD34 20
Russell Ct. SW1 BW40 41
Cleveland Row
Russell Gdns. N20 BU27 14
Russell Gdns. NW11 BR32 22
Russell Gdns. W14 BR41 40
Russell Gdns. Rich. BK48 57
Russell Gdns. Ms. W14 BR41 40
Russell Grn. Clo. Pur. BY58 78
Russell Gro. NW7 BO28 13
Russell Gro. SW9 BY43 51
Russell Hill, Pur. BX58 78
Russell Hill Pl. Pur. BY55 78
Russell Hill Pl. Pur. BY59 84
Russell Hill Rd. Pur. BY58 78
Russell La. N20 BU27 14
Russell Mead, Har. BH29 12
Russell Pl. SW1 BW42 41
Vauxhall Bridge Rd.
Russell Rd. E4 CD28 16
Russell Rd. E10 CE32 25
Russell Rd. E16 CH39 44
Russell Rd. E17 CD31 25
Russell Rd. N8 BW32 23
Russell Rd. N13 BX29 15
Russell Rd. N15 CA32 24
Culvert Rd.
Russell Rd. N20 BU27 14
Russell Rd. NW9 BO32 22
Russell Rd. SW19 BS50 59
Russell Rd. W14 BR41 40
Russell Rd. Buck. H. CH26 10
Russell Rd. Nthlt. BG35 29
Russell Rd. Twick. BH46 48
Russell Rd. Walt. BC53 65
Russell's Footpath SW16 BX49 60
Russell Sq. WC1 BW38 32

Russell St. E13 CH37 35
Russell St. WC2 BX40 42
Russel Rd. Mitch. BU52 68
Russet Cres. N7 BX35 33
Stockorchard Cres.
Russett Clo. Orp. CO56 81
Russett Way. Swan. CS51 73
Russett Way SE13 CE44 52
Conington Rd.
Russia La. E2 CC37 34
Russia Row EC2 BZ39 42
Milk St.
Rusthall Av. W4 BN42 40
Rusthall Clo. Croy. CC53 70
Rustic Av. SW16 BV50 59
Rustic Pl. Wem. BK35 30
Rustington Wk. Mord. BR54 67
Ruston Av. Surb. BM54 66
Ruston Clo. W11 BR39 40
St. Marks Rd.
Rushton Ms. W11 BR39 40
Ruston St. E3 CD37 34
Rust Sq. SE5 BZ43 51
Rutford Rd. SW16 BX49 60
Ruth Clo. Har. BL31 21
Ruthen Clo. Epsom BN60 82
Rutherford Clo. Sutt. BT57 77
Rutherford Way. Bush. BG26 4
Rutherglen Rd. SE2 CO43 54
Rutherland St. SW1 BW42 41
Rutherwick Rise. Couls. BX62 84
Rutherwyke Clo. Epsom BP57 76
Ruthin Rd. SE3 CH43 53
Ruthven St. E9 CC37 34
Lauriston Rd.
Rutland App. Horn. CW32 28
Rutland Av. Sid. CO47 63
Rutland Clo. SW14 BN45 49
Rutland Clo. SW19 BU50 59
Rutland Rd.
Rutland Clo. Bex. CP47 63
Rutland Clo. Chess. BL57 75
Rutland Clo. Dart. CV46 55
Rutland Clo. Epsom BN58 76
Rutland Clo. SE5 BZ45 51
Rutland Dr. Horn. CW32 28
Rutland Dr. Mord. BR53 67
Rutland Gdns. N4 BY32 24
Rutland Gdns. SW7 BU41 41
Rutland Gdns. W13 BJ39 39
Rutland Gdns. Croy. CA56 78
Rutland Gdns. Dag. CP35 36
Rutland Gdns. Rich. BK47 57
Rutland Gate. SW7 BU41 41
Rutland Gate. Belv. CR41 45
Rutland Gate. Brom. CG52 70
Rutland Gate Ms. SW7 BU41 41
Rutland Gro. W6 BP42 40
Rutland Ms. SW7 BS37 32
Rutland Ms. S. SW7 BU42 41
Rutland Rd.
Rutland Ms. St. SW7 BU42 41
Ennismore Ms. Gdns.
Rutland Pk. NW2 BQ36 31
Rutland Pk. SE6 CD48 61
Rutland Rd. E7 CJ36 35
Rutland Rd. E9 CC37 34
Rutland Rd. E11 CH32 26
Rutland Rd. E17 CE32 25
Rutland Rd. SW19 BU50 59
Rutland Rd. Har. BG32 20
Rutland Rd. Ilf. CL35 35
Rutland Rd. Sthl. BF38 29
Rutland Rd. Twick. BG48 56
Rutland St. SW7 BU41 41
Rutland Wk. SE6 CD48 61
Rutland Way. Orp. CP53 72
Rutlish Rd. SW19 BS51 68
Rutter Gdns. Mitch. BT52 68
Ruttesland St. N1 CA37 33
Hoxton St.
Rutts Ter. SE14 CC44 52
Rutts, The. Bush. BG26 4
Ruvigny Gdns. SW15 BQ45 49
Ruxley Clo. Epsom BM56 75
Ruxley Clo. Sid. CP50 63
Ruxley Cres. Esher BJ57 75
Ruxley La. Epsom BM57 75
Ruxley Ridge, Esher BJ57 75
Ryan Clo. SE3 CJ45 53
Ryarsh Cres. Orp. CN56 81
Rycott Path SE22 CB47 60
Lordship La.
Rycroft Cres. Barn. BP25 6
Rycliff Sq. SE3 CG44 52
Rydal Clo. SW16 BW49 59
Rydal Clo. Pur. BZ60 84
Rydal Ct. NW4 BR30 13
Rydal Cres. Grnf. BJ37 30
Rydal Dr. Bexh. CQ44 54
Rydal Gdns. NW9 BO32 22
Rydal Gdns. SW15 BO49 58
Rydal Gdns. Houns. BF46 47
Rydal Gdns. Wem. BK33 21
Rydal Rd. SW16 BW49 59
Rydal Way. Enf. CC25 9
Rydal Way. Ruis. BD35 29
Rydens Av. Walt. BC55 74

Rydens Av. Walt. BD55 74
Rydens Clo. Walt. BD55 74
Rydens Gro. Walt. BD56 74
Rydens Rd. Walt. BC55 74
Rydens Rd. Walt. BD54 65
Ryde Pl. Twick. BK46 48
Ryde Clo. Brom. CH49 62
Ryder Clo. Bush. BF25 4
Ryder Gdns. E6 CL39 44
Ryder Gdns. Rain. CT36 37
Ryder's Ter. NW8 BT37 32
Ryder St. SW1 BW40 41
Ryde Vale Rd. SW12 BV48 59
Rydons Clo. SE9 CK45 53
Rydons Pk. Walt. BD55 74
Rydon St. N1 BZ37 33
St. Paul St.
Rydston Clo. N7 BX36 33
Sutterton St.
Rye Clo. Bex. CR46 54
Ryecoates Mead SE21 CA47 60
Ryde Cres. Orp. CP54 72
Ryecroft Av. Ilf. CL30 17
Ryecroft Av. Twick. BF47 56
Ryecroft Rd. SE13 CF46 52
Ryecroft Rd. SW16 BY50 60
Ryecroft Rd. Orp. CM53 71
Ryecroft St. SW6 BS44 50
Ryecroft Way N17 CA31 24
Ryefield, Orp. CP55 81
Ryefield Cres. Pnr. BC30 11
Ryefield Rd. SE19 BZ50 60
Ryehill Ct. N. Mal. BO54 67
Rye Hill Est. SE15 CC45 52
Rye Hill Pk. SE15 CC45 52
Ryelands Cres. SE12 CJ46 53
Rye La. SE15 CB44 51
Rye Rd. SE15 CC45 52
Rye, The N14 BW26 7
Rye Wk. SW15 BQ46 49
Rye Way, Edg. BL29 12
Ryfold Rd. SW19 BS48 59
Ryhope Rd. N11 BV28 14
Rylandes Rd. NW2 BP34 22
Ryland Rd. NW5 BV36 32
Rylands Rd. S. Croy. CB58 78
Rylett Cres. W12 BO41 40
Rylett Rd. W12 BO41 40
Rylston Rd. N13 BZ27 15
Rylston Rd. SW6 BR43 49
Rymer Rd. SW18 BT45 50
Alma Rd.
Rymer Rd. Croy. CA54 69
Rymer St. SE24 BY46 51
Rymill St. E16 CL40 44
Rysbrack St. SW3 BU41 41
Rythe Ct. Surb. BJ54 66
Rythe Rd. Esher BH56 75
Sabbarton St. E16 CG39 43
Victoria Dock Rd.
Sabella Ct. E3 CE37 34
Mostyn Gro.
Sabella St. SE1 BY40 42
Sabine Rd. SW11 BU45 50
Sable Clo. Houns. BD45 47
Sable St. N1 BY36 33
Sach Rd. E5 CB34 24
Sackville Av. Brom. CH54 71
Sackville Clo. Har. BG34 20
Sackville Cres. Rom. CW30 19
Sackville Est. SW16 BX48 60
Sackville Gdns. Ilf. CK33 26
Sackville Rd. Dart. CV48 64
Sackville Rd. Sutt. BS57 77
Sackville St. W1 BW40 41
Saddlescombe Way N12 BS28 14
Saddleworth Rd. Rom. CV29 19
Saddleworth Sq. Rom. CV29 19
Sadie St. SE5 BZ44 51
Orpheus St.
Sadler Clo. Mitch. BU51 68
Sadlers Ride, E. Mol. BG51 65
Saffron Clo. NW11 BR32 22
Saffron Hill EC1 BY38 33
Saffron Rd. Rom. CS30 19
Saffron St. EC1 BY39 42
Saffron Hill
Saftesbury Rd. Cars. BT54 68
Sail St. SE11 BX42 42
Sainsbury Rd. SE19 BZ49 60
St. Agatha's Dr.
Kings. On T.
Latchmere Rd.
St. Agathas Gro. Cars. BU54 68
St. Agatha's Wk. BM50 57
Kings. On T.
Alexandra Rd.
St. Agnes Clo. E9 CC37 34
Gore Rd.
St. Agnes Pl. SE11 CC37 34
Pennethorpe Clo.
St. Agnes Rd. E11 BY43 51
St. Aidan's Rd. SE22 CB46 51
St. Aidan's Rd. W13 BJ41 39
St. Albans Av. E6 CK38 35
St. Albans Av. W4 BN42 40
St. Albans Av. Felt. BD49 56

St. Albans Clo. NW11 BS33 23
North End Rd.
St. Alban's Cres. N22 BY30 15
St. Alban's Cres. CH29 17
Wdf. Grn.
St. Albans Gdns. Tedd. BJ49 57
St. Alban's Gro. W8 BS41 41
St. Alban's Gro. Cars. BU54 68
St. Albans La. NW11 BS33 23
West Heath Br.
St. Alban's Ms. W2 BT39 41
Edgware Rd.
St. Albans Pl. W1 BY37 33
St. Alban's Rd. NW5 BV34 23
St. Alban's Rd. NW10 BO37 31
St. Albans Rd. Dart. CW47 64
St. Alban's Rd. Ilf. CN33 27
St. Alban's Rd. BL50 57
Kings. On T.
St. Alban's Rd. Sutt. BR56 76
St. Albans Rd. Wat. BC23 4
St. Alban's Rd. Wdf. Grn. CH29 17
St. Alban's St. SW1 BW40 41
Jermyn St.
St. Alban's Ter. W6 BR43 49
St. Alfege Pass. SE10 CF43 52
Roan St.
St. Alfege Rd. SE7 CJ43 53
St. Alphage Ter. NW9 BN31 22
St. Alphages Gdns. EC2 BZ39 42
St. Alphege Rd. N9 CC26 9
St. Alphonsus Rd. SW4 BW45 50
St. Andrew's Av. Horn. CU35 37
St. Andrew's Av. Wem. BJ35 30
St. Andrews Clo. NW2 BP34 22
St. Andrew's Clo. Islw. BH44 48
St. Andrew's Clo. Ruis. BD34 20
St. Andrew's Clo. Stan. BK30 12
St. Andrew's Clo. N12 BT28 14
St. Andrews Ct. SW18 BT48 59
Waynflete St.
St. Andrews Dr. Orp. CO53 72
St. Andrews Dr. Stan. BK30 12
St. Andrew's Gro. N16 BZ33 24
St. Andrew's Hill EC4 BY39 42
St. Andrew's Ms. N16 CA33 24
Dunsmore Rd.
St. Andrews Pl. NW1 BW38 32
St. Andrew's Rd. E11 CG32 25
St. Andrew's Rd. E13 CH38 35
St. Andrew's Rd. N9 CC26 9
St. Andrew's Rd. NW9 BN33 22
St. Andrew's Rd. NW10 BP36 31
St. Andrew's Rd. NW11 BR32 22
St. Andrew's Rd. W3 BO40 40
St. Andrews Rd. W14 BR43 49
St. Andrews Rd. Cars. BU55 77
St. Andrew's Rd. Couls. BV61 83
St. Andrew's Rd. Croy. BZ56 78
St. Andrew's Rd. Enf. BZ24 8
St. Andrew's Rd. Ilf. CK33 26
St. Andrew's Rd. Rom. CS32 28
St. Andrew's Rd. Sid. CP48 63
St. Andrews Rd. Surb. BK53 66
St. Andrews Rd. Surb. BK53 66
St. Andrews Sq. W11 BQ39 40
Lancaster Rd.
St. Andrew St. EC4 BY39 42
St. Andrews Rd. W14 BR43 49
St. Anne's Clo. N6 BV34 23
St. Anne's Ct. W1 BW39 41
Wardour St.
St. Anne's Gdns. NW10 BL38 30
St. Anne's Rd. E11 CF34 25
St. Anne's Rd. Har. BH32 21
St. Anne's Rd. Wem. BK35 30
St. Ann's, Bark. CM37 35
St. Ann's Cres. SW18 BT48 59
St. Ann's Gdns. NW5 BV36 32
Queen's Cres.
St. Ann's Hl. SW18 BS46 50
St. Ann's La. SW1 BW41 41
Old Pye St.
St. Ann's Pk. Rd. SW18 BT46 50
St. Ann's Pass. SW13 BO45 49
Cross St.
St. Ann's Rd. N9 CA27 15
St. Ann's Rd. N15 BY32 24
St. Anns Rd. SW13 BO44 49
St. Ann's Rd. W11 BQ40 40
St. Ann's Rd. SW1 BW41 41
St. Anns Rd. Bark. CM37 35
Morley Rd.
St. Ann's Ter. NW8 BT37 32
St. Ann's Vill. W11 BQ40 40
St. Ann's Way. S. Croy. BY57 78
St. Anselms Ct. SW16 BX49 60
St. Anselms Pl. W1 BV40 41
Davies St.
St. Anthony's Av. CJ29 17
Wdf. Grn.
St. Anthonys Clo. E1 CB40 42
St. Anthonys Clo. SW17 BU48 59
College Gdns.
St. Anthony's La. Swan. CU51 73
St. Anthony's Rd. E7 CH36 35

St. Arvans Clo. Croy. CA55 78
St. Asaph Rd. SE4 CC45 52
St. Aubyns Av. SW19 BR49 58
St. Aubyn's Av. Houns. BF46 47
St. Aubyn's Clo. Orp. CN55 81
St. Aubyn's Gdns. Orp. CN55 81
St. Aubyn's Rd. SE19 CA50 60
St. Audrey Av. Bexh. CR44 54
St. Augustine's Av. W5 BL37 30
St. Augustines Av. Brom. CK53 71
St. Augustine's Av. BZ57 78
S. Croy.
St. Augustine's Av. Wem. BL34 21
St. Augustine's Rd. NW1 BW36 32
St. Augustines Rd. Belv. CQ42 45
St. Austell Clo. Edg. BL30 12
St. Austell Rd. SE13 CF44 52
St. Awdry's Rd. Bark. CM36 35
St. Awdry's Wk. Bark. CM36 35
St. Barnabas Clo. Beck. CF51 70
St. Barnabas Rd. E17 CE32 25
St. Barnabas Rd. Mitch. BV50 59
St. Barnabas Rd. Sutt. BT56 77
St. Barnabas Rd. CH30 17
Wood. Grn.
St. Barnabas St. SW1 BV42 41
St. Barnabas Vil. SW8 BX44 51
Guildford Rd.
St. Bartholomew's Rd. E6 CK37 35
St. Benet Clo. SW17 BU48 59
St. Benet Pl. EC3 BZ40 42
Gracechurch St.
St. Benets Gro. Cars. BT54 68
St. Bernard's, Croy. CA55 78
St. Bernard's Rd. E6 CJ37 35
St. Blaise Av. Brom. CH51 71
St. Botolph Row EC3 CA39 42
Houndsditch
St. Botolph St. EC3 CA39 42
St. Bride's Av. Edg. BL30 12
St. Brides Clo. Belv. CP41 45
St. Katherines Rd.
St. Bride's Pass. EC4 BY39 42
Dorset Rise
St. Bride St. EC4 BY39 42
St. Catherines Clo. SW17 BU48 59
College Gdns.
St. Catherines Ct. W4 BO42 40
Newton Gro.
St. Catherine's Dr. SE14 CC44 52
Kitto Rd.
St. Catherine's Rd. E4 CE27 16
St. Chads Gdns. Rom. CO33 27
St. Chads Pl. WC1 BX38 33
St. Chads Rd. Rom. CO32 27
St. Chad's St. WC1 BX38 33
St. Charles' Pl. W10 BR39 40
Chesterton Rd.
St. Charles Sq. W10 BQ39 40
St. Christophers Clo. Islw. BH44 48
St. Christophers Clo. Islw. BG44 47
Thornbury Rd.
St. Christopher's Pl. W1 BV39 41
Barrett St.
St. Clair Clo. Ilf. CK30 17
St. Clair Dr. Wor. Pk. BP56 76
St. Clair Rd. E13 CH37 35
St. Clair's Rd. Croy. CA55 78
St. Clare St. EC3 CA39 42
Minories
St. Clement's La. WC2 BX39 42
Portugal St.
St. Clement St. N7 BY36 33
St. Cloud Rd. SE27 BZ49 60
St. Crispin's Clo. Sthl. BE39 38
St. Cross St. EC1 BY39 42
St. Cuthbert's NW3 BS34 23
St. Cuthberts Gdns. Pnr. BE29 11
Westfield Pk.
St. Cuthbert's Rd. NW2 BR36 31
St. Cyprian's St. SW17 BU49 59
St. David's Clo. Wem. BN34 22
St. Davids Clo. W. Wick. CE54 70
St. Davids, Couls. BX62 84
St. David's Dr. Edg. BL30 12
St. Davids Pl. NW4 BP33 22
St. David's Rd. Swan. CT50 64
St. Denis Rd. SE27 BZ49 60
St. Dionis Rd. SW6 BR44 49
St. Donatt's Rd. SE14 CD44 52
St. Dunstan's Alley EC3 CA40 42
Idol La.
St. Dunstan's Av. W3 BN40 40
St. Dunstan's Clo. Hayes BC42 38
St. Dunstan's Gdns. W3 BN40 40
St. Dunstan's Av.
St. Dunstan's Hill EC3 CA40 42
Idol La.
St. Dunstan's Hill. Sutt. BR57 76
St. Dunstan's La. EC3 CA40 42
St. Dunstans La. Beck. CF53 70
St. Dusntans Rd. E7 CJ36 35
St. Dunstan's Rd. SE25 CA52 69
St. Dunstan's Rd. W6 BQ42 40
St. Dunstan's Rd. W7 BH41 39

St. Mary's Cotts. SW19 BS51 68
St. Mary's Rd.
St. Mary's Ct. E6 CK38 35
St. Mary's Ct. W5 BK41 39
St. Mary's Rd.
St. Mary's Cres. NW4 BP31 22
St. Mary's Cres. Hayes BC40 38
St. Mary's Rd.
St. Marys Est. E2 CA37 33
Weymouth Ter.
St. Mary's Gdns. SE11 BY42 42
St. Mary's Gro. N1 BY36 33
St. Mary's Gro. SW13 BP45 49
St. Mary's Gro. W4 BM43 48
St. Mary's Gro. Rich. BL45 48
St. Mary's Ms. W2 BT39 41
St. Mary's Rd. E10 CF34 25
St. Mary's Rd. E13 CH37 35
St. Mary's Rd. N8 BX31 24
High St.
St. Mary's Rd. N9 CB26 8
St. Mary's Rd. NW10 BO37 31
St. Mary's Rd. NW11 BR33 22
St. Mary's Rd. SE25 CC44 52
St. Mary's Rd. SE25 CA52 69
St. Mary's Rd. SW19 BR49 58
St. Mary's Rd. W5 BK40 39
St. Mary's Rd. Barn. BU26 7
St. Mary's Rd. Bex. CS47 64
St. Marys Rd. E. Mol. BG53 65
St. Mary's Rd. Ilf. CM34 26
St. Mary's Rd. Merton BS51 68
SW19
St. Mary's Rd. S. Croy. BZ58 78
St. Mary's Rd. Surb. BK53 66
St. Mary's Rd. Surb. BK54 66
St. Mary's Rd. Swan. CS52 73
St. Mary's Rd. Wat. BC24 4
St. Mary's Rd. Wimb. BR49 58
SW19
St. Marys Rd. Wor. Pk. BO55 76
St. Mary's Sq. W2 BT39 41
St. Mary's Ter. Gdns.
St. Mary's Ter. W2 BT39 41
St. Mary St. SE18 CK42 44
St. Marys Way, Wdf. Grn. CL28 17
St. Mary's Wk. SE11 BY42 42
St. Mathew's Rd. SW2 BX45 51
St. Matthew's Av. Surb. BL54 66
St. Matthew's Clo. Rain. CU36 37
St. Matthew's Dr. Brom. CK52 71
St. Matthews Rd. W5 BL40 39
St. Matthews Row E2 CB38 33
St. Matthews St. SW1 BW41 41
Old Pye St.
St. Matthias Clo. BO32 22
St. Maur Rd. SW6 BR44 49
St. Merryn Clo. SE18 CM43 53
St. Meryl Est. Wat. BE27 11
St. Michael's Alley EC3 BZ39 42
Cornhill
St. Michael's Av. N9 CC26 9
St. Michael's Av. N22 BQ35 31
St. Michael's Av. Wem. BM36 30
St. Michael's Clo. N3 BR30 13
Hendon La.
St. Michaels Clo. N12 BU28 14
St. Michaels Clo. Belv. CP41 45
St. Helens Rd.
St. Michael's Clo. Brom. CK52 71
St. Michael's Clo. Walt. BD55 74
St. Michael's Cres. Pnr. BE32 20
St. Michael's Gdns. W10 BR39 40
Ladbroke Clo.
St. Michael's Rise, Well. CO44 54
St. Michael's Rd. SW9 BX44 51
St. Michael's Rd. Croy. BZ54 69
St. Michael's Rd. Wall. BW56 77
St. Michael's Rd. Well. CO45 54
St. Michael's St. SE8 CD44 52
Tanner's Hill
St. Michael's Ter. W2 BT39 41
St. Michael's Ter. N22 BX30 15
St. Mildred's Ct. EC2 BZ39 42
Poultry
St. Mildred's Rd. SE12 CG47 61
St. Neots Rd. Rom. CW29 19
St. Nicholas Av. Horn. CU34 28
St. Nicholas Clo. Borwd. BK25 5
Elstree Hill N.
St. Nicholas Glebe SW17 BV49 59
St. Nicholas La. Chis. CK51 71
St. Nicholas Rd. SE18 CN42 45
St. Nicholas Rd. Surb. BH53 66
St. Nichola's Rd. Sutt. BS56 77
St. Nicholas Rd. SE8 CE44 52
Lucas St.
St. Nichola's Way, Sutt. BS56 77
St. Nicholas Way, Sutt. BS56 77
Robin Hood La.
St. Norbert Grn. SE4 CD45 52
St. Norbert Rd. SE4 CC46 52
St. Norman's Way, Epsom BP58 76
St. Olaf's Rd. SW6 BR43 49
St. Olave's Est. SE1 CA41 42
St. Olave's Rd. E6 CL37 35

St. Olaves Wk. SW16 BW51 68
St. Oswalds Pl. SE11 BX42 42
St. Oswald's Rd. SW16 BY51 69
St. Oswulf St. SW1 BW42 41
Erasmus St.
St. Pancras Ct. N2 BT30 14
St. Pancras Way NW1 BW36 32
St. Pancras Way Est. BW36 32
NW1
St. Pancras Way
St. Paul's Alley EC4 BY39 42
St. Paul's Churchyard
St. Paul's Av. NW2 BP36 31
St. Paul's Av. SE16 CC40 43
St. Pauls Av. Har. BL32 21
St. Pauls Chuchyard EC4 BY39 42
St. Paul's Clo. Chess. BK56 75
St. Pauls Clo. Houns. BE44 47
St. Pauls Clo. W5 BL41 39
St. Paul's Cray Est. Chis. CN52 72
St. Paul's Cray Rd. Chis. CM50 62
St. Pauls Cres. NW1 BW36 32
St. Paul's Dr. E15 CF35 34
St. Paul's Pl. N1 BZ36 33
St. Paul's Rd. N1 BY36 33
St. Paul's Rd. N17 CB29 15
St. Paul's Rd. Bark. CM37 35
St. Paul's Rd. Brent. BK43 48
St. Paul's Rd. Erith CS43 55
St. Pauls Rd. Rich. BL45 48
St. Paul's Rd. Th. Hth. BZ52 69
St. Paul's Shrubbery N1 BZ36 33
St. Paul's
St. Paul's Sq. Brom. CG51 70
Church Rd.
St. Paul's St. E3 CD39 43
St. Pauls Ter. SE17 BY43 51
St. Paul St. N1 BZ37 33
St. Paul's Way E3 CD39 43
St. Paul's Way N3 BS29 14
St. Pauls Wood Hill, Orp. CN51 72
St. Peter's Av. E17 CG31 25
St. Peter's Av. N18 CB28 15
St. Petersburgh Ms. W2 BS40 41
St. Petersburgh Pl. W2 BS40 41
St. Peters Clo. E2 CB38 33
St. Peters Clo. Barn. BP25 6
St. Peters Clo. Bush. BG26 4
St. Peters Clo. Chis. CM50 62
St. Peter's Clo. Ilf. CN31 27
St. Peter's Clo. Ruis. BD34 20
St. Peter's Clo. SW17 BV48 59
College Gdns.
St. Peters Clo. W5 BK39 39
Regal Clo.
St. Peter's Ct. SE3 CG45 52
St. Peter's Gro. W6 BP42 40
St. Peter's Rd. N9 CB26 8
St. Peter's Rd. SW6 BR44 49
Filmer Rd.
St. Peter's Rd. W6 BP42 40
St. Peter's Rd. Croy. BZ56 78
St. Peters Rd. E. Mol. BF52 65
St. Peters Rd. BM51 66
Kings. On T.
Cambridge Rd.
St. Peter's Rd. Sthl. BF39 38
St. Peter's Rd. Twick. BJ46 48
St. Peter's Sq. W6 BP42 40
St. Peters St. N1 BY37 33
St. Peter's St. S. Croy. BZ56 78
St. Peter's Ter. SW6 BR43 49
Reporton Rd.
St. Peter's Vill. W6 BP42 40
St. Peter's Way N1 CA36 33
De Beauvoir Sq.
St. Peters Way W5 BK39 39
St. Philip's Av. Wor. Pk. BP55 76
St. Philip Sq. SW8 BV44 50
St. Philip's Rd. E8 CB36 33
St. Philips Rd. Surb. BK53 66
St. Philips St. SW8 BV44 50
St. Philips Way N1 BZ37 33
Linton St.
St. Quintin Rd. Well. CN45 54
St. Quintin Av. W10 BQ39 40
St. Quintin Gdns. W10 BQ39 40
St. Quintin Rd. E13 CH37 35
St. Raphael's Way NW10 BM36 31
St. Regis Clo. N10 BV30 14
St. Ronans Res. Wdf. CH29 17
Grn.
St. Rule St. SW8 BW44 50
St. Saviours Av. Dart. CW46 55
St. Saviours Est. SE1 CA41 42
St. Saviour's Rd. SW2 BX46 51
St. Saviour's Rd. Croy. BY53 69
Saints Dr. E7 CJ35 35
St. Silas Pl. NW5 BV36 32
St. Silas St. NW5 BV36 32
St. Simons Av. SW15 BQ46 49
St. Stephens Av. E17 CF32 25
St. Stephens Av. W12 BP40 40
St. Stephens Av. W13 BJ39 39
St. Stephens Clo. E17 CE32 25
St. Stephen's Clo. NW8 BU37 32
St. Stephen's Clo. Sthl. BF39 38

St. Stephens Cres. W2 BS39 41
St. Stephen's Cres. BY52 69
Th. Hth.
St. Stephens Gdn. Est. BS39 41
W2
St. Stephens Gdns. W2 BS39 41
St. Stephens Gdns. BK46 48
Twick.
St. Stephens Ms. W2 BS39 41
Chepstow Rd.
St. Stephens Pl. Twick. BK46 48
St. Stephen's Rd. E3 CD37 34
St. Stephen's Rd. E6 CJ36 35
St. Stephens Rd. E17 CE32 25
St. Stephens Rd. W13 BJ39 39
St. Stephen's Rd. Barn. BQ25 6
St. Stephen's Rd. Houns. BF46 47
St. Stephen's Row EC4 BZ39 42
Walbrook
St. Stephen's Ter. SW8 BX43 51
St. Swithin's La. EC4 BZ40 42
St. Swithun's Rd. SE13 CF46 52
St. Thomas Ct. Bex. CR47 63
St. Thomas Dr. Orp. CM54 71
St. Thomas Dr. Pnr. BE30 11
St. Thomas Rd. E16 CH39 44
St. Thomas Rd. N14 BW26 7
St. Thomas Rd. Belv. CS41 46
St. Thomas Gdns. NW5 BV36 32
Queen's Cres.
St. Thomas's Gdns. Ilf. CM36 35
St. Thomas's Pl. E9 CC36 34
St. Thomas's Rd. N4 BY34 24
St. Thomas's Rd. NW10 BO37 31
St. Thomas's Rd. W4 BN43 49
St. Thomas's Sq. E9 CB36 33
St. Thomas St. SE1 BZ40 42
St. Thomas's Way SW6 BR43 49
St. Ursula Rd. Sthl. BF39 38
St. Ursulas Gro. Pnr. BD32 20
St. Vincent Est. E14 CD40 43
St. Vincent Rd. Twick. BG46 47
St. Vincent Rd. Walt. BC55 74
St. Vincent's Rd. Dart. CW46 55
St. Vincent St. W1 BV39 41
Aybrook St.
Saintway, The, Stan. BH28 12
St. Wilfreds Clo. Barn. BT25 7
St. Wilfreds Rd.
St. Wilfred's Rd. Barn. BT25 7
St. Winifred's Av. E12 CK35 35
St. Winifreds Clo. Chig. CM28 17
St. Winifred's Rd. Tedd. BJ50 57
St. Winifrid's Wk. SE17 BY43 51
Lorrimore Rd.
Saigasso Clo. E16 CJ39 44
Royal Rd.
Salamanca Pl. SE1 BX42 42
Salamanca St.
Salamanca St. SE1 BX42 42
Salamanca St. SE11 BX42 42
Salamons Way, Rain. CT39 46
Salcombe Dr. Mord. BQ54 67
Salcombe Dr. Rom. CR32 27
Salcombe Gdns. NW7 BQ29 13
Salcombe Rd. E17 CD33 25
Salcombe Waye, Ruis. BC34 20
Salcot Cres. Croy. CF58 79
Salcott Rd. SW11 BU46 50
Salcott Rd. Croy. BX55 78
Salehurst Clo. Har. BL32 21
Salehurst Rd. SE4 CD46 52
Salem Pl. Croy. BZ55 78
Salem Rd. W2 BS40 41
Porchester Gdns.
Sale Pl. W2 BU39 41
Sale, The E4 CG28 16
Salford Rd. SW2 BW47 59
Salisbury Av. N3 BR31 22
Salisbury Av. Bark. CM36 35
Salisbury Av. Sutt. BR57 76
Salisbury Av. Swan. CU52 73
Salisbury Ct. EC4 BY39 42
Salisbury Gdns. SW19 BR50 58
Salisbury Pl. W1 BU39 41
Salisbury Plain NW4 BQ32 22
Brent St.
Salisbury Rd. E4 CE27 16
Salisbury Rd. E7 CH36 35
Salisbury Rd. E10 CF34 25
Salisbury Rd. E12 CJ35 35
Salisbury Rd. E17 CF32 25
Salisbury Rd. N4 BY32 24
Salisbury Rd. N9 CB27 15
Salisbury Rd. N22 BY30 15
Salisbury Rd. SE25 CB53 69
Salisbury Rd. SW19 BR50 58
Salisbury Rd. W13 BJ41 39
Salisbury Rd. Bans. BS60 83
Salisbury Rd. Barn. BR24 6
Salisbury Rd. Bex. CR47 63
Salisbury Rd. Brom. CK53 71
Salisbury Rd. Cars. CB56 77
Salisbury Rd. Dag. CR36 36
Salisbury Rd. Felt. BD47 56
Salisbury Rd. Har. BG32 20
Salisbury Rd. Houns. BD45 47
Salisbury Rd. Ilf. CN34 27

Salisbury Rd. N. Mal. BN52 67
Salisbury Rd. Pnr. BC31 20
Salisbury Rd. Rich. BL45 48
Salisbury Rd. Rom. CU32 28
Salisbury Rd. Sthl. BE42 38
Salisbury Rd. Wat. BC23 4
Salisbury Rd. Wor. Pk. BN56 76
Salisbury Sq. EC4 BY39 42
Salisbury St. NW8 BU38 32
Salisbury St. W3 BN41 40
Salisbury Walk N19 BW34 23
Girdlestone Est.
Salix Clo. Sun. BC50 56
Salmen Rd. E13 CG37 34
Salmond Clo. Stan. BJ29 12
Salmon La. E14 CD39 43
Salmon Rd. Belv. CR42 45
Salmons Rd. N9 CB26 8
Salmons Rd. Chess. BL57 75
Salmon St. NW9 BM33 21
Salomons Rd. E13 CJ39 44
Salop Rd. E17 CC32 25
Saltash Clo. Sutt. BR56 76
Saltash Rd. Ilf. CM29 17
Saltash Rd. Well. CP44 54
Saltcoats Rd. W4 BO41 40
Greenend Rd.
Salterford Rd. SW17 BV50 59
Salter Rd. SE16 CD40 43
Salter's Hill SE19 BZ49 60
Salters Rd. W10 BQ38 31
Salters Rd. E17 CF31 25
Salter St. E14 CE40 43
Salter St. NW10 BO38 31
Salterton Rd. N7 BX34 24
Saltford Clo. Erith CT42 46
Saltoun Rd. SW2 BY45 51
Saltram Clo. N15 CA31 24
Saltram Cres. W9 BR38 31
Saltwell St. E14 CE40 43
Saltwood Clo. Orp. CP56 81
Salusbury Rd. NW6 BR37 31
Salvador SW17 BU49 59
Salvia Gdns. Grnf. BJ37 30
Salvin Rd. SW15 BQ45 49
Salway Clo. Wdf. Grn. BG29 16
Salway Pl. E15 CF36 34
Broadway
Salway Rd. E15 CF36 34
Samantha Clo. E17 CD33 25
Samantha Mews, Hav. CT27 19
Sam Bartram Clo. SE7 CJ42 44
Sampson Av. Barn. BQ25 6
Samels Ct. W6 BP42 40
S. Black Lion La.
Samos Clo. SE3 CG43 52
Samos Rd. SE20 CB51 69
Sampson St. E1 CB40 42
Samson St. E13 CJ37 35
Samuel Lewis Bldgs. SE5 BZ44 51
Samuel Lewis Dws. SW6 BS43 50
Vanston Pl.
Samuel Lewis Trust CB35 33
Bldgs. E8
Samuel Lewis Trust BY36 33
Dws. N1
Samuel St. SE18 CK42 44
Sancroft Clo. NW2 BP34 22
Sancroft St. Stan. BJ30 12
Sancroft St. SE11 BX42 42
Sanctuary Clo. Dart. CV46 55
Sanctuary St. SE1 BZ41 42
Sanctuary, The, Bex. CP46 54
Sanctuary, The, Mord. BS53 68
Sandall Clo. W5 BL38 30
Sandall Rd. NW5 BW36 32
Sandall Rd. W5 BL38 30
Sandal Rd. N18 CB28 15
Sandal Rd. N. Mal. BN53 67
Sandal St. E15 CG37 34
Sandalwood Rd. Felt. BC48 56
Sandbach Pl. SE18 CM42 44
Sandbourne Av. SW19 BS51 68
Sandbourne Rd. SE4 CD44 52
Sandbrook Clo. NW7 BN29 13
Sunnydale Gro.
Sandbrook Rd. N16 CA34 24
Sandby Grn. SE9 CK45 53
Sandcliff Rd. Erith CS42 46
Sandell St. SE1 BY41 42
Sanders Clo. Hamptn. BG49 56
Sandersfield Gdns. BS61 83
Bans.
Sandersfield Rd. Bans. BS61 83
Sanders La. NW7 BQ29 13
Sanderson Clo. SW4 BW47 59
Sanderstead Av. NW2 BR34 22
Sanderstead Clo. SW4 BW47 59
Atkins Rd.
Sanderstead Ct. Av. CB60 84
S. Croy.
Sanderstead Hill S. CA59 84
Croy.
Sanderstead Rd. E10 CD33 25
Sanderstead Rd. Orp. CO53 72
Sanderstead Rd. S. BZ57 78
Croy.

Sanders Way N19 BX33 24
Hornsey Rise
Sanders Way N19 BW33 23
Sussex Way
Sandfield Gdns. Th. Hth. BY52 69
Sandfield Pass. Th. Hth. BZ52 69
Sandfield Rd. Th. Hth. BY52 69
Sandford Av. N22 BY29 15
Sandford Av. Loug. CL24 10
Sandford Clo. E6 CK38 35
Sandford Rd.
Sandford Ct. N16 CA33 24
Sandford Rd. E6 CK38 35
Sandford Rd. Bexh. CQ45 54
Sandford Rd. Brom. CH52 71
Sandford Row SE17 BZ42 42
Sandford St. SW6 BS43 50
Sandford St. SE14 CD43 52
Sandgate Rd. N18 CB28 15
Sandgate Rd. Well. CP43 54
Sandgate St. SE15 CB43 51
Sandhills Wall. BW56 77
Sandhurst Av. Har. BF32 20
Sandhurst Av. Surb. BM54 66
Sandhurst Clo. NW9 BM31 21
Sandhurst Clo. S. Croy. CA58 78
Sandhurst Ct. SW2 BX45 51
Sandhurst Dr. Ilf. CN35 36
Sandhurst Rd. N9 CC25 9
Sandhurst Rd. NW9 BM31 21
Sandhurst Rd. SE6 CF47 61
Sandhurst Rd. Bex. CP46 54
Sandhurst Rd. Orp. CO55 81
Sandhurst Rd. Sid. CN48 63
Sandhurst Way S. Croy. CA57 78
Sandiford Rd. Sutt. BR55 76
Sandiland Cres. Brom. CG55 79
Sandilands Croy. CB55 78
Sandilands La. SW6 BS44 50
Sandilands Rd. SW6 BS44 50
Sandison St. SE15 CA45 51
Sandland St. WC1 BX39 42
Sandling Rise SE9 CL48 62
Sandlings, The, N22 BY30 15
Sandmere Rd. SW4 BX45 51
Sandon Clo. Esher BG54 65
Sandow Av. Horn. CV34 28
Sandown Av. Dag. CS36 37
Sandown Av. Esher BG56 74
Sandown Clo. Houns. BC44 47
Sandown Ct. Esher BF56 74
Sandown Rd. SE25 CB53 69
Sandown Rd. Couls. BV61 83
Sandown Rd. Esher BG56 74
Sandown Way Nthlt. BE36 29
Sandpiper Rd. S. Croy. CC58 79
Sandpit Rd. Brom. CG49 61
Sandpit Rd. Dart. CV45 55
Sandpits Rd. Croy. CC56 79
Sandpits Rd. Rich. BK48 57
Sandra Clo. N22 BZ30 15
New Rd.
Sandra Clo. Houns. BF46 47
Sandridge Clo. Har. BH31 21
Sandridge Ct. N16 BZ34 24
Kings Crescent Est.
Sandringham Av. SW20 BR51 67
Sandringham Clo. Enf. CA23 8
Sandringham Clo. Ilf. CM31 26
Sandringham Gdns.
Sandringham Ct. W9 BT38 32
Sandringham Ct. Har. BF34 20
Sandringham Cres. Har. BF34 20
Sandringham Dr. Well. CN44 54
Sandringham Gdns. N8 BX32 24
Sandringham Gdns. N12 BT29 14
Sandringham Gdns. Houns. BC43 47
Sandringham Gdns. Ilf. CM31 26
Sandringham Ms. W5 BK40 39
High St.
Sandringham Rd. E7 CJ35 35
Sandringham Rd. E8 CA35 33
Sandringham Rd. E10 CF32 25
Sandringham Rd. N22 BZ31 24
Sandringham Rd. NW2 BP36 31
Sandringham Rd. NW11 BR33 22
Sandringham Rd. Bark. CN35 36
Sandringham Rd. Brom. CH49 62
Sandringham Rd. Nthlt. BF36 29
Sandringham Rd. Th. Hth. BZ53 69
Sandringham Rd. Wor. Pk. BP55 76
Sandrock Pl. Croy. CC56 79
Sandrock Rd. SE13 CE45 52
Sandsend La. SW6 BS44 50
Sandstone Pl. N6 BV34 23
Sandstone Rd. SE12 CH48 62
Sands Way IG8 CK29 17
Sandtoft Rd. SE7 CH43 53
Sandwell Cres. NW6 BS35 32
Sumatra Rd.
Sandwich St. WC1 BX38 33
Sandy Bury Orp. CM55 80
Sandy Clo. Twick. BJ47 57
Sandycombe Rd. Felt. BC47 56
Sandycombe Rd. Rich. BL45 48
Sandycombe Rd. Twick. BK46 48
Sandycroft SE2 CO43 54
Sandy Hill Av. SE18 CL42 44
Sandyhill Rd. Ilf. CL35 35
Sandy Hill Rd. Wall. BW58 77
Sandy La. Bush. BG24 4
Sandy La. Kings. On T. BK48 57
Sandy La. Mitch. BV51 68
Sandy La. Nthwd. BC27 11
Sandy La. Orp. CO54 72
Sandy La. Orp. CP51 72
Sandy La. Rain. CW40 46
Sandy La. Rich. BK48 57
Sandy La. Sutt. BR58 76
Sandy La. Tedd. BJ50 57
Sandy La. Walt. BC53 65
Sandy La. Est. Rich. BK48 57
Sandy La. N. Wall. BW56 77
Sandy La. S. Wall. BW58 77
Sandymount Av. Stan. BK28 12
Sandy Ridge Chis. CK50 62
Sandy Rd. NW3 BS34 23
Sandy's Row E1 CA39 42
Sandy Way Croy. CD55 79
Sanford St. SE14 CD43 52
Sanford Ter. N16 CA34 24
Sanford Wk. N16 CA34 24
Smalley Rd.
Sanford Wk. SE14 CD43 52
Chubworthy St.
Sanger Av. Chess. BL56 75
Sangley Rd. SE6 CE47 61
Sangley Rd. SE25 CA52 69
Sangora Rd. SW11 BT45 50
Strathblaine Rd.
Sansom Rd. E11 CG34 25
Sanson St. SE5 BZ43 51
Sans Wk. EC1 BY38 33
Woodbridge St.
Santley St. SW4 BX45 51
Santos Rd. SW18 BS46 50
Saphora Clo. Orp. CM56 80
Sapphire Rd. SE14 CD42 43
Saracen Clo. Croy. BZ53 69
Crescent, The
Saracen St. E14 CE39 43
Sarah St. N1 CA38 33
Saratoga Rd. E5 CC35 34
Sardinia St. WC2 BX39 42
Kingsway
Sarita Clo. Har. BG30 11
Sarjant Path SW19 BR48 58
Queensmere Rd.
Sark Clo. Houns. BF43 47
Sark Clo. Houns. BF44 47
Sutton Rd.
Sark Wk. E16 CH39 44
Sarnsfield Rd. Enf. BZ24 8
Sarre Rd. NW2 BR35 31
Sarre Rd. Orp. CP53 72
Sarsen Av. Houns. BE44 47
Sarsfield Rd. SW12 BU47 59
Sarsfield Rd. Grnf. BJ37 30
Sartor Rd. SE15 CC45 52
Satchwell Rd. E2 CA38 33
Bethnal Green Rd.
Sauls Grn. E11 CG34 25
Napier Rd.
Saunders Ness Rd. E14 CF42 43
Saunders Rd. SE18 CN42 45
Saunder's Way SE28 CO40 45
Saunderton Rd. Wem. BJ35 30
Saunton Rd. Horn. CU34 28
Savage Gdns. E6 CK39 44
Savage Gdns. EC3 CA40 42
Savernake Rd. N9 CB25 8
Savernake Rd. NW3 BU35 32
Savile Gdns. Croy. CA55 78
Savile Row W1 BW40 41
Saville Clo. N. Mal. BO53 67
Saville Gdns. Croy. CA55 78
Saville Rd. E16 CK40 44
Saville Rd. W4 BN41 40
Saville Rd. Rom. CQ32 27
Saville Rd. Twick. BH47 57
Saville Row Wdf. Grn. CH29 17
Savill Gdns. SW20 BP52 67
Savona Clo. SW19 BQ50 58
Savona Est. SW8 BW43 50
Savona St. SW8 BW43 50
Savoy Clo. Edg. BM28 12
Savoy Ct. WC2 BX40 42
Strand
Savoy Pl. WC2 BX40 42
Savoy Rd. Dart. CV46 55
Savoy St. WC2 BX40 42
Savoy Way WC2 BX40 42
Carting La.
Sawkins Cl. SW19 BR48 58
Quensmere Rd.
Sawley Rd. W12 BP40 40
Sawtry Clo. Cars. BT54 68
Sawyers Clo. Dag. CS36 37
Sawyers La. Borwd. BJ23 5
Sawyers Lawn W13 BJ39 39
Sawyer St. SE1 BZ41 42
Saxby St. SW2 BX47 60
Saxham Rd. Bark. CN37 36
Saxlingham Rd. E4 CF27 16
Saxon Av. Felt. BE48 56
Saxonbury Av. Sun. BC51 65
Saxonbury Clo. Mitch. BT52 68
Saxonbury Gdns. Surb. BK54 66
Saxon Clo. Rom. CW30 19
Saxon Dr. W3 BM39 39
Saxon Gdns. Sthl. BE40 38
Saxon Ho. Felt. BE48 56
Saxon Rd. E6 CK38 35
Saxon Rd. N22 BY30 15
Saxon Rd. SE25 BZ53 69
Saxon Rd. Brom. CG50 61
Saxon Rd. Dart. CW49 64
Saxon Rd. Ilf. CL36 35
Saxon Rd. Sthl. BE40 38
Saxon Rd. Walt. BD55 74
Saxon Rd. Wem. BM34 21
Saxon Rd. KT12 BD55 74
Linley Dr.
Saxon Wk. Sid. CP50 63
Cray Rd.
Saxon Way N14 BW25 7
Saxton Clo. SE13 CF45 52
Saxville Rd. Orp. CO52 72
Sayer's Wk. TW10 BL47 57
Stafford Pla.
Sayes Ct. Est. SE8 CD42 43
Sayes Ct. Gdns. SE8 CD42 43
Sayes Ct. Rd. Orp. CO52 72
Sayes Ct. St. SE8 CD43 52
Scad's Hill Clo. Orp. CN53 72
Scales Rd. N17 CA31 24
Scampston Ms. W10 BQ39 40
Scandrett St. E1 CB40 42
Wapping High St.
Scarba Wk. N1 BZ36 33
Scarba Wk. N1 BZ36 33
Marquess Est.
Scarborough Clo. Sutt. BR59 82
Scarborough Rd. E11 CF33 25
Scarborough Rd. N4 BY33 24
Scarborough Rd. N9 CC26 9
Scarbrook Rd. Croy. BZ55 78
Scarle Rd. Wem. BK36 30
Scarlet Rd. SE6 CG48 61
Scarsbrook Rd. SE3 CJ45 53
Scarbrook St. Croy. BZ55 78
Scarbrook Rd.
Scarsdale Gro. SE5 CA43 51
Neate St.
Scarsdale Rd. Har. BG34 20
Scarsdale Vill. W8 BS41 41
Scarth Rd. SW13 BO45 49
Scawen Rd. SE8 CD42 43
Scawfell St. E2 CA37 33
Scaynes Link N12 BS28 14
Sceaux Est. SE5 CA44 51
Sceptre Rd. E2 CC38 34
Sceyness Link N12 BS28 14
Chanctonbury Way
Schofield Rd. SE3 CH43 53
Scholars Rd. E4 CF26 9
Scholars Rd. SW12 BW47 59
Scholefield Rd. N19 BW33 23
Scholfield Rd. N19 BW34 23
School Alley, Twick. BJ47 57
School Approach N1 CA38 33
Kingsland Rd.
Schoolhouse La. E1 CC40 43
School Ho. La. Tedd. BJ50 57
School La. Bush. BF26 4
School la. Kings. On T. BK51 66
Park Rd.
School La. Pnr. BE31 20
School La. Surb. BM54 66
School La. Swan. CU51 73
School La. Well. CO45 54
School Lane SE23 CB48 60
School Pass. Kings. On T. BL51 66
School Pass. Sthl. BE40 38
School Pl. E1 CB38 33
Buckhurst St.
School Rd. E12 CK35 35
School Rd. NW10 BN38 31
School Rd. W4 BN42 40
Belmont Rd.
School Rd. Chis. CM51 71
School Rd. Dag. CR37 36
School Rd. E. Mol. BG52 65
School Rd. Hamptn. BG50 56
School Rd. Houns. BG45 47
School Rd. Kings. On T. BK51 66
Park Rd.
School Rd. Av. Hamptn. BG50 56
School Wk. Sun. BC52 65
School Way N12 BT28 14
School Way N12 BT29 14
School Way, Dag. CP34 27
Schubert Clo. Borwd. BK25 5
Schubert Rd. SW15 BR46 49
Sclater St. E1 CA38 33
Scobie Pl. N16 CA35 33
Amhurst Rd.
Scoresby St. SE1 BY40 42
Scorton Av. Grnf. BJ37 30
Scoter Clo. IG8 CH29 17
Scotch Common W13 BJ39 39
Scot Gro. Pnr. BE29 11
Scotland Gn. N17 CA30 15
Scotland Grn. Rd. Enf. CC25 9
Scotland Grn. Rd. N. Enf. CC24 9
Scotland Rd. Buck. H. CJ26 10
Scotney Wk. Horn. CV36 37
Scotsdale Clo. Orp. CN52 72
Scotsdale Clo. Sutt. BR57 76
Scotsdale Rd. SE12 CH46 53
Scotswold Wk. N17 CB29 15
Waverley Rd.
Scotswood Wk. N17 CB29 15
Waverley Rd.
Scotswood Wk. N17 CB29 15
Northumberland Pk.
Scott Clo. Epsom. BN56 76
Scott Cres. Erith CT44 55
Scott Cres. Har. BF33 20
Scott Ellis Gdns. NW8 BT38 32
Scottes La. Dag. CP33 27
Valence Av.
Scott Lidgett Cres. SE16 CB41 42
Scotts Av. Brom. CF51 70
Scotts Dr. Hamptn. BF50 56
Scotts Fm. Rd. Epsom BN57 76
Scott's La. Brom. CF52 70
Scott's Rd. E10 CF33 25
Scotts Rd. W12 BP41 40
Scott's Rd. Brom. CH50 62
Scott's Rd. Sthl. BD41 38
Scott St. E1 CB38 33
Scotts Wol. Clo. Bush. BE23 4
Scotts Wol. Rd. Bush. BE23 4
Scoulding Rd. E16 CG39 43
Roger's Rd.
Scouler St. E14 CF40 43
Harrap St.
Scout App. NW10 BN35 31
Scout La. SW4 BW45 50
Scout Way NW7 BN28 13
Scovell Rd. SE1 BZ41 42
Scrafton Rd. Ilf. CL34 26
Scrattons Ter. Bark. CP37 36
Scriven St. E8 CA37 33
Scrooby St. SE6 CE46 52
Scrubs La. NW10 BP38 31
Scrubs La. W10 BP38 31
Scrutton Clo. SW12 BW47 59
Scrutton St. EC2 CA38 33
Scudamore La. NW9 BN31 22
Scutari Rd. SE22 CB46 51
Scylla Rd. SE15 CB45 51
Seabrook Dr. W. Wick. CF55 79
Seabrook Gdns. Rom. CR33 27
Seabrook Rd. Dag. CP34 27
Seaburn Clo. Rain. CT38 37
Seacoal La. EC4 BY39 42
Seacourt Rd. SE2 CP41 45
Seacroft Gdns. Wat. BD27 11
Seafield Rd. N11 BW28 14
Seaford Rd. E17 CE31 25
Seaford Rd. N15 BZ31 24
Seaford Rd. W13 BJ40 39
Seaford Rd. Enf. CA24 8
Seaford St. WC1 BX38 33
Seaforth Av. N. Mal. BP53 67
Seaforth Clo. Rom. CT29 19
Seaforth Cres. N1 BZ35 33
Seaforth Gdns. N21 BX26 8
Seaforth Gdns. Epsom BO56 76
Seaforth Gdns. Wdf. Grn. CJ28 17
Seaforth Pl. SW1 BW41 41
Buckingham Gate
Seager Pl. E3 CD39 43
Burdett Rd.
Seagrave Rd. SW6 BS43 50
Seagry Rd. E11 CH32 26
Sealand Wk. Nthlt. BD38 29
Wayfarer Rd.
Seal St. E8 CA35 33
Searchwood Rd. Warl. CB62 84
Searle Clo. SW11 BU43 50
Searle Pl. N19 BX33 24
Searles Rd. SE1 BZ42 42
Sears St. SE5 BZ43 51
Seasprite Clo. Nthlt. BD38 29
Seaton Av. Ilf. CN35 36
Seaton Clo. E13 CH38 35
New Barn St.
Seaton Clo. Twick. BG46 47
Seaton Gdns. Ruis. BC34 20
Seaton Pl. NW1 BW38 32
Seaton Point N16 CB34 24
Downs Est.
Seaton Rd. Dart. CU47 64
Seaton Rd. Mitch. BU51 68
Seaton Rd. Twick. BG46 47
Seaton Rd. Well. CP43 54
Seaton Rd. Wem. BL37 30
Seaton St. N18 CB28 15
Sebastian St. EC1 BY38 33
Sebastopol Rd. N9 CB28 15
Sebbon St. N1 BY36 33
Sebert Rd. E7 CH35 35
Sebright Pass. E2 CB37 33

Sheldwick Ter. Brom. CK53 71
Shelford Pl. N16 BZ34 24
Shelford Rise SE19 CA50 60
Shelford Rd. Barn. BQ25 6
Shelgate Rd. SW11 BU46 50
Shellgrove Est. N16 CA35 33
Shelley Av. E12 CK36 35
Shelley Av. Grnf. BG38 29
Shelley Av. Horn. CT34 28
Shelley Clo. Bans. BQ61 82
Shelley Clo. Edg. BM28 12
Shelley Clo. Grnf. BG38 29
Shelley Clo. Hayes BC39 38
Shelley Clo. Orp. CN55 81
Shelley Cres. Houns. BD44 47
Shelley Cres. Sthl. BE39 38
Shelley Dr. Well. CN44 54
Shelley Gdns. Wem. BK34 21
Shelley Gro. Loug. CK24 10
Shelley Rd. NW10 BN37 31
Shelley Rd. Har. BH31 21
Shellgrove Ms. N16 CA35 33
Shellgrove Rd.
Shellness Rd. N16 CB35 33
Pembury Est.
Shell Rd. SE13 CE45 52
Shellwood Rd. SW11 BU44 50
Shelton Rd. SW19 BS51 68
Shelton St. WC2 BX39 42
Shemerdine Clo. E3 CD39 43
Shenfield Rd. Wdf. Grn. CH29 17
Shenfield St. N1 CA37 33
Shenley Rd. SE5 CA44 51
Shenley Rd. Borwd. BM24 5
Shenley Rd. Houns. BE44 47
Shenstone Clo. Dart. CS45 55
Shenstone Gdns. Rom. CV30 19
Shepherdess Pl. N1 BZ38 33
Shepherdess Wk.
Shepherdess Wk. N1 BZ37 33
Shepherd Mkt. W1 BV40 41
Shepherd's Bush Grn. W12 BQ41 40
Shepherd's Bush Market W12 BQ41 40
Shepherds Bush Pl. W12 BQ41 40
Shepherd's Bush Rd. W6 BQ42 40
Shepherd's Clo. N6 BV32 23
Shepherd's Clo. Rom. CP32 27
Shepherds Grn. Chis. CM50 62
Shepherd's Hill N6 BV32 23
Shepherd's La. E9 CC35 34
Shepherds La. Dart. CU47 64
Shepherds Path NW3 BT35 32
Lyndhurst Ter.
Shepherds Path Nthlt. BE36 29
Ridgeway Wk.
Shepherds Pl. W1 BV40 41
Lees Pl.
Shepherd St. E3 CE38 34
Shepherd St. W1 BV40 41
Shepherd's Wk. NW3 BT35 32
Hampstead High St.
Shepherds Way S. Croy. CC57 79
Shepley Clo. Cars. BV55 77
Shepley Clo. Horn. CV35 37
Sheppard St. E16 CG38 34
Shepperton Rd. N1 BZ37 33
Shepperton Rd. Orp. CM53 71
Sheppey Clo. Erith CU43 55
Sheppey Gdns. Dag. CP36 36
Sheppey Rd.
Sheppey Rd. Dag. CO36 36
Sheppey Wk. N1 BZ36 33
Marquess Est.
Shepstone St. E6 CL40 44
Sherard Rd. SE9 CK46 53
Sheraton Dr. Epsom BN59 82
Sheraton St. W1 BW39 41
Wardour St.
Sherborne Av. Enf. CC23 9
Sherborne Av. Sthl. BF42 38
Sherborne Cres. Cars. BU54 68
Sherborne Gdns. W13 BJ39 39
Sherborne Gdns. Rom. CR28 18
Sherborne La. EC4 BZ40 42
King William St.
Sherborne Rd. Orp. CN53 72
Sherborne Rd. Sutt. BS55 77
Sherborne St. N1 BZ37 33
Sherbourne Clo. Epsom BP62 82
Sherborne Gdns. NW9 BM31 21
Sherbourne Rd. Chess. BL56 75
Sherbrooke Clo. Bexh. CQ45 54
Sherbrooke Clo. Bexh. CR45 54
Sherbrooke Gdns. E6 CK39 44
Sherbrooke Rd. SW6 BR43 49
Sherbrook Gdns. N21 BY26 8
Shere Av. Sutt. BQ58 76
Shere Clo. Chess. BK56 75
Shereden Rd. E4 CF28 16
Shere Rd. Ilf. CL32 26
Sherfield Gdns. SW15 BO46 49
Sheridan Clo. Rom. CV29 19
Sheridan Cres. Chis. CL51 71
Penn Gdns.
Sheridan Gdns. Har. BK32 21
Sheridan Rd. E7 CG34 25
Sheridan Rd. E12 CK35 35
Sheridan Rd. Bel. CR42 45
Sheridan Rd. Bexh. CQ45 54
Sheridan Rd. Rich. BK48 57
Sheridan Ter. Nthlt. BF35 29
Sheridan Walk Cars. BU56 77
Park Hill
Sheridan Wk. NW11 BS32 23
Sheriden Rd. SW19 BR51 67
Sheringham Av. E12 CK35 35
Sheringham Av. N14 BW25 7
Sheringham Av. Felt. BC48 56
Sheringham Av. Rom. CS32 28
Sheringham Av. Twick. BE47 56
Sheringham Dr. Bark. CN35 36
Sheringham Rd. N7 BX36 33
Sheringham Rd. SE10 CC51 70
Sherington Av. Pnr. BF29 11
Sherington Rd. SE7 CH43 53
Sherland Rd. Twick. BH47 57
Sherlies Av. Orp. CN55 81
Shermanbdry Pl. Erith CT43 55
Shermanbury Pl. Erith CT43 55
Sherman Rd. Brom. CH51 71
Shernells Way SE2 CO42 45
Shernhall St. E17 CF31 25
Sherrard Rd. E7 CJ36 35
Sherrard Rd. E12 CJ36 35
Sherrards Way, Barn. BS25 7
Sherrick Gn. Rd. NW10 BP35 31
Sherriff Rd. NW6 BS36 32
Sherringham Av. N17 CB30 15
Sherrock Gdns. NW4 BP31 22
Sherwin Rd. SE14 CC44 52
Sherwood Av. E18 CH31 26
Sherwood Av. SW16 BW50 59
Sherwood Av. Grnf. BH36 30
Sherwood Av. Hayes BC38 29
Sherwood Clo. SW15 BP45 49
Sherwood Clo. W13 BJ40 39
Sherwood Gdns. Bark. CM36 35
Sherwood Park Av. Sid. CO47 63
Sherwood Pk. Rd. Mitch. BW52 68
Sherwood Pk. Rd. Sutt. BS56 77
Sherwood Rd. NW4 BQ31 22
Sherwood Rd. SW19 BR50 58
Sherwood Rd. Couls. BW61 83
Sherwood Rd. Croy. CB54 69
Sherwood Rd. Har. BG34 20
Sherwood Rd. Hamptn. BG49 56
Sherwood Rd. Ilf. CM31 26
Sherwood Rd. Well. CO44 54
Sherwoods Rd. Wat. BE26 4
Sherwood St. N20 BT27 14
Sherwood St. W1 BW40 41
Brewer St.
Sherwood Ter. N20 BT27 14
Sherwood Way W. Wick. CE55 79
Shetland Rd. E3 CD37 34
Shield Dr. Brent. BJ43 48
Shieldhall St. SE2 CP42 45
Shiliber Wk. Chig. CN27 18
Shilling St. N1 BY36 33
Cross St.
Shillington St. SW11 BU44 50
Shillitoe Rd. N13 BY28 15
Shinfield St. W12 BQ39 40
Shinglewell Rd. Erith CR43 54
Ship All. E1 CB40 42
Wellclose Sq.
Shipford Path SE23 CC48 61
Ship & Half Moon Pass. SE18 CL41 44
Shipka Rd. SW12 BV47 59
Ship La. SW14 BN45 49
Ship La. S. At H. CV51 73
Shipman Rd. E16 CH39 44
Shipman Rd. SE23 CC48 61
Ship & Mermaid Row SE1 BZ41 42
Weston St.
Ship St. SE8 CE44 52
Ship Tavern Pass. EC3 CA39 42
Lime St.
Shipton Clo. Dag. CP34 27
Shipton St. E2 CA38 33
Shipway Ter. N16 CA34 24
Victorian Rd.
Shirburn Clo. SE23 CC47 61
Shirbutt St. E14 CE40 43
Shirebrook Rd. SE3 CJ45 53
Shirehall Clo. NW4 BQ32 22
Shirehall Gdns. NW4 BQ32 22
Shirehall La. NW4 BQ32 22
Shirehall Pk. NW4 BQ32 22
Shirehall Rd. Dart. CV49 64
Shire La. Orp. CK57 80
Shire La. Orp. CN56 81
Shire Meade Borwd. BL25 5
Shires, The, Rich. BL49 57
Shirland Ms. W9 BR38 31
Shirland Rd. W9 BR38 31
Shirley Av. Bex. CP47 63
Shirley Av. Croy. CC54 70
Shirley Av. Sutt. BR58 76
Shirley Av. Sutt. BT56 77
Shirley Ch. Rd. Croy. CC55 79
Shirley Clo. Dart. CV45 55
Shirley Clo. Houns. BG46 47
Shirley Clo. E17 CE32 25
Addison Rd.
Shirley Cres. Beck. CC52 70
Shirley Dr. Houns. BG46 47
Shirley Gdns. W7 BH40 39
Shirley Gdns. Bark. CN36 36
Shirley Gdns. Horn. CV34 28
Shirley Gro. N9 CC26 9
Shirley Hills Rd. Croy. CC56 79
Shirley Ho. Dr. SE7 CJ43 53
Shirley Oak Rd. Croy. CC54 70
Shirley Pk. Rd. Croy. CB54 69
Shirley Rd. E15 CG36 34
Shirley Rd. W4 BN41 40
Shirley Rd. Croy. CB54 69
Shirley Rd. Enf. BZ24 8
Shirley Rd. Sid. CN48 63
Shirley Rd. Wall. BW58 77
Shirley St. E16 CG39 43
Shirley St. N1 BX37 33
Shirley Way, Croy. CD55 79
Shirlock Rd. NW3 BU35 32
Mansfield Rd.
Shobden Rd. N17 BZ30 15
Shoebury Rd. E6 CK36 35
Shoe La. EC4 BY39 42
Sholdon Gdns. Orp. CP53 72
Shooters Av. Har. BK31 21
Shooters Hill, SE18 CK44 53
Shooters Hill, Well. CK44 53
Shooter's Hill Rd. SE3 CF44 52
Shooter's Hill Rd. SE18 CF44 52
Shootery Clo. SE9 CK48 62
Shoot Up Hl. NW2 BR35 31
Shord Hill Ken. BZ57 78
Shord Hill Ken. BZ61 84
Shoreditch High St. E1 CA38 33
Shore Est. E9 CC36 34
Shore Gro. Felt. BF48 56
Shoreham Clo. Croy. CC53 70
Shoreham Clo. SW18 BS46 50
York Rd.
Shoreham Clo. Bexh. CP47 63
Stansted Cres.
Shoreham La. Orp. CR57 81
Shoreham Rd. Orp. CO51 72
Shoreham St. SW18 BS46 50
Barchard St.
Shoreham Way, Brom. CH53 71
Shore Pl. E9 CC36 34
Shore Rd. E9 CC36 34
Shore St. SW19 BT51 68
Shorncliffe Rd. SE1 CA42 42
Shorndean St. SE6 CF47 61
Shorne Clo. Sid. CO46 54
Park Mead
Shornefield Clo. Brom. CL52 71
Shorrolds Rd. SW6 BR43 49
Shortcroft Rd. Epsom BO57 76
Shortcrofts Rd. Dag. CQ36 36
Short Gate N12 BR28 13
Shortland Rd. E10 CE33 25
Shortlands W6 BQ42 40
Shortlands Clo. N18 BZ27 15
Shortlands Gdns. Brom CG51 70
Shortlands Gro. Brom. CF51 70
Shortlands Ms. W6 BQ42 40
Shortlands Rd. Brom. CF52 70
Shortlands Rd. Kings. On T. BL50 57
Short Rd. E11 CG34 25
Short Rd. E15 CF37 34
Short Rd. W4 BO43 49
Shorts Croft NW9 BM31 21
Shorts Gdns. WC2 BX39 42
Shorts Rd. Cars. BU56 77
Short St. NW4 BQ31 22
Short St. SE1 BY41 42
Short Way N12 BU29 14
Short Way SE9 CK45 53
Short Way Twick. BG47 56
Shotfield Wall. BV57 77
Shottendane Rd. SW6 BS44 50
Shottfield Av. SW14 BO45 49
Shoulder Of Mutton All. E14 CD40 43
Narrow St.
Shouldham St. W1 BU39 41
Showers Way, Hayes BC40 38
Shrapnel Rd. SE9 CK45 53
Shrewsbury Av. SW14 BN45 49
Shrewsbury Av. Har. BL31 21
Shrewsbury Clo. Surb. BK55 75
Shrewsbury Cres. NW10 BN37 31
Shrewsbury Ho. SW3 BU43 50
Shrewsbury La. SE18 CL44 53
Shrewsbury Ms. W2 BS39 41
Chepstow Rd.
Shrewsbury Rd. E7 CJ35 35
Shrewsbury Rd. N11 BW29 14
Shrewsbury Rd. NW10 BO37 31
Shrewsbury Rd. W2 BS39 41
Shrewsbury Rd. Beck. CD52 70
Shrewsbury Rd. Cars. BU53 68
Shrewsbury Wk. Islw. BJ45 48
South St.
Shrewton Rd. SW17 BU50 59
Shroffold Rd. Brom. CG49 61
Shropshire Clo. Mitch. BX52 69
Shropshire Rd. N22 BX29 15
Shroton St. NW1 BU39 41
Shrubberies, The, Chig. CM28 17
Shrubberies, The, E18 CH30 17
Shrubbery Gdns. N21 BY26 8
Shrubbery Rd. N9 CB27 15
Shrubbery Rd. SW16 BX49 60
Shrubbery Rd. Sthl. BF40 38
Shrubland Est. E8 CA36 33
Shrubland Gro. Wor. Pk. BQ55 76
Shrubland Rd. E8 CA37 33
Shrubland Rd. E10 CE33 25
Shrubland Rd. E17 CE32 25
Shrubland Rd. Bans. BR61 82
Shrubland Av. Croy. CE55 79
Shrubland Clo. N20 BT26 7
Shurbland Clo. Chig. CM29 17
Shuna Wk. N1 BZ36 33
Marquess Est.
Shurland Av. Barn. BT25 7
Shurland Gdns. SE15 CA43 51
Rosemary Rd.
Shurlock Av. Swan. CS51 73
Shurlock Dr. Orp. CM56 80
Broadwater Gdns.
Shuttle Clo. Sid. CN47 63
Shuttle Mead, Bex. CQ47 63
Shuttle Rd. Dart. CU45 55
Shuttle St. E1 CA38 33
Buxton St.
Shuttleworth Rd. SW11 BT44 50
Sibella Rd. SW4 BW44 50
Sibley Gro. E12 CK36 35
Sibthorpe Rd. SE12 CH47 62
Sibthorp Rd. Mitch. BU51 68
Sibton Rd. Cars. BU54 68
Sicilian Av. WC1 BX39 42
Bloomsbury Way
Sickert Ct. N1 BZ36 33
Sickle Cnr. Dag. CR38 36
Sidbury Av. SW6 BR45 49
Sidbury St. SW6 BR44 49
Sidcup By-pass, Sid. CM48 62
Sidcup High St. Sid. CO49 63
Sidcup Hill, Sid. CO49 63
Sidcup Hill Gdns. Sid. CP49 63
Sidcup Rd. SE9 CJ47 62
Sidcup Rd. SE12 CJ47 62
Siddons La. NW1 BU38 32
Siddons Rd. N17 CB30 15
Siddons Rd. SE23 CC48 61
Siddons Pl. Croy. BY55 78
Side Rd. E17 CD32 25
South Gro.
Sidewood Rd. SE9 CM47 62
Sidford Pl. SE1 BX41 42
Sidings, The E11 CF33 25
Sidmouth Av. Islw. BH44 48
Sidmouth Clo. Wat. BC27 11
Sidmouth Dr. Ruis. BC34 20
Sidmouth Rd. E10 CF34 25
Sidmouth Rd. NW2 BQ36 31
Sidmouth Rd. SE15 CA44 51
Sumner Est.
Sidmouth Rd. Orp. CO53 72
Sidmouth Rd. Well. CP43 54
Sidmouth St. WC1 BX38 33
Sidney Av. N13 BX28 15
Sidney Gdns. Brent. BK43 48
Boston Manor Rd.
Sidney Gro. N1 BY38 33
Wakley St.
Sidney Est. E1 CC39 43
Sidney Rd. E7 CH34 26
Sidney Rd. N22 BX29 15
Sidney Rd. SE25 CB53 69
Sidney Rd. SW9 BX44 51
Sidney Rd. Beck. CD51 70
Sidney Rd. Har. BG31 20
Sidney Rd. Sutt. BS56 77
Sidney Rd. Twick. BJ46 48
Sidney Rd. Walt. BC54 65
Sidney Sq. E1 CC39 43
Sidney St. E1 CB39 42
Sidworth St. E8 CB36 33
Siemens Rd. SE18 CJ41 44
Sigdon Rd. E8 CB35 33
Sigers, The, Pnr. BC32 20
Sigismund St. SE10 CG41 43
Silas St. Est. NW5 BV36 32
Silbury St. N1 BZ38 33
East Rd.
Silcester Ct. Th. Hth. BY52 69
Silchester Ms. W10 BQ40 40
Walmer Rd.
Silchester Rd. W10 BQ39 40
Silcote Rd. SE5 CA42 42
Albany Rd.
Silecroft Rd. Bexh. CR44 54
Silesia Bldgs. E8 CB36 33
London La.
Silex St. SE1 BY41 42
Silkfield Rd. NW9 BO32 22
Silkins, The, Rom. CT30 19
Silk Clo. SE12 CH46 53
Silk Mill Rd. Wat. BC26 4

Silk Mills Path SE13 CF44 52
Silkstream Rd. Edg. BN30 13
Silk St. EC2 BZ39 42
Silsoe Rd. N22 BX30 15
Silver Birch Av. E4 CD29 16
Silver Birch Clo. Dart. CT49 64
Silverbirch Wk. NW3 BU36 32
  Maitland Park Vw.
Silvercliffe Gdns. Barn. BU24 7
Silver Clo. Har. BG29 11
Silver Clo. Sutt. BR56 76
Silver Cres. W4 BM42 39
Silverdale SE26 CC49 61
Silverdale, Enf. BX24 8
Silverdale Av. Ilf. CN32 27
Silverdale Clo. W7 BH40 39
  Cherington Rd.
Silverdale Clo. W13 BH40 39
Silverdale Clo. Har. BE35 29
Silverdale Dr. Horn. CU35 37
Silverdale Dr. Sun. BC51 65
Silverdale Gdns. Hayes BC41 38
Silverdale Rd. E4 CF29 16
Silverdale Rd·Bexh. CR44 54
Silverdale Rd. Bush. BE25 4
Silverdale Rd. Orp. CM52 71
Silverdale Rd. Orp. CO52 72
Silverhall St. Islw. BJ45 48
Silverholme, Har. BK33 21
Silver Jubilee Way, BC45 47
  Houns.
Silverland St. E16 CK40 44
Silver La. Pur. BW59 83
Silver La. W. Wick. CF55 79
Silverleigh Rd. Th. Hth. BX52 69
Silvermere Av. Rom. CR28 18
Silvermere Rd. SE6 CE47 61
Silver Pl. W1 BW40 41
Silver Rd. W12 BQ40 40
  Lexington St.
Silver Spring Clo. Erith CR43 54
Silverston Way, Stan. BK29 12
Silver St. EC2 BZ39 42
  Wood St.
Silver St. N18 BZ28 15
Silver St. Enf. BZ24 8
Silverthorne Gdns. E4 CE27 16
Silverthorne Rd. SW8 BV44 50
Silverton Rd. W6 BQ43 49
Silvertown By-pass E16 CJ40 44
Silvertown Way E16 CG39 43
Silvertree Clo. Wat. BC55 74
Silvertree La. Grnf. BG37 29
Silver Wk. SE16 CD40 43
Silver Way, Rom. CR31 27
Silverwood Clo. Beck. CE50 61
  Brackley Rd.
Silvester Rd. SE22 CA46 51
Silvester St. SE1 BZ41 42
Silwood Est. SE16 CC42 43
Silwood St. SE16 CC42 43
Simla Clo. SE14 CD43 52
  Chubworthy St.
Simone Clo. Brom. CJ51 71
Simmil Rd. Esher BH56 75
Simmons Clo. N20 BU27 14
Simmons La. E4 CF27 16
Simmons Rd. SE18 -CL42 44
  Brookhill Rd.
Simmons Wk. E15 CF35 34
  Waddington Rd.
Simmons Way N20 BU27 14
Simm's Clo. Cars. BU55 77
Simms Rd. SE1 CB42 42
Simnel Rd. SE12 CH47 62
Simon Clo. W11 BS40 41
  Portobello Rd.
Simonds Rd. E10 CE34 25
Simone Dr. Ken. BZ62 84
Simons Wk. E15 CF35 34
Simons Wk. E15 CF36 34
Simpson Rd. Houns. BE46 47
Simpson Rd. Rain. CT36 37
Simpson Rd. Rich. BK49 57
Simpsons Rd. E14 CE40 43
Simpsons Rd. Brom. CH52 71
Simpson St. SW11 BU44 50
Simrose Ct. SW18 BS46 50
Sims Clo. Rom. CT31 28
  Junction Rd.
Sims Wk. SE3 CG45 52
  Lee Rd.
Sinclaire Clo. Enf. CA23 8
Sinclair Gdns. W14 BQ41 40
Sinclair Gro. NW4 BQ32 22
Sinclair Rd. E4 CD28 16
Sinclair Rd. W14 BQ41 40
Sinclare Clo. Enf. CA23 8
  Carterhatch La.
Sindall Rd. Grnf. BJ37 30
Sinderby Clo. Borwd. BL23 5
Singapore Rd. W13 BJ40 39
Singer St. EC2 BZ38 33
  Cowper St.
Singleton Clo. Horn. CT35 37
  Cowdray Way
Singleton Clo. Mitch. BU50 59
Singleton Rd. Dag. CQ35 36

Singleton Scarp N12 BS28 14
Sinnott Rd. E17 CC30 16
Sion Rd. Twick. BJ47 57
Sir Alexander Clo. W3 BO40 40
  Sir Alexander Rd.
Sir Alexander Rd. W3 BO40 40
Sirdar Rd. N22 BY31 24
Sirdar Rd. W11 BO40 40
Sirdar Rd. Mitch. BU50 59
Sir Thom. More Est. SW3 BT43 50
Sirus Rd. Nthwd. BC28 11
Sisley Rd. Nthwd. BC28 11
Sisley Rd. Bark. CN37 36
Sispara Gdns. SW18 BR46 49
Sissinghurst Rd. Croy. CB54 69
Sisters Av. SW11 BU45 50
Sistova Rd. SW12 BV47 59
Sittingbourne Av. Enf. BZ25 8
Sitwell Gro. Stan. BH28 12
Siverst Clo. Nthlt. BF36 29
Siviter Way, Dag. CR36 36
Siward Rd. N17 BZ30 15
Siward Rd. SW17 BT48 59
Siward Rd. Brom. CH52 71
Sixth Av. E12 CK35 35
Sixth Av. W10 BR38 31
Sixth Av. Enf. CA25 8
Sixth Cross Rd. Twick. BG48 56
Skardu Rd. NW2 BR35 31
Skeena Hl. SW18 BR47 58
Skeet Hill La. Orp. CO54 72
Skeffington Rd. E6 CK37 35
Skelbrook St. SW18 BS48 59
Skelgill Rd. SW15 BR45 49
Skelton Rd. E7 CH36 35
Skeltons La. E10 CE33 25
Skelwith Rd. W6 BQ43 49
Sketty Rd. Enf. CA24 8
Skibbs La. Orp. CQ56 81
Skiers St. E15 CF37 34
Skiffington Clo. SW2 BY47 60
Skomer Wk. N1 BZ36 33
Sky Peals Rd. Wdf. Grn. CF30 16
Sladebrook Rd. SE3 CJ45 53
Sladedale Rd. SE18 CN42 45
Slades Clo. Enf. BX44 51
Slade Gdns. Erith CT44 55
Slade Grn. Rd. Erith CT44 55
Slade Grn. Rd. Erith CU43 55
Slades Clo. Enf. BY24 8
Slade's Cotts, Chis. CL49 62
Slades Dr. Chis. CM48 62
Slades Gdns. Enf. BY23 8
Slades Hill, Enf. BY24 8
Slades Rise, Enf. BY24 8
Slade, The SE18 CN43 54
Slade Wk. SE5 BZ43 51
Slagrove Pl. SE13 CE46 52
Slaidburn St. SW10 BT43 50
Slaithwaite Rd. SE13 CF45 52
Sleaford Grn. Wat. BD27 11
Sleaford St. SW8 BW43 50
Sleath Walk SW19 BT51 68
  Brangwyn Cres.
Slewins La. Horn. CV32 28
Slingsby Pl. WC2 BX40 42
  Long Acre
Slippers Pl. SE16 CB41 42
Sloane Av. SW3 BU42 41
Sloane Ct. E. SW3 BV42 41
Sloane Ct. W. SW3 BV42 41
Sloane Gdns. SW1 BV42 41
Sloane Gdns. Orp. CM55 80
Sloane Sq. SW1 BU42 41
Sloane St. SW1 BU41 41
Sloane Ter. SW1 BV42 41
Sloane Wk. Croy. CD53 70
Slough La. NW9 BN32 22
Sly St. E1 CB39 42
  Cannon St. Rd.
Smallbrook Ms. W2 BT39 41
  Craven Rd.
Smallbury Av. Islw. BH44 48
Smalley Clo. N16 CA34 24
  Smalley Rd.
Smalley Rd. N16 CA34 24
Smallwood Rd. SW17 BT49 59
Smardale Rd. SW18 BT46 50
  Alma Rd.
Smarden Clo. Belv. CR42 45
  Essenden Rd.
Smarden Gro. SE9 CK49 62
  Prestbury Sq.
Smart Clo. Rom. CU30 19

Smart's La. Loug. CJ24 10
Smart's Pl. N18 CB28 15
  Fore St.
Smarts Pl. WC2 BX39 42
  Stukeley St.
Smart St. E2 CC38 34
Smeaton Rd. SW18 BS47 59
Smeaton Rd. Wdf. Grn. CK28 17
Smedley St. SW4 BW44 50
Smedley St. SW8 BW44 50
Smeed Rd. E3 CE36 34
Smithambottom La. Pur. BW59 83
Smithambowns Rd. Pur. BW60 83
Smithfield St. EC1 BY39 42
Smithies Ct. E15 CF35 34
Smithies Rd. SE2 CO42 45
Smithson Rd. N17 BZ30 15
Smith Sq. SW1 BX41 42
Smith St. SW3 BU42 41
Smith St. SW3 BU42 41
Smith St. Wat. BD24 4
Smith Ter. SW3 BU42 41
Smithwood Clo. SW19 BR47 58
Smithy St. E1 CC39 43
Smooth Field Est. Houns. BF45 47
Smyrks Rd. SE17 CA42 42
Smyrna Rd. NW6 BS36 32
Smythe St. E14 CE40 43
Snag La. Sev. CN68 81
Snakes La. Wdf. Grn. CH28 17
Snaresbrook Dr. Stan. BK28 12
Snaresbrook Rd. E11 CG31 25
Snarsgate St. W10 BO39 40
Snead St. SE14 CD43 52
Sneath Av. NW11 BR33 22
Snellings Rd. Walt. BD56 74
Snell's Pk. N18 CA29 15
Sneyd Av. NW2 BQ35 31
Snodland Clo. Orp. CL58 80
Snowbury Rd. SW6 BS44 50
Snowdon St. EC2 CA38 33
Snowdrop Path. Rom. CV29 19
Snows Hill EC1 BY39 42
Snows Fields SE1 BZ41 42
Snowshill Rd. E12 CK35 35
Soames St. SE15 CA45 51
Soames Wk. N. Mal. BO51 67
Socket La. Brom. CH53 71
Soho Sq. W1 BW39 41
Soho St. W1 BW39 41
  Soho Sq.
Solander Gdns. Est. E1 CC40 43
Solebay St. E1 CD38 34
Solebay St. E3 CD38 34
Solent Rd. NW6 BS35 32
Soley Ms. N1 BY38 33
  Percy St.
Solna Av. SW15 BQ46 49
Solna Rd. N21 BZ26 8
Solomon's Pass. SE15 CB45 51
Solomons Ter. N20 BT26 7
Solom's Ct. Rd. Bans. BT62 83
Solon Rd. SW2 BX45 51
Solway Clo. Houns. BE45 47
Solway Rd. N22 BY30 15
Solway Rd. SE22 CA45 51
Somaford Gro. Barn. BT25 7
Somali Rd. NW2 BR35 31
Somerby Rd. Bark. CM36 35
Somerden Rd. Orp. CP54 72
Somerfield Rd. N4 BY34 24
Somerford Clo. Pnr. BC31 20
Somerford Est. N16 CA35 33
Somerford Gro. N16 CA35 33
Somerford Gro. N17 CB29 15
Somerford St. E1 CB38 33
  Brady St.
Somerhill Av. Sid. CO47 63
Somerhill Rd. Well. CO44 54
Somerleyton Pass. SW9 BY45 51
  Mayall Rd.
Somerleyton Rd. SW9 BY45 51
Somersby Gdns. Ilf. CK32 26
Somerset Av. SW20 BP51 67
Somerset Av. Chess. BK56 75
Somerset Av. Well. CN46 54
Somerset Clo. N. Mal. BO53 67
Somerset Clo. Wdf. Grn. CH30 17
  Harold Rd.
Somerset Clo. Epsom BN58 76
  Hollymoor La.
Somerset Clo. Walt. BC57 74
  Queens Rd.
Somerset Est. SW11 BT44 50
Somerset Gdns. N6 BV33 23
Somerset Gdns. SE13 CE44 52
Somerset Gdns. SW16 BX52 69
Somerset Gdns. Tedd. BH49 57
Somerset Rd. E17 CE32 25
Somerset Rd. N18 CA28 15
Somerset Rd. N17 CA31 24
Somerset Rd. NW4 BQ32 22
Somerset Rd. SW19 BQ48 58
Somerset Rd. W4 BN41 40
Somerset Rd. W13 BJ40 39
Somerset Rd. Barn. BS25 7

Somerset Rd. Brent. BK43 48
Somerset Rd. Dart. CU46 55
Somerset Rd. Har. BG32 20
Somerset Rd. Sthl. BE51 66
Somerset Rd. Orp. CO54 72
Somerset Rd. Sthl. BE39 38
Somerset Rd. Tedd. BH49 57
Somerset Sq. W14 BR41 40
Somerset Way, Houns. BE43 47
Somersham Rd. Bexh. CQ44 54
Somers Ms. W2 BU39 41
Somers Pl. SW2 BX47 60
Somers Rd. E17 CD31 25
Somers Way, Bush. BG26 4
Somerton Av. Rich. BM45 48
Somerton Clo. Pur. BY61 84
Somerton Rd. NW2 BQ34 22
Somerton Rd. SE15 CB45 51
Somertrees Av. SE12 CH48 62
Somervell Ct. Har. BF35 29
Somervell Rd. Har. BH35 30
Somerville Av. SW13 CC44 52
Somerville Rd. SE20 CC50 61
Somerville Rd. Rom. CP32 27
Sounds Lodge, Swan. CS53 73
Sondes St. SE17 BZ43 51
Sophia Clo. N7 BX36 33
  Mackenzie Rd.
Sonia Ct. Har. BH32 21
Sonia Gdns. N12 BT28 14
Sonia Gdns. NW10 BO35 31
Sonia Gdns. Houns. BF43 47
Sonning Rd. SE25 CB53 69
Sophia Rd. E10 CE33 25
Sophia Rd. E16 CH39 44
Sopwith Rd. Houns. BD43 47
Sorrel Bank, Croy. CD58 79
Sorrel Clo. SE28 CQ40 45
Sorrel Wk. Rom. CT31 28
Sorrento St. SUtt. BS55 77
Sotheby Rd. N4 BY34 24
Sotheron Rd. SW6 BS43 50
Sotheron Rd. Wat. BD24 4
Soudan Rd. SW11 BU44 50
Souldern Rd. W14 BQ41 40
Souldern St. Wat. BC25 4
South Access Rd. E17 CD33 25
South Acre NW9 BO30 13
South Africa Rd. W12 BP40 40
Southall La. Houns. BC43 47
Southampton Bldgs. WC2 BY39 42
Southampton Gdns. BX53 69
  Mitch.
Southampton Pl. WC1 BX39 42
Southampton Rd. NW5 BU35 32
Southampton Row WC1 BX39 42
Southampton St. WC2 BX40 42
Southampton Way SE5 BZ43 51
Southacre Way, Pnr. BD30 11
South Audley St. W1 BV40 41
South Av. E4 CE26 9
South Av. Cars. BU57 77
South Av. Rich. BM44 48
  Sandycombe Rd.
South Av. Sthl. BE40 38
South Av. Gdns. Sthl. BE40 38
South Bank, Chis. CM49 62
Southbank, Surb. BJ54 66
South Bank, Surb. BL53 66
South Bank Lo. Surb. BL53 66
South Bank Ter. Surb. BL53 66
South Black Lion La. W6 BP42 49
South Boltons Gdns. BS42 41
  SW10
South Borders, The, Pur. BW59 83
Southborough Clo. Surb. BK54 66
Southborough La. Brom. CK53 71
Southborough Rd. Brom. CK53 71
Southborough Rd. Surb. BL54 66
Southbourne, Brom. CH54 71
Southbourne Av. NW9 BO30 13
Southbourne Clo. Pnr. BE33 29
Southbourne Cres. NW4 BR31 22
Southbourne Gdns. SE12 CH46 53
Southbourne Gdns. Ilf. CM35 35
Southbourne Gdns. Ruis. BC33 20
Southbridge Pl. Croy. BZ56 78
Southbridge Rd. Croy. BZ55 78
Southbridge Way, Sthl. BE41 38
Southbrook Rd. SE12 CG46 52
Southbrook Rd. SW16 BX51 69
Southbury Av. Enf. CB25 8
Southbury Clo. Horn. CU35 37
Southbury Rd. Enf. CA24 8
S. C. G. Smallholdings BQ60 82
  Rd. Epsom
South Church La. N13 BX28 15
  Palmerston Cres.
Southchurch Rd. E6 CK37 35
South Circular Rd. SE23 CA47 61
South Circular Rd. Rich. BM43 48
South Clo. N6 BV32 23
South Clo. Barn. BR24 6
South Clo. Bexh. CP45 54

South Clo. Dag. CR37 36
South Clo. Mord. BS53 68
South Clo. Pnr. BD30 11
South Clo. Pnr. BE33 20
South Clo. Twick. BF48 56
Southcombe St. W14 BR42 40
Southcote Av. Surb. BM54 66
Southcote Rd. E17 CC32 25
Southcote Rd. N19 BW35 32
Southcote Rd. SE25 CB53 69
Southcote Rd. S. Croy. CA58 78
South Cres. WC1 BW39 41
Southcroft Av. Well. CN45 54
Southcroft Av. W. Wick. CF55 79
Southcroft Rd. SW16 BV50 59
Southcroft Rd. SW17 BV50 59
Southcroft Rd. Orp. CN55 81
South Cross Rd. Ilf. CL32 26
Southdale, Chig. CM29 17
Southdean Gdns. SW19 BR48 58
South Dene NW7 BN27 13
Southdown Av. W7 BJ41 39
Southdown Clo. SW20 BQ50 58
Crescent Rd.
Southdown Cres. Har. BF33 20
South Down Cres. Ilf. CN32 27
Southdown Rd. SW20 BQ51 67
Southdown Rd. Cars. BV58 77
Southdown Rd. Horn. CU33 28
Southdown Rd. Walt. BE56 74
South Dr. SW11 BU44 50
South Dr. Bans. BU60 83
South Dr. Couls. BW61 83
South Dr. Orp. CN56 81
South Dr. Rom. CV31 28
South Dr. Sutt. BR58 76
South Ealing Rd. W5 BK41 39
South Eastern Av. N9 CA27 15
South Eaton Pl. SW1 BV42 41
South Eden Pk. Rd. Beck. CE53 70
South Edwardes Sq. W8 BR41 40
South End W8 BS41 41
St. Alban's Gro.
South End, Croy. BZ56 78
Southend Arterial Rd. Rom. CV30 19
South End Clo. NW3 BU35 32
South End Rd.
Southend Clo. SE9 CL46 53
Southend Cres. SE9 CL46 53
South End Grn. NW3 BU35 32
Southend La. SE6 CD49 61
Southend La. SE26 CD49 61
Southend Rd. E6 CK36 35
Southend Rd. E18 CH30 17
Southend Rd. E17 CF30 16
South End Rd. NW3 BU35 32
Southend Rd. Beck. CE50 61
South End Rd. Rain. CU37 37
Southend Rd. Wdf. Grn. CH30 17
South End Row W8 BS41 41
Southerby Mews N5 BY34 24
Southerby Rd.
Southerland Rd. N9 CB26 8
Southern Av. SE25 CA52 69
Southern Av. Felt. BC47 56
Southern Dr. Loug. CK25 10
Southern Gro. E3 CD38 34
Southernhay, Loug. CJ24 10
Southern Pl. Swan. CS52 73
Southern Rd. E13 CH37 35
Southern Rd. N2 BU31 23
Southern Row W10 BR38 31
Southern St. N1 BX37 33
Southern Way, Rom. CR32 27
Southerton Rd. W6 BQ41 40
South Esk Rd. E7 CJ36 35
Southey Rd. N15 CA32 24
Southey Rd. SW9 BY44 51
Southey Rd. SW19 BS50 59
Southey St. SE20 CC50 61
Southfield, Barn. BQ25 6
Southfield Pk. Har. BF31 20
Southfield Rd. N17 CA30 15
Avenue, The
Southfield Rd. W4 BO41 40
Southfield Rd. Chis. CN52 72
Southfield Rd. Enf. CB25 8
Southfields NW4 BP30 13
Southfields, E. Mol. BH53 66
Southfields Cotts. W7 BH41 39
Southfields Gdns. Twick. BH49 57
Southfields Rd. SW18 BS46 50
Southfleet Rd. Orp. CN55 81
South Gdns. SW19 BT50 59
Southgate Circle N14 BW26 7
Southgate Gro. N1 BZ36 33
Southgate Rd. N1 BZ37 33
South Gro. E17 CD32 25
South Gro. N6 BV33 23
South Gro. N15 BZ32 24
South Grove Ho. N6 BV33 23
South Hall Dr. Rain. CU39 46
South Hill, Chis. CK50 62
South Hill Av. Har. BG34 20
South Hill Gro. Har. BH35 30
South Hill Rd. Brom. CG52 70
South Hl. Pk. NW3 BU35 32
South Hl. Pk. Gdns. NW3 BU35 32
Southill La. Pnr. BC31 20
Southill Rd. Chis. CK50 62
South Island Pl. SW9 BX43 51
South Lambeth Pl. SW8 BX43 51
South Lambeth Rd.
South Lambeth Rd. SW8 BX43 51
Southland Rd. SE18 CN43 54
Southlands Av. Orp. CM56 80
Southlands Clo. Couls. BX62 84
Southlands Gro. Brom. CK52 71
Southlands Rd. Brom. CJ53 71
Southland Way, Houns. BG46 47
South La. Kings. On T. BK52 66
South La. N. Mal. BN52 67
South La. W. N. Mal. BO54 67
South La. N. Mal. BN52 67
South Lodge Av. Mitch. BX52 69
South Lodge Cres. Enf. BW24 7
South Lodge Dr. N14 BW24 7
Southly Clo. Sutt. BS55 77
South Mall N9 CB27 15
Southmead NW9 BO30 13
Southmead, Epsom BO57 76
South Meadows, Wem. BL35 30
Park Lawns
Southmead Rd. SW19 BR47 58
South Molton La. W1 BV39 41
South Molton St. W1 BV39 41
Southmont Rd. Esher BH55 75
Southmoor Wk. E9 CD36 34
Trowbridge Est.
South Norwood Hill SE19 CA50 60
South Norwood Hill SE25 CA51 69
Southold Rise SE9 CK48 62
Southolm St. SW11 BV44 50
Southover N12 BS27 14
Southover, Brom. CH49 62
South Pde. SW3 BT42 41
South Parade W4 BN42 40
South Pk. Clo. N. Mal. BN52 67
South Pk. Ct. Beck. CE50 61
South Pk. Cres. SE6 CG47 61
South Pk. Cres. Ilf. CM34 26
South Pk. Dr. Ilf. CN34 27
South Pk Est. SE16 CB41 42
South Pk. Hill Rd. S. Croy. BZ56 78
South Pk. Rd. SW19 BR50 58
South Pk. Rd. Ilf. CM34 26
South Pk. Ter. Ilf. CM34 26
South Pk. Way, Ruis. BD36 29
South Pl. SW19 BQ50 58
Thornton Rd.
South Pl. Enf. CC25 9
South St.
South Pl. Surb. BL54 66
South Pl. Ms. EC2 BZ39 42
South Pl.
Southport Rd. SE18 CM42 44
South Rise, Cars. BU58 77
South Rd. N9 CB26 8
South Rd. SE23 CC48 61
South Rd. SW19 BT50 59
South Rd. W5 BK42 39
South Rd. Edg. BM30 12
South Rd. Erith CT43 55
South Rd. Felt. BD49 56
South Rd. Hamptn. BE50 56
South Rd. Rom. CQ32 27
South Rd. Sthl. BE40 38
South Rd. Twick. BG48 56
South Row SE3 CG44 52
Southsea Av. Wat. BC24 4
Southsea Rd. Kings. On T. BL52 66
South Side W6 BO41 40
South Side Com. SW19 BQ50 58
Southspring, Sid. CM47 62
South Sq. NW11 BS32 23
South St. W1 BV40 41
South St. Brom. CH51 71
South St. Enf. CC25 9
South St. Epsom BN60 82
South St. Islw. BJ45 48
South St. Rain. CS37 37
South St. Rom. CT32 28
South Ter. SW7 BU42 41
South Ter. Surb. BL53 66
South, The, Glade, Bex. CQ47 63
South Vale SE19 CA50 60
South Vale, Har. BH35 30
Southvale Rd. SE3 CG44 52
South View, Brom. CJ51 71
South View, Dart. CT46 55
South View Av. NW10 BO35 31
Southview Clo. Swan. CT52 73
Willow Av.
South View Clo. Bex. CQ46 54
Southview Cres. Ilf. CL32 26
South View Dr. E18 CH31 26
Southview Gdns. Wall. BW57 77
South View Rd. N8 BW31 23
Southview Rd. Brom. CF49 61
South Vw. Rd. Dart. CV48 64
South Vw. Rd. Loug. CK25 10
South Vw. Rd. Pnr. BC29 11
Southviews, S. Croy. CC58 79
South Vill. NW1 BW36 32
Southville Clo. Epsom BN57 76
Southville Rd. Surb. BJ54 66
South Wk. W. Wick. CG55 79
Southwark Br. EC4 BZ40 42
Southwark Br. SE1 BZ40 42
Southwark Br. Rd. SE1 BY41 42
Southwark Gro. SE1 BZ40 42
Southwark Park Rd. SE16 CA42 42
Southwark St. SE1 BY40 42
Southwater Clo. E14 CD39 43
South Way N9 CC27 16
South Way N11 BW29 14
Ringway
Southway N20 BS27 14
Southway NW11 BS32 23
Southway SW20 BQ52 67
South Way, Brom. CH54 71
South Way, Cars. BT58 77
South Way, Croy. CD55 79
South Way, Har. BF31 20
Southway, Wall. BW56 77
South Way, Wem. BL35 30
Southway Clo. W12 BP41 40
Scotts Rd.
Southwell Av. Nthlt. BF36 29
Southwell Gdns. SW7 BT41 41
Southwell Grove Rd. E11 CG34 25
Southwell Rd. SE5 BZ45 51
Southwell Rd. Croy. BY53 69
Southwell Rd. Har. BK32 21
South Western Rd. Twick. BJ46 48
Southwest Rd. E11 CF33 25
South Wharf Rd. W2 BT39 41
Southwick Ms. W2 BT39 41
Southwick St.
Southwick St. W2 BU39 41
Southwick Pl. W2 BU39 41
Southwold Dr. Bark. CO35 36
Southwold Rd. E5 CB34 24
Southwold Rd. Bex. CR47 63
Southwood Av. N6 BV33 23
Southwood Av. Couls. BW61 83
Southwood Av. Kings. On T. BN51 67
Southwood Clo. Brom. CK52 71
Southwood Clo. Wor. Pk. BQ54 67
Southwood Ct. NW11 BS32 23
Southwood Dr. Surb. BN54 67
Southwood Gdns. Esher BJ55 75
Southwood Gdns. Ilf. CL31 26
Southwood La. N6 BV33 23
Southwood Lawn Rd. N6 BV33 23
Southwood Rd. SE9 CL48 62
Southwood Rd. SE28 CO40 45
South Worple Av. SW14 BO45 49
South Worple Way
South Worple Way SW14 BN45 49
Sovereign Clo. W5 BK39 39
Sowerby Clo. SE9 CK46 53
Sowrey Av. Rain. CT36 37
Spacewaye, Felt. BC46 47
Spa Clo. SE25 CA51 69
Spa Green Est. EC1 BY38 33
Spa Hill SE19 BZ51 69
Spalding Rd. SW17 BV49 59
Spalding Rd. NW4 BQ32 22
Spanby Rd. E3 CE38 34
Spangate SE3 CG45 52
Spaniards Clo. NW11 BT33 23
Spaniards End NW3 BT33 23
Spaniards Rd. NW3 BT34 23
Spanish Pl. W1 BV39 41
Spanish Rd. SW18 BT46 50
Spareleaze Hill, Loug. CK25 10
Sparepenny La. Eyns. CV54 73
Sparkbridge Rd. Har. BH31 21
Spa Rd. SE16 CA41 42
Sparrow Dr. Orp. CM54 71
Sparrow Farm Rd. Epsom BP56 76
Sparrow Fm. Dr. Felt. BD47 56
Sparrow Grn. Dag. CR34 27
Sparrows Herne, Bush. BF26 4
Sparrows La. SE9 CM47 62
Sparrows Way, Bush. BG26 4
Sparrows Herne
Sparsholt Rd. N19 BX33 24
Sparsholt Rd. Bark. CN37 36
Sparta St. SE10 CE44 52
Spear Ms. SW5 BS42 41
Spearman St. SE18 CL43 53
Spearpoint Gdns. Ilf. CN31 27
Spears Rd. N19 BX33 24
Speart La. Houns. BE43 47
Spedan Clo. NW3 BS34 23
Spedan Tower NW3 BS34 23
Speedwell St. SE8 CE43 52
Speer Rd. Surb. BH54 66
Speke Hill SE9 CK48 62
Speke Rd. Th. Hth. BZ51 69
Speldhurst Clo. Brom. CG53 70
Speldhurst Rd. E9 CC36 34
Speldhurst Rd. W4 BN41 40
Spelman St. E1 CA39 42
Spencer Av. N13 BX29 15
Spencer Av. Hayes BC39 38
Spencer Clo. Croy. BZ54 69
Spencer Clo. NW10 BL38 30
Spencer Clo. Orp. CN55 81
Spencer Clo. Wdf. Grn. CJ28 17
Spencer Ct. NW8 BT37 32
Spencer Ct. SW20 BP51 67
Spencer Rd.
Spencer Ct. Rich. BK49 57
Spencer Dr. N2 BT32 23
Spencer Gdns. SE9 CK46 53
Spencer Gdns. SW14 BN46 49
Spencer Hill SW19 BR50 58
Spencer Hl. Rd. SW19 BR50 58
Spencer Pk. SW18 BT46 50
Spencer Pass. E3 CB37 33
Dinmont St.
Spencer Pl. SW1 BW41 41
Greycoat Pl.
Spencer Rise NW5 BV34 23
Spencer Rd. E6 CJ37 35
Spencer Rd. E17 CF30 16
Spencer Rd. N11 BV28 14
Spencer Rd. N17 CB30 15
Spencer Rd. SW18 BT45 50
Spencer Rd. SW20 BP51 67
Spencer Rd. W3 BN40 40
Spencer Rd. W4 BN43 49
Spencer Rd. Brom. CG50 61
Spencer Rd. E. Mol. BG53 65
Spencer Rd. Har. BH30 12
Spencer Rd. Ilf. CN33 27
Spencer Rd. Islw. BG44 47
Spencer Rd. Mitch. BU54 68
Spencer Rd. Mitch. BV52 68
Spencer Rd. Rain. CS38 37
Spencer Rd. S. Croy. CA56 78
Spencer Rd. Twick. BH48 57
Spencer Rd. Wem. BK34 21
Spencer St. EC1 BY38 33
Spencer St. Sthl. BD41 38
Spencer Wk. SW15 BQ45 49
Spenser Gro. N16 CA35 33
Spenser Rd. SE24 BY46 51
Spenser St. SW1 BW41 41
Spensley Wk. N16 BZ34 24
Speranza St. SE18 CN42 45
Sperling Rd. N17 CA30 15
Spert St. E14 CD40 43
Spey Side N14 BW25 7
Spey Way, Rom. CT29 19
Spezia Rd. NW10 BP37 31
Spices Yd. Croy. BZ56 78
South End
Spicer Clo. SE5 BY44 51
Spielman Rd. Dart. CW44 55
Spiers Clo. N. Mal. BO53 67
Spigurnell Rd. N17 BZ30 15
Spikes Br. Rd. Sthl. BE39 38
Spilsby Rd. Rom. CV29 19
Spilsey Clo. NW9 BO30 13
Spindlewood Gdns. Croy. BZ56 78
Spinel Clo. SE18 CN42 45
Spingate Clo. Horn. CV35 37
Spinnells Rd. Har. BE33 20
Spinglewood Gdns. Croy. CA56 78
Spinney Clo. N. Mal. BO53 67
Spinney Gdns. Dag. CQ35 36
Spinney Oak, Brom. CK51 71
Spinneys, The, Brom. CK51 71
Spinney, The N21 BY26 8
Spinney, The SW16 BW48 59
Spinney, The, Barn. BS23 7
Spinney, The, Pur. BY59 84
Spinney, The, Sid. CQ49 63
Spinney, The, Stan. BL28 12
Spinney, The, Sun. BC51 65
Spinney, The, Sutt. BQ56 76
Spinney, The, Wat. BC23 4
Spinney, The, Wem. BJ34 21
Spires, The, Dart. CV48 64
Spital Sq. E1 CA39 42
Spital St. E1 CA39 42
Spital St. Dart. CV46 55
Spondon Rd. N15 CB31 24
Tynemouth Rd.
Spooner Wk. Wall. BW56 77
Sportsbank St. SE6 CF47 61
Spottons Gro. N17 BZ30 15
Spout Hill, Croy. CE56 79
Spratt Hall Rd. E11 CH32 26
Spray St. SE18 CL42 44
Spreighton Rd. E. Mol. BF53 65
Sprimont Pl. SW3 BU42 41
Springall St. SE15 CB43 51
Springbank N21 BX25 8
Springbank Av. Horn. CV35 37
Springbank Rd. SE13 CF46 52
Springbourne Ct. Beck. CF51 70
Spring Bridge Ms. W5 BK40 39
Spring Bridge Rd.
Spring Bridge Rd. W5 BK40 39
Spring Clo. Borwd. BM23 5
Spring Clo. Sutt. BR57 76
Spring Cott. Surb. BK53 66
St. Leonards Rd.
Spring Ct. Sid. CO48 63
Springcroft Av. N2 BU31 23
Spring Crofts, Bush. BF25 4
Springdale Rd. N16 BZ35 33
Spring Dr. Pnr. BC32 20
Springfield E5 CB33 24

Springfield, Bush. BG26 4
Springfield Av. N10 BW31 23
Springfield Av. SW20 BR52 67
Springfield Av. Hamptn. BF50 56
Springfield Av. Swan. CT52 73
Springfield Clo. Stan. BJ27 12
Springfield Dr. Ilf. CM32 26
Springfield Est. SW8 BW44 50
Springfield Gdns. E5 CB33 24
Springfield Gdns. NW9 BN32 22
Springfield Gdns. Brom. CK52 71
Springfield Gdns. Ruis. BC33 20
Springfield Gdns. Wdf. Grn. CJ29 17
Springfield Gdns. W. Wick. CE55 79
Springfield Gro. SE7 CJ43 53
Springfield Gro. Sun. BC51 65
Springfield La. NW6 BS37 32
Springfield Mt. NW9 BN32 22
Springfield Rd. E4 CG26 9
Springfield Rd. E6 CK36 35
Springfield Rd. E15 CG38 34
Springfield Rd. E17 CD32 25
Springfield Rd. N11 BV28 14
Springfield Rd. N15 CB31 24
Springfield Rd. NW8 BT37 32
Springfield Rd. SE26 CB49 60
Springfield Rd. SW19 BR49 58
Springfield Rd. W7 BH40 39
Springfield Rd. Bexh. CR45 54
Springfield Rd. Brom. CK52 71
Springfield Rd. Epsom BQ58 76
Springfield Rd. Har. BH32 21
Springfield Rd. Hayes BL52 66
Springfield Rd. Kings. On T. BL52 66
Springfield Rd. Tedd. BJ49 57
Springfield Rd. Th. Hth. BZ51 69
Springfield Rd. Twick. BF47 56
Springfield Rd. Wall. BV56 77
Springfield Rd. Well. CO45 54
Springfield Wk. NW6 BS37 32
Springfield Wk. Orp. CM54 71
Farm Av.
Spring Gdns. N5 BZ35 33
Spring Gdns. SE11 BX42 42
Spring Gdns. SW1 BW40 41
Spring Gdns. E. Mol. BG53 65
Spring Gdns. Horn. CU35 37
Spring Gdns. Orp. CO57 81
Spring Gdns. Rom. CS32 28
Spring Gdns. Wall. BW56 77
Spring Gdns. Wdf. Grn. CJ29 17
Goding St.
Spring Gro. TW12 BF51 65
Spring Gro. E3 CE37 34
Old Ford Rd.
Spring Gro. W4 BM42 39
Spring Gro. Loug. CJ25 10
Spring Gro. Cres. Houns. BG44 47
Spring Gro. Rd. Houns. BF44 47
Spring Gro. Rd. Rich. BL46 48
Springhead Rd. Erith CT43 55
Spring Hill SE26 CC49 61
Peak Hill
Spring Hill E5 CB33 24
Springhill Clo. SE5 BZ45 51
Spring Lake, Stan. BJ28 12
Spring La. E5 CB33 24
Spring La. SE25 CB53 69
Spring Pk. Av. Croy. CC55 79
Spring Pk. Dr. N4 BZ33 24
Spring Pk. Dr. Beck. CF51 70
Spring Pk. Rd. Croy. CC55 79
Spring Pass. SW15 BQ45 49
Embankment, The
Spring Pl. NW5 BV35 32
Springpond Rd. Dag. CQ35 36
Springrice Rd. SE13 CF46 52
Spring St. W2 BT39 41
Spring St. Epsom BO58 76
Springvale, Bexh. CR45 54
Spring Vale, Dart. CV47 64
Springvale Av. Brent. BL42 39
Spring Vale Clo. Swan. CT50 64
Springvale Est. W14 BR41 40
Springvale Ter. W14 BQ41 40
Spring Villa Rd. Edg. BM29 12
Springwater Clo. SE18 CL44 53
Springwell Av. NW10 BO37 31
Springwell Clo. SW16 BX49 60
Springwell Rd. SW16 BX49 60
Springwell Rd. Houns. BD44 47
Springwood Cres. Edg. BM27 12
Sprowston Ms. E7 CH35 35
Sprowston Rd. E7 CH35 35
Sprucedale Gdns. Croy. CC56 79
Spruce Dale Gdns. Wall. BX58 78
Spruce Hills Rd. E17 CE30 16
Sprules Rd. SE4 CD44 52
Spurfield, E. Mol. BF52 65
Spurgeon Av. SE19 BZ51 69
Spurgeon Rd. SE19 BZ51 69
Spurgeon St. SE1 BZ41 42
Spurling Rd. SE22 CA45 51
Spurling Rd. Dag. CQ36 36
Spurrell Av. Bex. CS49 64
Spur Rd. N15 BZ31 24
Philip La.
Spur Rd. SW1 BW41 41
Spur Rd. Bark. CM38 35
Spur Rd. Edg. BL28 12
Spur Rd. Felt. BC46 47
Spur Rd. Islw. BJ43 48
Spur Rd. Orp. CO55 81
Spur Rd. Est. Edg. BL28 12
Spurstowe Ter. E8 CB35 33
Square, The, Cars. BV56 77
Square, The, Ilf. CL33 26
Square, The, Rich. BL45 48
Quadrant, The
Square, The, Wdf. Grn. CH28 17
Squarey St. SW17 BT48 59
Squires La. N3 BS30 14
Squires Mt. NW3 BT34 23
East Heath Rd.
Squires Way, Dart. CS49 64
Squires Wood Dr. Chis. CK50 62
Bullerswood Dr.
Squirrel Clo. Houns. BD45 47
Squirrels Clo. N12 BT28 14
Woodside Av.
Squirrels Grn. Wor. Pk. BO55 76
Squirrels Heath Av. Rom. CU31 28
Squirrels Hth. La. Rom. CV31 28
Squirrel's Hth Rd. Rom. CW31 28
Squirrel's La. Buck. H. CJ27 17
Squirrel's, The, Bush. BG25 4
Squirrels, The SE13 CF45 52
Squirrels, The, Pnr. BE31 20
Squirrels Way, Epsom BN61 82
Squirries St. E2 CB38 33
Stable Clo. Nthlt. BF37 29
Stable End, Orp. CM55 80
Stable Inn Bldgs. WC2 BY39 42
Southampton Bldgs.
Stables, The, Buck. H. CJ26 10
Stable Way W10 BQ39 40
Latimer Rd.
Stable Wk. N2 BT30 14
Stables Way SE11 BY42 42
Stable Yd. Rd. SW1 BW40 41
Stacey Av. N18 CC28 16
Stacey Clo. E10 CF32 25
Stacey St. WC2 BW39 41
Stackhouse St. SW3 BU41 41
Pavilion Rd.
Stacy Path SE5 BZ43 51
Elmington Est.
Stadium Rd. SE18 CK43 53
Stadium Rd. NW4 BQ33 22
Stadium Rd. Dart. CT46 55
Crayford Rd.
Stadium St. SW10 BT43 50
Stadium Way, Wem. BL35 30
Staffa Rd. E10 CC33 25
Stafford Av. Horn. CV31 28
Stafford Av. Wall. BX56 78
Stafford Clo. NW6 BS38 32
Stafford Clo. Sutt. BR57 76
Stafford Ct. W8 BS41 41
Stafford Cripps Est. EC1 BZ38 33
Stafford Ms. NW6 BS38 32
Stafford Pl. SW1 BW41 41
Stafford Pl. Rich. BL47 57
Stafford Rd. E3 CD37 34
Stafford Rd. E7 CJ36 35
Stafford Rd. NW6 BS38 32
Stafford Rd. Croy. BX56 78
Stafford Rd. Har. BG30 11
Stafford Rd. N. Mal. BN52 67
Stafford Rd. Sid. CN49 63
Stafford Rd. Wall. BW57 77
Staffordshire St. SE15 CB43 51
Stafford St. W1 BW40 41
Stafford Ter. W8 BS41 41
Staff St. EC1 BZ38 33
Cranwood St.
Stagbury Av. Couls. BU62 83
Stag Clo. Edg. BM30 12
Staggart Grn. Chig. CO29 18
Stag La. NW9 BN30 13
Stag La. SW15 BO48 58
Stag La. Buck. H. CH27 17
Stag La. Edg. BN30 13
Stag Pl. SW1 BW41 41
Stag Ride SW19 BO49 58
Stainbank Rd. Mitch. BV52 68
Stainby Rd. N15 CA31 24
Colsterworth Rd.
Stainer Rd. Borwd. BK23 5
Stainer St. SE1 BZ40 42
Staines Av. Sutt. BQ55 76
Staines Rd. Felt. BC46 47
Staines Rd. Houns. BE45 47
Staines Rd. Ilf. CM35 35
Staines Rd. Twick. BF48 56
Staines Rd. E. Sun. BC50 56
Stainforth Rd. E17 CE31 25
Stainforth Rd. Ilf. CM33 26
Staining La. EC2 BZ39 42
Gresham St.
Stainmore Clo. Chis. CM50 62
Stainsbury St. E2 CC37 34
Royston St.
Stainsby Pl. E14 CE39 43
Stainsby Rd. E14 CE39 43
Stainton Rd. SE6 CF46 52
Stainton Rd. Enf. CC23 9
Stalham St. SE16 CB41 42
Stalisfield Pl. Orp. CL58 80
Stambourne Way SE19 CA50 60
Stambourne Way, W. Wick. CF55 79
Stamford Brook Av. W6 BO41 40
Stamford Brook Rd. W6 BO41 40
Stamford Clo. N15 CB31 24
Stamford Rd.
Stamford Clo. Har. BH29 12
Stamford Clo. Sthl. BF40 38
Stamford Ct. W6 BO42 40
Stamford Dr. Brom. CG52 70
Stamford Gdns. Dag. CP36 36
Stamford Gro. E. N16 CB33 24
Oldhill St.
Stamford Gro. W. N16 CB33 24
Oldhill St.
Stamford Hill N16 CA34 24
Stamford Hill Est. N16 CA33 24
Stamford Ho. W12 BP41 40
Stamford Rd. KT12 BD55 74
Colne Dr.
Stamford Rd. E6 CK37 35
Stamford Rd. N1 CA36 33
Stamford Rd. N15 CB32 24
Stamford Rd. Dag. CO37 36
Stamford Rd. Wat. BC23 4
Stamford St. NW8 BT38 32
Stamford St. SE1 BY40 42
Stamp Pl. E2 CA38 33
Stanborough Pass. E8 BT36 32
Nasland High St.
Stanborough Rd. Islw. BG45 47
Stanbridge Rd. SW15 BQ45 49
Stanbrook Rd. SE2 CO41 45
Stanbury Rd. SE15 CB44 51
Stancroft NW9 BO32 22
Standard Rd. NW10 BN38 31
Standard Rd. Belv. CR42 45
Standard Rd. Bexh. CQ45 54
Standard Rd. Houns. BE45 47
Standard Rd. Orp. CL58 80
Standen Av. Horn. CV34 28
Standen Rd. SW18 BR47 58
Standfield Gdns. Dag. CR36 36
Standfield Rd. Dag. CR35 36
Standish Rd. W6 BP42 40
Stand Rd. Wat. BD27 11
Stane Way SE18 CJ43 53
Staneway, Epsom BP58 76
Stanfield Rd. E3 CD37 34
Stanford Clo. Rom. CR32 27
Stanford Clo. Wdf. Grn. CK28 17
Stanford Clo. Hamptn. BE50 56
Stanford Pl. SE17 CA42 42
Old Kent Rd.
Stanford Rd. N11 BU28 14
Stanford Rd. SW16 BW51 68
Stanford Rd. W8 BS41 41
Stanford St. SW1 BW42 41
Vincent Sq.
Stanford Way SW16 BW51 68
Stangate Cres. Borwd. BN24 6
Stangate Gdns. Stan. BJ28 12
Stanger Rd. SE25 CB52 69
Stanham Pl. Dart. CU45 55
Stanham Rd. Dart. CU46 55
Stanhope Av. N3 BR31 22
Stanhope Av. Brom. CG54 70
Stanhope Av. Har. BG30 11
Stanhope Bldgs. SE1 BZ41 42
Redcross Way
Stanhope Gdns. N4 BY32 24
Stanhope Gdns. N6 BV32 23
Stanhope Gdns. NW7 BO28 13
Stanhope Gdns. SW7 BT42 41
Stanhope Gdns. Dag. CQ34 27
Stanhope Gdns. Ilf. CK33 26
Stanhope Gate W1 BV40 41
Stanhope Gro. Beck. CD53 70
Stanhope Ms. E. SW7 BT42 41
Stanhope Gdns.
Stanhope Ms. S. SW7 BT42 41
Gloucester Rd.
Stanhope Pk. Rd. Grnf. BG38 29
Stanhope Pl. W2 BU39 41
Stanhope Rd. E17 CE32 25
Stanhope Rd. N6 BW32 23
Stanhope Rd. N11 BV28 14
Stanhope Rd. N12 BT28 14
Stanhope Rd. Barn. BQ25 6
Stanhope Rd. Bexh. CQ44 54
Stanhope Rd. Cars. BV57 77
Stanhope Rd. Croy. CA55 78
Stanhope Rd. Dag. CQ34 27
Stanhope Rd. Grnf. BG39 38
Stanhope Rd. Rain. CU37 37
Stanhope Rd. Sid. CO49 63
Stanhope St. NW1 BW37 32
Stanhope Ter. W2 BT40 41
Stanier Clo. SW5 BR42 40
Aisgill Av.
Stanlake Rd. W12 BP40 40
Stanlake Vill. W12 BQ40 40
Stanley Av. Bark. CN37 36
Stanley Av. Beck. CF51 70
Stanley Av. Dag. CQ33 27
Stanley Av. Grnf. BG37 29
Stanley Av. N. Mal. BP53 67
Stanley Av. Rom. CU31 28
Stanley Av. Wem. BL36 30
Stanley Clo. Couls. BX62 84
Stanley Clo. Horn. CV34 28
Stanley Clo. Rom. CU31 28
Stanley Clo. Wem. BL36 30
Stanley Ct. Rd. Islw. BG44 47
Stanley Cres. W11 BR40 40
Stanley Gdns. NW2 BQ35 31
Stanley Gdns. SW17 BV50 59
Ashbourne Rd.
Stanley Gdns. W3 BO40 40
Stanley Gdns. W11 BR40 40
Stanley Gdns. Sth. Croy. CB59 84
Stanley Gdns. Wall. BW57 77
Stanley Gdns. Ms. W11 BR40 40
Kensington Pk. Rd.
Stanley Gdns. Rd. Tedd. BH49 57
Stanley Gro. SW8 BV44 50
Stanley Gro. Croy. BY53 69
Stanley Park Dr. Wem. BL36 30
Stanley Pk. Rd. Cars. BU57 77
Stanley Pass. NW1 BX37 33
Stanley Rd. E4 CF26 9
Stanley Rd. E10 CE32 25
Stanley Rd. E12 CK35 35
Stanley Rd. E15 CF37 34
Stanley Rd. E18 CG30 16
Stanley Rd. N2 BT31 23
Stanley Rd. N9 CA26 8
Stanley Rd. N10 BV29 14
Stanley Rd. N11 BW29 14
Stanley Rd. NW9 BP33 22
Broadway, The
Stanley Rd. SW14 BM45 48
Stanley Rd. SW19 BS50 59
Stanley Rd. W3 BN41 40
Stanley Rd. Brom. CH52 71
Stanley Rd. Cars. BV57 77
Stanley Rd. Croy. BY54 69
Stanley Rd. Enf. CA24 8
Stanley Rd. Enf. CC25 9
Stanley Rd. Har. BG34 20
Stanley Rd. Horn. CV34 28
Stanley Rd. Houns. BG45 47
Stanley Rd. Ilf. CM34 26
Stanley Rd. Mitch. BV50 59
Stanley Rd. Mord. BS52 68
Stanley Rd. Nthwd. BC30 11
Stanley Rd. Orp. CN54 72
Stanley Rd. Sid. CO48 63
Stanley Rd. Sthl. BE40 38
Stanley Rd. Sutt. BS57 77
Stanley Rd. Twick. BG48 56
Stanley Rd. Wat. BD24 4
Stanley Rd. Wem. BL36 30
Stanley Rd. N. Rain. CT37 37
Stanley Rd. S. Rain. CT37 37
Stanley Sq. Cars. BU58 77
Stanley St. E6 CL40 44
Stanley St. SE8 CD43 52
Stanley Way, Orp. CO53 72
Stanmer St. SW11 BU44 50
Stanmore Gdns. Rich. BL45 48
Stanmore Gdns. Sutt. BT55 77
Stanmore Hill, Stan. BJ27 12
Stanmore Pl. NW1 BV37 32
Arlington Rd.
Stanmore Rd. E11 CG33 25
Stanmore Rd. N15 BY31 24
Stanmore Rd. Belv. CS42 46
Stanmore Rd. Rich. BL45 48
Stanmore Rd. Wat. BC23 4
Stanmore St. N1 BX37 33
Bermerton St.
Stanmore Ter. Beck. CE51 70
Stanmore Way, Loug. CL23 10
Stannard Cres. E6 CL39 44
Stannard Rd. E8 CB36 33
Stannary St. SE11 BY43 51
Stansfield Rd. SW9 BX45 51
Stansfield Rd. Houns. BC44 47
Stansgate Rd. Dag. CR34 27
Stanstead Clo. Brom. CG53 70
Stanstead Rd. E11 CH32 26
Stanstead Rd. SE6 CC47 61
Stanstead Rd. SE23 CC47 61
Stansted Clo. Horn. CV36 37
Stansted Cres. Bex. CP47 63
Stanswood Gdns. SE5 CA43 51
Sedgmoor Pl.
Stanthorpe Rd. SW16 BX49 60
Stanton Av. Tedd. BH50 57
Stanton Clo. Epsom BM56 75
Stanton Clo. Wor. Pk. BQ55 76
Stanton Rd. SW13 BO44 49
Stanton Rd. SW20 BQ51 67
Stanton Rd. Croy. BZ54 69
Stanton Sq. SE26 CD49 61
Stanton St. SE15 CB44 51
Stanton Way SE26 CD49 61

Stanway Clo. Chig. CN28 18
Tine Rd.
Stanway Ct. N1 CA37 33
Hoxton St.
Stanway Gdns. W3 BM40 39
Stanway Gdns. Edg. BN29 13
Stanway St. N1 CA37 33
Stanwick Rd. W14 BR42 40
Stanworth St. SE1 CA41 42
Millstream Rd.
Stanwyck Dr. Chig. CM28 17
Stanwyck Gdns. Rom. CU28 19
Stapenhill Rd. Wem. BJ34 21
Staple Clo. Bex. CS48 64
Staplefield Clo. SW2 BX47 60
Staplefield Clo. Pnr. BE29 11
Stapleford Av. Ilf. CN32 27
Stapleford Clo. SW19 BR47 58
Beaumont Rd.
Stapleford Clo. BM52 66
Kings. On T.
Vincent Rd.
Stapleford Gdns. Rom. CR29 18
Staplefield Rd. Wem. BK36 30
Stapleford Way. Bark. CO38 36
Bastable Av.
Staplehurst Rd. SE13 CF46 52
Staplehurst Rd. Cars. BU57 77
Staple Inn Bldgs. WC1 BY39 42
Holborn
Staple's Rd. Loug. CJ24 10
Staple St. SE1 BZ41 42
Stapleton Cres. Rain. CU36 37
Stapleton Gdns. Croy. BY56 78
Stapleton Hall Rd. N4 BX33 24
Stapleton Rd. Orp. CN55 81
Stapleton Rd. SW17 BV48 59
Stapleton Rd. Bexh. CQ43 54
Stapley Rd. Belv. CR42 45
Stapylton Rd. Barn. BR24 6
Starboard Way E14 CE41 43
Tiller St.
Starch House La. Ilf. CM30 17
Starcross St. NW1 BW38 32
Starfield Rd. W12 BP41 40
Star & Garter Hl. Rich. BL47 57
Star Hill, Dart. CT46 55
Starkleigh Way SE16 CB42 42
Egan Way
Star La. E16 CG38 34
Star La. Orp. CP52 72
Starling Clo. Buck. H. CH26 10
Starling Clo. Pnr. BD31 20
Star Path, Nthlt. BF37 29
Leander Rd.
Star Rd. Islw. BG44 47
Star Rd. W14 BR43 49
Star St. E16 CG39 43
Star St. W2 BT39 41
Starts Clo. Orp. CL55 80
Starts Hill Av. Orp. CL56 80
Starts Hill Rd. Orp. CL56 80
Starts Rd. Orp. CL55 80
Star Yd. WC2 BY39 42
Staten Gdns. Twick. BH47 57
Statham Gro. N16 BZ34 24
Statham Gro. N18 CA28 15
Station App. E7 CH35 35
Woodford Rd.
Station App. N11 BV28 14
Station App. SE3 CH45 53
Station App. SE26 CC49 61
Station App. SW6 BR45 49
Station App. SW11 BU45 50
St. Johns Hill
Station App. Bark. CM36 35
Station App. Sid. CO48 63
Station App. W3 BN41 40
Kingswood Rd.
Station App. W7 BH40 39
Station St.
Station App. Bex. CR47 63
High St.
Station App. Bexh. CQ44 54
Pickford La.
Station App. Bexh. CS44 55
Barnehurst Rd.
Station App. Buck. H. CJ28 17
Station App. Chelsfield CO56 81
Station App. Chis. CK50 62
Elmstead Woods
Station App. Chis. CL51 71
Station App. Couls. BW61 83
Station App. Dart. CW46 55
Station App. Debden CM24 10
Station App. E. Croy. BZ55 78
Station App. Elt. SE9 CK46 53
Station App. Epsom BN60 82
Station App. Ewell E. BP58 76
Station App. Ewell W. BO58 76
Station App. Grnf. BJ36 30
Station App. Har. BH33 21
Station App. Hatch End BE29 11
Station App. Hayes CG54 70
Station App. BH55 75
Hinchley Wd.
Station App. Loug. CK25 10
Station App. Mott. SE9 CK47 62

Station App. Orp. CN55 81
Station App. Pnr. BE31 20
Station App. Pur. BY59 84
Whytecliffe Rd.
Station App. Rich. BM44 48
Station App. Ruis. BC35 29
Station App. St. Mary CO52 72
Cray
Station App. Sthl. BE41 38
Station App. Stoneleigh BO56 76
Station App. Sun. BC51 65
Station App. Swan. CT52 73
Station App. Upr. Warl. CA62 84
Station App. Well. CO44 54
Station App. (Cray.) Dart. CT46 55
Station Rd.
Station App. Elm Pk. CU35 37
Horn.
Station App. N. Sid. CO48 63
Station Pde.
Station App. Nth Sid. CO48 63
Station Rd.
Station App. Rd. W4 BM43 48
Grove Park Rd.
Station App. Southgate BW26 7
N14
Station App. Sth. Sid. CO48 63
Station Rd.
Station Av. Ewell W. BO58 76
Station Av. N. Mal. BO52 67
Station Av. Rich. BM44 48
Station Pde.
Station Bldgs. Hayes CG54 70
Station Clo. N3 BS30 14
Station Clo. Hmptn. BF51 65
Station Cres. N15 BZ31 24
Station Cres. SE3 CH42 44
Station Cres. Wem. BJ36 30
Station Dr. NW1 BW38 32
Station Est. Elmers End CC52 70
Station Est. Rd. Felt. BC47 56
Station Garage Mews BW50 59
SW16
Estreham Rd.
Station Gdns. W4 BN43 49
Station Gro. Wem. BL36 30
Station Hill, Hayes CH55 80
Station La. Edg. BM29 12
Station La. Horn. CV34 28
Station Pde. E11 CH32 26
High St.
Station Pde. NW2 BQ36 31
Station Pde. W3 BM39 39
Station Pde. Chipstead BU62 83
Station Pass. E18 CH30 17
Maybank Rd.
Station Pl. N4 BY34 24
Station Rise SE27 BY48 60
Norwood Rd.
Station Rd. E4 CF26 9
Station Rd. E7 CH35 35
Station Rd. E10 CF34 25
Station Rd. E12 CJ35 35
Station Rd. E17 CD32 25
Station Rd. N3 BS30 14
Station Rd. N11 BV28 14
Station Rd. N17 CB31 24
Station Rd. N18 CA28 15
Silver St.
Station Rd. N19 BW34 23
Junction Rd.
Station Rd. N22 BX30 15
Station Rd. NW4 BP32 22
Station Rd. NW7 BO28 13
Station Rd. NW10 BO37 31
Station Rd. SE20 CC50 61
Station Rd. SW13 BO44 49
Station Rd. W5 BL39 39
Station Rd. Barkingside CM31 26
Station Rd. Barn. BS25 7
Station Rd. Belmont BS58 77
Station Rd. Belv. CR41 45
Station Rd. Bexh. CQ45 54
Station Rd. Borwd. BM24 5
Station Rd. Brom. CH51 71
Station Rd. Cars. BU56 77
Station Rd. Chess. BL56 75
Station Rd. Chig. CL27 17
Station Rd. Chingford E4 CF26 9
Station Rd. Claygate BH56 75
Station Rd. Crayford CT47 64
Station Rd. Dag. CP33 27
Station Rd. Edg. BM29 12
Station Rd. Esher BG55 74
Station Rd. Hamptn. BK51 66
Station Rd. Hanwell W7 BH40 39
Station Rd. Har. BH31 21
Station Rd. Harold Wd. CW30 19
Station Rd. Hmptn. BF51 65
Station Rd. Houns. BF45 47
Station Rd. Ilf. CL34 26
Station Rd. Ken. BZ60 84
Station Rd. Kings. On T. BM51 66
Station Rd. Knockholt CO58 81
Station Rd. Loug. CK24 10
Station Rd. Motspur Pk. BP53 67

Station Rd. Norwood CA52 69
Junc. SE25
Station Rd. Orp. CN55 81
Station Rd. Pnr. BE31 20
Station Rd. St. Mary, CP52 72
Cray.
Station Rd. Shortlands CG51 70
Station Rd. Sid. CO49 63
Station Rd. Slou. BO53 67
Station Rd. Sun. BC50 56
Station Rd. Swan. CT52 73
Station Rd. Tedd. BH49 57
Station Rd. Thames BH54 66
Ditton
Station Rd. Twick. BH47 57
Station Rd. Upr. Warl. CA62 84
Station Rd. Wat. BC23 4
Station Rd. W. Croy. BZ54 69
Station Rd. W. Wick. CF54 70
Station Rd. Winchmore BY26 8
Hill N21
Station Rd. Gidea Pk. CU31 28
Rom.
Station Rd. N. Belv. CR41 45
Station Rd. N. Har. BF32 20
Station Sq. Pett's Wd. CM53 71
Station Sq. St. Mary CO52 72
Cray
Station St. E15 CF36 34
Station St. E16 CL40 44
Station Ter. NW10 BQ37 31
Station Ter. SE5 BZ44 51
Station Vw. Grnf. BG37 29
Station Vill. NW7 BQ29 13
Bittacy Hill
Station Way, Cheam BR57 76
Station Way, Claygate BH57 75
Station Way, Roding Vall. CJ28 17
Station Yd. Twick. BJ47 57
Staunton Rd. BL50 57
Kings. On T.
Staunton St. SE8 CD43 52
Staveley Clo. N7 BX35 33
Penn Rd.
Staveley Clo. SE15 CC44 52
Staveley Clo. E9 CC35 34
Churchill Wk.
Staveley Gdns. W4 BN44 49
Staveley Rd. W4 BN43 49
Staverton Rd. NW2 BQ36 31
Staverton Rd. Horn. CV32 28
Stavordale Rd. N5 BY35 33
Stavordale Rd. Cars. BT54 68
Stayners Rd. E1 CC38 34
Stayton Rd. Sutt. BS55 77
Steadman Clo. Bex. CT48 64
Stead St. SE17 BZ42 42
Stean St. E8 CA37 33
Stebbing Way, Bark. CO37 36
Stebondale St. E14 CF42 43
Steeds Rd. N10 BU30 14
Steeds Way, Loug. CK24 10
Steele Rd. E11 CG35 34
Steele Rd. N17 CA31 24
Steele Rd. NW10 BN37 31
Steele Rd. W4 BN41 40
Steele Rd. Islw. BJ45 48
Steele's Ms. NW3 BU36 32
Steele's Rd. NW3 BU36 32
Steel's La. E1 CC39 43
Devonport St.
Steen Way SE22 CA46 51
Dulwich Gro.
Steep Clo. Orp. CN57 81
Steep Hill SW16 BW48 59
Steeple Clo. SW6 BR45 49
Steeple Clo. SW19 BR49 58
Steeplestone Clo. N18 BZ28 15
Steeprth St. SW18 BS48 59
Steerlands, Bush. BF26 4
Steer's Mead, Mitch. BU51 68
Stella Rd. SW17 BU50 68
Stelling Rd. Erith CS43 55
Stembridge Rd. SE19 CB51 69
Stephen Av. Rain. CU36 37
Stephen Clo. BR6 CN55 81
Stephendale Rd. SW6 BS45 50
Stephen Rd. Bexh. CS45 55
Stephens Clo. Pnr. BD32 20
Stephens Gro. SE13 CF45 52
Stephen's Ms. W1 BW39 41
Gresse St.
Stephenson Rd. W7 BH39 39
Stephenson Rd. Houns. BF47 56
Stephenson St. E16 CG38 34
Stephenson St. NW10 BO38 31
Stephenson Way NW1 BW38 32
Stephenson Rd. E15 CG37 34
Stephen St. W1 BW39 41
Stepney Causeway E1 CC39 43
Stepney Green E1 CC39 43
Stepney Grn. Dws. E1 CC39 43
Hayfield Pass
Stepney High St. E1 CC39 43
Stepney Way E1 CB39 42
Sterling Av. Edg. BL28 12
Sterling Rd. Enf. BZ23 8

Sterling St. SW7 BU41 41
Montpelier Pl.
Sterndale Rd. W14 BQ41 40
Sterndale Rd. Dart. CW47 64
Sterne St. W12 BQ41 40
Sternhall La. SE15 CB44 51
Sternhold Av. SW2 BW48 59
Sterry Cres. Dag. CR35 36
Sterry Dr. E. Mol. BH53 66
Sterry Dr. Epsom BO56 76
Sterry Gdns. Dag. CR36 36
Sterry Rd. Bark. CN37 36
Sterry Rd. Dag. CR35 36
Sterry St. SE1 BZ41 42
Steucers La. SE23 CD47 61
Stevedale Rd. Well. CP44 54
Stevenage Cres. Borwd. BL23 5
Stevenage Rd. E12 CL36 35
Stevenage Rd. SW6 BQ43 49
Stevens Av. E9 CC36 34
Stevens Clo. Beck. CE50 61
Stumps Hill La.
Stevens Clo. Bex. CS49 64
Stevens Clo. Hamptn. BE50 56
Stevens Cott. NW2 BP36 31
High Rd.
Stevens Grn. Bush. BG26 4
Stevenson Clo. Erith CU43 55
Stevens Rd. Dag. CO34 27
Stevens's La. Esher BJ57 75
Steventon Rd. W12 BO40 40
Steward St. E1 CA39 42
Steward Wk. Rom. CT32 28
Liberty, The
Stewart Clo. NW9 BN32 22
Stewart Clo. Hamptn. BE50 56
Stewart Clo. Chis. CK55 71
Stewart's La. SW8 BV43 50
Stewarts Rd. SW8 BW44 50
Stewart St. E14 CF41 43
Stewart's Wk. SW3 BU42 41
Stew La. EC4 BZ40 42
Broken Wharf
Steyne Rd. W3 BM40 39
Steyning Gro. Ken. BY61 84
Steyning Gro. SE9 CK49 62
Steynings Way N12 BS28 14
Steyning Way, Houns. BD45 47
Sth. Birkbeck Rd. E11 CF34 25
Sth. Black Lion La. W6 BP42 40
Sth. Countess Rd. E17 CD31 25
Sth. Croxted Rd. SE21 BZ48 60
Sth. Eldon Pl. EC2 BZ39 42
Sth. Esk Rd. E7 CJ36 35
Sth. Lodge W5 BK40 39
Webster Gdns.
Sth. Molten Rd. E16 CH39 44
Sth. Norwood Hl. SE25 CA51 69
Sth. Tenter St. E1 CA40 42
Sth. Western Rd. Twick. BJ46 48
Stickland Rd. Belv. CR42 45
Stickleton Clo. Grnf. BF38 29
Stifford Est. E1 CC39 43
Stilecroft Gdns. Wem. BJ34 21
Stile Hall Gdns. W4 BM42 39
Stile Path, Sun. BC52 65
Stiles Clo. Brom. CK53 71
Stillingfleet Rd. SW13 BP43 49
Stillington St. SW1 BW42 41
Stillness Rd. SE23 CD46 52
Stipulakis Dr. Hayes BD39 29
Stilton Cres. NW10 BN36 31
Stirling Clo. Bans. BR62 82
Stirling Clo. Rain. CU38 37
Stirling Dr. Orp. CO56 81
Stirling Rd. E13 CH37 35
Stirling Rd. E17 CD31 25
Stirling Rd. N17 CB30 15
Stirling Rd. N22 BY30 15
Stirling Rd. SW9 BX44 51
Stirling Rd. W3 BM41 39
Stirling Rd. Har. BH31 21
Stirling Rd. Hayes BC40 38
Stirling Rd. Twick. BF47 56
Stirling Rd. Pth. E17 CD31 25
Stirling Rd. SW7 BU41 41
Stirling Wk. Surb. BM53 66
Stirling Way N18 BZ28 15
Stirling Way, Borwd. BN25 6
Stirling Way, Croy. BX54 69
Stiven Cres. Har. BE34 20
Stoats Nest Rd. Couls. BX60 84
Stoats Nest Vill. Couls. BX61 84
Stockbury Rd. Croy. CC53 70
Stockdale Rd. Dag. CQ34 27
Stockdove Way, Grnf. BH38 30
Stockfield Rd. SW16 BX48 60
Stockfield Rd. Esher BH56 75
Stockingswater La. Enf. CD23 9
Stockland Rd. Rom. CS32 28
Stock La. Dart. CV49 64
Stock Orchard Cres. N7 BX35 33
Stock Orchard St. N7 BX35 33
Stockport Rd. SW16 BW51 68
Stocksfield Rd. E17 CF31 25

Stock St. E13 CH37 35
Stockton Gdns. N17 BZ29 15
Stockton Gdns. NW7 BN27 13
Stockton Rd. N17 BZ29 15
Stockton Rd. N18 CB29 15
Stockwell Av. SW9 BX45 51
Stockwell Gdns. SW9 BX44 51
Stockwell Grn. SW9 BX44 51
Stockwell La. SW9 BX44 51
Stockwell Pk. SW9 BX45 51
Stockwell Park Cres. SW9 BX44 51
Stockwell Park Est. SW9 BX44 51
Stockwell Park Rd. SW9 BX44 51
Stockwell Rd. SW9 BX44 51
Stockwell St. SE10 CF43 52
Stockwood St. SW11 BT45 50
Plough Rd.
Stodart Rd. SE20 CC51 70
Stofield Gdns. SE9 CJ48 62
Stoford Clo. SW19 BR47 58
Southmead Rd.
Stoke Av. Ilf. CO29 18
Stokenchurch St. SW6 BS44 50
Stoke Newington Church St. N16 BZ34 24
Stoke Newington Common N16 CA34 24
Stoke Newington High St. N16 CA34 24
Stoke Newington Rd. N16 CA35 33
Stoke Pl. NW10 BO38 31
Stoke Rd. Kings. On T. BN50 58
Stoke Rd. Rain. CV37 37
Stoke Rd. Walt. BD55 74
Stokesby Rd. Chess. BL57 75
Stokesley St. W12 BO39 40
Stokes Rd. E6 CK38 35
Stokes Rd. Croy. CC53 70
Stompond La. Walt. BC55 74
Stonard Rd. N13 BY27 15
Stonard Rd. Dag. CO35 36
Stonards Hill, Loug. CK25 10
Stondon Pk. SE23 CD47 61
Stondon Wk. E6 CJ37 35
Stonebridge Est. E8 CA37 33
Stonebridge Pk. NW10 BN36 31
Stonebridge Rd. N15 CA32 24
Stonebridge Way, Wem. BM36 30
Stone Bldgs. WC2 BX39 42
Stone Clo. Dag. CQ34 27
Stonecot Clo. Sutt. BR54 67
Stonecot Hill, Sutt. BR54 67
Stonecroft Rd. Erith CS43 55
Stonecroft Way, Croy. BX54 69
Stone Cross Rd. Swan. CS53 73
Stonecutter St. EC4 BY39 42
Shoe La.
Stonefield Clo. Bexh. CR45 54
Stonefield Clo. Ruis. BE35 29
Stonefield Way SE7 CJ43 53
Green Bay Rd.
Stonefield Way, Ruis. BE35 29
Stonegrove, Edg. BL28 12
Stone Gro. Ct. Edg. BL28 12
Stone Gro. Gdns. Edg. BL28 12
Stonehall Av. Ilf. CK32 26
Stone Hall Rd. N21 BX26 8
Stoneham Rd. E5 CB34 24
Stoneham's Hill, Dart. CT45 55
Stonehill Clo. SW14 BN46 49
Stonehill Grn. Rd. Dart. CS50 64
Stonehill Rd. SW14 BN46 49
Stonehill Rd. W4 BM42 39
Wellesley Rd.
Stonehills Ct. SE21 CA48 60
Stone Ho. Ct. EC3 CA39 42
Houndsditch
Stonehouse La. Sev. CP58 81
Stonehouse Rd. Enf. CC25 9
Stonehouse Rd. N11 BW28 14
Stonehouse Rd. Sev. CP58 81
Stonehouse St. SW4 BW45 50
Stoneleigh Av. Wor. Pk. BP55 76
Stoneleigh Ct. Ilf. CK31 26
Stoneleigh Cres. Epsom BO56 76
Stoneleigh Pk. Av. Croy. CC53 70
Stoneleigh Pk. Rd. Epsom BO57 76
Stoneleigh Pl. W11 BQ40 40
Stoneleigh Rd. N17 CA30 15
Chestnut Rd.
Stoneleigh Rd. N17 CA31 24
Stoneleigh Rd. Cars. BU54 68
Stoneleigh Rd. Ilf. CK31 26
Stoneleigh St. W11 BQ40 40
Stonell's Rd. SW11 BU46 50
Chatham Rd.
Stonenest St. N4 BX33 24
Evershot Rd.
Stone Pk. Av. Beck. CE52 70
Stone Pl. Wor. Pk. BP55 76
Stone Rd. Brom. CG53 70
Stones Alley, Wat. BC24 4
Stones End St. SE1 BZ41 42
Stone's Rd. Epsom BO59 82
Stone St. Croy. BY56 78
Stonewood Rd. Erith CT42 46
Stone Yard La. E14 CF40 43
Stoneycroft Clo. SE12 CG47 61
Stoneycroft Rd. Wdf. Grn. CK29 17
Stoneydown Av. E17 CD31 25
Stoneyfield Rd. Couls. BX62 84
Stoneyfields Gdns. Edg. BN28 13
Stoneyfields La. Edg. BN28 13
Stoney La. E1 CA39 42
Stoney La. SE19 CA50 60
Stoney St. SE1 BZ40 42
Stonhouse St. SW4 BW45 50
Stonor Rd. W14 BR42 40
Stony Hill, Esher BE57 74
Stony Path, Loug. CK23 10
Stopford Rd. E13 CH37 35
Store Rd. E16 CL41 44
Store St. E15 CF35 34
Store St. WC1 BW39 41
Storey Rd. E17 CD32 25
Storey Rd. N6 BU32 23
Storey's Gate SW1 BW41 41
Storey St. E16 CL40 44
Stories Ms. SE5 CA44 51
Stories Rd. SE5 CA45 51
Stork Rd. E7 CG36 34
Storksmead Rd. Edg. BO29 13
Stormont Av. N6 BU33 23
Stormont Rd. SW11 BV45 50
Stormont Way, Chess. BK56 75
Storrington Rd. Croy. CA54 69
Stothard St. E1 CC38 34
Colebert Av.
Stoughton Av. Sutt. BQ56 76
Stour Av. Sthl. BF41 38
Stourcliffe St. W1 BU39 41
Stour Clo. Kes. CJ56 80
Stourhead SW19 BQ47 58
Stourhead Clo. SW19 BQ47 58
Castlecombe Dr.
Stourhead Gdns. SW20 BP51 67
Stour Rd. E3 CE36 34
Stour Rd. Dag. CR34 27
Stour Rd. Dart. CU45 55
Stourton Av. Felt. BE49 56
Stowage, The SE8 CE43 52
Stowe Gdns. N9 CA26 8
Stowell Av. Croy. CF58 79
Stowe Pl. N15 CA31 24
Stowe Rd. W12 BP41 40
Stowe Rd. Orp. CO56 81
Stowting Rd. Orp. CN56 81
Stox Mead, Har. BG30 11
Stracey Rd. E7 CH35 35
Stracey Rd. NW10 BN37 31
Strachen Pl. SW19 BQ50 58
Stradbroke Dr. Chig. CL29 17
Stradbroke Gro. Ilf. CK31 26
Stradbroke Rd. N5 BZ35 33
Balfour Rd.
Stradbrooke Gro. Buck. H. CJ26 10
Stradella Rd. SE24 BZ46 51
Strafford Av. Ilf. CL30 17
Strafford Rd. W3 BN41 40
Bollo Bridge Rd.
Strafford Rd. Barn. BR24 6
Strafford Rd. Houns. BE45 47
Strafford Rd. Twick. BJ47 57
Strahan Rd. E3 CD38 34
Straight Rd. Rom. CU28 19
Straight, The, Sthl. BE41 38
Straitsmouth SE10 CF43 52
Strakers Rd. SE22 CB45 51
Strand WC2 BX40 42
Strandfield Clo. SE18 CN42 45
Strand La. WC2 BX40 42
Temple Pl.
Strand Pl. N18 CA28 15
Strand On The Green W4 BM43 48
Stranraer Way N1 BX36 33
Gifford St.
Stratfield Rd. Borwd. BL24 5
Stratford Av. W8 BS41 41
Stratford Clo. Bark. CO36 36
Stratford Clo. Dag. CS36 37
Stratford Gro. SW15 BQ45 49
Stratford Pl. W1 BV39 41
Stratford Rd. W3 BN41 40
Stratford Rd. E13 CG37 34
Stratford Rd. NW4 BQ31 22
Stratford Rd. W8 BS41 41
Stratford Rd. Hayes BC38 29
Stratford Rd. Sthl. BE42 38
Stratford Rd. Th. Hth. BY52 69
Stratford Rd. Wat. BC23 4
Stratford St. E14 CE41 43
Stratford Vill. NW1 BW36 32
Strathan Clo. SW18 BR46 49
Strathaven Rd. SE12 CH46 53
Strathblaine Rd. SW11 BT45 50
Strathbrook Rd. SW16 BX50 60
Strathcona Rd. Wem. BK34 21
Strathdale SW16 BX49 60
Strathdon Dr. SW17 BT48 59
Strathearn Av. Twick. BF47 56
Strathearn Pl. W2 BU40 41
Strathearn Rd. SW19 BS49 59
Strathearn Rd. Sutt. BS56 77
Stratheden Rd. SE3 CH44 53
Strathfield Gdns. Bark. CM36 35
Strathleven Rd. SW2 BX45 51
Strathmore Gdns. N3 BS30 14
Hervey Clo.
Strathmore Gdns. W8 BS40 41
Strathmore Gdns. Edg. BM30 12
Strathmore Gdns. Horn. CT33 28
Strathmore Rd. SW19 BS48 59
Strathmore Rd. Croy. BZ54 69
Strathnairn St. SE1 CB42 42
Strathray Gdns. NW3 BU36 32
Strath Ter. SW11 BU45 50
Strathville Rd. SW18 BS48 59
Strathyre Av. SW16 BX52 69
Stratmore Rd. Tedd. BH49 57
Stratton Av. Wall. BW58 77
Stratton Clo. SW19 BS51 68
Stratton Clo. Bexh. CQ45 54
Stratton Clo. Edg. BL29 12
Stratton Clo. Houns. BF44 47
Stratton Clo. KT12 BD54 68
St. Johns Dr.
Strattondale St. E14 CF41 43
Stratton Dr. Bark. CN35 36
Stratton Gdns. Sthl. BE39 38
Stratton Rd. SW19 BS51 68
Stratton Rd. Bexh. CQ45 54
Stratton St. W1 BV40 41
Strauss Rd. W4 BN41 40
Strawberry Hill Clo. Twick. BH48 57
Strawberry Hill Rd. Twick. BH48 57
Strawberry La. Cars. BU55 77
Strawberry Vale N2 BT30 14
Strawberry Vale, Twick. BJ48 57
Streamdale SE2 CO43 54
Stream La. Edg. BM28 12
Streamway, Belv. CQ43 54
Streatfield Av. E6 CK37 35
Streatfield Rd. Har. BK31 21
Streatham Clo. SW16 BX48 60
Leigham Ct. Rd.
Streatham Com. N. SW16 BX49 60
Streatham Com. S. SW16 BX50 60
Streatham Ct. SW16 BX48 60
Streatham High Rd. SW16 BX49 60
Streatham Hill SW2 BX48 60
Streatham Hl. Est. SW16 BX48 60
Streatham Pl. SW2 BX47 60
Streatham Rd. SW16 BU51 68
Streatham Rd. Mitch. BU51 68
Streatham St. WC1 BX39 42
Streatham Vale SW16 BW51 68
Streathbourne Rd. SW17 BV48 59
Streatley Pl. NW3 BT35 32
Streatley Rd. NW6 BR36 31
Streimer Rd. E15 CF37 34
Strelley Way W3 BO40 40
Stretton Rd. Croy. CA54 69
Stretton Rd. Rich. BK48 57
Strickland Av. Dart. CW45 55
Strickland Row SW18 BT47 59
Strickland St. SE8 CE44 52
Stride Rd. E13 CG37 34
Strode Clo. N10 BV29 14
Strode Rd. E7 CH35 35
Strode Rd. N17 CA30 15
Strode Rd. NW10 BP36 31
Strode Rd. SW6 BR43 49
Strone Rd. E7 CJ36 35
Strone Rd. E12 CJ36 35
Strongbow Cres. SE9 CK46 53
Strongbow Rd. SE9 CK46 53
Strongbridge Clo. Har. BF33 20
Stronsa Rd. W12 BO41 40
Strood Av. Rom. CS33 28
Stroud Cres. SW15 BP48 58
Stroudes Clo. Wor. Pk. BO54 67
Stroud Field, Nthlt. BE36 29
Arnold Rd.
Stroud Gate, Har. BF35 29
Stroud Grn. Gdns. Croy. CC53 70
Stroud Green Rd. N4 BX33 24
Stroud Grn. Way, Croy. CB54 69
Stroud Rd. SE25 CB53 69
Stroud Rd. SW19 BS48 59
Strutton Rd. SW19 BS48 59
Strutton Ground SW1 BW41 41
Strype St. E1 CA39 42
Leyden St.
Stuart Av. NW4 BP33 22
Stuart Av. W5 BL40 39
Stuart Av. Brom. CH54 71
Stuart Av. Har. BE34 20
Stuart Av. Walt. BC54 65
Stuart Cres. N22 BX30 15
Stuart Cres. Croy. CD55 79
Stuart Evans Clo. Well. CP45 54
Stuart Gdns. Tedd. BH49 57
Stuart Mantle Way, Erith CS43 55
Stuart Pl. Mitch. BU51 68
Stuart Rd. NW6 BS38 32
Stuart Rd. SE15 CC45 52
Stuart Rd. SW19 BS48 59
Stuart Rd. W3 BN40 40
Stuart Rd. Bark. CN36 36
Dawson Av.
Stuart Rd. Barn. BU26 7
Stuart Rd. Har. BH31 21
Stuart Rd. Rich. BJ48 57
Stuart Rd. Th. Hth. BZ52 69
Stuart Rd. Well. CO44 54
Stubbs Way SW19 BT51 68
Brangwyn Cres.
Stucley Pl. NW1 BV37 32
Camden High St.
Stucley Rd. Houns. BG43 47
Studdridge St. SW6 BS44 50
Studholme Ct. NW6 BS35 32
Studholme St. SE15 CB43 51
Studio Dr. Wem. BM34 21
Empire Way
Studio, The, Bush. BF25 4
Studland Clo. Sid. CN48 63
Studland Rd. SE26 CC49 61
Studland Rd. W7 BG39 38
Studland Rd. Kings. On T. BL50 57
Studland St. W6 BP42 40
Studley Av. E4 CF29 16
Studley Ct. Sid. CO49 63
Studley Dr. Ilf. CJ32 26
Studley Gra. Rd. W7 BH41 39
Studley Rd. E7 CH36 35
Studley Rd. SW4 BX44 51
Studley Rd. Dag. CP36 36
Stukeley Rd. E7 CH36 35
Stukeley St. WC2 BX39 42
Stumps Hl. La. Beck. CE50 61
Stumps La. Whyt. CA62 84
Sturdy Rd. SE15 CB44 51
Sturge Av. E17 CE30 16
Sturges Field, Chis. CM50 62
Sturgess Av. NW4 BP33 22
Sturge St. SE1 BZ41 42
Sturmer Way N7 BX35 33
Stockorhcard Cres.
Sturrock Clo. N15 BZ31 24
Ida Rd.
Sturry St. E14 CE39 43
Sturt St. N1 BZ37 33
Stutfield St. E1 CB39 42
Styles Gdns. SW9 BY45 51
Styles Way Beck. CF52 70
Sudbourne Rd. SW2 BX46 51
Sudbrooke Rd. SW12 BU46 50
Sudbrook Gdns. Rich. BK48 57
Sudbrook La. Rich. BL47 57
Sudbury Av. Wem. BK34 21
Sudbury Ct. E5 CC35 34
Clapton Park Est.
Sudbury Ct. Dr. Har. BH34 21
Sudbury Ct. Rd. Har. BH34 21
Sudbury Cres. Brom. CH50 62
Sudbury Cres. Wem. BJ35 30
Sudbury Cft. Wem. BH35 30
Sudbury Gdns. Croy. CA55 78
Sudbury Heights Av. Grnf. BH35 30
Sudbury Hill Har. BH34 21
Sudbuy Hill Clo. Wem. BH35 30
Sudbury Par. Wem. BH35 30
Sudbury Par. Wem. BJ35 30
Sudbury Rd. Bark. CN35 36
Sudeley St. N1 BY37 33
Sudlow Rd. SW18 BS46 50
Sudrey St. SE1 BZ41 42
Suez Av. Grnf. BH37 30
Suez Rd. Enf. CD24 9
Suffield Rd. E4 CE28 16
Suffield Rd. N15 CA32 24
Suffield Rd. SE20 CC51 70
Suffolk Ct. Ilf. CN32 27
Suffock Ct. E10 CE33 25
Suffolk Gro. SE16 CC42 43
Hawkstone Rd.
Suffolk La. EC4 BZ40 42
Suffolk Pl. E17 CD31 25
Suffolk Pl. SW1 BW40 41
Suffolk St.
Suffolk Rd. E13 CG38 34
Suffolk Rd. N15 BZ32 24
Suffolk Rd. NW10 BO36 31
Suffolk Rd. SE25 CA52 69
Suffolk Rd. SW13 BO43 49
Suffolk Rd. Bark. CM36 35
Suffolk Rd. Dag. CS35 37
Suffolk Rd. Dart. CW46 55
Suffolk Rd. Enf. CB25 8
Suffolk Rd. Har. BE32 20
Suffolk Rd. Ilf. CN32 27
Suffolk Rd. Sid. CP50 63
Suffolk Rd. Wor. Pk. BO55 76
Suffolk St. E7 CH35 35
Suffolk St. SW1 BW40 41
Sugar House La. E15 CF37 34
Sugar Loaf Wk. E2 CC38 34
Sugden Rd. SW11 BV45 50
Sugden Rd. Surb. BJ54 66
Sugden St. SE5 BZ43 51
Sugden Way, Bark. CN37 36
Sulgrave Rd. W6 BQ41 40
Sulina Rd. SW2 BX47 60
Sulivan Ct. SW6 BS44 50
Sulivan Rd. SW6 BS45 50
Sullivan Av. E16 CJ39 44
Sullivan Clo. SW11 BU45 50

Sullivan Ct. SW6 BS45 50
Sullivan Rd. E. Mol. BF52 65
Sullivan Way, Borwd. BK25 5
Sultan Rd. E11 CH31 26
Sultan St. SE5 BZ43 51
Sultan St. Beck. CC51 70
Sumatra Rd. NW6 BS35 32
Sumburgh Rd. SW12 BV46 50
Summer Av. E. Mol. BH53 66
Summer Ct. Rd. E1 CC39 43
W. Arbour St.
Summerfield Av. NW6 BR37 31
Summerfield La. Surb. BK55 75
Summerfield Rd. W5 BJ38 30
Summerfield Rd. Loug. CJ25 10
Summerfield St. SE12 CG47 61
Summer Gdns. E. Mol. BH53 66
Summer Hill, Chis. CL51 71
Summerhill Clo. Orp. CN56 81
Summerhill Gro. Enf. CA25 8
Summerhill Rd. N15 BZ31 24
Summerhill Rd. Dart. CV47 64
Summer Hill Vill. Chis. CL51 71
Summerhouse Av. Houns. BE44 47
Summerhouse Dr. Bex. CS49 64
Summerhouse Dr. Dart. CS48 64
Summerhouse La. Wat. BS23 4
Summer House Rd. N16 CA34 24
Summerland Gdns. N10 BV31 23
Muswell Hill Broadway
Summerlands Av. W3 BN40 40
Summerlee Av. N2 BU31 23
Summerlee Gdns. N2 BU31 23
Summerley St. SW18 BS48 59
Summer Rd. E. Mol. BG53 65
Summersby Rd. N6 BV32 23
Summersby Rd. N6 BV32 23
Summers La. N12 BT29 14
Summers Row. N12 BU29 14
Summers St. EC1 BY38 33
Back Hill
Summerstown SW17 BT48 59
Summertrees Sun. BC51 65
Summerville Gdns. Sutt. BR57 76
Summerville Rd. Dart. CW46 55
Summerwood Rd. Islw. BH46 48
Summit Av. NW9 BN32 22
Summit Clo. N14 BW27 14
Summit Clo. NW2 BR35 31
Summit Clo. NW9 BN31 22
Summit Clo. Edg. BM29 12
Summit Dr. Wdf. Grn. CJ30 17
Summit Est. N16 CB33 24
Summit Rd. E17 CE31 25
Summit Rd. Nthlt. BF36 29
Summit, The Loug. CK23 10
Summit Way N14 BV27 14
Summit Way, SE19 CA50 60
Sumner Av. SE15 CA44 51
Sumner Rd.
Sumner Bldgs. SE1 BZ40 42
Sumner St.
Sumner Clo. Orp. CM56 80
Isabell Dr.
Sumner Gdns. Croy. BY54 69
Sumner Est. SE15 CA44 51
Sumner Pl. SW7 BT42 41
Sumner Pl. Ms. SW7 BT42 41
Sumner Pl.
Sumner Rd. SE15 CA43 51
Sumner Rd. Croy. BY54 69
Sumner Rd. Har. BG33 20
Sumner Rd. S. Croy. BY54 69
Sumner St. SE1 BY40 42
Sumpter Clo. NW3 BT36 32
Sun All. Rich. BL45 48
Sunbeam Rd. NW10 BN38 31
Sunbridge Rd. Croy. CA54 69
Sunbury Av. NW7 BN28 13
Sunbury Av. SW14 BN45 49
Sunbury Ct. Rd. Sun. BD51 65
Sunbury Gdns. NW7 BN28 13
Sunbury La. SW11 BT44 50
Sunbury La. Walt. BC53 65
Sunbury Rd. Sutt. BQ55 76
Sunbury St. SE18 CK41 44
Suncourt Erith CT44 55
Suncroft Pl. SE26 CC48 61
Sundale Av. Sth. Croy. CC58 79
Sunderland Ct. SE22 CB47 60
Sunderland Mt. SE23 CC48 61
Sunderland Rd. SE23 CC48 61
Sunderland Rd. W5 BK41 39
Sunderland Ter. W2 BS39 41
Sunderland Way E12 CJ34 26
Sundew Av. W12 BP40 40
Sundial Av. SE25 CA52 69
Sundorne Rd. SE7 CH42 44
Sundown Av. Sth. Croy. CA59 84
Sundown Pl. Ilf. CL34 26
Ilford Hill
Sundridge Av. Brom. CJ51 71
Sundridge Av. Well. CM44 53
Sundridge Ho. Brom. CH49 62
Sunfields Pl. SE3 CH43 53
Sunflower Way, Rom. CV30 19
Sunkist Way Wall. BX58 78
Sunland Av. Bexh. CQ45 54
Sun La. SE3 CH43 53
Sunleigh Rd. Wem. BL37 30
Sunley Gdns. Grnf. BJ37 30
Sunmead Rd. Sun. BC52 65
Sunna Gdns. Sun. BC51 65
Sunningdale Av. W3 BO39 40
Sunningdale Av. Bark. CM37 35
Sunningdale Av. Felt. BE48 56
Sunningdale Av. Rain. CU38 37
Sunningdale Av. Ruis. BD33 20
Sunningdale Clo. Stan. BJ29 12
Sunningdale Gdns. NW9 BN32 22
Sunningdale Rd. Brom. CK52 71
Sunningdale Rd. Rain. CU36 37
Sunningdale Rd. Sutt. BR55 76
Sunningfields Cres. NW4 BP30 13
Sunningfields Rd. NW4 BP30 13
Sunninghill Rd. SE13 CE44 52
Sunny Bank SE25 CB52 69
Sunny Bank, Epsom BN61 82
Sunny Cres. NW10 BN36 31
Sunnycroft Rd. SE25 CB52 69
Sunnycroft Rd. Houns. BF44 47
Sunnycroft Rd. Sthl. BF39 38
Sunnydale, SE12 CH46 53
Sunnydale Orp. CL55 80
Sunnydale Gro. NW7 BN29 13
Sunnydene Av. E4 CF28 16
Sunnydene Av. Ruis. BC33 20
Sunnydene Clo. Rom. CW29 19
Sunnydene Gdns. Wem. BK36 30
Sunnydene Rd. Pur. BY60 84
Sunnydene St. SE26 CD49 61
Sunnyfield NW7 BO27 13
Sunnyfield Rd. Chis. CO52 72
Sunny Gdns. Rd. NW4 BP30 13
Sunnyhill Rd. SW16 BX49 60
Sunny Hl. NW4 BP31 22
Sunnyhurst Clo. Sutt. BS55 77
Sunnymead Av. Mitch. BW52 68
Sunnymead Rd. NW9 BN33 22
Sunnymead Rd. SW15 BP46 49
Sunnymede Chig. CO27 18
Sunnymede Av. Cars. BT59 83
Sunnymede Av. Epsom BO58 76
Sunnymede Dr. Ilf. CL32 26
Sunny Nook Gdns. Sth. Croy BZ57 78
Selsdon Rd.
Sunnyside SW19 BR50 58
Sunny Side Walt. BD53 65
Sunnyside Dr. E4 CF26 9
Sunnyside Pass. SW19 BR50 58
Sunnyside Rd. E10 CE33 25
Sunnyside Rd. N19 BW33 23
Sunnyside Rd. NW2 BR34 22
Sunnyside Rd. W5 BK40 39
Sunnyside Rd. Ilf. CM34 26
Sunnyside Rd. Tedd. BG49 56
Sunnyside Rd. E. N9 CB27 15
Sunnyside Rd. N. N9 CB27 15
Sunnyside Rd. S. N9 CA27 15
Sunny, The, Rd. Enf. CC23 9
Sunny Vw. NW9 BN32 22
Sunny Way N12 BU29 14
Sunray Av. SE24 BZ45 51
Sunray Av. Brom. CK53 71
Sunray Av. Surb. BM55 75
Sunrise Av. Horn. CV34 28
Sunrise Clo. Felt. BE48 56
Sun St. Pass. EC2 CA39 42
Sunset Av. E4 CE26 9
Sunset Av. Wdf. Grn. CG28 16
Sunset Dr. Hav. CU28 19
Sunset Gdns. SE25 CA51 69
Sunset Rd. SE5 BZ45 51
Sunset Vw. Barn. BR23 6
Sunshine Way, Mitch. BU51 68
Sun St. EC2 BZ39 42
Sun Rd. W14 BR42 40
Sunwell St. SE15 CB44 51
Surbiton Ct. Surb. BK53 66
Surbiton Cres. Surb. BL53 66
Surbiton Hall Clo. Kings. On T. BL52 66
Surbiton Hill Pk. Surb. 'BL53 66
Surbiton Hill Rd. Surb. BL52 66
Surbition Pk. Ter. Kings. On T. BL52 66
Surbiton Rd. Kings. On T. BK52 66
Surgeon St. SE18 CL41 44
Surrendale Pl. W9 BS38 32
Surrey Canal Rd. SE15 CC43 52
Surrey Cres. W4 BM42 39
Chiswick High Rd.
Surrey Dr. Horn. CW31 28
Sussex Gdns. N6 BU32 23
Surrey Gdns. W4 BM42 39
Chiswick High Rd.
Surrey Gro. SE17 CA42 42
Surrey La. SW11 BU44 50
Surrey Ms. SE27 CA49 60
Hamilton Rd.
Surrey Mt. SE23 CB47 60
Surrey Rd. SE15 CC46 52
Surrey Rd. Bark. CN36 36
Surrey Rd. Dag. CR35 36
Surrey Rd. Har. BG32 20
Surrey Rd. W. Wick. CE54 70
Surrey Row, SE1 BY41 42
Surrey Sq. SE17 CA42 42
Surrey St. E13 CH38 35
Surrey St. WC2 BX40 42
Temple Pl.
Surrey St. Croy. BZ55 78
Surrey Ter. SE17 CA42 42
Surrey Sq.
Surridge Clo. Rain. CV38 37
Surridge Gdns. SE19 BZ50 60
Surr St. N7 BX35 33
Surry Gro. Sutt. BT55 77
Susan Clo. Rom. CR31 27
Susannah St. E14 CE39 43
Susan Rd. SE3 CH44 53
Susan Wood Chis. CL51 71
Susex Av. Islw. BH45 48
Sussex Av. Rom. CW29 19
Sussex Clo. Ilf. CK32 26
Radnor Cres.
Sussex Clo. N. Mal. BO52 67
Sussex Clo. Twick. BJ46 48
Westmorland Clo.
Sussex Cres. Nthlt. BF36 29
Sussex Gdns. N4 BZ32 24
Rosebery Gdns.
Sussex Gdns. W2 BT39 41
Sussex Gdns. Chess. BK57 75
Sussex Ms. NW1 BU38 32
Sussex Place
Sussex Ms. E. W2 BT40 41
Clifton Pl.
Sussex Pla. Erith CR43 54
Sussex Pl. W2 BT39 41
Sussex Pl. W6 BQ42 40
Sussex Rd. E6 CL37 35
Sussex Rd. Cars. BU57 77
Sussex Rd. Erith CR43 54
Sussex Rd. Har. BF32 20
Sussex Rd. N. Mal. BO52 67
Sussex Rd. Orp. CP53 72
Sussex Rd. Sid. CO49 63
Sussex Rd. Sth. Croy. BZ57 78
Sussex Rd. Sthl. BD41 38
Sussex Rd. W. Wick. CE54 70
Sussex Sq. W2 BT40 41
Sussex St. E13 CH38 35
Sussex St. SW1 BV42 41
Sussex Way N7 BW33 23
Sussex Way N19 BW33 23
Sussex Way, Barn. BV25 7
Sutcliffe Clo. NW11 BS32 23
Sutcliffe Clo. Bush. BG24 4
Sutcliffe Rd. SE18 CN43 54
Sutcliffe Rd. Well. CP44 54
Sutherland Av. W9 BS38 32
Sutherland Av. W13 BJ39 39
Sutherland Av. Hayes BC42 38
Sutherland Av. Orp. CN53 72
Sutherland Av. Well. CN45 54
Sutherland Clo. Barn. BR24 6
Sutherland Ct. NW9 BM32 21
Sutherland Dr. SW19 BT51 68
Brangwyn Cres.
Sutherland Gdns. SW14 BO45 49
Sutherland Gdns. Wor. Pk. BP54 67
Sutherland Gro. SW18 BR46 49
Sutherland Gro. Tedd. BH49 57
Sutherland Pl. W2 BS39 41
Sutherland Point. N16 CB34 24
Sutherland Rd. E3 CD37 34
Sutherland Rd. E17 CC30 16
Sutherland Rd. N17 CB29 15
Sutherland Rd. W4 BO43 49
Sutherland Rd. W13 BJ39 39
Sutherland Rd. Belv. CR41 45
Sutherland Rd. Croy. BY54 69
Sutherland Rd. Enf. CC25 9
Sutherland Rd. Sthl. BE39 38
Sutherland Rd. Path. E17 CC31 25
Sutherland Row SW1 BV42 41
Sutherland St.
Sutherlands Av. Sun. BC51 65
Sutherland Sq. SE17 BZ42 42
Sutherland St. E3 CD37 34
Sutherland St. SW1 BV42 41
Sutherland Wk. SE17 BZ42 42
Sutlej Rd. SE7 CJ43 53
Sutterton St. N7 BV36 32
Blundell St.
Sutton Clo. Loug. CK26 10
Sutton Clo. Pnr. BC32 20
Sutton Common Rd. Sutt. BR54 67
Sutton Ct. W4 BN43 49
Sutton Ct. Rd. E13 CJ38 35
Sutton Ct. Rd. W4 BN43 49
Sutton Ct. Rd. Sutt. BT57 77
Sutton Cres. Barn. BQ25 6
Sutton Dene Houns. BF44 47
Sutton Dws. N1 BY36 33
Sutton Dws. SE8 CC42 43
Sutton Dws. SW3 BU42 41
Sutton Dws. W10 BQ39 40
Sutton Gdns. Croy. CA53 69
Sutton Grn. Bark. CN37 36
Saxham Rd.
Sutton Gro. Sutt. BT56 77
Sutton Hall Rd. Houns. BF43 47
Sutton La. W4 BN42 40
Sutton La. Houns. BE45 47
Sutton La. Sutt. BS59 83
Sutton La. S. W4 BN43 49
Sutton Pk. Rd. Sutt. BS57 77
Sutton Path Borwd. BM23 5
Sutton Pl. E9 CC35 34
Sutton Rd. E13 CG38 34
Sutton Rd. E17 CC30 16
Sutton Rd. N10 BV30 14
Sutton Rd. Bark. CN37 36
Sutton Rd. Houns. BF44 47
Sutton Rd. Wat. BD24 4
Sutton Row, W1 BW39 41
Suttons Av. Horn. CV34 28
Suttons Gdns. Horn. CV34 28
Suttons La. Horn. CV35 37
Suttons Parkway Upmin. CW35 37
Sutton Sq. Houns. BE44 47
Sutton St. E1 CC40 43
Sutton Way W10 BQ38 31
Sutton Way Houns. BE44 47
Swaby Rd. SW18 BT47 59
Swaffham Way N22 BY29 15
Swaffield Rd. SW18 BS47 59
Swain Rd. Th. Hth. BZ53 69
Swain's La. N6 BV34 23
Swainson Rd. W3 BO41 40
Swain's Rd. SW17 BU50 59
Swaisland Rd. Dart. CU46 55
Swaislands Dr. Dart. CT46 55
Crayford Way
Swaislands Dr. Dart. CT46 55
Crayford Way
Swale Rd. Dart. CU45 55
Swallands Rd. SE6 CE48 61
Swallow Clo. SE14 CC44 52
Swallow Clo. Bush. BF26 4
Swallow Dr. Grnf. BE37 29
Hazelmere Rd.
Swallowdale Sth. Croy. CC58 79
Swallowfield Rd. SE7 CH42 44
Swallow St. W1 BW40 41
Piccadilly
Swallow Wk. Rain. CU36 37
Swanage Rd. E4 CF29 16
Swanage Rd. SW18 BT46 50
Swanage Waye, Hayes BD39 38
Swanbourne Dr. Horn. CV35 37
Swanbourne Dr. Horn. CV35 37
Swanbridge Rd. Bexh. CR44 54
Swan Clo. Felt. BE49 56
Swan Clo. Orp. CO52 72
Swan Ct. N20 BT27 14
Swan Ct. SW3 BU42 41
Swanfield St. E2 CA38 33
Swan La. EC4 BZ40 42
Wharfside
Swan La. N20 BT27 14
Swan La. Dart. CT47 64
Swan Pl. SW13 BO44 49
Swanley By-pass Swan. CS52 73
Swanley La. Swan. CT52 73
Swanley Rd. Well. CN44 54
Swanley Village Rd. Swan. CU51 73
SWan Mead, SE1 CA41 42
Swan Ms. SW9 BX44 51
Stockwell Rd.
Swan Rd. SE7 CJ41 44
Swan Rd. SE16 CC41 43
Swan Rd. Felt. BE49 56
Swan Rd. Sthl. BF39 38
Swanscombe Rd. W4 BO42 40
Swanscombe Rd. W11 BQ40 40
Swansea Rd. Enf. CC24 9
Swansland Gdns. E17 CD30 16
Mcentee Av.
Swanston Path. Wat. BD27 11
Swanston Path. Wat. BD27 11
Swan St. SE1 BZ41 42
SWan St. Islw. BJ45 48
Swanton Gdns. SW19 BQ47 58
Swanton Rd. Erith CR43 54
Swan Wk. SW3 BU43 50
Swan Wk. Rom. CT32 28
Swan Way, Enf. CC23 9
Swan Wf. EC4 BZ40 42
Wharfside
Swanwick Clo. SW15 BO47 58
Swan Yd. N1 BY36 33
Highbury Stn. Rd.
Sward Rd. Orp. CO53 72
Swaton Rd. E3 CE38 34
Swaylands Rd. Belv. CR43 54
Swedenborg Gdns. E1 CB40 42
Sweeney Cres. SE1 CA41 42
Sweeps La. Orp. CP53 72
Sweet Briar Grn. N9 CA27 15
Briary La.
Sweet Briar Gro. N9 CA27 15
Sweetbriar La. Epsom BN60 82
Sweet Briar Wk. N18 CA28 15

Sweetenham Wk. SE18	CM42	44
Sandbach Pl.		
Sweetmans Av. Pnr.	BD31	20
Sweets Way N20	BT27	14
Swete St. E13	CH37	35
Sweyn Pl. SE3	CH44	53
Swift Clo. Har.	BF34	20
Swift Rd. Felt.	BD49	56
Swiftsden Way, Brom.	CG50	61
Swift St. SW6	BR44	49
Swift Rd. Sthl.	BF41	38
Swinborn Ct. SE5	BZ45	51
Basingdon Way		
Swinbrook Rd. W10	BR39	40
Swinburne Cres. Croy.	CC53	70
Swinburne Rd. SW15	BP45	49
Swinderby Rd. Wem.	BL36	30
Swindon Clo. Ilf.	CN34	27
Salisbury Rd.		
Swindon Clo. Rom.	CW28	19
Swindon Gdns. Rom.	CW28	19
Swindon La. Rom.	CW28	19
Swindon St. W12	BP40	40
Swinfield Clo. Felt.	BE48	56
Swinford Gdns. SW9	BY45	51
Swingate La. SE18	CN43	54
Swinnerton St. E9	CD35	34
Swinton Clo. Wem.	BM33	21
Swinton Pl. WC1	BX38	33
Swinton St.		
Swinton St. WC1	BX38	33
Swires Shaw Kes.	CJ57	80
Swithland Gdns. SE9	CK49	69
Swyncombe Av. W5	BJ42	39
Sybourn St. E17	CD33	25
Sycamore Av. W5	BK41	39
Sycamore Av. Sid.	CN46	54
Sycamore Clo. Bush.	BE23	4
Sycamore Clo. Nthlt.	BE37	29
Sycamore Dr. Swan.	CT52	73
Sycamore Gdns. Mitch.	BT51	68
Bank Av.		
Sycamore Gdns. W6	BP41	40
Sycamore Gro. NW9	BN33	22
Sycamore Gro. N. Mal.	BN52	67
Sycamore Gro. SE20	CB51	69
Sycamore Rd. SW19	BQ50	58
Sycamore Rd. Dart.	CV47	64
Sycamore Wk. Ilf.	CM31	26
Civic Way		
Sycamore Way, Th. Hth.	BY53	69
Sydenham Av. SE26	CB49	60
Sydenham Hill. SE23	CB47	60
Sydenham Hill, SE26	CB47	60
Sydenham Pk. SE26	CC49	61
Sydenham Pk. Rd. SE26	CC48	61
Sydenham Rise SE23	CB48	60
Sydenham Rd. SE26	CC49	61
Sydenham Rd. Croy.	BZ54	69
Sydmons St. SE23	CC47	61
Sydner Rd. N16	CA35	33
Sydney Av. Pur.	BX59	84
Sydney Cl. SW3	BT42	41
Sydney Gro. NW4	BQ32	22
Sydney Ms. SW3	BT42	41
Sydney Ms. Clo. SW7	BT42	41
Sydney Pl. SW7	BU42	41
Sydney Rd. E11	CH32	26
Sydney Rd. N8	BY31	24
Sydney Rd. N10	BV30	14
Sydney Rd. SE2	CP41	45
Sydney Rd. SW20	BQ51	67
Sydney Rd. W13	BJ40	39
Sydney Rd. Bexh.	CP45	54
Sydney Rd. Enf.	BZ24	8
Sydney Rd. Felt.	BC47	56
Sydney Rd. Ilf.	CM30	17
Sydney Rd. Rich.	BL45	48
Sydney Rd. Sid.	CN49	63
Sydney Rd. Tedd.	BH49	57
Sydney Rd. Wdf. Grn.	CH28	17
Sydney Sq. SE15	CB43	51
Latona Rd.		
Sydney St. SW3	BU42	41
Sylvan Av. N3	BS30	14
Sylvan Av. N22	BX29	15
Sylvan Av. NW7	BO29	13
Sylvan Av. Horn.	CW32	28
Sylvan Av. Rom.	CQ32	27
Sylvan Clo. Sth. Croy.	CB58	78
Sylvan Gdns. Surb.	BK54	66
Sylvan Gro. SE15	CB43	51
Sylvan Hl. SE19	CA51	69
Sylvan Rd. E7	CH36	35
Sylvan Rd. E11	CH32	26
Sylvan Rd. E17	CE32	25
Sylvan Rd. SE19	CA51	69
Sylvan Rd. Ilf.	CM34	26
Sylvan Wy. Chig.	CO27	18
Sylvan Way. Dag.	CO34	27
Sylvan Way W. Wick.	CG56	79
Sylverdale Rd. Croy.	BY55	78
Sylverdale Rd. Ken.	BY60	84
Sylvester Av. Chis.	CK50	62
Syvester Gdns. Ilf.	CO28	18
Sylvester Rd. E8	CB36	33
Sylvester Rd. E17	CD33	25
Sylvester Rd. Wem.	BK35	30

Sylvia Av. Pnr.	BE29	11
Sylvia Ct. Wem.	BM36	30
Sylvia Gdns. Wem.	BM36	30
Symes Ms. NW1	BW37	32
Symons St. SW3	BU42	41
Syon La. Islw.	BH43	48
Syon Pk. Gdns. Islw.	BH43	48
Syracuse Av. Rain.	CV38	37
Tabard Garden Est. SE1	BZ41	42
Tabard St. SE1	BZ41	42
Tabarin Way, Epsom	BQ61	82
Tabernacle Av. E13	CH38	35
Tabernacle St. EC2	BZ38	33
Tableer Av. SW4	BW46	50
Tabley Rd. N7	BX35	33
Tabor Gdns. Sutt.	BR57	76
Tabor Gro. SW19	BR50	58
Tabor Rd. W6	BP41	40
Tachbrook Est. SW1	BW42	41
Tachbrook Rd. Sthl.	BD41	38
Tachbrook Rd. St. SW1	BW42	41
Tackbrook Ms. SW1	BW42	41
Longmore St.		
Tadema Rd. SW10	BT43	50
Tadmor St. W12	BQ40	40
Tadworth Av. N. Mal.	BO53	67
Tadworth Pde. Horn.	CU35	37
Tadworth Rd. NW2	BP34	22
Taffy's How, Mitch.	BU51	68
Tait Rd. Croy.	CA54	69
Takeley Clo. Rom.	CS30	19
Talacre Rd. NW5	BV36	32
Talbot Av. N2	BT31	23
Talbot Av. Wat.	BE26	4
Talbot Clo. N15	CA31	24
Talbot Rd.		
Talbot Ct. EC3	BZ40	42
Gracechurch St.		
Talbot Cres. NW4	BP32	22
Talbot Gdns. Ilf.	CO34	27
Talbot Houses SW3	CG44	52
Talbot Pl. SE3	CG44	52
Talbot Rd. W13	BJ40	39
Talbot Rd. E6	CK37	35
Talbot Rd. E7	CH35	35
Talbot Rd. N6	BV32	23
Talbot Rd. N15	CA31	24
Talbot Rd. N22	BW30	14
Talbot Rd. W11	BR39	40
Talbot Rd. W13	BJ40	39
Talbot Rd. Brom.	CH52	71
Talbot Rd. Cars.	BV56	77
Talbot Rd. Dag.	CQ36	36
Talbot Rd. Har.	BH30	12
Talbot Rd. Islw.	BJ45	48
Talbot Rd. Sthl.	BE42	38
Talbot Rd. Th. Hth.	BZ52	69
Talbot Rd. Twick.	BH47	57
Talbot Rd. Th. Hth.	BZ52	69
Talbot Rd. Twick.	BH47	57
Talbot Rd. Wem.	BK36	30
Talbot Sq. W2	BT39	41
Talbot Yd. SE1	BZ40	42
Borough High St.		
Talfourd Pl. SE15	CA44	51
Talfourd Rd. SE15	CA44	51
Talgarth Rd. W6	BQ42	40
Talgarth Rd. W14	BQ42	40
Talisman Clo. SE26	CB49	60
Talisman Way, Epsom	BQ61	82
Talisman Way, Wem.	BL34	21
Tallack Clo. Har.	BH29	12
Tallack Rd. E10	CD33	25
Tall Elms Clo. Brom.	CG53	70
Tallis Gro. SE7	CH43	53
Tallis St. EC4	BY40	42
Tallis Way, Borwd.	BK23	5
Tally Ho. Cor. N12	BT28	14
Talma Gdns. Twick.	BH46	48
Talmage Clo. SE23	CC47	61
Talma Rd. SW2	BY45	51
Talwin St. E3	CE38	34
Tamarisk Sq. W12	BO40	40
Tamar Sq. Wdf. Grn.	CH29	17
Tamar St. SE7	CJ42	44
Tamar Way N17	CA31	24
Tamian Way, Houns.	BD45	47
Tamplin Ms. W9	BS38	32
Warlock Rd.		
Tamworth Av. Wdf. Grn.	CG29	16
Tamworth Gdns. Pnr.	BD30	11
Tamworth La. Mitch.	BV51	68
Tamworth Pk. Mitch.	BV52	68
Tamwroth Pl. Croy.	BZ55	78
Tamworth Rd. Croy.	BY55	78
Tamworth St. SW6	BS43	50
Tancred Rd. N4	BY32	24
Tandridge Dr. Orp.	CM54	71
Tandridge Gdns. Sth. Croy.	CA60	84
Tanfield Av. NW2	BO35	31
Tanfield Rd. Croy.	BZ56	78
Tangent Rd. Rom.	CV30	19
Tangier Rd. Rich.	BM45	48
Tangier Way, Tad.	BR62	82
Tangier Wd. Tad.	BR62	82
Tanglewood Clo. Croy.	CC55	79
Tanglewood Clo. Stan.	BH27	12

Tangley Gro. SW15	BO46	49
Tangmere Cres. Horn.	CU36	37
Tangmere Gdns. Nthlt.	BD37	29
Tangmere Way NW9	BO30	13
Tanhurst Wk. SE2	CP41	45
Alsike Rd.		
Tanhurst Walk, Belv.	CP41	45
Lanridge Rd.		
Tankerton Rd. Surb.	BL55	75
Tankerton St. WC1	BX38	33
Argyle Wk.		
Tankerville Rd. SW16	BW50	59
Tankridge Rd. NW2	BP34	22
Tanneries, The E1	CC38	34
Cephas Av.		
Tanners End La. N18	CA28	15
Tanner's Hl. SE8	CD44	52
Tanners La. Ilf.	CM31	26
Tanner La. Ilf.	CM31	26
Tanner St. SE1	CA41	42
Tanner St. Bark.	CM36	35
Tannery Clo. Dag.	CR34	27
Tannsfield Rd. SE26	CC49	61
Tanrides Rd. Orp.	CM54	71
Tanswell Est. SE1	BY41	42
Tanswell St. SE1	BY41	42
Tansy Clo. Rom.	CV29	19
Tantallon Rd. SW12	BV47	59
Tant Av. E16	CG39	43
Tantony Gro. Rom.	CP31	27
Tanza Rd. NW3	BU35	32
Tapestry Clo. Sutt.	BS57	77
Tapiar Way N17	CA31	24
Taplow St. N1	BZ37	33
Tappesfield Rd. SE15	CC45	52
Tapster St. Barn.	BR24	6
Tarbert Rd. SE22	CA46	51
Tarbert Wk. E1	CC39	43
Juniper St.		
Tariff Rd. N17	CB29	15
Tarleton Gdns. SE23	CB48	60
Tarling Clo. Sid.	CO48	63
Tarling Rd. E16	CG39	43
Tarling Rd. N2	BT30	14
Oak La.		
Tarling St. E1	CB39	42
Tarling Est. E1	CC39	43
Tarling Rd. N2	BT30	14
Tarnwood Pk. SE9	CK47	62
Tarnwood Pk. Est SE9	CK47	62
Tarrington Rd. SW16	BW49	59
Tarry La. SE8	CD42	43
Yeoman St.		
Tarver Rd. SE17	BZ42	42
Tarves Way SE10	CE43	52
Norman Rd.		
Tash Pl. N11	BV28	14
Woodland Rd.		
Tasker Rd. NW3	BU35	32
Tasmania Ter. N18	BZ29	15
Tasman Rd. SW9	BX45	51
Tasman Wk. E16	CJ39	44
Royal Rd.		
Tasso Rd. W6	BR43	49
Tatam Rd. NW10	BN36	31
Tate Rd. E16	CK40	44
Tate Rd. Sutt.	BS56	77
Tatnell Rd. SE23	CD46	52
Tattenham Cnr. Rd. Epsom	BO62	82
Tattenham Cres. Epsom	BP62	82
Tattenham Gro. Epsom	BP62	82
Tattenham Way, Tad.	BQ62	82
Tattersall Clo. SE9	CK46	53
Tatum St. SE17	BZ42	42
Taunton Av. SW20	BP51	67
Taunton Av. Houns.	BG44	47
Taonton Clo. Bexh.	CS44	55
Taunton Clo. Sutt.	BS54	68
Taunton Dr. Enf.	BY24	8
Slades Rise		
Taunton Pl. NW1	BU38	32
Taunton Rd. SE12	CG46	52
Taunton Rd. Grnfd.	BF37	29
Taunton Rd. Rom.	CV28	19
Taunton Way, Stan.	BL31	21
Taunton Dr. Enf.	BY24	8
Taverners Clo. W11	BR40	40
Addison Av.		
Taverner Sq. N5	BZ35	33
Taverners Way E4	CG26	9
Tavistock Av. Grnf.	BJ37	30
Tavistock Clo. Rom.	CV30	19
Tavistock Cres. W11	BR39	40
Tavistock Cres. Mitch.	BX52	69
Tavistock Gdns. Ilf.	CN35	36
Tavistock Gro. Croy.	BZ54	69
Tavistock Pl. N14	BV25	7
Tavistock Pl. WC1	BW38	32
Tavistock Pl. E18	CH31	26
Tavistock Rd. E7	CG35	34
Tavistock Rd. E15	CG36	34
Tavistock Rd. E18	CH31	26
Tavistock Rd. N4	BZ32	24
Tavistock Rd. NW10	BO37	31
Tavistock Rd. W11	BR39	40
Tavistock Rd. Brom.	CG52	70
Tavistock Rd. Cars.	BT54	68

Tavistock Rd. Croy.	BZ54	69
Tavistock Rd. Edg.	BL30	12
Tavistock Rd. Wat.	BD23	4
Tavistock Rd. Well.	CP44	54
Tavistock Sq. WC1	BW38	32
Tavistock St. WC2	BX40	42
Tavistock Ter. N19	BW34	23
Tavistock Wk. Cars.	BT54	68
Taviton St. WC1	BW38	32
Tavy Br. SE2	CP41	45
Tawney Rd. SE28	CO40	45
Tawny Way, SE16	CC42	43
Tayben Av. Twick.	BH46	48
Taybridge Rd. SW11	BV45	50
Taylor Av. Rich.	BM44	48
Taylor Clo. N17	CB29	15
Northumberland Pk.		
Taylor Clo. Hamptn.	BG49	56
Taylor Clo. Rom.	CR29	18
Taylor Rd. Mitch.	BU50	59
Taylor Rd. Wall.	BV56	77
Taylors Bldgs. SE18	CL42	44
Spray St.		
Taylors Clo. DA15	CN48	63
Taylors Ct. E15	CF35	34
Taylors Grn. W3	BO39	40
Long Dr.		
Taylors La. NW10	BO36	31
Taylors La. SE26	CB49	60
Taylors La. Barn.	BR23	6
Taylor St. SE18	CL42	44
Taymount Grange, SE26	CC48	61
Taymount Ri. SE23	CC48	61
Tayport Cl. N1	BX36	33
Tayport Clo. N1	BX36	33
Gifford St.		
Tay Way Rom.	CT30	19
Taywood Rd. Nthlt.	BE38	29
Teale St. E2	CB37	33
Tealing Dr. Epsom	BN56	76
Teasel Way E15	CG38	34
Memorial Av.		
Teather St. SE5	CA43	51
Southampton Way		
Tebworth Rd. N17	CA29	15
Church Rd.		
Tedder Clo. Chess.	BK56	75
Mansfield		
Tedder Clo. Ruis.	BC35	29
Tedder Rd. Sth. Croy.	CC57	79
Teddington Clo. Epsom	BN58	76
Teddington Pk. Tedd.	BH49	57
Teddington Pk. Rd. Tedd.	BH49	57
Tedworth Gdns. SW3	BU42	41
Tedworth Sq. SW3	BU42	41
Tess Av. Grnf.	BH37	30
Teesdale Av. Islw.	BJ44	48
Teesdale Est. E2	CB38	33
Teesdale Gdns. Islw.	BJ44	48
Teesdale Rd. E11	CG32	25
Teesdale St. E2	CB37	33
Tees Rd. Rom.	CV27	19
Tee, The, W3	BO39	40
Teevan Clo. Croy.	CB54	69
Teevan Rd. Croy.	CB54	69
Teignmouth Clo. SW4	BW45	50
Teignmouth Clo. Edg.	BL30	12
Teignmouth Gdns. Grnf.	BJ37	30
Teignmouth Rd. NW2	BQ35	31
Teignmouth Rd. Well.	CP44	54
Telcote Way, Ruis.	BD33	20
Telegraph Hill NW3	BS34	23
Telegraph La. Esher	BJ57	75
Telegraph Ms. Ilf.	CO33	27
Eastwood Rd.		
Telegraph Rd. SW15	BP46	49
Telegraph St. EC2	BZ39	42
Telemann Sq. SE3	CH45	53
Telephone Pl. SW6	BR43	49
Telferscot Rd. SW12	BW47	59
Telford Av. SW2	BW47	59
Telford Clo. SE19	CA50	60
Telford Rd. Houns.	BF47	56
Telford Rd. N11	BW29	14
Telford Rd. NW9	BO33	22
Broadway, The,		
Telford Rd. SE9	CM48	62
Telford Rd. W10	BR39	40
Telford Rd. Sthl.	BF40	38
Telford Ter. SW1	BW43	50
Churchill Gdns. Rd.		
Telford Way, W3	BO39	40
Telham Rd. E6	CL37	35
Tell Gro. SE22	CA45	51
Tellisford, Esher	BF56	74
Tellson Av. SE18	CJ44	53
Telscombe Clo. Orp.	CN55	81
Temme Av. E15	CG36	34
Temperley Rd. SW12	BV47	59
Tempest Way, Rain.	CU36	37
Templar Ho. NW2	BR36	31
Templar Pl. Hamptn.	BF50	56
Templars Av. NW11	BR32	22
Templars Cres. N3	BS30	14
Templar St. SE5	BY44	51
Temple Av. EC4	BY40	42
Temple Av. N20	BT26	7

Temple Av. Croy. CD55 79
Temple Av. Dag. CR33 27
Temple Clo. N3 BR30 13
Templecombe Rd. E9 CC37 34
Templecombe Way Mord. BR53 67
Temple Fortune Hl. NW11 BS32 23
Temple Fortune la. NW11 BR32 22
Temple Gdns. NW11 BR32 22
Temple Gdns. Dag. CP34 27
Bennetts Cas. La.
Temple Gro. NW11 BS32 23
Temple Gro. Enf. BY24 8
Templehof Av. NW4 BQ33 22
Temple Hill Dart. CW46 55
Temple La. EC4 BY39 42
Templeman Clo. Pur. BY57 78
Templeman Clo. Pur. BY61 84
Templeman Rd. W7 BH39 39
Templemead Clo. W3 BO39 40
Carlisle Av.
Temple Mead Clo. Stan. BJ29 12
Temple Mill Rd. E15 CE35 34
Temple Mills La. E15 CE35 34
Temple Pl. WC2 BX40 42
Temple Rd. E6 CK37 35
Temple Rd. N8 BX31 24
Temple Rd. NW2 BQ35 31
Temple Rd. W4 BN41 40
Temple Rd. W5 BK41 39
Temple Rd. Croy. BZ56 78
Temple Rd. Epsom BN59 82
Temple Rd. Houns. BF45 47
Temple Rd. Rich. BL45 48
Temple Sheen SW14 BN45 49
Temple Sheen Rd. SW14 BM45 48
Temple St. E2 CB37 33
Temple, The SW1 BW40 41
Templeton Av. E4 CE27 16
Templeton Clo. SE19 BZ51 69
Templeton Pl. SW5 BS42 41
Templeton Rd. N15 BZ32 24
Temple Way Sutt. BT55 77
Templewood W13 BJ39 39
Templewood Av. NW3 BS34 23
Templewood Gdns. NW3 BS34 23
Tempsford Av. Borwd. BN24 6
Temsford Clo. Har. BG30 11
Tenbury Clo. E7 CJ35 35
Romford Rd.
Tenby Av. Har. BJ30 12
Tenby Clo. N15 CA31 24
Tenby Clo. Rom. CQ32 27
Tenby Gdns. Nthlt. BF36 29
Tenby Rd. E17 CD32 25
Tenby Rd. Edg. BL30 12
Tenby Rd. Enf. CC24 9
Tenby Rd. Rom. CQ32 27
Tenby Rd. Well. CP44 54
Tench St. E1 CB40 42
Tenda Rd. SE16 CB42 42
Tendring Way, Rom. CP32 27
Tenham Av. SW2 BW48 59
Tenham Ter. SW3 BU42 41
Tenison Ct. W1 BW40 41
Kingly St.
Tenison Way, SE1 BX40 42
Tenniel Ct. W2 CJ40 44
Porchester Gdns.
Tennison Av. Borwd. BM25 5
Tennison Rd. SE25 CA52 69
Tennis St. SE1 BZ41 42
Tenniswood Rd. Enf. CA23 8
Tennyson Av. E11 CH33 26
Tennyson Av. E12 CK36 35
Tennyson Av. NW9 BN31 22
Tennyson Av. N. Mal. BP53 67
Tennyson Av. Twick. BH47 57
Tennyson Clo. DA16 CN44 54
Keats Rd.
Tennyson Clo. Felt. BC46 47
Tennyson Ct. Rich. BK49 57
Tennyson Rd. E10 CE34 25
Tennyson Rd. E15 CG36 34
Tennyson Rd. E17 CD32 25
Tennyson Rd. NW6 BR37 31
Tennyson Rd. NW7 BP28 13
Tennyson Rd. SE20 CC50 61
Tennyson Rd. SW19 BT50 59
Tennyson Rd. W7 BH40 39
Tennyson Rd. Houns. BG44 47
Tennyson Rd. Rom. CU29 19
Tennyson Rd. Well. CN44 54
Shelley Dri.
Tennyson St. SW8 BV44 50
Tennyson Way, Horn. CU34 28
Tenplars Dr. Har. BG29 11
Tensing Rd. Sthl. BF41 38
Ten St. EC2 BZ39 42
Tentelow La. Sthl. BF42 38
Tenterden Clo. NW4 BQ31 22
Tenterden Dr. NW4 BQ31 22
Tenterden Gdns. NW4 BQ31 22
Tenterden Gdns. Croy. CB54 69
Tenterden Gro. NW4 BQ31 22
Tenterden Rd. N17 CA29 15
Tenterden Rd. Croy. CB54 69
Tenterden Rd. Dag. CQ34 27
Tenterden St. W1 BV39 41
Tenter Gro. E1 CA39 42
Brune St.
Tenter St. EC2 BZ39 42
Moor La.
Tent St. E1 CB38 33
Terborch Way, SE22 CA46 51
Dulwich Gro.
Tercel Path Chig. CO28 18
Terling Clo. E11 CG34 25
Terling Rd. Dag. CR34 27
Terminus Pl. SW1 BV41 41
Terrace Gdns. SW13 BO44 49
Terrace Gdns. Wat. BC23 4
Terrace Hill, Croy. BY55 78
Terrace La. Rich. BL46 48
Friars Stile Rd.
Terrace Rd. E9 CC36 34
Terrace Rd. E13 CH37 35
Terrace Rd. Walt. BC54 64
Terrace, The, N3 BR30 13
Hendon La.
Terrace, The, NW6 BS37 32
Terace, The, NW6 BS37 32
Terrace, The, SW13 BO44 49
Terrace, The, W14 BR42 40
Terrace, The, Kings. on T. BK51 66
Church Gro.
Terrace, The, Rich. BL46 48
Terrace Wk. Dag. CQ35 36
Terrapin Rd. SW17 BV48 59
Terrick Rd. N22 BX30 15
Terrick St. W12 BP39 40
Terrilands, Pnr. BE31 20
Terront Rd. N15 BZ32 24
Testerton St. W11 BQ40 40
Tetcott Rd. SW10 BT43 50
Kings Rd.
Tetherdown, N10 BV30 14
Tetterby Way SE16 CB42 42
Bonamy Estate West, The
Teversham La. SW8 BX44 51
Teviot Clo. Well. CO44 54
Stuart Rd.
Teviot St. E14 CF38 34
Tewkesbury Av. SE23 CB47 60
Tewkesbury Av. Pnr. BE32 20
Tewkesbury Clo. N15 BZ32 24
Tewkesbury Rd.
Tewkesbury Gdns. NW9 BM31 21
Tewkesbury Rd. N15 BZ32 24
Tewkesbury Rd. Cars. BT54 68
Tewkesbury Ter. N11 BW29 14
Tewson Rd. SE18 CN42 45
Teynham Av. Enf. BZ25 8
Teynahm Grn. Brom. CH53 71
Teynton Ter. N17 BZ30 15
Thackeray Av. N17 CB30 15
Thackeray Clo. SW19 BQ50 58
Thackeray Dr. Rom. CO33 27
Thackeray Rd. E6 CJ37 35
Thackeray Rd. SW8 BV44 50
Thackeray St. W8 BS41 41
Thackery Clo. Har. BF33 20
Thakeham Clo. SE26 CB49 60
Thalia Clo. SE10 CF43 52
Feathers Pl.
Thames Av. Dag. CS38 37
Thames Av. Grnf. BH37 30
Thames Bank SW14 BN44 49
Thames Clo. Hamptn. BF51 65
Thames Clo. Rain. CU39 46
Thamesgate Clo. Rich. BJ49 57
Thameshill Av. Rom. CS30 19
Thame Side Kings. On T. BK51 66
Thameside, Tedd. BK50 57
Thames Mead Walt. BC53 65
Thamesmead Spine Rd. DA18 CQ40 45
Thames Meadow E. Mol. BF51 65
Thames Pl. E14 CD40 43
Thames Prom. Twick. BJ45 48
Thames Rd. E16 CJ40 44
Thames Rd. W4 BM43 48
Thames Rd. Bark. CN38 36
Thames Rd. Dart. CT44 55
Thames Side. Tedd. BK50 57
Thames St. SE10 CE43 52
Thames St. Hamptn. BF51 65
Thames St. Kings. On T. BK51 66
Thames St. Sun. BC52 65
Thames View Est. Bark. CN37 36
Thames Village W4 BN44 49
Thanescroft Gdns. Croy. CA55 78
Thanet Pl. Croy. BZ56 78
Thanet Rd. Bex. CR47 63
Thanet Rd. Erith CT43 55
Thanet St. WC1 BX38 33
Thane Vill. N7 BX34 24
Thanington Ct. SE9 CN46 54
Tharp Rd. Wall. BW56 77
Thatcham Gdns. N20 BT26 7
Thatches Gro. Rom. CQ31 27
Thatchers Clo. Loug. CM23 10
Mannock Dr.
Thatchers Way Islw. BG46 47
Reapers Way
Thavies Inn, EC1 BY39 42
St. Andrew St.
Thaxted Rd. SE9 CM48 62
Thaxted Rd. Buck. H. CJ26 10
Thaxted Wk. Rain. CT37 37
Thaxton Rd. W14 BR43 49
Thayers Farm Rd. Beck. CD51 70
Thayer St. W1 BV39 41
Theatre St. SW11 BU45 50
Theberton St. N1 BY37 33
The Courtyard N1 BY36 33
Barnsbury Terr.
Theed St. SE1 BY40 42
Thelma Gdns. SE3 CK44 53
Thelma Gdns. Felt. BE48 56
Thelma Gro. Tedd. BJ50 57
Blackmores Gro.
Theobald Clo. Borwd. BL23 5
Theobald St.
Theobalds Ct. N16 BZ34 24
Kings Crescent Est.
Theobald Cres. Har. BF30 11
Theobald Rd. E17 CD33 25
Theobald Rd. Croy. BY55 78
Theobalds Av. N12 BT28 14
Theobalds Rd. WC1 BX39 42
Theobald St. SE1 BZ41 42
Theobald St. Borwd. BL23 5
Theodore Rd. SE13 CF46 52
Therapia La. Croy. BW54 68
Therapia La. Croy. BX53 69
Therapia Rd. SE22 CB46 51
Theresa Rd. W6 BP42 40
Theresa St. W6 BP42 40
Therfield Ct. N16 BZ34 24
Kings Crescent Est.
Thermopylae Gte. E14 CE42 43
The Sandlings N22 BY30 15
Theseus Walk N1 BY37 33
Nelson Pl.
Thesiger Rd. SE20 CC50 61
Thessaly Rd. SW8 BW43 50
Thetford Cl. N22 BY29 15
Thetford Clo. N15 CA32 24
Norfolk Av.
Thetford Gdns. Dag. CQ37 36
Thetford Rd. Dag. CP37 36
Thetford Rd. N. Mal. BN53 67
Thetford Rd. N. Mal. BO53 67
Theydon Gdns. Rain. CT36 37
Theydon Gro. Wdf. Grn. CJ29 17
Theydon Rd. E5 CC34 25
Theydon St. E17 CD33 25
Thicket Cres. Sutt. BT56 77
Thicket Gro. Dag. CP36 36
Thicket Rd. SE20 CB50 60
Thicket Rd. Sutt. BT56 77
Third Av. E12 CK35 35
Third Av. E13 CH38 35
Third Av. E17 CE32 25
Third Av. W3 BO40 40
Third Av. W10 BR38 31
Third Av. Dag. CR37 36
Third Av. Enf. CA25 8
Third Av. Rom. CP32 27
Third Av. Wem. BK34 21
Third Clo. E. Mol. BG52 65
Third Cross Rd. Twick. BG48 56
Third Way, Wem. BM35 30
Thirlby Rd. Edg. BN30 13
Thirleby Rd. SW1 BW41 41
Thirlmere Av. Grnf. BJ38 30
Thirlmere Gdns. Wem. BK33 21
Thirlmere Ri. Brom. CG50 61
Thirlmere Rd. N10 BV30 14
Thirlmere Rd. SW16 BW49 59
Thirlmere Rd. Bexh. CS44 55
Thirsk Clo. Nthlt. BF36 29
Thirsk Rd. SE25 BZ52 69
Thirsk Rd. SW11 BV45 50
Thirsk Rd. Mitch. BV50 59
Thirza Rd. Dart. CW46 55
Thistlecroft Gdns. Stan. BK30 12
Thistlecroft Rd. Walt. BD56 74
Thistledene E. Mol. BH53 66
Thistledene Av. Har. BE34 20
Thistleden Av. Rom. CR28 18
Thistle Gro. SW10 BT42 41
Thistlemead Chis. CL51 71
Thistle Mead Loug. CL24 10
Thistlewaite Rd. E5 CB34 24
Thistleworth Clo. Islw. BG43 47
Thomas A'becket Clo. Wem. BH35 30
Thomas Baines Rd. SW11 BT45 50
Thomas Derby Ct. W11 BR39 40
Thomas Doyle St. SE1 BY41 42
Thomas La. SE6 CE47 61
Thomas More St. E1 CB40 42
Thomas More Way N2 BT31 23
Thomas Rd. E14 CD39 43
Thomas Sims Ct. Rom. CV36 37
South End. Rd.
Thomas St. SE18 CL42 44
Thompson Avenue SE5 BZ43 51
Thompson Av. Rich. BM45 48
Thompson Rd. SE22 CA46 51
Thompson Rd. Dag. CQ34 27
Thomson Cres. Croy. BY54 69
Thomson Rd. Har. BH31 21
Thorburn Way SW19 BT51 68
Brangwyn Cres.
Thoresby St. N1 BZ38 33
Thorkhill Gdns. Surb. BJ54 66
Thorkhill Rd. Surb. BJ54 66
Thornaby Gdns. N18 CB29 15
Thorn Av. Bush. BG26 4
Thorn Bank, Edg. BM29 12
Thornbury Av. Islw. BG43 47
Thornbury Gdns. Borwd. BN24 6
Thornbury Rd. SW2 BX46 51
Thornbury Rd. Islw. BG43 47
Thornby Rd. E5 CC34 25
Thorncliffe Rd. SW2 BX46 51
Thorncliffe Rd. Sthl. BE42 38
Thorn Clo. Brom. CL53 71
Thorn Clo. Nthlt. BE38 29
Thorncombe Rd. SE22 CA46 51
Thorncroft Horn. CU32 28
Thorncroft Rd. Sutt. BS56 77
Thorncroft St. SW8 BX43 51
Thorndean St. SW18 BT48 59
Thorndene Av. N11 BV26 7
Thorndike Clo. SW10 BT43 50
Thorndon Clo. Orp. CN51 72
Thorndon Gdns. Epsom BO56 76
Thorndon Rd. Orp. CN51 72
Thorndyke Av. Nthlt. BD37 29
Thorne Clo. E11 CG35 34
Thorne Clo. E16 CG39 43
Thorne Clo. Erith CS43 55
Thorne Clo. N. Mal. BN52 67
Thorneloe Gdns. Croy. BY56 78
Thorne Pass. SW13 BO44 49
White Hart La.
Thorne Rd. SW8 BX43 51
Thorne Rd. N. Mal. BN52 67
Thornes Clo. Beck. CF52 70
Thorne St. E16 CG39 43
Thorne St. SW13 BO45 49
Thornet Wd. Rd. Brom. CL52 71
Thorney Hedge Rd. W4 BM42 39
Thorney St. SW1 BW42 41
Thornfield Av. NW7 BR30 13
Thornfield Rd. W12 BP41 40
Thornfield Rd. Bans. BS62 83
Thornford Rd. SE13 CF46 52
Thorngate Rd. W9 BS38 32
Thorngorve Rd. E13 CH37 35
Thornham Gro. E15 CF35 34
Thornham St. SE10 CE43 52
Thornhaugh St. WC1 BW38 32
Russell Sq.
Thornhill Av. SE18 CN43 54
Thornhill Av. Surb. BL55 75
Thornhill Cres. N1 BX36 33
Thornhill Gdns. E10 CE34 25
Thornhill Gdns. Bark. CN36 36
Thornhill Rd. E10 CE34 25
Thornhill Rd. N1 BY36 33
Thornhill Rd. Croy. BZ54 69
Thonrhill Rd. Surb. BL55 75
Thornhill Sq. N1 BX36 33
Thorn La. Rain. CV37 37
Thornlaw Rd. SE27 BY49 60
Thornley Clo. N17 CB29 15
Thornley Dr. Har. BF34 20
Thornley Pl. SE10 CG42 43
Caradoc St.
Thornsbeach Rd. SE6 CF47 61
Thornsett Pl. SE20 CB51 69
Thornsett Rd. SE20 CB51 69
Thornsett Rd. SW18 BS47 59
Thornton Av. SW2 BW47 59
Thornton Av. W4 BO42 40
Thornton Av. Croy. BX53 69
Thornton Ct. SW20 BQ53 67
Thornton Dene Beck. CE51 70
Thornton Gdns. SW12 BW47 59
Thonton Gro. Pnr. BF29 11
Thornton Hl. SW19 BR50 58
Thornton Pl. W1 BU39 41
Thornton Rd. E11 CF34 25
Thornton Rd. SW12 BW47 59
Thornton Rd. SW14 BN45 49
Thornton Rd. SW19 BQ50 58
Thornton Rd. Barn. BR24 6
Thornton Rd. Belv. CR42 45
Thornton Rd. Brom. CH49 62
Thornton Rd. Cars. BT54 68
Thonrton Rd. Cars. BU54 68
Thornton Rd. Croy. BX54 69
Thornton Rd. Ilf. CL35 35
Thornton Rd. E. SW19 BQ50 58
Thornton Rd.
Thornton Row Th. Hth. BY53 69
Thornton's Fm. Av. Rom. CS33 28
Thornton Way NW11 BS32 23
Thorntree Rd. SE7 CJ42 44
Thornville St. SE8 CF44 52
Thornwood Clo. E18 CH30 17
Thornwood Rd. SE13 CG46 52
Thorogood Gdns. E15 CG35 34

Thorogood Way, Rain.	CT37	37		

Thorogood Way, Rain. CT37 37
Thorold Clo. Sth. Croy. CC58 79
Thorold Rd. N22 BX29 15
Thorold Rd. Ilf. CL34 26
Thoparch Rd. SW8 BW44 50
Thorpebank Rd. W12 BP40 40
Thorpe Clo. Nthlt. BE36 29
Thorpe Clo. Orp. CN55 81
Thorpe Cres. E17 CD30 16
Thorpe Cres. Wat. BC26 4
Thorpedale Gdns. Ilf. CL31 26
Thorpedale Rd. N4 BX34 24
Thorpe Hall Ms. W5 BK39 39
Eaton Rise
Thorpe Lodge, Horn. CW32 28
Thorpehall Rd. E17 CF30 16
Thorpe Rd. E6 CK37 35
Thorpe Rd. E7 CG35 34
Thorpe Rd. E17 CF30 16
Thorpe Rd. N15 CA32 24
Thorpe Rd. Bark. CM36 35
Thorpe Rd. Kings. on T. BL50 57
Thorpe St. E1 CA39 42
Wentworth St.
Thorpewood Av. SE26 CB48 60
Thorsden Way SE19 CA49 60
Oaks Av.
Thorverton Rd. NW2 BR34 22
Thoydon Rd. E3 CD37 34
Thrale Rd. SW16 BW49 59
Thrale St. SE1 BZ40 42
Thrawl St. E1 CA39 42
Thraxed Pl. SW20 BQ50 58
Threadneedle St. EC2 BZ39 42
Three Colts La. E2 CB38 33
Three Colt St. E14 CD39 43
Three Corners, Bexh. CR44 54
Three Kings Rd. E16 BY39 42
Gough Sq.
Three Kings Rd. Mitch. BU52 68
Three Kings Yd. W1 BV40 41
Three Mill La. E3 CF38 34
Three Nun Ct. EC2 BZ39 42
Aldermanbury
Three Oak La. SE1 CA41 42
Threshers Pl. W11 BR40 40
Thriftwood SE23 CC48 61
Thrift Fm. Rd. Borwd. BN23 6
Thrigby Rd. Chess. BL57 75
Throckmorten Rd. E16 CH39 44
Throgmorton Av. EC2 BZ39 42
Throgmorton St. EC2 BZ39 42
Throwley Clo. SE2 CP41 45
Throwley Rd. Sutt. BS56 77
Throwley Way, Sutt. BS56 77
Thrupp Clo. Mitch. BV51 68
Thrupp's Av. Walt. BD56 74
Thrupp's La. Walt. BD56 74
Thrush St. SE17 BZ42 42
Thruxton Way SE15 CA43 51
Thurbarn Rd. SE6 CE49 61
Thurland Rd. SE16 CB41 42
Thurlby Clo. Har. BJ32 21
Gayton Rd.
Thurlby Clo. Wdf. Grn. CK28 17
Thurlby Rd. SE27 BY49 60
Thurlby Rd. Wem. BK36 30
Thurleigh Av. SW12 BV46 50
Thurleigh Rd. SW12 BU46 50
Thurleston Av. Mord. BR53 67
Thurlestone Av. N12 BU29 14
Thurloe Clo. SW7 BU42 41
Thurloe Ct. SW3 BU42 41
Fulham Rd.
Thurloe Gdns. Rom. CT32 28
Thurloe Pl. SW7 BT42 41
Thurloe Pl. Ms. SW7 BT42 41
Thurloe Pl.
Thurloe Sq. SW7 BU42 41
Thurloe St. SW7 BT42 41
Thurlow Gdns. Ilf. CM29 17
Thurlow Gdns. Wem. BK35 30
Thurlow Hl. SE21 BZ47 60
Thurlow Pk. Rd. SE21 BY47 60
Thurlow Rd. NW3 BT35 32
Thurlow Rd. W7 BJ41 39
Elthorne Pk. Rd.
Thurlow Rd. W13 BJ41 39
Park Rd.
Thurlow St. SE17 BZ42 42
Thurlow Ter. NW5 BU35 32
Thurlstone Av. Ilf. CN35 36
Thurlstone Rd. SE27 BY48 60
Thurlton Rd. Ruis. BC34 20
Thurnby Ct. Twick. BH48 57
Thursland Rd. Sid. CQ49 63
Thursley Cres. Croy. CF57 79
Thursley Gdns. SW19 BQ48 58
Thursley Rd. SE9 CK48 62
Thurso St. SW17 BT49 59
Thurston Rd. SE13 CE44 52
Thurston Rd. SW20 BP50 58
Thurston Rd. Sthl. BE39 38
Thurtle Rd. E2 CA37 33
Thwaite Clo. Erith CS43 55
Thyra Rd. N12 BS29 14
Thyer Clo. Orp. CM56 80
Tibbatts Rd. E3 CE38 34

Tibbenham Wk. E13 CG37 34
Tibberton Sq. N1 BZ37 33
Popham Rd.
Tibbet's Ride SW15 BQ47 58
Tibbetts Clo. SW19 BQ47 58
Ticehurst Rd. SE23 CD48 61
Tichmarsh, Epsom BN58 76
Tickford Clo. SE2 CP41 45
Tidal Basin Rd. E16 CG40 43
Tidenham Gdns. Croy. CA55 78
Tideway Clo. Rich. BJ49 57
Tideswell Rd. SW15 BQ45 49
Tideswell Rd. Croy. CE55 79
Tidey St. E3 CE39 43
Tidford Rd. Well. CN44 54
Tidsworth Rd. E3 CE38 34
Tidworth Rd. E3 CE38 34
Tiepigs La. W. Wick. CG55 79
Tierney Rd. SW2 BX47 60
Tiger Way E5 CB35 33
Tilbrook Rd. SE3 CJ45 53
Tilbury Clo. SE15 CA43 51
Willowbrook Rd.
Tilbury Clo. Orp. CO51 72
Tilbury Rd. E6 CK37 35
Tilbury Rd. E10 CF33 25
Tilbury Sq. Hayes BD39 38
Tildesley Rd. SW15 BQ46 49
Tile Fm. Rd. Orp. CM55 80
Tilehurst Rd. SW18 BT47 59
Tilehurst Rd. Sutt. BR56 76
Tile Kiln La. N6 BW33 23
Tile Kiln La. N13 BZ28 15
Tile Kiln La. Bex. CS48 64
Tileyard Rd. N7 BX36 33
Tilford Av. Croy. CF58 79
Tilford Gdns. SW19 BQ47 58
Tilia Rd. E5 CB35 33
Till Av. Farn. CW54 73
Tiller Rd. E14 CE41 43
Tillett Clo. NW10 BN36 31
Tillingbourne Gdns. N3 BR31 22
Tillingbourne Grn. Orp. CN52 72
Tillingbourne Way N3 BR31 22
Tillingbourne Gdns.
Tillingham Way N12 BS28 14
Tillotson Rd. N9 CA27 15
Tillotson Rd. Har. BF29 11
Tillotson Rd. Ilf. CL33 26
Tilmans Mead, Farn. CW54 73
Tilman St. E1 CB39 42
Tilney Ct. EC1 BZ38 33
Old St.
Tilney Dr. Buck. H. CH27 17
Tilney Gdns. N1 CA36 33
Baxter Rd.
Tilney Rd. Dag. CQ36 36
Tilney Rd. Sthl. BD42 38
Tilney St. W1 BV40 41
Tilson Gdns. SW2 BW47 59
Tilson Ho. SW2 BX47 60
Tilson Rd. N17 CB30 15
Tilton St. SW6 BR43 49
Tiltwood, The W3 BN40 40
Acacia Rd.
Tiltyard App. SE9 CK46 53
Timber Clo. Chis. CL51 71
Timbercroft, Epsom BN56 76
Timbercroft La. SE18 CN43 54
Timber Dene NW4 BQ30 13
Timber Slip Dr. Wall. BW58 77
Timber St. EC1 BZ38 33
Baltic St.
Timberwharf Rd. N16 CB32 24
Times Sq. Sutt. BS56 77
Throwley Way
Timothy Rd. E3 CD39 43
Timsbury Wk. SW15 BP47 58
Foxcombe Rd.
Tindal St. SW9 BY44 51
Tindale Clo. Sth. Croy. BZ59 84
Tinderbox Alley SW14 BN45 49
North Worple Way
Tine Rd. Chig. CN28 18
Tinsley Rd. E1 CC39 43
Tintagel Clo. Epsom BO60 82
Tintagel Cres. SE22 CA45 51
Tintagel Dr. Stan. BK28 12
Tintagel Gdns. SE22 CA45 51
Oxonian St.
Tintagel Rd. Orp. CP55 81
Tintern Av. NW9 BM31 21
Tintern Clo. SW15 BR46 49
Tintern Clo. SW19 BT50 59
Tintern Ct. W13 BJ40 39
Tintern Gdns. N14 BX26 8
Tintern Rd. N22 BZ30 15
Tintern Rd. Cars. BT54 68
Tintern St. SW4 BX45 51
Tintern Way, Har. BF33 20
Tinto Rd. E16 CH38 35
Tinworth St. SE11 BX42 42
Tippetts Clo. Enf. BZ23 8
Tipthorpe Rd. SW11 BV45 50
Tipton Dr. Croy. CA56 78
Tiptree Cres. Ilf. CL31 26
Tiptree Rd. Ruis. BC35 29
Tirlemont Rd. Sth. Croy. BZ57 78

Tirrell Rd. Croy. BZ53 69
Tisbury Rd. SW16 BX51 69
Tisdall Pl. SE17 BZ42 42
Titchborne Row W2 BU39 41
Hyde Park Cres.
Titchfield Rd. Cars. BT54 68
Titchfield Wk. Cars. BT54 68
Titchwell Rd. SW18 BT47 59
Tite St. SW3 BU42 41
Tithe Barn Way, Nthlt. BC37 29
Tithe Clo. NW7 BP30 13
Tithe Ct. NW4 BP30 13
Tithe Farm Av. Har. BF34 20
Tithe Farm Clo. Har. BF34 20
Tithepit Shaw La. Warl. CB62 84
Tithe Wk. NW7 BP30 13
Titian Av. Bush. BH26 5
Titley Clo. E4 CE28 16
Titmuss Av. SE28 CO40 45
Titmuss St. W12 BP41 40
Goldhawk Rd.
Tiverton Av. Ilf. CL31 26
Tiverton Dr. SE9 CM47 62
Tiverton Rd. N15 BZ32 24
Tiverton Rd. N18 CA28 15
Tiverton Rd. NW10 BQ37 31
Tiverton Rd. Edg. BL30 12
Tiverton Rd. Houns. BF44 47
Tiverton Rd. Ruis. BC34 20
Tiverton Rd. Wem. BL37 30
Tiverton St. SE1 BZ41 42
Tiverton Way, Chess. BK56 75
Tivoli Rd. N8 BW32 23
Tivoli Rd. SE27 BZ49 60
Tivoli Rd. Houns. BE45 47
Tivoli Way SE27 BZ49 60
Holderness Way
Tobin Clo. NW3 BU36 32
Todds Walk N7 BX34 24
Andover Est.
Tokenhouse Yd. EC2 BZ39 42
Tokyngton Av. Wem. BL36 30
Toland Sq. SW15 BP46 49
Tolcarne Dr. Pnr. BC30 11
Tolcarne Dr. Pnr. BC31 20
Toley Av. Wem. BL33 21
Tolhurst St. SE4 CD45 52
Foxwell St.
Tollesbury Gdns. Ilf. CM31 26
Tollet St. E1 CC38 34
Tollgate Dr. SE21 CA48 60
Tollgate Gdns. NW6 BS37 32
Tollgate Rd. E16 CJ39 44
Tollington Pk. N4 BX34 24
Tollington Pl. N4 BX34 24
Tollington Rd. N7 BX35 33
Tollington Way N7 BX34 24
Tolsford Rd. E5 CB35 33
Clarence Rd.
Tolson Rd. Islw. BJ45 48
Tolverne Rd. SW20 BQ51 67
Tolworth Clo. Surb. BM54 66
Tolworth Gdns. Rom. CP32 27
Tolworth Pk. Rd. Surb. BL55 75
Tolworth Rise N. Surb. BM54 66
Tolworth Rise S. Surb. BM54 66
Tolworth Rd. Surb. BL55 75
Tomahawk Gdns. Nthlt. BD38 29
Javelin Way
Tom Cribb Rd. SE18 CM41 44
Tomlins Gro. E3 CE38 34
Tomlinson Clo. E2 CA38 33
Tomlinson Clo. W4 BM42 39
Oxford Rd. N.
Tomlins Orchard, Bark. CM37 35
Tomlins Wk. N7 BX34 24
Andover Est.
Tom Mann Clo. Bark. CN37 36
Tomswood Hill, Ilf. CL29 17
Tomswood Rd. Chig. CL29 17
Tonbridge Clo. Bans. BU60 83
Tonbridge Rd. E. Mol. BE53 65
Tonbridge Rd. Har. BL31 21
Tonbridge Rd. WC1 BX38 33
Tonfield Rd. Sutt. BR54 67
Tonge Clo. Beck. CE53 70
Tonsley Hill SW18 BS46 50
Tonsley Pl. SW18 BS46 50
Tonsley Rd. SW18 BS46 50
Tonsley St. SW18 BS46 50
Tonstall Rd. Epsom BN59 82
Tonstall Rd. Mitch. BV51 68
Tooke Clo. Pnr. BE30 11
Took's Ct. EC4 BY39 42
Cursitor St.
Toolands Rd. Islw. BJ44 48
Tooley St. SE1 BZ40 42
Toorack Rd. Har. BG30 11
Tooting Bec Gdns. BW49 59
SW16
Tooting Bec Rd. SW17 BV48 59
Tooting Gro. SW17 BU49 59
Tooting High St. SW17 BU49 59
Toots Wd. Rd. Brom. CG53 70
Topcliffe Dr. Orp. CM56 80
Top Dartford Rd. Swan. CT50 64
Topham Sq. N17 BZ30 15

Topham St. EC1 BY38 33
Top Ho. Rise E4 CF26 9
Topiary Sq. Rich. BL45 48
Topley St. SE9 CJ45 53
Top Pk. Beck. CF53 70
Topsfield Rd. N8 BX32 24
Topsham Rd. SW17 BU48 59
Torbay Rd. NW6 BR36 31
Torbay Rd. Har. BE34 20
Torbridge Clo. Edg. BL29 12
Torbrook Clo. Bex. CQ46 54
Torcross Dr. SE23 CC48 61
Torcross Rd. Ruis. BC34 20
Tor Gdns. W8 BS41 41
Tormead Clo. Brom. CG50 61
Tormead Rd. Sutt. BS57 77
Tormount Rd. SE18 CN43 54
Toronto Av. E12 CK35 35
Toronto Rd. E11 CF35 34
Toronto Rd. Ilf. CL33 26
Torquay Gdns. Ilf. CJ31 26
Torquay St. W2 BS39 41
Harrow Rd.
Torrance Clo. Horn. CU33 28
Torrance Rd. SE3 CJ44 53
Torrens Rd. E15 CG36 34
Torrens Rd. SW2 BX46 51
Torrens Sq. E15 CG36 34
Torrens St. N1 BY37 33
Torre Wk. Cars. BU54 68
Torriano Av. NW5 BW35 32
Torriano Cotts. NW5 BW35 32
Torriano Av.
Torriano Est. NW1 BW35 32
Torridge Gdns. SE15 CC45 52
Torridge Rd. Th. Hth. BY53 69
Toridon Rd. SE6 CF47 61
Toridon Rd. SE13 CF47 61
Torrington Av. N12 BT28 14
Torrington Clo. N12 BT28 14
Torrington Pk.
Torrington Clo. Esher BH57 75
Torrington Dr. Har. BF35 29
Torrington Dr. Loug. CM24 10
Torrington Gdns. N11 BW29 14
Torrington Gdns. Grnf. BK37 30
Torrington Gdns. Loug. CM24 10
Torrington Gro. N12 BU28 14
Torrington Pk. N12 BT28 14
Torrington Pl. WC1 BW39 41
Torrington Rd. E18 CH31 26
Torrington Rd. Dag. CQ33 27
Torrington Rd. Esher BH57 75
Torrington Rd. Grnf. BK37 30
Torrington Way, Mord. BS54 68
Tor Rd. Well. CP45 54
Torwood Rd. SW15 BP46 49
Totham Lo. SW20 BP51 67
Richmond Rd.
Tothill St. SW1 BW41 41
Totnes Rd. Well. CO43 54
Totnes Wk. N2 BT31 23
Tottenhall Rd. N13 BY29 15
Tottenham Grn. E. N15 CA31 24
Tottenham Grn. W. N15 CA31 24
Town Hall App.
Tottenham La. N8 BX32 24
Tottenham Ms. W1 BW39 41
Tottenham St.
Tottenham Rd. N1 BZ36 33
Tottenham Rd. N1 CA36 33
Tottenham St. W1 BW39 41
Totterdge Com. N20 BP27 13
Totterdown St. SW17 BU49 59
Totteridge Clo. N20 BS27 14
Totteridge La. N20 BS27 14
Totternhoe Clo. Har. BK32 21
Totton Rd. Th. Hth. BY52 69
Totty St. E3 CD37 34
Toulmin St. SE1 BZ41 42
Toussaint Ms. Har. BW43 49
Tovil Clo. SE20 CB51 69
Towcester Rd. E3 CE38 34
Tower Bridge CA40 42
Tower Bridge App. E1 CA40 42
Tower Bridge Rd. SE1 CA41 42
Tower Clo. SE20 CB50 60
Tower Clo. Ilf. CM29 17
Tower Clo. Orp. CN55 81
Tower Ct. N16 CA33 24
Tower Gdns. Rd. N17 BZ30 15
Tower Hamlets Rd. E7 CG35 34
Tower Hamlets Rd. E17 CE31 25
Tower Hill EC3 CA40 42
Tower Ms. E17 CE31 25
High St.
Tower Pl. EC3 CA40 42
Tower Ri. Rich. BL45 48
Evelyn Ter.
Tower Rd. NW10 BP36 31
Tower Rd. Belv. CS42 46
Tower Rd. Bexh. CR45 54

Tower Rd. Dart.	CV47	64
Tower Rd. Orp.	CN55	81
Tower Rd. Twick.	BH48	57
Tower Royal EC4	BZ39	42
Cannon St.		
Towers Pl. Rich.	BL46	48
Eton St.		
Towers Rd. Pnr.	BE30	11
Towers Rd. Sthl.	BF38	29
Towers, The, Ken.	BZ61	84
Tower St. WC2	BW39	41
Tower Ter. N22	BX30	15
Mayes Rd.		
Tower Vw. Croy.	CD54	70
Towfield Rd. Felt.	BE48	56
Towncroft Cres. Orp.	CM53	71
Town Ct. La. Orp.	CM54	71
Town Court Path N4	BZ33	24
Towneymead, Nthlt.	BE37	29
Town Hall App. N15	CA31	24
Town Hall App. N16	CA35	33
Milton Bro.		
Town Hall Av. W4	BN42	40
Town Hall Rd. SW11	BU45	50
Townholm Cres. W7	BH41	39
Townley Ct. E15	CG36	34
Faraday Rd.		
Townley Rd. SE22	CA46	51
Townley Rd. Bexh.	CQ46	54
Townley St. SE17	BZ42	42
Townmeade Rd. Rich.	BM44	48
Townmead Est. SW6	BS45	50
Townmead Rd. SW6	BS45	50
Town Meadow, Brent.	BK43	48
Town Quay, Bark.	CL37	35
Town Rd. N9	CB27	15
Townsend Av. N14	BW28	14
Townsend La. NW9	BN33	22
Townsend Rd. N15	CA32	24
Townsend Rd. Sthl.	BE40	38
Townsend St. SE17	BZ42	42
Townsends Yd. N6	BV33	23
Townshend Est. NW8	BU37	32
Townshend Rd. NW8	BU37	32
Townshend Rd. Chis.	CL49	62
Townshend Rd. Rich.	BL45	48
Townshend Ter. Rich.	BL45	48
Townson Av. Nthlt.	BC37	29
Townson Way, Nthlt.	BC37	29
Town, The, Enf.	BZ24	8
Town Wf. Islw.	BJ45	48
Towton Rd. SE27	BZ48	60
Toxsowa Ho. SE21	CA47	60
Toynbee Rd. SW20	BQ51	67
Toynbee St. E1	CA39	42
Toyne Way N6	BU32	23
Kenwood Rd.		
Tracery, The, Bans.	BS61	83
Tracy Av. NW2	BP35	31
Tracy Ct. Stan.	BK29	12
Tradescant Rd. SW8	BX43	51
Trading Est. W3	BM38	30
Trading Est. Rd. NW10	BN38	31
Trafalgar Av. N17	CA29	15
Trafalgar Av. SE15	CA42	42
Trafalgar Av. Wor. Pk.	BQ54	67
Trafalgar Dr. Walt.	BC55	74
Trafalgar Gdns. E1	CC39	43
Trafalgar Gro. SE10	CF43	52
Trafalgar Pl. N18	CB28	15
Trafalgar Rd. SE10	CF43	52
Trafalgar Rd. SW19	BS50	59
Trafalgar Rd. Dart.	CW48	64
Trafalgar Rd. Rain.	CT37	37
Trafalgar Rd. Twick.	BG48	56
Trafalgar Sq. WC2	BW40	41
Trafalgar St. SE17	BZ42	42
Trafalgar Ter. Har.	BG33	20
Nelson Rd.		
Trafford Clo. E15	CE35	34
Trafford Rd. Th. Hth.	BX53	69
Tramway Av. E15	CF36	34
Broadway		
Tramway Av. N9	CB26	8
Tramway Path, Mitch.	BU53	68
Tranby Pl. E9	CC35	34
Tranmere Rd. N9	CA26	8
Tranmere Rd. SW18	BT47	59
Tranmere Rd. Twick.	BF47	56
Tranquil Rise, Erith	CT42	46
Tranquil Vale SE3	CG44	52
Transay Wk. N1	BZ36	33
Clephane Rd.		
Transept St. NW1	BU39	41
Transmere Clo. Orp.	CM53	71
Transport Av. Brent.	BJ43	48
Trap's Hill, Loug.	CK24	10
Traps La. N. Mal.	BO51	67
Travanion Rd. W14	BR43	49
Travellers Way, Houns.	BD44	47
Travers Rd. N7	BY34	24
Trawley Rd. SW6	BS45	50
Traynor Pl. N1	BZ37	33
Sherborne St.		
Treacy Clo. Bush.	BG27	11
Treadgold St. W11	BQ40	40
Treadway St. E2	CB37	33
Tredwell Rd. SE27	BY49	60
Treadwell Rd. Epsom	BO61	82
Treaty Rd. Houns.	BF45	47
Treaty St. N1	BX37	33
Trebeck St. W1	BV40	41
Curzon St.		
Trebovir Rd. SW5	BS42	41
Treby St. E3	CD38	34
Trecastle Way N7	BW35	32
Carleton Rd.		
Tredegar Rd. E3	CD37	34
Tredegar Rd. N11	BW29	14
Tredegar Rd. Dart.	CU48	64
Tredegar Sq. E3	CD38	34
Tredegar Ter. E3	CD38	34
Trederwen Rd. E8	CB37	33
Tredown Rd. SE26	CC49	61
Tree Clo. Rich.	BK47	57
Tree Mt. Ct. Epsom	BO60	82
Treen Av. SW13	BO45	49
Tree Rd. E16	CJ39	44
Tree Tops Clo. Belv.	CQ42	45
Treewall Gdns. Brom.	CH49	62
Trefgarne Rd. Dag.	CR34	27
Trefil Walk N7	BX35	33
Warlters Rd.		
Trefoil Rd. SW18	BT46	50
Tregaron Av. N8	BX32	24
Tregarvon Rd. SW11	BV45	50
Tregenna Av. Har.	BE35	29
Tregenna Clo. N14	BW25	7
Trego Rd. E3	CD36	34
Tregothnan Rd. SW9	BX45	51
Tregunter Rd. SW10	BS43	50
Trehearn Rd. Ilf.	CM29	17
Treherne Ct. SW17	BV49	59
Treherne Rd. SW9	BY44	51
Trehern Rd. SW14	BN45	49
Trehurst St. E5	CD35	34
Trelawney Est. E9	CC36	34
Trelawney Rd. Ilf.	CM29	17
Trelawn Rd. E10	CF34	25
Trelawn Rd. SW2	BY46	51
Treloar Gdns. SE19	BZ50	60
Tremadoc Rd. SW4	BW45	50
Tremaine Clo. SE4	CE44	52
Tremaine Rd. SE20	CB51	69
Trematon, Tedd.	BK50	57
Tremlett Gro. N19	BW34	23
Tremlett Mews N19	BW34	23
Junction Rd.		
Trenanace Gdns. Ilf.	CO34	27
Trenchard Av. Ruis.	BC35	29
Trenchard Clo. Stan.	BJ29	12
Trenchard Clo. Walt.	BD56	74
Trenchard Ct. Mord.	BS53	68
Trenchard St. SE10	CF42	43
Trenholme Clo. SE20	CB50	60
Trenholme Rd. SE20	CB50	60
Trenmar Gdns. NW10	BP38	31
Trent Av. W5	BK41	39
Trent Gdns. N14	BV25	7
Trentham Dr. Orp.	CO53	72
Trentham St. SW18	BS47	59
Trent Rd. SW2	BX46	51
Trent Rd. Buck. H.	CH26	10
Trent Way, Wor. Pk.	BQ55	76
Trentwood Side, Enf.	BX24	8
Treport St. SW18	BS47	59
Tresco Clo. Brom.	CG50	61
Hillbrow Rd.		
Trescoe Gdns. Har.	BE33	20
Trescoe Gdns. Rom.	CS28	19
Tresco Gdns. Ilf.	CO34	27
Tresco Rd. SE15	CB45	51
Tresham Cres. W1	BU38	32
Tresham Rd. Bark.	CN36	36
Tresham Wk. E9	CC35	34
Churchill Wk.		
Tressillian Cres. SE4	CE45	52
Tressillian Rd. SE4	CD45	52
Trestis Clo. Hayes	BD39	38
Treswell Rd. Dag.	CQ37	36
Tretawn Gdns. NW7	BO28	13
Tretawn Pk. NW7	BO28	13
Trevanion Rd. W14	BR42	40
Treve Av. Har.	BG33	20
Trevelyan Av. E12	CK35	35
Trevelyan Clo. Dart.	CW45	55
Trevelyan Cres. Har.	BL33	21
Trevelyan Gdns. NW10	BQ37	31
Trevelyan Rd. E15	CG35	34
Trevelyan Rd. SW17	BU49	59
Trevelyan Ct. N. Mal.	BO54	67
Treveris St. SE1	BY40	42
Bear Lane		
Treverton Est. W10	BQ39	40
Treverton St. W10	BQ38	31
Treville St. SW15	BP47	58
Treviso Rd. SE23	CC47	61
Farren Rd.		
Trevithick Dr. Dart.	CW45	55
Trevithick Rd. Dart.	CW46	55
Trevithick St. SE8	CE42	43
Watergate St.		
Trevone Gdns. Pnr.	BE32	20
Trevor Clo. Barn.	BT25	7
Trevor Clo. Brom.	CG54	70
Trevor Clo. Islw.	BH46	48
Trevor Clo. Nthlt.	BD37	29
Trevor Clo. Stan.	BH29	12
Trevor Gdns. Edg.	BN30	13
Trevor Gdns. Nthlt.	BD37	29
Trevor Pl. SW7	BU41	41
Trevor Rd. SW19	BR50	58
Trevor Rd. Edg.	BN30	13
Trevor Rd. Wdf. Grn.	CH29	17
Trevor Sq. SW7	BU41	41
Trevor St. SW7	BU41	41
Trevose Rd. E17	CF30	16
Trevose Way, Wat.	BD27	11
Trewince Rd. SW20	BQ51	67
Trewint St. SW18	BT48	59
Trewsbury Rd. SE26	CC49	61
Triandra Way, Hayes	BD39	38
Triangle Ct. E16	CJ39	44
Tollgate Rd.		
Triangle Pl. SW4	BW45	50
Triangle Rd. E8	CB37	33
Triangle, The, Bark.	CM36	35
Park Av.		
Triangle, The, Hamptn.	BG51	65
Triangle, The, Kings. On T.	BN51	67
Trident Gdns. Nthlt.	BD38	29
Jetstar Way		
Trident Way, Sthl.	BC41	38
Trigon Rd. SW8	BX43	51
Trilby Rd. SE23	CC48	61
Trimmer Wk. Brent.	BL43	48
Netley Rd.		
Trinder Gdns. N19	BX33	24
Trinder Rd. N19	BX33	24
Trinder Rd. Barn.	BQ25	6
Tring Av. W5	BL40	39
Tring Av. Sthl.	BE39	38
Tring Av. Wem.	BM36	30
Tring Clo. Ilf.	CM32	26
Tring Clo. Rom.	CW28	19
Tring Gdns. Rom.	CW28	19
Tring Grn. Rom.	CW28	19
Tring Wk. Rom.	CW28	19
Trinidad Gdns. Dag.	CS36	37
Trinidad St. E14	CD40	43
Trinity Av. N2	BT31	23
Trinity Av. Enf.	CA25	8
Trinity Ch. Rd. SW13	BP43	49
Trinity Church Sq. SE1	BZ41	42
Trinity Clo. E11	CG34	25
Trinity Clo. SE7	CJ42	44
Trinity Clo. SE13	CF45	52
Wisteria Rd.		
Trinity Clo. NW3	BT35	32
Hampstead High St.		
Trinity Clo. Brom.	CK54	71
Trinity Clo. Houns.	BE45	47
Trinity Clo. Sth. Croy.	CA58	78
Trinity Cotts. Rich.	BL45	48
Trinity Rd.		
Trinity Ct. N1	CA37	33
Trinity Cres. SW17	BU48	59
Trinity Est. SE8	CD42	43
Trinity Gdns. E16	CG39	43
Trinity Gdns. SW9	BX45	51
Trinity Grn. E1	CC38	34
Trinity Gro. SE10	CF44	52
Trinity Pl. Bexh.	CQ45	54
Trinity Ri. SW2	BY47	60
Trinity Rd. N2	BT31	23
Trinity Rd. N22	BX29	15
Trinity Rd. SW17	BT46	50
Trinity Rd. SW18	BT46	50
Trinity Rd. SW19	BS50	59
Trinity Rd. Ilf.	CM31	26
Trinity Rd. Rich.	BL45	48
Trinity Rd. Sthl.	BE40	38
Trinity Sq. EC3	CA40	42
Trinity St. E16	CG39	43
Trinity St. SE1	BZ41	42
Trinity St. Enf.	BZ23	8
Trinity Way W3	BO40	40
Trio Pl. SE1	BZ41	42
Trisian Sq. SE3	CG45	52
Tristram Clo. E17	CF31	25
Tristram Rd. Brom.	CG49	61
Triton Sq. NW1	BW38	32
Tritton Av. Croy.	BX56	78
Tritton Rd. SE21	BZ48	60
Trojan Way, Croy.	BX55	78
Tronsay Wk. N1	BZ36	33
Marquess Est.		
Troon St. E1	CD39	43
Trosley Rd. Belv.	CR43	54
Trossachs Rd. SE22	CA46	51
Trothy Rd. SE1	CB42	42
Trott St. SW11	BT44	50
Trott Rd. N10	BU29	14
Trotwood IG7	CM29	17
Troughton Rd. SE7	CH42	44
Troutbeck Rd. E14	CD44	52
Trouville Rd. SW4	BW46	50
Trowbridge Rd. E9	CD36	34
Trowbridge Rd. Rom.	CV29	19
Trowlock Av. Tedd.	BK50	57
Trowlock Way, Tedd.	BK50	57
Broom Rd.		
Troy Ct. W8	BS41	41
Troy Ct. SE18	CL42	44
Troy Rd. SE19	BZ50	60
Troy Town SE15	CB45	51
Nutbrook St.		
Trucks Alley, Swan.	CR51	72
Trulock Ct. N17	CB29	15
Trulock Rd. N17	CB29	15
Truman's Rd. N16	CA35	33
Truman St. SE16	CB41	42
Trumper's Way W7	BH41	39
Trumpington Rd. E7	CG35	34
Trump St. EC2	BZ39	42
King St.		
Trundlers Way, Bush.	BH26	5
Trundle St. SE1	BZ41	42
Weller St.		
Trundley's Rd. SE8	CC42	43
Trundley's Ter. SE8	CC42	43
Trunks Alley, Swan.	CR51	72
Truro Gdns. Ilf.	CK33	26
Truro Rd. E17	CD32	25
Truro Rd. N22	BX29	15
Truro St. NW3	BV36	32
Truro Wk. Rom.	CV29	19
Truslove Rd. SE27	BY49	60
Trussley Rd. W6	BQ41	40
Trustings Clo. Esher	BJ57	75
Trustons Gdns. Horn.	CU33	28
Tryfan Clo. Ilf.	CJ32	26
Tryon St. SW3	BU42	41
Tuam Rd. SE18	CM43	53
Tubbenden Clo. Orp.	CN55	81
Tubbenden Dr. Orp.	CM56	80
Tubbenden La. Orp.	CM56	80
Tubbenden La. Sth. Orp.	CM56	80
Tubbs Rd. NW10	BO37	31
Tucker St. Wat.	BD25	4
Tuck Rd. Rain.	CU36	37
Tudor Av. Hamptn.	BF50	56
Tudor Av. Rom.	CU31	28
Tudor Av. Wor. Pk.	BP55	76
Tudor Clo. NW3	BU35	32
Tudor Clo. NW7	BP29	13
Tudor Clo. NW9	BN34	22
Tudor Clo. SW2	BX46	51
Tudor Clo. Bans.	BR61	82
Tudor Clo. Chess.	BL56	75
Tudor Clo. Chig.	CL27	17
Tudor Clo. Chis.	CK51	71
Tudor Clo. Couls.	BY62	84
Tudor Clo. Dart.	CU46	55
Tudor Clo. Pnr.	BC32	20
Tudor Clo. Sth. Croy.	CB61	84
Tudor Clo. Sutt.	BQ57	76
Tudor Clo. Wall.	BW57	77
Tudor Clo. Wdf. Grn.	CH28	17
Tudor Ct. E17	CD32	25
Tudor Ct. SE9	CK45	53
Tudor Ct. N. Wem.	BM35	30
Tudor Ct. S. Wem.	BM35	30
Tudor Cres. Ilf.	CL29	17
Tudor Dr. Kings. On T.	BK49	57
Tudor Dr. Mord.	BQ53	67
Tudor Dr. Rom.	CU31	28
Tudor Dr. Walt.	BD54	65
Tudor Est. NW10	BM38	30
Tudor Gdns. NW9	BN34	22
Tudor Gdns. Twick.	BH47	57
Tudor Gdns. SW13	BO45	49
Treen Av.		
Tudor Gdns. W3	BM39	39
Tudor Gdns. Rom.	CU31	28
Tudor Gdns. W. Wick.	CF55	79
Tudor Gro. E9	CC36	34
Tudor Pl. W1	BW39	41
Gresse St.		
Tudor Pl. Mitch.	BU50	59
Tudor Rd. E4	CE29	16
Tudor Rd. E6	CJ37	35
Tudor Rd. E9	CB37	33
Tudor Rd. N9	CB26	8
Tudor Rd. SE19	CA50	60
Tudor Rd. SE25	CB53	69
Tudor Rd. Bark.	CN37	36
Tudor Rd. Barn.	BS24	7
Tudor Rd. Beck.	CE52	70
Tudor Rd. Hamptn.	BF50	56
Tudor Rd. Har.	BG30	11
Tudor Rd. Houns.	BG45	47
Tudor Rd. Kings. On T.	BM50	57
Tudor Rd. Pnr.	BD30	11
Tudor Rd. Sthl.	BE40	38
Tudor St. EC4	BY39	42
Tudor Wk. Bex.	CQ46	54
Tudor Way N14	BW26	7
Tudor Way W3	BL41	39
Tudor Way, Orp.	CM53	71
Tudor Well Clo. Stan.	BJ28	12
Tudway Rd. SE3	CJ45	53
Tudway Rd. SE9	CJ45	53
Tufnail Rd. Dart.	CW46	55
Tufnell Park Rd. N7	BW35	32
Tufter Rd. Chig.	CN28	18
Tufton Gdns. E. Mol.	BF51	65
Tufton Rd. E4	CE28	16
Tufton St. SW1	BW41	41
Tugela Rd. Croy.	BZ53	69
Tugela St. SE6	CD48	61

Tulip Clo. Rom.	CV29	19
Tulip Ct. Pnr.	BD31	20
Nursery Rd.		
Tullerie St. E2	CB37	33
Tulse Clo. Beck.	CF52	70
Tulse Hill SW2	BY46	51
Tulse Hl. Est. SW2	BY46	51
Tulsemere Rd. SE27	BZ48	60
Tumblewood Rd. Bans.	BR61	82
Tumbling Way, Walt.	BC53	65
Tuncombe Rd. N18	CA28	15
Tunis Rd. W12	BP40	40
Tunley Rd. NW10	BO37	31
Tunley Rd. SW17	BV47	59
Tunmarsh La. E13	CH38	35
Tunnel Av. SE10	CF40	43
Tunnel Gdns. N11	BW29	14
Tunstall Av. Ilf.	CO29	18
Tunstall Clo. Orp.	CN56	81
Tunstall Rd. SW9	BX45	51
Tunstall Rd. Croy.	CA54	69
Tunstall Wk. Brent.	BK42	39
Ealing Rd.		
Tunworth Clo. NW9	BN32	22
Tunworth Cres. SW15	BO46	49
Turin Rd. N9	CC26	9
Turin St. E2	CB38	33
Turkey Oak Clo. CA51	SE19	69
Hamlyn Gdns.		
Turk's Row SW3	BU42	41
Turle Rd. SW16	BW51	68
Turlewray Clo. EC1	BX34	24
Turley Clo. E15	CG37	34
Tunragain La. EC4	CM39	44
Farringdon St.		
Turnagain La. Dart.	CU48	64
Turnage Rd. Dag.	CQ33	27
Turnant Rd. N17	BZ30	15
Lordship La.		
Turnberry Way, Orp.	CM54	71
Turner Av. N15	CA31	24
Turner Av. Mitch.	BU51	68
Turner Av. Twick.	BG48	56
Turner Clo. NW11	BS32	23
Turner Dr. NW11	BS32	23
Turner Rd. E17	CF31	25
Turner Rd. Bush.	BG24	4
Turner Rd. Edg.	BL30	12
Turner Rd. N. Mal.	BN54	67
Turners La. Walt.	BC57	74
Turners Rd. E3	CD39	43
Turner St. E1	CB39	42
Turner St. E16	CG39	43
Turners Way NW11	BT33	23
Wildwood Rd.		
Turner's Wood NW11	BT33	23
Wildwood Rd.		
Turneville Rd. W14	BR43	49
Turney Rd. SE21	BZ47	60
Turney Rd. W14	BR43	49
Turnham Grn. Ter. W4	BO42	40
Turnham Rd. SE4	CD46	52
Turnmill St. EC1	BY38	33
Turnpike Ct. Bexh.	CP45	54
Crook Log		
Turnpike Dr. Orp.	CP58	81
Turnpike La. N8	BX31	24
Turnpike Link, Croy.	CA55	78
Turnpin La. SE10	CF43	52
King William Wk.		
Turnville Rd. W14	BR44	49
Turpentine La. SW1	BV42	41
Sutherland St.		
Turpin Av. Rom.	CR29	18
Turpington Clo. Brom.	CK54	71
Turpington La. Brom.	CK54	71
Turpin's La. Wdf. Grn.	CK28	17
Turpin Way, Wall.	BV57	77
Turquand St. SE17	BZ42	42
Turret Gro. SW4	BW45	50
Turnstone Clo. Sth. Croy.	CC58	79
Turtle Rd. SW17	BT48	59
Turton Mkt. Wem.	BL35	30
Turton Rd.		
Turton Rd. Wem.	BL35	30
Turville St. E2	CA38	33
Old Nichol St.		
Tuscan Rd. SE18	CM42	44
Tuskar St. SE10	CG42	43
Tuttlebee La. Buck. H.	CH27	17
Twedwell Clo. Brom.	CK52	71
Tweedale Ct. E15	CE35	34
Tweedale Rd. Cars.	BT54	68
Tweed Glen, Rom.	CS29	19
Tweed Grn. Rom.	CS29	19
Tweedmouth Rd. E13	CH37	35
Tweed Way, Rom.	CS29	19
Tweedy Rd. Brom.	CH51	71
Twentyman Clo.	CH28	17
Wdf. Grn.		
Twickenham Br. Rich.	BK46	48
Twickenham Br. Twick.	BK46	48
Twickenham Clo. Croy.	BX55	78
Twickenham Gdns. Grnf.	BJ35	30
Twickenham Gdns. Har.	BH29	12
Twickenham Rd. E11	CF34	25
Twickenham Rd. Felt.	BE49	56
Twickenham Rd. Islw.	BJ46	48
Twickenham Rd. Rich.	BK45	48
Twickenham Rd. Tedd.	BJ49	57
Twigs Clo. Erith	CT43	55
Twilley St. SW18	BS47	59
Twineham Grn. N12	BS28	14
Twining Av. Twick.	BG48	56
Twinn Rd. NW7	BR29	13
Twisden Rd. NW5	BV35	32
Twybridge Way NW10	BN36	31
Twyford Abbey Rd.	BL38	30
NW10		
Twyford Av. N2	BU31	23
Twyford Av. W3	BM40	39
Twyford Cres. W3	BM40	39
Twyford Pl. WC2	BX39	42
Kingsway		
Twyford Rd. Cars.	BT54	68
Twyford Rd. Har.	BF33	20
Twyford Rd. Ilf.	CM35	35
Twyford St. N1	BX37	33
Tyas Rd. E16	CG38	34
Tybenham Rd. SW19	BR52	67
Tyberry Rd. Enf.	CB24	8
Tyburn La. Har.	BH33	21
Tycehurst Gdns. Ilf.	CM35	35
Tycehurst Hill, Loug.	CK24	10
Tye La. Orp.	CM56	80
Tyers St. SE11	BX42	42
Tyers Ter. SE11	BX42	42
Tyeshurst Clo. SE2	CQ42	45
Tykeswater La. Borwd.	BK23	5
Tylecroft Rd. SW16	BX51	69
Tyle Grn. Horn.	CW31	28
Tyler Rd. Beck.	CE51	70
Tylers Clo. Loug.	CK26	10
Tylers Cres. Horn.	CV35	37
Tylers Est. SE1	CA41	42
Tylers Gate SE1	CA41	42
Tylers Gate, Har.	BL32	21
Tylers Grn. Rd. Swan.	CS53	73
Tyler St. SE10	CG42	43
Tylers Way, Wat.	BG24	4
Tylney Av. SE19	CA49	60
Tylney Rd. E7	CJ35	35
Tylney Rd. Brom.	CJ51	71
Tynan Clo. Felt.	BC47	56
Tyndall Rd. E10	CF34	25
Tyndall Rd. Well.	CN45	54
Tyneham Rd. SW11	BV44	50
Tynemouth Rd. N15	CA31	24
Tynemouth Rd. Mitch.	BV50	59
Tynemouth St. SW6	BT44	50
Tyne Rd. Ilf.	CM34	26
Type St. E2	CC37	34
Tyrawley Rd. SW6	BS44	50
Tyrone Rd. E6	CK37	35
Tyron Way, Sid.	CN49	63
Tyrrell Av. Well.	CO46	54
Tyrrell Clo. Har.	BH35	30
Tyrrell Rd. SE22	CB45	51
Tyrrel Way NW9	BO33	22
Tyrwhitt Rd. SE4	CE45	52
Tysoe St. EC1	BY38	33
Tyson Rd. SE23	CC47	61
Tyssen Pl. E8	BT36	32
Ramsgate St.		
Tyssen Rd. N16	CA34	24
Stoke Newington High St.		
Tyssen St. E8	CA36	33
Tytherton Rd. N19	BW34	23
Uamvar St. E14	CE39	43
Uckfield Gro. Mitch.	BV51	68
Udall Gdns. Rom.	CR29	18
Udall St. SW1	BW42	41
Vincent Sq.		
Udney Pk. Rd. Tedd.	BJ50	57
Uffington Rd. NW10	BP37	31
Uffington Rd. SE27	BY49	60
Ufford Clo. Har.	BF29	11
Ufford Rd. Har.	BF29	11
Ufford St. SE1	BY41	42
Ufton Gro. N1	BZ36	33
Ufton Rd. N1	BZ36	33
Ullathorne Rd. SW16	BW49	59
Ulleswater Rd. N14	BX28	15
Ullswater Clo. SW15	BN49	58
Ullswater Clo. Brom.	CG50	61
Ullswater Cres. SW15	BN49	58
Ullswater Cres. Couls.	BW61	83
Ullswater Cres. Couls.	BX57	78
Ullswater Rd. SE27	BY48	60
Ullswater Rd. SW13	BP43	49
Ullswater Way, Horn.	CU35	37
Ulster Gdns. N13	BZ28	15
Ulster Pl. NW1	BV38	32
Ulundi Rd. SE3	CG43	52
Ulva Rd. SW15	BQ45	49
Ravenna Rd.		
Ulverdale Rd. SW10	BT43	50
Ulverscroft Rd. SE22	CA46	51
Ulverstone Rd. SE27	BY48	60
Ulverstone Rd. E17	CF30	16
Ulysses Rd. NW6	BR35	31
Umberston St. E1	CB39	42
Umbria St. SW15	BP46	49
Umfreville Rd. N4	BY32	24
Ummer Gro. Borwd.	BK25	5
Underbridge Way, Enf.	CD24	9
Undercliff Rd. SE13	CE45	52
Underhill, Barn.	BS25	7
Underhill Rd. SE22	CB46	51
Underne Av. N14	BV27	14
Undershaft EC3	CA39	42
St. Mary Axe		
Undershaw Rd. Brom.	CG48	61
Underwood, Croy.	CF57	79
Underwood Rd. E1	CB38	33
Underwood Rd. E4	CE28	16
Underwood Row N1	BZ38	33
Underwood, The SE9	CK48	62
Undine St. SW17	BU49	59
Uneeda Dr. Grnf.	BG37	29
Union Cotts. E15	CG36	34
Union Ct. Rich.	BL46	48
Eton St.		
Union Ct. E15	CF37	34
Union Ct. EC2	CA39	42
Wormwood St.		
Union Ct. Ilf.	CL34	26
Ilford Hill		
Union Ct. Rich.	BL46	
Eton St.		
Union Dr. E1	CD38	34
Solebay St.		
Union Gro. SW8	BW44	50
Union La. Islw.	BJ44	48
Park Rd.		
Union Rd. N11	BW29	14
Union Rd. SW4	BW44	50
Union Rd. Nthlt.	BF37	29
Union Rd. SW8	BW44	50
Union Rd. Brom.	CJ53	71
Union Rd. Croy.	BZ54	69
Union Rd. Wem.	BL36	30
Union Row N18	CB29	15
Union Sq. N1	BZ37	33
Union St. E15	CF37	34
Union St. SE1	BY41	42
Union St. Barn.	BR24	6
Union St. Kings. On T.	BK51	66
Union Wk. E2	CA38	33
University Clo. NW9	BO29	13
Rivington Cres.		
University Pl. Erith	CR43	54
Belmont Rd.		
University Pl. Erith	CR44	54
Becton Pl.		
University Rd. SW19	BT50	59
University St. WC1	BW38	32
Unwin Clo. SE15	CB43	51
Unwin Rd. Islw.	BH45	48
Upbrook Ms. W2	BT39	41
Upcerne Rd. SW10	BT43	50
Burnaby St.		
Upchurch Clo. SE20	CB50	60
Woodbin Gro.		
Upcroft Av. Edg.	BN28	13
Updale Rd. Sid.	CN49	63
Upfield, Croy.	CB55	78
Upfield Rd. W7	BH38	30
Uphall Rd. Ilf.	CL34	26
Upham Pk. Rd. W4	BO42	40
Uphill Dr. NW7	BO28	13
Uphill Dr. NW9	BN32	22
Uphill Gro. NW7	BO28	13
Uphill Rd. NW7	BO28	13
Upland Ct. Rd. Rom.	CW30	19
Upland Rd. E13	CG38	34
Upland Rd. SE22	CB46	51
Upland Rd. Bexh.	CQ45	54
Upland Rd. Sth. Croy.	BZ56	78
Upland Rd. Sutt.	BT57	77
Uplands SW16	BY49	60
Uplands, Beck.	CE51	70
Uplands Av. E17	CC30	16
Uplands Clo. SW14	BM46	48
Uplands Ct. N21	BY26	8
Uplands End, Wdf. Grn.	CK29	17
Uplands Pk. Rd. Enf.	BY24	8
Uplands Rd. N8	BX32	24
Uplands Rd. Barn.	BV26	7
Uplands Rd. Ken.	BZ61	84
Uplands Rd. Orp.	CO54	72
Uplands Rd. Rom.	CP31	27
Uplands Rd. Wdf. Grn.	CK29	17
Uplands, The, Loug.	CK24	10
Uplands, The, Ruis.	BC33	20
Uplands Way N21	BY25	8
Upland Way, Epsom	BQ62	82
Upminster Rd. Horn.	CW34	28
Upminster Rd. N. Rain.	CV38	37
Upminster Rd. S. Rain.	CU38	37
Upney Clo. Horn.	CV35	37
Upney La. Bark.	CN35	36
Upnor Way SE17	CA42	42
Uppark Dr. Ilf.	CM32	26
Upr. Abbey Rd. Belv.	CQ42	45
Upper Addison Gdns.	BR41	40
W14		
Upr. Bardsey Wk. N1	BZ36	33
Marquess Est.		
Upr. Belgrave St. SW1	BV41	41
Upr. Berkeley St. W1	BU39	41
Upr. Beulah Hl. SE19	CA51	69
Upr. Brentwood Rd. Rom.	CV31	28
Upr. Brighton Rd. Surb.	BK53	66
Upper Brockley Rd. SE4	CD45	52
Upper Brook St. W1	BV40	41
Upper Butts Brent.	BK43	48
Upr. Caldy Wk. N1	BZ36	33
Marquess Est.		
Upr. Cavendish Av. N3	BS31	23
Upr. Cheyne Row SW3	BU43	50
Upr. Clapton Rd. E5	CB33	24
Upr. Court Rd. Epsom	BN59	82
Upper Elmers End Rd.	CD52	70
Beck.		
Upr. Farm Rd.	BE52	65
Upr. Fosters NW4	BQ32	22
Upr. Green E. Mitch.	BU51	68
Upr. Green W. Mitch.	BU51	68
Upper Grosvenor St. W1	BV40	41
Upr. Grotto Rd. Twick.	BH48	57
Upper Ground SE1	BY40	42
Upper Gro. SE25	CA52	69
Upr. Grove Rd. Belv.	CQ43	54
Upr. Gulland Wk. N1	BZ36	33
Marquess Est.		
Upr. Ham Rd. Rich.	BK49	57
Upper Handa Wk. N1	BZ36	33
Marquess Est.		
Upper Harley St. NW1	BV38	32
Upr. High St. Epsom	BO60	82
Upper Hill Vw. Rd. Pnr.	BE29	11
Upper Hitch, Wat.	BE27	11
Upr. Holly Hill Rd. Belv.	CR42	45
Upper James St. W1	BW40	41
Beak St.		
Upper John St. W1	BW40	41
Beak St.		
Upr. Lismore Wk. N1	BZ36	33
Marquess Est.		
Upper Mall W6	BP42	40
Upper Marsh SE1	BX41	42
Upr. Montagu St. W1	BU39	41
Upr. Mulgrave Rd. Sutt.	BR57	76
Upr. North St. E14	CE39	43
Upr. Paddock Rd. Wat.	BE25	4
Upr. Palace Rd. E. Mol.	BG52	65
Upr. Pk. Loug.	CJ24	10
Upr. Pk. Rd. N11	BV28	14
Upper Pk. Rd. NW3	BU35	32
Upr. Park Rd. Belv.	CR42	45
Upr. Park Rd. Brom.	CH51	71
Upr. Park Rd.	BM50	57
Kings. On T.		
Upper Phillimore Gdns.	BS41	41
W8		
Upr. Pillory Downs. Cars.	BV60	83
Upper Pines, Bans.	BU62	83
Upr. Rainham Rd. Horn.	CT33	28
Upr. Ramsey WK. N1	BZ36	33
Marquess Est.		
Upr. Richmond Rd. SW14	BM45	48
Upr. Richmond Rd. SW15	BM45	48
Upr. Richmond Rd. Rich.	BM45	48
Upper Rd. E13	CH38	35
Upper Rd. Wall.	BW56	77
Upr. St. Martin's La. WC2	BX40	42
Long Acre		
Upr. Selsdon Rd.	CA57	78
Sth. Croy.		
Upr. Sheppey Wk. N1	BZ36	33
Marquess Est.		
Upr. Sheridan Rd. Belv.	CR42	45
Coleman Rd.		
Upr. Shirley Rd. Croy.	CC55	79
Upper Sq. Islw.	BJ45	48
Upper Staithe W4	BN44	49
Upper St. N1	BY37	33
Upr. Sunbury Rd.	BE51	65
Hamptn.		
Upr. Sutton La. Houns.	BF44	47
Upper Tail. Wat.	BE27	11
Upr. Teddington Rd.	BK50	57
Kings. On T.		
Upr. Teddington Rd.	BK50	57
Tedd.		
Upper Ter. NW3	BT34	23
Windmill Hill		
Upper Thames St. EC4	BY40	42
Upr. Tollington Pk. N4	BY33	24
Upperton Rd. E13	CJ38	35
Upperton Rd. Sid.	CN49	63
Upr. Tooting Pk. SW17	BU48	59
Upr. Tooting Rd. SW17	BU49	59
Upr. Town Rd. Grnf.	BF38	29
Upr. Tulse Hill SW2	BX47	60
Upr. Walthamstow Rd.	CF31	25
E17		
Upr. Wickham La. Well.	CO45	54
Upr. Wimpole St. W1	BV39	41
Upper Woburn Pl. WC1	BW38	32
Uppingham Av. Stan.	BJ30	12
Upsdell Av. N13	BY29	15
Upstall St. SE5	BY44	51
Upton Av. E7	CH36	35
Upton Clo. Bex.	CQ46	54
Upton Dene, Sutt.	BS57	77
Upton Gdns. Har.	BJ32	21
Upton La. E7	CH37	35
Upton Lo. Clo. Bush.	BG26	4
Upton Park Rd. E7	CH36	35
Upton Rd. N18	CB28	15

Upton Rd. SE18 CM43 53
Upton Rd. Bexh. CQ45 54
Upton Rd. Houns. BF45 47
Upton Rd. Th. Hth. BZ51 69
Upton Rd. Wat. BC24 4
Upton Rd. S. Bex. CQ45 54
Upway N12 BU29 14
Upwood Rd. SE12 CG46 52
Upwood Rd. SW16 BX51 69
Urban Av. Horn. CV34 28
Urlwin St. SE5 BZ43 51
Urmston Dr. SW19 BR47 58
Ursula St. SW11 BU44 50
Urswick Gdns. Dag. CQ36 36
Urswick Rd. E9 CC35 34
Urswick Rd. Dag. CQ36 36
Usher Rd. E3 CD37 34
Usk Rd. SW11 BT45 50
Usk St. E2 CC38 34
Usk St. E16 CG40 43
Uvedale Rd. Dag. CR34 27
Uvedale Rd. Enf. BZ25 8
Uxbridge Gdns. Felt. BD48 56
Uxbridge Rd. W3 BJ40 39
Uxbridge Rd. W5 BJ40 39
Uxbridge Rd. W12 BP40 40
Uxbridge Rd. W13 BJ40 39
Uxbridge Rd. Felt. BD48 56
Uxbridge Rd. Hmptn. BF49 56
Uxbridge Rd. BK52 66
Kings. On T.
Uxbridge Rd. Pnr. BC30 11
Uxbridge Rd. Sthl. BF40 38
Uxbridge St. W8 BS40 41
Uxendon Cres. Wem. BL33 21
Uxendon Hill, Wem. BL33 21

Valance Av. E4 CG26 9
Valan Leas. Brom. CG52 70
Vale Av. Borwd. BM25 5
Vale Clo. W9 BT38 32
Vale Clo. Couls. BX60 84
Vale Clo. Twick. BJ48 57
Vale Cotts. Brom. CH52 71
Vale Cres. SW15 BO49 58
Valecroft, Pnr. BE32 20
Vale Dr. Barn. BR24 6
Vale End SE22 CA45 51
Grove Vale
Vale Gro. N4 BZ33 24
Vale Gro. W3 BN40 40
Vale La. W3 BM39 39
Valence Av. Dag. CP33 27
Valence Cir. Dag. CP34 27
Valence Ho. Dag. CQ34 27
Valence Rd. Erith CS43 55
Valence Wood Rd. Dag. CP34 27
Valencia Rd. Stan. BK28 12
Valentine Av. Bex. CQ48 63
Valentine Ct. SE23 CC48 61
Valentine Pl. SE1 BY41 42
Valentine Rd. E9 CC36 34
Valentine Rd. Har. BG34 29
Valentines Rd. Ilf. CL33 26
Valentines Way, Rom. CS34 28
Vale Of Heath NW3 BT34 23
East Heath Rd.
Vale Pl. W14 BQ41 40
Spring Vale Ter.
Valerie Ct. Bush. BG26 4
Vale Rise NW11 BR33 22
Vale Rd. E7 CH36 35
Vale Rd. N4 BZ33 24
Vale Rd. Brom. CL51 71
Vale Rd. Bush. BE25 4
Vale Rd. Dart. CU47 64
Vale Rd. Esher BH58 75
Vale Rd. Mitch. BW52 68
Vale Rd. Sutt. BS56 77
Vale Rd. Wor. Pk. BO55 76
Vale Rd. N. Surb. BL55 75
Vale Rd. S. Surb. BL55 75
Vale Row N5 BY34 24
Gillespie Rd.
Vale Royal N7 BX36 33
Valeswood Rd. Brom. CG49 61
Vale Ter. N4 BZ32 24
Vale, The N10 BV30 14
Vale, The N14 BW26 7
Vale, The NW11 BQ34 22
Vale, The SW3 BT43 50
Vale, The W3 BO40 40
Vale, The, Couls. BW60 83
Vale, The, Croy. CC55 79
Vale, The, Felt. BC46 47
Vale, The, Houns. BE43 47
Vale, The, Ruis. BD34 20
Vale, The, Sun. BC50 56
Vale, The, Wdf. Grn. CH29 17
Valerian Way E15 CG38 34
Valetta Gro. E13 CG37 34
Valetta Rd. W3 BO41 40
Valette St. E9 CB36 33
Vale Way, Horn. CU35 37
Valiant Clo. Nthlt. BD38 29
Ruislip Rd.
Valiant Clo. Rom. CR30 18
Vallance Rd. E1 CB38 33
Vallance Rd. E2 CB38 33

Vallance Rd. N22 BW30 14
Vallentin Ct. E17 CF31 25
Vallentin Rd.
Vallentin Rd. E17 CF31 25
Valley Av. N12 BT28 14
Valley Clo. Dart. CT46 55
Valley Clo. Loug. CK25 10
Valley Clo. Pnr. BC30 11
Valley Clo. Wdf. Grn. CH29 17
Valley Dr. NW9 BM32 21
Valleyfield Rd. SW16 BX49 60
Valley Flds. Cres. Enf. BY23 8
Valley Gdns. SW19 BT50 59
Valley Gdns. Wem. BL36 30
Valley Gro. SE7 CJ42 44
Valley Hill, Loug. CK26 10
Valley Ms. Twick. BJ48 57
Cross Deep
Valley Rd. N12 BT28 14
Valley Rd. SW16 BX49 60
Valley Rd. Belv. CR42 45
Valley Rd. Brom. CG51 70
Valley Rd. Dart. CT46 55
Valley Rd. Erith CS42 46
Valley Rd. Ken. BZ60 84
Valley Rd. Orp. CO51 72
Valley Side E4 CE27 16
Valley Vw. Barn. BR25 6
Valley Vw. Gdns. Ken. CA57 78
Valley Wv. Gdns. Ken. CA61 84
Valley Wk. Croy. CC55 79
Valliant Clo. Nthlt. BD38 29
Valliere Rd. NW10 BP38 31
Valliers Wood Rd. Sid. CM47 62
Vallis Way W13 BJ39 39
Vally Sq. Wat. BK56 75
Valmar Rd. SE5 BZ44 51
Valnay St. SW17 BU49 59
Valognes Av. E17 BX30 15
Valonia Gdns. SW18 BR46 49
Vambery Rd. SE18 CM43 53
Vanbrough Cres. Nthlt. BD37 29
Vanburgh Clo. Orp. CN54 72
Vanburgh Fields SE3 CG43 52
Vanbrugh Hill SE3 CG42 43
Vanbrugh Hill SE10 CG42 43
Vanbrugh Pk. SE3 CG43 52
Vanbrugh Pk. Rd. SE3 CG43 52
Vanbrugh Pk. Rd. W. SE3 CG43 52
Vanbrugh Rd. W4 BN41 40
Vanbrugh Ter. SE3 CG44 52
Vancouver Clo. Epsom BN59 82
Vancouver Rd. SE23 CD48 61
Vancouver Rd. Edg. BM29 12
Vancouver Rd. Hayes BC38 29
Vancouver Rd. Rich. BK49 57
Vanderbilt Rd. SW18 BS47 59
Vandome Clo. E16 CH39 44
Vandon Pass. SW1 BW41 41
Petty France
Vandon St. SW1 BW41 41
Van Dyck Av. N. Mal. BN54 67
Vandyke Clo. SW15 BQ46 49
Vandyke Cross SE9 CK46 53
Vandy St. EC2 CA38 33
Vane Clo. Har. BL32 21
Vane Clo. NW3 BT35 32
Hampstead High St.
Vane St. SW1 BV42 41
Vincent Sq.
Vanguard Clo. Croy. BY54 69
Vanguard Clo. Rom. CR30 18
Vanguard St. SE8 CE44 52
Vanguard Way, Wall. BX57 78
Vanoc Gdns. Brom. CG48 61
Vansittart Rd. E7 CG35 34
Vansittart St. SE14 CD43 52
Vanston Pl. SW6 BS43 50
Vant Rd. SW17 BU49 59
Varcoe Rd. SE16 CB42 42
Vardens Rd. SW11 BT45 50
Varden St. E1 CB39 42
Varley Rd. E16 CH39 44
Varna Rd. Hamptn. BF51 65
Varndell St. NW1 BW38 32
Vartry Rd. N15 BZ32 24
Vassall Rd. SW9 BY43 51
Vauban St. SE16 CA41 42
Vaudan St. SE16 CA41 42
Vaudrey Clo. E1 CC38 34
Vaughan Clo. Hamptn. BE50 56
Vaughan Av. NW4 BP32 22
Vaughan Av. W6 BO42 40
Vaughan Av. Horn. CV35 37
Vaughan Gdns. Ilf. CK33 26
Vaughan Rd. E15 CG36 34
Vaughan Rd. SE5 BZ44 51
Vaughan Rd. Har. BG32 20
Vaughan Rd. Surb. BJ54 66
Vaughan Rd. Well. CN44 54
Vaughan St. SE16 CC38 34
Vaughan Way E1 CB42 42
Vauxhall Bridge BX42 42
Vauxhall Bridge Rd. SW1 BW41 41
Vauxhall Cross SW8 BX43 51
Vauxhall Gdns. Sth. Croy. BZ57 78

Vauxhall Gdns. Est. SE11 BX42 42
Vauxhall Gro. SW8 BX43 51
Vauxhall Pl. Dart. CW47 64
Vauxhall St. SE11 BX42 42
Vauxhall Wk. SE11 BX42 42
Vawdrey Clo. SW17 BV50 59
Vectis Rd.
Vectis Rd. SW17 BV50 59
Veda Rd. SE13 CE45 52
Vega Rd. Bush. BG26 4
Velde Way SE22 CA46 51
Venables St. NW8 BT38 32
Vencourt Pl. W6 BP42 40
Dulwich Gro.
Venetia Rd. N4 BY32 24
Venetia Rd. W5 BK41 39
Venetian Rd. SE5 BZ44 51
Venette Clo. Rain. CU39 46
Venner Rd. SE26 CC49 61
Venners Clo. Bexh. CT44 55
Venn St. SW4 BW45 50
Venour Rd. E3 CD38 34
Ventnor Av. Stan. BJ30 12
Ventnor Dr. N20 BS27 14
Ventnor Gdns. Bark. CN36 36
Ventnor Rd. SE14 CC43 52
Ventnor Rd. Sutt. BS57 77
Venture Clo. Bex. CQ47 63
Venue St. E14 CF39 43
Venus Rd. SE18 CK41 44
Veny Cres. Horn. CV36 37
Vera Av. N21 BY25 8
Vera Rd. SW6 BR44 49
Verbena Gdns. W6 BP42 40
Verdant Rd. Bexh. CQ44 54
Verdant Ct. SE6 CG47 61
Verdant La. SE6 CG47 61
Verdayne Av. Croy. CC55 79
Verderers Rd. Chig. CO28 18
Verdun Rd. SE18 CO43 54
Verdun Rd. SW13 BP43 49
Vereker Dr. Sun. BC52 65
Vereker Rd. W14 BR42 40
Vere Rd. Loug. CM24 10
Vere St. W1 BV39 41
Vermont Rd. SE19 BZ50 60
Vermont Rd. SW18 BS46 50
Vermont Rd. Sutt. BS55 77
Verney Gdns. Dag. CQ35 36
Verney Rd. SE16 CB42 42
Verney Rd. Dag. CQ35 36
Verney St. NW10 BN34 22
Verney Way SE16 CB42 42
Vernham Rd. SE18 CM43 53
Vernon Av. E12 CK35 35
Vernon Av. SW20 BQ51 67
Vernon Av. Wdf. Grn. CH29 17
Vernon Clo. Epsom BN57 76
Vernon Clo. Orp. CO52 72
Vernon Cres. Barn. BV25 7
Vernon Dr. Stan. BJ30 12
Vernon Pl. WC1 BX39 42
Bloomsbury Way
Vernon Rise WC1 BX38 33
Percy Circus
Vernon Rise, Grnf. BG35 29
Vernon Rd. E3 CD37 34
Vernon Rd. E11 CG33 25
Vernon Rd. E15 CG36 34
Vernon Rd. E17 BX32 24
Vernon Rd. N8 BY31 24
Vernon Rd. SW14 BO45 49
Vernon Rd. Bush. BE25 4
Vernon Rd. Ilf. CN33 27
Vernon Rd. Rom. CS28 19
Vernon Rd. Sutt. BT56 77
Vernon Sq. WC1 BX38 33
Penton Rise
Vernon St. W14 BR42 40
Vernon Yd. W11 BR40 40
Portobello Rd.
Verona Dr. Surb. BL55 75
Verona Rd. E7 CH36 35
Upton La.
Veronica Clo. Rom. CV29 19
Veronica Rd. SW17 BV48 59
Veronique Gdns. Ilf. CM32 26
Verran Rd. SW12 BV47 59
Balham Gro.
Versailles Rd. SE20 CB50 60
Verulam Av. E17 CD32 25
Verulam Av. Pur. BW59 83
Verulam Rd. Grnf. BF38 29
Verulam St. EC1 BY39 42
Verwood Rd. Har. BG30 11
Vespan Rd. W12 BP41 40
Vesta Rd. SE4 CD44 52
Vestris Rd. SE23 CC48 61
Vestry Ms. SE5 CA44 51
Vestry Rd. E17 CE32 25
Vestry Rd. SE5 CA44 51
Vestry St. N1 BZ38 33
Vevey St. SE6 CD48 61
Veysey Gdns. Dag. CR34 27

Viaduct Bldgs. EC1 BY39 42
Saffron Hill
Viaduct Pl. E2 CB38 33
Viaduct St. E2 CB38 33
Viaduct, The E18 CH30 17
Vian St. SE13 CE45 52
Vibart Gdns. SW2 BX47 60
Vibart Wk. N1 BX37 33
Havelock St.
Vicarage Av. SE3 CH44 53
Vicarage Clo. Erith CS43 55
Vicarage Ct. W8 BS41 41
Vicarage Gate
Vicarage Cres. SW11 BT44 50
Vicarage Dr. SW14 BN46 49
Vicarage Dr. Bark. CM36 35
Vicarage Farm Rd. Houns. BE44 47
Vicarage Flds. Walt. BD53 65
Vicarage Gdns. SW14 BN46 49
Vicarage Gdns. W8 BS41 41
Vicarage Gdns. Mitch. BU52 68
Vicarage Gte. W8 BS41 41
Vicarage Gro. SE5 BZ44 51
Vicarage La. E6 CK38 35
Vicarage La. E15 CG36 34
Vicarage La. Chig. CM27 17
Vicarage La. Epsom BP58 76
Vicarage La. Ilf. CM33 26
Vicarage La. Walt. CM42 44
Vicarage Path N8 BW33 23
Vicarage Rd. E10 CE33 25
Vicarage Rd. E15 CG36 34
Vicarage Rd. N17 CB30 15
Vicarage Rd. SE18 CM42 44
Vicarage Rd. SW14 BN46 49
Vicarage Rd. Bex. CR47 63
Vicarage Rd. Croy. BY55 78
Vicarage Rd. Dag. CR36 36
Vicarage Rd. Horn. CU33 28
Vicarage Rd. Kings. On T. BK51 66
Vicarage Rd. Sutt. BS56 77
Vicarage Rd. Tedd. BJ49 57
Vicarage Rd. Twick. BG46 47
Vicarage Rd. Twick. BH48 57
Vicarage Rd. Wat. BC26 4
Vicarage Rd. Wdf. Grn. CK29 17
Vicarage Wk. Walt. BC54 65
Vicarage Way NW10 BN34 22
Vicarage Way, Har. BF33 20
Vicars Clo. E15 CH37 35
Vicars Clo. Enf. CA23 8
Vicars Clo. E2 CC37 34
Pennethorpe Clo.
Vicars Hl. SE13 CE45 52
Vicar's Moor La. N21 BY26 8
Vicars Oak Rd. SE19 CA50 60
Vicars Rd. NW5 BV35 32
Vicars Wk. Dag. CO34 27
Viceroy Clo. NW8 BU37 32
Viceroy Rd. SW8 BX41 51
Hartington Rd.
Vickers Rd. Erith CS42 46
Victor App. Horn. CV33 28
Victor Gdns. Horn. CV33 28
Victor Gro. Wem. BL36 30
Victoria Arc. Sthl. BE40 38
Victoria Av. E6 CJ37 35
Victoria Av. EC2 CA39 42
New St.
Victoria Av. N3 BR30 13
Victoria Av. Barn. BT24 7
Victoria Av. E. Mol. BF52 65
Victoria Av. Houns. BE46 47
Victoria Av. Sth. Croy. BZ58 78
Victoria Av. Surb. BK54 66
Victoria Av. Wall. BV55 77
Victoria Av. Wem. BM36 30
Victoria Clo. Barn. BT24 7
Victoria Cotts. N10 BV30 14
Victoria Ct. W3 BM41 39
Victoria Ct. Wem. BM36 30
Victoria Cres. N15 CA32 24
Victoria Cres. SE19 CA50 60
Victoria Cres. SW19 BR50 58
Victoria Dock Rd. E16 CG39 43
Victoria Dr. SW19 BQ47 58
Victoria Embank. SW1 BZ41 42
Victoria Gdns. W11 BS41 41
Church St.
Victoria Gdns. Houns. BE44 47
Victoria Gro. N12 BT28 14
Victoria Gro. W2 BS40 41
Victoria Gro. W8 BT41 41
Victoria Hill Rd. Swan. CT51 73
Victoria La. Barn. BR24 6
Victoria Ms. NW6 BS37 32
Victoria Ms. SW4 BV45 50
Victoria Rise
Victorian Gro. N16 CA34 24
Victorian Rd. N16 CA34 24
Victoria Pk. Rd. E9 CB37 33
Victoria Pk. Sq. E2 CC38 34
Victoria Pl. Epsom BO59 82
Victoria Pl. Rich. BK46 48
Victoria Rise SW8 BV45 50
Victoria Rd. E4 CG26 9
Victoria Rd. E13 CH37 35

Victoria Rd. E17 CF30 16
Victoria Rd. E18 CH30 17
Victoria Rd. N4 BX33 24
Victoria Rd. N9 CA27 15
Victoria Rd. N15 CB31 24
Victoria Rd. N18 CA28 15
Victoria Rd. N22 BW30 14
Victoria Rd. NW4 BQ31 22
Victoria Rd. NW6 BR37 31
Victoria Rd. NW7 BO28 13
Victoria Rd. NW10 BN39 40
Victoria Rd. SW14 BN45 49
Victoria Rd. W3 BN39 40
Victoria Rd. W5 BJ39 39
Victoria Rd. W8 BT41 41
Victoria Rd. Bark. CL36 35
Victoria Rd. Barn. BT24 7
Victoria Rd. Bexh. CR46 54
Victoria Rd. Brom. CJ53 71
Victoria Rd. Buck. H. CJ27 17
Victoria Rd. Bush. BF26 4
Victoria Rd. Chis. CL49 62
Victoria Rd. Couls. BW61 83
Victoria Rd. Dag. CR35 36
Victoria Rd. Dart. CV46 55
Victoria Rd. Erith CT43 55
Victoria Rd. Felt. BC47 56
Victoria Rd. Kings. On T. BL51 66
Victoria Rd. Mitch. BU50 59
Victoria Rd. Rich. BL44 48
Victoria Rd. Rom. CT32 28
Victoria Rd. Ruis. BC33 20
Victoria Rd. Sid. CN48 63
Victoria Rd. Sthl. BE41 38
Victoria Rd. Surb. BK53 66
Victoria Rd. Sutt. BT56 77
Victoria Rd. Twick. BJ47 57
Victoria Sq. SW1 BV41 41
Beeston Pl.
Victoria St. E15 CG36 34
Victoria St. SW1 BV41 41
Victoria St. Belv. CQ42 45
Victoria Ter. N4 BY33 24
Victoria Ter. Har. BG33 20
Victoria Vills. Rich. BL45 48
Victoria Way SE7 CH42 44
Victor Rd. N7 BX34 24
Victor Rd. NW10 BP38 31
Victor Rd. SE20 CC50 61
Victor Rd. Har. BG31 20
Victor Rd. Tedd. BH49 57
Victor Vill. N9 BZ27 15
Victor Wk. Horn. CV33 28
Victory Av. Mord. BT53 68
Victory Pl. SE17 BZ42 42
Victory Pl. SE19 CA50 60
Victory Rd. SW19 BT50 59
Victory Rd. Rain. CU37 37
Victory Sq. SE5 BZ43 51
Victory Way, Rom. CR30 18
View Clo. N6 BU33 23
View Clo. Chig. CM28 17
View Clo. Har. BG31 20
Viewfield Rd. SW18 BR46 49
Viewfield Rd. Sid. CP47 63
Viewland Rd. SE18 CN42 45
View Rd. N6 BU33 23
View, The SE2 CQ42 45
Viga Rd. N21 BY25 8
Vigilant Clo. SE26 CB49 60
Vigilant, The SE26 CB49 60
Vignoles Rd. Rom. CR33 27
Vigo St. W1 BW40 41
Viking Rd. Sthl. BE40 38
Villacourt Rd. SE18 CO43 54
Village Clo. E4 CF28 16
Village Grn. Rd. Dart. CU45 55
Village Rd. N3 BR29 13
Village Rd. Enf. BZ26 8
Village Rd. Enf. CA25 8
Village Row, Sutt. BS57 77
Village, The SE7 CJ43 53
Village Way NW10 BN35 31
Village Way SE21 BZ46 51
Village Way, Beck. CE52 70
Village Way, Pnr. BE33 20
Village Way, Sth. Croy. CB60 84
Village Way, E. Har. BE33 20
Villa Rd. SW9 BY45 51
Villas Rd. SE18 CM42 44
Villa St. SE17 BZ42 42
Villiers Av. Surb. BL53 66
Villiers Av. Twick. BE47 56
Villiers Clo. E10 CE34 25
Villiers Clo. Kings. On T. BL52 66
Villiers Ct. N20 BT26 7
Villiers Path, Surb. BL53 66
Villiers Rd. NW2 BP36 31
Villiers Rd. Beck. CC51 70
Villiers Rd. Islw. BH44 48
Villiers Rd. Kings. On T. BL52 66
Villiers Rd. Sthl. BE40 38
Villiers Rd. Wat. BE25 4
Villiers St. WC2 BX40 42
Vincam Clo. Twick. BF47 56
Vincent Av. Cars. BT59 83
Vincent Av. Surb. BM54 66
Vincent Clo. Barn. BS24 7
Vincent Clo. Brom. CH52 71
Vincent Clo. Esher BF55 74
Vincent Clo. Ilf. CM29 17
Vincent Ct. NW4 BQ31 22
Vincent Gdns. NW2 BO34 22
Vincent Rd. E4 CF29 16
Vincent Rd. N15 BZ31 24
Vincent Rd. N22 BY30 15
Vincent Rd. SE18 CL42 44
Vincent Rd. W3 BN41 40
Palmerston Rd.
Vincent Rd. Couls. BW61 83
Vincent Rd. Croy. CA54 69
Vincent Rd. Dag. CQ36 36
Vincent Rd. Houns. BD44 47
Vincent Rd. Islw. BG44 47
Vincent Rd. Kings. On T. BM52 66
Vincent Rd. Rain. CV38 37
Vincent Rd. Wem. BL36 30
Vincent Row, Hamptn. BG50 56
Vincents Path, Nthlt. BE36 29
Arnold Rd.
Vincent Sq. SW1 BW42 41
Vincent St. E16 CG39 43
Vincent St. SW1 BW42 41
Vincent Ter. N1 BY37 33
Vince St. EC1 BZ38 33
Vine Clo. Surb. BL53 66
Vine Clo. Sutt. BT55 77
Vine Ct. E1 CB39 42
Vine Ct. Har. BL32 21
Vinegar All. E17 CE31 25
Vine Gdns. Ilf. CM35 35
Vinegar Yd. SE1 CA41 42
St. Thomas St.
Vine Hill EC1 BY38 33
Vine La. SE1 CA40 42
Vine Pl. Houns. BF45 47
Vineries Bank NW7 BP28 13
Vineries, The N14 BW25 7
Vineries, The, Enf. CA24 8
Vine Rd. E15 CG36 34
Vine Rd. N17 CD29 16
Love La.
Vine Rd. SW13 BO45 49
Vine Rd. E. Mol. BG52 65
Vine Rd. Orp. CN57 81
Vinery Vill. NW8 BU38 32
Park Rd.
Vine St. Br. EC1 BY38 33
Farringdon Rd.
Vines Av. N3 BS30 14
Vine St. EC3 CA39 42
Vine St. W1 BW40 41
Swallow St.
Vine St. Rom. CS31 28
Vine Yd. SE1 BZ41 42
Sanctuary St.
Vineyard Av. NW7 BR29 13
Vine Yard Clo. SE6 CE47 61
Vineyard Clo. Kings. On T. BK51 66
Vineyard Hill Rd. SW19 BR49 58
Vineyard Pass. Rich. BL46 48
Vine Yard Path SW14 BN45 49
North Worple Way
Vineyard Rd. Felt. BC48 56
Vineyard Row KT1 BK51 66
Vineyard, The, Rich. BK46 48
Vineyard Wk. EC1 BY38 33
Pine St.
Viney Bank, Croy. CD58 79
Viney Rd. SE13 CE45 52
Vining St. SW9 BY45 51
Vinson Clo. Orp. CO54 72
Vinters Pl. EC4 BZ40 42
Viola Av. SE2 CO42 45
Viola Av. Felt. BD46 47
Viola Sq. W12 BO40 40
Violet Gdns. Croy. BY56 78
Violet Hill NW8 BT37 32
Violet La. Croy. BY56 78
Violet Rd. E3 CE38 34
Violet Rd. E17 CE32 25
Violet Rd. E18 CH30 17
Violet St. E2 CB38 33
Three Colts La.
Virgil St. SE1 BX41 42
Virginia Gdns. Ilf. CM30 17
Virginia Rd. E2 CA38 33
Virginia Rd. Th. Hth. BY51 69
Virginia St. E1 CB40 42
Viscount Gro. Nthlt. BD38 29
Wayfarer Rd.
Viscount St. EC1 BZ38 33
Vista Av. Enf. CC23 9
Vista Dr. Ilf. CJ32 26
Vista, The SE9 CJ46 53
Vista Way, Har. BL32 21
Vivian Av. NW4 BP32 22
Vivian Av. Wem. BM35 30
Vivian Clo. Wat. BC27 11
Vivian Gdns. Wat. BC26 4
Vivian Gdns. Wem. BM35 30
Vivian Sq. SE15 CB45 51
Vivian Rd. E3 CD37 34
Vivian Sq. SE15 CB45 51
Scylla Rd.
Vivian Way N2 BT32 23
Vivienne Clo. Twick. BK46 48
Voce Rd. SE18 CM43 53
Voewood Clo. N. Mal. BO53 67
Voltaire Rd. SW4 BW45 50
Voluntary Pl. E11 CH32 26
Vorley Rd. N19 BW34 23
Voss Ct. SW16 BX50 60
Voss St. E2 CB38 33
Vulcan Clo. Wall. BX57 78
Vulcan Gate, Enf. BY23 8
Uplands Park Rd.
Vulcan Rd. SE4 CD44 52
Vulcan Ter. SE4 CD44 52
Vulcan Way N7 BX36 33
Vulcan Way, Croy. CG58 79
Vyner Rd. W3 BN40 40
Vyner St. E2 CB37 33
Vyne, The, Bexh. CR45 54
Vyse Clo. Barn. BQ24 6
Wadding St. SE17 BZ42 42
Waddington Rd. E15 CF35 34
Waddington St. E15 CF36 34
Waddington Ter. Bexh. CR46 54
Waddington Way SE19 BZ50 60
Waddon Clo. Croy. BY55 78
Waddon Ct. Rd. Croy. BY56 78
Waddon Ind. Est. Croy. BX56 78
Waddon Marsh Way, Croy. BX54 69
Waddon New Rd. Croy. BY55 78
Waddon Pk. Av. Croy. BY56 78
Waddon Rd. Croy. BY55 78
Waddon Way, Croy. BY57 78
Wade Av. Orp. CP54 72
Wade Rd. E16 CJ39 44
Leyes Rd.
Wadeville Clo. Belv CR43 54
Wade's Gro. N21 BY26 8
Wade's Hill N21 BY26 8
Wades La. Tedd. BJ49 57
Wadeson St. E2 CB37 33
Wades Pl. E14 CE40 43
Wadeville Av. Rom. CQ32 27
Wadham Av. E17 CE29 16
Wadham Gdns. NW3 BU37 32
Wadham Gdns. Grnf. BG36 29
Wadham Rd. E17 CE30 16
Wadham Rd. SW15 BR45 49
Wadhurst Clo. SE20 CB51 69
Wadhurst Rd. SW8 BW44 50
Wadhurst Rd. W4 BN41 40
Wadley Rd. E11 CG33 25
Wadsworth Clo. Grnf. BK37 30
Wadsworth Clo. Enf. CC25 9
Church Rd.
Wadsworth Rd. Grnf. BJ37 30
Wager St. E3 CD38 34
Waggon La. N17 CB29 15
Waghorn Rd. E13 CJ37 35
Waghorn Rd. Har. BK31 21
Waghorn St. SE15 CB45 51
Wagner St. SE15 CC43 52
Ilderton Rd.
Wagtail Gdns. Sth. Croy. CC58 79
Waid Clo. Dart. CW46 55
Wainfleet Av. Rom. CS30 19
Wainford Clo. SW19 BQ47 58
Windlesham Rd.
Waite Daives Rd. SE12 CG47 61
Waite St. SE15 CA43 51
Waithman St. EC4 BY39 42
Pilgrim St.
Wakefield Gdns. SE19 CA50 60
Wakefield Gdns. Ilf. CK32 26
Wakefield Rd. N11 BW28 14
Wakefield Rd. N15 CA32 24
Wakefield Rd. Rich. BK46 48
Wakefield St. E6 CJ37 35
Wakefield St. N18 CB28 15
Wakefield St. WC1 BX38 33
Wakehams Hill, Pnr. BE31 20
Wakehams Hl. Pnr. BE31 20
Wakeham St. N1 BZ36 33
Wakehurst Rd. SW11 BU46 50
Wakeling Rd. W7 BH39 39
Wakeling St. E14 CD39 43
Wakelin Rd. E15 CG37 34
Wakeman Rd. NW10 BQ38 31
Wakemans Hill Av. NW9 BN32 22
Wakerfield Clo. Horn. CW32 28
Wakering Rd. Bark. CM36 35
Wakley St. EC1 BY38 33
Walberswick St. SW8 BX43 51
Walbrook EC4 BZ40 42
Walburgh St. E1 CB39 42
Walburton Rd. Pur. BW60 83
Walcorde Av. SE17 BZ42 42
Browning St.
Walcot Sq. SE11 BY42 42
Walcot Rd. Enf. CD23 9
Waldeck Gro. SE27 BY48 60
Waldeck Rd. N15 BY31 24
Waldeck Rd. SW14 BN45 49
Lower Richmond Rd.
Waldeck Rd. W4 BM43 48
Waldeck Rd. W13 BJ39 39
Waldeck Rd. Dart. CW47 64
Waldeck Ter. SW14 BN45 49
Lower Richmond Rd.
Waldegrave Av. Tedd. BH49 57
Waldegrave Gdns. Twick. BH48 57
Waldegrave Pk. Twick. BH49 57
Waldegrave Rd. N8 BY31 24
Waldegrave Rd. SE19 CA50 60
Waldegrave Rd. W5 BL39 39
Waldegrave Rd. Brom. CK52 71
Waldegrave Rd. Dag. CP34 27
Waldegrave Rd. Twick. BH49 57
Waldegrove, Croy. CA55 78
Waldemar Av. SW6 BR44 49
Waldemar Av. W5 BK40 39
Waldemar Rd. SW19 BS49 59
Walden Av. N13 BZ28 15
Walden Av. Chis. CK49 62
Walden Av. Rain. CS38 37
Walden Clo. Belv. CQ42 45
Walden Gdns. Th. Hth. BX52 69
Waldenhurst Rd. Orp. CP54 72
Walden Rd. N17 BZ30 15
Lordship La.
Walden Rd. Chis. CK50 62
Walden Rd. Horn. CV32 28
Waldens Clo. Orp. CP54 72
Waldenshaw Rd. SE23 CC47 61
Waldens Rd. Orp. CQ54 72
Walden St. E1 CB39 42
New Rd.
Walden Way NW7 BQ29 13
Walden Way, Horn. CV32 28
Walden Way, Ilf. CN29 18
Waldo Pl. Mitch. BU50 59
Waldo Rd. NW10 BP38 31
Waldo Rd. Brom. CJ52 71
Waldram Cres. SE23 CC47 61
Waldram Pk. Rd. SE23 CC47 61
Waldram Pl. SE23 CC47 61
Waldram Cres.
Waldron Gdns. Brom. CF52 70
Waldronhyrst, Croy. BY56 78
Waldron Ms. SW3 BU43 50
Old Church St.
Waldron Rd. SW18 BT48 59
Waldron Rd. Har. BH33 21
Waldrons Path, Croy. BZ56 78
Waldrons, The, Croy. BY56 78
Waldron Yd. Har. BG34 20
Waleran Bldgs. SE1 CA42 42
Old Kent Rd.
Walerand Rd. SE13 CF44 52
Wales Av. Cars. BU56 77
Wales Farm Rd. W3 BN39 40
Waley St. E1 CC39 43
Walfield Av. N20 BS26 7
Walford Rd. N16 CA35 33
Walfrey Gdns. Dag. CQ36 36
Walham Gro. SW6 BS43 50
Walham Rise SW19 BR49 58
Walham Yd. SW6 BS43 50
Walham Gro.
Walkden Rd. Chis. CL49 62
Walker Clo. Hamptn. BE50 56
Walker Clo. DA1 CT45 55
Walker Clo. SE18 CM42 44
Walkers Ct. W1 BW40 41
Peter St.
Walkerscroft Mead SE21 BZ47 60
Walkfield Dr. Epsom BP62 82
Walkford Way SE15 CA43 51
Walkley Rd. Dart. CU46 55
Walks, The N2 BT31 23
Walk, The, Horn. CW34 28
Wallace Cres. Cars. BU56 77
Wallace Flds. Epsom BO59 82
Wallace Rd. N1 BZ36 33
Wallbutton Rd. SE4 CD44 52
Wallcote Av. NW2 BQ33 22
Wall End Rd. E6 CK36 35
Wallenger Av. Rom. CU31 28
Waller Rd. SE14 CC44 52
Wallers Clo. Wdf. Grn. CK29 17
Wallers Hoppit, Loug. CK23 10
Wallflower St. W12 BO40 40
Wallgrave Rd. SW5 BS42 41
Wallgrave Rd. W8 BS43 50
Wallhouse Rd. Erith CU43 55
Wallingford Av. W10 BQ39 40
Wallington Ct. Wall. BV57 77
Wallington Rd. Ilf. CN33 27
Wallington Sq. Wall. BV57 77
Wallis Clo. SW11 BT45 50
Wayland Rd.
Wallis Clo. Dart. CT48 64
Wallis Rd. E9 CD36 34
Wallis Rd. Sthl. BF39 38
Wallis's Cotts. SW2 BX47 60
Wallorton Gdns. SW14 BN45 49
Wall St. N1 BZ36 33
Wallwood Rd. E11 CF33 25
Wallwood St. E14 CD39 43
Wally St. E1 CC39 43
Walmer Clo. Rom. CR30 18
Walmer Clo. Barn. BT23 7
Walmer Gdns. W13 BJ41 39
Walmer Rd. W11 BR40 40
Walmer Ter. SE18 CM42 44

Walmgate Rd. Grnf. BJ37 30
Walmington Fold N12 BS29 14
Walm La. NW2 BQ36 31
Walney Wk. N1 BZ36 33
Marquess Est.
Walnut Clo. Cars. BU56 77
Park Hill
Walnut Clo. Epsom BO61 82
Walnut Clo. Ilf. CM31 26
Civic Way
Walnut Grn. Bush. BE23 4
Walnut Gro. Enf. BZ25 8
Walnuts Rd. Orp. CO54 72
Walnut Tree Av. Dart. CW48 64
Walnut Tree Clo. Chis. CM51 71
Walnut Tree Cotts. SW19 BR49 58
Church Rd.
Walnut Tree Rd. SE10 CG42 43
Walnut Tree Rd. Brent. BL43 48
Walnut Tree Rd. Dag. CQ34 27
Walnut Tree Rd. Erith CT42 46
Walnut Tree Rd. Houns. BE43 47
Walnut Tree Wk. SE11 BY42 42
Walnut Way, Buck. H. CJ27 17
Walnut Way, Ruis. BD36 29
Walnut Way, Swan. CS51 73
Walpole Av. Couls. BU62 83
Walpole Av. Rich. BL44 48
Walpole Clo. W13 BK41 39
Walpole Clo. Pnr. BF29 11
Walpole Ct. Twick. BH48 57
Walpole Cres. Tedd. BH49 57
Walpole Gdns. W4 BN42 40
Walpole Gdns. Twick. BH48 57
Walpole Pl. Tedd. BH49 57
Walpole Rd. E6 CJ36 35
Walpole Rd. E17 CD31 25
Walpole Rd. E18 CG30 16
Walpole Rd. N17 BZ30 15
Walpole Rd. SW19 BT50 59
Walpole Rd. Brom. CJ53 71
Walpole Rd. Croy. BZ55 78
Walpole Rd. Surb. BL54 66
Walpole Rd. Tedd. BH49 57
Walpole Rd. Twick. BH48 57
Walpole St. SW3 BU42 41
Walpole Way, Barn. BQ25 6
Walrond Av. Wem. BL35 30
Walsham Rd. SE14 CC44 52
Walsham Rd. Felt. BC47 56
Walsingham Gdns. Epsom BO56 76
Walsingham Pk. Chis. CM51 71
Walsingham Rd. W13 BJ40 39
Walsingham Rd. E5 CB34 24
Walsingham Rd. Croy. CF58 79
Walsingham Rd. Enf. BZ24 8
Walsingham Rd. Mitch. CO51 72
Walsingham Walk DA17 CR43 54
Walsing Rd. Mitch. BU53 68
Walters Rd. SE25 CA52 69
Walters Rd. Enf. CC25 9
Walter's St. E2 CC38 34
Walter St. Kings. On T. BL51 66
Canbury Pass.
Walters Yd. Brom. CH51 71
Walter Ter. E1 CC39 43
Watertcon Rd. W9 BR38 31
Walter Wk. Edg. BN29 13
Waltham Av. NW9 BM32 21
Waltham Clo. Dart. CU46 65
Waltham Clo. Orp. CP54 72
Waltham Ct. Har. BF30 11
Waltham Dr. Edg. BM30 12
Waltham Pk. Wy. E17 CE30 16
Waltham Rd. Cars. BT54 68
Waltham Rd. Sthl. BE41 38
Waltham Rd. Wdf. Grn. CK29 17
Walthamstow Av. E4 CD29 16
Waltham Way E4 CD28 16
Waltheof Av. N17 BZ30 15
Waltheof Gdns. N17 BZ30 15
Walthorne Gdns. Dag. CR36 36
Acre Rd.
Walton Av. Har. BE35 29
Walton Av. N. Mal. BO52 67
Walton Av. Sutt. BR55 76
Walton Clo. NW2 BP34 22
Walton Clo. Har. BG31 20
Walton Cres. Har. BE35 29
Walton Dr. Har. BG31 20
Walton Gdns. W3 BM39 39
Walton Gdns. Grnf. BH35 30
Walton Gdns. Wem. BL34 21
Walton Grn. Croy. CE58 79
Walton Pk. Walt. BD55 74
Walton Pk. La. Walt. BD55 74
Walton Pl. SW3 BU41 41
Walton Rd. E12 CL34 26
Walton Rd. E13 CJ37 35
Walton Rd. Bush. BD24 4
Walton Rd. E. Mol. BE52 65
Walton Rd. Epsom BO62 82
Walton Rd. Har. BG31 20
Walton Rd. Rom. CQ29 18
Walton Rd. Sid. CO48 63
Walton Rd. Walt. BD53 65
Walton St. SW3 BU42 41
Walton St. Enf. BZ23 8

Walton Ter. SW8 BX43 51
South Lambeth Rd.
Walton Way W3 BM39 39
Walton Way, Mitch. BW52 68
Walverns Clo. Wat. BD25 4
Walworth Rd. SE17 BZ42 42
Walworth St. SE17 BZ42 42
Walwyn Av. Brom. CJ52 71
Wanborough Dr. SW15 BP47 58
Wandle Bank, SW19 BT50 59
Wandle Bank Croy. BX55 78
Wandle Ct. Epsom BN56 76
Wandle Ct. Gdns. Croy. BX55 78
Wandle Rd. SW17 BU48 59
Wandle Rd. Croy. BX55 78
Wandle Rd. Croy. BZ55 78
Wandle Rd. Mord. BT52 68
Wandle Rd. Wall. BV55 77
Wandle Side Croy. BX55 78
Wandle Side Wall. BV55 77
Wandle Way, Mitch. BU53 68
Wandon Rd. SW6 BS43 50
Wandsworth Br. Rd. SW6 BS44 50
Wandsworth High St. BS46 50
SW18
Wandsworth Plain, SW18 BS46 50
Wandsworth Rd. SW8 BV45 50
Wangey Rd. Rom. CP33 27
Wanless Rd. SE24 BZ45 51
Wanley Rd. SE5 BZ45 51
Wanlip Rd E13 CH38 35
Wannock Gdns. Ilf. CL29 17
Wansbeck Rd. E9 CC36 34
Wansbury Way, Swan. CU53 73
Wansey St. SE17 BZ42 42
Wansford Pk. Borwd. BN24 6
Wansford Rd. Wdf. Grn. CH30 17
Wanstead Clo. Brom. CJ51 71
Wanstead La. Ilf. CJ32 26
Wanstead Park Av. E12 CJ33 26
Wanstead Park Rd. Ilf. CJ32 26
Wanstead Pl. E11 CH32 26
Wanstead Rd. Brom. CJ51 71
Wansunt Rd. Bex. CS47 64
Wantage Rd. SE12 CG46 52
Wantz La. Rain. CU38 37
Wantz Rd. Dag. CR35 36
Wapping Dock St. E1 CB40 42
Cinnamon St.
Wapping High St. E1 CB40 42
Wapping La. E1 CB40 42
Wapping Wall. E1 CC40 43
Warbank Clo. Croy. CG58 79
Warbank Cres. Croy. CG58 79
Warbeck Rd. W12 BP40 40
Warberry Rd. N22 BX30 15
Warboys App. BM50 57
Kings. On T.
Warboys Cres. E4 CF28 16
Warboys Rd. BM50 57
Kings. On T.
Warburton Clo. Har. BG30 11
Warburton Rd. Twick. BF47 56
Warburton Ter. E17 CE30 16
Wardale Clo. SE16 CB41 42
Southwark Pk. Rd.
Ward Clo. Erith CS43 55
Wardell Clo. NW9 BO29 13
Wardell Field NW9 BO30 13
Warden Av. Har. BE33 20
Warden Av. Rom. CS28 19
Warden Ct. Har. BE33 20
Warden Rd. NW5 BV36 32
Wardens Gro. SE1 BZ40 42
Wardle St. E9 CC35 34
Wardley St. SW18 BS47 58
Wardo Av. SW6 BR44 49
Wardour St. W1 BW39 41
Ward Rd. E15 CF37 34
Ward Rd. N19 BX34 23
Wardrobe, The, Rich. BK46 48
Old Palace Yard
Wardove Rd. Sth. Croy. BY57 78
Wards La. Borwd. BH23 5
Wards Rd. Ilf. CM33 26
Wareham Clo. Houns. BF45 47
Waremead Rd. Ilf. CL32 26
Warenford Way, Borwd. BM23 5
Warfield Rd. NW10 BO39 40
Warfield Rd. Hampton. BF51 65
Wargrave Av. N15 CA32 24
Wargrave Rd. Har. BG34 20
Warham Rd. N4 BY32 24
Warham Rd. Har. BH30 12
Warham Rd. Sth. Croy. BY56 78
Waring Clo. Orp. CN57 81
Waring Dr. Orp. CN57 81
Waring Rd. Sid. CP50 63
Waring St. SE27 BZ49 60
Warkworth Gdns. Islw. BJ43 48
Warkworth Rd. N17 BZ29 15
Warland Rd. SE18 CM43 53
Warley Av. Dag. CQ33 27
Warley Av. Hayes BC39 38
Warley Clo. E10 CC33 24
Warley Rd. N9 CC27 16
Warley Rd. Hayes BC39 38
Warley Rd. Ilf. CL30 17

Warley Rd. Wdf. Grn. CH29 17
Warley St. E2 CC38 34
Warlingham Rd. Th. Hth. BY52 69
Warlock Rd. W9 BR38 31
Warlters Clo. N7 BX35 33
Warlters Mews N7 BX35 33
Warlters Pl. N7 BX35 33
Warlters Rd. N7 BX35 33
Warltersville Rd. N19 BX33 24
Warmington Rd. SE24 BZ46 51
Warmington St. E13 CH38 35
Warminster Gdns. SE25 CB51 69
Warminster Rd. SE25 CA51 69
Warminster Sq. SE25 CB51 69
Warminster Way, Mitch. BV51 68
Warndon St. SE16 CC42 43
Warneford Pl. Wat. BE25 4
Warneford Rd. Orp. CN56 81
Warneford Rd. Har. BK31 21
Warneford St. E9 CB37 33
Warner Av. Sutt. BR55 76
Warner Clo. E15 CG35 34
Warner Clo. NW9 BO33 22
Warner Pl. E2 CB37 33
Warner Rd. E17 CD31 25
Warner Rd. N8 BW31 23
Warner Rd. SE5 BZ44 51
Warner Rd. Brom. CG50 61
Warners Clo. Wdf. Grn. CH28 17
Warner's La. Kings. On T. BK49 57
Warner St. EC1 BY38 33
Warnford Rd. Orp. CN56 81
Warnham Court Rd. Cars. BU57 77
Warnham Rd. N12 BU28 14
Warple Way W3 BO40 40
Warren Av. E10 CF34 25
Warren Av. Brom. CG50 61
Warren Av. Orp. CN56 81
Warren Av. Rich. BM45 48
Warren Av. Sth. Croy. CC57 79
Warren Av. Sutt. BR58 76
Warren Clo. SE21 BZ46 51
Warren Clo. N9 CC26 9
Warren Clo. Bexh. CR46 54
Warren Clo. Esher BF56 74
Warren Ct. Beck. CD50 61
Warren Ct. Chig. CM28 17
Warren Cres. N9 CA26 8
Warren Cutting BN50 58
Kings. On T.
Warrender Way Ruis. BC33 20
Warren Dr. Grnf. BF38 29
Warren Dr. Horn. CU35 37
Warren Dr. Orp. CO56 81
Warren Dr. Ruis. BD33 20
Warren Dr. S. Surb. BN54 67
Warren Dr. N. Surb. BN54 66
Warren Farm Clo. E11 CJ32 26
Warren Rd.
Warren Flds. Stan. BK28 12
Warren Gdns. E15 CF35 34
Warren Gdns. Orp. CO56 81
Warren Gro. Borwd. BN24 6
Warren Hill Epsom BN61 82
Warren Hill Loug. CJ25 10
Warren Hill Rd. Loug. CH25 10
Warren Ho. Kings. On T. BN50 58
Warren La. SE18 CL41 44
Warren La. Stan. BH27 12
Warren Mead Bans. B061 82
Warren Pk. Kings. On T. BN50 58
Warren Pk. Rd. Sutt. BT57 77
Warren Pond Rd. E4 CG26 9
Warren Ri N. Mal. BN51 67
Warren Rd. E4 CF27 16
Warren Rd. E10 CF34 25
Warren Rd. E11 CJ32 26
Warren Rd. NW2 BN37 22
Warren Rd. SW19 BU50 59
Warren Rd. Bans. BQ60 82
Warren Rd. Bexh. CR46 54
Warren Rd. Brom. CH55 80
Warren Rd. Bush. BG26 4
Warren Rd. Croy. CA54 69
Warren Rd. Dart. CW48 64
Warren Rd. Ilf. CM32 26
Warren Rd. Kings. On T. BN50 58
Warren Rd. Orp. CN56 81
Warren Rd. Pur. BY59 84
Warren Rd. Sid. CP48 63
Warren Rd. Twick. BG46 47
Warren St. Ms. W1 BW38 32
Warren St.
Warren St. W1 BW38 32
Warren Ter. Rom. CP31 27
Warren, The, E12 CK35 35
Warren, The, Cars. BT58 77
Warren, The, Hayes BC39 38
Warren, The, Houns. BE43 47
Warren, The, Wor. Pk. BN56 76
Warren, The, Dr. E11 CJ33 26
Warren Wk. SE7 CJ43 53
Warren Way NW7 BR29 13
Warren Wd. Rd. Brom. CG55 79
Warren Wood Clo. Brom. CG55 79
Warrick Gro. Croy. CA55 78
Warriner Av. Horn. CV34 28

Warriner Gdns. SW11 BU44 50
Warrington Cres. W9 BT38 32
Warrington Gdns. W9 BS38 32
Warwick Av.
Warrington Gdns. Horn. CV32 28
Warrington Rd. Croy. BY55 78
Warrington Rd. Dag. CP34 27
Warrington Rd. Har. BH32 21
Warrington Rd. Rich. BL46 48
Hermitage, The,
Warrington Sq. Dag. CP34 27
Warrior Rd. SE5 BY43 51
Warrior Sq. E12 CL35 35
Warsaw Clo. Ruis. BC36 29
Warspite Rd. SE18 CK41 44
Warton Rd. E15 CF37 34
Warwick Av. W2 BS38 32
Warwick Av. W9 BS38 32
Warwick Av. Edg. BM27 12
Warwick Av. Har. BE35 29
Warwick Clo. Bush. BH26 5
Warwick Clo. Hamptn. BG50 56
Warwick Clo. Orp. CO55 81
Warwick Cotts. Barn. BT25 7
Warwick Ct. WC1 BX39 42
Warwick Ct. Surb. BL55 75
Warwick Cres. W2 BT39 41
Warwick Dene W5 BL40 39
Warwick Dr. SW15 BP45 49
Warwick Est. W2 BS39 41
Warwick Gdns. N4 BZ32 24
Warwick Gdns. W14 BR41 40
Warwick Gdns. Ilf. CL33 26
Warwick Gdns. Rom. CV31 29
Warwick Gdns. Surb. BH53 66
Warwick Gro. E5 CB33 24
Warwick Gro. Surb. BL54 66
Warwick Ho. St. SW1 BW40 41
Warwick La. EC4 BY39 42
Warwick La. Upmin. BX39 42
Warwick Pl. W5 BK41 39
Warwick Rd.
Warwick Pl. W9 BT39 41
Warwick Pl. Ms. W9 BT39 41
Warwick Pl. N. SW1 BW42 41
Warwick Rd. E4 CE28 16
Warwick Rd. E11 CH32 26
Warwick Rd. E12 CK35 35
Warwick Rd. E15 CG36 34
Warwick Rd. E17 CD30 16
Warwick Rd. N11 BW29 14
Warwick Rd. N18 CA28 15
Warwick Rd. SE20 CB52 69
Warwick Rd. SW5 BR42 40
Warwick Rd. W5 BK41 39
Warwick Rd. W14 BR42 40
Warwick Rd. Barn. BS24 7
Warwick Rd. Borwd. BN24 6
Warwick Rd. Couls. BW60 83
Warwick Rd. Houns. BC45 47
Warwick Rd. Kings. On T. BK51 66
Warwick Rd. N. Mal. BN52 67
Warwick Rd. Rain. CV38 37
Warwick Rd. Sid. CO49 63
Warwick Rd. Sthl. BE41 38
Warwick Rd. Surb. BH53 66
Warwick Rd. Sutt. BS56 77
Warwick Rd. Th. Hth. BY52 69
Warwick Rd. Twick. BH47 57
Warwick Rd. Well. CP45 54
Warwick Row EC4 BV41 41
Warwick Row SW1 BY39 42
Warwick Sq. SW1 BW42 41
Warwick Sq. EC4 BY39 42
Warwick St. W1 BW40 41
Warwick Ter. SE18 CM43 53
Warwick Way SW1 BW42 41
Washington Av. E12 CK35 35
Washington Rd. E6 CJ36 35
Washington Rd. E18 CG30 16
Washington Rd. SW13 BP43 49
Washington Rd. BM51
Kings. On T.
Wastdale Rd. SE23 CC47 61
Watchfield Ct. W4 BN42 40
Watcombe Cotts. Rich. BM43 48
Bushwood Rd.
Watcombe Rd. SE25 CB53 69
Waterbank Rd. SE6 CE48 61
Waterbeach Rd. Dag. CP35 36
Waterbrook La. NW4 BO32 22
Watercroft Rd. Sev. CO58 81
Waterdale Rd. SE2 CO43 54
Waterden Rd. E15 CE35 34
Waterer Rise, Wall. BW57 77
Waterer Rd. N20 BT27 14
Waterfall Rd.
Waterfall Cotts. SW19 BT50 59
Waterfall Rd. N11 BV28 14
Waterfall Rd. N14 BV28 14
Waterfall Rd. SW19 BT50 59
Waterfall Ter. SW17 BU50 59
Waterfield Clo. SE28 CO40 45
Waterfield Dr. Warl. CB62 84
Waterford Rd. SW6 BS43 50

Water Gdns. Stan.	BJ29	12
Gordon Av.		
Watergate EC4	BY40	42
Tudor St.		
Watergate, Wat.	BD27	11
Watergate St. SE8	CE43	52
Watergate Wk. WC2	BX40	42
Waterhall Av. E4	CG27	16
Waterhead Clo. Erith	CT43	55
Waterhouse Clo. W6	BQ42	40
Gt. Church La.		
Water La. E15	CG36	34
Water La. Ilf.	CN34	27
Water La. Kings. On T.	BK51	66
Water La. Rich.	BK46	48
Water La. Sid.	CQ48	63
Water La. Twick.	BJ47	57
Water La. Wat.	BD24	4
Waterloo Br. SE1	BX40	42
Waterloo Br. WC2	BX40	42
Waterloo Est. E2	CB37	33
Waterloo Gdns. E2	CC37	34
Waterloo Gdns. Rom.	CS32	28
St. Andrews Rd.		
Waterloo Ms. SE5	BZ43	51
Elmington Rd.		
Waterloo Pl. NW6	BR36	31
Willesdon La.		
Waterloo Pl. SW1	BW40	41
Waterloo Pl. Rich.	BL45	48
Quadrant, The		
Waterloo Pl. Rich.	BM43	48
Waterloo Rd. E6	CJ36	35
Waterloo Rd. E7	CG35	34
Wellington Rd.		
Waterloo Rd. E10	CE33	25
Waterloo Rd. NW2	BP33	22
Waterloo Rd. SE1	BX40	42
Waterloo Rd. Epsom	BN59	82
Waterloo Rd. Ilf.	CM30	17
Waterloo Rd. Rom.	CT32	28
Waterloo Rd. Sutt.	BT56	77
Waterloo St. EC1	BZ38	33
Lever St.		
Waterloo Ter. N1	BY36	33
Waterloo Wk. E9	CC35	34
Churchill Wk.		
Waterlow Ct. NW11	BS33	23
Waterlow Rd. N19	BW33	23
Waterman Clo. Wat.	BC25	4
Waterman St. SW15	BQ45	49
Watermead La. Cars.	BU53	68
Watermead Rd. SE6	CE49	61
Water Mill Clo. Rich.	BK48	57
Water Mill La. N18	CA28	15
Water Mill Way, Felt.	BE48	56
Water Rd. Wem.	BL37	30
Watersedge, Epsom	BN56	76
Watersfield Way, Edg.	BK29	12
Waters Gdns. Dag.	CR35	36
Waterside, Beck.	CD51	70
Waterside, Dart.	CT46	55
Waterside Clo. SE16	CB41	42
Bevington St.		
Waterside Pl. NW1	BV37	32
Princess Rd.		
Waterside Rd. Sthl.	BF41	38
Waterson St. E2	CA38	33
Watersplash La. Hayes	BC42	38
Waters Rd. SE6	CG48	61
Waters Rd. Kings. On T.	BM51	66
Waters Sq. Kings. On T.	BM52	66
Water St. WC2	BY40	42
Maltravers St.		
Water St. Kings. On T.	BL51	66
Canbury Pass.		
Water Tower Hill, Croy.	BZ56	78
Waterville Rd. N17	BZ30	15
Waterworks La. E5	CC34	25
Lea Bridge Rd.		
Waterworks Rd. SW2	BX46	51
Watery La. SW20	BR51	67
Watery La. Nthlt.	BD37	29
Watery La. Sid.	CO50	63
Wates Way, Mitch.	BU53	68
Watford By-pass, Borwd.	BJ26	5
Watford Clo. SW11	BU44	50
Petworth St.		
Watford Flds. Rd. Wat.	BD25	4
Watford Rd. E16	CH39	44
Watford Rd. Borwd.	BK25	5
Watford Rd. Har.	BJ33	21
Watford Rd. Wat.	BF23	4
Watford Way NW4	BN28	13
Watford Way NW7	BN28	13
Watkin Rd. Wem.	BM34	21
Watkinson Rd. N7	BX36	33
Watling Av. Edg.	BN30	13
Watling Ct. Bush.	BK25	5
Watling Ct. EC4	BZ39	42
Watling St.		
Watling Farm Clo. Stan.	BJ26	5
Watling Gdns. NW2	BR36	31
Watling St. EC4	BZ39	42
Watling St. Bexh.	CR45	54
Watling St. Borwd.	BK23	5
Watlington Gro. SE26	CD49	61
Watney Rd. SW14	BN45	49
Watney's Rd. Mitch.	BW53	68
Watson Av. E6	CL36	35
Watson Av. Sutt.	BR55	76
Watson Clo. N16	BZ35	33
Matthias Rd.		
Watson Clo. SW19	BU50	59
Watsons Ms. W1	BU39	41
Crawford Pl.		
Watson's Rd. N22	BX30	15
Watson's St. SE8	CD43	52
Watson St. E13	CH37	35
Watsons Yd. NW2	BO34	22
Wattendon Rd. Ken.	BY61	84
Wattisfield Rd. E5	CC34	25
Watt's Br. Rd. Erith	CT43	55
Watts Gro. E3	CE39	43
Watt's La. Chis.	CL51	71
Watts La. Tedd.	BJ49	57
Watts Rd. Surb.	BJ54	66
Watts St. E1	CB40	42
Wat Tyler Rd. SE10	CF44	52
Wauthier Clo. N13	BY28	15
Wavell Dr. Sid.	CN46	54
Wavel Ms. NW6	BS36	32
Wavendon Av. W4	BN42	40
Waveney Av. SE15	CB45	51
Waverley Av. E4	CD28	16
Waverley Av. Ken.	CA61	84
Waverley Av. Surb.	BM53	66
Waverley Av. Sutt.	BS55	77
Waverley Av. Twick.	BE47	56
Waverley Av. Wem.	BL36	30
Waverley Clo. E18	CJ30	17
Waverley Clo. Brom.	CJ53	71
Waverley Cres. SE18	CM42	44
Waverley Cres. Rom.	CV29	19
Waverley Gdns. NW10	BL37	30
Waverley Gdns. Bark.	CN37	36
Waverley Gdns. Ilf.	CM30	17
Waverley Gdns. Nthwd.	BC30	11
Waverley Gro. N3	BQ31	22
Waverley Pl. NW8	BT37	32
Waverley Rd. E17	CF31	25
Waverley Rd. E18	CJ30	17
Waverley Rd. N8	BW32	23
Waverley Rd. N17	CA30	15
Waverley Rd. N17	CB29	15
Waverley Rd. SE18	CM42	44
Waverley Rd. SE25	CB52	69
Waverley Rd. Enf.	BY24	8
Waverley Rd. Epsom	BP57	76
Waverley Rd. Har.	BE33	20
Waverley Rd. Rain.	CU38	37
Waverley Rd. Sthl.	BF40	38
Waverley Wk. W2	BS39	41
Waverley Way, Cars.	BU57	77
Waverton Rd. SW18	BT47	59
Waverton St. W1	BV41	41
Wavertree Ct. SW2	BX47	60
Wavertree Rd. E18	CH30	17
Wavertree Rd. SW2	BX47	60
Waxlow Cres. Sthl.	BF39	38
Waxlow Rd. NW10	BN37	31
Waxwell Clo. Pnr.	BD30	11
Waxwell La. Pnr.	BD30	11
Waye Av. Houns.	BC44	47
Wayfarer Rd. Nthlt.	BD38	29
Wayford St. SW11	BU44	50
Wayland Av. E8	CB35	33
Waylands, Swan.	CT52	73
Waylett Pl. Wem.	BK34	21
Wayne Clo. Orp.	CN55	81
Waynefleet Tower Av. Esher	BF55	74
Wayneflete Sq. W10	BQ40	40
Waynfleet Av. Croy.	BY55	78
Waynflete St. SW18	BT48	59
Wayside NW11	BR33	22
Wayside SW14	BN46	49
Wayside Av. Bush.	BG25	4
Wayside Av. Horn.	CV34	28
Wayside Clo. N14	BW25	7
Wayside Clo. Rom.	CT31	28
Wayside Ct. Twick.	BK46	48
Arlington Rd.		
Wayside Gdns. SE9	CK49	62
Wayside Gro.		
Wayside Gdns. Dag.	CR35	36
Wayside Gro. SE9	CK49	62
Wayside Mews, Ilf.	CL32	26
Gaysham Av.		
Weald Clo. Brom.	CK55	80
Weald La. Har.	BG30	11
Weald Rise, Har.	BH29	12
Weald Sq. E5	CB34	24
Rossington St.		
Wealdstone Rd. Sutt.	BR55	76
Weald, The, Chis.	CK50	62
Weald Way, Rom.	CR32	27
Wealdwood Gdns. Pnr.	BF29	11
High Banks Rd.		
Weale Rd. E4	CF27	16
Weatherley Clo. E3	CD39	43
Weardale Gdns. Enf.	BZ23	8
Weardale Rd. SE13	CF45	52
Wear Pl. E2	CB38	33
Wearside Rd. SE13	CE45	52
Wear St. E2	CB38	33
Teesdale St.		
Weavers La. SE1	CA40	42
Weaver St. E1	CB38	33
Weavers Ter. SW6	BS43	50
Micklethwaite Rd.		
Weaver Wk. SE27	BY49	60
Webber Clo. Erith	CU43	55
Webber Row SE1	BY41	42
Webber St. SE1	BY41	42
Webb Est. E5	CB33	24
Webb Gdns. E13	CH38	35
Kelland Rd.		
Webb Rd. SE3	CG43	52
Webbs Rd. SW11	BU45	50
Webbs Rd. Hayes	BC38	29
Webb St. SE1	CA41	42
Webster Gdns. W5	BK40	39
Webster Rd. E11	CF34	25
Webster Rd. SE16	CB41	42
Webster Vill. W5	BK40	39
Webster Gdns.		
Wedderburn Rd. NW3	BT35	32
Wedderburn Rd. Bark.	CM37	35
Wedgewood Way SE19	BZ50	60
Beulah Hill		
Wedlake Clo. Horn.	CW33	28
Wedlake St. W10	BR38	31
Kensal Rd.		
Wedmore Av. Ilf.	CL30	17
Wedmore Gdns. N19	BW34	23
Holloway Rd.		
Wedmore Ms. N19	BW34	23
Wedmore St.		
Wedmore Rd. Grnf.	BG38	29
Wedmore St. N19	BW34	23
Wednesbury Gdns. Rom.	CW29	19
Wednesbury Grn. Rom.	CW29	19
Wednesbury Rd. Rom.	CW29	19
Weech Rd. NW6	BS35	32
Weedington Rd. NW5	BV35	32
Weekley Sq. SW11	BT45	50
Th. Baines Rd.		
Weigall Rd. SE12	CH46	53
Weighhouse St. W1	BV40	41
Weighton Ms. SE20	CB51	69
Weighton Rd.		
Weighton Rd. SE20	CB51	69
Weighton Rd. Har.	BG30	11
Weimar St. SW15	BR45	49
Weinhurst Gdns. Sutt.	BT56	77
Weir Clo. Bex.	CR47	63
Weirdale Av. N20	BU27	14
Weir Hall Av. N18	BZ29	15
Weir Hall Clo. N17	BZ29	15
Weir Hall Gdns. N18	BZ28	15
Weir Hall Rd. N17	BZ29	15
Weir Hall Rd. N18	BZ29	15
Weir Rd. SW12	BW47	59
Weir Rd. SW19	BS49	59
Weir Rd. Walt.	BC53	65
Weirs Pass. NW1	BW38	32
Chalton St.		
Weiss Rd. SW15	BQ45	49
Welbeck Av. Brom.	CH49	62
Welbeck Av. Hayes	BC38	29
Welbeck Av. Sid.	CO47	63
Welbeck Clo. N12	BT28	14
Welbeck Clo. Borwd.	BM24	5
Welbeck Clo. Epsom	BP57	76
Welbeck Clo. N. Mal.	BO53	67
Welbeck Rd. E6	CJ38	35
Welbeck Rd. Barn.	BT25	7
Welbeck Rd. Cars.	BU54	68
Welbeck Rd. Har.	BF33	20
Welbeck Rd. Sutt.	BT55	77
Welbeck St. W1	BV39	41
Welbeck Wk. Cars.	BT54	68
Welbeck Rd.		
Welbeck Way W1	BV39	41
Welby St. SE5	BY44	51
Welch Pl. Pnr.	BD30	11
Welcomes Rd. Ken.	BZ61	84
Weldon Clo. Ruis.	BC36	29
Weld Pl. N11	BV28	14
Welfare Rd. E15	CG36	34
Welford Pl. SW19	BR49	58
Welham Rd. SW17	BV49	59
Welhouse Rd. Cars.	BU54	68
Wellacre Rd. Har.	BJ32	21
Wellan Clo. Well.	CO46	54
Welland Gdns. Grnf.	BH37	30
Wellands Clo. Brom.	CK51	71
Welland St. SE10	CF43	52
Well App. Barn.	BQ25	5
Wellbrook Rd. Orp.	CL56	80
Well Clo. E1	CB40	42
Well Clo. Ruis.	BE34	20
Wellclose Sq. E1	CB40	42
Wellclose St. E1	CB40	42
Wellclose Sq.		
Wellcome Av. Dart.	CW45	55
Well Cottage Clo. E11	CJ33	26
Well Ct. EC4	BZ39	42
Queen St.		
Well Ct. NW8	BT37	32
Well Ct. NW8	BT38	32
Welldale Rd. SE16	CC42	43
Welldon Cres. Har.	BH32	21
Wellers Cres. Har.	BH32	21
Wellers Ct. NW1	BX37	33
Weller St. SE1	BZ41	42
Wellesford Clo. Bans.	BR62	82
Wellesley Av. W6	BP41	40
Wellesley Av. Nthwd.	BC28	11
Wellesley Ct. W9	BT38	32
Wellesley Ct. Rd. Croy.	BZ55	78
Wellesley Gro.		
Wellesley Cres. Twick.	BH48	57
Wellesley Gro. Croy.	BZ55	78
Wellesley Pl. NW5	BV35	32
Wellesley Rd. E11	CH32	26
Wellesley Rd. E17	CE32	25
Wellesley Rd. N22	BY30	15
Redvers Rd.		
Wellesley Rd. NW5	BV35	32
Wellesley Rd. W4	BM42	39
Wellesley Rd. Croy.	BZ55	78
Wellesley Rd. Har.	BH32	21
Wellesley Rd. Ilf.	CL34	26
Wellesley Rd. Sutt.	BT57	77
Wellesley Rd. Twick.	BG48	56
Wellesley St. E1	CC39	43
Wellfield Av. N10	BV31	23
Wellfield Rd. SW16	BX49	60
Wellfields Rd. Loug.	CL24	10
Wellfield Wk. SW16	BX49	60
Wellfit Rd. SE24	BY45	51
Hinton Rd.		
Wellgarth Gdns. Grnf.	BJ36	30
Wellgarth Rd. NW11	BS33	23
Well Hall Rd. SE9	CK45	53
Well Hill, Orp.	CR57	81
Well Hill La. Orp.	CR57	81
Wellhouse La. Barn.	BQ24	6
Wellhouse Rd. Beck.	CD52	70
Wellington Arch SW1	BV42	41
Wellington Av. E4	CE27	16
Wellington Av. N9	CB27	15
Wellington Av. N15	CA32	24
Wellington Av. Houns.	BF46	47
Wellington Av. Pnr.	BE30	11
Wellington Av. Sid.	CO46	54
Wellington Av. Wor. Pk.	BQ56	76
Wellington Clo. E4	CE27	16
Wellington Av.		
Wellington Clo. W11	BS39	41
Ledbury Rd.		
Wellington Clo. Dag.	CS36	37
Wellington Cres. N. Mal.	BN52	67
Wellington Dr. Dag.	CS36	37
Wellington Gdns. SE7	CJ42	44
Wellington Gdns. Hmptn.	BG49	56
Wellingtonia Av. Hav.	CS27	19
Wellington Pass. E11	CH32	26
Wellington Rd.		
Wellington Pl. N2	BU32	23
Wellington Pl. NW8	BT38	32
Wellington Rd. E6	CK37	35
Wellington Rd. E7	CB35	34
Wellington Rd. E10	CD33	25
Wellington Rd. E11	CH32	26
Wellington Rd. E17	CD31	25
Wellington Rd. NW8	BT37	32
Wellington Rd. NW10	BQ39	40
Wellington Rd. SW19	BS48	59
Wellington Rd. W5	BK41	39
Wellington Rd. Belv.	CQ42	45
Wellington Rd. Bex.	CP46	54
Wellington Rd. Brom.	CJ52	71
Wellington Rd. Croy.	BY54	69
Wellington Rd. Dart.	CV46	55
Wellington Rd. Enf.	CA26	8
Wellington Rd. Har.	BH31	21
Wellington Rd. Hmptn.	BG49	56
Wellington Rd. Orp.	CO53	72
Wellington Rd. Pnr.	BE29	11
Wellington Rd. Wat.	BC23	4
Wellington Rd. N. Houns.	BE45	47
Wellington Rd. S. Houns.	BE45	47
Wellington Row E2	CA38	33
Wellington Sq. SW3	BU42	41
Wellington St. E5	CB34	24
Wellington St. SE18	CL42	44
Wellington St. WC2	BX40	42
Wellington Ter. Har.	BG33	20
Wellington Way E3	CE38	34
Wellington Yd. Rich.	BK46	48
George St.		
Welling Way SE9	CM45	53
Welling Way, Well.	CM45	53
Well La. SW14	BN46	49
Wellmeadow Rd. SE6	CG46	52
Wellmeadow Rd. SE13	CG46	52
Wellmeadow Rd. W7	BJ42	39
Wellow Wk. Cars.	BT54	68
Well Pass. NW3	BT34	23
Well Rd. NW3	BT34	23
Well Rd. NW10	BQ38	31
Well Rd. Barn.	BQ25	6
Wells Clo. Nthlt.	BD38	29
Wells Dr. NW9	BN33	22
Wells Gdns. Dag.	CR35	36
Wells Gdns. Ilf.	CK33	26
Wells Gdns. Rain.	CT36	37
Wellshouse Rd. NW10	BQ39	40

Wellside Clo. Barn. BQ24 6
Wellside Gdns. SW14 BN46 49
Wells Ms. W1 BW49 41
Wellsmoor Gdns. Brom. CL52 71
Wells Pk. Rd. SE26 CB48 60
Wells Pl. W5 BK40 39
Wellsprings Cres. Wem. BM34 21
Wells Rise NW8 BU37 32
Wells Rd. W12 BQ41 40
Wells Rd. Brom. CK51 71
Wells St. W1 BW39 41
Wellstead Av. N9 CC26 9
Wellstead Rd. E6 CL37 35
Wells Ter. N4 BY34 24
Wells, The N14 BW26 7
Well St. E9 CB36 33
Well St. E15 CG36 34
Wells Way SE5 BZ43 51
Wellwood Clo. Couls. BX60 84
Wellwood Rd. Ilf. CO33 27
Welsford St. SE1 CB42 42
Welsh Clo. E13 CH38 35
Welshpool St. E8 CB37 33
Weltje Rd. W6 BP42 40
Welton Rd. SE18 CN43 54
Welwyn St. E2 CC38 24
Globe Rd.
Wembley Hill Rd. Wem. BL34 21
Wembley Park Dr. Wem. BL35 30
Wembley Rd. Hmptn. BF51 65
Wembley Way. Wem. BM36 30
Wemborough Rd. Stan. BK30 12
Wembury Rd. N6 BV33 23
Wemyss Rd. SE3 CG44 52
Wendela Ct. Har. BH34 21
Wendell Rd. W12 BO41 40
Wendling Rd. Sutt. BT55 77
Wendon St. E3 CD37 34
Wendover Dr. N. Mal. BO53 67
Wendover Rd. NW10 BO37 31
Wendover Rd. SE9 CJ45 53
Wendover Rd. Brom. CH52 71
Wendover Way, Bush. BG25 4
Wendover Way, Horn. CV35 37
Springbank Av.
Wendover Way, Orp. CO53 72
Wendover Way, Well. CO46 54
Wend, The, Couls. BW60 83
Wendy Clo. Enf. CA25 8
First Av.
Wendy Way, Wem. BL37 30
Weneth Hall Rd. Ilf. CK31 26
Wenlock Rd. N1 BZ37 33
Wenlock Rd. Edg. BM29 12
Wenlock St. N1 BZ37 33
Wennington Rd. E3 CC37 34
Grove Rd.
Wennington Rd. Rain. CU38 37
Wensley Av. Wdf. Grn. CG29 16
Wensley Clo. Rom. CR28 18
Wensleydale Av. Ilf. CK30 17
Wensleydale Gdns. Hmptn. BF50 56
Wensleydale Pass. Hmptn. BF51 65
Wensleydale Rd. Hamptn. BF50 56
Wensley Rd. N18 CB29 15
Wentland Rd. SE6 CF48 61
Wentworth Av. N3 BS29 14
Wentworth Clo. N3 BS29 14
Wentworth Clo. Mord. BS54 68
Wentworth Clo. Orp. CN56 81
Wentworth Clo. Surb. BK55 75
Wentworth Dr. Dart. CU46 56
Wentworth Dr. Pnr. BC32 20
Wentworth Gdns. N13 BY28 15
Wentworth Hill, Wem. BL33 21
Wentworth Ms. E3 CD38 34
Eric St.
Wentworth Pk. N3 BS29 14
Wentworth Pl. Stan. BJ29 12
Wentworth Rd. E12 CJ35 35
Wentworth Rd. NW11 BR32 22
Wentworth Rd. Barn. BQ24 6
Wentworth Rd. Croy. BY54 69
Wentworth Rd. Sthl. BD42 38
Wentworth St. E1 CA39 42
Wentworth Way, Pnr. BD31 20
Wentworth Way, Rain. CU38 37
Wentworth Way, Sth. Croy. CB60 84
Wenvoe Av. Bexh. CR44 54
Wernbrook St. SE18 CM43 53
Werndee Rd. SE25 CB52 69
Werrington St. NW1 BW37 32
Werter Rd. SW15 BR45 49
Wesley Av. NW10 BN38 31
Wesley Av. Houns. BE44 47
Wesley Clo. Har. BG34 20
Wesley Clo. Orp. CP52 72
Wesley Rd. E10 CF33 25
Wesley Rd. NW10 BN37 31
Hillside
Wesley Rd. Hayes BC39 38
Wesley St. W1 BV39 41
Weymouth St.
Wessex Av. SW19 BS52 68

Wessex Bldgs. N19 BW34 23
Wedmore St.
Wessex Clo. Ilf. CN32 27
Wessex Clo. Kings. On T. BM51 66
Gloucester Rd.
Wessex Dr. Erith CT44 55
Wessex Dr. Pnr. BE29 11
Wessex Gdns. NW11 BR33 22
Wessex La. Grnf. BG27 29
Wessex St. E2 CC38 34
Wessex Way NW11 BR33 22
Westacres, Esher BE57 74
Westall Rd. Loug. CL24 10
West App. Orp. CM53 71
West Arbour St. E1 CC39 43
West Av. E17 CE31 25
West Av. N3 BS29 14
West Av. NW4 BQ32 22
West Av. Hayes BC40 38
West Av. Pnr. BE32 20
West Av. Sthl. BE40 38
West Av. Wall. BX56 78
West Av. Rd. E17 CE31 25
West Bank N16 CA33 24
West Bank, Bark. CL37 35
West Bank, Enf. BZ23 8
Westbank Rd. Hamptn. BG50 56
W. Barnes La. N. Mal. BP52 67
West Barnes La. N. Mal. BP53 67
West Beech Rd. N22 BY31 24
Westbere Dr. Stan. BK28 12
Westbere Rd. NW2 BR35 31
Westbourne Av. N9 CB27 15
Eastbournia Av.
Westbourne Av. W3 BN39 40
Westbourne Av. Sutt. BR55 76
Westbourne Br. W2 BT39 41
Westbourne Clo. Hayes BD38 29
Westbourne Cres. W2 BT40 41
Westbourne Dr. SE23 CC48 61
Westbourne Gdns. W2 BS39 41
Westbourne Gte. W2 BS39 41
Westbourne Gro. W11 BR40 40
Westbourne Gro. Ms. BS39 41
W11
Westbourne Gro.
Westbourne Gro. Ter. W2 BS39 41
Westbourne Pk. Ms. W2 BS39 41
Westbourne Gdns.
Westbourne Pk. Pass. W2 BS39 41
Westbourne Pk. Rd. W2 BR39 40
Westbourne Pk. Rd. W11 BR39 40
Westbourne Pk. Vill. W2 BS39 41
Westbourne Rd. N7 BY36 33
Eastbournia Av.
Arundel Sq.
Westbourne Rd. SE26 CC50 61
Westbourne Rd. Bexh. CP43 54
Westbourne Rd. Croy. CA53 69
Westbourne St. W2 BT40 41
Westbourne Ter. W2 BT39 41
Westbourne Ter. Ms. W2 BT39 41
Westbourne Ter. Rd. W2 BT39 41
Westbridge Rd. SW11 BT44 50
Westbrook Av. Hamptn. BE50 56
Westbrook Clo. Barn. BT24 7
Westbrook Cres. Barn. BT24 7
Westbrook Dr. Orp. CP54 72
Westbrooke Cres. Well. CP45 54
Westbrooke Rd. Sid. CM48 62
Westbrooke Rd. Well. CP45 54
Westbrook Rd. SE3 CH44 53
Westbrook Rd. Houns. BE43 47
Westbrook Rd. Th. Hth. BZ51 69
Westbrook Sq. Barn. BT24 7
Westbury Av. N22 BY31 24
Westbury Av. Esher BH57 75
Westbury Av. Sthl. BF38 29
Westbury Av. Wem. BL36 30
Westbury Clo. Ruis. BC33 20
Westbury Gro. N3 BS29 14
Westbury La. Buck. H. CH27 17
Westbury Lodge Clo. Pnr. BD31 20
Westbury Pl. Brent. BK43 48
Hamilton Rd.
Westbury Rd. E7 CH35 35
Westbury Rd. E17 CD31 25
Westbury Rd. N11 BX29 15
Westbury Rd. N12 BS29 14
Westbury Rd. SE20 CC51 70
Westbury Rd. W5 BL39 39
Westbury Rd. Bark. CM37 35
Westbury Rd. Beck. CD52 70
Westbury Rd. Brom. CJ51 71
Westbury Rd. Buck. H. CH27 17
Westbury Rd. Buck. H. CJ27 17
Westbury Rd. Croy. BZ53 69
Westbury Rd. Felt. BD47 56
Westbury Rd. Ilf. CK34 26
Westbury Rd. N. Mal. BN52 67
Westbury Rd. Wat. BC25 4
Westbury Rd. Wem. BL36 30
Westbury St. SW8 BW44 50
Westbury Ter. E7 CH36 35
Westcar La. Walt. BC57 74
West Central St. WC1 BX39 42
New Oxford St.

Westchester Dr. NW4 BQ31 22
West Clo. N9 CA27 15
West Clo. Barn. BP25 6
West Clo. Barn. BV24 7
West Clo. Grnf. BG37 29
West Clo. Rain. CU38 37
West Clo. Wem. BL33 21
Westcombe Av. Croy. BX53 69
Westcombe Ct. SE3 CG43 52
Westcombe Dr. Barn. BS25 7
Westcombe Hill SE3 CH42 44
Westcombe Pk. Rd. SE3 CG43 52
West Common Rd. Brom. CH55 80
Westcoombe Av. SW20 BO51 67
Westcote Rd. SW16 BW49 59
West Cotts. NW6 BS35 32
Westcott Clo. Croy. CE58 79
Westcott Cres. W7 BH39 39
Westcott Rd. SE17 BY43 51
Westcott Way, Sutt. BQ58 76
West Ct. SE18 CL43 53
Prince Imperial Way
West Ct. Wem. BK34 21
Westcroft Clo. NW2 BR35 31
Westcroft Est. NW2 BR35 31
Westcroft Gdns. Mord. BR52 67
Westcroft Rd. Cars. BV56 77
Westcroft Sq. W6 BO42 40
Westcroft Wy. NW2 BR35 31
West Cromwell Rd. SW5 BR42 40
West Cromwell Rd. W14 BR42 40
West Cross Way, Brent. BJ43 48
Westdale Rd. SE18 CL43 53
Westdean Av. SE12 CH47 62
Westdean Clo. SW18 BS46 50
Denton St.
West Dene, Sutt. BR57 76
Park La.
West End Ct. Pnr. BD31 20
Westdene Dr. Rom. CV28 19
Westdown Rd. E15 CF35 34
Westdown Rd. SE6 CE47 61
West Dr. N8 BW31 23
Redston Rd.
West Dr. SW11 BU43 50
West Dr. SW16 BW49 59
West Dr. Cars. BT58 77
West Dr. Har. BG29 11
West Dr. Sutt. BO58 76
West Dr. Tad. BQ62 82
West Dr. Gdns. Har. BG29 11
West Eaton Pl. SW1 BV42 41
Wested La. Swan. CU54 73
West Ella Rd. NW10 BO36 31
West End Av. E10 CF32 25
West End Av. Pnr. BD31 20
West End Gdns. Esher BE56 74
West End Gdns. Nthlt. BD37 29
West End La. NW6 BS35 32
West End La. Barn. BQ24 6
West End La. Esher BE57 74
West End La. Esher BF56 74
West End La. Pnr. BD31 20
West End La. Nthlt. BD36 29
West End Rd. Nthlt. BE40 38
Westerdale Ct. N5 BY35 33
Leigh Rd.
Westerdale Rd. SE10 CH42 44
Westerfield Rd. N15 CA32 24
Westergate Rd. SE2 CQ43 54
Westerham Av. N9 BZ27 15
Westerham Dr. Sid. CO46 54
Westerham Rd. E10 CE32 25
Westerham Rd. Kes. CJ58 80
Westerley Cres. SE26 CD49 61
Western Av. NW9 BO32 22
Western Av. W3 BJ38 30
Western Av. W5 BJ38 30
Western Av. Dag. CS36 37
Western Av. Grnf. BG38 30
Western Av. Rom. CV30 19
Western Av. Ruis. BC36 29
Western Circus W3 BO40 40
Western Ct. N3 BS29 14
Western Ct. W3 BN39 40
York Rd.
Western Gdns. W5 BM40 39
Western La. SW12 BV47 59
Western Pde. Barn. BS25 7
Western Rd. E13 CJ37 35
Western Rd. E17 CF32 25
Western Rd. N2 BU31 23
Western Rd. N22 BX30 15
Western Rd. SW9 BY45 51
Western Rd. SW19 BT51 68
Western Rd. W5 BK40 39
Western Rd. Mitch. BT51 68
Western Rd. Rom. CT32 28
Western Rd. Sthl. BD42 38
Western Rd. Sutt. BS56 77
Western Rd. NW10 BN38 31
Western St. E15 CF36 34
Western Trading Est. BN38 31
NW10
Westernville Gdns. Ilf. CM33 26
Western Way, Barn. BS25 7
Western Way SE28 CM41 44
West Ferry Est. E14 CE42 43

Westferry Rd. E14 CE42 43
Westfield N6 BV34 23
West Field SW13 BO45 49
Cross St.
Westfield Av. SW13 BO45 49
Westfield Av. S. Croy. BZ60 84
Westfield Clo. Enf. CD24 9
Westfield Clo. Sutt. BR56 76
Westfield Dr. Har. BK32 21
Westfield Gdns. Har. BK31 21
Westfield La. Har. BK31 21
Westfield, Loug. CJ25 10
Longfield
Westfield Pk. Pnr. BE29 11
Westfield Rd. N8 BX31 24
Westfield Rd. NW7 BM27 13
Westfield Rd. SE18 CJ41 44
Westfield Rd. W13 BJ40 39
Westfield Rd. Beck. CD51 70
Westfield Rd. Bexh. CS44 55
Westfield Rd. Croy. BY55 78
Westfield Rd. Dag. CQ35 36
Westfield Rd. Mitch. BU51 68
Westfield Rd. Surb. BK53 66
Westfield Rd. Sutt. BR56 76
Westfield Rd. Walt. BE54 65
Westfields SW13 BO45 49
Railway Side
Westfields Rd. W3 BM39 39
West Gdns. E1 CB40 42
West Gdns. SW17 BU50 59
West Gdns. Epsom BO58 76
West Gate Ms. W10 BQ38 31
West Row
Westgate Rd. SE25 CB52 69
Westgate Rd. Beck. CE51 70
Westgate Rd. Dart. CV46 55
Westgate St. E8 CB37 33
Westgate Ter. SW10 BS42 41
Westglade Ct. Har. BK32 21
West Green Rd. N15 BY31 24
West Gro. SE10 CF44 52
West Gro. Walt. BC56 74
West Gro. Wdf. Grn. CJ29 17
Westgrove La. SE10 CF44 52
West Halkin St. SW1 BV41 41
West Hall Ct. N6 BV34 23
West Hallows SE9 CJ47 62
Westhall Rd. SE5 BY43 51
West Hall Rd. Rich. BM44 48
Westhall Rd. Warl. CB62 84
West Ham La. E15 CE36 34
West Hampstead Ms. SE36 32
NW6
West Harding St. EC4 BY39 42
Fetter La.
Westharold, Swan. CS52 73
Westhay Gdns. SW14 BM46 48
West Heath Av. NW11 BS33 23
West Hth. Clo. NW3 BS34 23
West Heath Clo. Dart. CT46 55
West Heath Ct. NW11 BS33 23
West Heath Dr. NW11 BS33 23
West Hth. Gdns. NW3 BS34 23
West Heath Rd. NW3 BS34 23
West Heath Rd. SE2 CP43 54
West Heath Rd. Dart. CT46 55
West Hill SW15 BO47 58
West Hill SW18 BO47 58
West Hill, Dart. CV46 55
West Hill, Epsom BN60 82
West Hill, Har. BH34 21
West Hill, Sth. Croy. CA58 78
West Hill, Wem. BL33 21
West Hill Ct. SW18 BR46 49
West Hill Rd.
West Hill Dr. Dart. CV46 55
West Hill Rise, Dart. CV46 55
West Hill Rd. SW18 BR46 49
West Hill Way N20 BS26 7
Westholm NW11 BS31 23
Westholme, Orp. CN54 72
Westholme Gdns. Ruis. BC33 20
Westhorne Av. SE9 CH47 62
Westhorpe Av. SE12 CH47 62
Westhorpe Gdns. NW4 BQ31 22
Westhorpe Rd. SW15 BQ45 49
Westhouse Clo. SW19 BR47 58
Westhurst Dr. Chis. CL49 62
West India Dock Rd. E14 CD39 43
West Kentish Town Est. BV35 32
NW5
Westlake Clo. N13 BY27 15
Westlake Rd. Bex. CP46 54
Blendon Rd.
Westland Av. Horn. CW33 28
Westland Ct. Epsom BN60 82
Westland Dr. Brom. CG55 79
Westland Pl. N1 BZ38 33
Westland Rd. Wat. BC23 4
Westlands Clo. Hayes BE42 38
Westlands Ct. Epsom BN61 82
Westlands Ter. SW12 BW46 50
Gaskarth Rd.
West La. SE16 CB41 42
Westlea Rd. W7 BJ41 39
Westleigh Av. SW15 BP46 49

White Kennett St. E1 CA39 42
Whiteledges W13 BK39 39
Whitelegg Rd. E13 CG37 34
Whiteley Rd. SE19 BZ49 60
Whiteley's Cotts. W14 BR42 40
Whiteley Way, Felt. BF48 56
White Lion Hl. EC4 BY40 42
White Lion St. N1 BY37 33
White Lodge Clo. NW3 BT33 23
White Lodge Clo. SE19 BZ50 60
White Lodge Est. SE19 BZ50 60
Whiteoak Dr. Beck. CF51 70
Whiteoaks La. Grnf. BG38 29
White Orchards, Stan. BJ28 12
White Orchard N20 BR26 6
White Post La. E9 CD36 34
White Post St. SE15 CC43 52
White Rd. E15 CG36 34
Whites Av. Ilf. CN32 27
Whites Grounds SE1 CA41 42
Whites Grounds Est. SE1 CA41 42
White's Row E1 CA39 42
Whites Sq. SW4 BW45 50
 Nelson's Row
Whitestile Rd. Brent. BK42 39
Whitestone La. NW3 BT34 23
 Heath St.
Whitestone Wk. NW3 BT34 23
 North End Way
White St. Sthl. BD41 38
Whitethorn Av. Couls. BV61 83
Whitethorn Gdns. Croy. CB55 78
Whitethorn Gdns. Enf. BZ25 8
Whitethorn Gdns. Horn. CV32 28
Whitethorn St. E3 CE38 34
Whitewebbs Way, Orp. CN51 72
Whitfield Pl. W1 BW38 32
Whitfield Rd. E6 CJ36 35
Whitfield Rd. SE3 CF44 52
Whitfield Rd. Bexh. CQ43 54
Whitfield St. W1 BW38 32
Whitford Gdns. Mitch. BU52 68
Whitgift Av. Sth. Croy. BY56 78
Whitgift St. SE11 BX42 41
Whitgift St. Croy. BZ55 78
Whiting Av. Bark. CL36 35
Whiting Hill Est. Barn. BP25 6
Whitings Barn. BQ25 6
Whitings Way E6 CL39 44
Whitland Rd. Cars. BT54 68
Whitlock Dr. SW19 BR47 58
Whitley Rd. N17 CA30 15
Whitlock Dr. SW18 BR47 58
Whitman Rd. E3 CD38 34
Whitmore Clo. N11 BV42 38
 High Rd.
Whitmore Gdns. NW10 BQ37 31
Whitmore Rd. N1 CA37 33
Whitmore Rd. Beck. CD52 70
Whitmore Rd. Har. BF33 20
Whitmores Clo. Epsom BN61 82
Whitnell Way SW15 BQ46 49
 Chartfield Av.
Whitnell Way SW15 BQ46 49
Whitney Av. Ilf. CJ31 26
Whitney Rd. E10 CE33 25
Whitney Wk. Sid. CQ50 63
 Maidstone Rd.
Whitstable Clo. Beck. CD51 70
Whittaker Av. Rich. BK46 48
Whittaker Rd. E6 CJ36 35
Whittaker Rd. Sutt. BR55 76
Whittaker St. SW1 BV42 41
 Bourne St.
Whitta Rd. E12 CJ35 35
Whittel Gdns. SE26 CC48 61
Whitting Av. Bark. CL36 35
Whittingstall Rd. SW6 BR44 49
Whittington Av. EC3 CA39 42
 Leadenhall St.
Whittington Ct. N2 BU32 23
Whittington Rd. N22 BX29 15
Whittington Way, Pnr. BE32 20
Whittlebury Clo. Cars. BU57 77
Whittle Clo. Sthl. BF39 38
Whittle Rd. Houns. BD43 47
Whittlesea Path, Har. BG30 11
Whittlesea Rd. Har. BG29 11
Whittlesey St. SE1 BY40 42
Whitton Av. E. Grnf. BH35 30
Whitton Av. W. Grnf. BF35 29
Whitton Av. W. Nthlt. BF35 29
Whitton Clo. Grnf. BJ36 30
Whitton Dene, Houns. BF46 47
Whitton Dr. Grnf. BJ36 30
Whitton Manor Rd. Islw. BF46 47
Whitton Rd. Houns. BF45 47
Whitton Rd. Twick. BH46 48
Whitton Waye, Houns. BF46 47
Whitwell Rd. E13 CH38 35
Whitworth Rd. SE18 CL43 53
Whitworth Rd. SE25 CA52 69
Whitworth St. SE10 CG42 43
Whorlton Rd. SE15 CB45 51
Whybridge Clo. Rain. CT37 37
Whymark Av. N22 BY31 24
Whytecroft, Houns. DD43 47

Whyteville Rd. E7 CH36 35
Wickersley Rd. SW11 BV44 50
 St. John's Hill
Wicker St. E1 CB39 42
Wicket, The, Croy. CE56 79
Wickford Clo. Rom. CW28 19
Wickford Dr. Rom. CW28 19
Wickford St. E1 CC38 34
Wickford Way E17 CC31 25
Wickham Av. Croy. CD55 79
Wickham Av. Sutt. BQ56 76
Wickham Chase, CF54 70
 W. Wick.
Wickham Clo. Enf. CB24 8
Wickham Clo. N. Mal. BO53 67
Wickham Cres. W. Wick. CF55 79
Wickham Gdns. SE4 CD45 52
Wickham La. SE2 CO42 45
Wickham Rd E4 CF29 16
Wickham Rd. SE4 CD45 52
Wickham Rd. Beck. CE51 70
Wickham Rd. Croy. CC55 79
Wickham Rd. Har. BG30 11
Wickham St. SE11 BX42 41
Wickham St. Well. CN44 54
Wickham Way, Beck. CF52 70
Wick La. E3 CD36 34
Wickliffe Av. N3 BR30 13
Wicklow St. WC1 BX38 33
Wick Rd. E9 CC36 34
Wick Rd. Tedd. BJ50 57
Wicksteed Clo. Bex. CS48 64
Wickwood St. SE5 BY44 51
Widdenham Rd. N7 BX35 33
Widdicombe Av. Har. BE34 20
Widdin St. E15 CF36 34
Widecombe Clo. Rom. CV30 19
Widecombe Clo. N2 BT32 23
Widecombe Gdns. Ilf. CK31 26
Widecombe Rd. SE9 CK48 62
Widecombe Way N2 BT32 23
Widegate St. E1 CA39 42
 Sandy's Row
Wide Way, Mitch. BW52 68
Widmore Lo. Rd. Brom. CJ51 71
Widmore Rd. Brom. CH51 71
Wieland Rd. Nthwd. BC29 11
Wigan Ho. E5 CB33 24
Wiggenhall Rd. Wat. BC24 4
Wiggington Av. Wem. BM36 30
Wiggins La. Rich. BK48 57
Wiggins Mead NW9 BO30 13
Wiggins & Pointers BK48 57
 Cotts. Rich.
 Ham St.
Wightman Rd. N4 BY31 24
Wightman Rd. N8 BY31 24
Wigley Rd. Felt. BD48 56
Wigmore Pl. W1 BV39 41
Wigmore Rd. Cars. BT55 77
Wigmore St. W1 BV39 41
Wigmore Wk. Cars. BT55 77
Wigram Rd. E11 CJ32 26
Wigram Sq. E17 CF31 25
Wigston Rd. E13 CH38 35
Wigton Gdns. Stan. BL30 12
Wigton Pl. E17 CD30 16
Wigton Rd. Rom. CW28 19
Wilberforce Rd. N4 BY34 24
Wilberforce Rd. NW9 BP32 22
Wilberforce Way SW19 BQ50 58
Wilbraham Pl. SW1 BV42 41
Wilbury Av. Sutt. BR58 76
Wilbury Way N18 BZ28 15
Wilby Ms. N11 BR40 40
Wilby Rd. SE5 BZ44 51
 Grove La.
Wilcot Av. Wat. BE26 4
Wilcox Clo. Borwd. BN23 6
Wilcox Rd. SW8 BX43 51
Wilcox Rd. Sutt. BS56 77
Wilcox Rd. Tedd. BG49 56
Wild Ct. WC2 BX39 42
Wildcroft Gdns. Edg. BK29 12
Wildcroft Manor SW15 BQ47 58
 Bristol Gdns.
Wildcroft Rd. SW15 BQ47 58
Wilderness Rd. Chis. CL50 62
Wilderness, The, Hamptn. BF49 56
 Park Rd.
Wilderton Rd. N16 CA33 24
Wildfell Rd. SE6 CE47 61
Wild Goose Dr. SE14 CC44 52
Wild Hatch NW11 BS32 23
Wild's Rents SE1 CA41 42
Wild St. WC2 BX39 42
Wildwood Clo. SE12 CG47 61
Wildwood Rise NW11 BT33 23
Wildwood Rd. NW11 BT33 23
Wilfred Av. Rain. CU39 46
Wilfred St. SW1 BW41 41
Wilfred Turney Est. W6 BQ41 40
Wilfrid Gdns. W3 BN39 40

Wilkers Oak SE19 CA49 60
 Wood Park
Wilkinson Clo. Dart. CW45 55
Wilkinson Rd. E16 CJ39 44
Wilkinson St. SW8 BX43 51
Wilkinson Way W4 BN41 40
Wilk Pl. N13 BY29 15
Willan Rd. N17 BZ30 15
Willard Est. SW8 BV45 50
Willocks Clo. Chess. BL55 75
Willcott Rd. W3 BM40 39
Will Crook's Gdns. SE9 CJ45 53
Willenhall Av. Barn. BS25 7
Willenhall Rd. SE18 CL42 44
Willersley Av. Orp. CM55 80
Willersley Av. Sid. CN47 63
Willersley Clo. Sid. CN47 63
Willesden Grn. NW2 BQ36 31
Willesden La. NW2 BQ36 31
Willesden La. NW6 BQ36 31
Willes Rd. NW5 BV36 32
Willett Clo. Nthlt. BD38 29
 Broomcroft Av.
Willett Clo. Orp. CN53 72
Willett Clo. Th. Hth. BY53 69
Willett Rd. Th. Hth. BY53 69
Willett Way, Orp. CM53 71
Willet Way SE16 CB42 42
 Egan Way
William Booth Dr. SE9 CK49 62
William Booth Rd. SE20 CB51 69
William Bonney Est. SW4 BW45 50
William Clo. Rom. CS30 19
William Fourth St. WC2 BX40 42
William Guy Gdns. E3 CF38 34
 Talwin St.
William Hayne Gdns. BQ55 76
 Wor. Pk.
William Ms. SW1 BU41 41
William Morley Clo. E6 CV37 35
William Morris Clo. E17 CD31 25
William Morris Ho. W6 BQ41 49
William Parnell Ho. SW6 BS44 50
William Rd. NW1 BW38 32
William Rd. SW19 BR50 58
William Rd. Sutt. BT56 77
William's Av. E17 CD30 16
William St. E10 CE32 25
William St. E15 CF36 34
William St. N12 BT28 14
 Lodge La.
William St. N17 CA29 15
William St. SW1 BU41 41
William St. Bark. CM36 35
William St. Bush. BD24 4
William St. Cars. BU55 77
William Ter. Croy. BY57 78
William Willisom Est. BR47 58
 SW19
Willifield Way NW11 BS31 23
Willingale Clo. Loug. CM23 10
Willingale Clo. Wdf. Grn. CJ29 17
Willingale Rd. Loug. CM24 10
Willingdon Rd. N22 BY30 15
Willingham Clo. NW5 BW35 32
 Leighton Rd.
Willingham Ter. NW5 BW35 32
Willingham Way, BM52 66
 Kings. on T.
Willington Ct. E5 CC35 34
 Clapton Park Est.
Willington Rd. SW9 BX45 51
Willis Av. Sutt. BU57 77
Willis Rd. E15 CG37 34
Willis Rd. Croy. BZ54 69
Willis Rd. Erith CS42 46
Willis St. E14 CE39 43
Willmore End SW19 BS51 68
Willoughby Av. Croy. BX56 78
Willoughby Dr. Rain. CT36 37
Willoughby Gro. N17 CB29 15
Willoughby La. N17 CB29 15
Willoughby Pk. Rd. N17 CB29 15
Willoughby Rd. N8 BY31 24
Willoughby Rd. NW3 BT35 32
Willoughby Rd. BL51 66
 Kings. on T.
Willoughby Rd. Twick. BK46 48
Willow Av. SW13 BO44 49
Willow Av. Sid. CO46 54
Willow Av. Swan. CT52 73
Willow Bank SW6 BR45 49
Willowbank, Rich. BJ48 57
Willow Bridge Rd. N1 BZ36 33
Willowbrook Clo. Sthl. BF41 38
Willowbrook Gro. SE15 CA43 51
Willowbrook Rd. SE15 CA43 51
Willow Clo. W5 BK39 39

Willow Clo. Bex. CQ46 54
Willow Clo. Brent. BK43 48
Willow Clo. Brom. CK53 71
Willow Clo. Buck. H. CJ27 17
Willow Clo. Erith CU44 55
Willow Clo. Horn. CU34 28
Willow Clo. Orp. CO54 72
Willow Clo. Th. Hth. BY53 69
Willow Cotts. Rich. BM43 48
 Waterloo Pl.
Willow Cott. Rd. Cars. BU53 68
Willow Ct. Edg. BL28 12
Willowcourt Av. Har. BJ32 21
Willowdene N6 BU33 23
 Denewood Rd.
Willowdene, Bush. BH26 5
Willow Dene, Pnr. BD30 11
Willowdene Clo. Twick. BG47 56
Willow Dr. Barn. BR24 6
Willow End N20 BS27 14
Willow End, Surb. BL54 66
Willow Gdns. N16 CA34 24
 Cazenove Rd.
Willow Gdns. Houns. BF44 47
Willow Gro. E13 CH37 35
Willow Gro. SE1 CA42 42
 Curtis St.
Willow Gro. Chis. CL50 62
Willow Hayne Dr. Walt. BC54 65
Willowherb Wk. Rom. CV29 19
Willow La. Mitch. BU53 68
Willow La. Wat. BC25 4
Willow Mead, Chig. CO27 18
Willowmead Clo. W5 BK38 30
 Brenthan Way
Willowmere, Esher BG56 84
Willow Pl. SW1 BW42 41
 Francis St.
Willow Rd. NW3 BT35 32
Willow Rd. W5 BL41 39
Willow Rd. Dart. CV47 64
Willow Rd. Enf. CA24 8
Willow Rd. Erith CU44 55
Willow Rd. N. Mal. BN52 67
Willow Rd. Rom. CO32 27
Willow Rd. Wall. BV57 77
Willows Av. Mord. BS53 68
Willows Clo. Pnr. BD30 11
Willow St. E4 CF26 9
Willow St. EC2 CA38 33
Willow St. Rom. CS31 28
Willow Tree Clo. Hayes BD38 29
Willow Tree Clo. SW18 BS47 59
Willowtree La. Nthlt. BE38 29
Willow Val. Chis. CL50 62
Willow Vale W12 BP40 40
Willow View SW19 BT51 68
 Phipps Bridge Rd.
Willow View SW19 BT51 68
Willow Wk. E17 CD32 25
Willow Wk. N2 BT30 14
Willow Wk. Dart. CV45 55
Willow Wk. N15 BY31 24
Willow Wk. N21 BX25 8
Willow Wk. SE1 CA42 42
Willow Wk. Croy. CE55 79
Willow Wk. Dart. CV45 55
Willow Wk. Ilf. CL34 26
 Station Rd.
Willow Wk. Orp. CL55 80
Willow Wk. Sutt. BR55 76
Willow Way N3 BS29 14
Willow Way SE26 CB48 60
Willow Way, Epsom BN57 76
Willow Way, Sun. BC52 65
Willow Way, Twick. BF48 56
Willow Way, Wem. BJ34 21
Willow Wd. Cres. SE25 CA53 69
Wills Cres. Houns. BF46 47
Wills Gro. NW7 BP28 13
Willshaw St. SE14 CE44 52
Willwood Way NW3 BT33 23
 North End Rd.
Willy St. WC1 BX39 42
 Gt. Russell St.
Wilman Gro. E8 CB36 33
Wilmar Gdns. W. Wick. CE54 70
Wilmer Clo. Kings. On T. BL49 57
Wilmer Cres. Kings. On T. BL49 57
Wilmer Gdns. N1 CA37 33
Wilmer Ho. Kings. On T. BL49 57
Wilmerlee Clo. E15 CF36 34
Wilmer Way N14 BW28 14
Wilmington Av. W4 BN43 49
Wilmington Av. Orp. CP55 81
Wilmington Ct. Rd. Dart. CU46 64
Wilmington Gdns. Bark. CM36 35
Wilmington Sq. WC1 BY38 33
Wilmington St. WC1 BY38 33
 Sylvester Rd.
Wilmot Clo. N3 BT30 14
Wilmot Pl. NW1 BW36 32
Wilmot Pl. W7 BH40 39
Wilmot Rd. E10 CE34 25
Wilmot Rd. N17 BZ31 24
Wilmot Rd. Dart. CU46 55
Wilmot Rd. Pur. BY59 84
Wilmot St. E2 CB38 33

Wilmot Way, Bans. BS60 83
Wilmount St. SE18 CL42 44
Wilna Rd. SW18 BS47 59
Wilna Yd. SW18 BS47 59
Wilna Rd.
Wilrose Cres. SE2 CO42 45
Wilsham St. W11 BQ40 40
Wilsmere Dr. Har. BH29 12
Wilsmere Dr. Nthlt. BE35 29
Wilsmere Dr. Nthlt. BE36 29
Wilson Av. Mitch. BU50 59
Wilson Clo. Dag. CS36 37
Wilson Gdns. Har. BG33 20
Wilson Gro. SE16 CB41 42
Wilson Rd. E6 CJ38 35
Wilson Rd. SE5 BZ44 51
Wilson Rd. Chess. BL57 75
Wilson Rd. Har. BG33 20
Wilson Rd. Ilf. CK33 26
Wilson's Pl. E14 CD39 43
Salmon La.
Wilsons Rd. W6 BQ42 40
Wilson St. E13 CH37 35
Wilson St. E17 CF32 25
Wilson St. EC2 BZ39 42
Wilson St. N21 BY26 8
Wilton Av. W4 BO42 40
Wilton Ct. N10 BV30 14
Wilton Cres. SW1 BV41 41
Wilton Cres. SW19 BR51 67
Wilton Dr. Rom. CS29 19
Wilton Gdns. Walt. BD54 65
Wilton Gro. SW19 BR50 58
Wilton Gro. N. Mal. BO53 67
Wilton Ms. SW1 BV41 41
Wilton Pde. Felt. BC48 56
Wilton Pk. Ct. SE18 CL43 53
Prince Imperial Rd.
Wilton Pl. SW1 BV41 41
Wilton Rd. N10 BV30 14
Wilton Rd. SW1 BV41 41
Wilton Rd. SW19 BU50 59
Wilton Rd. Barn. BU24 7
Wilton Rd. Houns. BD45 47
Wilton Rd. Ilf. CL35 35
Wilton Row SW1 BV41 41
Wilton Sq. N1 BZ37 33
Wilton St. SW1 BV41 41
Wilton Ter. SW1 BV41 41
Wilton Way E8 CB36 33
Wilton Yd. W10 BQ40 40
Bard Rd.
Wiltshire Av. Horn. CW31 28
Wiltshire Clo. SW3 BU42 41
Wiltshire Gdns. Twick. BG47 56
Wiltshire Rd. N1 BZ37 33
Wiltshire Rd. SW9 BY45 51
Wiltshire Rd. Orp. CO54 72
Wiltshire Rd. Th. Hth. BY52 69
Wilverley Cres. N. Mal. BO53 67
Wimbart Rd. SW2 BX47 60
Wimbledon Clo. SW20 BQ50 58
Wimbledon Hl. Rd. SW19 BR50 58
Wimbledon Pk. Est. SW19 BR47 58
Wimbledon Parkside SW19 BQ47 58
Wimbledon Pk. Rd. SW18 BR48 58
Wimbledon Pk. Rd. SW19 BR48 58
Wimbledon Rd. SW17 BT49 59
Wimbolt St. E2 CB38 33
Wimborne Av. Hayes BC39 38
Wimborne Av. Orp. CN52 72
Wimborne Av. Sthl. BF42 38
Wimborne Clo. SE12 CG46 52
Wimborne Clo. Epsom BO60 82
Wimborne Dr. NW9 BM31 21
Wimborne Dr. Pnr. BD33 20
Wimborne Gdns. W13 BJ39 39
Wimborne Rd. N9 CB27 15
Wimborne Way, Beck. CC52 70
Wimbourne Clo. Buck. H. CH27 17
Wimbourne Ct. N1 BZ37 33
Wimbourne Rd. N17 CA30 15
Wimbourne St. N1 BZ37 33
Wimbrel Clo. Pur. BZ59 84
Wimpole Clo. Kings. On T. BL51 66
Wimpole Ms. W1 BV39 41
Wimpole St. W1 BV39 41
Winans Wk. SW9 BY44 51
Wincanton Cres. Nthlt. BF35 29
Wincanton Gdns. Ilf. CL30 17
Wincanton Rd. SW18 BR47 58
Wincanton Rd. Rom. CV27 19
Winchcombe Rd. Cars. BT54 68
Winchcomb Gdns. SE9 CJ45 53
Winchelsea Av. Bexh. CQ43 54
Winchelsea Clo. SW15 BQ46 49
Winchelsea Clo. SW15 BQ46 49
Chartfield Av.
Winchelsea Cres. E. Mol. BG51 65
Winchelsea Rd. E7 CH34 26
Winchelsea Rd. N17 CA31 24
Winchelsea Rd. NW10 BN37 31
Winchelsey Rd. Sth. Croy. CA57 78
Winchendon Rd. SW6 BR44 49
Winchendon Rd. Tedd. BG49 56
Winchester Av. NW6 BR37 31
Winchester Av. NW9 BM31 21
Winchester Av. Houns. BE43 47
Winchester Clo. Brom. CG52 70
Winchester Clo. Enf. CA25 8
Winchester Clo. Esher BF56 74
Winchester Clo. Kings. On T. BM50 57
Winchester Dr. Pnr. BD32 20
Winchester Mews NW3 BT36 32
Winchester Rd.
Winchester Ms. NW3 BT36 32
Winchester Rd.
Winchester Pk. Brom. CG52 70
Winchester Pl. E8 CA35 33
Kingsland High St.
Winchester Pl. N6 BV33 23
Winchester Pl. W3 BN41 40
Park Rd. E.
Winchester Rd. E4 CF29 16
Winchester Rd. N6 BV33 23
Winchester Rd. N9 CA26 8
Winchester Rd. NW3 BT36 32
Winchester Rd. Bexh. CP44 54
Winchester Rd. Brom. CG52 70
Winchester Rd. Felt. BE48 56
Winchester Rd. Har. BL31 21
Winchester Rd. Ilf. CM34 26
Winchester Rd. Nthwd. BC30 11
Winchester Rd. Orp. CO56 81
Winchester Rd. Twick. BJ46 48
Winchester Rd. Walt. BC54 65
Winchester Sq. SE1 BZ40 42
Winchester Wk.
Winchester St. SW1 BV42 41
Winchester St. W3 BN41 40
Winchester Wk. SE1 BZ40 42
Winchet Wk. Croy. CC53 69
Long Lane
Winchfield Clo. Har. CK32 21
Winchfield Rd. SE26 CD49 61
Winchmore Hill Rd. N14 BW26 7
Winchmore Hill Rd. N21 BW26 7
Winckley Clo. Har. BL32 21
Wincott St. SE11 BY42 42
Wincrofts Dr. SE9 CM45 53
Windborough Rd. Cars. BV57 77
Windermer Av. N3 BS31 23
Windermere Av. NW6 BR37 31
Windermere Av. SW19 BS52 68
Windermere Av. Horn. CU35 37
Windermere Av. Ruis. BD33 20
Windermere Av. Wem. BK33 21
Windermere Clo. Dart. CU47 64
Windermere Clo. Orp. CL55 80
Windermere Ct. SW13 BO43 49
Windermere Gdns. Ilf. CK32 26
Windermere Gro. Wem. BK33 21
Windermere Av.
Windermere Rd. N10 BV30 14
Windermere Rd. N19 BW34 23
Holloway Rd.
Windermere Rd. SW15 BO49 58
Windermere Rd. SW16 BW51 68
Windermere Rd. W5 BK41 39
Windermere Rd. Bexh. CR44 54
Windermere Rd. Bexh. CS44 55
Windermere Rd. Couls. BX61 84
Windermere Rd. Croy. CA54 69
Windermere Rd. Sthl. BE39 38
Windermere Rd. W. Wick. CG55 79
Winders Rd. SW11 BU44 50
Windfield Clo. SE26 CC49 61
Windham Av. Croy. CF58 79
Windham Rd. Rich. BL45 48
Windings, The, Sth. Croy. CA59 84
Winding Way, Dag. CP34 27
Winding Way, Har. BH35 30
Windlass Pl. SE8 CD42 43
Windlesham Rd. SW19 BQ47 58
Windley Clo. SE23 CC48 61
Windmill Av. Epsom BO59 82
Windmill Clo. Surb. BK54 66
Windmill Clo. Epsom BO59 82
Windmill Dr. SW4 BV46 50
Windmill End, Epsom BO59 82
Windmill Gdns. Enf. BY23 8
Windmill Hill, Enf. BY24 8
Windmill La. E15 CF36 34
Windmill La. Barn. BO25 6
Windmill La. Bush. BG26 4
Windmill La. Epsom BO59 82
Windmill La. Grnf. BG38 29
Windmill La. Sthl. BG40 38
Windmill La. Surb. BJ53 66
Windmill Rd. N18 BZ28 15
Windmill Rd. SW18 BT46 50
Windmill Rd. SW19 BP47 58
Windmill Rd. W4 BO42 40
Windmill Rd. W5 BK42 39
Windmill Rd. Brent. BK42 39
Windmill Rd. Croy. BZ54 69
Windmill Rd. Dag. CQ34 27
Windmill Rd. Hamptn. BF49 56
Windmill Rd. Mitch. BW53 68
Windmill Row SE11 BY42 42
Windmill St. W1 BW39 41
Windmill St. Bush. BH26 5
Windmill Wk. SE1 BY40 42
Windover Av. NW9 BN31 22
Windrush Clo. W4 BN43 49
Bolton Rd.
Windrush La. SE23 CC48 61
Windsland Ms. W2 BT39 41
London St.
Windsor Av. E17 CD30 16
Windsor Av. SW19 BT51 68
Windsor Av. E. Mol. BF52 65
Windsor Av. Edg. BM28 12
Windsor Av. N. Mal. BN53 67
Windsor Av. Sutt. BR55 76
Windsor Clo. N3 BR30 13
Windsor Rd.
Windsor Clo. Borwd. BM23 5
Windsor Clo. Har. BF34 20
Windsor Clo. Nthwd. BC30 11
Windsor Clo. Brent. BJ43 48
Amalgamated Dr.
Windsor Ct. N14 BW26 7
Windsor Cres. Har. BF35 29
Windsor Cres. Wem. BM34 21
Windsor Dr. Barn. BU25 7
Windsor Dr. Dart. CU46 55
Windsor Dr. Orp. CO57 81
Windsor Gro. SE27 BZ49 60
Windsor Pl. SW1 BW42 41
Francis St.
Windsor Rd. E4 CE28 16
Chivers Rd.
Windsor Rd. E7 CH35 35
Windsor Rd. E10 CE34 25
Windsor Rd. E11 CH33 26
Windsor Rd. N3 BR30 13
Windsor Rd. N7 BX34 24
Windsor Rd. N13 BY27 15
Windsor Rd. N17 CB30 15
Windsor Rd. NW2 BP36 31
Windsor Rd. W5 BK40 39
Windsor Rd. Barn. BQ25 6
Windsor Rd. Bexh. CQ45 54
Windsor Rd. Dag. CQ34 27
Windsor Rd. Har. BG30 11
Windsor Rd. Horn. CV33 28
Windsor Rd. Houns. BC44 47
Windsor Rd. Ilf. CL35 35
Windsor Rd. Kings. On T. BL50 57
Windsor Rd. Rich. BL44 48
Windsor Rd. Sid. CO50 63
Windsor Rd. Sthl. BE41 38
Windsor Rd. Sun. BC50 56
Windsor Rd. Tedd. BG49 56
Windsor Rd. Th. Hth. BY51 69
Windsor Rd. Wor. Pk. BP55 76
Windsor St. N1 BY37 33
Windsor Ter. N1 BZ38 33
Windsor Wk. SE5 BZ44 51
Windus Rd. N16 CA33 24
Windus Wk. N16 CA33 24
Alkham Rd.
Windy Ridge, Brom. CK51 71
Windyridge Clo. SW19 BQ49 58
Wine Clo. E1 CC40 43
Wine Office Ct. EC4 BY39 42
Winforton St. SE10 CF44 52
Winfrith Rd. SW18 BT47 59
Wingate Cres. Croy. BW53 68
Wingate Rd. W6 BP41 40
Wingate Rd. Ilf. CL35 35
Wingate Rd. Sid. CP49 63
Sidcup-hill
Wingate Rd. Sid. CP50 63
Sidcup Hill
Wingfield Ms. SE15 CB45 51
Wingfield St.
Wingfield Pl. Sid. CN48 63
Wingfield Rd. E15 CF35 34
Wingfield Rd. E17 CE32 25
Wingfield Rd. Kings. On T. BL50 57
Wingfield St. SE15 CB45 51
Wingfield Way, Ruis. BC35 29
Wingford Rd. SW2 BX46 51
Wingletye La. Horn. CW31 28
Wingmore Rd. SE24 BZ45 51
Wingrave Rd. W6 BQ43 49
Wingrove Rd. SE6 CG48 61
Winifred Av. Horn. CV35 37
Winifred Gro. SW11 BU45 50
Marjorie Gro.
Winifred Rd. SW19 BS51 68
Winifred Rd. Couls. BV61 83
Winifred Rd. Dag. CQ34 27
Winifred Rd. Dart. CU46 55
Winifred Rd. Erith CT42 46
Winifred Rd. Hamptn. BF49 56
Winifred St. E16 CK40 44
Winifred Ter. Enf. CA26 8
Winkfield Rd. E13 CH37 35
Winkfield Rd. N22 BY30 15
Winkley St. E2 CB38 33
Canrobert St.
Winkworth Pl. Bans. BR60 82
Winkworth Pl. Bans. BS60 83
Winkworth Rd. Bans. BS60 83
Winlaton Rd. Brom. CF49 61
Winn Common Rd. SE18 CN43 54
Winnings Wk. Nthlt. BE36 29
Arnold Rd.
Winnington Clo. N2 BT32 23
Winnington Rd. N2 BT32 23
Winn Rd. SE12 CH47 62
Winns Av. E17 CD31 25
Winns Ms. N15 CD31 25
Grove Park Rd.
Winns Ter. E17 CE30 16
Winsbeach E17 CF31 25
Winscombe Cres. W5 BK38 30
Winscombe St. N19 BV34 23
Winscombe Way, Stan. BJ28 12
Winsford Rd. SE6 CD48 61
Winsford Ter. N18 BZ28 15
Winsham Gro. SW11 BV46 50
Winslade Rd. SW2 BX46 51
Winslade Way SE6 CE47 61
Winsland Ms. W2 BT39 41
Winsland St.
Winsland St. W2 BT39 41
Winsley St. W1 BW39 41
Winslow Clo. Pnr. BC32 20
Winslow Gro. E4 CG27 16
Hoppett Rd.
Winslow Rd. W6 BQ43 49
Winslow Way, Felt. BE48 56
Winslow Way, Walt. BD55 74
Winsor Est. W9 BS39 41
Winsor Ter. E5 CL39 44
Winstanley Rd. SW11 BT45 50
Winstead Gdns. Dag. CS35 37
Winston Av. NW9 BO33 22
Winston Clo. Har. BH29 12
Winston Clo. Rom. CR31 27
Winston Ct. Har. BF29 11
Winstone Clo. Rom. CR31 27
Marlborough Rd.
Winstone Wk. W4 BN41 40
Winston Rd. N16 BZ35 33
Winstre Rd. Borwd. BM23 5
Winter Av. E6 CK37 35
Winterborne Av. Orp. CM55 80
Winterbourne Rd. SE6 CD47 61
Winterbourne Rd. Dag. CP34 27
Winterbourne Rd. Th. Hth. BY52 69
Winter Box Wk. Rich. BL46 48
Kings Rd.
Winterbrook Rd. SE24 BZ46 51
Winterdown Gdns. Esher BE57 74
Winterdown Rd. Esher BE57 74
Winterfold Clo. SW19 BR48 58
Winters Rd. Surb. BJ54 66
Winterstoke Gdns. NW7 BP28 13
Winterstoke Rd. SE6 CD47 61
Winterton Pl. SW10 BT43 50
Winterwell Rd. SW2 BX46 51
Winthorpe Rd. SW15 BR45 49
Winthrop Pl. E1 CB39 42
Winthrop St.
Winthrop St. E1 CB39 42
Winton Av. N11 BW29 14
Winton Clo. N9 CC26 9
Winton Gdns. Edg. BL29 12
Whitchurch La.
Winton Rd. Orp. CL56 80
Winton Way SW16 BY49 60
Wisbeach Rd. Croy. BZ53 69
Wisborough Rd. Sth. Croy. CA58 78
Wisdons Clo. Dag. CR34 27
Wise La. NW7 BP28 13
Wiseman Rd. E10 CE34 25
Wise Rd. E15 CF37 34
Wiseton Rd. SW17 BU47 59
Wishart Rd. SE3 CJ44 53
Wisley Rd. SW11 BU46 50
Wisley Rd. Orp. CO50 63
Wisteria Clo. Orp. CL55 80
Wisteria Gdns. Swan. CS51 73
Wisteria Rd. SE13 CF45 52
Witan St. E2 CB38 33
Coventry Rd.
Witham Rd. W13 BJ40 39
Uxbridge Rd.
Witham Clo. Loug. CK25 10
Witham Rd. SE20 CC52 70
Witham Rd. Dag. CR35 36
Witham Rd. Horn. CU32 28
Witham Rd. Islw. BG44 47
Withens Clo. Orp. CP52 72
Witherby Clo. Croy. CA56 78
Witherfield Way SE16 CB42 42
Egan Way
Witherings, The, Horn. CW32 28
Witherington Rd. N5 BY35 33
Withers Mead NW9 BO30 13
Withers Pl. EC1 BZ38 33
Old St.
Witherston Way SE9 CL48 62
Withycombe Rd. SW19 BQ47 58
Victoria Dr.
Withy Mead E4 CF27 16
Witley Ct. Sthl. BE32 20
Witley Cres. Croy. CF57 79
Witley Gdns. Sthl. BE42 38

Witley Rd. N19 BW34 23
Holloway Rd.
Witney Clo. Pnr. BE29 11
Witney Path SE23 CC48 61
Wittenham Way E4 CF27 16
Wittering Wk. Horn. CV36 37
Wittersham Rd. Brom. CG49 61
Witton Gdns. E. Mol. BF52 65
Wivenhoe Clo. SE15 CB45 61
Wivenhoe Rd. Bark. CO37 36
Wiverton Rd. SE26 CC50 61
Wix Rd. Dag. CP37 36
Wix's La. SW4 BV45 50
Woburn Av. Horn. CU35 37
Woburn Av. Pur. BY59 84
Woburn Clo. Bush. BG25 4
Woburn Clo. SW16 BT50 59
Woburn Pl. WC1 BW38 32
Woburn Rd. Cars. BU54 68
Woburn Rd. Croy. BZ54 69
Woburn Sq. WC1 BW38 32
Woburn Wk. WC1 BW38 32
Wodeham St. E1 CB39 42
Woffington Clo.
Kings. On T. BK51 66
Wofington Clo. Tedd. BK50 57
Upr. Teddington Rd.
Woking Clo. SW15 BO45 49
Woldham Rd. Brom. CJ52 71
Wolds Dr. Orp. CL56 80
Wolfe Clo. Brom. CH53 71
Wolfe Clo. Hayes BC38 29
Ayles Rd.
Wolfe Cres. SE7 CJ42 44
Wolftencroft Clo. SW11 BU45 50
Wolferton Rd. E12 CK35 35
Wolffe Gdns. E15 CG36 34
Wolfington Rd. SE27 BY49 60
Wolfram Clo. SE13 CG46 52
Wolfstan St. W12 BO40 40
Wolmer Clo. Edg. BM28 12
Wolmer Gdns. Edg. BM27 12
Wolseley Av. SW19 BS48 59
Wolseley Gdns. W4 BM43 48
Wolseley Rd. E7 CH36 35
Wolseley Rd. N8 BW32 23
Wolseley Rd. N22 BX30 15
Wolseley Rd. W4 BN42 40
Wolseley Rd. Har. BH31 21
Wolseley Rd. Mitch. BV54 68
Wolseley Rd. Rom. CS33 28
Wolseley St. SE1 CA41 42
Wolsen St. E1 CC39 43
Sidney St.
Wolsey Av. E6 CL38 35
Wolsey Av. E17 CD31 25
Wolsey Av. Surb. BH53 66
Wolsey Clo. Houns. BG46 47
Wolsey Clo. Kings. On T. BM51 66
Wolsey Clo. Sthl. BG41 38
Wolsey Clo. SW20 BP50 58
Wolsey Clo. Wor. Pk. BO56 76
Wolsey Cres. Croy. CF58 79
Wolsey Cres. Mord. BR54 67
Wolsey Dr. Kings. On T. BL49 57
Wolsey Dr. Walt. BD54 65
Wolsey Gdns. Ilf. CL29 17
Wolsey Gro. Edg. BN29 13
Wolsey Gro. Esher BF56 74
Wolsey Ms. NW5 BW36 32
Caversham Rd.
Wolsey Rd. N1 BZ35 33
Wolsey Rd. E. Mol. BG52 65
Wolsey Rd. Enf. CB23 8
Wolsey Rd. Esher BF56 74
Wolsey Rd. Hmptn. BF50 56
Wolsey Rd. Sun. BC50 56
Wolsey St. E1 CC39 43
Sidney St.
Wolsley Clo. Dart. CT46 55
Wolsey Way, Chess. BM56 75
Wolstonbury N12 BR28 13
Wolvercote Rd. SE2 CP41 45
Wolverley St. E2 CB38 33
Bethnal Green Rd.
Wolverton Av.
Kings. On T. BM51 66
Wolverton Gdns. W5 BL40 39
Wolverton Gdns. W6 BQ42 40
Wolverton Rd. Stan. BK29 12
Wolverton Way N14 BW25 7
Wolves La. N13 BY29 15
Wolves La. N22 BY29 15
Womersley Rd. N8 BX32 24
Wonersh Way, Sutt. BQ58 76
Wonford Clo. N. Mal. BO51 67
Wontford Rd. Pur. BY61 84
Wontner Rd. SW17 BU48 59
Woodall Cres. Horn. CW33 28
Woodbank N12 BS28 14
Woodbank Rd. Brom. CG48 61
Woodbastwick Rd. SE26 CC50 61
Woodberry Av. N12 BT29 14
Woodberry Av. N21 BY27 15
Woodberry Av. Har. BF31 20
Woodberry Clo. Sun. BC50 56
Woodberry Cres. N10 BV31 23
Woodberry Down N4 BZ33 24

Woodberry Down Est. N4 BZ33 24
Woodberry Gdns. N12 BT29 14
Woodberry Gro. N4 BZ33 24
Woodberry Gro. N12 BT29 14
Woodberry Way E4 CF26 9
Woodberry Way N12 BT29 14
Woodbine Clo. Twick. BG48 56
Woodbine Gro. SE20 CB50 60
Woodbine La. Wor. Pk. BP55 76
Woodbine Pl. E11 CH32 26
Woodbine Rd. Sid. CN47 63
Woodbines Av.
Kings. On T. BK52 66
Woodbine Ter. E9 CC36 34
Homerton Rd.
Woodborough Rd. SW15 BP45 49
Woodbourne Av. SW16 BW48 59
Woodbourne Clo. SW16 BW48 59
Woodbourne Dr. Esher BH57 75
Woodbourne Gdns. Wall. BV57 77
Woodbridge Clo. Rom. CV28 19
Woodbridge Ct. CK29 17
Woodbridge La. Rom. CV27 19
Woodbridge Rd. Bark. CN35 36
Woodbridge St. EC1 BY38 33
Woodbrook Rd. SE2 CO43 54
Woodburn Clo. NW4 BQ32 22
Woodbury Clo. E11 CH31 26
Woodbury Clo. Croy. CA55 78
Woodbury Dr. Sutt. BT58 77
Woodbury Hill, Loug. CK23 10
Woodbury Pk. Rd. W13 BJ38 30
Woodbury Rd. E17 CE31 25
Woodbury St. SW17 BU49 59
Woodchester Sq. W2 BS39 41
Woodchurch Clo. Sid. CM48 62
Woodchurch Dr. Brom. CJ50 62
Woodchurch Rd. NW6 BS36 32
Wood Clo. E2 CB38 33
Wood Clo. NW9 BN33 22
Wood Clo. Bex. CT48 64
Wood Clo. Har. BG33 20
Woodclyffe Dr. Chis. CL51 71
Woodcock Dell Av. Har. BK33 21
Woodcock Hill, Borwd. BM25 5
Woodcombe Cres. SE23 CC47 61
Woodcote Av. NW7 BQ29 13
Woodcote Av. Horn. CU35 37
Woodcote Av. Th. Hth. BY52 69
Woodcote Av. Wall. BV58 77
Woodcote Clo. Enf. CC25 9
Woodcote Clo. Epsom BN60 82
Woodcote Clo.
Kings. On T. BL50 57
Woodcote Dr. Pur. BW58 77
Woodcote End, Epsom BN61 82
Woodcote Grn. Rd.
Epsom BN61 82
Woodcote Grn. Rd. Wall. BW58 77
Woodcote Gro. Cars. BV59 83
Woodcote Gro. Rd. BW61 83
Couls.
Woodcote Hurst, Epsom BN61 82
Woodcote La. Pur. BW59 83
Woodcote Pk. Av. Pur. BW59 83
Woodcote Pk. Rd. Epsom BN61 82
Woodcote Pl. SE27 BY49 60
Woodcote Rd. E11 CH33 26
Woodcote Rd. Epsom BN60 82
Woodcote Rd. Orp. CM54 71
Woodcote Valley Rd. Pur. BW60 83
Woodcote Vill. Pur. BW59 83
Woodcrest Rd. Pur. BX60 84
Woodcroft N21 BX26 8
Woodcroft SE9 CK48 62
Woodcroft Av. NW7 BO29 13
Woodcroft Av. Stan. BJ30 12
Woodcroft Cres. Grnf. BJ36 30
Woodcroft Rd. Th. Hth. BY53 69
Wood Dene, Leath. BG58 74
Wood Dr. Chis. CK50 62
Woodedge Clo. E4 CG26 9
Forest Side
Woodend SE19 BZ50 60
Wood End, Esher BG55 74
Woodend, Sutt. BT55 77
Wood End Av. Har. BF35 29
Wood End Clo. Nthlt. BG35 29
Wood End Gdns. Nthlt. BG35 29
Wood End Gdns. Enf. BX24 8
Wood End La. Nthlt. BF36 29
Woodend Rd. E17 CF30 16
Wood End Rd. Har. BG35 29
Woodend, The, Wall. BV58 77
Wood End Way, Nthlt. BG35 29
Wooder Gdns. E7 CG35 34
Woodfall Av. Barn. BR25 6
Woodfall Rd. N4 BY33 24
Woodfall St. SW3 BU42 41
Woodfarrs SE5 BZ45 51
Woodfield Av. NW9 BO31 22
Woodfield Av. SW16 BW48 59
Woodfield Av. W5 BK38 30
Woodfield Av. Cars. BV57 77

Woodfield Av. Wem. BK34 21
Woodfield Clo. SE19 BZ50 60
Woodfield Cres. W5 BK38 30
Woodfield Dr. Barn. BV26 7
Woodfield Dr. Rom. CU31 28
Woodfield Gdns. W9 BR39 40
Woodfield Rd.
Woodfield Gdns. N. Mal. BO53 67
Woodfield Gro. SW16 BW48 59
Woodfield La. SW16 BW48 59
Woodfield Pl. W9 BR39 40
Woodfield Rise, Bush. BG26 4
Woodfield Rd. W5 BK38 30
Woodfield Rd. W9 BR39 40
Woodfield Rd. Surb. BH55 75
Woodfield Rd. Houns. BC44 47
Woodfields, The, CA59 84
Sth. Croy.
Woodfield Way N11 BW29 14
Woodfield Way, Horn. CV33 28
Woodford Av. Ilf. CJ31 26
Woodford Bridge Rd. Ilf. CJ31 26
Woodford Cres. Pnr. BC30 11
Woodford New Rd. N17 CG31 25
Woodford New Rd. E18 CG31 25
Woodford New Rd. CG31 25
Wdf. Grn.
Woodford Pl. Wem. BL33 21
Woodford Rd. E7 CH35 35
Woodford Rd. E18 CG33 25
Woodford Rd. Wat. BC23 4
Woodford Trading Est. CJ30 17
E18
Woodgate Av. Chess. BK56 75
Woodgate Cres. Nthwd. BC29 11
Woodgavil, Bans. BR61 82
Woodger Rd. W12 BQ41 40
Goldhawk Rd.
Woodgrange Av. N12 BT29 14
Woodgrange Av. W5 BL40 39
Woodgrange Av. Enf. CB25 8
Woodgrange Av. Har. BK32 21
Woodgrange Clo. Har. BK32 21
Woodgrange Gdns. Enf. CB25 8
Woodgrange Rd. E7 CH35 35
Woodhall Av. SE21 CA48 60
Woodhall Dr. SE21 CA48 60
Woodhall Dr. Pnr. BE30 11
Woodhall Gate, Pnr. BD29 11
Woodhall La. Wat. BD27 11
Woodhall Rd. Pnr. BD29 11
Woodham Ct. E18 CG31 25
Woodham Rd. SE6 CF48 61
Woodhatch Spinney, BX57 78
Couls.
Woodhatch Spinney, BX61 84
Couls.
Woodhaven Gdns. Ilf. CL31 26
Brandville Gdns.
Woodhaven Gdns. Ilf. CM31 26
Brandville Gdns.
Woodhayes Rd. SW19 BQ50 58
Woodhead Dr. Orp. CN55 81
Woodheyes Rd. NW10 BN35 31
Woodhill SE18 CK41 44
Woodhill Cres. Har. BK32 21
Woodhouse Av. Grnf. BH37 30
Woodhouse Clo. Grnf. BH37 30
Woodhouse Gro. E12 CK36 35
Woodhouse Rd. E11 CG34 25
Woodhouse Rd. N12 BT29 14
Woodhurst Av. Orp. CL53 71
Woodhurst Rd. SE2 CO42 45
Woodhurst Rd. W3 BN40 40
Woodhyrst Gdns. Ken. BY57 78
Woodhyrst Gdns. Ken. BY61 84
Woodington Clo. SE9 CL46 53
Woodison St. E3 CD38 34
Woodknoll Dr. Brom. CK51 71
Woodland App. Grnf. BJ36 30
Whitton Dr.
Woodland Clo. NW9 BN32 22
Woodland Clo. SE19 CA50 60
Woodland Hl.
Woodland Clo. Epsom BO57 76
Woodland Clo. Wdf. Grn. CH27 17
Woodland Cres. SE10 CG43 52
Woodland Gdns. N10 BV32 23
Woodland Gdns. Islw. BH45 48
Woodland Gdns. CC58 79
Sth. Croy.
Woodland Gro. SE10 CG42 43
Woodland Hl. SE19 CA50 60
Woodland Rise N10 BV31 23
Woodland Rise, Grnf. BJ36 30
Woodland Rd. E4 CF26 9
Woodland Rd. N11 BV28 14
Woodland Rd. SE19 CA49 60
Woodland Rd. Loug. CK24 10
Woodland Rd. Th. Hth. BY52 69
Woodlands NW11 BR32 22
Woodlands SW20 BO52 67
Woodlands, Har. BF31 20
Woodlands Av. E11 CH33 26
Woodlands Av. N12 BT29 14

Woodlands Av. W3 BM40 39
High St.
Woodlands Av. Horn. CV32 28
Woodlands Av. N. Mal. BN51 67
Woodlands Av. Rom. CQ32 27
Woodlands Av. Ruis. BD33 20
Woodlands Av. Sid. CN47 63
Woodlands Av. Wor. Pk. BO55 76
Woodlands Clo. NW11 BR32 22
Woodlands Clo. Borwd. BM24 5
Woodlands Clo. Brom. CK51 71
Woodlands Clo. Esher BH57 75
Woodlands Clo. Swan. CT52 73
Woodlands Dr. Har. BH29 12
Woodlands Dr. Sun. BD51 65
Woodlands Dr. Wat. BC23 4
Woodlands Gdns. E17 CG31 25
Woodford New Rd.
Woodlands Gro. Couls. BV62 83
Woodlands Gro. Islw. BH44 48
Woodlands Pk. Bex. CS49 64
Woodlands Pk. Rd. N15 BY32 24
Woodlands Pk. Rd. SE10 CG43 52
Woodlands Rise, Swan. CT51 73
Woodlands Rd. E11 CG34 25
Woodlands Rd. E17 CF31 25
Woodlands Rd. N9 CC26 9
Woodlands Rd. SW13 BO45 49
Woodlands Rd. Bexh. CQ45 54
Woodlands Rd. Brom. CK51 71
Woodlands Rd. Bush. BE24 4
Woodlands Rd. Enf. BZ23 8
Woodlands Rd. Har. BH32 21
Woodlands Rd. Ilf. CM34 26
Woodlands Rd. Islw. BG45 47
Woodlands Rd. Islw. BH44 48
Woodlands Rd. Orp. CO57 81
Woodlands Rd. Rom. CT31 28
Woodlands Rd. Sthl. BD40 38
Woodlands Rd. Surb. BK54 66
Woodlands St. SE13 CF47 61
Woodlands, The N14 BV26 7
Woodlands, The SE13 CF47 61
Woodlands, The SE19 BZ50 60
Woodlands, The, Esher BG55 74
Woodlands, The, Islw. BH44 48
Woodlands, The, Orp. CO57 81
Woodlands, The, Wall. BV58 77
Woodlands Way SW15 BR46 49
Woodland Ter. SE7 CK42 44
Woodland Ter. SE18 CK42 44
Woodland Wk. SE10 CG42 43
Woodland Way N21 BY27 15
Woodland Way NW7 BO29 13
Woodland Way NW11 BS32 23
Woodland Way SE2 CP42 45
Woodland Way, Croy. CD54 70
Woodland Way, Mitch. BV50 59
Woodland Way, Mord. BR52 67
Woodland Way, Orp. CM52 71
Woodland Way, Pur. BY60 84
Woodland Way, Surb. BM55 75
Woodland Way, CH27 17
Wdf. Grn.
Woodland Way, W. Wick. CE56 79
Wood La. N6 BV32 23
Wood La. NW9 BN33 22
Wood La. W12 BQ39 40
Wood La. Dag. CP35 36
Wood La. Horn. CU35 37
Wood La. Islw. BH43 48
Wood La. Stan. BJ27 12
Wood La. Tad. BR62 82
Wood La. Wdf. Grn. CG28 16
Woodlawn Cres. Twick. BF48 56
Woodlawn Dr. Felt. BD48 56
Woodlawn Rd. SW6 BQ43 49
Woodlea Dr. Brom. CG53 70
Woodlea Est. Brom. CG53 70
Woodlea Rd. N16 CA34 24
Woodleigh Av. N12 BU29 14
Woodleigh Gdns. SW16 BX48 60
Woodley Clo. SW17 BU50 59
Woodley Rd. Orp. CP55 81
Wood Lodge La. CF55 79
W. Wick.
Woodman La. E4 CG25 9
Woodman Path, Chig. CN29 18
Woodman Rd. Couls. BW61 83
Woodman St. E16 CJ40 44
Woodmans Ct. Stan. BL30 12
Woodmansterne La. BS61 83
Bans.
Woodmansterne La. Cars. BU59 83
Woodmansterne Rd. BW50 59
SW16
Woodmansterne Rd. BU58 77
Cars.
Woodmansterne Rd. BW61 83
Couls.
Woodmansterne St. Bans. BU61 83
Woodmans Mews W12 BP39 40
Woodmere SE9 CK47 62
Woodmere Av. Croy. CC54 70
Woodmere Clo. Croy. CC54 70
Woodmere Gdns. Croy. CC54 70
Woodmere Way, Beck. CF53 70
Woodmount, Swan. CS54 73
Woodnook Rd. SW16 BV49 59

Woodpecker Clo. N9 CB25 8
Woodpecker Clo. Bush. BG26 4
Woodpecker Mt. Croy. CD58 79
Woodpecker Rd. SE14 CD43 52
Woodpecker Rd. SE28 CP40 45
Woodplade La. Couls. BW62 83
Woodquest Av. SE24 BZ46 51
Woodride, Barn. BT23 7
Wood Ride, Orp. CM52 71
Woodridings Av. Pnr. BE30 11
Woodridings Clo. Pnr. BE29 11
Woodriffe Rd. E11 CF33 25
Wood Rise, Pnr. BC32 20
Woodrow SE18 CK42 44
Woodrow Clo. Grnf. BJ36 30
Woodrush Way, Rom. CP31 27
Woods Bldgs. E1 CB39 42
Durward St.
Woodseer St. E1 CA39 42
Woodsford Sq. W14 BR41 40
Woodshire Rd. Dag. CR34 27
Woodside NW11 BS32 23
Woodside SW19 BR50 58
Woodside, Borwd. BL24 5
Woodside, Buck. H. CJ27 17
Woodside, Orp. CO56 81
Woodside Av. N6 BU32 23
Woodside Av. N12 BS28 14
Woodside Av. SE25 CB53 69
Woodside Av. Chis. CL49 62
Woodside Av. Esher BH54 66
Woodside Av. Walt. BC56 74
Woodside Av. Wem. BL37 30
Woodside Clo. Bexh. CS45 55
Woodside Clo. Rain. CV38 37
Woodside Clo. Stan. BJ28 12
Woodside Clo. Surb. BN54 67
Woodside Clo. Wem. BL37 30
Woodside Ct. Croy. CB54 69
Woodside Ct. Rd. Croy. CB54 69
Woodside Cres. Sid. CN48 63
Woodside Dr. Dart. CT49 64
Woodside End, Wem. BL37 30
Woodside Gdns. E4 CE28 16
Woodside Gdns. N17 CA30 15
Woodside Grange Rd. N12 BS28 14
Woodside Grn. SE25 CB53 69
Woodside Gro. N12 BT27 14
Woodside La. N12 BS27 14
Woodside La. Bex. CP46 54
Woodside Pk. SE25 CB53 69
Woodside Pk. Av. E17 CF32 25
Woodside Pk. Rd. N12 BS28 14
Woodside Pl. Wem. BL37 30
Woodside Rd. E13 CJ38 35
Woodside Rd. N22 BX29 15
Woodside Rd. SE25 CB53 69
Woodside Rd. Bexh. CS45 55
Woodside Rd. Brom. CK53 71
Woodside Rd. Kings. On T. BL50 57
Woodside Rd. N. Mal. BN51 67
Woodside Rd. Nthwd. BC29 11
Woodside Rd. Pur. BW60 83
Woodside Rd. Sid. CN48 63
Woodside Rd. Sutt. BT55 77
Woodside Rd. Wdf. Grn. CH28 17
Woodside Vw. SE25 CB53 69
Woodside Vw. Sev. CR58 81
Woodside Way, Croy. CC53 70
Woodside Way, Mitch. BV51 68
Woods Ms. W1 BV40 41
Woodsome Rd. NW5 BV34 23
Woods Pl. SE1 CA41 42
Woodspring Rd. SW19 BR48 58
Woods Rd. SE15 CB44 51
Woodstead Gro. Edg. BL29 12
Woods, The, Nthwd. BC28 11
Woodstock, Surb. BK55 75
Woodstock Av. NW11 BR33 22
Woodstock Av. W13 BJ41 39
Woodstock Av. Islw. BJ46 48
Woodstock Av. Sthl. BE38 29
Woodstock Av. Sutt. BR54 67
Woodstock Clo. Bex. CQ47 63
Woodstock Clo. Stan. BL30 12
Woodstock Ct. SE12 CH46 53
Woodstock Cres. N9 CA25 8
Woodstock Gdns. Beck. CE50 61
Woodstock Gdns. Ilf. CO34 27
Woodstock Gro. W12 BQ41 40
Woodstock La. Esher BJ57 75
Woodstock Rise, Sutt. BR54 67
Woodstock Rd. E7 CJ36 35
Woodstock Rd. E17 CF30 16
Woodstock Rd. N4 BY33 24
Woodstock Rd. NW11 BR33 22
Woodstock Rd. W4 BO42 40
Woodstock Rd. Bush. BH26 5
Woodstock Rd. Cars. BV56 77
Woodstock Rd. Couls. BV61 83
Woodstock Rd. Croy. BZ55 78
Woodstock Rd. Wem. BL37 30
Woodstock St. E16 CG39 43
Woodstock St. W1 BV39 41
Oxford St.
Woodstock Ter. E14 CE40 43
Woodstock Way, Mitch. BV51 68
Woodstone Av. Epsom BP56 76
Wood St. E16 CH39 44
Ethel Rd.
Wood St. E17 CF31 25
Wood St. EC2 BZ39 42
Wood St. W4 BO42 40
Wood St. Barn. BQ24 6
Wood St. Kings. On T. BK51 66
Wood St. Mitch. BU54 68
Wood St. Swan. CV51 73
Woodsyre Est. SE26 CA49 60
Wood, The, Surb. BL53 66
Woodthorpe Rd. SW15 BP45 49
Wood Vale N10 BW32 23
Wood Vale SE23 CB47 60
Woodvale Av. SE25 CA52 69
Wood Vale Est. SE23 CB47 60
Woodview Av. E4 CF28 16
Woodview Clo. Sth. Croy. CB60 84
Woodview Rd. Swan. CS51 73
Woodville SE3 CH44 53
Woodville Clo. Tedd. BJ49 57
Woodville Clo. SE12 CH46 53
Woodville Ct. Wat. BC23 4
Woodville Gdns. NW11 BQ33 22
Hendon Way
Woodville Gdns. W5 BL39 39
Woodville Gdns. Ilf. CL31 26
Woodville Gro. N16 CA35 33
Woodville Rd.
Woodville Rd. E11 CG33 25
Woodville Rd. E17 CD32 25
Woodville Rd. E18 CH30 17
Woodville Rd. N16 CA35 33
Woodville Rd. NW6 BR37 31
Woodville Rd. NW11 BQ33 22
Woodville Rd. W5 BK39 39
Woodville Rd. Barn. BS24 7
Woodville Rd. Mord. BS52 68
Woodville Rd. Rich. BJ48 57
Woodville Rd. Th. Hth. BZ52 69
Woodward Av. NW4 BP32 22
Woodwardee Rd. SE22 CA46 51
Woodward Gdns. Dag. CO36 36
Woodward Rd.
Woodward Rd. Dag. CO36 36
Wood Way, Orp. CL55 80
Woodway Cres. Har. BJ32 21
Woodwaye, Wat. BD26 4
Woodwell St. SW18 BT46 50
North Side
Woodyard La. SE21 CA47 60
Woodyates Rd. SE12 CG46 52
Woolacombe Rd. SE3 CH44 53
Woolaston Rd. N4 BY32 24
Umfreville Rd.
Wooler St. SE17 BZ42 42
Woolf Clo. SE28 CO40 45
Woolhampton Way, Chig. CO27 18
Woolmead Av. NW4 BP33 22
Woolmer Gdns. N18 CB29 15
Woolmer Rd. N18 CB28 15
Woolmore St. E14 CF40 43
Woolneigh St. SW6 BS45 50
Wool Rd. SW20 BP50 58
Woolston Clo. E17 CC30 16
Woolstone Rd. SE23 CD48 61
Woolwich Church St. SE18 CK41 44
Woolwich Common SE18 CL43 53
Woolwich High St. SE18 CK41 44
Woolwich Ind. Est. SE28 CN41 45
Woolwich Manor Way E16 CL40 44
Woolwich New Rd. SE18 CL42 44
Woolwich Rd. SE2 CP43 54
Woolwich Rd. SE7 CG42 43
Woolwich Rd. SE10 CG42 43
Woolwich Rd. Belv. CP43 54
Woolwich Rd. Bexh. CR45 54
Wooster Gdns. E14 CF39 43
Wootton Clo. Horn. CV32 28
Wootton Gro. N3 BS30 14
Station Rd.
Wootton St. SE1 BY41 42
Worbeck Rd. SE20 CB51 69
Worcester Av. N17 CB29 15
Worcester Clo. Croy. CD55 79
Worcester Clo. Mitch. BV51 68
Worcester Ct. KT12 BD54 65
Rodney Rd.
Worcester Cres. NW7 BO27 13
Worcester Cres. Wdf. Grn. CH28 17
Worcester Gdns. Grnf. BG36 29
Worcester Gdns. Ilf. CK33 26
Worcester Gdns. Sutt. BS57 77
Worcester Gdns. Wor. Pk. BO55 76
Worcester Pk. Rd. Wor. Pk. BN55 76
Worcester Rd. E12 CK35 35
Worcester Rd. SW19 BR49 58
Worcester Rd. Sutt. BS57 77
Wordsworth Av. E12 CK36 35
Wordsworth Av. E18 CG31 25
Wordsworth Av. Grnf. BG38 29
Wordsworth Clo. Rom. CV30 19
Wordsworth Dr. Sutt. BQ56 76
Wordsworth Pde. N8 BY31 24
Alfoxton Av.
Wordsworth Rd. N16 CA35 33
Wordsworth Rd. SE20 CC50 61
Wordsworth Rd. Hamptn. BE49 56
Wordsworth Rd. Har. BH31 21
Wordsworth Rd. Wall. BW57 77
Wordsworth Rd. Well. CN44 54
Wordsworth Wk. NW11 BS31 23
Worfield St. SW11 BU43 50
Worgan St. SE11 BX42 42
Worland Rd. E15 CG36 34
Worlds End SW10 BT43 50
World End La. N21 BX25 8
Worlds End, Epsom BN61 82
Worlds End La. Orp. CN57 81
Worlds End Pass. SW10 BT43 50
Worlidge St. W6 BQ42 40
Worlingham Rd. SE22 CA45 51
Wormholt Est. W12 BO40 40
Wormholt Rd. W12 BP40 40
Wormwood St. EC2 CA39 42
Wornington Rd. W10 BR38 31
Wornington Yd. W10 BR38 31
Wornington Rd.
Woronzow Rd. NW8 BT37 32
Worple Av. SW19 BQ50 58
Worple Av. Islw. BJ46 48
Worple Clo. Har. BE33 20
Worple Rd. SW19 BQ51 67
Worple Rd. SW20 BQ51 67
Worple Rd. Epsom BN61 82
Worple Rd. Islw. BJ45 48
Worple Rd. Ms. SW19 BR50 58
Worple St. SW14 BN45 49
Worple Way, Har. BE33 20
Worple Way, Rich. BL46 48
Worship St. EC2 BZ38 33
Worslade Rd. SW17 BT49 59
Worsley Br. Rd. SE26 CD49 61
Worsley Br. Rd. Beck. CD49 61
Worsley Rd. E11 CG35 34
Worsopp Dr. SW4 BW45 50
Worth Clo. Orp. CN56 81
Worthfield Clo. Epsom BN57 76
Worthing Clo. E15 CG37 34
Worthing Rd. Houns. BE43 47
Worthington Rd. Surb. BL54 66
Worthydown Ct. SE18 CL43 53
Prince Imperial Rd.
Wortley Rd. E6 CJ36 35
Wortley Rd. Croy. BY54 69
Worton Gdns. Islw. BG44 47
Worton Rd. Islw. BG45 47
Worton Way, Islw. BG44 47
Wotton Grn. Orp. CP52 72
Wotton Rd. NW2 BQ35 31
Wotton Way, Sutt. BQ58 76
Wouldham Rd. E16 CG39 43
Wragby Rd. E11 CG34 25
Wrampling Pl. N9 CB26 8
Croyland Rd.
Wray Av. Ilf. CL31 26
Wray Clo. Horn. CV33 28
Wray Cres. N4 BX34 24
Wrayfield Rd. Sutt. BQ55 76
Wray Rd. Sutt. BR58 76
Wrekin Rd. SE18 CM43 53
Wren Av. NW2 BQ35 31
Wren Av. Sthl. BE42 38
Wren Clo. Sth. Croy. CC58 79
Wren Cres. Bush. BG26 4
Wren Gdns. Dag. CP35 36
Wren Gdns. Horn. CT33 28
Wren Rd. SE5 BZ44 51
Wren Rd. Dag. CP35 36
Wren Rd. Sid. CP49 63
Wren St. WC1 BX38 33
Gray's Inn Rd.
Wrentham Av. NW10 BQ37 31
Wrenthorpe Rd. Brom. CG49 61
Wrenwood Way, Pnr. BC31 20
Wrexham Rd. E3 CE37 34
Wrexham Rd. Rom. CV27 19
Wricklemarsh Rd. SE3 CH44 53
Wrigglesworth St. SE14 CC43 52
Wright Rd. Houns. BC43 47
Wright Rd. NW4 BQ33 22
Wright's Alley SW19 BQ50 58
Wrightsbridge Rd. Rom. CW27 19
Wrights Bldgs. SE1 CA41 42
Grange Rd.
Wright's Clo. SE13 CF45 52
Wisteria Rd.
Wright's La. W8 BS41 41
Wrights Pl. NW10 BN36 31
Mitchell Way
Wright's Rd. E3 CD37 34
Wright's Rd. SE25 CA52 69
Wrights Row, Wall. BV56 77
Wrights Wk. SW14 BN45 49
North Worple Way
Wrigley Clo. E4 CF28 16
Avenue, The
Writtle Wk. Rain. CT37 37
Wrotham Rd. NW1 BW36 32
Wrotham Rd. W13 BJ40 39
Mattock La.
Wrotham Rd. Barn. BR23 6
Wrotham Rd. Well. CP44 54
Wroths Path, Loug. CK23 10
Wrottesley Rd. NW10 BP37 31
Wrottesley Rd. SE18 CL43 53
Wroughton Rd. SW11 BU46 50
Wroughton Ter. NW4 BP31 22
Babington Rd.
Wroxall Rd. Dag. CP36 36
Wroxham Gdns. N11 BW29 14
Wroxton Rd. SE15 CB44 51
Wrythe Grn. Rd. Cars. BU55 77
Wrythe La. Cars. BT54 68
Wulfstan St. W12 BO39 40
Wyatt Clo. Hayes BC39 38
Wyatt Clo. Nthlt. BE36 29
Wyatt Pk. Rd. SW2 BX48 60
Wyatt Rd. DA1 CT45 55
Wyatt Rd. E7 CH36 35
Wyatt Rd. N5 BZ34 24
Wyatts La. E17 CF31 25
Wybert St. NW1 BV38 32
Munster Sq.
Wyborne Way NW10 BN36 31
Wyburn Av. Barn. BR24 6
Wyche Gro. Sth. Croy. BZ57 78
Wych Elm Pass. Kings. On T. BL50 57
Wycherley Cres. Barn. BS25 7
Wychling Clo. Orp. CP54 72
Wychwood Av. Edg. BK29 12
Wychwood Av. Th. Hth. BZ52 69
Wychwood Clo. Edg. BK29 12
Wychwood End N6 BW33 23
Wychwood Gdns. Ilf. CK31 26
Wychwood Wk. Edg. BK29 12
Wycliffe Clo. Well. CN44 54
Wycliffe Rd. SW19 BS50 59
Wyclif St. EC1 BY38 33
Wycombe Gdns. NW11 BS34 23
Wycombe Rd. N17 CB30 15
Wycombe Rd. Ilf. CK32 26
Wycombe Rd. Wem. BM37 30
Wydehurst Rd. Croy. CB54 69
Wydell Clo. Mord. BQ53 67
Wydeville Manor Rd. SE12 CH49 62
Wye Clo. Orp. CN54 72
Wye St. SW11 BT44 50
Wyeth's Rd. Epsom BO60 82
Wyevale Clo. Pnr. BC31 20
Wyfields, Ilf. CL30 17
Wyfold Rd. SW6 BR44 49
Wyhill Wk. Dag. CS36 37
Wyke Gdns. W7 BJ41 39
Wykeham Av. Dag. CP36 36
Wykeham Av. Horn. CV32 28
Wykeham Grn. Dag. CP36 36
Wykeham Hill, Wem. BL33 21
Wykenham Rd. NW4 BQ31 22
Wykeham Rd. Har. BJ31 21
Wyke Rd. E3 CE36 34
Wyke Rd. SW20 BQ51 67
Wyldfield Gdns. N9 CA27 15
Latymer Rd.
Wyld Way, Wem. BM36 30
Wyleu St. SE23 CD47 61
Wylie Rd. Sthl. BF41 38
Wyllen Clo. E1 CC38 34
Wylo Dr. Barn. BO25 6
Wymering Rd. W9 BS38 32
Wymond St. SW15 BQ45 49
Wynaud Ct. N22 BX29 15
Palmerston Rd.
Wyncham Av. Sid. CN47 63
Wynchgate N14 BW26 7
Wynchgate, Har. BH29 12
Wyncote Way, Sth. Croy. CC58 79
Wyncroft Clo. Brom. CK52 71
Wyndale Av. NW9 BM32 21
Wyndcliff Rd. SE7 CH42 44
Wyndcroft Clo. Enf. BY24 8
Wyndham Av. Barn. BU26 7
Wyndham Clo. Orp. CM54 71
Wyndham Clo. Sutt. BS57 77
Sackville Rd.
Wyndham Cres. N19 BW34 23
Wyndham Cres. Houns. BF46 47
Wyndham Est. SE5 BZ43 51
Wyndham Ms. W1 BU39 41
Upr. Montagu St.
Wyndham Pl. W1 BU39 41
Wyndham Rd. E6 CJ36 35
Wyndham Rd. SE5 BY43 51
Wyndham Rd. W13 BJ41 39
Wyndham Rd. Kings. On T. BL50 57
Wyndham St. W1 BU39 41
Wyneham Rd. SE24 BZ46 51
Wynell Rd. SE23 CC48 61
Wynford Gro. Orp. CO52 72
Wynford Rd. N1 BX37 33
Wynford Way SE9 CK48 62
Wynlie Gdns. Pnr. BC30 11
Wynndale Rd. E18 CH30 17
Wynne Rd. SW9 BY44 51

CENTRAL LONDON

CENTRAL LONDON AREA

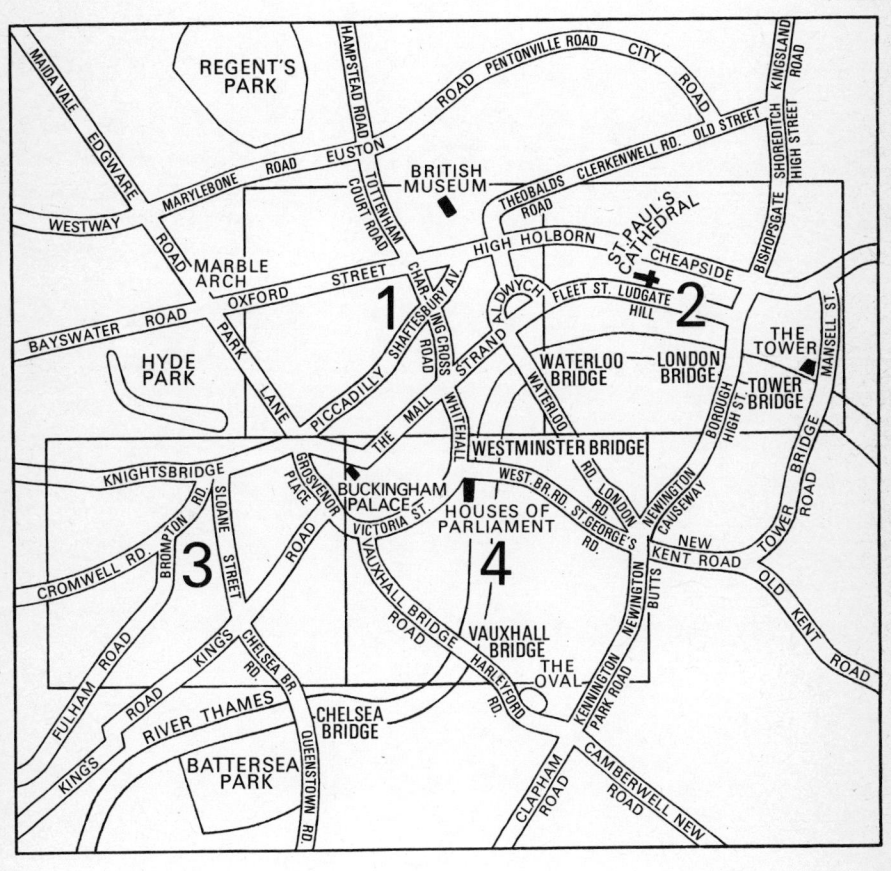

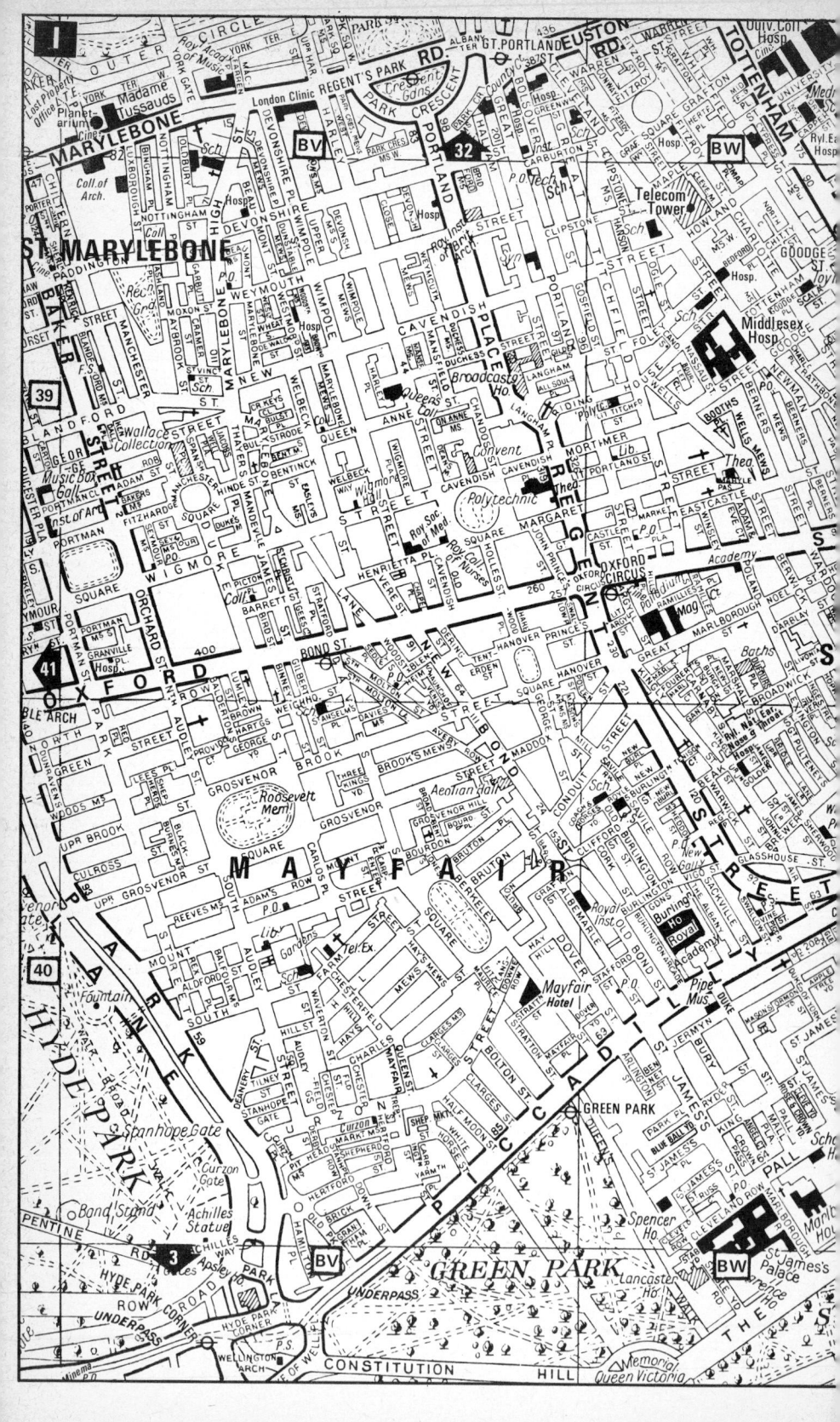

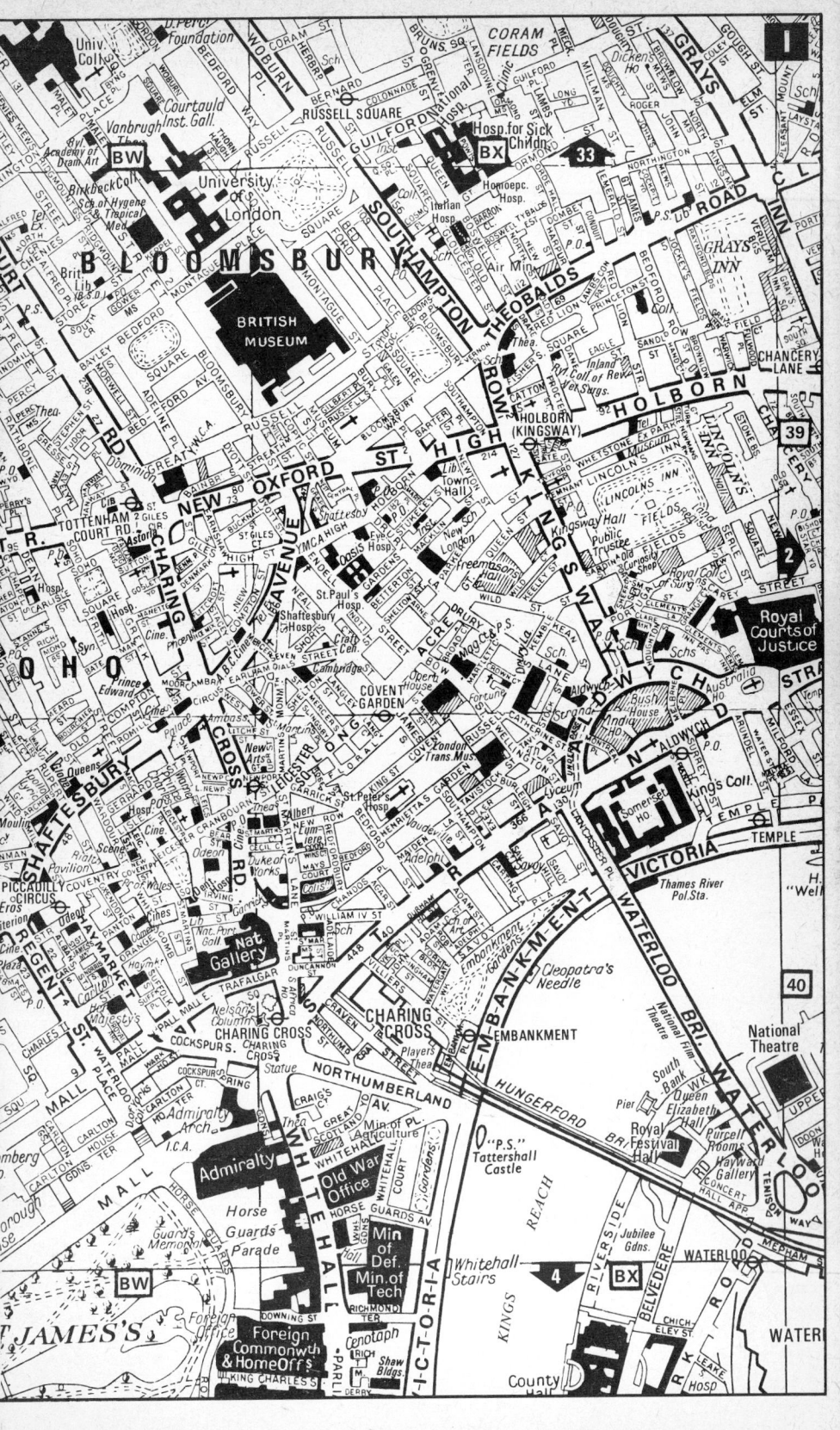

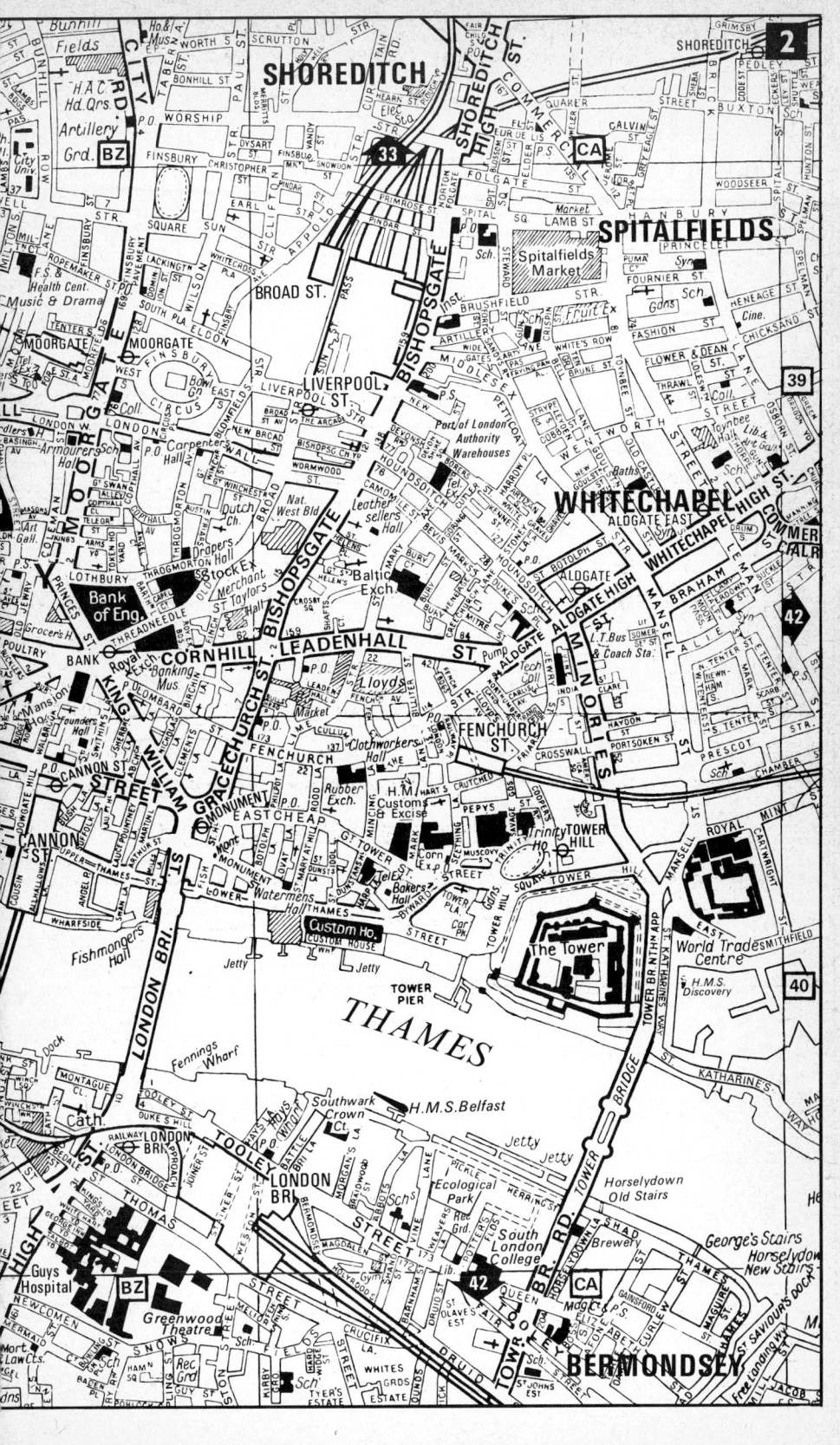

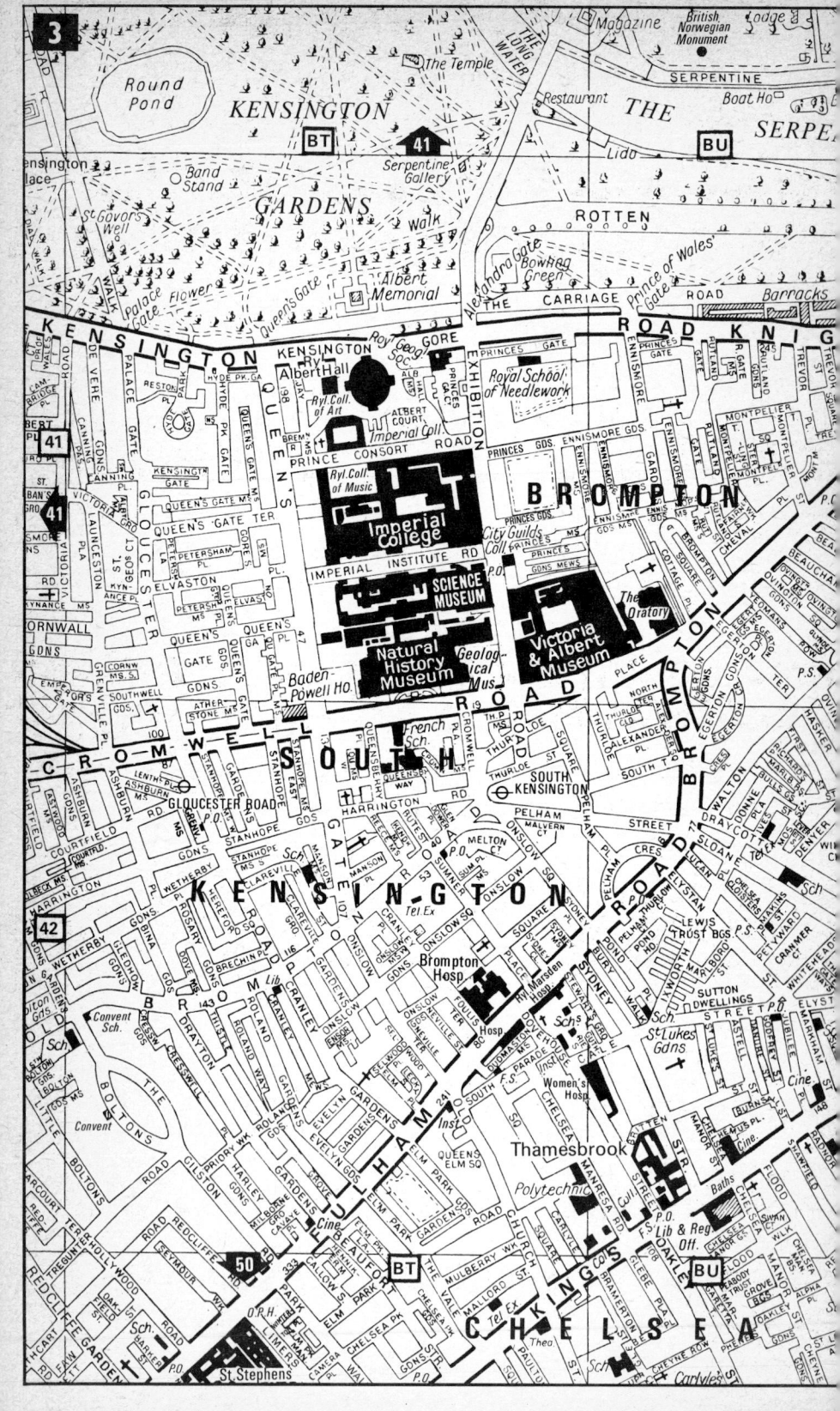

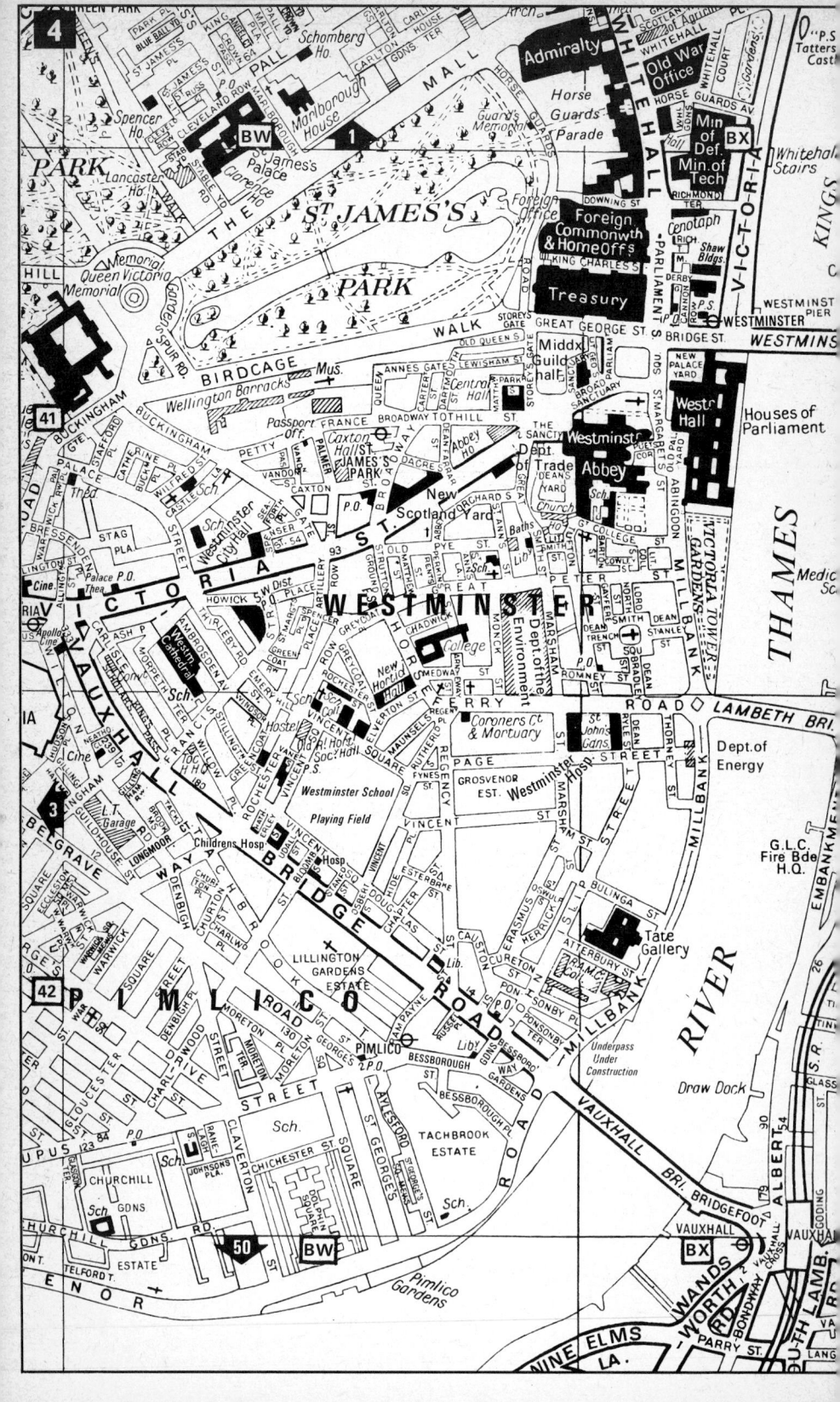

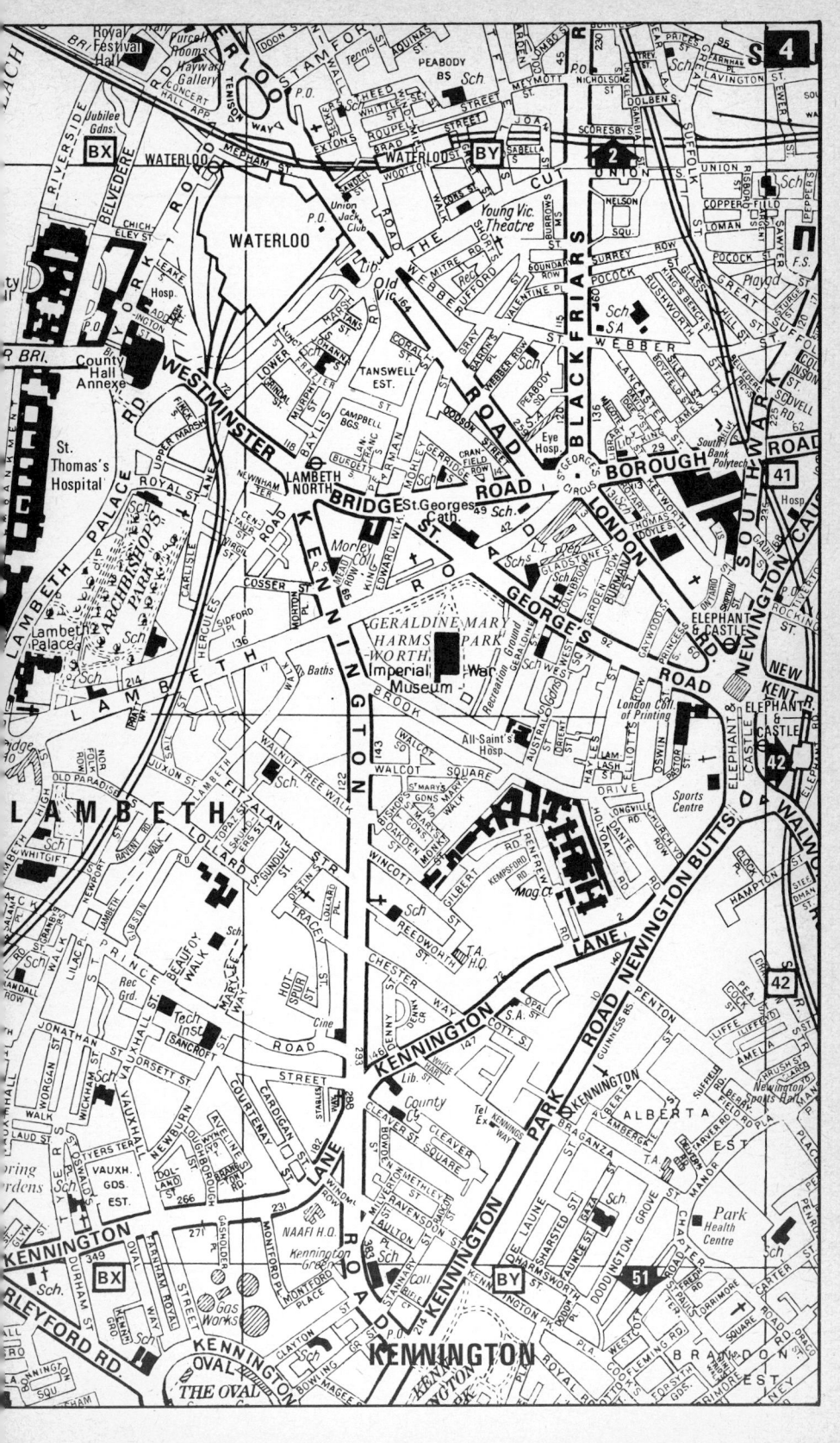

CENTRAL LONDON

Index to Streets

Clements La. EC4	BZ40	2
Cleveland Ms. W1	BW39	1
Cleveland Row, SW1	BW40	1
Cleveland Yd. SW1	BW40	1
Clifford St. W1	BW40	1
Clifton St. EC2	CA39	2
Clink St. SE1	BZ40	2
Cliveden Pl. SW1	BV42	3
Cloak La. EC4	BZ40	2
Clock Pl. SE1	BY42	4
Cloth Fair, EC1	BY39	2
Coach & Horses Yd. W1	BW40	1
Cobb St. E1	CA39	2
Cock La. EC1	BY39	2
Cockpit Yd. WC1	BX39	1
Cockspur Ct. SW1	BW40	1
Cockspur St. SW1	BW40	1
Coleman St. EC2	BZ39	2
Colin St. SE1	BY40	2
College Hill, EC4	BZ40	2
College St. EC4	BZ40	2
Colnbrook St. SE1	BY41	4
Colombo St. SE1	BY40	2
Colville Pl. W1	BW39	1
Concert Hall App. SE1	BX40	1
Conduit St. W1	BV40	1
Cons St. SE1	BY41	4
Constitution Hill, SW1	BV41	3
Cooper's Row, EC3	CA40	2
Copperfield St. SE1	BY41	4
Copthall Av. EC2	BZ39	2
Copthall Cl. EC2	BZ39	2
Coptic St. WC1	BX39	1
Coral St. SE1	BY41	4
Cork St. W1	BW40	1
Cornhill, EC3	BZ39	2
Cornwall Mews St. SW7	BT41	3
Cornwall Rd. SE1	BY40	2
Cosmo Pl. WC1	BX39	1
Cosser St. SE1	BY41	4
Cottage Pl. SW3	BU41	3
Cottage Wk. SW1	BU41	3
Cottington St. SE11	BY42	4
Coulson St. SW3	BU42	3
Courtenay St. SE11	BX42	4
Courtfield Ms. SW7	BT42	3
Cousin La. EC4	BZ40	2
Covent Gdn. WC2	BX40	1
Coventry St. W1	BW40	1
Cowcross St. EC1	BY39	2
Cowley St. SW1	BX41	4
Cox's Ct. EC1	BZ39	2
Craig's Ct. SW1	BX40	1
Cramer St. W1	BV39	1
Cranbourn St. WC2	BW40	1
Cranfield Row, SE1	BY41	4
Cranley Gdns. SW7	BT42	3
Cranley Ms. SW7	BT42	3
Cranley Pl. SW7	BT42	3
Cranmer Ct. SW3	BU42	3
Craven Pl. WC2	BX40	1
Craven St. WC2	BX40	1
Creechurch La. EC3	CA39	2
Creed La. EC4	BY39	2
Crescent Pl. SW3	BU42	3
Crescent, The EC3	CA40	2
Cresswell Gdns. SW5	BT42	3
Cresswell Pl. SW10	BT42	3
Cripplegate St. EC1	BZ39	2
Crispin St. E1	CA39	2
Cromwell Ms. SW7	BT42	3
Cromwell Pl. SW7	BT42	3
Crosby Sq. EC3	CA39	2
Cross Keys Clo. W1	BV39	1
Crosswall, EC3	CA40	2
Crown Ct. WC2	BX39	1
Crown Office Row, EC4	BY40	2
Crown Pass, SW1	BW40	1
Crutched Friars, EC3	CA40	2
Culford Gdns. SW3	BU42	3
Cullum St. EC3	CA40	2
Culross St. W1	BV40	1
Cumberland St. SW1	BV42	3
Cundy St. Est. SW1	BV42	3
Cundy St. SW1	BV42	3
Cureton St. SW1	BW42	4
Cursitor St. EC4	BY39	2
Curzon Pl. W1	BV40	1
Curzon St. W1	BV40	1
Custom House Wf. EC3	CA40	2
Cutler St. E1	CA39	2
Cut, The SE1	BY41	4
Dacre St. SW1	BW41	4
Dane St. WC1	BX39	1
Dante Rd. SE11	BY42	4
Danube St. SW3	BU42	3
D'arblay St. W1	BW39	1
Dartmouth St. SW1	BW41	4
Davidge St. SE1	BY41	4
Davies Ms. W1	BV40	1
Davies St. W1	BV39	1
Davies St. W1	BV40	1
Dean Bradley St.	BX41	4
Deanery St. W1	BV40	1
Dean Farrar St. SW1	BW41	4
Dean Ryle St. SW1	BX42	4
Dean's Ct. EC4	BY39	2
Deans Ms. W1	BV39	1
Dean Stanley St. SW1	BX41	4
Dean St. W1	BW39	1
Dean Trench St. SW1	BX41	4
Delverton Rd. SE17	BY42	4
Denbigh Pl. SW1	BW42	4
Denbigh St. SW1	BW42	4
Denman St. W1	BW40	1
Denmark Pl. WC2	BW39	1
Denmark St. WC2	BW39	1
Denny Cres. SE11	BY42	4
Denny St. SE11	BY42	4
Denyer St. SW3	BU42	3
Derby Gate, SW1	BX41	4
Derby St. W1	BV40	1
Dering St. W1	BV39	1
De Vere Gdns. W8	BT41	3
Devonshire Cl. W1	BV39	1
Devonshire Ms. S. W1	BV39	1
Devonshire Row, EC2	CA39	2
Devonshire Sq. EC2	CA39	2
Devonshire St. W1	BV39	1
De Walden St. W1	BV39	1
Distaff La. EC4	BZ40	2
Distin St. SE11	BY42	4
Dodson St. SE1	BY41	4
Dolben St. SE1	BY40	2
Dolland St. SE11	BX42	4
Dolphin Sq. SW1	BW42	4
Dombey St. WC1	BX39	1
Dominion St. EC2	BZ39	2
Donne Pl. SW3	BU42	3
Doon St. SE1	BY40	2
Dorrington St. EC1	BY39	2
Dorset Ms. SW1	BV41	3
Dorset Rise, EC4	BY39	2
Douglas St. SW1	BW42	4
Dovehouse St. SW3	BT42	3
Dove Ms. SW5	BT42	3
Dover St. W1	BV40	1
Dover Yd. W1	BW40	1
Dowgate Hill, EC4	BZ40	2
Downing St. SW1	BX41	4
Down St. W1	BV40	1
Doyley St. SW1	BV42	3
Drake St. WC1	BX39	1
Draycott Av. SW3	BU42	3
Draycott Pl. SW3	BU42	3
Draycott Ter. SW3	BU42	3
Drayton Gdns. SW10	BT42	3
Drum St. E1	CA39	2
Drury La. WC2	BX39	1
Dryden St. WC2	BX39	1
Duchess Ms. W1	BV39	1
Duchess St. W1	BV39	1
Duchy St. SE1	BY40	2
Duck La. W1	BW39	1
Dudmaston Ms. SW3	BT42	3
Dufours Pl. W1	BW39	1
Duke of Wellington Pl. SW1	BV41	3
Duke of York St. SW1	BW40	1
Duke St. Hill, SE1	BZ40	2
Duke St. St. James's SW1	BW40	1
Duke's Ms. W1	BV39	1
Duke's Pl. EC3	CA39	2
Duke St. SW1	BW40	1
Duke St. W1	BV39	1
Duncannon St. WC2	BX40	1
Dunraven St. W1	BV40	1
Dunstable Ms. W1	BV39	1
Durham House St. WC2	BX40	1
Durham Pl. SW3	BU42	3
Dyott St. WC1	BW39	1
Dyott St. WC1	BX39	1
Eagle Ct. EC1	BY39	2
Eagle Pl. SW1	BW40	1
Eagle St. WC1	BX39	1
Earlham St. WC2	BW39	1
Earlham St. WC2	BX39	1
Earl St. EC2	BZ39	2
Earnshaw St. WC2	BW39	1
Easleys Ms. W1	BV39	1
Eastcastle St. W1	BW39	1
Eastcheap, EC3	BZ40	2
East Harding St. EC4	BY39	2
East Poultry Av. EC1	BY39	2
East Smithfield, E1	CA40	2
East St. EC2	BZ39	2
East Tenter St. E1	CA39	2
Eaton Cl. SW1	BV42	3
Eaton Gate, SW1	BV42	3
Eaton La. SW1	BV41	3
Eaton Ms. SW1	BV42	3
Eaton Ms. N. SW1	BV42	3
Eaton Ms. S. SW1	BV42	3
Eaton Ms. W. SW1	BV42	3
Eaton Pl. SW1	BV41	3
Eaton Row, SW1	BV41	3
Eaton Sq. SW1	BV42	3
Eaton Ter. SW1	BV42	3
Ebury Bridge, SW1	BV42	3
Ebury Bridge Est. SW1	BV42	3
Ebury Bridge Rd. SW1	BV42	3
Ebury Ms. SW1	BV42	3
Ebury Ms. E. SW1	BV41	3
Ebury Sq. SW1	BV42	3
Ebury St. SW1	BV42	3
Eccleston Bridge, SW1	BV42	3
Eccleston Ms. SW1	BV41	3
Eccleston Ms. SW1	BV42	3
Eccleston Pl. SW1	BV42	3
Eccleston Sq. SW1	BV42	3
Eccleston St. SW1	BV41	3
Edinburgh Gate, SW1	BU41	3
Egerton Cres. SW3	BU42	3
Egerton Gdns. SW3	BU41	3
Egerton Ms. SW3	BU41	3
Egerton Pl. SW3	BU41	3
Egerton Ter. SW3	BU41	3
Elder St. E1	CA39	2
Eldon St. EC2	BZ39	2
Elephant & Castle, SE1	BY42	4
Elizabeth Bridge, SW1	BV42	3
Elizabeth St. SW1	BV42	3
Elliott's Row, SE11	BY42	4
Ellis St. SW1	BU42	3
Elm Park Gdns. SW10	BT42	3
Elm Park La. SW3	BT42	3
Elm Pl. SW7	BT42	3
Elvaston Ms. SW7	BT41	3
Elvaston Pl. SW7	BT41	3
Elverton St. SW1	BW42	4
Ely Pl. EC1	BY39	2
Elystan Pl. SW3	BU42	3
Elystan St. SW3	BU42	3
Embankment Pl. WC2	BX40	1
Emerald St. WC1	BX39	1
Emerson Pl. SE1	BZ40	2
Emerson St. SE1	BZ40	2
Emery Hill St. SW1	BW41	4
Endell St. WC2	BX39	1
English Gdns. SE1	CA40	2
Ennismore Gdns. SW7	BT41	3
Ennismore Gdns. Ms. SW7	BU41	3
Ennismore Ms. SW7	BT41	3
Ennismore Ms. SW7	BU41	3
Ensor Ms. SW7	BT42	3
Erasmus St. SW1	BW42	4
Essex Ct. EC4	BY39	2
Essex St. EC4	BY39	2
Essex St. WC2	BY40	2
Essex Ter. EC4	BY40	2
Esterbrooke St. SW1	BW42	4
Evelyn Gdns. SW7	BT42	3
Evelyn Yd. W1	BW39	1
Ewer St. SE1	BZ40	2
Exeter St. WC2	BX40	1
Exhibition Rd. SW7	BT41	3
Exton St. SE1	BY40	2
Fairholt St. SW7	BU41	3
Falconberg Ms. W1	BW39	1
Falcon Cl. SE1	BZ40	2
Farm St. W1	BV40	1
Farnham Pl. SE1	BY40	2
Farnham Royal, SE11	BX42	4
Farringdon St. EC1	BY39	2
Farringdon St. EC4	BY39	2
Fashion St. E1	CA39	2
Fenchurch Av. EC3	CA39	2
Fenchurch Bldgs. EC3	CA39	2
Fenchurch Ct. EC3	CA39	2
Fenchurch St. EC3	CA40	2
Fetter La. EC4	BY39	2
Field Ct. WC1	BX39	1
Finch La. EC3	BZ39	2
Finck St. SE1	BX41	4
Finsbury Av. EC2	BZ39	2
Finsbury Cir. EC2	BZ39	2
Finsbury Pavement, EC2	BZ39	2
Finsbury St. EC2	BZ39	2
First St. SW3	BU42	3
Fisher St. WC1	BX39	1
Fish St. Hill, EC3	BZ40	2
Fitzalan St. SE11	BX42	4
Fitzhardinge St. W1	BV39	1
Fitzmaurice Pl. W1	BV40	1
Fleet La. EC4	BY39	2
Fleet St. EC4	BY39	2
Flitcroft St. WC2	BW39	1
Flood St. SW3	BU42	3
Floral St. WC2	BX40	1
Flower And Dean St. E1	CA39	2
Foley St. W1	BW39	1
Folgate St. E1	CA39	2
Fore St. EC2	BZ39	2
Fore Street Av. EC2	BZ39	2
Fort St. E1	CA39	2
Foster La. EC2	BZ39	2
Foubert's Pl. W1	BW39	1
Foulis Ter. SW7	BT42	3
Fountain Ct. EC4	BY40	2
Fournier St. E1	CA39	2
Fox Ct. EC1	BY39	2
Francis St. SW1	BW42	4
Franklin's Row, SW3	BU42	3
Frazier St. SE1	BY41	4
Friar St. EC4	BY39	2
Friday St. EC4	BZ40	2
Frith St. W1	BW39	1
Frostic Pl. E1	CA39	2
Frying Pan All. E1	CA39	2
Fulham Rd. SW10	BT42	3
Fulwood Pl. WC1	BX39	1
Furnival St. EC4	BY39	2
Fynes St. SW1	BW42	4
Galen Pl. WC1	BX39	1
Gambia St. SE1	BY40	2
Ganton St. W1	BW40	1
Garbutt Pl. W1	BV39	1
Garden Row, SE1	BY41	4
Gardners La. EC4	BZ40	2
Garlick Hill, EC4	BZ40	2
Garrick St. WC2	BX40	1
Gasholder Pl. SE11	BX42	4
Gate St. WC2	BX39	1
Gateways, The SW3	BU42	3
Gatliff Rd. SW1	BV42	3
Gaunt St. SE1	BY41	4
Gayfere St. SW1	BX41	4
Gaywood St. SE1	BY41	4
Gaza St. SE17	BY42	4
Gees Ct. W1	BV39	1
George Inn Yd. SE1	BZ40	2
George Yd. W1	BV40	1
Geraldine St. SE11	BY41	4
Gerald Rd. SW1	BV42	3
Gerrard St. W1	BW40	1
Gerridge St. SE1	BY41	4
Gibson Rd. SE11	BX42	4
Gilbert Pl. WC1	BX39	1
Gilbert Rd. SE11	BY42	4
Gilbert St. W1	BV39	1
Gildea St. W1	BV39	1
Gillingham Ms. SW1	BV42	3
Gillingham Row, SW1	BW42	4
Gillingham St. SW1	BV42	3
Gilston Rd. SW10	BT42	3
Giltspur St. EC1	BY39	2
Gladstone St. SE1	BY41	4
Glasgow Ter. SW1	BW42	4
Glass Hill St. SE1	BY41	4
Glasshouse St. W1	BW40	1
Glasshouse Wk. SE11	BX42	4
Glass House Yd. EC1	BZ39	2
Gledhow Gdns. SW5	BT42	3
Glendower Pl. SW7	BT42	3
Gloucester Rd. SW7	BT41	3
Gloucester St. SW1	BW42	4
Glynde Ms. SW3	BU41	3
Glyn St. SE11	BX42	4
Godfrey St. SW3	BU42	3
Goding St. SE11	BX42	4
Godliman St. EC4	BY39	2
Golden Sq. W1	BW40	1
Goodge Pl. W1	BW39	1
Goodge St. W1	BW39	1
Goodwins Ct. WC2	BX40	1
Gore St. SW7	BT41	3
Goring St. EC3	CA39	2
Gosfield St. W1	BV39	1
Goslett Yd. WC2	BW39	1
Gough Sq. EC4	BY39	2
Goulston St. E1	CA39	2
Gower Ms. WC1	BW39	1
Gracechurch St. EC3	BZ40	2
Grafton St. W1	BV40	1
Graham Ter. SW1	BV42	3
Granby's Bldgs. SE11	BX42	4
Grand Av. EC1	BY39	2
Grange Ct. WC2	BX39	1
Grantham Pl. W1	BV40	1
Granville Pl. W1	BV39	1
Gravel La. E1	CA39	2
Gray's Inn Sq. WC1	BX39	1
Gray St. SE1	BY41	4
Gt. Castle St. W1	BV39	1
Gt. Chapel St. W1	BW39	1
Gt. College St. SW1	BX41	4
Gt. George St. SW1	BW41	4
Gt. Guildford St. SE1	BZ40	2
Gt. James St. WC1	BX39	1
Gt. Marlborough St. W1	BW39	1
Gt. New St. EC4	BY39	2
Gt. Ormond St. WC1	BX39	1
Gt. Peter St. SW1	BW41	4
Gt. Pulteney St. W1	BW40	1
Gt. Queen St. WC2	BX39	1
Gt. Russell St. WC1	BW39	1
Gt. Russell St. WC1	BX39	1
Gt. St. Helen's, EC3	CA39	2
Gt. St. Thomas Apostle, EC4	BZ40	2
Gt. Scotland Yd. SW1	BX40	1
Gt. Smith St. SW1	BW41	4
Gt. Suffolk St. SE1	BY40	2
Gt. Swan All. EC2	BZ39	2
Gt. Trinity La. EC4	BZ40	2
Gt. Turnstile, WC2	BX39	1
Gt. Winchester St. EC2	BZ39	2
Gt. Windmill St. W1	BW40	1
Greek St. W1	BW39	1
Greencoat Pl. SW1	BW42	4
Greencoat Row, SW1	BW41	4
Greenhills Rents, EC1	BY39	2
Green St. W1	BV40	1
Greet St. SE1	BY40	2
Grenville Ms. SW7	BT42	3
Grenville Pl. SW7	BT41	3
Gresham St. EC2	BZ39	2
Gresse St. W1	BW39	1
Greville St. EC1	BY39	2
Greycoat Pl. SW1	BW41	4
Greycoat St. SW1	BW41	4
Grey Eagle St. E1	CA39	2
Greyfriars Pass. EC1	BY39	2

Morpeth Ter. SW1	BW41	4
Mortimer St. W1	BV39	1
Morton Pl. SE1	BY41	4
Morwell St. WC1	BW39	1
Mossop St. SW3	BU42	3
Motcomb St. SW1	BU41	3
Mount Row, W1	BV40	1
Mount St. W1	BV40	1
Moxon St. W1	BV39	1
Murphy St. SE1	BY41	4
Muscovy St. EC3	CA40	2
Museum St. WC1	BX39	1
Nassau St. W1	BW39	1
Neal St. WC2	BX39	1
Neathouse Cl. SW1	BW42	4
Nelson Sq. SE1	BY41	4
Neville St. SW7	BT42	3
Neville Ter. SW7	BT42	3
New Bond St. W1	BV39	1
New Bond St. W1	BV40	1
New Bridge St. EC4	BY39	2
New Broad St. EC2	BZ39	2
Newburgh St. W1	BW39	1
New Burlington Ms. W1	BW40	1
New Burlington Pl. W1	BW40	1
New Burlington St. W1	BW40	1
Newburn St. SE11	BX42	4
Newbury St. EC1	BZ39	2
Newcastle Ct. EC4	BY39	2
New Cavendish St. W1	BV39	1
New Change, EC4	BZ39	2
New Compton St. WC2	BW39	1
New Ct. WC2	BX39	1
New Coventry St. W1	BW40	1
New Fetter La. EC4	BY39	2
Newgate St. EC1	BY39	2
New Goulston St. E1	CA39	2
Newington Butts, SE1	BY42	4
Newington Causeway, SE1	BY41	4
Newman Pass. W1	BW39	1
Newmans Row, WC2	BX39	1
Newman St. W1	BW39	1
Newman Yd. W1	BW39	1
Newnham St. E1	CA39	2
Newnham Ter. SE1	BY41	4
New North St. WC1	BX39	1
New Oxford St. WC1	BW39	1
New Oxford St. WC1	BX39	1
New Palace Yd. SW1	BX41	4
Newport Ct. WC2	BW40	1
Newport Pl. WC2	BW40	1
Newport St. SE11	BX42	4
Newport St. WC2	BW40	1
New Row, WC2	BX40	1
New Sq. WC2	BX39	1
New St. EC2	CA39	2
New Street Sq. EC4	BY39	2
Newton St. WC2	BX39	1
New Turnstile, WC2	BX39	1
Nicholas La. EC4	BZ40	2
Nicholson St. SE1	BY40	2
Noble St. EC2	BZ39	2
Noel St. W1	BW39	1
Norfolk Row, SE11	BX42	4
Norris St. SW1	BW40	1
North Audley St. W1	BV40	1
North Ct. W1	BW39	1
North Cres. WC1	BW39	1
North Tenter St. E1	CA39	2
North Ter. SW3	BU41	3
Northumberland All. EC3	CA39	2
Northumberland Av. WC2	BX40	1
Northumberland St. WC2	BX40	1
Norton Folgate, E1	CA39	2
Norwich St. EC4	BY39	2
Nottingham Ct. WC2	BX39	1
Nottingham St. W1	BV39	1
Oakden St. SE11	BY42	4
Oakley St. SW3	BU42	3
Oat La. EC2	BZ39	2
Ogle St. W1	BW39	1
Old Bailey, EC4	BY39	2
Old Barrack Yd. SW1	BV41	3
Old Bond St. W1	BW40	1
Old Broad St. EC2	BZ39	2
Old Burlington St. W1	BW40	1
Old Castle St. E1	CA39	2
Old Cavendish St. W1	BV39	1
Old Change Ct. EC4	BZ39	2
Old Church St. SW3	BT42	3
Old Compton St. W1	BW40	1
Old Gloucester St. WC1	BX39	1
Old Jewry, EC2	BZ39	2
Old North St. WC1	BX39	1
Old Palace Yd. SW1	BX41	4
Old Paradise St. SE11	BX42	4
Old Park La. W1	BV40	1
Old Pye St. SW1	BW41	4
Old Queen St. SW1	BW41	4
Old Seacoal La. EC4	BY39	2
Old Sq. WC2	BX39	1
O'meara St. SE1	BZ40	2
Onslow Gdns. SW7	BT42	3
Onslow Ms. E. SW7	BT42	3
Onslow Ms. W. SW7	BT42	3
Onslow Sq. SW7	BT42	3
Onslow St. EC1	BY39	2
Ontario St. SE1	BY41	4
Opal St. SE11	BY42	4
Orange St. WC2	BW40	1
Orchard St. W1	BV39	1
Orde Hall St. WC1	BX39	1
Orient St. SE11	BY42	4
Ormonde Gate, SW3	BU42	3
Ormond Yd. SW1	BW40	1
Orsett St. SE11	BX42	4
Osbert St. SW1	BW42	4
Osborn St. E1	CA39	2
Oswin St. SE11	BY42	4
Oval Way, SE11	BX42	4
Ovington Gdns. SW3	BU41	3
Ovington Ms. SW3	BU41	3
Ovington Sq. SW3	BU41	3
Ovington St. SW3	BU42	3
Oxendon St. SW1	BW40	1
Oxford Circus, W1	BV39	1
Oxford St. W1	BV39	1
Paddington St. W1	BV39	1
Page St. SW1	BW42	4
Palace Gate, W8	BT41	3
Palace Pl. SW1	BV41	3
Palace St. SW1	BW41	4
Pall Mall, SW1	BW40	1
Pall Mall E. SW1	BW40	1
Pall Mall Pl. SW1	BW40	1
Palmer St. SW1	BW41	4
Pancras La. EC4	BZ39	2
Panton St. SW1	BW40	1
Panyer All. EC4	BZ39	2
Paris Gdn. SE1	BY40	2
Park Cl. SW1	BU41	3
Parker St. WC2	BX39	1
Park La. W1	BV40	1
Park Pl. SW1	BW40	1
Park St. W1	BV40	1
Parliament Sq. SW1	BX41	4
Parliament St. SW1	BX41	4
Passmore St. SW1	BV42	3
Pastor St. SE11	BY42	4
Paternoster Row, EC4	BZ39	2
Paternoster Sq. EC4	BY39	2
Pavilion Rd. SW1	BU41	3
Pavilion St. SW1	BU41	3
Peabody Av. SW1	BV42	3
Peabody Bldgs. SE1	BY40	2
Peabody Sq. SE1	BY41	4
Peacock St. SE17	BY42	4
Pearman St. SE1	BY41	4
Pelham Ct. SW3	BU42	3
Pelham Cres. SW7	BU42	3
Pelham Pl. SW7	BT42	3
Pelham St. SW7	BT42	3
Pembroke Cl. SW1	BV41	3
Penton Pl. SE17	BY42	4
Pepys St. EC3	CA40	2
Percy Ms. W1	BW39	1
Percy St. W1	BW39	1
Perkins Rents, SW1	BW41	4
Perry's Pl. W1	BW39	1
Petersham La. SW7	BT41	3
Petersham Ms. SW7	BT41	3
Petersham Pl. SW7	BT41	3
Peter's La. EC1	BY39	2
Peter St. W1	BW40	1
Petticoat La. E1	CA39	2
Petty France, SW1	BW41	4
Petyward, SW3	BU42	3
Philpot La. EC3	CA40	2
Phoenix St. WC2	BW39	1
Piccadilly, W1	BV40	1
Piccadilly Circus, W1	BW40	1
Piccadilly Pl. W1	BW40	1
Pickle Herring St. SE1	CA40	2
Picton Pl. W1	BV39	1
Pike Gdns. SE1	BZ40	2
Pilgrim St. EC4	BY39	2
Pimlico Rd. SW1	BV42	3
Pindar St. EC2	CA39	2
Pit Head Ms. W1	BV40	1
Playhouse Yd. EC4	BY39	2
Plough Pl. EC4	BY39	2
Plumtree Ct. EC4	BY39	2
Pocock St. SE1	BY41	4
Poets Corner, SW1	BX41	4
Poland St. W1	BW39	1
Pollen St. W1	BV39	1
Pond House, SW3	BU42	3
Pond Pl. SW3	BU42	3
Ponsonby Pl. SW1	BW42	4
Ponsonby Ter. SW1	BW42	4
Pont St. Ms. SW1	BU41	3
Pont St. SW1	BU41	3
Poppins Ct. EC4	BY39	2
Portman Ms. S. W1	BV39	1
Portman Sq. W1	BV39	1
Portman St. W1	BV39	1
Portpool La. EC1	BY39	2
Portsoken St. E1	CA40	2
Portugal St. WC2	BX39	1
Potter's Flds. SE1	CA40	2
Poultry, EC2	BZ39	2
Poultry Av. EC1	BY39	2
Pratt Wk. SE11	BX42	4
Prescot St. E1	CA40	2
Prices St. SE1	BY40	2
Primrose Hill, EC4	BY39	2
Primrose St. EC2	CA39	2
Prince Consort Rd. SW7	BT41	3
Princelet St. E1	CA39	2
Princes Gdns. SW7	BT41	3
Princes Gdns. Ms. SW7	BT41	3
Princes Gate, SW7	BT41	3
Princes Gate Ct. SW7	BT41	3
Princes Ms. SW7	BT41	3
Princess St. SE1	BY41	4
Princes St. EC2	BZ39	2
Prince's St. W1	BV39	1
Princeton St. WC1	BX39	1
Priory Wk. SW10	BT42	3
Procter St. WC1	BX39	1
Providence Ct. W1	BV40	1
Puddle Dock, EC4	BY40	2
Puma Ct. E1	CA39	2
Quality Ct. WC2	BY39	2
Queen Anne Mews W1	BV39	1
Queen Anne's Gate, SW1	BW41	4
Queen Anne St. W1	BV39	1
Queenhithe, EC4	BZ40	2
Queen St. Pl. EC4	BZ40	2
Queensbury Ms. W. SW7	BT42	3
Queensbury Pl. SW7	BT42	3
Queensbury Way, SW7	BT42	3
Queens Elm Sq. SW3	BT42	3
Queen's Gate Gdns. SW7	BT41	3
Queen's Gate Ms. SW7	BT41	3
Queen's Gate Pl. Ms. SW7	BT41	3
Queen's Gate Pl. SW7	BT41	3
Queen's Gate, SW7	BT41	3
Queen's Gate Ter. SW7	BT41	3
Queens Sq. WC1	BX39	1
Gt. Ormond St.		
Queen St. EC4	BZ40	2
Queen St. Mayfair W1	BV40	1
Queen's Walk, SW1	BW40	1
Queen Victoria St. EC4	BY40	2
Radcot St. SE11	BY42	4
Radnor Wk. SW3	BU42	3
Railway App. SE1	BZ40	2
Railway Pl. EC3	CA40	2
Ralston St. SW3	BU42	3
Ramillies Pl. W1	BW39	1
Ramillies St. W1	BW39	1
Rampayne St. SW1	BW42	4
Randall Rd. SE11	BX42	4
Randall Row, SE11	BX42	4
Ranelagh Gro. SW1	BV42	3
Ranelagh St. SW1	BW42	4
Rangoon St. EC3	CA39	2
Raphael St. SW7	BU41	3
Rathbone Pl. W1	BW39	1
Rathbone St. W1	BW39	1
Ravensdon St. SE11	BY42	4
Ravent Rd. SE11	BX42	4
Rawlings St. SW3	BU42	3
Raymond Bldgs. WC1	BX39	1
Redcliffe Rd. SW10	BT42	3
Redesdale St. SW3	BU42	3
Red Lion Ct. EC4	BY39	2
Red Lion Sq. WC1	BX39	1
Red Lion St. WC1	BX39	1
Red Pl. W1	BV40	1
Reece Ms. SW7	BT42	3
Reedworth St. SE11	BY42	4
Reeves Ms. W1	BV40	1
Regency St. SW1	BW42	4
Regent Pl. SW1	BW42	4
Regent Pl. W1	BW40	1
Regent St. SW1	BW40	1
Regent St. W1	BV39	1
Remnant St. WC2	BX39	1
Renfrew Rd. SE11	BY42	4
Rennie St. SE1	BY40	2
Reston Pl. SW7	BT41	3
Rex Pl. W1	BV40	1
Richard's Pl. SW3	BU42	3
Richmond Bldgs. W1	BW39	1
Richmond Ter. SW1	BX41	4
Richmond Ter. Ms. SW1	BX41	4
Ridgmount Gdns. WC1	BW39	1
Ridgmount Pl. WC1	BW39	1
Riding House St. W1	BV39	1
Risborough St. SE1	BY41	4
Riverside Wk. SE1	BX41	4
Robert St. WC2	BX40	1
Rochester Row, SW1	BW42	4
Rochester St. SW1	BW41	4
Roland Gdns. SW10	BT42	3
Roland Way, SW10	BT42	3
Rolls Bldgs. EC4	BY39	2
Rolls Pass. EC4	BY39	2
Romilly St. W1	BW40	1
Romney St. SW1	BW41	4
Rood La. EC3	CA40	2
Rosary Gdns. SW7	BT42	3
Rose Alley, SE1	BZ40	2
Rose & Crown Yd. SW1	BW40	1
Rosemaker St. EC2	BZ39	2
Rosemoor St. SW3	BU42	3
Rose St. WC2	BX40	1
Rotary St. SE1	BY41	4
Rotten Row, SW7	BT41	3
Roupell St. SE1	BY40	2
Royal Arc. SW1	BW40	1
Royal Av. SW3	BU42	3
Royal Exchange Bldgs. EC3	BZ39	2
Royal Mint St. E1	CA40	2
Royal St. SE1	BX41	4
Rupert St. W1	BW40	1
Rushworth St. SE1	BY41	4
Russell Ct. SW1	BW40	1
Russell Pl. SW1	BW42	4
Russell St. WC2	BX40	1
Russia Row, EC2	BZ39	2
Rutherford St. SW1	BW42	4
Rutland Gdns. SW7	BU41	3
Rutland Gate, SW7	BU41	3
Rutland Gate Ms. SW7	BU41	3
Rutland Ms. St. SW7	BU41	3
Rutland St. SW7	BU41	3
Ryder St. SW1	BW40	1
Rysbrook St. SW3	BU41	3
Sabella St. SE1	BY40	2
Sackville St. W1	BW40	1
Saffron Hill, EC1	BY39	2
Sail St. SE11	BX42	4
St. Albans St. SW1	BW40	1
St. Alphages Gdns. EC2	BZ39	2
St. Andrews Hill, EC4	BY39	2
St. Andrews Hill, EC4	BY40	2
St. Andrew St. EC4	BY39	2
St. Anne's Ct. W1	BW39	1
St. Ann's La. SW1	BW41	4
St. Ann's St. SW1	BW41	4
St. Anselm's Pl. W1	BV40	1
St. Barnabas St. SW1	BV42	3
St. Botolph St. EC3	CA39	2
St. Bride St. EC4	BY39	2
St. Christopher's Pl. W1	BV39	1
St. Clare St. EC3	CA39	2
St. Clement's La. WC2	BX39	1
St. Cross St. EC1	BY39	2
St. Dunstans Hill, EC3	CA40	2
St. Dunstan's La. EC3	CA40	2
St. George's Cir. SE1	BY41	4
St. George's Ct. SW7	BT41	3
St. George's Dr. SW1	BV42	3
St. George's Rd. SE1	BY41	4
St. George's Sq. Ms. SW1	BW42	4
St. George's Sq. SW1	BW42	4
St. George St. W1	BV40	1
St. Giles Circus, WC1	BW39	1
St. Giles Ct. WC2	BW39	1
St. Giles Ct. WC2	BX39	1
St. Giles High St. WC2	BW39	1
St. Helen's Pl. EC3	CA39	2
St. James's Mkt. SW1	BW40	1
St. James's Pl. SW1	BW40	1
St. James's Sq. SW1	BW40	1
St. John's La. EC1	BY39	2
St. Katharines Way, E1	CA40	2
St. Leonard's. Ter. SW3	BU42	3
St. Luke's St. SW3	BU42	3
St. Margarets Ms. WC2	BX40	1
St. Margarets Pl. SW1	BW41	4
St. Margaret St. SW1	BX41	4
St. Mark St. E1	CA39	2
St. Martin's La. WC2	BX40	1
St. Martins Le Grand EC1	BZ39	2
St. Martin's Pl. WC2	BX40	1
St. Martins St. WC2	BW40	1
St. Mary at Hill, EC3	CA40	2
St. Mary Axe, EC3	CA39	2
St. Mary's Gdns. SE11	BY42	4
St. Mary's Wk. SE11	BY42	4
St. Matthew St. SW1	BW41	4
St. Olaves Ct. EC2	BZ39	2
St. Oswald's Pl. SE11	BX42	4
St. Oswulf St. SW1	BW42	4
St. Paul's Churchyard, EC4	BY39	2
St. Swithin's La. EC4	BZ40	2
St. Thomas St. SE1	BZ40	2
St. Vincent St. W1	BV39	1
Salamanca Pl. SE11	BX42	4
Salamanca St. SE11	BX42	4
Salisbury Ct. EC4	BY39	2
Salisbury Sq. EC4	BY39	2
Sancroft St. SE11	BX42	4
Sanctuary, The SW1	BW41	4
Sandell St. SE1	BY41	4
Sandland St. WC1	BX39	1
Sandys Row, E1	CA39	2
Sardinia St. WC2	BX39	1
Saunders St. SE11	BX42	4
Savage Gdns. EC3	CA40	2
Saville Row, W1	BW40	1
Savoy Ct. WC2	BX40	1
Savoy Hill, WC2	BX40	1
Savoy Pl. WC2	BX40	1
Savoy St. WC2	BX40	1
Scala St. W1	BW39	1
Scarborough St. E1	CA39	2
Scoresby St. SE1	BY40	2
Seacoal La. EC4	BY39	2
Seaforth Pl. SW1	BW41	4
Secker St. SE1	BY40	2
Sedding St. SW1	BV42	3
Sedley Pl. W1	BV39	1
Seething La. EC3	CA40	2
Selwood Pl. SW7	BT42	3
Selwood Ter. SW7	BT42	3
Semley Pl. SW1	BV42	3
Serle St. WC2	BX39	1

NOTES

To help you record specific details, the following pages can be used to add your own notes.

NOTES

NOTES

NOTES

NOTES

Titles in the A1 Street Atlas series

Title	Inches to mile	Size	ISBN
Birmingham & West Midlands	4	$8\frac{1}{2}''$ X $5\frac{1}{2}''$	0 09 219680 2
Edinburgh	4	$8\frac{1}{2}''$ X $5\frac{1}{2}''$	0 09 219360 9
Glasgow	4	$8\frac{1}{2}''$ X $5\frac{1}{2}''$	0 09 202130 1
Greater Bristol & Bath	4	$8\frac{1}{2}''$ X $5\frac{1}{2}''$	0 09 202280 4
Leeds & Bradford	4	$8\frac{1}{2}''$ X $5\frac{1}{2}''$	0 09 219040 5
London	3	$7''$ X $4\frac{1}{2}''$	0 09 202200 6
Manchester	4	$8\frac{1}{2}''$ X $5\frac{1}{2}''$	0 09 202140 9
Merseyside	4	$8\frac{1}{2}''$ X $5\frac{1}{2}''$	0 09 202270 7
Sheffield & Rotherham	4	$8\frac{1}{2}''$ X $5\frac{1}{2}''$	0 09 202100 X
Stoke on Trent	4	$8\frac{1}{2}''$ X $5\frac{1}{2}''$	0 09 202180 8

Printed in Great Britain at the University Press, Oxford